Ideas and Methods of Supersymmetry and Supergravity

or

A Walk Through Superspace

Tiger, Tiger, Burning Bright
In the Forest of the Night
What Immortal Hand or Eye
Could Frame Thy Fearful
Symmetry

William Blake:
Tiger

Ideas and Methods of Supersymmetry and Supergravity

or

A Walk Through Superspace

Ioseph L Buchbinder and Sergei M Kuzenko

Tomsk State University, Russia

Revised Edition

Published in 1998 by
Taylor & Francis Group
6000 Broken Sound Parkway NW, Suite 300
Boca Raton, FL 33487-2742

Published in Great Britain by
Taylor & Francis Group
2 Park Square
Milton Park, Abingdon
Oxon OX14 4RN

First hardback edition 1995, Revised (paperback) edition 1998

ISBN-13: 978-0-7503-0506-8 (hbk)
ISBN-13: 978-0-367-80253-0 (ebk)
DOI: 10.1201/9780367802530

Library of Congress Cataloging-in-Publication Data

Catalog record is available from the Library of Congress

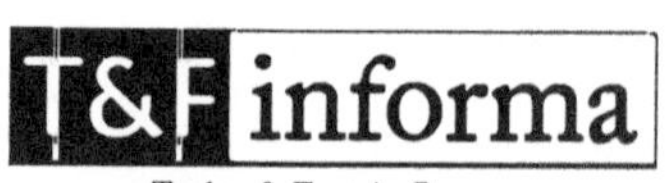

Taylor & Francis Group
is the Academic Division of T&F Informa plc.

Visit the Taylor & Francis Web site at
http://www.taylorandfrancis.com

Contents

Preface to the First Edition

The discovery of supersymmetry is one of the most distinguished achievements of theoretical physics in the second half of the twentieth century. The fundamental significance of supersymmetry was realized almost immediately after the pioneering papers by Gol'fand and Likhtman, Volkov and Akulov and Wess and Zumino. Within a short space of time the supersymmetric generalization of the standard model was found, the supersymmetric theory of gravity constructed, and the approaches to supersymmetric string theory developed. It has also emerged that supersymmetry is of interest in certain quantum mechanical problems and even some classical ones.

In essence, supersymmetry is the extension of space–time symmetry (Galileo symmetry, Poincaré symmetry, conformal symmetry, etc.) by fermionic generators and it serves as the theoretical scheme which naturally unifies bosons and fermions. Therefore, it is natural that supersymmetry should form the basis of most modern approaches to finding a unified theory of all fundamental interactions.

At present, the ideas and methods of supersymmetry are widely used by specialists in high-energy theoretical physics. The conceptions of supersymmetry play an important role in quantum field theory, in the theory of elementary particles, in gravity theory and in many aspects of mathematical physics. It is evident that supersymmetry must be an element of the basic education of the modern theoretical physicist. Hence there is a need for a textbook intended as an introduction to the subject, in which the fundamentals of supersymmetry and supergravity are expounded in detail.

The present book is just such a textbook and it aims to acquaint readers with the fundamental concepts, ideas and methods of supersymmetry in field theory and gravity. It is written for students specializing in quantum field theory, gravity theory and general mathematical physics, and also for young researchers in these areas. However, we hope that experts will find something of interest too.

The main problem which tormented us constantly while working on the book was the choice of material. We based our decisions on the following principles.

(i) The account must be closed and complete. The book should contain everything necessary for the material to be understood. Therefore we included in the book a series of mathematical sections concerning group theory, differential geometry and the foundations of algebra and analysis with anticommuting elements. We also included some material on classical and quantum field theory.

(ii) A detailed exposition. Since the book is intended as an introduction to the subject we tried to set forth the material in detail, with all corresponding calculations, and with discussions of the initial motivations and ideas.

(iii) The basic content of the subject. We proceeded from the fact that a book that aspires to the role of a textbook should, in the main, contain only completed material, the scientific significance of which will not change in the coming years, which has received recognition, and has already found some applications. These requirements essentially restrict the choice of material. In principle, this means that the basic content of the book should be devoted to four-dimensional $N = 1$ supersymmetry.

It is possible that such a point of view will provoke feelings of disappointment in some readers who would like to study extended supersymmetry, higher-dimensional supersymmetry and two-dimensional supersymmetry. Although realizing the considerable significance of these subjects, we consider, nevertheless, that the corresponding material either does not satisfy the completeness criterion, is too specialized, or, because of its complexity, is far beyond the scope of the present book. As far as two-dimensional supersymmetry is concerned, its detailed exposition, in our view, should be carried out in the context of superstring theory. However, we are sure that the reader, having studied this book, will be well prepared for independent research in any directions of supersymmetry.

(iv) The superfield point of view. The natural formulation of four-dimensional $N = 1$ supersymmetry is realized in a language of superspace and superfields—that is, this formulation is accepted for the material's account in the book. Initially we did not want to discuss a component formulation at all, considering that it is only of very narrow interest. But then we came to the conclusion that the component language is useful for illustrations, and for analogies and comparisons with conventional field theory; hence we have included it.

At present there exists an extensive literature on the problems of supersymmetry and supergravity. Since we give a complete account of material here, we decided that it was unnecessary to provide an exhaustive list of references. Instead we use footnotes which direct the attention of the reader to a few papers and books. Moreover, we give a list of pioneer papers, basic reviews and books, and also the fundamental papers whose results we have used.

The book consists of seven chapters. In Chapter 1 the mathematical backgrounds are considered. This material is used widely in the remaining chapters of the book and is standard for understanding supersymmetry. Chapter 2 is devoted to algebraic aspects of supersymmetry and the concepts of superspace and superfield. In Chapter 3 the classical superfield theory is given. Chapter 4 is devoted to the quantum superfield theory. In Chapters 5 and 6 the superfield formulation of supergravity is studied. Chapter 7 is devoted to the theory of quantized superfields in curved superspace. This chapter can be considered as a synthesis of the results and methods developed in all previous chapters. The material of Chapter 7 allows us to see how the geometrical structure of curved superspace displays itself by studying quantum aspects. The subject of this chapter is more complicated and, to a certain extent, reflects the authors' interests.

The book has pedagogical character and is based on lectures given by the authors at the Department of Quantum Field Theory in Tomsk State University. Certainly, it cannot be considered as an encyclopedia of supersymmetry. As with any book of such extensive volume, this one may not be free of misprints. References to them will be met with gratitude by the authors. We are grateful to Institute of Physics Publishing for its constant support during work on the book and its patience concerning our inability to finish the work on time. We are especially grateful to V N Romanenko for her invaluable help in preparation of the manuscript. One of us (SMK) wishes to express his deep gratitude to J V Yarevskaya, the first reader of this book, and also to the Alexander von Humboldt Foundation for financial support in the final stage of preparing this text.

We hope that the book will be useful for the young generation of theoretical physicists and perhaps will influence the formation of their interest in the investigation of supersymmetry.

Ioseph L Buchbinder
Sergei M Kuzenko

Tomsk and Hannover,
1994

Preface to the Revised Edition

The first edition of *Ideas and Methods of Supersymmetry and Supergravity* was published in 1995 and received positive reviews. It demonstrated the necessity for a detailed, closed and complete account of the fundamentals of supersymmetric field theory and its most important applications including discussion of motivations and nuances.

We are happy that the first edition proved to be a success and are grateful to Institute of Physics Publishing for publishing this revised edition. The general structure of the book remains without change. $N = 1$ supersymmetry, which is the main content of the book, is still the basic subject of many modern research papers devoted to supersymmetry. Although field theories possessing extended supersymmetry attract great attention due to their remarkable properties and prospects, their description both at the classical and quantum levels is realized, in many cases, in terms of $N = 1$ superfields with the help of the methods given in this book.

For the revised edition we have corrected a number of misprints and minor errors and supplemented the text with new material that fits naturally into the original content of the book. We have added a new subsection (3.4.6) which is devoted to the component structure of the massive vector multiplet. Of course, this model can be investigated by purely superfield methods given in the book, but as the component approach is more familiar to a great many practitioners of supersymmetry, we decided to present such a consideration for the massive vector multiplet; this point was not given in the first edition. We also included a new section (3.8) which deals with the non-minimal scalar multiplet being a variant realization of the superspin-0 multiplet. The non-minimal scalar multiplet possesses remarkable properties, as a supersymmetric field theory, and is expected to be important for phenomenological applications, in particular, for constructing supersymmetric extensions of the low-energy QCD effective action. Since those supersymmetric theories involving non-minimal scalar multiplets are still under investigation in the research literature, we have restricted ourselves in section (3.8) to the description of a few relevant models and provided references to recent research papers of interest.

One of the difficulties we faced while writing the book was associated with the practical impossibility of composing an adequate bibliography of the works on supersymmetry and supergravity related to our book. A complete list of such works seems to contain more than ten thousand titles and evena shorter list demands an essential increase in the book's volume. The references given in the book include only the monographs and review papers and also those papers which, in our opinion, have exerted influence on the development of the material under consideration as a whole. Of course, the choice of these papers is rather subjective and reflects our scientific interests.

After the first edition was published we received a number of responses from many of our colleagues to whom we are sincerely grateful. While working on the book, we took pleasure in communicating and discussing various aspects of supersymmetry and field theory with N. Dragon, E. S. Fradkin, S. J. Gates, E. A. Ivanov, O. Lechtenfeld, D. Lüst, V. I. Ogievetsky, H. Osborn, B. A. Ovrut, A. A. Tseytlin, I. V. Tyutin, M. A. Vasiliev, P. West and B. M. Zupnik. We express our deepest gratitude to all of them. We thank our editor at Institute of Physics Publishing, Jim Revill, for his friendly support and encouragement on this project. One of us (SMK) is grateful to the Alexander von Humboldt Foundation for financial support and to the Institute for Theoretical Physics at the University of Hannover for kind hospitality during the preparation of this edition.

Ioseph L Buchbinder
Sergei M Kuzenko

Tomsk
March 1998

1. Mathematical Background

Oh my children! my poor children!
Listen to the words of wisdom,
Listen to the words of warning,
From the lips of the Great Spirit,
From the Master of Life, who made you!

Henry Longfellow:
The Song of Hiawatha

1.1. The Poincaré group, the Lorentz group

1.1.1. Definitions

Space–time structure in special relativity is determined by the set of general principles:

1. Space and time are homogeneous.
2. Space is isotropic.
3. In all inertial reference systems the speed of light has the same value c.

From the mathematical point of view, these principles mean that the space–time coordinates x^m and x'^m of two arbitrary inertial reference systems are related by a linear non-homogeneous transformation

$$x'^m = \Lambda^m{}_n x^n + b^m \tag{1.1.1}$$

leaving the metric

$$\mathrm{d}s^2 = \eta_{mn}\,\mathrm{d}x^m\,\mathrm{d}x^n \tag{1.1.2}$$

invariant. We have used the following notation: $x^m = (x^0, x^1, x^2, x^3)$, $x^0 = ct$, where t is the time coordinate (in what follows, we set $c = 1$), $\vec{x} = (x^1, x^2, x^3)$ are the space coordinates and η_{mn} is the Minkowski metric

$$\eta_{mn} = \begin{pmatrix} -1 & 0 & 0 & 0 \\ 0 & 1 & 0 & 0 \\ 0 & 0 & 1 & 0 \\ 0 & 0 & 0 & 1 \end{pmatrix}.$$

DOI: 10.1201/9780367802530-1

The requirement for invariance of the metric ds^2 is equivalent to the equation

$$\Lambda^{\mathrm{T}}\eta\Lambda=\eta. \tag{1.1.3}$$

Here Λ^{T} is the matrix transpose of Λ.

Equation (1.1.3) is not the only necessary restriction on the parameters $\Lambda^m{}_n$. On physical grounds, two additional conditions are to be taken into account. Using (1.1.3), we find

$$\det\Lambda=\pm 1$$

$$(\Lambda^0{}_0)^2-(\Lambda^1{}_0)^2-(\Lambda^2{}_0)^2-(\Lambda^3{}_0)^2=1.$$

It is now easy to show that in order to preserve the direction of time one must demand

$$\Lambda^0{}_0\geqslant 1 \tag{1.1.4a}$$

and to preserve parity (spatial orientation), one has to choose the branch

$$\det\Lambda=1. \tag{1.1.4b}$$

The transformations (1.1.1) with parameters $\Lambda^m{}_n$ constrained by the relations (1.1.3, 1.1.4) are called the 'Poincaré transformations'. In the homogeneous case, when $b^m=0$, they are called the 'Lorentz transformations'. We shall denote the Poincaré transformations symbolically as (Λ, b) and the Lorentz transformations simply as Λ.

The union of all Poincaré transformations forms a real Lie group under the multiplication law

$$(\Lambda_2, b_2)\times(\Lambda_1, b_1)=(\Lambda_2\Lambda_1, b_2+\Lambda_2 b_1). \tag{1.1.5}$$

This is the 'Poincaré group', denoted below the symbol Π. Analogously, the union of all Lorentz transformations forms a real (semisimple) Lie group. This is called the (proper orthochroneous) 'Lorentz group'. We denote it by $SO(3,1)^{\uparrow}$.

Note that the set of all homogeneous transformations (1.1.1) constrained by the relation (1.1.3) forms a real Lie group denoted by $O(3,1)$ which consists of four disconnected pieces

$$O(3,1)=\{SO(3,1)^{\uparrow}, \Lambda_{\mathrm{P}}SO(3,1)^{\uparrow}, \Lambda_{\mathrm{T}}SO(3,1)^{\uparrow}, \Lambda_{\mathrm{PT}}SO(3,1)^{\uparrow}\},$$

where

$$\Lambda_{\mathrm{P}}=\begin{pmatrix}1&0&0&0\\0&-1&0&0\\0&0&-1&0\\0&0&0&-1\end{pmatrix}\qquad \Lambda_{\mathrm{T}}=\begin{pmatrix}-1&0&0&0\\0&1&0&0\\0&0&1&0\\0&0&0&1\end{pmatrix}\qquad \Lambda_{\mathrm{PT}}=\Lambda_{\mathrm{P}}\Lambda_{\mathrm{T}}$$

hence it includes the transformations of spatial reflection Λ_{P}, time reversal Λ_{T} and total reflection Λ_{PT}. The set of all homogeneous transformations

(1.1.1) constrained by the equations (1.1.3) and (1.1.4*b*) also forms a Lie group, $SO(3,1)$, which consists of two disconnected pieces

$$SO(3,1)=\{SO(3,1)^{\uparrow},\ \Lambda_{PT}SO(3,1)^{\uparrow}\}.$$

1.1.2. Useful decomposition in $SO(3,1)^{\uparrow}$

Now we shall give a deeper insight into the structure of the Lorentz group. To begin with, note that an arbitrary element $\Lambda\in SO(3,1)^{\uparrow}$ may be represented as

$$\Lambda=R\Lambda_x(\psi)\tilde{R} \tag{1.1.6}$$

where R and $\tilde{R}$ are space rotations:

$$R=\begin{pmatrix}1 & 0 & 0 & 0\\ 0 & R^1{}_1 & R^1{}_2 & R^1{}_3\\ 0 & R^2{}_1 & R^2{}_2 & R^2{}_3\\ 0 & R^3{}_1 & R^3{}_2 & R^3{}_3\end{pmatrix}\qquad R^{\mathrm{T}}R=\mathbb{1}\qquad \det R=1 \tag{1.1.7}$$

and Λ_x is a standard Lorentz boost in the x^0, x^1 plane

$$\Lambda_x(\psi)=\begin{pmatrix}\cosh\psi & \sinh\psi & 0 & 0\\ \sinh\psi & \cosh\psi & 0 & 0\\ 0 & 0 & 1 & 0\\ 0 & 0 & 0 & 1\end{pmatrix}. \tag{1.1.8}$$

There is a strict mathematical proof of equation (1.1.6), but it can be most easily seen by considering the following physical argument. The transformation (1.1.8) corresponds to the situation where the x^2- and x^3-directions of two inertial systems K and K' coincide, and the system K' moves along the x^1-axis of the system K. If $\Lambda\neq\Lambda_x(\psi)$, we can rotate the spatial axes of systems K and K' at $t=0$ so as to obtain just such a situation. This gives equation (1.1.6).

Let us also recall a well-known fact about the rotation group $SO(3)$. Namely that an arbitrary element $R\in SO(3)$ may be represented as a product of rotations around the coordinate axes:

$$R=R_x(\varphi_1)R_y(\varphi_2)R_z(\varphi_3)$$

$$R_x(\varphi)=\begin{pmatrix}1 & 0 & 0 & 0\\ 0 & 1 & 0 & 0\\ 0 & 0 & \cos\varphi & \sin\varphi\\ 0 & 0 & -\sin\varphi & \cos\varphi\end{pmatrix}\qquad R_y(\varphi)=\begin{pmatrix}1 & 0 & 0 & 0\\ 0 & \cos\varphi & 0 & -\sin\varphi\\ 0 & 0 & 1 & 0\\ 0 & \sin\varphi & 0 & \cos\varphi\end{pmatrix}$$

$$R_z(\varphi)=\begin{pmatrix}1 & 0 & 0 & 0\\ 0 & \cos\varphi & \sin\varphi & 0\\ 0 & -\sin\varphi & \cos\varphi & 0\\ 0 & 0 & 0 & 1\end{pmatrix}. \tag{1.1.9}$$

1.1.3. Universal covering group of the Lorentz group

From equations (1.1.6, 8, 9) it follows that the Lorentz group is connected. But $SO(3,1)^{\uparrow}$ is not simply connected. Let us recall that a connected Lie group G is termed simply connected if any closed path in G can be shrunk down to a single point. The main property of simply connected groups is a one-to-one correspondence between representations of the group and the corresponding Lie algebra. Namely, any representation of the Lie algebra $\mathscr{G}$ of a simply connected Lie group G is the differential of some representation of G. But this is not true for non-simply connected Lie groups.

However, to any connected Lie group G one can relate a (unique up to isomorphism) 'universal covering group' $\tilde{G}$ with the following properties:

1. $\tilde{G}$ is simply connected.
2. There exists an analytic homomorphism $\rho: \tilde{G}\to G$ such that $G\simeq\tilde{G}/\mathrm{Ker}\,\rho$, where $\mathrm{Ker}\,\rho$ is a discrete subgroup of the centre of $\tilde{G}$. Since the homomorphism ρ is locally one-to-one, the group G and its universal covering group $\tilde{G}$ have isomorphic Lie algebras.

In quantum field theory, one needs to know representations of the Lie algebra $so(3,1)$ associated with $SO(3,1)^{\uparrow}$ rather than representations of $SO(3,1)^{\uparrow}$ itself. As is seen from the discussion above, to construct representations of the Lorentz algebra $so(3,1)$, it is sufficient to find a universal covering group for $SO(3,1)^{\uparrow}$, denoted by $Spin(3,1)$, and to determine its representations.

The main property of $Spin(3,1)$ is given by the following theorem.

Theorem. $Spin(3,1)\cong SL(2,\mathbb{C})$, where $SL(2,\mathbb{C})$ is the Lie group of 2×2 complex unimodular matrices.

Proof. Introduce in the linear space of complex 2×2 matrices the basis σ_m, $m=0,1,2,3$,

$$\sigma_0=\begin{pmatrix}1 & 0\\ 0 & 1\end{pmatrix}\qquad \sigma_1=\begin{pmatrix}0 & 1\\ 1 & 0\end{pmatrix}\qquad \sigma_2=\begin{pmatrix}0 & -\mathrm{i}\\ \mathrm{i} & 0\end{pmatrix}\qquad \sigma_3=\begin{pmatrix}1 & 0\\ 0 & -1\end{pmatrix}$$

where $\vec{\sigma}$ are the Pauli matrices. It is convenient to define the one-to-one map of the Minkowski space on the set of 2×2 Hermitian matrices

$$x^m\to\mathbf{x}=x^m\sigma_m=\begin{pmatrix}x^0+x^3 & x^1-\mathrm{i}x^2\\ x^1+\mathrm{i}x^2 & x^0-x^3\end{pmatrix} \tag{1.1.10}$$

$$\mathbf{x}^{+}=\mathbf{x}\qquad \det\mathbf{x}=-\eta_{mn}x^m x^n.$$

Here η_{mn} is the Minkowski metric.

Let us now consider a transformation of the form

$$\mathbf{x} \rightarrow \mathbf{x}' = x'^m \sigma_m = N\mathbf{x}N^+ \qquad N \in SL(2, \mathbb{C}). \tag{1.1.11}$$

Since det $N = 1$, this transformation preserves the interval $\eta_{mn}x^m x^n = \eta_{mn}x'^m x'^n$. Thus $x'^m = (\Lambda(N))^m{}_n x^n$, where $\Lambda(N) \in O(3, 1)$.

From relation (1.1.11) we see that the map

$$\pi: SL(2, \mathbb{C}) \rightarrow O(3, 1)$$

defined by the rule

$$N \rightarrow \Lambda(N)$$

is an analytic homomorphism. In fact, we shall show that $\pi(SL(2, \mathbb{C})) = SO(3, 1)^{\uparrow}$.

First, note that

$$\text{Ker}\,\pi = \pm\begin{pmatrix} 1 & 0 \\ 0 & 1 \end{pmatrix}. \tag{1.1.12}$$

Indeed, $N \in \text{Ker}\,\pi$ if and only if $N\mathbf{x}N^+ = \mathbf{x}$ for any 2×2 matrix $\mathbf{x}$. In particular, the choice $\mathbf{x} = \mathbb{1}$ gives $NN^+ = \mathbb{1}$, hence $N^+ = N^{-1}$. So, our condition takes the form $N\mathbf{x}N^{-1} = \mathbf{x}$ for any $\mathbf{x}$. This is possible if and only if $N \sim \mathbb{1}$.

As the next step, we reconstruct elements of $SL(2, \mathbb{C})$, which are mapped into rotations $R_x(\varphi)$, $R_y(\varphi)$, $R_z(\varphi)$ defined in expressions (1.1.9). It is a simple exercise to check that

$$\begin{aligned}
\pi^{-1}(R_x(\varphi)) &= \pm\exp\left[\mathrm{i}\frac{\varphi}{2}\sigma_1\right] \equiv \pm N_1(\varphi) \\
\pi^{-1}(R_y(\varphi)) &= \pm\exp\left[\mathrm{i}\frac{\varphi}{2}\sigma_2\right] \equiv \pm N_2(\varphi) \\
\pi^{-1}(R_z(\varphi)) &= \pm\exp\left[\mathrm{i}\frac{\varphi}{2}\sigma_3\right] \equiv \pm N_3(\varphi).
\end{aligned} \tag{1.1.13}$$

Analogously, the Lorentz boost $\Lambda_x(\psi)$ defined in equation (1.1.8) and the Lorentz boosts $\Lambda_y(\psi)$, $\Lambda_z(\psi)$ (i.e. Lorentz transformations in the planes x^0, x^2 and x^0, x^3, correspondingly) are generated by

$$\begin{aligned}
\pi^{-1}(\Lambda_x(\psi)) &= \pm\exp\left[\frac{\psi}{2}\sigma_1\right] \equiv \pm M_1(\psi) \\
\pi^{-1}(\Lambda_y(\psi)) &= \pm\exp\left[\frac{\psi}{2}\sigma_2\right] \equiv \pm M_2(\psi) \\
\pi^{-1}(\Lambda_z(\psi)) &= \pm\exp\left[\frac{\psi}{2}\sigma_3\right] \equiv \pm M_3(\psi).
\end{aligned} \tag{1.1.14}$$

The expressions (1.1.13, 14) show that the one-parameter subgroups $N_i(\varphi)$, $M_i(\psi)$, $i=1, 2, 3$, in $SL(2, \mathbb{C})$ are mapped into $SO(3, 1)^{\uparrow}$. But these subgroups generate $SL(2, \mathbb{C})$. Hence, all elements of $SL(2, \mathbb{C})$ are mapped into $SO(3, 1)^{\uparrow}$, so $\pi(SL(2, \mathbb{C})) \subset SO(3, 1)^{\uparrow}$. On the other hand, the identities (1.1.6, 9) mean that the rotations $R_x(\varphi)$, $R_y(\varphi)$, $R_z(\varphi)$ and the boost $\Lambda_x(\psi)$ generate the Lorentz group. Then equations (1.1.13, 14) tell us that $\pi(SL(2, \mathbb{C})) = SO(3, 1)^{\uparrow}$. From this and equation (1.1.12), we see that

$$SO(3, 1)^{\uparrow} \cong SL(2, \mathbb{C})/\mathbb{Z}_2$$

$$\mathbb{Z}_2 = \left\{ \pm \begin{pmatrix} 1 & 0 \\ 0 & 1 \end{pmatrix} \right\}. \tag{1.1.15}$$

So, $SL(2, \mathbb{C})$ is a double-covering group of $SO(3, 1)^{\uparrow}$.

Finally, we briefly prove that $SL(2, \mathbb{C})$ is simply connected. Every element $N \in SL(2, \mathbb{C})$ can be represented uniquely in the form

$$N = g\mathbf{z} \tag{1.1.16}$$

where g is a unimodular unitary matrix and $\mathbf{z}$ is a unimodular Hermitian matrix with positive trace:

$$g \in SU(2) \quad \mathbf{z}^{+} = \mathbf{z} \quad \det \mathbf{z} = 1 \quad \operatorname{Tr} \mathbf{z} > 0. \tag{1.1.17}$$

The group $SU(2)$ is simply connected. Indeed, any $g \in SU(2)$ can be written as

$$g = \begin{pmatrix} p & q \\ -q^* & p^* \end{pmatrix} \qquad |p|^2 + |q|^2 = 1.$$

Thus topologically, $SU(2)$ is a three-sphere S^3 ($(u^1)^2 + (u^2)^2 + (u^3)^2 + (u^4)^2 = 1$, where $p = u^1 + \mathrm{i}u^2$, $q = u^3 + \mathrm{i}u^4$), which is a simply connected manifold.

Now consider the manifold of Hermitian 2×2 matrices $\mathbf{z}$ constrained by (1.1.17). If we parameterize $\mathbf{z}$ as $\mathbf{z} = z^m \sigma_m$, $(z^m)^* = z^m$, then the constraints (1.1.17) imply that

$$(z^0)^2 - (z^1)^2 - (z^2)^2 - (z^3)^2 = 1 \qquad z^0 > 0.$$

This manifold is evidently simply connected.

So, we may represent $SL(2, \mathbb{C})$ as a product of two simply connected manifolds, and hence it is also simply connected. Due to relation (1.1.15), $SL(2, \mathbb{C})$ is the universal covering group of the Lorenz group. This completes the proof of the theorem.

1.1.4. Universal covering group of the Poincaré group

Our goal now is to construct a universal covering group of the Poincaré group. For this purpose, we return once again to the space of 2×2 Hermitian matrices (1.1.10) and consider a new class of linear transformations over it by adding a non-homogeneous term in the right-hand side of equation (1.1.11):

$$\mathbf{x} \to \mathbf{x}' = x'^m \sigma_m = N \mathbf{x} N^+ + \mathbf{b}$$
$$N \in SL(2, \mathbb{C}) \qquad \mathbf{b} = \mathbf{b}^+ = b^m \sigma_m. \tag{1.1.18}$$

Such a transformation associated with a pair $(N, \mathbf{b})$ looks in components like $x'^m = (\Lambda(N))^m{}_n x^n + b^m$, i.e. it coincides with the Poincaré transformation $(\Lambda(N), b)$.

The set $\tilde{\Pi}$ of all pairs $(N, \mathbf{b})$, N and $\mathbf{b}$ being as in relations (1.1.18), forms a ten-dimensional real Lie group with respect to the multiplication law

$$(N_2, \mathbf{b}_2) \times (N_1, \mathbf{b}_1) = (N_2 N_1, N_2 \mathbf{b}_1 N_2^+ + \mathbf{b}_2). \tag{1.1.19}$$

Evidently, this group is simply connected. The above correspondence (1.1.18) constitutes the covering mapping

$$\varphi\colon \tilde{\Pi} \to \Pi$$
$$\varphi((N, \mathbf{b})) = (\Lambda(N), b) \tag{1.1.20}$$

of the simply connected group $\tilde{\Pi}$ on to the Poincaré group. Since the correspondence $N \to \Lambda(N)$ is a group homomorphism, and by virtue of equations (1.1.5) and (1.1.19), the mapping (1.1.20) is an analytic holomorphism of $\tilde{\Pi}$ on to the Poincaré group. Its kernel consists of two elements, $\mathrm{Ker}\,\varphi = \{(\pm \mathbb{1}, 0)\}$. The above arguments show that $\tilde{\Pi}$ is the universal covering group of the Poincaré group.

1.2. Finite-dimensional representations of *Spin*(3, 1)

1.2.1. Connection between representations of $SO(3,1)^\uparrow$ and $SL(2,\mathbb{C})$

A linear represenation T of a Lie group G in an n-dimensional vector space $\mathbf{V}_T$ is defined as a homomorphism of G into the Lie group of non-singular linear transformations acting on this vector space,

$$T\colon g \to T(g) \qquad g \in G$$
$$T(g_1)T(g_2) = T(g_1 g_2) \qquad g_1, g_2 \in G.$$

Let $T\colon \Lambda \to T(\Lambda)$ be a representation of the Lorentz group $SO(3,1)^\uparrow$. Then we automatically obtain a representation $\tilde{T}$ of its universal covering group $SL(2, \mathbb{C})$ by the rule

$$\tilde{T}\colon N \to T(\pi(N)) \tag{1.2.1}$$

where π is the covering mapping constructed in subsection 1.1.3. For example, the vector representation

$$T_v\colon \Lambda \to \Lambda$$
$$V^m \to V'^m = \Lambda^m{}_n V^n \tag{1.2.2}$$

or the covector representation

$$T_{cv}\colon\ \Lambda \to (\Lambda^{T})^{-1}$$
$$V_m \to V'_m = \Lambda_m{}^n V_n \qquad (1.2.3)$$
$$\Lambda_m{}^n = \eta_{mk}\Lambda^k{}_l\eta^{ln} \qquad \eta_{mk}\eta^{kn} = \delta_m{}^n$$

of the Lorentz group immediately generate representations of $SL(2, \mathbb{C})$. In equations (1.2.2, 3), V^m and V_m are the components of some Lorentz vector and covector correspondingly.

Any representation $\tilde{T}$ of $SL(2, \mathbb{C})$, associated with a representation T of $SO(3, 1)^{\uparrow}$ according to expression (1.2.1), satisfies the property

$$\tilde{T}(N) = \tilde{T}(-N) \qquad N \in SL(2, \mathbb{C}). \qquad (1.2.4)$$

But there exist representations of $SL(2, \mathbb{C})$ for which this property is not true. As we shall see, one can construct irreducible representations (irreps) of $SL(2, \mathbb{C})$ for which

$$\tilde{T}(N) = -\tilde{T}(-N). \qquad (1.2.5)$$

Any such $SL(2, \mathbb{C})$ irrep is not a representation of $SO(3, 1)^{\uparrow}$. However, it may be treated as a double-valued representation of the Lorentz group.

1.2.2. Construction of SL(2, C) irreducible representations

In what follows, we denote $SL(2, \mathbb{C})$ indices by small Greek letters. In particular, components of a matrix $N \in SL(2, \mathbb{C})$ are $N_\alpha{}^\beta$, $\alpha, \beta = 1, 2$. Components of the complex conjugate matrix N^* are denoted by dotted indices, $N^*{}_{\dot{\alpha}}{}^{\dot{\beta}}$.

Define the fundamental representation of $SL(2, \mathbb{C})$

$$T_s\colon\ N \to N$$
$$\psi_\alpha \to \psi'_\alpha = N_\alpha{}^\beta \psi_\beta. \qquad (1.2.6)$$

An object ψ_α transforming according to this representation is called a 'two-component left-handed Weyl spinor'. The representation T_s is called the (left-handed) Weyl spinor representation of the Lorentz group. It is denoted by symbol $(\frac{1}{2}, 0)$.

Taking an n-fold tensor product of T_s, $n = 2, 3, \ldots$, $T_s \otimes T_s \otimes \ldots \otimes T_s$, we obtain new representations of $SL(2, \mathbb{C})$ of the form

$$\psi_{\alpha_1\alpha_2\ldots\alpha_n} \to \psi'_{\alpha_1\alpha_2\ldots\alpha_n} = N_{\alpha_1}{}^{\beta_1} N_{\alpha_2}{}^{\beta_2} \ldots N_{\alpha_n}{}^{\beta_n} \psi_{\beta_1\beta_2\ldots\beta_n}. \qquad (1.2.7)$$

The representation contragradient to the representation T_s (1.2.6) is given by

$$T_{cs}\colon\ N \to (N^{T})^{-1}$$
$$\psi^\alpha \to \psi'^\alpha = \psi^\beta (N^{-1})_\beta{}^\alpha. \qquad (1.2.8)$$

This representation is equivalent to T_s. Indeed, the condition of unimodularity for $N \in SL(2, \mathbb{C})$ can be written in the form

$$\varepsilon_{\alpha\beta} = N_\alpha{}^\gamma N_\beta{}^\delta \varepsilon_{\gamma\delta}$$

or (1.2.9)

$$\varepsilon^{\alpha\beta} = \varepsilon^{\gamma\delta}(N^{-1})_\gamma{}^\alpha (N^{-1})_\delta{}^\beta$$

where $\varepsilon_{\alpha\beta}$ and $\varepsilon^{\alpha\beta}$ are antisymmetric tensors defined by

$$\begin{aligned} &\varepsilon_{\alpha\beta} = -\varepsilon_{\beta\alpha} \qquad \varepsilon_{12} = -1 \\ &\varepsilon^{\alpha\beta} = -\varepsilon^{\beta\alpha} \qquad \varepsilon^{12} = 1 \\ &\varepsilon^{\alpha\beta}\varepsilon_{\beta\gamma} = \delta^\alpha{}_\gamma . \end{aligned} \tag{1.2.10}$$

Equation (1.2.9) means that $(N^{-1})_\beta{}^\alpha = \varepsilon^{\alpha\gamma} N_\gamma{}^\delta \varepsilon_{\delta\beta}$, and hence the representations T_s and T_{cs} are equivalent.

The identities (1.2.9) imply that $\varepsilon_{\alpha\beta}$ and $\varepsilon^{\alpha\beta}$ are invariant tensors of the Lorentz group. Hence we can use them for lowering or raising spinor indices, which will be done in this text according to the rules

$$\psi^\alpha = \varepsilon^{\alpha\beta}\psi_\beta \qquad \psi_\alpha = \varepsilon_{\alpha\beta}\psi^\beta . \tag{1.2.11}$$

Now consider the complex conjugate representation of the representation T_s:

$$\begin{aligned} &T_{\bar{s}}\colon\ N \to N^* \\ &\psi_{\dot{\alpha}} \to \psi'_{\dot{\alpha}} = N^*{}_{\dot{\alpha}}{}^{\dot{\beta}}\psi_{\dot{\beta}} . \end{aligned} \tag{1.2.12}$$

An object $\psi_{\dot{\alpha}}$ transforming according to this representation is said to be a 'two-component right-handed Weyl spinor'. The representation $T_{\bar{s}}$ is called the (right-handed) Weyl spinor representation of the Lorentz group. It is denoted by the symbol $(0, \frac{1}{2})$.

Taking the tensor product of $T_{\bar{s}}$ with itself m times, $T_{\bar{s}} \otimes \ldots \otimes T_{\bar{s}}$, $m = 2, 3, \ldots$, one obtains new representations of $SL(2, \mathbb{C})$ of the form

$$\psi_{\dot{\alpha}_1\dot{\alpha}_2\ldots\dot{\alpha}_m} \to \psi'_{\dot{\alpha}_1\dot{\alpha}_2\ldots\dot{\alpha}_m} = N^*{}_{\dot{\alpha}_1}{}^{\dot{\beta}_1} N^*{}_{\dot{\alpha}_2}{}^{\dot{\beta}_2} \ldots N^*{}_{\dot{\alpha}_m}{}^{\dot{\beta}_m} \psi_{\dot{\beta}_1\dot{\beta}_2\ldots\dot{\beta}_m} . \tag{1.2.13}$$

We can also consider the more general situation

$$\underbrace{T_s \otimes \ldots \otimes T_s}_{n} \otimes \underbrace{T_{\bar{s}} \otimes \ldots \otimes T_{\bar{s}}}_{m}$$

obtaining, as a result, Lorentz spin-tensors with dotted and undotted spinor indices

$$\begin{aligned} &\psi_{\alpha_1\alpha_2\ldots\alpha_n\dot{\beta}_1\dot{\beta}_2\ldots\dot{\beta}_m} \to \psi'_{\alpha_1\alpha_2\ldots\alpha_n\dot{\beta}_1\dot{\beta}_2\ldots\dot{\beta}_m} \\ &\quad = N_{\alpha_1}{}^{\gamma_1} N_{\alpha_2}{}^{\gamma_2} \ldots N_{\alpha_n}{}^{\gamma_n} N^*{}_{\dot{\beta}_1}{}^{\dot{\delta}_1} N^*{}_{\dot{\beta}_2}{}^{\dot{\delta}_2} \ldots N^*{}_{\dot{\beta}_m}{}^{\dot{\delta}_m} \psi_{\gamma_1\gamma_2\ldots\gamma_n\dot{\delta}_1\dot{\delta}_2\ldots\dot{\delta}_m} . \end{aligned} \tag{1.2.14}$$

The representation contragradient to the representation $T_{\bar{s}}$ (1.2.12) is

$$T_{c\bar{s}}:\ N \to (N^{+})^{-1}$$
$$\psi^{\dot{\alpha}} \to \psi'^{\dot{\alpha}} = \psi^{\dot{\beta}}(N^{-1})^{*}{}_{\dot{\beta}}{}^{\dot{\alpha}} \tag{1.2.15}$$

This representation is equivalent to $T_{\bar{s}}$ since antisymmetric tensors $\varepsilon_{\dot{\alpha}\dot{\beta}}$ and $\varepsilon^{\dot{\alpha}\dot{\beta}}$, where

$$\varepsilon_{\dot{\alpha}\dot{\beta}} = -\varepsilon_{\dot{\beta}\dot{\alpha}} \qquad \varepsilon_{\dot{1}\dot{2}} = -1$$
$$\varepsilon^{\dot{\alpha}\dot{\beta}} = -\varepsilon^{\dot{\beta}\dot{\alpha}} \qquad \varepsilon^{\dot{1}\dot{2}} = 1 \tag{1.2.16}$$

are invariant Lorentz tensors and can be used for lowering or raising dotted spinor indices by the rules

$$\psi^{\dot{\alpha}} = \varepsilon^{\dot{\alpha}\dot{\beta}}\psi_{\dot{\beta}} \qquad \psi_{\dot{\alpha}} = \varepsilon_{\dot{\alpha}\dot{\beta}}\psi^{\dot{\beta}} \tag{1.2.17}$$

Representations of the form (1.2.14) with unconstrained tensors $\psi_{\alpha_1\alpha_2\ldots\alpha_n\dot{\beta}_1\dot{\beta}_2\ldots\dot{\beta}_m}$ are reducible when $n>1$ or $m>1$. For example, an arbitrary second-rank tensor with undotted indices $\psi_{\alpha\beta}$ may be decomposed in a Lorentz invariant way as

$$\psi_{\alpha\beta} = \frac{1}{2}(\psi_{\alpha\beta} + \psi_{\beta\alpha}) + \frac{1}{2}(\psi_{\alpha\beta} - \psi_{\beta\alpha}) \equiv \psi_{(\alpha\beta)} - \frac{1}{2}\varepsilon_{\alpha\beta}(\varepsilon^{\gamma\delta}\psi_{\gamma\delta})$$

where $(\alpha\,\beta\ldots)$ denotes symmetrization in indices α, β, ... In general, a tensor of the form (1.2.14) turns out to be irreducible if $\psi_{\alpha_1\alpha_2\cdots\alpha_n\dot{\beta}_1\dot{\beta}_2\ldots\dot{\beta}_m}$ is totally symmetric in its undotted indices and independently in its dotted indices. Thus irreducible representations are realized on tensors

$$\psi_{\alpha_1\alpha_2\ldots\alpha_n\dot{\beta}_1\dot{\beta}_2\ldots\dot{\beta}_m} = \psi_{(\alpha_1\alpha_2\ldots\alpha_n)(\dot{\beta}_1\dot{\beta}_2\ldots\dot{\beta}_m)}. \tag{1.2.18}$$

The corresponding irrep is denoted as $(n/2, m/2)$. Its dimension is $(n+1)(m+1)$. Note that $(n/2, m/2)$ is a single-valued representation of the Lorentz group if $(n+m)$ is even, otherwise, it is double-valued.

Let $\psi_{\alpha_1\alpha_2\ldots\alpha_n\dot{\beta}_1\dot{\beta}_2\ldots\dot{\beta}_m}$ be a $(n/2, m/2)$-type tensor. Taking the complex conjugate of equation (1.2.14), we find that $(\psi_{\alpha_1\alpha_2\ldots\alpha_n\dot{\beta}_1\dot{\beta}_2\ldots\dot{\beta}_m})^*$ transforms as a $(m/2, n/2)$-type tensor. Therefore, the following mapping

$$*:\ \mathbf{V}_{(n/2,m/2)} \to \mathbf{V}_{(m/2,n/2)}$$
$$\psi_{\alpha_1\alpha_2\ldots\alpha_n\dot{\beta}_1\dot{\beta}_2\ldots\dot{\beta}_m} \to \bar{\psi}_{\beta_1\beta_2\ldots\beta_m\dot{\alpha}_1\dot{\alpha}_2\ldots\dot{\alpha}_n} \equiv (\psi_{\alpha_1\alpha_2\ldots\alpha_n\dot{\beta}_1\dot{\beta}_2\ldots\dot{\beta}_m})^* \tag{1.2.19}$$

is defined and $\bar{\psi}$ is said to be the complex conjugate spin-tensor of ψ. Evidently, its square coincides with the identity operator. If $n \neq m$, the $(n/2, m/2)$-representation is complex. But with every $(n/2, m/2)$ representation, $n \neq m$, one may associate a real representation of the Lorentz group in the following way. Taking the direct sum representation $(n/2, m/2) \oplus (n/2, m/2)$ we consider within the space of this representation $\mathbf{V}_{(n/2,m/2)} \oplus \mathbf{V}_{(m/2,n/2)}$ (which is mapped on to itself by *) a Lorentz invariant subspace denoted by $\mathbf{V}^R_{(n/2,m/2)}$ and selected by the condition that * coincides with the identity operator on this subspace. Arbitrary pairs

$$(\psi_{\alpha_1\ldots\alpha_n\dot\beta_1\ldots\dot\beta_m}, \bar\psi_{\alpha_1\ldots\alpha_m\dot\beta_1\ldots\dot\beta_n})$$

$$\psi_{\alpha_1\ldots\alpha_n\dot\beta_1\ldots\dot\beta_m}=\psi_{(\alpha_1\ldots\alpha_n)(\dot\beta_1\ldots\dot\beta_m)}$$

span $\mathbf{V}^R_{(n/2,m/2)}$.

In the case $n=m$, we can define real tensors. By definition, they satisfy the equation

$$\psi_{\alpha_1\alpha_2\ldots\alpha_n\dot\beta_1\dot\beta_2\ldots\dot\beta_m}=\bar\psi_{\alpha_1\alpha_2\ldots\alpha_n\dot\beta_1\dot\beta_2\ldots\dot\beta_m}. \tag{1.2.20}$$

1.2.3. Invariant Lorentz tensors

Invariant tensors of the Lorentz group are useful for lowering, raising or covariant contraction of indices. Up until now, we have found the following invariant tensors: the Minkowski metric η_{mn} and its inverse η^{mn}, the spinor metrics $\varepsilon_{\alpha\beta}$, $\varepsilon_{\dot\alpha\dot\beta}$ and their inverse $\varepsilon^{\alpha\beta}$, $\varepsilon^{\dot\alpha\dot\beta}$, and the Levi–Civita totally antisymmetric tensor ε_{abcd} ($\varepsilon_{0123}=-1$). Now we find one more invariant tensor carrying Lorentz as well as spinor indices.

Let us rewrite the relation (1.1.11) in the form

$$(\Lambda(N))^m{}_n x^n\sigma_m=x^nN\sigma_nN^+$$

or

$$\sigma_m=N\sigma_nN^+(\Lambda(N)^{-1})^n{}_m. \tag{1.2.21}$$

Here $\Lambda(N)$ is the Lorentz transformation corresponding to an element $N\in SL(2,\mathbb{C})$. The identify (1.2.21) shows that, denoting components of σ_m as

$$(\sigma_m)_{\alpha\dot\alpha} \tag{1.2.22}$$

we obtain a Lorentz invariant tensor with one space–time index m, one undotted spinor index α and one dotted spinor index $\dot\alpha$. We may also introduce σ-matrices with upper spinor indices

$$(\tilde\sigma_m)^{\dot\alpha\alpha}\equiv\varepsilon^{\dot\alpha\dot\beta}\varepsilon^{\alpha\beta}(\sigma_m)_{\beta\dot\beta}$$

$$\tilde\sigma_m=(+\mathbb{1},-\vec\sigma). \tag{1.2.23}$$

One can check that the σ-matrices satisfy the useful identities

$$(\sigma_a\tilde\sigma_b+\sigma_b\tilde\sigma_a)_\alpha{}^\beta=-2\eta_{ab}\delta_\alpha{}^\beta \tag{1.2.24}$$

$$(\tilde\sigma_a\sigma_b+\tilde\sigma_b\sigma_a)^{\dot\alpha}{}_{\dot\beta}=-2\eta_{ab}\delta^{\dot\alpha}{}_{\dot\beta} \tag{1.2.25}$$

$$\mathrm{Tr}(\sigma_a\tilde\sigma_b)=-2\eta_{ab} \tag{1.2.26}$$

$$(\sigma^a)_{\alpha\dot\alpha}(\tilde\sigma_a)^{\dot\beta\beta}=-2\delta_\alpha^\beta\delta_{\dot\alpha}^{\dot\beta} \tag{1.2.27}$$

$$\sigma_a \tilde{\sigma}_b \sigma_c = (\eta_{ac}\sigma_b - \eta_{bc}\sigma_a - \eta_{ab}\sigma_c) + i\varepsilon_{abcd}\sigma^d \tag{1.2.28}$$

$$\tilde{\sigma}_a \sigma_b \tilde{\sigma}_c = (\eta_{ac}\tilde{\sigma}_b - \eta_{bc}\tilde{\sigma}_a - \eta_{ab}\tilde{\sigma}_c) - i\varepsilon_{abcd}\tilde{\sigma}^d. \tag{1.2.29}$$

It is seen from these relations that the Minkowski metric and the Levi-Civita tensor are expressible in terms of the σ-matrices.

The σ-matrices are usually used to convert space-time indices into spinor ones and vice versa according to the general rule: one vector index is equivalent to a pair of spinor indices, dotted and undotted,

$$V_{\alpha\dot{\alpha}} = (\sigma^a)_{\alpha\dot{\alpha}} V_a \qquad V_a = -\frac{1}{2}(\tilde{\sigma}_a)^{\dot{\alpha}\alpha} V_{\alpha\dot{\alpha}}. \tag{1.2.30}$$

In many cases, the conversion of vector indices into spinor ones leads to some technical advantages. Moreover, conditions of irreducibility are much simpler when working with two-component objects.

For example, let us consider a second-rank tensor X_{ab}. Convert the space-time indices a, b into spinor ones according to relation (1.2.30):

$$X_{ab} \to X_{\alpha\beta\dot{\alpha}\dot{\beta}} = (\sigma^a)_{\alpha\dot{\alpha}}(\sigma^b)_{\beta\dot{\beta}} X_{ab}.$$

The spin-tensor $X_{\alpha\beta\dot{\alpha}\dot{\beta}}$ can be decomposed into irreducible components as follows

$$\begin{aligned} X_{\alpha\beta\dot{\alpha}\dot{\beta}} &= X_{[\alpha\beta](\dot{\alpha}\dot{\beta})} + X_{(\alpha\beta)[\dot{\alpha}\dot{\beta}]} + X_{[\alpha\beta][\dot{\alpha}\dot{\beta}]} + X_{(\alpha\beta)(\dot{\alpha}\dot{\beta})} \\ &= \varepsilon_{\alpha\beta} X_{(\dot{\alpha}\dot{\beta})} + \varepsilon_{\dot{\alpha}\dot{\beta}} X_{(\alpha\beta)} + \varepsilon_{\alpha\beta}\varepsilon_{\dot{\alpha}\dot{\beta}} X + X_{(\alpha\beta)(\dot{\alpha}\dot{\beta})} \end{aligned} \tag{1.2.31}$$

where $[\alpha\beta]$ denotes antisymmetrization in indices α, β. The irreducible components of $X_{\alpha\beta\dot{\alpha}\dot{\beta}}$ are $X_{(\alpha\beta)(\dot{\alpha}\dot{\beta})}$, $X = \frac{1}{4}\varepsilon^{\alpha\beta}\varepsilon^{\dot{\alpha}\dot{\beta}} X_{\alpha\beta\dot{\alpha}\dot{\beta}}$, $X_{(\alpha\beta)} = -\frac{1}{2}\varepsilon^{\dot{\alpha}\dot{\beta}} X_{(\alpha\beta)\dot{\alpha}\dot{\beta}}$, $X_{(\dot{\alpha}\dot{\beta})} = -\frac{1}{2}\varepsilon^{\alpha\beta} X_{\alpha\beta(\dot{\alpha}\dot{\beta})}$. Here we have used the identities $\varepsilon^{\alpha\beta}\varepsilon_{\alpha\beta} = \varepsilon^{\dot{\alpha}\dot{\beta}}\varepsilon_{\dot{\alpha}\dot{\beta}} = -2$. If X_{ab} is a symmetric tensor, $X_{ab} = X_{ba}$, then the first and second terms in equation (1.2.31) vanish. If, in addition, X_{ab} is traceless, $X^a_a = 0$, then the third term in equation (1.2.31) is also zero. In the case when $X_{ab} = -X_{ba}$, only the first and the second terms in equation (1.2.31) are non-zero. Hence, an arbitrary antisymmetric tensor X_{ab} is equivalent to a pair of symmetric bi-spinors. To write down explicitly this correspondence, introduce the matrices

$$\begin{aligned} (\sigma_{ab})_\alpha{}^\beta &= -\frac{1}{4}(\sigma_a\tilde{\sigma}_b - \sigma_b\tilde{\sigma}_a)_\alpha{}^\beta \\ (\tilde{\sigma}_{ab})^{\dot{\alpha}}{}_{\dot{\beta}} &= -\frac{1}{4}(\tilde{\sigma}_a\sigma_b - \tilde{\sigma}_b\sigma_a)^{\dot{\alpha}}{}_{\dot{\beta}} \\ (\sigma_{ab})_{\alpha\beta} &\equiv \varepsilon_{\beta\gamma}(\sigma_{ab})_\alpha{}^\gamma = (\sigma_{ab})_{\beta\alpha} \\ (\tilde{\sigma}_{ab})_{\dot{\alpha}\dot{\beta}} &\equiv \varepsilon_{\dot{\alpha}\dot{\gamma}}(\tilde{\sigma}_{ab})^{\dot{\gamma}}{}_{\dot{\beta}} = (\tilde{\sigma}_{ab})_{\dot{\beta}\dot{\alpha}}. \end{aligned} \tag{1.2.32}$$

Then we obtain

$$X_{ab} = (\sigma_{ab})_{\alpha\beta} X^{\alpha\beta} - (\tilde{\sigma}_{ab})_{\dot{\alpha}\dot{\beta}} X^{\dot{\alpha}\dot{\beta}}$$

$$X_{\alpha\dot{\alpha}\beta\dot{\beta}} \equiv (\sigma^a)_{\alpha\dot{\alpha}}(\sigma^b)_{\alpha\beta}X_{ab} = 2\varepsilon_{\alpha\beta}X_{\dot{\alpha}\dot{\beta}} + 2\varepsilon_{\dot{\alpha}\dot{\beta}}X_{\alpha\beta}$$

$$X_{\alpha\beta} = \frac{1}{2}(\sigma^{ab})_{\alpha\beta}X_{ab} \qquad X_{\dot{\alpha}\dot{\beta}} = -\frac{1}{2}(\tilde{\sigma}^{ab})_{\dot{\alpha}\dot{\beta}}X_{ab} \tag{1.2.33}$$

$$X_{ab} = -X_{ba} \qquad X_{\alpha\beta} = X_{\beta\alpha} \qquad X_{\dot{\alpha}\dot{\beta}} = X_{\dot{\beta}\dot{\alpha}}.$$

If X_{ab} is a real tensor, then $X_{\alpha\beta}$ and $X_{\dot{\alpha}\dot{\beta}}$ are complex conjugates of each other, $\bar{X}_{\dot{\alpha}\dot{\beta}} = X_{\dot{\alpha}\dot{\beta}}$.

Now consider another example. Let C_{abcd} be a tensor subjected to the constraints

$$C_{abcd} = -C_{bacd} = -C_{abdc} = C_{cdab} \tag{1.2.34a}$$

$$C^a{}_{bac} = 0 \tag{1.2.34b}$$

$$\varepsilon^{abcd}C_{fbcd} = 0. \tag{1.2.34c}$$

The corresponding spin-tensor is seen to be

$$C_{\alpha\beta\gamma\delta\dot{\alpha}\dot{\beta}\dot{\gamma}\dot{\delta}} = (\sigma^a)_{\alpha\dot{\alpha}}(\sigma^b)_{\beta\dot{\beta}}(\sigma^c)_{\gamma\dot{\gamma}}(\sigma^d)_{\delta\dot{\delta}}C_{abcd} = \varepsilon_{\dot{\alpha}\dot{\beta}}\varepsilon_{\dot{\gamma}\dot{\delta}}C_{\alpha\beta\gamma\delta} + \varepsilon_{\alpha\beta}\varepsilon_{\gamma\delta}C_{\dot{\alpha}\dot{\beta}\dot{\gamma}\dot{\delta}}$$

where tensors $C_{\alpha\beta\gamma\delta}$ and $C_{\dot{\alpha}\dot{\beta}\dot{\gamma}\dot{\delta}}$ are symmetric in all their indices. Note that equations (1.2.34) are the algebraic constraints on the Weyl tensor in general relativity.

Remark. In what follows, by representations of the Lorentz group we mean both the single- and double-valued representations, i.e. arbitrary $SL(2, \mathbb{C})$-representations. Analogously, by representations of the Poincaré group we mean its single- and double-valued representations (both being infinite-dimensional, in general), i.e. arbitrary $\tilde{\Pi}$-representations.

1.3. The Lorentz algebra

The Lorentz group and its universal covering group $SL(2, \mathbb{C})$ are locally isomorphic. So, the Lie algebra $so(3, 1)$ of the Lorentz group (the 'Lorentz algebra') and the Lie algebra $sl(2, \mathbb{C})$ of the Lie group $SL(2, \mathbb{C})$ are isomorphic.

Nearly all elements $N \in SL(2, \mathbb{C})$ can be expressed in the exponential form

$$N = \exp(\vec{z}\vec{\sigma}) \equiv \exp(z) \qquad z \in sl(2, \mathbb{C}) \tag{1.3.1}$$

where $\vec{z} = (z_1, z_2, z_3)$ is a complex three-vector, and $\vec{\sigma}$ are the Pauli matrices. The exponential from $sl(2, \mathbb{C})$ covers $SL(2, \mathbb{C})$ apart from only a (complex) two-dimensional surface in $SL(2, \mathbb{C})$ consisting of elements

$$\begin{pmatrix} -1-ab & a^2 \\ -b^2 & -1+ab \end{pmatrix} = g\begin{pmatrix} -1 & 1 \\ 0 & -1 \end{pmatrix}g^{-1} \qquad g \in SL(2, \mathbb{C})$$

where (a, b) is a non-vanishing complex two-vector. The union of all $SL(2, \mathbb{C})$

points, given in the form (1.3.1), is an everywhere dense set in $SL(2, \mathbb{C})$. On these grounds, we shall later write $SL(2, \mathbb{C})$ elements in the form (1.3.1) without comment. Note that if $N \in SL(2, \mathbb{C})$, then at least one of the matrices N and $(-N)$ admits the exponential form (1.3.1).

Considering $SL(2, \mathbb{C})$ as a complex Lie group, the parameters $\vec{z}$ from equation (1.3.1) play the role of local complex coordinates, and the Pauli matrices form a basis in the corresponding Lie algebra. Since we treat $SL(2, \mathbb{C})$ as the universal covering group of $SO(3, 1)^{\uparrow}$, a real Lie group, we must understand $SL(2, \mathbb{C})$ as a real six-dimensional Lie group. Introduce real local coordinates in $SL(2, \mathbb{C})$ using the following parametrization:

$$N = \exp\left(\frac{1}{2} K^{ab} \sigma_{ab}\right) \qquad N \in SL(2, \mathbb{C})$$

$$(K^{ab})^* = K^{ab} \qquad K^{ab} = -K^{ba}. \tag{1.3.2}$$

The matrices σ^{ab} were defined in equation (1.2.32). The complex ($\vec{z}$) and real (K^{ab}) coordinates in $SL(2, \mathbb{C})$ are related by the rule

$$z_1 = \frac{1}{2}(K^{01} + \mathrm{i}K^{23}) \qquad z_2 = \frac{1}{2}(K^{02} + \mathrm{i}K^{31}) \qquad z_3 = \frac{1}{2}(K^{03} + \mathrm{i}K^{12}).$$

By virtue of equation (1.1.11), the infinitesimal Lorentz transformation corresponding to an infinitesimal $SL(2, \mathbb{C})$ transformation

$$N = \mathbb{1} + \frac{1}{2} K^{ab} \sigma_{ab} \tag{1.3.3}$$

is given by the expression

$$x^m \to x'^m = x^m + K^m{}_n x^n . \tag{1.3.4}$$

When deriving equation (1.3.4), we have used the identities (1.2.28, 29).

Recall one important result from group theory. If $f: G \to H$ is a homomorphism of Lie groups, $\mathrm{d}f: \mathscr{G} \to \mathscr{H}$ is the corresponding homomorphism of Lie algebras, then $f(\exp X) = \exp \mathrm{d}f(X)$, where $X \in \mathscr{G}$ and $\exp X$ is the exponential from $\mathscr{G}$ into G.

In accordance with equation (1.3.2), the matrices σ_{ab} form a basis of the real Lie algebra of $SL(2, \mathbb{C})$. Let T be a representation of $SL(2, \mathbb{C})$. Then, using equation (1.3.2) and the above result, we deduce that

$$T(N) = \exp\left(\frac{1}{2} K^{ab} M_{ab}\right) \qquad N \in SL(2, \mathbb{C}). \tag{1.3.5}$$

Here the 'generators' $M_{ab} \equiv \mathrm{d}T(\sigma_{ab})$ define a representation of the Lorentz algebra. Note that $M_{ab} = -M_{ba}$.

The commutation relations for the generators can be easily obtained by noting that $\mathrm{d}T$ is a Lie algebra homomorphism, so that

$$[M_{ab}, M_{cd}] = [\mathrm{d}T(\sigma_{ab}), \mathrm{d}T(\sigma_{cd})] = \mathrm{d}T([\sigma_{ab}, \sigma_{cd}]).$$

Recalling the explicit form (1.2.32) for the σ-matrices, one finds

$$[\sigma_{ab}, \sigma_{cd}] = \eta_{ad}\sigma_{bc} - \eta_{ac}\sigma_{bd} + \eta_{bc}\sigma_{ad} - \eta_{bd}\sigma_{ac}. \tag{1.3.6}$$

This gives

$$[M_{ab}, M_{cd}] = \eta_{ad}M_{bc} - \eta_{ac}M_{bd} + \eta_{bc}M_{ad} - \eta_{bd}M_{ac}. \tag{1.3.7}$$

These commutation relations define the Lorentz algebra and its representations.

In the cases of the representations $(\frac{1}{2}, 0)$, $(0, \frac{1}{2})$ and $(\frac{1}{2}, \frac{1}{2})$, the Lorentz generators are

$${}^{s}M_{ab}(\psi_\alpha) = (\sigma_{ab})_\alpha{}^\beta \psi_\beta \tag{1.3.8a}$$

$${}^{\bar{s}}M_{ab}(\bar{\psi}^{\dot{\alpha}}) = (\tilde{\sigma}_{ab})^{\dot{\alpha}}{}_{\dot{\beta}} \bar{\psi}^{\dot{\beta}} \tag{1.3.8b}$$

$${}^{v}M_{ab}(V^c) = \delta^c_a V_b - \delta^c_b V_a. \tag{1.3.8c}$$

The matrices ${}^{v}M_{ab}$ form a basis of the Lie algebra $so(3, 1)$ of the Lorentz group $SO(3, 1)^{\uparrow}$. Recall that any element $\Lambda \in SO(3, 1)^{\uparrow}$ can be written as

$$\Lambda = \Lambda(N) = \Lambda(-N)$$

for some $N \in SL(2, \mathbb{C})$, with respect to the covering map $\pi: SL(2, \mathbb{C}) \to SO(3, 1)^{\uparrow}$. As was pointed out above, N or $(-N)$ or both can be represented in the exponential form (1.3.2), therefore we deduce that

$$\Lambda = \exp\left(\frac{1}{2} K^{ab}\, {}^{v}M_{ab}\right)$$

for every Lorentz matrix $\Lambda \in SO(3, 1)^{\uparrow}$.

Let us introduce a new basis for the Lorentz algebra, replacing the generators M_{ab} with vector indices by operators $M_{\alpha\beta}$ and $\bar{M}_{\dot{\alpha}\dot{\beta}}$ with spinor indices defined as follows

$$\begin{aligned} M_{\alpha\beta} &= \frac{1}{2}(\sigma^{ab})_{\alpha\beta} M_{ab} \\ \bar{M}_{\dot{\alpha}\dot{\beta}} &= -\frac{1}{2}(\tilde{\sigma}^{ab})_{\dot{\alpha}\dot{\beta}} M_{ab}. \end{aligned} \tag{1.3.9}$$

After this redefinition, equation (1.3.5) takes the form

$$\begin{aligned} T(N) &= \exp(K^{\alpha\beta} M_{\alpha\beta} + \bar{K}^{\dot{\alpha}\dot{\beta}} \bar{M}_{\dot{\alpha}\dot{\beta}}) \\ K_{\alpha\beta} &= K_{\beta\alpha} = \frac{1}{2}(\sigma^{ab})_{\alpha\beta} K_{ab}. \end{aligned} \tag{1.3.10}$$

Using the commutation relations (1.3.7), it is not difficult to obtain the commutation relations for generators $M_{\alpha\beta}$ and $\bar{M}_{\dot{\alpha}\dot{\beta}}$. These are:

$$[M_{\alpha\beta}, M_{\gamma\delta}] = \frac{1}{2}\{\varepsilon_{\alpha\gamma}M_{\beta\delta} + \varepsilon_{\alpha\delta}M_{\beta\gamma} + \varepsilon_{\beta\gamma}M_{\alpha\delta} + \varepsilon_{\beta\delta}M_{\alpha\gamma}\}$$

$$[M_{\alpha\beta}, \bar{M}_{\dot{\alpha}\dot{\beta}}] = 0 \tag{1.3.11}$$

$$[\bar{M}_{\dot{\alpha}\dot{\beta}}, \bar{M}_{\dot{\gamma}\dot{\delta}}] = \frac{1}{2}\{\varepsilon_{\dot{\alpha}\dot{\gamma}}\bar{M}_{\dot{\beta}\dot{\delta}} + \varepsilon_{\dot{\alpha}\dot{\delta}}\bar{M}_{\dot{\beta}\dot{\gamma}} + \varepsilon_{\dot{\beta}\dot{\gamma}}\bar{M}_{\dot{\alpha}\dot{\delta}} + \varepsilon_{\dot{\beta}\dot{\delta}}\bar{M}_{\dot{\alpha}\dot{\gamma}}\}.$$

It is now a simple exercise to check that the commutation relations of the operators $M_{\alpha\beta}$ (M_{11}, $M_{12} = M_{21}$ and M_{22}) coincides with those for standard generators of $sl(2, \mathbb{C})$. The same assertion is true for the algebra of generators $\bar{M}_{\dot{\alpha}\dot{\beta}}$. So, the Lorentz algebra is isomorphic to a direct sum of two (mutually conjugate) $sl(2, \mathbb{C})$ algebras.

By virtue of equations (1.3.8) and (1.3.9), the generators $M_{\alpha\beta}$ and $\bar{M}_{\dot{\alpha}\dot{\beta}}$ act on undotted and dotted spinors according to the rules

$$M_{\alpha\beta}(\psi_\gamma) = \frac{1}{2}(\varepsilon_{\gamma\alpha}\psi_\beta + \varepsilon_{\gamma\beta}\psi_\alpha)$$

$$M_{\alpha\beta}(\bar{\psi}_{\dot{\gamma}}) = 0 \qquad \bar{M}_{\dot{\alpha}\dot{\beta}}(\psi_\gamma) = 0 \tag{1.3.12}$$

$$\bar{M}_{\dot{\alpha}\dot{\beta}}(\bar{\psi}_{\dot{\gamma}}) = \frac{1}{2}(\varepsilon_{\dot{\gamma}\dot{\alpha}}\bar{\psi}_{\dot{\beta}} + \varepsilon_{\dot{\gamma}\dot{\beta}}\bar{\psi}_{\dot{\alpha}}).$$

These relations serve as the main motivation for the introduction of generators (1.3.9). We see that the undotted generators $M_{\alpha\beta}$ move undotted spinor indices only, while the dotted generators $\bar{M}_{\dot{\alpha}\dot{\beta}}$ give non-vanishing results only when acting on dotted spinor indices.

Let us consider operators

$$C_1 = M^{\alpha\beta}M_{\alpha\beta} \qquad C_2 = \bar{M}^{\dot{\alpha}\dot{\beta}}\bar{M}_{\dot{\alpha}\dot{\beta}}. \tag{1.3.13}$$

Using equation (1.3.11), we find $[C_1, M_{\alpha\beta}] = 0$ and $[C_2, \bar{M}_{\dot{\alpha}\dot{\beta}}] = 0$. On the other hand, C_1 trivially commutes with $\bar{M}_{\dot{\alpha}\dot{\beta}}$ and C_2 commutes with $M_{\alpha\beta}$. So, C_1 and C_2 are the Casimir operators of the Lorentz group. We are going to calculate values of these operators in the $(n/2, m/2)$ representations. It is sufficient to find C_1 in the case of some $(n/2, 0)$ representation. Given a totally symmetric tensor $\psi_{\alpha_1\alpha_2\ldots\alpha_n}$, from equation (1.3.12) we obtain

$$M_{\gamma\delta}\psi_{\alpha_1\alpha_2\ldots\alpha_n} = \frac{1}{2}\sum_{i=1}^{n}(\varepsilon_{\alpha_i\gamma}\psi_{\delta\alpha_1\ldots\hat{\alpha}_i\ldots\alpha_n} + \varepsilon_{\alpha_i\delta}\psi_{\gamma\alpha_1\ldots\hat{\alpha}_i\ldots\alpha_n})$$

where $\hat{\alpha}_i$ means that the corresponding index is omitted. Raising the indices α and β in the first relation (1.3.12) gives

$$M^{\alpha\beta}\psi_\gamma = -\frac{1}{2}(\delta^\alpha_\gamma\psi^\beta + \delta^\beta_\gamma\psi^\alpha).$$

Taking into account that $\varepsilon_{\alpha\beta}$ is a Lorentz invariant tensor, we then have

$$M^{\gamma\delta}M_{\gamma\delta}\psi_{\alpha_1\alpha_2\ldots\alpha_n}=\frac{1}{2}(\varepsilon_{\alpha_i\gamma}M^{\gamma\delta}\psi_{\delta\alpha_1\ldots\hat{\alpha}_i\ldots\alpha_n}+\varepsilon_{\alpha_i\delta}M^{\gamma\delta}\psi_{\gamma\alpha_1\ldots\hat{\alpha}_i\ldots\alpha_n})$$

$$=\sum_{i=1}^{n}\varepsilon_{\alpha_i\gamma}M^{\gamma\delta}\psi_{\delta\alpha_1\ldots\hat{\alpha}_i\ldots\alpha_n}$$

$$=\sum_{i=1}^{n}\varepsilon_{\alpha_i\gamma}\left\{-\frac{3}{2}\psi^{\gamma}{}_{\alpha_1\ldots\hat{\alpha}_i\ldots\alpha_n}-\frac{1}{2}\sum_{j\neq i}(\delta^{\delta}_{\alpha_j}\psi^{\gamma}{}_{\delta\alpha_1\ldots\hat{\alpha}_i\ldots\hat{\alpha}_j\ldots\alpha_n}\right.$$

$$\left.+\,\delta^{\gamma}_{\alpha_j}\psi^{\delta}{}_{\delta\alpha_1\ldots\hat{\alpha}_i\ldots\hat{\alpha}_j\ldots\alpha_n})\right\}.$$

The last term here vanishes since $\psi_{\alpha_1\ldots\alpha_n}$ is totally symmetric. On the same grounds, the first two terms give the following contribution

$$-\left(\frac{3}{2}n+\frac{1}{2}(n^2-n)\right)\psi_{\alpha_1\ldots\alpha_n}=-\frac{1}{2}n(n+2)\psi_{\alpha_1\ldots\alpha_n}.$$

The above calculation leads to the final results:

$$M^{\alpha\beta}M_{\alpha\beta}|_{(n/2,m/2)}=-\frac{1}{2}n(n+2)\mathbb{1}$$

$$\bar{M}^{\dot{\alpha}\dot{\beta}}\bar{M}_{\dot{\alpha}\dot{\beta}}|_{(n/2,m/2)}=-\frac{1}{2}m(m+2)\mathbb{1}. \tag{1.3.14}$$

Our final comment concerns the fact that the set of $(n/2, m/2)$ representations, where n, $m=0$, 1, 2, ..., actually embraces all irreducible finite-dimensional representations of the Lorentz group. As is well known, finite-dimensional representations of $sl(2, \mathbb{C})$ are parametrized by an integer $n=0$, 1, 2, ... The dimension of the corresponding representation is equal to $(n+1)$. Since the Lorentz algebra is the direct sum of two conjugate $sl(2, \mathbb{C})$ copies (generated by $M_{\alpha\beta}$ and $\bar{M}_{\dot{\alpha}\dot{\beta}}$), its irreducible representations are parametrized by pair of integers $n, m=0$, 1, 2, ..., and have the dimensions $(n+1)(m+1)$. But the $(n/2, m/2)$-sequence is characterized by just the same property.

1.4. Two-component and four-component spinors

Following P. Dirac, half-integer spin particles are usually described in terms of four-component spinor fields. However, when working with field theories in a superspace, one has inevitably to break the habit of using the four-component spinor formalism and to develop the habit of using the two-component spinor formalism. The reason is in the fact that all operations with two-component spinors are, as usual, much simpler than with the corresponding four-component ones. Now, we are going to show how two-component and four-component spinors are related, as well as how to translate two-component expressions into four-component language. To begin with, we list the main facts about two-component spinors. In accordance with the spin-statistics theorem, we shall assume all spinors to be anticommuting (odd elements of some Grassmann algebra), i.e.

$$\psi_\alpha\chi_\beta = -\chi_\beta\psi_\alpha \qquad \bar{\psi}_{\dot{\alpha}}\bar{\chi}_\beta = -\bar{\chi}_\beta\bar{\psi}_{\dot{\alpha}} \qquad \psi_\alpha\bar{\chi}_\beta = -\bar{\chi}_\beta\psi_\alpha \tag{1.4.1}$$

for any spinors ψ_α and χ_α. The detailed treatment of 'anticommuting numbers' will be given in Section 1.9.

1.4.1. Two-component spinors

Let ψ_α be an arbitrary undotted spinor, and $\bar{\chi}^{\dot{\alpha}}$ be an arbitrary dotted spinor. Infinitesimal Lorentz transformations act on them by the rule:

$$\delta\psi_\alpha = \frac{1}{2}K^{ab}(\sigma_{ab})_\alpha{}^\beta\psi_\beta = K_\alpha{}^\beta\psi_\beta$$

$$\delta\bar{\chi}^{\dot{\alpha}} = \frac{1}{2}K^{ab}(\tilde{\sigma}_{ab})^{\dot{\alpha}}{}_{\dot{\beta}}\bar{\chi}^{\dot{\beta}} = -\bar{K}^{\dot{\alpha}}{}_{\dot{\beta}}\bar{\chi}^{\dot{\beta}}. \tag{1.4.2a}$$

Index raising and lowering gives

$$\delta\psi^\alpha = -\psi^\beta\left(\frac{1}{2}K^{ab}\sigma_{ab}\right)_\beta{}^\alpha = -\psi^\beta K_\beta{}^\alpha = -K^\alpha{}_\beta\psi^\beta$$

$$\delta\bar{\chi}_{\dot{\alpha}} = -\bar{\chi}_{\dot{\beta}}\left(\frac{1}{2}K^{ab}\tilde{\sigma}_{ab}\right)^{\dot{\beta}}{}_{\dot{\alpha}} = \bar{\chi}_{\dot{\beta}}\bar{K}^{\dot{\beta}}{}_{\dot{\alpha}} = \bar{K}_{\dot{\alpha}}{}^{\dot{\beta}}\bar{\chi}_{\dot{\beta}} \tag{1.4.2b}$$

where we have used the observation that $(\sigma_{ab})_{\alpha\beta}$ and $(\tilde{\sigma}_{ab})_{\dot{\alpha}\dot{\beta}}$ are symmetric in their spinor indices. Finite Lorentz transformations are given by

$$\psi'_\alpha = (\exp K)_\alpha{}^\beta\psi_\beta \qquad \bar{\chi}'_{\dot{\alpha}} = (\exp\bar{K})_{\dot{\alpha}}{}^{\dot{\beta}}\bar{\chi}_{\dot{\beta}}. \tag{1.4.2c}$$

We define spinor bi-linear scalar and vector combinations

$$\psi\chi \equiv \psi^\alpha\chi_\alpha = -\psi_\alpha\chi^\alpha \qquad \psi^2 \equiv \psi\psi$$

$$\bar{\psi}\bar{\chi} \equiv \bar{\psi}_{\dot{\alpha}}\bar{\chi}^{\dot{\alpha}} = -\bar{\psi}^{\dot{\alpha}}\bar{\chi}_{\dot{\alpha}} \qquad \bar{\psi}^2 \equiv \bar{\psi}\bar{\psi} \tag{1.4.3}$$

$$\psi\sigma_a\bar{\chi} \equiv (\sigma_a)_{\alpha\dot{\alpha}}\psi^\alpha\bar{\chi}^{\dot{\alpha}} \qquad \bar{\chi}\tilde{\sigma}_a\psi \equiv \bar{\chi}_{\dot{\alpha}}(\tilde{\sigma}_a)^{\dot{\alpha}\alpha}\psi_\alpha.$$

By virtue of equation (1.4.1), the following identities

$$\psi\chi = \chi\psi \qquad \bar{\psi}\bar{\chi} = \bar{\chi}\bar{\psi} \qquad \psi\sigma_a\bar{\chi} = -\bar{\chi}\tilde{\sigma}_a\psi \tag{1.4.4}$$

hold. The difference in the definitions of $\psi\chi$ and $\overline{\psi\chi}$ is reasonable since we wish to have a conjugation rule of the form: $(\psi\chi)^* = \bar{\psi}\bar{\chi}$. But spinor conjugation should be understood as Hermitian conjugation (conjugation and transposition of places), so $(\psi\chi)^* = (\psi^\alpha\chi_\alpha)^* = (\chi_\alpha)^*(\psi^\alpha)^* = \bar{\chi}_{\dot{\alpha}}\bar{\psi}^{\dot{\alpha}}$. Due to this difference, we have

$$\psi_\alpha\chi_\beta = \psi_\beta\chi_\alpha + \varepsilon_{\alpha\beta}\psi\chi$$

$$\bar{\psi}_{\dot{\alpha}}\bar{\chi}_{\dot{\beta}} = \bar{\psi}_{\dot{\beta}}\bar{\chi}_{\dot{\alpha}} - \varepsilon_{\dot{\alpha}\dot{\beta}}\bar{\psi}\bar{\chi}. \tag{1.4.5}$$

The product of two spinors can be reduced by the rules:

$$\psi_\alpha\chi_\beta = \frac{1}{2}\varepsilon_{\alpha\beta}\psi\chi - \frac{1}{2}(\sigma^{ab})_{\alpha\beta}\psi\sigma_{ab}\chi$$

$$\psi\sigma_{ab}\chi \equiv \psi^\alpha(\sigma_{ab})_\alpha{}^\beta\chi_\beta \tag{1.4.6a}$$

$$\bar{\psi}_{\dot{\alpha}}\bar{\chi}_{\dot{\beta}} = -\frac{1}{2}\varepsilon_{\dot{\alpha}\dot{\beta}}\bar{\psi}\bar{\chi} - \frac{1}{2}(\tilde{\sigma}^{ab})_{\dot{\alpha}\dot{\beta}}\bar{\psi}\tilde{\sigma}_{ab}\bar{\chi} \tag{1.4.6b}$$

$$\bar{\psi}\tilde{\sigma}_{ab}\bar{\chi} \equiv \bar{\psi}_{\dot{\alpha}}(\tilde{\sigma}_{ab})^{\dot{\alpha}}{}_{\dot{\beta}}\bar{\chi}^{\dot{\beta}}$$

$$\psi_\alpha\bar{\chi}_{\dot{\alpha}} = -\frac{1}{2}(\sigma^a)_{\alpha\dot{\alpha}}\psi\sigma_a\bar{\chi}. \tag{1.4.6c}$$

To derive equation (1.4.6*a*), one has simply to write

$$\psi_\alpha\chi_\beta = \frac{1}{2}(\psi_\alpha\chi_\beta - \psi_\beta\chi_\alpha) + \frac{1}{2}(\psi_\alpha\chi_\beta + \psi_\beta\chi_\alpha)$$

and then use equation (1.2.33). Equation (1.4.6*c*) is a trivial consequence of equation (1.2.27). In group theoretical language, equation (1.4.5*a*) means that $(\frac{1}{2},0)\otimes(\frac{1}{2},0) = (0,0)\otimes(1,0)$, while equation (1.4.5*c*) means that $(\frac{1}{2},0)\otimes(0,\frac{1}{2}) = (\frac{1}{2},\frac{1}{2})$.

Finally, we list the Fierz rearrangement rules. For arbitrary spinors ψ_1^α, ψ_2^α, ψ_3^α and ψ_4^α we have

$$(\psi_1\psi_2)(\psi_3\psi_4) = -(\psi_1\psi_3)(\psi_2\psi_4) - (\psi_1\psi_4)(\psi_2\psi_3)$$

$$(\psi_1\psi_2)(\bar{\psi}_3\bar{\psi}_4) = -\frac{1}{2}(\psi_1\sigma^a\bar{\psi}_4)(\psi_2\sigma_a\bar{\psi}_3). \tag{1.4.7}$$

One can easily prove these identities with the help of equations (1.4.5) and (1.4.6*c*).

1.4.2. Dirac spinors

Let ψ_α be an undotted spinor, and $\bar{\chi}^{\dot{\alpha}}$ be a dotted spinor. We incorporate these spinors into a four-component column

$$\mathbf{\Psi} = \begin{pmatrix} \psi_\alpha \\ \bar{\chi}^{\dot{\alpha}} \end{pmatrix}. \tag{1.4.8}$$

This double spinor is called a 'Dirac spinor' (here and later, four-component spinor objects are denoted by boldface letters). By virtue of equation (1.4.2*a*), an infinitesimal Lorentz transformation acts on Ψ by the rule

$$\delta\boldsymbol{\Psi}=\frac{1}{2}K^{ab}\begin{pmatrix}\sigma_{ab} & 0\\ 0 & \tilde{\sigma}_{ab}\end{pmatrix}\boldsymbol{\Psi}. \tag{1.4.9}$$

Now, recalling the explicit form of σ_{ab} and $\tilde{\sigma}_{ab}$ (equation (1.2.32)), it is natural to introduce 4×4 matrices

$$\gamma_a=\begin{pmatrix}0 & \sigma_a\\ \tilde{\sigma}_a & 0\end{pmatrix}. \tag{1.4.10}$$

Then, the transformation law (1.4.9) takes the form

$$\delta\boldsymbol{\Psi}=\frac{1}{2}K^{ab}\Sigma_{ab}\boldsymbol{\Psi}$$

$$\Sigma_{ab}=-\frac{1}{4}[\gamma_a,\gamma_b]. \tag{1.4.11}$$

The matrices (1.4.10) satisfy the algebra

$$\{\gamma_a,\gamma_b\}=-2\eta_{ab}\mathbb{1}_4 \tag{1.4.12}$$

so γ_a are the usual Dirac matrices (in the special representation). Equation (1.4.11) is the standard transformation law of Dirac spinors.

Given a Dirac spinor (1.4.8), we conjugate ψ_α to obtain $\bar{\psi}_{\dot{\alpha}}$ and conjugate $\bar{\chi}^{\dot{\alpha}}$ to obtain χ^α. Let us combine the resulting two-component spinors in a four-component row

$$\bar{\boldsymbol{\Psi}}\equiv(\chi^\alpha,\bar{\psi}_{\dot{\alpha}}). \tag{1.4.13}$$

Its transformation law, using equations (1.4.2*b*) and (1.4.10, 11), is

$$\delta\bar{\boldsymbol{\Psi}}=-\bar{\boldsymbol{\Psi}}\left(\frac{1}{2}K^{ab}\Sigma_{ab}\right). \tag{1.4.14}$$

What is more, $\bar{\boldsymbol{\Psi}}$ satisfies the equation: $\bar{\boldsymbol{\Psi}}=\boldsymbol{\Psi}^+\gamma_0$. So, $\bar{\boldsymbol{\Psi}}$ is the ordinary Dirac conjugate spinor of $\boldsymbol{\Psi}$.

One more four-component object is obtained from $\boldsymbol{\Psi}$ by transposition of ψ and χ. Namely, let us consider the spinor

$$\boldsymbol{\Psi}_C=\begin{pmatrix}\chi_\alpha\\ \bar{\psi}^{\dot{\alpha}}\end{pmatrix}.$$

Evidently, it transforms as a Dirac spinor,

$$\delta\boldsymbol{\Psi}_C=\frac{1}{2}K^{ab}\Sigma_{ab}\boldsymbol{\Psi}_C. \tag{1.4.15}$$

What is more, it satisfies the equation

$$\Psi_C = C\bar{\Psi}^{\mathrm{T}} \tag{1.4.16}$$

where C and its inverse matrix C^{-1} have the form

$$C = \begin{pmatrix} \varepsilon_{\alpha\beta} & 0 \\ 0 & \varepsilon^{\dot{\alpha}\dot{\beta}} \end{pmatrix} \qquad C^{-1} = \begin{pmatrix} \varepsilon^{\alpha\beta} & 0 \\ 0 & \varepsilon_{\dot{\alpha}\dot{\beta}} \end{pmatrix}.$$

It is easy to see that C is a unitary antisymmetric matrix obeying the identity

$$C^{-1}\gamma_m C = -\gamma_m^{\mathrm{T}}.$$

So, C is the 'charge conjugation matrix', and Ψ_C is the 'charge conjugate spinor' of Ψ.

1.4.3. Weyl spinors

We introduce the Lorentz invariant matrix $\gamma_5 = -\mathrm{i}\gamma_0\gamma_1\gamma_2\gamma_3$ with the properties

$$\gamma_5 = \begin{pmatrix} \mathbb{1}_2 & 0 \\ 0 & -\mathbb{1}_2 \end{pmatrix} \qquad \gamma_5^2 = \mathbb{1}_4 \qquad \gamma_5\gamma_m + \gamma_m\gamma_5 = 0$$

and define operators P_{L} and P_{R} as follows

$$P_{\mathrm{L}} = \frac{1}{2}(\mathbb{1}_4 + \gamma_5) \qquad P_{\mathrm{R}} = \frac{1}{2}(\mathbb{1}_4 - \gamma_5).$$

Here $\mathbb{1}_2$ is the 2×2 unit matrix and $\mathbb{1}_4$ is the 4×4 unit matrix.

Due to the identities

$$P_{\mathrm{L}}^2 = P_{\mathrm{L}} \qquad P_{\mathrm{R}}^2 = P_{\mathrm{R}} \qquad P_{\mathrm{L}}P_{\mathrm{R}} = P_{\mathrm{R}}P_{\mathrm{L}} = 0$$

these operators are Lorentz invariant projectors. Acting with P_{L} and P_{R} on an arbitrary Dirac spinor Ψ, one obtains four-component objects

$$\Psi_{\mathrm{L}} \equiv P_{\mathrm{L}}\Psi = \begin{pmatrix} \psi_\alpha \\ 0 \end{pmatrix} \qquad \Psi_{\mathrm{R}} \equiv P_{\mathrm{R}}\Psi = \begin{pmatrix} 0 \\ \bar{\chi}^{\dot{\alpha}} \end{pmatrix} \tag{1.4.17}$$

which transform as Dirac spinors and satisfy the Lorentz covariant constraints

$$\gamma_5\Psi_{\mathrm{L}} = \Psi_{\mathrm{L}} \qquad \gamma_5\Psi_{\mathrm{R}} = -\Psi_{\mathrm{R}}. \tag{1.4.18}$$

The relations (1.4.17) show that Ψ_{L} and Ψ_{R} are the Dirac forms of two-component left-handed and right-handed Weyl spinors ψ_α and $\bar{\chi}^{\dot{\alpha}}$ correspondingly. Very often, the spinors Ψ_{L} and Ψ_{R} are called Weyl spinors too.

1.4.4. Majorana spinors

In our combinatorical exercises of subsection 1.4.2, we have not considered one interesting possibility: $\psi_\alpha = \chi_\alpha$. If ψ_α is an undotted two-component spinor, then the four-component object

$$\Psi_{M} = \begin{pmatrix} \psi_{\alpha} \\ \bar{\psi}^{\dot{\alpha}} \end{pmatrix} \qquad (1.4.19)$$

transforms as a Dirac spinor,

$$\delta\Psi_{M} = \frac{1}{2} K^{ab}\Sigma_{ab}\Psi_{M}.$$

Its main property is that Ψ_{M} coincides with its charge conjugate spinor,

$$\Psi_{M} = C\bar{\Psi}^{T}_{M}. \qquad (1.4.20)$$

So Ψ_{M} is called a 'real' (or 'Majorana') four-component spinor.

Any Dirac spinor (1.4.8) can be represented as a sum of two Majorana spinors Φ_{M} and Λ_{M} by the rule:

$$\Psi = \Phi_{M} + \mathrm{i}\Lambda_{M} \Rightarrow \psi_{\alpha} = \varphi_{\alpha} + \mathrm{i}\lambda_{\alpha} \quad \chi_{\alpha} = \varphi_{\alpha} - \mathrm{i}\lambda_{\alpha}. \qquad (1.4.21)$$

1.4.5. The reduction rule and the Fierz identity

The set of 4×4 matrices

$$\gamma_{A} = \{\mathbb{1}_{4}, \mathrm{i}\gamma_{a}, 2\mathrm{i}\Sigma_{ab}, \gamma_{a}\gamma_{5}, \gamma_{5}\}$$

forms a basis in the linear space of 4×4 matrices. Defining the corresponding set with upper indices,

$$\gamma^{A} = \{\mathbb{1}_{4}, \mathrm{i}\gamma^{a}, 2\mathrm{i}\Sigma^{ab}, \gamma^{a}\gamma_{5}, \gamma^{5}\}$$

we have the identities

$$\frac{1}{4}\mathrm{Tr}(\gamma^{A}\gamma_{B}) = \delta^{A}{}_{B} \qquad \gamma^{A}\gamma_{A} = \mathbb{1}_{4} \text{ (no sum)}.$$

In accordance with these identities, if Γ is a 4×4 matrix, then

$$\Gamma = \sum_{A} C^{A}\gamma_{A} \qquad C^{A} = \frac{1}{4}\mathrm{Tr}(\gamma^{A}\Gamma)$$

or, in components,

$$\Gamma_{ij} = \frac{1}{4}\sum_{A}(\gamma^{A})_{kl}\Gamma_{lk}(\gamma_{A})_{ij}.$$

Since Γ is arbitrary, one obtains the following completeness relation

$$\delta_{il}\delta_{jk} = \frac{1}{4}\sum_{A}(\gamma^{A})_{kl}(\gamma_{A})_{ij}. \qquad (1.4.22)$$

This relation is a four-component version of equation (1.2.27).

Now let Ψ_{1}, Ψ_{2}, Ψ_{3} and Ψ_{4} be arbitrary Dirac spinors. Using equation

(1.4.22), one can easily show that

$$\bar{\Psi}_{1i}\Psi_{2j} = \frac{1}{4}\sum_A \bar{\Psi}_1\gamma^A\Psi_2(\gamma_A)_{ji} \tag{1.4.23}$$

$$(\bar{\Psi}_1\Psi_2)(\bar{\Psi}_3\Psi_4) = -\frac{1}{4}\sum_A (\bar{\Psi}_1\gamma^A\Psi_4)(\bar{\Psi}_3\gamma_A\Psi_2). \tag{1.4.24}$$

Equation (1.4.23) is a four-component generalization of the two-component reduction rules (1.4.6). Equation (1.4.24) is a four-component version of the Fierz identities (1.4.7). After some practice, one can find that working with equations (1.4.6, 7) is much simpler than with equations (1.4.23, 24).

1.4.6. Two-component and four-component bi-linear combinations
In conclusion, we consider the connection between two-component and four-component bi-linear combinations. Given two Dirac spinors

$$\Psi_1 = \begin{pmatrix} \psi_{1\alpha} \\ \bar{\chi}_1{}^{\dot{\alpha}} \end{pmatrix} \qquad \Psi_2 = \begin{pmatrix} \psi_{2\alpha} \\ \bar{\chi}_2{}^{\dot{\alpha}} \end{pmatrix}$$

we have

$$\begin{aligned}
\bar{\Psi}_1\Psi_2 &= \chi_1\psi_2 + \bar{\psi}_1\bar{\chi}_2 \\
\bar{\Psi}_1\gamma_5\Psi_2 &= \chi_1\psi_2 - \bar{\psi}_1\bar{\chi}_2 \\
\bar{\Psi}_1\gamma_a\Psi_2 &= \chi_1\sigma_a\bar{\chi}_2 - \psi_2\sigma_a\bar{\psi}_1 \\
\bar{\Psi}_1\gamma_a\gamma_5\Psi_2 &= -\chi_1\sigma_a\bar{\chi}_2 - \psi_2\sigma_a\bar{\psi}_1 \\
\bar{\Psi}_1\Sigma_{ab}\Psi_2 &= \chi_1\sigma_{ab}\psi_2 + \bar{\psi}_1\tilde{\sigma}_{ab}\bar{\chi}_2 \\
\bar{\Psi}_1\Sigma_{ab}\gamma_5\Psi_2 &= \chi_1\sigma_{ab}\psi_2 - \bar{\psi}_1\tilde{\sigma}_{ab}\bar{\chi}_2.
\end{aligned}$$

These relations show how to express an arbitrary bi-linear combination of Dirac spinors in terms of two-component bi-linear combinations and vice versa.

1.5. Representations of the Poincaré group

1.5.1. The Poincaré algebra
To begin analysis of the Poincaré group, we give its realization as a matrix Lie group. Let us associate with every element (Λ, b) of the Poincaré group a 5×5 matrix $\langle\Lambda, b\rangle$ defined as follows:

$$\langle\Lambda, b\rangle = \left(\begin{array}{c|c} \Lambda^m{}_n & b^m \\ \hline 0 & 1 \end{array}\right). \tag{1.5.1}$$

Since the multiplication law in the Poincaré group is given as $(\Lambda_2, b_2) \times (\Lambda_1, b_1) = (\Lambda_2\Lambda_1, b_2 + \Lambda_2 b_1)$, the correspondence $(\Lambda, b) \to \langle \Lambda, b \rangle$ is an isomoprhism of the Poincaré group on the subgroup of $GL(5, \mathbb{R})$ consisting of matrices of the form (1.5.1), where $\Lambda^m{}_n$ is restricted by equations (1.1.3,4) and b^m is an arbitrary four-vector. So, the Poincaré group can be identified with this matrix group. We denote it by $\tilde{\Pi}$. Minkowski space can be identified with the hyperplane $x^4 = 1$ in $\mathbb{R}^5$ (with coordinates $x^0, x^1, \ldots, x^4$). Then, the linear operator (1.5.1) acts on this hyperplane as the corresponding Poincaré transformation (Λ, b).

The Lie algebra of the Lie group $\tilde{\Pi}$ consists of 5×5 matrices of the form

$$\langle K, \mathfrak{b} \rangle = \left(\begin{array}{c|c} K^m{}_n & \mathfrak{b}^m \\ \hline 0 & 0 \end{array} \right) \qquad K_{mn} = -K_{nm} \tag{1.5.2}$$

where K_{mn} is an arbitrary antisymmetric Lorentz tensor and $\mathfrak{b}^m$ is an arbitrary four-vector. We identify this matrix Lie algebra with the Lie algebra of the Poincaré group (the 'Poincaré algebra' will be denoted by $\mathscr{P}$). The multiplication law in the Poincaré algebra is

$$[\langle K_2, \mathfrak{b}_2 \rangle, \langle K_1, \mathfrak{b}_1 \rangle] = \langle [K_2, K_1], K_2\mathfrak{b}_1 - K_1\mathfrak{b}_2 \rangle \qquad (K_2\mathfrak{b}_1)^m = K_2{}^m{}_n \mathfrak{b}_1^n. \tag{1.5.3}$$

It is useful for later applications to introduce a basis $(\mathbf{j}_{ab}, \boldsymbol{p}_a)$, $\mathbf{j}_{ab} = -\mathbf{j}_{ba}$, for the Poincaré algebra by the rule

$$\langle K, \mathfrak{b} \rangle = \frac{\mathrm{i}}{2} K^{ab} \mathbf{j}_{ab} - \mathrm{i}\mathfrak{b}^a \boldsymbol{p}_a$$

$$\mathbf{j}_{ab} = \left(\begin{array}{c|c} -\mathrm{i}\,{}^{\mathrm{v}}M_{ab} & 0 \\ \hline 0 & 0 \end{array} \right) \tag{1.5.4}$$

$$\boldsymbol{p}_0 = \left(\begin{array}{c|c} & \mathrm{i} \\ & 0 \\ 0 & 0 \\ & 0 \\ \hline 0 & 0 \end{array} \right), \quad \ldots, \quad \boldsymbol{p}_3 = \left(\begin{array}{c|c} & 0 \\ & 0 \\ 0 & 0 \\ & \mathrm{i} \\ \hline 0 & 0 \end{array} \right).$$

Here ${}^{\mathrm{v}}M_{ab}$ are the Lorentz generators in the vector representation (equation (1.3.8c)). Using rule (1.5.4), we obtain the commutation relations in the

Poincaré algebra:

$$[\rho_a, \rho_b] = 0$$

$$[\mathrm{j}_{ab}, \rho_c] = \mathrm{i}\eta_{ac}\rho_b - \mathrm{i}\eta_{bc}\rho_a \tag{1.5.5}$$

$$[\mathrm{j}_{ab}, \mathrm{j}_{cd}] = \mathrm{i}\eta_{ac}\mathrm{j}_{bd} - \mathrm{i}\eta_{ad}\mathrm{j}_{bc} + \mathrm{i}\eta_{bd}\mathrm{j}_{ac} - \mathrm{i}\eta_{bc}\mathrm{j}_{ad}.$$

Note that every element $\langle \Lambda, b \rangle$ from Π can be represented as

$$\langle \Lambda, b \rangle = \exp\left(\frac{\mathrm{i}}{2} K^{ab}\mathrm{j}_{ab} - \mathrm{i}\hat{b}^a \rho_a\right)$$

$$\Rightarrow \Lambda^m{}_n = (\exp K)^m{}_n \qquad b^m = \left(\frac{\mathrm{e}^K - \mathbb{1}_4}{K}\right)^m{}_n \hat{b}^n. \tag{1.5.6}$$

Now, let T be a representation of the Poincaré group, and $T(\Lambda, b)$ be some representation operator. In accordance with equation (1.5.6), we have

$$T(\Lambda, b) = \exp\left(\frac{\mathrm{i}}{2} K^{ab}\mathbf{J}_{ab} - \mathrm{i}\hat{b}^a \mathbf{P}_a\right) \tag{1.5.7}$$

where the Poincaré generators $\mathbf{J}_{ab} = \mathrm{d}T(\mathrm{j}_{ab})$ and $\mathbf{P}_a = \mathrm{d}T(\rho_a)$ provide a representation of the Poincaré algebra. Here $\mathbf{P}_a$ are the generators of translations and $\mathbf{J}_{ab}$ are the generators of Lorentz transformations. As a result of equation (1.5.5), the generators satisfy the following commutation relations:

$$[\mathbf{P}_a, \mathbf{P}_b] = 0$$

$$[\mathbf{J}_{ab}, \mathbf{P}_c] = \mathrm{i}\eta_{ac}\mathbf{P}_b - \mathrm{i}\eta_{bc}\mathbf{P}_a \tag{1.5.8}$$

$$[\mathbf{J}_{ab}, \mathbf{J}_{cd}] = \mathrm{i}\eta_{ac}\mathbf{J}_{bd} - \mathrm{i}\eta_{ad}\mathbf{J}_{bc} + \mathrm{i}\eta_{bd}\mathbf{J}_{ac} - \mathrm{i}\eta_{bc}\mathbf{J}_{ad}.$$

These commutation relations define an arbitrary representation of the Poincaré algebra.

The Poincaré group has two Casimiar operators (commuting with $\mathbf{P}_a$ and $\mathbf{J}_{ab}$)

$$C_1 = -\mathbf{P}^a\mathbf{P}_a \qquad C_2 = \mathbf{W}^a\mathbf{W}_a \tag{1.5.9}$$

where $\mathbf{W}_a$ is the Pauli–Lubanski vector

$$\mathbf{W}_a = \frac{1}{2}\varepsilon_{abcd}\mathbf{J}^{bc}\mathbf{P}^d. \tag{1.5.10}$$

Using equation (1.5.8), one can prove the following properties of the Pauli–Lubanski vector

$$\mathbf{W}^a\mathbf{P}_a = 0 \qquad [\mathbf{W}_a, \mathbf{P}_b] = 0 \tag{1.5.11a}$$

$$[\mathbf{J}_{ab}, \mathbf{W}_c] = \mathrm{i}\eta_{ac}\mathbf{W}_b - \mathrm{i}\eta_{bc}\mathbf{W}_a \tag{1.5.11b}$$

$$[\mathbf{W}_a, \mathbf{W}_b] = \mathrm{i}\varepsilon_{abcd}\mathbf{W}^c\mathbf{P}^d. \tag{1.5.11c}$$

1.5.2. Field representations

Let $\Phi(x)$ be an arbitrary Lorentz tensor field on Minkowski space (indices are suppressed). By definition, its components in two different coordinate systems, related by some Poincaré transformation (1.1.1), are connected by the rule

$$\Phi'(x') = T(\Lambda)\Phi(x) \qquad (1.5.12)$$

where $T(\Lambda)$ is a finite-dimensional representation of the Lorentz group. Proceeding from the transformation law (1.5.12), we construct a representation of the Poincaré group acting in a linear space of tensor fields as follows

$$\begin{gathered}\Phi(x) \to \Phi_{(\Lambda,b)}(x) \\ \Phi_{(\Lambda,b)}(x) = T(\Lambda)\Phi(\Lambda^{-1}(x-b)).\end{gathered} \qquad (1.5.13)$$

This representation is infinite dimensional. The corresponding Poincaré generators are

$$\begin{gathered}\mathbf{P}_a = -\mathrm{i}\partial_a \\ \mathbf{J}_{ab} = \mathrm{i}(x_b\partial_a - x_a\partial_b) - \mathrm{i}M_{ab}\end{gathered} \qquad (1.5.14)$$

where M_{ab} are the generators of the Lorentz representation $T(\Lambda)$, $T(\Lambda) = \exp(\frac{1}{2}K^{ab}M_{ab})$.

1.5.3. Unitary representations

In quantum field theory, Poincaré invariance means that any Poincaré transformation (Λ, b) induces a unitary transformation

$$U(\Lambda, b) = \exp\left[\mathrm{i}\left(-b^a\mathbb{P}_a + \frac{1}{2}K^{ab}\mathbb{J}_{ab}\right)\right] \qquad (1.5.15)$$

acting in a Hilbert space of particle states. The union of operators $U(\Lambda, b)$ provides us with a unitary representation of the Poincaré group. Operators $\mathbb{P}_a$ and $\mathbb{J}_{ab}$ are the corresponding generators. The generators $\mathbb{P}_a$ are identified with the energy–momentum operator, $\mathbb{P}_a = (-\mathbb{H}, \vec{\mathbb{P}})$, where $\mathbb{H}$ and $\vec{\mathbb{P}}$ are the Hamiltonian and the momentum of some quantized field. Therefore, the first Casimir operator (1.5.9) coincides with the squared mass operator. In any irreducible representation of the Poincaré group, all particle states are eigenstates of this operator with the same mass, $\mathbb{P}^a\mathbb{P}_a = -m^2\mathbb{1}$, $m^2 \geqslant 0$. Note that there exist Poincaré representations with negative mass squared. However, physically, such representations are not admissible. Under the supposition of non-negativity of mass, the Poincaré group has one more Casimiar operator in addition to the operators (1.5.9), namely sign $(\mathbb{P}_0)$. From a physical point of view, we should consider only positive-energy representations, characterized by the condition

$$\langle\Psi|\mathbb{P}_0|\Psi\rangle<0 \tag{1.5.16}$$

for any non-zero state $|\Psi\rangle$.

The second Casimir operator (1.5.9) describes the spin of the particles. To clarify this assertion, we introduce one auxiliary notion which will be useful also when constructing irreducible representations of the (super) Poincaré group.

1.5.4. Stability subgroup

In a Hilbert space of one-particle states with a given mass m, we consider the substance V_q of particle states having a given four-momentum q_a,

$$\mathbb{P}_a|q\rangle=q_a|q\rangle$$

for any state $|q\rangle\in V_q$. The four-vector q_a lies on the 'mass-shell' surface

$$p^a p_a=-m^2 \qquad p_0<0 \tag{1.5.17}$$

in momentum space. We define the set H_q of group elements (Λ, b) such that the corresponding operators $U(\Lambda, b)$ transform V_q onto itself. Evidently, H_q forms a subgroup of the Poincaré group. It will be called the stability subgroup for V_q.

To find H_q, note that, according to (1.5.8) an operator $\exp[(i/2)K^{ab}\mathbb{J}_{ab}]$ transforms some state $|q\rangle$ into another state $|q'\rangle$, where $q'^a=(\exp K)^a{}_b q^b$. Demanding $q'=q$, we obtain the equation $K^a{}_b q^b=0$, which has the following general solution:

$$K_{ab}=\varepsilon_{abcd}q^c n^d$$

where n^d is an arbitrary vector. Therefore, the stability subgroup consists of elements of the form

$$\exp\left[\mathrm{i}\left(-b^a\mathbb{P}_a+\frac{1}{2}\varepsilon^{abcd}q_c n_d \mathbb{J}_{ab}\right)\right] \tag{1.5.18}$$

for arbitrary vectors b, n. Being restricted to V_q, elements of H_q can be expressed in terms of the Pauli–Lubanski vector, since operator (1.5.18) coincides (on V_q) with

$$\exp(-\mathrm{i}\alpha)\exp(-\mathrm{i}n_a\mathbb{W}^a) \tag{1.5.19}$$

where $\alpha=b^a q_a$. It is worth now recalling the identify (1.5.11*c*). This indicates that components of the Pauli–Lubanski vector form a Lie algebra ($so(3)$ in the massive case and $so(2)$ in the massless case, see below) restricted to V_q. So, the stability subgroup acts on V_q as the $U(1)\times SO(3)$ group in the massive case and as the $U(1)\times U(1)$ group in the massless case.

All vectors from V_q describe particle states with the same momentum. On physical grounds, any two linearly independent states $|1\rangle$, $|2\rangle\in V_q$ should correspond to different spin polarizations to be transforming into each other under the action of H_q. In accordance with expression (1.5.19), the stability subgroup is generated (up to phase factors) on V_q by the Pauli–Lubanski vector. Hence, the spin properties of the particles are characterized by the Pauli–Lubanski vector.

For a given particle, the spectrum of its spin polarizations contains only a finite number of values. This means that in any physically admissible irreducible representation of the Poincaré group all subspaces V_q are finite dimensional and H_q acts irreducibly on V_q.

Now one could ask about the coset space Π/H_q of the Poincaré group over some stability subgroup H_q. It is clear that Π/H_q may be parameterized by a set of Lorentz transformations $\Omega[p]$ transferring the four-momentum q^a into another four-momentum p^a from the same 'mass-shell' surface (1.5.17), $p^a=(\Omega[p])^a{}_b q^b$. Given some such family $\{\Omega[p]\}$, we find that a unitary operator $U(\Omega[p],0)$ transforms V_q on to V_p. In the massive case, the most useful choice for the test momentum q^a is the momentum of a particle at rest. Then, a useful candidate for the role of $\{\Omega[p]\}$ is the family of Lorentz transformations

$$\Omega[p]=\left(\begin{array}{c|ccc} E_p/m & p^1/m & p^2/m & p^3/m \\ \hline p^1/m & & & \\ p^2/m & & \delta^{IJ}+\dfrac{p^I p^J}{m(E_p+m)} & \\ p^3/m & & & \end{array}\right) \tag{1.5.20}$$

$$p^a=(E_p, p^1, p^2, p^3) \qquad E_p=\sqrt{\vec{p}^2+m^2}.$$

In the massless case, the test momentum is usually chosen in the form $p^a=(E,0,0,E)$. The corresponding family $\{\Omega[p]\}$ is

$$\Omega[p]=\begin{pmatrix} \frac{1}{2}\frac{|\vec{p}|}{E}(1+\alpha) & 0 & 0 & \frac{1}{2}\frac{|\vec{p}|}{E}(1-\alpha) \\ \frac{1}{2}\frac{p^1}{E}(1-\alpha) & n^1(p) & m^1(p) & \frac{1}{2}\frac{p^1}{E}(1+\alpha) \\ \frac{1}{2}\frac{p^2}{E}(1-\alpha) & n^2(p) & m^2(p) & \frac{1}{2}\frac{p^2}{E}(1+\alpha) \\ \frac{1}{2}\frac{p^3}{E}(1-\alpha) & n^3(p) & m^3(p) & \frac{1}{2}\frac{p^3}{E}(1+\alpha) \end{pmatrix} \tag{1.5.21}$$

$$p^a=(|\vec{p}|, p^1, p^2, p^3) \qquad \alpha=E^2/\vec{p}^2.$$

Here $\vec{n}(p)$ and $\vec{m}(p)$ are three-vectors such that the set $\{\vec{n}, \vec{m}, \vec{p}/|\vec{p}|\}$ forms an orthonormal basis in the space.

Now we are in a position to describe the main steps in the construction of unitary irreducible Poincaré representations. Since $\mathbb{P}^a\mathbb{P}_a$ is the Casimir operator, irreducible Poincaré representations are classified by mass. Let us fix some mass value m and choose a test momentum q^a from the 'mass-shell' surface (1.5.17). Then one has to find all (finite-dimensional) unitary irreducible representations of H_q, characterized by the condition that any operator (1.5.18) acts in accordance with rule (1.5.19). Suppose that T_q is a given representation (with the described properties) of H_q acting on a space V_q, and let $|q, \sigma; m^2\rangle$ be an orthonormal basis in V_q; here the variable σ runs over a finite number of values (spin variable). Further, we formally associate with any point $p^a \neq q^a$ on the 'mass-shell' surface (1.5.17) a vector space V_p of the same dimension as V_q. Let $|p, \sigma; m^2\rangle$ be a basis in V_p. As the next step, we define a Hilbert space of one-particle states as a formal direct sum $\mathscr{H} = \oplus_p V_p$ (for all p^a, including q^a) with the following inner product

$$\langle p, \sigma; m^2 | p', \sigma'; m^2\rangle = E_p\delta_{\sigma\sigma'}\delta(\vec{p}-\vec{p}'). \tag{1.5.22}$$

Finally, let us define a unitary Poincaré representation on $\mathscr{H}$ by four requirements: (1) for any $(\Lambda, b) \in H_q$, the unitary operator $U(\Lambda, b)$ restricted to V_q coincides with the operator $T_q(\Lambda, b)$; (2) each subspace V_p carries the momentum p^a,

$$\mathbb{P}_a |p, \sigma; m^2\rangle = p_a |p, \sigma; m^2\rangle \tag{1.5.23}$$

(3) for a given family of Lorentz transformations $\Omega[p]$, which parameterizes the coset space Π/H_q, we have

$$|p, \sigma; m^2\rangle = U(\Omega[p], 0)\, |q, \sigma; m^2\rangle \tag{1.5.24}$$

(4) for every Lorentz transformation $(\Lambda, 0)$, we have

$$U(\Lambda, 0)\, |p, \sigma; m^2\rangle = U(\Omega[\Lambda p], 0)U(\Omega^{-1}[\Lambda p]\Lambda\Omega[p], 0)\, |q, \sigma; m^2\rangle \tag{1.5.25}$$

where $(\Lambda p)^a = \Lambda^a{}_b p^b$. Note that $\Omega^{-1}[\Lambda p]\Lambda\Omega[p] \in H_q$, so the action of the operator $U(\Omega^{-1}[\Lambda p]\Lambda\Omega[p], 0)$ on the subspace V_q is given by requirement (1). It is a technical matter to check that we have indeed obtained some unitary representation of the Poincaré group. Evidently, this representation is irreducible.

What we have described above is simply Wigner's method of induced representations as applied to the Poincaré group.

1.5.5. Massive irreducible representations

We proceed by finding massive irreducible representations of the Poincaré group. As explained above, it is sufficient to construct all (unitary) irreducible finite-dimensional representations of the stability subgroup H_q corresponding to an arbitrary four-momentum on the 'mass-shell' $p^2 = -m^2$,

$p_0<0$. For simplicity, we choose the momentum

$$q_a=(-m, 0, 0, 0) \tag{1.5.26}$$

of a particle at rest.

Being restricted to the subspace V_q, components of $\mathbb{W}_a=\frac{1}{2}\varepsilon_{abcd}\mathbb{J}^{bc}\mathbb{P}^d$ should have the form

$$\mathbb{W}_0=0 \qquad \mathbb{W}_I=m\mathbb{S}_I \qquad I=1, 2, 3$$
$$\mathbb{S}_I=\frac{1}{2}\varepsilon_{IJK}\mathbb{J}^{JK} \tag{1.5.27}$$

and satisfy the algebra

$$[\mathbb{S}_I, \mathbb{S}_J]=\mathrm{i}\varepsilon_{IJK}\mathbb{S}_K. \tag{1.5.28}$$

The algebra (1.5.28) coincides with the angular momentum algebra $su(2)$ (or $so(3)$). As is well known, finite-dimensional irreducible representations of $su(2)$ are characterized by the condition

$$(\mathbb{S}_1)^2+(\mathbb{S}_2)^2+(\mathbb{S}_3)^2=s(s+1)\mathbb{1} \tag{1.5.29}$$

where possible values of s are: $s=0, \frac{1}{2}, 1, \frac{3}{2}, \ldots$. For a given s, the dimension of the representation is $(2s+1)$. Equations (1.5.27, 29) give

$$\mathbb{W}^a\mathbb{W}_a=m^2s(s+1)\mathbb{1} \qquad s=0, 1/2, 1, 3/2, \ldots. \tag{1.5.30}$$

This relation determines the spectrum of values which the spin operator can take in unitary irreducible massive Poincaré representations. The quantum number s is called a spin. Hence, in the massive case, irreducible representations of the Poincaré group are classified by mass m and spin s.

1.5.6. Massless irreducible representations

Now we are going to discuss the massless case, where

$$\mathbb{P}^a\mathbb{P}_a=0.$$

On the upper light-cone surface

$$p^ap_a=0 \qquad p_0<0$$

we choose the four-momentum

$$q_a=(-E, 0, 0, E) \qquad E\neq 0 \tag{1.5.31}$$

and analyse unitary finite-dimensional representations of the stationary subgroup H_q.

When acting on the subspace V_q, the components of the Pauli–Lubanski vector are

$$\mathbb{W}_0 = -E\mathbb{J}_{12} \qquad \mathbb{W}_3 = E\mathbb{J}_{12}$$
$$\mathbb{W}_1 = E(\mathbb{J}_{23} + \mathbb{J}_{20}) \equiv E\mathbb{R}_1 \tag{1.5.32}$$
$$\mathbb{W}_2 = E(\mathbb{J}_{13} + \mathbb{J}_{10}) \equiv E\mathbb{R}_2$$

where we have used equations (1.5.10) and (1.5.11*a*). As a result of equation (1.5.11*c*), the operators $\mathbb{J}_{12}$, $\mathbb{R}_1$ and $\mathbb{R}_2$ satisfy the algebra

$$[\mathbb{R}_1, \mathbb{R}_2] = 0$$
$$[\mathbb{J}_{12}, \mathbb{R}_1] = -i\mathbb{R}_2 \qquad [\mathbb{J}_{12}, \mathbb{R}_2] = i\mathbb{R}_1 \tag{1.5.33}$$

which is simply the Lie algebra of the group E_2 of translations and rotations on a two-dimensional plane. The Casimir operator of E_2 is

$$(\mathbb{R}_1)^2 + (\mathbb{R}_2)^2$$

and in any unitary irreducible representation this operator looks like

$$(\mathbb{R}_1)^2 + (\mathbb{R}_2)^2 = \mu^2 \mathbb{I} \qquad \mu^2 \geqslant 0. \tag{1.5.34}$$

If $\mu^2 > 0$, a basis in the space of representation consists of states

$$|\vec{r}\rangle \qquad \mathbb{R}_1|\vec{r}\rangle = r_1|\vec{r}\rangle \qquad \mathbb{R}_2|\vec{r}\rangle = r_2|\vec{r}\rangle$$

labelled by points $\vec{r} = (r_1, r_2)$ of the one-sphere

$$(r_1)^2 + (r_2)^2 = \mu^2.$$

Thus, if $\mu^2 > 0$, the operators $\mathbb{R}_1$ and $\mathbb{R}_2$ have a continuous spectrum, and the representation is infinite-dimensional.

On the other hand, the subspace V_q should have a finite dimension. Therefore, no other possibility is available, but $\mu^2 = 0$ and $\mathbb{R}_1$, $\mathbb{R}_2$ are trivial on V_q, hence

$$\mathbb{W}_1 = \mathbb{W}_2 = 0. \tag{1.5.35}$$

Recalling equation (1.5.19), we see now that H_q acts on V_q as the product of a $U(1)$ phase group and an Abelian group generated by the only operator $\mathbb{J}_{12}$. Since the action of H_q on V_q should be irreducible, this space includes only one non-trivial state,

$$\mathbb{J}_{12}|\lambda\rangle = \lambda|\lambda\rangle \tag{1.5.36}$$

where λ can takes values

$$\lambda = 0, \pm 1/2, \pm 1, \pm 3/2, \ldots \tag{1.5.37}$$

The quantum number λ is called 'helicity'. Sometimes, the quantity $|\lambda|$ is called the 'spin of a massless particle'.

The restriction (1.5.37) is quite understandable. Indeed, operators $\exp(i\varphi\mathbb{J}_{12})$ describe space rotations along the direction of particle motion.

Making the rotation on $\varphi = 2\pi$, we obtain

$$e^{2\pi i \mathbb{J}_{12}}|\lambda\rangle = e^{2\pi i\lambda}|\lambda\rangle.$$

The resulting phase factor must be equal to ± 1 depending on whether our representation of the Poincaré group is single- or two-valued.

The set of relations (1.5.32, 35, 36) means that, in the reference system (1.5.31) the equality

$$\mathbb{W}_a = \lambda \mathbb{P}_a \tag{1.5.38}$$

is fulfilled. Since $\mathbb{W}_a$ and $\mathbb{P}_a$ transform as Lorentz vectors, equation (1.5.38) is satisfied in any coordinate system. Hence, the helicity is a Poincaré invariant characteristic of massless particles. Our conclusion is that massless irreducible representations of the Poincaré group are classified by helicity.

1.6. Elements of differential geometry and gravity

1.6.1. Lorentz manifolds

In general relativity, space-time is a four-dimensional connected smooth manifold, i.e. a connected topological space M covered by a set of open charts $\{U_i\}_{i \in A}$, for each of which a homeomorphism

$$x_i\colon\ U_i \to \mathbb{R}^4$$

of U_i onto an open subset of $\mathbb{R}^4$ is defined, such that the transition functions

$$f_{ij} = x_i \circ x_j^{-1}\colon\ x_j(U_i \cap U_j) \to x_i(U_i \cap U_j)$$

are smooth whenever $U_i \cap U_j \neq \varnothing$. The functions $x_i^m(p) = \{x_i^0, x_i^1, x_i^2, x_i^3\}$, $p \in U_i$, are called the local coordinates of p in the chart U_i. Orientedness of the manifold M means that local coordinates may be chosen in such a way that the transition functions satisfy the condition

$$\det(\partial x_i^{m}/\partial x_j^{n}) > 0$$

for any two charts U_i and U_j with non-empty overlap. The orientedness is needed to define an integration over the manifold.

We anticipate that the reader is familiar with the concept of tensors and tensor fields on manifolds. In particular, a vector field is defined in any chart U by smooth functions $v^m(x)$,such that the operator $v(p) = v^m \partial/\partial x^m|_p, p \in M$, does not depend on the choice of chart. The set of vectors $\{\partial/\partial x^m|_p\}$ forms a basis (holonomic basis) in the tangent space $T_p(M)$ at $p \in M$. A covector field is given in any coordinate chart U by smooth functions $w_m(x)$, such that the one-form $w(p) = w_m(x)\, dx^m|_p$ does not depend on the choice of chart. The set of one-forms $\{dx^m|_p\}$ is a basis in the cotangent space $T_p^*(M)$ at point $p \in M$.

Any manifold always admits a globally well-defined smooth Riemannian metric but not, in general, a pseudo-Riemannian metric. It can be shown that a manifold admits a metric of Lorentzian signature $(-, +, +, +)$ if and only if there exists a global vector field, non-vanishing at each point of the manifold (global line field). In fact, any non-compact manifold admits a global line field and, hence, a Lorentzian metric. For a compact orientable manifold, existence of a Lorentzian metric is equivalent to the fact that the manifold has zero Euler–Poincaré characteristic.

For definiteness, we shall assume through this text that our space-time manifold is a non-compact, topologically Euclidean manifold. This means that the manifold M, considered as a topological space, is homeomorphic to Euclidean space $\mathbb{R}^4$. In particular, M may be covered by a single chart with local coordinates x^m, $m=0, 1, 2, 3$.

The choice of local coordinates on M is usually a matter of convenience. So, the group of (invertible) general coordinate transformations

$$x^m \to x'^m = f^m(x) \qquad \det(\partial f^m/\partial x_n) \neq 0 \tag{1.6.1a}$$

or, in the infinitesimal form,

$$x^m \to x'^m = x^m - K^m(x) \tag{1.6.1b}$$

naturally acts on M. Every general coordinate transformation changes components of tensor fields according to tensorial laws. For example, given a vector field $v = v^m(x)\partial_m$, $\partial_m = \partial/\partial x^m$, its components in new local coordinates are

$$v'^m(x') = \frac{\partial x'^m}{\partial x^n} v^n(x).$$

In the infinitesimal case, we have

$$\delta v^m(x) = v'^m(x) - v^m(x) = K^n \partial_n v^m - v^n \partial_n K^m \tag{1.6.2}$$

or, in terms of a Lie bracket,

$$\delta v^m \partial_m = [K^n \partial_n, v^m \partial_m]. \tag{1.6.3}$$

Analogously, the transformation law

$$\Phi'(x') = \Phi(x)$$

of a scalar field $\Phi(x)$ can be written in the infinitesimal case as

$$\delta\Phi(x) = \Phi'(x) - \Phi(x) = [K^n \partial_n, \Phi]. \tag{1.6.4}$$

When a space–time metric $ds^2 = g_{mn}(x)\, dx^m\, dx^n$ has been introduced, one can define (in addition to the general coordinate transformation group) an action on M of the so called local Lorentz group by the following rule. In the tangent space $T_p(M)$ at some point $p \in M$, we consider an orthonormal frame $\{e_a\}$, $e_a = e_a{}^m \partial_m|_p$, where $a=0, 1, 2, 3$,

$$\langle e_a, e_b \rangle \equiv g_{mn} e_a{}^m e_b{}^n = \eta_{ab}.$$

Then, an infinitesimal transformation

$$e_a \to e'_a = e_a + K_a{}^b e_b \qquad K_{ab} \equiv K_a{}^c \eta_{cb} = -K_{ba}$$

transfers the frame $\{e_a\}$ to another orthonormal frame $\{e'_a\}$, and therefore represents some infinitesimal Lorentz transformation in $T_p(M)$. Now let $\{e_a(p)\}$, $e_a(p) = e_a{}^m(x)\partial_m$, be a set of smooth vector fields on M, forming an orthonormal frame at each point $p \in M$,

$$g_{mn}(x) e_a{}^m(x) e_b{}^n(x) = \eta_{ab}. \tag{1.6.5}$$

The set $\{e_a(p)\}$ is called a 'vierbein'. Then, the relation

$$e_a{}^m(x) \to e'_a{}^m(x) = e_a{}^m(x) + K_a{}^b(x) e_b{}^m(x) \tag{1.6.6}$$
$$K_{ab} = -K_{ba}$$

defines an infinitesimal local Lorentz transformation. Here K_{ab} are scalar fields on M. Exponentiating equation (1.6.6), we obtain finite local Lorentz transformations.

In contrast to the holonomic basis $\{\partial/\partial x^m\}$, the vierbein forms, as usual, an anholonomic basis in the sense that a commutator of basis fields does not vanish

$$\begin{gathered}[e_a, e_b] = \mathscr{C}_{ab}{}^c(x) e_c \\ \mathscr{C}_{ab}{}^c = \{(e_a e_b{}^m) - (e_b e_a{}^m)\} e_m{}^c\end{gathered} \tag{1.6.7}$$

where $\mathscr{C}_{ab}{}^c$ are the 'anholonomy coefficients' and $e_m{}^c(x)$ is the inverse vierbein,

$$e_m{}^a e_a{}^n = \delta_m{}^n \qquad e_a{}^m e_m{}^b = \delta_a{}^b. \tag{1.6.8}$$

By virtue of equations (1.6.6, 8), the local Lorentz transformations act on the inverse vierbein as follows

$$e_m{}^a(x) \to e'_m{}^a(x) = e_m{}^a(x) + K^a{}_b(x) e_m{}^b(x). \tag{1.6.9}$$

Owing to (1.6.5), the inverse vierbein satisfies the relation

$$g_{mn}(x) = \eta_{ab} e_m{}^a(x) e_n{}^b(x). \tag{1.6.10}$$

It is clear that the local Lorentz transformations do not change the metric. So any two vierbeins connected by some local Lorentz transformation are physically equivalent. Note that the smooth one-forms $\{e^a = e_m{}^a \, dx^m\}$ represent a basis in the cotangent space $T^*_p(M)$ of any point $p \in M$.

Let $v(p)$ be a vector field. In the tangent spaces $T_p(M)$ there are two natural frames: one with curved-space indices $\{\partial/\partial x^m\}$ and another with flat-space indices $\{e_a\}$. Decomposing the vector field with respect to these two frames, one obtains

$$v(p) = v^m \, \partial/\partial x^m = v^a e_a$$
$$v^a = e_m{}^a v^m \qquad v^m = e_a{}^m v^a.$$

The components v^m with curved-space indices transform by the vector law (1.6.2) with respect to the general coordinate group and stay invariant with respect to the local Lorentz group. On the other hand, the components v^a with flat-space indices are scalar fields with respect to the general coordinate group and transform by the vector law with respect to the local Lorentz group,

$$v^a(x) \to v'^a(x) = v^a(x) + K^a{}_b(x)v^b(x).$$

This example illustrates the general situation. Namely, starting from a world tensor (a tensor with curved-space indices only) and using the vierbein and its inverse, one can convert all curved-space indices into flat-space ones, obtaining as a result an object which is a world scalar field and a Lorentz tensor field. Of course, we can also consider tensor fields which carry curved-space indices at the same time. The frame $e_a{}^m$ gives such an example.

Note that in this text we use the following notational conventions. Small letters from the beginning of the Latin alphabet are used for flat-space vector indices and small letters from the middle are used for curved-space indices.

In principle, there is no great advantage in working with Lorentz tensors instead of world tensors. The main importance of the local Lorentz group structure on a space-time manifold lies in the fact that spinor fields, which are used for describing half-integer spin particles, can be defined only as (linear) representations of the Lorentz group. There is no way to realize spinors as linear representations of the general coordinate group.

In conclusion, we write down the transformation law of a spin-tensor field $\Phi_{\alpha_1 \ldots \alpha_A \dot{\alpha}_1 \ldots \dot{\alpha}_B}$ with A undotted indices and B dotted indices with respect to infinitesimal general coordinate (1.6.1b) and local Lorentz (1.6.6) transformations:

$$\begin{aligned} \delta\Phi_{\alpha_1 \ldots \alpha_A \dot{\alpha}_1 \ldots \dot{\alpha}_B}(x) &= \Phi'_{\alpha_1 \ldots \alpha_A \dot{\alpha}_1 \ldots \dot{\alpha}_B}(x) - \Phi_{\alpha_1 \ldots \alpha_A \dot{\alpha}_1 \ldots \dot{\alpha}_B}(x) \\ &= \left(K^n \partial_n + \frac{1}{2} K^{ab} M_{ab} \right) \Phi_{\alpha_1 \ldots \alpha_A \dot{\alpha}_1 \ldots \dot{\alpha}_B}(x) \end{aligned} \tag{1.6.11}$$

where M_{ab} are the Lorentz generators.

1.6.2. Covariant differentiation of world tensors

The prescription of how to covariantly differentiate world tensors is well known. One simply has to introduce the Christoffel symbols

$$\Gamma^k{}_{mn} = \frac{1}{2} g^{kp}(\partial_m g_{pn} + \partial_n g_{pm} - \partial_p g_{mn}) \tag{1.6.12}$$

and to replace operators ∂_m by 'covariant derivatives' ∇_m defined as follows:

$$\begin{aligned} \nabla_m v^n &= \partial_m v^n + \Gamma^n{}_{mp} v^p \\ \nabla_m w_n &= \partial_m w_n - \Gamma^p{}_{mn} w_p \end{aligned} \tag{1.6.13}$$

and so on. Given some world tensor, its covariant derivative is also a world tensor. The metric g_{mn} is covariantly constant

$$\nabla_p g_{mn} = 0.$$

The covariant derivatives commute only when acting on a scalar field. In the tensor cases, we have

$$[\nabla_m, \nabla_n] v^k = \mathscr{R}^k{}_{pmn} v^p$$

$$[\nabla_m, \nabla_n] w_k = -\mathscr{R}^p{}_{kmn} w_p$$

and so on, where $\mathscr{R}^k{}_{pmn}$ is the 'curvature tensor':

$$\mathscr{R}^k{}_{pmn} = \partial_m \Gamma^k{}_{np} - \partial_n \Gamma^k{}_{mp} + \Gamma^k{}_{mr} \Gamma^r{}_{np} - \Gamma^k{}_{nr} \Gamma^r{}_{mp}. \tag{1.6.14}$$

and has the following algebraic properties:

$$\mathscr{R}_{kpmn} = -\mathscr{R}_{pkmn} = -\mathscr{R}_{kpnm} = \mathscr{R}_{mnkp}$$
$$\mathscr{R}^k{}_{pmn} + \mathscr{R}^k{}_{mnp} + \mathscr{R}^k{}_{npm} = 0 \tag{1.6.15}$$

where $\mathscr{R}_{kpmn} = g_{kl} \mathscr{R}^l{}_{pmn}$, as well as satisfying the Bianchi identities

$$\nabla_r \mathscr{R}^k{}_{pmn} + \nabla_m \mathscr{R}^k{}_{pnr} + \nabla_n \mathscr{R}^k{}_{prm} = 0. \tag{1.6.16}$$

Extracting from $\mathscr{R}^k{}_{pmn}$ the Ricci tensor $\mathscr{R}_{mn} = \mathscr{R}^k{}_{mkn}$, $\mathscr{R}_{mn} = \mathscr{R}_{nm}$, and the scalar curvature $\mathscr{R} = g^{mn} \mathscr{R}_{mn}$, one obtains the Weyl tensor

$$C^k{}_{pmn} = \mathscr{R}^k{}_{pmn} + \frac{1}{2}(\mathscr{R}_{mp}\delta^k_n - \mathscr{R}_{np}\delta^k_m + g_{mp}\mathscr{R}^k_n - g_{np}\mathscr{R}^k_m) + \frac{1}{6}(\delta^k_m g_{np} - \delta^k_n g_{mp})\mathscr{R}. \tag{1.6.17}$$

The Weyl tensor is traceless in any pair of its indices and has the same algebraic properties (1.6.15) as the curvature.

1.6.3 Covariant differentiation of the Lorentz tensor

To obtain a covariant differentiation rule moving some Lorentz tensor to another one, it is sufficient to introduce a spin connection $\omega_{mab}(x)$, $\omega_{mab} = -\omega_{mba}$, taking its values in the Lorentz algebra and transforming by the law

$$\delta\omega_{mab} = K^n \partial_n \omega_{mab} + (\partial_m K^n)\omega_{nab} - \partial_m K_{ab} + K_a{}^c \omega_{mcb} + K_b{}^c \omega_{mac} \tag{1.6.18}$$

with respect to the general coordinate and local Lorentz transformations. Then the operators

$$\hat{\nabla}_a = e_a{}^m \partial_m + \frac{1}{2}\omega_{abc} M^{bc} \equiv e_a{}^m \hat{\nabla}_m$$
$$\omega_{abc} = e_a{}^m \omega_{mbc} \tag{1.6.19}$$

have the following transformation law

$$\delta\hat{\nabla}_a = \left[K^m \partial_m + \frac{1}{2} K^{bc} M_{bc}, \hat{\nabla}_a \right] \tag{1.6.20}$$

when acting on tensors with flat-space indices only. Recalling equation (1.6.11), we see that the operators $\hat{\nabla}_a$ transfer any Lorentz tensor to another one (with an additional vector index). So, $\hat{\nabla}_a$ are 'Lorentz covariant derivatives'.

The covariant derivatives satisfy the algebra

$$\begin{gathered} [\hat{\nabla}_a, \hat{\nabla}_b] = \mathscr{T}_{ab}{}^c \hat{\nabla}_c + \frac{1}{2} \mathscr{R}_{ab}{}^{cd} M_{cd} \\ \mathscr{T}_{ab}{}^c = \mathscr{C}_{ab}{}^c + \omega_{ab}{}^c - \omega_{ba}{}^c \\ \mathscr{R}_{ab}{}^{cd} = e_a \omega_b{}^{cd} - e_b \omega_a{}^{cd} - \mathscr{C}_{ab}{}^f \omega_f{}^{cd} + \omega_a{}^{cf} \omega_{bf}{}^d - \omega_b{}^{cf} \omega_{af}{}^d \end{gathered} \tag{1.6.21}$$

where the anholonomy coefficients $\mathscr{C}_{ab}{}^c$ were defined in equation (1.6.7). The field strengths $\mathscr{T}_{ab}{}^c$ and $\mathscr{R}_{ab}{}^{cd}$ are called the 'torsion tensor' and the 'curvature tensor', respectively. In general, the vierbein and the spin connection are completely independent fields. This is clear from their physical interpretation: the vierbein is a gauge field for the general coordinate group, while the spin connection is a gauge field for the local Lorentz group. However, one can consider a geometry in which the vierbein and the spin connection are related to each other in a covariant way, due to some constraints on the torsion. For example, the torsion-free condition

$$\mathscr{T}_{ab}{}^c = 0 \tag{1.6.22}$$

determines, by virtue of (1.6.21), the spin connection in terms of vierbein as follows

$$\omega_{abc} = \frac{1}{2}(\mathscr{C}_{bca} + \mathscr{C}_{acb} - \mathscr{C}_{abc}). \tag{1.6.23}$$

In this case, the curvatures (1.6.14) and (1.6.21) appear to be the curved-space form and the flat-space form of the same tensor,

$$\mathscr{R}^a{}_{bcd} = e_k{}^a e_b{}^p e_c{}^m e_d{}^n \mathscr{R}^k{}_{pmn}. \tag{1.6.24}$$

Sometimes, it seems reasonable to consider tensors with both flat-space and cuved-space indices. Then one has to modify the definition of covariant derivatives (1.6.19) by including terms with the Christoffel symbols acting on curved-space indices. For example, if Ψ^a_m is a Lorentz vector and world covector field, then

$$\hat{\nabla}_m \Psi^a_n = \partial_m \Psi^a_n + \omega_m{}^a{}_b \Psi^b_n - \Gamma^p{}_{mn} \Psi^a_p.$$

In the torsion-free case (1.6.22), the vierbein is covariantly constant,

$$\hat{\nabla}_m e_n{}^a = 0$$

as a consequence of equations (1.6.12, 23). Now, the derivatives ∇_m (1.6.13) and $\hat{\nabla}_a$ (1.6.19) are consistent in the sense that they represent the curved-space form and the flat-space form of the same operator. For example, for a vector field $v^a = e_m{}^a v^m$, we have

$$\nabla_m v^n = e_m{}^a e_b{}^n \hat{\nabla}_a v^b.$$

Equation (1.6.24) is a consequence of the last assertion.

From now on, we consider only torsion-free covariant derivatives and denote operators ∇ and $\hat{\nabla}$ by the same symbol ∇.

1.6.4. Frame deformations

We are going to discuss a rather technical question—how to transform geometrical objects (the covariant derivatives and curvature) with respect to an arbitrary variation of the vierbein

$$e_a{}^m \to e_a{}^m + \delta e_a{}^m \qquad \delta e_a{}^m(x) = H_a{}^b(x) e_b{}^m(x) \tag{1.6.25}$$

where H is a second rank Lorentz tensor field. For this purpose, it is useful to decompose H into its symmetric and antisymmetric parts:

$$\begin{gathered} H_a{}^b(x) = K_a{}^b(x) + \mathcal{H}_a{}^b(x) \\ K_{ab} = -K_{ba} \qquad \mathcal{H}_{ab} = \mathcal{H}_{ba}. \end{gathered} \tag{1.6.26}$$

The frame deformation $\delta e_a{}^m = K_a{}^b e_b{}^m$ corresponds to a local Lorentz transformation, under which the covariant derivatives change as in equation (1.6.20), or, expanding the commutator, as

$$\delta \nabla_a = K_a{}^b \nabla_b - \frac{1}{2}(\nabla_a K^{bc}) M_{bc}$$

and the curvature changes as a fourth-rank Lorentz tensor. We must now study the case of a symmetric H.

Let us consider the frame deformation

$$\delta e_a{}^m = \mathcal{H}_a{}^b e_b{}^m \tag{1.6.27}$$

$\mathcal{H}$ being a symmetric tensor field. In accordance with equation (1.6.19), we can represent the corresponding changing of covariant derivatives in the form

$$\nabla_a \to \nabla'_a = \nabla_a + \mathcal{H}_a{}^b \nabla_b + \frac{1}{2} \hat{\omega}_a{}^{bc} M_{bc} \tag{1.6.28}$$

where $\hat{\omega}_a{}^{bc}$ is a 'connection deformation'. To determine it, one must impose the torsion-free condition on the derivatives ∇'_a. This leads to

$$\hat{\omega}_{abc} = \nabla_b \mathcal{H}_{ac} - \nabla_c \mathcal{H}_{ab}. \tag{1.6.29}$$

Commuting the derivatives ∇'_a, we find the change in the curvature, the Ricci tensor and the scalar curvature:

$$\delta \mathscr{R}_{abcd} = \nabla_a(\nabla_c \hat{H}_{bd} - \nabla_d \hat{H}_{bc}) - \nabla_b(\nabla_c \hat{H}_{ad} - \nabla_d \hat{H}_{ac}) + \hat{H}_{ak}\mathscr{R}^k{}_{bcd} - \hat{H}_{bk}\mathscr{R}^k{}_{acd} \Rightarrow$$

$$\delta \mathscr{R}_{ab} = \nabla^c \nabla_c \hat{H}_{ab} - \nabla_a \nabla^c \hat{H}_{bc} - \nabla_b \nabla^c \hat{H}_{ac} + \nabla_a \nabla_b \hat{H}^c_c + 2\hat{H}^{cd}\mathscr{R}_{cadb} \quad (1.6.30)$$

$$\delta \mathscr{R} = 2\nabla^c \nabla_c \hat{H}^a_a - 2\nabla^a \nabla^b \hat{H}_{ab} + 2\hat{H}^{ab}\mathscr{R}_{ab}.$$

Note that the vierbein deformation (1.6.27) induces the following (in fact, arbitrary) metric variation

$$\delta g_{mn} = -2\hat{H}_{mn} = -2e_m{}^a e_n{}^b \hat{H}_{ab}. \quad (1.6.31)$$

A particular transformation of the type (1.6.27)

$$e_a{}^m \to e_a{}^m + \sigma e_a{}^m \Rightarrow$$
$$g_{mn} \to g_{mn} - 2\sigma g_{mn} \quad (1.6.32)$$

is known as a 'Weyl transformation'. Making the specialization of equations (1.6.28–30) to the case $\hat{H}_{ab} = \sigma \eta_{ab}$, we find how the Weyl transformations change all geometrical objects:

$$\nabla_a \to \nabla_a + \sigma \nabla_a - (\nabla^b \sigma) M_{ab}$$

$$\delta \mathscr{R}_{abcd} = \eta_{bd} \nabla_a \nabla_c \sigma - \eta_{bc} \nabla_a \nabla_d \sigma + \eta_{ac} \nabla_b \nabla_d \sigma - \eta_{ad} \nabla_b \nabla_c \sigma + 2\sigma \mathscr{R}_{abcd} \quad (1.6.33)$$

$$\delta \mathscr{R}_{ab} = \eta_{ab} \nabla^c \nabla_c \sigma + 2\nabla_a \nabla_b \sigma + 2\sigma \mathscr{R}_{ab}$$

$$\delta \mathscr{R} = 6\nabla^c \nabla_c \sigma + 2\sigma \mathscr{R}.$$

The Weyl tensor (1.6.17) is seen to transform homogeneously:

$$\delta C_{abcd} = 2\sigma C_{abcd}. \quad (1.6.34)$$

1.6.5. The Weyl tensor

The Weyl tensor is an important characteristic of space–time. Namely the Weyl tensor measures whether our space–time is conformally flat or not. Recall that a space–time M is called 'conformally flat' if there exists a coordinate system on M in which the metric has the form

$$g_{mn}(x) = \varphi(x)\eta_{mn} \quad (1.6.35)$$

for some positive-definite scalar function φ on M. It can be shown that a space-time is conformally flat if and only if the Weyl tensor vanishes, i.e.

$$C_{abcd} = 0. \quad (1.6.36)$$

Now we give a deeper insight into the structure of the Weyl tensor. Recall that it is traceless in any pair of its indices and has all the algebraic properties (1.2.34*a*,*c*) of the curvature. Let us decompose C_{abcd} into its self-dual and antiself-dual components using the Levi–Civita tensor:

$$C_{(\pm)}{}^{ab}{}_{cd}=\frac{1}{2}C^{ab}{}_{cd}\mp\frac{i}{4}\varepsilon^{abef}C_{efcd}$$

$$\frac{1}{2}\varepsilon^{ab}{}_{ef}C_{(\pm)}{}^{ef}{}_{cd}=\pm iC_{(\pm)}{}^{ab}{}_{cd} \tag{1.6.37}$$

$$C_{abcd}=C_{(+)abcd}+C_{(-)abcd}.$$

Using equations (1.2.34*b*,*c*), one finds

$$C_{(\pm)}{}^{a}{}_{bad}=0 \qquad \varepsilon^{abcd}C_{(\pm)fbcd}=0 \tag{1.6.38}$$

and therefore

$$C_{(\pm)abcd}=C_{(\pm)cdab}. \tag{1.6.39}$$

We see that $C_{(\pm)abcd}$ is (anti) self-dual in the first and the second pairs of its indices. Further, making use of equation (1.6.37) and the properties of the Levi–Civita tensor, one can prove the identities

$$C_{(+)abcd}C_{(-)}{}^{abef}=0$$

$$C_{(\pm)abcd}C_{(\pm)}{}^{abcf}=\frac{1}{4}C^{2}_{(\pm)}\delta_{d}{}^{f} \tag{1.6.40}$$

where

$$C^{2}_{(\pm)}\equiv C_{(\pm)abcd}C_{(\pm)}{}^{abcd}.$$

Algebraically, the Weyl tensor and the Ricci curvature are independent. But they are connected by some differential relations. Indeed, based on the identities

$$\nabla^{d}\mathscr{R}_{dabc}=\nabla_{b}\mathscr{R}_{ac}-\nabla_{c}\mathscr{R}_{ab}$$

$$\nabla^{b}\mathscr{R}_{ab}=\frac{1}{2}\nabla_{a}\mathscr{R} \tag{1.6.41}$$

which are consequences of the Bianchi identities (1.6.16), one obtains

$$\nabla^{d}C_{dabc}=\frac{1}{2}(\nabla_{b}\mathscr{R}_{ac}-\nabla_{c}\mathscr{R}_{ab})+\frac{1}{12}(\eta_{ab}\nabla_{c}\mathscr{R}-\eta_{ac}\nabla_{b}\mathscr{R}). \tag{1.6.42}$$

1.6.6. Four-dimensional topological invariants

In four dimensions, there exist two functionals, quadratic in curvature, with purely topological origin: the Pontrjagin invariant

$$P=\int d^{4}x\, e^{-1}(C^{2}_{(+)}-C^{2}_{(-)}) \qquad e=\det(e_{a}{}^{m}) \tag{1.6.43}$$

and the Euler invariant

$$\chi=\int \mathrm{d}x\, e^{-1}\left(C^2_{(+)}+C^2_{(-)}-2\mathscr{R}^{ab}\mathscr{R}_{ab}+\frac{2}{3}\mathscr{R}^2 \right). \qquad (1.6.44)$$

Being explicitly constructed from a metric g_{mn}, P and χ do not change under its arbitrary variations

$$\frac{\delta}{\delta g_{mn}}P=0 \qquad \frac{\delta}{\delta g_{mn}}\chi=0. \qquad (1.6.45)$$

Therefore, the Pontrjagin invariant and the Euler invariant depend only on the topological structure of space–time. To prove relations (1.6.45), we employ the results of two previous subsections.

As usual, we represent the metric in the form (1.6.10) and consider an arbitrary vierbein variation (1.6.25). The functionals P and χ are scalars with respect to the general coordinate and local Lorentz transformations. Thus, they are evidently invariant under the deformations (1.6.25) with any antisymmetric H_{ab}. It is convenient to start with the Weyl transformations (1.6.32) which leave invariant the functionals

$$I_{(\pm)}=\int \mathrm{d}^4x\, e^{-1}\, C^2_{(\pm)}$$

by virtue of the transformation laws (1.6.34) and $\delta e=4\sigma e$. Analogously, the functional

$$J=\int \mathrm{d}^4x\, e^{-1}\left(\mathscr{R}^{ab}\mathscr{R}_{ab}+\frac{1}{3}\mathscr{R}^2 \right)$$

is Weyl invariant as a consequence of equation (1.6.33). We see that the Pontrjagin invariant and the Euler invariant do not react to the Weyl deformations. On these grounds, it is sufficient to consider only the deformations (1.6.25) with a traceless symmetric H_{ab}. This is a tedious exercise, involving employment of the relations (1.6.30, 37–42), to show that

$$\begin{aligned}\delta I_{(+)}&=\delta I_{(-)}=\delta J\\ &=2\int \mathrm{d}^4x\, e^{-1}\, \hat{H}^{ab}\left\{ \nabla^c\nabla_c\mathscr{R}_{ab}-\frac{1}{3}\nabla_a\nabla_b\mathscr{R}-\frac{2}{3}\mathscr{R}\mathscr{R}_{ab}+2\mathscr{R}^{cd}\mathscr{R}_{cadb} \right\} \qquad (1.6.46)\\ &\hat{H}_{ab}=\hat{H}_{ba} \qquad \hat{H}^a_a=0.\end{aligned}$$

This completes the proof.

1.6.7. Einstein gravity and conformal gravity

We now recall two gravity models based on different gauge groups. The first one is Einstein gravity describing (as a field theory) propagation of a spin-two massless particle (the gravition). The theory is characterized by the action

$$S_G = \frac{1}{2\kappa^2}\int d^4x\, e^{-1}\mathscr{R} \tag{1.6.47}$$

where κ is a gravitational coupling constant. The gravitational field can be treated in terms of the vierbein or the metric. In the vierbein approach, the symmetry group of Einstein gravity is a product of the general coordinate group and the local Lorentz group. The vierbein transformation law is

$$\delta e_a{}^m = K^n \partial_n e_a{}^m - e_a K^m + K_a{}^b e_b{}^m. \tag{1.6.48}$$

It is instructive to rewrite this deformation in the form (1.6.25). Using the torsion-free condition, one finds

$$\begin{gathered}\delta e_a{}^m = H_a{}^b e_b{}^m + \tilde{K}_a{}^b e_b{}^m \\ H_{ab} = -\nabla_{(a}K_{b)} \qquad K^a = K^m e_m{}^a \\ \tilde{K}_{ab} = K_{ab} - K^c \omega_{cab} - \nabla_{[a}K_{b]}.\end{gathered} \tag{1.6.49}$$

In the metric approach, the symmetry group of Einstein gravity is reduced to the general coordinate group. The metric transformation law can be easily obtained from equations (1.6.31, 49):

$$\delta g_{mn} = \nabla_m K_n + \nabla_n K_m. \tag{1.6.50}$$

The equations of motion for S_G are

$$\mathscr{R}_{ab} - \frac{1}{2}\eta_{ab}\mathscr{R} = 0, \quad \Rightarrow \quad \mathscr{R}_{ab} = 0. \tag{1.6.51}$$

To derive them, one must use equation (1.6.30). Recalling equation (1.6.42), we see that the Weyl tensor satisfies the on-shell equations

$$\nabla^d C_{dabc} = 0. \tag{1.6.52}$$

The second gravity model we would like to discuss is conformal gravity. It is characterized by a larger gauge group with respect to Einstein gravity since the corresponding action

$$S_C = \frac{1}{\gamma}\int d^4x\, e^{-1}\, C^{abcd}C_{abcd} \tag{1.6.53}$$

is invariant, as has been shown above, under the Weyl transformations. The price for this additional symmetry is that S_C is a higher-derivative model. In the vierbein approach, S_C is invariant under the general coordinate, local Lorentz and Weyl transformations:

$$\delta e_a{}^m = K^n \partial_n e_a{}^m - e_a K^m + K_a{}^b e_b{}^m + \sigma e_a{}^m. \tag{1.6.54}$$

In the metric approach, the transformation law is given in the form:

$$\delta g_{mn} = \nabla_m K_n + \nabla_n K_m - 2\sigma g_{mn}. \tag{1.6.55}$$

The equations of motion for S_C are

$$\begin{gathered} \nabla^c \nabla_c \mathscr{R}_{ab} - \frac{1}{3} \nabla_a \nabla_b \mathscr{R} - \frac{2}{3} \mathscr{R} \mathscr{R}_{ab} + 2 \mathscr{R}^{cd} \mathscr{R}_{cadb} \\ - \frac{1}{6} \eta_{ab} (\nabla^c \nabla_c \mathscr{R} - \mathscr{R}^2 + 3 \mathscr{R}^{cd} \mathscr{R}_{cd}) = 0 \end{gathered} \tag{1.6.56}$$

as a consequence of equation (1.6.46). Any conformally flat metric is a solution of these equations.

1.6.8. Energy–momentum tensor

Let $S[\Phi, e_a{}^m]$ be a model of some field $\Phi = \{\Phi^i\}$ coupled to a gravity background. All information about coupling to gravity is encoded in the variational derivative of the action with respect to the vierbein

$$T_m{}^a(x) = -\delta S / \delta e_a{}^m(x). \tag{1.6.57}$$

Note that the variational derivative is defined as follows

$$\delta S = \int \mathrm{d}^4 x \, e^{-1} \, \delta e_a{}^m \frac{\delta S}{\delta e_a{}^m}.$$

The symmetric part of $T_m{}^a$

$$T^{(ab)} = \frac{1}{2} (e^{am} T_m{}^b + e^{bm} T_m{}^a) = 2 e_m{}^a e_n{}^b \frac{\delta S}{\delta g_{mn}} \tag{1.6.58}$$

is known as the 'energy–momentum tensor'.

We anticipate the model to be invariant under the transformations (1.6.48) supplemented by general coordinate and local Lorentz transformations of matter fields Φ. What are consequences this invariance leads to? Let the matter fields be on-shell,

$$\delta S / \delta \Phi = 0. \tag{1.6.59}$$

Then, the Lorentz invariance means that

$$\int \mathrm{d}^4 x \, e^{-1} \, K_a{}^b e_b{}^m T_m{}^a = 0$$

for any antisymmetric field $K_{ab}(x)$. So on-shell, T^{ab} has no antisymmetric part, and hence T^{ab} coincides with the energy–momentum tensor. The invariance under the general coordinate transformations, by virtue of equations (1.6.48, 49), means

$$\int \mathrm{d}^4x\, e^{-1}\, \nabla_{(a}K_{b)}T^{ab}=0$$

for any vector field $K^a(x)$. Therefore, the on-shell energy–momentum tensor satisfies the equation

$$\nabla_a T^{ab}=0. \tag{1.6.60}$$

Suppose, in addition, that the action $S[\Phi, e_a{}^m]$ is invariant under the Weyl transformations

$$\delta e_a{}^m=\sigma e_a{}^m \qquad \delta\Phi^i=\sigma d_{(i)}\Phi^i\equiv\sigma(d\Phi)^i \tag{1.6.61}$$

where $d_{(i)}$ are constants. Then, the on-shell energy–momentum tensor is traceless,

$$T^a_a=0 \tag{1.6.62}$$

as a consequence of conditions (1.6.61).

1.6.9. The covariant derivatives algebra in spinor notation

For completeness, we now rewrite the algebra (1.6.21) of the torsion-free covariant derivatives in two-component $SL(2,\mathbb{C})$ notation. This notation happens to be very useful when working with field theories in superspace which will be analysed in detail later.

Recalling the one-to-one correspondence (1.2.30) between Lorentz vectors and $SL(2,\mathbb{C})$ bi-spinors with one dotted and one undotted index, we introduce the bi-spinor derivatives

$$\nabla_{\alpha\dot\alpha}=(\sigma^a)_{\alpha\dot\alpha}\nabla_a. \tag{1.6.63}$$

Then the algebra (1.6.21) under the constraint $\mathscr{T}_{ab}{}^c=0$ takes the form

$$[\nabla_{\alpha\dot\alpha}, \nabla_{\beta\dot\beta}]=\frac{1}{2}\mathscr{R}_{\alpha\dot\alpha,\beta\dot\beta}{}^{cd}M_{cd}=\mathscr{R}_{\alpha\dot\alpha,\beta\dot\beta,\gamma\delta}M^{\gamma\delta}+\bar{\mathscr{R}}_{\alpha\dot\alpha,\beta\dot\beta,\dot\gamma\dot\delta}\bar{M}^{\dot\gamma\dot\delta} \tag{1.6.64}$$

$$\mathscr{R}_{\alpha\dot\alpha,\beta\dot\beta,\gamma\delta}=\frac{1}{2}(\sigma^a)_{\alpha\dot\alpha}(\sigma^b)_{\beta\dot\beta}(\sigma^{cd})_{\gamma\delta}\mathscr{R}_{abcd}$$

where we have used equation (1.2.33). Because of algebraic curvature constraints (1.6.15), $\mathscr{R}_{\alpha\dot\alpha,\beta\dot\beta,\gamma\delta}$ and $\bar{\mathscr{R}}_{\alpha\dot\alpha,\beta\dot\beta,\dot\gamma\dot\delta}$ can be decomposed in terms of their irreducible components as follows:

$$\begin{aligned}\mathscr{R}_{\alpha\dot\alpha,\beta\dot\beta,\gamma\delta}&=\varepsilon_{\dot\alpha\dot\beta}C_{\alpha\beta\gamma\delta}+\varepsilon_{\alpha\beta}E_{\gamma\delta\dot\alpha\dot\beta}+\varepsilon_{\dot\alpha\dot\beta}(\varepsilon_{\alpha\gamma}\varepsilon_{\beta\delta}+\varepsilon_{\alpha\delta}\varepsilon_{\beta\gamma})F\\ \bar{\mathscr{R}}_{\alpha\dot\alpha,\beta\dot\beta,\dot\gamma\dot\delta}&=\varepsilon_{\alpha\beta}\bar{C}_{\dot\alpha\dot\beta\dot\gamma\dot\delta}+\varepsilon_{\dot\alpha\dot\beta}\bar{E}_{\alpha\beta\dot\gamma\dot\delta}+\varepsilon_{\alpha\beta}(\varepsilon_{\dot\alpha\dot\gamma}\varepsilon_{\dot\beta\dot\delta}+\varepsilon_{\dot\alpha\dot\delta}\varepsilon_{\dot\beta\dot\gamma})\bar{F}.\end{aligned} \tag{1.6.65}$$

Here $C_{\alpha\beta\gamma\delta}$ is a completely symmetric tensor, and $E_{\gamma\delta\dot\alpha\dot\beta}$ is a tensor symmetric in its dotted and in its undotted indices. The spin-tensors from equations

(1.6.65) are connected with the curvature components by the rules:

$$F=\bar{F}=\frac{1}{12}\mathcal{R}$$

$$E_{\alpha\beta\dot{\alpha}\dot{\beta}}=\bar{E}_{\alpha\beta\dot{\alpha}\dot{\beta}}=\frac{1}{2}(\sigma^a)_{\alpha\dot{\alpha}}(\sigma^b)_{\beta\dot{\beta}}\left(\mathcal{R}_{ab}-\frac{1}{4}\eta_{ab}\mathcal{R}\right)$$

$$C_{\alpha\beta\gamma\delta}=\frac{1}{6}\{(\sigma^{ab})_{\alpha\beta}(\sigma^{cd})_{\gamma\delta}+(\sigma^{ab})_{\alpha\gamma}(\sigma^{cd})_{\delta\beta}+(\sigma^{ab})_{\alpha\delta}(\sigma^{cd})_{\beta\gamma}\}C_{(-)abcd} \quad (1.6.66)$$

$$\bar{C}_{\dot{\alpha}\dot{\beta}\dot{\gamma}\dot{\delta}}=\frac{1}{6}\{(\tilde{\sigma}^{ab})_{\dot{\alpha}\dot{\beta}}(\tilde{\sigma}^{cd})_{\dot{\gamma}\dot{\delta}}+(\tilde{\sigma}^{ab})_{\dot{\alpha}\dot{\gamma}}(\tilde{\sigma}^{cd})_{\dot{\delta}\dot{\beta}}+(\tilde{\sigma}^{ab})_{\dot{\alpha}\dot{\delta}}(\tilde{\sigma}^{cd})_{\dot{\beta}\dot{\gamma}}\}C_{(+)abcd}.$$

To derive the last two relations, we have used the observation that the matrices σ^{ab} and $\tilde{\sigma}^{ab}$ are (anti) self-dual:

$$\begin{aligned}\frac{1}{2}\varepsilon^{abcd}\sigma_{cd}&=-\mathrm{i}\sigma^{ab}\\ \frac{1}{2}\varepsilon^{abcd}\tilde{\sigma}_{cd}&=\mathrm{i}\tilde{\sigma}^{ab}.\end{aligned} \quad (1.6.67)$$

Finally, the Bianchi identities (1.6.41, 42) take the form

$$\nabla^{\alpha\dot{\alpha}}E_{\alpha\beta\dot{\alpha}\dot{\beta}}=-\frac{1}{2}\nabla_{\beta\dot{\beta}}\mathcal{R} \quad (1.6.68)$$

$$3\nabla^{\rho}{}_{\dot{\rho}}C_{\alpha\beta\gamma\rho}=\nabla_{\alpha}{}^{\dot{\alpha}}E_{\gamma\beta\dot{\alpha}\dot{\rho}}+\nabla_{\beta}{}^{\dot{\alpha}}E_{\alpha\gamma\dot{\alpha}\dot{\rho}}+\nabla_{\gamma}{}^{\dot{\alpha}}E_{\beta\alpha\dot{\alpha}\dot{\rho}}.$$

1.7. The conformal group

1.7.1. Conformal Killing vectors

Let M be a space–time manifold with local coordinates x^m and metric $ds^2=g_{mn}(x)\,dx^m\,dx^n$ (of Lorentzian signature). Given a vector field $\xi=\xi^m(x)\partial_m$ we can define an infinitesimal general coordinate transformation

$$x^m\to x'^m=x^m+\xi^m(x) \quad (1.7.1)$$

which changes the metric as follows

$$\delta g_{mn}(x)=g'_{mn}(x)-g_{mn}(x)=-\nabla_m\xi_n-\nabla_n\xi_m. \quad (1.7.2)$$

A vector field $\xi^m(x)$ is called a 'conformal Killing' vector if it satisfies the equation

$$\nabla_m\xi_n+\nabla_n\xi_m=\frac{1}{2}g_{mn}(\nabla_k\xi^k). \quad (1.7.3)$$

Any conformal Killing vector induces the general coordinate transformation which locally scales the metric

$$\delta g_{mn} = -2\sigma[\xi]g_{mn} \qquad \sigma[\xi] = \frac{1}{4}\nabla_k\xi^k. \tag{1.7.4}$$

It is convenient to introduce some vierbein $e_a{}^m(x)$ for the metric and to convert curved-space indices in equation (1.7.3) into flat-space ones in standard fashion: one has to change ξ^m to $\xi^a = \xi^m e_m{}^a$ and ∇_m to the Lorentz covariant derivatives ∇_a. Then equation (1.7.3) takes the form

$$\nabla_a\xi_b + \nabla_b\xi_a = \frac{1}{2}\eta_{ab}(\nabla_c\xi^c). \tag{1.7.5}$$

Equation (1.7.4) is now equivalent to the identity

$$-\left[\xi^c\nabla_c + \frac{1}{2}K[\xi]^{bc}M_{bc}, \nabla_a\right] = \sigma[\xi]\nabla_a + (\nabla^c\sigma[\xi])M_{ca}$$

$$K[\xi]^{bc} = \frac{1}{2}\nabla^b\xi^c - \frac{1}{2}\nabla^c\xi^b \tag{1.7.6}$$

where M_{bc} are the Lorentz generators. This identity means that the composition of the general coordinate transformation

$$x^m \to x'^m = x^m - \xi^m(x)$$
$$\delta e_a{}^m = \xi e_a{}^m - e_a\xi^m \tag{1.7.7a}$$

the local Lorentz transformation

$$\delta e_a{}^m = K[\xi]_a{}^b e_b{}^m - \xi^c\omega_{ca}{}^b e_b{}^m \tag{1.7.7b}$$

and the Weyl transformation

$$\delta e_a{}^m = \sigma[\xi]e_a{}^m \tag{1.7.7c}$$

where ξ^m is a conformal Killing vector, do not change the vierbein, and hence the metric.

1.7.2. Conformal Killing vectors in Minkowski space

Not every manifold admits conformal Killing vectors because of global restrictions on the curvature and topology of the manifold. Minkowski space admits non-trivial solutions of equation (1.7.5). Now we find all conformal Killing vectors in Minkowski space, where equation (1.7.5) takes the form

$$\partial_a\xi_b + \partial_b\xi_a = \frac{1}{2}\eta_{ab}(\partial_c\xi^c). \tag{1.7.8}$$

It has the evident consequences:

$$\Box\xi_a = -\frac{1}{2}\partial_a(\partial_c\xi^c) \qquad \Box \equiv \partial^a\partial_a \tag{1.7.9}$$

$$\Box(\partial_c\xi^c) = 0.$$

After applying the master equation (1.7.8) once more, we find

$$\begin{gathered}\Box(\partial_a\xi_b + \partial_b\xi_a) = 0 \\ \partial_a\partial_b(\partial_c\xi^c) = 0.\end{gathered} \tag{1.7.10}$$

Keeping in mind the last identity, we differentiate the master equation twice, resulting in

$$\partial_c\partial_d\partial_a\xi_b + \partial_c\partial_d\partial_b\xi_a = 0 \quad \Rightarrow$$

$$\partial_c\partial_a\partial_b\xi_d + \partial_c\partial_a\partial_d\xi_b = 0$$

$$\partial_c\partial_b\partial_d\xi_a + \partial_c\partial_b\partial_a\xi_d = 0.$$

These three equations have the general solution

$$\partial_c\partial_d\partial_a\xi_b = 0. \tag{1.7.11}$$

We see that conformal Killing vectors in Minkowski space are at most quadratic in space–time coordinates. This observation along with the master equation, leads to the final expression for conformal Killing vectors:

$$\begin{gathered}\xi^a(x) = b^a + \Delta x^a + K^a{}_b x^b + f^a x^2 - 2x^a(x, f) \\ K_{ab} = -K_{ba} \qquad (x, f) \equiv x^a f_a\end{gathered} \tag{1.7.12}$$

where b^a, Δ, K_{ab}, f^a are arbitrary real constant parameters. Evidently, the parameters b^a and K_{ab} correspond to the Poincaré transformations. The parameters Δ and f^a induce infinitesimal space–time transformations of the form (1.7.1) known as 'dilatations' (or scaling transformations) and 'special conformal boosts', respectively.

1.7.3. The conformal algebra

Let $\xi_{(1)} = \xi^a_{(1)}\partial_a$ and $\xi_{(2)} = \xi^a_{(2)}\partial_a$ be two conformal Killing vectors in Minkowski space. Their Lie bracket

$$[\xi_{(1)}, \xi_{(2)}] = \xi_{(3)} = \xi^a_{(3)}\partial_a$$

$$\xi^a_{(3)} = \xi^b_{(1)}\partial_b\xi^a_{(2)} - \xi^b_{(2)}\partial_b\xi^a_{(1)}$$

gives the vector field $\xi_{(3)}$ which also satisfies the master equation (1.7.8). Therefore, the set of all conformal Killing vectors forms a Lie algebra called the 'conformal algebra'.

We introduce a basis $\{P_a, J_{ab}, D, V_a\}$ for the conformal algebra by the rule:

$$\xi = \xi^a(x)\partial_a = \mathrm{i}\left\{b^a P_a - \frac{1}{2}K^{ab}J_{ab} - \Delta D + f^a V_a\right\}$$

for any conformal Killing vector $\xi^a(x)$ of the form (1.7.12), where

$$\begin{aligned} P_a &= -\mathrm{i}\partial_a \qquad J_{ab} = \mathrm{i}(x_b\partial_a - x_a\partial_b) \\ D &= \mathrm{i}x^a\partial_a \qquad V_a = \mathrm{i}(2x_a x^b\partial_b - x^2\partial_a). \end{aligned} \tag{1.7.13}$$

The basis vector fields satisfy the algebra

$$\begin{aligned} &[D, P_a] = -\mathrm{i}P_a \qquad [D, V_a] = \mathrm{i}V_a \\ &[V_a, P_b] = 2\mathrm{i}\eta_{ab}D - 2\mathrm{i}J_{ab} \\ &[J_{ab}, P_c] = \mathrm{i}\eta_{ac}P_b - \mathrm{i}\eta_{bc}P_a \\ &[J_{ab}, V_c] = \mathrm{i}\eta_{ac}V_b - \mathrm{i}\eta_{bc}V_a \\ &[J_{ab}, J_{cd}] = \mathrm{i}\eta_{ac}J_{bd} - \mathrm{i}\eta_{ad}J_{bc} + \mathrm{i}\eta_{bd}J_{ac} - \mathrm{i}\eta_{bc}J_{ad}. \end{aligned} \tag{1.7.14}$$

All other commutators vanish. These commutation relations define the conformal algebra. So we can forget about its explicit realization (1.7.13) and postulate the conformal algebra as an abstract real Lie algebra subject to two requirements: (1) it has a basis $\{\boldsymbol{\rho}_a, \mathbf{j}_{ab} = -\mathbf{j}_{ba}, \mathbf{d}, \mathbf{v}_a\}$ with multiplication law as in (1.7.14); (2) its general element X is of the form

$$X = \mathrm{i}\left(-b^a\boldsymbol{\rho}_a + \frac{1}{2}K^{ab}\mathbf{j}_{ab} + \Delta\mathbf{d} - f^a\mathbf{v}_a\right) \tag{1.7.15}$$

where b^a, $K^{ab} = -K^{ba}$, Δ, f^a are real parameters.

The 'conformal group' is formally obtained by exponentiation of the conformal algebra. So, nearly all its elements can be written as

$$g = \exp\left[\mathrm{i}\left(-b^a\boldsymbol{\rho}_a + \frac{1}{2}K^{ab}\mathbf{j}_{ab} + \Delta\mathbf{d} - f^a\mathbf{v}_a\right)\right]. \tag{1.7.16}$$

This definition is rather formal. Now we give two realizations of the conformal group.

1.7.4. Conformal transformations

The conformal group can be formally realized as a group of nonlinear transformations in Minkowski space

$$x^a \to x'^a = g \cdot x^a.$$

For infinitesimal group elements

$$g \approx 1 + \mathrm{i}\left(-b^a\boldsymbol{\rho}_a + \frac{1}{2}K^{ab}\mathbf{j}_{ab} + \Delta\mathbf{d} - f^a\mathbf{v}_a\right)$$

we define the corresponding transformations as follows:

$$x^a \to g \cdot x^a = x^a + \xi^a(x) \tag{1.7.17}$$

where $\xi^a(x)$ is the conformal Killing vector (1.7.12). Exponentiation leads to the transformations:

1. *Translations*

$$\mathrm{e}^{-\mathrm{i}b^c \rho_c} \cdot x^a = x^a + b^a \tag{1.7.18a}$$

2. *Lorentz transformations*

$$\mathrm{e}^{\mathrm{i}(1/2)K^{bc}\mathbf{j}_{bc}} \cdot x^a = (\exp K)^a{}_b x^b \tag{1.7.18b}$$

3. *Dilatations*

$$\mathrm{e}^{\mathrm{i}\Delta \mathbf{d}} \cdot x^a = \mathrm{e}^{\Delta} x^a \tag{1.7.18c}$$

4. *Special conformal transformations*

$$\mathrm{e}^{-\mathrm{i}f^c \mathbf{v}_c} \cdot x^a = \frac{x^a + f^a x^2}{1 + 2(f, x) + f^2 x^2}. \tag{1.7.18d}$$

The only comment required concerns the derivation of equation (1.7.18*d*). For this purpose, let us consider the inversion transformation defined on a domain of Minkowski space:

$$R\colon\ x^a \to x'^a = x^a/x^2 \qquad R^2 = \mathbb{1}. \tag{1.7.19}$$

It is an easy exercise to show that the inversion locally scales the metric $\mathrm{d}s^2 = \eta_{ab}\,\mathrm{d}x^a\,\mathrm{d}x^b$,

$$\mathrm{d}s^2 \to \mathrm{d}s'^2 = \mathrm{d}s^2/(x^2)^2.$$

Hence, the inversion is a discrete conformal transformation. Now, one can check that the transformation

$$R\,\mathrm{e}^{-\mathrm{i}f^c \rho_c} R \cdot x^a$$

coincides with the right-hand side of equation (1.7.18*d*). On the other hand, in the case of infinitesimal parameters f^a, we have

$$R(1 - \mathrm{i}f^c \rho_c) R \cdot x^a = x^a + (f^a x^2 - 2x^a (f, x)) = (1 - \mathrm{i}f^c \mathbf{v}_c) \cdot x^a.$$

So, we obtain the identity

$$R\,\mathrm{e}^{-\mathrm{i}f^a \rho_a} R = \mathrm{e}^{-\mathrm{i}f^a \mathbf{v}_a} \tag{1.7.20}$$

which proves the relation (1.7.18*d*). Note that the following identities

$$R\,\mathrm{e}^{\mathrm{i}\Delta\mathbf{d}} R = \mathrm{e}^{-\mathrm{i}\Delta\mathbf{d}} \qquad R\,\mathrm{e}^{\mathrm{i}(1/2)K^{ab}\mathbf{j}_{ab}} R = \mathrm{e}^{\mathrm{i}(1/2)K^{ab}\mathbf{j}_{ab}}$$

also hold.

Remark. The inversion transformation (1.7.19) turns out to be indeterminate for the points of the light-cone surface $x^2=0$ in Minkowski space. This leads to the fact that the special conformal transformations (1.7.18*d*) are not defined globally on Minkowski space. Thus, the conformal group is not a true transformation group of Minkowski space; its global action can be constituted only on a compactified version of Minkowski space. When considering below finite conformal transformations on Minkowski space (or finite superconformal transformations on a superspace, see Chapters 2 and 6), our discussion will be rather formal. The reader should keep in mind that only infinitesimal conformal transformations prove to be well defined on Minkowski space; finite conformal transformations are well defined in general on some domains in Minkowski space.

1.7.5. Matrix realization of the conformal group

We now give an exact realization for the conformal group as a group of linear transformations acting in a six-dimensional space $\mathbb{R}^6$ with coordinates $y^I = y^0, y^1, \ldots, y^5$ and the metric

$$dS^2 = -(y^0)^2+(y^1)^2+\ldots+(y^4)^2-(y^5)^2 = y^a y_a+(y^4)^2-(y^5)^2 \qquad (1.7.21)$$

Let us consider the group $O(4,2)$ of linear homogeneous transformations $y^I \to y'^I = A^I{}_J y^J$ preserving the above metric. All $O(4,2)$-transformations move the 'light-cone' surface

$$y^a y_a+(y^4)^2-(y^5)^2=0 \qquad (1.7.22)$$

onto itself. $O(4,2)$ is a 15-dimensional Lie group consisting of four connected components. We identify the conformal group with the component of unit in $O(4,2)$ denoted by $SO(4,2)^{\uparrow}$ (the matrices from $SO(4,2)^{\uparrow}$ are specified by the requirements that their diagonal 2×2 and 4×4 blocks labelled by the indices 0, 5 and 1, 2, 3, 4, respectively, have positive determinants). Let us comment upon this definition.

It is useful to redefine the variables y^4 and y^5 by the rule

$$y^4=\frac{1}{2}(\alpha-\beta) \qquad y^5=\frac{1}{2}(\alpha+\beta).$$

Then equation (1.7.22) takes the form

$$y^a y_a = \alpha\beta.$$

We parametrize locally the 'light-cone' as follows

$$\begin{pmatrix} y^a \\ \alpha \\ \beta \end{pmatrix} = \begin{pmatrix} \alpha x^a \\ \alpha \\ \alpha x^2 \end{pmatrix} \qquad (1.7.23)$$

where x^a will be identified with the coordinates of Minkowski space. Now consider particular $SO(4,2)^{\uparrow}$ transformations given with respect to the variables (y^a, α, β) by the matrices

$$A_1 = \begin{pmatrix} \Lambda^a{}_d & b^a & 0 \\ 0 & 1 & 0 \\ 2b_c\Lambda^c{}_d & b^2 & 1 \end{pmatrix} \tag{1.7.24a}$$

$$A_2 = \begin{pmatrix} \delta^a{}_d & 0 & 0 \\ 0 & e^{-\Delta} & 0 \\ 0 & 0 & e^{\Delta} \end{pmatrix} \tag{1.7.24b}$$

$$A_3 = \begin{pmatrix} \delta^a{}_d & 0 & f^a \\ 2f_d & 1 & f^2 \\ 0 & 0 & 1 \end{pmatrix} \tag{1.7.24c}$$

where Λ is an element of the Lorentz group. The transformation (1.7.24a) acts on the surface (1.7.23) by the rule

$$\begin{pmatrix} \alpha' x'^a \\ \alpha' \\ a' x'^2 \end{pmatrix} = \begin{pmatrix} \alpha(\Lambda^a{}_b x^b + b^a) \\ \alpha \\ \alpha(\Lambda x + b)^2 \end{pmatrix}$$

so we recognize here some ordinary Poincaré transformation in Minkowski space. Analogously, the operators A_2 and A_3 act on the space–time as a dilatation and a special conformal transformation, respectively. Matrices of the form (1.7.24a–c) generate $SO(4,2)^{\uparrow}$. Note also that the six-dimensional transformation

$$y^a \to y^a \qquad \alpha \to \beta \qquad \beta \to \alpha$$

corresponds to the space–time inversion (1.7.19).

1.7.6. Conformal invariance

The conformal group turns our to be the space–time symmetry group of some massless field theories. One possible way to understand this assertion is as follows.

Consider a field theory in Minkowski space and suppose that it can be extended to a curved space so that its curved-space action is invariant under Weyl transformations of the metric (1.6.32) supplemented by some σ-dependent transformations of the matter fields. In detail, let $S[\Phi]=\int d^4x\mathscr{L}(\Phi)$ be an action describing the dynamics of some fields $\Phi=\{\Phi^i\}$ in Minkowski space. Let $g_{mn}(x)$ be a curved space–time metric. If $S[\Phi]$ is a massless field theory, it is possible, as a rule, to define a generally covariant action

$$S[\Phi, g_{mn}]=\int d^4x\sqrt{-g}\,\mathscr{L}(\Phi, g) \tag{1.7.25}$$

where $g=\det(g_{mn})$, such that:

1. In the flat-space limit, $S[\Phi, g_{mn}]$ reduces to the original action $S[\Phi]$,
$$S[\Phi, g_{mn}=\eta_{mn}]=S[\Phi] \tag{1.7.26}$$
where η_{mn} is the Minkowski metric.
2. The action $S[\Phi, g_{mn}]$ is invariant under the following transformations
$$\delta g_{mn}=-2\sigma g_{mn} \qquad \delta\Phi^i=\sigma d_{(i)}\Phi^i=\sigma(d\Phi)^i \tag{1.7.27}$$
which have the form (1.6.61) in the vierbein approach. The constants $d_{(i)}$ are known as 'conformal weights' of massless fields Φ^i.

We shall argue later that the local invariance (1.7.27) is possible in the massless case only.

Due to the Weyl invariance and general covariance, the action $S[\Phi, g_{mn}]$ does not change under the transformations

$$\delta\Phi=\xi^a\nabla_a\Phi+\frac{1}{2}K^{ab}M_{ab}\Phi+\sigma d\Phi=\xi^n\partial_n\Phi+\frac{1}{2}(K^{ab}+\xi^n\omega_n{}^{ab})M_{ab}\Phi+\sigma d\Phi$$

$$\delta e_a{}^m=-(\nabla_a\xi^b)e_b{}^m+K_a{}^be_b{}^m+\sigma e_a{}^m=\xi e_a{}^m-e_a\xi^m+(K_a{}^b+\xi^n\omega_{na}{}^b)e_b{}^m+\sigma e_a{}^m$$

where $\xi^m(x)$, $K^{ab}(x)=-K^{ba}$ and $\sigma(x)$ are arbitrary parameters (it has been supposed above that all Φ^i are Lorentz tensor fields). Now let $\xi^m(x)$ be a conformal Killing vector with respect to the metric g_{mn}. Then the composition of the general coordinate transformation (1.7.7*a*), the local Lorentz transformation (1.7.7*b*) and the Weyl transformation (1.7.7*c*) do not change the vierbein and, hence, the metric

$$\delta e_a{}^m=-(\nabla_a\xi^b)e_b{}^m+K[\xi]_a{}^be_b{}^m+\sigma[\xi]e_a{}^m=0.$$

So, the Weyl invariance (1.7.27) of $S[\Phi, g_{mn}]$ leads to the fact that every transformation of the form

$$\begin{gathered}-\delta\Phi=\xi^a\nabla_a\Phi+\frac{1}{2}K[\xi]^{ab}M_{ab}\Phi+\sigma[\xi]d\Phi\\ \delta e_a{}^m=0 \quad\Rightarrow\quad \delta g_{mn}=0\end{gathered} \tag{1.7.28}$$

where ξ^a is a conformal Killing vector, preserves the action $S[\Phi, g_{mn}]$.

Taking now the flat-space limit in equation (1.7.28) ($g_{mn} \to \eta_{mn}$, $\nabla_a \to \partial_a$ and so on) and using the boundary condition (1.7.26), one finds the action $S[\Phi]$ in Minkowski space to be invariant under conformal transformations

$$
\begin{gathered}
-\delta\Phi = \xi^a \partial_a \Phi + \frac{1}{2} K[\xi]^{bc} M_{bc} \Phi + \sigma[\xi] d\Phi \\
K[\xi]^{bc} = \frac{1}{2}(\partial^b \xi^c - \partial^c \xi^b) \qquad \sigma[\xi] = \frac{1}{4} \partial_c \xi^c
\end{gathered}
\tag{1.7.29}
$$

where ξ^a is an arbitrary conformal Killing vector (1.7.12).

Consider scale transformations with $\xi^a = \Delta x^a$ in equation (1.7.29):

$$\delta\Phi = -\Delta x^a \partial_a \Phi - \Delta d\Phi$$

or in the case of finite transformations,

$$\Phi'(x) = \mathrm{e}^{-\Delta d} \Phi(\mathrm{e}^{-\Delta} x). \tag{1.7.30}$$

It is clear now that the conformal weight $d_{(i)}$ coincides with the dimension of Φ^i. The action $S[\Phi]$ is invariant under (1.7.30). This invariance makes it possible to change arbitrary scales in the theory under consideration, and therefore anticipates the absence of fixed dimensional constants (otherwise, it implies an infinite number of dimensional constants). On these grounds, the conformal symmetry is admissible for massless theories only. To be more exact, a theory may admit conformal symmetry if it is massless or describes particles of all possible positive masses $0 < m < \infty$. Indeed, at the quantum level, the conformal symmetry means that in a Hilbert space of physical states we have a representation of the conformal algebra (1.7.14) by Hermitian operators $\{\mathbb{P}_a, \mathbb{J}_{ab}, \mathbb{D}, \mathbb{V}_a\}$, where $\mathbb{P}_a$ and $\mathbb{J}_{ab}$ are the Poincaré generators, $\mathbb{D}$ is the dilatation generator and $\mathbb{V}_a$ are the generators of special conformal transformations. Due to the identity

$$[\mathbb{D}, \mathbb{P}_a] = -\mathrm{i}\mathbb{P}_a$$

the squared mass operator $(-\mathbb{P}^a\mathbb{P}_a)$ is characterized by the following transformation law

$$\mathrm{e}^{-\mathrm{i}\Delta\mathbb{D}}(-\mathbb{P}^a\mathbb{P}_a)\,\mathrm{e}^{\mathrm{i}\Delta\mathbb{D}} = \mathrm{e}^{-2\Delta}(-\mathbb{P}^a\mathbb{P}_a)$$

with respect to the dilatations. So, if $|\Psi\rangle$ is some state of given positive mass m, then the state $|\Psi'\rangle = \mathrm{e}^{\mathrm{i}\Delta\mathbb{D}}|\Psi\rangle$ carries the mass $m' = \mathrm{e}^{-2\Delta} m$. Both states $|\Psi'\rangle$ and $|\Psi\rangle$ must belong, by virtue of the conformal invariance, to the same Hilbert space. Hence, conformally invariant theories describe only massless particles or particles with all possible positive masses.

1.7.7. Examples of conformally invariant theories

Our general consideration will be accompanied by three examples. The first example is the theory of a massless self-interacting scalar field $\eta(x)$ with the action

$$S[\eta] = -\frac{1}{2}\int d^4x \left\{\partial^a\eta\,\partial_a\eta + \frac{\lambda}{4!}\eta^4\right\} \tag{1.7.31}$$

where λ is a dimensionless coupling constant. The unique continuation of $S[\eta]$ in a curved space–time, consistent with the above given requirements, is

$$S[\eta, g_{mn}] = -\frac{1}{2}\int d^4x \sqrt{-g}\left\{g^{mn}\,\partial_m\eta\,\partial_n\eta + \frac{1}{6}\mathcal{R}\eta^2 + \frac{\lambda}{4!}\eta^4\right\} \tag{1.7.32}$$

where $\mathcal{R}$ is the scalar curvature. Using the transformation law (1.6.33) of the scalar curvature, one can prove that the action $S[\eta, g_{mn}]$ is invariant under the Weyl transformations

$$\delta g_{mn} = -2\sigma g_{mn} \qquad \delta\eta = \sigma\eta. \tag{1.7.33}$$

Now equation (1.7.29) says that the action $S[\eta]$ is invariant under the following conformal transformations

$$-\delta\eta = \xi^a\partial_a\eta + \frac{1}{4}(\partial_a\xi^a)\eta \tag{1.7.34}$$

where ξ^a is a conformal Killing vector. Making use of the master equation (1.7.8), one can explicitly check the invariance (1.7.34).

Our second example is the theory of a free massless Majorana spinor field $\Psi(x)$ with the action

$$S[\Psi] = -\frac{i}{2}\int d^4x \bar{\Psi}\gamma^a\,\partial_a\Psi. \tag{1.7.35}$$

Its continuation to a curved space–time is given by

$$S[\Psi, e_a{}^m] = -\frac{i}{2}\int d^4x\, e^{-1}\,\bar{\Psi}\gamma^a\,\nabla_a\Psi \tag{1.7.36}$$

where $e_a{}^m$ is the vierbein. This action is invariant under the Weyl transformations

$$\delta e_a{}^m = \sigma e_a{}^m \qquad \delta\Psi = \frac{3}{2}\sigma\Psi. \tag{1.7.37}$$

Then equation (1.7.29) shows that $S[\Psi]$ is invariant under the conformal transformations

$$-\delta\Psi = \xi^a\,\partial_a\Psi + \frac{1}{2}K^{bc}[\xi]\Sigma_{bc}\Psi + \frac{3}{8}(\partial_a\xi^a)\Psi. \tag{1.7.38}$$

Here we have used the Lorentz transformation law (1.4.11) of four-component spinors.

The last example is the Yang–Mills theory describing the dynamics of vector fields A_a taking values in a compact Lie algebra, $A_a = \mathrm{i} A_a{}^I T^I$ and T^I are the generators of the algebra, $\mathrm{tr}(T^I T^J) = \delta^{IJ}$. The action is

$$S[A] = \frac{1}{4g^2} \,\mathrm{tr} \int \mathrm{d}^4x \, F_{ab} F^{ab}. \tag{1.7.39}$$

The only continuation of this action to a curved space–time, consistent with the gauge invariance of $S[A]$, is

$$\begin{gathered} S[A, g_{mn}] = \frac{1}{4g^2} \,\mathrm{tr} \int \mathrm{d}^4x \sqrt{-g} \, F_{mn} F_{kl} g^{mk} g^{nl} \\ F_{mn} = \nabla_m A_n - \nabla_n A_m - [A_m, A_n]. \end{gathered} \tag{1.7.40}$$

Here A_m carries a curved-space index. The action $S[A, g_{mn}]$ is evidently invariant under the Weyl rescalings

$$\begin{cases} \delta g_{mn} = -2\sigma g_{mn} \\ \delta A_m = 0 \end{cases} \Leftrightarrow \begin{cases} \delta e_a{}^m = \sigma e_a{}^m \\ \delta A_a = \sigma A_a \end{cases} \tag{1.7.41}$$

where $A_a = e_a{}^m A_m$, and $e_a{}^m$ is some vierbein for the metric. Recalling equation (1.7.29), we conclude that the action $S[A]$ is invariant under the conformal transformations

$$-\delta A_a = \xi^c \partial_c A_a + \frac{1}{2} (\partial_a \xi^c - \partial^c \xi_a) A_c + \frac{1}{4} (\partial_c \xi^c) A_a. \tag{1.7.42}$$

1.7.8. Example of a non-conformal massless theory

Now we present an example of conformally non-invariant massless field theory. This is the model of a second-rank antisymmetric tensor field $B^{cd}(x)$ with the action

$$\begin{gathered} S(B] = \frac{1}{2} \int \mathrm{d}^4x L^a(B) L_a(B) \\ L_a(B) = \frac{1}{2} \varepsilon_{abcd} \, \partial^b B^{cd}. \end{gathered} \tag{1.7.43}$$

It is a simple exercise to show that the theory $S[B]$ cannot be continued to curved space–time in a way consistent with Weyl invariance. Moreover, in the next section, we give a direct proof of the fact that the Poincaré group is the maximal space–time symmetry group of $S[B]$.

1.8. The mass-shell field representations

In Section 1.5, the irreducible unitary representations (massive and massless) of the Poincaré group have been described. Now, we give their realizations in terms of tensor fields restricted by some supplementary conditions. We shall also consider irreducible massless field representations of the conformal group.

1.8.1. Massive field representations of the Poincaré group

Recall that the irreducible massive Poincaré representations are classified by mass and spin. For definiteness, we fix some mass value $m>0$. The irreducible massive spin-zero representation is seen to admit the only realization in terms of a scalar field $\Phi(x)$ under the mass-shell equation

$$\mathbf{P}^a\mathbf{P}_a\Phi=-m^2\Phi \qquad \mathbf{P}_a=-\mathrm{i}\partial_a. \tag{1.8.1}$$

However, in the case of some non-vanishing spin s, the irreducible massive spin-s representation admits several realizations in terms of fields.

Let us consider the linear space $\mathscr{H}_{(A,B)}$ of $(A/2, B/2)$-type spin-tensor fields $\Phi_{\alpha_1\alpha_2\ldots\alpha_A\dot{\alpha}_1\dot{\alpha}_2\ldots\dot{\alpha}_B}(x)$ totally symmetric in their A undotted indices and independently in their B dotted indices, $A+B=2s$, and satisfying the following supplementary condition

$$\partial^{\alpha\dot{\alpha}}\Phi_{\alpha\alpha_1\ldots\alpha_{A-1}\dot{\alpha}\dot{\alpha}_1\ldots\dot{\alpha}_{B-1}}(x)=0, \tag{1.8.2}$$

and the Klein–Gordon equation

$$(\Box-m^2)\Phi_{\alpha_1\alpha_2\ldots\alpha_A\dot{\alpha}_1\dot{\alpha}_2\ldots\dot{\alpha}_B}(x)=0. \tag{1.8.3}$$

Here $\partial_{\alpha\dot{\alpha}}=(\sigma^a)_{\alpha\dot{\alpha}}\partial_a$. As we shall show, the supplementary condition (1.8.2) is needed to select the spin-s representation. Note that this condition is absent in the cases $A=2s$, $B=0$ and $A=0$, $B=2s$.

One more restriction should be added to obtain a positive energy representation. Namely that in the 'momentum space' decomposition for $\Phi_{\alpha_1\alpha_2\ldots\alpha_A\dot{\alpha}_1\dot{\alpha}_2\ldots\dot{\alpha}_B}(x)$, only positive frequency plane waves should be kept

$$\Phi_{\alpha_1\alpha_2\ldots\alpha_A\dot{\alpha}_1\dot{\alpha}_2\ldots\dot{\alpha}_B}(x)=\int\frac{\mathrm{d}^3\vec{p}}{p^0}\,\mathrm{e}^{\mathrm{i}px}\,\Phi_{\alpha_1\alpha_2\ldots\alpha_A\dot{\alpha}_1\dot{\alpha}_2\ldots\dot{\alpha}_B}(p)$$
$$p^a=(p^0,\vec{p}) \qquad p^0=\sqrt{m^2+\vec{p}^{\,2}}. \tag{1.8.4}$$

This decomposition should be extended to include negative frequency modes when the discrete space–time symmetries (time reversal and parity) are taken into account. In momentum space, the restriction (1.8.2) takes the form

$$p^{\alpha\dot{\alpha}}\Phi_{\alpha\alpha_1\ldots\alpha_{A-1}\dot{\alpha}\dot{\alpha}_1\ldots\dot{\alpha}_{B-1}}(p)=0. \tag{1.8.5}$$

Note that the supplementary conditions (1.8.2) and (1.8.4) and the mass-shell equation (1.8.3) are invariant under Poincaré transformations.

Any field from $\mathscr{H}_{(A,B)}$ has $(A+B+1)$ independent components. To prove this, it is useful to employ the momentum space decomposition and to make the transition into the rest frame, in which $p_{\alpha\dot{\alpha}}=p^a(\sigma_a)_{\alpha\dot{\alpha}}=m\delta_{\alpha\dot{\alpha}}$; then equation (1.8.5) means that the rest-frame field components are totally symmetric in all their indices. There is another, more elegant, way to prove the above statement. Let us consider the following operator

$$\Delta_{\alpha\dot{\alpha}}=\frac{\mathrm{i}}{m}\mathbf{P}_{\alpha\dot{\alpha}}=\frac{1}{m}\partial_{\alpha\dot{\alpha}} \tag{1.8.6}$$

invertible through the equation (1.8.3),

$$\Delta_\alpha{}^{\dot{\alpha}}\Delta^\beta{}_{\dot{\alpha}}=\delta_\alpha{}^\beta \qquad \Delta^\alpha{}_{\dot{\alpha}}\Delta_\alpha{}^{\dot{\beta}}=\delta_{\dot{\alpha}}{}^{\dot{\beta}}. \tag{1.8.7}$$

Now, if $B\neq 0$, we can define a one-to-one map of $\mathscr{H}_{(A,B)}$ on $\mathscr{H}_{(A+1,B-1)}$ by the rule:

$$\Phi_{\alpha_1\ldots\alpha_A\dot{\alpha}_1\ldots\dot{\alpha}_B}(x)\rightarrow\Phi_{\alpha_1\ldots\alpha_A\alpha_{A+1}\dot{\alpha}_1\ldots\dot{\alpha}_{B-1}}(x)$$
$$=\Delta_{\alpha_{A+1}}{}^{\dot{\alpha}_B}\Phi_{\alpha_1\ldots\alpha_A\dot{\alpha}_1\ldots\dot{\alpha}_B}(x)\in\mathscr{H}_{(A+1,B-1)}. \tag{1.8.8}$$

Applying this operation B times, we obtain the one-to-one map of $\mathscr{H}_{(A,B)}$ on $\mathscr{H}_{(2s,0)}$. But all fields $\Phi_{\alpha_1\ldots\alpha_{2s}}(x)$ from $\mathscr{H}_{(2s,0)}$ are totally symmetric in their indices, so they have $(2s+1)$ independent components. Rather beautifully, this shows that all spaces $\mathscr{H}_{(2s,0)}$, $\mathscr{H}_{(2s-1,1)}$, $\ldots$, $\mathscr{H}_{(0,2s)}$ describe equivalent representations of the Poincaré group.

Now we are going to demonstrate that the spin operator $\mathbf{W}^a\mathbf{W}_a$, where $\mathbf{W}_a$ is the Pauli–Lubanski vector, is a multiple of the identity operator on $\mathscr{H}_{(A,B)}$. Owing to the explicit expressions (1.5.14) for the Poincaré generators, the Pauli–Lubanski vector is

$$\mathbf{W}_a=-\frac{1}{2}\varepsilon_{abcd}M^{bc}\partial^d. \tag{1.8.9}$$

It is worth changing here the Lorentz generators with vector indices to generators with spinor indices according to the rules (1.3.9) and (1.2.33):

$$\mathbf{W}_a=+\frac{1}{2}\varepsilon_{abcd}\partial^b\{-(\sigma^{cd})_{\alpha\beta}M^{\alpha\beta}+(\tilde{\sigma}^{cd})_{\dot{\alpha}\dot{\beta}}\bar{M}^{\dot{\alpha}\dot{\beta}}\}.$$

Making use of properties (1.6.67) for the σ-matrices, we obtain

$$\mathbf{W}_a=\mathrm{i}(\sigma_{ab})_{\alpha\beta}M^{\alpha\beta}\partial^b+\mathrm{i}(\tilde{\sigma}_{ab})_{\dot{\alpha}\dot{\beta}}\bar{M}^{\dot{\alpha}\dot{\beta}}\partial^b.$$

Finally, we convert the vector index 'a' into a pair of spinor indices, dotted and undotted, resulting in

$$\mathbf{W}_{\alpha\dot{\alpha}}=-\mathrm{i}\partial^\beta{}_{\dot{\alpha}}M_{\beta\alpha}+\mathrm{i}\partial_\alpha{}^{\dot{\beta}}\bar{M}_{\dot{\beta}\dot{\alpha}}. \tag{1.8.10}$$

Then, the spin operator takes the form

$$\mathbf{W}_a\mathbf{W}_a = -\frac{1}{2}\mathbf{W}^{\alpha\dot{\alpha}}\mathbf{W}_{\alpha\dot{\alpha}} = -\frac{m^2}{2}(M^{\alpha\beta}M_{\alpha\beta} + \bar{M}^{\dot{\alpha}\dot{\beta}}\bar{M}_{\dot{\alpha}\dot{\beta}}) + M_{\alpha\beta}\bar{M}_{\dot{\alpha}\dot{\beta}}\partial^{\alpha\dot{\alpha}}\partial^{\beta\dot{\beta}}. \tag{1.8.11}$$

Here we have used the mass-shell equation (1.8.3). The values of $M^{\alpha\beta}M_{\alpha\beta}$ and $\bar{M}^{\dot{\alpha}\dot{\beta}}\bar{M}_{\dot{\alpha}\dot{\beta}}$ in the $(n/2, m/2)$ representation series have been calculated in Section 1.3 (see equation (1.3.14)). So, we only need to determine the last term in (1.8.11). Recalling how $M_{\alpha\beta}$ and $\bar{M}_{\dot{\alpha}\dot{\beta}}$ act on spinor indices (equation (1.3.12)), one obtains

$$\begin{aligned}\partial^{\alpha\dot{\alpha}}\partial^{\beta\dot{\beta}}M_{\alpha\beta}\bar{M}_{\dot{\alpha}\dot{\beta}}\Phi_{\gamma_1\ldots\gamma_A\dot{\gamma}_1\ldots\dot{\gamma}_B} &= \sum_{k=1}^{B}\partial^{\alpha}{}_{\dot{\gamma}_k}\partial^{\beta\dot{\beta}}M_{\alpha\beta}\Phi_{\gamma_1\ldots\gamma_A\dot{\beta}\dot{\gamma}_1\ldots\hat{\dot{\gamma}}_k\ldots\dot{\gamma}_B}\\ &= \frac{1}{2}\sum_{p=1}^{A}\sum_{k=1}^{B}\partial_{\gamma_p\dot{\gamma}_k}\partial^{\beta\dot{\beta}}\Phi_{\beta\gamma_1\ldots\hat{\gamma}_p\ldots\gamma_A\dot{\beta}\dot{\gamma}_1\ldots\hat{\dot{\gamma}}_k\ldots\dot{\gamma}_B}\\ &\quad + \frac{1}{2}\sum_{p=1}^{A}\sum_{k=1}^{B}\partial^{\alpha}{}_{\dot{\gamma}_k}\partial_{\gamma_p}{}^{\dot{\beta}}\Phi_{\alpha\gamma_1\ldots\hat{\gamma}_p\ldots\gamma_A\dot{\beta}\dot{\gamma}_1\ldots\hat{\dot{\gamma}}_k\ldots\dot{\gamma}_B}.\end{aligned}$$

Here the first term vanishes due to the supplementary condition (1.8.2). For the same reason, the second term can be rewritten as

$$\frac{1}{2}\sum_{p=1}^{A}\sum_{k=1}^{B}\partial^{\alpha}{}_{\dot{\gamma}_k}\partial_{\alpha}{}^{\dot{\beta}}\Phi_{\gamma_1\ldots\gamma_A\dot{\beta}\dot{\gamma}_1\ldots\hat{\dot{\gamma}}_k\ldots\dot{\gamma}_B} = \frac{1}{2}ABm^2\Phi_{\gamma_1\ldots\gamma_A\dot{\gamma}_1\ldots\dot{\gamma}_B}$$

where we have used equation (1.8.3). The relation obtained together with equation (1.3.14) lead to the final result

$$\mathbf{W}^a\mathbf{W}_a|_{\mathscr{H}_{(A,B)}} = m^2 s(s+1)\mathbb{1} \qquad s = A/2 + B/2. \tag{1.8.12}$$

To summarize, we have shown that the massive spin-s Poincaré representation can be described in terms of $(A/2, B/2)$-type fields, $A + B = 2s$, restricted by equations (1.8.2, 3). Note that the irreducible integer spin representations are usually described by choosing $A = B$ and considering in $\mathscr{H}_{(s,s)}$ the subspace of real spin-tensor fields

$$\Phi_{\alpha_1\alpha_2\ldots\alpha_s\dot{\alpha}_1\dot{\alpha}_2\ldots\dot{\alpha}_s}(x) = \bar{\Phi}_{\alpha_1\alpha_2\ldots\alpha_s\dot{\alpha}_1\dot{\alpha}_2\ldots\dot{\alpha}_s}(x). \tag{1.8.13}$$

All spinor indices here can be converted into vector ones

$$\Phi_{a_1a_2\ldots a_s}(x) \equiv (-1)^s\frac{1}{2^s}(\tilde{\sigma}_{a_1})^{\dot{\alpha}_1\alpha_1}(\tilde{\sigma}_{a_2})^{\dot{\alpha}_2\alpha_2}\ldots(\tilde{\sigma}_{a_s})^{\dot{\alpha}_s\alpha_s}\Phi_{\alpha_1\alpha_2\ldots\alpha_s\dot{\alpha}_1\dot{\alpha}_2\ldots\dot{\alpha}_s}(x) \tag{1.8.14}$$

obtaining a real tensor field which is totally symmetric and traceless:

$$\Phi_{a_1a_2\ldots a_s} = \Phi_{(a_1a_2\ldots a_s)} \qquad \Phi^a{}_{aa_1a_2\ldots a_{s-2}} = 0. \tag{1.8.15}$$

The supplementary condition (1.8.2) is now

$$\partial^a\Phi_{aa_1a_2\ldots a_{s-1}}(x) = 0. \tag{1.8.16}$$

The irreducible half-integer spin representations are usually described by choosing $A-1=B$ or $B-1=A$. In the first case, we write any field from $\mathcal{H}_{(B+1,B)}$ as

$$\Psi_{\alpha\alpha_1\ldots\alpha_B\dot{\alpha}_1\ldots\dot{\alpha}_B}(x)$$

and then convert each pair $(\alpha_i,\dot{\alpha}_i)$, $i=1,\ldots,B$, into a vector index,

$$\Psi_{a_1\ldots a_B\alpha}(x)\equiv(-1)^B\frac{1}{2^B}(\tilde{\sigma}_{a_1})^{\dot{\alpha}_1\alpha_1}\ldots(\tilde{\sigma}_{a_B})^{\dot{\alpha}_B\alpha_B}\Psi_{\alpha\alpha_1\ldots\alpha_B\dot{\alpha}_1\ldots\dot{\alpha}_B}(x) \quad (1.8.17)$$

obtaining a spin-tensor which is totally symmetric, traceless and σ-traceless:

$$\Psi_{a_1a_2\ldots a_B\alpha}=\Psi_{(a_1a_2\ldots a_B)\alpha} \qquad \Psi^a{}_{aa_1\ldots a_{B-2}\alpha}=0$$
$$(\tilde{\sigma}^a)^{\dot{\alpha}\alpha}\Psi_{aa_1\ldots a_{B-1}\alpha}=0. \quad (1.8.18)$$

The supplementary condition (1.8.2) can now be written as

$$\partial^a\Psi_{aa_1\ldots a_{B-1}\alpha}(x)=0. \quad (1.8.19)$$

The second case, $B-1=A$, is treated analogously. Any field from $\mathcal{H}_{(A,A+1)}$ should be represented in the form

$$\Psi_{\alpha_1\ldots\alpha_A\dot{\alpha}_1\ldots\dot{\alpha}_A\dot{\alpha}}(x)$$

and then each pair $(\alpha_i,\dot{\alpha}_i)$, $i=1,\ldots,A$, transformed into a vector index,

$$\Psi_{a_1\ldots a_A\dot{\alpha}}\equiv(-1)^A\frac{1}{2^A}(\tilde{\sigma}_{a_1})^{\dot{\alpha}_1\alpha_1}\ldots(\tilde{\sigma}_{a_A})^{\dot{\alpha}_A\alpha_A}\Psi_{\alpha_1\ldots\alpha_A\dot{\alpha}\dot{\alpha}_1\ldots\dot{\alpha}_A} \quad (1.8.20)$$

The resultant spin-tensor satisfies the following algebraic

$$\Psi_{a_1a_2\ldots a_A\dot{\alpha}}=\Psi_{(a_1a_2\ldots a_A)\dot{\alpha}} \qquad \Psi^a{}_{aa_1\ldots a_{A-2}\dot{\alpha}}=0$$
$$(\sigma^a)_{\alpha\dot{\alpha}}\Psi_{aa_1\ldots a_{A-1}}{}^{\dot{\alpha}}=0 \quad (1.8.21)$$

and differential

$$\partial^a\Psi_{aa_1\ldots a_{A-1}\dot{\alpha}}(x)=0. \quad (1.8.22)$$

constraints.

In conclusion, note that, if $A\neq B$, the Poincaré representation on $\mathcal{H}_{(A,B)}$ is irreducible until considering the operation of complex conjugation.

1.8.2. Real massive field representations

We would like to continue analysis of the massive case and to consider real field representations. Each space $\mathcal{H}_{(A,B)}$ is assumed here to describe massive fields possessing both positive and negative frequency modes in their Fourier decomposition. Recall that the operation of complex conjugation provides us with a one-to-one mapping * (1.2.19) of the linear space of

$(A/2, B/2)$ tensors on the space of $(B/2, A/2)$ tensors and, as a consequence, of the mass-shell space $\mathscr{H}_{(A,B)}$ on $\mathscr{H}_{(B,A)}$. On the other hand, making use of operator $\Delta_{\alpha\dot{\alpha}}$ (1.8.6), we can define, by analogy with (1.8.8), the one-to-one mapping Δ of $\mathscr{H}_{(A,B)}$ on $\mathscr{H}_{(B,A)}$ as follows

$$\Phi_{\alpha_1\ldots\alpha_A\dot{\alpha}_1\ldots\dot{\alpha}_B}(x)\to\Phi_{\alpha_1\ldots\alpha_B\dot{\alpha}_1\ldots\dot{\alpha}_A}(x)=\Delta_{\alpha_1}{}^{\dot{\gamma}_1}\ldots\Delta_{\alpha_B}{}^{\dot{\gamma}_B}\Delta^{\gamma_1}{}_{\dot{\alpha}_1}\ldots\Delta^{\gamma_A}{}_{\dot{\alpha}_A}\Phi_{\gamma_1\ldots\gamma_A\dot{\gamma}_1\ldots\dot{\gamma}_B}(x).$$

Now we select a Poincaré invariant subspace $\mathscr{H}^R_{(A,B)}$ in $\mathscr{H}_{(A,B)}$, which is characterized by the coincidence of the maps * and Δ on it. Any field from $\mathscr{H}^R_{(A,B)}$ satisfies the equation

$$\bar{\Phi}_{\alpha_1\ldots\alpha_B\dot{\alpha}_1\ldots\dot{\alpha}_A}(x)=\Delta_{\alpha_1}{}^{\dot{\gamma}_1}\ldots\Delta_{\alpha_B}{}^{\dot{\alpha}_B}\Delta^{\gamma_1}{}_{\dot{\alpha}_1}\ldots\Delta^{\gamma_A}{}_{\dot{\alpha}_A}\Phi_{\gamma_1\ldots\gamma_A\dot{\gamma}_1\ldots\dot{\gamma}_B}(x). \quad (1.8.23)$$

We shall call such fields 'real massive fields'.

To clarify what equation (1.8.23) means, let us consider some particular cases. In the case $A=B=s$, we have

$$\begin{aligned}\bar{\Phi}_{\alpha_1\ldots\alpha_s\dot{\alpha}_1\ldots\dot{\alpha}_s}(x)&=\Delta_{\alpha_1}{}^{\dot{\gamma}_1}\ldots\Delta_{\alpha_s}{}^{\dot{\gamma}_s}\Delta^{\gamma_1}{}_{\dot{\alpha}_1}\ldots\Delta^{\gamma_s}{}_{\dot{\alpha}_s}\Phi_{\gamma_1\ldots\gamma_s\dot{\gamma}_1\ldots\dot{\gamma}_s}(x)\\&=\Delta_{\gamma_1}{}^{\dot{\gamma}_1}\ldots\Delta_{\gamma_s}{}^{\dot{\gamma}_s}\Delta^{\gamma_1}{}_{\dot{\alpha}_1}\ldots\Delta^{\gamma_s}{}_{\dot{\alpha}_s}\Phi_{\alpha_1\ldots\alpha_s\dot{\gamma}_1\ldots\dot{\gamma}_s}(x)=\Phi_{\alpha_1\ldots\alpha_s\dot{\alpha}_1\ldots\dot{\alpha}_s}(x)\end{aligned}$$

where we have used the supplementary condition (1.8.2) and the mass-shell equation (1.8.3). So, if $A=B$, equation (1.8.23) gives nothing more than the reality condition (1.8.13).

In the case $A-1=B$, equation (1.8.23) together with equations (1.8.2, 3) lead to

$$\begin{aligned}\bar{\Psi}_{\alpha_1\ldots\alpha_B\dot{\alpha}\dot{\alpha}_1\ldots\dot{\alpha}_B}(x)&=\Delta_{\alpha_1}{}^{\dot{\gamma}_1}\ldots\Delta_{\alpha_B}{}^{\dot{\gamma}_B}\Delta^{\delta}{}_{\dot{\alpha}}\Delta^{\gamma_1}{}_{\dot{\alpha}_1}\ldots\Delta^{\gamma_B}{}_{\dot{\alpha}_B}\Psi_{\delta\gamma_1\ldots\gamma_B\dot{\gamma}_1\ldots\dot{\gamma}_B}(x)\\&=\Delta_{\gamma_1}{}^{\dot{\gamma}_1}\ldots\Delta_{\gamma_B}{}^{\dot{\gamma}_B}\Delta^{\delta}{}_{\dot{\alpha}}\Delta^{\gamma_1}{}_{\dot{\alpha}_1}\ldots\Delta^{\gamma_B}{}_{\dot{\alpha}_B}\Psi_{\delta\alpha_1\ldots\alpha_B\dot{\gamma}_1\ldots\dot{\gamma}_B}(x)\\&=\Delta^{\delta}{}_{\dot{\alpha}}\Psi_{\delta\alpha_1\ldots\alpha_B\dot{\alpha}_1\ldots\dot{\alpha}_B}(x).\end{aligned}$$

In terms of the spin-tensor field (1.8.17), this result is

$$\begin{aligned}m\Psi_{a_1\ldots a_B}{}^{\dot{\alpha}}(x)&=\partial^{\dot{\alpha}\gamma}\Psi_{a_1\ldots a_B\gamma}(x)\quad\Rightarrow\\-m\Psi_{a_1\ldots a_B\alpha}(x)&=\partial_{\alpha\dot{\gamma}}\Psi_{a_1\ldots a_B}{}^{\dot{\gamma}}(x).\end{aligned} \quad (1.8.24)$$

The equations (1.8.24) are elegant in three respects. First, the Klein–Gordon equation (1.8.3) is a consequence of the first-order differential equations (1.8.24). Secondly, the fields $\Psi_{a_1\ldots a_B\alpha}$ and $\Psi_{a_1\ldots a_B\dot{\alpha}}$ under the Klein–Gordon equation, describe two different massive particles having spin $(B+\frac{1}{2})$. When equations (1.8.24) are imposed, however, we have only the independent field $\Psi_{a_1\ldots a_B\alpha}$, which corresponds to a single massive particle with spin $(B+\frac{1}{2})$. Thirdly, let us make the replacement

$$\Psi_{a_1\ldots a_B\alpha}\to e^{(3/4)\pi i}\Psi_{a_1\ldots a_B\alpha},\quad \Psi_{a_1\ldots a_B\dot{\alpha}}\to e^{-(3/4)\pi i}\Psi_{a_1\ldots a_B\dot{\alpha}}$$

and incorporate the resultant fields into a four-component column

$$\Psi_{a_1\ldots a_B}=\begin{pmatrix}\Psi_{a_1\ldots a_B\alpha}\\ \Psi_{a_1\ldots a_B}{}^{\dot{\alpha}}\end{pmatrix} \quad (1.8.25)$$

which is a Majorana spinor and rank-B Lorentz tensor. Then equations (1.8.24) coincide with the Dirac equation

$$(\mathrm{i}\gamma^b\partial_b + m)\Psi_{a_1\dots a_B}(x) = 0 \tag{1.8.26}$$

where the γ-matrices were defined in (1.4.10).

Finally, if $|A-B| \neq 0$, 1, the reality condition (1.8.23) can be represented in the form of a differential equation independent of the Klein–Gordon equation but having the same or higher-order $|A-B|$.

1.8.3. Massless field representations of the Poincaré group

To describe the massless case, we consider a $(A/2, B/2)$-type spin-tensor field $G_{\alpha_1\dots\alpha_A\dot{\alpha}_1\dots\dot{\alpha}_B}(x)$, totally symmetric in its A undotted indices and in its B dotted indices, satisfying the following supplementary conditions

$$\begin{aligned}\partial^{\gamma\dot{\gamma}}G_{\gamma\alpha_1\dots\alpha_{A-1}\dot{\alpha}_1\dots\dot{\alpha}_B}(x) &= 0\\ \partial^{\gamma\dot{\gamma}}G_{\alpha_1\dots\alpha_A\dot{\gamma}\dot{\alpha}_1\dots\dot{\alpha}_{B-1}}(x) &= 0.\end{aligned} \tag{1.8.27}$$

As will be shown, the supplementary conditions are sufficient to select a single helicity state. The equations (1.8.27) can be rewritten in the following equivalent form

$$\partial_{\gamma\dot{\gamma}}G_{\alpha_1\dots\alpha_A\dot{\alpha}_1\dots\dot{\alpha}_B}(x) = \partial_{\alpha_k\dot{\gamma}}G_{\gamma\alpha_1\dots\hat{\alpha}_k\dots\alpha_A\dot{\alpha}_1\dots\dot{\alpha}_B}(x) = \partial_{\gamma\dot{\alpha}_l}G_{\alpha_1\dots\alpha_A\dot{\gamma}\dot{\alpha}_1\dots\hat{\dot{\alpha}}_l\dots\dot{\alpha}_B}(x) \tag{1.8.28}$$

where $k = 1, \dots, A$ and $l = 1, \dots, B$. If $A \neq 0$ or $B \neq 0$, the on-shell equation

$$\Box G_{\alpha_1\dots\alpha_A\dot{\alpha}_1\dots\dot{\alpha}_B}(x) = 0 \tag{1.8.29}$$

follows from the supplementary conditions. If $A = B = 0$, there are no supplementary conditions, and we have only the on-shell equation

$$\Box\Phi(x) = 0. \tag{1.8.30}$$

Note that the supplementary conditions (1.8.27) are invariant with respect to the Poincaré transformations.

The massless analogue of decomposition (1.8.4) is

$$\begin{aligned}G_{\alpha_1\dots\alpha_A\dot{\alpha}_1\dots\dot{\alpha}_B}(x) &= \int \frac{\mathrm{d}^3\vec{p}}{|\vec{p}|}\,\mathrm{e}^{\mathrm{i}px}\,G_{\alpha_1\dots\alpha_A\dot{\alpha}_1\dots\dot{\alpha}_B}(p)\\ p^a = (p^0, \vec{p}) &\qquad p^0 = |\vec{p}|.\end{aligned} \tag{1.8.31}$$

In momentum space, the conditions (1.8.27) take the form

$$\begin{aligned}p^{\gamma\dot{\gamma}}G_{\gamma\alpha_1\dots\alpha_{A-1}\dot{\alpha}_1\dots\dot{\alpha}_B}(p) &= 0\\ p^{\gamma\dot{\gamma}}G_{\alpha_1\dots\alpha_A\dot{\gamma}\dot{\alpha}_1\dots\dot{\alpha}_{B-1}}(p) &= 0.\end{aligned} \tag{1.8.32}$$

It is now an easy task to prove that the fields under consideration have only one independent component. Indeed, choosing a reference system, in which $p^a=(E, 0, 0, E)$, we have

$$p^{\dot{\alpha}\alpha}=p^a(\tilde{\sigma}_a)^{\dot{\alpha}\alpha}=2E\begin{pmatrix}0 & 0\\ 0 & 1\end{pmatrix}.$$

Then equations (1.8.32) mean that the only non-vanishing component is

$$G_{\underbrace{1\ldots1}_{A}\underbrace{\dot{1}\ldots\dot{1}}_{B}}(p).$$

We now show that any of the fields under consideration carry a definite helicity. Using the Pauli–Lubanski vector (1.8.10) and recalling definition (1.3.12), one obtains

$$\begin{aligned}
&\mathbf{W}_{\beta\dot{\beta}}G_{\alpha_1\ldots\alpha_A\dot{\alpha}_1\ldots\dot{\alpha}_B}\\
&\quad=-\frac{\mathrm{i}}{2}\partial^{\gamma}{}_{\dot{\beta}}\sum_{k=1}^{A}(\varepsilon_{\alpha_k\gamma}G_{\beta\alpha_1\ldots\hat{\alpha}_k\ldots\alpha_A\dot{\alpha}_1\ldots\dot{\alpha}_B}+\varepsilon_{\alpha_k\beta}G_{\gamma\alpha_1\ldots\hat{\alpha}_k\ldots\alpha_A\dot{\alpha}_1\ldots\dot{\alpha}_B})\\
&\qquad+\frac{\mathrm{i}}{2}\partial_{\beta}{}^{\dot{\gamma}}\sum_{k=1}^{B}(\varepsilon_{\dot{\alpha}_k\dot{\gamma}}G_{\alpha_1\ldots\alpha_A\dot{\beta}\dot{\alpha}_1\ldots\hat{\dot{\alpha}}_k\ldots\dot{\alpha}_B}+\varepsilon_{\dot{\alpha}_k\dot{\beta}}G_{\alpha_1\ldots\alpha_A\dot{\gamma}\dot{\alpha}_1\ldots\hat{\dot{\alpha}}_k\ldots\dot{\alpha}_B})\\
&\quad=-\frac{\mathrm{i}}{2}\sum_{k=1}^{A}\partial_{\alpha_k\dot{\beta}}G_{\beta\alpha_1\ldots\hat{\alpha}_k\ldots\alpha_A\dot{\alpha}_1\ldots\dot{\alpha}_B}+\frac{\mathrm{i}}{2}\sum_{k=1}^{B}\partial_{\beta\dot{\alpha}_k}G_{\alpha_1\ldots\alpha_A\dot{\beta}\dot{\alpha}_1\ldots\hat{\dot{\alpha}}_k\ldots\dot{\alpha}_B}
\end{aligned}$$

where we have used equations (1.8.27). Applying the second form (1.8.28) of the supplementary conditions, one obtains

$$\begin{aligned}
\mathbf{W}_{\beta\dot{\beta}}G_{\alpha_1\ldots\alpha_A\dot{\alpha}_1\ldots\dot{\alpha}_B}&=\frac{1}{2}(A-B)\mathbf{P}_{\beta\dot{\beta}}G_{\alpha_1\ldots\alpha_A\dot{\alpha}_1\ldots\dot{\alpha}_B}\\
\mathbf{P}_{\beta\dot{\beta}}&=-\mathrm{i}\partial_{\beta\dot{\beta}}.
\end{aligned}\tag{1.8.33}$$

Therefore, we have found that any $(A/2, B/2)$-type field under the supplementary conditions (1.8.27) (or under the on-shell equation (1.8.30) when $A=B=0$) describes a massless particle having helicity $\lambda=(A-B)/2$.

1.8.4. Examples of massless fields

It is our purpose now to demonstrate how massless fields, described in the previous subsection, arise in field theories. To start with, we consider two different field realizations of a spin-zero massless particle. The first realization is built in terms of a scalar field $\varphi(x)$. The action

$$S[\varphi]=-\frac{1}{2}\int \mathrm{d}^4x\,\partial^a\varphi\,\partial_a\varphi \tag{1.8.34}$$

leads to the on-shell equation (1.8.30). So, the field $\varphi(x)$ carries helicity $\lambda=0$.

The second realization is described in terms of a second-rank antisymmetric tensor field $B_{cd}(x)$. Its classical dynamics are dictated by the action (1.7.43) being invariant under the following gauge transformations

$$B_{cd}(x) \rightarrow B'_{cd}(x) = B_{cd} + \partial_{[c}\lambda_{d]} \tag{1.8.35}$$

where $\lambda_d(x)$ is an arbitrary vector field. It is the field strength $L_a(B)$ that is invariant under these transformations. The equation of motion

$$\frac{1}{2}\varepsilon_{abcd}\,\partial^c L^d(B) = 0 \tag{1.8.36a}$$

and its consequence

$$\partial_{[c}L_{d]}(B) = 0 \tag{1.8.36b}$$

can be rewritten in two-component spinor notation:

$$\partial^{\gamma\dot{\gamma}}L_{\gamma\dot{\alpha}}(B) = 0 \qquad \partial^{\gamma\dot{\gamma}}L_{\alpha\dot{\gamma}}(B) = 0. \tag{1.8.37}$$

To derive equations (1.8.37) from (1.8.36), one has to do the same steps as in deriving equation (1.8.10) from (1.8.9). The equations (1.8.37) mean that $L_{\alpha\dot{\alpha}}(B)$ is a massless field. In accordance with (1.8.33), $L_{\alpha\dot{\alpha}}(B)$ carries helicity $\lambda = 0$. So we may conclude that the field models (1.8.34) and (1.7.43) are equivalent, since they describe the same Poincaré representation. The equivalence can be seen also as follows. The equation (1.8.36*b*) means that $L_a(B) = \partial_a\varphi$, where $\varphi(x)$ is a scalar field satisfying, by virtue of (1.8.37), the on-shell equation (1.8.30). Therefore, the models (1.7.43) and (1.8.34) lead to the same dynamics.

Now consider the model (1.7.35) describing the dynamics of a massless Majorana spinor field Ψ. It is useful to rewrite the Dirac equation

$$\gamma^a\partial_a\Psi = 0$$

in two-component spinor notation:

$$\partial^{\alpha\dot{\alpha}}\Psi_\alpha = 0 \qquad \partial^{\alpha\dot{\alpha}}\bar{\Psi}_{\dot{\alpha}} = 0. \tag{1.8.38}$$

It is seen that $\Psi_\alpha(x)$ and $\bar{\Psi}_{\dot{\alpha}}(x)$ are massless fields. Recalling equation (1.8.33), we conclude that our model describes two massless particles having helicities $\lambda = \pm\frac{1}{2}$.

One more example is given by electrodynamics. The action

$$S[A] = -\frac{1}{4}\int d^4x\, F^{ab}F_{ab}$$
$$F_{ab} = \partial_a A_b - \partial_b A_a \tag{1.8.39}$$

is invariant under the gauge transformations

$$\delta A_b = \partial_b\lambda$$

where $\lambda(x)$ is an arbitrary scalar field. So the only physical observable is the field strength F_{ab}. We rewrite the first

$$\frac{1}{2}\varepsilon^{abcd}\,\partial_b F_{cd}=0$$

and the second

$$\partial^b F_{ab}=0$$

Maxwell's equations in two-component spinor notation:

$$\partial^{\gamma\dot\gamma}F_{\alpha\gamma}=0 \qquad \partial^{\gamma\dot\gamma}\bar F_{\dot\alpha\dot\gamma}=0 \tag{1.8.40}$$

where

$$F_{\alpha\dot\alpha\beta\dot\beta}=(\sigma^a)_{\alpha\dot\alpha}(\sigma^b)_{\beta\dot\beta}F_{ab}=2\varepsilon_{\alpha\beta}\bar F_{\dot\alpha\dot\beta}+2\varepsilon_{\dot\alpha\dot\beta}F_{\alpha\beta}.$$

The equations (1.8.40) imply that $F_{\alpha\beta}(x)$ and $\bar F_{\dot\alpha\dot\beta}(x)$ are massless fields carrying helicities $\lambda=\pm 1$, respectively.

Our next example is the Rarita–Schwinger model

$$S[\Psi_a]=\frac{1}{2}\int \mathrm{d}^4x\;\varepsilon^{abcd}\bar\Psi_a\gamma_b\gamma_5\partial_c\Psi_d \tag{1.8.41}$$

where $\Psi_a(x)$ is a Majorana spinor and Lorentz vector field,

$$\Psi_a=\begin{pmatrix}\Psi_{a\gamma}\\ \bar\Psi_a^{\dot\gamma}\end{pmatrix}.$$

In terms of the two-component spinors, the action is

$$S[\Psi_a]=\frac{1}{2}\int \mathrm{d}^4x\;\varepsilon^{abcd}\bar\Psi_a\tilde\sigma_b\Psi_{cd}$$
$$\Psi_{cd\alpha}\equiv\partial_c\Psi_{d\alpha}-\partial_d\Psi_{c\alpha}. \tag{1.8.42}$$

This model is a gauge theory since $S[\Psi_a]$ is invariant under the following transformations

$$\Psi_{a\alpha}(x)\to\Psi'_{a\alpha}(x)=\Psi_{a\alpha}(x)+\partial_a\epsilon_\alpha(x)$$

where $\epsilon_\alpha(x)$ is an arbitrary spinor field. Evidently, these transformations do not change the field strength $\Psi_{ab\alpha}$. To analyse the equations of motion

$$\varepsilon^{abcd}(\tilde\sigma_b)^{\dot\beta\beta}\Psi_{cd\beta}=0 \tag{1.8.43}$$

it is useful to rewrite $\Psi_{ab\alpha}$ in spinor notation introducing the spin-tensor

$$\Psi_{\gamma\dot\gamma\delta\dot\delta\alpha}=(\sigma^c)_{\gamma\dot\gamma}(\sigma^d)_{\delta\dot\delta}\Psi_{cd\alpha}=2\varepsilon_{\gamma\delta}\Psi_{\dot\gamma\dot\delta\alpha}+2\varepsilon_{\dot\gamma\dot\delta}\Psi_{\gamma\delta\alpha}$$
$$\Psi_{\dot\gamma\dot\delta\alpha}=\Psi_{\dot\delta\dot\gamma\alpha} \qquad \Psi_{\gamma\delta\alpha}=\Psi_{\delta\gamma\alpha} \tag{1.8.44}$$

Then equation (1.8.43) is nothing more than the following two equations

$$\Psi_{\dot{\delta}\dot{\gamma}\alpha} = 0 \qquad \Psi^{\alpha}{}_{\dot{\gamma}\alpha} = 0 \quad \Rightarrow$$
$$\Psi_{\alpha\beta\gamma} = \Psi_{(\alpha\beta\gamma)}. \tag{1.8.45}$$

Therefore, $\Psi_{\dot{\delta}\dot{\gamma}\alpha}$ vanishes on-shell, and $\Psi_{\alpha\beta\gamma}$ is totally symmetric under the equations of motion. What is more, $\Psi_{\alpha\beta\gamma}$ satisfies some differential constraints. Indeed, starting from the obvious relation

$$\varepsilon^{abcd}\,\partial_b\Psi_{cd\gamma} = 0$$

and using (1.8.44, 45), one obtains

$$\partial^{\gamma\dot{\gamma}}\Psi_{\alpha\beta\gamma}(x) = 0 \quad \Rightarrow$$
$$\partial^{\gamma\dot{\gamma}}\bar{\Psi}_{\dot{\alpha}\dot{\beta}\dot{\gamma}}(x) = 0. \tag{1.8.46}$$

Therefore, $\Psi_{\alpha\beta\gamma}(x)$ and $\bar{\Psi}_{\dot{\alpha}\dot{\beta}\dot{\gamma}}(x)$ are massless fields and, in accordance with (1.8.33), their helicities are $\pm\frac{3}{2}$, respectively.

In conclusion, let us consider linearized gravity. Its action is obtained from the Einstein gravity action (1.6.47) by representing the metric in the form

$$g_{mn}(x) = \eta_{mn} - \frac{\kappa}{2}H_{mn}(x)$$

where H_{mn} is a small fluctuation, and keeping in S_G only terms quadratic in H_{mn}. This gives

$$S[H] = -\frac{1}{2}\int \mathrm{d}^4x\{\partial^c H^{ab}\,\partial_c H_{ab} - \partial^c H^a_a \partial_c H^b_b + 2\partial_b H^{ab}(\partial_a H^c_c - \partial^c H_{ac})\}. \tag{1.8.47}$$

The action does not change under the linearized gauge transformations

$$H_{ab}(x) \to H'_{ab}(x) = H_{ab}(x) + \partial_{(a}\xi_{b)}(x)$$

where $\xi_a(x)$ is an arbitrary vector field. Linearized field strengths, invariant under these transformations, are obtained from (1.6.30) by keeping only H_{ab} terms and setting $\nabla_a = \partial_a$. Then one obtains

$$\check{\mathcal{R}}_{abcd} = \partial_a\partial_c H_{bd} - \partial_b\partial_c H_{ad} + \partial_b\partial_d H_{ac} - \partial_a\partial_d H_{bc}$$
$$\check{\mathcal{R}}_{ab} = \Box H_{ab} - \partial^c(\partial_a H_{bc} + \partial_b H_{ac}) + \partial_a\partial_b H^c_c \tag{1.8.48}$$
$$\check{\mathcal{R}} = 2\Box H^a_a - 2\partial^a\partial^b H_{ab}.$$

Imposing the equations of motion

$$\check{\mathcal{R}}_{ab} = 0 \tag{1.8.49}$$

we work with the linearized Weyl tensor $\check{C}_{abcd}$. In two-component spinor notation, this is described by totally symmetric spin-tensors $\check{C}_{\alpha\beta\gamma\delta}$ and $\bar{\check{C}}_{\dot{\alpha}\dot{\beta}\dot{\gamma}\dot{\delta}}$

which satisfy, by virtue of equations (1.8.49) and (1.6.68), the constraints

$$\partial^{\delta\dot{\delta}}\check{C}_{\alpha\beta\gamma\delta}=0 \qquad \partial^{\delta\dot{\delta}}\check{\bar{C}}_{\dot{\alpha}\dot{\beta}\dot{\gamma}\dot{\delta}}=0.$$

Therefore, $\check{C}_{\alpha\beta\gamma\delta}$ and $\check{\bar{C}}_{\dot{\alpha}\dot{\beta}\dot{\gamma}\dot{\delta}}$ are massless fields describing, due to equation (1.8.33), the helicity states ± 2, respectively.

The given examples illustrate the general situation: having some gauge theory $S[\Phi]$, the dynamical fields Φ do not belong, as is usual, to the family of massless fields described in Section 1.8.3. What is more, the original fields Φ cannot be treated as physical observables because of gauge arbitrariness in their choice. Rather, it is the gauge invariant field strengths $G(\Phi)$ built from the dynamical fields which play the role of physical observables. Namely these objects satisfy the usual criteria imposed on massless fields. To clarify the spin content of a theory, one must analyse the corresponding field strengths.

1.8.5. Massless field representations of the conformal group

Let $S[\Phi]$ be a massless field theory invariant with respect to conformal transformations (1.7.29). This transformation law defines a field representation of the conformal group. The corresponding generators $\{\mathbf{P}_a, \mathbf{J}_{ab}, \mathbf{D}, \mathbf{V}_a\}$ are introduced by the rule

$$\delta\Phi(x)=\mathrm{i}\left\{-b^a\mathbf{P}_a+\frac{1}{2}K^{ab}\mathbf{J}_{ab}+\Delta\mathbf{D}-f^a\mathbf{V}_a\right\}\Phi(x)$$

and their explicit expressions are

$$\begin{gathered}\mathbf{P}_a=-\mathrm{i}\partial_a \qquad \mathbf{J}_{ab}=\mathrm{i}(x_b\partial_a-x_a\partial_b)-\mathrm{i}M_{ab}\\ \mathbf{D}=\mathrm{i}x^a\partial_a+\mathrm{i}d \\ \mathbf{V}_a=\mathrm{i}(2x_ax^b\partial_b-x^2\partial_a)+2\mathrm{i}x^bM_{ab}+2\mathrm{i}x_ad.\end{gathered} \tag{1.8.50}$$

They satisfy the commutation relations (1.7.14) with arbitrary conformal weights d.

One can construct, with the help of the dynamical fields Φ, different secondary fields: $\partial_a\Phi$, $\partial_a\partial_b\Phi$ and so on. In general, their transformation laws have a structure other than that of equation (1.7.29). For example, considering $\partial_a\Phi$, one finds

$$\begin{aligned}-\delta(\partial_a\Phi)=\xi^b\partial_b(\partial_a\Phi)+\frac{1}{2}K[\xi]^{bc}M_{bc}(\partial_a\Phi)+\sigma[\xi]d'(\partial_a\Phi)\\ +(\partial^b\sigma[\xi])\{M_{ba}\Phi+\eta_{ab}d\Phi\}\end{aligned} \tag{1.8.51}$$

where $d'=d+1$. The difference from equation (1.7.29) is the presence of some non-homogeneous terms. But gauge invariant field strengths $G(\Phi)$, being secondary fields, should transform, at least on-shell, homogeneously

$$-\delta G(\Phi)=\xi^a\partial_a G(\Phi)+\frac{1}{2}K[\xi]^{bc}M_{bc}G(\Phi)+\sigma[\xi]d_G G(\Phi) \qquad (1.8.52)$$

where d_G are the scale dimensions of $G(\Phi)$. This assertion may be considered as a principle. But there are some obvious physical arguments. Indeed, let us anticipate that the gauge transformations (internal, as a rule) in the theory $S[\Phi]$ commute with transformations from the conformal group (space–time group). Then any conformal transformation should transform $G(\Phi)$ into a gauge invariant object. So, the right-hand side of equation (1.8.52) should be represented in the form of some operator acting on $G(\Phi)$. Under this assertion, the right-hand side of equation (1.8.52) is restored uniquely.

Now one has to clarify what the requirements are for the off-shell transformation law (1.8.52) to be consistent with the on-shell equations (1.8.27). The consistency conditions are

$$\begin{aligned}\partial^{\alpha_1\dot{\gamma}}\,\delta G_{\alpha_1\alpha_2\ldots\alpha_A\dot{\alpha}_1\ldots\dot{\alpha}_B}(x)&=0\\ \partial^{\gamma\dot{\alpha}_1}\,\delta G_{\alpha_1\ldots\alpha_A\dot{\alpha}_1\dot{\alpha}_2\ldots\dot{\alpha}_B}(x)&=0\end{aligned} \qquad (1.8.53)$$

for any conformal transformation (1.8.52). Note that the Poincaré transformations satisfy these requirements. So we are to study the dilatations and the special conformal transformations. As applied to these two cases, the conditions (1.8.53) can be rewritten in the form:

$$\begin{aligned}\partial^{\alpha_1\dot{\gamma}}\mathbf{D}G_{\alpha_1\alpha_2\ldots\alpha_A\dot{\alpha}_1\ldots\dot{\alpha}_B}(x)&=0\\ \partial^{\gamma\dot{\alpha}_1}\mathbf{D}G_{\alpha_1\ldots\alpha_A\dot{\alpha}_1\dot{\alpha}_2\ldots\dot{\alpha}_B}(x)&=0\end{aligned} \qquad (1.8.54a)$$

$$\begin{aligned}\partial^{\alpha_1\dot{\gamma}}\mathbf{V}_{\beta\dot{\beta}}G_{\alpha_1\alpha_2\ldots\alpha_A\dot{\alpha}_1\ldots\dot{\alpha}_B}(x)&=0\\ \partial^{\gamma\dot{\alpha}_1}\mathbf{V}_{\beta\dot{\beta}}G_{\alpha_1\ldots\alpha_A\dot{\alpha}_1\dot{\alpha}_2\ldots\dot{\alpha}_B}(x)&=0\end{aligned} \qquad (1.8.54b)$$

where the generators $\mathbf{D}$ and $\mathbf{V}_{\beta\dot{\beta}}$ are the same as in (1.8.50) but with conformal weight d_G:

$$\begin{aligned}\mathbf{D}&=-\frac{1}{2}x^{\alpha\dot{\alpha}}\partial_{\alpha\dot{\alpha}}+\mathrm{i}d_G\\ \mathbf{V}_{\beta\dot{\beta}}&=-\mathrm{i}x_{\beta}{}^{\dot{\delta}}x^{\delta}{}_{\dot{\beta}}\partial_{\delta\dot{\delta}}+2\mathrm{i}d_G x_{\beta\dot{\beta}}-2\mathrm{i}(x^{\delta}{}_{\dot{\beta}}M_{\beta\delta}+x_{\beta}{}^{\dot{\delta}}\bar{M}_{\dot{\beta}\dot{\delta}}).\end{aligned} \qquad (1.8.55)$$

Here we have converted all vector indices into spinor ones.

It is not difficult to check that equations (1.8.54*a*) are fulfilled identically (under equations (1.8.27)). A more interesting situation arises when considering the restrictions (1.8.54*b*). To start the analysis, let us suppose that $A \neq 0$. Using equations (1.8.55), one obtains

$$
\begin{aligned}
&\mathbf{V}_{\beta\dot\beta} G_{\alpha_1\ldots\alpha_A\dot\alpha_1\ldots\dot\alpha_B} \\
&\quad = -ix_{\beta\dot\delta}x_{\delta\dot\beta}\partial^{\delta\dot\delta} G_{\alpha_1\ldots\alpha_A\dot\alpha_1\ldots\dot\alpha_B} + 2id_G x_{\beta\dot\beta} G_{\alpha_1\ldots\alpha_A\dot\alpha_1\ldots\dot\alpha_A} \\
&\qquad - \mathrm{i}\sum_{k=1}^{A} x_{\alpha_k\dot\beta} G_{\beta\alpha_1\ldots\hat\alpha_k\ldots\alpha_A\dot\alpha_1\ldots\dot\alpha_B} + \mathrm{i}x_{\delta\dot\beta}\sum_{k=1}^{A}\varepsilon_{\alpha_k\beta} G^{\delta}{}_{\alpha_1\ldots\hat\alpha_k\ldots\alpha_A\dot\alpha_1\ldots\dot\alpha_B} \\
&\qquad - \mathrm{i}\sum_{k=1}^{B} x_{\beta\dot\alpha_k} G_{\alpha_1\ldots\alpha_A\dot\beta\dot\alpha_1\ldots\hat{\dot\alpha}_k\ldots\dot\alpha_B} + \mathrm{i}x_{\beta\dot\delta}\sum_{k=1}^{B}\varepsilon_{\dot\alpha_k\dot\beta} G_{\alpha_1\ldots\alpha_A\dot\alpha_1\ldots\hat{\dot\alpha}_k\ldots\dot\alpha_B}{}^{\dot\delta}.
\end{aligned}
$$

Taking into account equations (1.8.27), one then finds

$$
\begin{aligned}
\partial^{\alpha_1\dot\gamma}\mathbf{V}_{\beta\dot\beta} G_{\alpha_1\alpha_2\ldots\alpha_A\dot\alpha_1\ldots\dot\alpha_B} = {}& 4\mathrm{i}(A/2+1-d_G)\delta^{\dot\gamma}_{\dot\beta} G_{\beta\alpha_2\ldots\alpha_A\dot\alpha_1\ldots\dot\alpha_B} \\
& + 2\mathrm{i}\sum_{k=1}^{B}\delta^{\dot\gamma}_{\dot\alpha_k} G_{\beta\alpha_2\ldots\alpha_A\dot\beta\dot\alpha_1\ldots\hat{\dot\alpha}_k\ldots\dot\alpha_B} \\
& - 2\mathrm{i}\sum_{k=1}^{B}\varepsilon_{\dot\alpha_k\dot\beta} G_{\beta\alpha_2\ldots\alpha_A\dot\alpha_1\ldots\hat{\dot\alpha}_k\ldots\dot\alpha_B}{}^{\dot\gamma}
\end{aligned}
$$

where we have used the identity $\partial_{\beta\dot\beta}x^{\alpha\dot\alpha} = -2\delta^{\alpha}_{\beta}\delta^{\dot\alpha}_{\dot\beta}$. Since $A \neq 0$, this expression vanishes if and only if

$$d_G = 1 + A/2 \qquad B = 0. \tag{1.8.56}$$

Then, the second equation (1.8.54*b*) is absent since $B = 0$. Otherwise, in the case $B \neq 0$, one obtains the following restrictions

$$d_G = 1 + B/2 \qquad A = 0. \tag{1.8.57}$$

Finally, in the case where $A = B = 0$ one can find $d_G = 1$.

We conclude that admissible tensor types for field strength, arising in a conformal field theory, are $(A/2, 0)$ or $(0, B/2)$ only. The corresponding scale dimensions (1.8.56) and (1.8.57) are known as 'canonical dimensions'.

The main results (1.8.56, 57) show, in particular, that the antisymmetric tensor field model $S[B]$ (1.7.43) is not a conformal theory, while the model $S[\Phi]$ (1.8.34), classically equivalent to the first one, is a conformal theory. The statement about equivalence of these theories does not contradict the fact that $S[\Phi]$ possesses conformal symmetry but $S[B]$ does not. The equivalence means coincidence of the two dynamics. But we have seen that $L_a(B) = \partial_a\Phi$ on-shell. Hence, since Φ is a conformally covariant field with the transformation law (1.7.34), $L_a(B)$ transforms in a non-covariant way (equation (1.8.51)).

1.9. Elements of algebra with supernumbers

Supersymmetric field theories (as well as all fermionic field theories) are formulated most naturally in the language of supermathematics based on the concept of Grassmann algebra. Now we are going to describe the main ideas of supermathematics and obtain some results which will be explored in the following chapters.

To make our consideration self-contained, it is worth starting by recalling some trivial definitions from algebra theory. A linear space $\mathscr{A}$ (complex or real) is said to be an 'algebra' (complex or real) if $\mathscr{A}$ is provided with a binary operation of multiplication $(\cdot)$

$$(a, b) \rightarrow ab \in \mathscr{A} \qquad \forall\, a, b \in \mathscr{A}$$

which satisfy the axioms

$$\left.\begin{aligned} a(\alpha b + \beta c) &= \alpha ab + \beta ac \\ (\alpha b + \beta c)a &= \alpha ba + \beta ca \end{aligned}\right. \qquad \forall\, a, b, c \in \mathscr{A}$$

where α and β are arbitrary numbers (complex or real; for the time being, we restrict outself to the complex case only). If the multiplication law is characterized by the property

$$ab = ba \qquad \forall\, a, b \in \mathscr{A}$$

the algebra is called commutative. If the multiplication law is characterized by the property

$$a(bc) = (ab)c \qquad \forall\, a, b, c \in \mathscr{A}$$

the algebra is called associative. If $\mathscr{A}$ contains a unit e with the property

$$ea = ae = a \qquad \forall\, a \in \mathscr{A}$$

the algebra is called a unital algebra (or algebra with unit).

An example of associative commutative algebra with unit is the algebra $C^\infty(M)$ of smooth functions on a manifold M. The multiplication law is defined as the product of functions: if $f, \varphi \in C^\infty(M)$, then $(f \cdot \varphi)(p) = f(p)\varphi(p)$, $p \in M$. An example of an associative non-commutative unital algebra is the algebra $Mat_n(\mathbb{C})$ of $n \times n$ complex matrices. The multiplication law in $Mat_n(\mathbb{C})$ is the matrix multiplication. All Lie algebras are non-associative algebras (without unit).

Let $\mathscr{A}$ be an associative algebra with unit e and $B \subset \mathscr{A}$ be some set of elements. The algebra is said to be generated by B if every element $a \in \mathscr{A}$ can be represented as a finite-order polynomial of elements from B:

$$a = \alpha e + \sum_{k=1}^{p} \sum_{i_1, i_2, \ldots, i_k} C_{i_1 i_2 \ldots i_k} b^{i_1} b^{i_2} \ldots b^{i_k}$$

where α and $C_{i_1 i_2 \ldots i_k}$ are complex numbers, and all b^i lie in B. Then, B is called a 'system of generating elements' for $\mathscr{A}$.

Now we are in a position to define Grassmann algebras.

1.9.1. Grassmann algebras Λ_N and Λ_∞

The 'Grassmann algebra' Λ_N is an associative unital algebra generated by a set of N linearly independent elements ζ^i, $i=1, 2, \ldots, N$, which anticommute with each other:

$$\zeta^i\zeta^j+\zeta^j\zeta^i=0 \qquad i, j=1, \ldots, N. \tag{1.9.1}$$

In particular, $(\zeta^i)^2=0$.

Every element $a\in\Lambda_N$ can be represented in the form

$$a=\alpha+\sum_{k=1}^{N}\frac{1}{k!}C_{i_1i_2\ldots i_k}\zeta^{i_1}\zeta^{i_2}\ldots\zeta^{i_k} \tag{1.9.2}$$

where summation over all repeated indices is to be understood, α and $C_{i_1i_2\ldots i_k}$ are complex numbers, and the Cs are totally antisymmetric in their indices (it has been supposed that the unit element coincides with the number $1\in\mathbb{C}$). It is clear that elements

$$\begin{array}{ll} 1 & \\ \zeta^i & \\ \zeta^{i_1}\zeta^{i_2} & i_1<i_2 \\ \vdots & \\ \zeta^1\zeta^2\ldots\zeta^N & \end{array}$$

form a basis for Λ_N. So, the algebras Λ_N are finite-dimensional, $\dim\Lambda_N=2^N$.

We shall mainly be interested in an 'infinite-dimensional Grassmann algebra' Λ_∞. By definition, this is an associative unital algebra generated by an infinite set $\{\zeta^i\}$, $i=1, 2, \ldots$, of linearly independent, anticommuting elements ζ^i

$$\zeta^i\zeta^j+\zeta^j\zeta^i=0 \qquad i, j=1, 2, \ldots \tag{1.9.3}$$

Elements of Λ_∞ are called 'supernumbers' and Λ_∞ is called the 'space of supernumbers'. Every supernumber $z\in\Lambda_\infty$ can be represented in the form

$$z=z_{\mathrm{B}}+z_{\mathrm{S}}$$

$$z_{\mathrm{S}}=\sum_{k=1}^{\infty}\frac{1}{k!}C_{i_1i_2\ldots i_k}\zeta^{i_1}\zeta^{i_2}\ldots\zeta^{i_k} \tag{1.9.4}$$

$$z_{\mathrm{B}},\ C_{i_1i_2\ldots i_k}\in\mathbb{C}.$$

Here the coefficients $C_{i_1i_2\ldots i_k}$ are totally antisymmetric in their indices, and only a finite number of them do not vanish. Following B. De Witt, z_{B} will be called the 'body' and z_{S} the 'soul' of z. Because of equation (1.9.3), $(z_{\mathrm{S}})^n=0$ for some integer n. If $\zeta^i z=0$, for all ζ^i, then $z=0$.

Every supernumber z can be decomposed into the sum of its 'even' z_c and 'odd' z_a parts defined by

$$z = z_c + z_a$$

$$z_c = z_B + \sum_{k=1}^{\infty} \frac{1}{(2k)!} C_{i_1 i_2 \ldots i_{2k}} \zeta^{i_1} \zeta^{i_2} \ldots \zeta^{i_{2k}} \tag{1.9.5}$$

$$z_a = \sum_{k=0}^{\infty} \frac{1}{(2k+1)!} C_{i_1 i_2 \ldots i_{2k-1}} \zeta^{i_1} \zeta^{i_2} \ldots \zeta^{i_{2k+1}}.$$

If $z_a = 0$, z is called a 'c-number'; if $z_c = 0$, z is called an 'a-number'. By virtue of equation (1.9.3), c-numbers commute with all supernumbers. The set of all c-numbers is denoted as $\mathbb{C}_c$ and forms a commutative sub-algebra in Λ_∞. By virtue of equation (1.9.3), a-numbers anticommute among themselves and commute with c-numbers. Given two a-numbers, their product is a bodiless c-number. For every a-number z, we have $z^2 = 0$. The set of all a-numbers is denoted as $\mathbb{C}_a$. We shall call c-numbers and a-numbers pure supernumbers. Our consideration shows that pure supernumbers are characterized by the following properties:

$$\mathbb{C}_c \cdot \mathbb{C}_c = \mathbb{C}_c \qquad \mathbb{C}_a \cdot \mathbb{C}_a \subset \mathbb{C}_c$$

$$\mathbb{C}_a \cdot \mathbb{C}_c = \mathbb{C}_c \cdot \mathbb{C}_a = \mathbb{C}_a.$$

These properties mean that Λ_∞ is a $\mathbb{Z}_2$-graded associative algebra (see Section 2.1).

Together with the algebra Λ_∞, one can consider its extension (closure) $\boldsymbol{\Lambda}_\infty$ with respect to the norm $\| \ \|$ defined as

$$\|z\|^2 = |z_B|^2 + \sum_{k=1}^{\infty} \sum_{i_1, i_2, \ldots, i_k} \frac{1}{k!} |C_{i_1 i_2 \ldots i_k}|^2. \tag{1.9.6}$$

The elements of $\boldsymbol{\Lambda}_\infty$ are arbitrary linear combinations of the form (1.9.4) having finite norms. Since

$$\|z + \omega\| \leqslant \|z\| + \|\omega\|$$

and

$$\|\zeta_{i_1} \zeta_{i_2} \ldots \zeta_{i_k} z\| \leqslant \|z\| \quad \Rightarrow$$

$$\Rightarrow \quad \|z \cdot \omega\| \leqslant \|z\| \cdot \|\omega\|$$

for any elements $z, \omega \in \boldsymbol{\Lambda}_\infty$, the set $\boldsymbol{\Lambda}_\infty$ is an algebra. It is clear that $\Lambda_\infty \subset \boldsymbol{\Lambda}_\infty$, and every element $z \in \boldsymbol{\Lambda}_\infty$ may be given as the limit of a sequence $\{z_a\}$, $a = 1, 2, \ldots$, of supernumbers $z_a \in \Lambda_\infty$ such that $\lim_{a \to \infty} \|z - z_a\| = 0$. The elements of $\boldsymbol{\Lambda}_\infty$ are also called supernumbers, and $\boldsymbol{\Lambda}_\infty$ is said to be the 'full space of supernumbers'. For every $z \in \boldsymbol{\Lambda}_\infty$, one finds $\lim_{n \to \infty} (z_S)^n = 0$. In this book we deal with Λ_∞ and formally extend results to the case of $\boldsymbol{\Lambda}_\infty$.

It is useful to connect with every pure supernumber z its 'Grassmann parity' $\varepsilon(z)$ by the rule

$$\varepsilon(z)=\begin{cases}0 & \text{if } z\in\mathbb{C}_c\\ 1 & \text{if } z\in\mathbb{C}_a.\end{cases} \tag{1.9.7}$$

Then, the properties of pure supernumbers described above are encoded in the relations

$$\begin{aligned}&1.\ \varepsilon(z\cdot\omega)=\varepsilon(z)+\varepsilon(\omega)\quad(\text{mod } 2)\\ &2.\ z\cdot\omega=(-1)^{\varepsilon(z)\varepsilon(\omega)}\omega\cdot z.\end{aligned} \tag{1.9.8}$$

To define real supernumbers, we introduce the operation (*) of 'complex' conjugation (involution) in Λ_∞ as follows:

$$\begin{gathered}(\zeta^i)^*=\zeta^i \qquad i=1,2,\ldots\\ (\alpha z)=\alpha^* z^* \qquad \alpha\in\mathbb{C}\\ (z+\omega)^*=z^*+\omega^* \qquad (z\omega)^*=\omega^* z^*\end{gathered} \tag{1.9.9}$$

where z and ω are arbitrary supernumbers. Then for every supernumber (1.9.4), we have

$$\begin{aligned}z^* &= z_{\mathrm{B}}^*+\sum_{k=1}^{\infty}\frac{1}{k!}C^*_{i_1 i_2\ldots i_k}\zeta^{i_k}\ldots\zeta^{i_2}\zeta^{i_1}\\ &= z_{\mathrm{B}}^*+\sum_{k=1}^{\infty}(-1)^{k(k-1)/2}\frac{1}{k!}C^*_{i_1 i_2\ldots i_k}\zeta^{i_1}\zeta^{i_2}\ldots\zeta^{i_k}.\end{aligned} \tag{1.9.10}$$

A supernumber z is said to be 'real' if $z^*=z$, 'imaginary' if $z^*=-z$, and 'complex' otherwise. The set of all real supernumbers in $\mathbb{C}_c$ will be denoted by $\mathbb{R}_c$. The set of all real supernumbers in $\mathbb{C}_a$ will be denoted by $\mathbb{R}_a$. Having two real c-numbers, their product is a real c-number. The product of a real c-number and a real a-number is a real a-number. The product of two a-numbers is a bodiless imaginary c-number.

1.9.2. Supervector spaces

Supervector spaces are linear spaces in the usual sense, but supplied with the additional operations of left and right multiplication by supernumbers. Following B. De Witt, we define a 'supervector (superlinear) space' as a set $\mathscr{L}$ of elements, called 'supervectors', together with a binary operation of addition (+), operations of left and right multiplication by supernumbers, and a mapping of complex conjugation (*), which satisfy the axioms:

1. $\vec{X}+\vec{Y}=\vec{Y}+\vec{X} \quad \forall\, \vec{X},\vec{Y}\in\mathscr{L}$
2. $(\vec{X}+\vec{Y})+\vec{Z}=\vec{X}+(\vec{Y}+\vec{Z}) \quad \forall\, \vec{X},\vec{Y},\vec{Z}\in\mathscr{L}$
3. There exists an element $\vec{0}\in\mathscr{L}$ such that

$$\vec{X}+\vec{0}=\vec{X} \qquad \forall\, \vec{X}\in\mathscr{L}$$

4. For every $\vec{X} \in \mathscr{L}$, there exists an element $\vec{Y} \in \mathscr{L}$ such that

$$\vec{X} + \vec{Y} = \vec{0}$$

5. For every $\alpha, \beta \in \Lambda_\infty$ and every $\vec{X}, \vec{Y} \in \mathscr{L}$, we have

$$(\alpha+\beta)\vec{X} = \alpha\vec{X} + \beta\vec{X} \qquad \vec{X}(\alpha+\beta) = \vec{X}\alpha + \vec{X}\beta$$

$$\alpha(\vec{X} + \vec{Y}) = \alpha\vec{X} + \alpha\vec{Y} \quad (\vec{X}+\vec{Y})\alpha = \vec{X}\alpha + \vec{Y}\alpha$$

$$(\alpha\beta)\vec{X} = \alpha(\beta\vec{X}) \qquad \vec{X}(\alpha\beta) = (\vec{X}\alpha)\beta$$

$$1\vec{X} = \vec{X} \qquad \vec{X}1 = \vec{X}.$$

Then one finds

$$0\vec{X} = \vec{X}0 = \vec{0} \qquad \alpha\vec{0} = \vec{0}\alpha = \vec{0}$$

$$\vec{X} + (-1)\vec{X} = \vec{X} + \vec{X}(-1) = \vec{0}.$$

6. Left and right multiplications are related as follows:
   (a) $(\alpha\vec{X})\beta = \alpha(\vec{X}\beta) \qquad \forall\ \alpha, \beta \in \Lambda_\infty \qquad \forall\ \vec{X} \in \mathscr{L}$
   (b) $\alpha\vec{X} = \vec{X}\alpha \qquad \forall\ \alpha \in \mathbb{C}_c \qquad \forall\ \vec{X} \in \mathscr{L}$
   (c) For every $\vec{X} \in \mathscr{L}$, there exist unique supervectors ${}^0\vec{X}$, ${}^1\vec{X} \in \mathscr{L}$ such that

$$\vec{X} = {}^0\vec{X} + {}^1\vec{X}$$

$$\alpha\,{}^0\vec{X} = {}^0\vec{X}\alpha \qquad \alpha\,{}^1\vec{X} = -{}^1\vec{X}\alpha \qquad \forall\ \alpha \in \mathbb{C}_a.$$

The ${}^0\vec{X}$ is called the 'even' part of $\vec{X}$, and ${}^1\vec{X}$ is called the 'odd' part of $\vec{X}$. If $\vec{X} = {}^0\vec{X}$ or $\vec{X} = {}^1\vec{X}$, it is called a 'pure' supervector, of '*c*-type' in the first case and of '*a*-type' in the second case. Associated with every pure supervector $\vec{X}$ is its Grassmann parity $\varepsilon(\vec{X})$ defined by

$$\varepsilon(\vec{X}) = \begin{cases} 0 & \text{if } \vec{X} \text{ even} \\ 1 & \text{if } \vec{X} \text{ odd.} \end{cases} \tag{1.9.11}$$

For any pure α and $\vec{X}$, we have

$$\alpha\vec{X} = (-1)^{\varepsilon(\alpha)\varepsilon(\vec{X})}\vec{X}\alpha. \tag{1.9.12}$$

7. For arbitrary $\alpha \in \Lambda_\infty$ and $\vec{X}, \vec{Y} \in \mathscr{L}$, we have

$$\vec{X}^{**} = \vec{X}$$

$$(\vec{X} + \vec{Y})^* = \vec{X}^* + \vec{Y}^*$$

$$(\alpha\vec{X})^* = \vec{X}^*\alpha^* \qquad (\vec{X}\alpha)^* = \alpha^*\vec{X}^*.$$

It is not difficult to see that if $\vec{X}$ is an even or odd supervector then $\vec{X}^*$ is also an even or odd supervector, respectively. A supervector $\vec{X}$ is called 'real' if $\vec{X}^* = \vec{X}$, 'imaginary' if $\vec{X}^* = -\vec{X}$ and 'complex' otherwise. The product of a real c-number and a real supervector is a real supervector. The product of a real a-number and a real c-type supervector is a real a-type supervector. Finally, the product of a real a-number and a real a-type supervector is an imaginary c-type supervector.

The set of all c-type complex (real) supervectors in $\mathscr{L}$ will be denoted by ${}^0\mathscr{L}$ (${}^0\mathscr{L}_R$), and the set of all a-type complex (real) supervectors by ${}^1\mathscr{L}$ (${}^1\mathscr{L}_R$). Each of ${}^0\mathscr{L}$ and ${}^1\mathscr{L}$ is a linear space in the usual sense, but supplied with the operation of multiplication by c-numbers and the mapping of complex conjugation. Each of ${}^0\mathscr{L}_R$ and ${}^1\mathscr{L}_R$ is a linear space in the usual sense, but supplied with the operation of multiplication by real c-numbers.

Associated with a supervector space $\mathscr{L}$ is the $\mathscr{L}$-parity mapping $\mathscr{P}$: $\mathscr{L} \rightarrow \mathscr{L}$ acting on an arbitrary supervector $\vec{X} = {}^0\vec{X} + {}^1\vec{X}$ by the law

$$\mathscr{P}({}^0\vec{X} + {}^1\vec{X}) = {}^0\vec{X} - {}^1\vec{X}. \tag{1.9.13a}$$

In the case of a pure supervector $\vec{X}$, the formula reads

$$\mathscr{P}(\vec{X}) = (-1)^{\varepsilon(\vec{X})}\vec{X}. \tag{1.9.13b}$$

Basic properties of the $\mathscr{L}$-parity mapping are:

$$\mathscr{P}^2 = \mathbb{1} \tag{1.9.14a}$$

$$\mathscr{P}(\vec{X} + \vec{Y}) = \mathscr{P}(\vec{X}) + \mathscr{P}(\vec{Y}) \tag{1.9.14b}$$

$$\mathscr{P}(\alpha\vec{X}) = (-1)^{\varepsilon(\alpha)}\alpha\mathscr{P}(\vec{X}) \qquad \mathscr{P}(\vec{X}\alpha) = (-1)^{\varepsilon(\alpha)}\mathscr{P}(\vec{X})\alpha. \tag{1.9.14c}$$

Here $\vec{X}$ and $\vec{Y}$ are arbitrary supervectors, α is a pure supernumber.

1.9.3. Finite-dimensional supervector spaces

Let $\{\vec{e}_M\}$ be a set of elements of a supervector space $\mathscr{L}$. These supervectors are said to be linearly independent if and only if the requirement that a finite linear combination $X^M\vec{e}_M$ (or $\vec{e}_M X^M$), $X^M \in \Lambda_\infty$, vanishes is equivalent to $X^M = 0$, for all X^M. The space $\mathscr{L}$ is said to be finite dimensional if it possesses a finite system of linearly independent supervectors $\{\vec{e}_M\}$, where $M = 1, 2, \ldots, d$, such that every $\vec{X} \in \mathscr{L}$ can be expressed in the form

$$\vec{X} = \vec{e}_M X^M_{(+)} = X^M_{(-)}\vec{e}_M \qquad X^M_{(\pm)} \in \Lambda_\infty. \tag{1.9.15}$$

Then, the system $\{\vec{e}_M\}$ is called a 'basis' for $\mathscr{L}$. Clearly, each of the decompositions in equation (1.9.15) is unique. The $X^M_{(+)}$ ($X^M_{(-)}$) are called the 'left (right) components' of $\vec{X}$ with respect to the basis $\{\vec{e}_M\}$.

Any two bases of a finite-dimensional supervector space $\mathscr{L}$ have an equal number of elements, which is called the 'total dimension' of $\mathscr{L}$. To ground the assertion, it is useful to introduce the notion of supermatrices. A 'supermatrix' is a matrix with elements being supernumbers. Given a supermatrix F, one can decompose its matrix elements onto their bodies and souls, resulting with the body F_B and the soul F_S of F, $F = F_B + F_S$. F_B is an ordinary complex matrix. The 'rank' of F is defined to be the rank of its body F_B. If F and G are $m \times n$ and $n \times p$ supermatrices, then the body of their product FG coincides with the product of their bodies, $(F \cdot G)_B = F_B \cdot G_B$. This implies that all statements about matrix rank transfer to the case of supermatrices. In particular, if F and G are $m \times n$ and $n \times m$ supermatrices such that

$$FG = \mathbb{1}_m \qquad GF = \mathbb{1}_n$$

$\mathbb{1}_m$ ($\mathbb{1}_n$) being the unit $m \times m$ ($n \times n$) matrix, then $m = n$ and $G = F^{-1}$. Obviously the former relation can be rewritten as

$$F_B G_B + F_S G_B + F_B G_S + F_S G_S = \mathbb{1}_m.$$

Since $\mathbb{1}_m$ has no soul, we deduce that $F_B G_B = \mathbb{1}_m$. Similarly, the latter relation leads to $F_B G_B = \mathbb{1}_n$, which confirms our assertion. An $n \times n$ supermatrix F is said to be 'non-singular' if its body F_B is non-singular. Every non-singular supermatrix has a unique inverse. We can write

$$F = F_B(\mathbb{1}_n + F_B^{-1} F_S)$$

where supermatrix $F_B^{-1} F_S$ has no body, hence $(F_B^{-1} F_S)^p = 0$ for some integer p (recall, for every supernumber z there exists an integer q such that $(z_S)^q = 0$). Now one obtains

$$F^{-1} = F_B^{-1} + \sum_{k=1}^{\infty} (-1)^k (F_B^{-1} F_S)^k F_B^{-1}$$

the power series being terminated at some finite order. It is clear that *a square supermatrix has an inverse if and only if its body is non-singular.* The inverse supermatrix is unique.

Let $\{\vec{e}_M\}$, $M = 1, 2, \ldots, d$, be a basis in $\mathscr{L}$. It is readily seen that supervectors

$$\vec{X}_a = G_a{}^M \vec{e}_M \qquad a = 1, 2, \ldots, d'$$

are linearly independent if and only if the supermatrix $G_a{}^M$ has rank d', in particular, $d' \leqslant d$. As a result, any two bases in $\mathscr{L}$ have the same number of elements.

Remark. Λ_∞ is a one-dimensional supervector space. Every supernumber with non-vanishing body can be taken in the role of a basis for Λ_∞.

Remark. In contrast with ordinary vector spaces, not every subspace of a finite-dimensional supervector space $\mathscr{L}$ has definite dimension. As an example, one can consider a subspace $z\mathscr{L}$, where $z \in \mathbb{C}_a$.

Given a d-dimensional supervector space $\mathscr{L}$, we may always choose a basis $\{\vec{E}_M\}$ consisting of pure supervectors only. Indeed, let $\{\vec{e}_M\}$ be some basis.

We decompose every $\vec{e}_M$ into the sum of its even and odd parts

$$\vec{e}_M = {}^0\vec{e}_M + {}^1\vec{e}_M.$$

On the other hand, each of ${}^0\vec{e}_M$ and ${}^1\vec{e}_M$ can be decomposed with respect to $\{\vec{e}_M\}$

$${}^0\vec{e}_M = f_M{}^N \vec{e}_N \qquad {}^1\vec{e}_M = \varphi_M{}^N \vec{e}_N \qquad f_M{}^N, \varphi_M{}^N \in \Lambda_\infty.$$

One can easily see that the rank of $2d \times d$ supermatrix $(f_M{}^N, \varphi_M{}^N)$ is equal to d. Therefore, there are p supervectors from the set $\{{}^0\vec{e}_M\}$ and q supervectors from the set $\{{}^1\vec{e}_m\}$, $p+q=d$, that form a basis which will be denoted by $\{\vec{E}_m\}$ and called a 'pure' basis. It is convenient to label c-type elements of the basis by small Latin letters and a-type elements by small Greek letters, $\vec{E}_m = (\vec{E}_m, \vec{E}_\mu)$, where $m = 1, 2, \ldots, p$ and $\mu = 1, 2, \ldots, q$. Together with the total dimension, the numbers p and q turn out to be invariant characteristics of the supervector space and are called the 'even' and 'odd dimensions' of $\mathscr{L}$, respectively. Then $\mathscr{L}$ is said to have dimension (p, q).

Remark. Λ_∞ is a supervector space of dimension $(1, 0)$. Any c-number with non-zero body can be taken in the role of a pure basis.

Every c-type supervector $\vec{X} \in \mathscr{L}$ can now be written as

$$\vec{X} = y^m \vec{E}_m + \theta^\mu \vec{E}_\mu \qquad y^m \in \mathbb{C}_c \qquad \theta^\mu \in \mathbb{C}_a. \tag{1.9.16}$$

Therefore, the set ${}^0\mathscr{L}$ of c-type supervectors in $\mathscr{L}$ is in one-to-one correspondence with points of the space $\mathbb{C}^{p|q} \equiv \mathbb{C}_c^p \times \mathbb{C}_a^q$:

$$\mathbb{C}^{p|q} = \{(y^1, y^2, \ldots, y^p, \theta^1, \theta^2, \ldots, \theta^q), y^m \in \mathbb{C}_c, \theta^\mu \in \mathbb{C}_a\}. \tag{1.9.17}$$

This implies, in particular, that the numbers p and q are invariants of the supervector space.

Every finite-dimensional supervector space $\mathscr{L}$ proves to have a pure basis $\{\vec{E}_M\}$, which at the same time is real, $(\vec{E}_M)^* = \vec{E}_M$. With respect to such a basis, real c-type supervectors are given in the form

$$\vec{X} = x^m \vec{E}_m + \mathrm{i}\theta^\mu \vec{E}_\mu \qquad x^m \in \mathbb{R}_c \qquad \theta^\mu \in \mathbb{R}_a. \tag{1.9.18}$$

Hence, the set ${}^0\mathscr{L}_R$ of real c-type supervectors in $\mathscr{L}$ is in one-to-one correspondence with points of the space $\mathbb{R}^{p|p} \equiv \mathbb{R}_c^p \times \mathbb{R}_a^q$:

$$\mathbb{R}^{p|q} = \{z^M = (x^1, x^2, \ldots, x^p, \theta^1, \theta^2, \ldots, \theta^q), x^m \in \mathbb{R}_c, \theta^\mu \in \mathbb{R}_a\}. \tag{1.9.19}$$

Looking at equation (1.9.18), it would appear more convenient to exchange the pure real basis $\{\vec{E}_M\}$ with the basis $\{\vec{\mathscr{E}}_M\}$, where $\vec{\mathscr{E}}_m = \vec{E}_m$ and $\vec{\mathscr{E}}_\mu = \mathrm{i}\vec{E}_\mu$. Every pure basis $\{\vec{\mathscr{E}}_M\}$ with the properties $(\vec{\mathscr{E}}_m)^* = \vec{\mathscr{E}}_m$ and $(\vec{\mathscr{E}}_\mu)^* = -\vec{\mathscr{E}}_\mu$ is known as a 'standard basis'.

Supernumber space $\mathbb{C}^{p|q}$, defined by (1.9.16), is known as a 'complex superspace of dimension' (p, q), supernumber space $\mathbb{R}^{p|q}$, defined by equation (1.9.18), is known as a 'real superspace of dimension' (p, q).

Pure supervectors from $\mathscr{L}$ are characterized by the following properties

$$\vec{X} = \vec{E}_M X^M_{(+)} = X^M_{(-)} \vec{E}_M$$

$$\varepsilon(X^M_{(+)}) = \varepsilon(X^M_{(-)}) = \varepsilon(\vec{X}) + \varepsilon_M \tag{1.9.20}$$

$$X^M_{(-)} = (-1)^{\varepsilon_M(1+\varepsilon(\vec{X}))} X^M_{(+)}$$

where we have introduced the notation

$$\varepsilon_M = \varepsilon(\vec{E}_M) = \begin{cases} 0 & \text{if } M = m \\ 1 & \text{if } M = \mu. \end{cases} \tag{1.9.21}$$

As may be seen, the right components of every a-type supervector coincide with the left ones, but this is not the case for c-type supervectors.

Let $\{\vec{E}_M\}$ and $\{\vec{E}'_M\}$ be two pure bases. Each supervector $\vec{E}'_M$ may be decomposed with respect to $\{\vec{E}_M\}$:

$$\vec{E}'_M = G_M{}^N \vec{E}_N \tag{1.9.22}$$

where

$$G_M{}^N = \begin{pmatrix} A_m{}^n & B_m{}^\nu \\ C_\mu{}^n & D_\mu{}^\nu \end{pmatrix} \qquad \begin{matrix} A_m{}^n, D_\mu{}^\nu \in \mathbb{C}_c \\ B_m{}^\nu, C_\mu{}^n \in \mathbb{C}_a. \end{matrix}$$

Supermatrix $G_M{}^N$ should be non-singular, that is its body

$$F_B = \begin{pmatrix} A_B & 0 \\ 0 & D_B \end{pmatrix}$$

is a non-singular matrix, hence A_B and D_B are invertible matrices. As a result, the supermatrices A and D are invertible, and the unique inverse of G can be expressed in the form

$$G^{-1} = \begin{pmatrix} (A - BD^{-1}C)^{-1} & -A^{-1}B(D - CA^{-1}B)^{-1} \\ -D^{-1}C(A - BD^{-1}C)^{-1} & (D - CA^{-1}B)^{-1} \end{pmatrix}. \tag{1.9.23}$$

It is worth pointing out that, because of relation (1.9.22), the supermatrices $BD^{-1}C$ and $CA^{-1}B$ are bodiless, therefore $(A - BD^{-1}C)$ and $(D - CA^{-1}B)$ are invertible.

1.9.4. Linear operators and supermatrices

Let $\mathscr{L}$ be some supervector space. A mapping $\mathscr{F}_l \colon \mathscr{L} \to \mathscr{L}$ is called a 'left linear operator' on $\mathscr{L}$ if

$$\begin{aligned} &1.\ \mathscr{F}_l(\vec{X} + \vec{Y}) = \mathscr{F}_l(\vec{X}) + \mathscr{F}_l(\vec{Y}) \qquad \forall\, \vec{X}, \vec{Y} \in \mathscr{L} \\ &2.\ \mathscr{F}_l(\vec{X}\alpha) = \mathscr{F}_l(\vec{X})\alpha \qquad \forall\, \alpha \in \Lambda_\infty \qquad \forall\, \vec{X} \in \mathscr{L}. \end{aligned} \tag{1.9.24a}$$

A mapping $\mathscr{F}_r$: $\mathscr{L} \to \mathscr{L}$ is called a 'right linear operator' on $\mathscr{L}$ if

$$\begin{aligned} &1.\ (\vec{X}+\vec{Y})\mathscr{F}_r = (\vec{X})\mathscr{F}_r + (\vec{Y})\mathscr{F}_r \qquad \forall\, \vec{X}, \vec{Y} \in \mathscr{L} \\ &2.\ (\alpha\vec{X})\mathscr{F}_r = \alpha(\vec{X})\mathscr{F}_r \qquad \forall\, \alpha \in \Lambda_\infty \qquad \forall\, \vec{X} \in \mathscr{L}. \end{aligned} \tag{1.9.24b}$$

The set of all left (right) linear operators on $\mathscr{L}$ will be denoted by $End^{(+)}\mathscr{L}$ ($End^{(-)}\mathscr{L}$). As will be shown, the spaces $End^{(+)}\mathscr{L}$ and $End^{(-)}\mathscr{L}$ have non-empty overlap but do not coincide in general. Now, we are going to discuss in detail the properties of $End^{(+)}\mathscr{L}$. Specific features of $End^{(-)}\mathscr{L}$ will be commented upon later.

The simplest elements in $End^{(+)}\mathscr{L}$ are linear operators of left multiplication by supernumbers: given some supernumber z, the corresponding operator $\hat{z} \in End^{(+)}\mathscr{L}$ is defined as follows

$$\hat{z}(\vec{X}) = z\vec{X} \qquad \forall\, \vec{X} \in \mathscr{L}. \tag{1.9.25}$$

$End^{(+)}\mathscr{L}$ is naturally provided with the operations of

1. addition

$$(\mathscr{F}_1 + \mathscr{F}_2)(\vec{X}) = \mathscr{F}_1(\vec{X}) + \mathscr{F}_2(\vec{X}) \tag{1.9.26a}$$

2. multiplication

$$(\mathscr{F}_1 \cdot \mathscr{F}_2)(\vec{X}) = \mathscr{F}_1(\mathscr{F}_2(\vec{X})) \tag{1.9.26b}$$

3. left multiplication by supernumbers

$$z\mathscr{F} \equiv \hat{z} \cdot \mathscr{F} \Leftrightarrow (z\mathscr{F})(\vec{X}) = z\mathscr{F}(\vec{X}) \tag{1.9.26c}$$

4. right multiplication by supernumbers

$$\mathscr{F}z \equiv \mathscr{F} \cdot \hat{z} \Leftrightarrow (\mathscr{F}z)(\vec{X}) = \mathscr{F}(z\vec{X}) \tag{1.9.36d}$$

Introduce a pure basis $\{\vec{E}_M\}$ in $\mathscr{L}$ and decompose every supervector $\vec{X} \in \mathscr{L}$ with respect to this basis in two different forms

$$\vec{X} = \begin{cases} \vec{E}_M X^M_{(+)} & (1.9.27a) \\ X^M_{(-)} \vec{E}_M & (1.9.27b) \end{cases} \qquad X^M_{(\pm)} \in \Lambda_\infty$$

introducing the left and the right components of $\vec{X}$ with respect to $\{\vec{E}_M\}$. The former decomposition in expression (1.9.27) will be used when dealing with left linear operators on $\mathscr{L}$, the latter when dealing with right linear operators.

To any left linear operator $\mathscr{F}$ on $\mathscr{L}$ we associate a supermatrix F^M_N by the prescription

$$\begin{aligned} &\mathscr{F}(\vec{E}_N) = \vec{E}_M F^M{}_N \qquad (\neq F^M{}_N \vec{E}_M!) \\ &F^M{}_N = \begin{pmatrix} A^m{}_n & B^m{}_\nu \\ C^\mu{}_n & D^\mu{}_\nu \end{pmatrix} \qquad F^M{}_N \in \Lambda_\infty. \end{aligned} \tag{1.9.28}$$

Then, in accordance with (1.9.24*a*), the operator $\mathscr{F}$ acts on every supervector (1.9.27*a*) according to the rule

$$\mathscr{F}(\vec{X})=\vec{X}'=\vec{E}_M X'^M_{(+)} \qquad X'^M_{(+)}=F^M{}_N X^N_{(+)}. \tag{1.9.29}$$

Note that the operators (1.9.25) of multiplication by pure supernumbers are characterized by supermatrices

$$(\hat{z})^M{}_N=\begin{pmatrix} z\mathbb{1}_p & 0 \\ 0 & z\mathbb{1}_q \end{pmatrix}=z\mathbb{1}_{(p,q)} \qquad z\in\mathbb{C}_c \tag{1.9.30a}$$

$$(\hat{z})^M{}_N=\begin{pmatrix} z\mathbb{1}_p & 0 \\ 0 & -z\mathbb{1}_q \end{pmatrix}\equiv z\hat{\mathbb{1}}_{(p,q)} \qquad z\in\mathbb{C}_a \tag{1.9.30b}$$

The supermatrix $\hat{\mathbb{1}}_{(p,q)}$ will be called the 'graded unit supermatrix'. The above results can be combined in a compact expression,

$$(\hat{z})^M{}_N=z(-1)^{\varepsilon(z)\varepsilon_M}\delta^M{}_N. \tag{1.9.30c}$$

Our consideration shows that every left linear operator on $\mathscr{L}$ is uniquely determined by its supermatrix with respect to a given pure basis in $\mathscr{L}$. Conversely, having a supermatrix $F^M{}_N$, the prescription (1.9.29) provides us with a left linear operator on $\mathscr{L}$. Thus $End^{(+)}\mathscr{L}$ is in one-to-one correspondence with the set of all supermatrices of the type (1.9.28), denoted by $Mat^{(+)}(p,q|\Lambda_\infty)$. If $\mathscr{F}_1$ and $\mathscr{F}_2$ are left linear operators on $\mathscr{L}$ and F_1 and F_2 are the corresponding supermatrices, then the operators $(\mathscr{F}_1+\mathscr{F}_2)$ and $\mathscr{F}_1\cdot\mathscr{F}_2$ lead to the supermatrices

$$F_1+F_2=\begin{pmatrix} A_1+A_2 & B_1+B_2 \\ C_1+C_2 & D_1+D_2 \end{pmatrix} \tag{1.9.31a}$$

and

$$F_1\cdot F_2=\begin{pmatrix} A_1A_2+B_1C_2 & A_1B_2+B_1D_2 \\ C_1A_2+D_1C_2 & C_1B_2+D_1D_2 \end{pmatrix} \tag{1.9.31b}$$

respectively. If a left linear operator $\mathscr{F}$ on $\mathscr{L}$ is characterized by the supermatrix F, then the operators $z\mathscr{F}$ and $\mathscr{F}z$, with z being some pure supernumber, lead to the supermatrices $\hat{z}F$ and $F\hat{z}$, respectively, where the supermatrix $\hat{z}$ is defined by equations (1.9.30). Explicitly, we have

$$\hat{z}F=\begin{pmatrix} zA & zB \\ (-1)^{\varepsilon(z)}zC & (-1)^{\varepsilon(z)}zD \end{pmatrix} \tag{1.9.31c}$$

and

$$F\hat{z}=\begin{pmatrix} Az & (-1)^{\varepsilon(z)}Bz \\ Cz & (-1)^{\varepsilon(z)}Dz \end{pmatrix}. \tag{1.9.31d}$$

These relations determine the operations of left and right multiplication by supernumbers in $Mat^{(+)}(p,q|\Lambda_\infty)$.

Now, we are going to describe even and odd elements in $End^{(+)}\mathscr{L}$ and $Mat^{(+)}(p, q|\Lambda_\infty)$. Given a left linear operator $\mathscr{F}$ on $\mathscr{L}$, we represent it as

$$\mathscr{F} = {}^0\mathscr{F} + {}^1\mathscr{F}$$
$$ {}^0\mathscr{F} \equiv \frac{1}{2}(\mathscr{F} + \mathscr{P}\mathscr{F}\mathscr{P}) \qquad {}^1\mathscr{F} \equiv \frac{1}{2}(\mathscr{F} - \mathscr{P}\mathscr{F}\mathscr{P}) \tag{1.9.32}$$

$\mathscr{P}$ being the $\mathscr{L}$-parity mapping (1.9.13), and show that ${}^0\mathscr{F}$ (${}^1\mathscr{F}$) is a left linear operator commuting (anticommuting) with all a-numbers. Choosing an arbitrary $\vec{X} \in \mathscr{L}$ and pure $\alpha \in \Lambda_\infty$ and using equation (1.9.14c), one obtains

$$\mathscr{P}\mathscr{F}\mathscr{P}(\vec{X}\alpha) = (-1)^{\varepsilon(\alpha)}\mathscr{P}\mathscr{F}(\mathscr{P}(\vec{X})\alpha) = (-1)^{\varepsilon(\alpha)}\mathscr{P}(\mathscr{F}(\mathscr{P}(\vec{X}))\alpha) = (\mathscr{P}\mathscr{F}\mathscr{P}(\vec{X}))\alpha$$

which implies that ${}^0\mathscr{F}, {}^1\mathscr{F} \in End^{(+)}\mathscr{L}$. Next, ${}^0\mathscr{F}$ and ${}^1\mathscr{F}$ possess the properties:

$${}^0\mathscr{F}: \begin{cases} {}^0\mathscr{L} \to {}^0\mathscr{L} \\ {}^1\mathscr{L} \to {}^1\mathscr{L} \end{cases} \qquad {}^1\mathscr{F}: \begin{cases} {}^0\mathscr{L} \to {}^1\mathscr{L} \\ {}^1\mathscr{L} \to {}^0\mathscr{L} \end{cases}. \tag{1.9.33}$$

Let us prove, as an example, the former relation. If ${}^0\vec{X} \in {}^0\mathscr{L}$ and ${}^1\vec{X} \in {}^1\mathscr{L}$, then

$${}^0\mathscr{F}({}^0\vec{X}) = \frac{1}{2}\mathscr{F}({}^0\vec{X}) + \frac{1}{2}\mathscr{P}\mathscr{F}({}^0\vec{X}) = {}^0(\mathscr{F}({}^0\vec{X})) \in {}^0\mathscr{L}$$

$${}^0\mathscr{F}({}^1\vec{X}) = \frac{1}{2}\mathscr{F}({}^1\vec{X}) - \frac{1}{2}\mathscr{P}\mathscr{F}({}^1\vec{X}) = {}^1(\mathscr{F}({}^0\vec{X})) \in {}^1\mathscr{L}.$$

Using relations (1.9.33), it is easy to prove that

$$\alpha\,{}^0\mathscr{F} = {}^0\mathscr{F}\alpha \qquad \alpha\,{}^1\mathscr{F} = (-1)^{\varepsilon(\alpha)}\,{}^1\mathscr{F}\alpha \tag{1.9.34}$$

where α is an arbitrary pure supernumber. In the case of ${}^1\mathscr{F}$, for example, we have

$$({}^1\mathscr{F}\alpha)(\vec{X}) = {}^1\mathscr{F}(\alpha\vec{X}) = (-1)^{\varepsilon(\alpha)\varepsilon(\vec{X})}\,{}^1\mathscr{F}(\vec{X})\alpha = (-1)^{\varepsilon(\alpha)}\alpha\,{}^1\mathscr{F}(\vec{X})$$

$\vec{X}$ being a pure supervector.

In accordance with the relations (1.9.33, 34), a linear operator $\mathscr{F} \in End^{(+)}\mathscr{L}$ is said to be of c-type, if it does not change supervector type, and a-type, if it changes supervector type:

$$\begin{aligned} c\text{-type:} &\qquad \varepsilon(\mathscr{F}(\vec{X})) = \varepsilon(\vec{X}) \\ a\text{-type:} &\qquad \varepsilon(\mathscr{F}(\vec{X})) = 1 + \varepsilon(\vec{X}) \pmod 2 \end{aligned} \tag{1.9.35}$$

for every pure $\vec{X} \in \mathscr{L}$. Linear operators of definite types will be called 'pure' and endowed with the Grassmann parity

$$\varepsilon(\mathscr{F}) = \begin{cases} 0 & \text{for } c\text{-type } \mathscr{F} \\ 1 & \text{for } a\text{-type } \mathscr{F}. \end{cases} \tag{1.9.36}$$

Every left linear operator on $\mathscr{L}$ is uniquely represented as the sum of its c-type and a-type parts by the law (1.9.32).

Equations (1.9.28) and (1.9.35) show that a pure operator $\mathscr{F}$ on $\mathscr{L}$ is characterized by a supermatrix F with the following entries:

$$c\text{-type: } \begin{cases} A^m{}_n, D^\mu{}_\nu \in \mathbb{C}_c \\ B^m{}_\nu, C^\mu{}_n \in \mathbb{C}_a \end{cases} \tag{1.9.37a}$$

$$a\text{-type: } \begin{cases} A^m{}_n, D^\mu{}_\nu \in \mathbb{C}_a \\ B^m{}_\nu, C^\mu{}_n \in \mathbb{C}_c. \end{cases} \tag{1.9.37b}$$

Making use of the notation (1.9.21) and the Grassmann parity function (1.9.36), these properties may be collected in a compact expression:

$$\varepsilon(F^M{}_N) = \varepsilon(\mathscr{F}) + \varepsilon_M + \varepsilon_N \ (\text{mod } 2). \tag{1.9.37c}$$

It is instructive to check explicitly that equation (1.9.37*c*) guarantees the desired properties (1.9.35) of $\mathscr{F}$. Let $\vec{X} = \vec{E}_M X^M_{(+)}$ be a pure supervector, $X^M_{(+)}$ as in equation (1.9.20). Then, from equation (1.9.37*c*), we obtain

$$\begin{aligned} \varepsilon(X'^M_{(+)}) &= \varepsilon(F^M{}_N X^N_{(+)}) = \{\varepsilon(\mathscr{F}) + \varepsilon_M + \varepsilon_N + \varepsilon_N + \varepsilon(\vec{X})\} \quad (\text{mod } 2) \\ &= \{\varepsilon(\mathscr{F}) + \varepsilon_M + \varepsilon(\vec{X})\} (\text{mod } 2) = \{\varepsilon(\mathscr{F}) + \varepsilon(X^M_{(+)})\} \quad (\text{mod } 2) \end{aligned}$$

which is in agreement with relations (1.9.35).

A supermatrix $F \in Mat^{(+)}(p, q | \Lambda_\infty)$ is said to be of *c*-type (*a*-type) if it satisfies the conditions (1.9.37*a*) ((1.9.37*b*)). Supermatrices of definite types are called pure. A left linear operator $\mathscr{F}$ on $\mathscr{L}$ is pure if and only if the corresponding supermatrix F is pure and of the same type as $\mathscr{F}$. The Grassmann parity $\varepsilon(F)$ of pure supermatrix F is defined to be equal to the Grassmann parity $\varepsilon(\mathscr{F})$ of the corresponding linear operator $\mathscr{F}$. From equation (1.9.34) we obtain the following important identities:

$$\hat{z}F = (-1)^{\varepsilon(z)\varepsilon(F)} F\hat{z} \tag{1.9.39a}$$

$$z\mathscr{F} = (-1)^{\varepsilon(z)\varepsilon(\mathscr{F})} \mathscr{F} z \Leftrightarrow \mathscr{F}(z\vec{X}) = (-1)^{\varepsilon(z)\varepsilon(\mathscr{F})} z\mathscr{F}(\vec{X}) \tag{1.9.39b}$$

for any pure $z \in \Lambda_\infty$, $F \in Mat^{(+)}(p, q | \Lambda_\infty)$ and $\mathscr{F} \in End^{(+)}\mathscr{L}$. So, *c-numbers commute with all supermatrices (linear operators), while a-numbers commute with c-type supermatrices (linear operators) and anticommute with a-types.*

The equations (1.9.26*a,c,d*) and (1.9.39*b*) show that $End^{(+)}\mathscr{L}$ is a supervector space (with an undefined operation of complex conjugation); the set ${}^0End^{(+)}\mathscr{L}$ (${}^1End^{(+)}\mathscr{L}$) of even (odd) supervectors in $End^{(+)}\mathscr{L}$ coincides with the set of *c*-type (*a*-type) linear operators defined by relations (1.9.35). Analogously, equations (1.9.31*a,c,d*)and (1.9.39*a*) show that $Mat^{(+)}(p, q | \Lambda_\infty)$ is a supervector space (with an undefined operation of complex conjugation); the set ${}^0Mat^{(+)}(p, q | \Lambda_\infty)$ (${}^1Mat^{(+)}(p, q | \Lambda_\infty)$) of even (odd) supervectors in $Mat^{(+)}(p, q | \Lambda_\infty)$ coincides with the set of *c*-type (*a*-type) supermatrices defined by relations (1.9.37). Each of $End^{(+)}\mathscr{L}$ and $Mat^{(+)}(p, q | \Lambda_\infty)$ has dimension $(p^2 + q^2, 2pq)$. Finally, $End^{(+)}\mathscr{L}$ and $Mat^{(+)}(p, q | \Lambda_\infty)$ form a unital associative algebra with respect to the operations of multiplications (1.9.26*b*) and (1.9.31*b*), respectively. The product of two *c*-type supermatrices is a *c*-type supermatrix, the product of c-type and a-type supermatrices is an a-type supermatrix, the product of two a-type supermatrices is a c-type supermatrix. Analogous results take place for linear operatons on $\mathscr{L}$.

In fact, there is a natural way to introduce operations of complex conjugation in supervector spaces $End^{(+)}\mathscr{L}$ and $Mat^{(+)}(p,q|\Lambda_\infty)$. For the former space this is done as follows. To every pure operator $\mathscr{F}\in End^{(+)}\mathscr{L}$ we associate the mapping $\mathscr{F}^*\colon \mathscr{L}\to\mathscr{L}$ acting on pure supervectors $\vec{X}\in\mathscr{L}$ as

$$\mathscr{F}^*(\vec{X})=(-1)^{\varepsilon(\mathscr{F})\varepsilon(\vec{X})}(\mathscr{F}(\vec{X}^*))^*. \tag{1.9.40a}$$

In the case of arbitrary $\mathscr{F}\in End^{(+)}\mathscr{L}$ and $\vec{X}\in\mathscr{L}$, one is to represent $\mathscr{F}$ and $\vec{X}$ as superpositions of their c-type and a-type parts and then to apply the above definition, assuming that

$$\begin{aligned}(\mathscr{F}_1+\mathscr{F}_2)^* &= \mathscr{F}_1^*+\mathscr{F}_2^*\\ \mathscr{F}^*(\vec{X}_1+\vec{X}_2) &= \mathscr{F}^*(\vec{X}_1)+\mathscr{F}^*(\vec{X}_2).\end{aligned} \tag{1.9.40b}$$

Let us show that the operation '*' is a complex conjugation in $End^{(+)}\mathscr{L}$. If α is a pure supernumber, then

$$\begin{aligned}\mathscr{F}^*(\vec{X}\alpha) &= (-1)^{\varepsilon(\mathscr{F})[\varepsilon(\vec{X})+\varepsilon(\alpha)]}(\mathscr{F}(\alpha^*\vec{X}^*))^*\\ &= (-1)^{\varepsilon(\mathscr{F})\varepsilon(\vec{X})}(\alpha^*\mathscr{F}(\vec{X}^*))^* = \mathscr{F}^*(\vec{X})\alpha\end{aligned}$$

therefore $\mathscr{F}^*\in End^{(+)}\mathscr{L}$. Next, one finds

$$\begin{aligned}(\mathscr{F}\alpha)^*(\vec{X}) &= (-1)^{[\varepsilon(\mathscr{F})+\varepsilon(\alpha)]\varepsilon(\vec{X})}(\mathscr{F}\alpha(\vec{X}^*))^*\\ &= (-1)^{\varepsilon(\mathscr{F})\varepsilon(\vec{X})}(\mathscr{F}(\vec{X}^*)\alpha)^* = \alpha^*\mathscr{F}^*(\vec{X})\end{aligned}$$

which implies

$$(\mathscr{F}\alpha)^* = \alpha^*\mathscr{F}^*$$

for every $\mathscr{F}\in End^{(+)}\mathscr{L}$ and every $\alpha\in\Lambda_\infty$. It is also easily verified that $(\mathscr{F}^*)^*=\mathscr{F}$, for every left linear operator on $\mathscr{L}$. As a result, the operation (1.9.40) satisfies all the properties of complex conjugation. We take this operation in the role of complex conjugation in $End^{(+)}\mathscr{L}$. The linear operator $\mathscr{F}\in End^{(+)}\mathscr{L}$ is said to be real if $\mathscr{F}^*=\mathscr{F}$, imaginary if $\mathscr{F}^*=-\mathscr{F}$, and complex otherwise. As is seen from equation (1.9.40a), these are real c-type operators which map every real supervector on to a real one.

Operation (1.9.40) induces a lot of complex conjugations in $Mat^{(+)}(p,q|\Lambda_\infty)$, depending on the pure basis chosen in $\mathscr{L}$ to identify $End^{(+)}\mathscr{L}$ and $Mat^{(+)}(p,q|\Lambda_\infty)$. Let us choose, for convenience, a standard basis $\{\vec{\mathscr{E}}_M\}$ in $\mathscr{L}$, $\vec{\mathscr{E}}_M^*=(-1)^{\varepsilon_M}\vec{\mathscr{E}}_M$. If F is the supermatrix of a pure operator $\mathscr{F}$, with respect to $\{\vec{\mathscr{E}}_M\}$, then the supermatrix of $\mathscr{F}^*$ will be denoted as F^{s*} and called the supercomplex conjugate of F. From the definition, we have

$$\mathscr{F}(\vec{\mathscr{E}}_M)=\vec{\mathscr{E}}_N F^N{}_M \qquad \mathscr{F}^*(\vec{\mathscr{E}}_M)=\vec{\mathscr{E}}_N(F^{s*})^N{}_M.$$

On the other hand, the use of equation (1.9.40) gives

$$\mathcal{F}^*(\vec{\mathscr{E}}_M)=(-1)^{\varepsilon(\mathcal{F})\varepsilon_M}(\mathcal{F}(\vec{\mathscr{E}}^*_M))^*=(-1)^{\varepsilon_M(1+\varepsilon(\mathcal{F}))}(\mathcal{F}(\vec{\mathscr{E}}_M))^*$$
$$=(-1)^{\varepsilon_M(1+\varepsilon(\mathcal{F}))}(\vec{\mathscr{E}}_N F^N{}_M)^*=(-1)^{\varepsilon(\mathcal{F})(\varepsilon_M+\varepsilon_N)+\varepsilon_M+\varepsilon_M\varepsilon_N}\vec{\mathscr{E}}_N(F^N{}_M)^*.$$

Therefore, we obtain

$$(F^{s*})^M{}_N=(-1)^{\varepsilon(F)(\varepsilon_M+\varepsilon_N)+\varepsilon_N+\varepsilon_M\varepsilon_N}(F^M{}_N)^*. \tag{1.9.41a}$$

Suppressing indices, this definition can be rewritten in the form:

c-type

$$F^{s*}=\begin{pmatrix} A^* & -B^* \\ C^* & D^* \end{pmatrix} \tag{1.9.41b}$$

a-type

$$F^{s*}=\begin{pmatrix} A^* & B^* \\ -C^* & D^* \end{pmatrix} \tag{1.9.41c}$$

A supermatrix F is said to be real if $F^{s*}=F$, imaginary if $F^{s*}=-F$, and complex otherwise. Pure real supermatrices satisfy the equation

$$(F^M{}_N)^*=(-1)^{\varepsilon(F)(\varepsilon_M+\varepsilon_N)+\varepsilon_N+\varepsilon_M\varepsilon_N}F^M{}_N \tag{1.9.42a}$$

In the c-type case, this means

$$(F^M{}_N)^*=(-1)^{\varepsilon_N+\varepsilon_M\varepsilon_N}F^M{}_N$$
$$F^M{}_N=\begin{pmatrix} A^m{}_n & \mathrm{i}B^m{}_\nu \\ C^\mu{}_n & D^\mu{}_\nu \end{pmatrix} \qquad \begin{matrix} A^m{}_n, D^\mu{}_\nu \in \mathbb{R}_c \\ B^m{}_\nu, C^\mu{}_n \in \mathbb{R}_a. \end{matrix} \tag{1.9.42b}$$

In the a-type case, we have

$$(F^M{}_N)^*=(-1)^{\varepsilon_M+\varepsilon_M\varepsilon_N}F^M{}_N$$
$$F^M{}_N=\begin{pmatrix} A^m{}_n & B^m{}_\nu \\ \mathrm{i}C^\mu{}_n & D^\mu{}_\nu \end{pmatrix} \qquad \begin{matrix} A^m{}_n, D^\mu{}_\nu \in \mathbb{R}_a \\ B^m{}_\nu, C^\mu{}_n \in \mathbb{R}_c. \end{matrix} \tag{1.9.42c}$$

Operation '$s*$', being rather unusual, proves more acceptable than the naive operation '$*$' defined by $(F^*)^M{}_N=(F^M{}_N)^*$, since the latter does not respect axiom 7 of supervector spaces while the former does: one finds

$$(\hat{z}F)^{s*}=F^{s*}\hat{z}^* \qquad (F\hat{z})^{s*}=\hat{z}^*F^{s*} \tag{1.9.43}$$

for arbitrary $z\in\Lambda_\infty$ and $F\in Mat^{(+)}(p,q|\Lambda_\infty)$. An important property of supercomplex conjugation is that

$$(F_1F_2)^{s*}=(-1)^{\varepsilon(F_1)\varepsilon(F_2)}F_1^{s*}F_2^{s*}. \tag{1.9.44}$$

This relation shows, in particular, that the product of two real c-type supermatrices is a real c-type supermatrix, the product of real c-type and

a-type supermatrices is a real type a-supermatrix, and the product of two a-type supermatrices is an imaginary c-type supermatrix. Analogous results hold for left linear operators on $\mathscr{L}$. The set of all real c-type (a-type) elements in $End^{(+)}\mathscr{L}$ and $Mat^{(+)}(p, q | \Lambda_\infty)$ will be denoted by ${}^0End_{\mathbf{R}}^{(+)}\mathscr{L}$ (${}^1End_{\mathbf{R}}^{(+)}\mathscr{L}$) and ${}^0Mat_{\mathbf{R}}^{(+)}(p, q | \Lambda_\infty)$ (${}^1Mat_{\mathbf{R}}^{(+)}(p, q | \Lambda_\infty)$), respectively.

Now, we briefly comment upon $End^{(-)}\mathscr{L}$ and its connection to $End^{(+)}\mathscr{L}$. Let $\mathscr{G}$ be a right linear operator on $\mathscr{L}$. In contrast to equation (1.9.28), its action on the basis is represented as

$$(\vec{E}_M)\mathscr{G} = G_M{}^N \vec{E}_N$$

$$G_M{}^N = \begin{pmatrix} A_m{}^n & B_m{}^\nu \\ C_\mu{}^n & D_\mu{}^\nu \end{pmatrix} \qquad G_M{}^N \in \Lambda_\infty. \tag{1.9.45}$$

Then $\mathscr{G}$ acts on every $\vec{X} \in \mathscr{L}$, written in the form (1.9.27b), by the rule

$$(\vec{X})\mathscr{G} = \vec{X}' = X'^M_{(-)} \vec{E}_M \qquad X'^M_{(-)} = X^N_{(-)} G_N{}^M. \tag{1.9.46}$$

Then $G_M{}^N$ is said to be the supermatrix of $\mathscr{G}$ in the basis $\{\vec{E}_M\}$. It is clear that every right linear operator is uniquely determined by its supermatrix. So, $End^{(-)}\mathscr{L}$ is in one-to-one correspondence with the set of all supermatrices of the type (1.9.45), the latter being denoted by $Mat^{(-)}(p, q | \Lambda_\infty)$. Supermatrices from $Mat^{(+)}(p, q | \Lambda_\infty)$ and $Mat^{(-)}(p, q | \Lambda_\infty)$ differ in the position of indices enumerating columns and rows. The supervector space structure on $End^{(-)}\mathscr{L}$ or $Mat^{(-)}(p, q | \Lambda_\infty)$ is introduced in perfect analogy with the cases of $End^{(+)}\mathscr{L}$ or $Mat^{(+)}(p, q | \Lambda_\infty)$. In particular, c-type supermatrices from $Mat^{(-)}(p, q | \Lambda_\infty)$ look like those in equation (1.9.22).

It follows from equation (1.9.39b) and its right analogue that every left c-type linear operator on $\mathscr{L}$ turns out to be a right c-type one and vice versa. But this is not the case for a-type operators. Nevertheless, there exists a simple way to associate with any left a-type linear operator a right a-type operator. The point is that every operator of the form

$$\mathscr{F}\mathscr{P} = -\mathscr{P}\mathscr{F} \qquad \mathscr{F} \in {}^1End^{(+)}\mathscr{L} \tag{1.9.47}$$

$\mathscr{P}$ being the $\mathscr{L}$-parity mapping (1.9.13), proves to be right linear,

$$\mathscr{F}\mathscr{P}(\alpha\vec{X}) = \alpha\mathscr{F}\mathscr{P}(\vec{X}) \qquad \forall\, \alpha \in \Lambda_\infty \qquad \forall\, \vec{X} \in \mathscr{L}. \tag{1.9.48}$$

Setting α and $\vec{X}$ to be pure, one obtains

$$\mathscr{F}\mathscr{P}(\alpha\vec{X}) = (-1)^{\varepsilon(\alpha)+\varepsilon(\vec{X})}\mathscr{F}(\alpha\vec{X}) = (-1)^{\varepsilon(\vec{X})}\alpha\mathscr{F}(\vec{X}) = \alpha\mathscr{F}\mathscr{P}(\vec{X}).$$

It is evident that $\mathscr{F}\mathscr{P} \in {}^1End^{(-)}\mathscr{L}$. As a result, the correspondence Σ: $End^{(+)}\mathscr{L} \to End^{(-)}\mathscr{L}$ defined by

$$\mathscr{F} = {}^0\mathscr{F} + {}^1\mathscr{F} \to {}^0\mathscr{F} + {}^1\mathscr{F}\mathscr{P} \tag{1.9.49}$$

${}^0\mathscr{F}$ and ${}^1\mathscr{F}$ being the c-type and a-type components of $\mathscr{F}$, determines a one-to-one mapping of $End^{(+)}\mathscr{L}$ on to $End^{(-)}\mathscr{L}$. Now, for every left linear operator $\mathscr{F} = {}^0\mathscr{F} + {}^1\mathscr{F}$ we can define its right action on $\mathscr{L}$ by the rule

$$(\vec{X})\mathscr{F} \equiv {}^0\mathscr{F}(\vec{X}) + {}^1\mathscr{F}\mathscr{P}(\vec{X}) \qquad \forall\, \vec{X} \in \mathscr{L}. \tag{1.9.50a}$$

When $\mathscr{F}$ and $\vec{X}$ are pure, this definition reads as

$$(\vec{X})\mathscr{F} \equiv (-1)^{\varepsilon(\mathscr{F})\varepsilon(\vec{X})}\mathscr{F}(\vec{X}). \tag{1.9.50b}$$

Since every element of ${}^0End^{(+)}\mathscr{L}$ is at the same time an element of ${}^0End^{(-)}\mathscr{L}$ and vice versa, these spaces coincide and can be denoted simply as ${}^0End\mathscr{L}$. However, we shall not identify the c-type supermatrix spaces ${}^0Mat^{(+)}(p, q \,|\, \Lambda_\infty)$ and ${}^0Mat^{(-)}(p, q \,|\, \Lambda_\infty)$ used to represent operators from ${}^0End^{(+)}\mathscr{L}$ and ${}^0End^{(-)}\mathscr{L}$. The reason for keeping this difference will be described in the next subsection.

Up to now we have studied the operator spaces $End^{(\pm)}\mathscr{L}$ associated with some supervector space $\mathscr{L}$. Given two supervector spaces $\mathscr{L}$ and $\tilde{\mathscr{L}}$, one can associate with them not only the spaces $End^{(\pm)}\mathscr{L}$ and $End^{(\pm)}\tilde{\mathscr{L}}$, but also the spaces $End^{(\pm)}(\mathscr{L}, \tilde{\mathscr{L}})$ of all left (right) linear operators from $\mathscr{L}$ to $\tilde{\mathscr{L}}$ and the spaces $End^{(\pm)}(\tilde{\mathscr{L}}, \mathscr{L})$ of all left (right) linear operators from $\tilde{\mathscr{L}}$ to $\mathscr{L}$. Let us briefly discuss, as an example, $End^{(+)}(\mathscr{L}, \tilde{\mathscr{L}})$. By definition, a mapping $\mathscr{F}_1 : \mathscr{L} \to \tilde{\mathscr{L}}$ is called a left linear operator acting from $\mathscr{L}$ to $\tilde{\mathscr{L}}$ if it satisfies the properties (1.9.24*a*). A supervector space structure on $End^{(+)}(\mathscr{L}, \tilde{\mathscr{L}})$ is introduced with the help of the operations (1.9.26*a,c,d*). In particular, pure linear operators from $End^{(+)}(\mathscr{L}, \tilde{\mathscr{L}})$ are characterized as follows:

$$\begin{array}{cc} c\text{-type} & a\text{-type} \\ \mathscr{F}: \begin{cases} {}^0\mathscr{L} \to {}^0\tilde{\mathscr{L}} \\ {}^1\mathscr{L} \to {}^1\tilde{\mathscr{L}} \end{cases} & \mathscr{F}: \begin{cases} {}^0\mathscr{L} \to {}^1\tilde{\mathscr{L}} \\ {}^1\mathscr{L} \to {}^0\tilde{\mathscr{L}}. \end{cases} \end{array} \tag{1.9.50}$$

The operation of complex conjugation in $End^{(+)}(\mathscr{L}, \tilde{\mathscr{L}})$ can be introduced through rule (1.9.40). Finally, the transformation (1.9.49) maps $End^{(+)}(\mathscr{L}, \tilde{\mathscr{L}})$ on to $End^{(-)}(\mathscr{L}, \tilde{\mathscr{L}})$. The overlap of $End^{(+)}(\mathscr{L}, \tilde{\mathscr{L}})$ with $End^{(-)}(\mathscr{L}, \tilde{\mathscr{L}})$ is given by the set of all c-type linear operators from $\mathscr{L}$ to $\tilde{\mathscr{L}}$, denoted by ${}^0End(\mathscr{L}, \tilde{\mathscr{L}})$.

1.9.5. Dual supervector spaces, supertransposition

Consider some supervector space $\mathscr{L}$. Supervector spaces ${}^*\mathscr{L} \equiv End^{(+)}(\mathscr{L}, \Lambda_\infty)$ and $\mathscr{L}^* \equiv End^{(-)}(\mathscr{L}, \Lambda_\infty)$ are said to be the 'left dual' and the 'right dual' of $\mathscr{L}$, respectively. Elements of ${}^*\mathscr{L}$ ($\mathscr{L}^*$) are called 'left (right) super 1-forms' on $\mathscr{L}$ (or 'supercovectors'). In the remainder of this section, we label super 1-forms by capital Latin letters in boldface and use the notation

$$\begin{array}{lll} \mathbf{J}_l(\vec{X}) \equiv \langle \mathbf{J} \,|\, \vec{X} \rangle \in \Lambda_\infty & (\forall\, \mathbf{J}_l \in {}^*\mathscr{L} & \forall\, \vec{X} \in \mathscr{L}) \\ (\vec{X})\mathbf{J}_r \equiv \langle \vec{X} \,|\, \mathbf{J} \rangle \in \Lambda_\infty & (\forall\, \mathbf{J}_r \in \mathscr{L}^* & \forall\, \vec{X} \in \mathscr{L}). \end{array}$$

Basic properties of left super 1-forms are:

$$\langle \mathbf{J} | \vec{X} + \vec{Y} \rangle = \langle \mathbf{J} | \vec{X} \rangle + \langle \mathbf{J} | \vec{Y} \rangle \qquad \langle \mathbf{J} | \vec{X}\alpha \rangle = \langle \mathbf{J} | \vec{X} \rangle \alpha$$
$$\langle \mathbf{J} + \mathbf{L} | \vec{X} \rangle = \langle \mathbf{J} | \vec{X} \rangle + \langle \mathbf{L} | \vec{X} \rangle \qquad \langle \alpha \mathbf{J} | \vec{X} \rangle = \alpha \langle \mathbf{J} | \vec{X} \rangle \tag{1.9.51}$$
$$\langle J\alpha | \vec{X} \rangle = \langle \mathbf{J} | \alpha \vec{X} \rangle$$

for all $\mathbf{J}, \mathbf{L} \in {}^*\mathscr{L}$ and $\vec{X}, \vec{Y} \in \mathscr{L}$ and $\alpha \in \Lambda_\infty$. To obtain their right analogues, one replaces supervectors by right super 1-forms and left super 1-forms by supervectors. Pure super 1-forms are characterized as follows:

$$\begin{array}{cc} c\text{-type} & a\text{-type} \\ \mathbf{J}: \begin{cases} {}^0\mathscr{L} \to \mathbb{C}_c \\ {}^1\mathscr{L} \to \mathbb{C}_a \end{cases} & \mathbf{J}: \begin{cases} {}^0\mathscr{L} \to \mathbb{C}_a \\ {}^1\mathscr{L} \to \mathbb{C}_c \end{cases}. \end{array} \tag{1.9.52}$$

Similarly to (1.9.40), complex conjugation in ${}^*\mathscr{L}$ and $\mathscr{L}^*$ is defined by

$$\langle \mathbf{J}^* | \vec{X} \rangle \equiv (-1)^{\varepsilon(\mathbf{J})\varepsilon(\vec{X})} \langle \mathbf{J} | \vec{X}^* \rangle^* \tag{1.9.53a}$$
$$\langle \vec{X} | \mathbf{J}^* \rangle \equiv (-1)^{\varepsilon(\mathbf{J})\varepsilon(\vec{X})} \langle \vec{X}^* | \mathbf{J} \rangle^*. \tag{1.9.53b}$$

Given a pure left super 1-form $\mathbf{J}$, one can obtain a right one through the prescription

$$\langle \vec{X} | \mathbf{J} \rangle \equiv (-1)^{\varepsilon(\mathbf{J})\varepsilon(\vec{X})} \langle \mathbf{J} | \vec{X} \rangle. \tag{1.9.54}$$

The set of all c-type super 1-forms

$${}^0\mathscr{L}^* \equiv {}^0({}^*\mathscr{L}) = {}^0(\mathscr{L}^*)$$

is said to be the 'dual' of ${}^0\mathscr{L}$. The set of all real c-type super 1-forms

$${}^0\mathscr{L}^*_{\mathbf{R}} \equiv {}^0({}^*\mathscr{L})_{\mathbf{R}} = {}^0(\mathscr{L}^*)_{\mathbf{R}}$$

is said to be the dual of ${}^0\mathscr{L}_{\mathbf{R}}$. Every $\mathbf{J} \in {}^0\mathscr{L}^*$ is a c-number-valued function on ${}^0\mathscr{L}$, every $\mathbf{J} \in {}^0\mathscr{L}^*_{\mathbf{R}}$ is a real c-number-valued function on ${}^0\mathscr{L}_{\mathbf{R}}$.

When $\mathscr{L}$ has dimension (p, q), the same is true for both ${}^*\mathscr{L}$ and $\mathscr{L}^*$. If we introduce a pure basis in $\mathscr{L}$, $\{\vec{E}_M\}$, then every super 1-form is completely specified by its values on $\vec{E}_M$:

$$\langle \mathbf{J} | \vec{X} \rangle = \langle \mathbf{J} | \vec{E}_M \rangle X^M_{(+)} \qquad \langle \vec{X} | \mathbf{J} \rangle = X^M_{(-)} \langle \vec{E}_M | \mathbf{J} \rangle.$$

In particular, there exist a unique set of left super 1-forms $\{\langle \mathbf{E}^M |\}$ and a unique set of right super 1-forms $\{| \mathbf{E}^M \rangle\}$ such that

$$\langle \mathbf{E}^M | \vec{E}_N \rangle = \delta^M{}_N \qquad \langle \vec{E}_M | \mathbf{E}^N \rangle = \delta_M{}^N. \tag{1.9.55}$$

The $\{\langle \mathbf{E}^M |\}$ prove to form a pure basis in ${}^*\mathscr{L}$ called the 'left dual' of $\{\vec{E}_M\}$. Similarly, the set $\{| \mathbf{E}^M \rangle\}$ forms a pure bases in $\mathscr{L}^*$ known as the 'right dual' of $\{\vec{E}_M\}$. It follows from equation (1.9.53) that the left (right) dual of a real pure basis in $\mathscr{L}$ is a standard basis in ${}^*\mathscr{L}$ ($\mathscr{L}^*$) and vice versa.

Let $\mathscr{F}$ be a pure left linear operator on $\mathscr{L}$. It can be considered as a right linear operator $\mathscr{F}_{\mathrm{r}}$ on $^*\mathscr{L}$ defined as

$$\langle(\mathbf{J})\mathscr{F}_{\mathrm{r}} \mid \vec{X}\rangle \equiv \langle\mathbf{J} \mid \mathscr{F}(\vec{X})\rangle$$

or as a left linear operator $\mathscr{F}^{\mathrm{sT}}$ on $^*\mathscr{L}$ defined as

$$\langle\mathscr{F}^{\mathrm{sT}}(\mathbf{J}) \mid \vec{X}\rangle \equiv (-1)^{\varepsilon(\mathscr{F})\varepsilon(\mathbf{J})}\langle\mathbf{J} \mid \mathscr{F}(\vec{X})\rangle. \tag{1.9.56}$$

Evidently, if α is a pure supernumber, one obtains

$$\begin{aligned}\langle\mathscr{F}^{\mathrm{sT}}(\mathbf{J}\alpha) \mid \vec{X}\rangle &= (-1)^{\varepsilon(\mathscr{F})[\varepsilon(\mathbf{J})+\varepsilon(\alpha)]}\langle\mathbf{J}\alpha \mid \mathscr{F}(\vec{X})\rangle \\ &= (-1)^{\varepsilon(\mathscr{F})\varepsilon(\mathbf{J})}\langle\mathbf{J} \mid \mathscr{F}(\alpha\vec{X})\rangle = \langle\mathscr{F}^{\mathrm{sT}}(\mathbf{J})\alpha \mid \vec{X}\rangle.\end{aligned}$$

The operator $\mathscr{F}^{\mathrm{sT}}$ is said to be the 'supertranspose' of $\mathscr{F}$. Given an arbitrary left linear operator on $\mathscr{L}$, $\mathscr{F} = {}^0\mathscr{F} + {}^1\mathscr{F}$, its supertranspose is defined as

$$\mathscr{F}^{\mathrm{sT}} = ({}^0\mathscr{F})^{\mathrm{sT}} + ({}^1\mathscr{F})^{\mathrm{sT}}.$$

Now, the operation of supertransposition determines a one-to-one mapping of $End^{(+)}\mathscr{L}$ on to $End^{(+)}(^*\mathscr{L})$. Its inverse mapping is defined to be the supertransposition on $End^{(+)}(^*\mathscr{L})$. Basic properties of supertransposition are:

$$\begin{gathered}(\mathscr{F}_1 + \mathscr{F}_2)^{\mathrm{sT}} = \mathscr{F}_1^{\mathrm{sT}} + \mathscr{F}_2^{\mathrm{sT}} \\ (\mathscr{F}_1 \cdot \mathscr{F}_2)^{\mathrm{sT}} = (-1)^{\varepsilon(\mathscr{F}_1)\varepsilon(\mathscr{F}_2)}\mathscr{F}_1^{\mathrm{sT}} \cdot \mathscr{F}_2^{\mathrm{sT}} \\ (\alpha \cdot \mathscr{F})^{\mathrm{sT}} = \alpha \cdot \mathscr{F}^{\mathrm{sT}} \qquad (\mathscr{F} \cdot \alpha)^{\mathrm{sT}} = \mathscr{F}^{\mathrm{sT}} \cdot \alpha \qquad (\forall\, \alpha \in \Lambda_\infty) \\ (\mathscr{F}^{\mathrm{sT}})^* = (\mathscr{F}^*)^{\mathrm{sT}} \qquad (\mathscr{F}^{\mathrm{sT}})^{\mathrm{sT}} = \mathscr{F}.\end{gathered} \tag{1.9.57}$$

The correspondence $\mathscr{F} \to \mathscr{F}^{\mathrm{sT}}$ described induces an operation of supertransposition which maps every supermatrix F from $Mat^{(+)}(p, q \mid \Lambda_\infty)$ into the supermatrix F^{sT} in $Mat^{(-)}(p, q \mid \Lambda_\infty)$ and vice versa. Introduce the supermatrices of $\mathscr{F}$ and $\mathscr{F}^{\mathrm{sT}}$ with respect to a pure basis $\{\vec{E}_M\}$ in $\mathscr{L}$ and its dual $\{\mathbf{E}^M\}$ in $^*\mathscr{L}$, respectively,

$$\begin{gathered}\mathscr{F}(\vec{E}_M) = \vec{E}_N F^N{}_M \\ \mathscr{F}^{\mathrm{sT}}(\mathbf{E}^M) = \mathbf{E}^N (F^{\mathrm{sT}})_N{}^M.\end{gathered}$$

Note, supermatrices from $Mat^{(+)}(p, q \mid \Lambda_\infty)$ are used as left linear operators on $\mathscr{L}$, while supermatrices from $Mat^{(-)}(p, q \mid \Lambda_\infty)$ are used as left linear operators on $^*\mathscr{L}$. The use of equations (1.9.55, 56) leads to

$$(F^{\mathrm{sT}})_M{}^N = (-1)^{\varepsilon(F)(\varepsilon_M + \varepsilon_N) + \varepsilon_M + \varepsilon_M\varepsilon_N} F^N{}_M. \tag{1.9.58a}$$

It is instructive to rewrite this definition in more detail:

c-type

$$\begin{gathered}(F^{\mathrm{sT}})_M{}^N = (-1)^{\varepsilon_M + \varepsilon_M\varepsilon_N} F^N{}_M \\ F^{\mathrm{sT}} = \begin{pmatrix} A^{\mathrm{T}} & C^{\mathrm{T}} \\ -B^{\mathrm{T}} & D^{\mathrm{T}} \end{pmatrix}\end{gathered} \tag{1.9.58b}$$

a-type

$$(F^{sT})_M{}^N = (-1)^{\varepsilon_N + \varepsilon_M \varepsilon_N} F^N{}_M$$

$$F^{sT} = \begin{pmatrix} A^T & -C^T \\ B^T & D^T \end{pmatrix}. \tag{1.9.58c}$$

The supermatrix F^{sT} is said to be the 'supertranspose' of F. Supertransposition can be extended in an obvious way to the full $Mat^{(+)}(p, q|\Lambda_\infty)$ resulting in a one-to-one mapping of $Mat^{(+)}(p, q|\Lambda_\infty)$ on to $Mat^{(-)}(p, q|\Lambda_\infty)$. Its inverse mapping will be taken in the role of supertransposition on $Mat^{(-)}(p, q|\Lambda_\infty)$. One readily deduces from (1.9.58a) that the supertranspose of a pure supermatrix $G_M{}^N \in Mat^{(-)}(p, q|\Lambda_\infty)$ reads

$$(G^{sT})^M{}_N = (-1)^{\varepsilon(G)(\varepsilon_M + \varepsilon_N) + \varepsilon_N + \varepsilon_M \varepsilon_N} G_N{}^M. \tag{1.9.59}$$

Comparing equations (1.9.58) and (1.9.59), one finds that supertransposition is sensitive to the positions of indices! That is why we do not identify ${}^0Mat^{(+)}(p, q|\Lambda_\infty)$ and ${}^0Mat^{(-)}(p, q|\Lambda_\infty)$. Basic properties of supertransposition are:

$$(F_1 + F_2)^{sT} = F_1^{sT} + F_2^{sT} \qquad (F^{sT})^{sT} = F$$

$$(F_1 \cdot F_2)^{sT} = (-1)^{\varepsilon(F_1)\varepsilon(F_2)} F_1^{sT} \cdot F_2^{sT}$$

$$(\hat{z} \cdot F)^{sT} = \hat{z} \cdot F^{sT} \qquad (F \cdot \hat{z})^{sT} = F^{sT} \cdot \hat{z} \qquad (\forall\, z \in \Lambda_\infty) \tag{1.9.60}$$

$$\hat{z}^{sT} = \hat{z}.$$

This can be supplemented by the relation

$$(F^{sT})^{s*} = (F^{s*})^{sT} \tag{1.9.61}$$

provided complex conjugation in $Mat^{(-)}(p, q|\Lambda_\infty)$ is defined in the manner

$$(G^{s*})_M{}^N = (-1)^{\varepsilon(G)(\varepsilon_M + \varepsilon_N) + \varepsilon_M + \varepsilon_M \varepsilon_N} (G_M{}^N)^* \tag{1.9.62}$$

different from equation (1.9.41). The point is that the dual of a standard basis in $\mathscr{L}$ is a real pure basis in ${}^*\mathscr{L}$.

Let X^M be components of a c-type supervector, $\varepsilon(X^M) = \varepsilon_M$. One readily finds that

$$\begin{aligned} F^M{}_N X^N &= X^N (F^{sT})_N{}^M && \forall\, F \in {}^0Mat^{(+)}(p, q|\Lambda_\infty) \\ X^N G_N{}^M &= (G^{sT})^M{}_N X^N && \forall G \in {}^0Mat^{(-)}(p, q|\Lambda_\infty) \end{aligned} \tag{1.9.63}$$

These relations will often be used below.

1.9.6. Bi-linear forms

Bi-linear forms on supervector spaces may be introduced in different ways. For example, associated with every right linear operator $\mathscr{H}: \mathscr{L} \to {}^*\mathscr{L}$ is the following bi-linear form on $\mathscr{L}$

$$(\vec{X}, \vec{Y})_{\mathscr{H}} \equiv \langle (\vec{X})\mathscr{H} | \vec{Y} \rangle \tag{1.9.64}$$

which possesses the property

$$(\alpha\vec{X}, \vec{Y}\beta)_{\mathscr{H}} = \alpha \cdot (\vec{X}, \vec{Y})_{\mathscr{H}} \cdot \beta \tag{1.9.65}$$

for arbitrary supernumbers α and β. When $\mathscr{H}$ is pure, we easily construct its supertranspose $\mathscr{H}^{\mathrm{sT}}$ as the left linear operator $\mathscr{H}^{\mathrm{sT}}: \mathscr{L} \to \mathscr{L}^*$ defined by

$$\langle \vec{X} \,|\, \mathscr{H}^{\mathrm{sT}}(\vec{Y}) \rangle \equiv (-1)^{\varepsilon(\mathscr{H})[\varepsilon(\vec{X}) + \varepsilon(\vec{Y})] + \varepsilon(\vec{X})\varepsilon(\vec{Y})} \langle (\vec{Y})\mathscr{H} \,|\, \vec{X} \rangle \tag{1.9.66}$$

$\vec{X}$ and $\vec{Y}$ being pure supervectors, thus obtaining another bi-linear form on $\mathscr{L}$

$$(\vec{X}, \vec{Y})_{\mathscr{H}^{\mathrm{sT}}} \equiv \langle \vec{X} \,|\, \mathscr{H}^{\mathrm{sT}}(\vec{Y}) \rangle \tag{1.9.67}$$

possessing property (1.9.65). The latter bi-linear form is called the supertranspose of the former. Bi-linear forms (1.9.64) and (1.9.67) are said to be c-type or a-type or mixed depending on the type of $\mathscr{H}$. A c-type bi-linear form (1.9.64) is called 'supersymmetric' if it coincides with its supertranspose, that is

$$(\vec{X}, \vec{Y})_{\mathscr{H}} = (-1)^{\varepsilon(\vec{X})\varepsilon(\vec{Y})} (\vec{Y}, \vec{X})_{\mathscr{H}} \tag{1.9.68}$$

and 'antisupersymmetric' if

$$(\vec{X}, \vec{Y})_{\mathscr{H}} = -(-1)^{\varepsilon(\vec{X})\varepsilon(\vec{Y})} (\vec{Y}, \vec{X})_{\mathscr{H}}. \tag{1.9.69}$$

It is worth pointing out that every c-type bi-linear form takes c-number values on even supervectors,

$$(\vec{X}, \vec{Y})_{\mathscr{H}} \in \mathbb{C}_c \qquad \forall\, \vec{X}, \vec{Y} \in {}^0\mathscr{L}. \tag{1.9.70}$$

Next, if $\mathscr{H}: \mathscr{L} \to {}^*\mathscr{L}$ is a real c-type linear operator,

$$(\mathscr{H}(\vec{X}))^* = \mathscr{H}(\vec{X}^*) \qquad \forall\, \vec{X} \in \mathscr{L}$$

(see also equation (1.9.40)), then the corresponding bi-linear form takes real c-number values on real even supervectors,

$$(\vec{X}, \vec{Y})_{\mathscr{H}} \in \mathbb{R}_c \qquad \forall\, \vec{X}, \vec{Y} \in {}^0\mathscr{L}_{\mathbf{R}}. \tag{1.9.71}$$

A real c-type linear operator $\mathscr{H}: \mathscr{L} \to {}^*\mathscr{L}$ is said to be 'self-adjoint' if the corresponding bi-linear form is supersymmetric. Below we discuss only c-type bi-linear forms and consider their values only on even supervectors.

Let us fix some standard basis $\{\vec{\mathscr{E}}_M\}$ in $\mathscr{L}$ and its left dual $\{\mathbf{E}^M\}$, which is a real pure basis in ${}^*\mathscr{L}$, and introduce left components for supervectors in ${}^0\mathscr{L}$ and right components for super 1-forms in ${}^0\mathscr{L}^*$:

$$\vec{X} = \vec{\mathscr{E}}_M X^M \qquad \varepsilon(X^M) = \varepsilon_M$$

$$\mathbf{J} = J_M \mathbf{E}^M \qquad \varepsilon(J_M) = \varepsilon_M.$$

As a result, we have

$$\langle \mathbf{J} \,|\, \vec{X} \rangle = J_M X^M = (-1)^{\varepsilon_M} X^M J_M \in \mathbb{C}_c \tag{1.9.72}$$

for all $\mathbf{J} \in {}^0\mathscr{L}^*$ and $\vec{X} \in {}^0\mathscr{L}$. Real supervectors and super 1-forms satisfy the constraints

$$(X^M)^* = X^M \qquad (J_M)^* = (-1)^{\varepsilon_M} J_M. \tag{1.9.73}$$

Given a c-type linear operator $\mathscr{H}: \mathscr{L} \to {}^*\mathscr{L}$, we represent its action as

$$(\vec{E}_M)\mathscr{H} = (-1)^{\varepsilon_M} H_{MN} \mathbf{E}^N \tag{1.9.74a}$$

hence

$$(\vec{X})\mathscr{H} = X^M H_{MN} \mathbf{E}^N \tag{1.9.74b}$$

H_{MN} being some supermatrix under the conditions

$$\varepsilon(H_{MN}) = \varepsilon_M + \varepsilon_N. \tag{1.9.75}$$

Then, the induced bi-linear form reads

$$(\vec{X}, \vec{Y})_{\mathscr{H}} = X^M H_{MN} Y^N. \tag{1.9.76}$$

Is supertranspose (1.9.67) is given by

$$\begin{aligned} (\vec{X}, \vec{Y})_{\mathscr{H}^{sT}} &= X^M (H^{sT})_{MN} Y^N \\ (H^{sT})_{MN} &= (-1)^{\varepsilon_M + \varepsilon_N + \varepsilon_M \varepsilon_N} H_{NM}. \end{aligned} \tag{1.9.77}$$

If our bi-linear form is supersymmetric, then its supermatrix satisfies the equation

$$H_{MN} = (-1)^{\varepsilon_M + \varepsilon_N + \varepsilon_M \varepsilon_N} H_{NM}. \tag{1.9.78}$$

Such supermatrices are called 'supersymmetric'.

Finally, the reality condition (1.9.71) means that

$$(H_{MN})^* = (-1)^{\varepsilon_M + \varepsilon_N + \varepsilon_M \varepsilon_N} H_{MN}. \tag{1.9.79}$$

Suppose the operator $\mathscr{H}$ is invertible. Introducing its inverse $\mathscr{H}^{-1}$: ${}^*\mathscr{L} \to \mathscr{L}$, we define the bi-linear form on ${}^*\mathscr{L}$:

$$(\mathbf{J}, \mathbf{L})_{\mathscr{H}^{-1}} \equiv \langle \mathbf{J} | (\mathbf{L})\mathscr{H}^{-1} \rangle. \tag{1.9.80}$$

It is instructive to obtain its explicit form. Since the supermatrix H_{MN} is non-singular, there exists a unique inverse $(H^{-1})^{MN}$,

$$H^{MR}(H^{-1})_{RN} = \delta^M{}_N \qquad (H^{-1})_{MR} H^{RN} = \delta_M{}^N. \tag{1.9.81}$$

Resolving equations (1.9.74) gives

$$\begin{aligned} (\mathbf{E}^M)\mathscr{H}^{-1} &= (-1)^{\varepsilon_N} (H^{-1})^{MN} \vec{E}_N \\ (\mathbf{J})\mathscr{H}^{-1} &= \vec{E}_M J_N (H^{-1})^{NM} \end{aligned}$$

therefore the expression (1.9.80) takes the form

$$(\mathbf{J}, \mathbf{L})_{\mathscr{H}^{-1}} \equiv J_M L_N (\mathscr{H}^{-1})^{NM} \tag{1.9.82}$$

which should be compared with equation (1.9.76). In conclusion, we note that the inverse of supersymmetric supermatrix (1.9.78) is characterized by the property:

$$(H^{-1})^{MN} = (-1)^{\varepsilon_M \varepsilon_N} (H^{-1})^{NM}. \tag{1.9.83}$$

1.10. Elements of analysis with supernumbers

1.10.1. Superfunctions

We are going to introduce a generalization of the usual smooth and analytic functions in $\mathbb{R}^n$ or $\mathbb{C}^n$ to the case of supernumber-valued functions. Instead of $\mathbb{R}^n$, we consider a real superspace $\mathbb{R}^{p|q} \equiv \mathbb{R}^p_c \times \mathbb{R}^q_a$ (1.9.19) parametrized by p c-number coordinates x^m, $m = 1, 2, \ldots, p$ and q a-number coordinates θ^μ, $\mu = 1, 2, \ldots, q$.

We start by generalizing the concept of analytic functions of ordinary real variables. A supernumber-valued function on $\mathbb{R}^{p|q}$

$$f: \mathbb{R}^{p|q} \to \Lambda_\infty$$

is called 'superanalytic' if it can be expanded in a Taylor series in z^M:

$$\begin{aligned} f(z) &= f(x^1, \ldots, x^p, \theta^1, \ldots, \theta^q) \\ &= \sum_{k=0}^{\infty} f_{M_1 M_2 \ldots M_k} z^{M_1} z^{M_2} \ldots z^{M_k} \qquad f_{M_1 M_2 \ldots M_k} \in \Lambda_\infty. \end{aligned} \tag{1.10.1}$$

To see what this definition leads to, we consider some particular cases. First, when $p = 0$, $q = 1$, we deal with functions of one real odd variable θ. Since a-numbers anticommute between themselves, $\theta^2 = 0$, and equation (1.10.1) means

$$f(\theta) = \varphi + \psi\theta \qquad \varphi, \psi \in \Lambda_\infty. \tag{1.10.2}$$

On the same grounds, in the case of $q \neq 0$, 1 odd variables θ^μ only, every product of more than q θs vanishes, and the most general expression for a superanalytic functions on $\mathbb{R}^{0|q}$ is

$$f(\theta^1, \theta^2, \ldots, \theta^q) = f_0 + \sum_{k=1}^{q} \frac{1}{k!} f_{[\mu_1 \mu_2 \ldots \mu_k]} \theta^{\mu_1} \theta^{\mu_2} \ldots \theta^{\mu_k} \qquad f_0, f_{\mu_1 \mu_2 \ldots \mu_k} \in \Lambda_\infty \tag{1.10.3}$$

with $f_{\mu_1 \mu_2 \ldots \mu_k}$ being totally antisymmetric in its indices. It is clear that the set of all superanalytic functions on $\mathbb{R}^{0|q} = \mathbb{R}^q_a$ forms a supervector space of dimension $(2^{q-1}, 2^{q-1})$.

Life is not so simple when $p \neq 0$. In particular, in the simplest case $p = 1$,

$q=0$ all terms in the Taylor expression

$$f(x)=\sum_{k=0}^{\infty} f_k x^k \qquad f_k \in \Lambda_\infty \tag{1.10.4}$$

survive and we are faced with the problem of finding restrictions on f_k providing existence of the expansion. But for a wide class of superanalytic functions, it is really a problem of ordinary analysis. Indeed, let us split the variable x into its body and soul, $x=x_B+x_S$, and let $\tilde{f}\colon \mathbb{R}\to\mathbb{R}(\mathbb{C})$ be an analytic function of an ordinary real variable,

$$\tilde{f}(x_B)=\sum_{k=0}^{\infty} \tilde{f}_k (x_B)^k \qquad \tilde{f}_k \in \mathbb{R}(\mathbb{C}).$$

Then, the supernumber-valued function on $\mathbb{R}_c$

$$f(x)=f(x_B+x_S)\equiv \tilde{f}(x_B)+\sum_{k=1}^{\infty}\frac{1}{k!}\tilde{f}^{(k)}(x_B)(x_S)^k \tag{1.10.5}$$

where $\tilde{f}^{(k)}$ denotes the kth derivative of $\tilde{f}$, is superanalytic on $\mathbb{R}_c$. Since $(x_S)^n=0$ for some integer n, only a finite number of terms in the expansion (1.10.5) are non-vanishing. Then $f(x)$ is said to be the super extension of $\tilde{f}(x_B)$. A remarkable result is that the general form of a superanalytic function on $\mathbb{R}_c$ is

$$f(x)=\sum_{k=0}^{\infty} f_{i_1 i_2\ldots i_k}(x)\zeta^{i_1}\zeta^{i_2}\ldots\zeta^{i_k} \tag{1.10.6}$$

where ζ^i are the generating elements of Λ_∞ and all the $f_{i_1\ldots i_k}(x)$ are functions of the type (1.10.5).

Another important observation is that equation (1.10.5) may serve as a prescription for constructing 'supersmooth' functions on $\mathbb{R}_c$. Having a smooth function $\tilde{f}$ on $\mathbb{R}$, we introduce its super extension using rule (1.10.5) resulting in a supersmooth function on $\mathbb{R}_c$. The nth derivative of $f(x)$ with respect to x is defined to be

$$f^{(n)}(x)=\frac{\mathrm{d}^n f(x)}{\mathrm{d}x^n}=\tilde{f}^{(n)}(x_B)+\sum_{k=1}^{\infty}\frac{1}{k!}\tilde{f}^{(n+k)}(x_B)(x_S)^k. \tag{1.10.7}$$

By definition, the general form of a supersmooth function on $\mathbb{R}_c$ is given by equation (1.10.6), where all the $f_{i_1\ldots i_k}(x)$ are super extensions of a smooth function on $\mathbb{R}$. Every supersmooth function $f(x)$ is smooth on the body and analytic on the soul of the c-number variable x.

Our previous consideration is easily extended to the case of supernumber-valued functions of c-number variables. In particular, the most general form of a superanalytic function $f\colon \mathbb{R}_c^p\to\Lambda_\infty$ is

$$f(x^1,\ldots,x^p)=\sum_{k=0}^{\infty}\frac{1}{k!}f_{(m_1 m_2\ldots m_k)}x^{m_1}x^{m_2}\ldots x^{m_k} \qquad f_{m_1 m_2\ldots m_k}\in\Lambda_\infty \tag{1.10.8}$$

with the fs being totally symmetric in their indices. Finally, if $p \neq 0$ and $q \neq 0$, every supersmooth or superanalytic function f: $\mathbb{R}^{p|q} \to \Lambda_\infty$ can be expressed in the form

$$f(x^1, \ldots, x^p, \theta^1, \ldots, \theta^q) = f_0(x^1, \ldots, x^p) + \sum_{k=1}^{q} \frac{1}{k!} f_{[\mu_1\mu_2 \ldots \mu_k]}(x^1, \ldots, x^p)\theta^{\mu_1}\theta^{\mu_2} \ldots \theta^{\mu_k} \tag{1.10.9}$$

with the fs being supersmooth or superanalytic functions of $\mathbb{R}^p_c$. Note that all supersmooth functions f: $\mathbb{R}^{p|q} \to \Lambda_\infty$ are supersmooth with respect to their c-number variables and superanalytic with respect to their a-number variables.

Let f: $\mathbb{R}^{p|q} \to \Lambda_\infty$ be some superfunction. For every point $z^M \in \mathbb{R}^{p|q}$, the image $f(z)$ may be decomposed into its even and odd parts

$$f(z) = f_c(z) + f_a(z) \qquad f_c(z) \in \mathbb{C}_c \qquad f_a(z) \in \mathbb{C}_a.$$

Thus, it is reasonable to consider purely even

$$f\colon \mathbb{R}^{p|q} \to \mathbb{C}_c \tag{1.10.10a}$$

or purely odd

$$f\colon \mathbb{R}^{p|q} \to \mathbb{C}_a \tag{1.10.10b}$$

mappings. Superfunctions of the type (1.10.10*a*) will be called 'bosonic' or 'even', and superfunctions of the type (1.10.10*b*) will be called 'fermionic' or 'odd'. Bosonic and fermionic superfunctions will also be called 'pure'. It is useful to associate with any pure superfunction f its Grassmann parity $\varepsilon(f)$ by the rule

$$\varepsilon(f) = \begin{cases} 0 & \text{if } f \text{ bosonic} \\ 1 & \text{if } f \text{ fermionic.} \end{cases} \tag{1.10.11}$$

Having some pure supersmooth function (1.10.9), the coefficients in the right-hand side of (1.10.9) carry the following Grassmann parities

$$\varepsilon(f_0) = \varepsilon(f) \qquad \varepsilon(f_{\mu_1\mu_2 \ldots \mu_k}) = \varepsilon(f) + k \pmod 2. \tag{1.10.12}$$

From here we will consider pure superfunctions only.

Let $f(x^1, \ldots, x^p, \theta^1, \ldots, \theta^q)$ be a supersmooth or superanalytic function. We define 'partial derivatives' of f with respect to its variables as follows

$$\frac{\partial}{\partial x^m} f(z) = \frac{\partial}{\partial x^m} f_0(x) + \sum_{k=1}^{q} \frac{1}{k!} \left(\frac{\partial}{\partial x^m} f_{[\nu_1\nu_2 \ldots \nu_k]}(x) \right) \theta^{\nu_1}\theta^{\nu_2} \ldots \theta^{\nu_k}$$

$$\frac{\vec{\partial}}{\partial \theta^\mu} f(z) = (-1)^{\varepsilon(f)} \sum_{k=1}^{q} (-1)^k \frac{1}{(k-1)!} f_{[\mu\nu_1 \ldots \nu_{k-1}]}(x)\theta^{\nu_1} \ldots \theta^{\nu_{k-1}} \tag{1.10.13}$$

$$f(z)\frac{\overleftarrow{\partial}}{\partial\theta^{\mu}}=\sum_{k=1}^{q}\frac{1}{(k-1)!}f_{[\nu_1\dots\nu_{k-1}\mu]}(x)\theta^{\nu_1}\dots\theta^{\nu_{k-1}}$$

where partial derivatives of functions of the type (1.10.8) are defined in the ordinary way:

$$\frac{\partial}{\partial x^m}f(x^1,\dots,x^p)=\sum_{k=0}^{\infty}\frac{1}{k!}f_{(mn_1\dots n_k)}x^{n_1}\dots x^{n_k}.$$

We have introduced *left* $\overrightarrow{\partial}/\partial\theta^{\mu}$ and *right* $\overleftarrow{\partial}/\partial\theta^{\mu}$ partial derivatives with respect to the odd variables since any infinitesimal displacement $\theta^{\mu}\to\theta^{\mu}+\mathrm{d}\theta^{\mu}$ results in the changing of f

$$\mathrm{d}f(z)=f(x,\,\theta+\mathrm{d}\theta)-f(x,\,\theta)$$

which can be represented in two forms:

$$\mathrm{d}f(z)=\mathrm{d}\theta^{\mu}\left(\frac{\overrightarrow{\partial}}{\partial\theta^{\mu}}f(z)\right)=\left(f(z)\frac{\overleftarrow{\partial}}{\partial\theta^{\mu}}\right)\mathrm{d}\theta^{\mu}. \tag{1.10.14a}$$

The left and right derivatives are related:

$$f(z)\frac{\overleftarrow{\partial}}{\partial\theta^{\mu}}=-(-1)^{\varepsilon(f)}\frac{\overrightarrow{\partial}}{\partial\theta^{\mu}}f(z). \tag{1.10.14b}$$

So, one can work with the left or the right derivatives only. In what follows, we shall use mainly the left derivatives which will be denoted

$$\frac{\partial}{\partial\theta^{\mu}}\equiv\frac{\overrightarrow{\partial}}{\partial\theta^{\mu}}. \tag{1.10.15}$$

Looking at equations (1.10.13), one finds that differentiation with respect to odd variables (by virtue of (1.10.12)) changes the type of superfunction. Namely, the operator $\partial/\partial\theta^{\mu}$ transforms every bosonic superfunction into a fermionic one and vice versa,

$$\varepsilon\left(\frac{\partial f}{\partial\theta^{\mu}}\right)=1+\varepsilon(f)\pmod 2 \tag{1.10.16a}$$

In contrast, differentiation with respect to even variables leaves the superfunction type unchanged,

$$\varepsilon\left(\frac{\partial f}{\partial x^{m}}\right)=\varepsilon(f). \tag{1.10.16b}$$

Introducing the following useful notation

$$\varepsilon_M\equiv\varepsilon(z^M)=\begin{cases}0 & M=m\\ 1 & M=\mu\end{cases} \tag{1.10.17}$$

(see also (1.9.21), and

$$\partial_M = (\partial_m, \partial_\mu) \equiv \frac{\partial}{\partial z^M} = \left(\frac{\partial}{\partial x^m}, \frac{\partial}{\partial \theta^\mu}\right) \tag{1.10.18}$$

one can unify equations (1.10.16*a*) and (1.10.16*b*) into a single equation:

$$\varepsilon(\partial_M f) = \varepsilon_M + \varepsilon(f) \qquad (\text{mod } 2). \tag{1.10.16c}$$

Further properties of the partial derivatives are:

$$1.\ \partial_M z^N = \delta_M{}^N \Leftrightarrow \begin{cases} \partial_m x^n = \delta_m{}^n \\ \partial_\mu \theta^\nu = \delta_\mu{}^\nu \\ \partial_m \theta^\nu = \partial_\mu x^n = 0 \end{cases} \tag{1.10.19a}$$

$$2.\ \partial_M \partial_N = (-1)^{\varepsilon_M \varepsilon_N} \partial_N \partial_M \Leftrightarrow \begin{cases} \partial_m \partial_n = \partial_n \partial_m \\ \partial_m \partial_\nu = \partial_\nu \partial_m \\ \partial_\mu \partial_\nu = -\partial_\nu \partial_\mu \end{cases} \tag{1.10.19b}$$

$$3.\ \partial_M (f\varphi) = (\partial_M f)\varphi + (-1)^{\varepsilon_M \varepsilon(f)} f(\partial_M \varphi) \tag{1.10.19c}$$

where f and φ are superfunctions. We see that the odd derivatives anticommute between themselves and commute with the even derivatives. What is more, pulling the operator $\partial/\partial\theta^\mu$ through a fermionic superfunction gives a minus sign. The equation (1.10.19*c*) can also be rewritten as

$$\partial_M \cdot f - (-1)^{\varepsilon_M \varepsilon(f)} f \cdot \partial_M = (\partial_M f). \tag{1.10.20}$$

Finally, we clarify how the operation of complex conjugation (1.9.10) acts on partial derivatives. Given a superfunction of the form (1.10.9), one obtains

$$\begin{aligned} f^* &= f_0^* + \sum_{k=1}^{q} \frac{1}{k!} \theta^{\nu_k} \ldots \theta^{\nu_2} \theta^{\nu_1} f^*_{[\nu_1 \nu_2 \ldots \nu_k]} \\ &= f_0^* + \sum_{k=1}^{q} (-1)^{k(k-1)/2} \frac{1}{k!} \theta^{\nu_1} \theta^{\nu_2} \ldots \theta^{\nu_k} f^*_{[\nu_1 \nu_2 \ldots \nu_k]} \\ &= f_0^* + \sum_{k=1}^{q} (-1)^{\{k(k-1)/2 + k + k\varepsilon(f)\}} \frac{1}{k!} f^*_{[\nu_1 \nu_2 \ldots \nu_k]} \theta^{\nu_1} \theta^{\nu_2} \ldots \theta^{\nu_k} \end{aligned}$$

where we have used equation (1.10.12). Making use of equation (1.10.13), we then obtain

$$\partial_\mu f^* = (-1)^{\varepsilon(f)} \sum_{k=1}^{q} (-1)^{\{k(k-1)/2 + k\varepsilon(f)\}} \frac{1}{(k-1)!} f^*_{[\mu\nu_1 \ldots \nu_{k-1}]} \theta^{\nu_1} \ldots \theta^{\nu_{k-1}}.$$

Now we change the order of operations,

$$(\partial_\mu f)^* = (-1)^{\varepsilon(f)} \sum_{k=1}^{q} (-1)^k \frac{1}{(k-1)!} \theta^{\nu_{k-1}} \dots \theta^{\nu_1} f^*_{[\mu\nu_1 \dots \nu_{k-1}]}$$

$$= (-1)^{\varepsilon(f)} \sum_{k=1}^{q} (-1)^{k+(k-1)(k-2)/2} \frac{1}{(k-1)!} \theta^{\nu_1} \dots \theta^{\nu_{k-1}} f^*_{[\mu\nu_1 \dots \nu_{k-1}]}$$

$$= (-1)^{\varepsilon(f)} \sum_{k=1}^{q} (-1)^{\{(k-1)(k-2)/2 + k + (k-1)(\varepsilon(f)+k)\}} \frac{1}{(k-1)!}$$

$$\times f^*_{[\mu\nu_1 \dots \nu_{k-1}]} \theta^{\nu_1} \dots \theta^{\nu_{k-1}}.$$

Comparing $\partial_\mu f^*$ and $(\partial_\mu f)^*$, we find

$$(\partial_\mu f)^* = -(-1)^{\varepsilon(f)} \partial_\mu f^*. \qquad (1.10.21)$$

This beautiful result can be unified with the trivial relation

$$(\partial_m f)^* = \partial_m f^*$$

as follows

$$(\partial_M f)^* = (-1)^{\varepsilon_M(1+\varepsilon(f))} \partial_M f^*. \qquad (1.10.22)$$

What does equation (1.10.22) mean? Let us consider 'real' bosonic

$$f: \mathbb{R}^{p|q} \to \mathbb{R}_c \qquad (1.10.23)$$

and fermionic

$$f: \mathbb{R}^{p|q} \to \mathbb{R}_a \qquad (1.10.24)$$

superfunctions. Then equation(1.10.22) tells us that an odd partial derivative of a real bosonic (fermionic) superfunction is an imaginary fermionic (real bosonic) superfunction.

1.10.2. Integration over $\mathbb{R}^{p|q}$

Our goal is to develop an integration theory on superspaces $\mathbb{R}^{p|q}$. Of course, the theory of integration over $\mathbb{R}^n$ will guide us. We shall consider in detail the simplest cases $p=1$, $q=0$ and $p=0$, $q=1$, and then generalize the results to the general case.

Let $f(x)$: $\mathbb{R}_c \to \Lambda_\infty$ be a supersmooth function. The main question is how to define the integral

$$\int_{x_1}^{x_2} f(x)\,\mathrm{d}x \qquad x_1, x_2 \in \mathbb{R}_c \qquad (1.10.25)$$

where the integration is performed over a contour

$$x(t): [0, 1] \to \mathbb{R}_c \qquad x(0) = x_1,\ x(1) = x_2.$$

The problem we are faced with is that, in contrast to $\mathbb{R}$ where integrals like

(1.10.25) are usually performed over a line

$$x_{\mathrm{B}}(t)=x_{\mathrm{B},1}+t(x_{\mathrm{B},2}-x_{\mathrm{B},1}) \tag{1.10.26}$$

there are now many more possibilities in the choice of integration contour. Indeed, splitting $x(t)$ into its body and soul,

$$x(t)=x_{\mathrm{B}}(t)+x_{\mathrm{S}}(t)$$

and fixing the body, for example, as in equation (1.10.26), we still have wide freedom in our choice of the soul. Therefore, it would be desirable to take as our integration prescription one which will guarantee that the integral (1.10.25) will not depend on the particular choice of integration contour (which implies that the soul is measureless). The second requirement of integration theory over $\mathbb{R}_c$ is a correspondence with the integration theory over $\mathbb{R}$. Namely, if $f(x)$ is the super extension of a smooth function $\tilde{f}(x_{\mathrm{B}})$ on $\mathbb{R}$ (equation (1.10.5)) and the end points in equation (1.10.25) are real numbers, then the integral should coincide with the ordinary integral

$$\int_{x_1}^{x_2} \tilde{f}(x_{\mathrm{B}})\,\mathrm{d}x_{\mathrm{B}}.$$

The above requirements are satisfied by an integration theory which is characterized by the property

$$\int_{x_1}^{x_2} \mathrm{d}\varphi=\int_{x_1}^{x_2} \varphi'(x)\,\mathrm{d}x=\varphi(x_2)-\varphi(x_1) \tag{1.10.27}$$

where $\varphi(x)$ is an arbitrary supersmooth function. Let us prove this statement for the example of a superfunction $f(x)$ which is the super extension of a smooth function $\tilde{f}(x_{\mathrm{B}})$ on $\mathbb{R}$. We introduce another smooth function $\tilde{F}(x_{\mathrm{B}})$ on $\mathbb{R}$,

$$\tilde{F}(x_{\mathrm{B}})=\int_0^{x_B} \tilde{f}(x'_{\mathrm{B}})\,\mathrm{d}x'_{\mathrm{B}} \qquad \tilde{F}'(x_{\mathrm{B}})=\tilde{f}(x_{\mathrm{B}})$$

and construct its extension $F(x)$ on $\mathbb{R}_c$ by the rule (1.10.5):

$$F(x)=\tilde{F}(x_{\mathrm{B}})+\sum_{k=1}^{\infty}\frac{1}{k!}\tilde{F}^{(k)}(x_{\mathrm{B}})(x_{\mathrm{S}})^k$$

$$=\tilde{F}(x_{\mathrm{B}})+\sum_{k=1}^{\infty}\frac{1}{k!}\tilde{f}^{(k-1)}(x_{\mathrm{B}})(x_{\mathrm{S}})^k.$$

As a result of(1.10.7), we have

$$F'(x)=f(x).$$

Applying equation (1.10.27), we have

$$\int_{x_1}^{x_2} f(x)\,\mathrm{d}x = F(x_2) - F(x_1).$$

This completes the proof. Note that our consideration happened also to be successful for the following reason: every supersmooth function $f(x)$ is analytic with respect to the soul of x.

Since the integrals (1.10.25) depend only on the end points of the integration contour, one may immediately extend distributions on $\mathbb{R}$ (like the δ-function) to distributions on $\mathbb{R}_c$, to develop Fourier transform theory on $\mathbb{R}_c$, and so on. In particular, the δ-function on $\mathbb{R}_c$ can be represented in a standard way, namely

$$\delta(x) = \frac{1}{2\pi}\int_{-\infty}^{\infty} \mathrm{e}^{\mathrm{i}px}\,\mathrm{d}p = \lim_{\varepsilon\to+0}\frac{1}{2\pi}\int_{-\infty}^{\infty} \mathrm{e}^{\mathrm{i}px-\varepsilon p^2}\,\mathrm{d}p \tag{1.10.28}$$

where the integrals here can be performed over every contour $p(t)$: $\mathbb{R}\to\mathbb{R}_c$ such that the body of $p(t)$ goes to $\pm\infty$ when t goes to $\pm\infty$, respectively,

$$\lim_{t\to\pm\infty} p_{\mathrm{B}}(t) = \pm\infty.$$

For every supersmooth function $f(x)$: $\mathbb{R}_c\to\Lambda_\infty$, we have

$$f(x) = \int_{-\infty}^{\infty} f(y)\delta(x-y)\,\mathrm{d}y. \tag{1.10.29}$$

In perfect analogy with integration over $\mathbb{R}^p$, the integration over $\mathbb{R}^{p|0} = \mathbb{R}_c^p$ will be understood as multiple integration consisting of sequences of ordinary integrations over $\mathbb{R}_c$

$$\int \mathrm{d}^p x\, f(x^1, x^2, \ldots, x^p) = \int_{-\infty}^{\infty}\mathrm{d}x^1 \int_{-\infty}^{\infty}\mathrm{d}x^2 \ldots \int_{-\infty}^{\infty}\mathrm{d}x^p\, f(x^1, x^2, \ldots, x^p)$$

$$\mathrm{d}^p x \equiv \mathrm{d}x^1\,\mathrm{d}x^2 \ldots \mathrm{d}x^p \tag{1.10.30}$$

where the supersmooth function $f(x^1, x^2, \ldots, x^p)$ should satisfy some obvious restrictions to ensure existence of the integral. We do not explicitly define integrals over finite domains in $\mathbb{R}_c^p$ since every such integral can be extended to an integral of the type (1.10.30). The δ-function on $\mathbb{R}_c^p$ is defined by the rule

$$\delta^p(x^m) \equiv \delta(x^1)\delta(x^2)\ldots\delta(x^p)$$

and has the same properties as the δ-function in $\mathbb{R}^p$.

We proceed to integrate over $\mathbb{R}_a$. Given some superfunction $f(\theta)$ on $\mathbb{R}_a$, which is expressed in the form (1.10.2), one must define an integral

$$\int f(\theta)\,\mathrm{d}\theta.$$

How do we do this? When considering integration over $\mathbb{R}_c$, it was possible, since $\mathbb{R}_c$ contains $\mathbb{R}$, to make use of our experience of integration over $\mathbb{R}$. Now any experience is absent. All we are able to do is to postulate some reasonable axioms determining the properties of integrals. Following F. Berezin, we postulate

$$\int [f(\theta)+h(\theta)]\,\mathrm{d}\theta = \int f(\theta)\,\mathrm{d}\theta + \int h(\theta)\,\mathrm{d}\theta \tag{1.10.31a}$$

$$\int z\cdot f(\theta)\,\mathrm{d}\theta = z\int f(\theta)\,\mathrm{d}z \qquad z\in\Lambda_\infty \tag{1.10.31b}$$

$$\int \frac{\mathrm{d}}{\mathrm{d}\theta} f(\theta)\,\mathrm{d}\theta = 0 \tag{1.10.31c}$$

$$\int \theta\,\mathrm{d}\theta = -1. \tag{1.10.31d}$$

Let us comment on the axioms. The equations (1.10.31*a,b*) express standard properties of ordinary integrals. Equation (1.10.31*c*) is a generalization of the well-known integral property (1.10.27). Since $\int f(\theta)\,\mathrm{d}\theta$ is understood as an integral over the whole space $\mathbb{R}_a$ and $\mathbb{R}_a$ is boundless, the right-hand side of property (1.10.31*c*) should vanish. Equivalently, the axiom demands that every integral be invariant with respect to translations on $\mathbb{R}_a$

$$\int f(\theta+\gamma)\,\mathrm{d}\theta = \int f(\theta)\,\mathrm{d}\theta \qquad \forall\,\gamma\in\mathbb{R}_a. \tag{1.10.32}$$

This axiom can also be treated as the rule for integration by parts:

$$\int \frac{\mathrm{d}f(\theta)}{\mathrm{d}\theta}\cdot h(\theta)\,\mathrm{d}\theta = -(-1)^{\varepsilon(f)}\int f(\theta)\frac{\mathrm{d}h(\theta)}{\mathrm{d}\theta}\,\mathrm{d}\theta. \tag{1.10.33}$$

Recalling the most general form (1.10.2) of superfunctions on $\mathbb{R}_a$, equation (1.10.31*c*) can also be expressed as the statement that $\mathbb{R}_a$ has vanishing 'volume'.

$$\int \mathrm{d}\theta \equiv 0. \tag{1.10.34}$$

After imposing the axioms (1.10.31*a–c*), the remaining task is to normalize the integral $\int \theta\,\mathrm{d}\theta$. It is the last axiom (1.10.31*d*) which plays the role of the normalization condition. Since the right-hand side in axiom (1.10.31*d*) is a *c*-number and the integration variable θ is an *a*-number, the measure $\mathrm{d}\theta$

should be treated as an a-number object anticommuting with all a-numbers,

$$d\theta z = -z\, d\theta \qquad (\forall\ z \in \mathbb{C}_a) \tag{1.10.35}$$

$$\Rightarrow \quad \int d\theta\theta = -\int \theta\, d\theta.$$

Then equation (1.10.31d) can be expressed in the form

$$\int d\theta\, f(\theta) = \frac{d}{d\theta} f(\theta). \tag{1.10.36}$$

Therefore, the integration over $\mathbb{R}_a$ is equivalent to differentiation.

We introduce the δ-function on $\mathbb{R}_a$:

$$\delta(\theta) \equiv \theta. \tag{1.10.37}$$

It possesses the standard δ-function properties

$$\begin{aligned} &\int d\theta\, \delta(\theta)\, f(\theta) = f(0) \\ &\delta(\theta - \theta')\, f(\theta') = \delta(\theta - \theta') f(\theta) \end{aligned} \tag{1.10.38}$$

as well as some unusual properties:

$$\begin{aligned} &\delta(\theta)\delta(\theta) \equiv 0, \\ &\delta(-\theta) = -\delta(\theta) \qquad \delta(0) = 0 \\ &\delta(\theta)\, f(\theta) = (-1)^{\varepsilon(f)} f(\theta)\, \delta(\theta). \end{aligned} \tag{1.10.39}$$

It is not difficult to obtain an analogue of the representation (1.10.28) for the δ-function on $\mathbb{R}_a$. It is

$$\delta(\theta) = \int d\lambda\, e^{\lambda\theta} \tag{1.10.40}$$

where the integral is performed over $\mathbb{R}_a$.

By analogy with integration over $\mathbb{R}_c^p$, integration over $\mathbb{R}_a^q$ will be understood as multiple integration:

$$\int d^q\theta\, f(\theta^1, \ldots, \theta^q) \equiv \int d\theta_q \int d\theta_{q-1} \cdots \int d\theta_1\, f(\theta^1, \ldots, \theta^q) \tag{1.10.41}$$

where $d\theta_\mu$ is the measure on $\mathbb{R}_a$ with coordinate θ^μ,

$$\int d\theta_\mu\, \theta^\nu = \delta_\mu{}^\nu. \tag{1.10.42}$$

The measures $d\theta_\mu$ are assumed to anticommute with all a-numbers and between themselves

$$d\theta_\mu \, d\theta_\nu + d\theta_\nu \, d\theta_\mu = 0. \tag{1.10.43}$$

For any superfunction $f(\theta^1, \ldots, \theta^q)$, which is expressed in the form (1.10.3), one obtains

$$\int d^q\theta \, f(\theta^1, \ldots, \theta^q) = (-1)^{q(1+\varepsilon(f))} f_{12\ldots q}. \tag{1.10.44}$$

Therefore, only the last term in the power series (1.10.3) gives a non-vanishing contribution to the integral.

The δ-function on $\mathbb{R}^q_a$ is

$$\delta^q(\theta^1, \theta^2, \ldots, \theta^q) \equiv \delta(\theta^1)\,\delta(\theta^2)\ldots\delta(\theta^q). \tag{1.10.45}$$

Its properties can be easily read off from expressions (1.10.38, 39).

Finally, integration over $\mathbb{R}^{p|q}$ is defined as multiple integration over $\mathbb{R}^p_c$ and $\mathbb{R}^q_a$:

$$\int d^{p+q}z \, f(z^M) \equiv \int d^px \int d^q\theta \, f(x^1, \ldots, x^p, \theta^1, \ldots, \theta^q). \tag{1.10.46}$$

1.10.3. Linear replacements of variables on $\mathbb{R}^{p|q}$

Consider a linear one-to-one mapping of $\mathbb{R}^{p|q}$ on itself

$$z^M \to z'^M = F^M{}_N z^N$$
$$F^M{}_N = \begin{pmatrix} A^m{}_n & B^m{}_\nu \\ C^\mu{}_n & D^\mu{}_\nu \end{pmatrix} \tag{1.10.47}$$

where F is a c-type non-singular supermatrix (recall that a supermatrix is non-singular if and only if it has non-singular body); F has to be a real supermatrix of the form (1.9.42b) for the z'^M to be real. Indeed, rewriting the transformation explicitly as

$$x'^m = A^m{}_n x^n + B^m{}_\nu \theta^\nu$$
$$\theta'^\mu = C^\mu{}_n x^n + D^\mu{}_\nu \theta^\nu$$

and demanding $(x'^m)^* = x'^m$ and $(\theta'^\mu)^* = \theta'^\mu$, one finds that the matrices A, D and C have real entries while the matrix B is purely imaginary. Evidently, the set of all invertible transformations (1.10.47) forms a group.

Let us treat the transformation (1.10.47) as a replacement of variables in integrals like (1.10.46). It is clear then that the measures $d^{p+q}z$ and $d^{p+q}z'$ in the old and new variables, respectively, differ by the transformation Jacobian $J(F)$

$$d^{p+q}z' = J(F)\, d^{p+q}z \tag{1.10.48}$$

which is some function of the supermatrix F. Our main goal is to determine

$J(F)$. We divide the solution of the problem into parts, first calculating the Jacobians of some particular transformations and returning to the general case at the end.

1. We start with the replacement

$$\begin{matrix} x'^m = A^m{}_n x^n \\ \theta'^\mu = \theta^\mu \end{matrix} \Rightarrow F = \begin{pmatrix} A & 0 \\ 0 & \mathbb{I}_q \end{pmatrix}. \tag{1.10.49}$$

This is a linear change of variables in $\mathbb{R}^p_c$. The Jacobian can be calculated in the same way as for $\mathbb{R}^p$. The result is

$$J(F) = \det A. \tag{1.10.50}$$

2. Now consider the replacement

$$\begin{matrix} x'^m = x^m \\ \theta'^\mu = D^\mu{}_\nu \theta^\nu \end{matrix} \Rightarrow F = \begin{pmatrix} \mathbb{I}_p & 0 \\ 0 & D \end{pmatrix}. \tag{1.10.51}$$

To find the Jacobian, note that the variables θ'^μ and the corresponding measures $d\theta'^\mu$ should satisfy the same basic identity (1.10.42) as in the former variables. So, we have

$$\begin{gathered} \delta_\mu{}^\nu = \int d\theta'_\mu\, \theta'^\nu = D^\nu{}_\lambda \int d\theta'_\mu\, \theta^\lambda \Rightarrow \\ d\theta'_\mu = d\theta_\sigma (D^{-1})^\sigma{}_\mu. \end{gathered} \tag{1.10.52}$$

By virtue of equation (1.10.43), one then has

$$d\theta'_1\, d\theta'_2 \ldots d\theta'_q = (\det D)^{-1}\, d\theta_1\, d\theta_2 \ldots d\theta_q.$$

Therefore, the transformation (1.10.51) is characterized by the Jacobian

$$J(F) = (\det D)^{-1}. \tag{1.10.53}$$

3. Consider the replacement

$$\begin{matrix} x'^m = x^m + B^m{}_\nu \theta^\nu \\ \theta'^\mu = \theta^\mu \end{matrix} \Rightarrow F = \begin{pmatrix} \mathbb{I}_p & B \\ 0 & \mathbb{I}_q \end{pmatrix}. \tag{1.10.54}$$

In this case we have

$$J(F) = 1. \tag{1.10.55}$$

Indeed, let us understand the integral (1.10.46) as follows

$$\int d^{p+q}z\, f(z^M) = \int d^p\theta \int d^q x\, f(x^1, \ldots, x^p, \theta^1, \ldots, \theta^q).$$

For fixed θs, the transformation (1.10.54) is a shift in $\mathbb{R}_c^q$, which keeps the measure $d^p x$ invariant.

4. Finally, consider the transformation

$$\begin{aligned} x'^m &= x^m \\ \theta'^\mu &= \theta^\mu + C^\mu{}_n x^n \end{aligned} \Rightarrow F = \begin{pmatrix} \mathbb{1}_p & 0 \\ C & \mathbb{1}_q \end{pmatrix}. \tag{1.10.56}$$

Now we understand the integration over $\mathbb{R}^{p|q}$ as in (1.10.46). For fixed xs, the replacement (1.10.56) is a shift on $\mathbb{R}_a^q$. But in accordance with equation (1.10.32), the measure $d^q\theta$ is translationally invariant. Therefore, the Jacobian is unity,

$$J(F) = 1. \tag{1.10.57}$$

Now we are in a position to find the Jacobian of the transformation (1.10.47). To do this, we represent the transformation as a product of some replacements, each of which belongs to one of the four types considered above. The matrix F can be expressed as

$$F = \begin{pmatrix} \mathbb{1}_p & 0 \\ 0 & D \end{pmatrix} \begin{pmatrix} \mathbb{1}_p & B \\ 0 & \mathbb{1}_q \end{pmatrix} \begin{pmatrix} A - BD^{-1}C & 0 \\ 0 & \mathbb{1}_q \end{pmatrix} \begin{pmatrix} \mathbb{1}_p & 0 \\ D^{-1}C & \mathbb{1}_q \end{pmatrix} \tag{1.10.58a}$$

or as

$$F = \begin{pmatrix} A & 0 \\ 0 & \mathbb{1}_q \end{pmatrix} \begin{pmatrix} \mathbb{1}_p & 0 \\ C & \mathbb{1}_q \end{pmatrix} \begin{pmatrix} \mathbb{1}_p & 0 \\ 0 & D - CA^{-1}B \end{pmatrix} \begin{pmatrix} \mathbb{1}_p & A^{-1}B \\ 0 & \mathbb{1}_q \end{pmatrix}. \tag{1.10.58b}$$

Therefore, the replacement (1.10.47) is characterized by the Jacobian

$$J(F) = \det(A - BD^{-1}C)\det{}^{-1}(D) = \det A \det{}^{-1}(D - CA^{-1}B) \tag{1.10.59}$$

where we have used equations (1.10.49–58). The Jacobian $J(F)$ proves to be real, $J(F) \in \mathbb{R}_c$, because F is a real supermatrix of the form (1.9.42*b*).

1.10.4. c-type supermatrices revisited

After obtaining some experience in analysis with supernumbers, it is worth returning again to an algebraic subject. We find it convenient to collect here the main operations with *c*-type $(p+q)\times(p+q)$ supermatrices, elements of ${}^0Mat^{(+)}(p, q|\Lambda_\infty)$ or ${}^0Mat^{(-)}(p, q|\Lambda_\infty)$. Every supermatrix $F \in {}^0Mat^{(+)}(p, q|\Lambda_\infty)$ is represented in the form (1.9.28), where the matrix element $F^M{}_N$ are pure supernumbers with the following Grassmann parities

$$\varepsilon(F^M{}_N) = \varepsilon_M + \varepsilon_N \pmod 2 \tag{1.10.60}$$

(see also equation (1.9.37*a*)). Every supermatrix $G \in {}^0Mat^{(-)}(p, q|\Lambda_\infty)$ looks as in equation (1.9.22). Supermatrices from $Mat^{(+)}(p,q\,|\,\Lambda_\infty)$ and $Mat^{(-)}(p,q\,|\,\Lambda_\infty)$ differ in the positions of indices.

Both spaces ${}^0Mat^{(+)}(p,q\,|\,\Lambda_\infty)$ and ${}^0Mat^{(-)}(p,q\,|\,\Lambda_\infty)$ are closed with respect to:

1. the operation of addition (1.9.31*a*);
2. the operation of multiplication (1.9.31*b*);
3. the operation of multiplication by *c*-numbers (1.9.31*c*);
4. the operation of supercomplex conjugation defined on ${}^0Mat^{(+)}(p,q\,|\,\Lambda_\infty)$ by equation (1.9.41*b*) and on ${}^0Mat^{(-)}(p,q\,|\,\Lambda_\infty)$ by

$$G^{s*}=\begin{pmatrix} A^* & B^* \\ -C^* & D^* \end{pmatrix} \tag{1.10.61}$$

   the specialization of equation (1.9.62) to the *c*-type case;
5. the operation of taking the inverse (1.9.23), which is defined for non-singular *c*-type supermatrices only.

The spaces are connected to each other by the operation of supertransposition:

$$\begin{gathered} \mathrm{sT}:\quad {}^0Mat^{(+)}(p,q\,|\,\Lambda_\infty)\to {}^0Mat_{\mathbb{R}}^{(+)}(p,q\,|\,\Lambda_\infty) \\ (F^{\mathrm{sT}})_M{}^N=(-1)^{\varepsilon_M+\varepsilon_M\varepsilon_N}F^N{}_M \end{gathered} \tag{1.10.62a}$$

$$\begin{gathered} \mathrm{sT}:\ {}^0Mat^{(-)}(p,q\,|\,\Lambda_\infty)\to {}^0Mat^{(+)}(p,q\,|\,\Lambda_\infty) \\ (G^{\mathrm{sT}})^M{}_N=(-1)^{\varepsilon_N+\varepsilon_M\varepsilon_N}G_N{}^M \end{gathered} \tag{1.10.62b}$$

Recall, supertransposition is characterized by the property (1.9.63).

The operation of supercomplex conjugation selects in ${}^0Mat_{\mathbb{R}}^{(+)}(p,q\,|\,\Lambda_\infty)$ the subset ${}^0Mat_{\mathbb{R}}^{(+)}(p,q\,|\,\Lambda_\infty)$ of real supermatrices which satisfy the equation $F^{s*}=F$ and have the form (1.9.42*b*). ${}^0Mat_{\mathbb{R}}^{(+)}(p,q\,|\,\Lambda_\infty)$ is closed with respect to the operations of addition and multiplication of its elements and the operation of multiplication by real *c*-numbers. The supertransposition maps ${}^0Mat_{\mathbb{R}}^{(+)}(p,q\,|\,\Lambda_\infty)$ onto the real subset ${}^0Mat_{\mathbb{R}}^{(-)}(p,q\,|\,\Lambda_\infty)$ in ${}^0Mat^{(-)}(p,q\,|\,\Lambda_\infty)$.

The set of non-singular supermatrices in ${}^0Mat^{(+)}(p,q\,|\,\Lambda_\infty)$, evidently, forms a group denoted by $GL^{(+)}(p,q\,|\,\Lambda_\infty)$. If $F\in {}^0Mat_{\mathbb{R}}^{(+)}(p,q\,|\,\Lambda_\infty)$ is a non-singular supermatrix, then, due to (1.9.23), its inverse supermatrix F^{-1} is real. Since the product of real supermatrices is a real supermatrix, the set of all non-singular supermatrices in ${}^0Mat_{\mathbb{R}}^{(+)}(p,q\,|\,\Lambda_\infty)$ forms a group denoted by $GL_{\mathbb{R}}^{(+)}(p,q\,|\,\Lambda_\infty)$. Supertransposition maps $GL^{(+)}(p,q\,|\,\Lambda_\infty)$ $(GL_{\mathbb{R}}^{(+)}(p,q\,|\,\Lambda_\infty))$ onto the group $GL^{(-)}(p,q\,|\,\Lambda_\infty)$ $(GL_{\mathbb{R}}^{(-)}(p,q\,|\,\Lambda_\infty))$ of non-singular (real non-singular) supermatrices in ${}^0Mat^{(-)}(p,q\,|\,\Lambda_\infty)$.

Now we are going to introduce a supergeneralization of the determinant. Recall that the determinant of an ordinary $n\times n$ matrix A can be defined as the Jacobian of the replacement of variables $x'=Ax$ on $\mathbb{R}^n$. It is this definition which will be generalized to the super-case. In the previous subsection, we found the Jacobian of the linear replacement of variables (1.10.47) in $\mathbb{R}^{p|q}$

associated with a non-singular supermatrix F. The Jacobian was given by equation (1.10.59). So it would be desirable to postulate the 'superdeterminant' or the 'Berezinian' of a non-singular supermatrix $F \in GL^{(+)}(p, q \mid \Lambda_\infty)$ as

$$\text{sdet}\, F \equiv \text{Ber}\, F = \det(A - BD^{-1}C) \det{}^{-1} D = \det A \det{}^{-1}(D - CA^{-1}B). \tag{1.10.63}$$

and similarly for supermatrices from $GL^{(-)}(p, q \mid \Lambda_\infty)$. The notion of superdeterminant was introduced by F. Berezin. In accordance with (1.10.63), the Berezinian can be calculated using two different prescriptions. The statement that the first and the second lead to the same final result is a consequence of our consideration in subsection 1.10.3. (In fact, the supermatrix F in (1.10.47) was real. To generalize equation (1.10.59) to the case of complex supermatrices, one has simply to consider linear replacements of variables in the complex superspace $\mathbb{C}^{p|q}$.) On the same grounds, since the Jacobian for the composition of two replacements of variables is equal to the product of the Jacobians, one immediately obtains

$$\begin{gathered} \text{Ber}(F_1 F_2) = \text{Ber}\, F_1\, \text{Ber}\, F_2 \quad \Rightarrow \\ \text{Ber}(F^{-1}) = (\text{Ber}\, F)^{-1}. \end{gathered} \tag{1.10.64}$$

We also have

$$\text{Ber}(F^{\text{sT}}) = \text{Ber}\, F. \tag{1.10.65}$$

Indeed, due to (1.9.58*b*) and (1.10.63), one can write

$$\begin{aligned} \text{Ber}(F^{\text{sT}}) &= \det[A^{\text{T}} + C^{\text{T}}(D^{\text{T}})^{-1} B^{\text{T}}] \det{}^{-1} D^{\text{T}} \\ &= \det[(A - BD^{-1}C)^{\text{T}}] \det{}^{-1} D^{\text{T}} = \text{Ber}\, F \end{aligned}$$

where we have used the fact that the entries of the submatrices B and C are a-numbers. Note that relation (1.10.65) is in complete agreement with the fact that the transformation (1.10.47) can be written in two equivalent forms

$$z'^M = F^M{}_N z^N = z^N (F^{\text{sT}})_N{}^M$$

in accordance with (1.9.63).

Finally, we introduce a supergeneralization of the trace of a matrix. Recall that $\det(\mathbb{1} + A) = 1 + \text{tr}\, A$, for every ordinary matrix A with infinitesimal elements. We take this identity as the starting point in defining a supertrace. Let F be an infinitesimal supermatrix (this means that the matrix elements F^M_N are infinitesimal supernumbers with respect to the norm (1.9.6)). Then the supermatrix $(\mathbb{1}_{(p,q)} + F)$ is non-singular and, due to (1.10.63), we have

$$\text{Ber}(\mathbb{1}_{(p,q)} + F) = 1 + \text{tr}\, A - \text{tr}\, D.$$

So we postulate the 'supertrace' of a supermatrix $F \in Mat^{(+)}(p, q \mid \Lambda_\infty)$ to be

$$\text{str}\, F = (-1)^{\varepsilon_M} F^M{}_M \tag{1.10.66}$$

and similarly for supermatrices from ${}^0Mat^{(-)}(p,q|\Lambda_\infty)$. The main supertrace properties are

$$\begin{aligned} \mathrm{str}(F^{sT}) &= \mathrm{str}\, F \\ \mathrm{str}(F_1F_2) &= \mathrm{str}(F_2F_1). \end{aligned} \tag{1.10.67}$$

The proof is quite trivial

$$\mathrm{str}(F^{sT}) = (-1)^{\varepsilon_M}(F^{sT})_M{}^M = (-1)^{\varepsilon_M}(-1)^{\varepsilon_M+\varepsilon_M\varepsilon_M}F^M{}_M = (-1)^{\varepsilon_M}F^M{}_M$$

$$\begin{aligned} \mathrm{str}(F_1F_2) &= (-1)^{\varepsilon_M}F_{1N}^M F_{2M}^N = (-1)^{\varepsilon_M}(-1)^{(\varepsilon_M+\varepsilon_N)(\varepsilon_M+\varepsilon_N)}F_{2M}^N F_{1N}^M \\ &= (-1)^N F_{2M}^N F_{1N}^M. \end{aligned}$$

The superdeterminant and supertrace are related as follows:

$$\mathrm{Ber}(\mathrm{e}^F) = \mathrm{e}^{\mathrm{str} F} \tag{1.10.68}$$

where F is an arbitrary supermatrix. To prove (1.10.68), one has simply to follow the ordinary proof. One introduces the c-number-valued function on $\mathbb{R}$

$$f(t) = \mathrm{Ber}(\mathrm{e}^{tF}) \qquad f(0) = 1.$$

For an infinitesimal displacement $\mathrm{d}t$, we have

$$f(t+\mathrm{d}t) = \mathrm{Ber}(\mathrm{e}^{tF}\,\mathrm{e}^{\mathrm{d}tF}) = \mathrm{Ber}(\mathrm{e}^{tF})\,\mathrm{Ber}(\mathbb{1}_{(p,q)} + \mathrm{d}tF) = f(t)\{1 + \mathrm{d}t\cdot\mathrm{str}\, F\}.$$

Then,

$$\mathrm{d}f(t)/\mathrm{d}t = (\mathrm{str}\, F)f(t) \qquad \Rightarrow f(t) = c\,\mathrm{e}^{t(\mathrm{str} F)}.$$

Taking into account the initial condition, one arrives at (1.10.68).

1.11. The supergroup of general coordinate transformations on $\mathbb{R}^{p|q}$

We would like to continue the study of real superspaces $\mathbb{R}^{p|q}$ parametrized by p even coordinates x^m and q odd coordinates θ^μ. As the next step it would be very interesting to generalize the concept of the general coordinate transformation group on manifolds which, as is well known, plays an important role in all gravity models.

Let us consider a one-to-one mapping of $\mathbb{R}^{p|q}$ on itself. In the coordinates on $\mathbb{R}^{p|q}$, this is

$$z^M \to z'^M = f^M(z) \Leftrightarrow \begin{cases} x^m \to x'^m = f^m(x,\theta) \\ \theta^\mu \to \theta'^\mu = f^\mu(x,\theta). \end{cases} \tag{1.11.1}$$

We restrict ourselves to the consideration of supersmooth transformations with f^m and f^μ being supersmooth functions. The necessary condition for the transformation being invertible is that the supermatrix of partial derivatives

$$G_M{}^N(z) \equiv \partial_M f^N(z)$$

is non-singular at each point $z^M \in \mathbb{R}^{p|q}$. Indeed, infinitesimally, every displacement $\mathrm{d}z'^M$ should be the image of a unique displacement $\mathrm{d}z^N$ related to each other by the rule

$$\mathrm{d}z'^M = \mathrm{d}z^N(\partial_N f^M).$$

The set of all invertible supersmooth transformations (1.11.1) forms a group. It will be called the 'general coordinate transformation supergroup' (the term 'supergroup' will be described properly in the next chapter).

In the present section, our treatment is based mainly on special technique developed by S. J. Gates and W. Siegel in their study of superfield supergravity[1].

1.11.1. The exponential form for general coordinate transformations

An infinitesimal coordinate transformation on $\mathbb{R}^{p|q}$ can be written as

$$z'^M = z^M - \tau K^M(z). \tag{1.11.2}$$

Here τ is an infinitesimal real parameter, $K^M(z) = (K^m(x, \theta), K^\mu(x, \theta))$, where $K^m(z)$ are real bosonic superfunctions and $K^\mu(z)$ are real fermionic superfunctions. We introduce the first-order differential operator

$$\begin{gathered} K = K^M(z)\partial_M = K^m(x, \theta)\partial_m + K^\mu(x, \theta)\partial_\mu \\ \varepsilon(K^M) = \varepsilon_M \qquad (K^M)^* = K^M \end{gathered} \tag{1.11.3}$$

which will be called a 'real supervector field'. Then equation (1.11.2) takes the form

$$z'^M = (1 - \tau K) z^M.$$

Exponentiating this infinitesimal transformation, one obtains a finite coordinate transformation,

$$z'^M = \mathrm{e}^{-K} z^M \tag{1.11.4}$$

or, more explicitly,

$$z'^M = z^M - K^M(z) + \sum_{r=2}^{\infty} \frac{(-1)^r}{r!} K^{N_1}(z)\partial_{N_1} \ldots K^{N_{r-1}}(z)\partial_{N_{r-1}} K^M(z).$$

Evidently, the transformation (1.11.4) is one-to-one. A sufficient condition for it to map $\mathbb{R}^{p|q}$ onto $\mathbb{R}^{p|q}$ is that the components of K^M vanish when the bodies of the c-number coordinates x^m go to infinity,

$$K^M \xrightarrow{|x_B| \to \infty} 0.$$

[1] S. J. Gates and W. Siegel, *Nucl. Phys.* **B 163** 519, 1980.

Therefore, existence of the exponential form (1.11.4) for a transformation (1.11.1) guarantees that the transformation is invertible.

Let us give two simple examples of coordinate transformations which admit the exponential form. First, consider a translation on $\mathbb{R}^{p|q}$:

$$z^M \to z'^M = z^M + b^M \qquad \partial_N b^M = 0.$$

This is expressed in the form (1.11.4) with $K = -b^M \partial_M$. Secondly, consider a linear transformation of the type (1.10.47). This admits the exponential form if $F = \mathrm{e}^R$ for some supermatrix R. Under this condition, one can prove that

$$\begin{aligned} z'^M &= F^M{}_N z^N = \mathrm{e}^K z^M \\ K &= R^M{}_N z^N \partial_M. \end{aligned} \tag{1.11.5}$$

Not all general coordinate transformations (1.11.1) possess the exponential form (1.11.4), only contractible ones. A coordinate transformation $z'^M = f^M(z)$ is said to be 'contractible' if it may be deformed continuously to the identity transformation, i.e. there exists a sequence of transformations

$$z'^M = f^M(z, t) \qquad t \in [0, 1]$$

such that $f^M(z, t)$ are continuous functions of a real parameter t, and $f^M(z, 1) = f^M(z)$, $f^M(z, 0) = z^M$. The set of all contractible transformations on $\mathbb{R}^{p|q}$ forms a subgroup of the general coordinate transformation supergroup. From there we shall consider only the contractible coordinate transformations.

The representation (1.11.4) is very useful in practice. Indeed, let $\Phi(z)$ be a supersmooth function on $\mathbb{R}^{p|q}$. In the variables z'^M, it takes the form

$$\Phi'(z') = \Phi(z). \tag{1.11.6}$$

A natural question to ask is: how does the function $\Phi'(z)$ differ from the former one $\Phi(z)$? To answer this we use the observation that, by virtue of equation (1.10.20),

$$[K, \Psi(z)] = (K\Psi(z))$$

where $\Psi(z)$ is an arbitrary superfunction, so

$$\begin{aligned} \mathrm{e}^{-K}\Psi(z)\,\mathrm{e}^{K} &= (\mathrm{e}^{-K}\Psi(z)) \quad \Rightarrow \\ \mathrm{e}^{-K} z^M\,\mathrm{e}^{K} &= (\mathrm{e}^{-K} z^M). \end{aligned} \tag{1.11.7}$$

Then, the left-hand side of (1.11.6) can be written as

$$\Phi'(\mathrm{e}^{-K} z\,\mathrm{e}^{K}) = \mathrm{e}^{-K}\Phi'(z).$$

As a result, we have

$$\Phi'(z) = \mathrm{e}^{K}\Phi(z). \tag{1.11.8}$$

To see further properties of the representation (1.11.4), we now discuss in detail supervector fields.

1.11.2. The operators K and $\overleftarrow{K}$

Let $K_1 = K_1^M(z)\partial_m$ and $K_2 = K_2^M(z)\partial_m$ be real supervector fields. In accordance with equations (1.10.19, 20, 22), their commutator

$$[K_1, K_2] = K_3 = K_3^M(z)\partial_M$$
$$K_3^M = K_1^N \partial_N K_2^M - K_2^N \partial_N K_1^M \tag{1.11.9}$$

is a real supervector field. Therefore, the set of all real supervector fields forms a Lie algebra. According to the Baker–Hausdorff formula, this implies that the transformation

$$z'^M = e^{K_1} e^{K_2} z^M$$

may be represented in the form (1.11.4). Another consequence is that, if K and V are real supervector fields, the operator

$$e^K V e^{-K} = \sum_{n=0}^{\infty} \frac{1}{n!} (L_K)^n V \qquad L_K V \equiv [K, V] \tag{1.11.10}$$

is a real supervector field.

For every supervector field K, we define the first-order differential operator $\overleftarrow{K}$ by the rule:

$$\overleftarrow{K} \equiv K^M \overleftarrow{\partial}_M = (-1)^{\varepsilon_M} \overleftarrow{\partial}_M K^M + (-1)^{\varepsilon_M} (\partial_M K^M) \tag{1.11.11}$$

where $\overleftarrow{\partial}_M$ is defined to be related to the right partial derivative $\overleftarrow{\partial}/\partial z^M$ in z^M by

$$\overleftarrow{\partial}_M \equiv (-1)^{\varepsilon_M} \frac{\overleftarrow{\partial}}{\partial z^M}. \tag{1.11.12a}$$

The $\overleftarrow{K}$ acts on a superfunction $\Phi(z)$ as follows

$$\Phi \overleftarrow{K} = K\Phi + (-1)^{\varepsilon_M} (\partial_M K^M)\Phi = (-1)^{\varepsilon_M} \partial_M (K^M \Phi). \tag{1.11.12b}$$

Lemma 1.

$$(1 \cdot e^{\overleftarrow{K}})(e^K \Phi) = (\Phi \cdot e^{\overleftarrow{K}}) \tag{1.11.13}$$

where $\Phi(z)$ is an arbitrary superfunction.

Proof. Define the superfunction $\Phi(z, \tau)$

$$\Phi(z, \tau) = \Phi(z) e^{\tau \overleftarrow{K}} \qquad \Phi(z, 0) = \Phi(z)$$

depending on a real parameter τ. $\Phi(\tau)$ satisfies the equation

$$d\Phi(\tau)/d\tau = \Phi(\tau)\overleftarrow{K}.$$

Define another superfunction $\Psi(z, \tau)$ by the rule

$$\Psi(z, \tau) = (1 \cdot e^{\tau \overleftarrow{K}})(e^{\tau K} \Phi(z)) \qquad \Psi(z, 0) = \Phi(z).$$

We have

$$\frac{\mathrm{d}\Psi(\tau)}{\mathrm{d}\tau}=((1\cdot \mathrm{e}^{\tau\overleftarrow{K}})\overleftarrow{K})(\mathrm{e}^{\tau K}\Phi)+(1\cdot \mathrm{e}^{\tau\overleftarrow{K}})K(\mathrm{e}^{\tau K}\Phi)$$

$$=[K(1\cdot \mathrm{e}^{\tau\overleftarrow{K}})+(-1)^{\varepsilon_M}(\partial_M K^M)(1\cdot \mathrm{e}^{\tau\overleftarrow{K}})](\mathrm{e}^{\tau K}\Phi)+(1\cdot \mathrm{e}^{\tau\overleftarrow{K}})K(\mathrm{e}^{\tau K}\Phi)$$

$$=K\Psi(\tau)+(-1)^{\varepsilon_M}(\partial_M K^M)\Psi(\tau)=\Psi(\tau)\overleftarrow{K}.$$

Therefore, $\Phi(\tau)$ and $\Psi(\tau)$ satisfy the same equation and the same boundary condition, hence they coincide.

Lemma 2.

$$(1\cdot \mathrm{e}^{\overleftarrow{K}})^{-1}=\mathrm{e}^{K}(1\cdot \mathrm{e}^{-\overleftarrow{K}}). \tag{1.11.14}$$

Proof. Choose $\Phi=(1\cdot \mathrm{e}^{-\overleftarrow{K}})$ in equation (1.11.13).

Lemma 3.

$$[\overleftarrow{K},\overleftarrow{V}]=-\overleftarrow{[K,V]}\equiv-\overleftarrow{L_K V} \tag{1.11.15}$$

where K and V are arbitrary supervector fields.

Proof.

$$[\overleftarrow{K},\overleftarrow{V}]=K^N\overleftarrow{\partial}_N V^M\overleftarrow{\partial}_M-V^N\overleftarrow{\partial}_N K^M\overleftarrow{\partial}_M$$

$$=(-1)^{\varepsilon_M\varepsilon_N}K^N V^M\overleftarrow{\partial}_N\overleftarrow{\partial}_M-(K^N\partial_N V^M)\overleftarrow{\partial}_M$$

$$-(-1)^{\varepsilon_M\varepsilon_N}V^N K^M\overleftarrow{\partial}_N\overleftarrow{\partial}_M+(V^N\partial_N K^M)\overleftarrow{\partial}_M$$

$$=V^N K^M\{\overleftarrow{\partial}_M\overleftarrow{\partial}_N-(-1)^{\varepsilon_M\varepsilon_N}\overleftarrow{\partial}_N\overleftarrow{\partial}_M\}-\{(KV^M)-(VK^M)\}\overleftarrow{\partial}_M=-(L_K V)^M\overleftarrow{\partial}_M.$$

Here we have used the fact that the derivatives $\overleftarrow{\partial}_M$ satisfy the same identity (1.10.19*b*) which was fulfiled by the derivatives ∂_M.

The equation (1.11.15) shows that the set of all operators $\overleftarrow{K}$, where K is a real supervector field, forms a Lie algebra, and the correspondence $K\rightarrow\overleftarrow{K}$ is an anti-isomorphism of the Lie algebras.

Lemma 4.

$$\overleftarrow{\mathrm{e}^{K}V\mathrm{e}^{-K}}=\mathrm{e}^{-\overleftarrow{K}}\,\overleftarrow{V}\,\mathrm{e}^{\overleftarrow{K}}. \tag{1.11.16}$$

Proof. The statement follows from equations (1.11.10) and (1.11.15).

Now we are able to prove an important result.

1.11.3. Theorem

$$z'^M=\mathrm{e}^{-K}z^M\quad\Rightarrow\quad \mathrm{Ber}(\partial_M z'^N)=(1\cdot \mathrm{e}^{-\overleftarrow{K}}). \tag{1.11.17}$$

Proof. Introduce the partial derivatives ∂'_M corresponding to the variables z'^M:

$$\partial'_M = \frac{\partial z^N}{\partial z'^M}\partial_N \equiv G_M{}^N(z)\partial_N \equiv G_M. \tag{1.11.18}$$

In accordance with equations (1.10.19*a*,*b*), the operators ∂'_M should satisfy the equations

$$\partial'_M z'^N = \delta_M{}^N$$
$$\partial'_M\partial'_N - (-1)^{\varepsilon_M\varepsilon_N}\partial'_N\partial'_M = 0.$$

The solution is unique and has the form

$$\partial'_M = \mathrm{e}^{-K}\,\partial_M\,\mathrm{e}^K. \tag{1.11.19}$$

On the other hand, in terms of the supervector fields G_M (1.11.18), we have

$$\begin{aligned}
0 &= \partial'_M\partial'_N - (-1)^{\varepsilon_M\varepsilon_N}\partial'_N\partial'_M = G_M{}^R\partial_R G_N{}^Q\partial_Q - (-1)^{\varepsilon_M\varepsilon_N}G_N{}^R\partial_R G_M{}^Q\partial_Q \\
&= (G_M G_N{}^Q)\partial_Q + (-1)^{\varepsilon_R(\varepsilon_N+\varepsilon_Q)}G_M{}^R G_N{}^Q\,\partial_R\partial_Q - (-1)^{\varepsilon_M\varepsilon_N}(G_N G_M{}^Q)\partial_Q \\
&\quad - (-1)^{\varepsilon_M\varepsilon_N}(-1)^{\varepsilon_R(\varepsilon_M+\varepsilon_Q)}G_N{}^R G_M{}^Q\,\partial_R\partial_Q \\
&= \{(G_M G_N{}^Q) - (-1)^{\varepsilon_M\varepsilon_N}(G_N G_M{}^Q)\}\,\partial_Q \\
&\quad + (-1)^{\varepsilon_M(\varepsilon_N+\varepsilon_R)}G_N{}^R G_M{}^Q\{\partial_Q\partial_R - (-1)^{\varepsilon_R\varepsilon_Q}\partial_R\partial_Q\} \\
&= \{(G_M G_N{}^Q) - (-1)^{\varepsilon_M\varepsilon_N}(G_N G_M{}^Q)\}\,\partial_Q \\
&= \{(G_M G_N{}^Q) - (-1)^{\varepsilon_M\varepsilon_N}(G_N G_M{}^Q)\}(G^{-1})_Q{}^R G_R.
\end{aligned}$$

Therefore,

$$\{(G_M G_N{}^Q) - (-1)^{\varepsilon_M\varepsilon_N}(G_N G_M{}^Q)\}(G^{-1})_Q{}^R = 0.$$

In this relation, we take the supertrace over the indices N and R (contract both sides with $(-1)^{\varepsilon_N}\delta_R{}^N$). This gives

$$\mathrm{str}\{(G_M G)G^{-1}\} = (-1)^{\varepsilon_N(1+\varepsilon_M)}(\partial_N G_M{}^N).$$

Evidently, the left-hand side coincides with $G_M\{\ln(\mathrm{Ber}\,G)\}$. We recognize the right-hand side as $(1\cdot\overleftarrow{G}_M)$. So, we obtain

$$G_M\{\ln(\mathrm{Ber}\,G)\} = (1\cdot\overleftarrow{G}_M). \tag{1.11.20}$$

But $(1\cdot\overleftarrow{G}_M)$ may be transformed further making use of equations (1.11.19) and (1.11.16):

$$(1\cdot\overleftarrow{G}_M) = (1\cdot\overleftarrow{\mathrm{e}^{-K}\partial_M\mathrm{e}^K}) = (1\cdot\mathrm{e}^{\overleftarrow{K}}\,\overleftarrow{\partial}_M\,\mathrm{e}^{-\overleftarrow{K}}) = (\partial_M(1\cdot\mathrm{e}^{\overleftarrow{K}}))\,\mathrm{e}^{-\overleftarrow{K}}.$$

Now apply equation (1.11.13)

$$(1\cdot\overleftarrow{G}_M) = (1\cdot\mathrm{e}^{-\overleftarrow{K}})\mathrm{e}^{-K}\,\partial_M(1\cdot\mathrm{e}^{-\overleftarrow{K}}) = (1\cdot\mathrm{e}^{-\overleftarrow{K}})G_M(\mathrm{e}^{-K}(1\cdot\mathrm{e}^{\overleftarrow{K}})).$$

All that remains is to apply equation (1.11.14). This gives

$$(1\cdot\overleftarrow{G}_M) = -G_M \ln(1\cdot \mathrm{e}^{-\overleftarrow{K}}).$$

Comparing with (1.11.20) gives

$$\mathrm{Ber}\, G = C(1\cdot \mathrm{e}^{-\overleftarrow{K}})^{-1}$$

where C is a universal constant which can be calculated by considering some special transformation. In the case of the identity transformation, Ber $G=1$ and $K=0$, so $C=1$. Finally, since $\mathrm{Ber}\, G = \mathrm{Ber}^{-1}(\partial_M z'^N)$, one obtains (1.11.17).

Remark. Making use of equations (1.11.15) and (1.11.17), one can prove the following statement:

$$z'^M = \mathrm{e}^{-K}\,\mathrm{e}^{-L} z^M \qquad \Rightarrow \mathrm{Ber}(\partial_M z'^N) = (1\cdot \overleftarrow{\mathrm{e}^{-K}\,\mathrm{e}^{-L}}) = (1\cdot \mathrm{e}^{-\overleftarrow{L}}\,\mathrm{e}^{-\overleftarrow{K}}). \qquad (1.11.21)$$

1.11.4. The transformation law for the volume element on $\mathbb{R}^{p|q}$

Let us treat the general coordinate transformation as a replacement of variables in an integral over $\mathbb{R}^{p|q}$

$$\int \mathrm{d}^{p+q} z'\, \Phi(z')$$

of some superfunction under reasonable boundary conditions. The replacement (1.11.4) changes the integration measure leading to the appearance of the Jacobian $J(z)$

$$\int \mathrm{d}^{p+q} z'\, \Phi(z') = \int \mathrm{d}^{p+q} z\, J(z)\Phi(z'(z)). \qquad (1.11.22)$$

The technique developed above makes it possible to determine the Jacobian of the transformation.

Theorem.

$$z'^M = \mathrm{e}^{-K} z^M$$

$$\int \mathrm{d}^{p+q} z'\, \Phi(z') = \int \mathrm{d}^{p+q} z\, \mathrm{Ber}(\partial_M z'^N)\Phi(z'(z)). \qquad (1.11.23)$$

Proof. Recall that $\Phi(\mathrm{e}^{-K} z) = \mathrm{e}^{-K}\Phi(z)$. So, the right-hand side of equation (1.11.22) is

$$\int \mathrm{d}^{p+q} z\, J(z)\, \mathrm{e}^{-K}\Phi(z) = \int \mathrm{d}^{p+q} z\, \mathrm{e}^{-K}[(\mathrm{e}^{K} J(z))\Phi(z)]. \qquad (1.11.24)$$

We insert into the integral the decomposition of unity

$$1 = (1\cdot \mathrm{e}^{-\overleftarrow{K}})(1\cdot \mathrm{e}^{-\overleftarrow{K}})^{-1} = (1\cdot \mathrm{e}^{-\overleftarrow{K}})\,\mathrm{e}^{-K}\,\mathrm{e}^{K}\,(1\cdot \mathrm{e}^{-\overleftarrow{K}})^{-1}.$$

Then, (1.11.24) takes the form

$$\int \mathrm{d}^{p+q}z(1\cdot \mathrm{e}^{-\overleftarrow{K}})\,\mathrm{e}^{-K}[\Phi(z)\,\mathrm{e}^{K}\{J(z)(1\cdot \mathrm{e}^{-\overleftarrow{K}})^{-1}\}]$$

$$= \int \mathrm{d}^{p+q}z[\Phi(z)\,\mathrm{e}^{K}\{J(z)(1\cdot \mathrm{e}^{-\overleftarrow{K}})^{-1}\}]\,\mathrm{e}^{-\overleftarrow{K}} \quad (1.11.25)$$

where we have used equation (1.11.13). By virtue of (1.11.12), $\mathrm{e}^{-\overleftarrow{K}} = 1 + \ldots$, where dots mean terms giving rise to total derivatives in the integral (1.11.25). Therefore, equation (1.11.25) can be equivalently rewriten as

$$\int \mathrm{d}^{p+q}z\,\Phi(z)\,\mathrm{e}^{K}\{J(z)(1\cdot \mathrm{e}^{-\overleftarrow{K}})^{-1}\}.$$

This integral should coincide with the left-hand side of expression (1.11.22). Since $\Phi(z)$ is a completely arbitrary superfunction, one demands that

$$\mathrm{e}^{K}\{J(z)(1\cdot \mathrm{e}^{-\overleftarrow{K}})^{-1}\} = 1 \qquad \Rightarrow J(z) = (1\cdot \mathrm{e}^{-\overleftarrow{K}}).$$

The equation (1.11.17) then leads to equation (1.11.23).

Remark. The Jacobian in (1.11.23) presents the Berezinian of the supermatrix

$$(F^{\mathrm{sT}})_M{}^N = \partial_M z'^N \equiv \frac{\vec{\partial}}{\partial z^M} z'^N$$

which is the supertranspose of

$$F^M{}_N = (-1)^{\varepsilon_N + \varepsilon_M\varepsilon_N}(F^{\mathrm{sT}})_N{}^M = z'^M \frac{\overleftarrow{\partial}}{\partial z^N}.$$

Here we have used the identity

$$f(z)\frac{\overleftarrow{\partial}}{\partial z^M} = (-1)^{\varepsilon_M + \varepsilon_M\varepsilon(f)}\frac{\vec{\partial}}{\partial z^M} f(z)$$

which generalizes equation (1.10.14*b*). Because of the relation $\mathrm{Ber}\,F = \mathrm{Ber}\,(F^{\mathrm{sT}})$ and equation (1.11.17), we can write the transformation Jacobian in three equivalent forms:

$$\mathrm{Ber}\left(\frac{\vec{\partial}}{\partial z^M} z'^N\right) = \mathrm{Ber}\left(z'^M \frac{\overleftarrow{\partial}}{\partial z^N}\right) = (1\cdot \mathrm{e}^{-\overleftarrow{K}}). \qquad (1.11.26)$$

1.11.5 Basic properties of integration theory over $\mathbb{R}^{p|2q}$

We would like to finish this section by listing the properties of integration theory over $\mathbb{R}^{p|2q}$, which are assumed to be valid (properly generalized) for functional integration in quantum field theory. A chief reason why we choose the number of a-type coordinates to be even here is the existence of non-singular supersymmetric bi-linear forms. Explicitly, consider a real supersymmetric bi-linear form on $\mathbb{R}^{p|q}$ (see subsection 1.9.6)

$$s^2 = z^M \eta_{MN} z^N \tag{1.11.27}$$

where

$$\begin{gathered} \varepsilon(\eta_{MN}) = \varepsilon_M + \varepsilon_N \\ \eta_{MN} = (-1)^{\varepsilon_M + \varepsilon_N + \varepsilon_M \varepsilon_N} \eta_{NM} \qquad (\eta_{MN})^* = \eta_{NM}. \end{gathered} \tag{1.11.28}$$

The requirement for non-singularity

$$\eta_{MN} z^N = 0 \Rightarrow z^N = 0 \tag{1.11.29}$$

implies that the bodies of the even–even and odd–odd blocks of η are invertible matrices. Since the odd–odd block is antisymmetric, owing to equation (1.11.28), its body can be invertible only if q is even. Note, every bi-linear form (1.11.27) under equations (1.11.28, 29) can be taken in the role of a supermetric on the superspace.

Integration over $\mathbb{R}^{p|2q}$ is characterized by the following properties:

1. *Linearity*

$$\int \mathrm{d}^{p+2q}z \{\alpha f_1(z) + \beta f_2(z)\} = \alpha \int \mathrm{d}^{p+2q}z f_1(z) + \beta \int \mathrm{d}^{p+2q}z f_2(z) \tag{1.11.30}$$

where $f_1(z)$ and $f_2(z)$ are arbitrary functions on $\mathbb{R}^{p|2q}$, α and β are arbitrary supernumbers. The measure $d^{p+2q}z$ is an even object commuting with all supernumbers.

2. *Integration by parts* (three equivalent forms)

$$\int \mathrm{d}^{p+2q}z \partial_M f(z) = 0 \tag{1.11.31a}$$

$$\int \mathrm{d}^{p+2q}z f(z+\xi) = \int \mathrm{d}^{p+2q}z f(z) \qquad \partial_M \xi^N = 0 \tag{1.11.31b}$$

$$\int \mathrm{d}^{p+2q}z (\partial_M f_1(z)) f_2(z) = -(-1)^{\varepsilon_M \varepsilon(f_1)} \int \mathrm{d}^{p+2q}z f_1(z) (\partial_M f_2(z)). \tag{1.11.31c}$$

3. *Replacement of variables*

Making a replacement of variables in integrals over $\mathbb{R}^{p|2q}$ is accompanied by the appearance of the transformation Jacobian:

$$\begin{gathered} \int \mathrm{d}^{p+2q}z' f(z') \overset{z'=z'(z)}{=\!=\!=} \int \mathrm{d}^{p+2q}z J(z) f(z'(z)) \\ J(z) = \mathrm{Ber}\left(\frac{\vec{\partial}}{\partial z^M} z'^N\right) = \mathrm{Ber}\left(z'^M \frac{\overleftarrow{\partial}}{\partial z^N}\right). \end{gathered} \tag{1.11.32}$$

4. *Existence of delta function*
There exists the unique delta function $\delta^{p+2q}(z)$ on $\mathbb{R}^{p|2q}$ possessing the property

$$\int \mathrm{d}^{p+2q}z\,\delta^{p+2q}(z)f(z) = f(0) \tag{1.11.33}$$

f being an arbitrary function on $\mathbb{R}^{p|2q}$. The $\delta^{p+2q}(z)$ can be expressed in the form

$$\delta^{p+2q}(z) = N\int \mathrm{d}^{p+2q}\omega\,\mathrm{e}^{\mathrm{i}\omega_M z^M} \tag{1.11.34}$$

where N is a normalization constant and the integral is over $\mathbb{R}^{p|2q}$ parametrized by real bosonic and imaginary fermionic coordinates,

$$(\omega_M)^* = (-1)^{\varepsilon_M}\omega_M.$$

Remark. A formal proof of equation (1.11.34) is as follows. Consider a superanalytic function $f(z)$ on $\mathbb{R}^{p|2q}$,

$$f(z) = \sum_{k=0}^{\infty} z^{M_1}z^{M_2}\dots z^{M_k} f_{M_1M_2\dots M_k} \qquad f_{M_1M_2\dots M_k}\in\Lambda_\infty,$$

and integrate it over $\mathbb{R}^{p|2q}$ with the weight (1.11.34):

$$\int \mathrm{d}^{p+2q}z\,\delta^{p+2q}(z)f(z)$$

$$= N\int \mathrm{d}^{p+2q}z\;\mathrm{d}^{p+2q}\omega \sum_{k=0}^{\infty}(-\mathrm{i})^k\frac{\vec{\partial}}{\partial\omega_{M_1}}\dots\frac{\vec{\partial}}{\partial\omega_{M_k}}\mathrm{e}^{\mathrm{i}\omega_M z^M} f_{M_1M_2\dots M_k}$$

$$= N\int \mathrm{d}^{p+2q}z\;\mathrm{d}^{p+2q}\omega\,\mathrm{e}^{\mathrm{i}\omega_M z^M} f(0)$$

hence

$$N^{-1} = \int \mathrm{d}^{p+2q}z\;\mathrm{d}^{p+2q}\omega\,\mathrm{e}^{\mathrm{i}\omega_M z^M}.$$

In the above, validity of integration by parts has been assumed.

5. *Gaussian integrals*
Choose a fixed supermatrix η_{NM} under the requirements (1.11.28) and (1.11.29). For every supermatrix H_{MN} under the same requirements, we have

$$N'\int \mathrm{d}^{p+2q}z\,\exp\left[\frac{\mathrm{i}}{2}z^M H_{MN}z^N\right] = \mathrm{Ber}^{-1}((\eta^{-1})^{MP}H_{PN}) \tag{1.11.35}$$

where N' is the universal normalization constant defined by

$$N' \int d^{p+2q}z \exp\left[\frac{i}{2} z^M \eta_{MN} z^N\right] = 1. \tag{1.11.36}$$

Remark. Relation (1.11.35) generalizes the well-known integral formula

$$\int_{-\infty}^{\infty} dt e^{\frac{i}{2}\lambda t^2} = \left(\frac{2\pi i}{\lambda}\right)^{1/2} \tag{1.11.37}$$

λ being a non-zero real number, and its a-number analogue

$$\int d\theta^2 d\theta^1 \, e^{\frac{i}{2}\lambda(\theta^1\theta^2 - \theta^2\theta^1)} = \lambda = \det{}^{1/2}\begin{pmatrix} 0 & \lambda \\ -\lambda & 0 \end{pmatrix} \tag{1.11.38}$$

where θ^1 and θ^2 are real a-number coordinates on $\mathbb{R}^2_a$.

To prove (1.11.35), let us represent H_{MN} in the manner

$$H_{MN} = (F^{sT})_M{}^K \eta_{KL} F^L{}_N \tag{1.11.39}$$

for some supermatrix $F \in GL^{(+)}_{\mathbb{R}}(p, 2q|\Lambda_\infty)$. Changing the integration variables in (1.11.35) according to

$$z^M \rightarrow z'^M = F^M{}_N z^N$$

and using equation (1.11.36), one obtains

$$N' \int d^{p+2q}z \exp\left[\frac{i}{2} z^M H_{MN} z^N\right] = \text{Ber}^{-1}(F). \tag{1.11.40}$$

On the other hand, equation (1.11.39) can be rewritten in the form

$$\tilde{H}^M{}_N \equiv (\eta^{-1})^{MP} H_{PN} = (\eta^{-1})^{MP}(F^{sT})_P{}^K \eta_{KL} F^L{}_N$$

or, in indexless notation,

$$\tilde{H} = \eta^{-1} F^{sT} \eta F$$

hence

$$\text{Ber}\, \tilde{H} = \text{Ber}^2(F).$$

This relation and equation (1.11.40) leads to equation (1.11.35).

2 Supersymmetry and Superspace

Within this restless, hurried, modern world
We took our hearts' full pleasure – You and I,
And now the white sails of our ship are furled,
And spent the lading of our argosy.

Oscar Wilde:
My Voice

2.0. Introduction: from $\mathbb{R}^{p|q}$ to supersymmetry

We have spent a lot of time studying supernumber spaces $\mathbb{R}^{p|q}$. It was beautiful mathematics; and now let us reveal its relation to physics.

A set of dynamical fields $\{\varphi^i\}$, arising in some classical field theory, should be treated as mappings from Minkowski space into a superspace

$$\varphi^i(x): \mathbb{R}^4 \rightarrow \mathbb{R}^{p|q} \tag{2.0.1}$$

where p is the number of bosonic fields and q is the number of fermionic fields. The situation is quite clear in the fermionic case. In quantum theory half-integer spin particles are described by Hermitian operators $\hat{\varphi}^i(x)$ on an appropriate Fock space which obey some anticommutation relations

$$\{\hat{\varphi}^i(x), \hat{\varphi}^j(y)\} = O(\hbar)$$

with $\hbar$ being the Planck constant. Taking the point of view that a classical theory corresponds to the limit $\hbar \rightarrow 0$, classical fermionic fields are anti-commuting. Anticipating fermionic fields to be valued in the odd part of a finite-dimensional Grassmann algebra Λ_N, one finds that all combinations of the form

$$(\bar{\Psi}_1 A_1 \Psi_2)(\bar{\Psi}_3 A_2 \Psi_4) \ldots (\bar{\Psi}_{2k-1} A_k \Psi_{2k})$$

DOI: 10.1201/9780367802530-2

vanish for sufficiently large k; here $A_1, A_2, \ldots, A_k$ are arbitrary operators constructed from the γ-matrices and space–time partial derivatives. Therefore, we have to look on fermionic fields as a-number-valued functions on Minkowski space. For example, the undotted spinor field $\psi_\alpha(x)$ is a mapping

$$\psi_\alpha(x): \mathbb{R}^4 \to \mathbb{C}_a^2$$

The Majorana spinor field Ψ_{M} is a mapping

$$\Psi_{\mathrm{M}}(x): \mathbb{R}^4 \to \mathbb{R}_a^4;$$

The Dirac spinor field $\Psi(x)$ is a mapping

$$\Psi(x): \mathbb{R}^4 \to \mathbb{C}_a^4 \cong \mathbb{R}_a^8.$$

Classical bosonic fields are usually considered as real number-valued functions on Minkowski space. This description is correct only in the case of a purely bosonic theory. However, having a system of interacting bosonic and fermionic fields, we must inevitably think of bosonic fields as c-number-valued functions on space–time. For example, in electrodynamics

$$S = -\int \mathrm{d}^4x \left\{ \frac{1}{4} F^{mn} F_{mn} + \bar{\Psi}\gamma^m(\mathrm{i}\partial_m + eA_m)\Psi \right\}$$

the vector field equation of motion

$$\partial^n F_{mn} = -e\bar{\Psi}\gamma_m\Psi$$

tells us that the body of $A_m(x)$ has free Maxwell dynamics and the soul of $A_m(x)$ is propagated by the charged bodiless current $j_m(x) = e\bar{\Psi}\gamma_m\Psi$. As a result, ordinary electromagnetic radiation is absent in this classical picture; radiation is a purely quantum effect (of course, one may generate non-trivial dynamics by switching on some external source). This may look rather unusual from an ordinary point of view. Nevertheless, the treatment of classical fields as mappings of the form (2.0.1) is most convenient when understanding the classical theory as a starting point for quantum field theory. According to B. De Witt, it would be more precise to speak about a 'superclassical' starting point.

Now let us return once more to equation (2.0.1). Note that every set $\{\varphi^i(x)\}$ of smooth fields on Minkowski space can be continued uniquely to a set $\{\tilde{\varphi}^i(x)\}$ of supersmooth fields on the c-number space $\mathbb{R}_c^4$. The resultant mapping

$$\{\tilde{\varphi}^i: \mathbb{R}_c^4 \to \mathbb{R}^{p|q} = \mathbb{R}_c^p \times \mathbb{R}_a^q\} \tag{2.0.2}$$

looks more symmetrical than equation (2.0.1). Of course, Minkowski space should be identified with the soulless surface in $\mathbb{R}_c^4$,

$$\mathbb{R}^4 = \{(x^m) \in \mathbb{R}_c^4,\ x_{\mathrm{S}}^m = 0,\ m = 0, 1, 2, 3\}.$$

After mapping (2.0.2), it remains to perform one more step — to consider supersmooth mappings of the form

$$\Phi^i : \mathbb{R}^{4|k} \to \mathbb{R}^{p|q}. \tag{2.0.3}$$

Can such generalized fields, which it is natural to call 'superfields', be interesting from the physical point of view? To answer this question, let us consider one particular case $k = 4$.

We parametrize $\mathbb{R}^{4|4}$ by four real c-number variables x^m, two complex a-number variables θ^α and their complex conjugate $(\theta^\alpha)^* \equiv \bar{\theta}^{\dot{\alpha}}$. Consider a superfield

$$V: \mathbb{R}^{4|4} \to \mathbb{R}_c. \tag{2.0.4}$$

This can be written explicitly as

$$\begin{aligned} V(x,\theta,\bar{\theta}) = A(x) + \theta^\alpha\psi_\alpha(x) + \bar{\theta}_{\dot{\alpha}}\bar{\psi}^{\dot{\alpha}}(x) + \theta^2 F(x) + \bar{\theta}^2\bar{F}(x) + \theta\sigma^a\bar{\theta}V_a(x) \\ + \bar{\theta}^2\theta^\alpha\lambda_\alpha(x) + \theta^2\bar{\theta}_{\dot{\alpha}}\bar{\lambda}^{\dot{\alpha}}(x) + \theta^2\bar{\theta}^2 G(x) \end{aligned} \tag{2.0.5}$$

where A, F, $\bar{F}$, V_a and G are bosonic fields and ψ_α, $\bar{\psi}^{\dot{\alpha}}$, λ_α and $\bar{\lambda}^{\dot{\alpha}}$ are fermionic fields. Interpreting odd coordinates θ^α and $\bar{\theta}^{\dot{\alpha}}$ on $\mathbb{R}^{4|4}$ as two-component Weyl spinors makes it possible to define an action of the Poincaré group on $\mathbb{R}^{4|4}$ as follows:

$$\begin{aligned} x^m \to x'^m &= (\exp K)^m{}_n x^n + b^m \\ \theta_\alpha \to \theta'_\alpha &= \left[\exp\left(\frac{1}{2}K^{ab}\sigma_{ab}\right)\right]_\alpha{}^\beta \theta_\beta \\ \bar{\theta}^{\dot{\alpha}} \to \bar{\theta}'^{\dot{\alpha}} &= \left[\exp\left(\frac{1}{2}K^{ab}\tilde{\sigma}_{ab}\right)\right]^{\dot{\alpha}}{}_{\dot{\beta}} \bar{\theta}^{\dot{\beta}}. \end{aligned} \tag{2.0.6}$$

Finally, demanding the Poincaré transformation law

$$V'(x',\theta',\bar{\theta}') = V(x,\theta,\bar{\theta}) \tag{2.0.7}$$

one finds that A, F, $\bar{F}$ and G are scalar fields, V_a is a vector field, and

$$\Psi_{\mathrm{M}} = \begin{pmatrix} \psi_\alpha \\ \bar{\psi}^{\dot{\alpha}} \end{pmatrix} \qquad \Lambda_{\mathrm{M}} = \begin{pmatrix} \lambda_\alpha \\ \bar{\lambda}^{\dot{\alpha}} \end{pmatrix}$$

are Majorana spinor fields. Therefore, every superfield of the form (2.0.4) is equivalent to a whole family of ordinary bosonic and fermionic fields.

The above example can be further generalized. Namely, every superspace $\mathbb{R}^{4|4N}$, $N = 1, 2, \ldots$, admits a natural action on it of the Poincaré group. So in these cases as well, superfields (2.0.3) are equivalent to a whole set of ordinary fields. We are faced with a very interesting situation. We have found a mechanism to unify bosonic and fermionic fields in a multiplet. But our consideration seems rather formal. It would be seen as less formal if an underlying principle leading to such unification existed. Remarkably, the key to the problem lies in the above example. The equation (2.0.5) tells us that $V(x, \theta, \bar{\theta})$ contains an equal number of bosonic and fermionic fields. This clearly applies in the general case of superfields on $\mathbb{R}^{4|4N}$, $N = 1, 2, \ldots$. This observation becomes a simple physical consequence under supposition of the existence of some symmetry between bosons and fermions. Indeed, imagine some hypothetical field theory possesses an invariance 'rotating' bosonic fields into fermionic ones and vice versa. Then the number of bosonic and fermionic partners should, in general, be equal. At first sight, such a symmetry may look very strange and unusual, since it changes particle statistics (the reason it is called a 'supersymmetry'). On the other hand, if God created two types of particles, He could also create a mechanism to transform one type into another, couldn't He? Namely supersymmetry is able to serve as the desirable underlying principle.

One can easily invent supersymmetry transformations having made some reasonable assumptions. To this end recall that the Poincaré group acts on $\mathbb{R}^{4|4}$ by means of the linear transformations (2.0.6). They include the Lorentz rotations of superspace coordinates and the x^m-translations (since the set of translations forms a commutative group, it is not possible to supplement space–time shifts $\delta x^m = b^m$ by linear transformations of θ, $\bar{\theta}$ of the type: $\delta\theta_\alpha = \kappa b_{\alpha\dot{\alpha}}\bar{\theta}^{\dot{\alpha}}$, $\delta\bar{\theta}_{\dot{\alpha}} = \bar{\kappa} b_{\alpha\dot{\alpha}}\theta^\alpha$, where κ is a complex constant; the only possible choice is $\kappa = 0$). However, while the even and odd coordinates of $\mathbb{R}^{4|4}$ are on an equal footing, θ-translations are not present in equation (2.0.6). To overcome this discrepancy, it would be natural to consider in superspace not only the Poincaré transformations but also shifts in θ, $\bar{\theta}$ supplemented by some transformations of x^m. Assuming linearity, the most general form for such a displacement, consistent with the needed Lorentz transformation laws for x^m and θ^α, $\bar{\theta}^{\dot{\alpha}}$, reads

$$\delta\theta^\alpha = \epsilon^\alpha \qquad \delta\bar{\theta}^{\dot{\alpha}} = \bar{\epsilon}^{\dot{\alpha}}$$

$$\delta x^m = k\theta\sigma^m\bar{\epsilon} + \bar{k}\epsilon\sigma^m\bar{\theta}$$

where ϵ^α is an arbitrary undotted spinor, and k is a complex parameter. The case $k \neq 0$ is most interesting, since it provides us with non-trivial mixing between even and odd superspace coordinates. Without loss of generality, it is possible to fix k as $k = i$ obtaining the supersymmetry transformation

$$\begin{aligned} \delta\theta^\alpha &= \epsilon^\alpha \qquad \delta\bar{\theta}^{\dot{\alpha}} = \bar{\epsilon}^{\dot{\alpha}} \\ \delta x^m &= \mathrm{i}(\theta\sigma^m\bar{\epsilon} - \epsilon\sigma^m\bar{\theta}). \end{aligned} \tag{2.0.8}$$

Finally, if one adapts the transformation law (2.0.7) to supersymmetry, this gives

$$\begin{aligned} \delta V(x, \theta, \bar{\theta}) &= V'(x, \theta, \bar{\theta}) - V(x, \theta, \bar{\theta}) \\ &= \mathrm{i}(\epsilon\sigma^m\bar{\theta} - \theta\sigma^m\bar{\epsilon})\partial_m V(x, \theta, \bar{\theta}) - (\epsilon^\alpha\partial_\alpha + \bar{\epsilon}^{\dot{\alpha}}\bar{\partial}_{\dot{\alpha}})V(x, \theta, \bar{\theta}) \end{aligned}$$

or, in terms of the fields from which $V(x, \theta, \bar{\theta})$ is constructed,

$$\delta A(x) = -\epsilon\psi(x) - \overline{\epsilon\psi}(x)$$

$$\delta\psi_\alpha(x) = -2\epsilon_\alpha F(x) - (\sigma^a\bar{\epsilon})_\alpha\{V_a(x) + \partial_a A(x)\}$$

$$\delta F(x) = -\bar{\epsilon}\bar{\lambda}(x) + \frac{i}{2}\partial_a\psi(x)\sigma^a\bar{\epsilon}$$

$$\delta V_a(x) = \epsilon\sigma_a\bar{\lambda}(x) + \lambda(x)\sigma_a\bar{\epsilon} + i\epsilon\sigma^b\tilde{\sigma}_a\partial_b\psi(x) - i\partial_b\bar{\psi}(x)\tilde{\sigma}_a\sigma^b\bar{\epsilon}$$

$$\delta\lambda_\alpha(x) = -2\epsilon_\alpha G(x) + \frac{i}{2}(\sigma^a\tilde{\sigma}^b\epsilon)_\alpha\partial_b V_a(x) - i(\sigma^a\bar{\epsilon})_\alpha\partial_a\bar{F}(x)$$

$$\delta G(x) = \frac{i}{2}\partial_a\{\lambda(x)\sigma^a\bar{\epsilon} - \epsilon\sigma^a\bar{\lambda}(x)\}.$$

As a result, we have constructed a symmetry mixing bosonic and fermionic fields. Note that if some component field from $V(x, \theta, \bar{\theta})$ was zero at the beginning, it becomes non-vanishing after making a supersymmetry transformation.

The set of Poincaré transformations (2.0.6) and supersymmetry transformations (2.0.8) form a group. But it is a rather unusual group, since the supersymmetry transformations are built of anticommuting parameters ϵ^α and $\bar{\epsilon}^{\dot{\alpha}}$. Before studying supersymmetry it would be instructive to familiarize the reader with such algebraic objects.

2.1. Superalgebras, Grassmann shells and super Lie groups

Superalgebras present a natural generalization of the concept of Lie algebras.

Every Lie algebra $\mathcal{G}$ is a non-associative algebra in which the multiplication law $[\ldots, \ldots]$: $\mathcal{G} \times \mathcal{G} \to \mathcal{G}$ is characterized by two basic properties:

1. $[a, b] = -[b, a]$ (2.1.1)
2. $[a, [b, c]] + [b, [c, a]] + [c, [a, b]] = 0$ (2.1.2)

for all elements, a, b, $c \in \mathcal{G}$. The bi-linear operation $[\ldots, \ldots]$ is usually called a Lie bracket.

There is a universal way to construct Lie algebras. Namely, given an associative algebra $\mathcal{U}$, one can introduce a Lie algebra structure on $\mathcal{U}$ by taking a commutator of elements:

$$[x, y] \equiv x \cdot y - y \cdot x \qquad \forall x, y \in \mathcal{U}. \tag{2.1.3}$$

Here the Jacobi identities (2.1.2) are evidently satisfied. In particular, if $\mathcal{U}$ is some matrix algebra, then the prescription (2.1.3) provides us with a matrix

Lie algebra. In fact, every finite-dimensional Lie algebra proves to be isomorphic to some matrix Lie algebra. So, we may look at Lie algebra as a set of linear operators closed under commutation.

Another algebraic operation, often used in applications, is anticommutation of operators. For example, to formulate particle dynamics in quantum field theory, one has to choose commutation relations for integer spin particles and anticommutation relations for half-integer spin particles. Naturally, the question arises: is it possible to generalize the Lie algebra multiplication law to include commutators and anticommutators as well?

2.1.1. Superalgebras

To define superalgebras, we will need one more auxiliary notion. A vector space (real or complex) L is said to be '$\mathbb{Z}_2$-graded' if it is decomposed into a direct sum of two subspaces ${}^0L$ and ${}^1L$, $L = {}^0L \oplus {}^1L$. It is useful to call ${}^0L$-elements and ${}^1L$-elements 'even' and 'odd vectors', respectively. Even and odd vectors in L are also called 'pure'. By analogy with the Grassmann algebra Λ_∞, we introduce a 'parity function' $\kappa(u)$ defined on the set of pure vectors in L as follows:

$$\kappa(u) = \begin{cases} 0 & \text{if } u \in {}^0L \\ 1 & \text{if } u \in {}^1L. \end{cases} \tag{2.1.4}$$

Every vector $u \in L$ may be decomposed uniquely into the sum of its even and odd parts,

$$u = {}^0u + {}^1u.$$

Using the parity function makes it possible to define a one-to-one linear mapping $\mathscr{P}$: $L \to L$ by the rule:

$$\mathscr{P}(u) = (-1)^{\kappa(u)}u \qquad \mathscr{P}^2 = \mathbb{I}, \tag{2.1.5}$$

for every pure vector $u \in L$. We will refer to $\mathscr{P}$ as the 'L-parity mapping'.

A superalgebra is a $\mathbb{Z}_2$-graded linear space $\mathscr{G} = {}^0\mathscr{G} \oplus {}^1\mathscr{G}$ and an algebra with respect to a multiplication $[\ldots,\ldots\}$: $\mathscr{G} \times \mathscr{G} \to \mathscr{G}$ which is characterized by three basic properties:

1. $[a, b\} = -(-1)^{\kappa(a)\kappa(b)}[b, a\}$ (2.1.6)
2. $\kappa([a, b\}) = \kappa(a) + \kappa(b) \pmod 2$ (2.1.7)
3. Super-Jacobi identities

$$(-1)^{\kappa(a)\kappa(c)}[a, [b, c\}\} + (-1)^{\kappa(b)\kappa(a)}[b, [c, a\}\} + (-1)^{\kappa(c)\kappa(b)}[c, [a, b\}\} = 0. \tag{2.1.8}$$

Here a, b and c are arbitrary pure elements from $\mathscr{G}$. The bilinear operation $[\ldots,\ldots\}$ is called a 'Lie superbracket'.

Before considering properties of superalgebras, it is instructive to describe a universal rule by which they may be constructed. Let $\mathscr{U}$ be a $\mathbb{Z}_2$-graded associative algebra, i.e. (1) $\mathscr{U}$ is an associative algebra; (2) $\mathscr{U}$ is a $\mathbb{Z}_2$-graded vector space, $\mathscr{U} = {}^0\mathscr{U} \oplus {}^1\mathscr{U}$; (3) the $\mathscr{U}$-parity mapping $\mathscr{P}$ is an automorphism of the algebra, $\mathscr{P}(x \cdot y) = \mathscr{P}(x) \cdot \mathscr{P}(y)$. Then $\mathscr{U}$ becomes a superalgebra with respect to the multiplication law

$$[x, y\} \equiv x \cdot y - (-1)^{\kappa(x)\kappa(y)} y \cdot x, \tag{2.1.9}$$

where x and y are arbitrary pure elements. The super-Jacobi identities (2.1.8) turn out to be satisfied identically. In particular, having some $\mathbb{Z}_2$-graded matrix algebra, the prescription (2.1.9) will provide us with a matrix superalgebra. Examples are given in the next subsection. So, one can imagine a superalgebra as a set of linear operators closed under (anti)commutation.

From here on our consideration will be confined to the case of superalgebras obtained in accordance with the prescription (2.1.9). So, we have

$$\begin{aligned} &[a, b\} = -[b, a\} = [a, b] \qquad && a, b \in {}^0\mathscr{G} \\ &[a, b\} = -[b, a\} = [a, b] \qquad && a \in {}^0\mathscr{G}, b \in {}^1\mathscr{G} \\ &[a, b\} = [b, a\} = \{a, b\} \qquad && a, b \in {}^1\mathscr{G}. \end{aligned} \tag{2.1.10}$$

For arbitrary elements $a, b \in \mathscr{G}$, we have

$$[a, b\} = [{}^0a, {}^0b] + [{}^0a, {}^1b] + [{}^1a, {}^0b] + \{{}^1a, {}^1b\}.$$

Now let us analyse the axioms. Equation (2.1.7) tells us that

$$[{}^0\mathscr{G}, {}^0\mathscr{G}] \subset {}^0\mathscr{G}, \tag{2.1.11a}$$

$$[{}^0\mathscr{G}, {}^1\mathscr{G}] \subset {}^1\mathscr{G} \tag{2.1.11b}$$

$$\{{}^1\mathscr{G}, {}^1\mathscr{G}\} \subset {}^0\mathscr{G}. \tag{2.1.11c}$$

Therefore, ${}^0\mathscr{G}$ is a Lie algebra. Equation (2.1.11*b*) shows that with every element ${}^0a \in {}^0\mathscr{G}$ we can associate a linear operator $\mathscr{F}({}^0a)$: ${}^1\mathscr{G} \to {}^1\mathscr{G}$ on ${}^1\mathscr{G}$ defined by

$$\mathscr{F}({}^0a){}^1c \equiv [{}^0a, {}^1c] \qquad \forall {}^1c \in {}^1\mathscr{G}. \tag{2.1.12}$$

Choosing in relation (2.1.8) $a = {}^0a$, $b = {}^0b$ and $c = {}^1c$ leads to

$$\mathscr{F}({}^0a)\mathscr{F}({}^0b){}^1c - \mathscr{F}({}^0b)\mathscr{F}({}^0a){}^1c = \mathscr{F}([{}^0a, {}^0b]){}^1c. \tag{2.1.13}$$

As a result, the set of all operators $\mathscr{F}({}^0a)$ forms a representation of the Lie algebra ${}^0\mathscr{G}$ in the linear space ${}^1\mathscr{G}$. The other super-Jacobi identities (2.1.8) give some restrictions on the possible type of this representation. To reveal them, we anticipate the superalgebra $\mathscr{G}$ to be finite dimensional, and let $\{e_i\}$ be a basis in ${}^0\mathscr{G}$, $i = 1, 2, \ldots, \dim {}^0\mathscr{G}$, and $\{e_\alpha\}$ be a basis in ${}^1\mathscr{G}$, $\alpha = 1, 2, \ldots, \dim {}^1\mathscr{G}$. Then, the set $e_I = \{e_i, e_\alpha\}$ is a 'pure basis' in $\mathscr{G}$. Every pure basis in $\mathscr{G}$ is said to be a 'set of generators' of the superalgebra.

Introduce 'structure constants' of the superalgebra defined by

$$[e_I, e_J\} = f_{IJ}{}^L e_L \qquad f_{IJ}{}^L \in \mathbb{C} \text{ (or } \mathbb{R}). \tag{2.1.14}$$

They are (anti)symmetric in their lower indices,

$$f_{JI}{}^L = -(-1)^{\kappa_I \kappa_J} f_{IJ}{}^L \tag{2.1.15}$$

in agreement with equation (2.1.6), where we have denoted $\kappa_I \equiv \kappa(e_I)$. By virtue of equations (2.1.11), the only non-vanishing components are

$$f_{ij}{}^l = -f_{ji}{}^l \qquad f_{i\alpha}{}^\beta = -f_{\alpha i}{}^\beta \qquad f_{\alpha\beta}{}^i = f_{\beta\alpha}{}^i. \tag{2.1.16}$$

The super-Jacobi identifies lead to the bilinear equations in the structure constants:

$$(-1)^{\kappa_J \kappa_L} f_{IJ}{}^K f_{LK}{}^M + (-1)^{\kappa_I \kappa_L} f_{JL}{}^K f_{IK}{}^M + (-1)^{\kappa_I \kappa_J} f_{LI}{}^K f_{JK}{}^M = 0 \tag{2.1.17}$$

From these equations, those essential for us are the following

$$f_{ij}{}^l f_{\alpha\beta}{}^j - f_{i\alpha}{}^\gamma f_{\gamma\beta}{}^l - f_{i\beta}{}^\gamma f_{\alpha\gamma}{}^l = 0. \tag{2.1.18}$$

$$f_{\alpha\beta}{}^i f_{i\gamma}{}^\delta + f_{\beta\gamma}{}^i f_{i\alpha}{}^\delta + f_{\gamma\alpha}{}^i f_{i\beta}{}^\delta = 0. \tag{2.1.19}$$

What do these identities mean?

First, note the generators $\mathscr{F}(e_i)$ of the ${}^0\mathscr{G}$-representation (2.1.12) are given by matrices $[\mathscr{F}(e_i)]^\alpha{}_\beta = f_{i\beta}{}^\alpha$. Secondly, the generators $\mathrm{ad}(e_i)$ of the ${}^0\mathscr{G}$-adjoint representation are given by matrices $[\mathrm{ad}(e_i)]^l{}_j = f_{ij}{}^l$. So, equation (2.1.18) means nothing more than that $f_{\alpha\beta}{}^i$ is an invariant tensor of the Lie algebra ${}^0\mathscr{G}$ transforming in the adjoint representation with respect to the index i and in the representation being contragredient to $\mathscr{F}$ with respect to each of the indices α and β. What is more, not every invariant tensor with such a structure may be chosen, only that which satisfies equation (2.1.19).

The above consideration makes it clear that not every Lie algebra admits an extension to a superalgebra; admissible Lie algebra should have an invariant tensor $f_{\alpha\beta}{}^i = f_{\beta\alpha}{}^i$ subject to constraint (2.1.19).

2.1.2. Examples of superalgebras

Let us consider the set of all $(p+q) \times (p+q)$ complex matrices. It is a $\mathbb{Z}_2$-graded associative algebra with respect to matrix multiplication and the grading defined as follows. For every matrix

$$R = \begin{pmatrix} A & B \\ C & D \end{pmatrix} \tag{2.1.20}$$

where A, B, C and D are $p \times p$, $p \times q$, $q \times p$ and $q \times q$ complex matrices, respectively, the decomposition into its even and odd parts reads

$$R = {}^0R + {}^1R$$
$${}^0R = \begin{pmatrix} A & 0 \\ 0 & D \end{pmatrix} \qquad {}^1R = \begin{pmatrix} 0 & B \\ C & 0 \end{pmatrix}. \tag{2.1.21}$$

This algebra is denoted usually by $Mat(p, q|\mathbb{C})$. In accordance with (2.1.9), $Mat(p, q|\mathbb{C})$ becomes a superalgebra with respect to the Lie superbracket

$$[R, S\} = [{}^0R, {}^0S] + [{}^0R, {}^1S] + [{}^1R, {}^0S] + \{{}^1R, {}^1S\}.$$

The resultant superalgebra is denoted by $gl(p, q|\mathbb{C})$. Restriction to the case of real $(p + q) \times (p + q)$ matrices gives the real superalgebra $gl(p, q|\mathbb{R})$.

By analogy with supermatrices, one can define in $Mat(p, q|\mathbb{C})$ the operation of supertrace by the rule:

$$\mathrm{str}\, R = \mathrm{tr}\, A - \mathrm{tr}\, D.$$

It is a simple exercise to check that

$$\mathrm{str}\,([R, S\}) = 0$$

for arbitrary matrices R, $S \in Mat(p, q|\mathbb{C})$. As a result, the set of all supertraceless matrices,

$$\mathrm{str}\, R = 0 \tag{2.1.22}$$

forms a superalgebra, denoted by $sl(p, q|\mathbb{C})$, which is a subalgebra in $gl(p, q|\mathbb{C})$. The set of all real matrices in $sl(p, q|\mathbb{C})$ presents the real superalgebra $sl(p, q|\mathbb{R})$.

The last example is the orthosymplectic superalgebra $osp(p, 2q|\mathbb{C})$. It is a subalgebra of $sl(p, 2q|\mathbb{C})$ selected by the conditions

$$A + A^{\mathrm{T}} = 0 \qquad JD + D^{\mathrm{T}}J = 0 \qquad B = \mathrm{i}C^{\mathrm{T}}J \tag{2.1.23}$$

where J is the $2q \times 2q$ symplectic matrix

$$J = \begin{pmatrix} 0 & \mathbb{1}_q \\ -\mathbb{1}_q & 0 \end{pmatrix}. \tag{2.1.24}$$

Equation (2.1.23) means that $A \in so(p, \mathbb{C})$ and $D \in sp(2q, \mathbb{C})$, so that

$${}^0osp(p, 2q|\mathbb{C}) = so(p, \mathbb{C}) \oplus sp(2q, \mathbb{C}).$$

2.1.3. The Grassmann shell of a superalgebra

Every complex $\mathbb{Z}_2$-graded vector space $L(\mathbb{C}) = {}^0L(\mathbb{C}) \oplus {}^1L(\mathbb{C})$ of dimension $(p + q)$ can be made into a supervector space $L(\Lambda_\infty)$ of dimension (p, q) defined as follows. Let $e_I = \{e_i, e_\alpha\}$, where $i = 1, 2, \ldots, p$ and $\alpha = 1, 2, \ldots, q$, be a pure basis in $L(\mathbb{C})$,

$$u = {}^0u + {}^1u = x^ie_i + x^\alpha e_\alpha = x^Ie_I \qquad x^I \in \mathbb{C}$$

for every $u \in L(\mathbb{C})$. Consider the set $L(\Lambda_\infty)$ of all formal linear combinations

$$\mathbf{u} = z^ie_i + z^\alpha e_\alpha = z^Ie_I \qquad z^I \in \Lambda_\infty \tag{2.1.25}$$

with z^I, $I = 1, 2, \ldots, p + q$, being arbitrary supernumbers. In $L(\Lambda_\infty)$ we define the operations

1. *addition*

$$\mathbf{u}_1 + \mathbf{u}_2 = (z_1^I + z_2^I)e_I \tag{2.1.26a}$$

2. *left multiplication by supernumbers*

$$\alpha \cdot \mathbf{u} = (\alpha z^I)e_I \qquad \alpha \in \Lambda_\infty \tag{2.1.26b}$$

3. *right multiplication by supernumbers*

$$\mathbf{u} \cdot \alpha = (-1)^{\kappa_I \varepsilon(\alpha)}(z^I \alpha)e_I \tag{2.1.26c}$$

where α is a pure supernumber. As a result, every $\mathbf{u} \in L(\Lambda_\infty)$ is decomposed into its even and odd parts by the rule:

$$\begin{aligned} \mathbf{u} &= {}^0\mathbf{u} + {}^1\mathbf{u} \\ {}^0\mathbf{u} &= \xi^i e_i + \xi^\alpha e_\alpha \qquad \xi^i \in \mathbb{C}_c, \quad \xi^\alpha \in \mathbb{C}_a \\ {}^1\mathbf{u} &= \eta^i e_i + \eta^\alpha e_\alpha \qquad \eta^i \in \mathbb{C}_a, \quad \eta^\alpha \in \mathbb{C}_c. \end{aligned} \tag{2.1.27}$$

It is readily seen that $L(\Lambda_\infty)$ becomes a supervector space of dimension (p, q), in which the operation of complex conjugation is undefined. Every even (odd) vector in $L(\mathbb{C})$ is, at the same time, an even (odd) supervector in $L(\Lambda_\infty)$. Following F. Berezin, the supervector space $L(\Lambda_\infty)$ is said to be the 'Grassmann shell' of the $\mathbb{Z}_2$-graded vector space $L(\mathbb{C})$ (originally, Berezin used the term 'Grassmann shell' for the set ${}^0L(\Lambda_\infty)$ of the even supervectors in $L(\Lambda_\infty)$).

From now on, we make no distinction between the parity functions ε on $L(\Lambda_\infty)$ and κ on $L(\mathbb{C})$, using the same notation ε for both.

Now let $\mathscr{G}(\mathbb{C})$ be a complex superalgebra with pure basis $\{e_I\}$. We take the Grassmann shell $\mathscr{G}(\Lambda_\infty)$ of the $\mathbb{Z}_2$-graded vector space $\mathscr{G}(\mathbb{C})$. Then, the operations of multiplication (2.1.10) in $\mathscr{G}(\mathbb{C})$ and multiplication by supernumbers (2.1.26*b*, *c*) in $\mathscr{G}(\Lambda_\infty)$ induce a multiplication in $\mathscr{G}(\Lambda_\infty)$ defined by

$$[\mathbf{u}, \mathbf{v}\} = [{}^0\mathbf{u}, {}^0\mathbf{v}] + [{}^0\mathbf{u}, {}^1\mathbf{v}] + [{}^1\mathbf{u}, {}^0\mathbf{v}] + \{{}^1\mathbf{u}, {}^1\mathbf{v}\} \tag{2.1.28}$$

for any supervectors $\mathbf{u}$, $\mathbf{v} \in \mathscr{G}(\Lambda_\infty)$. Choosing in (2.1.28) the supervectors $\mathbf{u} = \xi^I e_I$ and $\mathbf{v} = \zeta^J e_J$ to be pure, $\varepsilon(\xi^I) = \varepsilon(\mathbf{u}) + \varepsilon_I$ and $\varepsilon(\zeta^J) = \varepsilon(\mathbf{v}) + \varepsilon_J$, one obtains

$$\begin{aligned} [\mathbf{u}, \mathbf{v}\} &= \mathbf{u}\mathbf{v} - (-1)^{\varepsilon(\mathbf{u})\varepsilon(\mathbf{v})}\mathbf{v}\mathbf{u} = (-1)^{\varepsilon(\mathbf{u})(\varepsilon(\mathbf{v}) + \varepsilon_J)}\zeta^J \xi^I [e_I, e_J\} \\ &= (-1)^{\varepsilon(\mathbf{u})(\varepsilon(\mathbf{v}) + \varepsilon_J)}\zeta^J \xi^I f_{IJ}{}^L e_L \end{aligned} \tag{2.1.29}$$

where equation (2.1.14) has been used. The main properties of the multiplication (2.1.28) are:

1. $[\mathbf{u}, \mathbf{v}\} = -(-1)^{\varepsilon(\mathbf{u})\varepsilon(\mathbf{v})}[\mathbf{v}, \mathbf{u}\}$
2. $\varepsilon([\mathbf{u}, \mathbf{v}\}) = \varepsilon(\mathbf{u}) + \varepsilon(\mathbf{v}) \pmod 2$

3. $(-1)^{\varepsilon(\mathbf{u})\varepsilon(\mathbf{w})}[\mathbf{u},[\mathbf{v},\mathbf{w}\}\} + (-1)^{\varepsilon(\mathbf{v})\varepsilon(\mathbf{u})}[\mathbf{v},[\mathbf{w},\mathbf{u}\}\} + (-1)^{\varepsilon(\mathbf{w})\varepsilon(\mathbf{v})}[\mathbf{w},[\mathbf{u},\mathbf{v}\}\} = 0$ (2.1.30)

   for any supervectors $\mathbf{u}, \mathbf{v}, \mathbf{w} \in \mathscr{G}(\Lambda_\infty)$

4. $[\alpha\mathbf{u} + \beta\mathbf{v}, \mathbf{w}\} = \alpha[\mathbf{u},\mathbf{w}\} + \beta[\mathbf{v},\mathbf{w}\} \qquad \forall\alpha,\beta \in \Lambda_\infty \qquad \forall\mathbf{u},\mathbf{v},\mathbf{w} \in \mathscr{G}(\Lambda_\infty)$.

To prove property 2, one has to use equations (2.1.7) and (2.1.29). The properties 1–3 show that $\mathscr{G}(\Lambda_\infty)$ forms a superalgebra. Property 4 means that the Lie superbracket $[\ldots,\ldots\}$ of $\mathscr{G}(\Lambda_\infty)$ is linear with respect to multiplication by supernumbers. The supervector space $\mathscr{G}(\Lambda_\infty)$ with the multiplication (2.1.28) is said to be the 'Grassmann shell of the superalgebra' $\mathscr{G}(\mathbb{C})$. We shall also call objects like $\mathscr{G}(\Lambda_\infty)$ 'Berezin superalgebras' to distinguish them from ordinary superalgebras over $\mathbb{C}$ or $\mathbb{R}$, which have been considered in subsection 2.1.1.

By definition, a Berezin superalgebra $\mathscr{G}$ is a supervector space $\mathscr{G}$ together with an operation of multiplication $[\ldots,\ldots\}$: $\mathscr{G} \times \mathscr{G} \to \mathscr{G}$ of elements from $\mathscr{G}$ such that the axioms (2.1.30) are satisfied. When $\mathscr{G}$ has finite dimension, all algebraic information is contained in the 'structure constants' defined by

$$[e_I, e_J\} = f_{IJ}{}^L e_L \qquad f_{IJ}{}^L \in \Lambda_\infty$$
$$\varepsilon(f_{IJ}{}^L) = \varepsilon_I + \varepsilon_J + \varepsilon_L \pmod 2 \tag{2.1.31}$$

with $\{e_I\}$ being a pure basis of $\mathscr{G}$. A Berezin superalgebra is said to be 'conventional' if it admits a pure basis in which the structure constants are ordinary numbers, i.e. as in equation (2.1.14); otherwise it is called 'unconventional'. Every conventional Berezin superalgebra is the Grassmann shell of some complex superalgebra. From now on we will discuss only conventional Berezin superalgebras.

Consider the set ${}^0\mathscr{G}(\Lambda_\infty)$ of even supervectors in $\mathscr{G}(\Lambda_\infty)$. Property 2 from (2.1.30) tells us that ${}^0\mathscr{G}(\Lambda_\infty)$ forms a subalgebra in the Berezin superalgebra $\mathscr{G}(\Lambda_\infty)$. In accordance with equations (2.1.28, 29), the multiplication on ${}^0\mathscr{G}(\Lambda_\infty)$ is given by

$$[\mathbf{u},\mathbf{v}] = \zeta^J \xi^I f_{IJ}{}^L e_L \tag{2.1.32}$$

for any supervectors $\mathbf{u} = \xi^I e_I$ and $\mathbf{v} = \zeta^J e_J$ from ${}^0\mathscr{G}(\Lambda_\infty)$, and due to (2.1.30), it is characterized by the properties

$$\left.\begin{array}{ll} 1.\ [\mathbf{u},\mathbf{v}] = -[\mathbf{v},\mathbf{u}] & \\ 2.\ [\mathbf{u},[\mathbf{v},\mathbf{w}]] + [\mathbf{v},[\mathbf{w},\mathbf{u}]] + [\mathbf{w},[\mathbf{u},\mathbf{v}]] = 0 & \\ 3.\ [\alpha\mathbf{u} + \beta\mathbf{v}, \mathbf{w}] = \alpha[\mathbf{u},\mathbf{w}] + \beta[\mathbf{v},\mathbf{w}] \qquad \forall\alpha,\beta \in \mathbb{C}_c & \forall\mathbf{u},\mathbf{v},\mathbf{w} \in {}^0\mathscr{G}(\Lambda_\infty) \end{array}\right\} \tag{2.1.33}$$

Therefore, we obtain a Lie algebra structure on ${}^0\mathscr{G}(\Lambda_\infty)$ which is induced by the superalgebra structure on $\mathscr{G}(\mathbb{C})$. Of course, the reason for this transmutation lies in the heterotic nature of supernumbers: c-numbers are commuting objects while a-numbers are anticommuting. The algebra ${}^0\mathscr{G}(\Lambda_\infty)$ is said to be the 'complex super Lie algebra associated to the superalgebra' $\mathscr{G}(\mathbb{C})$. More generally, a complex super Lie algebra is defined to be the even part ${}^0\mathscr{G}$ of a Berezin superalgebra $\mathscr{G}$; the multiplication on ${}^0\mathscr{G}$ is characterized by the properties (2.1.33).

There are three basic differences between ordinary Lie algebras and super Lie algebras. First, Lie algebras are endowed with the operation of multiplication by ordinary numbers only, while for super Lie algebras it is defined for arbitrary c-numbers. Secondly, generators of Lie algebras satisfy commutation relations, while generators of super Lie algebras are subject to (anti)commutation relations like (2.1.31). Finally, generators of a Lie algebra are elements of the algebra. But not all generators of a super Lie algebra belong to the algebra; the even generators belong, whereas the odd generators do not.

Let $\mathscr{G}$ be a Berezin superalgebra and ${}^0\mathscr{G}$ be the corresponding super Lie algebra. Suppose that the supervector space $\mathscr{G}$ can be provided by an operation of complex conjugation such that the subset ${}^0\mathscr{G}_{\mathbf{R}}$ of real supervectors in ${}^0\mathscr{G}$ forms a subalgebra of the super Lie algebra ${}^0\mathscr{G}$. Then ${}^0\mathscr{G}_{\mathbf{R}}$ is said to be a 'real super Lie algebra'. The operation of multiplication in ${}^0\mathscr{G}_{\mathbf{R}}$ respects the properties 1, 2 (2.1.33), but the third property should be substituted by

$$[\alpha\mathbf{u} + \beta\mathbf{v}, \mathbf{w}] = \alpha[\mathbf{u}, \mathbf{w}] + \beta[\mathbf{v}, \mathbf{w}] \qquad \forall\alpha, \beta \in \mathbb{R}_c, \quad \forall\mathbf{u}, \mathbf{v}, \mathbf{w} \in {}^0\mathscr{G}_{\mathbf{R}}. \tag{2.1.34}$$

It is not difficult to find restrictions on the structure constants of $\mathscr{G}$ (2.1.31) (with respect to a standard basis $\{e_I\}$, $e_I^* = (-1)^{\varepsilon_I}e_I$) to ensure that the subset ${}^0\mathscr{G}_{\mathbf{R}}$ is subalgebra of ${}^0\mathscr{G}$. The result is

$$(f_{IJ}{}^L)^* = (-1)^{(\varepsilon_I + \varepsilon_J)(1 + \varepsilon_L) + \varepsilon_I\varepsilon_J} f_{IJ}{}^L. \tag{2.1.35}$$

When the structure constants are ordinary numbers, equation (2.1.35) takes a simpler form

$$(f_{IJ}{}^L)^* = (-1)^{\varepsilon_I\varepsilon_J} f_{IJ}{}^L. \tag{2.1.36}$$

2.1.4. Examples of Berezin superalgebras and super Lie algebras

First, let us construct Grassmann shells of the complex superalgebras presented in subsection 2.1.2. We begin with the superalgebra $gl(p, q|\mathbb{C})$. Taking the convention that matrix indices are positioned in the manner

$$R = (R^M{}_N)$$

one easily finds that the Grassmann shell of the $\mathbb{Z}_2$-graded matrix space $Mat(p, q|\mathbb{C})$ coincides with the supermatrix space $Mat^{(+)}(p, q|\Lambda_\infty)$ studied in Section 1.9. $Mat^{(+)}(p, q|\Lambda_\infty)$ becomes the Grassmann shell of the superalgebra $gl(p, q|\mathbb{C})$ with respect to the superbracket defined by

$$[F_1, F_2\} = F_1F_2 - (-1)^{\varepsilon(F_1)\varepsilon(F_2)}F_2F_1 \tag{2.1.37}$$

where F_1 and F_2 are arbitrary pure supermatrices. The Berezin superalgebra obtained is denoted by $gl^{(+)}(p, q|\Lambda_\infty)$. It is clear that the set ${}^0Mat^{(+)}(p, q|\Lambda_\infty)$ of c-type supermatrices in $Mat^{(+)}(p, q|\Lambda_\infty)$ (see equation (1.9.37a)) forms the super Lie algebra ${}^0gl^{(+)}(p, q|\Lambda_\infty)$ with respect to the ordinary Lie bracket. Finally, recall that we introduced in $Mat^{(+)}(p, q|\Lambda_\infty)$ the operation '$s*$' of supercomplex conjugation (1.9.41), which is characterized by property (1.9.44). Therefore, if F_1 and F_2 are real c-type supermatrices (see equation (1.9.42b)), then F_1F_2 and hence $[F_1F_2]$ are real. As a result, the set ${}^0Mat^{(+)}_{\mathbb{R}}(p, q|\Lambda_\infty)$ of real c-type supermatrices forms a subalgebra in ${}^0gl^{(+)}(p, q|\Lambda_\infty)$ denoted by ${}^0gl^{(+)}_{\mathbb{R}}(p, q|\Lambda_\infty)$. The algebra ${}^0gl^{(+)}_{\mathbb{R}}(p, q|\Lambda_\infty)$ is a real super Lie algebra.

To construct the Grassmann shell of the superalgebra $sl(p, q|\mathbb{C})$, we introduce the operation supertrace in the supermatrix algebra $Mat^{(+)}(p, q|\Lambda_\infty)$. Recall that supertrace has been defined for c-type supermatrices only (see equation (1.10.66)). We generalize this definition to the whole $Mat^{(+)}(p, q|\Lambda_\infty)$ demanding two basic postulates:

$$\left.\begin{array}{ll} 1. & \mathrm{str}\, F = (-1)^{\varepsilon_M + \varepsilon(F)\varepsilon_M} F^M{}_M, \\ & \text{for every pure supermatrix } F \\ 2. & \mathrm{str}\,(F_1 + F_2) = \mathrm{str}\, F_1 + \mathrm{str}\, F_2, \\ & \text{for any supermatrices } F_1 \text{ and } F_2. \end{array}\right\} \qquad (2.1.38)$$

So, the supertrace of a-type supermatrices coincides with the ordinary trace. The operation of supertrace is characterized by the following properties:

$$\left.\begin{array}{ll} 1. & \mathrm{str}\,(\hat{z}F) = z\, \mathrm{str}\, F, \qquad \mathrm{str}\,(F\hat{z}) = (\mathrm{str}\, F)z, \\ 2. & \mathrm{str}\, \hat{z} = z(p - q), \\ 3. & \mathrm{str}\,(F^{s*}) = (\mathrm{str}\, F)^*, \end{array}\right\} \qquad (2.1.39)$$

where z and F are an arbitrary supernumber and supermatrix, respectively;

4. $\mathrm{str}\,(F_1F_2) = (-1)^{\varepsilon(F_1)\varepsilon(F_2)}\, \mathrm{str}\,(F_2F_1)$,

for arbitrary pure supermatrices F_1 and F_2.

Now we consider in $Mat^{(+)}(p, q|\Lambda_\infty)$ the subset $sl^{(+)}(p, q|\Lambda_\infty)$ of supertraceless supermatrices,

$$\mathrm{str}\, F = 0 \qquad \forall F \in sl(p, q|\Lambda_\infty). \qquad (2.1.40)$$

Property 2 in (2.1.38) and property 1 in (2.1.39) show that $sl^{(+)}(p, q|\Lambda_\infty)$ is a supervector space. Recalling the operation of multiplication by supernumbers in $Mat^{(+)}(p, q|\Lambda_\infty)$ (see equation (1.9.31)), one finds $sl^{(+)}(p, q|\Lambda_\infty)$ to be the Grassmann shell of the $\mathbb{Z}_2$-graded vector space $sl(p, q|\mathbb{C})$. Due to the property 4 in (2.1.39), $sl^{(+)}(p, q|\Lambda_\infty)$ becomes the Berezin superalgebra with respect to the superbracket (2.1.37). The subset of c-type supermatrices in $sl^{(+)}(p, q|\Lambda_\infty)$ forms a super Lie algebra denoted by ${}^0sl^{(+)}(p, q|\Lambda_\infty)$. Finally, the subset of real c-type supermatrices in ${}^0sl^{(+)}(p, q|\Lambda_\infty)$ forms a real super Lie algebra denoted by ${}^0sl_{\mathbb{R}}^{(+)}(p, q|\Lambda_\infty)$.

The Grassmann shell of the orthosymplectic superalgebra $osp(p,2q|\mathbb{C})$ turns out to coincide with the subset of $Mat^{(+)}(p,2q|\Lambda_\infty)$ denoted by $osp^{(+)}(p, 2q|\Lambda_\infty)$ and consisting of all supermatrices under the equation

$$\eta_{MP}F^P{}_N + (F^{sT})_M{}^P\eta_{PN} = 0 \tag{2.1.41a}$$

or, in index notation,

$$\eta F + F^{sT}\eta = 0 \tag{2.1.41b}$$

where

$$\eta = \begin{pmatrix} \mathbb{1}_p & 0 \\ 0 & \mathrm{i}J \end{pmatrix}, \tag{2.1.42}$$

J being as in equation (2.1.24). The matrix η_{MN} is a supersymmetric real supermatrix in the sense of the definition (1.9.78). The operation of supertransposition for $Mat^{(+)}(p, q|\Lambda_\infty)$ is given by equations (1.9.58) (for $Mat^{(-)}(p, q|\Lambda_\infty)$ see equation (1.9.59)). Equation (2.1.41) is evidently consistent. Indeed let us extend the operation of supertransposition (1.9.77), defined originally for c-type supermatrices only, to the case of pure arbitrary supermatrices with lower positioned indices:

$$(H^{sT})_{MN} = (-1)^{\varepsilon(H)(\varepsilon_M + \varepsilon_N) + \varepsilon_M + \varepsilon_N + \varepsilon_M\varepsilon_N}H_{NM}. \tag{2.1.43}$$

Then, the identities

$$\begin{aligned} (HF)^{sT} &= (-1)^{\varepsilon(H)\varepsilon(F)}F^{sT}H^{sT} \qquad \forall F \in Mat^{(+)}(p, q|\Lambda_\infty) \\ (GH)^{sT} &= (-1)^{\varepsilon(G)\varepsilon(H)}H^{sT}G^{sT} \qquad \forall G \in Mat^{(-)}(p, q|\Lambda_\infty) \end{aligned} \tag{2.1.44}$$

imply that the supertranspose of equation (2.1.41) is just the same equation. Using the relations (1.9.60) and (1.9.61), one can readily verify that (1) $osp^{(+)}(p, 2q|\Lambda_\infty)$ is a supervector space and a Berezin superalgebra with respect to the superbracket (2.1.37); (2) with respect to the conjugation (1.9.42), the real even subset ${}^0osp_{\mathbb{R}}^{(+)}(p, 2q|\Lambda_\infty)$ of $osp^{(+)}(p, 2q|\Lambda_\infty)$ forms a real super Lie algebra.

Our last example, very important for the subsequent discussion, is the 'Berezin superalgebra of supervector fields' on a real superspace $\mathbb{R}^{p|q}$ parametrized by p real c-number coordinates x^m and q real a-number coordinates θ^μ. We will now use the notation adopted in Section 1.10; in particular, $z^M = (x^m, \theta^\mu)$ and $\partial_M = (\partial_m, \partial_\mu) \equiv \partial/\partial z^M$. By definition, a 'supervector field' on $\mathbb{R}^{p|q}$ is a first-order differential operator,

$$K = K^M(z)\partial_M = K^m(x, \theta)\partial_m + K^\mu(x, \theta)\partial_\mu \tag{2.1.45}$$

with $K^M(z)$ being supersmooth functions on $\mathbb{R}^{p|q}$. The set of all supervector fields on $\mathbb{R}^{p|q}$ will be denoted by SVF(p, q). A supervector field K is said to be c-type (a-type) if its components $K(z)$ are pure superfunctions on $\mathbb{R}^{p|q}$ with the following Grassmann parities:

$$\begin{array}{ll} c\text{-type} & a\text{-type} \\ \varepsilon(K^M(z)) = \varepsilon_M & \varepsilon(K^M(z)) = 1 + \varepsilon_M \quad (\text{mod } 2) \end{array} \tag{2.1.46}$$

Supervector fields of definite type are called pure. Every pure supervector K is provided with the Grassmann parity $\varepsilon(K)$ defined by

$$\varepsilon(K) = \begin{cases} 0 & \text{for } c\text{-type } K \\ 1 & \text{for } a\text{-type } K. \end{cases} \tag{2.1.47}$$

Note that all supervector fields considered in Section 1.11, were of c-type.

We define in SVF(p, q) the operations

$$\left.\begin{array}{l} \text{1. } \textit{addition} \\ \quad K_1 + K_2 = (K_1^M + K_2^M)\partial_M \\ \text{2. } \textit{left multiplication by supernumbers} \\ \quad \alpha K = (\alpha K^M)\partial_M \\ \text{3. } \textit{right multiplication by supernumbers} \\ \quad K\alpha = (-1)^{\varepsilon(\alpha)\varepsilon_M}(K^M\alpha)\partial_M \\ \text{4. } \textit{complex conjugation} \\ \quad K \rightarrow \bar{K} \equiv \bar{K}^M\partial_M = (-1)^{\varepsilon(K)\varepsilon_M}(K^M)^*\partial_M. \end{array}\right\} \tag{2.1.48}$$

Then, all the supervector space axioms prove to be satisfied, and SVF(p, q) becomes an infinite-dimensional supervector space. It should be pointed out that, due to operations (2.1.48), real c-type supervector fields, $\bar{K} = K$, are characterized by real components, $(K^M)^* = K^M$. So, the above definition of complex conjugation is consistent with the notion of real supervector fields introduced in Section 1.11.

Let us consider the set of all supersmooth functions on $\mathbb{R}^{p|q}$, denoted by $C^\infty(\mathbb{R}^{p|q})$. $C^\infty(\mathbb{R}^{p|q})$ is endowed, in an obvious way, with the structure of an infinite-dimensional supervector space. Every supervector field K defines a left linear operator on $C^\infty(\mathbb{R}^{p|q})$ by the rule:

$$\Phi(z) \rightarrow (K\Phi)(z) = K^M(z)\partial_M\Phi(z) \qquad \forall\Phi(z) \in C^\infty(\mathbb{R}^{p|q}). \tag{2.1.49}$$

Every c-type supervector field is a c-type linear operator on $C^\infty(\mathbb{R}^{p|q})$ since it does not change superfunction types. Analogously, every a-type supervector field is a left a-type linear operator on $C^\infty(\mathbb{R}^{p|q})$. Using equation (1.10.22), one can prove the identity

$$(K\Phi)^* = (-1)^{\varepsilon(K)\varepsilon(\Phi)}\bar{K}\Phi^* \tag{2.1.50}$$

for arbitrary pure supervector field K and superfunction Φ. Replacing here $\Phi \rightarrow \Phi^*$ brings this relation into the general form (1.9.40).

Given arbitrary pure supervector fields $K = K^M\partial_M$ and $L = L^M\partial_M$, we define their superbracket by the rule

$$[K, L\} = K \cdot L - (-1)^{\varepsilon(K)\varepsilon(L)} L \cdot K \tag{2.1.51}$$

which, due to equations (1.10.19, 20) and (2.1.46), is also a pure supervector field,

$$[K, L\} = ((KL^M) - (-1)^{\varepsilon(K)\varepsilon(L)}(LK^M))\partial_M. \tag{2.1.52}$$

The superbracket (2.1.51) proves to satisfy all the axioms (2.1.30) for a Berezin superalgebra. As a result, SVF(p, q) obtains the structure of a Berezin superalgebra. Then, the subset ^{0}SVF(p, q) of c-type supervector fields in SVF(p, q) forms a super Lie algebra. Finally, since any pure supervector fields K and L satisfy the identity

$$\overline{[K, L\}} = (-1)^{\varepsilon(K)\varepsilon(L)}[\bar{L}, \bar{K}\} \tag{2.1.53}$$

the subset ^{0}SVF$_{\mathbf{R}}(p, q)$ of real c-type supervector fields $^0SVF(p, q)$ forms a real super Lie algebra. This super Lie algebra has been studied in Section 1.11.

2.1.5. Representations of (Berezin) superalgebras and super Lie algebras

For our discussion of representation theory it seems convenient to start with Berezin superalgebra representations.

Let $\mathscr{G}(\Lambda_\infty)$ be a Berezin superalgebra and $\mathscr{L}(\Lambda_\infty)$ be a supervector space. A mapping T: $\mathscr{G}(\Lambda_\infty) \to End^{(+)}\mathscr{L}(\Lambda_\infty)$ from $\mathscr{G}(\Lambda_\infty)$ into the algebra of left linear operators on $\mathscr{L}(\Lambda_\infty)$ is said to be a 'linear representation of the Berezin superalgebra' in $\mathscr{L}(\Lambda_\infty)$ under fulfilment of the conditions:

$$\left.\begin{aligned} &1.\ \varepsilon(T(\mathbf{u})) = \varepsilon(\mathbf{u}) \\ &\quad \text{for every pure } \mathbf{u} \in \mathscr{G}(\Lambda_\infty) \\ &2.\ T(\alpha\mathbf{u} + \beta\mathbf{u}) = \alpha T(\mathbf{u}) + \beta T(\mathbf{v}) \\ &\quad \text{for } \forall\alpha, \beta \in \Lambda_\infty \text{ and } \forall\mathbf{u}, \mathbf{v} \in \mathscr{G}(\Lambda_\infty) \\ &3.\ T([\mathbf{u}, \mathbf{v}\}) = [T(\mathbf{u}), T(\mathbf{v})\} \\ &\quad \text{for any pure elements } \mathbf{u}, \mathbf{v} \in \mathscr{G}(\Lambda_\infty). \end{aligned}\right\} \tag{2.1.54}$$

$End^{(+)}\mathscr{L}(\Lambda_\infty)$ can be provided with a Berezin superalgebra structure with respect to the superbracket:

$$[\mathscr{F}_1, \mathscr{F}_2\} = \mathscr{F}_1\mathscr{F}_2 - (-1)^{\varepsilon(\mathscr{F}_1)\varepsilon(\mathscr{F}_2)}\mathscr{F}_2\mathscr{F}_1 \tag{2.1.55}$$

where $\mathscr{F}_1$ and $\mathscr{F}_2$ are arbitrary left linear operators on $\mathscr{L}(\Lambda_\infty)$. Thus, every linear representation T: $\mathscr{G}(\Lambda_\infty) \to End^{(+)}\mathscr{L}(\Lambda_\infty)$ is a homomorphism of the Berezin superalgebras.

In accordance with the first requirement (2.1.54), every even (odd) element of $\mathscr{G}(\Lambda_\infty)$ maps to a c-type (a-type) linear operator on $\mathscr{L}(\Lambda_\infty)$, respectively.

As a result, the $\mathscr{G}(\Lambda_\infty)$-representation in $\mathscr{L}(\Lambda_\infty)$ induces a 'linear representation of the super Lie algebra' ${}^0\mathscr{G}(\Lambda_\infty)$ in $\mathscr{L}(\Lambda_\infty)$, T: ${}^0\mathscr{G}(\Lambda_\infty) \to {}^0End\, \mathscr{L}(\Lambda_\infty)$, which is characterized by the properties:

$$\left.\begin{array}{l} 1.\ \ T(\alpha\mathbf{u} + \beta\mathbf{v}) = \alpha T(\mathbf{u}) + \beta T(\mathbf{v}) \\ \quad\ \forall\alpha, \beta \in \mathbb{C}_c \text{ and } \forall\mathbf{u}, \mathbf{v} \in {}^0\mathscr{G}(\Lambda_\infty) \\ 2.\ \ T([\mathbf{u}, \mathbf{v}]) = [T(\mathbf{u}), \mathbf{T}(\mathbf{v})] \\ \quad\ \forall\mathbf{u}, \mathbf{v} \in {}^0\mathscr{G}(\Lambda_\infty). \end{array}\right\} \tag{2.1.56}$$

For every $\mathbf{u} \in {}^0\mathscr{G}(\Lambda_\infty)$, the operator $T(\mathbf{u})$ maps the subset ${}^0\mathscr{L}(\Lambda_\infty)$ of even supervectors in $\mathscr{L}(\Lambda_\infty)$ into itself and the subset ${}^1\mathscr{L}(\Lambda_\infty)$ of odd supervectors in $\mathscr{L}(\Lambda_\infty)$ into itself also. So, while the representation of the Berezin superalgebra $\mathscr{G}(\Lambda_\infty)$ acts on the whole supervector space $\mathscr{L}(\Lambda_\infty)$, the corresponding representation of the super Lie algebra ${}^0\mathscr{G}(\Lambda_\infty)$ acts independently in each of the subspaces ${}^0\mathscr{L}(\Lambda_\infty)$ and ${}^1\mathscr{L}(\Lambda_\infty)$. This is very important for the following reasons. First, suppose that $\mathscr{L}(\Lambda_\infty)$ has a finite dimension (p, q). Then, the subspace ${}^0\mathscr{L}(\Lambda_\infty)$ can be identified with the complex superspace $\mathbb{C}^{p|q}$ and the subspace ${}^1\mathscr{L}(\Lambda_\infty)$ can be identified with the superspace $\mathbb{C}^{q|p}$ (see subsection 1.9.2). Therefore, every finite-dimensional representation of the super Lie algebra ${}^0\mathscr{G}(\Lambda_\infty)$ leads to linear (supermatrix) representations in superspace and superspace is the main subject of our book. Secondly, suppose we have a linear representation T of $\mathscr{G}(\Lambda_\infty)$ in the supervector space $C^\infty(\mathbb{R}^{p|q})$ of supersmooth functions on $\mathbb{R}^{p|q}$. Then it induces representations of the super Lie algebra ${}^0\mathscr{G}(\Lambda_\infty)$ in the space ${}^0C^\infty(\mathbb{R}^{p|q})$ of bosonic fields on $\mathbb{R}^{p|q}$ and in the space ${}^1C^\infty(\mathbb{R}^{p|q})$ of fermionic fields on $\mathbb{R}^{p|q}$. That is why super Lie algebras are of great importance.

When $\mathscr{G}(\Lambda_\infty)$ is endowed with an operation of complex conjugation such that the subset ${}^0\mathscr{G}_{\mathbb{R}}(\Lambda_\infty)$ of real even elements in $\mathscr{G}(\Lambda_\infty)$ forms a real super Lie algebra, then every linear representation T: $\mathscr{G}(\Lambda_\infty) \to End\, \mathscr{L}(\Lambda_\infty)$ induces a linear representation of the real super Lie algebra ${}^0\mathscr{G}_{\mathbb{R}}(\Lambda_\infty)$ in $\mathscr{L}(\Lambda_\infty)$, T: ${}^0\mathscr{G}_{\mathbb{R}}(\Lambda_\infty) \to {}^0End\, \mathscr{L}(\Lambda_\infty)$, for which the first requirement of (2.1.56) should be substituted by

$$T(\alpha\mathbf{u} + \beta\mathbf{v}) = \alpha T(\mathbf{u}) + \beta T(\mathbf{v}), \qquad \forall\alpha, \beta \in \mathbb{R}_c \qquad \forall\mathbf{u}, \mathbf{v} \in {}^0\mathscr{G}_{\mathbb{R}}(\Lambda_\infty). \tag{2.1.57}$$

Now we proceed to consideration of superalgebra representations. For the time being, we restrict ourselves to the case of finite-dimensional representations.

Let $\mathscr{G}(\mathbb{C})$ be a complex superalgebra. A linear mapping T: $\mathscr{G}(\mathbb{C}) \to Mat(p, q|\mathbb{C})$ from $\mathscr{G}(\mathbb{C})$ into a $\mathbb{Z}_2$-graded matrix algebra is said to be a 'finite-dimensional representation of the superalgebra' $\mathscr{G}(\mathbb{C})$ under fulfilment of the following conditions

$$\left.\begin{array}{l} 1.\ \ \kappa(T(a)) = \kappa(a) \text{ for every pure } a \in \mathscr{G}(\mathbb{C}) \\ 2.\ \ T([a, b\}) = [T(a), T(b)\} \text{ for any pure } a, b \in \mathscr{G}(\mathbb{C}). \end{array}\right\} \tag{2.1.58}$$

The first requirement means that, for every $a \in \mathscr{G}(\mathbb{C})$, we have

$$T(a) = T({}^0a) + T({}^1a)$$

$$T({}^0a) = \begin{pmatrix} A(a) & 0 \\ 0 & D(a) \end{pmatrix} \qquad T({}^1a) = \begin{pmatrix} 0 & B(a) \\ C(a) & 0 \end{pmatrix}. \tag{2.1.59}$$

Therefore, a representation of $\mathscr{G}(\mathbb{C})$ gives us a homomorphism T: $\mathscr{G}(\mathbb{C}) \to gl(p, q|\mathbb{C})$ of the superalgebras.

It is convenient to identify $Mat(p, q|\mathbb{C})$ with the algebra $End\, \mathscr{L}(\mathbb{C})$ of all linear operators acting in a $\mathbb{Z}_2$-graded vector space $\mathscr{L}(\mathbb{C})$ of dimension $(p + q)$ by the standard rule: every linear operator $\mathscr{R}$ on $\mathscr{L}(\mathbb{C})$ is represented with respect to a given pure basis $E_M = \{E_m, E_\mu\}$ in $\mathscr{L}(\mathbb{C})$, $m = 1, 2, \ldots, p$ and $\mu = 1, 2, \ldots, q$, by the matrix $R \in Mat(p, q|\mathbb{C})$ defined as follows

$$\mathscr{R}(E_M) = E_N R^N{}_M.$$

Then the operator $\mathscr{R}$ acts in $\mathscr{L}(\mathbb{C})$ as

$$X = x^M E_M \to \mathscr{R}(X) = X'$$

$$X' = x'^M E_M \qquad x'^M = R^M{}_N x^N$$

for every $X \in \mathscr{L}(\mathbb{C})$. After this identification, $End\, \mathscr{L}(\mathbb{C})$ inherits the structure of a $\mathbb{Z}_2$-graded associative algebra and becomes a superalgebra with respect to the superbracket (2.1.9). So, every representation T: $\mathscr{G}(\mathbb{C}) \to Mat(p, q|\mathbb{C})$ may be understood as a representation T: $\mathscr{G}(\mathbb{C}) \to End\, \mathscr{L}(\mathbb{C})$ in a $\mathbb{Z}_2$-graded vector space.

Now we show that any linear representation T of some finite-dimensional superalgebra $\mathscr{G}(\mathbb{C})$ in a finite-dimensional $\mathbb{Z}_2$-graded vector space $\mathscr{L}(\mathbb{C})$ induces a linear representation $\hat{T}$ of the Grassmann shell $\mathscr{G}(\Lambda_\infty)$ of the superalgebra $\mathscr{G}(\mathbb{C})$ on the Grassmann shell $\mathscr{L}(\Lambda_\infty)$ of the $\mathbb{Z}_2$-graded vector space $\mathscr{L}(\mathbb{C})$.

First, we construct a one-to-one linear mapping $\hat{}$: $End\, \mathscr{L}(\mathbb{C}) \to End^{(+)} \mathscr{L}(\Lambda_\infty)$, preserving the grading and the multiplication. Elements of the supervector space $\mathscr{L}(\Lambda_\infty)$ will be denoted by an arrow, and they will be written in the form (1.9.27*a*) placing basis supervectors on the left. Then, given a linear operator $\mathscr{R}$ on $\mathscr{L}(\mathbb{C})$, it induces the left linear operator $\hat{\mathscr{R}}$ *on* $\mathscr{L}(\Lambda_\infty)$ defined as follows

$$\vec{X} = \vec{E}_M X^M \to \hat{\mathscr{R}}(\vec{X}) = \vec{X}'$$

$$\vec{X}' = \vec{E}_M X'^M \qquad X'^M = R^M{}_N X^N$$

for every $\vec{X} \in \mathscr{L}(\Lambda_\infty)$. Hence, the operators $\hat{\mathscr{R}}$ and $\mathscr{R}$ are characterized by the same (super)matrix R with respect to the basis $\{\vec{E}_M\}$ in $\mathscr{L}(\Lambda_\infty)$ and the basis $\{E_M\}$ in $\mathscr{L}(\mathbb{C})$. If $\mathscr{R}$ is a pure linear operator on $\mathscr{L}(\mathbb{C})$, then $\hat{\mathscr{R}}$ is a pure left linear operator on $\mathscr{L}(\Lambda_\infty)$ of the same type, $\kappa(\mathscr{R}) = \varepsilon(\hat{\mathscr{R}})$. One can readily check also that $\widehat{\mathscr{R}_1\mathscr{R}_2} = \hat{\mathscr{R}}_1\hat{\mathscr{R}}_2$, $\forall \mathscr{R}_1, \mathscr{R}_2 \in End\, \mathscr{L}(\mathbb{C})$. So, in accordance with

(2.1.58), we have

$$\hat{T}([a,b\}) = [\hat{T}(a), \hat{T}(b)\} \tag{2.1.60}$$

for any pure elements a, $b \in \mathscr{G}(\mathbb{C})$.

As the next step, we construct a linear mapping $\hat{T}$: $\mathscr{G}(\Lambda_\infty) \to End^{(+)}\mathscr{L}(\Lambda_\infty)$ as follows. Let $\{e_I\}$ be a pure basis for $\mathscr{G}(\mathbb{C})$; every element $a \in \mathscr{G}(\mathbb{C})$ has the form

$$a = y^I e_I = y^i e_i + y^\alpha e_\alpha \qquad y^I \in \mathbb{C}$$

and every element $\mathbf{a} \in \mathscr{G}(\Lambda_\infty)$ has the form

$$\mathbf{a} = \xi^I e_I = \xi^i e_i + \xi^\alpha e_\alpha \qquad \xi^I \in \Lambda_\infty.$$

We postulate

$$\hat{T}(\mathbf{a}) = \xi^I \hat{T}(e_I).$$

Then, using equations (1.9.35b) and (2.1.60), one finds

$$\hat{T}([\mathbf{a},\mathbf{b}\}) = [\hat{T}(\mathbf{a}), \hat{T}(\mathbf{b})\}$$

for any pure elements $\mathbf{a}$, $\mathbf{b} \in \mathscr{G}(\Lambda_\infty)$. Therefore, the correspondence $\hat{T}$: $\mathscr{G}(\Lambda_\infty) \to End^{(+)}\mathscr{L}(\Lambda_\infty)$ is a representation of the Berezin superalgebra $\mathscr{G}(\Lambda_\infty)$.

2.1.6. Super Lie groups

As is well known, Lie groups and Lie algebras are related by means of the exponential mapping. Given a connected Lie group G and the corresponding Lie algebra $\mathscr{G}_G$, (almost all) group elements can be represented as exponentials of Lie algebra elements,

$$\begin{aligned} g &= \exp a \\ a &= \xi^i e_i \qquad \xi^i \in \mathbb{C} \text{ (or } \mathbb{R}) \end{aligned} \tag{2.1.61}$$

where $\{e_i\}$ is a basis for $\mathscr{G}_G$. The components ξ^i, $i = 1, 2, \ldots, \dim G$, of Lie algebra elements play the role of local coordinates (in a neighbourhood of the identity) of the group manifold. The fact that the union of elements of the form (2.1.61) presents a group is a consequence of the Baker–Hausdorff formula:

$$\exp(a)\exp(b) = \exp\left(a + b + \frac{1}{2}[a,b] + \frac{1}{12}[a,[a,b]] + \frac{1}{12}[b,[b,a]] + \ldots\right) \tag{2.1.62}$$

where dots mean commutators of third order and higher.

Now one can ask whether it is possible, using the exponential mapping, to construct some group-like objects starting from superalgebras, Berezin superalgebras or super Lie algebras. The answer is negative in the case of superalgebras or Berezin superalgebras, because the multiplication law in

(Berezin) superalgebras is a graded commutator, including both commutators and anticommutators. But there is no graded generalization of the Baker–Hausdorff formula. The situation is different in the case of super Lie algebras, since the multiplication operation is now an ordinary Lie bracket. As a result, one can duplicate the ansatz (2.1.61). Namely, given a super Lie algebra ${}^0\mathscr{G}(\Lambda_\infty)$ (complex or real), we associate formally with every element $\mathbf{a}$,

$$\mathbf{a} = \xi^I e_I = \xi^i e_i + \xi^\alpha e_\alpha \qquad \begin{cases} \xi^i \in \mathbb{C}_c & (\text{or } \mathbb{R}_c) \\ \xi^\alpha \in \mathbb{C}_a & (\text{or } \mathbb{R}_a) \end{cases}$$

the symbol

$$\mathbf{g}(\xi^I) = \exp \mathbf{a}. \tag{2.1.63}$$

In the set of all symbols, denoted by G, we introduce a multiplication operation using the Baker–Hausdorff formula. The group obtained is said to be a 'super Lie group' (or 'supergroup'), since its elements are parametrized by supernumbers. Super lie groups provide us with an example of 'supermanifolds' which present a generalization of ordinary manifolds. A supermanifold is a space which looks locally like a domain in $\mathbb{R}^{p|q}$. We are not able to give here a detailed treatment of supermanifolds and super Lie groups, but refer the reader to the book by B. De Witt. We content ourselves with giving some examples of super Lie groups.

The first example is the super Lie group $GL^{(+)}(p, q|\Lambda_\infty)$ of non-singular c-type supermatrices in $Mat^{(+)}(p, q|\Lambda_\infty)$ which was introduced in Section 1.10. Every supermatrix $P \in GL^{(+)}(p, q|\Lambda_\infty)$ from a small neighbourhood of the unit supermatrix can be represented in the form

$$P = \exp F \qquad F \in {}^0gl^{(+)}(p, q|\Lambda_\infty).$$

It is clear then that the supermatrix multiplication in $GL^{(+)}(p, q|\Lambda_\infty)$ coincides with the definition given above of multiplication in super Lie groups. ${}^0gl^{(+)}(p, q|\Lambda_\infty)$ is the super Lie algebra of the super Lie group $GL(p, q|\Lambda_\infty)$.

A super Lie group which corresponds to the real super Lie algebra ${}^0gl^{(+)}_{\mathbb{R}}(p, q|\Lambda_\infty)$ is the set $GL^{(+)}_{\mathbb{R}}(p, q|\Lambda_\infty)$ of non-singular real c-type supermatrices in $Mat^{(+)}(p, q|\Lambda_\infty)$. Recall that this super Lie group has been realized in subsection 1.10.3 as the group of invertible linear transformations in the superspace $\mathbb{R}^{p|q}$.

A super Lie group which corresponds to the real super Lie algebra ${}^0sl^{(+)}_{\mathbb{R}}(p, q|\Lambda_\infty)$ is the set $SL^{(+)}_{\mathbb{R}}(p, q|\Lambda_\infty)$ of unimodular real c-type supermatrices in $Mat(p, q|\Lambda_\infty)$. Indeed, every supermatrix P expressed in the form

$$P = \exp F \qquad F \in {}^0sl^{(+)}_{\mathbb{R}}(p, q|\Lambda_\infty)$$

has unit Berezinian,

$$\mathrm{Ber}\, P = 1$$

in accordance with equation (1.9.68). The super Lie group $SL^{(+)}_{\mathbb{R}}(p, q|\Lambda_\infty)$

can be realized as the group of unimodular linear transformations of the form (1.10.47) acting in real superspace $\mathbb{R}^{p|q}$.

A super Lie group, corresponding to the real super Lie algebra ${}^0osp^{(+)}_{\mathbb{R}}(p, 2q|\Lambda_\infty)$ is the set $OSP^{(+)}_{\mathbb{R}}(p, 2q|\Lambda_\infty)$ of unimodular real c-type supermatrices in $Mat^{(+)}(p, 2q|\Lambda_\infty)$ which satisfy the equation

$$F^{sT}\eta F = \eta \tag{2.1.64}$$

where the supermatrix η was defined in (2.1.42). The super Lie group $OSP^{(+)}_{\mathbb{R}}(p, 2q|\Lambda_\infty)$ has a deep geometrical origin. To see this, let us introduce in the real superspace $\mathbb{R}^{p|2q}$ with coordinates $z^M = (x^m, \theta^\mu)$, where $m = , 2, \ldots, p$ and $\mu = 1, 2, \ldots, 2q$, the non-singular supermetric

$$s^2 = x^m\delta_{mn}x^n + \mathrm{i}\theta^\mu J_{\mu\nu}\theta^\nu = z^{\mathrm{T}}\eta z. \tag{2.1.65}$$

Now consider the set of all linear transformations on $\mathbb{R}^{p|2q}$ of the form $z \to z' = Fz$ preserving the supermetric (2.1.65). Then every such transformation proves to be given by a supermatrix F under equation (2.1.64).

Finally, a super Lie group corresponding to the super Lie algebra ${}^0\mathrm{SVF}_{\mathbb{R}}(p, q)$ of real c-type supervector fields on $\mathbb{R}^{p|q}$, is the supergroup of general coordinate transformations on $\mathbb{R}^{p|q}$, which has been studied in Section 1.11.

To summarize, super Lie algebras and super Lie groups play the same role on superspaces $\mathbb{R}^{p|q}$ which Lie algebras and Lie groups play on ordinary spaces $\mathbb{R}^n$.

2.1.7. *Unitary representations of real superalgebras*

Let $\mathscr{G}$ be a real superalgebra and $\mathscr{H}$ be a Hilbert space. A linear mapping $T\colon \mathscr{G} \to End\,\mathscr{H}$ from $\mathscr{G}$ into the algebra $End\,\mathscr{H}$ of linear operators on $\mathscr{H}$ is said to be a 'unitary representation of the superalgebra' $\mathscr{G}$ in $\mathscr{H}$ under the following conditions:

$$\left.\begin{array}{l} 1.\ (T(a))^+ = -(\mathrm{i})^{\kappa(a)}T(a) \\ 2.\ T([a, b\}) = [T(a), T(b)\} \end{array}\right\} \tag{2.1.66}$$

for any pure elements $a, b \in \mathscr{G}$.

The first requirement seems rather unusual but it is necessary to reconcile the main property of Hermitian conjugation $(AB)^+ = B^+A^+$ with the superalgebra multiplication. Let us comment further. First, the demand that $(T({}^0a))^+ = -T({}^0a)$, $\forall {}^0a \in {}^0\mathscr{G}$, is standard. Then every operator $\exp(T({}^0a))$ is unitary and the union of all such operators forms a unitary representation of a Lie group related to the Lie algebra ${}^0\mathscr{G}$. Further, for all odd elements ${}^1a, {}^1b \in {}^1\mathscr{G}$, we have

$$\{T({}^1a), T({}^1b)\} = T({}^0a)$$

for some ${}^0a \in {}^0\mathscr{G}$, hence the operator on the right is antiHermitian. So, the demand $(T({}^1a))^+ = -\mathrm{i}T({}^1a)$ is quite correct.

Let $\{e_I\}$ be a pure basis for the superalgebra,

$$\begin{gathered} a = x^I e_I = x^i e_i + x^\alpha e_\alpha \qquad x^I \in \mathbb{R} \qquad \forall a \in \mathscr{G} \\ [e_i, e_j] = f_{ij}{}^k e_k \qquad [e_i, e_\alpha] = f_{i\alpha}{}^\beta e_\beta \qquad \{e_\alpha, e_\beta\} = f_{\alpha\beta}{}^i e_i \end{gathered} \tag{2.1.67}$$

where the structure constants are real. In accordance with (2.1.66), the representation generators $T(e_I)$ are not Hermitian. It is useful to change the real basis $\{e_I\}$ to another basis $\{\hat{e}_I\}$ defined as follows

$$\begin{gathered} \hat{e}_i = \frac{1}{\mathrm{i}} e_i \qquad \hat{e}_\alpha = \frac{1}{\sqrt{\mathrm{i}}} e_\alpha \\ [\hat{e}_i, \hat{e}_j] = -\mathrm{i} f_{ij}{}^k \hat{e}_k \qquad [\hat{e}_i, \hat{e}_\alpha] = -\mathrm{i} f_{i\alpha}{}^\beta \hat{e}_\beta \\ \{\hat{e}_\alpha, \hat{e}_\beta\} = f^i{}_{\alpha\beta} \hat{e}_i. \end{gathered} \tag{2.1.68}$$

Then the representation generators $T(\hat{e}_I)$ are Hermitian. In the parametrization (2.1.68), elements of the superalgebra are

$$a = \mathrm{i}x^i \hat{e}_i + \sqrt{\mathrm{i}}x^\alpha \hat{e}_\alpha \qquad x^I \in \mathbb{R}, \quad \forall a \in \mathscr{G}. \tag{2.1.69}$$

2.2. The Poincaré superalgebra

2.2.1. Uniqueness of the N = 1 Poincaré superalgebra

Now all the facilities which we require to introduce a basic object of our book — the Poincaré superalgebra, which presents an extension of the Poincaré algebra $\mathscr{P}$ — are at our disposal. The main motivation for the appearance of this object in theoretical physics was a very old theorists' dream to find a non-trivial extension (other than the direct sum) of the Poincaré algebra, i.e. of the Lie algebra of the space–time symmetry group of any relativistic quantum field theory. The problem has proved to have no affirmative solution in the class of Lie algebras. The well-known theorem of S. Coleman and J. Mandula states that, due to assumptions of the S-matrix approach, the most general Lie group of symmetries in a quantum field theory (whose spectrum contains massive particles) is a direct product of the Poincaré group and some internal group,

$$\Pi \times G \qquad G = G_1 \times [U(1)]^n \tag{2.2.1}$$

where G_1 is a semi-simple compact group. The corresponding Lie algebra has the direct sum structure

$$\mathscr{P} \oplus \mathscr{G} \qquad \mathscr{G} = \mathscr{G}_1 \oplus \mathscr{G}_2 \tag{2.2.2}$$

where $\mathscr{G}_1$ is the semi-simple Lie algebra of G_1 and $\mathscr{G}_2$ is an Abelian algebra.

Since elements of $\mathscr{G}$ commute with elements of $\mathscr{P}$, the only generators with Lorentz indices present in the symmetry Lie algebra are the Poincaré generators $\{\boldsymbol{p}_a, \mathbf{j}_{ab}\}$ of space–time translations and Lorentz rotations. After suffering a setback in the class of Lie algebras, it was natural to look for success in the class of superalgebras (in fact, this is the way superalgebras were created). Fortunately, it turned out that it was possible to extend the Poincaré algebra by several sets of fermionic generators with spinor indices thus obtaining superalgebras.

Let us show that there exists a unique superalgebra extension of the Poincaré algebra by four a-type generators $(\mathbf{q}_\alpha, \bar{\mathbf{q}}^{\dot\alpha})$, with $\mathbf{q}_\alpha$ carrying an undotted spinor index and $\bar{\mathbf{q}}^{\dot\alpha}$ a dotted spinor index. Recall, in accordance with the results of subsection 2.1.1, a-type elements of a superalgebra $\mathscr{G}$ transform in some representation of the Lie algebra ${}^0\mathscr{G}$. Since we have chosen the representations $(\frac{1}{2}, 0)$ and $(0, \frac{1}{2})$ to act on the subspaces generated by $\mathbf{q}_\alpha$ and $\bar{\mathbf{q}}^{\dot\alpha}$, respectively, these generators must commute with the Poincaré generators as follows

$$\begin{aligned} [\mathbf{j}_{ab}, \mathbf{q}_\alpha] &= \mathrm{i}(\sigma_{ab})_\alpha{}^\beta \mathbf{q}_\beta \qquad & [\boldsymbol{p}_a, \mathbf{q}_\alpha] &= 0 \\ [\mathbf{j}_{ab}, \bar{\mathbf{q}}^{\dot\alpha}] &= \mathrm{i}(\tilde\sigma_{ab})^{\dot\alpha}{}_{\dot\beta} \bar{\mathbf{q}}^{\dot\beta} \qquad & [\boldsymbol{p}_a, \bar{\mathbf{q}}^{\dot\alpha}] &= 0 \end{aligned} \tag{2.2.3}$$

where we have used the fact that space–time translations do not act on spinor or vector indices.

Now we are going to analyse anticommutation relations in the assumed superalgebra. From equation (2.1.11c), we can write

$$\begin{gathered} \{\mathbf{q}_\alpha, \mathbf{q}_\beta\} = f_{\alpha\beta}{}^a \boldsymbol{p}_a + \frac{1}{2} f_{\alpha\beta}{}^{ab} \mathbf{j}_{ab} \\ f_{\alpha\beta}{}^a = f_{\beta\alpha}{}^a \qquad f_{\alpha\beta}{}^{ab} = f_{\beta\alpha}{}^{ab} = -f_{\alpha\beta}{}^{ba} \end{gathered} \tag{2.2.4a}$$

and

$$\begin{gathered} \{\bar{\mathbf{q}}_{\dot\alpha}, \bar{\mathbf{q}}_{\dot\beta}\} = f_{\dot\alpha\dot\beta}{}^a \boldsymbol{p}_a + \frac{1}{2} f_{\dot\alpha\dot\beta}{}^{ab} \mathbf{j}_{ab} \\ f_{\dot\alpha\dot\beta}{}^a = f_{\dot\beta\dot\alpha}{}^a \qquad f_{\dot\alpha\dot\beta}{}^{ab} = f_{\dot\beta\dot\alpha}{}^{ab} = -f_{\dot\alpha\dot\beta}{}^{ba} \end{gathered} \tag{2.2.4b}$$

and

$$\begin{gathered} \{\mathbf{q}_\alpha, \bar{\mathbf{q}}_{\dot\alpha}\} = f_{\alpha\dot\alpha}{}^a \boldsymbol{p}_a + \frac{1}{2} f_{\alpha\dot\alpha}{}^{ab} \mathbf{j}_{ab} \\ f_{\alpha\dot\alpha}{}^{ab} = -f_{\alpha\dot\alpha}{}^{ba}. \end{gathered} \tag{2.2.4c}$$

Recall that spinor indices are raised and lowered with the help of the spinor metric $\varepsilon_{\alpha\beta}$ and $\varepsilon_{\dot\alpha\dot\beta}$ (see equations (1.2.11, 17)). In accordance with the results of subsection 2.1.1, the set of structure constants $\{f_{\alpha\beta}{}^a, f_{\alpha\beta}{}^{ab}, f_{\dot\alpha\dot\beta}{}^a, f_{\dot\alpha\dot\beta}{}^{ab}, f_{\alpha\dot\alpha}{}^a, f_{\alpha\dot\alpha}{}^{ab}\}$ should form an invariant tensor of the Poincaré algebra, i.e. it must satisfy equation (2.1.18). The indices i, j and l from (2.1.18) are now Poincaré indices, and the indices α, β and γ from (2.1.18) are now spinor indices. The non-vanishing structure constants of the Poincaré algebra are $f_{a,cd}{}^{b} = -f_{cd,a}{}^{b}$ and $f_{ab,cd}{}^{fl} = -f_{cd,ab}{}^{fl}$, and their explicit values can be readily found from (1.5.5). Then, choosing in equation (2.1.18) $i = a$ and $l = b$ and using equation (2.2.3), one obtains

$$f_{\alpha\beta}{}^{ab} = f_{\dot{\alpha}\dot{\beta}}{}^{ab} = f_{\alpha\dot{\alpha}}{}^{ab} = 0.$$

The relations (2.1.18) tell us that the other structure constants $f_{\alpha\beta}{}^{a}$, $f_{\dot{\alpha}\dot{\beta}}{}^{a}$ and $f_{\alpha\dot{\alpha}}{}^{a}$ are invariant tensors of the Lorentz algebra. But the Lorentz group has no invariant tensors like $f_{\alpha\beta}{}^{a}$ or $f_{\dot{\alpha}\dot{\beta}}{}^{a}$. Further, the only candidate for the role of $f_{\alpha\dot{\alpha}}{}^{a}$ is the invariant tensor $(\sigma^{a})_{\alpha\dot{\alpha}}$. As a result, the anticommutation relations (2.2.4) are simplified drastically:

$$\begin{gathered}\{\mathbf{q}_{\alpha}, \mathbf{q}_{\beta}\} = 0 \qquad \{\bar{\mathbf{q}}_{\dot{\alpha}}, \bar{\mathbf{q}}_{\dot{\beta}}\} = 0 \\ \{\mathbf{q}_{\alpha}, \bar{\mathbf{q}}_{\dot{\alpha}}\} = 2k(\sigma^{a})_{\alpha\dot{\alpha}}\boldsymbol{p}_{a}\end{gathered} \tag{2.2.5}$$

with k being some constant. It is a simple exercise to check that the second equation (2.1.19) for the structure constants is satisfied now identically. So, we have obtained a superalgebra.

Finally, it would be desirable to demand that in any unitary representation T of the obtained superalgebra (from the physical point of view, such representations are of the greatest importance) the generators $\mathbb{Q}_{\alpha} = T(\mathbf{q}_{\alpha})$ and $\bar{\mathbb{Q}}_{\dot{\alpha}} = T(\bar{\mathbf{q}}_{\dot{\alpha}})$ were Hermitian conjugate to each other,

$$\bar{\mathbb{Q}}_{\dot{\alpha}} = (\mathbb{Q}_{\alpha})^{+}. \tag{2.2.6}$$

In other words, we wish to treat the pair $(\mathbf{q}_{\alpha}, \bar{\mathbf{q}}^{\dot{\alpha}})$ as a Majorana spinor. Then, the constant k in (2.2.5) should be real and positive. Indeed, equation (2.2.5) leads to

$$-k\mathbb{P}_{a} = \frac{1}{4}(\tilde{\sigma}_{a})^{\dot{\beta}\alpha}\{\mathbb{Q}_{\alpha}, \bar{\mathbb{Q}}_{\dot{\beta}}\} = \frac{1}{4}(\tilde{\sigma}_{a})^{\dot{\beta}\alpha}\{\mathbb{Q}_{\alpha}, (\mathbb{Q}_{\beta})^{+}\} \tag{2.2.7}$$

where $\mathbb{P}_{a} = T(\boldsymbol{p}_{a}) = (-\mathbb{E}, \vec{\mathbb{P}})$ is the (Hermitian) energy-momentum operator, hence k is real. Choosing here $a = 0$ gives

$$k\mathbb{E} = \frac{1}{4}\mathbb{Q}_{1}(\mathbb{Q}_{1})^{+} + \frac{1}{4}\mathbb{Q}_{2}(\mathbb{Q}_{2})^{+} + \frac{1}{4}(\mathbb{Q}_{1})^{+}\mathbb{Q}_{1} + \frac{1}{4}(\mathbb{Q}_{2})^{+}\mathbb{Q}_{2}. \tag{2.2.8}$$

Since physically acceptable unitary Poincaré representations are characterized by condition (1.5.16) (positivity of energy) and due to positive definiteness of the operator in the right-hand side of equation (2.2.8), we must choose $k > 0$. Hence one can set $k = 1$ by making a simple rescaling of $\mathbf{q}_{\alpha}$ and $\bar{\mathbf{q}}_{\dot{\alpha}}$.

Let us now write down the complete list of (anti)commutation relations of the superalgebra:

$$
\begin{gathered}
[\mathbf{p}_a, \mathbf{p}_b] = 0 \qquad [\mathbf{j}_{ab}, \mathbf{p}_c] = \mathrm{i}\eta_{ac}\mathbf{p}_b - \mathrm{i}\eta_{bc}\mathbf{p}_a \\
[\mathbf{j}_{ab}, \mathbf{j}_{cd}] = \mathrm{i}\eta_{ac}\mathbf{j}_{bd} - \mathrm{i}\eta_{ad}\mathbf{j}_{bc} + \mathrm{i}\eta_{bd}\mathbf{j}_{ac} - \mathrm{i}\eta_{bc}\mathbf{j}_{ad} \\
[\mathbf{j}_{ab}, \mathbf{q}_\alpha] = \mathrm{i}(\sigma_{ab})_\alpha{}^\beta \mathbf{q}_\beta \qquad [\mathbf{p}_a, \mathbf{q}_\alpha] = 0 \\
[\mathbf{j}_{ab}, \bar{\mathbf{q}}^{\dot\alpha}] = \mathrm{i}(\tilde\sigma_{ab})^{\dot\alpha}{}_{\dot\beta}\bar{\mathbf{q}}^{\dot\beta} \qquad [\mathbf{p}_a, \bar{\mathbf{q}}^{\dot\alpha}] = 0 \\
\{\mathbf{q}_\alpha, \mathbf{q}_\beta\} = 0 \qquad \{\bar{\mathbf{q}}_{\dot\alpha}, \bar{\mathbf{q}}_{\dot\beta}\} = 0 \\
\{\mathbf{q}_\alpha, \bar{\mathbf{q}}_{\dot\alpha}\} = 2(\sigma^a)_{\alpha\dot\alpha}\mathbf{p}_a
\end{gathered}
\tag{2.2.9}
$$

This real superalgebra is known as the 'Poincaré superalgebra'. It will be denoted by $S\mathscr{P}$. Its a-type generators $\mathbf{q}_\alpha$ and $\bar{\mathbf{q}}_{\dot\alpha}$ are called 'supersymmetry generators'. Every element X of the Poincaré superalgebra can be represented, in agreement with subsection 2.1.7, as follows:

$$
\begin{gathered}
X = \mathrm{i}\left(-b^a\mathbf{p}_a + \frac{1}{2}K^{ab}\mathbf{j}_{ab}\right) + \sqrt{\mathrm{i}}(\kappa^\alpha\mathbf{q}_\alpha + \bar\kappa^{\dot\alpha}\bar{\mathbf{q}}_{\dot\alpha}) \\
b^a, K^{ab} = -K^{ba} \in \mathbb{R} \qquad \kappa^\alpha, \bar\kappa^{\dot\alpha} = (\kappa^\alpha)^* \in \mathbb{C}.
\end{gathered}
\tag{2.2.10}
$$

For later use, we rewrite the Poincaré superalgebra in spinor notation, converting every vector index into a pair of dotted and undotted indices. When applied to the Poincaré generators, this operation leads to the change $\{\mathbf{p}_a, \mathbf{j}_{ab}\} \to \{\mathbf{p}_{\alpha\dot\alpha}, \mathbf{j}_{\alpha\beta}, \bar{\mathbf{j}}_{\dot\alpha\dot\beta}\}$. Then one finds

$$
\begin{gathered}
[\mathbf{j}_{\alpha\beta}, \mathbf{p}_{\gamma\dot\gamma}] = \frac{\mathrm{i}}{2}\varepsilon_{\gamma\alpha}\mathbf{p}_{\beta\dot\gamma} + \frac{\mathrm{i}}{2}\varepsilon_{\gamma\beta}\mathbf{p}_{\alpha\dot\gamma} \\
[\mathbf{j}_{\alpha\beta}, \mathbf{j}_{\gamma\delta}] = \frac{\mathrm{i}}{2}(\varepsilon_{\gamma\alpha}\mathbf{j}_{\beta\delta} + \varepsilon_{\gamma\beta}\mathbf{j}_{\alpha\delta} + \varepsilon_{\delta\alpha}\mathbf{j}_{\gamma\beta} + \varepsilon_{\delta\beta}\mathbf{j}_{\gamma\alpha}) \\
[\mathbf{j}_{\alpha\beta}, \mathbf{q}_\gamma] = \frac{\mathrm{i}}{2}\varepsilon_{\gamma\alpha}\mathbf{q}_\beta + \frac{\mathrm{i}}{2}\varepsilon_{\gamma\beta}\mathbf{q}_\alpha \\
\{\mathbf{q}_\alpha, \bar{\mathbf{q}}_{\dot\alpha}\} = 2\mathbf{p}_{\alpha\dot\alpha}.
\end{gathered}
\tag{2.2.11}
$$

Other (anti)commutators vanish or may be found by Hermitian conjugation.

2.2.2. *Extended Poincaré superalgebras*

It was shown above that the Poincaré superalgebra is the only possible superalgebra such that its even part coincides with $\mathscr{P}$ and the odd part transforms in the real (Majorana) representation $(\frac{1}{2}, 0) \oplus (0, \frac{1}{2})$ of the Lorentz group. However, one can consider a more general problem: to find all possible superalgebras $\mathscr{M}$ with ${}^0\mathscr{M}$ being of the form (2.2.2) and ${}^1\mathscr{M}$ being generated by a set of elements, each of which carries at least one spinor index. Then, since $\{{}^1\mathscr{M}, {}^1\mathscr{M}\} \subset {}^0\mathscr{M}$, it follows from the Coleman–Mandula theorem that every generator of ${}^1\mathscr{M}$ should carry exactly one spinor index (otherwise, ${}^0\mathscr{M}$

will contain a Lorentz tensor generator different from the Poincaré generators).

Of course, we wish to have a real superalgebra $\mathcal{M}$. Therefore, ${}^1\mathcal{M}$ is a direct sum of several real $(\frac{1}{2}, 0) \oplus (0, \frac{1}{2})$ representations, i.e. ${}^1\mathcal{M}$ is generated by elements $\mathbf{q}_\alpha{}^A$ and $\bar{\mathbf{q}}_{\dot{\alpha}B}$, where $A, B = 1, 2, \ldots, N$, with $\bar{\mathbf{q}}_{\dot{\alpha}A}$ being Hermitian conjugate of $\mathbf{q}_\alpha{}^A$. Further, one must satisfy equations (2.1.18) and (2.1.19). These equations give very strong restrictions on the structure constants. An analysis similar to that of subsection 2.2.1 leads to the following (anti)commutation relations:

$$\begin{gathered}
[\mathbf{j}_{ab}, \mathbf{q}_\alpha{}^A] = \mathrm{i}(\sigma_{ab})_\alpha{}^\beta \mathbf{q}_\beta{}^A \qquad [\boldsymbol{p}_a, \mathbf{q}_\alpha{}^A] = 0 \\
[\mathbf{j}_{ab}, \bar{\mathbf{q}}^{\dot{\alpha}}{}_A] = \mathrm{i}(\tilde{\sigma}_{ab})^{\dot{\alpha}}{}_{\dot{\beta}} \bar{\mathbf{q}}^{\dot{\beta}}{}_A \qquad [\boldsymbol{p}_a, \bar{\mathbf{q}}^{\dot{\alpha}}{}_A] = 0 \\
\{\mathbf{q}_\alpha{}^A, \mathbf{q}_\beta{}^B\} = \varepsilon_{\alpha\beta}\mathbf{c}^{AB} \qquad \{\bar{\mathbf{q}}_{\dot{\alpha}A}, \bar{\mathbf{q}}_{\dot{\beta}B}\} = \varepsilon_{\dot{\alpha}\dot{\beta}}\bar{\mathbf{c}}_{AB} \\
\{\mathbf{q}_\alpha{}^A, \bar{\mathbf{q}}_{\dot{\alpha}B}\} = 2(\sigma^a)_{\alpha\dot{\alpha}}\delta^A_B \boldsymbol{p}_a \\
[\mathbf{c}^{AB}, \mathbf{q}_\alpha{}^D] = [\mathbf{c}^{AB}, \bar{\mathbf{q}}_{\dot{\alpha}D}] = 0 \\
[\bar{\mathbf{c}}_{AB}, \mathbf{q}_\alpha{}^D] = [\bar{\mathbf{c}}_{AB}, \bar{\mathbf{q}}_{\dot{\alpha}D}] = 0 \\
[\mathbf{t}_i, \mathbf{q}_\alpha{}^A] = -(S_i)^A{}_B \mathbf{q}_\alpha{}^B \\
[\mathbf{t}_i, \bar{\mathbf{q}}_{\dot{\alpha}A}] = (S_i)^B{}_A \bar{\mathbf{q}}_{\dot{\alpha}B} \\
[\mathbf{t}_i, \mathbf{t}_j] = \mathrm{i} f_{ij}{}^k \mathbf{t}_k
\end{gathered} \tag{2.2.12}$$

together with the Poincaré commutation relations. Here $\{\mathbf{t}_i\}$ are generators of the semi-simple Lie algebra $\mathcal{G}_1$, $\{\mathbf{c}^{AB} = -\mathbf{c}^{BA}, \bar{\mathbf{c}}_{AB} = -\bar{\mathbf{c}}_{BA}\}$ are generators of the Abelian algebra $\mathcal{G}_2$, see (2.2.2), with $\bar{\mathbf{c}}_{AB}$ the Hermitian conjugate of $\mathbf{c}^{AB}$. The matrices $(S_i)^A{}_B$ are Hermitian, $S_i^+ = S_i$, and they form a representation of the algebra $\mathcal{G}_1$, $[S_i, S_j] = \mathrm{i} f_{ij}{}^k S_k$. Finally, the generators $\mathbf{c}^{AB}$ and $\bar{\mathbf{c}}_{AB}$ of $\mathcal{G}_2$ should be 'invariant tensors' of $\mathcal{G}_1$, i.e. one has

$$\begin{gathered}
(S_i)^A{}_D \mathbf{c}^{DB} + (S_i)^B{}_D \mathbf{c}^{AD} = 0 \\
\bar{\mathbf{c}}_{DB}(S_i)^D{}_A + \bar{\mathbf{c}}_{AD}(S_i)^D{}_B = 0.
\end{gathered} \tag{2.2.13}$$

As may be seen, the generators $\mathbf{c}^{AB}$ and $\bar{\mathbf{c}}_{AB}$ commute with every element of the superalgebra. On these grounds, they are called 'central charges'. Note that central charges may exist when $N \geqslant 2$. For a more detailed derivation of equations (2.2.12) and (2.2.13), see the book by J. Wess and J. Bagger.

The superalgebra (2.2.12) is known as an 'N-extended Poincaré superalgebra', depending on the number of spinor generators. The Poincaré superalgebra (2.2.9) is called the $N = 1$ (or 'simple') Poincaré superalgebra. It follows from the above consideration that the superalgebra (2.2.12) is the most general (finite-dimensional) extension of the Poincaré algebra, consistent with axioms of quantum field theory. This assertion is known as the Haag, Lopuszanski and Sohnius theorem.

From now on, we shall study the $N = 1$ Poincaré superalgebra only; this case, being simple, is worked out in detail and contains the main ingredients of all supersymmetric theories.

2.2.3. *Matrix realization of the Poincaré superalgebra*

The Poincaré superalgebra has been introduced above as an abstract superalgebra. Now we give its realization in terms of matrices. Let us consider in $Mat(4, 1 \mid \mathbb{C})$ matrices $\{\rho_a, \mathbf{j}_{ab}, \mathbf{q}_\alpha, \bar{\mathbf{q}}_{\dot{\alpha}}\}$ defined as follows

$$\rho_a = \left(\begin{array}{c|c} -\frac{1}{8}(\mathbb{1}_4 + \gamma_5)\gamma_a & \begin{matrix}0\\0\\0\\0\end{matrix} \\ \hline \begin{matrix}0 & 0 & 0 & 0\end{matrix} & 0 \end{array}\right) = \left(\begin{array}{cc|c} \mathbf{0}_2 & -\frac{1}{4}\sigma_a & \begin{matrix}0\\0\end{matrix} \\ \mathbf{0}_2 & \mathbf{0}_2 & \begin{matrix}0\\0\end{matrix} \\ \hline \begin{matrix}0 & 0\end{matrix} & \begin{matrix}0 & 0\end{matrix} & 0 \end{array}\right)$$

$$\mathbf{j}_{ab} = \left(\begin{array}{c|c} -\mathrm{i}\Sigma_{ab} & \begin{matrix}0\\0\\0\\0\end{matrix} \\ \hline \begin{matrix}0 & 0 & 0 & 0\end{matrix} & 0 \end{array}\right) = \left(\begin{array}{cc|c} -\mathrm{i}\sigma_{ab} & \mathbf{0}_2 & \begin{matrix}0\\0\end{matrix} \\ \mathbf{0}_2 & -\mathrm{i}\tilde{\sigma}_{ab} & \begin{matrix}0\\0\end{matrix} \\ \hline \begin{matrix}0 & 0\end{matrix} & \begin{matrix}0 & 0\end{matrix} & 0 \end{array}\right) \tag{2.2.14}$$

$$\mathbf{q}_1 = \left(\begin{array}{c|c} \mathbf{0}_4 & \begin{matrix}0\\-1\\0\\0\end{matrix} \\ \hline \begin{matrix}0 & 0 & 0 & 0\end{matrix} & 0 \end{array}\right) \qquad \mathbf{q}_2 = \left(\begin{array}{c|c} \mathbf{0}_4 & \begin{matrix}1\\0\\0\\0\end{matrix} \\ \hline \begin{matrix}0 & 0 & 0 & 0\end{matrix} & 0 \end{array}\right)$$

$$\bar{\mathbf{q}}_{\dot{1}} = \left(\begin{array}{c|c} \mathbf{0}_4 & \begin{matrix}0\\0\\0\\0\end{matrix} \\ \hline \begin{matrix}0 & 0 & -1 & 0\end{matrix} & 0 \end{array}\right) \qquad \bar{\mathbf{q}}_{\dot{2}} = \left(\begin{array}{c|c} \mathbf{0}_4 & \begin{matrix}0\\0\\0\\0\end{matrix} \\ \hline \begin{matrix}0 & 0 & 0 & 1\end{matrix} & 0 \end{array}\right)$$

where $\mathbf{0}_n$ means the zero $n \times n$ matrix. The matrices γ_a, Σ_{ab} and γ_5 were introduced in Section 1.4. One can readily check that the matrices (2.2.14) satisfy the (anti)commutation relations (2.2.9). Note that the superalgebra, generated by the matrices (2.2.14), is a subalgebra of $sl(4, 1|\mathbb{C})$.

2.2.4. Grassmann shell of the Poincaré superalgebra

As we know, with every complex superalgebra $\mathscr{G}(\mathbb{C})$ one can relate a Berezin superalgebra $\mathscr{G}(\Lambda_\infty)$ (the Grassmann shell of $\mathscr{G}(\mathbb{C})$) and a super Lie algebra ${}^0\mathscr{G}(\Lambda_\infty)$ (the even part of $\mathscr{G}(\Lambda_\infty)$). We now construct these objects for the complex shell $S\mathscr{P}(\mathbb{C})$ of the Poincaré superalgebra. It is a complex superalgebra of dimension $(10 + 4)$ with a general element $X \in S\mathscr{P}(\mathbb{C})$ of the form

$$X = x^a \rho_a + \frac{1}{2} x^{ab} \mathbf{j}_{ab} + x^\alpha \mathbf{q}_\alpha + x^{\dot{\alpha}} \bar{\mathbf{q}}_{\dot{\alpha}}$$
$$x^a, x^{ab} = -x^{ba}, x^\alpha, x^{\dot{\alpha}} \in \mathbb{C}. \tag{2.2.15}$$

Recall that elements of the Poincaré superalgebra have the form (2.2.10). Now, introduce the Grassmann shell $S\mathscr{P}(\Lambda_\infty)$ of $S\mathscr{P}(\mathbb{C})$. It is a Berezin superalgebra of dimension $(10, 4)$ with a pure basis $\{\rho_a, \mathbf{j}_{ab}, \mathbf{q}_\alpha, \bar{\mathbf{q}}_{\dot{\alpha}}\}$ such that

$$z\rho_a = \rho_a z \qquad z\mathbf{j}_{ab} = \mathbf{j}_{ab} z$$
$$z\mathbf{q}_\alpha = (-1)^{\varepsilon(z)} \mathbf{q}_\alpha z \qquad z\bar{\mathbf{q}}_{\dot{\alpha}} = (-1)^{\varepsilon(z)} \bar{\mathbf{q}}_{\dot{\alpha}} z \tag{2.2.16}$$

for any pure supernumber $z \in \Lambda_\infty$. Every element $\vec{X} \in S\mathscr{P}(\Lambda_\infty)$ can be represented as follows

$$\vec{X} = \xi^a \rho_a + \frac{1}{2} \xi^{ab} \mathbf{j}_{ab} + \xi^\alpha \mathbf{q}_\alpha + \xi^{\dot{\alpha}} \bar{\mathbf{q}}_{\dot{\alpha}}$$
$$\xi^a, \xi^{ab} = -\xi^{ba}, \xi^\alpha, \xi^{\dot{\alpha}} \in \Lambda_\infty. \tag{2.2.17}$$

Using the matrix realization (2.2.14) of the Poincaré superalgebra, it is not difficult to obtain a supermatrix realization of the Berezin superalgebra $S\mathscr{P}(\Lambda_\infty)$.

Let us endow the supervector space $S\mathscr{P}(\Lambda_\infty)$ with an operation of complex conjugation according to the rule

$$(\rho_a)^* \equiv -\rho_a \qquad (\mathbf{j}_{ab})^* \equiv -\mathbf{j}_{ab} \qquad (\mathbf{q}_\alpha)^* \equiv -\bar{\mathbf{q}}_{\dot{\alpha}}.$$

Then every real c-type supervector $\vec{X} \in {}^0S\mathscr{P}(\Lambda_\infty)$ is of the form

$$\vec{X} = \mathrm{i}\left(-b^a \rho_a + \frac{1}{2} K^{ab} \mathbf{j}_{ab} + \epsilon^\alpha \mathbf{q}_\alpha + \bar{\epsilon}_{\dot{\alpha}} \bar{\mathbf{q}}^{\dot{\alpha}} \right) \qquad \bar{\epsilon}_{\dot{\alpha}} \equiv (\epsilon_\alpha)^*$$
$$b^a, K^{ab} = -K^{ba} \in \mathbb{R}_c \qquad \epsilon^\alpha \in \mathbb{C}_a. \tag{2.2.18}$$

Due to the (anti)commutation relations (2.2.9), one easily finds that the subset ${}^0S\mathscr{P}(\Lambda_\infty)$ of real c-type supervectors forms a subalgebra of the super Lie algebra ${}^0S\mathscr{P}(\Lambda_\infty)$. So ${}^0S\mathscr{P}_{\mathbb{R}}(\Lambda_\infty)$ is a real super Lie algebra. It is called the '($N = 1$) super Poincaré algebra'.

Recalling the spinor notation (1.4.3), the last two terms in (2.2.18) can be rewritten as follows:

$$\epsilon^\alpha \mathbf{q}_\alpha + \bar{\epsilon}_{\dot{\alpha}} \bar{\mathbf{q}}^{\dot{\alpha}} = \epsilon\mathbf{q} + \overline{\epsilon\mathbf{q}} = \mathbf{q}\epsilon + \overline{\mathbf{q}\epsilon} \qquad (2.2.19)$$

where we have used equation (2.2.16).

2.2.5. *The super Poincaré group*

A super Lie group corresponding to the super Poincaré algebra is known as the ($N = 1$) 'super Poincaré group'. It is denoted by $S\Pi$. Every element of $S\Pi$ is of the form

$$\begin{aligned} g(b, \epsilon, \bar{\epsilon}, K) &= \exp\left[\mathrm{i}\left(-b^a p_a + \frac{1}{2}K^{ab}\mathbf{j}_{ab} + \epsilon\mathbf{q} + \overline{\epsilon\mathbf{q}}\right)\right] \\ &= \exp\left[\mathrm{i}\left(\frac{1}{2}b^{\alpha\dot{\alpha}}p_{\alpha\dot{\alpha}} + K^{\alpha\beta}\mathbf{j}_{\alpha\beta} + \bar{K}^{\dot{\alpha}\dot{\beta}}\mathbf{j}_{\dot{\alpha}\dot{\beta}} + \epsilon\mathbf{q} + \overline{\epsilon\mathbf{q}}\right)\right]. \end{aligned} \qquad (2.2.20)$$

So, points of the super Poincaré group are parametrized by real c-number variables b^a and $K^{ab} = -K^{ba}$ as well as by a-number variables $(\epsilon_\alpha, \bar{\epsilon}^{\dot{\alpha}})$ forming a Majorana spinor. Similarly to the Poincaré group, supergroup elements

$$g(b) = \exp(-\mathrm{i}b^a p_a) \qquad b^a \in \mathbb{R}_c \qquad (2.2.21)$$

will be called 'space–time translations' and elements

$$g(K) = \exp\left(\frac{1}{2}K^{ab}\mathbf{j}_{ab}\right) \qquad K^{ab} \in \mathbb{R}_c \qquad (2.2.22)$$

will be called 'Lorentz transformations'. Elements

$$g(\epsilon, \bar{\epsilon}) = \exp[\mathrm{i}(\epsilon\mathbf{q} + \overline{\epsilon\mathbf{q}})] \qquad \epsilon^\alpha \in \mathbb{C}_a \qquad (2.2.23)$$

are said to be 'supersymmetry transformations'. The union of all elements (2.2.22), denoted by $SO(3, 1|\mathbb{R}_c)^\uparrow$, forms a super Lie group, which represents a c-number shell of the Lorentz group $SO(3, 1)^\uparrow$. Ordinary translations and ordinary Lorentz transformations correspond to soulless parameters in (2.2.21) and (2.2.22). We will refer to $SO(3, 1|\mathbb{R}_c)^\uparrow$ as the Lorentz group over $\mathbb{R}_c$.

It should be noticed that, since $[\mathbf{j}, p] \sim p$ and $[\mathbf{j}, \mathbf{q}] \sim \mathbf{q}$, the set of elements (2.2.20) with $b^a \in \mathbb{R}_c$, $\epsilon^\alpha \in \mathbb{C}_a$, but $K^{ab} \in \mathbb{R}$, forms a subgroup of the super Poincaré group. Therefore, one may restrict the parameters K^{ab} in (2.2.18) and (2.2.20) to be ordinary real numbers. However, since $\{\mathbf{q}, \bar{\mathbf{q}}\} \sim p$,

we have

$$[\epsilon_1\mathbf{q} + \bar{\epsilon}_1\bar{\mathbf{q}}, \epsilon_2\mathbf{q} + \bar{\epsilon}_2\bar{\mathbf{q}}] = 2(\epsilon_1\sigma^a\bar{\epsilon}_2 - \epsilon_2\sigma^a\bar{\epsilon}_1)\boldsymbol{\rho}_a$$

and hence

$$\exp[\mathrm{i}(\epsilon_1\mathbf{q} + \bar{\epsilon}_1\bar{\mathbf{q}})]\exp[\mathrm{i}(\epsilon_2\mathbf{q} + \bar{\epsilon}_2\bar{\mathbf{q}})]$$
$$= \exp[\mathrm{i}\{(\epsilon_1 + \epsilon_2)\mathbf{q} + (\bar{\epsilon}_1 + \bar{\epsilon}_2)\bar{\mathbf{q}} + \mathrm{i}(\epsilon_1\sigma^a\bar{\epsilon}_2 - \epsilon_2\sigma^a\bar{\epsilon}_1)\boldsymbol{\rho}_a\}] \quad (2.2.25)$$

as a consequence of the Baker–Hausdorff formula, so the parameters b^a in (2.2.18) and (2.2.20) should be arbitrary real c-numbers.

An invariance in quantum field theory with respect to the super Poincaré group is called '$N = 1$ (or simple) supersymmetry'.

2.3. Unitary representations of the Poincaré superalgebra

We proceed by studying one-particle unitary representations of the Poincaré superalgebra. The main goal will be to classify irreducible representations T of $S\mathscr{P}$ each of which acts in a Hilbert space $\mathscr{H}$ of one-particle states. The representation generators will be denoted by

$$\mathbb{P}_a = T(\boldsymbol{\rho}_a) \qquad \mathbb{J}_{ab} = T(\mathbf{j}_{ab}) \qquad \mathbb{Q}_\alpha = T(\mathbf{q}_\alpha) \qquad \bar{\mathbb{Q}}_{\dot{\alpha}} = T(\bar{\mathbf{q}}_{\dot{\alpha}}) \quad (2.3.1)$$

where $\mathbb{P}_a$ and $\mathbb{J}_{ab}$ are the Hermitian generators of space–time translations and Lorentz transformations, and $\mathbb{Q}_\alpha$ and $\bar{\mathbb{Q}}_{\dot{\alpha}}$ are the supersymmetry generators which satisfy equation (2.2.6). The generators $\mathbb{P}_a$ and $\mathbb{J}_{ab}$ form a representation of the Poincaré algebra and the set of all unitary operators

$$\exp\left[\mathrm{i}\left(-b^a\mathbb{P}_a + \frac{1}{2}K^{ab}\mathbb{J}_{ab}\right)\right]$$

where b^a and $K^{ab} = -K^{ba}$ are real number parameters, gives a unitary representation of the Poincaré group.

2.3.1. *Positivity of energy*

As is known, the Poincaré group has irreducible unitary representations of two types: positive-energy representations, characterized by the condition (1.5.14), and negative-energy representations. Only the positive-energy representations are physically admissible. Thus the negative-energy representations, being possible only in principle, must be discarded.

Supersymmetry changes the situation radically. The Poincaré superalgebra (2.2.9) leads to equation (2.2.8) with $k = 1$, therefore every state $|\Psi\rangle$ of the Hilbert (or Fock) space is characterized by a non-negative average energy

$$E_\Psi = \langle\Psi|\mathbb{E}|\Psi\rangle$$
$$= \frac{1}{4}(\langle\mathbb{Q}_1^+\Psi|\mathbb{Q}_1^+\Psi\rangle + \langle\mathbb{Q}_2^+\Psi|\mathbb{Q}_2^+\Psi\rangle + \langle\mathbb{Q}_1\Psi|\mathbb{Q}_1\Psi\rangle + \langle\mathbb{Q}_2\Psi|\mathbb{Q}_2\Psi\rangle) \geqslant 0. \quad (2.3.2)$$

As a result, all unitary representations of the Poincaré superalgebra are positive-energy representations of the Poincaré group.

Let us suppose the Poincaré superalgebra acts in a Fock space of some field theory, and let $|\text{vac}\rangle$ be a Poincaré invariant vacuum state, $\mathbb{P}_a|\text{vac}\rangle = 0$. Then, due to equation (2.3.2), this state is annihilated by the supersymmetry generators:

$$\mathbb{Q}_\alpha|\text{vac}\rangle = \bar{\mathbb{Q}}_{\dot{\alpha}}|\text{vac}\rangle = 0. \tag{2.3.3}$$

So, the vacuum state is a supersymmetrically invariant state. Conversely, having a supersymmetric state, which satisfied equation (2.3.3), it is inevitably a Poincaré invariant vacuum state. Every non-supersymmetric state has positive energy.

2.3.2. *Casimir operators of the Poincaré superalgebra*

In order to classify irreducible unitary representations of the Poincaré superalgebra, it is worth finding its Casimir operators — that is, polynomials in the $S\mathscr{P}$ generators commuting with each element of $S\mathscr{P}$. Recall that the Poincaré algebra has two Casimir operators: the squared mass operator $C_1 = -\mathbb{P}^a\mathbb{P}_a$ and the spin operator $C_2 = \mathbb{W}^a\mathbb{W}_a$, where $\mathbb{W}_a$ is the Pauli–Lubanski vector (1.5.10). The Poincaré representations are classified by mass and spin.

Due to the supersymmetry algebra (2.2.9), the energy–momentum operator $\mathbb{P}_a$ commutes with the supersymmetry generators $\mathbb{Q}_\alpha$ and $\bar{\mathbb{Q}}_{\dot{\alpha}}$, hence the squared mass operator also commutes. So, $C_1 = -\mathbb{P}^a\mathbb{P}_a$ is the Casimir operator of the Poincaré superalgebra. However, the spin operator does not commute with the supersymmetry generators. This follows from the identity

$$[\mathbb{W}_{\alpha\dot{\alpha}}, \mathbb{Q}_\beta] = \frac{1}{2}\varepsilon_{\alpha\beta}\mathbb{Q}_\gamma\mathbb{P}^\gamma{}_{\dot{\alpha}} - \frac{1}{2}\mathbb{Q}_\alpha\mathbb{P}_{\beta\dot{\alpha}} = \frac{1}{2}\mathbb{Q}_\beta\mathbb{P}_{\alpha\dot{\alpha}} - \mathbb{Q}_\alpha\mathbb{P}_{\beta\dot{\alpha}} \tag{2.3.4}$$

where we have rewritten the Pauli–Lubanski vector and the Poincaré generators in the spinor notations ($\mathbb{W}_{\alpha\dot{\alpha}} = (\sigma^a)_{\alpha\dot{\alpha}}\mathbb{W}_a$, $\mathbb{P}_{\alpha\dot{\alpha}} = (\sigma^a)_{\alpha\dot{\alpha}}\mathbb{P}_a$ and $\mathbb{J}_{\alpha\beta} = \frac{1}{2}(\sigma^{ab})_{\alpha\beta}\mathbb{J}_{ab}$); in particular, one has

$$\begin{aligned} \mathbb{W}_{\alpha\dot{\alpha}} &= \mathrm{i}\mathbb{J}_{\alpha\beta}\mathbb{P}^\beta{}_{\dot{\alpha}} - \mathrm{i}\bar{\mathbb{J}}_{\dot{\alpha}\dot{\beta}}\mathbb{P}_\alpha{}^{\dot{\beta}} \\ [\mathbb{W}_{\alpha\dot{\alpha}}, \mathbb{W}_{\beta\dot{\beta}}] &= -\mathbb{W}_{\alpha\dot{\beta}}\mathbb{P}_{\beta\dot{\alpha}} + \mathbb{W}_{\beta\dot{\alpha}}\mathbb{P}_{\alpha\dot{\beta}}. \end{aligned} \tag{2.3.5}$$

Equation (2.3.4) can be readily obtained from (2.2.11).

The fact that the spin operator does not commute with the supersymmetry generators means nothing more than that irreducible representations of the Poincaré superalgebra contain particles of different spins.

To find a supersymmetric invariant generalization of the spin operator, let us consider the operator

$$\mathbb{Z}_a = \mathbb{W}_a - \frac{1}{8}(\tilde{\sigma}_a)^{\dot{\alpha}\alpha}[\mathbb{Q}_\alpha, \bar{\mathbb{Q}}_{\dot{\alpha}}] \tag{2.3.6}$$

which is a generalization of the Pauli–Lubanski vector. Using equations (2.3.4) and the identity

$$[[\mathbb{Q}_\alpha, \bar{\mathbb{Q}}_{\dot{\alpha}}], \mathbb{Q}_\beta] = 4\mathbb{Q}_\alpha \mathbb{P}_{\beta\dot{\alpha}} \tag{2.3.7}$$

one finds

$$[\mathbb{Z}_a, \mathbb{Q}_\alpha] = \frac{1}{2}\mathbb{Q}_\alpha \mathbb{P}_a. \tag{2.3.8}$$

One can also prove the relations

$$\begin{aligned} [\mathbb{Z}_a, \mathbb{P}_b] &= 0 \\ [\mathbb{Z}_a, \mathbb{Z}_b] &= \mathrm{i}\varepsilon_{abcd}\mathbb{Z}^c\mathbb{P}^d. \end{aligned} \tag{2.3.9}$$

Note that the Pauli–Lubanski vector obeys such relations also (see equation (1.5.11)).

In accordance with equations (2.3.8) and (2.3.9), the operator $(\mathbb{Z}_a\mathbb{P}_b - \mathbb{Z}_b\mathbb{P}_a)$ commutes with the momentum generators and the supersymmetry generators,

$$[\mathbb{Z}_{[a}\mathbb{P}_{b]}, \mathbb{P}_c] = [\mathbb{Z}_{[a}\mathbb{P}_{b]}, \mathbb{Q}_\alpha] = 0. \tag{2.3.10}$$

Then, the scalar operator

$$-\frac{1}{2}(\mathbb{Z}_a\mathbb{P}_b - \mathbb{Z}_b\mathbb{P}_a)(\mathbb{Z}^a\mathbb{P}^b - \mathbb{Z}^b\mathbb{P}^a)$$

commutes with $\mathbb{J}_{ab}$ and also, due to (2.3.10), with $\mathbb{P}_a$ and $\mathbb{Q}_\alpha$, $\bar{\mathbb{Q}}_{\dot{\alpha}}$. As a result, the fourth-order polynomial

$$C = (\mathbb{Z}, \mathbb{P})^2 - \mathbb{Z}^2\mathbb{P}^2 \tag{2.3.11}$$

is a Casimir operator of the Poincaré superalgebra. It is called the 'superspin operator'.

Let us analyse the spectrum of eigenvalues the superspin operator can take. Given an irreducible massive representation of the Poincaré superalgebra in a Hilbert space $\mathscr{H}$, we consider in $\mathscr{H}$ the subspace V_q of particle states having a given four-momentum q_a (see also subsection 1.5.4). As usual, it is useful to choose the momentum (1.5.26) of a particle at rest. Recall that the Pauli–Lubanski vector transforms V_q into itself. Since the supersymmetry generators commute with $\mathbb{P}_a$, each of $\mathbb{Q}_\alpha$ and $\bar{\mathbb{Q}}_{\dot{\alpha}}$ transforms V_q into itself. Therefore, the operator Z_a transforms V_q into itself. When restricted to V_q, the superspin operator (2.3.11) takes the form

$$m^2((\mathbb{Z}_1)^2 + (\mathbb{Z}_2)^2 + (\mathbb{Z}_3)^2) \tag{2.3.12}$$

and the operators $(1/m)\ \mathbb{Z}_I \equiv \mathbb{S}_I$, $I = 1, 2, 3$, proportional to the space components of $\mathbb{Z}_a$ satisfy, by virtue of (2.3.9), the commutation relations

(1.5.28) of the algebra $su(2)$. Then equations (1.5.29) and (2.3.12) lead to

$$(\mathbb{Z}, \mathbb{P})^2 - \mathbb{Z}^2\mathbb{P}^2 = m^4 Y(Y+1)\mathbb{I} \qquad Y = 0, \frac{1}{2}, 1, \frac{3}{2}, \ldots \tag{2.3.13}$$

The quantum number Y is called 'superspin'.

The above consideration makes clear that massive representations of the Poincaré superalgebra are classified by mass and superspin.

2.3.3. *Massive irreducible representations*

Let us fix some mass (m) and superspin (Y). The corresponding irreducible representation T of the Poincaré superalgebra provides us with a reducible representation of the Poincaré algebra. The main problem now is to decompose T into a direct sum of irreducible Poincaré representations. This may be done as follows.

In the Hilbert space $\mathscr{H}$ of particle states we consider subspaces $\mathscr{H}_{(+)}$, $\mathscr{H}_{(-)}$ and $\mathscr{H}_{(0)}$ defined by the rule:

$$\begin{aligned} &\mathscr{H}_{(+)}: \{|\Psi\rangle \in \mathscr{H}, \bar{\mathbb{Q}}_{\dot{\alpha}}|\Psi\rangle = 0\} \\ &\mathscr{H}_{(-)}: \{|\Psi\rangle \in \mathscr{H}, \mathbb{Q}_{\alpha}|\Psi\rangle = 0\} \\ &\mathscr{H}_{(0)}: \{|\Psi\rangle \in \mathscr{H}, \mathbb{Q}^2|\Psi\rangle = \bar{\mathbb{Q}}^2|\Psi\rangle = 0\} \end{aligned} \tag{2.3.14}$$

where $\mathbb{Q}^2 = \mathbb{Q}^\alpha\mathbb{Q}_\alpha$, $\bar{\mathbb{Q}}^2 = \bar{\mathbb{Q}}_{\dot{\alpha}}\bar{\mathbb{Q}}^{\dot{\alpha}}$. Each of the subspaces $\mathscr{H}_{(+)}$, $\mathscr{H}_{(-)}$ and $\mathscr{H}_{(0)}$ is invariant with respect to the Poincaré group. Indeed, the Poincaré transformations $U(b, K) = \exp[\mathrm{i}(b^a\mathbb{P}_a + (1/2)K^{ab}\mathbb{J}_{ab})]$ act in $\mathscr{H}$ by the law:

$$|\Psi\rangle \to |\Psi'\rangle = U(b, K)|\Psi\rangle.$$

So, if $|\Psi\rangle \in \mathscr{H}_{(+)}$, $\bar{\mathbb{Q}}_{\dot{\alpha}}|\Psi\rangle = 0$, then we deduce from the supersymmetry algebra (2.2.9) that

$$\bar{\mathbb{Q}}^{\dot{\alpha}}|\Psi'\rangle = \left(\exp\left(\frac{1}{2}K^{ab}\tilde{\sigma}_{ab}\right)\right)^{\dot{\alpha}}{}_{\dot{\beta}}U(b, K)\bar{\mathbb{Q}}^{\dot{\beta}}|\Psi\rangle = 0.$$

Analogous arguments are applicable to $\mathscr{H}_{(-)}$ and $\mathscr{H}_{(0)}$.

It is not difficult to construct projectors for the subspaces (2.3.14). Using the supersymmetry algebra (2.2.9), one readily obtains the identities

$$[\mathbb{Q}^2, \bar{\mathbb{Q}}_{\dot{\alpha}}] = 4\mathbb{P}_{\alpha\dot{\alpha}}\mathbb{Q}^\alpha \qquad [\bar{\mathbb{Q}}^2, \mathbb{Q}_\alpha] = -4\mathbb{P}_{\alpha\dot{\alpha}}\bar{\mathbb{Q}}^{\dot{\alpha}} \tag{2.3.15a}$$

$$\bar{\mathbb{Q}}^2\mathbb{Q}_\alpha\bar{\mathbb{Q}}^2 = 0 \qquad \mathbb{Q}^2\bar{\mathbb{Q}}_{\dot{\alpha}}\mathbb{Q}^2 = 0 \tag{2.3.15b}$$

$$\mathbb{Q}^2\bar{\mathbb{Q}}^2\mathbb{Q}^2 = -16\mathbb{P}^2\mathbb{Q}^2 \qquad \bar{\mathbb{Q}}^2\mathbb{Q}^2\bar{\mathbb{Q}}^2 = -16\mathbb{P}^2\bar{\mathbb{Q}}^2. \tag{2.3.15c}$$

To prove equations (2.3.15*b*, *c*), one has simply to notice that any product like $\mathbb{Q}_{\alpha_1}\mathbb{Q}_{\alpha_2}\ldots\mathbb{Q}_{\alpha_n}$ or $\bar{\mathbb{Q}}_{\dot{\alpha}_1}\bar{\mathbb{Q}}_{\dot{\alpha}_2}\ldots\bar{\mathbb{Q}}_{\dot{\alpha}_n}$, $n > 2$, vanishes. Then, projectors on the subspaces $\mathscr{H}_{(+)}$, $\mathscr{H}_{(-)}$ and $\mathscr{H}_{(0)}$ turn out to be

$$P_{(+)} = \frac{1}{16m^2}\bar{\mathbb{Q}}^2\mathbb{Q}^2 \qquad P_{(-)} = \frac{1}{16m^2}\mathbb{Q}^2\bar{\mathbb{Q}}^2$$

$$P_{(0)} = -\frac{1}{8m^2}\mathbb{Q}^\alpha\bar{\mathbb{Q}}^2\mathbb{Q}_\alpha = -\frac{1}{8m^2}\bar{\mathbb{Q}}_{\dot{\alpha}}\mathbb{Q}^2\bar{\mathbb{Q}}^{\dot{\alpha}} \tag{2.3.16}$$

$$P_{(i)}P_{(j)} = \delta_{ij}P_{(i)}$$

where the indices i, j take values $+$, $-$, 0. Further, one can prove the relation

$$P_{(+)} + P_{(-)} + P_{(0)} = \mathbb{1} \tag{2.3.17}$$

so the decomposition

$$\mathscr{H} = \mathscr{H}_{(+)} \oplus \mathscr{H}_{(-)} \oplus \mathscr{H}_{(0)} \tag{2.3.18}$$

takes place. What is more, the subspaces $\mathscr{H}_{(+)}$, $\mathscr{H}_{(-)}$ and $\mathscr{H}_{(0)}$ are orthogonal to each other.

As the following step, we point out that the operator (2.3.6), when restricted to $\mathscr{H}_{(+)}$ or $\mathscr{H}_{(-)}$, is of the form

$$\mathbb{Z}_a\Big|_{\mathscr{H}_{(\pm)}} = \mathbb{W}_a \mp \frac{1}{2}\mathbb{P}_a$$

hence

$$\mathbb{Z}_{[a}\mathbb{P}_{b]}\Big|_{\mathscr{H}_{(\pm)}} = \mathbb{W}_{[a}\mathbb{P}_{b]}.$$

Then, due to equation (2.3.13), we have

$$\mathbb{W}^2\Big|_{\mathscr{H}_{(\pm)}} = m^2Y(Y+1)\mathbb{1}. \tag{2.3.19}$$

Therefore, each of the subspaces $\mathscr{H}_{(+)}$ and $\mathscr{H}_{(-)}$ carries the spin Y.

It is worth pointing out one more important observation. Namely, any state $|\Psi\rangle \in \mathscr{H}$ may be obtained by acting with the supersymmetry generators on states from one of the subspaces $\mathscr{H}_{(+)}$, $\mathscr{H}_{(-)}$ and $\mathscr{H}_{(0)}$. For example, starting from $\mathscr{H}_{(+)}$, every state $|\Psi\rangle \in \mathscr{H}_{(-)}$ can be represented as

$$|\Psi\rangle = \frac{1}{16m^2}\mathbb{Q}^2\bar{\mathbb{Q}}^2|\Psi\rangle = \mathbb{Q}^2|\Phi\rangle \qquad |\Phi\rangle \in \mathscr{H}_{(+)}$$

every state $|\Psi\rangle \in \mathscr{H}_{(0)}$ is represented as

$$|\Psi\rangle = -\frac{1}{8m^2}\mathbb{Q}^\alpha\bar{\mathbb{Q}}^2\mathbb{Q}_\alpha|\Psi\rangle = \mathbb{Q}^\alpha|\Phi_\alpha\rangle \qquad |\Phi_\alpha\rangle \in \mathscr{H}_{(+)}.$$

Together with equations (2.3.19), this observation means that a representation of the Poincaré superalgebra in a Hilbert space $\mathscr{H}$ is irreducible if and only if the corresponding Poincaré representations in the subspaces $\mathscr{H}_{(+)}$ and $\mathscr{H}_{(-)}$ are irreducible.

The only remaining task is to decompose $\mathscr{H}_{(0)}$ into irreducible Poincaré representations. For this purpose, let us once more consider in $\mathscr{H}$ the subspace V_q, where, as usual, q_a is the momentum of a particle at rest. The Pauli–Lubanski vector reduces on V_q to the form (1.5.27), where the operators $\mathbb{S}_I$ satisfy the commutation relations (1.5.28) of the algebra $su(2)$. The equation (2.3.4) reduces on V_q to the form

$$[\mathbb{S}_I, \mathbb{Q}_\alpha] = -\frac{1}{2}(\sigma_I)_\alpha{}^\beta \mathbb{Q}_\beta \tag{2.3.20}$$

where $(\sigma_I)_\alpha{}^\beta$, $I = 1, 2, 3$, are the ordinary Pauli matrices. So, the supersymmetry generators $\mathbb{Q}_\alpha$ form an $SU(2)$-spinor on the subspace V_q.

The subspace V_q is decomposed into the direct sum

$$V_q = V_{q(+)} \oplus V_{q(-)} \oplus V_{q(0)} \tag{2.3.21}$$

where $V_{q(+)} = V_q \cap \mathscr{H}_{(+)}$ and so on. Since $\mathscr{H}_{(+)}$ describes the spin-Y Poincaré representation, one can choose a basis $\{|\Psi_{\alpha_1\alpha_2\ldots\alpha_n}\rangle\}$ in $V_{q(+)}$ to be totally symmetric $SU(2)$-tensor of rank $n = 2Y$,

$$|\Psi_{\alpha_1\alpha_2\cdots\alpha_n}\rangle = |\Psi_{(\alpha_1\alpha_2\ldots\alpha_n)}\rangle \in V_{q(+)}.$$

Then, the states

$$\mathbb{Q}_{\alpha_1}|\Psi_{\alpha_1\alpha_2\ldots\alpha_n}\rangle \in V_{q(0)}$$

generate $V_{q(0)}$. In accordance with equation (2.3.20), these states represent an $SU(2)$-tensor. It contains two irreducible (totally symmetric) $SU(2)$-tensors:

$$|\Psi_{\alpha_1\alpha_2\ldots\alpha_{n+1}}\rangle = \mathbb{Q}_{\alpha_1}|\Psi_{\alpha_2\ldots\alpha_{n+1}}\rangle - \frac{1}{n+1}\sum_{k=2}^{n+1} \varepsilon_{\alpha_1\alpha_k}\mathbb{Q}^\gamma|\Psi_{\gamma\alpha_2\ldots\hat{\alpha}_k\ldots\alpha_{n+1}}\rangle$$

and

$$|\Psi_{\alpha_1\alpha_2\ldots\alpha_{n-1}}\rangle = \mathbb{Q}^\gamma|\Psi_{\gamma\alpha_1\alpha_2\ldots\alpha_{n-1}}\rangle.$$

Evidently, we have

$$\mathbb{W}^2|\Psi_{\alpha_1\alpha_2\ldots\alpha_{n+1}}\rangle = ms'(s'+1)|\Psi_{\alpha_1\alpha_2\ldots\alpha_{n+1}}\rangle \qquad s' = Y + 1/2$$

$$\mathbb{W}^2|\Psi_{\alpha_1\alpha_2\ldots\alpha_{n-1}}\rangle = ms''(s''+1)|\Psi_{\alpha_1\alpha_2\ldots\alpha_{n-1}}\rangle \qquad s'' = Y - 1/2.$$

As a result, $\mathscr{H}_{(0)}$ contains two irreducible Poincaré representations: of spin $(Y+\frac{1}{2})$ and $(Y-\frac{1}{2})$, respectively, when $Y>0$. In the case $Y=0$, $\mathscr{H}_{(0)}$ describes only one Poincaré representation, of spin $\frac{1}{2}$.

To summarize, the unitary representations of the Poincaré superalgebra are classified by mass and superspin. In the case $Y \neq 0$, the corresponding irreducible representation describes four particles of spin $Y-\frac{1}{2}, Y, Y, Y+\frac{1}{2}$ with the same mass $m \neq 0$. When $Y=0$, the representation describes two scalar particles and a spin-$\frac{1}{2}$ particle with the same mass.

2.3.4. *Massless irreducible representations*

Now we are going to study massless unitary representations of the Poincaré superalgebra. They are characterized by the massless equation

$$\mathbb{P}^a\mathbb{P}_a = 0. \tag{2.3.22}$$

This equation gives very strong constraints on the supersymmetry generators. Using the supersymmetry algebra (2.2.9), one can prove the identity

$$\{\mathbb{P}_{\alpha\dot\beta}\bar{\mathbb{Q}}^{\dot\beta}, \mathbb{P}_{\beta\dot\alpha}\mathbb{Q}^{\beta}\} = -2\mathbb{P}_{\alpha\dot\alpha}\mathbb{P}^2.$$

In accordance with equation (2.3.22), this operator should vanish. Then, for every state $|\Psi\rangle$ from a Hilbert space $\mathscr{H}$ of particle states, we have

$$\langle A_\alpha\Psi|A_\alpha\Psi\rangle + \langle (A_\alpha)^+\Psi|(A_\alpha)^+\Psi\rangle = 0$$

where $A_\alpha = \mathbb{P}_{\alpha\dot\beta}\bar{\mathbb{Q}}^{\dot\beta}$. As a result, one must impose the operational constraints

$$\mathbb{P}_{\alpha\dot\beta}\bar{\mathbb{Q}}^{\dot\beta} = \mathbb{P}_{\beta\dot\alpha}\mathbb{Q}^{\beta} = 0. \tag{2.3.23}$$

Further, recall the identities (2.3.15*a*). Then, due to equation (2.3.23), one must also demand

$$[\mathbb{Q}^2, \bar{\mathbb{Q}}_{\dot\alpha}] = [\bar{\mathbb{Q}}^2, \mathbb{Q}_\alpha] = 0. \tag{2.3.24}$$

Finally, for any states $|\Psi\rangle \in \mathscr{H}$, we can write

$$\mathbb{P}_{\alpha\dot\alpha}\mathbb{Q}^2|\Psi\rangle = \frac{1}{2}(\mathbb{Q}_\alpha\bar{\mathbb{Q}}_{\dot\alpha} + \bar{\mathbb{Q}}_{\dot\alpha}\mathbb{Q}_\alpha)\mathbb{Q}^2|\Psi\rangle = \frac{1}{2}\mathbb{Q}_\alpha\bar{\mathbb{Q}}_{\dot\alpha}\mathbb{Q}^2|\Psi\rangle$$

$$= \frac{1}{2}\mathbb{Q}_\alpha[\bar{\mathbb{Q}}_{\dot\alpha}, \mathbb{Q}^2]|\Psi\rangle = 0$$

where we have used equation (2.3.24). Therefore, we have

$$\mathbb{P}_a\mathbb{Q}^2|\Psi\rangle = 0 \qquad \forall|\Psi\rangle \in \mathscr{H}. \tag{2.3.25}$$

It is necessary to point out that the space $\mathscr{H}$ must be understood as a space of one-particle states (not a Fock space). Hence if $|\Psi\rangle$ is a non-vanishing state, its average energy must be positive, $\langle\Psi|\mathbb{E}|\Psi\rangle > 0$. In other words, the energy operator $\mathbb{E}$ must be invertible. So, equation (2.3.25) forces us to demand the constraints

$$\mathbb{Q}^2 = \bar{\mathbb{Q}}^2 = 0 \tag{2.3.26}$$

in addition to equations (2.3.23, 24). In fact, equation (2.3.24) is now a consequence of constraints (2.3.23) and (2.3.26).

By analogy with the massive case, we can introduce Poincaré invariant subspaces in $\mathscr{H}$:

$$\begin{aligned}\mathscr{H}_{(+)}&: \{|\Psi\rangle \in \mathscr{H}, \bar{\mathbb{Q}}_{\dot{\alpha}}|\Psi\rangle = 0\} \\ \mathscr{H}_{(-)}&: \{|\Psi\rangle \in \mathscr{H}, \mathbb{Q}_{\alpha}|\Psi\rangle = 0\}.\end{aligned} \tag{2.3.27}$$

In contrast to the massive case, the subspace $\mathscr{H}_{(0)}$ is now a trivial one, due to constraint (2.3.26). Each of the subspaces $\mathscr{H}_{(+)}$ and $\mathscr{H}_{(-)}$ describes some massless representation of the Poincaré group. The supersymmetry generators transform $\mathscr{H}_{(+)}$ onto $\mathscr{H}_{(-)}$ and vice versa,

$$\mathbb{Q}_{\alpha}\mathscr{H}_{(+)} \sim \mathscr{H}_{(-)} \qquad \bar{\mathbb{Q}}_{\dot{\alpha}}\mathscr{H}_{(-)} \sim \mathscr{H}_{(+)} \tag{2.3.28}$$

in accordance with the identity $\mathbb{P}_{\alpha\dot{\alpha}} = \frac{1}{2}\{\mathbb{Q}_{\alpha}, \bar{\mathbb{Q}}_{\dot{\alpha}}\}$.

To make our discussion complete, we must clarify possible helicity eigenvalues $\lambda_{(+)}$ and $\lambda_{(-)}$, which the Poincaré representations on $\mathscr{H}_{(+)}$ and $\mathscr{H}_{(-)}$ may take.

2.3.5. Superhelicity

In the massless case, the relation (2.3.4) is simplified to the form

$$[\mathbb{W}_{\alpha\dot{\alpha}}, \mathbb{Q}_{\beta}] = -\frac{1}{2}\mathbb{Q}_{\alpha}\mathbb{P}_{\beta\dot{\alpha}} = -\frac{1}{2}\mathbb{Q}_{\beta}\mathbb{P}_{\alpha\dot{\alpha}} \tag{2.3.29}$$

as a consequence of equation (2.3.23). Hence, instead of the operator $\mathbb{Z}_a$ (equation (2.3.6)), it is worth introducing another operator

$$\mathbb{L}_a = \mathbb{W}_a - \frac{1}{16}(\tilde{\sigma}_a)^{\dot{\alpha}\alpha}[\mathbb{Q}_{\alpha}, \bar{\mathbb{Q}}_{\dot{\alpha}}] \tag{2.3.30}$$

which commutes with the momentum generators and the supersymmetry generators

$$[\mathbb{L}_a, \mathbb{P}_b] = 0 \qquad [\mathbb{L}_a, \mathbb{Q}_{\alpha}] = 0. \tag{2.3.31}$$

Using equations (1.5.11) and (2.3.23, 29), one can also prove

$$\begin{aligned}\mathbb{L}^a\mathbb{P}_a &= 0 \\ [\mathbb{L}_a, \mathbb{L}_b] &= \mathrm{i}\varepsilon_{abcd}\mathbb{L}^c\mathbb{P}^d .\end{aligned} \tag{2.3.32}$$

Expressions (2.3.31) and (2.3.32) show that the operator $\mathbb{L}_a$, being supersymmetric invariant, possesses all the properties of the Pauli–Lubanski vector. Then, applying the same line of argument that was used in subsection 1.5.6 to deduce the massless equation $\mathbb{W}_a = \lambda\mathbb{P}_a$, λ being the helicity, one finds that the operators $\mathbb{L}_a$ and $\mathbb{P}_a$ must be proportional to each other, $\mathbb{L}_a \propto \mathbb{P}_a$, in every irreducible massless representation of the Poincaré superalgebra. So, we introduce a new quantum number κ defined by

$$\mathbb{L}_a = \left(\kappa + \frac{1}{4}\right)\mathbb{P}_a. \tag{2.3.33}$$

This quantum number is called 'superhelicity'. Superhelicity characterizes massless representations of the Poincaré superalgebra.

Given some representation of superhelicity κ, let us calculate helicity values on the subspaces $\mathscr{H}_{(+)}$ and $\mathscr{H}_{(-)}$ defined above. For any state $|\Psi\rangle \in \mathscr{H}_{(+)}$, we have

$$\mathbb{L}_{\alpha\dot{\alpha}}|\Psi\rangle = \left(\mathbb{W}_{\alpha\dot{\alpha}} + \frac{1}{8}\left[\mathbb{Q}_\alpha, \bar{\mathbb{Q}}_{\dot{\alpha}}\right]\right)|\Psi\rangle = \left(\mathbb{W}_{\alpha\dot{\alpha}} - \frac{1}{8}\bar{\mathbb{Q}}_{\dot{\alpha}}\mathbb{Q}_\alpha\right)|\Psi\rangle$$

$$= \left(\mathbb{W}_{\alpha\dot{\alpha}} - \frac{1}{4}\mathbb{P}_{\alpha\dot{\alpha}}\right)|\Psi\rangle = \left(\kappa + \frac{1}{4}\right)\mathbb{P}_{\alpha\dot{\alpha}}|\Psi\rangle.$$

This gives

$$\mathbb{W}_{\alpha\dot{\alpha}}|\Psi\rangle = \left(\kappa + \frac{1}{2}\right)\mathbb{P}_{\alpha\dot{\alpha}}|\Psi\rangle \qquad \forall|\Psi\rangle \in \mathscr{H}_{(+)}. \tag{2.3.34}$$

Analogously, one finds

$$\mathbb{W}_{\alpha\dot{\alpha}}|\Psi\rangle = \kappa\mathbb{P}_{\alpha\dot{\alpha}}|\Psi\rangle \qquad \forall|\Psi\rangle \in \mathscr{H}_{(-)}. \tag{2.3.35}$$

Therefore, the helicities of the $\mathscr{H}_{(+)}$- and $\mathscr{H}_{(-)}$-Poincaré representations are equal to $(\kappa + \frac{1}{2})$ and κ, respectively.

To summarize, the massless unitary representations of the Poincaré superalgebra are classified by superhelicity $\kappa, \kappa = 0, \pm\frac{1}{2}, \pm 1, \pm\frac{3}{2}, \ldots$. For a given superhelicity κ, the corresponding representation describes two massless particles of helicities κ and $(\kappa + \frac{1}{2})$.

2.3.6. *Equality of bosonic and fermionic degrees of freedom*

Our discussion of unitary representations of the Poincaré superalgebra would be incomplete without pointing out one important consequence of the above results — that is, each $S\mathscr{P}$-representation describes an equal number of bosonic and fermionic degrees of freedom. It is worth recalling what is usually understood by the notion 'number of degrees of freedom'. Given a massive spin-s particle, its number of degrees of freedom, denoted by N_s, is defined to be equal to the number of different spin polarizations of the particle at rest. In other words, N_s coincides with the dimension of every subspace V_q in a Hilbert space of one-particle spin-s states. In the massive case we have $N_s = \dim V_q = (2s + 1)$. A massless helicity-$\lambda$ particle has only one degree of freedom. Since the relation $\mathbb{W}_a = \lambda\mathbb{P}_a$ is Poincaré covariant, every massless particle of definite helicity possesses one and only one spin polarization.

We have seen above that the massive $S\mathscr{P}$-representation of superspin Y describes four particles with spins $Y-\frac{1}{2}$, Y, Y, $Y+\frac{1}{2}$, i.e. two bosonic and two fermionic particles. Therefore, the complete number of bosonic (fermionic) degrees of freedom is equal to $2(2Y+1)$.

For a given superhelicity κ, the corresponding massless $S\mathscr{P}$-representation describes two massless particles having helicities κ and $(\kappa+\frac{1}{2})$, i.e. one bosonic and one fermionic particle.

2.4. Real superspace $\mathbb{R}^{4|4}$ and superfields

As is known, the unitary Poincaré representations can be realized in terms of fields on Minkowski space (see Section 1.8.) Undoubtedly, it would be interesting to obtain analogous realizations for the unitary representations of the Poincaré superalgebra. How is this done? Clearly, the existence of field Poincaré representations was possible merely due to the fact that from the very beginning the Poincaré group was introduced as the group of transformations in Minkowski space. As for the Poincaré superalgebra, one cannot relate it to some ordinary Lie group but only to a super Lie group —the super Poincaré group. A super Lie group cannot be realized as a group of transformations acting in some ordinary space $\mathbb{R}^n$, only in some superspace $\mathbb{R}^{p|q}$. So, to achieve the above aim, one may choose some reasonable superspace, then define on it some reasonable action of the super Poincaré group and so on, and so forth. But what principle do we have to be guided by? Evidently, we cannot make use of the principle leading to the Poincaré transformations. Recall, they are those transformations of $\mathbb{R}^4$ which preserve the Minkowski metric. In our case, we know nothing about either superspace or its metric. We have only the super Poincaré group at our disposal. Fortunately, there exists a purely algebraic way to introduce Minkowski space and Poincaré transformations starting from the Poincaré group. It is this approach which may be generalized to the super case.

2.4.1. Minkowski space as the coset space $\Pi/SO(3,1)^{\uparrow}$

Let us consider the left coset space $\Pi/SO(3,1)^{\uparrow}$, where Π is the Poincaré group and $SO(3,1)^{\uparrow}$ is the Lorentz group. We are going to show that this coset space can be identified with Minkowski space.

Points of the coset space $SO(3,1)^{\uparrow}$ are equivalence classes. For any group element $g\in\Pi$, its equivalence class $\bar{g}$ is defined to be the following set of group elements:

$$\bar{g}=\{gh,\, h\in SO(3,1)^{\uparrow}\}. \tag{2.4.1}$$

Recall, every equivalence class $\bar{g}$ is uniquely determined by its arbitrary element, i.e. $\bar{g} = \overline{gh}$, $\forall h \in SO(3,1)^{\uparrow}$. How can one parametrize points of the coset space?

Elements of the Poincaré group are parametrized by ten real variables b^a, $K^{ab} = -K^{ba}$, via the exponential mapping,

$$g(b, K) = \exp\left[\mathrm{i}\left(-b^a \mathrm{p}_a + \frac{1}{2} K^{ab} \mathrm{j}_{ab}\right)\right].$$

In particular, translations and Lorentz transformations look like $g(b, 0)$ and $g(0, K)$, respectively. Since $[\mathrm{j}, \mathrm{p}] \sim \mathrm{p}$, making use of the Baker–Hausdorff formula gives

$$g(b, K) = g(x, 0) g(0, K) \tag{2.4.2a}$$

$$g(b, K) = g(0, K) g(y, 0) \tag{2.4.2b}$$

where x^a and y^a are functions of b^a and K^{ab}. One can take the variables $\{x^a, K^{ab}\}$ or $\{y^a, K^{ab}\}$ in the role of local coordinates on Π. The variables $\{x^a, K^{ab}\}$ prove to be best adapted to the coset space $\Pi/SO(3,1)^{\uparrow}$, since for any elements $g(b_1, K_1)$ and $g(b_2, K_2)$ from the same equivalence class $\bar{g} \in \Pi/SO(3,1)^{\uparrow}$ we have

$$g(b_1, K_1) = g(x, 0) g(0, K_1) \qquad g(b_2, K_2) = g(x, 0) g(0, K_2).$$

As a result, every equivalence class $\overline{g(b, K)}$ is uniquely determined by the translation $g(x, 0)$ defined by equation (2.4.2*a*). Since translations $g(x, 0) = \exp(-\mathrm{i}x^a \mathrm{p}_a)$ are uniquely determined by four real numbers x^a, we obtain a one-to-one correspondence between $\Pi/SO(3,1)^{\uparrow}$ and $\mathbb{R}^4$. Thus, we can identify $\Pi/SO(3,1)^{\uparrow}$ and $\mathbb{R}^4$ by the rule:

$$\overline{g(b, K)} = \overline{g(x, 0)} \equiv \overline{\exp[-\mathrm{i}x^a \mathrm{p}_a]}. \tag{2.4.3}$$

Let us consider the left action of the Poincaré group on the coset space $\Pi/SO(3,1)^{\uparrow}$. Namely, with every group element $g_0 \in \Pi$ we relate a mapping $\hat{g}_0$: $\Pi/SO(3,1)^{\uparrow} \to \Pi/SO(3,1)^{\uparrow}$ of the coset space to itself, defined as follows:

$$\bar{g} \to \bar{g}' = \hat{g}_0(\bar{g}) = \overline{g_0 g} \qquad \forall \bar{g} \in \Pi/SO(3,1)^{\uparrow}. \tag{2.4.4}$$

Clearly, we have $\hat{g}_1 \circ \hat{g}_2 = \widehat{g_1 g_2}$. Further, every transformation (2.4.4) induces some mapping of $\mathbb{R}_4$ to itself, due to the identification (2.4.3). Let us analyse these transformations. It is sufficient to study two particular cases: the translations $g_0 = g(b, 0)$, and the Lorentz transformations $g_0 = g(0, K)$.

Translations

In accordance with equation (2.4.4), one has

$$\overline{g(x, 0)} \to \overline{g(b, 0) g(x, 0)} = \overline{g(x + b, 0)}.$$

Therefore, the corresponding transformation in $\mathbb{R}^4$ is

$$x^a \to x'^a = x^a + b^a. \tag{2.4.5}$$

Lorentz transformations

In accordance with equation (2.4.4), one has

$$\overline{g(x,0)} \to \overline{g(0,K)g(x,0)}.$$

Since $\bar{g} = \overline{gh}$, $\forall h \in SO(3,1)^{\uparrow}$, we can write

$$\overline{g(0,K)g(x,0)} = \overline{g(0,K)g(x,0)g(0,-K)}.$$

The expression under the bar can be rewritten as

$$g(0,K)g(x,0)g(0,-K) = \mathrm{e}^{\frac{1}{2}K^{bc}\mathbf{j}_{bc}}\mathrm{e}^{-\mathrm{i}x^a\mathbf{p}_a}\mathrm{e}^{-\frac{1}{2}K^{bc}\mathbf{j}_{bc}} = \exp\left(-\mathrm{i}x^a\mathrm{e}^{\frac{1}{2}K^{bc}\mathbf{j}_{bc}}\mathbf{p}_a\mathrm{e}^{-\frac{1}{2}K^{bc}\mathbf{j}_{bc}}\right).$$

Making use of the Poincaré algebra gives

$$\mathrm{e}^{\frac{1}{2}K^{bc}\mathbf{j}_{bc}}\mathbf{p}_a\mathrm{e}^{-\frac{1}{2}K^{bc}\mathbf{j}_{bc}} = (\mathrm{e}^{-K})_a{}^b\mathbf{p}_b.$$

Finally, since $K_{ab} = -K_{ba}$, we obtain

$$\overline{g(0,K)g(x,0)} = \overline{g(x',0)}$$

where

$$x'^a = (\mathrm{e}^K)^a{}_b x^b. \tag{2.4.6}$$

The expressions (2.4.5) and (2.4.6) reproduce ordinary Poincaré transformations.

The above discussion shows that Minkowski space can be identified with the coset space $\Pi/SO(3,1)^{\uparrow}$.

2.4.2. *Real superspace* $\mathbb{R}^{4|4}$

Let us try to generalize the left coset construction described above to the super case. Consider the coset space $S\Pi/SO(3,1|\mathbb{R}_c)^{\uparrow}$, where $S\Pi$ is the super Poincaré group and $SO(3,1|\mathbb{R}_c)^{\uparrow}$ is the Lorentz group over $\mathbb{R}_c$. Its elements are equivalence classes

$$\bar{g} = \{gh, h \in SO(3,1|\mathbb{R}_c)^{\uparrow}\}$$

for any $g \in S\Pi$. Elements of the Poincaré supergroup are parametrized as in equation (2.2.20). Then, since $[\mathbf{j}, \mathbf{p}] \sim \mathbf{p}$ and $[\mathbf{j}, \mathbf{q}] \sim \mathbf{q}$, the Baker–Hausdorff formula gives

$$g(b,\epsilon,\bar{\epsilon},K) = g(x,\theta,\bar{\theta},0)\,g(0,0,0,K) \tag{2.4.7}$$

where real c-numbers x^a and complex a-numbers θ^α and $\bar{\theta}^{\dot{\alpha}}$ (conjugate to each other) are functions of the supergroup coordinates. Evidently, equation (2.4.7) generalizes equation (2.4.2*a*). The variables $\{x^a, \theta^\alpha, \bar{\theta}^{\dot{\alpha}}, \mathrm{K}^{ab}\}$ may be used to parametrize $S\Pi$, instead of the original coordinates. Now, it is easy to prove that points of the coset space $S\Pi/SO(3,1|\mathbb{R}_c)^{\uparrow}$ can be identified with points of real superspace $\mathbb{R}^{4|4}$ parametrized by the rule

$$\mathbb{R}^{4|4} = \{(z^A) = (x^a, \theta^\alpha, \bar{\theta}_{\dot{\alpha}}), \bar{\theta}_{\dot{\alpha}} = (\theta_\alpha)^*, x^a \in \mathbb{R}_c, \theta_\alpha \in \mathbb{C}_a\}. \tag{2.4.8}$$

The identification is as follows

$$\overline{g(b,\epsilon,\bar{\epsilon},\mathrm{K})} = \overline{g(z,0)} \equiv \exp\left[\mathrm{i}(-x^a p_a + \theta\mathbf{q} + \bar{\theta}\bar{\mathbf{q}})\right]. \tag{2.4.9}$$

Furthermore, one may define the left action of the super Poincaré group on the coset space $S\Pi/SO(3,1|\mathbb{R}_c)^\dagger$ in the same fashion as in (2.4.4). This induces some action of the super Poincaré group on the superspace $\mathbb{R}^{4|4}$ due to the identification (2.4.9). It is sufficient to find transformations of $\mathbb{R}^{4|4}$ corresponding to the supergroup elements (2.2.21–23). Using the Poincaré superalgebra (2.2.9), one obtains:

Translations

$$x'^a = x^a + b^a \qquad \theta'^\alpha = \theta^\alpha \qquad \bar{\theta}'_{\dot{\alpha}} = \bar{\theta}_{\dot{\alpha}}. \tag{2.4.10}$$

Lorentz transformations

$$\begin{aligned} x'^a &= (\mathrm{e}^K)^a{}_b x^b \\ \theta'_\alpha &= \left(\exp\left(\frac{1}{2}K^{ab}\sigma_{ab}\right)\right)_\alpha{}^\beta \theta_\beta \\ \bar{\theta}'^{\dot{\alpha}} &= \left(\exp\left(\frac{1}{2}K^{ab}\tilde{\sigma}_{ab}\right)\right)^{\dot{\alpha}}{}_{\dot{\beta}}\bar{\theta}^{\dot{\beta}}. \end{aligned} \tag{2.4.11}$$

We see that odd superspace coordinates θ_α and $\bar{\theta}^{\dot{\alpha}}$ transform as (un)dotted spinors. To derive equations (2.4.10, 11), one has to perform the same steps as in deriving equations (2.4.5, 6).

Supersymmetry transformations

$$\begin{aligned} x'^a &= x^a - \mathrm{i}\epsilon\sigma^a\bar{\theta} + \mathrm{i}\theta\sigma^a\bar{\epsilon} \\ \theta'^\alpha &= \theta^\alpha + \epsilon^\alpha \qquad \bar{\theta}'_{\dot{\alpha}} = \bar{\theta}_{\dot{\alpha}} + \bar{\epsilon}_{\dot{\alpha}}. \end{aligned} \tag{2.4.12}$$

Let us comment on equation (2.4.12). In accordance with prescription (2.4.4), the element $g(\epsilon,\bar{\epsilon})$ (2.2.23) of the super Poincaré group acts on the coset space by the rule

$$\overline{g(z,0)} \to \overline{g(\epsilon,\bar{\epsilon})g(z,0)}.$$

One can write

$$\begin{aligned} g(\epsilon,\bar{\epsilon})\,g(z,0) &= \exp\left[\mathrm{i}(\epsilon\mathbf{q} + \bar{\epsilon}\bar{\mathbf{q}})\right]\exp\left[\mathrm{i}(-x^a p_a + \theta\mathbf{q} + \bar{\theta}\bar{\mathbf{q}})\right] \\ &= \exp(-\mathrm{i}x^a p_a)\exp\left[\mathrm{i}(\epsilon\mathbf{q} + \bar{\epsilon}\bar{\mathbf{q}})\right]\exp\left[\mathrm{i}(\theta\mathbf{q} + \bar{\theta}\bar{\mathbf{q}})\right] \end{aligned}$$

because $[p_a, \mathbf{q}_\alpha] = 0$. Then equation (2.2.25) leads to equation (2.4.12).

Combining equations (2.4.10–12) leads to the most general form of super Poincaré transformations on $\mathbb{R}^{4|4}$:

$$\begin{aligned} x'^a &= (\mathrm{e}^K)^a{}_b x^b + \mathrm{i}(\theta\sigma^a\bar{\epsilon} - \epsilon\sigma^a\bar{\theta}) + b^a \\ \theta'^\alpha &= (\mathrm{e}^{-K})^\alpha{}_\beta\theta^\beta + \epsilon^\alpha \Leftrightarrow \theta'_\alpha = (\mathrm{e}^K)_\alpha{}^\beta\theta_\beta + \epsilon_\alpha \\ \bar{\theta}_{\dot{\alpha}} &= (\mathrm{e}^{\bar{K}})_{\dot{\alpha}}{}^{\dot{\beta}}\bar{\theta}_{\dot{\beta}} + \bar{\epsilon}_{\dot{\alpha}} \Leftrightarrow \bar{\theta}'^{\dot{\alpha}} = (\mathrm{e}^{-\bar{K}})^{\dot{\alpha}}{}_{\dot{\beta}}\bar{\theta}^{\dot{\beta}} + \bar{\epsilon}^{\dot{\alpha}} \end{aligned} \tag{2.4.13}$$

where

$$K_{ab} = -K_{ba} = (K_{ab})^* \qquad K_{\alpha\beta} = K_{\beta\alpha} = \frac{1}{2}(\sigma^{ab})_{\alpha\beta}K_{ab} \qquad \bar{K}_{\dot{\alpha}\dot{\beta}} = (K_{\alpha\beta})^*.$$

So, the super Poincaré group acts on the superspace $\mathbb{R}^{4|4}$ as a group of linear inhomogeneous transformations. Using expressions (2.4.13), it is not difficult to obtain a supermatrix realization of $S\Pi$ analogous to the matrix realization of Π given in subsection 1.5.1.

Looking at equation (2.4.12), we see that the supersymmetry transformations represent z-independent shifts of the odd superspace coordinates together with θ-dependent shifts of the even superspace coordinates. More precisely, decomposing each even coordinate x^a, $a = 0, 1, 2, 3$, into its body and soul,

$$x^a = (x^a)_B + (x^a)_S \tag{2.4.14}$$

the supersymmetry transformations change the soul leaving the body invariant,

$$(x^a)_B \to (x^a)_B \qquad (x^a)_S \to (x^a)_S + i(\theta\sigma^a\bar{\epsilon} - \epsilon\sigma^a\bar{\theta}).$$

Even if all x^a were soulless before making a supersymmetry transformation, they acquire some soul afterwards. Supersymmetry requires soul.

Recall that the transformation (2.4.12) corresponds to the supergroup element $g(\epsilon, \bar{\epsilon})$ (2.2.23). Let us consider an element

$$g = (g(\epsilon_1, \bar{\epsilon}_1))^{-1}(g(\epsilon_2, \bar{\epsilon}_2))^{-1}g(\epsilon_1, \bar{\epsilon}_1)g(\epsilon_2, \bar{\epsilon}_2) \tag{2.4.15}$$

where ϵ_1^α and ϵ_2^α are arbitrary undotted spinors. Due to equation (2.2.25), we have $(g(\epsilon, \bar{\epsilon}))^{-1} = g(-\epsilon, -\bar{\epsilon})$. On the same grounds, one finds

$$g = e^{-ib^a p_a} \qquad b^a = 2i(\epsilon_2\sigma^a\bar{\epsilon}_1 - \epsilon_1\sigma^a\bar{\epsilon}_2). \tag{2.4.16}$$

Therefore, the sequence (2.4.15) of supersymmetry transformations on $\mathbb{R}^{4|4}$ presents a bodiless translation of the even superspace coordinates. In fact, every bodiless translation (2.4.10) may be represented as a sequence of supersymmetry transformations.

It is useful to treat the superspace $\mathbb{R}^{4|4}$ as a trivial fibre bundle over Minkowski space such that the projection π: $\mathbb{R}^{4|4} \to \mathbb{R}^4$ from the fibre bundle into the base manifold—Minkowski space—is given by

$$\pi((x^a, \theta^\alpha, \bar{\theta}_{\dot{\alpha}})) = ((x^a)_B)$$

for any superspace point $(x^a, \theta^\alpha, \theta_{\dot{\alpha}})$. Every Poincaré transformation (2.4.5) or (2.4.6) on the base space can be extended to a transformation (2.4.10) or (2.4.11), respectively, on the fibre bundle. Every super Poincaré transformation (2.4.13) on the fibre bundle is projected into a Poincaré transformation

$$(x'^a)_B = ((e^K)^a{}_b)_B(x^b)_B + (b^a)_B$$

on the base manifold, where $(K^{ab})_B$ and $(b^a)_B$ are the bodies of the parameters K^{ab} and b^a.

2.4.3. Supersymmetric interval

As is known, the Poincaré transformations leave invariant the interval (1.1.2). It is not difficult to find its superspace generalization invariant under all super Poincaré transformations. Consider a superspace two-point function $w^a(z_1, z_2)$, where z_1^A and z_2^A are arbitrary points of $\mathbb{R}^{4|4}$, defined by

$$w^a(z_1, z_2) = (x_2 - x_1)^a + \mathrm{i}\theta_1\sigma^a(\bar{\theta}_2 - \bar{\theta}_1) - \mathrm{i}(\theta_2 - \theta_1)\sigma^a\bar{\theta}_1. \quad (2.4.17)$$

This two-point function proves to be invariant under supersymmetry transformations (2.4.12). Clearly, it is also invariant under translations (2.4.10). Hence every super Poincaré transformation (2.4.13) leaves invariant the two-point function

$$\mathrm{d}s^2 = w^a w_a. \quad (2.4.18)$$

This is called a 'supersymmetric interval'.

Let us recall also that every space–time transformation $x^a \to x'^a = f^a(x)$ preserving the interval (1.1.2) is a Poincaré transformation. For this reason, flat space–time admits a preferable class of reference systems — inertial systems, in which the space–time metric has flat form (1.1.2). Below we shall show that every superspace transformation $z^A \to z'^A = f^A(z)$ preserving the supersymmetric interval (2.4.17) presents some super Poincaré transformation. So, by analogy with Minkowski space, one can speak about super-inertial systems — that is, reference systems (z^A) on $\mathbb{R}^{4|4}$ in which the supersymmetric interval has the form (2.4.17). Given two super-inertial systems, their coordinates are related by a super Poincaré transformation.

2.4.4. Superfields

A supersmooth function $V: \mathbb{R}^{4|4} \to \Lambda_\infty$ on real superspace $\mathbb{R}^{4|4}$ is said to be a 'superfield' (recall, supersmooth functions are defined to be smooth with respect to even superspace coordinates and analytic with respect to odd superspace coordinates). Since superspace is parametrized by rule (2.4.8) and spinor indices α or $\dot{\alpha}$ take only two values, the odd superspace coordinates satisfy the identities

$$\theta_\alpha\theta_\beta\theta_\gamma = 0 \qquad \bar{\theta}_{\dot{\alpha}}\bar{\theta}_{\dot{\beta}}\bar{\theta}_{\dot{\gamma}} = 0. \quad (2.4.19)$$

Due to the reduction rules (1.4.6), we also have

$$\begin{aligned} \theta_\alpha\theta_\beta &= \frac{1}{2}\varepsilon_{\alpha\beta}\theta^2 \qquad \theta^\alpha\theta^\beta = -\frac{1}{2}\varepsilon^{\alpha\beta}\theta^2 \\ \bar{\theta}_{\dot{\alpha}}\bar{\theta}_{\dot{\beta}} &= -\frac{1}{2}\varepsilon_{\dot{\alpha}\dot{\beta}}\bar{\theta}^2 \qquad \bar{\theta}^{\dot{\alpha}}\bar{\theta}^{\dot{\beta}} = \frac{1}{2}\varepsilon^{\dot{\alpha}\dot{\beta}}\bar{\theta}^2 \\ \theta_\alpha\bar{\theta}_{\dot{\alpha}} &= -\frac{1}{2}(\sigma_a)_{\alpha\dot{\alpha}}\theta\sigma^a\bar{\theta}. \end{aligned} \quad (2.4.20)$$

Therefore, an expansion of a superfield $V(z) = V(x, \theta, \bar{\theta})$ in a power series in θ^α and $\bar{\theta}_{\dot{\alpha}}$ reads

$$\begin{aligned} V(x, \theta, \bar{\theta}) = {} & A(x) + \theta^\alpha \psi_\alpha(x) + \bar{\theta}_{\dot{\alpha}} \bar{\varphi}^{\dot{\alpha}}(x) + \theta^2 F(x) + \bar{\theta}^2 G(x) \\ & + \theta \sigma^a \bar{\theta} C_a(x) + \bar{\theta}^2 \theta^\alpha \lambda_\alpha(x) + \theta^2 \bar{\theta}_{\dot{\alpha}} \bar{\eta}^{\dot{\alpha}}(x) + \theta^2 \bar{\theta}^2 D(x). \end{aligned} \quad (2.4.21)$$

The coefficients in this expansion are said to be 'component fields of the superfield'.

From now on we restrict ourselves to the consideration of 'bosonic superfields'

$$V: \mathbb{R}^{4|4} \to \mathbb{C}_c$$

and 'fermionic superfields'

$$V: \mathbb{R}^{4|4} \to \mathbb{C}_a$$

only. In these cases component fields are ordinary bosonic and fermionic fields over Minkowski space. This should be understood as follows. Let $V(z)$ be, for example, a bosonic superfield. Then, component fields $A(x)$, $F(x)$, $G(x)$, $C_a(x)$ and $D(x)$ are bosonic supersmooth functions on the c-number space $\mathbb{R}^4_c$, the other component fields are fermionic supersmooth functions on $\mathbb{R}^4_c$. Their restrictions from $\mathbb{R}^4_c$ to $\mathbb{R}^4$ represent smooth bosonic and fermionic fields on $\mathbb{R}^4$ identified with Minkowski space.

Given a superfield $V(z)$, we define its complex conjugate superfield $V^*(z)$ by the rule

$$V^*(z) \equiv (V(z))^* \qquad \forall z \in \mathbb{R}^{4|4}. \quad (2.4.22)$$

Since conjugation rules for the superspace coordinates have the form

$$\begin{aligned} & (x^a)^* = x^a \qquad (\theta^\alpha)^* = \bar{\theta}^{\dot{\alpha}} \qquad (\bar{\theta}_{\dot{\alpha}})^* = \theta_\alpha \\ & (\theta_\alpha \theta_\beta)^* = \bar{\theta}_{\dot{\beta}} \bar{\theta}_{\dot{\alpha}} \Rightarrow (\theta^2)^* = \bar{\theta}^2 \\ & (\bar{\theta}_{\dot{\beta}} \bar{\theta}_{\dot{\alpha}})^* = \theta_\alpha \theta_\beta \Rightarrow (\bar{\theta}^2)^* = \theta^2 \\ & (\theta_\alpha \bar{\theta}_{\dot{\beta}})^* = \theta_\beta \bar{\theta}_{\dot{\alpha}} \Rightarrow (\theta \sigma^a \bar{\theta})^* = \theta \sigma^a \bar{\theta} \end{aligned} \quad (2.4.23)$$

the component fields of $V^*(z)$ are related to the component fields of $V(z)$ in the following way

$$\begin{aligned} V^*(x, \theta, \bar{\theta}) = {} & A^*(x) + (-1)^{\varepsilon(V)} \theta^\alpha \varphi_\alpha(x) + (-1)^{\varepsilon(V)} \bar{\theta}_{\dot{\alpha}} \bar{\psi}^{\dot{\alpha}}(x) + \theta^2 G^*(x) \\ & + \bar{\theta}^2 F^*(x) + \theta \sigma^a \bar{\theta} C^*_a(x) + (-1)^{\varepsilon(V)} \bar{\theta}^2 \theta^\alpha \eta_\alpha(x) \\ & + (-1)^{\varepsilon(V)} \theta^2 \bar{\theta}_{\dot{\alpha}} \bar{\lambda}^{\dot{\alpha}}(x) + \theta^2 \bar{\theta}^2 D^*(x) \end{aligned} \quad (2.4.24)$$

where

$$\begin{aligned} & \varphi_\alpha(x) = (\bar{\varphi}_{\dot{\alpha}}(x))^* \qquad \bar{\psi}^{\dot{\alpha}}(x) = (\psi^\alpha(x))^* \\ & \eta_\alpha(x) = (\bar{\eta}_{\dot{\alpha}}(x))^* \qquad \bar{\lambda}^{\dot{\alpha}}(x) = (\lambda^\alpha(x))^* \end{aligned}$$

and $\varepsilon(V)$ is the Grassmann parity of $V(z)$.

Introduce partial derivatives of the superspace coordinates

$$\partial_A = (\partial_a, \partial_\alpha, \bar{\partial}^{\dot\alpha}) \equiv \frac{\partial}{\partial z^A} = \left(\frac{\partial}{\partial x^a}, \frac{\partial}{\partial \theta^\alpha}, \frac{\partial}{\partial \bar\theta_{\dot\alpha}}\right) \tag{2.4.25}$$

$$\partial_A z^B = \delta_A{}^B \Leftrightarrow \begin{cases} \partial_a x^b = \delta_a{}^b & \partial_a \theta^\alpha = \partial_a \bar\theta_{\dot\alpha} = 0 \\ \partial_\alpha \theta^\beta = \delta_\alpha{}^\beta & \partial_\alpha x^b = \partial_\alpha \bar\theta_{\dot\beta} = 0 \\ \bar\partial^{\dot\alpha} \bar\theta_{\dot\beta} = \delta^{\dot\alpha}{}_{\dot\beta} & \bar\partial^{\dot\alpha} x^b = \bar\partial^{\dot\alpha} \theta^\beta = 0. \end{cases}$$

We also define partial derivatives ∂^A corresponding to the variables $z_A = (x_a, \theta_\alpha, \bar\theta^{\dot\alpha})$,

$$\partial^A = (\partial^a, \partial^\alpha, \bar\partial_{\dot\alpha}) = \left(\frac{\partial}{\partial x_a}, \frac{\partial}{\partial \theta_\alpha}, \frac{\partial}{\partial \bar\theta^{\dot\alpha}}\right) \tag{2.4.26}$$

$$\partial^A z_B = \delta^A{}_B \Leftrightarrow \begin{cases} \partial^a x_b = \delta^a{}_b & \partial^a \theta_\beta = \partial^a \bar\theta^{\dot\beta} = 0 \\ \partial^\alpha \theta_\beta = \delta^\alpha{}_\beta & \partial^\alpha x_b = \partial^\alpha \bar\theta^{\dot\beta} = 0 \\ \bar\partial_{\dot\alpha} \bar\theta^{\dot\beta} = \delta_{\dot\alpha}{}^{\dot\beta} & \bar\partial_{\dot\alpha} x_b = \bar\partial_{\dot\alpha} \theta_\beta = 0. \end{cases}$$

The derivatives ∂_A and ∂^A are related as follows

$$\partial^a = \eta^{ab}\partial_b \qquad \partial^\alpha = -\varepsilon^{\alpha\beta}\partial_\beta \qquad \bar\partial_{\dot\alpha} = -\varepsilon_{\dot\alpha\dot\beta}\bar\partial^{\dot\beta}. \tag{2.4.27}$$

Recall that partial derivatives of even superspace coordinates are always left ones.

Basic properties of the partial derivatives are:

$$\begin{aligned} &1.\ [\partial_A, \partial_B\} \equiv \partial_A\partial_B - (-1)^{\varepsilon_A\varepsilon_B}\partial_B\partial_A = 0; \\ &2.\ \partial_A(V\cdot U) = (\partial_A V)U + (-1)^{\varepsilon_A\varepsilon(V)} V\partial_A U; \\ &3.\ \varepsilon(\partial_A V) = \varepsilon_A + \varepsilon(V) \qquad (\text{mod } 2); \\ &4.\ (\partial_a V)^* = \partial_a V^*, \quad (\partial_\alpha V)^* = -(-1)^{\varepsilon(V)}\bar\partial_{\dot\alpha} V^* \quad (\bar\partial^{\dot\alpha} V)^* = -(-1)^{\varepsilon(V)}\partial^\alpha V^*. \end{aligned} \tag{2.4.28}$$

Their derivation duplicates the general analysis of Section 1.10 with one modification: in Section 1.10 we parametrized superspaces $\mathbb{R}^{p|q}$ by real even and odd variables, $\mathbb{R}^{4|4}$ is parametrized by four real *c*-number variables x^a and four complex *a*-number variables θ^α and $\bar\theta^{\dot\alpha}$ conjugate to each other.

2.4.5. *Superfield representations of the super Poincaré group*

The notion of tensor fields is easily generalized to the superspace. For example, a tensor superfield of Lorentz type $(n/2, m/2)$ is defined by two requirements: (1) in every super-inertial system, it is determined by a set of $(n+1)(m+1)$ superfields $V_{\alpha_1\alpha_2\cdots\alpha_n\dot\alpha_1\dot\alpha_2\cdots\dot\alpha_m}(z)$ (component superfields), totally symmetric in n undotted indices and m dotted indices; (2) a super Poincaré transformation (2.4.13) changes the component superfields according to the law

$$V'_{\alpha_1\alpha_2\ldots\alpha_n\dot{\alpha}_1\dot{\alpha}_2\ldots\dot{\alpha}_m}(z') = e^{\frac{1}{2}K^{ab}M_{ab}}V_{\alpha_1\alpha_2\ldots\alpha_n\dot{\alpha}_1\dot{\alpha}_2\ldots\dot{\alpha}_m}(z) \tag{2.4.29}$$

where the Lorentz generators M_{ab} act on the external superfield indices only. Removing the restriction of $V_{\alpha_1\alpha_2\ldots\alpha_n\dot{\alpha}_1\dot{\alpha}_2\ldots\dot{\alpha}_m}(z)$ being totally symmetric in its undotted indices and dotted indices, the transformation law (2.4.29) defines an arbitrary Lorentz tensor superfield.

The operation of complex conjugation maps every tensor superfield $V_{\alpha_1\alpha_2\ldots\alpha_n\beta_1\beta_2\ldots\beta_m}(z)$ into

$$\bar{V}_{\beta_1\beta_2\ldots\beta_m\dot{\alpha}_1\dot{\alpha}_2\ldots\dot{\alpha}_n}(z) \equiv (V_{\alpha_1\alpha_2\ldots\alpha_n\beta_1\beta_2\ldots\beta_m}(z))^* \tag{2.4.30}$$

which is also a tensor superfield, in accordance with equation (2.4.29). Given a tensor superfield of Lorentz type $(n/2,m/2)$, its complex conjugate tensor superfield has Lorentz type $(m/2,n/2)$. In the case $n=m$, we can consider real tensor superfields defined by the equation

$$\bar{V}_{\alpha_1\alpha_2\ldots\alpha_n\dot{\alpha}_1\dot{\alpha}_2\ldots\dot{\alpha}_m}(z) = V_{\alpha_1\alpha_2\ldots\alpha_n\dot{\alpha}_1\dot{\alpha}_2\ldots\dot{\alpha}_m}(z) \tag{2.4.31}$$

As a rule, we will assume every tensor superfield carrying an even (odd) total number of indices to be bosonic (fermionic),

$$\varepsilon(V_{\alpha_1\alpha_2\ldots\alpha_n\dot{\alpha}_1\dot{\alpha}_2\ldots\dot{\alpha}_m}(z)) = n+m \qquad (\text{mod } 2). \tag{2.4.32}$$

The notion of tensor superfields given above will play a most important role in subsequent chapters for two basic reasons. First, in the case of Poincaré transformations (2.4.10, 11), equation (2.4.29) means nothing more than the fact that the component fields of $V_{\alpha_1\alpha_2\ldots\alpha_n\dot{\alpha}_1\dot{\alpha}_2\ldots\dot{\alpha}_m}(z)$ are ordinary bosonic and fermionic tensor fields. For example, let $V(z)$ be a scalar superfield transforming according to the law

$$V'(z') = V(z) \tag{2.4.33}$$

with respect to the super Poincaré group. Choosing a space–time translation (2.4.10) gives

$$\begin{aligned} &A'(x') + \theta'^{\alpha}\psi'_{\alpha}(x') + \bar{\theta}'_{\dot{\alpha}}\bar{\varphi}'^{\dot{\alpha}}(x') + \ldots \\ &= A'(x') + \theta^{\alpha}\psi'_{\alpha}(x') + \bar{\theta}_{\dot{\alpha}}\bar{\varphi}'^{\dot{\alpha}}(x') + \\ &= A(x) + \theta^{\alpha}\psi_{\alpha}(x) + \bar{\theta}_{\dot{\alpha}}\bar{\varphi}^{\dot{\alpha}}(x) + \ldots . \end{aligned}$$

In the case of Lorentz transformations (2.4.11) equation (2.4.31) leads to

$$\begin{aligned} &A'(x') + \theta'^{\alpha}\psi'_{\alpha}(x') + \bar{\theta}'_{\dot{\alpha}}\bar{\varphi}'^{\dot{\alpha}}(x') + \ldots \\ &= A'(x') + \theta^{\alpha}(e^{-\frac{1}{2}K^{ab}\sigma_{ab}})_{\alpha}{}^{\beta}\psi'_{\beta}(x') + \bar{\theta}_{\dot{\alpha}}(e^{-\frac{1}{2}K^{ab}\tilde{\sigma}_{ab}})^{\dot{\alpha}}{}_{\dot{\beta}}\bar{\varphi}'^{\dot{\beta}}(x') + \ldots \\ &= A(x) + \theta^{\alpha}\psi_{\alpha}(x) + \bar{\theta}_{\dot{\alpha}}\bar{\varphi}^{\dot{\alpha}}(x) + \ldots \end{aligned}$$

Therefore, in expansion (2.4.21), component fields $A(x)$, $F(x)$, $G(x)$ and $D(x)$ are scalar fields, $C_a(x)$ is a vector field, $\psi_{\alpha}(x)$ and $\lambda_{\alpha}(x)$ are undotted spinor fields and $\bar{\varphi}^{\dot{\alpha}}(x)$ and $\bar{\eta}^{\dot{\alpha}}(x)$ are dotted spinor fields. The second reason why tensor superfields are of primary importance is that the transformation law (2.4.29) automatically provides us with a realization of supersymmetry transformations on component tensor fields. Therefore, if we work out a technique to handle superfields, in particular to construct super Poincaré invariant functionals of superfields, we shall arrive at a supersymmetric field theory.

Supersymmetry transformation laws of component fields will be analysed below.

Tensor superfields provide us with representations of the super Poincaré group. For example, consider the space $\mathscr{L}_{(n,m)}$ of tensor superfields of Lorentz type $(n/2,m/2)$. With every element $g=g(b,\epsilon,\bar{\epsilon},K)$ of the super Poincaré group we relate a one-to-one mapping

$$T(g)\colon \mathscr{L}_{(n,m)} \rightarrow \mathscr{L}_{(n,m)}$$

which is given by rewriting the transformation law (2.4.29) in the form:

$$T(g)\colon V(z) \rightarrow V'(z)=\mathrm{e}^{\frac{i}{2}K^{ab}M_{ab}}V(g^{-1}\cdot z) \qquad \forall\, V(z)\in \mathscr{L}_{(n,m)} \tag{2.4.34}$$

where we have suppressed indices. Here $z'^A=g^{-1}\cdot z^A$ is the super Poincaré transformation corresponding to the element g^{-1}. Evidently, the correspondence $g \rightarrow T(g)$ determines a representation of the super Poincaré group. Therefore, every representation operator $T(g)$ can be expressed as

$$\begin{aligned} T(g(b,\epsilon,\bar{\epsilon},K)) &= \exp[\mathrm{i}(-b^a\mathbf{P}_a+\tfrac{1}{2}K^{ab}\mathbf{J}_{an}+\epsilon\mathbf{Q}+\overline{\epsilon\mathbf{Q}})] \\ &= \exp[\mathrm{i}(\tfrac{1}{2}b^{\alpha\dot{\alpha}}\mathbf{P}_{\alpha\dot{\alpha}}+K^{\alpha\beta}\mathbf{J}_{\alpha\beta}+\bar{K}^{\dot{\alpha}\dot{\beta}}\bar{\mathbf{J}}_{\dot{\alpha}\dot{\beta}}+\epsilon\mathbf{Q}+\overline{\epsilon\mathbf{Q}})] \end{aligned} \tag{2.4.35}$$

where

$$\mathbf{P}_a=\mathrm{d}T(p_a) \qquad \mathbf{J}_{ab}=\mathrm{d}T(\mathbf{j}_{ab}) \qquad \mathbf{Q}_\alpha=\mathrm{d}T(\mathbf{q}_\alpha) \qquad \bar{\mathbf{Q}}^{\dot{\alpha}}=\mathrm{d}T(\bar{\mathbf{q}}^{\dot{\alpha}})$$

are generators of the representation. One can easily find the generators by recalling the explicit form for the super Poincaré transformations. The result is

$$\begin{aligned} \mathbf{P}_a &= -\mathrm{i}\partial_a \\ \mathbf{J}_{ab} &= \mathrm{i}(x_b\partial_a-x_a\partial_b+(\sigma_{ab})^{\alpha\beta}\theta_\alpha\partial_\beta-(\tilde{\sigma}_{ab})^{\dot{\alpha}\dot{\beta}}\bar{\theta}_{\dot{\alpha}}\bar{\partial}_{\dot{\beta}}-M_{ab}) \\ \mathbf{Q}_\alpha &= \mathrm{i}\partial_\alpha+(\sigma^a)_{\alpha\dot{\alpha}}\bar{\theta}^{\dot{\alpha}}\partial_a \\ \bar{\mathbf{Q}}_{\dot{\alpha}} &= -\mathrm{i}\bar{\partial}_{\dot{\alpha}}-\theta^\alpha(\sigma^a)_{\alpha\dot{\alpha}}\partial_a \end{aligned} \tag{2.4.36}$$

or, in spinor notation,

$$\begin{aligned} \mathbf{P}_{\alpha\dot{\alpha}} &= -\mathrm{i}\partial_{\alpha\dot{\alpha}} \\ \mathbf{J}_{\alpha\beta} &= -\tfrac{\mathrm{i}}{4}(x_\alpha{}^{\dot{\alpha}}\partial_{\beta\dot{\alpha}}+x_\beta{}^{\dot{\alpha}}\partial_{\alpha\dot{\alpha}})+\tfrac{1}{2}(\theta_\alpha\partial_\beta+\theta_\beta\partial_\alpha)-\mathrm{i}M_{\alpha\beta} \\ \bar{\mathbf{J}}_{\dot{\alpha}\dot{\beta}} &= -\tfrac{\mathrm{i}}{4}(x^\alpha{}_{\dot{\alpha}}\partial_{\alpha\dot{\beta}}+x^\alpha{}_{\dot{\beta}}\partial_{\alpha\dot{\alpha}})+\tfrac{1}{2}(\bar{\theta}_{\dot{\alpha}}\bar{\partial}_{\dot{\beta}}+\bar{\theta}_{\dot{\beta}}\bar{\partial}_{\dot{\alpha}})-\mathrm{i}\bar{M}_{\dot{\alpha}\dot{\beta}} \\ \mathbf{Q}_\alpha &= \mathrm{i}\partial_\alpha+\bar{\theta}^{\dot{\alpha}}\partial_{\alpha\dot{\alpha}} \qquad \bar{\mathbf{Q}}_{\dot{\alpha}} = -\mathrm{i}\bar{\partial}_{\dot{\alpha}}-\theta^\alpha\partial_{\alpha\dot{\alpha}} \end{aligned} \tag{2.4.37}$$

where

$$\partial_{\alpha\dot{\alpha}}=(\sigma^a)_{\alpha\dot{\alpha}}\partial_a \qquad x_{\alpha\dot{\alpha}}=(\sigma^a)_{\alpha\dot{\alpha}}x_a \qquad \partial_{\alpha\dot{\alpha}}x^{\beta\dot{\beta}}=-2\delta_\alpha^\beta\delta_{\dot{\alpha}}^{\dot{\beta}} \tag{2.4.38}$$

As an exercise, one can check that the operators (2.4.36) really form a representation of the Poincaré superalgebra. The reader should always keep in mind that the Lorentz generators M_{ab} in equations (2.4.34) and (2.4.36) act on external superfield indices.

Infinitesimal supersymmetry transformations act on any tensor superfields by the law

$$\delta V_{\alpha_1\ldots\alpha_n\dot{\alpha}_1\ldots\dot{\alpha}_m}(z)=\mathrm{i}(\varepsilon\mathbf{Q}+\bar{\varepsilon}\bar{\mathbf{Q}})V_{\alpha_1\ldots\alpha_n\dot{\alpha}_1\ldots\dot{\alpha}_m}(z). \tag{2.4.39}$$

As for the superspace supersymmetry transformations (2.4.12), they can also be written in terms of supersymmetry generators $\mathbf{Q}_\alpha$ and $\bar{\mathbf{Q}}_{\dot{\alpha}}$ as follows

$$z'^A=z^A-\mathrm{i}(\varepsilon\mathbf{Q}+\bar{\varepsilon}\bar{\mathbf{Q}})z^A. \tag{2.4.40}$$

2.4.6. Mass dimensions

In conclusion, let us discuss a technical question regarding dimensions (in units of mass) of the superspace coordinates. Evidently, for the even superspace coordinates we have

$$[x^a]=-1 \qquad [\partial_a]=1. \tag{2.4.41}$$

To determine the dimension of θ^α and $\bar{\theta}^{\dot{\alpha}}$, it is necessary to recall equation (2.4.12), which forces us to demand

$$[\theta^\alpha]=[\bar{\theta}^{\dot{\alpha}}]=-1/2 \qquad [\partial_\alpha]=[\bar{\partial}_{\dot{\alpha}}]=1/2 \tag{2.4.42}$$

After this, having a superfield of some fixed dimension, one can easily determine dimensions of its component fields.

The concept of superspace and superfields was first introduced by A. Salam and J. Strathdee.

2.5. Complex superspace $\mathbb{C}^{4|2}$, chiral superfields and covariant derivatives

We are now going to describe one more realization of the super Poincaré group as a group of transformations in superspace. Using this realization will turn out to be very helpful in two respects. In the present section it will facilitate an introduction of chiral superfields—important representations of supersymmetry. Later, in Chapter 5, this realization will serve as a starting point for constructing supergravity—that is, a gauge theory of supersymmetry.

The point of departure of our discussion is the observation that the set $\{x^a_{(+)}, \theta^\alpha\}$ of complex variables $x^a_{(+)}\equiv x^a+\mathrm{i}\theta\sigma^a\bar{\theta}$ and θ^α is closed with respect to the super Poincaré group, because the supersymmetry transformations act on these variables as follows

$$x^a_{(+)} \to x'^a_{(+)} = x^a_{(+)} + 2\mathrm{i}\theta\sigma^a\bar{\epsilon} + \mathrm{i}\epsilon\sigma^a\bar{\epsilon} \qquad \theta^\alpha \to \theta'^\alpha = \theta^\alpha + \epsilon^\alpha \tag{2.5.1}$$

All the other results of this present section are in a sense consequences of this observation.

2.5.1. Complex superspace $\mathbb{C}^{4|2}$

Consider a complex superspace $\mathbb{C}^{4|2}$ spanned by four complex c-number coordinates y^a and two complex a-number coordinates θ^α. We introduce super Poincaré transformations on $\mathbb{C}^{4|2}$ by associating with every element $g \in S\Pi$ a one-to-one mapping $y^a \to y'^a = g\cdot y^a$, $\theta^\alpha \to \theta'^\alpha = g\cdot\theta^\alpha$ of $\mathbb{C}^{4|2}$ to itself, defined as follows:

Translations

$$g(b)\cdot y^a = y^a + b^a \qquad g(b)\cdot\theta^\alpha = \theta^\alpha \tag{2.5.2a}$$

Lorentz transformations

$$g(K)\cdot y^a = (\mathrm{e}^K)^a{}_b y^b \qquad g\cdot\theta_\alpha = (\exp(\tfrac{1}{2}K^{ab}\sigma_{ab}))_\alpha{}^\beta\theta_\beta \tag{2.5.2b}$$

Supersymmetry transformations

$$g(\epsilon,\bar{\epsilon})\cdot y^a = y^a + 2\mathrm{i}\theta\sigma^a\bar{\epsilon} + \mathrm{i}\epsilon\sigma^a\bar{\epsilon} \qquad g(\epsilon,\bar{\epsilon})\cdot\theta^\alpha = \theta^\alpha + \epsilon^\alpha \tag{2.5.2c}$$

The most general form of super Poincaré transformations on $\mathbb{C}^{4|2}$ reads

$$\begin{aligned} y'^a &= (\exp K)^a{}_b y^b + 2\mathrm{i}\theta\sigma^a\bar{\epsilon} + b^a + \mathrm{i}\epsilon\sigma^a\bar{\epsilon} \\ \theta'^\alpha &= (\exp(-K))^\alpha{}_\beta\theta^\beta + \epsilon^\alpha \end{aligned} \tag{2.5.3}$$

Evidently, expressions (2.5.2) define a group of transformations on $\mathbb{C}^{4|2}$, i.e. $g_1(g_2\cdot y^a) = (g_1g_2)\cdot y^a$ and $g_1(g_2\cdot\theta^\alpha) = (g_1g_2)\cdot\theta^\alpha$, for any elements g_1 and g_2 of the super Poincaré group.

Beautifully, since $\mathbb{C}^{4|2}$ is complex, one can define an action on $\mathbb{C}^{4|2}$ of a complex super Lie group corresponding to the complex shell (2.2.17) of the super Poincaré algebra. Elements of this super Lie group are parametrized by complex c-number variables f^a, $K^{\alpha\beta} = K^{\beta\alpha}$ and $\bar{L}^{\dot\alpha\dot\beta} = \bar{L}^{\dot\beta\dot\alpha}$ and complex a-number variables ϵ^a and $\bar{\xi}_{\dot\alpha}$,

$$g = \exp[\mathrm{i}(-f^a p_a + K^{\alpha\beta}\mathbf{j}_{\alpha\beta} + \bar{L}^{\dot\alpha\dot\beta}\bar{\mathbf{j}}_{\dot\alpha\dot\beta} + \epsilon\mathbf{q} + \bar{\xi}\bar{\mathbf{q}})]$$

where, in contrast to the super Poincaré group, $\bar{L}^{\dot\alpha\dot\beta} \neq (K^{\alpha\beta})^*$ and $\bar{\xi}_{\dot\alpha} \neq (\varepsilon_\alpha)^*$. Transformations of $\mathbb{C}^{4|2}$ absent in expressions (2.5.2) have the form

$$\mathrm{e}^{\mathrm{i}K^{\beta\gamma}\mathbf{j}_{\beta\gamma}}\cdot y^{\alpha\dot\alpha} = (\mathrm{e}^{-K})^\alpha{}_\beta y^{\beta\dot\alpha} \qquad \mathrm{e}^{\mathrm{i}K^{\beta\gamma}\mathbf{j}_{\beta\gamma}}\cdot\theta^\alpha = (\mathrm{e}^{-K})^\alpha{}_\beta\theta^\beta \tag{2.5.4a}$$

$$\mathrm{e}^{\mathrm{i}\bar{L}^{\dot\beta\dot\gamma}\bar{\mathbf{j}}_{\dot\beta\dot\gamma}}\cdot y^{\alpha\dot\alpha} = (\mathrm{e}^{-\bar{L}})^{\dot\alpha}{}_{\dot\beta} y^{\alpha\dot\beta} \qquad \mathrm{e}^{\mathrm{i}\bar{L}^{\dot\beta\dot\gamma}\bar{\mathbf{j}}_{\dot\beta\dot\gamma}}\cdot\theta^\alpha = \theta^\alpha \tag{2.5.4b}$$

$$e^{i\epsilon\mathbf{q}}\cdot y^a = y^a \qquad e^{i\epsilon\mathbf{q}}\cdot\theta^\alpha = \theta^\alpha + \epsilon^\alpha \tag{2.5.4c}$$

$$e^{i\bar{\xi}\bar{\mathbf{q}}}\cdot y^a = y^a + 2i\theta\sigma^a\bar{\xi} \qquad e^{i\bar{\xi}\bar{\mathbf{q}}}\cdot\theta^\alpha = \theta^\alpha \tag{2.5.4d}$$

The reader may forget, for a time, transformations (2.5.4) because they will be used only in Section 2.9 when studying the superconformal group.

2.5.2. Holomorphic superfields

Complex superspace $\mathbb{C}^{4|2}$ can equivalently be considered as a real superspace $\mathbb{R}^{8|4}$ with coordinates y^a, $\bar{y}^a = (y^a)^*$, θ^α and $\bar{\theta}^{\dot{\alpha}} = (\theta^\alpha)^*$. In general, a superfield on $\mathbb{C}^{4|2}$ is a supersmooth function U: $\mathbb{C}^{4|2} \to \Lambda_\infty$ of all these variables, $U = U(y,\bar{y},\theta,\bar{\theta})$. However, objects best adapted to the complex structure on $\mathbb{C}^{4|2}$ are 'holomorphic superfields' depending on the variables y^a and θ^α only,

$$\Phi = \Phi(y,\theta) \Leftrightarrow \partial\Phi/\partial\bar{y}^a = \partial\Phi/\partial\bar{\theta}^{\dot{\alpha}} = 0 \tag{2.5.5}$$

and antiholomorphic superfields defined by

$$\Psi = \Psi(\bar{y},\bar{\theta}) \Leftrightarrow \partial\Psi/\partial y^a = \partial\Psi/\partial\theta^\alpha = 0. \tag{2.5.6}$$

Clearly, an antiholomorphic superfield is complex conjugate to a holomorphic superfield. Every holomorphic superfield can be expanded in a power series in θ^α:

$$\Phi(y,\theta) = A(y) + \theta^\alpha\psi_\alpha(y) + \theta^2 F(y). \tag{2.5.7}$$

In accordance with equation (2.5.3), the super Poincaré transformations represent holomorphic mappings of $\mathbb{C}^{4|2}$ to itself. Therefore, it is possible to introduce into consideration tensor holomorphic superfields like $\Phi_{\alpha_1\alpha_2\ldots\alpha_n\dot{\alpha}_1\dot{\alpha}_2\ldots\dot{\alpha}_m}(y,\theta)$ defined by the transformation law

$$\Phi'_{\alpha_1\alpha_2\ldots\alpha_n\dot{\alpha}_1\dot{\alpha}_2\ldots\dot{\alpha}_m}(y',\theta') = e^{\frac{1}{2}K^{ab}M_{ab}}\Phi_{\alpha_1\alpha_2\ldots\alpha_n\dot{\alpha}_1\dot{\alpha}_2\ldots\dot{\alpha}_m}(y,\theta) \tag{2.5.8}$$

with respect to the super Poincaré transformations, where the Lorentz generators M_{ab} act, as usual, on external superfield indices only.

Similarly to ordinary holomorphic functions on $\mathbb{C}^n$, a holomorphic superfield $\Phi(y,\theta)$ proves to be an analytic function of complex c-number variables y^a. This means that $\Phi(y,\theta)$ possesses a convergent Taylor expansion in y^a near each point of the superspace. In particular, holomorphic superfields are not localizable in y^a, i.e. one cannot make a holomorphic superfield on $\mathbb{C}^{4|2}$ non-vanishing in a small neighbourhood of some point y_0^a only. Evidently, this is a very severe restriction from the point of view of field theory. Nevertheless, there is a way to obtain localizable holomorphic superfields. One should simply restrict the region of the superfield definition from the whole superspace $\mathbb{C}^{4|2}$ to a surface in $\mathbb{C}^{4|2}$ such that, for every point (y^a,θ^α) on the surface, the imaginary part of y^a is bodiless,

$$(y^a - \bar{y}^a)_B = 0. \tag{2.5.9}$$

This surface will be denoted $\mathbb{C}_s^{4|2}$ and termed the 'complex truncated superspace'.

The surface $\mathbb{C}_s^{4|2}$ is interesting for two reasons. First, due to equation (2.5.3), every super Poincaré transformation maps $\mathbb{C}_s^{4|2}$ on to itself. Therefore, the notion of tensor holomorphic superfields can be transferred from $\mathbb{C}^{4|2}$ to $\mathbb{C}_s^{4|2}$. Secondly, every holomorphic superfield on $\mathbb{C}_s^{4|2}$ may be represented in the form

$$\Phi(y,\theta) = \sum \frac{1}{n!} \frac{\partial^n \Phi(x,\theta)}{\partial x^{a_1} \ldots \partial x^{a_n}}\bigg|_{x=\frac{1}{2}(y+\bar{y})} \frac{(y-\bar{y})^{a_1}}{2} \cdots \frac{(y-\bar{y})^{a_n}}{2} \tag{2.5.10}$$

where $\Phi(x,\theta)$ is a supersmooth function of four real c-number variables x^a and two complex a-number variables θ^α. Conversely, for every such superfunction $\Phi(x,\theta)$, equation (2.5.10) defines a holomorphic superfield on $\mathbb{C}_s^{4|2}$ (but not, in general, on $\mathbb{C}^{4|2}$). To confirm these assertions, the reader may recall the discussion in Section 1.10 concerning supersmooth functions. We adopt the convention that every holomorphic superfield appearing below is defined on $\mathbb{C}_s^{4|2}$.

2.5.3. $\mathbb{R}^{4|4}$ *as a surface in* $\mathbb{C}^{4|2}$

Let us introduce a family of surfaces in $\mathbb{C}^{4|2}$ determined by

$$y^a - \bar{y}^a = 2\mathrm{i}\mathscr{H}^a(\tfrac{1}{2}y + \tfrac{1}{2}\bar{y},\, \theta,\, \bar{\theta}) \tag{2.5.11}$$

Here $\mathscr{H}^a$, $a = 0, 1, 2, 3$, are real superfields on $\mathbb{R}^{4|4}$. Every surface (2.5.11) can be parametrized by four real c-number variables x^a, given by

$$x^a = \tfrac{1}{2}(y + \bar{y})^a \tag{2.5.12}$$

and four complex a-number variables θ^α and $\bar{\theta}^{\dot{\alpha}}$ conjugate to each other. Therefore, we may look on this surface as a real superspace $\mathbb{R}^{4|4}$ equipped with a set of four real superfields $\mathscr{H}^a(x,\theta,\bar{\theta})$ on $\mathbb{R}^{4|4}$. So, it seems natural to use the notation $\mathbb{R}^{4|4}(\mathscr{H})$ for such a surface.

Now, let us try to find a super Poincaré invariant surface $\mathbb{R}^{4|4}(\mathscr{H})$—that is, a surface which moves into itself when applying an arbitrary super Poincaré transformation. First, invariance with respect to the translations (2.5.2*a*) imposes the restriction

$$\mathscr{H}^a(x+b,\theta,\bar{\theta}) = \mathscr{H}^a(x,\theta,\bar{\theta})$$

because the combination $(y-\bar{y})$ is translationally invariant. We see that $\mathscr{H}^a$ is a function of θ and $\bar{\theta}$ only. Secondly, the invariance with respect to the Lorentz transformations (2.5.2*b*) leads to

$$\mathscr{H}^a(\theta',\bar{\theta}') = (\mathrm{e}^K)^a{}_b \mathscr{H}^b(\theta,\bar{\theta}).$$

The most general solution of this equation proves to be $\mathscr{H}^a = k\theta\sigma^a\bar{\theta}$, where

k is a constant real c-number. Finally, the invariance with respect to the supersymmetry transformations (2.5.2c) fixes this constant resulting in

$$\mathscr{H}^a = \theta\sigma^a\bar{\theta}. \tag{2.5.13}$$

As a result, we arrive at a beautiful conclusion. Namely, the family $\{\mathbb{R}^{4|4}(\mathscr{H})\}$ of surfaces in $\mathbb{C}^{4|2}$ includes a unique super Poincaré invariant surface—$\mathbb{R}^{4|4}(\theta\sigma\bar{\theta})$. But this is not the whole story. Restricting transformations (2.5.2) from $\mathbb{C}^{4|2}$ to $\mathbb{R}^{4|4}(\theta\sigma\bar{\theta})$ one finds they coincide with the super-Poincaré transformations on $\mathbb{R}^{4|4}$ (see equation (2.4.10–12)). Therefore, one can identify the superspace $\mathbb{R}^{4|4}$ with the surface $\mathbb{R}^{4|4}(\theta\sigma\bar{\theta})$ in $\mathbb{C}^{4|2}$. The identification works as follows

$$(x^a,\ \theta^\alpha,\ \bar{\theta}_{\dot{\alpha}}) \leftrightarrow (x^a + \mathrm{i}\theta\sigma^a\bar{\theta},\ \theta^\alpha) \tag{2.5.14}$$

2.5.4. *Chiral superfields*

Let $\Phi_{\alpha_1\ldots\alpha_n\dot{\alpha}_1\ldots\dot{\alpha}_m}(y,\theta)$ be a tensor holomorphic superfield. Restricting it to the surface $\mathbb{R}^{4|4}(\theta\sigma\bar{\theta})$ and applying the prescription (2.5.14) leads to the tensor superfield

$$\Phi_{\alpha_1\ldots\alpha_n\dot{\alpha}_1\ldots\dot{\alpha}_m}(x,\theta,\bar{\theta}) = \Phi_{\alpha_1\ldots\alpha_n\dot{\alpha}_1\ldots\dot{\alpha}_m}(x + \mathrm{i}\theta\sigma\bar{\theta},\ \theta) \tag{2.5.15}$$

on real superspace $\mathbb{R}^{4|4}$. Evidently, it is a quite trivial fact that we have really obtained a tensor superfield on $\mathbb{R}^{4|4}$. However, due to its importance, let us comment on it in detail. Under a super Poincaré transformation, $\Phi_{\alpha_1\ldots\alpha_n\dot{\alpha}_1\ldots\dot{\alpha}_m}(y,\theta)$ changes according to the law (2.5.8). For every point (y^a,θ^α) from the surface $\mathbb{R}^{4|4}(\theta\sigma\bar{\theta})$, the transformed point (y'^a,θ'^α) will also lie on $\mathbb{R}^{4|4}(\theta\sigma\bar{\theta})$. Therefore, we have

$$\Phi'_{\alpha_1\ldots\alpha_n\dot{\alpha}_1\ldots\dot{\alpha}_m}(x' + \mathrm{i}\theta'\sigma\bar{\theta}',\theta') = \mathrm{e}^{\frac{1}{2}K^{ab}M_{ab}}\Phi_{\alpha_1\ldots\alpha_n\dot{\alpha}_1\ldots\dot{\alpha}_m}(x + \mathrm{i}\theta\sigma\bar{\theta},\theta).$$

This gives

$$\Phi'_{\alpha_1\ldots\alpha_n\dot{\alpha}_1\ldots\dot{\alpha}_m}(x',\theta',\bar{\theta}') = \mathrm{e}^{\frac{1}{2}K^{ab}M_{ab}}\Phi_{\alpha_1\ldots\alpha_n\dot{\alpha}_1\ldots\dot{\alpha}_m}(x,\theta,\bar{\theta})$$

which represents a tensor transformation law. Note that the transformed superfield depends on the superspace coordinates in the same fashion as in superfield (2.5.15),

$$\Phi'_{\alpha_1\ldots\alpha_n\dot{\alpha}_1\ldots\dot{\alpha}_m}(x,\theta,\bar{\theta}) = \Phi'_{\alpha_1\ldots\alpha_n\dot{\alpha}_1\ldots\dot{\alpha}_m}(x + \mathrm{i}\theta\sigma\bar{\theta},\theta). \tag{2.5.16}$$

A tensor superfield on $\mathbb{R}^{4|4}$ defined by equation (2.5.15) is said to be a 'tensor chiral superfield'. Its conjugate tensor superfield

$$\bar{\Phi}_{\alpha_1\ldots\alpha_m\dot{\alpha}_1\ldots\dot{\alpha}_n}(x,\theta,\bar{\theta}) = \bar{\Phi}_{\alpha_1\ldots\alpha_m\dot{\alpha}_1\ldots\dot{\alpha}_n}(x - \mathrm{i}\theta\sigma\bar{\theta},\bar{\theta}) \tag{2.5.17}$$

is said to be a 'tensor antichiral superfield'. As may be seen, chiral superfields depend on x and $\bar{\theta}$ only through the combination $(x + \mathrm{i}\theta\sigma\bar{\theta})$, antichiral superfields depend on x and θ only through the combination $(x - \mathrm{i}\theta\sigma\bar{\theta})$.

Obviously, the set of tensor chiral (or antichiral) superfields of Lorentz type $(n/2,m/2)$ forms a super Poincaré invariant subspace in the space of tensor superfields of Lorentz type $(n/2,m/2)$. So it would be desirable to obtain a covariant constraint selecting this subspace. This is an easy problem upon taking into account the observation that equation (2.5.15) can be rewritten in the form

$$\Phi_{\alpha_1\ldots\alpha_n\dot{\alpha}_1\ldots\dot{\alpha}_m}(x,\theta,\bar{\theta})=e^{i\mathscr{H}}\Phi_{\alpha_1\ldots\alpha_n\dot{\alpha}_1\ldots\dot{\alpha}_m}(x,\theta) \qquad (2.5.18)$$
$$\mathscr{H}=\theta\sigma^a\bar{\theta}\partial_a$$

Therefore, every tensor chiral superfield satisfies the equation

$$\bar{D}_{\dot{\alpha}}\Phi_{\alpha_1\ldots\alpha_n\dot{\alpha}_1\ldots\dot{\alpha}_m}(x,\theta,\bar{\theta})=0$$
$$\bar{D}_{\dot{\alpha}}\equiv e^{i\mathscr{H}}(-\bar{\partial}_{\dot{\alpha}})e^{-i\mathscr{H}}=-\bar{\partial}_{\dot{\alpha}}-i\theta^{\alpha}\partial_{\alpha\dot{\alpha}}. \qquad (2.5.19)$$

Analogously, every tensor antichiral superfield can be represented in the form

$$\bar{\Phi}_{\alpha_1\ldots\alpha_n\dot{\alpha}_1\ldots\dot{\alpha}_m}(x,\theta,\bar{\theta})=e^{-i\mathscr{H}}\bar{\Phi}_{\alpha_1\ldots\alpha_n\dot{\alpha}_1\ldots\dot{\alpha}_m}(x,\bar{\theta}) \qquad (2.5.20)$$

and hence it satisfies the equation

$$D_{\alpha}\bar{\Phi}_{\alpha_1\ldots\alpha_n\dot{\alpha}_1\ldots\dot{\alpha}_m}(x,\theta,\bar{\theta})=0$$
$$D_{\alpha}\equiv e^{-i\mathscr{H}}\partial_{\alpha}e^{i\mathscr{H}}=\partial_{\alpha}+i\bar{\theta}^{\dot{\alpha}}\partial_{\alpha\dot{\alpha}}. \qquad (2.5.21)$$

2.5.5. Covariant derivatives

The differential operators D_{α} and $\bar{D}_{\dot{\alpha}}$ introduced in the previous subsection anticommute with the supersymmetry generators

$$\{D_{\alpha},Q_{\beta}\}=\{D_{\alpha},\bar{Q}_{\dot{\beta}}\}=0$$
$$\{\bar{D}_{\dot{\beta}},Q_{\beta}\}=\{\bar{D}_{\dot{\alpha}},\bar{Q}_{\dot{\beta}}\}=0. \qquad (2.5.22)$$

These identities can be checked explicitly. However, there are two different ways to understand relations (2.5.22). First, consider a tensor chiral superfield $\Phi(z)$ (indices are suppressed). After applying an infinitesimal supersymmetry transformation, it takes the form

$$\Phi'(z)=\Phi(z)+\delta\Phi(z) \qquad \delta\Phi(z)=i(\epsilon Q+\overline{\epsilon Q})\Phi(z).$$

Superfields $\Phi(z)$ and $\Phi'(z)$ are chiral $\bar{D}_{\dot{\alpha}}\Phi=\bar{D}_{\dot{\alpha}}\Phi'=0$. Therefore, we must have

$$[\bar{D}_{\dot{\alpha}},\ \epsilon Q+\overline{\epsilon Q}]=0$$

at least when acting on chiral superfields. Secondly, let us recall how the coset space $S\Pi/SO(3,1|\mathbb{R}_c)^{\uparrow}$ has been identified above with real superspace $\mathbb{R}^{4|4}$. Namely, points of the coset space are in one-to-one correspondence with elements

$$g(z)=e^{i(-x^a p_a+\theta q+\overline{\theta q})}$$

of the super Poincaré group. The supersymmetry transformations act on these elements by left shifts $l_{g(\varepsilon,\bar{\varepsilon})}$

$$g(z) \rightarrow g(z') = l_{g(\varepsilon,\bar{\varepsilon})} g(z) = e^{i(\varepsilon q + \overline{\varepsilon q})} g(z)$$

and lead to the superspace transformation

$$z^A \rightarrow z'^A = z^A - i(\varepsilon Q + \overline{\varepsilon Q}) z^A \tag{2.3.23}$$

Now, consider right shifts $r_{g(\eta,\bar{\eta})}$

$$g(z) \rightarrow g(z') = r_{g(\eta,\bar{\eta})} g(z) = g(z) e^{i(\eta q + \overline{\eta q})}$$

which induce the superspace transformation

$$z^A \rightarrow z'^A = z^A + (\eta D + \overline{\eta D}) z^A \tag{2.5.24}$$

Since the left and right shifts commute, $l_{g(\varepsilon,\bar{\varepsilon})} r_{g(\eta,\bar{\eta})} = r_{g(\eta,\bar{\eta})} l_{g(\varepsilon,\bar{\varepsilon})}$ and from their explicit expressions (2.5.23) and (2.5.24), one obtains expression (2.5.22).

Derivatives

$$D_A = (\partial_a, D_\alpha, \bar{D}^{\dot{\alpha}}) \tag{2.5.25}$$

are seen to form a complete set of first-order differential operators commuting with the supersymmetry transformations,

$$[D_A, \varepsilon Q + \overline{\varepsilon Q}] = 0 \tag{2.5.26}$$

This is the first reason why D_A are called 'covariant derivatives'. The second reason is that for every tensor superfield $V_{\alpha_1 \ldots \alpha_n \dot{\alpha}_1 \ldots \dot{\alpha}_m}(z)$, the superfield

$$U_{A\alpha_1 \ldots \alpha_n \dot{\alpha}_1 \ldots \dot{\alpha}_m}(z) \equiv D_A V_{\alpha_1 \ldots \alpha_n \dot{\alpha}_1 \ldots \dot{\alpha}_m}(z)$$

turns out also to be a tensor superfield, with the transformation law

$$U'^A{}_{\alpha_1 \ldots \alpha_n \dot{\alpha}_1 \ldots \dot{\alpha}_m}(z') = e^{\frac{1}{2} K^{bc} M_{bc}} U_{A\alpha_1 \ldots \alpha_n \dot{\alpha}_1 \ldots \dot{\alpha}_m}(z)$$

with respect to the super Poincaré group. Here the Lorentz generators M_{bc} act on all external indices including index A. In other words, a covariant derivative moves every tensor superfield into a tensor superfield. The operators D_α and $\bar{D}_{\dot{\alpha}}$ are said to be 'spinor covariant derivatives'.

2.5.6. *Properties of covariant derivatives*

The covariant derivatives satisfy the algebra

$$\begin{gathered} \{D_\alpha, D_\beta\} = \{\bar{D}_{\dot{\alpha}}, \bar{D}_{\dot{\beta}}\} = [D_\alpha, \partial_a] = [\bar{D}_{\dot{\alpha}}, \partial_a] = 0 \\ \{D_\alpha, \bar{D}_{\dot{\alpha}}\} = -2i\partial_{\alpha\dot{\alpha}} = 2P_{\alpha\dot{\alpha}}. \end{gathered} \tag{2.5.27}$$

We see that the spinor covariant derivatives and the supersymmetry generators satisfy similar anticommutation relations. Furthermore, the dotted and undotted covariant derivatives are related by complex conjugation as follows

$$(D_\alpha V)^* = (-1)^{\varepsilon(V)} \bar{D}_{\dot\alpha} V^* \qquad (D^2 V)^* = \bar{D}^2 V^* \tag{2.5.28}$$

for an arbitrary superfield $V(z)$ (bosonic or fermionic), where we have introduced the notation

$$\begin{aligned} D^2 &= D^\alpha D_\alpha \qquad D^\alpha = \varepsilon^{\alpha\beta} D_\beta \\ \bar{D}^2 &= \bar{D}_{\dot\alpha} \bar{D}^{\dot\alpha} \qquad \bar{D}^{\dot\alpha} = \varepsilon^{\dot\alpha\dot\beta} \bar{D}_{\dot\beta}. \end{aligned} \tag{2.5.29}$$

Let us list the identities involving the covariant derivatives which are known to be most relevant for practical superfield calculations:

$$D_\alpha D_\beta = \tfrac{1}{2}\varepsilon_{\alpha\beta} D^2 \qquad \bar{D}_{\dot\alpha} \bar{D}_{\dot\beta} = -\tfrac{1}{2}\varepsilon_{\dot\alpha\dot\beta} \bar{D}^2 \tag{2.5.30a}$$

$$D_\alpha D_\beta D_\gamma = 0 \qquad \bar{D}_{\dot\alpha} \bar{D}_{\dot\beta} \bar{D}_{\dot\gamma} = 0 \tag{2.5.30b}$$

$$[D^2, \bar{D}_{\dot\alpha}] = -4i\partial_{\alpha\dot\alpha} D^\alpha \qquad [\bar{D}^2, D_\alpha] = 4i\partial_{\alpha\dot\alpha} \bar{D}^{\dot\alpha} \tag{2.5.30c}$$

$$D^\alpha \bar{D}^2 D_\alpha = \bar{D}_{\dot\alpha} D^2 \bar{D}^{\dot\alpha} \tag{2.5.30d}$$

$$D^2 \bar{D}^2 + \bar{D}^2 D^2 - 2D^\alpha \bar{D}^2 D_\alpha = 16\Box \tag{2.5.30e}$$

$$D^2 \bar{D}_{\dot\alpha} D^2 = 0 \qquad \bar{D}^2 D_\alpha \bar{D}^2 = 0 \tag{2.5.30f}$$

$$D^2 \bar{D}^2 D^2 = 16 D^2 \Box \qquad \bar{D}^2 D^2 \bar{D}^2 = 16 \bar{D}^2 \Box \tag{2.5.30g}$$

All the identities can be readily proven with the help of relations (2.5.27). On the same grounds, one can see that a product of $n \geqslant 5$ spinor covariant derivatives may be reduced to an expression containing terms with at most four D and $\bar{D}$ factors.

It is worth pointing out two simple applications of the above identities. Given a tensor superfield $V_{\alpha_1 \ldots \alpha_n \dot\alpha_1 \ldots \dot\alpha_m}(z)$, the object

$$\bar{D}^2 V_{\alpha_1 \ldots \alpha_n \dot\alpha_1 \ldots \dot\alpha_m}(z) \tag{2.5.31}$$

is a tensor chiral superfield, and the object

$$D^2 V_{\alpha_1 \ldots \alpha_n \dot\alpha_1 \ldots \dot\alpha_m}(z) \tag{2.5.32}$$

is a tensor antichiral superfield. The reader may check that every tensor chiral superfield can be represented in the form (2.5.31). Furthermore, for every chiral superfield $\Phi(z)$, $\bar{D}_{\dot\alpha}\Phi = 0$, we have

$$\bar{D}^2 D_\alpha \Phi = 0 \tag{2.5.33}$$

as a consequence of equation (2.5.30*c*).

2.6. The on-shell massive superfield representations

The main goal of the present section is to give a realization in terms of superfields for the massive super Poincaré representations described in Section 2.3.

2.6.1. On-shell massive superfields

To begin with, we must formulate what is to be understood by the notion 'on-shell massive superfield'. By analogy with the Poincaré case, every tensor superfield of Lorentz type $(A/2,B/2)$,

$$V_{\alpha_1\ldots\alpha_A\dot\alpha_1\ldots\dot\alpha_B}(z)=V_{(\alpha_1\ldots\alpha_A)(\dot\alpha_1\ldots\dot\alpha_B)}(z) \tag{2.6.1}$$

satisfying the mass-shell equation

$$(\Box-m^2)V_{\alpha_1\ldots\alpha_A\dot\alpha_1\ldots\dot\alpha_B}(z)=0 \qquad m^2>0 \tag{2.6.2}$$

and the supplementary condition

$$\mathbf{P}^{\alpha\dot\alpha}V_{\alpha\alpha_1\ldots\alpha_{A-1}\dot\alpha\dot\alpha_1\ldots\dot\alpha_{B-1}}(z)=0 \qquad \mathbf{P}_{\alpha\dot\alpha}=-\mathrm{i}\partial_{\alpha\dot\alpha} \tag{2.6.3}$$

imposed when $A\neq0$ and $B\neq0$, is said to be an on-shell massive superfield of Lorentz type $(A/2,B/2)$. For arbitrary non-negative integers A and B, $A+B\equiv2Y$, the space of all on-shell massive $(A/2,B/2)$-type superfields, denoted by $\mathscr{H}_{(A,B)}$, forms a representation of the super Poincaré group.

Given a non-negative (half-)integer Y, spaces $\mathscr{H}_{(2Y,0)}, \mathscr{H}_{(2Y-1,1)}, \ldots, \mathscr{H}_{(0,2Y)}$ describe equivalent representations of the super Poincaré group. This assertion can be proved in the same fashion as was done in the Poincaré case. Namely, the operator $\Delta_{\alpha\dot\alpha}$ (1.8.6) is invertible under the fulfilment of equation (2.6.2) and provides us with a one-to-one mapping of $\mathscr{H}_{(2Y,0)}$ on $\mathscr{H}_{(A,B)}$, where $B\neq0$ and $A+B=2Y$, defined as follows

$$V_{\alpha_1\ldots\alpha_A\dot\alpha_1\ldots\dot\alpha_B}=\Delta^{\gamma_1}{}_{\dot\alpha_1}\ldots\Delta^{\gamma_B}{}_{\dot\alpha_B}V_{\alpha_1\ldots\alpha_A\gamma_1\ldots\gamma_B} \tag{2.6.4}$$

where $V_{\alpha_1\ldots\alpha_{A+B}}(z)$ is an arbitrary element of $\mathscr{H}_{(2Y,0)}$.

In contrast to the Poincaré case, every space $\mathscr{H}_{(A,B)}$ constitutes a reducible representation of the super Poincaré group, because $\mathscr{H}_{(A,B)}$ contains at least three super Poincaré invariant subspaces:

$$\begin{aligned}
&\mathscr{H}^{(+)}_{(A,B)}\colon \{V(z)\in\mathscr{H}_{(A,B)} \qquad \bar{\mathrm{D}}_{\dot\alpha}V(z)=0\}\\
&\mathscr{H}^{(-)}_{(A,B)}\colon \{V(z)\in\mathscr{H}_{(A,B)} \qquad \mathrm{D}_{\alpha}V(z)=0\}\\
&\mathscr{H}^{(0)}_{(A,B)}\colon \{V(z)\in\mathscr{H}_{(A,B)} \qquad \mathrm{D}^2V(z)=\bar{\mathrm{D}}^2V(z)=0\}
\end{aligned} \tag{2.6.5}$$

where superfield indices have been suppressed. It is seen that $\mathscr{H}^{(+)}_{(A,B)}$ consists of chiral on-shell superfields and $\mathscr{H}^{(-)}_{(A,B)}$ includes antichiral on-shell superfields only. Every element of $\mathscr{H}^{(0)}_{(A,B)}$ is said to be a 'linear superfield'.

One can easily find projection operators for the subspaces $\mathscr{H}^{(+)}_{(A,B)}$, $\mathscr{H}^{(-)}_{(A,B)}$ and $\mathscr{H}^{(0)}_{(A,B)}$. Making use of identities (2.5.30) gives

$$\begin{aligned}
&\mathscr{P}_{(+)}=\frac{1}{16}\frac{\bar{\mathrm{D}}^2\mathrm{D}^2}{\Box}=\frac{1}{16m^2}\bar{\mathrm{D}}^2\mathrm{D}^2\\
&\mathscr{P}_{(-)}=\frac{1}{16}\frac{\mathrm{D}^2\bar{\mathrm{D}}^2}{\Box}=\frac{1}{16m^2}\mathrm{D}^2\bar{\mathrm{D}}^2
\end{aligned} \tag{2.6.6}$$

$$\mathscr{P}_{(0)} = -\frac{1}{8}\frac{D^\alpha \bar{D}^2 D_\alpha}{\Box} = -\frac{1}{8m^2} D^\alpha \bar{D}^2 D_\alpha = -\frac{1}{8m^2} \bar{D}_{\dot\alpha} D^2 \bar{D}^{\dot\alpha}$$

$$\mathscr{P}_{(i)}\mathscr{P}_{(j)} = \delta_{ij}\mathscr{P}_{(i)}$$

where $\mathscr{P}_{(i)} = (\mathscr{P}_{(+)}, \mathscr{P}_{(-)}, \mathscr{P}_{(0)})$. The equation (2.5.30*e*) leads to

$$\mathscr{P}_{(+)} + \mathscr{P}_{(-)} + \mathscr{P}_{(0)} = \mathbb{1} \tag{2.6.7}$$

therefore we have the decomposition

$$\mathscr{H}_{(A,B)} = \mathscr{H}^{(+)}_{(A,B)} \oplus \mathscr{H}^{(-)}_{(A,B)} \oplus \mathscr{H}^{(0)}_{(A,B)}. \tag{2.6.8}$$

It is not difficult to see that the super Poincaré representations on the spaces $\mathscr{H}^{(\pm)}_{(A,B)}$ (but not on $\mathscr{H}^{(0)}_{(A,B)}$) are irreducible. Now, we have to decompose $\mathscr{H}^{(0)}_{(A,B)}$ into a direct sum of invariant subspaces and then determine superspin values corresponding to each of the irreducible representations under consideration. But before doing this, it is worth discussing the question of where the difference between the Poincaré and super Poincaré cases lies. Why were restrictions (2.6.1–3) sufficient in the Poincaré case to select out irreducible representations and yet they have proved to be incomplete in order to play the same role in the super Poincaré case? The point is that in superspace, in contrast to Minkowski space, we have at our disposal not only the super Poincaré generators acting on superfields but also the spinor covariant derivatives which possess the property of preserving tensor structure when acting on superfields. As a result, each space $\mathscr{H}_{(A,B)}$ is endowed with the action of a super Lie algebra including the super Poincaré algebra as a subalgebra.

2.6.2. *Extended super-Poincaré algebra*

Following E. Sokatchev, let us extend the set of super Poincaré generators (2.4.37) by adding the spinor covariant derivatives and consider the linear differential operators on $\mathscr{H}_{(A,B)}$:

$$\mathrm{i}(\tfrac{1}{2}b^{\alpha\dot\alpha}\mathbf{P}_{\alpha\dot\alpha} + K^{\alpha\beta}\mathbf{J}_{\alpha\beta} + \bar{K}^{\dot\alpha\dot\beta}\bar{\mathbf{J}}_{\dot\alpha\dot\beta} + \varepsilon\mathbf{Q} + \overline{\varepsilon\mathbf{Q}}) + \eta\mathrm{D} + \overline{\eta\mathrm{D}} \tag{2.6.9}$$

where the super Poincaré parameter are defined in the standard way and $(\eta^\alpha, \bar\eta_{\dot\alpha})$ are *a*-numbers forming a Majorana spinor. The set of all operators (2.6.9) is seen to form a super Lie algebra with respect to the ordinary Lie bracket, and the generators of the algebra satisfy the following (anti)commutation relations:

$$\begin{aligned}
&[\mathbf{J}_{\alpha\beta}, \mathbf{P}_{\gamma\dot\gamma}] = \tfrac{1}{2}\varepsilon_{\gamma\alpha}\mathbf{P}_{\beta\dot\gamma} + \tfrac{1}{2}\varepsilon_{\gamma\beta}\mathbf{P}_{\alpha\dot\gamma}\\
&[\mathbf{J}_{\alpha\beta}, \mathbf{J}_{\gamma\delta}] = \tfrac{1}{2}(\varepsilon_{\gamma\alpha}\mathbf{J}_{\beta\delta} + \varepsilon_{\gamma\beta}\mathbf{J}_{\alpha\delta} + \varepsilon_{\delta\alpha}\mathbf{J}_{\gamma\beta} + \varepsilon_{\delta\beta}\mathbf{J}_{\gamma\alpha})\\
&[\mathbf{J}_{\alpha\beta}, \mathbf{Q}_\gamma] = \tfrac{1}{2}\varepsilon_{\gamma\alpha}\mathbf{Q}_\beta + \tfrac{1}{2}\varepsilon_{\gamma\beta}\mathbf{Q}_\alpha\\
&[\mathbf{J}_{\alpha\beta}, \mathrm{D}_\gamma] = \tfrac{1}{2}\varepsilon_{\gamma\alpha}\mathrm{D}_\beta + \tfrac{1}{2}\varepsilon_{\gamma\beta}\mathrm{D}_\alpha\\
&\{\mathbf{Q}_\alpha, \bar{\mathbf{Q}}_{\dot\alpha}\} = \{\mathrm{D}_\alpha, \bar{\mathrm{D}}_{\dot\alpha}\} = 2\mathbf{P}_{\alpha\dot\alpha}.
\end{aligned} \tag{2.6.10}$$

The remaining (anti)commutators vanish or may be found by Hermitian conjugation (using the rule $(\mathbf{P}_{\alpha\dot{\alpha}})^{+}=\mathbf{P}_{\alpha\dot{\alpha}}$, $(\mathbf{J}_{\alpha\beta})^{+}=\bar{\mathbf{J}}_{\dot{\alpha}\dot{\beta}}$, $(\mathbf{Q}_{\alpha})^{+}=\bar{\mathbf{Q}}_{\dot{\alpha}}$ and $(\mathrm{D}_{\alpha})^{+}=\bar{\mathrm{D}}_{\dot{\alpha}}$). The algebra (2.6.10) presents nothing more than the $N=2$ Poincaré superalgebra without central charges (see Section 2.2).

It is a simple exercise to check that every space $\mathscr{H}_{(A,B)}$ has no non-trivial subspaces invariant under all of the operators (2.6.9). It is the spinor covariant derivatives which mix superfields from $\mathscr{H}^{(+)}_{(A,B)}$, $\mathscr{H}^{(-)}_{(A,B)}$ and $\mathscr{H}^{(0)}_{(A,B)}$.

2.6.3. *The superspin operator*

As the next step, we express the superspin operator (see subsection 2.3.2), corresponding to superfield representations,

$$C=(\mathbf{Z},\,\mathbf{P})^{2}-\mathbf{Z}^{2}\mathbf{P}^{2} \tag{2.6.11}$$

where

$$\begin{aligned}\mathbf{Z}_{a}&=\mathbf{W}_{a}-\tfrac{1}{8}(\tilde{\sigma}_{a})^{\dot{\alpha}\alpha}[\mathbf{Q}_{\alpha},\,\bar{\mathbf{Q}}_{\dot{\alpha}}]\\ \mathbf{W}_{a}&=\tfrac{1}{2}\varepsilon_{abcd}\mathbf{J}^{bc}\mathbf{P}^{d}\end{aligned} \tag{2.6.12}$$

in an explicitly supersymmetrically invariant form. The point is that superspace is endowed with not only the supersymmetry generators but the covariant derivatives also. The covariant derivatives, being supersymmetric invariant objects, have a structure similar to the structure of the supersymmetry generators. On these grounds, it is worth expecting that the Casimir operator C can be re-expressed in terms of D_{α} and $\bar{\mathrm{D}}_{\dot{\alpha}}$.

The basic observation is that the operator Z_{a} can be rewritten, after some algebra with the super Poincaré generators (2.4.36), using the rule

$$\begin{aligned}\mathbf{Z}_{a}&=\tilde{\mathbf{Z}}_{a}-\mathbf{A}\cdot\mathbf{P}_{a}\\ \tilde{\mathbf{Z}}_{a}&=-\tfrac{1}{2}\varepsilon_{abcd}M^{bc}\mathbf{P}^{d}+\tfrac{1}{8}(\tilde{\sigma}_{a})^{\dot{\alpha}\alpha}[\mathrm{D}_{\alpha},\,\bar{\mathrm{D}}_{\dot{\alpha}}]\\ \mathbf{A}&=\tfrac{1}{2}(\theta^{\beta}\partial_{\beta}-\bar{\theta}^{\dot{\beta}}\bar{\partial}_{\dot{\beta}}).\end{aligned} \tag{2.6.13}$$

Here $\mathbf{A}$ is the generator of γ_{5}-rotations

$$V(x,\theta,\bar{\theta})\rightarrow V'(x,\theta,\bar{\theta})=V(x,\mathrm{e}^{\frac{1}{2}\varphi}\theta,\mathrm{e}^{-\frac{1}{2}\varphi}\bar{\theta}).$$

It commutes with $\mathbf{P}_{a}$, so we have

$$\mathbf{Z}_{[a}\mathbf{P}_{b]}=\tilde{\mathbf{Z}}_{[a}\mathbf{P}_{b]}$$

and the superspin operator takes the form

$$C=(\tilde{\mathbf{Z}},\,\mathbf{P})^{2}-\tilde{\mathbf{Z}}^{2}\mathbf{P}^{2}. \tag{2.6.14}$$

Obviously, this expression is explicitly supersymmetrically invariant. The operator $\tilde{\mathbf{Z}}_{a}$ consists of two terms. The first one,

$$\tilde{\mathbf{W}}_{a}=-\tfrac{1}{2}\varepsilon_{abcd}M^{bc}\mathbf{P}^{d}\qquad \tilde{\mathbf{W}}_{\alpha\dot{\alpha}}=M_{\alpha\beta}\mathbf{P}^{\beta}{}_{\dot{\alpha}}-\bar{M}_{\dot{\alpha}\dot{\beta}}\mathbf{P}_{\alpha}{}^{\dot{\beta}} \tag{2.6.15}$$

is the 'space–time' Pauli–Lubanski vector. It is sensitive to superfield tensor types and x-dependence but not to θ-dependence. The second one, $\frac{1}{8}\tilde{\sigma}_a^{\dot{\alpha}\alpha}[\mathrm{D}_\alpha, \bar{\mathrm{D}}_{\dot{\alpha}}]$, is sensitive to superfield x- and θ-dependence and is inert with respect to superfield tensor type.

After using the identity $\hat{\mathbf{W}}^a\mathbf{P}_a=0$ and the mass-shell equation $\mathbf{P}^2=-m^2$, one obtains

$$C=\frac{1}{64}(\mathbf{P}^{\dot{\alpha}\alpha}[\mathrm{D}_\alpha, \bar{\mathrm{D}}_{\dot{\alpha}}])^2-\frac{m^2}{32}[\mathrm{D}^\alpha, \bar{\mathrm{D}}^{\dot{\alpha}}][\mathrm{D}_\alpha, \bar{\mathrm{D}}_{\dot{\alpha}}]+m^2\hat{\mathbf{W}}^2+\frac{m^2}{4}\hat{\mathbf{W}}^{\alpha\dot{\alpha}}[\mathrm{D}_\alpha, \bar{\mathrm{D}}_{\dot{\alpha}}]. \tag{2.6.16}$$

As for the third term here, we can profit from our old result (1.8.12), which leads to

$$\hat{\mathbf{W}}^2|_{\mathscr{H}_{(A,B)}}=m^2Y(Y+1)\mathbb{1} \qquad Y=(A+B)/2 \tag{2.6.17}$$

The first and second terms in expression (2.6.16) can be simplified with the help of equation (2.5.27). The final expression for the superspin operator is

$$\begin{gathered} C|_{\mathscr{H}_{(A,B)}}=m^4\{Y(Y+1)\mathbb{1}+\tfrac{3}{4}\mathscr{P}_{(0)}+\mathbf{B}\} \\ \mathbf{B}=\frac{1}{4m^2}\hat{\mathbf{W}}^{\alpha\dot{\alpha}}[\mathrm{D}_\alpha, \bar{\mathrm{D}}_{\dot{\alpha}}] \end{gathered} \tag{2.6.18}$$

where $\mathscr{P}_{(0)}$ is the projector onto the subspace of linear superfields in $\mathscr{H}_{(A,B)}$. The operator $\mathbf{B}$ turns out to have the following interesting properties:

$$\begin{gathered} \mathbf{B}\mathscr{P}_{(0)}=\mathscr{P}_{(0)}\mathbf{B}=\mathbf{B} \\ \mathbf{B}^2=Y(Y+1)\mathscr{P}_{(0)}-\mathbf{B} \end{gathered} \tag{2.6.19}$$

where $\mathbf{B}$ is assumed to act on $\mathscr{H}_{(A,B)}$. So, we can rewrite equation (2.6.18) in the form

$$C|_{\mathscr{H}^{(A,B)}}=m^4\{Y(Y+1)\mathbb{1}+(\tfrac{3}{4}+\mathbf{B})\mathscr{P}_{(0)}\} \tag{2.6.20}$$

Recalling equation (2.6.5), this immediately gives

$$C|_{\mathscr{H}^{(+)}_{(A,B)}}=C|_{\mathscr{H}^{(-)}_{(A,B)}}=m^4Y(Y+1)\mathbb{1} \tag{2.6.21}$$

Therefore, each of the spaces $\mathscr{H}^{(\mp)}_{(A,B)}$ of (anti)chiral on-shell superfields of Lorentz type $(A/2,B/2)$ provides us with the superspin-Y representation of the Poincaré superalgebra.

2.6.4. *Decomposition of $\mathscr{H}^{(0)}_{(A,B)}$ into irreducible representations*

We are going to show that every space $\mathscr{H}^{(0)}_{(A,B)}$, $A+B\neq 0$, describes two irreducible super Poincaré representations of superspins $(Y\pm\frac{1}{2})$, and the space $\mathscr{H}^{(0)}_{(0,0)}$ describes the super Poincaré representation of superspin $\frac{1}{2}$. Since all of the spaces $\mathscr{H}^{(0)}_{(2Y,0)}, \mathscr{H}^{(0)}_{(2Y-1,1)}, \ldots, \mathscr{H}^{(0)}_{(0,2Y)}$ realize equivalent representations of the super Poincaré group (see subsection 2.6.1), it is sufficient to restrict our consideration to the case $B=0$.

Let $V_{\alpha_1\ldots\alpha_A}\in\mathscr{H}^{(0)}_{(A,0)}$ be an arbitrary linear on-shell superfield,

$$\mathrm{D}^2V_{\alpha_1\ldots\alpha_A}=\bar{\mathrm{D}}^2V_{\alpha_1\ldots\alpha_A}=(\Box-m^2)V_{\alpha_1\ldots\alpha_A}=0 \tag{2.6.22}$$

When $A\neq 0$, we have the identity

$$V_{\alpha_1\ldots\alpha_A}=-\frac{1}{8m^2}\,\mathrm{D}^{\gamma}\bar{\mathrm{D}}^2\mathrm{D}_{(\gamma}V_{\alpha_1\ldots\alpha_A)}+\frac{1}{8m^2}\,\frac{A}{A+1}\,\mathrm{D}_{(\alpha_1}\bar{\mathrm{D}}^2\mathrm{D}^{\gamma}V_{\alpha_2\ldots\alpha_A)\gamma} \tag{2.6.23}$$

where parentheses $(\ldots)$ denote, as usual, the total symmetrization of indices, for example,

$$\mathrm{D}_{(\alpha_1}\bar{\mathrm{D}}^2\mathrm{D}^{\gamma}V_{\alpha_2\ldots\alpha_A)\gamma}=\frac{1}{A}\sum_{k=1}^{A}\mathrm{D}_{\alpha_k}\bar{\mathrm{D}}^2\mathrm{D}^{\gamma}V_{\alpha_1\ldots\hat{\alpha}_k\ldots\alpha_A\gamma}$$

and symbol $\hat{\alpha}_k$ means that index α_k is omitted. In accordance with equation (2.6.23), for every $V_{\alpha_1\ldots\alpha_A}\in\mathscr{H}^{(0)}_{(A,0)}$ there exist chiral on-shell superfields $\chi_{\alpha_1\ldots\alpha_{A+1}}\in\mathscr{H}^{(+)}_{(A+1,0)}$ and $\eta_{\alpha_1\ldots\alpha_{A-1}}\in\mathscr{H}^{(+)}_{(A-1,0)}$,

$$\bar{\mathrm{D}}_{\dot{\alpha}}\chi_{\alpha_1\ldots\alpha_{A+1}}=(\Box-m^2)\chi_{\alpha_1\ldots\alpha_{A+1}}=0$$

$$\bar{\mathrm{D}}_{\dot{\alpha}}\eta_{\alpha_1\ldots\alpha_{A-1}}=(\Box-m^2)\eta_{\alpha_1\ldots\alpha_{A-1}}=0$$

such that the following representation

$$V_{\alpha_1\ldots\alpha_A}=\mathrm{D}^{\gamma}\chi_{\gamma\alpha_1\ldots\alpha_A}+\mathrm{D}_{(\alpha_1}\eta_{\alpha_2\ldots\alpha_A)} \tag{2.6.24}$$

takes place. Conversely, for arbitrary chiral on-shell superfields $\chi_{\alpha_1\ldots\alpha_{A+1}}\in\mathscr{H}^{(+)}_{(A+1,0)}$ and $\eta_{\alpha_1\ldots\alpha_{A-1}}\in\mathscr{H}^{(+)}_{(A-1,0)}$, the superfield $V_{\alpha_1\ldots\alpha_A}$ constructed by the rule (2.6.24) belongs to $\mathscr{H}^{(0)}_{(A,0)}$. Moreover, the correspondence

$$(\chi_{\alpha_1\ldots\alpha_{A+1}},\eta_{\alpha_1\ldots\alpha_{A-1}})\rightarrow V_{\alpha_1\ldots\alpha_A}$$

is one-to-one, because equation (2.6.24) can be resolved as follows:

$$\begin{aligned}\chi_{\alpha_1\ldots\alpha_{A+1}}&=-\frac{1}{8m^2}\,\bar{\mathrm{D}}^2\mathrm{D}_{(\alpha_1}V_{\alpha_2\ldots\alpha_{A+1})}\\ \eta_{\alpha_1\ldots\alpha_{A-1}}&=\frac{1}{8m^2}\,\frac{A+1}{A}\,\bar{\mathrm{D}}^2\mathrm{D}^{\gamma}V_{\gamma\alpha_1\ldots\alpha_{A-1}}\end{aligned} \tag{2.6.25}$$

We come to the conclusion that the super Poincaré representation on the space $\mathscr{H}^{(0)}_{(A,0)}$ is equivalent to the super Poincaré representation on the direct sum space $\mathscr{H}^{(+)}_{(A+1,0)}\oplus\mathscr{H}^{(+)}_{(A-1,0)}$. For every positive (half-)integer Y, the space $\mathscr{H}^{(0)}_{(2Y,0)}$, provides us with two irreducible super Poincaré representations of superspins $(Y\pm\frac{1}{2})$. It is not difficult to find a suppplementary condition selecting out each superspin. Namely, the highest superspin, $Y+\frac{1}{2}$, is extracted by the condition

$$\mathrm{D}^{\gamma} V_{\gamma\alpha_1\ldots\alpha_{A-1}} = 0 \tag{2.6.26}$$

in accordance with equation (2.6.24). The lowest superspin, $Y-\frac{1}{2}$, is extracted by the condition

$$\mathrm{D}_{(\alpha_1} V_{\alpha_2\ldots\alpha_{A-1})} = 0. \tag{2.6.27}$$

Now, it is worth recalling the decomposition (2.6.8). In accordance with equation (2.6.24), we can write every on-shell superfield $U_{\alpha_1\ldots\alpha_A} \in \mathscr{H}_{(A,0)}$ in the form

$$\begin{gathered} U_{\alpha_1\ldots\alpha_A} = \Phi_{\alpha_1\ldots\alpha_A} + \Psi_{\alpha_1\ldots\alpha_A} + \mathrm{D}^{\gamma}\chi_{\gamma\alpha_1\ldots\alpha_A} + \mathrm{D}_{(\alpha_1}\eta_{\alpha_2\ldots\alpha_A)} \\ \Phi_{\alpha_1\ldots\alpha_A} \in \mathscr{H}^{(+)}_{(A,0)} \qquad \Psi_{\alpha_1\ldots\alpha_A} \in \mathscr{H}^{(-)}_{(A,0)} \\ \chi_{\alpha_1\ldots\alpha_{A+1}} \in \mathscr{H}^{(+)}_{(A+1,0)} \qquad \eta_{\alpha_1\ldots\alpha_{A-1}} \in \mathscr{H}^{(+)}_{(A-1,0)} \end{gathered} \tag{2.6.28}$$

which represents a decomposition into irreducible superfields.

The case $Y=0$ is treated similarly. Every superfield $V \in \mathscr{H}^{(0)}_{(0,0)}$ can be represented as

$$V = -\frac{1}{8m^2} \mathrm{D}^{\alpha}\bar{\mathrm{D}}^2\mathrm{D}_{\alpha} V$$

and hence

$$V = \mathrm{D}^{\alpha}\chi_{\alpha} \qquad \chi_{\alpha} = -\frac{1}{8m^2}\bar{\mathrm{D}}^2\mathrm{D}_{\alpha} V \qquad \chi_{\alpha} \in \mathscr{H}^{(+)}_{(1,0)}. \tag{2.6.29}$$

These relations establish the equivalence of the super Poincaré representations on $\mathscr{H}^{(0)}_{(0,0)}$ and $\mathscr{H}^{(+)}_{(1,0)}$, therefore the space $\mathscr{H}^{(0)}_{(0,0)}$ realizes the superspin-$\frac{1}{2}$ representation.

By analogy with equation (2.6.28), every scalar on-shell superfield $U \in \mathscr{H}_{(0,0)}$ can be written in the form

$$\begin{gathered} U = \Phi + \Psi + \mathrm{D}^{\alpha}\chi_{\alpha} \\ \bar{\mathrm{D}}_{\dot{\alpha}}\Phi = \bar{\mathrm{D}}_{\dot{\alpha}}\chi_{\alpha} = 0 \qquad \mathrm{D}_{\alpha}\Psi = 0. \end{gathered} \tag{2.6.30}$$

We summarize the results. If $A \neq 0$ or $B \neq 0$, the representation $\mathscr{H}_{(A,B)}$ is the direct sum of four irreducible super Poincaré representations with superspins $Y-\frac{1}{2}$, Y, Y, $Y+\frac{1}{2}$, where $2Y = A+B$. When $A=B=0$, the representation of $\mathscr{H}_{(0,0)}$ is the direct sum of three super Poincaré representations with superspin 0, 0, $\frac{1}{2}$. Every massive super Poincare representation can be realized in terms of (anti)chiral superfields. This is the reason why (anti)chiral superfields are objects of primary importance.

2.6.5. Projection operators

To complete the above consideration, it is worth finding projection operators extracting from $\mathscr{H}^{(0)}_{(A,B)}$ the subspaces of superspin $Y-\frac{1}{2}$ and $Y+\frac{1}{2}$, respectively. To do this, one can use the following simple observation. If a linear operator F on a vector space $\mathscr{L}$ takes eigenvalues $f_1, f_2, \ldots, f_n$ such that

$$\mathscr{L}=\mathscr{L}_1\oplus\mathscr{L}_2\oplus\ldots\oplus\mathscr{L}_n \qquad F|_{\mathscr{L}_i}=f_i\mathbb{1}$$

then projection operators on eigenspaces $\mathscr{L}_i$ are given in the form

$$\Pi_1=\frac{(F-f_2)\ldots(F-f_n)}{(f_1-f_2)\ldots(f_1-f_n)},\ldots,\Pi_n=\frac{(F-f_1)\ldots(F-f_{n-1})}{(f_n-f_1)\ldots(f_n-f_{n-1})}$$

$$\Pi_i\Pi_j=\delta_{ij}\Pi_i \qquad \Pi_1+\ldots+\Pi_n=\mathbb{1}.$$

In our case, the superspin operator acts on $\mathscr{H}_{(A,B)}$ and its eigenvalues are

$$m^4(Y-1/2)(Y+1/2) \qquad m^4Y(Y+1) \qquad m^4(Y+1/2)(Y+3/2)$$

Then equations (2.6.19, 20) and the prescription just given lead to

$$\begin{aligned}\Pi_{Y-1/2}&=\frac{1}{2Y+1}\{Y\mathbb{1}-\mathbf{B}\}\mathscr{P}_{(0)}\\ \Pi_{Y+1/2}&=\frac{1}{2Y+1}\{(Y+1)\mathbb{1}+\mathbf{B}\}\mathscr{P}_{(0)}.\end{aligned} \tag{2.6.31}$$

2.6.6. Real representations

The mapping of superfield complex conjugation defined by equation (2.4.30) converts a mass-shell space $\mathscr{H}_{(A,B)}$ to $\mathscr{H}_{(B,A)}$. Using this mapping and the operator $\Delta_{\alpha\dot{\alpha}}$ which changes tensor type, one can define real massive superfield representations in the same fashion as was done in Section 1.8 for the massive field representations. In particular, in the case $A=B$ one can impose the reality condition

$$V_{\alpha_1\ldots\alpha_A\dot{\alpha}_1\ldots\dot{\alpha}_A}(z)=\bar{V}_{\alpha_1\ldots\alpha_A\dot{\alpha}_1\ldots\dot{\alpha}_A}(z) \tag{2.6.32}$$

which defines a real tensor superfield.

Evidently, the set of real on-shell superfields of Lorentz type $(A/2,A/2)$ represents a super Poincaré invariant subspace in $\mathscr{H}_{(A,A)}$. To decompose a real on-shell superfield onto irreducible superfields, one can, as a first step, represent it in the form

$$V_{\alpha_1\ldots\alpha_A\dot{\alpha}_1\ldots\dot{\alpha}_A}=U_{\alpha_1\ldots\alpha_A\dot{\alpha}_1\ldots\dot{\alpha}_A}+\bar{U}_{\alpha_1\ldots\alpha_A\dot{\alpha}_1\ldots\dot{\alpha}_A} \tag{2.6.33}$$

with $U_{\alpha_1\ldots\alpha_A\dot{\alpha}_1\ldots\dot{\alpha}_A}$ being some complex on-shell superfield; after this, it is sufficient to apply to $U_{\alpha_1\ldots\alpha_A\dot{\alpha}_1\ldots\dot{\alpha}_A}$ a decomposition onto irreducible superfields as adopted in $\mathscr{H}_{(A,A)}$. For example, given a real scalar superfield

$V(z)$, we write it as

$$V = U + \bar{U}$$

and make use of equation (2.6.30). This gives

$$\begin{gathered} V = \Phi + \bar{\Phi} + \mathrm{D}^{\alpha}\eta_{\alpha} + \bar{\mathrm{D}}_{\dot{\alpha}}\bar{\eta}^{\dot{\alpha}} \equiv \Phi + \bar{\Phi} + V_{\mathrm{L}} \\ \bar{\mathrm{D}}_{\dot{\alpha}}\Phi = \bar{\mathrm{D}}_{\dot{\alpha}}\eta_{\alpha} = 0 \qquad \bar{\mathrm{D}}^2 V_{\mathrm{L}} = \mathrm{D}^2 V_{\mathrm{L}} = 0. \end{gathered} \tag{2.6.34}$$

It is possible to subject superfields to extraordinary reality conditions for the simple reason that we have at our disposal the spinor covariant derivatives in addition to the space–time derivatives. In particular, when $A = B + 1$, one can demand the equation

$$\mathrm{D}^{\gamma} U_{\gamma\alpha_1\ldots\alpha_B\dot{\alpha}_1\ldots\dot{\alpha}_B} = \bar{\mathrm{D}}_{\dot{\gamma}} \bar{U}_{\alpha_1\ldots\alpha_B\dot{\alpha}_1\ldots\dot{\alpha}_B}{}^{\dot{\gamma}}. \tag{2.6.35}$$

Let us analyse this equation in the simplest case $A = 1$ and $B = 0$. Consider an arbitrary on-shell superfield $U_{\alpha}(z)$. In accordance with equation (2.6.28), it can be represented in the form

$$\begin{gathered} U_{\alpha} = \Phi_{\alpha} + \Psi_{\alpha} + \mathrm{D}^{\beta}\chi_{\alpha\beta} + \mathrm{D}_{\alpha}\eta \\ \bar{\mathrm{D}}_{\dot{\alpha}}\Phi_{\alpha} = \bar{\mathrm{D}}_{\dot{\alpha}}\chi_{\alpha\beta} = \bar{\mathrm{D}}_{\dot{\alpha}}\eta = 0 \qquad \chi_{\alpha\beta} = \chi_{\beta\alpha} \qquad \mathrm{D}_{\alpha}\Psi_{\beta} = 0. \end{gathered} \tag{2.6.36}$$

Imposing the equation

$$\mathrm{D}^{\alpha} U_{\alpha} = \bar{\mathrm{D}}_{\dot{\alpha}} \bar{U}^{\dot{\alpha}} \tag{2.6.37}$$

leads to the equality

$$\mathrm{D}^{\alpha}\Phi_{\alpha} - \bar{\mathrm{D}}_{\dot{\alpha}}\bar{\Phi}^{\dot{\alpha}} = \bar{\mathrm{D}}^2\bar{\eta} - \mathrm{D}^2\eta.$$

Since Φ_{α} is a chiral superfield, the left-hand side is a linear superfield. The expression on the right is the sum of (anti)chiral superfields. Therefore, we have

$$\eta = 0 \qquad \mathrm{D}^{\alpha}\Phi_{\alpha} = \bar{\mathrm{D}}_{\dot{\alpha}}\bar{\Phi}^{\dot{\alpha}}. \tag{2.6.38}$$

Taking D_{α} and $\bar{\mathrm{D}}_{\dot{\alpha}}$ on both sides gives

$$-\tfrac{1}{4}\mathrm{D}^2\Phi_{\alpha} = \mathbf{P}_{\alpha\dot{\alpha}}\bar{\Phi}^{\dot{\alpha}} \qquad \tfrac{1}{4}\bar{\mathrm{D}}^2\bar{\Phi}_{\dot{\alpha}} = \mathbf{P}_{\alpha\dot{\alpha}}\Phi^{\alpha}. \tag{2.6.39}$$

These relations provide us with one more example of possible reality conditions.

A natural generalization of the final conditions to the scalar case reads

$$-\tfrac{1}{4}\mathrm{D}^2\Phi = \mu\bar{\Phi} \qquad -\tfrac{1}{4}\bar{\mathrm{D}}^2\bar{\Phi} = \bar{\mu}\Phi \tag{2.6.40}$$

where μ is a complex constant. Since here $\Phi(z)$ is a chiral superfield, $\bar{\mathrm{D}}_{\dot{\alpha}}\Phi = 0$, equation (2.6.40) leads to the mass-shell equations

$$(\Box - |\mu|^2)\Phi = 0 \qquad (\Box - |\mu|^2)\bar{\Phi} = 0. \tag{2.6.41}$$

So, one can look on the reality condition (2.6.40) as an equation of motion for a chiral scalar superfield. When imposing the mass-shell equations (2.6.41) only, (anti)chiral superfields $\bar{\Phi}$ and Φ are independent and describe two irreducible representations of superspin $Y=0$. However, choosing stronger equations (2.6.40) leads to dependence between $\bar{\Phi}$ and Φ, therefore the system is reduced to describing a single superspin $Y=0$.

In conclusion, let us point out that the constant μ in equation (2.6.40) can be made real after making a redefinition $\Phi(z) \to e^{i\alpha}\Phi(z)$.

2.7. The on-shell massless superfield representations

In this section, we give a realization in terms of superfields for the massless super Poincaré representations described in Section 2.3.

2.7.1. *Consistency conditions*

It has been shown in Section 2.3 that every massless unitary representation of the Poincaré superalgebra is characterized by the operatorial constraints

$$\mathbb{P}_{\alpha\dot{\alpha}}\mathbb{Q}^{\alpha} = \mathbb{P}_{\alpha\dot{\alpha}}\bar{\mathbb{Q}}^{\dot{\alpha}} = \mathbb{Q}^2 = \bar{\mathbb{Q}}^2 = 0.$$

Recall that these constraints were necessary to make the supersymmetry algebra consistent with the unitarity and the on-shell equation $\mathbb{P}^2 = 0$. Since we intend to realize in superspace the massless unitary representations, we subject massless superfields to the operatorial constraints

$$\mathbf{P}_{\alpha\dot{\alpha}}\mathbf{Q}^{\alpha} = \mathbf{P}_{\alpha\dot{\alpha}}\bar{\mathbf{Q}}^{\dot{\alpha}} = 0 \tag{2.7.1}$$

and

$$\mathbf{Q}^2 = \bar{\mathbf{Q}}^2 = 0 \tag{2.7.2}$$

in addition to the on-shell equation

$$\mathbf{P}^2 = 0. \tag{2.7.3}$$

Note that the supersymmetry generators in superspace can be written as

$$\mathbf{Q}_{\alpha} = \mathrm{i}(\partial_{\alpha} + \bar{\theta}^{\dot{\alpha}}\mathbf{P}_{\alpha\dot{\alpha}}) \qquad \bar{\mathbf{Q}}_{\dot{\alpha}} = -\mathrm{i}(\bar{\partial}_{\dot{\alpha}} + \theta^{\alpha}\mathbf{P}_{\alpha\dot{\alpha}})$$

which leads to

$$\mathbf{P}^{\alpha\dot{\alpha}}\mathbf{Q}_{\alpha} = \mathrm{i}\mathbf{P}^{\alpha\dot{\alpha}}\partial_{\alpha} - \mathrm{i}\bar{\theta}^{\dot{\alpha}}\mathbf{P}^2.$$

Then, equations (2.7.1) and (2.7.3) lead to

$$\mathbf{P}_{\alpha\dot{\alpha}}\partial^{\alpha} = \mathbf{P}_{\alpha\dot{\alpha}}\bar{\partial}^{\dot{\alpha}} = 0. \tag{2.7.4}$$

Analogously, equations (2.7.2–4) lead to

$$\partial^{\alpha}\partial_{\alpha} = \bar{\partial}_{\dot{\alpha}}\bar{\partial}^{\dot{\alpha}} = 0. \tag{2.7.5}$$

Furthermore, the covariant derivatives can be written in the form

$$\mathrm{D}_\alpha = \partial_\alpha - \bar\theta^{\dot\alpha}\mathbf{P}_{\alpha\dot\alpha} \qquad \bar{\mathrm{D}}_{\dot\alpha} = -\bar\partial_{\dot\alpha} + \theta^\alpha \mathbf{P}_{\alpha\dot\alpha}.$$

Then, equations (2.7.3, 4) give

$$\mathbf{P}_{\alpha\dot\alpha}\mathrm{D}^\alpha = \mathbf{P}_{\alpha\dot\alpha}\bar{\mathrm{D}}^{\dot\alpha} = 0. \tag{2.7.6}$$

Analogously, equations (2.7.3, 5, 6) give

$$\mathrm{D}^2 = \bar{\mathrm{D}}^2 = 0. \tag{2.7.7}$$

The above considerations show that the set of equations (2.7.1–3) is equivalent to the set of equations (2.7.3, 6, 7). However, the second set seems preferable to the first one, because the corresponding equations are explicitly supersymmetrically invariant.

2.7.2. *On-shell massless superfields*

A tensor superfield of Lorentz type $(A/2,B/2)$, $V_{\alpha_1\ldots\alpha_A\dot\alpha_1\ldots\dot\alpha_B}(z)$, is said to be an 'on-shell massless superfield' if it satisfies equations (2.7.3, 6, 7) and the supplementary conditions

$$\mathbf{P}^{\gamma\dot\gamma}V_{\gamma\alpha_1\ldots\alpha_{A-1}\dot\alpha_1\ldots\dot\alpha_B}(z) = 0 \tag{2.7.8a}$$

$$\mathbf{P}^{\gamma\dot\gamma}V_{\alpha_1\ldots\alpha_A\dot\gamma\dot\alpha_1\ldots\dot\alpha_{B-1}}(z) = 0 \tag{2.7.8b}$$

as well. In fact, when $A \neq 0$ or $B \neq 0$, the on-shell equation (2.7.3) is a consequence of the supplementary conditions.

We are going to classify on-shell massless superfields. First, consider the case $A \neq 0$, $B \neq 0$. Given a $(A/2,B/2)$-type superfield under the supplementary conditions, we impose the first equation (2.7.6),

$$\mathbf{P}_{\gamma\dot\gamma}\mathrm{D}^\gamma V_{\alpha_1\ldots\alpha_A\dot\alpha_1\ldots\dot\alpha_B}(z) = 0.$$

Then, making use of equation (2.7.8*a*), this leads to

$$\mathbf{P}_{\alpha_1\dot\gamma}\mathrm{D}^\gamma V_{\gamma\alpha_2\ldots\alpha_A\dot\alpha_1\ldots\dot\alpha_B}(z) = 0.$$

Analogously, imposing the second equation (2.7.6) and making use of equation (2.7.8*b*) gives

$$\mathbf{P}_{\gamma\dot\alpha_1}\bar{\mathrm{D}}^{\dot\gamma} V_{\alpha_1\ldots\alpha_A\dot\gamma\dot\alpha_2\ldots\dot\alpha_B}(z) = 0.$$

Therefore, we have

$$\mathrm{D}^\gamma V_{\gamma\alpha_1\ldots\alpha_{A-1}\dot\alpha_1\ldots\dot\alpha_B}(z) = 0 \tag{2.7.9a}$$

$$\bar{\mathrm{D}}^{\dot\gamma} V_{\alpha_1\ldots\alpha_A\dot\gamma\dot\alpha_1\ldots\dot\alpha_{B-1}}(z) = 0 \tag{2.7.9b}$$

since the superfields in the left-hand sides of equations (2.7.9) carry zero momentum. Now, both equations (2.7.7) are satisfied identically,

$$\mathrm{D}^2 V_{\alpha_1\ldots\alpha_A\dot\alpha_1\ldots\dot\alpha_B} = -2\mathrm{D}_{\alpha_1}\mathrm{D}^\gamma V_{\gamma\alpha_2\ldots\alpha_A\dot\alpha_1\ldots\dot\alpha_B} = 0.$$

It is seen that the set of equations (2.7.6–8) is equivalent to equations (2.7.8, 9).

Furthermore, we are to consider three possible superfield types to which $V_{\alpha_1\ldots\alpha_A\dot\alpha_1\ldots\dot\alpha_B}$ may belong: (a) chiral; (b) antichiral; (c) neither chiral nor antichiral. In the first case, the total set of massless constraints (equivalent to equations (2.7.8, 9)) is

$$\begin{aligned} \bar{\mathrm{D}}_{\dot\gamma}\Phi_{\alpha_1\ldots\alpha_A\dot\alpha_1\ldots\dot\alpha_B}(z)&=0\\ \mathrm{D}^{\gamma}\Phi_{\gamma\alpha_1\ldots\alpha_{A-1}\dot\alpha_1\ldots\dot\alpha_B}(z)&=0\\ \mathbf{P}^{\gamma\dot\gamma}\Phi_{\alpha_1\ldots\alpha_A\dot\gamma\dot\alpha_1\ldots\dot\alpha_{B-1}}(z)&=0. \end{aligned} \tag{2.7.10}$$

Similarly, in the second case we have

$$\begin{aligned} \mathrm{D}_{\gamma}\bar\Phi_{\alpha_1\ldots\alpha_A\dot\alpha_1\ldots\dot\alpha_B}(z)&=0\\ \bar{\mathrm{D}}^{\dot\gamma}\bar\Phi_{\alpha_1\ldots\alpha_A\dot\gamma\dot\alpha_1\ldots\dot\alpha_{B-1}}(z)&=0\\ \mathbf{P}^{\gamma\dot\gamma}\bar\Phi_{\gamma\alpha_1\ldots\alpha_{A-1}\dot\alpha_1\ldots\dot\alpha_B}(z)&=0. \end{aligned} \tag{2.7.11}$$

Equations (2.7.10, 11) determine massless (anti)chiral superfields. Finally, let us consider the last case. Now, we can construct, starting from $V_{\alpha_1\ldots\alpha_A\dot\alpha_1\ldots\dot\alpha_B}$, two secondary superfields:

$$\Phi_{\alpha_1\ldots\alpha_A\dot\alpha_1\ldots\dot\alpha_{B+1}}(z)=\bar{\mathrm{D}}_{\dot\alpha_1}V_{\alpha_1\ldots\alpha_A\dot\alpha_2\ldots\dot\alpha_{B+1}}(z) \tag{2.7.12}$$

and

$$\bar\Phi_{\alpha_1\ldots\alpha_{A+1}\dot\alpha_1\ldots\dot\alpha_B}(z)=\mathrm{D}_{\alpha_1}V_{\alpha_2\ldots\alpha_{A+1}\dot\alpha_1\ldots\dot\alpha_B}(z). \tag{2.7.13}$$

The first superfield is chiral and symmetric (due to equation (2.7.9*b*)) in its dotted indices, hence it belongs to Lorentz type $(A/2,(B+1)/2)$. What is more, it satisfies all the constraints (2.7.10). Similarly, $\bar\Phi_{\alpha_1\ldots\alpha_{A+1}\dot\alpha_1\ldots\dot\alpha_B}(z)$ is an antichiral superfield of Lorentz type $((A+1)/2,B/2)$ under constraints (2.7.11). Therefore, the (anti)chiral secondary superfields are also on-shell massless superfields. One can look on a general massless superfield as a superposition of massless (anti)chiral superfields.

As the next step, let us treat the case $A\neq 0$, $B=0$. Now, one can readily see that the total set of massless constraints is given by the equations

$$\begin{aligned} \mathrm{D}^{\gamma}V_{\gamma\alpha_1\ldots\alpha_{A-1}}(z)&=0\\ \mathbf{P}^{\gamma\dot\gamma}V_{\gamma\alpha_1\ldots\alpha_{A-1}}(z)&=0\\ \bar{\mathrm{D}}^2V_{\alpha_1\ldots\alpha_A}(z)=\mathbf{P}_{\gamma\dot\gamma}\bar{\mathrm{D}}^{\dot\gamma}V_{\alpha_1\ldots\alpha_A}(z)&=0. \end{aligned} \tag{2.7.14}$$

When taking the massless superfield to be chiral, equations (2.7.14) are simplified drastically. Namely, each $(A/2,0)$-type superfield $\Phi_{\alpha_1\ldots\alpha_A}(z)$ under the constraints

$$\begin{aligned} \bar{\mathrm{D}}_{\dot\gamma}\Phi_{\alpha_1\ldots\alpha_A}(z)&=0\\ \mathrm{D}^{\gamma}\Phi_{\gamma\alpha_1\ldots\alpha_{A-1}}(z)&=0 \end{aligned} \tag{2.7.15}$$

proves to be massless. One more solution of equations (2.7.14) reads

$$\begin{aligned} \mathrm{D}_{\gamma}\bar{\Phi}_{\alpha_1\ldots\alpha_A}(z)&=0 \\ \mathbf{P}^{\dot{\gamma}\gamma}\bar{\Phi}_{\gamma\alpha_1\ldots\alpha_{A-1}}(z)&=0 \end{aligned} \tag{2.7.16}$$

$$\bar{\mathrm{D}}^2\bar{\Phi}_{\alpha_1\ldots\alpha_A}(z)=0$$

which defines an antichiral massless superfield of Lorentz type $(A/2, 0)$. Finally, in the case of a general massless superfield, neither chiral nor antichiral, one can construct two secondary superfields:

$$\Phi_{\alpha_1\ldots\alpha_A\dot{\alpha}}(z)=\bar{\mathrm{D}}_{\dot{\alpha}}V_{\alpha_1\ldots\alpha_A}(z) \tag{2.7.17}$$

and

$$\bar{\Phi}_{\alpha_1\ldots\alpha_{A-1}}(z)=\mathrm{D}_{\alpha_1}V_{\alpha_2\ldots\alpha_{A+1}}(z) \tag{2.7.18}$$

which are chiral and antichiral massless superfields, respectively.

The case $A=0$ and $B\neq 0$ is treated in complete analogy with the previous one. So, we investigate the last possibility, $A=B=0$. Now, the total set of massless constraints can be represented in the form

$$\begin{aligned} \bar{\mathrm{D}}^2V(z)&=\bar{\mathrm{D}}^2\mathrm{D}_{\alpha}V(z)=0 \\ \mathrm{D}^2V(z)&=\mathrm{D}^2\bar{\mathrm{D}}_{\dot{\alpha}}V(z)=0 \end{aligned} \tag{2.7.19}$$

due to the identities

$$[\mathrm{D}^2,\bar{\mathrm{D}}_{\dot{\alpha}}]=4\mathbf{P}_{\alpha\dot{\alpha}}\mathrm{D}^{\alpha} \qquad [\bar{\mathrm{D}}^2,\mathrm{D}_{\alpha}]=-4\mathbf{P}_{\alpha\dot{\alpha}}\bar{\mathrm{D}}^{\dot{\alpha}}.$$

There is no need to impose the on-shell equation

$$\Box V(z)=0$$

since it follows by virtue of equation (2.5.30*e*), from equations (2.7.19). In accordance with constraints (2.7.19), a massless chiral superfield is defined by

$$\bar{\mathrm{D}}_{\dot{\alpha}}\Phi(z)=0 \qquad \mathrm{D}^2\Phi(z)=0 \tag{2.7.20}$$

and a massless antichiral scalar superfield is defined by

$$\mathrm{D}_{\alpha}\bar{\Phi}(z)=0 \qquad \bar{\mathrm{D}}^2\bar{\Phi}(z)=0. \tag{2.7.21}$$

In the case of a general massless scalar superfield $V(z)$, one can construct two secondary massless superfields: chiral

$$\Phi_{\dot{\alpha}}(z)=\bar{\mathrm{D}}_{\dot{\alpha}}V(z) \tag{2.7.22}$$

and antichiral

$$\bar{\Phi}_{\alpha}(z)=\mathrm{D}_{\alpha}V(z). \tag{2.7.23}$$

In the following chapters of our book it will be shown how massless superfields, described in this section, arise in supersymmetric field theories.

2.7.3. Superhelicity

As is known, massless super Poincaré representations are classified by superhelicity κ (see subsection 2.3.5). We are going to determine superhelicity values corresponding to all of the massless superfields considered earlier. For this purpose, we rewrite the superhelicity operator

$$\mathbf{L}_a = \mathbf{W}_a - \frac{1}{16}(\tilde{\sigma}_a)^{\dot{\alpha}\alpha}[\mathbf{Q}_\alpha, \bar{\mathbf{Q}}_{\dot{\alpha}}]$$

$$\mathbf{W}_a = \frac{1}{2}\varepsilon_{abcd}\mathbf{J}^{bc}\mathbf{P}^d$$

in an explicitly supersymmetric invariant form. Recalling expressions for the super-Poincaré generators (2.4.36) and taking into account massless constraints (2.7.4), one finds

$$\mathbf{L}_{\alpha\dot{\alpha}} = \tilde{\mathbf{W}}_{\alpha\dot{\alpha}} - \frac{1}{8}[\mathrm{D}_\alpha, \bar{\mathrm{D}}_{\dot{\alpha}}] \tag{2.7.24}$$

where $\tilde{\mathbf{W}}_{\alpha\dot{\alpha}}$ is the 'space–time' Pauli–Lubanski vector (2.6.15). It acts on a superfield of Lorentz type $(A/2, B/2)$ subject to the supplementary conditions (2.7.8) as follows

$$\tilde{\mathbf{W}}_{\gamma\dot{\gamma}}V_{\alpha_1\ldots\alpha_A\dot{\alpha}_1\ldots\dot{\alpha}_B}(z) = \frac{1}{2}(A - B)\mathbf{P}_{\gamma\dot{\gamma}}V_{\alpha_1\ldots\alpha_A\dot{\alpha}_1\ldots\dot{\alpha}_B}(z) \tag{2.7.25}$$

(see subsection 1.8.3). We say that a massless superfield has a superhelicity κ if it satisfies the equation

$$\mathbf{L}_{\alpha\dot{\alpha}} = \left(\kappa + \frac{1}{4}\right)\mathbf{P}_{\alpha\dot{\alpha}}. \tag{2.7.26}$$

Only (anti)chiral massless superfields have definite superhelicities. Every chiral massless superfield of Lorentz type $(A/2, B/2)$ proves to have superhelicity

$$\kappa\Big|_{(A/2,\, B/2),\ \mathrm{chiral}} = \frac{1}{2}(A - B). \tag{2.7.27}$$

Every antichiral massless superfield of Lorentz type $(A/2, B/2)$ proves to have superhelicity

$$\kappa\Big|_{(A/2,\, B/2),\ \mathrm{antichiral}} = \frac{1}{2}(A - B) - \frac{1}{2}. \tag{2.7.28}$$

Given a massless chiral superfield $\Phi_{\alpha_1\ldots\alpha_A\dot{\alpha}_1\ldots\dot{\alpha}_B}$, its secondary antichiral superfield

$$\bar{\Phi}_{\alpha_1\ldots\alpha_{A+1}\dot{\alpha}_1\ldots\dot{\alpha}_B}(z) = \mathrm{D}_{\alpha_1}\Phi_{\alpha_2\ldots\alpha_{A+1}\dot{\alpha}_1\ldots\dot{\alpha}_B}(z)$$

is also massless, and both superfields have the same superhelicity.

2.8. From superfields to component fields

In our opinion, the reader has had the opportunity to become convinced that working with superfields is not much harder than with fields in space–time. The formalism developed in Sections 2.4–2.7 makes it possible to handle a superfield as a simple indivisible object such that at any stage of practical calculations there is no need to think about its explicit construction in terms of the components fields. Unfortunately, at present it has not been possible to extract all the necessary physical information directly from superfields. For example, so far no-one has a clear unerstanding how to formulate canonical quantization on superfield language. It is a quite general situation that in order to do reasonable physical analysis of a supersymmetric theory one must re-express the theory in terms of components fields. Hence one should carry out the reduction from superfields to components in an optimum way. In this section we describe the reduction technique invented by J. Wess and B. Zumino.

2.8.1. *Chiral scalar superfield*

To fix the ideas let us start with a chiral scalar superfield $\Phi(z) = \exp[\mathrm{i}\theta\sigma^a\bar{\theta}\partial_a]\Phi(x,\theta)$. Its component expansion is

$$\Phi(x,\theta,\bar{\theta}) = A(x) + \theta^\alpha\psi_\alpha(x) + \theta^2 F(x) + \mathrm{i}\theta\sigma^a\bar{\theta}\partial_a A(x) + \frac{\mathrm{i}}{2}\theta^2\bar{\theta}\tilde{\sigma}^a\partial_a\psi(x) + \frac{1}{4}\theta^2\bar{\theta}^2\Box A(x). \tag{2.8.1}$$

In this expansion only the first three components are independent fields, the rest are secondary fields. Expression (2.8.1) determines the chiral superfield in terms of the component fields. Now we want to discuss the task of determining the component fields from the superfield.

Introduce a mapping projecting every superfield $V(x,\theta,\bar{\theta})$ into its zeroth-order (in θ^α and $\bar{\theta}^{\dot{\alpha}}$) component field:

$$V\Big| \equiv V(x,\theta,\bar{\theta})\Big|_{\theta=0,\,\bar{\theta}=0}. \tag{2.8.2}$$

We will refer to this mapping as the 'space projection'. Obviously, component fields of any superfield can be obtained by first taking some partial derivatives in θ and $\bar{\theta}$ and then applying the space projection. In the chiral scalar case we have

$$A(x) = \Phi| \qquad \psi_\alpha(x) = \partial_\alpha\Phi| \qquad F(x) = \frac{1}{4}\partial^\alpha\partial_\alpha\Phi| \qquad \partial_{\alpha\dot{\alpha}}A(x) = \frac{\mathrm{i}}{2}[\partial_\alpha, \bar{\partial}_{\dot{\alpha}}]\Phi|$$

and so on. But proceeding in such a way, we have no explicit realization of how, for example, $[\partial_\alpha, \bar{\partial}_{\dot\alpha}]\Phi|$ is expressed through $\Phi|$. How can one implement this? Recalling expressions for the covariant derivatives, one finds the identities

$$\begin{aligned} &\mathrm{D}_\alpha V| = \partial_\alpha V| \qquad \mathrm{D}^2 V| = -\partial^\alpha \partial_\alpha V| \\ &\bar{\mathrm{D}}_{\dot\alpha} V| = -\bar{\partial}_{\dot\alpha} V| \quad \bar{\mathrm{D}}^2 V| = -\bar{\partial}_{\dot\alpha}\bar{\partial}^{\dot\alpha} V| \\ &[\mathrm{D}_\alpha, \bar{\mathrm{D}}_{\dot\alpha}]V = -[\partial_\alpha, \bar{\partial}_{\dot\alpha}]V| \end{aligned} \tag{2.8.3}$$

where $V(z)$ is an arbitrary superfield. After this, the independent component fields of our chiral superfield can be defined as follows

$$A(x) = \Phi| \qquad \psi_\alpha(x) = \mathrm{D}_\alpha \Phi| \qquad F(x) = -\frac{1}{4}\mathrm{D}^2\Phi|. \tag{2.8.4}$$

It is now evident that, due to the chirality constraint $\bar{\mathrm{D}}_{\dot\alpha}\Phi = 0$, the space projection of any number of Ds and $\bar{\mathrm{D}}$s applied to $\Phi(z)$ is expressed in terms of the above fields.

The given definition is very useful for obtaining supersymmetric transformation laws of component fields. Namely, in accordance with the identity

$$\mathrm{i}(\epsilon\mathbf{Q} + \overline{\epsilon\mathbf{Q}}) = -(\epsilon\mathrm{D} + \overline{\epsilon\mathrm{D}}) + 2\mathrm{i}(\epsilon\sigma^a\bar\theta - \theta\sigma^a\bar\epsilon)\partial_a$$

we have

$$\mathrm{i}(\epsilon\mathbf{Q} + \overline{\epsilon\mathbf{Q}})V| = -(\epsilon\mathrm{D} + \overline{\epsilon\mathrm{D}})V| \tag{2.8.5}$$

for every superfield $V(z)$. Since the covariant derivatives anticommute with the supersymmetry generators, fields (2.8.4) transform according to the rule

$$\begin{aligned} \delta A(x) &= \mathrm{i}(\epsilon\mathbf{Q} + \overline{\epsilon\mathbf{Q}})\Phi| = -\epsilon\mathrm{D}\Phi| = -\epsilon\psi(x) \\ \delta\psi_\alpha(x) &= \mathrm{D}_\alpha\{\mathrm{i}(\epsilon\mathbf{Q} + \overline{\epsilon\mathbf{Q}})\Phi\}| = \mathrm{i}(\epsilon\mathbf{Q} + \overline{\epsilon\mathbf{Q}})\mathrm{D}_\alpha\Phi| \\ &= -\epsilon\mathrm{D}\mathrm{D}_\alpha\Phi| - \overline{\epsilon\mathrm{D}}\mathrm{D}_\alpha\Phi| = -2\epsilon_\alpha F(x) - 2\mathrm{i}\bar\epsilon^{\dot\alpha}\partial_{\alpha\dot\alpha}A(x) \\ \delta F(x) &= -\frac{1}{4}\mathrm{D}^2\{\mathrm{i}(\epsilon\mathbf{Q} + \overline{\epsilon\mathbf{Q}})\Phi\}| = -\frac{\mathrm{i}}{4}(\epsilon\mathbf{Q} + \overline{\epsilon\mathbf{Q}})\mathrm{D}^2\Phi| \\ &= \frac{1}{4}\overline{\epsilon\mathrm{D}}\mathrm{D}^2\Phi| = -\mathrm{i}\bar\epsilon\tilde\sigma^a\partial_a\psi(x). \end{aligned} \tag{2.8.6}$$

Therefore, in this approach the determination of transformation laws consists of simple manipulations with the covariant derivatives.

Consider the antichiral superfield $\bar\Phi(z)$ conjugate to $\Phi(z)$. It is characterized by the component fields

$$\bar{A}(x) = \bar\Phi| \qquad \bar\psi_{\dot\alpha}(x) = \bar{\mathrm{D}}_{\dot\alpha}\bar\Phi| \qquad \bar{F}(x) = -\frac{1}{4}\bar{\mathrm{D}}^2\bar\Phi|. \tag{2.8.7}$$

The supersymmetry transformations act on them according to the law

$$\begin{aligned}
\delta\bar{A}(x) &= -\bar{\epsilon}\bar{\psi}(x)\\
\delta\bar{\psi}_{\dot{\alpha}}(x) &= -2\bar{\epsilon}_{\dot{\alpha}}\bar{F}(x) + 2\mathrm{i}\epsilon^{\alpha}\partial_{\alpha\dot{\alpha}}\bar{A}(x)\\
\delta\bar{F}(x) &= -\mathrm{i}\epsilon\sigma^{a}\partial_{a}\bar{\psi}(x).
\end{aligned} \tag{2.8.8}$$

Rather beautifully, using definition (2.8.4), one can immediately obtain component equations which follow from the massive superfield equations

$$-\frac{1}{4}\bar{\mathrm{D}}^2\bar{\Phi} + m\Phi = 0 \tag{2.8.9a}$$

$$-\frac{1}{4}\mathrm{D}^2\Phi + m\bar{\Phi} = 0 \tag{2.8.9b}$$

defining the superspin-0 representation. Namely, taking the space projection of equation (2.8.9) leads to

$$\bar{F} + mA = 0 \qquad F + m\bar{A} = 0. \tag{2.8.10}$$

Then, taking D_{α} from equation (2.8.9*a*) and $\bar{\mathrm{D}}_{\dot{\alpha}}$ from equation (2.8.9*b*) and making the space projection gives

$$\mathrm{i}\partial_{\alpha\dot{\alpha}}\bar{\psi}^{\dot{\alpha}} + m\psi_{\alpha} = 0 \qquad -\mathrm{i}\partial_{\alpha\dot{\alpha}}\psi^{\alpha} + m\bar{\psi}_{\dot{\alpha}} = 0. \tag{2.8.11}$$

Finally, taking $(-\frac{1}{4}\mathrm{D}^2)$ from equation (2.8.9*a*) and $(-\frac{1}{4}\bar{\mathrm{D}}^2)$ from equation (2.8.9*b*) and making the space projection we obtain

$$\Box\bar{A} + mF = 0 \qquad \Box A + m\bar{F} = 0. \tag{2.8.12}$$

It is seen that the complex scalar field $F(x)$ does not have independent dynamics, since it is expressed in terms of $A(x)$, while the complex scalar field $A(x)$ and the Majorana spinor field

$$\boldsymbol{\Psi}(x) = \begin{pmatrix}\psi_{\alpha}(x)\\ \bar{\psi}^{\dot{\alpha}}(x)\end{pmatrix}$$

satisfy the Klein–Gordon equation

$$(\Box - m^2)A = 0$$

and Dirac equation

$$(\mathrm{i}\gamma^{a}\partial_{a} + m)\boldsymbol{\Psi} = 0$$

respectively.

2.8.2. *Chiral tensor superfield of Lorentz type $(n/2, 0)$*

Consider a chiral tensor superfield $\Phi_{\alpha_1\alpha_2\ldots\alpha_n}(z)$ totally symmetric in its indices.

By analogy with the scalar case, we define its component fields according to the rule

$$
\begin{aligned}
A_{\alpha_1\alpha_2\ldots\alpha_n}(x) &= \Phi_{\alpha_1\alpha_2\ldots\alpha_n}| \\
\psi_{\beta(\alpha_1\ldots\alpha_n)}(x) &= D_\beta\Phi_{\alpha_1\ldots\alpha_n}| \\
F_{\alpha_1\alpha_2\ldots\alpha_n}(x) &= -\frac{1}{4}D^2\Phi_{\alpha_1\ldots\alpha_n}|.
\end{aligned}
\tag{2.8.13}
$$

Obviously, the first and third fields are totally symmetric in their indices. As for the second field, it can be decomposed into two Lorentz irreducible fields, of Lorentz types $((n-1)/2, 0)$ and $((n+1)/2, 0)$, as follows

$$
\begin{aligned}
\psi_{\beta\alpha_1\ldots\alpha_n}(x) &= \eta_{\beta\alpha_1\ldots\alpha_n}(x) + \frac{1}{n+1}\sum_{k=1}^{n}\varepsilon_{\beta\alpha_k}\lambda_{\alpha_1\ldots\hat{\alpha}_k\ldots\alpha_n}(x) \\
\eta_{\alpha_1\alpha_2\ldots\alpha_{n+1}}(x) &= D_{(\alpha_1}\Phi_{\alpha_2\ldots\alpha_n)}\Big| \\
\lambda_{\alpha_1\alpha_2\cdots\alpha_{n-1}}(x) &= D^\beta\Phi_{\beta\alpha_1\alpha_2\ldots\alpha_{n-1}}\Big|.
\end{aligned}
\tag{2.8.14}
$$

Therefore, a chiral tensor superfield of Lorentz type $(n/2, 0)$ contains four irreducible tensor fields of Lorentz types $((n-1)/2, 0)$, $((n/2, 0)$, $(n/2, 0)$ and $((n+1)/2, 0)$, respectively.

Transformation laws of the component fields with respect to the supersymmtery transformations can be obtained in the same fashion as was done in the scalar case. The results are

$$
\begin{aligned}
\delta A_{\alpha_1\ldots\alpha_n}(x) &= -\epsilon^\beta\eta_{\beta\alpha_1\ldots\alpha_n}(x) + \frac{1}{n+1}\sum_{k=1}^{n}\epsilon_{\alpha_k}\lambda_{\alpha_1\ldots\hat{\alpha}_k\ldots\alpha_n}(x) \\
\delta\lambda_{\alpha_1\ldots\alpha_{n-1}}(x) &= -2\epsilon^\beta F_{\beta\alpha_1\ldots\alpha_{n-1}}(x) + 2i\bar{\epsilon}_{\dot\beta}\partial^{\beta\dot\beta}A_{\beta\alpha_1\ldots\alpha_n}(x) \\
\delta\eta_{\alpha_1\alpha_2\ldots\alpha_{n+1}}(x) &= -2\epsilon_{(\alpha_1}F_{\alpha_2\ldots\alpha_{n+1})}(x) + 2i\bar{\epsilon}_{\dot\alpha}\partial_{(\alpha_1}{}^{\dot\alpha}A_{\alpha_2\ldots\alpha_{n+1})}(x) \\
\delta F_{\alpha_1\ldots\alpha_n}(x) &= -i\bar{\epsilon}_{\dot\beta}\partial^{\beta\dot\beta}\eta_{\beta\alpha_1\ldots\alpha_n}(x) - \frac{i}{n+1}\sum_{k=1}^{n}\bar{\epsilon}^{\dot\beta}\partial_{\alpha_k\dot\beta}\lambda_{\alpha_1\ldots\hat{\alpha}_k\ldots\alpha_n}(x).
\end{aligned}
\tag{2.8.15}
$$

It is instructive to clarify the component form of the equation

$$D^\beta\Phi_{\beta\alpha_1\alpha_2\ldots\alpha_{n-1}} = 0$$

defining a massless superfield. This equation has the following consequences

$$D^2\Phi_{\alpha_1\ldots\alpha_n} = \partial^{\beta\dot\beta}\Phi_{\beta\alpha_1\ldots\alpha_{n-1}} = \partial^{\beta\dot\beta}D_{\alpha_1}\Phi_{\beta\alpha_2\ldots\alpha_n} = 0.$$

Taking the space projection of all the equations gives

$$\lambda_{\alpha_1\ldots\alpha_{n-1}}(x) = F_{\alpha_1\ldots\alpha_n}(x) = 0 \tag{2.8.16}$$

and

$$\partial^{\beta\dot\beta} A_{\beta\alpha_1\ldots\alpha_n\ 1}(x) = \partial^{\beta\dot\beta}\eta_{\beta\alpha_1\ldots\alpha_n}(x). \tag{2.8.17}$$

Therefore, component fields $\lambda_{\alpha_1\ldots\alpha_{n\ 1}}$ and $F_{\alpha_1\ldots\alpha_n}$ vanish on-shell, and the rest are on-shell massless fields. The on-shell form of transformation laws (2.8.15) read

$$\begin{aligned}\delta A_{\alpha_1\ldots\alpha_n}(x) &= -\epsilon^{\beta}\eta_{\beta\alpha_1\ldots\alpha_n}(x)\\ \delta\eta_{\alpha_1\alpha_2\ldots\alpha_{n+1}}(x) &= 2\mathrm{i}\bar\epsilon_{\dot\alpha}\partial_{(\alpha_1}{}^{\dot\alpha}A_{\alpha_2\ldots\alpha_{n+1})}(x).\end{aligned} \tag{2.8.18}$$

2.8.3. *Real scalar superfield*

Consider a real scalar superfield $V(z)$. We define its component fields as follows

$$\begin{gathered}A(x) = V| \qquad \psi_\alpha(x) = \mathrm{D}_\alpha V| \qquad \bar\psi_{\dot\alpha}(x) = \bar{\mathrm{D}}_{\dot\alpha}V|\\ F(x) = -\frac{1}{4}\mathrm{D}^2V| \qquad \bar F(x) = -\frac{1}{4}\bar{\mathrm{D}}^2V| \qquad V_{\alpha\dot\alpha} = \frac{1}{2}[\mathrm{D}_\alpha, \bar{\mathrm{D}}_{\dot\alpha}]V|\\ \lambda_\alpha = -\frac{1}{4}\mathrm{D}_\alpha\bar{\mathrm{D}}^2V| \qquad \bar\lambda_{\dot\alpha} = -\frac{1}{4}\bar{\mathrm{D}}_{\dot\alpha}\mathrm{D}^2V|\\ D(x) = \frac{1}{32}(\mathrm{D}^2\bar{\mathrm{D}}^2 + \bar{\mathrm{D}}^2\mathrm{D}^2)V|.\end{gathered} \tag{2.8.19}$$

Note that the last three component fields here do not coincide with the corresponding ones in the power series expansion (2.0.5) because, for example, we have

$$-\frac{1}{4}\mathrm{D}_\alpha\bar{\mathrm{D}}^2V| = \frac{1}{4}\partial_\alpha\bar\partial_{\dot\alpha}\bar\partial^{\dot\alpha}V| + \frac{\mathrm{i}}{2}\partial_{\alpha\dot\alpha}\bar\partial^{\dot\alpha}V|.$$

The supersymmetry transformations act on the component fields of $V(z)$ as follows

$$\begin{aligned}\delta A(x) &= -\epsilon\psi(x) - \overline{\epsilon\psi}(x)\\ \delta\psi_\alpha(x) &= -2\epsilon_\alpha F(x) - \bar\epsilon^{\dot\alpha}V_{\alpha\dot\alpha}(x) - \mathrm{i}\bar\epsilon^{\dot\alpha}\partial_{\alpha\dot\alpha}A(x)\\ \delta F(x) &= -\overline{\epsilon\lambda}(x)\\ \delta V_{\alpha\dot\alpha}(x) &= 2(\bar\epsilon_{\dot\alpha}\lambda_\alpha(x) - \epsilon_\alpha\bar\lambda_{\dot\alpha}(x)) - 2\mathrm{i}(\epsilon_\alpha\partial_{\beta\dot\alpha}\psi^\beta(x) + \bar\epsilon_{\dot\alpha}\partial_{\alpha\dot\beta}\bar\psi^{\dot\beta}(x))\\ &\quad + \mathrm{i}\partial_{\alpha\dot\alpha}(\overline{\epsilon\psi}(x) - \epsilon\psi(x))\\ \delta\lambda_\alpha(x) &= -2\epsilon_\alpha D(x) - \mathrm{i}\epsilon_\alpha\partial_a V^a(x) - 2\mathrm{i}\bar\epsilon^{\dot\alpha}\partial_{\alpha\dot\alpha}F(x)\\ \delta D(x) &= -\frac{\mathrm{i}}{2}\partial_{\alpha\dot\alpha}(\epsilon^\alpha\bar\lambda^{\dot\alpha}(x) + \bar\epsilon^{\dot\alpha}\lambda^\alpha(x)).\end{aligned} \tag{2.8.20}$$

2.8.4. Linear real scalar superfield

Finally, let us consider a real scalar superfield restricted by the linearity constraint

$$D^2 V = \bar{D}^2 V = 0. \tag{2.8.21}$$

In accordance with expressions (2.8.19), this superfield has the following component fields

$$A(x) = V| \qquad \psi_\alpha(x) = D_\alpha V| \qquad \bar{\psi}_{\dot{\alpha}}(x) = \bar{D}_{\dot{\alpha}} V| \qquad V_{\alpha\dot{\alpha}}(x) = \frac{1}{2}[D_\alpha, \bar{D}_{\dot{\alpha}}]V|. \tag{2.8.22}$$

Not all of these fields are unconstrained. Indeed, one can readily prove the identity

$$[D^2, \bar{D}^2] = -4i\partial^{\alpha\dot{\alpha}}[D_\alpha, \bar{D}_{\dot{\alpha}}].$$

This identity and equation (2.8.21) show that the vector field $V_a(x)$ is transverse,

$$\partial^a V_a(x) = 0. \tag{2.8.23}$$

Analogously, every linear tensor superfield contains a constrainted component field.

Transformation laws of component fields (2.8.21) follow from expressions (2.8.20):

$$\begin{aligned} \delta A(x) &= -\epsilon\psi(x) - \bar{\epsilon}\bar{\psi}(x) \\ \delta\psi_\alpha(x) &= -\bar{\epsilon}^{\dot{\alpha}} V_{\alpha\dot{\alpha}}(x) - i\bar{\epsilon}^{\dot{\alpha}}\partial_{\alpha\dot{\alpha}} A(x) \\ \delta V_{\alpha\dot{\alpha}}(x) &= -2i(\epsilon_\alpha \partial_{\beta\dot{\alpha}}\psi^\beta(x) + \bar{\epsilon}_{\dot{\alpha}}\partial_{\alpha\beta}\bar{\psi}^\beta(x)) + i\partial_{\alpha\dot{\alpha}}(\bar{\epsilon}\bar{\psi}(x) - \epsilon\psi(x)). \end{aligned} \tag{2.8.24}$$

It is not difficult to see that $\partial^a(\delta V_a(x)) = 0$.

To summarize, the reduction of superfields to component fields is done most effectively by means of covariant differentiation supplemented by space projection.

2.9. The superconformal group

We have seen in Section 2.5 that the super Poincaré transformations (2.5.3) acting on the complex superspace $\mathbb{C}^{4|2}$ leave invariant the surface $\mathbb{R}^{4|4}(\theta\sigma\bar{\theta})$ defined by

$$y^a - \bar{y}^a = 2i\theta\sigma^a\bar{\theta}. \tag{2.9.1}$$

This surface has been identified with the real superspace $\mathbb{R}^{4|4}$ by setting the variables

$$x^a = \frac{1}{2}(y^a + \bar{y}^a) \tag{2.9.2}$$

to be the c-number coordinates on $\mathbb{R}^{4|4}$. It seems reasonable to ask: do other holomorphic transformations of $\mathbb{C}^{4|2}$ exist leaving $\mathbb{R}^{4|4}(\theta\sigma\bar{\theta})$ invariant? The answer is yes. Before reading the present section, it is worth recalling the remark given at the end of subsection 1.7.4.

2.9.1. *Superconformal transformations*

A holomorphic mapping of $\mathbb{C}^{4|2}$ onto itself leaving $\mathbb{R}^{4|4}(\theta\sigma\bar{\theta})$ invariant is said to be 'superconformal'. Clearly, the set of all superconformal transformations forms a group. It is called the superconformal group. All the super Poincaré transformations are superconformal. The simplest superconformal transformations are:

Dilatations

$$e^{i\Delta \mathbf{d}} \cdot y^a = e^{\Delta} y^a \qquad e^{i\Delta \mathbf{d}} \cdot \theta^\alpha = e^{\Delta/2}\theta^\alpha \tag{2.9.3}$$

Axial or γ_5-rotations

$$e^{i\Omega \mathbf{a}} \cdot y^a = y^a \qquad e^{i\Omega \mathbf{a}} \cdot \theta^\alpha = e^{-i\Omega/2}\theta^\alpha. \tag{2.9.4}$$

Here Δ and Ω are real c-number parameters, $\mathbf{d}$ and $\mathbf{a}$ denote the generators of the scale and axial transformations, respectively. The above transformations act on $\mathbb{R}^{4|4}(\theta\sigma\bar{\theta})$ as follows:

Dilatations

$$e^{i\Delta \mathbf{d}} \cdot x^a = e^{\Delta} x^a \qquad e^{i\Delta \mathbf{d}} \cdot \theta^\alpha = e^{\Delta/2}\theta^\alpha \qquad e^{i\Delta \mathbf{d}} \cdot \bar{\theta}_{\dot{\alpha}} = e^{\Delta/2}\bar{\theta}_{\dot{\alpha}} \tag{2.9.5}$$

γ_5-rotations

$$e^{i\Omega \mathbf{a}} \cdot y^a = y^a \qquad e^{i\Omega \mathbf{a}} \cdot \theta^\alpha = e^{-i\Omega/2}\theta^\alpha \qquad e^{i\Omega \mathbf{a}} \cdot \bar{\theta}_{\dot{\alpha}} = e^{i\Omega/2}\bar{\theta}_{\dot{\alpha}}. \tag{2.9.6}$$

Combining θ_α and $\bar{\theta}^{\dot{\alpha}}$ in the four-component column

$$\Theta = \begin{pmatrix} \theta_\alpha \\ \bar{\theta}^{\dot{\alpha}} \end{pmatrix}$$

that is a Majorana spinor, the γ_5-rotations act on Θ by the rule:

$$e^{i\Omega \mathbf{a}} \cdot \Theta = e^{-\frac{i}{2}\Omega\gamma_5}\Theta.$$

Next, let us consider the antiholomorphic mapping defined on $\mathbb{C}^{4|2}$ as

$$y'^a = R \cdot y^a = \frac{\bar{y}^a}{\bar{y}^2} \qquad \theta'^\alpha = R \cdot \theta^\alpha = -\frac{\bar{y}^b(\bar{\theta}\tilde{\sigma}_b)^\alpha}{\bar{y}^2}. \tag{2.9.7}$$

This mapping is called the 'superinversion'. Obviously, it represents a superspace analogue of the space–time inversion (1.7.19). One can easily verify that the superinversion moves $\mathbb{R}^{4|4}(\theta\sigma\bar{\theta})$ into itself. In addition, it coincides with the inverse mapping,

$$R^2 = \mathbb{I} \tag{2.9.8}$$

Therefore, for every superconformal transformation S, the mapping RSR is superconformal. As a result, taking one of the known superconformal transformations (super Poincaré, scale or axial) in the role of S, we may obtain a new superconformal transformation. Recall that the analogous trick has been applied in Section 1.7 in obtaining the special conformal transformations in Minkowski space.

First, we choose space–time translations $\exp(-\mathrm{i}f^a\,\mathbf{p}_a)$ in the role of S and consider the transformations

$$\exp(-\mathrm{i}f^a\mathbf{v}_a) \equiv R\exp(-\mathrm{i}f^a\mathbf{p}_a)R. \tag{2.9.9}$$

Using the relation (2.5.2*a*) and (2.9.7), one then arrives at

Special conformal transformations

$$\begin{aligned}
\mathrm{e}^{-\mathrm{i}(f,\mathbf{v})}\cdot y^a &= \frac{y^a + f^a y^2}{1 + 2(f,y) + f^2y^2} \\
\mathrm{e}^{-\mathrm{i}(f,\mathbf{v})}\cdot \theta^\alpha &= \frac{\theta^\alpha - y^a f^b(\theta\sigma_a\tilde{\sigma}_b)^\alpha}{1 + 2(f,y) + f^2y^2}
\end{aligned} \tag{2.9.10}$$

generalizing the space–time transformations (1.7.18*d*). As an exercise, we suggest the reader find the restriction of (2.9.10) to $\mathbb{R}^{4|4}(\theta\sigma\bar{\theta})$.

Further, let us choose a supersymmetry transformation in the role of S. So we are to evaluate the mappings

$$\mathrm{e}^{\mathrm{i}(\eta^\alpha\mathbf{s}_\alpha + \bar{\eta}_{\dot{\alpha}}\bar{\mathbf{s}}^{\dot{\alpha}})} \equiv R\mathrm{e}^{\mathrm{i}(\eta^\alpha\mathbf{q}_\alpha + \bar{\eta}_{\dot{\alpha}}\bar{\mathbf{q}}^{\dot{\alpha}})}R. \tag{2.9.11}$$

Making use of the relations (2.5.2*c*) and (2.9.7) gives

S-supersymmetry transformations

$$\begin{aligned}
\mathrm{e}^{\mathrm{i}(\eta\mathbf{s} + \bar{\eta}\bar{\mathbf{s}})}\cdot y^a &= y^a - 2\theta\sigma^a\tilde{\sigma}_b\eta y^b + \mathrm{i}(2y^a y_b - y^2\delta^a_b)\eta\sigma^b\bar{\eta} + 4\eta^2\theta^2 y^a \\
&\quad + 4\mathrm{i}y^a y_b\theta\sigma^b\bar{\eta}\eta^2 - y^a y^2\bar{\eta}^2\eta^2 \\
\mathrm{e}^{\mathrm{i}(\eta\mathbf{s} + \bar{\eta}\bar{\mathbf{s}})}\cdot \theta^\alpha &= \theta^\alpha + 2\eta^\alpha\theta^2 - \mathrm{i}(\bar{\eta}\tilde{\sigma}_b)^\alpha y^b[1 - 2\theta\eta + \theta^2\eta^2] - \eta^\alpha\bar{\eta}^2 y^2 \\
&\quad + \theta^\alpha\eta^2\bar{\eta}^2 y^2.
\end{aligned} \tag{2.9.12}$$

Note, ordinary supersymmetry transformations (2.5.2*c*) are sometimes called *Q*-supersymmetric, to distinguish them from the *S*-supersymmetric ones.

Subsequent application of the above trick does not give new superconformal transformations, since one can readily prove that

$$
\begin{aligned}
Re^{\frac{i}{2}K^{ab}\mathbf{j}_{ab}}R &= e^{\frac{i}{2}K^{ab}\mathbf{j}_{ab}} \\
Re^{i\Delta\mathbf{d}}R &= e^{-i\Delta\mathbf{d}} \\
Re^{i\Omega\mathbf{a}}R &= e^{-i\Omega\mathbf{a}}.
\end{aligned}
\tag{2.9.13}
$$

However, it can be shown that the transformations described turn out to generate the superconformal group. Namely, in a neighbourhood of the identity of the superconformal group, every group element looks like

$$
g = \exp\left[i\left(-b^a\mathbf{p}_a - f^a\mathbf{v}_a + \frac{1}{2}K^{ab}\mathbf{j}_{ab} + \Delta\mathbf{d} + \Omega\mathbf{a} + + \epsilon\mathbf{q} + \overline{\epsilon\mathbf{q}} + \eta\mathbf{s} + \overline{\eta\mathbf{s}}\right)\right].
\tag{2.9.14}
$$

We shall argue this assertion in Chapter 6.

In the case of infinitesimal parameters in equation (2.9.14), the corresponding superconformal transformation of $\mathbb{C}^{4|2}$ reads

$$
y'^a = y^a + \lambda^a(y,\theta) \qquad \theta'^\alpha = \theta^\alpha + \lambda^\alpha(y,\theta)
\tag{2.9.15}
$$

where

$$
\begin{aligned}
\lambda^a(y,\theta) &= b^a + K^a{}_b y^b + \Delta y^a + f^a y^2 - 2y^a(f,y) + 2i\theta\sigma^a\bar{\epsilon} - 2\theta\sigma^a\tilde{\sigma}_b\eta y^b \\
\lambda^\alpha(y,\theta) &= \epsilon^\alpha - i(\bar{\eta}\tilde{\sigma}_b)^\alpha y^b + \frac{1}{2}(\Delta - i\Omega)\theta^\alpha + f^a y^b(\theta\sigma_a\tilde{\sigma}_b)^\alpha - K^\alpha{}_\beta\theta^\beta + 2\eta^\alpha\theta^2.
\end{aligned}
\tag{2.9.16}
$$

When restricted to $\mathbb{R}^{4|4}(\theta\sigma\bar{\theta})$ (defined by equations (2.9.1, 2)), this transformation acts as follows:

$$
\begin{aligned}
x'^a &= x^a + \frac{1}{2}\left(e^{i\mathscr{H}}\lambda^a(x,\theta) + e^{-i\mathscr{H}}\bar{\lambda}^a(x,\bar{\theta})\right) \\
\theta'^\alpha &= \theta^\alpha + e^{i\mathscr{H}}\lambda^\alpha(x,\theta) \\
\bar{\theta}'_{\dot{\alpha}} &= \bar{\theta}_{\dot{\alpha}} + e^{-i\mathscr{H}}\bar{\lambda}_{\dot{\alpha}}(x,\bar{\theta})
\end{aligned}
\tag{2.9.17}
$$

where $\mathscr{H} = \theta\sigma^a\bar{\theta}\partial_a$.

2.9.2. The supersymmetric interval and superconformal transformations

We have seen in Section 2.4 that the super Poincaré transformations leave invariant the supersymmetric interval

$$
ds^2 = \omega^a\omega_a \qquad \omega^a = dx^a + i\theta\sigma^a d\bar{\theta} - id\theta\sigma^a\bar{\theta}.
\tag{2.9.18}
$$

Let us now consider how superconformal transformations act on this object.

Evidently, the dilatations act on ds^2 in the manner

$$
e^{i\Delta\mathbf{d}}: ds^2 \to e^{2\Delta}ds^2
\tag{2.9.19}
$$

and the γ_5-rotation leave ds^2 unchanged. Next, the superinversion (2.9.7) changes ω^a according to the rule

$$\omega'^{\alpha\dot{\alpha}} = \frac{1}{x_{(+)}^2 x_{(-)}^2} x_{(+)\beta}{}^{\dot{\alpha}} x_{(-)}{}^{\alpha}{}_{\dot{\beta}} \omega^{\beta\dot{\beta}}$$

where we have introduced the notation

$$x_{(\pm)}^a = x^a \pm i\theta\sigma^a\bar{\theta}$$

As a result, the inversion locally rescales the interval

$$R: ds^2 \to \frac{1}{x_{(+)}^2 x_{(-)}^2} ds^2. \tag{2.9.20}$$

The same is true, owing to identities (2.9.9) and (2.9.11), for the special conformal and S-supersymmetry transformations. Therefore we come to the remarkable result: the peculiar feature of superconformal transformations in that they, at most, locally rescale the supersymmetric interval.

2.9.3. *The superconformal algebra*

The generators of the superconformal group obey the algebra:

$$\begin{aligned}
&\{\mathbf{q}_\alpha, \bar{\mathbf{q}}_{\dot{\alpha}}\} = 2\boldsymbol{p}_{\alpha\dot{\alpha}} \qquad \{\mathbf{s}_\alpha, \bar{\mathbf{s}}_{\dot{\alpha}}\} = 2\mathbf{v}_{\alpha\dot{\alpha}} \\
&[\mathbf{d}, \mathbf{q}_\alpha] = -\frac{i}{2}\mathbf{q}_\alpha \qquad [\mathbf{d}, \mathbf{s}_\alpha] = \frac{i}{2}\mathbf{s}_\alpha \\
&[\mathbf{a}, \mathbf{q}_\alpha] = -\frac{1}{2}\mathbf{q}_\alpha \qquad [\mathbf{a}, \mathbf{s}_\alpha] = \frac{1}{2}\mathbf{s}_\alpha \\
&[\mathbf{v}_{\alpha\dot{\alpha}}, \mathbf{q}_\beta] = 2\varepsilon_{\alpha\beta}\bar{\mathbf{s}}_{\dot{\alpha}} \qquad [\boldsymbol{p}_{\alpha\dot{\alpha}}, \mathbf{s}_\beta] = 2\varepsilon_{\alpha\beta}\bar{\mathbf{q}}_{\dot{\alpha}} \\
&\{\mathbf{s}_\alpha, \mathbf{q}_\beta\} = 4i\mathbf{j}_{\alpha\beta} + 2i\varepsilon_{\alpha\beta}\mathbf{d} + 6\varepsilon_{\alpha\beta}\mathbf{a}
\end{aligned} \tag{2.9.21a}$$

and

$$\begin{aligned}
&[\mathbf{d}, \boldsymbol{p}_{\alpha\dot{\alpha}}] = -i\boldsymbol{p}_{\alpha\dot{\alpha}} \qquad [\mathbf{d}, \mathbf{v}_{\alpha\dot{\alpha}}] = i\mathbf{v}_{\alpha\dot{\alpha}} \\
&\frac{i}{4}[\mathbf{v}_{\alpha\dot{\alpha}}, \boldsymbol{p}_{\beta\dot{\beta}}] = \varepsilon_{\alpha\beta}\bar{\mathbf{j}}_{\dot{\alpha}\dot{\beta}} + \varepsilon_{\dot{\alpha}\dot{\beta}}\mathbf{j}_{\alpha\beta} + \varepsilon_{\alpha\beta}\varepsilon_{\dot{\alpha}\dot{\beta}}\mathbf{d}
\end{aligned} \tag{2.9.21b}$$

and

$$\begin{aligned}
&[\mathbf{j}_{\alpha\beta}, \mathbf{q}_\gamma] = i\varepsilon_{\gamma(\alpha}\mathbf{q}_{\beta)} \qquad [\mathbf{j}_{\alpha\beta}, \mathbf{s}_\gamma] = i\varepsilon_{\gamma(\alpha}\mathbf{s}_{\beta)} \\
&[\mathbf{j}_{\alpha\beta}, \boldsymbol{p}_{\gamma\dot{\gamma}}] = i\varepsilon_{\gamma(\alpha}\boldsymbol{p}_{\beta)\dot{\gamma}} \qquad [\mathbf{j}_{\alpha\beta}, \mathbf{v}_{\gamma\dot{\gamma}}] = i\varepsilon_{\gamma(\alpha}\mathbf{v}_{\beta)\dot{\gamma}} \\
&[\mathbf{j}_{\alpha\beta}, \mathbf{j}_{\gamma\delta}] = i\varepsilon_{\gamma(\alpha}\mathbf{j}_{\beta)\delta} + i\varepsilon_{\delta(\alpha}\mathbf{j}_{\beta)\gamma}.
\end{aligned} \tag{2.9.21c}$$

The other (anti)commutators vanish or can be found by Hermitian conjugation (considering $\rho_a, \mathbf{v}_a, \mathbf{d}, \mathbf{a}$ as Hermitian operators and setting $(\mathbf{j}_{\alpha\beta})^+ = \bar{\mathbf{j}}_{\dot{\alpha}\dot{\beta}}, (\mathbf{q}_a)^+ = \bar{\mathbf{q}}_{\dot{\alpha}}$ and $(\mathbf{s}_\alpha)^+ = \bar{\mathbf{s}}_{\dot{\alpha}}$). The above (anti)commutation relations define the superconformal algebra.

To derive (2.9.21), one can consider a simple representation of the superconformal group acting on the space of scalar superfields by the rule

$$g: U(z) \to U_g(z) = U(g^{-1} \cdot z) \tag{2.9.22}$$

for every superfield U and every element g of the superconformal group. This representation is characterized by the super Poincaré generators (2.4.37) (with $M_{\alpha\beta} = \bar{M}_{\dot{\alpha}\dot{\beta}} = 0$), the dilatation generator

$$\mathrm{D} = \frac{\mathrm{i}}{2}\{(x^a_{(+)} + x^a_{(-)})\partial_a + \theta^\alpha\partial_\alpha + \bar{\theta}^{\dot{\alpha}}\bar{\partial}_{\dot{\alpha}}\} \tag{2.9.23a}$$

the axial generator

$$A = \frac{1}{2}(\theta^\alpha\partial_\alpha - \bar{\theta}^{\dot{\alpha}}\bar{\partial}_{\dot{\alpha}}) \tag{2.9.23b}$$

the special conformal generators

$$\begin{aligned} V_{\alpha\dot{\alpha}} = (\sigma^a)_{\alpha\dot{\alpha}} V_a = &-\frac{\mathrm{i}}{2}(x_{(+)\alpha}{}^{\dot{\beta}} x_{(+)}{}^{\beta}{}_{\dot{\alpha}} + x_{(-)\alpha}{}^{\dot{\beta}} x_{(-)}{}^{\beta}{}_{\dot{\alpha}})\partial_{\beta\dot{\beta}} \\ &+ 2\mathrm{i}(\theta_\alpha x_{(+)}{}^{\beta}{}_{\dot{\alpha}}\partial_\beta + \bar{\theta}_{\dot{\alpha}} x_{(-)\alpha}{}^{\dot{\beta}}\bar{\partial}_{\dot{\beta}}) \end{aligned} \tag{2.9.23c}$$

and the S-supersymmetry generators

$$\begin{aligned} S_\alpha &= -\mathrm{i}x_{(+)\alpha}{}^{\dot{\beta}}\theta^\beta\partial_{\beta\dot{\beta}} + 2\mathrm{i}\theta^2\partial_\alpha + x_{(-)\alpha}{}^{\dot{\beta}}\bar{\partial}_{\dot{\beta}} \\ \bar{S}_{\dot{\alpha}} &= -\mathrm{i}x_{(-)}{}^{\beta}{}_{\dot{\alpha}}\bar{\theta}^{\dot{\beta}}\partial_{\beta\dot{\beta}} + x_{(+)}{}^{\beta}{}_{\dot{\alpha}}\partial_\beta - 2\mathrm{i}\bar{\theta}^2\bar{\partial}_{\dot{\alpha}}. \end{aligned} \tag{2.9.23d}$$

Other superfield representations of the superconformal group will be discussed in Chapter 6.

In conclusion, we would like to describe one of the possible explanations of the symmetry between expressions in the l.h.s. and r.h.s. of (2.9.21). Consider the transformations (2.5.4). Together with transformation (2.5.2*a*), where b^a is complex, they define the action on $\mathbb{C}^{4|2}$ of the complex shell of the super Poincaré group. Let us introduce the holomorphic mappings

$$\begin{gathered} \exp(\mathrm{i}f^a\mathbf{v}_a) \equiv R\exp(\mathrm{i}\bar{f}^a\rho_a)R \\ \exp(\mathrm{i}\eta^\alpha\mathbf{s}_\alpha) \equiv R\exp(\mathrm{i}\bar{\eta}_{\dot{\alpha}}\bar{\mathbf{q}}^{\dot{\alpha}})R \qquad \exp(\mathrm{i}\bar{\kappa}_{\dot{\alpha}}\bar{\mathbf{s}}^{\dot{\alpha}}) = R\exp(\mathrm{i}\kappa^\alpha\mathbf{q}_\alpha)R. \end{gathered} \tag{2.9.24}$$

Now, the transformations (2.5.2*a*) and (2.9.3, 4), where b^a and Δ, Ω are complex, and the transformations (2.5.4) and (2.9.24) define the action on $\mathbb{C}^{4|2}$ of the complex shell of the superconformal group. It is a simple exercise to check that

$$R \exp(\mathrm{i}\Delta\mathbf{d})R = \exp(-\mathrm{i}\bar{\Delta}\mathbf{d}) \qquad R \exp(\mathrm{i}\Omega\mathbf{a})R = \exp(-\mathrm{i}\bar{\Omega}\mathbf{a})$$
$$R \exp(\mathrm{i}K^{\alpha\beta}\mathbf{j}_{\alpha\beta})R = \exp(\mathrm{i}\bar{K}^{\dot{\alpha}\dot{\beta}}\bar{\mathbf{j}}_{\dot{\alpha}\dot{\beta}}). \tag{2.9.25}$$

The relations (2.9.24, 25) make it possible to reconstruct the expression on the r.h.s. of (2.9.21) starting from the corresponding ones on the l.h.s.

3 Field Theory in Superspace

Gaily bedight,
A gallant knight,
In sunshine and in shadow,
Had journeyed long,
Singing a song.
In search of Eldorado.

Edgar Allan Poe:
Eldorado

3.1. Supersymmetric field theory

We proceed to a systematic study of supersymmetric field theories. By definition, a field theory is said to be 'supersymmetric' if its symmetry group coincides with the super Poincaré group or includes this supergroup as a subgroup. The family of all supersymmetric field theories forms a subclass in the class of all relativistic (or Poincaré invariant) field theories. Our primary goal is to define requirements on a field theory in order for it to be supersymmetric. Then, we are to clarify dynamical properties of supersymmetric field theories, both at the classical and quantum levels. This chapter is devoted to consideration of classical aspects. Quantum theory will be discussed in the next chapter.

3.1.1. Quick review of field theory

To begin with, it is worth restoring in mind basic principles of classical field theory; for a more detailed treatment see, for example, the lectures of B. De Witt*. As is well known, any dynamical system is determined by specifying

*B.S. De Witt, The Space–time Approach to Quantum Field Theory, in: *Relativity, Groups and Topology*, eds B.S. De Witt and R. Stora (Elsevier Science Publishers B.V., 1984)

DOI: 10.1201/9780367802530-3

a 'space of dynamical variables' (or 'space of histories') $\mathbf{\Phi}$ and an 'action functional' $S[\varphi]$. Each point $\varphi = \{\varphi^i(x)\}$ of $\mathbf{\Phi}$ is said to be a 'field history'. Here $\varphi^i(x)$ are smooth bosonic and fermionic fields on a space–time, where index 'i' (labelling field statistics types and tensor types) takes a number of values fixed for the given system. Fields φ^i forming histories possess an arbitrary behaviour in finite regions of the space–time but obey certain boundary conditions at infinity, appropriate for the system. The action functional $S[\varphi]$, which is a mapping

$$S\colon \mathbf{\Phi} \to \mathbb{R}_c \tag{3.1.1}$$

determines the dynamical field equations of the system

$$_{i,}S[\varphi] \equiv \delta S[\varphi]/\delta\varphi^i = 0 \tag{3.1.2}$$

where left functional (or variational) derivatives are defined as follows

$$\delta S[\varphi] = S[\varphi + \delta\varphi] - S[\varphi] = \int \mathrm{d}^4x\,\delta\varphi^i(x)\frac{\delta S[\varphi]}{\delta\varphi^i(x)} \tag{3.1.3}$$

with $\delta\varphi^i(x)$ being arbitrary field variations. In obtaining the dynamical equations, the following identity

$$\frac{\delta\varphi^j(x')}{\delta\varphi^i(x)} = \delta_i{}^j\delta^4(x - x') \tag{3.1.4}$$

is often helpful. Every solution $\varphi_0 = \{\varphi_0^i(x)\}$ of the dynamical equations is said to be a 'dynamical field history'. The set of all dynamical histories forms a subspace in $\mathbf{\Phi}$, called the 'dynamical subspace' (or 'mass shell surface') and is denoted by $\mathbf{\Phi}_0$. Throughout this book we adopt the following convention with regard to the global structure of $\mathbf{\Phi}$: for every history $\varphi \in \mathbf{\Phi}$, there exists $\varphi_0 \in \mathbf{\Phi}_0$ such that the displacements

$$\Delta\varphi^i(x) = \varphi^i(x) - \varphi_0^i(x)$$

have compact supports in the space–time.

In the case of a dynamical system in Minkowski space, the Poincaré group is assumed to act on the space of histories $\mathbf{\Phi}$ by means of some transformations

$$\varphi^i(x) \to \varphi'^i(x) = \varphi_g^i(x) \tag{3.1.5}$$

defined for every group element $g \in \Pi$. The dynamical system is said to be a 'relativistic field theory' if the dynamical subspace $\mathbf{\Phi}_0$ is a Poincaré invariant surface in $\mathbf{\Phi}$,

$$_{i,}S[\varphi] = 0 \Rightarrow {}_{i,}S[\varphi_g] = 0. \tag{3.1.6}$$

This requirement is satisfied automatically when the action functional is chosen to be a scalar with respect to the Poincaré group

$$S[\varphi] = S[\varphi_g] \tag{3.1.7}$$

for every field history $\varphi \in \mathbf{\Phi}$ and every group element $g \in \Pi$. One can look on equation (3.1.7) as a postulate of relativistic field theory. More generally, having a field theory with some invariance group (rigid or gauge), the action functional should remain unchanged under all transformations from the invariance group.

One more assumption of relativistic field theory is that dynamical variables $\varphi^i(x)$ may be chosen in such a way that Poincaré transformations (3.1.5) are linear and homogeneous. As a result, $\varphi^i(x)$ are tensor fields on Minkowski space, with the Poincaré transformation law (1.5.13).

The field theories usually considered are local ones. A field theory is said to be 'local' if the dynamical equations involve a finite number of time derivatives (and space derivatives, as a consequence of the Poincaré covariance). In local field theory the action functional has the form

$$S[\varphi] = \int \mathrm{d}^4x L(\varphi, \partial_a\varphi, \ldots, \partial_{a_1}\partial_{a_2}\ldots\partial_{a_k}\varphi) \tag{3.1.8}$$

where the integral is performed over Minkowski space. The integrand L is called the 'Lagrangian' (or the 'Lagrange function'). To guarantee equation (3.1.7), it is sufficient to choose the Lagrangian to be a scalar field,

$$L_g(x) = L(g^{-1}\cdot x) \tag{3.1.9}$$

with respect to the Poincaré group. Explicitly, since the Poincaré transformations preserve the space–time volume,

$$\det\left(\frac{\partial x'^a}{\partial x^b}\right) = 1 \qquad x'^a = g^{-1}\cdot x^a \tag{3.1.10}$$

we have

$$S[\varphi_g] = \int \mathrm{d}^4x L(g^{-1}\cdot x) = \int \mathrm{d}^4x' L(x') = S[\varphi].$$

In practice, one never needs to know the concrete values of $S[\varphi]$ on fixed histories, but only its functional structure. In this respect, a number of formal manipulations with the action functional, like integration by parts for instance, are available. These operations are justified by adding suitable boundary terms to $S[\varphi]$ or by imposing convenient restrictions on the dynamical variables, as will always be assumed below.

The most popular field theories are those in which the dynamical equations are at most second order in time derivatives. In this case the action functional can be represented in the form

$$S[\varphi] = \int \mathrm{d}^4x L(\varphi, \partial_a\varphi) \tag{3.1.11}$$

modulo boundary terms.

3.1.2. The space of superfield histories; the action superfunctional

In a supersymmetric field theory the space of field histories is a transformation space of the super Poincaré group. So it is worth finding a systematic way of obtaining representations of the super Poincaré group on functional spaces. At present, the only known systematic and most elegant way is based on the use of superspace and superfields. Indeed, we have seen that tensor superfields provide us with representations of the super Poincaré group. What is more, every unitary representation of the super Poincaré group admits a superfield realization. Furthermore, each superfield is determined in terms of its component fields and they automatically form a multiplet with respect to the super Poincaré group. Therefore, every space of tensor superfields leads to some space of tensor fields ('component field space') with defined action of the super Poincaré group.

Conversely, the question arises: does any space of fields with defined action of the super Poincaré group admit a structure of component field space for some space of superfields? We do not know the answer to this question. But all known $N = 1$ supersymmetric field theories admit superfields realizations, and we restrict our consideration to the case of such theories. In each theory from this class, there are two equivalent realizations of the space of dynamical variables: as the space of field histories $\mathbf{\Phi}$ and the 'space of superfield histories' which will be denoted by $\mathbb{V}$. Any point of $\mathbb{V}$ presents a set of tensor superfields $v = \{v^I(z)\}$ on $\mathbb{R}^{4|4}$ and the totality of all their component fields gives a field history $\{\varphi^i(x)\}$ from $\mathbf{\Phi}$. In other words, $\mathbf{\Phi}$ is the component field space for $\mathbb{V}$. Since the spaces $\mathbf{\Phi}$ and $\mathbb{V}$ are in one-to-one correspondence, the action functional $S[\varphi]$ can be re-expressed as a functional on the space of superfield histories, i.e. as a mapping

$$S\colon \mathbb{V} \to \mathbb{R}_c \tag{3.1.12}$$

which will be denoted by $S[v]$. It is useful to call $S[v]$ the 'action superfunctional' (more generally, any mapping from a space of superfunctions into the Grassmann algebra Λ_∞ is said to be a 'superfunctional'). If v is a superfield history and φ is the corresponding field history, then

$$S[v] = S[\varphi]. \tag{3.1.13}$$

The 'super Poincaré invariance' means that the action superfunctional should remain unchanged under the super Poincaré transformations (2.4.29),

$$S[v] = S[v_g] \tag{3.1.14}$$

for every supergroup element $g \in S\Pi$.

The dynamical subspace $\mathbf{\Phi}_0$ consists of field histories corresponding to stationary values of the action functional. When φ_0 is a dynamical history, we have

$$S[\varphi_0 + \delta\varphi] - S[\varphi_0] = 0 \qquad \forall \delta\varphi^i(x). \tag{3.1.15}$$

Let us denote by $\mathbb{V}_0$ the subset in $\mathbb{V}$, for which $\mathbf{\Phi}_0$ is the component field space. The set $\mathbb{V}_0$ is the superfield version of the dynamical subspace. Any $v_0 = \{v_0^I(z)\} \in \mathbb{V}_0$ will be called a 'dynamical superfield history'. Due to one-to-one correspondence between $\mathbf{\Phi}$ and $\mathbb{V}$, equation (3.1.15) leads to

$$S[v_0 + \delta v] - S[v_0] = 0 \qquad \forall \delta v^i(z) \tag{3.1.16}$$

for every dynamical superfield history $v_0(z)$. So, the action superfunctional $S[v]$ takes stationary values at each point of $\mathbb{V}_0$. We see that *the dynamical behaviour of the superfield system is determined by the stationary action principle.*

To rewrite equation (3.1.16) in a differential form, it is worth discussion superfield variational derivatives.

3.1.3. *Integration over $\mathbb{R}^{4|4}$ and superfunctional derivatives*

Now is the time to resort to the integration theory over $\mathbb{R}^{p|q}$ developed in Section 1.10. We will be mainly interested in integrals over $\mathbb{R}^{4|4}$ of the general form

$$\int \mathrm{d}^8 z \, v(z) = \int \mathrm{d}^4 x \int \mathrm{d}^2 \theta \int \mathrm{d}^2 \bar{\theta} \, v(x, \theta, \bar{\theta}) \tag{3.1.17}$$

where $v(x, \theta, \bar{\theta})$ is some superfield. The integrals over a-number variables θ^α and $\bar{\theta}_{\dot{\alpha}}$ are defined as follows

$$\int \mathrm{d}\theta_\alpha \, \theta^\beta = \delta_\alpha{}^\beta \qquad \int \mathrm{d}\bar{\theta}^{\dot{\alpha}} \, \bar{\theta}_{\dot{\beta}} = \delta^{\dot{\alpha}}{}_{\dot{\beta}} \tag{3.1.18}$$

and the multiple measures are

$$\mathrm{d}^2\theta = \frac{1}{4} \varepsilon^{\alpha\beta} \, \mathrm{d}\theta_\alpha \, \mathrm{d}\theta_\beta \qquad \mathrm{d}^2\bar{\theta} = \frac{1}{4} \varepsilon_{\dot{\alpha}\dot{\beta}} \, \mathrm{d}\bar{\theta}^{\dot{\alpha}} \, \mathrm{d}\bar{\theta}^{\dot{\beta}} \tag{3.1.19}$$

instead of equation (1.10.41), such that

$$\int \mathrm{d}^2\theta \, \theta^2 = 1 \qquad \int \mathrm{d}^2\bar{\theta} \, \bar{\theta}^2 = 1. \tag{3.1.20}$$

The difference in definition (by a constant) of multiple integrals (1.10.41) and (3.1.17, 19) is introduced to have the property

$$\int \mathrm{d}^8 z V(z) = \int \mathrm{d}^4 x D(x) \tag{3.1.21}$$

where $D(x)$ is the highest component field of $V(z)$ in the $\theta, \bar{\theta}$-expansion (2.4.21). Therefore, if $v(z)$ is a real superfield then integral $\int \mathrm{d}^8 z v(z)$ is a real supernumber. More generally, we have

$$\left(\int d^8 z v(z)\right)^* = \int d^8 z v^*(z). \tag{3.1.22}$$

Superfields arising as integrands will always be assumed to vanish when the bodies of x^m go to $(\pm\infty)$, which implies, together with equation (1.10.31*c*), the relation

$$\int d^8 z(\partial_A v(z)) = 0 \tag{3.1.23}$$

where ∂_A are the partial derivatives defined in equation (2.4.25). This result presents the superspace rule for integration by parts. It can be rewritten in terms of the covariant derivatives (2.5.25):

$$\int d^8 z(D_A v(z)) = 0 \tag{3.1.24}$$

which follows from

$$\int d^8 z D_\alpha v = \int d^8 z \partial_\alpha v + i\int d^8 z(\sigma^a\bar\theta)_\alpha \partial_a v = \int d^8 z \partial_\alpha v + i\int d^8 z \partial_a((\sigma^a\bar\theta)_\alpha v).$$

Recall that integration in *a*-number variables is equivalent to differentiation. Then, due to equation (3.1.20), one obtains

$$\int d^2\theta = \frac{1}{4}\partial^\alpha\partial_\alpha \qquad \int d^2\bar\theta = \frac{1}{4}\bar\partial_{\dot\alpha}\bar\partial^{\dot\alpha}. \tag{3.1.25}$$

By analogy with the derivation of equation (3.1.24) from equation (3.1.23), the final relations can be rewritten in terms of the spinor covariant derivatives:

$$\int d^8 z v(z) = -\frac{1}{4}\int d^4x\, d^2\theta \bar D^2 v(z) = \frac{1}{16}\int d^4x D^2\bar D^2 v(z) = \int d^4x\left(\frac{1}{16}D^2\bar D^2 v|\right) \tag{3.1.26a}$$

and

$$\int d^8 z v(z) = -\frac{1}{4}\int d^4x\, d^2\bar\theta D^2 v(z) = \frac{1}{16}\int d^4x \bar D^2 D^2 v(z) = \int d^4x\left(\frac{1}{16}\bar D^2 D^2 v|\right) \tag{3.1.26b}$$

where symbol '|' means space projection (see Section 2.8).

Introduce the δ-function on $\mathbb{R}^{4|4}$:

$$\begin{gathered}\delta^8(z) = \delta^4(x)\delta^2(\theta)\delta^2(\bar\theta) \\ \delta^2(\theta) = \theta^2 \qquad \delta^2(\bar\theta) = \bar\theta^2.\end{gathered} \tag{3.1.27}$$

It is characterized by the properties:

$$\int d^8z'\delta(z - z')v(z') = v(z)$$

$$\delta^8(z - z')v(z') = \delta^8(z - z')v(z) \tag{3.1.28}$$

$$\delta^8(z) = \delta^8(-z) \qquad (\delta^8(z))^2 = 0 \qquad \delta^8(0) = 0$$

for an arbitrary superfield $v(z)$ (compare with equations (1.10.38, 39)).

After the given excursion into integration theory, we are in a position to introduce variational derivatives associated with different superfield types.

Let us consider a dynamical system with the space of histories $\mathbb{V}$ being the set of all real scalar superfields $V(z)$, $V = \bar{V}$, under some boundary conditions. For any superfield $V \in \mathbb{V}$, we denote by $\{\varphi^i(x)\}$ the set of component fields of $V(z)$. Using the supernumber norm (1.9.6), $\mathbb{V}$ can be turned into a metric space with the distance function $\rho(V_1, V_2)$ between two arbitrary superfields $V_1, V_2 \in \mathbb{V}$ defined as follows

$$\rho(V_1, V_2) = \{\max_{i,x} \| \varphi^i_1(x) - \varphi^i_2(x) \|, \quad x \in \mathbb{R}^4\}. \tag{3.1.29}$$

Three basic properties

1. $\rho(V_1, V_2) = \rho(V_2, V_1) > 0, \quad \text{if } V_1 \neq V_2$
2. $\rho(V, V) = 0$
3. $\rho(V_1, V_2) \leqslant \rho(V_1, V_3) + \rho(V_2, V_3)$

are obvious. The mapping (3.1.12) defining the action superfunctional will be assumed to be continuous. Then, under reasonble assumptions about the form of $S[V]$, two values of the action values $S[V]$ and $S[V + \delta V]$ where $\delta V(z)$ is an infinitesimal displacement, $\rho(V, V + \delta V) \to 0$, are related by the rule

$$\delta S[V] = S[V + \delta V] - S[V] = \int d^8z\delta V(z)\frac{\delta S[V]}{\delta V(z)} + O((\delta V)^2). \tag{3.1.30}$$

The superfield $\delta S[V]/\delta V(z)$ will be called the 'superfunctional derivative' of $S[V]$ at $V[z]$.

In complete analogy with ordinary variational analysis, one can prove that

$$\int d^8z V(z)\Psi(z) = 0 \qquad \forall V(z) = \bar{V}(z) \Rightarrow \Psi(z) = 0. \tag{3.1.31}$$

Here V can be restricted to have compact support in the space–time. Therefore, due to the arbitrariness of the superfield variation $\delta V(z)$ in (3.1.30), the stationary action principle (3.1.16) leads to the equation

$$\delta S[V]/\delta V(z) = 0 \tag{3.1.32}$$

which is the 'dynamical superfield equation' of the system.

For calculating the superfunctional derivative of different possible actions $S[V]$, it is useful first to do this in the case of a special superfunctional defined by

$$F : \mathbb{V} \to \mathbb{R}_c \qquad F(V) = V(z')$$

where z' is a fixed point of $\mathbb{R}^{4|4}$. Making use of the superspace δ-function (3.1.27) gives

$$\delta F(V) = \delta V(z') = \int \mathrm{d}^8 z \delta V(z) \delta(z - z').$$

Therefore, we have the relation

$$\frac{\delta V(z')}{\delta V(z)} = \delta^8(z - z') \tag{3.1.33}$$

which can be used further in practical calculations as a formal rule.

When a dynamical system is described by p real scalar superfields $V^I(z)$, $V^I = \bar{V}^I$, completely arbitrary modulo certain boundry conditions, equation (3.1.30) should be substituted by

$$\delta S[V] = \int \mathrm{d}^8 z \delta V^I(z) \frac{\delta S[V]}{\delta V^I(z)} \tag{3.1.34}$$

and, instead of equation (3.1.33), we will have

$$\frac{\delta V^J(z')}{\delta V^I(z)} = \delta_I{}^J \delta^8(z - z'). \tag{3.1.35}$$

When the total number of V^I is even, it is sometimes useful to combine the superfields under considerations into complex scalar superfields $V^{\hat{I}}$ and their conjugate $\bar{V}^{\hat{I}}(z)$. Then equation (3.1.35) tales the form

$$\frac{\delta V(z')}{\delta \bar{V}(z)} = \frac{\delta \bar{V}(z')}{\delta V(z)} = 0$$

$$\frac{\delta V(z')}{\delta V(z)} = \frac{\delta \bar{V}(z')}{\delta \bar{V}(z)} = \delta^8(z - z') \tag{3.1.36}$$

where indices are suppressed.

Now consider a dynamical system, for which the space of histories $\mathbb{V}$ coincides with the set of all possible pairs $(\Phi, \bar{\Phi})$, where $\Phi(z)$ is an arbitrary chiral scalar superfield, $\bar{D}_{\dot{\alpha}} \Phi = 0$, modulo boundary conditions. How can one represent an infinitesimal variation of the action superfunctional $S[\Phi, \bar{\Phi}]$?

Recall, each chiral superfield $\Phi(z)$ is of the form

$$\Phi(z) = e^{i\mathscr{H}}\hat{\Phi}(x,\theta) \qquad \mathscr{H} = \theta\sigma^a\bar{\theta}\partial_a. \tag{3.1.37}$$

Hence $\Phi(z)$ is determined by superfield $\hat{\Phi}(z,\theta)$ which represents a superfunction on superspace $\mathbb{R}^4_c \times \mathbb{C}^2_a$. Therefore, one can look on the action superfunctional as a superfunctional of $\hat{\Phi}(x,\theta)$ and $\hat{\bar{\Phi}}(x,\bar{\theta}) = (\hat{\Phi}(x,\theta))^*$,

$$S[\Phi,\bar{\Phi}] = S[\hat{\Phi},\hat{\bar{\Phi}}]. \tag{3.1.38}$$

Obviously, $\hat{\Phi}(x,\theta)$ is an arbitrary (modulo the boundary conditions) bosonic superfunction on $\mathbb{R}^4_c \times \mathbb{C}^2_a$. Then we can write

$$\begin{aligned}\delta S[\Phi,\bar{\Phi}] &= S[\Phi+\delta\Phi,\bar{\Phi}+\delta\bar{\Phi}] - S[\Phi,\bar{\Phi}]\\ &= \int d^6z\,\delta\hat{\Phi}(x,\theta)\frac{\delta S}{\delta\hat{\Phi}(x,\theta)} + \int d^6\bar{z}\,\delta\hat{\bar{\Phi}}(x,\bar{\theta})\frac{\delta S}{\delta\hat{\bar{\Phi}}(x,\bar{\theta})}\end{aligned} \tag{3.1.39}$$

where we have introduced the notation:

$$d^6z = d^4x\,d^2\theta \qquad d^6\bar{z} = d^4x\,d^2\bar{\theta}. \tag{3.1.40}$$

Note that $\delta S/\delta\hat{\Phi}(x,\theta)$ is a superfunction on $\mathbb{R}^4_c \times \mathbb{C}^2_a$.

Furthermore, due to equation (3.1.37), the following identity

$$\int d^6z\hat{\Phi}(x,\theta) = \int d^6z\Phi(z) \tag{3.1.41}$$

holds if the component fields of $\hat{\Phi}$ vanish at infinity. Therefore, one can rewrite equation (3.1.39) in the form

$$\delta S[\Phi,\bar{\Phi}] = \int d^6z\,\delta\Phi(z)\frac{\delta S}{\delta\Phi(z)} + \int d^6\bar{z}\,\delta\bar{\Phi}(z)\frac{\delta S}{\delta\bar{\Phi}(z)} \tag{3.1.42}$$

where

$$\frac{\delta S}{\delta\Phi(z)} = e^{i\mathscr{H}}\frac{\delta S}{\delta\hat{\Phi}(x,\theta)} \qquad \frac{\delta S}{\delta\bar{\Phi}(z)} = e^{-i\mathscr{H}}\frac{\delta S}{\hat{\bar{\Phi}}(x,\bar{\theta})}.$$

Evidently, we have

$$\bar{D}_{\dot{\alpha}}\frac{\delta S}{\delta\Phi(z)} = 0 \qquad D_\alpha\frac{\delta S}{\delta\bar{\Phi}(z)} = 0. \tag{3.1.43}$$

Chiral superfield $\delta S/\delta\Phi(z)$ is called the 'superfunctional derivative' of $S[\Phi,\bar{\Phi}]$ at $\Phi(z)$.

By analogy with equation (3.1.31) one readily proves that

$$\int d^6z\,\hat{\Phi}(x,\theta)\hat{\Psi}(x,\theta) = 0 \qquad \forall\hat{\Phi}(x,\theta) \Rightarrow \hat{\Psi}(x,\theta) = 0. \tag{3.1.44}$$

Then the stationary action principle (3.1.16) and equation (3.1.42) give the dynamical superfield equations:

$$\frac{\delta S}{\delta \Phi(z)} = 0 \qquad \frac{\delta S}{\delta \bar{\Phi}(z)} = 0. \tag{3.1.45}$$

Since the action superfunctional is real, these equations are conjugate to each other.

For calculating the superfunctional derivatives, it is useful to operate with the identities:

$$\begin{aligned} \frac{\delta \Phi(z')}{\delta \Phi(z)} &= -\frac{1}{4}\bar{\mathrm{D}}^2 \delta^8(z - z') \equiv \delta_+(z, z') \\ \frac{\delta \bar{\Phi}(z')}{\delta \bar{\Phi}(z)} &= -\frac{1}{4}\mathrm{D}^2 \delta^8(z - z') \equiv \delta_-(z, z') \\ \frac{\delta \Phi(z')}{\delta \bar{\Phi}(z)} &= \frac{\delta \bar{\Phi}(z')}{\delta \Phi(z)} = 0. \end{aligned} \tag{3.1.46}$$

One can readily prove these identities by considering the superfunctional

$$F[\Phi, \bar{\Phi}] = \Phi(z')$$

with z' being a fixed point of $\mathbb{R}^{4|4}$. Its variation is represented as

$$\begin{aligned} \delta F[\Phi, \bar{\Phi}] = \delta \Phi(z') &= \int \mathrm{d}^8 z \, \delta^8(z - z') \delta \Phi(z) = -\frac{1}{4} \int \mathrm{d}^6 z \bar{\mathrm{D}}^2 (\delta^8(z - z') \delta \Phi(z)) \\ &= \int \mathrm{d}^6 z \, \delta \Phi(z) \delta_+(z, z') \end{aligned}$$

where equation (3.1.26*a*) has been used. The $\delta_+(z, z')$ will be called the 'chiral delta-function'. We list its basic properties:

$$\begin{gathered} \bar{\mathrm{D}}_{\dot{\alpha}} \delta_+(z, z') = 0 \qquad \delta_+(z, z') = \delta_+(z', z) \\ \int \mathrm{d}^6 z' \delta_+(z, z') \Phi(z') = \Phi(z) \\ (\delta_+(z, z'))^2 = 0 \qquad \delta_+(z, z) = 0 \end{gathered} \tag{3.1.47}$$

where $\Phi(z)$ is a chiral superfield. The $\delta_-(z, z')$ is called the 'antichiral delta-function'. Conjugating the above identities, one obtains the basic properties of $\delta_-(z, z')$.

Equations (3.1.33) and (3.1.46) show that there is no universal form for superfunctional derivatives, in contrast to the space–time case, where we had equation (3.1.4). The point is that different superfields are in fact, defined, on different superspaces (real superfield on $\mathbb{R}^{4|4}$, chiral superfield on $\mathbb{R}^4_c \times \mathbb{C}^2_a$).

We have discussed two possible realizations of dynamical variables: in terms of real scalar superfields and in terms of (anti)chiral scalar superfields.

Generally, throughout this book, we will consider only supersymmetric field theories described by unconstrained tensor superfields and unconstrained tensor (anti)chiral superfields. This means that the superfields variables $V^I(z)$, parametrizing the space of histories $\mathbb{V}$, may take arbitrary values modulo the boundary conditions and modulo the constraints of chirality and antichirality, which superfields $v^I(z)$ satisfy. For every such theory the stationary action principle (3.1.16) is equivalent to the dynamical superfield equations

$$\delta S[v]/\delta v^I(z) = 0 \tag{3.1.48}$$

and the superfunctional derivatives are calculated by applying identities like (3.1.33), (3.1.36) or (3.1.46), depending on superfield types.

3.1.4. Local supersymmetric field theories

We are going to describe a wide family of local supersymmetric field theories. It is useful to start with two simple theorems.

Theorem 1. All the super Poincaré transformations on $\mathbb{R}^{4|4}$ have unit Berezinian,

$$\begin{gathered} z^A \to z'^A = g \cdot z^A \qquad g \in S\Pi \\ \Rightarrow \mathrm{Ber}\left(\frac{\partial}{\partial z^A} z'^B\right) = 1. \end{gathered} \tag{3.1.49}$$

Proof. It is sufficient to prove the theorem for particular super Poincaré transformations (2.4.10–12). In the cases of translations (2.4.10) and supersymmetry transformations (2.4.12), the statement is evidently satisfied. In the case of Lorentz transformations (2.4.11), the statement is also satisfied since matrices $(e^K)^a{}_b$ and $(e^K)_\alpha{}^\beta$ are unimodular.

Theorem 2. Let $\mathscr{L}(z)$ be a scalar superfield, and $\mathscr{L}_c(z)$ be a chiral scalar superfield, $\bar{D}_{\dot{\alpha}}\mathscr{L}_c = 0$. Then the integrals

$$I = \int d^8z \mathscr{L}(z) \qquad I_c = \int d^6z \mathscr{L}_c(z) \tag{3.1.50}$$

are invariant under the super Poincaré transformations.

Proof. In accordance with equation (2.4.34), a supergroup element $g \in S\Pi$ moves $\mathscr{L}(z)$ into

$$\mathscr{L}'(z) = \mathscr{L}(g^{-1} \cdot z)$$

and hence

$$I' = \int d^8z \mathscr{L}'(z) = \int d^8z \mathscr{L}(g^{-1} \cdot z).$$

Introduce the new integration variables

$$z'^A = g^{-1}\cdot z \qquad \mathrm{d}^8 z = \mathrm{d}^8 z' \,\mathrm{Ber}^{-1}\left(\frac{\partial}{\partial z^A} z'^B\right).$$

Making use of (3.1.49) gives $I' = I$.

To prove the second part of the theorem, we note that $\mathscr{L}_c(z)$ can be represented in the form

$$\mathscr{L}_c(z) = -\frac{1}{4}\bar{\mathrm{D}}^2 U(z) \tag{3.1.51}$$

where $U(z)$ is a scalar superfield, see also subsection 2.5.6. Then, due to equation (3.1.26a), we have

$$I_c = \int \mathrm{d}^8 z U(z).$$

Therefore, the second part of the theorem is reduced to the first part, and the proof is complete.

Remark. In accordance with equation (3.1.26a), we have

$$\int \mathrm{d}^8 z \mathscr{L}_c(z) = 0 \qquad \bar{\mathrm{D}}_{\dot\alpha}\mathscr{L}_c = 0. \tag{3.1.52}$$

Now, consider a supersymmetric field theory with an action superfunctional of the general form

$$\begin{aligned} S[v] = &\int \mathrm{d}^8 z \mathscr{L}(v, \mathrm{D}_A v, \ldots, \mathrm{D}_{A_1}\ldots\mathrm{D}_{A_m} V) \\ &+ \left\{\int \mathrm{d}^6 z \mathscr{L}_c(v, \mathrm{D}_A v, \ldots, \mathrm{D}_{A_1}\ldots\mathrm{D}_{A_n} v) + \text{c.c.}\right\} \\ &\mathscr{L} = \bar{\mathscr{L}} \qquad \bar{\mathrm{D}}_{\dot\alpha}\mathscr{L}_c = 0 \end{aligned} \tag{3.1.53}$$

where $\mathscr{L}$ and $\mathscr{L}_c$ are supersmooth functions of their arguments; the symbol 'c.c.' means complex conjugation. The reality condition $\mathscr{L} = \bar{\mathscr{L}}$ is imposed in order that $S[v]$ be real. To guarantee the requirement of super Poincaré invariance (3.1.14), $\mathscr{L}$ and $\mathscr{L}_c$ should be scalar superfields constructed in terms of tensor superfields $v^I(z)$, $\mathrm{D}_A v^I(z)$, $\mathrm{D}_A \mathrm{D}_B v^I(z), \ldots$. The function $\mathscr{L}$ will be called the 'super Lagrangian' and $\mathscr{L}_c$ will be called the 'chiral super Lagrangian'. Using the technique of reduction from superfields to component fields (see Section 2.8) one can readily see that $S[v]$ leads to a local field theory at the component level, with the action functional

$$\begin{aligned} S[\varphi] &= \int \mathrm{d}^4 x L(\varphi, \partial_a\varphi, \ldots, \partial_{a_1}\ldots\partial_{a_k}\varphi) \\ L &= \frac{1}{32}\{\bar{\mathrm{D}}^2, \mathrm{D}^2\}\mathscr{L}\Big| - \frac{1}{4}\mathrm{D}^2\mathscr{L}_c\Big| - \frac{1}{4}\bar{\mathrm{D}}^2\bar{\mathscr{L}}_c\Big| \end{aligned} \tag{3.1.54a}$$

(see also equations (3.1.26)). Up to total derivative terms, the Lagrangian can be represented in two different forms:

$$L = \frac{1}{16}\bar{D}^2 D^2 \mathscr{L}\Big| - \frac{1}{4}D^2\mathscr{L}_c\Big| - \frac{1}{4}\bar{D}^2\bar{\mathscr{L}}_c\Big| \tag{3.1.54b}$$

or

$$L = \frac{1}{16}D^2\bar{D}^2\mathscr{L}\Big| - \frac{1}{4}D^2\mathscr{L}_c\Big| - \frac{1}{4}\bar{D}^2\bar{\mathscr{L}}_c\Big| \tag{3.1.54c}$$

Any action superfunctional of the form (3.1.53) will be called 'local'.

We have taken the super Lagrangians to be dependent on covariant derivatives D_A of v^I but not on partial derivatives ∂_A of v^I. The reason for this is that the spinor partial derivatives are not tensor operations; when $v^I(z)$ are tensor superfields, then $\partial_\alpha v^I(z)$ or $\bar{\partial}_{\dot{\alpha}} v^I(z)$ are not. It should be pointed out that in Minkowski space partial derivatives ∂_a are automatically covariant ones.

In accordance with (3.1.26), one can rewrite the action superfunctional (3.1.53) as follows

$$S[v] = \int d^6z\,\hat{\mathscr{L}}_c(z) + \text{c.c.}$$
$$\hat{\mathscr{L}}_c = \mathscr{L}_c - \frac{1}{8}\bar{D}^2\mathscr{L}. \tag{3.1.55}$$

Evidently, this action superfunctional is local. As a result, one can work with chiral super Lagrangians only. But a reverse transformation, being possible in principle, may result in a non-local superfunctional. We can represent $\mathscr{L}_c$ in the form (3.1.51), to obtain

$$S[v] = \int d^8z\,\hat{\mathscr{L}} \qquad \hat{\mathscr{L}} = L + U + \bar{U}. \tag{3.1.56}$$

As a rule, this superfunctional is non-local. For example, consider the chiral super Lagrangian

$$\mathscr{L}_c(\Phi) = \frac{m}{2}\Phi^2$$

where m is a constant, and $\Phi(z)$ is a chiral scalar superfield. Using identities (2.5.30e) and (2.5.33), we have

$$\Phi = \frac{1}{16}\frac{\bar{D}^2 D^2}{\Box}\Phi \qquad \Rightarrow \frac{m}{2}\Phi^2 = -\frac{1}{4}\bar{D}^2\left\{\left(-\frac{m}{8}\right)\Phi\frac{D^2}{\Box}\Phi\right\}.$$

This leads to

$$\frac{m}{2}\int d^6z\Phi^2 = -\frac{m}{8}\int d^8z\Phi\frac{D^2}{\Box}\Phi \tag{3.1.57}$$

and the obtained superfunctional is non-local.

In conclusion, we must answer the following question: what restrictions on the action superfunctional can we impose in order to make the dynamical field equations at most second order in time derivatives? Recall that the space–time derivatives can be expressed by the rule

$$\partial_a = -\frac{i}{4}(\tilde{\sigma}_a)^{\dot{\alpha}\alpha}\{D_\alpha, \bar{D}_{\dot{\alpha}}\}.$$

Therefore, the dynamical superfield equations should be at most of fourth order in the spinor covariant derivatives. Then, the action superfunctional can be represented in the general form

$$\begin{aligned}S[v] = &\int d^8z\mathscr{L}(v, D_Av, D_AD_Bv)\\ &+\left\{\int d^6z\mathscr{L}_c(v, D_Av, D_AD_Bv, D_AD_BD_Cv) + \text{c.c.}\right\}\end{aligned} \tag{3.1.58}$$

3.1.5. *Mass dimensions*

As is known, the action functional is dimensionless (in units of mass). The same is true for the action superfunctional. We wish to find mass dimensions of super Lagrangians. By virtue of equations (2.4.42) and (3.1.20) one obtains

$$[d^2\theta] = [d^2\bar{\theta}] = 1.$$

Since $[d^4x] = -4$, we have

$$[d^8z] = -2 \qquad [d^6z] = -3. \tag{3.1.59}$$

hence, the dimensions of super Lagrangians are

$$[\mathscr{L}] = 2 \qquad [\mathscr{L}_c] = 3. \tag{3.1.60}$$

3.1.6. *Chiral representation*

When investigating a dynamical system, it very often proves useful to consider different pictures (or representations) for dynamical variables. For example, the two most popular quantum mechanical pictures are the Schrödinger representation and the Heisenberg representation. In field theory one may choose the dynamical variables to be fields in Minkowski space or, after taking Fourier transforms, in momentum space. In the case of supersymmetric field theory, there is a representation for dynamical variables — the 'chiral representation'—which does not have an analogue in ordinary field theory. This representation is characterized by the requirement that, for every chiral superfield Φ, its chiral transform $\Phi^{(C)}$ coincides with the superfield $\hat{\Phi}$ defined by equation (3.1.37).

The chiral representation is introduced as follows. Consider a dynamical system with the action superfunctional $S[v]$ being of the form (3.1.53). One changes every superfield variable $V^I(z)$ to

$$v^{(C)I}(z) = e^{-i\mathscr{H}} v^I(z) \qquad \mathscr{H} = \theta\sigma^a\bar{\theta}\partial_a \tag{3.1.61a}$$

and every differential operator $\mathscr{F}$ to

$$\mathscr{F}^{(C)} = e^{-i\mathscr{H}} \mathscr{F} e^{i\mathscr{H}}. \tag{3.1.61b}$$

Obviously, the super Lagrangians can be represented as

$$\mathscr{L}(v, D_A v, \ldots) = e^{i\mathscr{H}} \mathscr{L}(v^{(C)}, D_A{}^{(C)} v^{(C)}, \ldots) \equiv e^{i\mathscr{H}} \mathscr{L}^{(C)}(v^{(C)}, D_A{}^{(C)} v^{(C)}, \ldots)$$
$$\mathscr{L}_c(v, D_A v, \ldots) = e^{i\mathscr{H}} \mathscr{L}_c(v^{(C)}, D_A{}^{(C)} v^{(C)}, \ldots) \equiv e^{i\mathscr{H}} \mathscr{L}_c{}^{(C)}(v^{(C)}, D_A{}^{(C)} v^{(C)}, \ldots).$$

Therefore, the action superfunctional does not change (modulo total derivative terms),

$$S[v] = S^{(C)}[v^{(C)}].$$

In accordance with equations (2.5.19, 21), the chiral transformed covariant derivatives are

$$\bar{D}^{(C)}_{\dot{\alpha}} = -\bar{\partial}_{\dot{\alpha}} \qquad D^{(C)}_{\alpha} = e^{-2i\mathscr{H}}\partial_\alpha e^{2i\mathscr{H}} = \partial_\alpha + 2i\bar{\theta}^{\dot{\alpha}}\partial_{\alpha\dot{\alpha}}. \tag{3.1.62}$$

Therefore, in this representation each chiral superfield depends only on x^a and θ^α,

$$\bar{D}_{\dot{\alpha}}\Phi = 0 \Leftrightarrow \Phi^{(C)}(z) = \hat{\Phi}(x, \theta). \tag{3.1.63}$$

Further, recalling the explicit form of the super Poincaré generators (2.4.37), their chiral transforms are

$$\begin{gathered} \mathbf{P}^{(C)}_a = \mathbf{P}_a \qquad \mathbf{J}^{(C)}_{ab} = \mathbf{J}_{ab} \\ \mathbf{Q}^{(C)}_\alpha = i\partial_\alpha \qquad \bar{\mathbf{Q}}^{(C)}_{\dot{\alpha}} = -ie^{-2i\mathscr{H}}\bar{\partial}_{\dot{\alpha}}e^{2i\mathscr{H}} = -i\bar{\partial}_{\dot{\alpha}} - 2\theta^\alpha\partial_{\alpha\dot{\alpha}} \end{gathered} \tag{3.1.64}$$

It is clear that all the super Poincaré transformations preserve the $\bar{\theta}$-independence of chiral superfields.

The chiral representation is best suited for operating with chiral superfields. For other superfield types, it leads to complications. For example, the operation of superfield complex conjugation in this representation is as follows

$$\overline{V^{(C)}} = e^{2i\mathscr{H}} \bar{V}^{(C)} \tag{3.1.65}$$

where $V(z)$ is an arbitrary superfield and $\bar{V}(z)$ is its conjugate. In particular, if $V(z)$ is a real superfield, $\bar{V} = V$, then

$$\overline{V^{(C)}} = e^{2i\mathscr{H}} V^{(C)}.$$

We see that the chiral representation modifies the reality condition. On these grounds, the previously used representation will be called 'real', to distinguish it from the chiral representation.

One can easily see that

$$e^{-i\mathscr{H}} e^{-i\mathscr{H}'} \delta^8(z - z') = \delta^8(z - z').$$

Therefore, the chiral transform of $\delta_+(z, z')$ is given, using equation (3.1.62), in the form

$$\delta_+^{(C)}(z, z') = \delta^4(z - z')\delta^2(\theta - \theta'). \tag{3.1.66}$$

By analogy with the chiral representation, one can introduce an 'antichiral representation' best suited to describing antichiral superfields. It is defined by replacing the operator $\mathscr{H}$ by $(-\mathscr{H})$ in equations (3.1.61).

3.2. Wess–Zumino model

We now review the most popular supersymmetric field theories. To maintain the tradition, we start with dynamical systems described by (anti)chiral superfields—supersymmetric analogues of scalar fields.

3.2.1. Massive chiral scalar superfield model

As has been shown in Section 2.6, the massive superspin-0 representation can be realized in terms of a chiral scalar superfield $\Phi(z)$ and its conjugate $\bar{\Phi}(z)$ using the equations

$$-\frac{1}{4}\bar{D}^2\bar{\Phi} + m\Phi = 0$$

$$-\frac{1}{4}D^2\Phi + m\bar{\Phi} = 0 \tag{3.2.1}$$

with m being the mass. We wish to find a dynamical system for which equations (3.2.1) are the dynamical superfield equations.

The above equations are linear in Φ and $\bar{\Phi}$, hence the action superfunctional $S[\Phi, \bar{\Phi}]$ must be quadratic in the superfields under consideration. Recall that the variational derivatives $\delta\Phi(z')/\delta\Phi(z)$ and $\delta\bar{\Phi}(z')/\delta\bar{\Phi}(z)$ involve the covariant derivatives (see equations (3.1.46)). Therefore it seems reasonble that the super Lagrangians will depend on Φ and $\bar{\Phi}$ only. Then there are three possible structures

$$I_1 = \int \mathrm{d}^8z\, \bar{\Phi}\Phi \qquad I_2 = \int \mathrm{d}^6z\, \Phi^2 \qquad I_3 = \int \mathrm{d}^6\bar{z}\, \bar{\Phi}^2$$

which may contribute to $S[\Phi, \bar{\Phi}]$. Two other structures $\int \mathrm{d}^8z\, \Phi^2$ and $\int \mathrm{d}^8z\, \bar{\Phi}^2$ vanish identically, in accordance with equation (3.1.52).

Making use of equations (3.1.46) gives

$$\frac{\delta I_1}{\delta\Phi(z)} = \int \mathrm{d}^8z'\, \bar{\Phi}(z')\frac{\delta\Phi(z')}{\delta\Phi(z)} = \int \mathrm{d}^8z'\, \bar{\Phi}(z')\left(-\frac{1}{4}\bar{D}^2\delta^8(z-z')\right)$$

$$= -\frac{1}{4}\bar{D}^2 \int \mathrm{d}^8z'\, \bar{\Phi}(z')\,\delta^8(z-z') = -\frac{1}{4}\bar{D}^2\bar{\Phi}(z)$$

$$\frac{\delta I_2}{\delta\Phi(z)} = 2\int \mathrm{d}^6z'\, \Phi(z')\frac{\delta\Phi(z')}{\delta\Phi(z)} = 2\int \mathrm{d}^6z'\, \Phi(z')\,\delta_+(z,z') = 2\int \mathrm{d}^6z'\, \Phi(z')\,\delta_+(z',z)$$

$$= 2\int \mathrm{d}^6z'\left(-\frac{1}{4}\bar{D}'^2\right)\left(\Phi(z')\,\delta^8(z'-z)\right)$$

$$= 2\int \mathrm{d}^8z'\, \Phi(z')\,\delta^8(z'-z) = 2\Phi(z)$$

$$\frac{\delta I_3}{\delta\Phi(z)} = 0.$$

It is instructive also to reobtain $\delta I_2/\delta\Phi(z)$ in the chiral representation. One finds, after using equation (3.1.66),

$$\frac{\delta I_2}{\delta\Phi^{(C)}(z)} = 2\int \mathrm{d}^6z'\, \Phi^{(C)}(z')\,\delta_+^{(C)}(z,z')$$

$$= 2\int \mathrm{d}^4x'\,\mathrm{d}^2\theta,\, \Phi^{(C)}(x',\theta')\,\delta^4(x-x')\,\delta^2(\theta-\theta') = 2\Phi^{(C)}(z).$$

The above identities show that the action superfunctional is of the form

$$S[\Phi,\bar{\Phi}] = \int \mathrm{d}^8z\bar{\Phi}\Phi + \frac{m}{2}\int \mathrm{d}^6z\Phi^2 + \frac{m}{2}\int \mathrm{d}^6\bar{z}\bar{\Phi}^2. \tag{3.2.2}$$

Here the first term is called the 'kinetic term' and the rest are said to be the 'mass terms'. Obviously, the action is invariant under super Poincaré transformations. Representing $\Phi(z) = \mathrm{e}^{\mathrm{i}\mathscr{H}}\Phi(x,\theta)$ and $\bar{\Phi}(z) = \mathrm{e}^{-\mathrm{i}\mathscr{H}}\bar{\Phi}(x,\bar{\theta})$ and integrating by parts, $S[\Phi,\bar{\Phi}]$ can be rewritten in the form

$$S[\Phi,\bar{\Phi}] = \int \mathrm{d}^8z\bar{\Phi}(x,\bar{\theta})\mathrm{e}^{2\mathrm{i}\mathscr{H}}\Phi(x,\theta) + \frac{m}{2}\int \mathrm{d}^6z\,\Phi^2(x,\theta) + \frac{m}{2}\int \mathrm{d}^6\bar{z}\,\bar{\Phi}^2(x,\bar{\theta}). \tag{3.2.3}$$

One can readily find the mass dimension of Φ and $\bar{\Phi}$. Explicitly, equations (3.1.59, 60) lead to

$$[\Phi(z)] = [\bar{\Phi}(z)] = 1. \tag{3.2.4}$$

3.2.2. Massless chiral scalar superfield model

A massless model is obtained by setting the mass parameter in (3.2.2) to be vanishing. The action superfunctional reads

$$S[\Phi, \bar{\Phi}] = \int d^8z \bar{\Phi}\Phi = \frac{1}{2}\int d^8z(\Phi + \bar{\Phi})^2 \tag{3.2.5}$$

and the dynamical equations are

$$-\frac{1}{4}\bar{D}^2\bar{\Phi} = 0 \qquad -\frac{1}{4}D^2\Phi = 0. \tag{3.2.6}$$

These equations determine on-shell massless (anti)chiral scalar superfields (see subsection 2.7.2), and the corresponding superhelicities are

$$\kappa(\Phi) = 0 \qquad \kappa(\bar{\Phi}) = -1/2. \tag{3.2.7}$$

3.2.3. Wess–Zumino model

We now consider the model of interacting (anti)chiral scalar superfields, proposed by J. Wess and B. Zumino, with the action

$$\begin{gathered} S[\Phi, \bar{\Phi}] = \int d^8z \bar{\Phi}\Phi + \int d^6z \mathscr{L}_c(\Phi) + \int d^6\bar{z} \bar{\mathscr{L}}_c(\bar{\Phi}) \\ \bar{D}_{\dot{\alpha}}\mathscr{L}_c(\Phi(z)) = 0. \end{gathered} \tag{3.2.8}$$

Here $\mathscr{L}_c$ is a holomorphic function of an ordinary complex variable, called the 'chiral superpotential'. Clearly, $\mathscr{L}_c(z) \equiv \mathscr{L}_c(\Phi(z))$ is a chiral scalar superfield, therefore the action superfunctional is super Poincaré invariant. The dynamical superfield equations have the form

$$-\frac{1}{4}\bar{D}^2\bar{\Phi} + \mathscr{L}'_c(\Phi) = 0 \qquad -\frac{1}{4}D^2\Phi + \bar{\mathscr{L}}'_c(\bar{\Phi}) = 0. \tag{3.2.9}$$

We will refer to the above theory as the 'general Wess–Zumino model'.

At the classical level, there is no restriction on the form of $\mathscr{L}_c(\Phi)$. The situation is different at the quantum level. As will be shown in Chapter 4, the theory with classical action (3.2.8) is renormalizable if the chiral superpotential is at most a third-order polynomial,

$$\mathscr{L}_c(\Phi) = \mu\Phi + \frac{m}{2}\Phi^2 + \frac{\lambda}{3!}\Phi^3 \tag{3.2.10}$$

where μ, m and λ are coupling constants. Note that the first term in $\mathscr{L}_c(\Phi)$

can be removed by the superfield redefinition $\Phi \to \Phi + \xi$, where ξ is a suitable constant. Then, one obtains the action

$$S[\Phi, \bar{\Phi}] = \int d^8z\bar{\Phi}\Phi + \left\{\int d^6z\left(\frac{m}{2}\Phi^2 + \frac{\lambda}{3!}\Phi^3\right) + \text{c.c.}\right\}. \tag{3.2.11}$$

This theory will be called the 'standard Wess–Zumino model'.

3.2.4. *Wess–Zumino model in component form*

Let us rewrite the action superfunctional (3.2.8) in terms of the component fields of Φ and $\bar{\Phi}$. We shall follow the prescription (3.1.54c) to obtain $S[\varphi]$. Recalling definition (2.8.4) of the component fields, we have

$$\begin{aligned}
\frac{1}{16}D^2\bar{D}^2(\bar{\Phi}\Phi)\Big| &= \frac{1}{16}D^2(\Phi\bar{D}^2\bar{\Phi})\Big| \\
&= \frac{1}{16}(D^2\Phi)|\bar{D}^2\bar{\Phi}| + \frac{1}{8}(D^\alpha\Phi)|D_\alpha\bar{D}^2\bar{\Phi}| + \frac{1}{16}\Phi|D^2\bar{D}^2\bar{\Phi}| \\
&= F(x)\bar{F}(x) - \frac{i}{2}(D^\alpha\Phi)|\partial_{\alpha\dot{\alpha}}\bar{D}^{\dot{\alpha}}\bar{\Phi}| + \Phi|\Box\bar{\Phi}| \\
&= F(x)\bar{F}(x) - \frac{i}{2}\psi^\alpha(x)\partial_{\alpha\dot{\alpha}}\bar{\psi}^{\dot{\alpha}}(x) + A(x)\Box\bar{A}(x)
\end{aligned}$$

the next contribution is

$$\begin{aligned}
-\frac{1}{4}D^2\mathscr{L}_c(\Phi)\Big| &= -\frac{1}{4}D^\alpha((D_\alpha\Phi)\mathscr{L}'_c(\Phi))\Big| \\
&= -\frac{1}{4}(D^2\Phi)|\mathscr{L}'_c(\Phi)| - \frac{1}{4}(D^\alpha\Phi)|(D_\alpha\Phi)|\mathscr{L}''_c(\Phi)| \\
&= F(x)\mathscr{L}'_c(A(x)) - \frac{1}{4}\psi^\alpha(x)\psi_\alpha(x)\mathscr{L}''_c(A(x))
\end{aligned}$$

the final contribution is

$$-\frac{1}{4}\bar{D}^2\bar{\mathscr{L}}_c(\bar{\Phi})| = \bar{F}(x)\bar{\mathscr{L}}'_c(\bar{A}(x)) - \frac{1}{4}\bar{\psi}_{\dot{\alpha}}(x)\bar{\psi}^{\dot{\alpha}}(x)\bar{\mathscr{L}}''_c(\bar{A}(x)).$$

Then, after introducing the notation

$$\mathscr{V}(A) = \mathscr{L}'_c(A) \tag{3.2.12}$$

the action functional reads

$$S[\varphi] = \int d^4x \left\{ \bar{F}F - \partial^a \bar{A} \partial_a A - \frac{i}{2} \psi \sigma^a \partial_a \bar{\psi} + \left(\left[F \mathscr{V}(A) - \frac{1}{4} \mathscr{V}'(A) \psi\psi \right] + \text{c.c.} \right) \right\} \tag{3.2.13}$$

where $\varphi = \{A, \bar{A}, \psi_\alpha, \bar{\psi}^{\dot{\alpha}}, F, \bar{F}\}$ represents the complete set of field variables.

It is instructive to rewrite the action in four-component spinor notation. Introducing the Majorana spinor

$$\Psi = \frac{1}{\sqrt{2}} \begin{pmatrix} \psi_\alpha \\ \bar{\psi}^{\dot{\alpha}} \end{pmatrix} \tag{3.2.14}$$

and recalling expressions for γ-matrices (see Section 1.4), one obtains

$$S[\varphi] = \int d^4x \left\{ \bar{F}F - \partial^a \bar{A} \partial_a A - \frac{i}{2} \bar{\Psi} \gamma^a \partial_a \Psi + \left(\left[F \mathscr{V}(A) - \frac{1}{4} \mathscr{V}'(A) \bar{\Psi}(\mathbb{1} + \gamma_5) \Psi \right] + \text{c.c.} \right) \right\}. \tag{3.2.15}$$

Not all of the component fields have non-trivial dynamics. The first dynamical equation (3.2.9) is equivalent to three component equations

$$\bar{F} + \mathscr{V}(A) = 0 \tag{3.2.16a}$$

$$i(\sigma^a \partial_a \bar{\psi})_\alpha - \mathscr{V}'(A) \psi_\alpha = 0 \tag{3.2.16b}$$

$$\Box \bar{A} + \mathscr{V}'(A) F - \frac{1}{4} \mathscr{V}''(A) \psi\psi = 0. \tag{3.2.16c}$$

Therefore, on-shell $\bar{F}(x)$ is expressed algebraically in terms of $A(x)$; $\bar{F}(x)$ does not have an independent dynamics. On these grounds, $F(x)$ and $\bar{F}(x)$ are said to be 'auxiliary fields'.

3.2.5. Auxiliary fields

Now, we would like to describe what the notion 'auxiliary fields' means in general. Let $\varphi = \{\hat{\varphi}, \check{\varphi}\}$ be field variables of a field theory with an action functional $S[\varphi] = S[\hat{\varphi}, \check{\varphi}]$. Suppose that the dynamical equations for fields $\check{\varphi}$ are such that: (1) they do not involve time derivatives of $\check{\varphi}$; (2) they can be uniquely resolved by expressing $\check{\varphi}$ in terms of the remaining fields,

$$\delta S[\varphi]/\delta \check{\varphi} = 0 \Leftrightarrow \check{\varphi} = f(\hat{\varphi}) \equiv \check{\varphi}_0.$$

Under these assumptions, $\check{\varphi}$ are said to be auxiliary fields and $\hat{\varphi}$ are said to be physical fields.

Instead of working with the total set of fields, one may eliminate from the very beginning the auxiliary fields, resulting in the 'physical' action

$$S[\hat{\varphi}] = S[\hat{\varphi}, \check{\varphi}]|_{\check{\varphi} = f(\hat{\varphi})}$$

Then, two actions $S[\varphi]$ and $S[\hat{\varphi}]$ lead to equivalent dynamical equations in $\hat{\varphi}$:

$$\frac{\delta S[\hat{\varphi}]}{\delta \hat{\varphi}} = \left.\frac{\delta S[\varphi]}{\delta \hat{\varphi}}\right|_{\check{\varphi}_0} + \left.\frac{\delta S[\varphi]}{\delta \check{\varphi}}\right|_{\check{\varphi}_0} \frac{\partial f(\hat{\varphi})}{\partial \hat{\varphi}}$$

therefore

$$\frac{\delta S[\hat{\varphi}]}{\delta \hat{\varphi}} = 0 \Leftrightarrow \left.\frac{\delta S[\varphi]}{\delta \hat{\varphi}}\right|_{\check{\varphi}_0} = 0.$$

However, very often it is convenient to keep the auxiliary fields, since their elimination may produce complications such as the loss of explicit covariance or the non-locality of the 'physical' action and so on. In particular, linear symmetry transformations may turn into nonlinear ones after elimination of the auxiliary fields. If $S[\varphi]$ possesses an invariance under linear transformations of the form

$$\delta \hat{\varphi} = R^{\wedge\wedge}\hat{\varphi} + R^{\wedge\vee}\check{\varphi} \qquad \delta \check{\varphi} = R^{\vee\wedge}\hat{\varphi} + R^{\vee\vee}\check{\varphi}$$

then $S[\hat{\varphi}]$ is invariant under transformations

$$\delta \hat{\varphi} = R^{\wedge\wedge}\hat{\varphi} + R^{\wedge\vee} f(\hat{\varphi}).$$

which are nonlinear, as a rule. The Wess–Zumino model provides us with an example.

3.2.6. Wess–Zumino model after auxiliary field elimination

Substituting equation (3.2.16*a*) into (3.2.15) expresses the Wess–Zumino model in terms of physical fields $\hat{\varphi} = \{A, \bar{A}, \psi_\alpha, \bar{\psi}^{\dot{\alpha}}\}$ as follows

$$S[\hat{\varphi}] = -\int d^4x \left\{ \partial^a \bar{A} \partial_a A + U(A, \bar{A}) + \frac{i}{2} \bar{\Psi} \gamma^a \partial_a \Psi \right.$$
$$\left. + \frac{1}{4} \mathscr{V}'(A) \bar{\Psi}(\mathbb{1} + \gamma_5)\Psi + \frac{1}{4} \bar{\mathscr{V}}'(A) \bar{\Psi}(\mathbb{1} - \gamma_5)\Psi \right\} \tag{3.2.17}$$

where

$$U(A, \bar{A}) = |\mathscr{V}(A)|^2 \tag{3.2.18}$$

is the potential of scalar fields. Note, the scalar potential is positive semidefinite, which is a direct consequence of the energy positivity in supersymmetric theories (see subsection 2.3.1).

In the case of the standard Wess–Zumino model (3.2.11), we have

$$\mathscr{V}(A) = mA + \frac{1}{2}\lambda A^2$$

and the action (3.2.17) takes the form

$$\begin{aligned} S[\hat{\varphi}] = -\int \mathrm{d}^4x \Big\{ &\bar{A}(-\Box + m^2)A + \frac{1}{2}\bar{\Psi}(\mathrm{i}\gamma^a\partial_a + m)\Psi \\ &+ \frac{1}{2}m\lambda(A + \bar{A})|A|^2 + \frac{1}{4}\lambda^2|A|^4 \\ &+ \frac{1}{4}\lambda(A + \bar{A})\bar{\Psi}\Psi + \frac{1}{4}\lambda(A - \bar{A})\bar{\Psi}\gamma_5\Psi \Big\} \end{aligned} \tag{3.2.19}$$

where we have supposed that the coupling constant λ is real. The resulting action describes cubic and quartic self-interaction of the scalar fields and Yukawa coupling between the scalar fields and the Majorana spinor field.

Now, let us discuss supersymmetric properties of the Wess–Zumino model, before and after elimination of the auxiliary fields. Since the action superfunctional (3.2.8) is super Poincaré invariant, it does not change under infinitesimal supersymmetry transformations

$$\delta_\epsilon\Phi = \mathrm{i}(\epsilon\mathbf{Q} + \overline{\epsilon\mathbf{Q}})\Phi \qquad \delta_\epsilon\bar{\Phi} = \mathrm{i}(\epsilon\mathbf{Q} + \overline{\epsilon\mathbf{Q}})\bar{\Phi}.$$

At the component level, they are given by equations (2.8.6, 8) (these transformations leave invariant the action functional (3.2.13)). The equation (2.8.6) defines a linear field representation of the supersymmetry algebra. In particular, commuting two supersymmetry transformations gives a space–time translation,

$$\begin{aligned} [\delta_{\epsilon_1}, \delta_{\epsilon_2}] &= b^a\partial_a \\ b^a &= 2\mathrm{i}(\epsilon_1\sigma^a\bar{\epsilon}_2 - \epsilon_2\sigma^a\bar{\epsilon}_1) \end{aligned} \tag{3.2.20}$$

in accordance with equation (2.2.24).

As for the action functional (3.2.17), it is invariant under the supersymmetry transformations

$$\begin{aligned} \delta_\epsilon A &= -\epsilon\psi \qquad & \delta_\epsilon\psi_\alpha &= 2\epsilon_\alpha\bar{\mathscr{V}}'(\bar{A}) - 2\mathrm{i}\bar{\epsilon}^{\dot{\alpha}}\partial_{\alpha\dot{\alpha}}A \\ \delta_\epsilon\bar{A} &= -\bar{\epsilon}\bar{\psi} \qquad & \delta_\epsilon\bar{\psi}_{\dot{\alpha}} &= 2\bar{\epsilon}_{\dot{\alpha}}\mathscr{V}'(A) + 2\mathrm{i}\epsilon^\alpha\partial_{\alpha\dot{\alpha}}\bar{A} \end{aligned} \tag{3.2.21}$$

which follow from equations (2.8.6, 8) by setting $F = -\bar{\mathscr{V}}'(\bar{A})$ and $\bar{F} = -\mathscr{V}'(A)$. We see that the spinor transformation laws become nonlinear after elimination of the auxiliary fields. This, however, is not the whole story. Taking the commutator of two supersymmetry transformations δ_{ϵ_1} and δ_{ϵ_2}

applied to A and ψ_α, one finds

$$[\delta_{\epsilon_1}, \delta_{\epsilon_2}]A = b^a\partial_a A$$
$$[\delta_{\epsilon_1}, \delta_{\epsilon_2}]\psi_\alpha = b^a\partial_a\psi_\alpha + \mathrm{i}b_{\alpha\dot\alpha}\delta S[\varphi]/\delta\bar\psi_{\dot\alpha}. \tag{3.2.22}$$

The second relation indicates that the supersymmetry algebra becomes broken when the auxiliary fields are eliminated! The supersymmetry algebra closes only modulo the dynamical equations of the spinor fields.

To summarize, auxiliary fields are needed to make the supersymmetry transformations linear as well as to make possible the very existence of the super Poincaré algebra off-shell. Without use of auxiliary fields, the super Poincaré algebra cannot in general be realized on the space of field histories but only on the mass shell surface.

It should be pointed out that the problems just discussed arise only when working with ordinary (component) fields. Living in superspace and operating with superfields, there is no need to worry about supersymmetry. Superspace makes supersymmetry manifest.

3.2.7. Generalization of the model

The Wess–Zumino model can be easily generalized to describe the dynamics of n chiral scalar superfields Φ^i and their conjugates $\bar\Phi^i_-$, $\bar\Phi^i(z) = (\Phi^i(z))^*$. An action superfunctional is taken to be

$$S[\Phi, \bar\Phi] = \int \mathrm{d}^8z\bar\Phi^i\Phi^i + \left\{\int \mathrm{d}^6z\mathscr{L}_c(\Phi^1, \Phi^2, \ldots, \Phi^n) + \text{c.c.}\right\} \tag{3.2.23}$$

where $\mathscr{L}_c$ is a holomorphic function of n complex variables. The action leads to the following equations of motion for Φ^i

$$-\frac{1}{4}\bar{\mathrm{D}}^2\bar\Phi^i + \mathscr{V}_i(\Phi) = 0 \qquad \mathscr{V}_i(\Phi) = \partial\mathscr{L}_c(\Phi)/\partial\Phi^i. \tag{3.2.24}$$

Turning from the superfields to the component fields

$$A^i(x) = \Phi^i| \qquad \psi^i_\alpha(x) = \mathrm{D}_\alpha\Phi^i| \qquad F^i(x) = -\frac{1}{4}\mathrm{D}^2\Phi^i| \tag{3.2.25}$$

the action reduces to the form

$$S = \int \mathrm{d}^4x\left\{\bar F^i F^i - \partial^a\bar A^i\partial_a A^i - \frac{\mathrm{i}}{2}\psi^i\sigma^a\partial_a\bar\psi^i\right.$$
$$\left. + \left[F^i\mathscr{V}_i(A) - \frac{1}{4}\frac{\partial^2\mathscr{L}_c(A)}{\partial A^i\partial A^j}\psi^i\psi^j + \text{c.c.}\right]\right\}. \tag{3.2.26}$$

If one eliminates the auxiliary fields, the purely scalar part of the action reads

$$S[A, \bar{A}] = -\int d^4x\{\partial^a \bar{A}^i \partial_a A^i + U(A, \bar{A})\}$$
$$U(A, \bar{A}) = \sum_{i=1}^{n} |\mathscr{V}^i(A)|^2. \tag{3.2.27}$$

The scalar potential is positive semidefinite.

From the expression (3.2.27) it is seen that not every scalar field theory of the general form

$$S[\varphi] = -\frac{1}{2}\int d^4x \partial^a \varphi^I \partial_a \varphi^I - \int d^4x U(\varphi) \tag{3.2.28}$$

where φ^I are real scalars can be supersymmetrized. An admissible model is one with an even number of real fields which can be combined into complex ones in such a way that the scalar potential has the structure (3.2.27).

3.3. Supersymmetric nonlinear sigma-models

In this section we endeavour to clarify the conditions under which a nonlinear scalar field theory of the general form

$$S[\varphi] = -\frac{1}{2}\int d^4x g_{IJ}(\varphi)\, \partial^a \varphi^I \partial_a \varphi^J \tag{3.3.1}$$

generalizing the kinetic term in equation (3.2.28), can be extended to a supersymmetric field theory. First, it is worth saying some words about models (3.3.1) known as nonlinear σ-models.

3.3.1. Four-dimensional σ-models

Nonlinear σ-models naturally arise in a geometrical framework as follows. Let $\mathscr{M}^n$ be an n-dimensional Riemann manifold with metric $g_{IJ}(\varphi)$, where φ^I are local coordinates on $\mathscr{M}^n$. The σ-model space of field histories coincides with the set of all smooth mappings

$$\varphi: \mathbb{R}^4 \to \mathscr{M}^n \tag{3.3.2}$$

from Minkowski space into $\mathscr{M}^n$ ('target space'). So, fields are coordinates on $\mathscr{M}^n$ depending on points of Minkowski space. An action functional is required to be invariant under the Poincaré transformations on Minkowski space and the general coordinate transformations

$$\varphi^I \to \varphi'^I = f^I(\varphi)$$
$$g_{IJ}(\varphi) \to g'_{IJ}(\varphi') = \frac{\partial \varphi^K}{\partial \varphi'^I} \frac{\partial \varphi^L}{\partial \varphi'^J} g_{KL}(\varphi) \tag{3.3.3}$$

of the target manifold. Evidently, the choice (3.3.1) is consistent with these requirements.

Of primary importance for us will be σ-models on complex manifolds. Recall, a manifold $\mathcal{M}^{2n}$ is said to be a complex n-dimensional manifold if local coordinates φ^I, $I = 1, 2, \ldots, 2n$, on the manifold can be chosen in such a way that, after introducing complex local coordinates u^i and $\bar{u}^{\underline{i}}$,

$$\varphi^i = \frac{1}{2}(u^i + \bar{u}^{\underline{i}}) \qquad \varphi^{i+n} = \frac{1}{2i}(u^i - \bar{u}^{\underline{i}}) \qquad i, \underline{i} = 1, \ldots, n \tag{3.3.4}$$

the transition functions are holomorphic,

$$u'^i = f^i(u) \tag{3.3.5}$$

in any non-empty overlap of two arbitrary charts. Given a positive metric $ds^2 = g_{IJ}(\varphi)\, d\varphi^I\, d\varphi^J$ on the manifold, in complex coordinates it takes the form

$$ds^2 = 2g_{i\underline{j}}\, du^i\, d\bar{u}^{\underline{j}} + g_{ij}\, du^i\, du^j + g_{\underline{i}\underline{j}}\, d\bar{u}^{\underline{i}}\, d\bar{u}^{\underline{j}}$$
$$g_{j\underline{i}} = (g_{i\underline{j}})^* \qquad g_{\underline{i}\underline{j}} = g_{\underline{j}\underline{i}} = (g_{ij})^*. \tag{3.3.6}$$

The metric is said to be Hermitian, if it is of the form

$$ds^2 = 2g_{i\underline{j}}(u, \bar{u})\, du^i\, d\bar{u}^{\underline{j}}. \tag{3.3.7}$$

Under the coordinate transformations (3.3.5), the components of Hermitian metric change according to the rule

$$g'_{i\underline{i}}(u', \bar{u}') = \frac{\partial u^j}{\partial u'^i} \frac{\partial \bar{u}^{\underline{j}}}{\partial \bar{u}'^{\underline{i}}} g_{j\underline{j}}(u, \bar{u}). \tag{3.3.8}$$

In the case of a complex target manifold with some Hermitian metric, the action (3.3.1) is rewritten as

$$S[u, \bar{u}] = -\int d^4x\, g_{i\underline{j}}(u, \bar{u}) \partial^a \bar{u}^{\underline{j}} \partial_a u^i. \tag{3.3.9}$$

One can look upon the actions (3.3.1) and (3.3.9) as nonlinear analogues of the kinetic terms in (3.2.28) and (3.2.27), respectively.

3.3.2. *Supersymmetric σ-models*

We are now going to generalize the superfield kinetic term in (3.2.23) to nonlinear case. Demanding the super Lagrangian to be independent of the covariant derivatives of (anti)chiral superfields $\bar{\Phi}^{\underline{i}}$ and Φ^i, the most general form of the action superfunctional reads

$$S[\Phi, \bar{\Phi}] = \int d^8z K(\Phi^i, \bar{\Phi}^{\underline{i}}) \tag{3.3.10}$$

where K is a real smooth function of n complex variables and their conjugates. Since the integral over $\mathbb{R}^{4|4}$ of a chiral superfield vanishes (see equation 3.1.52)), $K(\Phi, \bar{\Phi})$ is determined only modulo the transformations

$$K(\Phi, \bar{\Phi}) \rightarrow K(\Phi, \bar{\Phi}) + \Lambda(\Phi) + \bar{\Lambda}(\bar{\Phi}) \tag{3.3.11}$$

with Λ being an arbitrary holomorphic function of n complex variables. The action (3.3.10), which has been suggested by B. Zumino, defines a 'supersymmetric nonlinear σ-model'.

Evidently, the action is super Poincaré invariant. Furthermore, the chirality-preserving reparametrizations

$$\Phi^i \rightarrow \Phi'^i = f^i(\Phi) \qquad \bar{\Phi}^{\underline{i}} \rightarrow \bar{\Phi}'^{\underline{i}} = \bar{f}^{\underline{i}}(\bar{\Phi}) \tag{3.3.12}$$

leave the action invariant provided $K(\Phi, \bar{\Phi})$ changes as follows

$$K'(\Phi', \bar{\Phi}') = K(\Phi, \bar{\Phi}) \tag{3.3.13}$$

modulo a transformation (3.3.11).

Let us investigate the component structure of the theory (3.3.10). In accordance with the prescription (3.1.54*c*), the action functional is represented as

$$S = \frac{1}{16}\int d^4x D^2\bar{D}^2 K(\Phi, \bar{\Phi})| = \int d^4x L.$$

Defining the component fields of Φ^i by the rule (3.2.25), a simple calculation leads to

$$\begin{aligned} L = & -K_{i\underline{j}}\left(\partial^a \bar{A}^{\underline{j}} \partial_a A^i - \bar{F}^{\underline{j}} F^i + \frac{i}{4}\psi^i \sigma^a \overleftrightarrow{\partial}_a \bar{\psi}^{\underline{j}}\right) \\ & - \frac{1}{4} K_{ij\underline{l}}(\bar{F}^{\underline{l}}\psi^i\psi^j - i\partial_a A^i \psi^j \sigma^a \bar{\psi}^{\underline{l}}) \\ & - \frac{1}{4} K_{i\underline{j}\underline{l}}(F^i \bar{\psi}^{\underline{j}}\bar{\psi}^{\underline{l}} + i\partial_a \bar{A}^{\underline{j}}\psi^i\sigma^a\bar{\psi}^{\underline{l}}) + \frac{1}{16} K_{ij\underline{i}\underline{j}}\psi^i\psi^j\bar{\psi}^{\underline{i}}\bar{\psi}^{\underline{j}} \end{aligned} \tag{3.3.14}$$

where we have introduced the notation

$$K_{i_1 \ldots i_p \underline{i}_1 \ldots \underline{i}_q} \equiv \frac{\partial^{p+q} K(A, \bar{A})}{\partial A^{i_1} \ldots \partial A^{i_p} \partial \bar{A}^{\underline{i}_1} \ldots \partial \bar{A}^{\underline{i}_q}}.$$

By comparing the obtained supersymmetric action with the nonlinear σ-model action (3.3.9), it is seen that the dynamics of scalar fields A^i and $\bar{A}^{\underline{i}}$

is determined by a Hermitian metric of the special type

$$g_{i\underline{j}} = \frac{\partial^2 K(A, \bar{A})}{\partial A^i \partial \bar{A}^{\underline{j}}}. \tag{3.3.15}$$

Every such metric induces an interesting geometry known to mathematicians as Kähler geometry.

3.3.3. *Kähler manifolds*

Let $\mathcal{M}$ be a complex manifold with Hermitian metric $g_{i\underline{j}}(u, \bar{u})$. Using the metric one can construct the second-order differential form

$$\omega = \frac{\mathrm{i}}{2} g_{i\underline{j}} \, \mathrm{d}u^i \wedge \mathrm{d}\bar{u}^{\underline{j}} \tag{3.3.16}$$

which is called the 'Kähler form'. The exterior differential of ω is written as

$$\mathrm{d}\omega = \frac{\mathrm{i}}{4}\left(\frac{\partial g_{i\underline{j}}}{\partial u^k} - \frac{\partial g_{k\underline{j}}}{\partial u^i}\right) \mathrm{d}u^k \wedge \mathrm{d}u^i \wedge \mathrm{d}\bar{u}^{\underline{j}} + \frac{\mathrm{i}}{4}\left(\frac{\partial g_{i\underline{j}}}{\partial \bar{u}^{\underline{k}}} - \frac{\partial g_{i\underline{k}}}{\partial \bar{u}^{\underline{j}}}\right) \mathrm{d}u^i \wedge \mathrm{d}\bar{u}^{\underline{j}} \wedge \mathrm{d}\bar{u}^{\underline{k}}$$

The Hermitian metric is said to be 'Kählerian' if the corresponding Kähler form is closed,

$$\mathrm{d}\omega = 0. \tag{3.3.17}$$

Then, $\mathcal{M}$ is called a 'Kähler manifold'. The above requirement is equivalent to the equations

$$\frac{\partial g_{i\underline{j}}}{\partial u^k} = \frac{\partial g_{k\underline{j}}}{\partial u^i} \qquad \frac{\partial g_{i\underline{j}}}{\partial \bar{u}^{\underline{k}}} = \frac{\partial g_{i\underline{k}}}{\partial \bar{u}^{\underline{j}}}. \tag{3.3.18}$$

These imply that the metric can be locally represented in the form (3.1.15). The corresponding function $K(u, \bar{u})$ is called a 'Kähler potential'. It should be pointed that $K(u, \bar{u})$ is defined locally and only modulo the transformations

$$K(u, \bar{u}) \to K(u, \bar{u}) + \Lambda(u) + \Lambda(\bar{u}). \tag{3.3.19}$$

Let us calculate the Christoffel symbols and the curvature tensor of a Kähler manifold. For this purpose we introduce 'real' coordinates $\varphi^I = \{u^i, \bar{u}^{\underline{i}}\}$ labelled by capital Latin letters. In these coordinates the metric and its inverse are

$$g_{IJ} = \begin{pmatrix} 0 & g_{i\underline{j}} \\ g_{\underline{i}j} & 0 \end{pmatrix} \qquad g^{IJ} = \begin{pmatrix} 0 & g^{i\underline{j}} \\ g^{\underline{i}j} & 0 \end{pmatrix}. \tag{3.3.20}$$

Here $g_{\underline{i}j} = g_{j\underline{i}}$, $g^{\underline{i}j} = g^{j\underline{i}}$ and $g^{i\underline{j}} g_{\underline{j}k} = \delta^i_k$, $g^{\underline{i}j} g_{j\underline{k}} = \delta^{\underline{i}}_{\underline{k}}$. In accordance with equation (1.6.12), the Christoffel symbols are defined by the rule

$$\Gamma^I{}_{JK} = \frac{1}{2} g^{IL}\left(\frac{\partial g_{LK}}{\partial \varphi^J} + \frac{\partial g_{JL}}{\partial \varphi^K} - \frac{\partial g_{JK}}{\partial \varphi^L}\right).$$

Making use of the explicit form of the metric (3.3.20) and equations (3.3.18) gives

$$\begin{aligned}
\Gamma^i{}_{jk} &= \frac{1}{2} g^{i\underline{l}}\left(\frac{\partial g_{\underline{l}k}}{\partial u^j} + \frac{\partial g_{j\underline{l}}}{\partial u^k}\right) = g^{i\underline{l}}\frac{\partial g_{j\underline{l}}}{\partial u^k} \\
\Gamma^{\underline{i}}{}_{\underline{j}\underline{k}} &= g^{\underline{i}l}\frac{\partial g_{\underline{j}l}}{\partial \bar{u}^{\underline{k}}} = g^{l\underline{i}}\frac{\partial g_{l\underline{j}}}{\partial \bar{u}^k} = (\Gamma^i{}_{jk})^*
\end{aligned} \tag{3.3.21}$$

where equation (3.3.6) has been used. On the same grounds, the other components prove to be vanishing,

$$\Gamma^i{}_{\underline{j}k} = \Gamma^i{}_{\underline{k}\underline{j}} = \Gamma^{\underline{i}}{}_{j\underline{k}} = \Gamma^{\underline{i}}{}_{jk} = 0. \tag{3.3.22}$$

In accordance with equation (1.6.14), the curvature tensor is

$$\mathscr{R}^I{}_{JKL} = \partial_K \Gamma^I{}_{LJ} - \partial_L \Gamma^I{}_{KJ} + \Gamma^I{}_{KE}\Gamma^E{}_{LJ} - \Gamma^I{}_{LE}\Gamma^E{}_{KJ}.$$

Looking at (3.3.22), one reads off

$$\mathscr{R}^i{}_{\underline{j}KL} = 0 \qquad \mathscr{R}^{\underline{i}}{}_{jKL} = 0 \tag{3.3.23}$$

hence $\mathscr{R}_{ijKL} = g_{i\underline{i}}\mathscr{R}^{\underline{i}}{}_{jKL} = 0$ and $\mathscr{R}_{\underline{i}\underline{j}KL} = g_{\underline{i}i}\mathscr{R}^i{}_{\underline{j}KL} = 0$. Due to the properties of the curvature tensor, we also have $\mathscr{R}_{KLij} = 0$ and $\mathscr{R}_{KL\underline{i}\underline{j}} = 0$. Therefore, the only non-vanishing components are: $\mathscr{R}_{\underline{i}jk\underline{l}}$, $\mathscr{R}_{\underline{i}j\underline{k}l}$, $\mathscr{R}_{i\underline{j}k\underline{l}}$ and $\mathscr{R}_{i\underline{j}\underline{k}l}$, which are related to $\mathscr{R}^{\underline{i}}{}_{\underline{j}k\underline{l}}$, $\mathscr{R}^{\underline{i}}{}_{\underline{j}\underline{k}l}$, $\mathscr{R}^i{}_{jk\underline{l}}$ and $\mathscr{R}^i{}_{j\underline{k}l}$, respectively.

A straightforward calculation gives

$$\begin{aligned}
\mathscr{R}^i{}_{jk\underline{l}} &= -\frac{\partial \Gamma^i{}_{jk}}{\partial \bar{u}^{\underline{l}}} \qquad & \mathscr{R}^{\underline{i}}{}_{\underline{j}\underline{k}l} &= -\frac{\partial \Gamma^{\underline{i}}{}_{\underline{j}\underline{k}}}{du^l} \\
\mathscr{R}^i{}_{j\underline{k}l} &= \frac{\partial \Gamma^i{}_{lj}}{\partial \bar{u}^{\underline{k}}} \qquad & \mathscr{R}^{\underline{i}}{}_{\underline{j}k\underline{l}} &= \frac{\partial \Gamma^{\underline{i}}{}_{\underline{l}\underline{j}}}{du^k}.
\end{aligned} \tag{3.3.24}$$

We see $\mathscr{R}^{\underline{i}}{}_{\underline{j}\underline{k}l} = (\mathscr{R}^i{}_{jk\underline{l}})^*$ and $\mathscr{R}^{\underline{i}}{}_{\underline{j}k\underline{l}} = (\mathscr{R}^i{}_{j\underline{k}l})^*$. We also have $\mathscr{R}^i{}_{jk\underline{l}} = \mathscr{R}^i{}_{kj\underline{l}}$.

Expressions (3.3.21, 24) lead to

$$\mathscr{R}_{i\underline{j}k\underline{l}} = \frac{\partial^2 g_{i\underline{j}}}{\partial u^k \partial \bar{u}^{\underline{l}}} - g^{m\underline{n}}\frac{\partial g_{i\underline{n}}}{\partial u^k}\frac{\partial g_{m\underline{j}}}{\partial \bar{u}^{\underline{l}}}.$$

Then, taking into account equation (3.3.15), one obtains the result

$$\mathscr{R}_{i\underline{j}k\underline{l}} = K_{ik\underline{j}\underline{l}} - g^{m\underline{n}}K_{ik\underline{n}}K_{m\underline{j}\underline{l}} \tag{3.3.25}$$

where we have used the notation adoted in the previous subsection. Now, it is obvious that

$$\mathscr{R}_{i\underline{j}k\underline{l}} = \mathscr{R}_{k\underline{j}i\underline{l}} = \mathscr{R}_{i\underline{l}k\underline{j}}. \tag{3.3.26}$$

Relations (3.3.23, 25) determine the curvature tensor of a Kähler manifold.

3.3.4. Kähler geometry and supersymmetric σ-models

Above it has been shown that in supersymmetric σ-models, scalar fields couple via the Kählerian metric and the Kähler potential coincides with the super Lagrangian in (3.3.10). So, every supersymmetric σ-model defines some Kähler geometry. Remarkably, the converse is also true. Namely, with a given Kähler manifold one can associate a supersymmetric σ-model. However, in this situation (anti)chiral superfields occurring in the action (3.3.10) take their values not in the Kähler manifold but in a supermanifold, in which the Kähler manifold is embedded.

First, let us show how to embed a real n-dimensional manifold $\mathcal{M}^n$ in a real supermanifold of dimension $(n, 0)$ (a supermanifold of dimension (p, q) looks locally like a domain of $\mathbb{R}^{p|q}$). The basic observation is that every smooth function of n ordinary real variables $f(\varphi^I)$, defined on an open set $V \in \mathbb{R}^n$, can be extended to a supersmooth function of n real c-number variables $f(\omega^I)$, defined on the domain $\hat{V} = \pi^{-1}(V) \subset \mathbb{R}^n_c$, where π: $\mathbb{R}^n_c \to \mathbb{R}^n$ is the map projecting any point $(\omega^I) \in \mathbb{R}^n_c$ into its body,

$$\pi((\omega^I)) = (\omega^I)_B = \varphi^I.$$

The above mentioned supermanifold is a fibre bundle $\hat{\mathcal{M}}^n$ over base manifold $\mathcal{M}^n$ with fibre $F = \pi^{-1}(\vec{o})$, $\vec{o} \in \mathbb{R}^n$, and Π: $\hat{\mathcal{M}}^n \to \mathcal{M}^n$ being the projection mapping. It is endowed with the structure of a supermanifold as follows. Choose an arbitrary chart (U, φ) in $\mathcal{M}^n$, where U is an open set of $\mathcal{M}^n$ and φ: $U \to V \subset \mathbb{R}^n$ is a homeomorphism defining local coordinates φ^I on U. Next, we extend (U, φ) to a chart $(\hat{U}, \omega)$ in $\hat{\mathcal{M}}^n$, where $\hat{U} = \Pi^{-1}(U)$ and ω: $\hat{U} \to \hat{V} = \pi^{-1}(V) \subset \mathbb{R}^n_c$ is a homeomorphism, introducing local c-number coordinates ω^I on $\hat{U}$, which are taken to satisfy the relation $\varphi \circ \Pi = \pi \circ \omega$. If two charts U and U' have non-empty overlap with the transition functions being of the form (3.3.3), then their transition superfunctions for the corresponding charts U and U' are taken to be the supersmooth extensions of the function $f^I(\varphi)$,

$$\omega^I \to \omega'^I = f^I(\omega). \tag{3.3.27}$$

As a result, $\hat{\mathcal{M}}^n$ turns out to be a supermanifold.

Furthermore, given a manifold $\mathcal{M}^{2n}$ endowed with the structure of a complex manifold, the corresponding supermanifold $\hat{\mathcal{M}}^{2n}$ also admits a complex structure. Explicitly, if one parametrizes $\mathcal{M}^{2n}$ as in equations (3.3.4, 5) and introduces local complex coordinates Φ^i and $\bar{\Phi}^{\underline{i}}$ on $\hat{\mathcal{M}}^{2n}$

$$\omega^i = \frac{1}{2}(\Phi^i + \bar{\Phi}^{\underline{i}}) \qquad \omega^{i+n} = \frac{1}{2\mathrm{i}}(\Phi^i - \bar{\Phi}^{\underline{i}}) \qquad i, \underline{i} = 1, \ldots, n \tag{3.3.28}$$

then the transition superfunctions prove to be (anti)holomorphic,

$$\Phi'^i = f^i(\Phi) \qquad \bar{\Phi}'^{\underline{i}} = \bar{f}^{\underline{i}}(\bar{\Phi}).$$

Finally, if $\mathcal{M}^{2n}$ possesses a Kählerian metric, it can be extended to a Kählerian metric on $\hat{\mathcal{M}}^{2n}$.

Now, we are in a position to give the geometrical interpretation of supersymmetric nonlinear σ-models. Let $\mathcal{M}$ be a Kähler manifold and $\hat{\mathcal{M}}$ be the corresponding supermanifold. We introduce into consideration the family of superholomorphic mappings of the form

$$\varphi\colon \mathbb{C}_s^{4|2} \to \hat{\mathcal{M}} \tag{3.3.29}$$

In local complex coordinates on $\hat{\mathcal{M}}$ every such mapping looks like

$$\begin{gathered}\Phi^i(y,\theta) = A^i(y) + \theta^\alpha \psi^i_\alpha(y) + \theta^2 F^i(y) \\ \partial\Phi^i/\partial\bar{y}^a = \partial\Phi^i/\partial\bar{\theta}^{\dot{\alpha}} \equiv 0.\end{gathered} \tag{3.3.30}$$

When restricted from $\mathbb{C}_s^{4|2}$ to the surface $\mathbb{R}^{4|4}(\theta\sigma\bar{\theta})$ in $\mathbb{C}_s^{4|2}$, on which $y^a = x^a + \mathrm{i}\theta\sigma^a\theta$, Φ^i becomes a chiral superfield. So, we recover σ-model superfield variables. Therefore, one can take the following point of view. The set of all superholomorphic mappings (3.3.29) is identified with the σ-model space of superfield histories. The action superfunctional is taken in the form (3.3.10), where $K(\Phi,\bar{\Phi})$ is a Kähler potential related to the Kählerian metric on $\hat{\mathcal{M}}$ by the rule

$$g_{i\underline{j}}(\Phi,\bar{\Phi}) = \partial^2 K(\Phi,\bar{\Phi})/\partial\Phi^i\partial\bar{\Phi}^{\underline{j}}.$$

Note that $K(\Phi,\bar{\Phi})$ is locally defined on $\hat{\mathcal{M}}$. But the action (3.3.10) is well defined globally on $\mathbb{R}^{4|4}$, since the inherent arbitrariness (3.3.11) in the choice of the super Lagrangian coincides with the arbitrariness (3.3.19) in the choice of the Kähler potential.

In conclusion, let us investigate the action of the transformations (3.3.12) (coordinate transformations on $\hat{\mathcal{M}}$) on the component fields of $\Phi^i(z)$. One readily finds

$$\begin{gathered}A'^i = f^i(A) \\ \psi'^i_\alpha = \mathrm{D}_\alpha\Phi'^i| = \frac{\partial A'^i}{\partial A^j}\psi^j_\alpha \\ F'^i = -\frac{1}{4}\mathrm{D}^2\Phi'^i| = -\frac{1}{4}\frac{\partial^2 A'^i}{\partial A^k\partial A^j}\psi^k\psi^j + \frac{\partial A'^i}{\partial A^j}F^j.\end{gathered} \tag{3.3.31}$$

It is seen that ψ^i_α transforms as a vector field on $\hat{\mathcal{M}}$. The transformation law for F^i is not covariant. Redefining F^i by the rule

$$\mathscr{F}^i = F^i - \frac{1}{4}\Gamma^i{}_{jk}\psi^j\psi^k \tag{3.3.32}$$

where $\Gamma^i{}_{jk}$ are the Christoffel symbols (3.3.21), one obtains the vector transformation law

$$\mathscr{F}'^i = \frac{\partial A'^i}{\partial A^j} \mathscr{F}^j. \tag{3.3.33}$$

After the redefinition (3.3.32), the Lagrangian (3.3.14) can be rewritten in the form

$$L = -g_{i\underline{j}}\left(\partial^a \bar{A}^{\underline{j}} \partial_a A^i - \bar{\mathscr{F}}^{\underline{j}} \mathscr{F}^i + \frac{\mathrm{i}}{4} \psi^i \sigma^a \overleftrightarrow{\nabla}_a \bar{\psi}^{\underline{j}}\right) + \frac{1}{16} \mathscr{R}_{i\underline{j}k\underline{l}} \psi^i \psi^k \bar{\psi}^{\underline{j}} \bar{\psi}^{\underline{l}} \tag{3.3.34}$$

Here $\mathscr{R}_{i\underline{j}k\underline{l}}$ is the curvature tensor (3.3.25), and ∇_a denote the target space covariant derivatives,

$$\nabla_a \psi^i_\alpha = \partial_a \psi^i_\alpha + \Gamma^i{}_{jl}(\partial_a A^j) \psi^l_\alpha$$

$$\nabla_a \bar{\psi}^{\underline{i}}_{\dot{\alpha}} = \partial_a \bar{\psi}^{\underline{i}}_{\dot{\alpha}} + \Gamma^{\underline{i}}{}_{\underline{j}\underline{l}} \partial_a \bar{A}^{\underline{j}} \bar{\psi}^{\underline{l}}_{\dot{\alpha}}.$$

We see that the supersymmetric σ-model Lagrangian is expressible in geometrical terms.

3.4. Vector multiplet models

3.4.1. Massive vector multiplet model

In Section 2.6 it was shown that the superspin-$\frac{1}{2}$ representation of the Poincaré superalgebra can be described in terms of a real scalar superfield subject to the Klein–Gordon equation and a supplementary condition. Now, we present the supersymmetric model of an unconstrained real scalar superfield $V(z)$, $V = \bar{V}$, in which the Klein–Gordon equation and the supplementary condition arise as consequences of the dynamical equation.

Let us consider the dynamical system with action superfunctional

$$S[V] = \frac{1}{8} \int d^8z V D^\alpha \bar{D}^2 D_\alpha V + m^2 \int d^8z V^2. \tag{3.4.1}$$

The kinetic term is real due to the identity (2.5.30*d*). Integrating by parts and making use of equation (3.1.26*a*), one can represent the action in the form

$$S[V] = \frac{1}{2} \int d^6z W^\alpha W_\alpha + m^2 \int d^8z V^2 \tag{3.4.2}$$

where we have introduced the superfield

$$W_\alpha = -\frac{1}{4} \bar{D}^2 D_\alpha V \qquad \bar{D}_{\dot{\alpha}} W_\alpha = 0. \tag{3.4.3}$$

Due to equation (2.5.30*d*), this chiral spinor superfield satisfies the reality constraint

$$\mathrm{D}^{\alpha}W_{\alpha} = \bar{\mathrm{D}}_{\dot{\alpha}}\bar{W}^{\dot{\alpha}}. \tag{3.4.4}$$

Let us also point out the identity

$$\int \mathrm{d}^6 z W^{\alpha}W_{\alpha} = \int \mathrm{d}^6 \bar{z} \bar{W}_{\dot{\alpha}}\bar{W}^{\dot{\alpha}}. \tag{3.4.5}$$

To find the dynamical equation, one uses the relation (3.1.33). The result is

$$\frac{1}{8}\mathrm{D}^{\alpha}\bar{\mathrm{D}}^2\mathrm{D}_{\alpha}V + m^2 V = 0. \tag{3.4.6}$$

Acting by the operators D^2 or $\bar{\mathrm{D}}^2$ on the left side and using equations (2.5.30*f*), one obtains

$$\mathrm{D}^2 V = \bar{\mathrm{D}}^2 V = 0. \tag{3.4.7}$$

Therefore, on-shell the superfield V is linear. Then, by virtue of equation (2.5.30*e*), the equation of motion is equivalent to

$$(\Box - m^2)V = 0. \tag{3.4.8}$$

These are the equations (3.4.7, 8), which determine the irreducible superspin-$\frac{1}{2}$ representation.

Remark. Mass dimensions of $V(z)$ and $W_{\alpha}(z)$ are

$$[V] = 0 \qquad [W_{\alpha}] = 3/2.$$

Remark. The action superfunctional has been taken in the form (3.4.1) in order that the dynamical subspace was given by equations (3.4.7, 8) as well as to have a standard mass term for the component vector field of $V(z)$,

$$V(z) \sim \theta\sigma^a\bar{\theta}V_a \qquad m^2V^2 \sim -\frac{m^2}{2}\theta^2\bar{\theta}^2 V^a V_a.$$

On-shell, only four component fields of $V(z)$ are non-vanishing (equations (2.8.22)). The multiplet of fields includes a transverse vector field. On these grounds, the theory under consideration is called the massive vector multiplet model.

The superspin-$\frac{1}{2}$ representation describes four particles of spins 0, $\frac{1}{2}$, $\frac{1}{2}$ and 1. It should be pointed out that there are three super Poincaré representations, of superspins $\frac{1}{2}$, 1 and $\frac{3}{2}$, including vector particles. But only the former representation describes particles of spins not higher than one.

3.4.2. Massless vector multiplet model

Let us set the mass parameter in equation (3.4.1) to zero. The resultant action

$$S[V] = \frac{1}{8}\int d^8z V D^\alpha \bar{D}^2 D_\alpha V = \frac{1}{2}\int d^6z W^\alpha W_\alpha \tag{3.4.9}$$

turns out to be invariant under the following gauge transformations

$$V \to V' = V + \frac{i}{2}(\bar{\Lambda} - \Lambda) \qquad \bar{D}_{\dot{\alpha}}\Lambda = 0 \tag{3.4.10}$$

where $\Lambda(z)$ is an arbitrary chiral superfield. Making use of equation (2.5.30*c*) gives

$$W'_\alpha = W_\alpha + \frac{i}{8}\bar{D}^2 D_\alpha \Lambda = W_\alpha.$$

We see that (anti)chiral superfields $\bar{W}_{\dot{\alpha}}$ and W_α are gauge invariant objects. $\bar{W}_{\dot{\alpha}}$ and W_α will be called 'superfield strengths'.

The dynamical equation can be written in two equivalent forms

$$D^\alpha W_\alpha = 0 \Leftrightarrow \bar{D}_{\dot{\alpha}}\bar{W}^{\dot{\alpha}} = 0 \tag{3.4.11}$$

due to the identity (3.4.4). Now, it is time to recall the results of Section 2.7. Obviously, the above equations state that W_α and $\bar{W}_{\dot{\alpha}}$ are on-shell massless superfields and their superhelicities are

$$\kappa(W_\alpha) = 1/2 \qquad \kappa(\bar{W}_{\dot{\alpha}}) = -1. \tag{3.4.12}$$

Therefore, W_α describes two massless particles of helicities $\frac{1}{2}$ and 1, while $\bar{W}_{\dot{\alpha}}$ describes particles of helicities -1 and $-\frac{1}{2}$.

3.4.3. Wess–Zumino gauge

It is instructive to investigate the component structure of the theory. First, let us discuss gauge transformations. Since Λ is chiral, we can write

$$\Lambda(x, \theta, \bar{\theta}) = e^{i\theta\sigma^a\bar{\theta}\partial_a}(u(x) + \theta^\alpha \rho_\alpha(x) + \theta^2 f(x)). \tag{3.4.13}$$

Here $u(x)$ and $f(x)$ are arbitrary complex scalar fields and $\rho_\alpha(x)$ is an arbitrary undotted spinor field. Then the transformation law (3.4.10) takes the form

$$\begin{aligned}\delta V(x, \theta, \bar{\theta}) &= \frac{i}{2}(\bar{u}(x) - u(x)) - \frac{i}{2}\theta^\alpha \rho_\alpha(x) - \frac{i}{2}\theta^2 f(x) + \frac{i}{2}\bar{\theta}_{\dot{\alpha}}\bar{\rho}^{\dot{\alpha}}(x) \\ &\quad + \frac{i}{2}\bar{\theta}^2 \bar{f}(x) + \frac{1}{2}\partial_a(u(x) + \bar{u}(x))\theta\sigma^a\bar{\theta} + \ldots\end{aligned} \tag{3.4.14}$$

where dots mean terms involving partial derivatives and at least third order in θ, $\bar{\theta}$. Obviously, the component fields A, ψ_α and F of our superfield (see equation (2.8.19)) have arbitrary displacements, hence they can be gauged away. Then V is reduced to

$$V = \theta\sigma^a\bar{\theta}V_a + \bar{\theta}^2\theta^\alpha\lambda_\alpha + \theta^2\bar{\theta}_{\dot{\alpha}}\bar{\lambda}^{\dot{\alpha}} + \theta^2\bar{\theta}^2\mathscr{D}. \tag{3.4.15}$$

This gauge fixing, known as the 'Wess–Zumino gauge', can also be defined as follows

$$V| = \mathrm{D}_\alpha V| = \mathrm{D}^2 V| = 0. \tag{3.4.16a}$$

Under these requirements, the component fields in the expansion (3.4.15) are given by the same relations as in equation (2.8.19),

$$V_{\alpha\dot{\alpha}} = \frac{1}{2}[\mathrm{D}_\alpha, \bar{\mathrm{D}}_{\dot{\alpha}}]V| \qquad \lambda_\alpha = -\frac{1}{4}\bar{\mathrm{D}}^2\mathrm{D}_\alpha V|$$

$$D = \frac{1}{32}\{\mathrm{D}^2, \bar{\mathrm{D}}^2\}V|. \tag{3.4.16b}$$

Imposition of the Wess–Zumino gauge does not completely fix the invariance (3.4.10). The transformations preserving the above gauge correspond to the choice

$$\Lambda(x, \theta, \bar{\theta}) = \mathrm{e}^{\mathrm{i}\theta\sigma^a\bar{\theta}\partial_a}\xi(x) \qquad \xi = \bar{\xi} \tag{3.4.17}$$

with $\xi(x)$ being a real function, and their explicit form is

$$V'_a = V_a + \partial_a\xi \qquad \lambda'_\alpha = \lambda_\alpha \qquad D' = D. \tag{3.4.18}$$

Obviously, the gauge transformation of V_a is the same as in Maxwell electrodynamics.

As for the component fields of the superfield strength W_α, they can be easily determined using equations (3.4.16). One finds

$$\begin{gathered} W_\alpha| = \lambda_\alpha \\ \mathrm{D}^\alpha W_\alpha| = -\frac{1}{8}\{\mathrm{D}^2, \bar{\mathrm{D}}^2\}V| = -4D \\ \mathrm{D}_{(\alpha}W_{\beta)}| = 2\mathrm{i}F_{\alpha\beta} \\ -\frac{1}{4}\mathrm{D}^2W_\alpha| = \frac{1}{2}\mathrm{D}_\alpha\mathrm{D}^\beta W_\beta| = -\mathrm{i}\partial_{\alpha\dot{\alpha}}\bar{\lambda}^{\dot{\alpha}} \end{gathered} \tag{3.4.20}$$

where we have introduced the vector field strength

$$\begin{gathered} F_{ab} = \partial_a V_b - \partial_b V_a \\ F_{\alpha\beta} = \frac{1}{2}(\sigma^{ab})_{\alpha\beta}F_{ab} = -\frac{1}{2}\partial_{(\alpha\dot{\alpha}}V_{\beta)}{}^{\dot{\alpha}}. \end{gathered} \tag{3.4.21}$$

Therefore, the expansion of W_α as a power series in θ reads

$$W_\alpha(x, \theta) = \lambda_\alpha + 2D\theta_\alpha + 2iF_{\alpha\beta}\theta^\beta - i\partial_{\alpha\dot{\alpha}}\bar{\lambda}^{\dot{\alpha}}\theta^2. \qquad (3.4.22)$$

The Wess–Zumino gauge is especially helpful, due to the gauge invariance of $S[V]$, for calculating the action functional. Following the prescription (3.1.54) gives

$$S = -\frac{1}{8}\int d^4x D^2(W^\alpha W_\alpha)| = \frac{1}{4}\int d^4x(D^\beta W^\alpha| D_\beta W_\alpha| - W^\alpha| D^2 W_\alpha|).$$

Making use of equations (3.4.20) leads to

$$S = \int d^4x\{-F^{\alpha\beta}F_{\alpha\beta} - i\lambda^\alpha\partial_{\alpha\dot{\alpha}}\bar{\lambda}^{\dot{\alpha}} + 2D^2\}.$$

It only remains to recall the identities

$$\frac{1}{2}F^{ab}F_{ab} = F^{\alpha\beta}F_{\alpha\beta} + \bar{F}^{\dot{\alpha}\dot{\beta}}\bar{F}_{\dot{\alpha}\dot{\beta}}$$

$$\int d^4x F^{\alpha\beta}F_{\alpha\beta} = \int d^4x \bar{F}^{\dot{\alpha}\dot{\beta}}\bar{F}_{\dot{\alpha}\dot{\beta}}. \qquad (3.4.23)$$

Then, the action functional takes the final form

$$S = \int d^4x\left\{-\frac{1}{4}F^{ab}F_{ab} - i\lambda\sigma^a\partial_a\bar{\lambda} + 2D^2\right\}. \qquad (3.4.24)$$

Beautifully, the first term coincides with the action of Maxwell electrodynamics.

3.4.4. *Supersymmetry transformations*

The action (3.4.9) is invariant under the supersymmetry transformations

$$\delta_\epsilon V = i(\epsilon \mathbf{Q} + \bar{\epsilon}\bar{\mathbf{Q}})V \qquad (3.4.25)$$

having the form (2.8.20) at the component level. Every such transformation breaks down the Wess–Zumino gauge. Indeed, if V is under the constraints (3.4.16), then making a supersymmetry transformation generates non-vanishing component fields ψ_α and F,

$$\psi'_\alpha = -\bar{\epsilon}^{\dot{\alpha}}V_{\alpha\dot{\alpha}} \qquad F' = -\epsilon\lambda$$

in accordance with equation (2.8.20). So, the Wess–Zumino gauge is not supersymmetric.

Of course, one can restore the Wess–Zumino gauge by accompanying the variation $\delta_\epsilon V$ with the ϵ-dependent gauge transformation

$$\delta_{\Lambda(\epsilon)} V = \frac{i}{2}(\bar{\Lambda}(\epsilon) - \Lambda(\epsilon))$$
$$\Lambda(\epsilon) = e^{i\theta\sigma^a\bar{\theta}\partial_a}(2i\theta\sigma^b\bar{\epsilon}V_b + 2i\theta^2\overline{\epsilon\lambda}). \tag{3.4.26}$$

The resultant supersymmetry transformation

$$\hat{\delta}_\epsilon V = i(\epsilon\mathbf{Q} + \overline{\epsilon\mathbf{Q}})V + \frac{i}{2}(\bar{\Lambda}(\epsilon) - \Lambda(\epsilon)) \tag{3.4.27}$$

has not, however, the universal form (3.4.25). In other words, imposing the Wess–Zumino gauge modifies the superfield supersymmetry generators. As a result, the Wess–Zumino gauge, being quite useful for component calculations, does not fit into covariant (explicitly supersymmetric) superfield calculations.

The supersymmetry transformation (3.4.27) acts on the component fields (3.4.16*b*) in accordance with the rule

$$\hat{\delta}_\epsilon V_a = \delta_\epsilon V_a = \epsilon\sigma_a\bar{\lambda} + \lambda\sigma_a\bar{\epsilon}$$
$$\hat{\delta}_\epsilon\lambda_\alpha = \delta_\epsilon\lambda_\alpha - \frac{1}{8}D_\alpha\bar{D}^2\bar{\Lambda}(\epsilon)| = -2iF_{\alpha\beta}\epsilon^\beta - 2\epsilon_\alpha D \tag{3.4.28}$$
$$\hat{\delta}_\epsilon D = \delta_\epsilon D = \frac{i}{2}\partial_a(\lambda\sigma^a\bar{\epsilon} - \epsilon\sigma^a\bar{\lambda})$$

where we have used the relation (2.8.20). The last two expressions can also be obtained by applying results of subsection 2.8.2 to the chiral spinor superfield W_α.

3.4.5. Super Lorentz gauge

As we have seen, the Wess–Zumino gauge is not supersymmetric. One would like to have a covariant gauge fixing condition — that is, one commuting with the supersymmetry transformations (3.4.25). A useful gauge choice consists in imposing the linearity constraint (3.4.7), which restricts the component vector fields of $V(z)$ to be transverse (see subsection 2.8.4). On these grounds, this gauge is known as the super Lorentz gauge.

3.4.6. Massive vector multiplet model revisited

The massive vector multiplet model defined by action (3.4.1) possesses a hidden point: several component fields have non-canonical dimensions. Therefore, we should carefully define the component fields of V in order to avoid higher derivatives at the component level (in any case, such potential higher derivatives are not dangerous and can be eliminated by a field redefinition).

Instead of using the definition (2.8.19), now it is more reasonable to introduce the component fields of V by the rule

$$\begin{aligned}
&\frac{1}{m}B(x) = V| \qquad \frac{1}{m}\chi_\alpha(x) = \mathrm{D}_\alpha V| \qquad \frac{1}{m}G(x) = -\frac{1}{4}\mathrm{D}^2 V| \\
&V_{\alpha\dot\alpha}(x) = \frac{1}{2}[\mathrm{D}_\alpha, \bar{\mathrm{D}}_{\dot\alpha}]V| \qquad \lambda_\alpha(x) = -\frac{1}{4}\bar{\mathrm{D}}^2\mathrm{D}_\alpha V| \\
&D(x) = \frac{1}{16}\mathrm{D}^\alpha\bar{\mathrm{D}}^2\mathrm{D}_\alpha V| \,.
\end{aligned} \tag{3.4.29}$$

With such a definition the component fields of W_α remain unchanged, see relation (3.4.20). Hence, we can use our old result (3.4.24) establishing the component form of the first term in action (3.4.2). As to the mass term in action (3.4.2), its component structure is

$$\begin{aligned}
m^2\int \mathrm{d}^8 z V^2 = \int \mathrm{d}^4 x\Big\{ -\frac{1}{2}\partial^a B\partial_a B + 2m\,BD + 2\bar{G}G - \frac{1}{2}m^2 V^a V_a \\
- \mathrm{i}\chi\sigma^a\partial_a\bar\chi - m(\lambda\chi + \bar\lambda\bar\chi)\Big\}\,.
\end{aligned} \tag{3.4.30}$$

It is seen that the superfield mass term contains the kinetic terms of the component fields B and χ. The component Lagrangian of the massive vector multiplet reads

$$\begin{aligned}
L = &-\frac{1}{4}F^{ab}F_{ab} - \frac{1}{2}m^2 V^a V_a - \frac{1}{2}\partial^a B\partial_a B - \frac{1}{2}m^2 B^2 \\
&- \mathrm{i}\lambda\sigma^a\partial_a\bar\lambda - \mathrm{i}\chi\sigma^a\partial_a\bar\chi - m(\lambda\chi + \bar\lambda\bar\chi) \\
&+ 2\bar{G}G + 2(D + \frac{1}{2}mB)^2\,.
\end{aligned} \tag{3.4.31}$$

Eliminating the auxiliary fields, the Lagrangian turns into

$$\begin{aligned}
L = &-\frac{1}{4}F^{ab}F_{ab} - \frac{1}{2}m^2 V^a V_a - \frac{1}{2}\partial^a B\partial_a B - \frac{1}{2}m^2 B^2 \\
&- \overline{\Psi}_\mathrm{D}(\mathrm{i}\gamma^a\partial_a + m)\Psi_\mathrm{D}\,.
\end{aligned} \tag{3.4.32}$$

Here we have introduced the Dirac spinor

$$\Psi_\mathrm{D} = \begin{pmatrix} \chi_\alpha \\ \bar\lambda^{\dot\alpha} \end{pmatrix}. \tag{3.4.33}$$

The massive vector multiplet model can be reformulated as a gauge theory such that all component fields have canonical dimensions in a Wess-Zumino gauge. Such a formulation is based on a superfield generalization of the

Stueckelberg approach to the gauge covariant description of massive vector particles. We introduce a compensating chiral scalar superfield $\Phi(z)$ and its conjugate $\bar{\Phi}(z)$ by making the replacement

$$V \longrightarrow V + \frac{\mathrm{i}}{2m}(\Phi - \bar{\Phi}) \tag{3.4.34}$$

in action (3.4.2). The resulting action superfunctional

$$\begin{aligned} S[V, \Phi, \bar{\Phi}] = \frac{1}{2} \int \mathrm{d}^6 z\, W^\alpha W_\alpha + m^2 \int \mathrm{d}^8 z\, V^2 \\ + \mathrm{i} m \int \mathrm{d}^8 z\, V(\Phi - \bar{\Phi}) + \frac{1}{2} \int \mathrm{d}^8 z\, \bar{\Phi}\Phi \end{aligned} \tag{3.4.35}$$

is invariant under the gauge transformations

$$\delta V = \frac{\mathrm{i}}{2}(\bar{\Lambda} - \Lambda) \qquad \delta\Phi = m\Lambda \tag{3.4.36}$$

with Λ an arbitrary chiral scalar superfield. The gauge freedom can be used to completely gauge away Φ. In the gauge

$$\Phi = 0 \tag{3.4.37}$$

we return to the original model (3.4.2). Instead of imposing such a gauge fixing, however, we are now able to choose the Wess-Zumino gauge (3.4.15). Then, the residual gauge invariance given by equation (3.4.17) can be further used to eliminate the real part of $\Phi|$. As a consequence, it is in our power to impose the conditions

$$V| = \mathrm{D}_\alpha V| = \mathrm{D}^2 V| = 0 \qquad \Phi| + \bar{\Phi}| = 0 \tag{3.4.38}$$

which completely fix the gauge freedom.

Gauge fixing condition (3.4.38) is most useful for passing to components. From action (3.4.35) one readily obtains the component Lagrangian

$$\begin{aligned} L = & -\frac{1}{4}F^{ab}F_{ab} - \frac{1}{2}m^2 V^a V_a - \frac{1}{2}\partial^a A \partial_a A - \frac{1}{2}m^2 A^2 \\ & - \mathrm{i}\lambda\sigma^a\partial_a\bar{\lambda} - \frac{\mathrm{i}}{4}\psi\sigma^a\partial_a\bar{\psi} - \frac{\mathrm{i}}{2}m(\lambda\psi - \bar{\lambda}\bar{\psi}) \\ & + \frac{1}{2}\bar{F}F + 2(D - \frac{1}{2}mA)^2 \end{aligned} \tag{3.4.39}$$

where $A = \bar{A} = -\mathrm{i}\Phi|$. The component Lagrangians (3.4.31) and (3.4.39) are seen to coincide if we identify $B = -A$, $G = \frac{1}{2}F$ and $\chi = \frac{\mathrm{i}}{2}\psi$.

3.5. Supersymmetric Yang–Mills theories

The main goal of the present section is the construction of Yang–Mills theories possessing supersymmetry. Our strategy consists of finding superfield theories having the Yang–Mills form at the component level. The role of matter will be played by (anti)chiral scalar superfields since their component content is scalar and spinor fields only. The role of gauge object will be played by a multiplet of Lie-algebra-valued real scalar superfields $V^I(z)$ transforming by

$$\delta V^I = \frac{i}{2}(\bar{\Lambda}^I - \Lambda^I) + O(V) \qquad \bar{D}_{\dot{\alpha}}\Lambda^I = 0 \tag{3.5.1}$$

with $\Lambda^I(z)$ being arbitrary chiral superfield parameters. This transformation law is reasonable on two grounds: (1) when reduced to the linearized level, it describes a set of free vector multiplets (see equation (3.4.10); (2) it admits the Wess–Zumino gauge

$$V^I = \theta\sigma^a\bar{\theta}V_a{}^I + \bar{\theta}^2\theta^\alpha\lambda_\alpha{}^I + \theta^2\bar{\theta}_{\dot{\alpha}}\bar{\lambda}^{\dot{\alpha}I} + \theta^2\bar{\theta}^2D^I. \tag{3.5.2}$$

To start with, we consider an Abelian gauge model.

3.5.1. *Supersymmetric scalar electrodynamics*

As is well known, an Abelian gauge vector field $V_a(x)$, transforming by the law $\delta V_a(x) = \partial_a\xi(x)$, arises as a compensating field when trying to make local the rigid phase invariance

$$A(x) \rightarrow e^{i\Omega}A(x)$$

in the theory of a complex scalar field $A(x)$ with the action

$$S = -\int d^4x\partial^m\bar{A}\partial_m A.$$

A gauge real scalar superfield $V(z)$, transforming by the law (3.4.10), can be introduced in a similar fashion.

Consider the massless chiral scalar superfield model (3.2.5). The action is invariant under the rigid $U(1)$-transformations

$$\Phi(z) \rightarrow e^{i\Omega}\Phi(z) \qquad \Omega = \bar{\Omega}. \tag{3.5.3}$$

In trying to localize these transformations, one is faced with the following problem: the choice of $\Omega(z)$ as a real superfield is inconsistent with the chirality of Φ,

$$\Omega(z) = \bar{\Omega}(z) \Rightarrow \bar{D}_{\dot{\alpha}}(e^{i\Omega}\Phi) \neq 0.$$

Therefore, if one wishes to have a local transformation, the parameter should be chiral

$$\Phi \rightarrow \Phi' = e^{i\Lambda}\Phi \qquad \bar{D}_{\dot{\alpha}}\Lambda = 0. \tag{3.5.4}$$

But this breaks the invariance of the action:

$$\bar{\Phi}\Phi \rightarrow \bar{\Phi}e^{i(\Lambda - \bar{\Lambda})}\Phi.$$

Following the ideology of gauge field theory, we introduce a compensating real scalar superfield V, with the transformation law (3.4.10), and change the action to

$$\int d^8z\bar{\Phi}e^{2V}\Phi. \tag{3.5.5}$$

Obviously, this superfunctional is gauge invariant.

Note that the transformation law (3.5.3) corresponds to unit $U(1)$ charge. In the case of a chiral scalar superfield, having $U(1)$ charge q, the above law reads

$$\Phi(z) \rightarrow e^{iq\Omega}\Phi(z). \tag{3.5.6}$$

This leads to a slight modification of the gauge invariant action (3.5.5) consisting in the insertion of q into the exponential, resulting in

$$S[\Phi, \bar{\Phi}; V] = \int d^8z\bar{\Phi}e^{2qV}\Phi. \tag{3.5.7}$$

The action is invariant under the gauge transformations

$$\begin{aligned} \Phi' &= e^{iq\Lambda}\Phi \qquad \bar{D}_{\dot{\alpha}}\Lambda = 0 \\ e^{2qV'} &= e^{iq\bar{\Lambda}}e^{2qV}e^{-iq\Lambda}. \end{aligned} \tag{3.5.8}$$

Adding the free vector multiplet action (3.4.9), the total gauge invariant action takes the form

$$S = \int d^8z\bar{\Phi}e^{2qV}\Phi + \frac{1}{2}\int d^6zW^\alpha W_\alpha. \tag{3.5.9}$$

Unexpectedly, the action turns out to be non-polynomial. However, in the Wess–Zumino gauge (3.4.15), where $V^3 = 0$, the exponential terminates.

Let us determine the component form of the action (3.5.9) in the Wess–Zumino gauge. Using equations (3.4.16) and defining the component fields of Φ by rule (2.8.4), one obtains the Lagrangian

$$L = \left\{ -(\partial^m + iqV^m)\bar{A}(\partial_m - iqV_m)A + 2q\bar{A}AD - \frac{i}{2}\psi\sigma^a(\partial_a + iqV_a)\bar{\psi} \right.$$

$$\left. - qA\overline{\psi\lambda} - q\bar{A}\psi\lambda + \bar{F}F \right\} - \left\{ \tfrac{1}{4}F^{ab}F_{ab} + i\lambda\sigma^a\partial_a\bar{\lambda} - 2D^2 \right\}. \tag{3.5.10}$$

After eliminating the auxiliary fields F and D, this takes the form

$$L = -(\partial^m + iqV^m)\bar{A}(\partial_m - iqV_m)A - \frac{1}{2}q^2(\bar{A}A)^2 - \frac{i}{2}\psi\sigma^a(\partial_a + iqV_a)\bar{\psi}$$

$$- qA\overline{\psi\lambda} - q\bar{A}\psi\lambda - \frac{1}{4}F^{ab}F_{ab} - i\lambda\sigma^a\partial_a\bar{\lambda}. \tag{3.5.11}$$

One can look upon this model as a supersymmetric generalization of scalar electrodynamics.

The Lagrangian (3.5.11) is invariant under the gauge transformations

$$A' = e^{iq\xi}A \qquad \psi'_\alpha = e^{iq\xi}\psi_\alpha$$
$$V'_a = V_a + \partial_a\xi \qquad \lambda'_\alpha = \lambda_\alpha. \tag{3.5.12}$$

Using the spinor component field ψ_α of Φ, we can construct three four-component spinor objects:

(1) the Majorana spinor field

$$\Psi_M = \frac{1}{\sqrt{2}}\begin{pmatrix}\psi_\alpha \\ \bar{\psi}^{\dot{\alpha}}\end{pmatrix} \tag{3.5.13a}$$

(2) the left Weyl spinor

$$\Psi_L = \begin{pmatrix}\psi_\alpha \\ 0\end{pmatrix} \tag{3.5.13b}$$

(3) the right Weyl spinor

$$\Psi_R = \begin{pmatrix}0 \\ \bar{\psi}^{\dot{\alpha}}\end{pmatrix}. \tag{3.5.13c}$$

In accordance with equation (3.5.12), Ψ_M, Ψ_L and Ψ_R transform as follows

$$\Psi'_M = e^{iq\xi\gamma_5}\Psi_M \tag{3.5.14a}$$

$$\Psi'_L = e^{iq\xi}\Psi_L \tag{3.5.14b}$$

$$\Psi'_R = e^{-iq\xi}\Psi_R. \tag{3.5.14c}$$

Correspondingly, the gauge invariant spinor kinetic term in equations (3.5.10, 11) can be rewritten as

$$-\frac{i}{2}\psi\sigma^a(\partial_a + iqV_a)\bar{\psi} = \begin{cases} -\frac{i}{2}\bar{\Psi}_M\gamma^a(\partial_a - iq\gamma_5 V_a)\Psi_M & (3.5.15a) \\ -\frac{i}{2}\bar{\Psi}_L\gamma^a(\partial_a - iqV_a)\Psi_L & (3.5.15b) \\ -\frac{i}{2}\bar{\Psi}_R\gamma^a(\partial_a + iqV_a)\Psi_R & (3.5.15c) \end{cases}$$

Therefore equation (3.5.12) describes local chiral or γ_5-transformations. We see that the theory under consideration describes chiral fermions at the component level.

3.5.2. *Supersymmetric spinor electrodynamics*

To obtain a Dirac spinor field in component form, let us consider two chiral scalar superfields Φ_+ and Φ_- with $U(1)$ charges q and $-q$, respectively. Their spinor component fields

$$\psi_{+\alpha} = D_\alpha\Phi_+| \qquad \psi_{-\alpha} = D_\alpha\Phi_-|$$

transform according to the laws

$$\psi'_{+\alpha} = e^{iq\xi}\psi_{+\alpha} \qquad \psi'_{-\alpha} = e^{-iq\xi}\psi_{-\alpha}$$

in accordance with equation (3.5.12). Then, the Dirac spinor

$$\Psi_D = \begin{pmatrix} \psi_{+\alpha} \\ \bar{\psi}_-{}^{\dot{\alpha}} \end{pmatrix} = \Psi_{+L} + \Psi_{-R} \tag{3.5.16}$$

is characterized by the transformation law

$$\Psi'_D = e^{iq\xi}\Psi_D \tag{3.5.17}$$

inherent to spinor electrodynamics.

The most general (renormalizable) gauge invariant action for Φ_+ and Φ_-, coupled to a gauge superfield V, reads

$$\begin{aligned} S_{SED} &= \int d^8z\{\bar{\Phi}_+ e^{2qV}\Phi_+ + \bar{\Phi}_- e^{-2qV}\Phi_-\} \\ &\quad + m\int d^6z\Phi_+\Phi_- + m\int d^6\bar{z}\bar{\Phi}_+\bar{\Phi}_- + \frac{1}{2}\int d^6z W^\alpha W_\alpha. \end{aligned} \tag{3.5.18}$$

Note that a mass term is forbidden in the case of a single chiral scalar superfield with non-zero $U(1)$ charge. In components, the above action is

$$S_{SED} = \int d^4x L_{SED} \tag{3.5.19}$$

$$L_{SED} = -\frac{1}{4}F^{ab}F_{ab} - \frac{1}{2}\bar{\Psi}_D\{\gamma^a(i\partial_a + qV_a) + m\}\Psi_D$$

$$- (\partial^m + iqV^m)\bar{A}_+(\partial_m - iqV_m)A_+ - (\partial^m - iqV^m)\bar{A}_-(\partial_m + iqV_m)A_-$$

$$- m^2(\bar{A}_+A_+ + \bar{A}_-A_-) - \tfrac{1}{2}q^2(\bar{A}_+A_+ - \bar{A}_-A_-)^2 - \frac{i}{2}\bar{\Lambda}_M\gamma^a\partial_a\Lambda_M$$

$$-\frac{1}{2}q\left\{A_+\bar{\Psi}_D(\mathbb{I} - \gamma_5)\Lambda_M - A_-\bar{\Lambda}_M(\mathbb{I} - \gamma_5)\Psi_D + \text{c.c.}\right\}$$

$$+ (\bar{F}_+ + mA_-)(F_+ + m\bar{A}_-) + (\bar{F}_- + mA_+)(F_- + m\bar{A}_+)$$

$$+ 2\left(D + \tfrac{1}{2}q(\bar{A}_+A_+ - \bar{A}_-A_-)\right)^2$$

where we have introduced the Majorana spinor

$$\Lambda_M = \begin{pmatrix} \lambda_\alpha \\ \bar{\lambda}^{\dot{\alpha}} \end{pmatrix}.$$

Obviously, the model obtained represents a supersymmetric extension of spinor electrodynamics.

In conclusion, we rewrite the action (3.5.18) in a different form. We unify Φ_+ and Φ_- into a two-component column

$$\Phi = \begin{pmatrix} \Phi_+ \\ \Phi_- \end{pmatrix} \qquad \bar{\Phi} = (\bar{\Phi}_+, \bar{\Phi}_-) \tag{3.5.20}$$

and make the redefinition

$$\Phi = \frac{1}{\sqrt{2}}\begin{pmatrix} 1 & -i \\ 1 & i \end{pmatrix}\chi \qquad \chi = \begin{pmatrix} \chi_1 \\ \chi_2 \end{pmatrix}. \tag{3.5.21}$$

After this, the action takes the form

$$S_{\text{SED}} = \int d^8z\bar{\chi}e^{2qV\sigma_2}\chi + \left\{\int d^6z\left(\frac{m}{2}\chi^{\text{T}}\chi + \frac{1}{4}W^\alpha W_\alpha\right) + c.c.\right\} \tag{3.5.22}$$

where σ_2 is the second Pauli matrix. The transformation law of χ is

$$\chi' = e^{iq\Lambda\sigma_2}\chi. \tag{3.5.23}$$

In summary, we have constructed the supersymmetric generalization of spinor electrodynamics.

3.5.3. Non-Abelian gauge superfield

We now generalize the above results to the non-Abelian case.

Let G be a compact connected Lie group (in general, G is the product of a finite number of simple and $U(1)$ group factors). Given a finite-dimensional unitary G-representation, we denote by $(T^I)^i{}_j$, where $I = 1, 2, \ldots, \dim G$, its Hermitian generators:

$$[T^I, T^J] = \mathrm{i} f^{IJK} T^K \qquad (T^I)^+ = T^I \tag{3.5.24}$$

and the structure constants are assumed to be totally antisymmetric. Any operator of the representation is written in the form

$$\exp(\mathrm{i}\eta^I T^I) \qquad \eta^I \in \mathbb{R}$$

since every compact connected Lie group is covered by the exponential mapping.

Consider a system of free massless chiral scalar superfields $\Phi^i(z)$. The action

$$S = \int \mathrm{d}^8 z \bar{\Phi}_i \Phi^i = \int \mathrm{d}^8 z \bar{\Phi} \Phi \qquad \bar{\Phi}_i = (\Phi^i)^*$$

is invariant under the rigid transformations

$$\Phi' = \mathrm{e}^{\mathrm{i}\eta^I T^I} \Phi \qquad \eta^I \in \mathbb{R}_c.$$

As in the Abelian case, when attempting to localize these transformations, the parameters are to be taken chiral,

$$\begin{gathered} \Phi' = \mathrm{e}^{\mathrm{i}\Lambda} \Phi \qquad \bar{\Phi}' = \bar{\Phi} \mathrm{e}^{-\mathrm{i}\bar{\Lambda}} \\ \Lambda = \Lambda^I T^I \qquad \bar{\Lambda} = \bar{\Lambda}^I T^I \qquad \bar{D}_{\dot{\alpha}} \Lambda^I = 0. \end{gathered} \tag{3.5.25}$$

In addition, it is necessary to modify the action by introducing a compensating multiplet of real scalar superfields $V^I(z)$ transforming by the rule

$$\mathrm{e}^{2V'} = \mathrm{e}^{\mathrm{i}\bar{\Lambda}} \mathrm{e}^{2V} \mathrm{e}^{-\mathrm{i}\Lambda} \qquad V = V^+ = V^I T^I. \tag{3.5.26}$$

The gauge-invariant action is

$$S = \int \mathrm{d}^8 z \bar{\Phi}_i (\mathrm{e}^{2V})^i{}_j \Phi^j = \int \mathrm{d}^8 z \bar{\Phi} \mathrm{e}^{2V} \Phi. \tag{3.5.27}$$

Lie-algebra-valued Hermitian superfield V with the transformation law (3.5.26) is said to be a 'gauge superfield' (or 'Yang–Mills superfield'). The transformations (3.5.25, 26) will be called 'supergauge' transformations. The equation (3.5.26) shows that V' is Hermitian, $(V')^+ = V'$. Next, due to equation (3.5.26) and the Baker–Hausdorff formula (2.1.62), V' has the form $V' = V'^I T^I$. To see the explicit connection between V'^I and V^I, let us consider infinitesimal gauge transformations (3.5.26).

3.5.4. *Infinitesimal gauge transformations*

In the infinitesimal case, equation (3.5.26) takes the form

$$\delta \mathrm{e}^{2V} = \mathrm{e}^{2(V+\delta V)} - \mathrm{e}^{2V} = \mathrm{i}\bar{\Lambda}\mathrm{e}^{2V} - \mathrm{i}\mathrm{e}^{2V}\Lambda. \tag{3.5.28}$$

Our goal is to determine δV in terms of V, Λ and $\bar{\Lambda}$. For this purpose we use the identity:

$$e^{A+\varepsilon B} = e^{A}\left(1 + \int_0^1 d\tau\, e^{-\tau A}\varepsilon B\, e^{\tau A}\right) \tag{3.5.29}$$

where A and B are arbitrary operators and ε is an infinitesimal parameter. This identity can be easily proved with the help of the auxiliary function

$$K(\tau, \varepsilon) = e^{-\tau A} e^{\tau(A+\varepsilon B)}$$

which satisfies the equation

$$\frac{dK(\tau,\varepsilon)}{d\tau} = e^{-\tau A}\varepsilon B\, e^{\tau A} K(\tau,\varepsilon) \approx e^{-\tau A}\varepsilon B\, e^{\tau A}.$$

Defining operator L_A by the rule

$$L_A B \equiv [A, B]$$

the relation (3.5.29) can be rewritten as

$$e^{A+\varepsilon B} = e^{A}\{1 + (L_A)^{-1}(1 - e^{-L_A})\varepsilon B\}. \tag{3.5.30}$$

Now, making use of equations (3.5.28, 30) gives

$$(L_V)^{-1}(1 - e^{-2L_V})\delta V = ie^{-2L_V}\bar{\Lambda} - i\Lambda$$

and acting on both sides by $L_V e^{L_V}$

$$\begin{aligned} 2\sinh(L_V)\delta V &= iL_V(e^{-L_V}\bar{\Lambda} - e^{L_V}\Lambda) \\ &= iL_V\cosh(L_V)(\bar{\Lambda} - \Lambda) - iL_V\sinh(L_V)(\bar{\Lambda} + \Lambda). \end{aligned}$$

This leads to

$$\begin{aligned} \delta V &= -\frac{i}{2}L_V(\bar{\Lambda} + \Lambda) + \frac{i}{2}L_V\coth(L_V)(\bar{\Lambda} - \Lambda) \\ &= \frac{i}{2}(\bar{\Lambda} - \Lambda) - \frac{i}{2}[V, \bar{\Lambda} + \Lambda] + \tfrac{i}{6}[V, [V, \bar{\Lambda} - \Lambda]] + O(V^4). \end{aligned} \tag{3.5.31}$$

In the Abelian case, this transformation law reduces to expression (3.4.10).

We see that the supergauge transformations are highly nonlinear. However, since $\delta V = \frac{i}{2}(\bar{\Lambda} - \Lambda) + O(V)$, these transformations can be used to impose the Wess–Zumino gauge. Those supergauge transformations which preserve the Wess–Zumino gauge are described by the parameters

$$\Lambda(x, \theta, \bar{\theta}) = e^{i\theta\sigma^a\bar{\theta}\partial_a}\xi(x) \qquad \xi = \xi^I T^I = \xi^+ \tag{3.5.32}$$

They act on V under the Wess–Zumino gauge condition by the rule

$$\delta V = \frac{i}{2}(\bar{\Lambda} - \Lambda) - \frac{i}{2}[V, \bar{\Lambda} + \Lambda] \tag{3.5.33a}$$

or, in components,

$$\begin{aligned}\delta V_a{}^I &= \partial_a \xi^I - \xi^K f^{KJI} V_a{}^J \\ \delta \lambda_\alpha{}^I &= -\xi^K f^{KJI} \lambda_\alpha{}^J \\ \delta D^I &= -\xi^K f^{KJI} D^J.\end{aligned} \tag{3.5.33b}$$

As for the matter multiplet Φ, defined by equation (3.5.25), the transformation (3.5.25) acts on the corresponding component fields

$$A^i = \Phi^i| \qquad \psi_\alpha{}^i = \mathrm{D}_\alpha \Phi^i| \qquad F^i = -\frac{1}{4}\mathrm{D}^2\Phi^i| \tag{3.5.34}$$

as follows

$$\begin{gathered}\delta A^i = \mathrm{i}\xi^K (T^K)^i{}_j A^j \qquad \delta\psi_\alpha{}^i = \mathrm{i}\xi^K (T^K)^i{}_j \psi_\alpha{}^j \\ \delta F^i = \mathrm{i}\xi^K (T^K)^i{}_j F^j.\end{gathered} \tag{3.5.35}$$

Therefore, we recover ordinary Yang–Mills transformations. In accordance with equation (3.5.33*b*), V_a^I is a gauge Yang–Mills field. Its superpartners λ_α^I and D^I transform, due to (3.5.33*b*), in the adjoint representation of the gauge group.

3.5.5. *Super Yang–Mills action*

We proceed by finding the gauge invariant action for the gauge superfield. Let us introduce the following Lie-algebra-valued spinor superfields

$$\begin{aligned}W_\alpha &= -\frac{1}{8}\bar{\mathrm{D}}^2(\mathrm{e}^{-2V} D_\alpha \mathrm{e}^{2V}) = W_\alpha{}^I T^I \\ \bar{W}_{\dot\alpha} &= \frac{1}{8}\mathrm{D}^2(\mathrm{e}^{2V}\bar{\mathrm{D}}_{\dot\alpha}\mathrm{e}^{-2V}) = (W_\alpha)^+\end{aligned} \tag{3.5.36}$$

and study some their properties. First, in the Abelian limit W_α reduces to the form (3.4.3). Secondly, W_α is chiral and $\bar{W}_{\dot\alpha}$ is antichiral,

$$\bar{\mathrm{D}}_{\dot\alpha} W_\alpha = 0 \qquad \mathrm{D}_\alpha \bar{W}_{\dot\alpha} = 0. \tag{3.5.37}$$

Finally, W_α and $\bar{W}_{\dot\alpha}$ change covariantly under the supergauge transformations (3.5.26). Explicitly, we have

$$\begin{aligned}W'_\alpha &= -\frac{1}{8}\bar{\mathrm{D}}^2(\mathrm{e}^{\mathrm{i}\Lambda}\mathrm{e}^{-2V}\mathrm{e}^{-\mathrm{i}\bar\Lambda}\mathrm{D}_\alpha \mathrm{e}^{\mathrm{i}\bar\Lambda}\mathrm{e}^{2V}\mathrm{e}^{-\mathrm{i}\Lambda}) = -\frac{1}{8}\bar{\mathrm{D}}^2(\mathrm{e}^{\mathrm{i}\Lambda}\mathrm{e}^{-2V}\mathrm{D}_\alpha(\mathrm{e}^{2V}\mathrm{e}^{-\mathrm{i}\Lambda})) \\ &= \mathrm{e}^{\mathrm{i}\Lambda} W_\alpha \mathrm{e}^{-\mathrm{i}\Lambda} - \frac{1}{8}\mathrm{e}^{\mathrm{i}\Lambda}\bar{\mathrm{D}}^2\mathrm{D}_\alpha \mathrm{e}^{-\mathrm{i}\Lambda}.\end{aligned}$$

Since $\bar{\mathrm{D}}^2\mathrm{D}_\alpha\eta = 0$ for every chiral superfield η, we have the result

$$W'_\alpha = e^{i\Lambda} W_\alpha e^{-i\Lambda}$$
$$\bar{W}'_{\dot{\alpha}} = e^{i\bar{\Lambda}} \bar{W}_{\dot{\alpha}} e^{-i\bar{\Lambda}}. \tag{3.5.38}$$

W_α and $\bar{W}_{\dot{\alpha}}$ will be called the 'Yang–Mills superfield strengths'.

The relation (3.5.37) tells us that the (anti)chiral scalar superfields

$$\mathrm{tr}(\bar{W}_{\dot{\alpha}} \bar{W}^{\dot{\alpha}}) \qquad \mathrm{tr}(W^\alpha W_\alpha)$$

where tr denotes matrix trace, are invariant with respect to the supergauge transformations. Therefore, the superfunctional

$$S = \frac{1}{4g^2} \int d^6 z \, \mathrm{tr}\,(W^\alpha W_\alpha) + \text{c.c.} \tag{3.5.39}$$

where g is a coupling constant, can be taken in the role of the gauge invariant action. Above we have supposed that the generators T^I are normalized by

$$\mathrm{tr}\,(T^I T^J) = \delta^{IJ}. \tag{3.5.40}$$

Let us investigate the action at the component level. Imposing the Wess–Zumino gauge, W_α takes the form

$$W_\alpha = -\frac{1}{4}\bar{D}^2 D_\alpha V + \frac{1}{4}\bar{D}^2([V, D_\alpha V])$$

To calculate the component content of W_α, we point out that the Wess–Zumino gauge is characterized by the conditions (3.4.16). Then, one readily obtains

$$\begin{aligned} W_\alpha| &= \lambda_\alpha \\ D^\alpha W_\alpha| &= -4D \\ D_{(\alpha} W_{\beta)}| &= 2iG_{\alpha\beta} \\ -\frac{1}{4}D^2 W_\alpha| &= -i\nabla_{\alpha\dot{\alpha}}\bar{\lambda}^{\dot{\alpha}}. \end{aligned} \tag{3.5.41}$$

Here we have introduced the Yang–Mills field strength:

$$\begin{aligned} G_{ab} &= \partial_a V_b - \partial_b V_a - i[V_a, V_b] \\ G_{\alpha\beta} &= \frac{1}{2}(\sigma^{ab})_{\alpha\beta} G_{ab} = -\frac{1}{2}\partial_{(\alpha\dot{\alpha}} V_{\beta)}{}^{\dot{\alpha}} + \frac{i}{4}[V_{\alpha\dot{\alpha}}, V_\beta{}^{\dot{\alpha}}] \end{aligned} \tag{3.5.42}$$

and the Yang–Mills covariant derivatives

$$\begin{aligned} \nabla_a B^i &= \partial_a B^i - iV_a{}^I (T^I)^i{}_j B^j \\ \nabla_a \lambda^\alpha &= \partial_a \lambda^\alpha - i[V_a, \lambda^\alpha]. \end{aligned} \tag{3.5.43}$$

Based on the above expressions, one finds

$$\frac{1}{4}\int d^6z\, \mathrm{tr}\,(W^\alpha W_\alpha) = \int d^4x\, \mathrm{tr}\left\{-\frac{1}{2}G^{\alpha\beta}G_{\alpha\beta} - \frac{1}{2}\mathrm{i}\lambda^\alpha\nabla_{\alpha\dot\alpha}\bar\lambda^{\dot\alpha} + D^2\right\}.$$

Now, it is helpful to recall the identity

$$\int d^4x\, \mathrm{tr}\,(G^{\alpha\beta}G_{\alpha\beta}) = \int d^4x\, \mathrm{tr}\,(\bar G^{\dot\alpha\dot\beta}\bar G_{\dot\alpha\dot\beta})$$

modulo total derivatives. This means, in particular, that the relation

$$\int d^6z\, \mathrm{tr}\,(W^\alpha W_\alpha) = \int d^6\bar z\, \mathrm{tr}\,(\bar W_{\dot\alpha}\bar W^{\dot\alpha}) \tag{3.5.44}$$

holds in the Wess–Zumino gauge and, due to gauge invariance, in the general case. Finally, we can write the component action

$$S = \frac{1}{g^2}\int d^4x\, \mathrm{tr}\left\{-\frac{1}{4}G^{ab}G_{ab} - \mathrm{i}\lambda\sigma^a\nabla_a\bar\lambda + 2D^2\right\}. \tag{3.5.45}$$

As is seen, we have obtained the supersymmetric extension of the Yang–Mills action.

3.5.6. *Super Yang–Mills models*

Gauge invariant coupling of matter to a Yang–Mills superfield is described by an action superfunctional of the general form

$$S = \int d^8z \bar\Phi_i (e^{2V})^i{}_j \Phi^j + \left\{\int d^6z \mathscr{L}_c(\Phi^i) + \text{c.c.}\right\} + \frac{1}{2g^2}\int d^6z\, \mathrm{tr}\,(W^\alpha W_\alpha). \tag{3.5.46}$$

Here the chiral superpotential $\mathscr{L}_c(\Phi^i)$ should be a group-invariant function,

$$\frac{\partial\mathscr{L}_c}{\partial\Phi^i}(T^I)^i{}_j\Phi^j = 0. \tag{3.5.47}$$

In particular, if the chiral superpotential is of the form

$$\mathscr{L}_c(\Phi^i) = \frac{1}{2}m_{ij}\Phi^i\Phi^j + \frac{1}{3!}\lambda_{ijk}\Phi^i\Phi^j\Phi^k \tag{3.5.48}$$

the coupling constants m_{ij} and λ_{ijk} should be invariant tensors of the group under consideration.

It is instructive to analyse the component form of the action. Defining the component fields of Φ^i by rule (3.5.34) and imposing the Wess–Zumino gauge condition (3.4.16) on V, one arrives at the action functional

$$S = \int d^4x \Big\{ -\nabla^m \bar{A}_i \nabla_m A^i + 2D^I \bar{A}_i (T^I)^i{}_j A^j - \frac{i}{2} \psi^i \sigma^a \nabla_a \bar{\psi}_i - \bar{\psi}_i \bar{\lambda}^I (T^I)^i{}_j A^j$$

$$- \bar{A}_i (T^I)^i{}_j \lambda^I \psi^j + \bar{F}_i F^i + \left(\left[F^i \mathscr{V}_i(A) - \frac{1}{4} \frac{\partial^2 \mathscr{L}_c(A)}{\partial A^i \partial A^j} \psi^i \psi^j \right] + \text{c.c.} \right) \qquad (3.5.49)$$

$$- \frac{1}{g^2} \left(\tfrac{1}{4} G^{abI} G_{ab}{}^I + i \lambda^I \sigma^a \nabla_a \bar{\lambda}^I - 2D^I D^I \right) \Big\}$$

where $\mathscr{V}_i(A)$ is given by equation (3.2.24). After eliminating the auxiliary fields F^i, $\bar{F}_i$ and D^I, the scalar fields have the non-negative potential

$$U(A, \bar{A}) = \sum_i |\mathscr{V}_i(A)| + \frac{g^2}{2} \sum_I (\bar{A} T^I A)^2. \qquad (3.5.50)$$

3.5.7. Real representations

In conclusion, we would like to discuss one important question similar to that arising when constructing supersymmetric spinor electrodynamics. Namely, let $\mathbf{\Psi}_D{}^\mu$ be a multiplet of Dirac spinors transforming in some (complex) unitary representation of the gauge group,

$$\delta \mathbf{\Psi}_D{}^\mu = i\xi^I (\mathscr{T}^I)^\mu{}_\nu \mathbf{\Psi}_D{}^\nu \qquad (\mathscr{T}^I)^+ = \mathscr{T}^I \qquad (3.5.51)$$

coupled to the Yang–Mills field in the standard fashion:

$$S = -\frac{1}{2} \int d^4x \bar{\mathbf{\Psi}}_{D\mu} (i\gamma^a \nabla_a + m) \mathbf{\Psi}_D{}^\mu. \qquad (3.5.52)$$

Suppose, one would like to find a super Yang–Mills model of the form (3.5.46), in which, after reduction to components, the spinor fields from Φ^i and $\bar{\Phi}_j$ can be combined somehow into Dirac spinors such that the spinor part of the action (3.5.46) coincides with expression (3.5.52). The question is the following: what is a group representation by which the multiplet $\{\Phi^i\}$ must transform?

Let us decompose $\mathbf{\Psi}_D{}^\mu$ into a combination of two Majorana spinors:

$$\mathbf{\Psi}_D{}^\mu = \mathbf{\Psi}^{\underline{\mu}} + i\mathbf{\Psi}^{\bar{\mu}}$$

$$\mathbf{\Psi}^{\underline{\mu}} = \begin{pmatrix} \psi_\alpha{}^{\underline{\mu}} \\ \bar{\psi}^{\dot{\alpha}\underline{\mu}} \end{pmatrix} \qquad \mathbf{\Psi}^{\bar{\mu}} = \begin{pmatrix} \psi_\alpha{}^{\bar{\mu}} \\ \bar{\psi}^{\dot{\alpha}\bar{\mu}} \end{pmatrix}. \qquad (3.5.53)$$

Introducing a multiplet of Majorana spinors in the manner

$$(\mathbf{\Psi}^i) = \begin{pmatrix} \mathbf{\Psi}^{\underline{\mu}} \\ \mathbf{\Psi}^{\bar{\mu}} \end{pmatrix} \qquad \mathbf{\Psi}^i = \begin{pmatrix} \psi_\alpha{}^i \\ \bar{\psi}^{\dot{\alpha}i} \end{pmatrix} \qquad (3.5.54)$$

it transforms by the real representation:

$$\delta \mathbf{\Psi}^i = i\xi^I (T^I)^i{}_j \mathbf{\Psi}^j$$

$$T^I = \begin{pmatrix} i\,\mathrm{Im}\mathscr{T}^I & i\,\mathrm{Re}\mathscr{T}^I \\ -i\,\mathrm{Re}\mathscr{T}^I & i\,\mathrm{Im}\mathscr{T}^I \end{pmatrix}. \qquad (3.5.55)$$

Since $\mathscr{T}^I$ are Hermitian, the matrices $\mathrm{Re}\mathscr{T}^I$ are symmetric and the matrices $\mathrm{Im}\mathscr{T}^I$ are antisymmetric. As a result, T^I are antisymmetric Hermitian matrices,

$$(T^I)^+ = T^I \qquad (T^I)^T = -T^I. \tag{3.5.56}$$

Now, it is clear how to resolve the problem raised above. First, one must take a set of chiral scalar superfields $\{\Phi^i\} = \{\Phi^{\underline{\mu}}, \Phi^{\bar{\mu}}\}$ transforming by the introduced real representation. Secondly, the multiplet (3.5.54) should be identified with the spinor component fields of Φ^i and $\bar{\Phi}^i$ by the rule

$$\Psi^i = \begin{pmatrix} D_\alpha \Phi^i| \\ \bar{D}^{\dot{\alpha}} \bar{\Phi}^i| \end{pmatrix}.$$

Then, the gauge invariant matter action

$$S = \int d^8z \bar{\Phi} e^{2V} \Phi + \frac{m}{2}\left\{\int d^6z \Phi^T\Phi + \text{c.c.}\right\} \tag{3.5.57}$$

leads to the desired dynamics of spinor fields. Therefore, in order to have a complex representation in components, one must use its real realization in superfields. When taking a complex representation for superfields, there is room for chiral fermions at the component level and, hence, for chiral anomalies in the quantum theory.

In what follows, we will often restrict ourselves to the consideration of real group representations for a description of matter chiral superfield multiplets. As applied to the gauge superfield $V = V^I T^I$, this requirement means

$$V^+ = V \qquad V^T = -V. \tag{3.5.58}$$

3.6. Geometric approach to super Yang–Mills theories

In this section we intend to give a clear geometric interpretation of supergauge transformations and the gauge superfield. This will be done after reminding ourselves of some relevant mathematical notions.

3.6.1. Complex and c-number shells of compact Lie groups

In formulating super Yang–Mills theories, there arise several groups: the gauge group G, its complexification G^c and their c-number shells $\hat{G}$ and $\hat{G}^c$. We are going to describe these objects.

We begin with an excursion into the theory of Lie groups. Let G be a compact connected n-dimensional Lie group, and $\mathscr{G}$ be the corresponding Lie algebra. The exponential mapping exp: $\mathscr{G} \to G$ turns out to cover G (due to compactness and connectness), therefore real coordinates $\eta^I (I = 1, \ldots, n)$ in $\mathscr{G}$ with respect to a basis $\{E^I\}$,

$$X \in \mathscr{G} \qquad X = \mathrm{i}\eta^I E^I \qquad \eta^I \in \mathbb{R} \tag{3.6.1}$$

can be used to parametrize G,

$$g(\eta) = \exp(\mathrm{i}\eta^I E^I). \tag{3.6.2}$$

It is worth noting that the map exp is on to but not one-to-one. Generators E^I are assumed to satisfy the commutation relations

$$[E^I, E^J] = \mathrm{i} f^{IJK} E^K \tag{3.6.3}$$

with the structure constants f^{IJK} being real and totally antisymmetric.

Introduce the complex shell $\mathscr{G}^c$ of the algebra $\mathscr{G}$. The $\mathscr{G}^c$ is a complex n-dimensional Lie algebra arising from $\mathscr{G}$ by taking the coordinates to be complex,

$$X \in \mathscr{G}^c \qquad X = \mathrm{i}\omega^I E^I \qquad \omega^I \in \mathbb{C}. \tag{3.6.4}$$

Supplying $\mathscr{G}^c$ with the operation of complex conjugation (involution)

$$X \to X^* = \mathrm{i}\bar{\omega}^I E^I \qquad \bar{\omega}^I = (\omega^I)^* \tag{3.6.5}$$

makes it possible to identify the initial algebra $\mathscr{G}$ with the real subset of $\mathscr{G}^c$. The algebra $\mathscr{G}^c$ can be equivalently treated as a real $2n$-dimensional Lie algebra, denoted by $(\mathscr{G}^c)_\mathbb{R}$, with elements

$$\mathrm{i}(\mathrm{Re}\,\omega^I E^I + \mathrm{Im}\,\omega^I S^I)$$

and generators E^I and $S^I (= \mathrm{i}E^I)$ under the commutation relations

$$[E^I, E^J] = \mathrm{i} f^{IJK} E^K \qquad [E^I, S^J] = \mathrm{i} f^{IJK} S^K \qquad [S^I, S^J] = -\mathrm{i} f^{IJK} E^K.$$

Then, the conjugation map $*$: $\mathscr{G}^c \to \mathscr{G}^c$ acts as an automorphism on $(\mathscr{G}^c)_\mathbb{R}$. This automorphism has a transparent meaning in the basis

$$\begin{gathered} E^I_{(\pm)} = \frac{1}{2}(E^I \mp \mathrm{i}S^I) \\ [E^I_{(\pm)}, E^J_{(\pm)}] = \mathrm{i} f^{IJK} E_{(\pm)} \qquad [E^I_{(+)}, E^J_{(-)}] = 0 \end{gathered} \tag{3.6.6}$$

with respect to which every element of $(\mathscr{G}^c)_\mathbb{R}$ is written

$$\mathrm{i}(\omega^I E^I_{(+)} + \bar{\omega}^I E^I_{(-)}). \tag{3.6.7}$$

It is seen that $(\mathscr{G}^c)_\mathbb{R}$ has the direct sum structure, $(\mathscr{G}^c)_\mathbb{R} = \mathscr{G}^c_{(+)} \oplus \mathscr{G}^c_{(-)}$, and the complex conjugation maps $\mathscr{G}^c_{(+)}$ on to $\mathscr{G}^c_{(-)}$ and vice versa.

There exists a unique (up to isomorphism) complex connected Lie group $\mathscr{G}^c$ satisfying two requirements:

(1) $\mathscr{G}^c$ is the Lie algebra of G^c;
(2) considering G^c as a real $2n$-dimensional Lie group, G is isomorphic to the subgroup in G^c corresponding to the subalgebra $\mathscr{G}$ in $\mathscr{G}^c$.

G^c is said to be 'the complex shell' of G (for example, if $G = SO(n)$, then $G^c = SO(n, \mathbb{C})$; if $G = SU(n)$, then $G^c = SL(n, \mathbb{C})$). In a neighbourhood of the unit of G^c, every element can be represented in the exponential form

$$g(\omega) = \exp(\mathrm{i}\omega^I E^I) \tag{3.6.8}$$

The variables ω^I play the role of local complex coordinates on G^c, at least in a neighbourhood of the unit. When treating G^c as a real Lie group, it is parametrized by variables ω^I and their conjugated $\bar{\omega}^I$ as follows

$$\begin{aligned} g(\omega, \bar{\omega}) &= \exp[\mathrm{i}(\omega^I E^I_{(+)} + \bar{\omega}^I E^I_{(-)})] \\ &= \exp(\mathrm{i}\omega^I E^I_{(+)}) \exp(\mathrm{i}\bar{\omega}^I E^I_{(-)}). \end{aligned} \tag{3.6.9}$$

Therefore, G^c possesses the direct product structure, $G^c = G^c_{(+)} \times G^c_{(-)}$, where the subgroups $G^c_{(+)}$ and $G^c_{(-)}$ are mutually conjugated. The initial group can be identified with the diagonal in this product. Note that G^c is non-compact, in contrast to G.

Now, let us discuss c-number shells of the groups G and G^c. Starting from the group manifold G, we embed it in the supermanifold $\hat{G}$ using the rule described in subsection 3.3.4. $\hat{G}$ is naturally endowed with the structure of a real super Lie group which is induced by the Lie group structure on G (see also the discussion of supergroups in subsection 2.1.6). Elements of $\hat{G}$ are parametrized as in equation (3.6.2), but in contrast to G, the local coordinates η^I on $\hat{G}$ are real c-numbers. Similarly, the group manifold G^c is embedded in the complex supermanifold $\hat{G}^c$, which becomes a complex super Lie group after inducing the group structure from G^c. Elements of $\hat{G}^c$ can be represented in the form (3.6.8), where ω^I are complex c-numbers.

Every unitary finite-dimensional representation $\mathscr{T}$ of the compact group G with Hermitian generators $T^I (= \mathrm{d}\mathscr{T}(E^I))$,

$$\mathscr{T}(g(\eta)) = \mathrm{e}^{\mathrm{i}\eta^I T^I} \qquad (T^I)^+ = T^I \tag{3.6.10}$$

can be continued to a holomorphic representation

$$\mathscr{T}(g(\omega)) = \mathrm{e}^{\mathrm{i}\omega^I T^I} \tag{3.6.11}$$

or antiholomorphic representation

$$\bar{\mathscr{T}}(g(\omega)) = \mathrm{e}^{-\mathrm{i}\bar{\omega}^I (T^I)^*} \tag{3.6.12}$$

of the complex group G^c. Choosing in equations (3.6.10–12) parameters $\eta^I(\omega^I)$ to be real (complex) c-numbers, one obtains representations of $\hat{G}$ and $\hat{G}^c$, respectively.

3.6.2. K-supergroup and Λ-supergroup

The basic symmetry principle underlying Yang–Mills theories is invariance under local (or gauge) group transformations. Every gauge transformation is specified by a smooth mapping of Minkowski space into the group G,

$$\mathscr{K}: \mathbb{R}^4 \to G. \tag{3.6.13}$$

In classical field theory, when fields are treated as Grassmann-algebra-valued functions on a space–time, it is more correct to consider mappings from Minkowski space into the c-number shell $\hat{G}$ of G,

$$\mathscr{K}: \mathbb{R}^4 \to \hat{G} \tag{3.6.14}$$

in the role of gauge transformations. The set of all such mappings forms an infinite-dimensional group (with respect to the multiplication law $(\mathscr{K}_1 \cdot \mathscr{K}_2)(x) = \mathscr{K}_1(x)\mathscr{K}_2(x)$ called the local $\hat{G}$-group. It is the gauge group of classical Yang–Mills theory.

Trying to keep a literal analogy with ordinary Yang–Mills theories, it may seem natural that in the supersymmetric case the role of gauge group should be played by the supergroup consisting of supersmooth mappings from the superspace $\mathbb{R}^{4|4}$ in $\hat{G}$,

$$\mathscr{K}: \mathbb{R}^{4|4} \to \hat{G}. \tag{3.6.15}$$

This group will be called 'the superlocal' G-group (or the 'K-supergroup'). Every element of the K-group is determined by a multiplet of real superfields $K^I(x, \theta, \bar{\theta})$:

$$\mathscr{K}(x, \theta, \bar{\theta}) = \exp(\mathrm{i}K^I(x, \theta, \bar{\theta})E^I) \qquad \bar{K}^I(z) = K^I(z). \tag{3.6.16}$$

Unfortunately, the K-supergroup has not arisen in the previous section. The point is that in the supersymmetric case there exists another natural candidate for the role of gauge group.

Recall that the complex truncated superspace $\mathbb{C}_s^{4|2}$, rather than the real superspace $\mathbb{R}^{4|4}$, is the fundamental object in supersymmetry; $\mathbb{R}^{4|4}$ is embedded into $\mathbb{C}_s^{4|2}$ in the super Poincaré invariant fashion. Therefore, it seems reasonble that $\mathbb{C}_s^{4|2}$ should be taken in the role of the arena upon which rigid $\hat{G}$-invariance is promoted to a local invariance. Furthermore, localization should be done in a way consistent with the complex structure on $\mathbb{C}_s^{4|2}$. This means that we change $\hat{G}$ to its complexification $\hat{G}^c$ and consider superholomorphic mappings of the form

$$\lambda: \mathbb{C}_s^{4|2} \to \hat{G}^c. \tag{3.6.17}$$

Every such mapping is determined by a multiplet of holomorphic superfields $\Lambda^I(y, \theta)$,

$$\lambda(y, \theta) = \exp(i\Lambda^I(y, \theta)E^I) \tag{3.6.18}$$

which become chiral when restricted from $\mathbb{C}_s^{4|2}$ to $\mathbb{R}^{4|4}$ (setting $y^a = x^a + i\theta\sigma^a\bar{\theta}$)

$$\lambda(z) = \exp(i\Lambda^I(z)E^I) \qquad \bar{D}_{\dot{\alpha}}\Lambda^I(z) = 0. \tag{3.6.19}$$

Given two superholomorphic mappings λ_1 and λ_2, their product $(\lambda_1 \cdot \lambda_2)(y, \theta) = \lambda_1(y, \theta)\lambda_2(y, \theta)$ is also superholomorphic. Hence, the set of all mappings (3.6.17) forms a supergroup, which will be called the 'Λ-supergroup'. The Λ-supergroup is the invariance group of the super Yang–Mills action (3.5.46).

3.6.3. *Gauge superfield*

Now, we are in a position to demonstrate, following E. Ivanov, the geometrical origin of the gauge superfield.

Consider the complex supermanifold

$$\mathcal{M}^{4+n|2} = \mathbb{C}^{4|2} \times \hat{G}^c \tag{3.6.20}$$

which can be parametrized by the complex coordinates y^a, θ^α and ω^I corresponding to $\mathbb{C}^{4|2}$ and $\hat{G}^c$, respectively. The Λ-supergroup naturally acts on $\mathcal{M}^{4+n|2}$ by means of left nonlinear shifts of the coordinates ω^I:

$$y'^a = y^a \qquad \theta'^\alpha = \theta^\alpha \qquad g(\omega') = \exp(i\Lambda^I(y, \theta)E^I)g(\omega). \tag{3.6.21}$$

In addition, one can realize the K-supergroup as a transformation group of $\mathcal{M}^{4+n|2}$ acting by right nonlinear shifts of the coordinates ω^I:

$$\begin{gathered} y'^a = y^a \qquad \theta'^\alpha = \theta^\alpha \\ g(\omega') = g(\omega)\exp\left(-iK^I\left(\frac{1}{2}y + \frac{1}{2}\bar{y}, \theta, \bar{\theta}\right)E^I\right). \end{gathered} \tag{3.6.22}$$

It is clear that the Λ-transformations commute with the K-transformations.

Next, we embed the real superspace $\mathbb{R}^{4|4}$ into $\mathcal{M}^{4+n|2}$ as follows:

$$\begin{aligned} y^a &= x^a + i\theta\sigma^a\bar{\theta} \\ \omega^I &= \Omega^I(x, \theta, \bar{\theta}) \end{aligned} \tag{3.6.23}$$

where Ω^I is the complex scalar superfield which determines the embedding. In accordance with equations (3.6.21, 22), the Λ- and K-supergroups act on Ω^I according to the law

$$\exp(i\Omega'(z)) = \exp(i\Lambda(z))\exp(i\Omega(z))\exp(-iK(z)) \tag{3.6.24}$$

where we have introduced the shorthand notation: $\Omega = \Omega^I E^I$ and so on.

Let us discuss the transformation law (3.6.24). We decompose Ω^I into real and imaginary parts. Obviously, real superfields $\mathrm{Re}\,\Omega^I$ and $\mathrm{Im}\,\Omega^I$ live in the subgroup $\hat{G}$ of $\hat{G}^c$ and the coset space $\hat{G}^c/\hat{G}$, respectively. Since the

K-supergroup acts on $\exp(\mathrm{i}\Omega)$ by arbitrary local $\hat{G}$-transformations, one can gauge away $\mathrm{Re}\,\Omega^I$ by a proper choice of K-transformation. Therefore, the superfields $\mathrm{Re}\,\Omega^I(z)$ are purely gauge degrees of freedom. Imposing the gauge fixing condition

$$\mathrm{Re}\,\Omega^I(z) = 0 \tag{3.6.25}$$

we work only with the Λ-supergroup which acts on the multiplet of real scalar superfields

$$V^I(z) \equiv \mathrm{Im}\,\Omega^I(z) \tag{3.6.26}$$

spanning the coset space $\hat{G}^c/\hat{G}$. Hence, the K-invariance is auxiliary. Its meaning is that $\hat{G}$-directions of the embedding (3.6.23) are never observed dynamically.

To find the transformation law of $V^I(z)$, it is worth noting that, to preserve the gauge (3.6.25), every Λ-transformation should be supplemented by a Λ-dependent K-transformation:

$$\exp(-V') = \exp(\mathrm{i}\Lambda)\exp(-V)\exp(-\mathrm{i}K[V,\Lambda]) \tag{3.6.27}$$

where $K^I[V,\Lambda]$ are rather complicated functions. Instead of finding their explicit form, we take another course. In the gauge (3.6.25), we have

$$\exp(-2V) = \exp(\mathrm{i}\Omega)\exp(-\mathrm{i}\bar{\Omega}). \tag{3.6.28}$$

In the general case, this relation defines a set of real scalar superfields $V^I(z)$ (the reality of V^I follows from the fact that the conjugation map $f\colon g(\omega) \rightarrow g(\bar{\omega})$ represents an antiholomorphic isomorphism of $\hat{G}^c$ on itself, $f(g(\omega_1)g(\omega_2)) = g(\bar{\omega}_1)g(\bar{\omega}_2)$. Equivalently, in the expression of (3.6.28) the imaginary part of Ω is excluded by construction. Next, taking the conjugate of (3.6.24) gives

$$\exp(\mathrm{i}\bar{\Omega}') = \exp(\mathrm{i}\bar{\Lambda})\exp(\mathrm{i}\bar{\Omega})\exp(-\mathrm{i}K). \tag{3.6.29}$$

Finally, from the relations (3.6.24, 28, 29) one obtains

$$\exp(-2V') = \exp(\mathrm{i}\Lambda)\exp(-2V)\exp(-\mathrm{i}\bar{\Lambda}). \tag{3.6.30}$$

As may be seen, we have recovered the transformation law of the gauge superfield (3.5.26). Therefore, the Yang–Mills superfield can be identified with the multiplet of real scalar superfields $V^I(z)$ residing in the coset space $\hat{G}^c/\hat{G}$ and transforming by the law (3.6.30) under the Λ-supergroup (clearly, this interpretation differs drastically from the well-known geometrical interpretation of ordinary Yang–Mills fields). It is worth emphasizing, however, that in the geometrical approach developed the real multiplet V^I represents the gauge-fixed version of the initial complex multiplet Ω^I. This observation serves as the foundation for the following point of view of super Yang–Mills theory:

(1) the full invariance group is the direct product of the Λ- and K-supergroups;

(2) the true dynamical object is the multiplet of complex scalar superfields $\Omega^I(z)$ residing in $\hat{G}^c$ and transforming by the law (3.6.24) under the invariance group.

Taking the above point of view, one can see a close analogy between formulations of two quite different theories: the super Yang–Mills theory and gravity. Objects corresponding to each other in the super Yang–Mills theory and in gravity, are given below:

$$\Omega^I(z) \leftrightarrow e_m{}^a(x)$$

$$V^I(z) \leftrightarrow g_{mn}(x)$$

$$K\text{-supergroup} \leftrightarrow \text{local Lorentz group}$$

$$\Lambda\text{-supergroup} \leftrightarrow \text{general coordinate transformation group}$$

$$\mathrm{e}^{-2V} = \mathrm{e}^{\mathrm{i}\bar{\Omega}}\mathrm{e}^{-\mathrm{i}\Omega} \leftrightarrow g_{mn} = e_m{}^a e_n{}^b \eta_{ab}.$$

In fact, the analogy can be promoted further. As is known, the vierbein $e_m{}^a(x)$ plays the role of a converter which turns curved-space tensors (transforming under the general coordinate transformation group) into tangent-space ones (transforming under the local Lorentz group). The vierbein's superanalogue $\Omega^I(z)$ is seen to have a similar meaning. Consider a multiplet of chiral scalar superfields $\Phi^i(z)$ transforming according to the law (3.5.25) under the Λ-supergroup. We convert this multiplet into the nonchiral one

$$\Phi^i_{(r)} = (\mathrm{e}^{-\mathrm{i}\Omega})^i{}_j\Phi^j \qquad \Omega = \Omega^I T^I. \tag{3.6.31}$$

In accordance with equation (3.6.24), $\Phi_{(r)}$ changes according to the law

$$\Phi'_{(r)} = \mathrm{e}^{\mathrm{i}K}\Phi_{(r)} \qquad K = K^I T^I \tag{3.6.32}$$

under the K-supergroup and is unchanged with respect to the Λ-supergroup. Next, converting the antichiral multiplet $\bar{\Phi}_i$ in the manner

$$\bar{\Phi}_{(r)i} = \bar{\Phi}_j(\mathrm{e}^{\mathrm{i}\bar{\Omega}})^j{}_i \qquad \bar{\Omega} = \bar{\Omega}^I T^I \tag{3.6.33}$$

one arrives, due to equations (3.5.25) and (3.6.29), at the transformation law

$$\bar{\Phi}'_{(r)} = \bar{\Phi}_{(r)}\mathrm{e}^{-\mathrm{i}K}. \tag{3.6.34}$$

Surprisingly, it follows from equation (3.6.28) that the action (3.5.27) can be rewritten in the free form

$$S = \int \mathrm{d}^8 z \bar{\Phi}_{(r)}\Phi_{(r)} \tag{3.6.35}$$

as if the interaction were absent.

3.6.4. Gauge covariant derivatives

Gauge superfields Ω^I and $\bar{\Omega}^I$ can be used to construct gauge covariant derivatives—that is, first-order differential operators preserving the transformation laws of matter multiplets. In the rest of this section, we restrict ourselves to consideration of real G-representations in which the generators satisfy equation (3.5.56).

Introduce some matter multiplet $\Phi^i(z)$ and its conjugate $\bar{\Phi}^i(z)$ (the $\bar{\Phi}$ index is given in the upper position in accordance with the reality of the G-representation), transforming under the Λ-supergroup by the laws

$$\Phi \to \Phi' = e^{i\Lambda}\Phi \tag{3.6.36a}$$

$$\bar{\Phi}^T \to \bar{\Phi}^{T\,\prime} = e^{i\bar{\Lambda}}\bar{\Phi}^T \tag{3.6.36b}$$

Taking a covariant derivative of Φ breaks the initial transformation law,

$$D_A\Phi \to D_A\Phi' \neq e^{i\Lambda}D_A\Phi.$$

To preserve the transformation laws, it is worth changing D_A by new sets of operators $\mathscr{D}_A^{(+)}$ and $\mathscr{D}_A^{(-)}$ of the form

$$\begin{gathered} \mathscr{D}_A^{(\pm)} = (\mathscr{D}_a^{(\pm)}, \mathscr{D}_\alpha^{(\pm)}, \bar{\mathscr{D}}^{\dot{\alpha}(\pm)}) \\ \mathscr{D}_A^{(\pm)} = D_A + i\Gamma_A^{(\pm)} \qquad \Gamma_A^{(\pm)} = \Gamma_A^{I(\pm)}(z)T^I \end{gathered} \tag{3.6.37}$$

with $\Gamma_A^{I(\pm)}(z)$ being 'superfield connections'. One demands that the Λ-supergroup act on $\mathscr{D}_A^{(\pm)}$ in such a way that the following relations

$$\mathscr{D}_A^{(+)}\Phi \to \mathscr{D}_A'^{(+)}\Phi' = e^{i\Lambda}\mathscr{D}_A^{(+)}\Phi \tag{3.6.38a}$$

$$\mathscr{D}_A^{(-)}\bar{\Phi}^T \to \mathscr{D}_A'^{(-)}\bar{\Phi}^{T} = e^{i\bar{\Lambda}}\mathscr{D}_A^{(-)}\bar{\Phi}^T \tag{3.6.38b}$$

hold. This means

$$\mathscr{D}_A^{(+)} \to \mathscr{D}_A'^{(+)} = e^{i\Lambda}\mathscr{D}_A^{(+)}e^{-i\Lambda} \tag{3.6.39a}$$

$$\mathscr{D}_A^{(-)} \to \mathscr{D}_A'^{(-)} = e^{i\bar{\Lambda}}\mathscr{D}_A^{(-)}e^{-i\bar{\Lambda}}. \tag{3.6.39b}$$

The operators $\mathscr{D}_A^{(\pm)}$ are said to be 'gauge covariant derivatives'.

A particular solution of the above problem, based on the use of the gauge superfield V^I, is given as follows:

$$\begin{gathered} \mathscr{D}_\alpha^{(+)} = e^{-2V}D_\alpha e^{2V} \qquad \bar{\mathscr{D}}_{\dot{\alpha}}^{(+)} = \bar{D}_{\dot{\alpha}} \\ \mathscr{D}_{\alpha\dot{\alpha}}^{(+)} \equiv (\sigma^a)_{\alpha\dot{\alpha}}\mathscr{D}_a^{(+)} = \frac{i}{2}\{\mathscr{D}_\alpha^{(+)}, \bar{\mathscr{D}}_{\dot{\alpha}}^{(+)}\} \end{gathered} \tag{3.6.40a}$$

$$\begin{gathered} \mathscr{D}_\alpha^{(-)} = D_\alpha \qquad \bar{\mathscr{D}}_{\dot{\alpha}}^{(-)} = e^{2V}\bar{D}_{\dot{\alpha}}e^{-2V} \\ \mathscr{D}_{\alpha\dot{\alpha}}^{(-)} \equiv (\sigma^a)_{\alpha\dot{\alpha}}\mathscr{D}_a^{(-)} = \frac{i}{2}\{\mathscr{D}_\alpha^{(-)}, \bar{\mathscr{D}}_{\dot{\alpha}}^{(-)}\} \end{gathered} \tag{3.6.40b}$$

Since $D_\alpha\bar{\Lambda} = \bar{D}_{\dot{\alpha}}\Lambda = 0$, making use of equation (3.6.30) leads to

$$\mathscr{D}_\alpha'^{(+)} = e^{-2V'}D_\alpha e^{2V'} = e^{i\Lambda}e^{-2V}e^{-i\bar{\Lambda}}D_\alpha e^{i\bar{\Lambda}}e^{2V}e^{-i\Lambda}$$
$$= e^{i\Lambda}e^{-2V}D_\alpha e^{2V}e^{-i\Lambda} = e^{i\Lambda}\mathscr{D}_\alpha^{(+)}e^{-i\Lambda}$$
$$\bar{\mathscr{D}}_{\dot{\alpha}}'^{(+)} = \bar{D}_{\dot{\alpha}} = e^{i\Lambda}\bar{D}_{\dot{\alpha}}e^{-i\Lambda} = e^{i\Lambda}\bar{\mathscr{D}}_{\dot{\alpha}}^{(+)}e^{-i\Lambda}.$$

Note, if Φ is chiral, then it is gauge covariantly chiral with respect to $\mathscr{D}^{(+)}$. It is worth noting also that the two sets $\mathscr{D}_A^{(+)}$ and $\mathscr{D}_A^{(-)}$ are connected by

$$\mathscr{D}_A^{(+)} = e^{-2V}\mathscr{D}_A^{(-)}e^{2V}. \tag{3.6.41}$$

In the general case, the gauge covariant derivatives (3.6.37) satisfy the algebra

$$\begin{aligned}[][\mathscr{D}_A^{(\pm)}, \mathscr{D}_B^{(\pm)}\} &= \mathscr{D}_A^{(\pm)}\mathscr{D}_B^{(\pm)} - (-1)^{\varepsilon_A\varepsilon_B}\mathscr{D}_B^{(\pm)}\mathscr{D}_A^{(\pm)} \\ &= T_{AB}{}^C\mathscr{D}_C^{(\pm)} + iF_{AB}^{(\pm)} \\ F_{AB}^{(\pm)} &= F_{AB}^{I(\pm)}(z)T^I. \end{aligned} \tag{3.6.42}$$

Here $T_{AB}{}^C$ is the supertorsion tensor defined in terms of the covariant derivatives D_A:

$$[D_A, D_B\} = T_{AB}{}^C D_C.$$

$F_{AB}^{(\pm)}$ is said to be the 'supercurvature tensor'. It is expressed in terms of the connection superfields by the rule:

$$F_{AB}^{(\pm)} = D_A\Gamma_B^{(\pm)} - (-1)^{\varepsilon_A\varepsilon_B}D_B\Gamma_A^{(\pm)} + i[\Gamma_A^{(\pm)}, \Gamma_B^{(\pm)}\}. \tag{3.6.43}$$

It is evident that $F_{AB}^{(\pm)}$ transforms covariantly under the Λ-supergroup (in the adjoint G-representation):

$$F_{AB}^{(+)} \rightarrow F_{AB}'^{(+)} = e^{i\Lambda}F_{AB}^{(+)}e^{-i\Lambda} \tag{3.6.44a}$$

$$F_{AB}^{(-)} \rightarrow F_{AB}'^{(-)} = e^{i\bar{\Lambda}}F_{AB}^{(-)}e^{-i\bar{\Lambda}}. \tag{3.6.44b}$$

In the case of the derivatives (3.6.40), the algebra (3.6.42) is

$$\begin{aligned} \{\mathscr{D}_\alpha^{(\pm)}, \bar{\mathscr{D}}_{\dot{\alpha}}^{(\pm)}\} &= -2i\mathscr{D}_{\alpha\dot{\alpha}}^{(\pm)} \\ \{\mathscr{D}_\alpha^{(\pm)}, \mathscr{D}_\beta^{(\pm)}\} &= \{\bar{\mathscr{D}}_{\dot{\alpha}}^{(\pm)}, \bar{\mathscr{D}}_{\dot{\beta}}^{(\pm)}\} = 0 \\ [\bar{\mathscr{D}}_{\dot{\alpha}}^{(\pm)}, \mathscr{D}_{\beta\dot{\beta}}^{(\pm)}] &= 2i\varepsilon_{\dot{\alpha}\dot{\beta}}W_\beta^{(\pm)} \\ [\mathscr{D}_\alpha^{(\pm)}, \mathscr{D}_{\beta\dot{\beta}}^{(\pm)}] &= 2i\varepsilon_{\alpha\beta}\bar{W}_{\dot{\beta}}^{(\pm)} \\ [\mathscr{D}_{\alpha\dot{\alpha}}^{(\pm)}, \mathscr{D}_{\beta\dot{\beta}}^{(\pm)}] &= -\varepsilon_{\alpha\beta}(\bar{\mathscr{D}}_{\dot{\alpha}}^{(\pm)}\bar{W}_{\dot{\beta}}^{(\pm)}) - \varepsilon_{\dot{\alpha}\dot{\beta}}(\mathscr{D}_\alpha^{(\pm)}W_\beta^{(\pm)}). \end{aligned} \tag{3.6.45}$$

Here the supercurvature components $W_\alpha^{(\pm)}$ and $\bar{W}_{\dot{\alpha}}^{(\pm)}$ are connected with the superfield strengths (3.5.36) by the formulae

$$W_\alpha^{(+)} = W_\alpha \qquad \bar{W}_{\dot{\alpha}}^{(+)} = e^{-2V}\bar{W}_{\dot{\alpha}}e^{2V} \tag{3.6.46a}$$

$$W_\alpha^{(-)} = e^{2V} W_\alpha e^{-2V} \qquad \bar{W}_{\dot{\alpha}}^{(-)} = \bar{W}_{\dot{\alpha}}. \tag{3.6.46b}$$

Let us comment on the derivation of the relations (3.6.45). As a result of (3.6.40*a*), we have

$$\Gamma_\alpha^{(+)} = -\mathrm{i}(e^{-2V} D_\alpha e^{2V}) \qquad \Gamma_{\dot{\alpha}}^{(+)} = 0 \Rightarrow \Gamma_{\alpha\dot{\alpha}}^{(+)} = \frac{\mathrm{i}}{2} \bar{D}_{\dot{\alpha}} \Gamma_\alpha^{(+)}.$$

This immediately leads to

$$[\bar{\mathscr{D}}_{\dot{\alpha}}^{(+)}, \bar{\mathscr{D}}_{\beta\dot{\beta}}^{(+)}] = 2\mathrm{i}\varepsilon_{\dot{\alpha}\dot{\beta}} W_\beta.$$

Similarly, one readily obtains

$$[\mathscr{D}_\alpha^{(-)}, \mathscr{D}_{\beta\dot{\beta}}^{(-)}] = 2\mathrm{i}\varepsilon_{\alpha\beta} \bar{W}_{\dot{\beta}}.$$

Then, making use of the relation (3.6.41), one obtains the commutators $[\mathscr{D}_\alpha^{(+)}, \mathscr{D}_{\beta\dot{\beta}}^{(+)}]$ and $[\bar{\mathscr{D}}_{\dot{\alpha}}^{(-)}, \bar{\mathscr{D}}_{\beta\dot{\beta}}^{(-)}]$. Finally, a commutator of vector derivatives can be calculated by the rule

$$[\mathscr{D}_{\alpha\dot{\alpha}}^{(\pm)}, \mathscr{D}_{\beta\dot{\beta}}^{(\pm)}] = \frac{\mathrm{i}}{2}[\{\mathscr{D}_\alpha^{(\pm)}, \mathscr{D}_{\dot{\alpha}}^{(\pm)}\}, \mathscr{D}_{\beta\dot{\beta}}^{(\pm)}]$$

$$= \frac{\mathrm{i}}{2}\{[\mathscr{D}_\alpha^{(\pm)}, \mathscr{D}_{\beta\dot{\beta}}^{(\pm)}], \mathscr{D}_{\dot{\alpha}}^{(\pm)}\} + \frac{\mathrm{i}}{2}\{\mathscr{D}_\alpha^{(\pm)}, [\mathscr{D}_{\dot{\alpha}}^{(\pm)}, \mathscr{D}_{\beta\dot{\beta}}^{(\pm)}]\}.$$

It is worth pointing out that this expression should be antisymmetric with respect to the double transposition $\alpha \leftrightarrow \beta$ and $\dot{\alpha} \leftrightarrow \dot{\beta}$. On the other hand, from expressions (3.6.45) we see that

$$[\mathscr{D}_{\alpha\dot{\alpha}}^{(\pm)}, \mathscr{D}_{\beta\dot{\beta}}^{(\pm)}] = -\varepsilon_{\alpha\beta}(\bar{\mathscr{D}}_{(\dot{\alpha}}^{(\pm)} \bar{W}_{\dot{\beta})}^{(\pm)}) - \varepsilon_{\dot{\alpha}\dot{\beta}}(\mathscr{D}_{(\alpha}^{(\pm)} W_{\beta)}^{(\pm)})$$
$$+ \frac{1}{2}\varepsilon_{\alpha\beta}\varepsilon_{\dot{\alpha}\dot{\beta}}(\bar{\mathscr{D}}_{\dot{\gamma}}^{(\pm)} \bar{W}^{\dot{\gamma}(\pm)} - \mathscr{D}^{\gamma(\pm)} W_\gamma^{(\pm)}).$$

This requires the identity

$$\bar{\mathscr{D}}_{\dot{\alpha}}^{(\pm)} \bar{W}^{\dot{\alpha}(\pm)} = \mathscr{D}^{\alpha(\pm)} W_\alpha^{(\pm)}. \tag{3.6.47}$$

The reader can check this explicitly.

Remark. Since the superfield strengths W_α and $\bar{W}_{\dot{\alpha}}$ are chiral and antichiral, respectively, it follows from the relations (3.6.40, 41, 46) that the supercurvatures $W_\alpha^{(\pm)}$ and $\bar{W}_{\dot{\alpha}}^{(\pm)}$ are gauge covariantly chiral and antichiral, respectively,

$$\bar{\mathscr{D}}_{\dot{\alpha}}^{(\pm)} W_\alpha^{(\pm)} = 0 \qquad \mathscr{D}_\alpha^{(\pm)} \bar{W}_{\dot{\alpha}}^{(\pm)} = 0. \tag{3.6.48}$$

In conclusion, we introduce gauge covariant derivatives $\mathscr{D}_A$ for a matter multiplet transforming under the K-supergroup by the law (3.6.32). They are given in the form

$$\mathscr{D}_A = e^{-\mathrm{i}\Omega} \mathscr{D}_A^{(+)} e^{\mathrm{i}\Omega} = e^{-\mathrm{i}\bar{\Omega}} \mathscr{D}_A^{(-)} e^{\mathrm{i}\bar{\Omega}}$$
$$\mathscr{D}_\alpha = e^{-\mathrm{i}\bar{\Omega}} D_\alpha e^{\mathrm{i}\bar{\Omega}} \qquad \bar{\mathscr{D}}_{\dot{\alpha}} = e^{-\mathrm{i}\Omega} \bar{D}_{\dot{\alpha}} e^{\mathrm{i}\Omega} \tag{3.6.49}$$

$$\mathscr{D}_{\alpha\dot{\alpha}} = \frac{\mathrm{i}}{2}\{\mathscr{D}_\alpha, \bar{\mathscr{D}}_{\dot{\alpha}}\}$$

where we have used equations (3.6.28). In accordance with equation (3, 6, 24, 29), the operators $\mathscr{D}_A$ are characterized by the transformation law

$$\mathscr{D}_A \to \mathscr{D}'_A = \mathrm{e}^{\mathrm{i}K}\mathscr{D}_A\mathrm{e}^{-\mathrm{i}K}. \tag{3.6.50}$$

3.6.5. Matter equations of motion

Gauge covariant derivatives arise in dynamical equations for matter multiplets. For example, consider the dynamical system (3.5.57). It is characterized by the equation of motion

$$\begin{gathered} -\frac{1}{4}\mathrm{D}^2\mathrm{e}^{2V}\Phi + m\bar{\Phi}^{\mathrm{T}} = 0 \\ -\frac{1}{4}\bar{\mathrm{D}}^2\mathrm{e}^{-2V}\bar{\Phi}^{\mathrm{T}} + m\Phi = 0 \end{gathered} \tag{3.6.51}$$

From this equation one obtains

$$\begin{gathered} \frac{1}{16}\bar{\mathrm{D}}^2\mathrm{e}^{-2V}\mathrm{D}^2\mathrm{e}^{2V}\Phi - m^2\Phi = 0 \\ \frac{1}{16}\mathrm{D}^2\mathrm{e}^{2V}\bar{\mathrm{D}}^2\mathrm{e}^{-2V}\bar{\Phi}^{\mathrm{T}} - m^2\bar{\Phi}^{\mathrm{T}} = 0. \end{gathered}$$

Recalling definition (3.6.40), we can rewrite the equations in terms of the gauge-covariant derivatives:

$$\begin{gathered} \frac{1}{16}\bar{\mathscr{D}}^{2(+)}\mathscr{D}^{2(+)}\Phi - m^2\Phi = 0 \\ \frac{1}{16}\mathscr{D}^{2(-)}\bar{\mathscr{D}}^{2(-)}\bar{\Phi}^{\mathrm{T}} - m^2\bar{\Phi}^{\mathrm{T}} = 0. \end{gathered} \tag{3.6.52}$$

Now, since $\bar{\mathscr{D}}^{(+)}_{\dot{\alpha}}\Phi = 0$ and $\mathscr{D}^{(-)}_{\alpha}\bar{\Phi}^{\mathrm{T}} = 0$, these fourth-order equations can be easily reduced, after some work with the algebra (3.6.45), to the following second-order equations:

$$\begin{gathered} \left(\mathscr{D}^{a(+)}\mathscr{D}^{(+)}_a - W^\alpha\mathscr{D}^{(+)}_\alpha - \frac{1}{2}(\mathscr{D}^{\alpha(+)}W_\alpha) - m^2\right)\Phi = 0 \\ \left(\mathscr{D}^{a(-)}\mathscr{D}^{(-)}_a + \bar{W}_{\dot{\alpha}}\bar{\mathscr{D}}^{\dot{\alpha}(-)} + \frac{1}{2}(\bar{\mathscr{D}}^{(-)}_{\dot{\alpha}}\bar{W}^{\dot{\alpha}}) - m^2\right)\bar{\Phi}^{\mathrm{T}} = 0. \end{gathered} \tag{3.6.53}$$

Evidently, the equations obtained are explicitly gauge covariant.

3.6.6. Gauge superfield dynamical equations

In conclusion, we would like to obtain dynamical equations corresponding to the pure supergauge action (3.5.39). To this end, it is worth discussing how to do functional variation with respect to V in an optimum way.

Let $F[V]$ be a functional of the gauge superfield. Making an infinitesimal displacement in V changes the functional by

$$\delta F[V] = F[V+\delta V] - F[V] = \mathrm{tr}\int \mathrm{d}^8 z\, \delta V(z) \frac{\delta F[V]}{\delta V(z)}. \tag{3.6.54}$$

If $F[V]$ is gauge invariant, $F[V'] = F[V]$, V' being as in equation (3.5.26), then the equation

$$\frac{\delta F[V]}{\delta V} = 0 \tag{3.6.55}$$

proves to be gauge covariant,

$$\frac{\delta F[V]}{\delta V} = 0 \Rightarrow \frac{\delta F[V']}{\delta V'} = 0.$$

However, the variational derivatives $\delta F[V]/\delta V$ transform into a nonlinear representation of the Λ-supergroup, since the variation δV changes in a complicated nonlinear way:

$$\begin{gathered} \mathrm{e}^{2V'} = \mathrm{e}^{\mathrm{i}\bar{\Lambda}} \mathrm{e}^{2V} \mathrm{e}^{-\mathrm{i}\Lambda} \Rightarrow \\ \mathrm{e}^{2(V' + \delta V')} = \mathrm{e}^{\mathrm{i}\bar{\Lambda}} \mathrm{e}^{2(V+\delta V)} \mathrm{e}^{-\mathrm{i}\Lambda}. \end{gathered} \tag{3.6.56}$$

Instead of considering δV, one can introduce covariantized variations $\Delta^{(\pm)}V$ defined in the manner

$$\begin{aligned} 2\Delta^{(+)}V &\equiv \mathrm{e}^{-2V}\delta \mathrm{e}^{2V} = \mathrm{e}^{-2V}\mathrm{e}^{2(V+\delta V)} - 1 \\ 2\Delta^{(-)}V &\equiv \delta \mathrm{e}^{2V}\mathrm{e}^{-2V} = \mathrm{e}^{2(V+\delta V)}\mathrm{e}^{-2V} - 1 \end{aligned} \tag{3.6.57}$$

which transforms covariantly:

$$\begin{aligned} \Delta^{(+)}V' &= \mathrm{e}^{\mathrm{i}\Lambda}\Delta^{(+)}V\mathrm{e}^{-\mathrm{i}\Lambda} \\ \Delta^{(-)}V' &= \mathrm{e}^{\mathrm{i}\bar{\Lambda}}\Delta^{(-)}V\mathrm{e}^{-\mathrm{i}\bar{\Lambda}} \end{aligned} \tag{3.6.58}$$

in accordance with (3.6.56). The initial variation δV is in a one-to-one correspondence with $\Delta^{(+)}V$ or $\Delta^{(-)}V$, since it follows from (3.6.57) that

$$\begin{aligned} \Delta^{(+)}V &= \frac{1-\exp(-2L_V)}{2L_V}\delta V = \delta V + \ldots \\ \delta V &= L_V \coth(L_V)\Delta^{(+)}V + L_V\Delta^{(+)}V \end{aligned} \tag{3.6.59}$$

and similarly for $\Delta^{(-)}V$. Expressing δV in (3.6.54) in terms of $\Delta^{(+)}V$ or $\Delta^{(-)}V$, one obtains

$$\delta F[V] = \mathrm{tr}\int \mathrm{d}^8 z \Delta^{(\pm)} V(z) \frac{\Delta^{(\pm)} F[V]}{\Delta^{(\pm)} V(z)}. \tag{3.6.60}$$

If $F[V]$ is gauge invariant, then $\Delta^{(\pm)}F[V]/\Delta^{(\pm)}V$ transforms covariantly:

$$\begin{aligned} \frac{\Delta^{(+)}F[V']}{\Delta^{(+)}V'} &= \mathrm{e}^{\mathrm{i}\Lambda} \frac{\Delta^{(+)}F[V]}{\Delta^{(+)}V} \mathrm{e}^{-\mathrm{i}\Lambda} \\ \frac{\Delta^{(-)}F[V']}{\Delta^{(-)}V'} &= \mathrm{e}^{\mathrm{i}\bar{\Lambda}} \frac{\Delta^{(-)}F[V]}{\Delta^{(-)}V} \mathrm{e}^{-\mathrm{i}\bar{\Lambda}}. \end{aligned} \tag{3.6.61}$$

Finally, equation (3.6.55) is satisfied if and only if the covariantized forms

$$\frac{\Delta^{(\pm)}F[V]}{\Delta^{(\pm)}V} = 0 \tag{3.6.62}$$

hold.

Now, let us vary the action (3.5.39). Taking into account equations (3.5.36, 44), we obtain

$$\begin{aligned} \delta S &= \frac{1}{g^2}\mathrm{tr}\int \mathrm{d}^6 z \delta W^\alpha W_\alpha = \frac{1}{2g^2}\mathrm{tr}\int \mathrm{d}^8 z \delta(\mathrm{e}^{-2V}\mathrm{D}^\alpha \mathrm{e}^{2V}) W_\alpha \\ &= -\frac{1}{g^2}\mathrm{tr}\int \mathrm{d}^8 z \Delta^{(+)}V[\mathrm{D}^\alpha W_\alpha + \{\mathrm{i}\Gamma^{\alpha(+)}, W_\alpha\}] \\ &= -\frac{1}{g^2}\mathrm{tr}\int \mathrm{d}^8 z \Delta^{(+)}V(\mathscr{D}^{\alpha(+)}W_\alpha). \end{aligned}$$

Therefore, the dynamical equations are

$$\mathscr{D}^{\alpha(+)}W_\alpha = 0. \tag{3.6.63}$$

Using the identity (3.6.47), they can be written in the form

$$\mathscr{D}^{\alpha(\pm)}W^{(\pm)}_\alpha = \bar{\mathscr{D}}^{(\pm)}_{\dot\alpha}\bar{W}^{\dot\alpha(\pm)} = 0 \tag{3.6.64}$$

representing the non-Abelian generalization of the vector multiplet equation of motion (3.4.11).

3.7. Classically equivalent theories

It has been shown in Sections 2.6 and 2.7 that different superfields can be used to describe the same super Poincaré representation. In the present section we would like to acquaint the reader with some supersymmetric field theories, each of which is characterized by classical dynamics identical to that arising in one of the previously considered models but realized in terms of other superfields.

3.7.1. Massive chiral spinor superfield model

Consider the theory of a chiral spinor superfield $\chi_\alpha(z)$, $\bar{D}_{\dot\alpha}\chi_\alpha = 0$, and its conjugate $\bar{\chi}_{\dot\alpha}$ with action superfunctional

$$S[\chi, \bar{\chi}] = -\int d^8z G^2 - m^2 \int d^6z \chi^2 - m^2 \int d^6\bar{z} \bar{\chi}^2$$
$$G = \frac{1}{2}(D^\alpha \chi_\alpha + \bar{D}_{\dot\alpha}\bar{\chi}^{\dot\alpha}) = \bar{G}. \tag{3.7.1}$$

Here the real scalar superfield G is linear,

$$D^2 G = \bar{D}^2 G = 0 \tag{3.7.2}$$

in accordance with the identities (2.5.30*f*).To find the dynamical equations of the model, one can use the variational rule

$$\frac{\delta \chi^\beta(z')}{\delta \chi^\alpha(z)} = -\frac{1}{4}\delta_\alpha{}^\beta \bar{D}^2 \delta^8(z - z'). \tag{3.7.3}$$

This gives

$$\frac{1}{8}\bar{D}^2 D_\alpha G + m^2 \chi_\alpha = 0 \qquad \frac{1}{8} D^2 \bar{D}_{\dot\alpha} G + m^2 \bar{\chi}_{\dot\alpha} = 0. \tag{3.7.4}$$

Since $D^\alpha \bar{D}^2 D_\alpha G = \bar{D}_{\dot\alpha} D^2 \bar{D}^{\dot\alpha} G$, one obtains the on-shell constraint

$$D^\alpha \chi_\alpha = \bar{D}_{\dot\alpha}\bar{\chi}^{\dot\alpha}. \tag{3.7.5}$$

After this, the system (3.7.4) can be rewritten in the form

$$(\Box - m^2)\chi_\alpha = 0$$
$$\bar{\chi}_{\dot\alpha} = -\frac{i}{4m^2}\partial_{\alpha\dot\alpha} D^2 \chi^\alpha \tag{3.7.6}$$

where we have used the chirality of χ_α ($\bar{D}^2 D^2 \chi_\alpha = 16\Box\chi_\alpha$).

Recall that when imposing the first equation (3.7.6) only, the (anti)chiral superfields $\bar{\chi}_{\dot\alpha}$ and χ_α describe two independent superspin-$\frac{1}{2}$ states (see Section 2.6). Adding the second equation (3.7.6) expresses $\bar{\chi}_{\dot\alpha}$ in terms of χ_α. As a result, the theory (3.7.1) describes the irreducible superspin-$\frac{1}{2}$ representation, similarly to the massive vector multiplet model (3.4.1). Therefore the theories (3.4.1) and (3.7.1) are dynamically equivalent.

The equivalence of the two theories (3.4.1) and (3.7.1) can also be seen as follows. Let us introduce an auxiliary model with action

$$S[V, \chi, \bar{\chi}] = \int d^8z(V^2 - 2VG) - m^2 \int d^6z \chi^2 - m^2 \int d^6\bar{z} \bar{\chi}^2 \tag{3.7.7}$$

where $V(z)$ is a real scalar superfield. We vary the action with respect to V, obtaining the equation

$$V - G = 0. \tag{3.7.8}$$

Next, taking the variation with respect to χ_α gives

$$\frac{1}{8}\bar{D}^2 D_\alpha V + m^2 \chi_\alpha = 0. \tag{3.7.9}$$

Due to equation (3.7.8), V can be eliminated from the action. This produces exactly the action (3.7.1),

$$S[V, \chi, \bar{\chi}]|_{\delta S/\delta V = 0} = S[\chi, \bar{\chi}].$$

Similarly, using equations (3.7.9), one can eliminate χ_α from the action (3.7.7). This procedure (up to a factor) the action (3.4.1),

$$S[V, \chi, \bar{\chi}]|_{\delta S/\delta \chi = 0} = \frac{1}{m^2} S[V].$$

Hence, the action (3.7.7) plays the role of a converter from the massive vector multiplet model to the massive spinor superfield model and vice versa. On these grounds the models (3.4.1) and (3.4.2) are said to be dual to each other.

3.7.2. Massless chiral spinor superfield model

Let us set the mass parameter in (3.7.1) to zero. The action

$$S[\chi, \bar{\chi}] = -\frac{1}{4}\int d^8 z (D^\alpha \chi_\alpha + \bar{D}_{\dot{\alpha}} \bar{\chi}^{\dot{\alpha}})^2 \tag{3.7.10}$$

turns out to be invariant under the gauge transformations

$$\delta\chi_\alpha = i\bar{D}^2 D_\alpha K \qquad \delta\bar{\chi}_{\dot{\alpha}} = -iD^2\bar{D}_{\dot{\alpha}} K \qquad K = \bar{K} \tag{3.7.11}$$

where K is an arbitrary real scalar superfield. What are the superhelicity states that the model (3.7.10) describes?

Note that the linear superfield G (3.7.1) is gauge invariant. The equations of motion derived from the action (3.7.10)

$$\bar{D}^2 D_\alpha G = 0 \qquad D^2 \bar{D}_{\dot{\alpha}} G = 0 \tag{3.7.12}$$

together with the identities (3.7.2) show that G is an on-shell massless superfield (see Section 2.7). G has no definite superhelicity. However, one can construct from G two secondary superfields: antichiral $\bar{\lambda}_\alpha = D_\alpha G$ and chiral $\lambda_{\dot{\alpha}} = \bar{D}_{\dot{\alpha}} G$. These superfields prove also to be on-shell massless, in accordance with the analysis of Section 2.7. Their superhelicities are

$$\kappa(\bar{\lambda}_\alpha) = 0 \qquad \kappa(\lambda_{\dot{\alpha}}) = -1/2. \tag{3.7.13}$$

Recall that the same superhelicity content arose in the massless chiral scalar superfield model (3.2.5). Therefore, the theories (3.2.5) and (3.7.10) are dynamically equivalent.

In fact, the two theories (3.2.5) and (3.7.10) are dual to each other, because they can be obtained from the action

$$S[V, \chi, \bar{\chi}] = \int d^8z\{V^2 + V(D^\alpha\chi_\alpha + \bar{D}_{\dot{\alpha}}\bar{\chi}^{\dot{\alpha}})\}. \tag{3.7.14}$$

Indeed, eliminating V with the help of its equation of motion, one obtains the action (3.7.10). On the other hand, varying $S[V, \chi, \bar{\chi}]$ with respect to χ_α and $\bar{\chi}_{\dot{\alpha}}$ gives

$$\bar{D}^2 D_\alpha V = D^2\bar{D}_{\dot{\alpha}} V = 0 \tag{3.7.15}$$

which can be resolved as follows

$$V = \Phi + \bar{\Phi} \qquad \bar{D}_{\dot{\alpha}}\Phi = 0 \tag{3.7.16}$$

where Φ is an arbitrary chiral scalar superfield. Substituting this solution into (3.7.14) and using the observation that

$$\begin{gathered} \bar{D}_{\dot{\alpha}}\Phi = 0 \qquad \bar{D}^2 G = 0 \Rightarrow \\ \int d^8z\Phi G = -\frac{1}{4}\int d^6z\Phi\bar{D}^2 G = 0 \end{gathered} \tag{3.7.17}$$

one arrives at the action (3.2.5).

In conclusion, let us discuss the component structure of the model (3.7.10). Following the prescription of Section 2.8, we define the component fields of χ_α by

$$\begin{aligned} \lambda_\alpha(x) &= \chi_\alpha| \qquad & B_{\alpha\beta}(x) &= D_{(\alpha}\chi_{\beta)}| \\ A(x) &= D^\alpha\chi_\alpha| \qquad & \eta_\alpha(x) &= -\frac{1}{4}D^2\chi_\alpha|. \end{aligned} \tag{3.7.18}$$

From the symmetric spin-tensors $B_{\alpha\beta}$ and $\bar{B}_{\dot{\alpha}\dot{\beta}}$ one can construct an antisymmetric real tensor B_{ab} by the standard rule

$$B_{ab} = (\sigma_{ab})_{\alpha\beta}B^{\alpha\beta} - (\tilde{\sigma}_{ab})_{\dot{\alpha}\dot{\beta}}\bar{B}^{\dot{\alpha}\dot{\beta}}.$$

Next, the component fields of the linear superfield G are expressed in terms of the fields (3.7.18) and their conjugates in the manner:

$$\begin{aligned} G| &= \frac{1}{2}(\bar{A} + A) \equiv \varphi(x) \\ D_\alpha G| &= \eta_\alpha - i\partial_{\alpha\dot{\alpha}}\bar{\lambda}^{\dot{\alpha}} \equiv \psi_\alpha(x) \\ \frac{1}{2}[D_\alpha, \bar{D}_{\dot{\alpha}}]G| &= -i\partial^\beta{}_{\dot{\alpha}}B_{\alpha\beta} + i\partial_\alpha{}^{\dot{\beta}}\bar{B}_{\dot{\alpha}\dot{\beta}} = -(\sigma^a)_{\alpha\dot{\alpha}}L_a(B). \end{aligned} \tag{3.7.19}$$

Here $L_a(B)$ is the antisymmetric tensor field strength defined in (1.7.43). It is worth noting that the component fields (3.7.19) are invariant under the gauge transformations (3.7.11), due to the gauge invariance of G. Finally, the component form of the action (3.7.10) can be calculated by the rule:

$$S = \int \mathrm{d}^4x L$$

$$-L = \frac{1}{16} \mathrm{D}^2\bar{\mathrm{D}}^2 G^2 \Big| = \frac{1}{8} \mathrm{D}^2(\bar{\mathrm{D}}_{\dot{\alpha}} G \bar{\mathrm{D}}^{\dot{\alpha}} G) \Big| = \frac{1}{4} \mathrm{D}_\alpha \bar{\mathrm{D}}_{\dot{\alpha}} G | \mathrm{D}^\alpha \bar{\mathrm{D}}^{\dot{\alpha}} G | + \frac{1}{4} \mathrm{D}^2 \bar{\mathrm{D}}_{\dot{\alpha}} G | \bar{\mathrm{D}}^{\dot{\alpha}} G |$$

$$= \frac{1}{4}\left(\frac{1}{2}[\mathrm{D}_\alpha, \bar{\mathrm{D}}_{\dot{\alpha}}]G| - \mathrm{i}\partial_{\alpha\dot{\alpha}} G|\right)^2 - \mathrm{i}\partial_{\alpha\dot{\alpha}} \mathrm{D}^\alpha G | \bar{\mathrm{D}}^{\dot{\alpha}} G |$$

which gives

$$S = \int \mathrm{d}^4x \left\{ \frac{1}{2} L^a(B) L_a(B) - \frac{1}{2} \partial^a \varphi \partial_a \varphi - \mathrm{i}\psi \sigma^a \partial_a \bar{\psi} \right\}. \tag{3.7.20}$$

This action represents the supersymmetric extension of the antisymmetric tensor field theory (1.7.43).

In components, the equivalence of the theories (3.2.5) and (3.7.10) is trivial. The former theory describes two massless scalar fields and a massless Majorana spinor field. The latter theory describes three massless fields: one scalar, one antisymmetric tensor and a Majorana spinor. But free massless scalar and antisymmtric tensor fields are dynamically equivalent (see subsection 1.8.4).

It is worth pointing out that in Minkowski space bosonic field theories can never be equivalent to fermionic ones, due to the spin-statistics theorem. The situation is different in superspace $\mathbb{R}^{4|4}$. Here some bosonic superfield theory may appear to be equivalent to a theory of fermionic superfields and this fact does not contradict the spin-statistics theorem.

3.7.3. Superfield redefinitions

In both theories considered above, their equivalence to theories realized in terms of different superfields, occurred due to the existence of proper duality transformations. Another way of obtaining dynamically equivalent models is superfield redefinitions. For example, one can always express a chiral scalar superfield through a real scalar superfield by the rule

$$\Phi = -\frac{1}{4}\bar{\mathrm{D}}^2 V \qquad V = \bar{V}. \tag{3.7.21}$$

However, since V has more degrees of freedom than Φ, this redefinition is accompanied by the appearance of gauge invariance

$$V \rightarrow V' = V + \frac{1}{2}(\mathrm{D}^\alpha \psi_\alpha + \bar{\mathrm{D}}_{\dot{\alpha}} \bar{\psi}^{\dot{\alpha}}) \qquad \bar{\mathrm{D}}_{\dot{\alpha}} \psi_\alpha = 0. \tag{3.7.22}$$

Here $\psi_\alpha(z)$ is an arbitrary chiral spinor superfield. Defining component fields of V as in expression (2.8.19), one can see that the fields A, ψ_α, $\bar{\psi}_{\dot{\alpha}}$ and the transverse part of V_a are purely gauge degrees of freedom. The other fields in expression (2.8.19) are related to the component fields of Φ as follows:

$$\Phi| = F(x) \qquad \mathrm{D}_\alpha\Phi| = \lambda_\alpha(x)$$
$$-\frac{1}{4}\mathrm{D}^2\Phi| = D(x) + \frac{\mathrm{i}}{2}\partial^a V_a(x). \tag{3.7.23}$$

Remark. The chiral parameters ψ_α in expressions (3.7.22) are defined modulo transformations

$$\psi_\alpha \rightarrow \psi'_\alpha = \psi_\alpha + \mathrm{i}\bar{\mathrm{D}}^2\mathrm{D}_\alpha K \qquad K = \bar{K}.$$

After making the redefinition (3.7.21), every chiral scalar superfield theory transforms into a real scalar superfield theory. For instance, the model (3.2.2) is changed to

$$S[V] = \frac{1}{2}\int \mathrm{d}^8z V\left(\frac{1}{16}\{\mathrm{D}^2, \bar{\mathrm{D}}^2\} - \frac{m}{4}(\mathrm{D}^2 + \bar{\mathrm{D}}^2)\right)V. \tag{3.7.24}$$

Similarly to the initial model (3.2.2), its transformed version (3.7.24) turns out to describe one superspin-0 state. Surprisingly, it is sufficient to change the mass term in the manner

$$\frac{m}{4}(\mathrm{D}^2 + \bar{\mathrm{D}}^2) \rightarrow m^2$$

and the resultant action

$$S[V] = \frac{1}{2}\int \mathrm{d}^8z V\left(\frac{1}{16}\{\mathrm{D}^2, \bar{\mathrm{D}}^2\} - m^2\right)V \tag{3.7.25}$$

will describe two superspin-0 states. The reader can easily check this statement.

3.8. Non-minimal scalar multiplet

One of the historic traditions popular among the practitioners of supersymmetry is to use term 'scalar multiplet' for the set of dynamical (super)fields in a supersymmetric theory realizing the superspin-0 representation on the mass shell. More precisely, this term is literally used for those supermultiplets which involve only physical scalar fields to describe the spin-0 states. The supersymmetric model, which has been discussed in subsection 3.7.2, also describes the massless superspin-0 multiplet, but with one of the physical scalars being traded for a gauge antisymmetric tensor off shell. That is why this model is often called the 'tensor multiplet'. It is scalar multiplets which realize the matter sector in supersymmetric phenomenological models.

Another tradition is that the scalar multiplets are identified with chiral scalar superfields and their conjugates. It is quite natural to ask whether such an identification is obligatory or, to a greater extent, a matter of habit and convenience? Certainly, the chiral scalar superfield provides the simplest and very convenient description of a scalar multiplet ('minimal scalar multiplet'). But there exist alternative (or variant) scalar multiplet realizations like that given in subsection 3.7.3 (as compared to the minimal scalar multiplet, here one of the two auxiliary scalar fields is converted into the divergence of a vector field, see equation (3.7.23)). Therefore, such nonstandard multiplets provide alternative descriptions of supersymmetric matter and cannot be simply ignored.

In the present section we consider several models of the 'non-minimal scalar multiplet' which was introduced for the first time by S. J. Gates and W. Siegel. The non-minimal scalar multiplet is described by a 'complex linear scalar superfield' which we did not meet in our previous study of supersymmetry.

3.8.1. Complex linear scalar superfield

A complex linear scalar superfield $\Gamma(z)$ is subject to the single constraint

$$\bar{\mathrm{D}}^2\Gamma = 0\,. \tag{3.8.1}$$

In contrast to the real linear scalar, no reality condition on Γ is required and thus $\mathrm{D}^2\Gamma \neq 0$.

Constraint (3.8.1) defines a reducible representation of supersymmetry. For example, Γ can be expressed in the manner

$$\Gamma = \Omega + \mathrm{D}^\alpha\Omega_\alpha \qquad \bar{\mathrm{D}}_{\dot\alpha}\Omega = \bar{\mathrm{D}}_{\dot\alpha}\Omega_\alpha = 0 \tag{3.8.2}$$

with Ω and Ω_α some chiral scalar and spinor superfields, respectively.

Complex linear scalar Γ and antichiral scalar $\bar\Phi$ are complementary to each other in the sense that an unconstrained complex scalar U can be decomposed as follows

$$U = \bar\Phi + \Gamma \qquad \mathrm{D}_\alpha\bar\Phi = \bar{\mathrm{D}}^2\Gamma = 0\,. \tag{3.8.3}$$

Similarly, real linear scalar G and chiral scalar Φ are complementary to each other in the sense that an unconstrained real scalar V reads

$$V = \Phi + \bar\Phi + G \qquad \bar{\mathrm{D}}_{\dot\alpha}\Phi = \bar{\mathrm{D}}^2 G = 0 \qquad \bar G = G\,. \tag{3.8.4}$$

The general solution to equation (3.8.1) is

$$\Gamma = \bar{\mathrm{D}}_{\dot\alpha}\bar\Psi^{\dot\alpha} \qquad \bar\Gamma = \mathrm{D}^\alpha\Psi_\alpha \tag{3.8.5}$$

with $\Psi_\alpha(z)$ being an unconstrained spinor superfield defined modulo transformations of the form

$$\delta\Psi_\alpha = \mathrm{D}^\beta\xi_{(\alpha\beta)} \tag{3.8.6}$$

where $\xi_{(\alpha\beta)}(z)$ is an unconstrained symmetric bi-spinor superfield. Therefore, any supersymmetric model described by constrained superfields Γ and $\bar\Gamma$ can be treated as a theory of unconstrained potentials Ψ_α and $\bar{\Psi}_{\dot{\alpha}}$ with the gauge invariance (3.8.6).

Remark. The complex linear superfield naturally originates if we consider a more general superfield redefinition than that given by relation (3.7.21). Namely, let us express the chiral scalar superfield through a complex scalar superfield by the rule

$$\Phi = -\frac{1}{4}\bar{D}^2 U \,.$$

Since U possesses more degrees of freedom than Φ, this redefinition is accompanied by the appearance of gauge invariance

$$U \to U' = U + \Gamma \qquad \bar{D}^2\Gamma = 0$$

with an arbitrary complex linear superfield Γ.

3.8.2. *Free non-minimal scalar multiplet*

Let us consider a simple dynamical system described by the action superfunctional

$$S[\Phi, \bar{\Phi}, U, \bar{U}] = \int d^8z \, \{\bar{\Phi}\Phi - \bar{U}U\} \tag{3.8.7}$$

where Φ is a chiral scalar superfield, and U an unconstrained complex scalar superfield. Since the dynamics of U and $\bar{U}$ are trivial, this model is obviously equivalent to the massless scalar multiplet model (3.2.5). But as long as U is unconstrained and complex, it is in our power to make the superfield redefinition

$$U \longrightarrow U + \bar{\Phi}$$

which turns the action into

$$\tilde{S}[\Phi, \bar{\Phi}, U, \bar{U}] = -\int d^8z \, \{\bar{U}U + U\Phi + \bar{U}\bar{\Phi}\} \,. \tag{3.8.8}$$

The superfields U and $\bar{U}$ can be eliminated with the help of their equations of motion. As a result, one arrives exactly at the model (3.2.5). On the other hand, varying $\tilde{S}$ with respect to Φ gives

$$\bar{D}^2 U = 0$$

and, hence, U becomes a complex linear superfield. After that, the second and third terms in (3.8.8) drop out, since

$$\int d^8z \, \Phi U = -\frac{1}{4}\int d^6z \, \Phi\bar{D}^2 U = 0 \,.$$

The above analysis shows that the massless scalar multiplet model (3.2.5) is equivalent to special theory of a complex linear superfield Γ and its conjugate $\bar{\Gamma}$ with the action superfunctional

$$S[\Gamma, \bar{\Gamma}] = -\int \mathrm{d}^8 z\, \bar{\Gamma}\Gamma\,. \tag{3.8.9}$$

It is instructive also to re-establish the equivalence by considering the dynamical equations $S[\Gamma, \bar{\Gamma}]$ leads to, which can be easily found by representing Γ and $\bar{\Gamma}$ as in relation (3.8.5). The results are summarized in the following Table.

Table 3.8.1. Minimal and non-minimal scalar multiplets.

Superfield type	Basic constraint	Equation of motion
antichiral scalar	$\mathrm{D}_\alpha \bar{\Phi} = 0$	$\bar{\mathrm{D}}^2 \bar{\Phi} = 0$
complex linear scalar	$\bar{\mathrm{D}}^2 \Gamma = 0$	$\mathrm{D}_\alpha \Gamma = 0$

For completeness, we also present a similar table establishing the duality between the minimal scalar and tensor multiplets, the latter considered in subsection 3.7.2.

Table 3.8.2. Minimal scalar and tensor multiplets.

Superfield type	Basic constraint	Equation of motion
chiral & antichiral scalars	$\bar{\mathrm{D}}^2 \mathrm{D}_\alpha (\Phi + \bar{\Phi}) = 0$	$\bar{\mathrm{D}}^2 (\Phi + \bar{\Phi}) = 0$
real linear scalar	$\bar{\mathrm{D}}^2 G = 0$	$\bar{\mathrm{D}}^2 \mathrm{D}_\alpha G = 0$

Action (3.8.9) defines the free non-minimal scalar multiplet model. To determine its component form, it turns out to be reasonable to introduce the component fields of Γ in the following way:

$$\begin{aligned} B &= \Gamma| & \bar{\chi}_{\dot{\alpha}} &= \bar{\mathrm{D}}_{\dot{\alpha}} \Gamma| \\ H &= -\frac{1}{4}\mathrm{D}^2 \Gamma| & U_{\alpha\dot{\alpha}} &= \bar{\mathrm{D}}_{\dot{\alpha}} \mathrm{D}_\alpha \Gamma| \\ \rho_\alpha &= \mathrm{D}_\alpha \Gamma| & \bar{\omega}_{\dot{\alpha}} &= \frac{1}{4}\mathrm{D}^\alpha \bar{\mathrm{D}}_{\dot{\alpha}} \mathrm{D}_\alpha \Gamma|\,. \end{aligned} \tag{3.8.10}$$

Then, a simple calculation leads to the component Lagrangian

$$\begin{aligned} L = &-\partial^a \bar{B} \partial_a B - \frac{\mathrm{i}}{2} \chi \sigma^a \partial_a \bar{\chi} \\ &- \bar{H} H + \frac{1}{2} \bar{U}^a U_a + \frac{1}{2}(\omega\rho + \bar{\omega}\bar{\rho})\,. \end{aligned} \tag{3.8.11}$$

Comparing the relations (3.8.10) and (3.8.11) with similar ones for the minimal scalar multiplet model, see equations (2.8.4) and (3.2.13), we observe that the non-minimal scalar multiplet model involves extra auxiliary fields: the complex vector field U_a and the two spinor fields ω_α and ρ_α. Moreover, the auxiliary scalar field H enters the Lagrangian with opposite sign as compared to the auxiliary scalar F in (3.2.13).

Remark. The action $S[\Gamma,\bar{\Gamma}]$ is the simplest representative of action superfunctionals of the general form

$$S_\zeta[\Gamma,\bar{\Gamma}] = \int d^8z \left\{ -\bar{\Gamma}\Gamma + \frac{\zeta}{2}(\Gamma^2 + \bar{\Gamma}^2) \right\} \qquad (3.8.12)$$

with ζ a dimensionless parameter. It is a simple exercise to show that for all values of ζ, except $\zeta = \pm 1$, these actions describe the scalar multiplet on shell. For $\zeta = \pm 1$ the corresponding equations of motion imply nothing but $\Gamma = \pm\bar{\Gamma}$ modulo an irrelevant constant.

3.8.3. *Mass generation I*

The only way to introduce a mass for a single non-minimal scalar multiplet is to add to the massless action (3.8.9) new terms containing the bare potentials Ψ_α and $\bar{\Psi}_{\dot{\alpha}}$ (3.8.5). So, let us consider the following action

$$S[\Psi,\bar{\Psi}] = -\int d^8z \left\{ \bar{\Gamma}\Gamma + 2m(\Psi^\alpha \Psi_\alpha + \bar{\Psi}_{\dot{\alpha}}\bar{\Psi}^{\dot{\alpha}}) \right\} \qquad (3.8.13)$$

which does not possess the gauge invariance (3.8.6). The corresponding equations of motion

$$\begin{aligned} -\frac{1}{4}D_\alpha\Gamma + m\Psi_\alpha &= 0 \\ -\frac{1}{4}\bar{D}_{\dot{\alpha}}\bar{\Gamma} + m\bar{\Psi}_{\dot{\alpha}} &= 0 \end{aligned} \qquad (3.8.14)$$

determine Ψ_α and $\bar{\Psi}_{\dot{\alpha}}$ in terms of Γ and $\bar{\Gamma}$, respectively. Their obvious consequences are

$$\begin{aligned} -\frac{1}{4}D^2\Gamma + m\bar{\Gamma} &= 0 \\ -\frac{1}{4}\bar{D}^2\bar{\Gamma} + m\Gamma &= 0\,. \end{aligned} \qquad (3.8.15)$$

We see that Γ is chiral on the mass shell and satisfies the dynamical equation (3.2.1) of the massive superspin-0 multiplet.

In summary, the massive scalar multiplet can be described by an unconstrained spinor superfield.

3.8.4. Mass generation II

Non-minimal scalar multiplets can acquire masses in tandem with chiral superfields. Following B. B. Deo and S. J. Gates, let us couple a single non-minimal scalar multiplet to a chiral scalar superfield my means of deforming the kinematic constraint (3.8.1) to

$$-\frac{1}{4}\bar{D}^2\Sigma = \mathcal{Q}(\Phi) \tag{3.8.16}$$

with a holomorphic function $\mathcal{Q}$. A general solution of this constraint is

$$\Sigma = \Gamma + \mathcal{F}[\Phi] \tag{3.8.17}$$

for some local functional $\mathcal{F}[\Phi]$.

An action superfunctional of the joint chiral-non-minimal dynamical system under consideration can be chosen in the form

$$S = \int d^8z \, \{\bar{\Phi}\Phi - \bar{\Sigma}\Sigma\} + \Big(\int d^6z \, \mathcal{L}_c(\Phi) + \text{c.c.}\Big) \tag{3.8.18}$$

with $\mathcal{L}_c(\Phi)$ the chiral superpotential.

To pass to components, we define the component fields of Σ in complete analogy with prescription (3.8.10), simply by replacing there Γ by Σ. With standard definition (2.8.4) of the component fields of Φ, a simple calculation then gives the component Lagrangian

$$\begin{aligned} L = & -\partial^a\bar{A}\partial_a A - \frac{i}{2}\psi\sigma^a\partial_a\bar{\psi} - \partial^a\bar{B}\partial_a B - \frac{i}{2}\chi\sigma^a\partial_a\bar{\chi} \\ & + \bar{F}F - \bar{H}H + \frac{1}{2}\bar{U}^a U_a + \frac{1}{2}(\omega\rho + \bar{\omega}\bar{\rho}) \\ & + \Big(F\big\{\mathcal{V}(A) - \bar{B}\mathcal{Q}'(A)\big\} \ + \ \text{c.c.}\Big) - |\mathcal{Q}(A)|^2 \\ & + \frac{1}{2}\Big(\mathcal{Q}'(A)\psi\chi + \frac{1}{2}\big\{\bar{B}\mathcal{Q}''(A) - \mathcal{V}'(A)\big\}\psi\psi \ + \ \text{c.c.}\Big) \end{aligned} \tag{3.8.19}$$

where $\mathcal{V}(A)$ is defined by relation (3.2.12). Eliminating the auxiliary fields leads to the following scalar potential

$$U(A, \bar{A}, B, \bar{B}) = |\mathcal{Q}(A)|^2 + |\mathcal{V}(A) - \bar{B}\mathcal{Q}'(A)|^2 \, . \tag{3.8.20}$$

Looking at the Lagrangian obtained, we see that the auxiliary fields of Σ do not couple to the component fields of Φ. In particular, the auxiliary scalar H does not contribute to the potential. This is not accidental. The point is that the structures $\bar{F}F$ and $\bar{H}H$ enter the Lagrangian with opposite signs. If we had linear in H contributions to L, they would produce corrections to the potential with the wrong sign and, as a consequence, the potential would not be positive semi-definite. Another important property of the model in field is that one can generate a nontrivial interaction without having chiral superpotential at all! The deformed constraint (3.8.16) is responsible for such a coupling. In particular, if we choose $\mathcal{L}_c(\Phi) = 0$ and $\mathcal{Q}(\Phi) = -m\Phi$, with m a positive parameter, and eliminate all the auxiliary fields, the final Lagrangian takes the form

$$\begin{aligned} L = & -\partial^a \bar{A} \partial_a A - \partial^a \bar{B} \partial_a B - m^2(\bar{A}A + \bar{B}B) \\ & - \frac{1}{2} \overline{\Psi}_{\rm D}({\rm i}\gamma^a \partial_a + m)\Psi_{\rm D} \end{aligned} \tag{3.8.21}$$

with all fields being now massive! Here $\Psi_{\rm D}$ denotes the Dirac spinor constructed of the physical spinor fields

$$\Psi_{\rm D} = \begin{pmatrix} {\rm D}_\alpha \Phi| \\ \bar{\rm D}^{\dot\alpha} \Sigma| \end{pmatrix} . \tag{3.8.22}$$

3.8.5. Supersymmetric electrodynamics

In subsection 3.5.2, we have realized the matter sector of supersymmetric electrodynamics in terms of two chiral scalar superfields with opposite charges. Here we are going to present an alternative formulation for this theory . Supersymmetric matter will be described by a chiral scalar $\Phi(z)$ along with a generalized complex linear superfield $\Sigma(z)$ constrained by

$$-\frac{1}{4}\bar{\rm D}^2 \Sigma + m\Phi = 0 . \tag{3.8.23}$$

The gauge group is postulated to act on Φ by the law

$$\Phi' = {\rm e}^{{\rm i}q\Lambda}\Phi \qquad \bar{\rm D}_{\dot\alpha}\Lambda = 0 . \tag{3.8.24}$$

Then, constraint (3.8.23) *enforces* Σ to possess the same transformation law

$$\Sigma' = {\rm e}^{{\rm i}q\Lambda}\Sigma . \tag{3.8.25}$$

The unique renormalizable and gauge invariant action reads

$$S_{\rm SED} = \int {\rm d}^8 z \, \{\bar\Phi {\rm e}^{2qV}\Phi - \bar\Sigma {\rm e}^{2qV}\Sigma\} + \frac{1}{2}\int {\rm d}^6 z \, W^\alpha W_\alpha . \tag{3.8.26}$$

As compared to the non-gauge models (3.8.12) and (3.8.17), in the present case both the ξ term and chiral superpotential are forbidden by gauge invariance.

To pass to components in a manifestly gauge covariant form, now we have to be a bit more delicate than in the case of purely chiral matter. Let us start with introducing a generalized chiral representation (compare with subsection 3.1.6). In this representation one works with the matter superfields

$$\begin{aligned} \Phi^{(C)} &= \Phi \qquad & \bar\Phi^{(C)} &= {\rm e}^{2qV}\bar\Phi \\ \Sigma^{(C)} &= \Sigma \qquad & \bar\Sigma^{(C)} &= {\rm e}^{2qV}\bar\Sigma \end{aligned} \tag{3.8.27}$$

whose gauge transformations involve only the chiral parameter Λ

$$\delta\begin{pmatrix}\Phi^{(C)}\\ \bar{\Phi}^{(C)}\end{pmatrix} = \mathrm{i}\Lambda\begin{pmatrix}q\Phi^{(C)}\\ -q\bar{\Phi}^{(C)}\end{pmatrix} \equiv \mathrm{i}\Lambda\mathbf{q}\begin{pmatrix}\Phi^{(C)}\\ \bar{\Phi}^{(C)}\end{pmatrix} \tag{3.8.28}$$

and completely similar for the column of $\Sigma^{(C)}$ and $\bar{\Sigma}^{(C)}$; here $\mathbf{q}$ denotes the operator of charge. Therefore, adapted for handling such superfields are the gauge covariant derivatives (3.6.40*a*) in which V must be identified with $V\mathbf{q}$. Herewith, the superfields $\Phi^{(C)}$ and $\bar{\Phi}^{(C)}$ are covariantly chiral and antichiral, respectively,

$$\bar{\mathcal{D}}^{(+)}_{\dot{\alpha}}\Phi^{(C)} = 0 \qquad \mathcal{D}^{(+)}_{\alpha}\bar{\Phi}^{(C)} = 0\,. \tag{3.8.29}$$

Similarly, the superfields $\Sigma^{(C)}$ and $\bar{\Sigma}^{(C)}$ are covariantly 'linear' and 'antilinear', respectively,

$$-\frac{1}{4}(\bar{\mathcal{D}}^{(+)})^2\Sigma^{(C)} + m\Phi^{(C)} = 0 \qquad -\frac{1}{4}(\mathcal{D}^{(+)})^2\bar{\Sigma}^{(C)} + m\bar{\Phi}^{(C)} = 0\,. \tag{3.8.30}$$

The representation under consideration is chiral in the sense that the covarinat derivatives (3.6.40*a*) correspond to the chiral transform of those defined by relation (3.6.49).

We define the component fields of $\Phi^{(C)}$ as gauge covariant space projections

$$A = \Phi^{(C)}| \qquad \psi_\alpha = \mathcal{D}^{(+)}_\alpha\Phi^{(C)}| \qquad F = -\frac{1}{4}(\mathcal{D}^{(+)})^2\Phi^{(C)}| \tag{3.8.31}$$

and analogously for $\Sigma^{(C)}$

$$\begin{aligned} B &= \Sigma^{(C)}| & \bar{\chi}_{\dot{\alpha}} &= \bar{\mathcal{D}}^{(+)}_{\dot{\alpha}}\Sigma^{(C)}| \\ H &= -\frac{1}{4}(\mathcal{D}^{(+)})^2\Sigma^{(C)}| & U_{\alpha\dot{\alpha}} &= \bar{\mathcal{D}}^{(+)}_{\dot{\alpha}}\mathcal{D}^{(+)}_{\alpha}\Sigma^{(C)}| \\ \rho_\alpha &= \mathcal{D}^{(+)}_\alpha\Sigma^{(C)}| & \bar{\omega}_{\dot{\alpha}} &= \frac{1}{4}\mathcal{D}^{(+)\alpha}\bar{\mathcal{D}}^{(+)}_{\dot{\alpha}}\mathcal{D}^{(+)}_{\alpha}\Sigma^{(C)}|\,. \end{aligned} \tag{3.8.32}$$

In the Wess-Zumino gauge (3.4.16), definition (3.8.31) of the component fields of Φ coincides with relation (2.8.4). The component fields of $\bar{\Phi}^{(C)}$ and $\bar{\Sigma}^{(C)}$ should also be defined via computing their gauge covariant space projections. Since the chiral representation is characterized by the conjugation rule

$$\overline{U^{(C)}} = \mathrm{e}^{2V\mathbf{q}}\,\bar{U}^{(C)} \tag{3.8.33}$$

in the Wess-Zumino gauge the component fields of $\bar{\Phi}^{(C)}$ and $\bar{\Sigma}^{(C)}$ coincide with the complex conjugates of those given by relations (3.8.31) and (3.8.32), respectively.

To find the component form of the action (3.8.26), it only remains to rewrite it in the form

$$\begin{aligned} S_{\mathrm{SED}} &= \int \mathrm{d}^8z\,\mathcal{L}_{\mathrm{MAT}} + \frac{1}{2}\int \mathrm{d}^6z\,W^\alpha W_\alpha \\ \mathcal{L}_{\mathrm{MAT}} &= \bar{\Phi}^{(C)}\Phi^{(C)} - \bar{\Sigma}^{(C)}\Sigma^{(C)} \end{aligned} \tag{3.8.34}$$

and to note that $\mathcal{L}_{\rm MAT}$ is gauge invariant, hence

$$\int d^8z\, \mathcal{L}_{\rm MAT} = \frac{1}{16}\int d^4x\, D^2\bar{D}^2\mathcal{L}_{\rm MAT}|$$
$$= \frac{1}{16}\int d^4x\, (\mathcal{D}^{(+)})^2(\bar{\mathcal{D}}^{(+)})^2\mathcal{L}_{\rm MAT}| \,. \qquad (3.8.35)$$

When computing the latter projection, one should keep in mind the basic constraints (3.8.29) and (3.8.30) as well as make use of the covariant derivative algebra (3.6.45), in particular, its simple consequences

$$[\mathcal{D}_\alpha, \bar{\mathcal{D}}^2] = -4i\mathcal{D}_{\alpha\dot\alpha}\bar{\mathcal{D}}^{\dot\alpha} + 8W_\alpha\mathbf{q} \qquad [\bar{\mathcal{D}}_{\dot\alpha}, \mathcal{D}^2] = 4i\mathcal{D}_{\alpha\dot\alpha}\mathcal{D}^\alpha - 8\bar{W}_{\dot\alpha}\mathbf{q}$$
$$[\mathcal{D}^2, \bar{\mathcal{D}}^2] = -4i\mathcal{D}^{\alpha\dot\alpha}[\mathcal{D}_\alpha, \bar{\mathcal{D}}_{\dot\alpha}] + 16\big(W^\alpha\mathcal{D}_\alpha + \bar{W}_{\dot\alpha}\bar{\mathcal{D}}^{\dot\alpha}\big)\mathbf{q} + 8(\mathcal{D}^\alpha W_\alpha)\mathbf{q}\,. \quad (3.8.36)$$

Finally, one should make use of the projections (3.4.20) and the identity

$$(\mathcal{D}_a^{(C)}U^{(C)})| = \nabla_a U^{(C)}| \qquad \nabla_a \equiv \partial_a - i\,V_a\,\mathbf{q} \qquad (3.8.37)$$

which holds for any superfield U, with V_a the electromagnetic gauge potential, see relation (3.4.15). As a result, one obtains the component Lagrangian

$$\begin{aligned} L_{\rm SED} = & -\frac{1}{4}F^{ab}F_{ab} - \frac{1}{2}\overline{\Psi}_{\rm D}\{i\gamma^a\nabla_a + m\}\Psi_{\rm D} \\ & -\nabla^a\bar{A}\nabla_a A - m^2\bar{A}A - \nabla^a\bar{B}\nabla_a B - m^2\bar{B}B \\ & -\frac{1}{2}q(\bar{A}A - \bar{B}B)^2 - \frac{i}{2}\overline{\Lambda}_{\rm M}\gamma^a\partial_a\Lambda_{\rm M} \\ & -\frac{1}{2}q\Big\{A\overline{\Psi}_{\rm D}(\mathbb{I}-\gamma_5)\Lambda_{\rm M} - \bar{B}\overline{\Lambda}_{\rm M}(\mathbb{I}-\gamma_5)\Psi_{\rm D} + {\rm c.c.}\Big\} \\ & +\big(\bar{F}+m\bar{B}\big)\big(F+mB\big) + 2\Big(D + \frac{1}{2}q(\bar{A}A - \bar{B}B)\Big)^2 \\ & -\bar{H}H + \frac{1}{2}\bar{U}^a U_a + \frac{1}{2}(\rho\omega + \bar\rho\bar\omega)\,. \end{aligned} \qquad (3.8.38)$$

Here $\Psi_{\rm D}$ denotes the matter Dirac spinor (3.8.22), and $\Lambda_{\rm M}$ the Majorana spinor entering the massless vector multiplet. Comparing the two Lagrangians (3.5.19) and (3.8.38), we see that they differ only in their auxiliary field structure. Therefore, the formulation of supersymmetric electrodynamics given in the present section can be equally well used for the description of this theory.

When $m = 0$ the basic constraint (3.8.23) tells us that Σ coincides with the complex linear superfield Γ which should be expressed through its potential by the rule (3.8.5). The corresponding gauge invariance (3.8.6) defines a gauge theory of an infinite stage of reducibility (see subsection 4.5.1). Covariant quantization of such a gauge theory turns out to be a nontrivial technical problem, and a complete solution of this problem was given by M. Grisaru *et al* [1]. For

[1] M. Grisaru, A. Van Proyen and D. Zanon *Nucl. Phys.* **502B** 345, 1997.

$m \neq 0$ one can avoid any quantization problems by adopting the point of view that the constraint (3.8.23) is nothing else but the definition of Φ. Then, we work with a single unconstrained complex scalar superfield Σ and its conjugate $\bar{\Sigma}$, in the role of matter superfields, and there is no extra gauge invariance associated with the matter sector.

3.8.6. *Couplings to Yang Mills superfield*

Complex linear superfields can be coupled to a non-Abelian gauge superfield very much like to the chiral matter. A multiplet of complex scalar superfields Γ transforms with respect to the gauge group by the rule (compare with relation (3.5.25))

$$\begin{aligned} &\Gamma' = \mathrm{e}^{\mathrm{i}\Lambda}\Gamma \qquad \bar{\Gamma}' = \bar{\Gamma}\mathrm{e}^{-\mathrm{i}\bar{\Lambda}} \\ &\Lambda = \Lambda^I T^I \qquad \bar{\Lambda} = \bar{\Lambda}^I T^I \qquad \bar{\mathrm{D}}_{\dot{\alpha}}\Lambda^I = 0 \end{aligned} \tag{3.8.39}$$

and this transformation is obviously compartible with constraint (3.8.1). The massless gauge invariant action reads

$$S = -\int \mathrm{d}^8 z\, \bar{\Gamma}_i (\mathrm{e}^{2V})^i{}_j \Gamma^j = -\int \mathrm{d}^8 z\, \bar{\Gamma}\mathrm{e}^{2V}\Gamma\,. \tag{3.8.40}$$

When working with purely chiral matter, it is always a nontrivial question how to get Dirac fermions in components, see subsection 3.5.7. For such a realization of supersymmetric matter, the chiral superfields should transform in a real representation of the gauge group just in order to have Dirac fermions at the component level. Another situation takes place if chiral superfields are introduced in tandem with complex linear ones, the latter being constrained by equation (3.8.23), for example. Now, we have a natural prescription to construct Dirac fermions, which is given by relation (3.8.22). Remarkably, the left handed and right handed components of the Dirac fermions are automatically distributed between the different superfields, Φ and Σ respectively. Moreover, Φ transforms in the same complex representation of the gauge group to which the Dirac spinor fields belong.

3.8.7. *Nonlinear sigma models*

The duality transform, which we implemented in subsection 2.8.2 to convert the minimal description of the scalar multiplet into the non-minimal one, can be extended to the case of nonlinear sigma models. Let us rewrite the supersymmetric sigma model action (3.3.10) in the following equivalent form

$$S[U,\bar{U},\Gamma,\bar{\Gamma}] = \int \mathrm{d}^8 z \left\{K(U,\bar{U}) - U\Gamma - \bar{U}\bar{\Gamma}\right\}. \tag{3.8.41}$$

Here U are unconstrained complex superfields, Γ complex linear superfields. This dynamical system is equivalent to the supersymmetric nonlinear sigma model, since the Γ-equations of motion require U to be chiral, and hence reduce the above action to that given by relation (3.3.10). On the other hand, if we vary the action with respect to U and $\bar{U}$, we obtain the equations

$$\frac{\partial}{\partial U}K(U,\bar{U}) = \Gamma \qquad \frac{\partial}{\partial \bar{U}}K(U,\bar{U}) = \bar{\Gamma} \tag{3.8.42}$$

which have to be solved to express U and $\bar{U}$ through Γ and $\bar{\Gamma}$. As a result, one arrives at the dual action

$$S[\Gamma,\bar{\Gamma}] = \int d^8z\, \tilde{K}(\Gamma,\bar{\Gamma}) \tag{3.8.43}$$

where $\tilde{K}$ is the Legendre transform of K

$$\tilde{K}(\Gamma,\bar{\Gamma}) = \left\{K(U,\bar{U}) - U\Gamma - \bar{U}\bar{\Gamma}\right\}\Big|_{U=U(\Gamma,\bar{\Gamma}),\,\bar{U}=\bar{U}(\Gamma,\bar{\Gamma})} . \tag{3.8.44}$$

We see that the duality transformation is accompanied by the exchange of potentials

$$K(\Phi,\bar{\Phi}) \longrightarrow \tilde{K}(\Gamma,\bar{\Gamma}) . \tag{3.8.45}$$

In the non-minimal picture (3.8.43), the Kähler invariance (3.3.11) is hidden.

At the component level, the nonlinear sigma model described by action (3.8.43) is radically different from that realized in terms of chiral superfields (3.3.10). We refer the interested reader to the original publication by B. B. Deo and S. J. Gates for a detailed discussion of the component structure of model (3.8.43).

Very interesting sigma models arise if we consider, once again, chiral superfields in tandem with complex linear ones. Let us here briefly discuss so called 'chiral-non-minimal nonlinear sigma-model' which was proposed for the supersymmetric description of the low-energy QCD effective action [2]. The action superfunctional reads

$$S[\Phi,\bar{\Phi},\Gamma,\bar{\Gamma}] = \int d^8z \left\{K(\Phi,\bar{\Phi}) - \Gamma^i\bar{\Gamma}^{\underline{j}}\,\frac{\partial^2 K(\Phi,\bar{\Phi})}{\partial\Phi^i\partial\bar{\Phi}^{\underline{j}}}\right\} \tag{3.8.46}$$

and possesses the Kähler invariance (3.3.11). In contrast to the chiral sigma model (3.3.10), here the target manifold is the tangent bundle over the Kähler manifold where Φ^i and $\bar{\Phi}^{\underline{j}}$ take their values. The complex linear variables Γ^i form a tangent vector at the point of the Kähler manifold with coordinates Φ^i and $\bar{\Phi}^{\underline{j}}$, with the transformation law

$$(\Gamma^i)' = \frac{\partial f^i(\Phi)}{\partial\Phi^j}\,\Gamma^j \tag{3.8.47}$$

with respect to the holomorphic reparametrizations (3.3.12). This transformation law is nicely compartible with the basic constraint (3.8.1). A nontrivial exercise for the reader is to find the component form of action (3.8.46).

[2] S. J. Gates *Phys. Lett.* **365B** 132, 1996; S. J. Gates, M. T. Grisaru, M. E. Knutt-Wehlau, M. Roček and O. A. Soloviev *Phys. Lett.* **396B** 167, 1997.

4 Quantized Superfields

If you can make one heap of all your winnings
And risk it on one turn of pitch-and-toss,
And lose, and start again at your beginnings.

Rudyard Kipling:
If

4.1. Picture-change operators

In the previous chapter, the classical dynamics of the most popular $N=1$ supersymmetric field theories was described. Now we turn our attention to investigating their quantum properties. Our primary goal will be to develop the superfield extension of the Feynman path integral perturbation theory. We anticipate that the reader is familiar with the functional methods of quantum field theory, although an introduction to these methods will be given below.

As has been shown earlier, there are two equivalent formulations for supersymmetric theories: in terms of component fields and in terms of superfields. The only difference between these formulations is that dynamical variables exist on different spaces: in the former case—on Minkowski space, in the latter—on superspace $\mathbb{R}^{4|4}$ (or its chiral subspaces). Owing to the universality of the path integral quantization procedure, it can be formally applied in both cases. When using the component field formulation, the path integral rules turn out to be correct in the sense that they lead to the same physical results as canonical quantization. When working in the superfield approach, it is worth expecting that the superfield path integral rules will guarantee the supersymmetry to remain explicit at any stage of calculations. A natural question to ask is: What is the relation of the superfield path integral rules to physical scattering amplitudes?

An answer to this question will be given in the following section. The chief goal of the present section is to introduce an auxiliary formalism, which

DOI: 10.1201/9780367802530-4

proves useful hereafter. We start by discussing (super) functional supermatrices and (super) functional derivatives. Then we describe picture-change operators which carry out transforms from superfields to component fields and vice versa. These operators will play an important role below.

4.1.1. Functional supermatrices

Many differential operators arising in field theories become linear if one considers their action on some linear functional spaces associated with spaces of histories. Under reasonable assumptions, it is possible to handle such operators as one does with ordinary supermatrices (see Sections 1.9 and 1.10). Here we are going to present the corresponding technique.

There are several linear functional spaces related to the space of histories $\mathbf{\Phi}$ of a dynamical system, which describe bosonic and fermionic tensor fields φ^i on Minkowski space. Recall that $\mathbf{\Phi}$ is spanned by all possible sets $\{\varphi^i(x)\}$ of smooth tensor fields on Minkowski space such that: (1) the fields have fixed Grassmann parities

$$\varepsilon(\varphi^i(x)) = \varepsilon_i \qquad \varepsilon_i = 0,\ 1 \tag{4.1.1}$$

(2) every field history includes together with each field φ^i its complex conjugate, that is,

$$(\varphi^i(x))^* = C^i{}_j \varphi^j(x) \tag{4.1.2}$$

$C^i{}_j$ being a real matrix fixed for the system under consideration; (3) the fields may have an arbitrary behaviour in finite regions of space–time and obey boundary conditions determined by the dynamical subspace $\mathbf{\Phi}_0$ (owing to these boundary conditions, in nonlinear theories $\mathbf{\Phi}$ cannot be treated as a linear space and should be viewed as an infinite-dimensional supermanifold). Functional spaces discussed are the following. First is the 'extended space of histories' $\tilde{\mathbf{\Phi}}$ spanned by all possible sets $\{\varphi^i(x)\}$ of smooth fields on $\mathbb{R}^4$ under requirements (1) and (2) (and maybe restricted to be bounded only). Omitting also requirement (2) gives the complexification $\tilde{\mathbf{\Phi}}_c$ of $\tilde{\mathbf{\Phi}}$. Second is the subspace $\Delta\mathbf{\Phi}$ in $\tilde{\mathbf{\Phi}}$ consisting of all histories with compact support in space–time, that is, for every history $\{\delta\phi^i(x)\} \in \Delta\mathbf{\Phi}$ every field $\delta\varphi^i$ may take non-zero values on a compact set in $\mathbb{R}^4$. Sometimes it is also reasonable to consider its complexification $\Delta\mathbf{\Phi}_c$. The space $\Delta\mathbf{\Phi}$ can be called the space of field variations. Finally, for every point $\varphi = \{\varphi^i(x)\}$ of $\mathbf{\Phi}$, one can introduce the tangent space $\mathbb{T}_\varphi$ to $\mathbf{\Phi}$ at φ, its complexification $(\mathbb{T}_\varphi)_c$ and subspaces in these.

Each of the spaces presented is naturally identified with the even or real even subset of some infinite-dimensional supervector space. For example, consider the supervector space $\tilde{\mathbf{\Phi}}(\Lambda_\infty)$ of all smooth mappings

$$\varphi\colon \mathbb{R}^4 \to \mathscr{L}^{(p,q)}$$

where $\mathscr{L}^{(p,q)}$ is a (p, q)-dimensional supervector space, with p and q being the number of bosonic and fermionic fields, respectively, which $\mathbf{\Phi}$ describes. Choosing a pure basis $\{\vec{E}_i\}$ in $\mathscr{L}^{(p,q)}$, $\varepsilon(\vec{E}_i)=\varepsilon_i$, one finds that c-type elements in $\tilde{\mathbf{\Phi}}(\Lambda_\infty)$ are of the form

$$\varphi(x)=\vec{E}_i\varphi^i(x)$$

where the smooth functions $\varphi^i(x)$ satisfy equation (4.1.1). Therefore, the set $\{\varphi^i(x)\}$ lies in $\tilde{\mathbf{\Phi}}_c$. Next, imposing the complex conjugation rules

$$\vec{E}_i^*=(-1)^{\varepsilon_i}\vec{E}_jC^j{}_i$$

one observes that real c-type elements in $\tilde{\mathbf{\Phi}}(\Lambda_\infty)$ obey the equations (4.1.1) and (4.1.2). Therefore, $\mathbf{\Phi}$-points span the real event subset of $\tilde{\mathbf{\Phi}}(\Lambda_\infty)$. Recalling now the identification of $\mathbb{C}^{p|q}$ and $\mathbb{R}^{p|q}$ with the even and real even subsets of a (p, q)-dimensional supervector space (see Section 1.9), one can look on $\tilde{\mathbf{\Phi}}$ ($\tilde{\mathbf{\Phi}}_c$) as an infinite-dimensional real (complex) superspace.

Let F be a linear differential operator acting on $\tilde{\mathbf{\Phi}}_c$ (a c-type linear operator on $\tilde{\mathbf{\Phi}}(\Lambda_\infty)$),

$$\varepsilon(F\varphi^i(x))=\varepsilon_i. \tag{4.1.3}$$

Associated with F is its 'kernel' (or 'functional supermatrix') $F^i{}_j(x, x')$ and the 'supertranspose kernel' $F^{sT}{}_i{}^j(x, x')$ defined by

$$F\varphi^i(x)=\int d^4x'\, F^i{}_j(x, x')\varphi^j(x')=\int d^4x'\, \varphi^j(x')F^{sT}{}_j{}^i(x', x) \tag{4.1.4}$$

for all $\varphi\in\tilde{\mathbf{\Phi}}_c$. One easily finds:

$$\varepsilon(F^i{}_j(x, x'))=\varepsilon(F^{sT}{}_i{}^j(x, x'))=\varepsilon_i+\varepsilon_j \tag{4.1.5}$$

and

$$F^{sT}{}_i{}^j(x, x')=(-1)^{\varepsilon_i+\varepsilon_i\varepsilon_j}F^j{}_i(x', x). \tag{4.1.6}$$

It is instructive to compare the final relation with equation (1.9.58b) and equation (4.1.4) with equation (1.9.63). The kernel is evaluated by the rule

$$F^i{}_j(x, x')=F\delta^i{}_j\delta^4(x-x') \tag{4.1.7}$$

where F is to act on the indices i and x only. The 'functional supertrace' of F is defined as

$$\text{sTr}\, F=\int d^4x\,(-1)^{\varepsilon_i}F^i{}_i(x, x)=\int d^4x\,(-1)^{\varepsilon_i}F^{sT}{}_i{}^i(x, x). \tag{4.1.8}$$

If F is a non-singular operator possessing logarithm, its 'functional superdeterminant' is defined as

$$\text{sDet}\, F=\exp(\text{sTr}(\ln F)). \tag{4.1.9}$$

The last two definitions represent infinite-dimensional generalizations of those given in Section 1.10. It should be remarked also that introducing the even–even, even–odd, odd–even and odd–odd parts of $F^i{}_j(x, x')$, one can represent sDet F as in equation (1.10.63). Another remark is that not every linear operator possesses well-defined sTr and sDet. We restrict ourselves to consideration of operators for which the right-hand sides in equations (4.1.8) and (4.1.9) exist.

Given two linear operators F_1 and F_2, the following relations hold:

$$(F_1F_2)^i{}_j(x, x') = \int \mathrm{d}^4x''\, F^i_{1k}(x, x'')F^k_{2j}(x'', x') \tag{4.1.10a}$$

$$\mathrm{sTr}(F_1F_2) = \mathrm{sTr}(F_2F_1) \tag{4.1.10b}$$

$$\mathrm{sDet}(F_1F_2) = \mathrm{sDet}\, F_1 \cdot \mathrm{sDet}\, F_2. \tag{4.1.10c}$$

To prove the latter, we set $F_1 = \mathrm{e}^A$, $F_2 = \mathrm{e}^B$ and use the Baker–Hausdorff formula:

$$\begin{aligned}
\mathrm{sDet}(\mathrm{e}^A\mathrm{e}^B) &= \mathrm{sDet}\left(\exp\left(A + B + \frac{1}{2}[A, B] + \ldots\right)\right) \\
&= \exp\left(\mathrm{sTr}\left(A + B + \frac{1}{2}[A, B] + \ldots\right)\right) = \exp(\mathrm{sTr}(A + B)\Big) \\
&= \exp(\mathrm{sTr}\, A)\exp(\mathrm{sTr}\, B).
\end{aligned}$$

To simplify subsequent formulae, we introduce a condensed notation based on unifying the symbols i and x, which specify $\varphi^i(x)$, in the label $\hat{\imath}$. In such notation, we have

$$\begin{gathered}
\varphi^{\hat{\imath}} = \varphi^i(x) \qquad \delta^{\hat{\imath}}{}_{\hat{\jmath}} = \delta^i{}_j\delta^4(x - x') \qquad \delta_{\hat{\imath}}{}^{\hat{\jmath}} = \delta_i{}^j\delta^4(x - x') \\
F^{\hat{\imath}}{}_{\hat{\jmath}}\varphi^{\hat{\jmath}} = \int \mathrm{d}^4x'\, F^i{}_j(x, x')\varphi^j(x') \qquad \mathrm{sTr}\, F = (-1)^{\varepsilon_{\hat{\imath}}} F^{\hat{\imath}}{}_{\hat{\imath}}.
\end{gathered} \tag{4.1.11}$$

Tensor-like objects

$$T^{\hat{\imath}_1 \ldots \hat{\imath}_n}{}_{\hat{\imath}_{n+1} \ldots \hat{\imath}_{n+m}} \quad \text{or} \quad T_{\hat{\imath}_1 \ldots \hat{\imath}_n}{}^{\hat{\imath}_{n+1} \ldots \hat{\imath}_{n+m}}$$

mean $(n+m)$-point functions

$$\begin{aligned}
T^{\hat{\imath}_1 \ldots \hat{\imath}_n}{}_{\hat{\imath}_{n+1} \ldots \hat{\imath}_{n+m}} &= T^{i_1 \ldots i_n}{}_{i_{n+1} \ldots i_{n+m}}(x_1, \ldots, x_n, x_{n+1}, \ldots, x_{n+m}) \\
T_{\hat{\imath}_1 \ldots \hat{\imath}_n}{}^{\hat{\imath}_{n+1} \ldots \hat{\imath}_{n+m}} &= T_{i_1 \ldots i_n}{}^{i_{n+1} \ldots i_{n+m}}(x_1, \ldots, x_n, x_{n+1}, \ldots, x_{n+m}).
\end{aligned}$$

4.1.2. Superfunctional supermatrices

Consider a supersymmetric dynamical system. As is known, it can be described by the space of histories $\mathbf{\Phi}$ or by the space of superhistories $\mathbb{V}$, in relation to which $\mathbf{\Phi}$ is the space of component fields. In general, the superfield variables $v^I(z)$ include unconstrained tensor superfields $V^{\underline{a}}(z)$, chiral tensor superfields $\Phi^{\underline{m}}(z)$ and their conjugate antichiral ones $\bar{\Phi}^{\underline{\dot{m}}}(z)$,

$$v^I(z) = (V^{\underline{a}}(z), \Phi^{\underline{m}}(z), \bar{\Phi}^{\underline{\dot{m}}}(z))$$
$$\bar{D}_{\dot{\alpha}}\Phi^{\underline{m}} = D_\alpha \bar{\Phi}^{\underline{\dot{m}}} = 0. \tag{4.1.12}$$

Throughout Sections 4.1–4.3, we use lower-case underlined letters from the beginning (middle) of the Latin alphabet to label unconstrained (chiral and antichiral) superfields; in the antichiral case, indices are in addition dotted. Superfield variables can be bosonic and fermionic,

$$\varepsilon(v^I(z)) = \varepsilon_I \qquad \varepsilon_I = 0, 1 \tag{4.1.13}$$

and are real in the sense that any superhistory $v = \{v^I\} \in \mathbb{V}$ includes together with every superfield v^I its complex conjugate,

$$(v^I(z))^* = C^I{}_J v^J(z) \qquad C^I{}_J = \left(\begin{array}{c|c|c} C^{\underline{a}}{}_{\underline{b}} & 0 & 0 \\ \hline 0 & 0 & C^{\underline{m}}{}_{\underline{\dot{n}}} \\ \hline 0 & C^{\underline{\dot{m}}}{}_{\underline{n}} & 0 \end{array}\right) \tag{4.1.14}$$

$C^I{}_J$ being a fixed real matrix. Similarly to $\mathbf{\Phi}$, the space $\mathbb{V}$ is an infinite-dimensional supermanifold, whose structure is specified by boundary conditions imposed on superhistories at space–time infinity. In contrast with the field variables, which are functions on Minkowski space, the $V^{\underline{a}}(z)$, $\Phi^{\underline{m}}(z)$ and $\bar{\Phi}^{\underline{\dot{m}}}(z)$ exist on different superspaces (parameterized by $(x, \theta, \bar{\theta})$, (x, θ) and $(x, \bar{\theta})$, respectively), and this is expressed in the explicit form of the unit operator on $\mathbb{V}$:

$$v^I(z) = \int d^{(J)}z'\ \delta^I{}_J(z, z')v^J(z') = \int d^{(J)}z'\ v^J(z')\delta_J{}^I(z', z) \tag{4.1.15}$$

where

$$d^{(I)}z = \begin{cases} d^8 z & I = \underline{a} \\ d^6 z & I = \underline{m} \\ d^6 \bar{z} & I = \underline{\dot{m}} \end{cases} \tag{4.1.16}$$

and

$$\delta^I{}_J(z, z') = \delta_J{}^I(z', z) = \delta^I_J \delta^{(I)}(z, z')$$
$$\delta^{(I)}(z, z') = \begin{cases} \delta^8(z - z') = \delta^4(x - x')(\theta - \theta')^2(\bar{\theta} - \bar{\theta}')^2 & I = \underline{a} \\ \delta_+(z, z') = \delta^4(x_{(+)} - x'_{(+)})(\theta - \theta')^2 & I = \underline{m} \\ \delta_-(z, z') = \delta^4(x_{(-)} - x'_{(-)})(\bar{\theta} - \bar{\theta}')^2 & I = \underline{\dot{m}} \end{cases} \tag{4.1.17}$$

$$x^a_{(\pm)} = x^a \pm \mathrm{i}\theta\sigma^a\bar{\theta}.$$

The fact that the chiral delta-function (3.1.46) can be written as in (4.1.17) is a direct consequence of the relation (3.1.66) holding in the chiral representation.

Associated with $\mathbb{V}$ are several infinite-dimensional superspaces: the 'extended space of superhistories' $\tilde{\mathbb{V}}$, the space of superfield variations $\Delta\mathbb{V}$, the tangent spaces $\mathbb{T}_v$ to $\mathbb{V}$ at v, for every $v \in \mathbb{V}$, and their complexifications $\tilde{\mathbb{V}}_c$, $\Delta\mathbb{V}_c$, $(\mathbb{T}_v)_c$. These are introduced by analogy with $\tilde{\mathbf{\Phi}}$, $\Delta\tilde{\mathbf{\Phi}}$, and so on. Note also that chiral and antichiral superfields describing points in $\tilde{\mathbb{V}}_c$, $\Delta\mathbb{V}_c$ and $(\mathbb{T}_v)_c$ are not related by complex conjugation.

Let $\mathbf{F}$ be a linear differential operator acting on $\tilde{\mathbb{V}}_c$. We define the 'superkernel' (or 'superfunctional supermatrix') of $\mathbf{F}$ as follows:

$$\mathbf{F}v^I(z) = \int \mathrm{d}^{(J)}z'\, \mathbf{F}^I{}_J(z, z')v^J(z') = \int \mathrm{d}^8z'\, \mathbf{F}^I{}_{\underline{b}}(z, z')V^{\underline{b}}(z')$$
$$+ \int \mathrm{d}^6z'\, \mathbf{F}^I{}_{\underline{n}}(z, z')\Phi^{\underline{n}}(z') + \int \mathrm{d}^6\bar{z}'\, \mathbf{F}^I{}_{\underline{\dot{n}}}(z, z')\bar{\Phi}^{\underline{\dot{n}}}(z') \qquad (4.1.18a)$$

some components of the superkernel being chiral or antichiral:

$$\begin{aligned} \bar{\mathrm{D}}_{\dot{\alpha}}\mathbf{F}^{\underline{m}}{}_J(z, z') &= \bar{\mathrm{D}}'_{\dot{\alpha}}\mathbf{F}^I{}_{\underline{n}}(z, z') = 0 \\ \mathrm{D}_\alpha \mathbf{F}^{\underline{\dot{m}}}{}_J(z, z') &= \mathrm{D}'_\alpha \mathbf{F}^I{}_{\underline{\dot{n}}}(z, z') = 0. \end{aligned} \qquad (4.1.18b)$$

The superkernel is evaluated by the rule

$$\mathbf{F}^I{}_J(z, z') = \mathbf{F}\delta^I{}_J(z, z') \qquad (4.1.19)$$

where $\mathbf{F}$ acts on the indices I and z only. Its supertranspose is defined by

$$\begin{aligned} \mathbf{F}v^I(z) &= \int \mathrm{d}^{(J)}z'\, v^J(z')\mathbf{F}^{\mathrm{sT}}{}_J{}^I(z', z) \\ \mathbf{F}^{\mathrm{sT}}{}_I{}^J(z, z') &= (-1)^{\varepsilon_I + \varepsilon_I\varepsilon_J}\mathbf{F}^J{}_I(z', z). \end{aligned} \qquad (4.1.20)$$

Similarly to equation (4.1.5), one now finds

$$\varepsilon(\mathbf{F}^I{}_J(z, z')) = \varepsilon(\mathbf{F}^{\mathrm{sT}}{}_I{}^J(z, z')) = \varepsilon_I + \varepsilon_J. \qquad (4.1.21)$$

The 'superfunctional supertrace' and 'superdeterminant' of $\mathbf{F}$ are defined in the manner:

$$\mathbf{sTr}\,\mathbf{F} = \int \mathrm{d}^{(I)}z\,(-1)^{\varepsilon_I}\mathbf{F}^I{}_I(z, z) = \int \mathrm{d}^{(I)}z\,(-1)^{\varepsilon_I}\mathbf{F}^{\mathrm{sT}}{}_I{}^I(z, z)$$
$$= \int \mathrm{d}^8z\,(-1)^{\varepsilon_{\underline{a}}}\mathbf{F}^{\underline{a}}{}_{\underline{a}}(z, z) + \int \mathrm{d}^6z\,(-1)^{\varepsilon_{\underline{m}}}\mathbf{F}^{\underline{m}}{}_{\underline{m}}(z, z)$$

$$+\int d^6\bar{z}(-1)^{\varepsilon_{\dot{m}}}\mathbf{F}^{\dot{m}}{}_{\dot{m}}(z,z) \tag{4.1.22}$$

$$\mathbf{sDet}\,\mathbf{F}=\exp(\mathbf{sTr}(\ln\mathbf{F})). \tag{4.1.23}$$

The properties (4.1.10*a–c*) are trivially extended to the supercase.

It may seem, looking at equations (4.1.18) and (4.1.20) that operating with superfunction objects is much harder than with functional ones. Some real complications exist: first, when introducing the superkernel, one must succeed in satisfying the (anti)chirality requirements (4.1.18*b*) (note, however, they fix the superkernel uniquely); second, it is required to integrate equations (4.1.18, 20) over different superspaces. Nevertheless, it is worth keeping in mind that every superfield variable describes a large number of ordinary fields. So the shortcomings mentioned can be looked on as a reasonably small cost for the advantage of working with a multiplet of fields as a single object. In addition, the operations **sTr** and **sDet** possess remarkable properties (having no direct space–time analogues), which often greatly simplify superfield calculations. Some of such properties are formulated in two simple theorems.

Theorem 1. Let $\mathbf{F}$ be a linear operator of the general structure

$$\mathbf{F}^I{}_J(z,z')=\begin{cases}\mathscr{F}^{\underline{a}}{}_{\underline{b}}(x,x',\theta,\bar{\theta})(\theta-\theta')^2(\bar{\theta}-\bar{\theta}')^2\\ \mathscr{F}^{\underline{m}}{}_{\underline{n}}(x_{(+)},x'_{(+)},\theta)(\theta-\theta')^2\\ \mathscr{F}^{\underline{\dot{m}}}{}_{\underline{\dot{n}}}(x_{(-)},x'_{(-)},\bar{\theta})(\bar{\theta}-\bar{\theta}')^2.\end{cases} \tag{4.1.24}$$

Then, for every positive integer N, the operator $\mathbf{F}^N$ has the same structure and

$$\mathbf{sTr}(\mathbf{F}^N)=0.$$

If $\mathbf{F}$ possesses a logarithm, then

$$\mathbf{sDet}\,\mathbf{F}=1.$$

The meaning of equation (4.1.24) is quite simple. It tells us that $\mathbf{F}$, regarded as a differential operator, is of zeroth order in the partial derivatives in θ and $\bar{\theta}$. As examples of operators of the type (4.1.24), we can suggest z-dependent supermatrices

$$\mathbf{F}^I{}_J(z,z')=\begin{cases}\mathscr{F}^{\underline{a}}{}_{\underline{b}}(z)\delta^8(z-z')\\ \mathscr{F}^{\underline{m}}{}_{\underline{n}}(z)\delta_+(z,z')\\ \mathscr{F}^{\underline{\dot{m}}}{}_{\underline{\dot{n}}}(z)\delta_-(z,z')\end{cases} \tag{4.1.25}$$

$$\bar{D}_{\dot{\alpha}}\mathscr{F}^{\underline{m}}{}_{\underline{n}}=D_\alpha\mathscr{F}^{\underline{\dot{m}}}{}_{\underline{\dot{n}}}=0$$

or generalized d'Alembertians

$$\hat{\Box} = \begin{cases} \delta^{\underline{a}}{}_{\underline{b}}[G^{ab}(z)\partial_a\partial_b + L^a(z)\partial_a] + Q^{\underline{a}}{}_{\underline{b}}(z) \\ \delta^{\underline{m}}{}_{\underline{n}}\,[G^{ab}_{(+)}(z)\partial_a\partial_b + L^a_{(+)}(z)\partial_a] + Q^{\underline{m}}_{(+)\underline{n}}(z) \\ \delta^{\underline{\dot{m}}}{}_{\underline{\dot{n}}}\,[G^{ab}_{(-)}(z)\partial_a\partial_b + L^a_{(-)}(z)\partial_a] + Q^{\underline{\dot{m}}}_{(-)\underline{\dot{n}}}(z) \end{cases} \tag{4.1.26}$$

where $G_{(\pm)}$, $L_{(\pm)}$ and $Q_{(\pm)}$ are chiral (antichiral) superfields. Owing to Theorem 1, we have

$$\mathbf{sDet}(\lambda\mathbb{I}) = 1 \tag{4.1.27}$$

for an arbitrary c-number λ with non-zero body.

Theorem 2. Let $\mathbf{K}$ be a first-order differential operator of the general structure

$$\mathbf{K} = \begin{cases} \delta^{\underline{a}}{}_{\underline{b}}K^A(z)\partial_A + K^{\underline{a}}{}_{\underline{b}}(z) \\ \delta^{\underline{m}}{}_{\underline{n}}\,[\Lambda^a(z)\partial_a + \Lambda^\alpha(z)\,\mathrm{e}^{\mathrm{i}\mathscr{H}}\partial_\alpha\,\mathrm{e}^{-\mathrm{i}\mathscr{H}}] + \Lambda^{\underline{m}}{}_{\underline{n}}(z) \\ \delta^{\underline{\dot{m}}}{}_{\underline{\dot{n}}}\,[\Omega^a(z)\partial_a + \Omega_{\dot{\alpha}}(z)\,\mathrm{e}^{-\mathrm{i}\mathscr{H}}\bar{\partial}^{\dot{\alpha}}\mathrm{e}^{\mathrm{i}\mathscr{H}}] + \Omega^{\underline{\dot{m}}}{}_{\underline{\dot{n}}}(z) \end{cases} \tag{4.1.28}$$

$$\mathscr{H} = \theta\sigma^a\bar{\theta}\partial_a$$

where all Λ are chiral and all Ω are antichiral. Then

$$\mathbf{sDet}\,\mathrm{e}^{\mathbf{K}} = 1.$$

Proof. Since $\mathbf{sDet}(\exp\mathbf{K}) = \exp(\mathbf{sTr}\,\mathbf{K})$, we are to prove $\mathbf{sTr}\,\mathbf{K} = 0$. Introducing the superkernel of $\mathbf{K}$, one observes that each of $\mathbf{K}^{\underline{a}}{}_{\underline{b}}(z, z')$, $\mathbf{K}^{\underline{m}}{}_{\underline{n}}(z, z')$ and $\mathbf{K}^{\underline{\dot{m}}}{}_{\underline{\dot{n}}}(z, z')$ is proportional to $(\theta - \theta')$ or $(\bar{\theta} - \bar{\theta}')$ and vanishes at $z = z'$.

To understand the importance of the above statement, suppose we have a linear representation T of some supergroup G (for example, the super Poincaré group) on $\tilde{\mathbb{V}}$ or on $\tilde{\mathbb{V}}_c$, the generators being first-order differential operators of the form (4.1.28) (as is the case in practice, in particular, for the super Poincaré group). Then every operator of the representation is unimodular,

$$\mathbf{sDet}(T(g)) = 1 \qquad \forall\, g \in G.$$

In quantum theory this implies the supergroup invariance of the path-integral measure!

By analogy with equation (4.1.11), we introduce the superfield condensed notation:

$$\begin{gathered} v^{\mathit{I}} = v^I(z) \qquad \delta^{\mathit{I}}{}_{\mathit{J}} = \delta^I{}_J(z, z') \qquad \delta_{\mathit{I}}{}^{\mathit{J}} = \delta_I{}^J(z, z') \\ \mathbf{F}^{\mathit{I}}{}_{\mathit{J}}v^{\mathit{J}} = \int \mathrm{d}^{(J)}z'\,\mathbf{F}^I{}_J(z, z')v^J(z') \qquad \mathbf{sTr}\,\mathbf{F} = (-1)^{\varepsilon_I}F^{\mathit{I}}{}_{\mathit{I}}. \end{gathered} \tag{4.1.29}$$

In the case of tensor-like objects

$$T^{\mathit{I}_1\ldots\mathit{I}_n}{}_{\mathit{I}_{n+1}\ldots\mathit{I}_{n+m}} = T^{I_n\ldots I_n}{}_{I_{n+1}\ldots I_{n+m}}(z_1, \ldots, z_n, z_{n+1}, \ldots, z_{n+m})$$

or

$$T_{\hat{I}_1 \ldots \hat{I}_n}{}^{\hat{I}_{n+1} \ldots \hat{I}_{n+m}} = T_{I_1 \ldots I_n}{}^{I_{n+1} \ldots I_{n+m}}(z_1, \ldots, z_n, z_{n+1}, \ldots, z_{n+m})$$

it will always be assumed that for $I_r = \underline{m}_r(\underline{\dot{m}}_r)$ the expressions on the right are chiral (antichiral) with respect to the argument z_r.

4.1.3. (Super)functional derivatives

Up to now, (super)functional derivatives, as they were introduced in Chapter 3, have been left ones. Below it will be convenient to operate with both left and right (super)functional derivatives. Here we give the necessary definitions.

Let $\Psi: \mathbf{\Phi} \to \Lambda_\infty$ be a functional on $\mathbf{\Phi}$. Its left functional derivative

$$\frac{\vec{\delta}}{\delta\varphi^i(x)} \Psi[\varphi]$$

and its right functional derivative

$$\Psi[\varphi] \frac{\overleftarrow{\delta}}{\delta\varphi^i(x)}$$

are defined by

$$\delta\Psi[\varphi] = \int \mathrm{d}^4x \, \delta\varphi^i(x) \frac{\vec{\delta}}{\delta\varphi^i(x)} \Psi[\varphi] = \int \mathrm{d}^4x \, \Psi[\varphi] \frac{\overleftarrow{\delta}}{\delta\varphi^i(x)} \delta\varphi^i(x) \qquad (4.1.30a)$$

or, in condensed notation

$$\delta\Psi[\varphi] = \delta\varphi^{\hat{i}} \frac{\vec{\delta}}{\delta\varphi^{\hat{i}}} \Psi[\varphi] = \Psi[\varphi] \frac{\overleftarrow{\delta}}{\delta\varphi^{\hat{i}}} \delta\varphi^{\hat{i}} \qquad (4.1.30b)$$

$\delta\varphi^i(x)$ being an infinitesimal variation. Functionally differentiating φ gives the unit operator on $\mathbf{\Phi}$:

$$\frac{\vec{\delta}}{\delta\varphi^{\hat{i}}} \varphi^{\hat{j}} = \delta_{\hat{i}}{}^{\hat{j}} \qquad \varphi^{\hat{i}} \frac{\overleftarrow{\delta}}{\delta\varphi^{\hat{j}}} = \delta^{\hat{i}}{}_{\hat{j}}. \qquad (4.1.31)$$

To simplify expressions involving repeated functional derivatives, it is useful to introduce the following notation:

$$_{\hat{i}_1 \ldots \hat{i}_n,}\Psi_{,\hat{i}_{n+1} \ldots \hat{i}_{n+m}}[\varphi] = \frac{\vec{\delta}}{\delta\varphi^{\hat{i}_1}} \cdots \frac{\vec{\delta}}{\delta\varphi^{\hat{i}_n}} \Psi[\varphi] \frac{\overleftarrow{\delta}}{\delta\varphi^{\hat{i}_{n+1}}} \cdots \frac{\overleftarrow{\delta}}{\delta\varphi^{\hat{i}_{n+m}}}. \qquad (4.1.32)$$

Symmetry properties of such tensor-like objects can be readily found by considering the nth variation of $\Psi[\varphi]$

$$\delta^n\Psi[\varphi] = \frac{\mathrm{d}^n}{\mathrm{d}t^n} \Psi[\varphi + t\delta\varphi] \Big|_{t=0}$$

and then representing it in several different forms. Thus one finds

$$\begin{aligned} \delta^{n+m}\Psi[\varphi] &= \delta\varphi^{\hat{i}_1} \ldots \delta\varphi^{\hat{i}_{n+m}} {}_{\hat{i}_{n+m} \ldots \hat{i}_1,}\Psi[\varphi] \\ &= \delta\varphi^{\hat{i}_1} \ldots \delta\varphi^{\hat{i}_n} {}_{\hat{i}_n \ldots \hat{i}_1,}\Psi_{,\hat{i}_{n+m} \ldots \hat{i}_{n+1}}[\varphi] \delta\varphi^{\hat{i}_{n+1}} \ldots \delta\varphi^{\hat{i}_{n+m}} \\ &= \Psi_{,\hat{i}_{n+m} \ldots \hat{i}_1}[\varphi] \delta\varphi^{\hat{i}_1} \ldots \delta\varphi^{\hat{i}_{n+m}}. \end{aligned}$$

This leads to

$$_{i,}\Psi[\varphi]=(-1)^{\varepsilon_i(1+\varepsilon(\Psi))}\Psi_{,i}[\varphi] \tag{4.1.33a}$$

$$_{i,}\Psi_{,j}[\varphi]=(-1)^{(\varepsilon_i+\varepsilon_j)(1+\varepsilon(\Psi))+\varepsilon_i\varepsilon_j}\,_{j,}\Psi_{,i}[\varphi] \tag{4.1.33b}$$

$$_{ij,}\Psi[\varphi]=(-1)^{\varepsilon_i\varepsilon_j}\,_{ji,}\Psi[\varphi] \tag{4.1.33c}$$

$$\Psi_{,ij}[\varphi]=(-1)^{\varepsilon_i\varepsilon_j}\Psi_{,ji}[\varphi]. \tag{4.1.33d}$$

In the first two relations, Ψ is assumed to be a pure functional:

$$\begin{array}{cc}\text{bosonic} & \text{fermionic}\\ \Psi:\boldsymbol{\Phi}\rightarrow\mathbb{C}_c & \Psi:\boldsymbol{\Phi}\rightarrow\mathbb{C}_a.\end{array}$$

Let Ψ be a bosonic functional. Equation (4.1.33*b*) tells us that, for every $\varphi\in\boldsymbol{\Phi}$, the supermatrix $_{i,}\Psi_{,j}[\varphi]$ is supersymmetric in the sense of definition (1.9.78). Associated with such a supermatrix is the c-number-valued bi-linear form on $\Delta\boldsymbol{\Phi}$:

$$(\delta\varphi_1,\delta\varphi_2)=(\delta\varphi_2,\delta\varphi_1)=\delta\varphi_1^i\,_{i,}\Psi_{,j}[\varphi]\,\delta\varphi_2^j.$$

In the case of a real bosonic functional $\Psi:\boldsymbol{\Phi}\rightarrow\mathbb{R}_c$, the above bi-linear form turns out to be real c-number-valued, hence the operator $_{i,}\Psi_{,j}[\varphi]$ is self-adjoint on $\Delta\boldsymbol{\Phi}$.

Now, let $\Psi:\mathbb{V}\rightarrow\Lambda_\infty$ be a superfunctional on $\mathbb{V}$. Its left superfunctional derivative $\vec{\delta}/\delta v^I(z)\Psi[v]$ is defined as

$$\begin{aligned}\delta\Psi[v]&=\int \mathrm{d}^{(I)}z\,\mathrm{d}v^I(z)\frac{\vec{\delta}}{\delta v^I(z)}\Psi[v]\\&=\int \mathrm{d}^8z\,\delta V^{\underline{a}}(z)\frac{\vec{\delta}}{\delta V^{\underline{a}}(z)}\Psi[v]+\int \mathrm{d}^6z\,\delta\Phi^{\underline{m}}(z)\frac{\vec{\delta}}{\delta\Phi^{\underline{m}}(z)}\Psi[v]\\&\quad+\int \mathrm{d}^6\bar{z}\,\delta\bar{\Phi}^{\underline{\dot{m}}}(z)\frac{\vec{\delta}}{\delta\bar{\Phi}^{\underline{\dot{m}}}(z)}\Psi[v]\end{aligned} \tag{4.1.34}$$

where

$$\bar{\mathrm{D}}_{\dot{\alpha}}\frac{\vec{\delta}}{\delta\Phi^{\underline{m}}(z)}\Psi[v]=0 \qquad \mathrm{D}_\alpha\frac{\vec{\delta}}{\delta\bar{\Phi}^{\underline{\dot{m}}}(z)}\Psi[v]=0.$$

The right superfunctional derivative $\Psi[v]\overleftarrow{\delta}/\delta v^I(z)$ is read off from $\delta\Psi[v]$ in which $\delta v^I(z)$ is to be placed on the right. In condensed notation, we have

$$\delta\Psi[v]=\delta v^I\frac{\vec{\delta}}{\delta v^I}\Psi[v]=\Psi[v]\frac{\overleftarrow{\delta}}{\delta v^I}\delta v^I. \tag{4.1.35}$$

Superfunctionally differentiating v gives the unit operator on $\mathbb{V}$:

$$\frac{\vec{\delta}}{\delta v^I}v^J=\delta_I{}^J \qquad v^I\frac{\overleftarrow{\delta}}{\delta v^J}=\delta^I{}_J. \tag{4.1.36}$$

In perfect analogy with equation (4.1.32), we introduce the notation

$$
{}_{I_1\ldots I_p,}\Psi_{I_{p+1}\ldots I_{p+q}}[v]=\frac{\vec{\delta}}{\delta v^{I_1}}\cdots\frac{\vec{\delta}}{\delta v^{I_p}}\Psi[v]\frac{\overleftarrow{\delta}}{\delta v^{I_{p+1}}}\cdots\frac{\overleftarrow{\delta}}{\delta v^{I_{p+q}}} \quad (4.1.37)
$$

Relatitions (4.1.33) are trivially extended to the case under consideration.

4.1.4. Picture-change operators

In developing the general formalism for supersymmetric dynamical systems, it proves reasonable to have an analytic realization of the fact that $\mathbf{\Phi}$ is the component field space for $\mathbb{V}$. This is the purpose of the present subsection.

Associated with every superhistory $\{v^I(z)\}\in\mathbb{V}$ is the unique history $\{\varphi^i(x)\}\in\mathbf{\Phi}$ spanned by the component fields of $V^{\underline{a}}$, $\Phi_{\underline{m}}$ and $\bar{\Phi}^{\underline{\dot{m}}}$. This correspondence is one-to-one and can be uniquely extended up to a correspondence between $\tilde{\mathbb{V}}$ and $\tilde{\mathbf{\Phi}}$. So we have two mappings

$$
\begin{aligned}
&P\colon \tilde{\mathbb{V}}\to\tilde{\mathbf{\Phi}} \qquad Q\colon \tilde{\mathbf{\Phi}}\to\tilde{\mathbb{V}}\\
&QP=\mathbb{1}_{\tilde{\mathbb{V}}} \qquad PQ=\mathbb{1}_{\tilde{\mathbf{\Phi}}}
\end{aligned} \quad (4.1.38)
$$

carrying out transforms from superfields to their component fields and vice versa. In accordance with the results of Section 2.8, the operators P and Q are linear. Introducing their kernels, we can write

$$
\begin{aligned}
&\varphi^i(x)=\int \mathrm{d}^{(I)}z'\, P^i{}_I(x,z')v^I(z')\\
&\bar{\mathrm{D}}'_{\dot{\alpha}}P^i{}_{\underline{m}}(x,z')=\mathrm{D}'_{\alpha}P^i{}_{\underline{\dot{m}}}(x,z')=0
\end{aligned} \quad (4.1.39a)
$$

and

$$
\begin{aligned}
&v^I(z')=\int \mathrm{d}^4x'\, Q^I{}_i(z,x')\varphi^i(x')\\
&\bar{\mathrm{D}}_{\dot{\alpha}}Q^{\underline{m}}{}_i(z,x')=\mathrm{D}_{\alpha}P^{\underline{\dot{m}}}{}_i(z,x')=0
\end{aligned} \quad (4.1.39b)
$$

for every $v\in\tilde{\mathbb{V}}$ and the corresponding $\varphi\in\tilde{\mathbf{\Phi}}$. Using condensed notation, the above relations take the form:

$$
\begin{aligned}
&\varphi^{\hat{i}}=P^{\hat{i}}{}_{\hat{I}}v^{\hat{I}} \qquad v^{\hat{I}}=Q^{\hat{I}}{}_{\hat{i}}\varphi^{\hat{i}}\\
&Q^{\hat{I}}{}_{\hat{i}}P^{\hat{i}}{}_{\hat{J}}=\delta^{\hat{I}}{}_{\hat{J}} \qquad P^{\hat{i}}{}_{\hat{I}}Q^{\hat{I}}{}_{\hat{j}}=\delta^{\hat{i}}{}_{\hat{j}}.
\end{aligned} \quad (4.1.40)
$$

The operators P and Q will be called the 'picture-change' operators.

Kernels of the picture-change operators are easily found in practice. Consider, for example, a chiral scalar superfield $\Phi(z)$. Its component fields are given by equation (2.8.4). Then one acts in the manner:

$$
A(x)=\Phi|=\int \mathrm{d}^6z'\delta_+(z,z')\Phi(z')\bigg|_{\theta=\bar{\theta}=0}=\int \mathrm{d}^6z'\,\delta_+(z|,z')\Phi(z') \quad (4.1.41)
$$

$$\psi_\alpha(x)=\mathrm{D}_\alpha\Phi|=\int \mathrm{d}^8z'\,\delta^8(z|-z')\mathrm{D}'_\alpha\Phi(z')=-\int \mathrm{d}^8z'\{\mathrm{D}'_\alpha\delta^8(z|-z')\}\Phi(z')$$

$$=\frac{1}{4}\int \mathrm{d}^6z'\{\bar{\mathrm{D}}'^2\mathrm{D}'_\alpha\delta^8(z|-z')\}\Phi(z')$$

$$F(x)=-\frac{1}{4}\mathrm{D}^2\Phi|=-\frac{1}{4}\int \mathrm{d}^8z'\,\delta^8(z|-z')\mathrm{D}'^2\Phi(z')$$

$$=-\frac{1}{4}\int \mathrm{d}^8z'\{\mathrm{D}'^2\delta^8(z|-z')\}\Phi(z')=\frac{1}{16}\int \mathrm{d}^6z'\{\bar{\mathrm{D}}'^2\mathrm{D}'^2\delta^8(z|-z')\}\Phi(z')$$

where

$$z^A|\equiv(x^a,\theta^\alpha,\bar\theta_{\dot\alpha})|=(x^a,0,0).$$

It is an almost trivial task to extend this exercise to all admissible superfield types.

The simplest properties of the picture-change operators are:

$$\varepsilon(P^{\hat i}{}_{\hat I})=\varepsilon(Q^{\hat I}{}_{\hat i})=\varepsilon_i+\varepsilon_I \tag{4.1.42a}$$

$$(P^i{}_I(x,z'))^*=(-1)^{\varepsilon_I+\varepsilon_i\varepsilon_I}C^i{}_jP^j{}_J(x,z')C^J{}_I \tag{4.1.42b}$$

$$(Q^I{}_i(x,z'))^*=(-1)^{\varepsilon_i+\varepsilon_i\varepsilon_I}C^I{}_JQ^J{}_j(x,z')C^j{}_i \tag{4.1.42c}$$

They follow from equations (4.1.3, 14). Together with $P^{\hat i}{}_{\hat I}$ and $Q^{\hat I}{}_{\hat i}$, we introduce their supertranspose supermatrices by the rule

$$\begin{aligned}P^{\mathrm{sT}}{}_{\hat I}{}^{\hat i}&=(-1)^{\varepsilon_I+\varepsilon_i\varepsilon_I}P^{\hat i}{}_{\hat I}\\ Q^{\mathrm{sT}}{}_{\hat i}{}^{\hat I}&=(-1)^{\varepsilon_i+\varepsilon_i\varepsilon_I}Q^{\hat I}{}_{\hat i}.\end{aligned} \tag{4.1.43}$$

Then we have

$$\begin{aligned}\varphi^{\hat i}&=P^{\hat i}{}_{\hat I}v^{\hat I}=v^{\hat I}P^{\mathrm{sT}}{}_{\hat I}{}^{\hat i}\\ v^{\hat I}&=Q^{\hat I}{}_{\hat i}\varphi^{\hat i}=\varphi^{\hat i}Q^{\mathrm{sT}}{}_{\hat i}{}^{\hat I}.\end{aligned} \tag{4.1.44}$$

The fact that the picture-change operators are linear has many important consequences. First, note that the mappings (4.1.38) are uniquely extended up to the linear mappings

$$P\colon\ \tilde{\mathbb{V}}_c\to\mathbf{\Phi}_c\qquad Q\colon\ \mathbf{\Phi}_c\to\tilde{\mathbb{V}}_c$$

inverse to each other. Let $\mathbf{F}$ be a linear operator on $\tilde{\mathbb{V}}_c$. It induces the linear operators on $\mathbf{\Phi}_c$:

$$F\equiv P\mathbf{F}Q \tag{4.1.45a}$$

Obviously, we have

$$F^{\hat i}{}_{\hat j}=P^{\hat i}{}_{\hat I}\mathbf{F}^{\hat I}{}_{\hat J}Q^{\hat J}{}_{\hat j}. \tag{4.1.45b}$$

The use of equations (4.1.21) and (4.1.42*a*) leads to

$$\begin{aligned}\mathbf{sTr}\,F&=(-1)^{\varepsilon_{i}}F^{\hat{i}}{}_{\hat{i}}=(-1)^{\varepsilon_{i}}P^{\hat{i}}{}_{\hat{I}}\mathbf{F}^{\hat{I}}{}_{\hat{J}}Q^{\hat{J}}{}_{\hat{i}}\\&=(-1)^{\varepsilon_{I}}\mathbf{F}^{\hat{I}}{}_{\hat{J}}Q^{\hat{J}}{}_{\hat{i}}P^{\hat{i}}{}_{\hat{I}}=(-1)^{\varepsilon_{I}}\mathbf{F}^{\hat{I}}{}_{\hat{I}}\end{aligned}$$

therefore

$$\mathbf{sTr}\,\mathbf{F}=\mathrm{sTr}\,F. \tag{4.1.46}$$

Next, if $\mathbf{F}$ possesses a logarithm, the same is true for F and their superdeterminants coincide

$$\mathbf{sDet}\,\mathbf{F}=\mathrm{sDet}\,F. \tag{4.1.47}$$

As a result, it is inessential which picture, superfield or component field, is used for computing supertraces and superdeterminants.

Furthermore, consider a superfunctional $\Psi[v]$ on $\mathbb{V}$. Associated with $\Psi[v]$ is the functional $\Psi[\varphi]$ on $\mathbf{\Phi}$

$$\Psi[\varphi]\equiv\Psi[v(\varphi)]=\Psi[Q\varphi]. \tag{4.1.48}$$

Representing an arbitrary variation of Ψ in several forms,

$$\delta\Psi=\delta v^{\hat{I}}\frac{\vec{\delta}}{\delta v^{\hat{I}}}\Psi=\delta\varphi^{\hat{i}}\frac{\vec{\delta}}{\delta\varphi^{\hat{i}}}\Psi$$

or

$$\delta\Psi=\Psi\frac{\overleftarrow{\delta}}{\delta v^{\hat{I}}}\delta v^{\hat{I}}=\Psi\frac{\overleftarrow{\delta}}{\delta\varphi^{\hat{i}}}\delta\varphi^{\hat{i}}$$

and using equations (4.1.44), one obtains the relation between the functional and superfunctional derivatives:

$$\frac{\vec{\delta}}{\delta v^{\hat{I}}}=P^{\mathrm{sT}}{}_{\hat{I}}{}^{\hat{i}}\frac{\vec{\delta}}{\delta\varphi^{\hat{i}}}\qquad\frac{\vec{\delta}}{\delta\varphi^{\hat{i}}}=Q^{\mathrm{sT}}{}_{\hat{i}}{}^{\hat{I}}\frac{\vec{\delta}}{\delta v^{\hat{I}}} \tag{4.1.49a}$$

$$\frac{\overleftarrow{\delta}}{\delta v^{\hat{I}}}=\frac{\overleftarrow{\delta}}{\delta\varphi^{\hat{i}}}P^{\hat{i}}{}_{\hat{I}}\qquad\frac{\overleftarrow{\delta}}{\delta\varphi^{\hat{i}}}=\frac{\overleftarrow{\delta}}{\delta v^{\hat{I}}}Q^{\hat{I}}{}_{\hat{i}}. \tag{4.1.49b}$$

In particular, the following identities hold:

$$\Psi_{,\hat{I}}[v]=\Psi_{,\hat{i}}[\varphi]P^{\hat{i}}{}_{\hat{I}}\qquad\Psi_{,\hat{i}}[\varphi]=\Psi_{,\hat{I}}[v]Q^{\hat{I}}{}_{\hat{i}} \tag{4.1.50a}$$

$$\begin{aligned}{}_{\hat{I},}\Psi_{,\hat{J}}[v]&=P^{\mathrm{sT}}{}_{\hat{I}}{}^{\hat{i}}({}_{\hat{i},}\Psi_{,\hat{j}}[\varphi])P^{\hat{j}}{}_{\hat{J}}\\{}_{\hat{i},}\Psi_{,\hat{j}}[\varphi]&=Q^{\mathrm{sT}}{}_{\hat{i}}{}^{\hat{I}}({}_{\hat{I},}\Psi_{,\hat{J}}[v])Q^{\hat{J}}{}_{\hat{j}}\end{aligned} \tag{4.1.50b}$$

Ψ being arbitrary.

Provided one of the operators P or Q is at our disposal, its inverse can be obtained by the following scheme. Introduce a 'Poincaré invariant supermetric' on $\Delta\Phi$ of the form:

$$\mathrm{d}s^2 = \int \mathrm{d}^4x\, \delta\varphi^i(x)\eta_{ij}\, \delta\varphi^j(x) = \delta\varphi^{\underline{i}}\eta_{\underline{i}\underline{j}}\, \delta\varphi^{\underline{j}}$$
$$\eta_{\underline{i}\underline{j}} = \eta_{ij}\, \delta^4(x - x') \tag{4.1.51}$$

where η_{ij} is a constant invertible matrix subject to the requirements:

(1) η_{ij} is supersymmetric and does not possess the even–odd and odd–even pieces,

$$\eta_{ji} = (-1)^{\varepsilon_i + \varepsilon_j + \varepsilon_i\varepsilon_j}\eta_{ij}$$
$$(-1)^{\varepsilon_i + \varepsilon_j}\eta_{ij} = \eta_{ij} \tag{4.1.52}$$

(2) η_{ij} satisfies the reality condition

$$(\eta_{ij})^* = (-1)^{\varepsilon_i\varepsilon_j}\eta_{kl}C^k{}_iC^l{}_j \tag{4.1.53}$$

which implies that the supermetric takes real c-number values;

(3) η_{ij} is a Lorentz invariant tensor, that is, it is a descendant of the invariant tensors $(\sigma^a)_{\alpha\dot\alpha}$ and $(\tilde\sigma^a)^{\dot\alpha\alpha}$.

The inverse of η_{ij} will be denoted by η^{ij}, $\eta_{ik}\eta^{kj} = \delta_i{}^j$. Next, introduce a 'super Poincaré invariant supermetric' on $\Delta\mathbb{V}$ of the form

$$\mathrm{d}s^2 = \int \mathrm{d}^8z\, \delta V^{\underline{a}}(z)\eta_{\underline{ab}}\, \delta V^{\underline{b}}(z) + \left\{\int \mathrm{d}^6z\, \delta\Phi^{\underline{m}}(z)\eta_{\underline{mn}}\, \delta\Phi^{\underline{n}}(z) + \text{c.c.}\right\}$$
$$= \delta v^{\underline{I}}\eta_{\underline{IJ}}\, \delta v^{\underline{J}} \tag{4.1.54}$$
$$\eta_{\underline{IJ}} = \eta_{IJ}\, \delta^{(I)}(z, z').$$

Here

$$\eta_{IJ} = \begin{pmatrix} \eta_{\underline{ab}} & & 0 \\ & \eta_{\underline{mn}} & \\ 0 & & \eta_{\underline{\dot m\dot n}} \end{pmatrix} \tag{4.1.55}$$

is a constant invertible matrix subject to requirements completely similar to those imposed on η_{ij} (with obvious modifications). The inverse of η_{IJ} will be denoted by η^{IJ}, $\eta_{IK}\eta^{KJ} = \delta_I{}^J$. Since $\Delta\Phi$ is the component field space for $\Delta\mathbb{V}$, it seems reasonable to restrict the supermetric on $\Delta\Phi$ (4.1.51) to be induced by the supermetric on $\Delta\mathbb{V}$ (4.1.54),

$$\mathrm{d}s^2 = \delta\varphi^{\underline{i}}\, Q^{\mathrm{sT}}{}_{\underline{i}}{}^{\underline{I}}\eta_{\underline{IJ}}Q^{\underline{J}}{}_{\underline{j}}\, \delta\varphi^{\underline{j}}. \tag{4.1.56}$$

As a result, the supermetric on $\Delta\Phi$ proves to be super Poincaré invariant. It is easily seen that the induced supermetric satisfies all the criteria formulated above. Comparing equations (4.1.51) and (4.1.56) leads to

$$\eta_{\underline{i}\underline{j}} = Q^{\mathrm{sT}}{}_{\underline{i}}{}^{\underline{I}}\eta_{\underline{IJ}}Q^{\underline{J}}{}_{\underline{j}} \tag{4.1.57a}$$

hence

$$P^{\hat{i}}{}_{\hat{I}} = \eta^{\hat{i}\hat{I}} Q^{\mathrm{sT}}{}_{\hat{I}}{}^{\hat{J}} \eta_{\hat{J}\hat{I}}. \tag{4.1.57b}$$

Therefore, P is expressed via the supertranspose of Q. Supermetrics η_{ij}, η^{ij} and η_{IJ}, η^{IJ} may be used for lowering and raising indices of fields and superfields, respectively.

Let us give an application of equation (4.1.57). Consider a non-generate bi-linear form on $\Delta\mathbb{V}$

$$(\delta v_1, \delta v_2) = \delta v_1^{\hat{I}} \mathbf{H}_{\hat{I}\hat{J}} \delta v_2^{\hat{J}}$$

where $\mathbf{H}_{\hat{I}\hat{J}}$ is a c-type non-singular differential operator. It induces the bi-linear form on $\Delta\mathbf{\Phi}$

$$\begin{gathered}(\delta\varphi_1, \delta\varphi_2) = \delta\varphi_1^{\hat{i}} H_{\hat{i}\hat{j}} \delta\varphi_2^{\hat{j}} \\ H_{\hat{i}\hat{j}} = Q^{\mathrm{sT}}{}_{\hat{i}}{}^{\hat{I}} \mathbf{H}_{\hat{I}\hat{J}} Q^{\hat{J}}{}_{\hat{j}}.\end{gathered} \tag{4.1.58}$$

Now, defining

$$\mathbf{H}^{\hat{I}}{}_{\hat{J}} \equiv \eta^{\hat{I}\hat{K}} \mathbf{H}_{\hat{K}\hat{J}} \qquad H^{\hat{i}}{}_{\hat{j}} = \eta^{\hat{i}\hat{k}} H_{\hat{k}\hat{j}} \tag{4.1.59}$$

and using equation (4.1.57b), one finds

$$H^{\hat{i}}{}_{\hat{j}} = P^{\hat{i}}{}_{\hat{I}} \mathbf{H}^{\hat{I}}{}_{\hat{J}} Q^{\hat{J}}{}_{\hat{j}} \tag{4.1.60}$$

therefore

$$\mathrm{sDet}(\eta^{\hat{i}\hat{k}} H_{\hat{k}\hat{j}}) = \mathbf{sDet}(\eta^{\hat{I}\hat{K}} \mathbf{H}_{\hat{K}\hat{J}}). \tag{4.1.61}$$

This relation extends the identity (4.1.47) to the case of supermatrices with indices in the lower position.

4.2. Equivalence of component field and superfield perturbation theories

This section is devoted to demonstrating the equivalence of two perturbation theories for computing correlation functions, based on the use of the component field and superfield formulations, respectively, in a large class of non-gauge supersymmetric dynamical systems.

4.2.1. (Super)field Green's functions

Consider some relativistic field theory with a classical action $S[\varphi]$. In this section we make three basic assumptions about the structure of $S[\varphi]$. First, $S[\varphi]$ is a local functional of the form (3.1.11). Second, it is non-gauge, which means the dynamical subspace $\mathbf{\Phi}_0$ possesses a history φ_0

$$S_{,i}[\varphi_0] = 0 \tag{4.2.1}$$

such that the Hessian at this point

$$_{\hat{i},}S_{,\hat{j}}[\varphi_0] \tag{4.2.2}$$

is non-singular, that is, the equation

$$_{\hat{i},}S_{,\hat{j}}[\varphi_0]\,\delta\varphi^{\hat{j}}=0 \tag{4.2.3}$$

has no solutions with compact support in space–time,

$$\{_{\hat{i},}S_{,\hat{j}}[\varphi_0]\,\delta\varphi^{\hat{j}}\}\neq 0 \qquad \forall\ \delta\varphi\in\Delta\mathbf{\Phi}\backslash 0. \tag{4.2.4}$$

Third, $S[\varphi]$ is analytic in a neighbourhood of φ_0 in $\mathbf{\Phi}$:

$$S[\varphi]=S[\varphi_0]+\sum_{n=2}^{\infty}\frac{1}{n!}S_{,\hat{i}_1\ldots\hat{i}_n}[\varphi_0]\Delta\varphi^{\hat{i}_n}\ldots\Delta\varphi^{\hat{i}_1} \tag{4.2.5}$$

$$\Delta\varphi^{\hat{i}}=\varphi^{\hat{i}}-\varphi_0^{\hat{i}}.$$

Introducing the linearized action

$$S_0[\Delta\varphi;\,\varphi_0]=\frac{1}{2}\Delta\varphi^{\hat{i}}\,{}_{\hat{i},}S_{,\hat{j}}[\varphi_0]\Delta\varphi^{\hat{j}} \tag{4.2.6}$$

and the self-coupling

$$S_{\rm INT}[\Delta\varphi;\,\varphi_0]=\sum_{n=3}^{\infty}\frac{1}{n!}S_{,\hat{i}_1\ldots\hat{i}_n}[\varphi_0]\Delta\varphi^{\hat{i}_n}\ldots\Delta\varphi^{\hat{i}_1} \tag{4.2.7}$$

the action can be written in the form

$$S[\varphi]=S[\varphi_0]+S_0[\Delta\varphi;\,\varphi_0]+S_{\rm INT}[\Delta\varphi;\,\varphi_0]. \tag{4.2.8}$$

Let us remark that the theory with action (4.2.6) leads to the dynamical equations (4.2.3). The union of all sets of bounded fields $\{\Delta\varphi^{\hat{i}}(x)\}$, on which the linearized action takes finite values, defines the tangent space $\mathbb{T}_{\varphi_0}$ to $\mathbf{\Phi}$ at φ_0.

Because the Hessian (4.2.2) is non-singular it possesses Green's functions defined by

$$_{\hat{i},}S_{,\hat{j}}[\varphi_0]G^{\hat{j}\hat{k}}=-\delta_{\hat{i}}{}^{\hat{k}} \qquad G^{\hat{i}\hat{j}}{}_{\hat{j},}S_{,\hat{k}}[\varphi_0]=-\delta^{\hat{i}}{}_{\hat{k}}. \tag{4.2.9}$$

If $G^{\hat{i}\hat{j}}$ is a Green's function, its supertranspose

$$(G^{\rm sT})^{\hat{i}\hat{j}}\equiv(-1)^{\varepsilon_i\varepsilon_j}G^{\hat{j}\hat{i}} \tag{4.2.10}$$

turns out to be a Green's function too. To pick out a unique solution of equation (4.2.9), one must impose proper boundary conditions. The Green's functions which play a crucial role in classical theory are the retarded and advanced ones, $G_{\rm ret}$ and $G_{\rm adv}$. We anticipate that the reader is familiar with the boundary conditions appropriate for these Green's functions. Here it is worth recalling that $G_{\rm ret}$ is the supertranspose of $G_{\rm adv}$ and vice versa. The Green's function most important in quantum theory is known as the Feynman propagator G_F. Its defining property is that G_F propagates only positive (negative) frequency modes in the remote future (past). Next, G_F is known to coincide with its supertranspose

$$G_F{}^{\hat{j}\hat{i}} = (-1)^{\varepsilon_i \varepsilon_j} G_F{}^{\hat{i}\hat{j}}. \tag{4.2.11}$$

Both these properties make it possible to operate with the Feynman propagator as with an ordinary supermatrix. Let us recall that if H_{MN} is a supersymmetric supermatrix (1.9.78), as is the case for ${}_{\hat{i},}S_{,\hat{j}}[\varphi]$, then its unique inverse satisfies the supertransposition property (1.9.83), which is as equation (4.2.11). Next, if H_{MN} changes infinitesimally, $H \rightarrow H + \delta H$, then its inverse acquires the displacement

$$\delta H^{-1} = -H^{-1}(\delta H)H^{-1}.$$

Similarly, when the operator $H_{\hat{i}\hat{j}} \equiv {}_{\hat{i},}S_{,\hat{j}}[\varphi_0]$ suffers an infinitesimal disturbance,

$${}_{\hat{i},}S_{,\hat{j}}[\varphi_0] \rightarrow {}_{\hat{i},}S_{,\hat{j}}[\varphi_0 + \varepsilon\delta\varphi] \qquad \delta\varphi \in \Delta\Phi$$

where $\varepsilon \rightarrow 0$, its inverse $H^{-1} = -G_F$ changes by the above law.

Suppose the dynamical system under consideration is supersymmetric. Then it can be described in terms of the superfield variables $v^I(z)$, of which $\varphi^i(x)$ are the component fields, by re-expressing φ in $S[\varphi]$ via v in accordance with the instruction (4.1.39*a*):

$$S[v] = S[\varphi(v)] = S[Pv]. \tag{4.2.12}$$

Equation (4.1.50*a*) tells us that every stationary point of $S[\varphi]$ corresponds to a unique stationary point of $S[v]$ and vice versa,

$$S_{,\hat{i}}[\varphi_0] = 0 \Leftrightarrow S_{,\hat{I}}[v_0] = 0 \qquad v_0^{\hat{I}} = Q^{\hat{I}}{}_{\hat{i}}\varphi_0^{\hat{i}}. \tag{4.2.13}$$

In accordance with equation (4.1.50*b*), we have

$${}_{\hat{I},}S_{,\hat{J}}[v] = P^{\mathrm{sT}}{}_{\hat{I}}{}^{\hat{i}}\, {}_{\hat{i},}S_{,\hat{j}}[\varphi] P^{\hat{j}}{}_{\hat{J}} \tag{4.2.14}$$

therefore the condition (4.2.4) implies

$$\{{}_{\hat{I},}S_{,\hat{J}}[v_0]\delta v^{\hat{J}}\} \neq 0 \qquad \forall\ \delta v \in \Delta\mathbb{V}\backslash 0.$$

As a result, the Hessian (4.2.2) is non-singular if and only if the super Hessian (4.2.14) evaluated at $v_0 = Q\varphi_0$ is non-singular. Finally, the analyticity of $S[\varphi]$ implies the analyticity of $S[v]$, which is of the form

$$S[v] = S[v_0] + S_0[\Delta v; v_0] + S_{\mathrm{INT}}[\Delta v; v_0]$$

$$\Delta v^{\hat{I}} = v^{\hat{I}} - v_0^{\hat{I}} \tag{4.2.15}$$

where the superfunctional

$$S_0[\Delta v; v_0] = \frac{1}{2}\Delta v^{\hat{I}}\, {}_{\hat{I},}S_{,\hat{J}}[v_0]\Delta v^{\hat{J}} \tag{4.2.16}$$

represents the superfield version of the linearized action (4.2.6), and the superfunctional

$$S_{\mathrm{INT}}[\Delta v; v_0] = \sum_{n=3}^{\infty} \frac{1}{n!} S_{,I_1 \ldots I_n}[v_0] \Delta v^{I_n} \ldots \Delta v^{I_1} \tag{4.2.17}$$

is the superfield form of equation (4.2.7).

Let $G^{\hat{i}\hat{j}}$ be a Green's function of ${}_{\hat{i},}S_{,\hat{j}}[\varphi_0]$. Introduce the superfield two-point object

$$\mathbf{G}^{IJ} \equiv Q^I{}_{\hat{i}} G^{\hat{i}\hat{j}} Q^{\mathrm{sT}}{}_{\hat{j}}{}^J. \tag{4.2.18}$$

Because of relations (4.2.9) and (4.2.14) this satisfies the equations

$${}_{I,}S_{,J}[v_0]\mathbf{G}^{JK} = -\delta_I{}^K \qquad \mathbf{G}^{IJ}{}_{J,}S_{,K}[v_0] = -\delta^I{}_K. \tag{4.2.19}$$

As is seen, $\mathbf{G}^{IJ}$ is a Green's function of ${}_{I,}S_{,J}[v_0]$. When $G^{\hat{i}\hat{j}}$ is subjected to some boundary conditions, like the Feynman, retarded or advanced ones, the same boundary conditions are fulfilled by its superfield version $\mathbf{G}^{IJ}$. The superfield version $\mathbf{G}_{\mathrm{F}}{}^{IJ}$ of the Feynman propagator $G_{\mathrm{F}}{}^{\hat{i}\hat{j}}$ is called the 'Feynman superpropagator' and satisfies, due to equation (4.2.11), the relation

$$\mathbf{G}_{\mathrm{F}}{}^{JI} = (-1)^{\varepsilon_I \varepsilon_J} \mathbf{G}_{\mathrm{F}}{}^{IJ}. \tag{4.2.20}$$

Remark. Let $G^{\hat{i}\hat{j}}$ be a Green's function of ${}_{\hat{i},}S_{,\hat{j}}[\varphi_0]$ and $\mathbf{G}^{IJ}$ be its superfield version. Introducing the operators

$$\begin{aligned} H^{\hat{i}}{}_{\hat{j}} &= \eta^{\hat{i}\hat{k}}{}_{\hat{k},}S_{,\hat{j}}[\varphi_0] \qquad & \mathbf{H}^I{}_J &\equiv \eta^{IK}{}_{K,}S_{,J}[v_0] \\ G^{\hat{i}}{}_{\hat{j}} &= G^{\hat{i}\hat{k}}\eta_{\hat{k}\hat{j}} \qquad & \mathbf{G}^I{}_J &= \mathbf{G}^{IK}\eta_{KJ} \end{aligned} \tag{4.2.21}$$

one obtains

$$H^{\hat{i}}{}_{\hat{k}} G^{\hat{k}}{}_{\hat{j}} = -\delta^{\hat{i}}{}_{\hat{j}} \qquad \mathbf{H}^I{}_K \mathbf{G}^K{}_J = -\delta^I{}_J. \tag{4.2.22}$$

Since $H^{\hat{i}}{}_{\hat{j}}$ and $\mathbf{H}^I{}_J$ are connected by law (4.1.60), the last relations lead to

$$\mathbf{G}^I{}_J = Q^I{}_{\hat{i}} G^{\hat{i}}{}_{\hat{j}} P^{\hat{j}}{}_J. \tag{4.2.23}$$

4.2.2. Generating functional

Under the basic assumptions made in the previous subsection, it is well known that the in–out vacuum amplitude and the mean values of time-ordered functionals of quantized fields, like n-point Green's functions, are given by the Feynman functional integral:

$$\langle \mathrm{out} | \mathrm{in} \rangle = \mathcal{N}_{\Phi} \int \mathcal{D}\varphi\, \mathrm{e}^{(\mathrm{i}/\hbar)S[\varphi]} \tag{4.2.24}$$

$$\langle \Psi[\varphi] \rangle \equiv \frac{\langle \mathrm{out} | T(\Psi[\varphi]) | \mathrm{in} \rangle}{\langle \mathrm{out} | \mathrm{in} \rangle} = \frac{\int \mathcal{D}\varphi \Psi[\varphi] \exp\{(\mathrm{i}/\hbar)S[\varphi]\}}{\int \mathcal{D}\varphi \exp\{(\mathrm{i}/\hbar)S[\varphi]\}} \tag{4.2.25}$$

here $\hbar$ is the Planck constant, $\mathcal{N}_\Phi$ is a normalization constant, and the volume element is given by

$$\mathcal{D}\varphi = \prod_{i,x} \mathrm{d}\varphi^i(x).$$

The symbol φ in the first line of equation (4.2.25) is to be understood as a qunatum operator, in the second as an integration variable. The integrations in both expressions are over field histories satisfying boundary conditions appropriate to the in- and out-states chosen. In the WKB-approximation used in practice, the integrations are restricted to some complexified neighbourhood of a dynamical history $\varphi_0 \in \Phi_0$, which is taken in the role of the classical ground state, the neighbourhood being specified by the Feynman boundary conditions: pure positive (negative) frequency in the remote future (past). The Feynman propagator is the unique Green's function of ${}_{i,}S_{,j}[\varphi_0]$ corresponding to these boundary conditions. When operating with functional integrals (and later with superfunctional ones), we shall formally assume that all the properties of integration theory over $\mathbb{R}^{p|2q}$ listed in subsection 1.11.5 generalize to the case under consideration, simply replacing functions by (super)functionals, superdeterminants by (super)functional superdeterminants and so on.

As is well known, instead of computing chronological averages (4.2.25), for each $\Psi[\varphi]$, it proves sufficient to evaluate the generating functional for Green's functions

$$Z[J] = \mathcal{N}_\Phi \int \mathcal{D}\varphi \exp\left\{\frac{\mathrm{i}}{\hbar}(S[\varphi] + J_{\hat{i}}\varphi^{\hat{i}})\right\} \tag{4.2.26}$$

where $\{J_{\hat{i}}\}$ is an arbitrary external source under the requirement

$$\varepsilon(J_i(x)) = \varepsilon_i. \tag{4.2.27}$$

Because of the variational law

$$\frac{\vec{\delta}}{\delta J_{\hat{i}}} J_{\hat{j}} = \delta^{\hat{i}}{}_{\hat{j}}$$

one finds

$$\langle\Psi[\varphi]\rangle = (Z[0])^{-1}\Psi\left[\frac{\hbar}{\mathrm{i}}\frac{\vec{\delta}}{\delta J}\right] Z[J]\bigg|_{J=0} \tag{4.2.28}$$

for every functional Ψ of quantized fields. The standard way of perturbatively computing $Z[J]$ is to represent $S[\varphi]$ in the form (4.2.8) and to make use of the identity

$$Z[J] = \exp\left\{\frac{\mathrm{i}}{\hbar}(S[\varphi_0] + J_{\hat{i}}\varphi_0^{\hat{i}}\right) \exp\left\{\frac{\mathrm{i}}{\hbar} S_{\mathrm{INT}}\left[\frac{\hbar}{\mathrm{i}}\frac{\vec{\delta}}{\delta J}; \varphi_0\right]\right\} Z_0[J] \tag{4.2.29}$$

where

$$Z_0[J] = \mathcal{N}_\Phi \int \mathscr{D}\varphi \exp\left\{\frac{\mathrm{i}}{\hbar}(S[\Delta\varphi; \varphi_0] + J_{\hat{i}}\, \Delta\varphi^{\hat{i}})\right\} \tag{4.2.30}$$

is the generating functional of the linearized dynamical system (4.2.6). The final Gaussian integration can be easily performed resulting in

$$Z_0[J] = \mathrm{sDet}^{-1/2}\left(\frac{\mathrm{i}}{\hbar} H^{\hat{i}}{}_{\hat{j}}\right) \exp\left\{\frac{1}{2}\frac{\mathrm{i}}{\hbar} J_{\hat{i}} J_{\hat{j}} G_{\mathrm{F}}{}^{\hat{j}\hat{i}}\right\} \tag{4.2.31a}$$

or, equivalently

$$Z_0[J] = \mathrm{sDet}^{1/2}(\hbar G_{\mathrm{F}}{}^{\hat{i}}{}_{\hat{j}}) \exp\left\{\frac{1}{2}\frac{\mathrm{i}}{\hbar} J_{\hat{i}} J_{\hat{j}} G_{\mathrm{F}}{}^{\hat{j}\hat{i}}\right\} \tag{4.2.31b}$$

with $H^{\hat{i}}{}_{\hat{j}}$ and $G_{\mathrm{F}}{}^{\hat{i}}{}_{\hat{j}}$ being defined as in expressions (4.2.21). Here we have used the normalization condition

$$\mathcal{N}_\Phi \int \mathscr{D}\varphi \exp\left\{\frac{1}{2}\mathrm{i}(\varphi - \varphi_0)^{\hat{i}} \eta_{\hat{i}\hat{j}} (\varphi - \varphi_0)^{\hat{j}}\right\} = 1. \tag{4.2.32}$$

4.2.3. Generating superfunctional

By analogy with the standard Feynman rules formulated above, one is in a position to develop a superfield quantum theory. First, one can introduce quantized superfields $v^{\hat{I}}_{\mathrm{quant}}$ related to the quantized fields $\varphi^{\hat{i}}_{\mathrm{quant}}$ by the law

$$v^{\hat{I}}_{\mathrm{quant}} = Q^{\hat{I}}{}_{\hat{i}}\, \varphi^{\hat{i}}_{\mathrm{quant}}.$$

Next, given a superfunctional Ψ of quantized superfields, its quantum average can be defined by the superfunctional integral

$$\langle\langle \Psi[v] \rangle\rangle \equiv \frac{\int \mathscr{D}v \Psi[v] \exp\{(\mathrm{i}/\hbar) S[v]\}}{\int \mathscr{D}v \exp\{(\mathrm{i}/\hbar) S[v]\}} \tag{4.2.33}$$

where

$$\mathscr{D}v = \mathscr{D}V \mathscr{D}\Phi \mathscr{D}\bar{\Phi}$$

and superfunctional integrals, similarly to functional ones, should be understood as sequences of Gaussian integrations. In particular, the 'n-point superfield Green's function' is defined to be

$$\langle\langle v^{\hat{I}_1} v^{\hat{I}_2} \ldots v^{\hat{I}_n} \rangle\rangle \qquad n \geqslant 2. \tag{4.2.34}$$

Introduce an external 'supersource' $\{\mathbf{J}_{\hat{I}}\}$ representing a set of unconstrained, chiral and antichiral superfields

$$\mathbf{J}_{\mathcal{I}} = \mathbf{J}_{I}(z) = (\mathbf{J}_{\underline{a}}(z), \mathbf{J}_{\underline{m}}(z), \mathbf{J}_{\underline{\dot{m}}}(z))$$
$$\bar{\mathrm{D}}_{\dot{\alpha}}\mathbf{J}_{\underline{m}} = \mathrm{D}_{\alpha}\mathbf{J}_{\underline{\dot{m}}} = 0 \tag{4.2.35}$$
$$\varepsilon(\mathbf{J}_{I}(z)) = \varepsilon_{I}$$

and consider the 'generating superfunctional for Green's functions'

$$\mathbf{Z}[\mathbf{J}] = \mathcal{N}_{\mathrm{V}} \int \mathcal{D}v \exp\left\{\frac{\mathrm{i}}{\hbar}(S[v] + \mathbf{J}_{\mathcal{I}}v^{\mathcal{I}})\right\} \tag{4.2.36}$$

$\mathcal{N}_{\mathrm{V}}$ being a normalization constant. Then every average (4.2.33) can be represented in the form

$$\langle\langle \Psi[v] \rangle\rangle = (\mathbf{Z}[0])^{-1}\Psi\left[\frac{\hbar}{\mathrm{i}}\frac{\vec{\delta}}{\delta \mathbf{J}}\right]\mathbf{Z}[\mathbf{J}]\Big|_{\mathbf{J}=0} \tag{4.2.37}$$

with $\vec{\delta}/\delta J$ being defined as

$$\frac{\vec{\delta}}{\delta \mathbf{J}_{\mathcal{I}}}\mathbf{J}_{\mathcal{J}} = \delta^{\mathcal{I}}{}_{\mathcal{J}}. \tag{4.2.38}$$

Therefore, the generating superfunctional contains all necessary information about the superfield quantum theory. To compute $\mathbf{Z}[\mathbf{J}]$, we proceed in complete analogy with the calculation of $Z[J]$. Namely, we represent the classical action superfunctional in the form (4.2.15) and make use of the identity

$$\mathbf{Z}[\mathbf{J}] = \exp\left\{\frac{\mathrm{i}}{\hbar}(S[v_0] + \mathbf{J}_{\mathcal{I}}v_0^{\mathcal{I}})\right\}\exp\left\{\frac{\mathrm{i}}{\hbar}S_{\mathrm{INT}}\left[\frac{\hbar}{\mathrm{i}}\frac{\vec{\delta}}{\delta \mathbf{J}}; v_0\right]\right\}\mathbf{Z}_0[\mathbf{J}] \tag{4.2.39}$$

where

$$\mathbf{Z}_0[\mathbf{J}] = \mathcal{N}_{\mathrm{V}} \int \mathcal{D}v \exp\left\{\frac{\mathrm{i}}{\hbar}(S[\Delta v; v_0] + \mathbf{J}_{\mathcal{I}}\Delta v_0^{\mathcal{I}})\right\} \tag{4.2.40}$$

is the generating superfunctional of the linearized theory (4.2.16). Choosing the normalization condition

$$\mathcal{N}_{\mathrm{V}} \int \mathcal{D}v \exp\left\{\frac{1}{2}\mathrm{i}(v - v_0)^{\mathcal{I}}\eta_{\mathcal{I}\mathcal{J}}(v - v_0)^{\mathcal{J}}\right\} = 1. \tag{4.2.41}$$

the above Gaussian integral takes the form

$$\mathbf{Z}_0[\mathbf{J}] = \mathbf{sDet}^{-1/2}\left(\frac{\mathrm{i}}{\hbar}\mathbf{H}^{\mathcal{I}}{}_{\mathcal{J}}\right)\exp\left\{\frac{1}{2}\frac{\mathrm{i}}{\hbar}\mathbf{J}_{\mathcal{I}}\mathbf{J}_{\mathcal{J}}\mathbf{G}_{\mathrm{F}}{}^{\mathcal{J}\mathcal{I}}\right\} \tag{4.2.42a}$$

or, equivalently

$$\mathbf{Z}_0[\mathbf{J}]=\mathbf{sDet}^{1/2}\left(\hbar\mathbf{G}_{\mathrm{F}}{}^{\hat{I}}{}_{\hat{J}}\right)\exp\left\{\frac{1}{2}\frac{\mathrm{i}}{\hbar}\mathbf{J}_{\hat{I}}\mathbf{J}_{\hat{J}}\mathbf{G}_{\mathrm{F}}{}^{\hat{J}\hat{I}}\right\} \tag{4.2.42b}$$

with $\mathbf{H}^{\hat{I}}{}_{\hat{J}}$ and $\mathbf{G}_{\mathrm{F}}{}^{\hat{I}}{}_{\hat{J}}$ being defined as in expressions (4.2.21). Recalling the property (4.1.27) of the superfunctional superdeterminant, the final relation can be rewritten as

$$\mathbf{Z}_0[\mathbf{J}]=\mathbf{sDet}^{1/2}(\mathbf{G}_{\mathrm{F}}{}^{\hat{I}}{}_{\hat{J}})\exp\left\{\frac{1}{2}\frac{\mathrm{i}}{\hbar}\mathbf{J}_{\hat{I}}\mathbf{J}_{\hat{J}}\mathbf{G}_{\mathrm{F}}{}^{\hat{J}\hat{I}}\right\}. \tag{4.2.42c}$$

4.2.4. Coincidence of $Z[J]$ *and* $\mathbf{Z}[\mathbf{J}]$

We now show that $Z[J]$ coincides with $\mathbf{Z}[\mathbf{J}]$, the latter being considered as a functional of (properly defined) component sources of $\mathbf{J}_{\hat{I}}$.

To start with, let us define the component sources $J_{\hat{i}}$ of the supersources $\mathbf{J}_{\hat{I}}$ by the rule:

$$J_{\hat{i}}=\mathbf{J}_{\hat{I}}Q^{\hat{I}}{}_{\hat{i}}\Rightarrow\mathbf{J}_{\hat{I}}=J_{\hat{i}}P^{\hat{i}}{}_{\hat{I}}. \tag{4.2.43}$$

Such a choice of component sources provides us with the relation

$$\mathbf{J}_{\hat{I}}v^{\hat{I}}=J_{\hat{i}}\varphi^{\hat{i}}. \tag{4.2.44}$$

Considering now supersources and sources with upper indices defined as

$$\mathbf{J}^{\hat{I}}\equiv\mathbf{J}_{\hat{J}}\eta^{\hat{J}\hat{I}}\qquad J^{\hat{i}}\equiv J_{\hat{j}}\eta^{\hat{j}\hat{i}} \tag{4.2.45}$$

we note that they turn out, due to equation (4.1.57), to be connected to each other by the law

$$J^{\hat{i}}=P^{\hat{i}}{}_{\hat{I}}\mathbf{J}^{\hat{I}}\qquad\mathbf{J}^{\hat{I}}=Q^{\hat{I}}{}_{\hat{i}}J^{\hat{i}} \tag{4.2.46}$$

that is, in the same fashion as $v^{\hat{I}}$ and $\varphi^{\hat{i}}$ are connected.

Recalling the definition of $P^{s\mathrm{T}}$ and $Q^{s\mathrm{T}}$ (4.1.43), one finds

$$\begin{aligned}J_{\hat{i}}&=(-1)^{\varepsilon_{\hat{i}}+\varepsilon_{\hat{I}}}Q^{s\mathrm{T}}{}_{\hat{i}}{}^{\hat{I}}J_{\hat{I}}\\ \mathbf{J}_{\hat{I}}&=(-1)^{\varepsilon_{\hat{i}}+\varepsilon_{\hat{I}}}P^{s\mathrm{T}}{}_{\hat{I}}{}^{\hat{i}}J_{\hat{i}}.\end{aligned} \tag{4.2.47}$$

Now, the use of the relations (4.2.23) and (4.2.43, 47) gives

$$\mathbf{J}_{\hat{I}}\mathbf{J}_{\hat{J}}\mathbf{G}_{\mathrm{F}}{}^{\hat{J}\hat{I}}=J_{\hat{i}}J_{\hat{j}}G_{\mathrm{F}}{}^{\hat{j}\hat{i}}. \tag{4.2.48}$$

Furthermore, it is not difficult to prove that

$$S_{\mathrm{INT}}\left[\frac{\hbar}{\mathrm{i}}\frac{\vec{\delta}}{\delta\mathbf{J}};v_0\right]=S_{\mathrm{INT}}\left[\frac{\hbar}{\mathrm{i}}\frac{\vec{\delta}}{\delta J};\varphi_0\right]. \tag{4.2.49}$$

By definition, we have

$$S_{\mathrm{INT}}[\Delta\varphi;\varphi_0]=\sum_{q=3}^{\infty}\frac{1}{q!}S_{,\hat{i}_1\ldots\hat{i}_q}[\varphi_0]\Delta\varphi^{\hat{i}_q}\ldots\Delta\varphi^{\hat{i}_1}$$

and

$$S_{\rm INT}[\Delta v; v_0] = \sum_{q=3}^{\infty} \frac{1}{q!} S_{,\hat{i}_1 \ldots \hat{i}_q}[v_0] P^{\hat{i}_q}{}_{\hat{I}_q} \Delta v^{\hat{I}_q} \ldots P^{\hat{i}_1}{}_{\hat{I}_1} \Delta v^{\hat{I}_1}.$$

On the other hand, because of equation (4.2.43) we have

$$\frac{\vec{\delta}}{\delta J_{\hat{i}}} = P^{\hat{i}}{}_{\hat{I}} \frac{\vec{\delta}}{\delta \mathbf{J}_{\hat{I}}} \tag{4.2.50}$$

which confirms equation (4.2.49).

As the next step, it is worth recalling that the relation (4.1.45) implies (4.1.46). The Feynman propagator is connected to its superfield version by the law (4.2.18), which leads to relation (4.2.23), and hence

$$\mathbf{sDet}(\hbar \mathbf{G}_{\rm F}{}^{\hat{I}}{}_{\hat{J}}) = \mathrm{sDet}(\hbar G_{\rm F}{}^{\hat{i}}{}_{\hat{j}}). \tag{4.2.51}$$

Finally, equations (4.2.12) and (4.2.44) tell us that

$$S[v_0] + \mathbf{J}_{\hat{I}} v_0^{\hat{I}} = S[\varphi_0] + J_{\hat{i}} \varphi_0^{\hat{i}}. \tag{4.2.52}$$

Now, the relations (4.2.48, 49, 51, 52) mean

$$Z[J] = \mathbf{Z}[\mathbf{J}(J)] \qquad \mathbf{J}_{\hat{I}}(J) = J_{\hat{i}} P^{\hat{i}}{}_{\hat{I}}. \tag{4.2.53}$$

Our final result is seen to be of great importance. Its obvious consequence is the fact that any chronological average $\langle \Psi[\varphi] \rangle$, Ψ being some functional, can be read off from the corresponding superfield analogue $\langle\langle \Psi[v] \rangle\rangle$, where $\Psi[\varphi]$ and $\Psi[v]$ are connected as in equation (4.1.48). From equations (4.2.28, 37, 50, 53) one deduces

$$\langle \Psi[\varphi] \rangle = \langle\langle \Psi[\phi(v)] \rangle\rangle = \langle\langle \Psi[Pv] \rangle\rangle. \tag{4.2.54}$$

In particular, the n-point Green's functions $\langle \varphi^{\hat{i}_1} \varphi^{\hat{i}_2} \ldots \varphi^{\hat{i}_n} \rangle$, $n = 2, 3, \ldots$, are expressed via the superfield Green's functions (4.2.34) as follows:

$$\begin{gathered} \langle \varphi^{\hat{i}} \varphi^{\hat{j}} \rangle = P^{\hat{i}}{}_{\hat{I}} \langle\langle v^{\hat{I}} v^{\hat{J}} \rangle\rangle P^{\rm sT}{}_{\hat{J}}{}^{\hat{j}} \\ \langle \varphi^{\hat{i}} \varphi^{\hat{j}} \varphi^{\hat{k}} \rangle = (-1)^{\varepsilon_k(\varepsilon_I + \varepsilon_J) + \varepsilon_j \varepsilon_I} P^{\hat{i}}{}_{\hat{I}} P^{\hat{j}}{}_{\hat{J}} P^{\hat{k}}{}_{\hat{K}} \langle\langle v^{\hat{K}} v^{\hat{J}} v^{\hat{I}} \rangle\rangle \end{gathered} \tag{4.2.55}$$

and so on.

In conclusion, it is worth saying some words about the general structure of $Z[J]$ and $\mathbf{Z}[\mathbf{J}]$. Assuming that $Z[J]$ can be expanded in a functional Taylor series, one immediately finds

$$(Z[0])^{-1} Z[J] = 1 + \frac{\mathrm{i}}{\hbar} J_{\hat{i}} \langle \varphi^{\hat{i}} \rangle + \sum_{n=2}^{\infty} \frac{1}{n!} \left(\frac{\mathrm{i}}{\hbar} \right)^n J_{\hat{i}_n} \ldots J_{\hat{i}_2} J_{\hat{i}_1} \langle \varphi^{\hat{i}_1} \varphi^{\hat{i}_2} \ldots \varphi^{\hat{i}_n} \rangle \tag{4.2.56}$$

where the quantized field average

$$\langle \varphi^{\hat{i}} \rangle = \frac{\hbar}{\mathrm{i}} \frac{\vec{\delta}}{\delta J_{\hat{i}}} \ln Z[J] \bigg|_{J=0} \tag{4.2.57}$$

is said to be the 'mean field'. Analysing the relations (4.2.29) and (4.2.31), one readily notes that

$$\langle \varphi^{i} \rangle = \varphi_0^{i} + O(\hbar) \tag{4.2.58}$$

so the mean field differs from the classical field φ_0 by quantum corrections. Further, the two-point correlator $\langle \varphi\varphi \rangle$ turns out to be of the form

$$\langle \varphi^{i}\varphi^{j} \rangle = \langle \varphi^{i} \rangle \langle \varphi^{j} \rangle + \frac{\hbar}{i} G_F{}^{ij} + O(\hbar^2). \tag{4.2.59}$$

In perfect analogy with $Z[J]$, the generating superfunctional has the form

$$(\mathbf{Z}[0])^{-1}\mathbf{Z}[\mathbf{J}] = 1 + \frac{i}{\hbar}\mathbf{J}_{I}\langle\langle v^{I} \rangle\rangle + \sum_{n=2}^{\infty} \frac{1}{n!}\left(\frac{i}{\hbar}\right)^{n} \mathbf{J}_{I_n}\ldots\mathbf{J}_{I_2}\mathbf{J}_{I_1}\langle\langle v^{I_1}v^{I_2}\ldots v^{I_n}\rangle\rangle. \tag{4.2.60}$$

The quantized superfield average

$$\langle\langle v^{I} \rangle\rangle = \frac{\hbar}{i}\frac{\vec{\delta}}{\delta \mathbf{J}_{I}} \ln \mathbf{Z}[\mathbf{J}]\bigg|_{\mathbf{J}=0} \tag{4.2.61}$$

will be called the 'mean superfield'. The mean field and the mean superfield are connected as follows

$$\langle \varphi^{i} \rangle = P^{i}{}_{I}\langle\langle v^{I} \rangle\rangle. \tag{4.2.62}$$

In other words, the union $\{\langle \varphi^{i} \rangle\}$ presents the full set of component fields for $\{\langle\langle v^{I} \rangle\rangle\}$. Similarly to equation (4.2.59), the two-point superfield correlator is of the form

$$\langle\langle v^{I}v^{J} \rangle\rangle = \langle\langle v^{I} \rangle\rangle\langle\langle v^{J} \rangle\rangle + \frac{\hbar}{i}\mathbf{G}_F{}^{IJ} + O(\hbar^2). \tag{4.2.63}$$

In summary, we have shown that ordinary perturbation theory and the superfield version defined above are equivalent for non-gauge supersymmetric systems. From now on, all computation work can be done using superfields, without any reference to components. Return to component fields should be carried out only at the final stage, with the help of the projection rule (4.2.54).

4.3. Effective action (super)functional

In this section we introduce a central object of modern quantum field theory—the effective action, in both component field and superfield approaches. A brief review of the renormalization theory will be given. We also present a wide class of (pathological) supersymmetric models finite at each order of perturbation theory, with trivial S-matrix.

4.3.1. Effective action

We begin by recalling that associated with the generating functional for Green's functions $Z[J]$ is the generating functional for connected Green's functions $W[J]$ defined by

$$Z[J]=e^{(i/\hbar)W[J]}. \tag{4.3.1}$$

The expansion of $W[J]$ in a power series in J reads

$$W[J]=W[0]+J_{\hat{i}}\langle\varphi^{\hat{i}}\rangle+\sum_{n=2}^{\infty}\frac{1}{n!}J_{\hat{i}_n}\ldots J_{\hat{i}_2}J_{\hat{i}_1}G_c{}^{\hat{i}_1\hat{i}_2\ldots\hat{i}_n} \tag{4.3.2}$$

$G_c{}^{\hat{i}_1\hat{i}_2\ldots\hat{i}_n}$ being the n-point connected Green's function. $G_c{}^{\hat{i}\hat{j}}$ is usually called the full propagator and is related to the two-point correlator (4.2.59) in the manner:

$$\begin{gathered}\langle\varphi^{\hat{i}}\varphi^{\hat{j}}\rangle=\langle\varphi^{\hat{i}}\rangle\langle\varphi^{\hat{j}}\rangle+\frac{\hbar}{i}G_c{}^{\hat{i}\hat{j}}\\ G_c{}^{\hat{i}\hat{j}}=G_F{}^{\hat{i}\hat{j}}+O(\hbar).\end{gathered} \tag{4.3.3}$$

As is seen, the full propagator coincides with the Feynman propagator in the limit $\hbar\to 0$; on these grounds, G_F is often called the tree-level propagator. By functionally differentiating $W[J]$, one obtains the functionals

$$\tilde{\varphi}^{\hat{i}}=\tilde{\varphi}^{\hat{i}}[J]\equiv\frac{\vec{\delta}}{\delta J_{\hat{i}}}W[J]=\langle\varphi^{\hat{i}}\rangle+J_{\hat{j}}G_c{}^{\hat{j}\hat{i}}+\sum_{n=2}^{\infty}\frac{1}{n!}J_{\hat{j}_n}\ldots J_{\hat{j}_1}G_c{}^{\hat{j}_1\ldots\hat{j}_n\hat{i}} \tag{4.3.4}$$

known as the mean field in the presence of sources. This coincides with the mean field when the sources are switched off

$$\tilde{\varphi}|_{J=0}=\langle\varphi\rangle. \tag{4.3.5}$$

Owing to the non-singularity of $G_F{}^{\hat{i}\hat{j}}$, the full propagator considered as a power series in $\hbar$ turns out to be perturbatively non-singular. Then relation (4.3.4) can be uniquely resolved with respect to J resulting in a functional of $\tilde{\varphi}$, $J_{\hat{i}}=J_{\hat{i}}[\tilde{\varphi}]$. Now, making the Legendre transform of $W[J]$, one arrives at the following functional

$$\Gamma[\tilde{\varphi}]=W[J[\tilde{\varphi}]]-J_{\hat{i}}[\tilde{\varphi}]\tilde{\varphi}^{\hat{i}} \tag{4.3.6}$$

which is called the effective action. It possesses the obvious property

$$\Gamma[\tilde{\varphi}]\frac{\overleftarrow{\delta}}{\delta\tilde{\varphi}^{\hat{i}}}=-J_{\hat{i}}. \tag{4.3.7}$$

Setting here $J=0$, hence $\tilde{\varphi}=\langle\varphi\rangle$, gives

$$\Gamma_{,\hat{i}}[\langle\varphi\rangle]=0 \tag{4.3.8}$$

therefore the mean field is a stationary point of $\Gamma[\tilde{\varphi}]$. Next, by functionally differentiating equation (4.3.7) and using the identity

$$\frac{\vec{\delta}}{\delta J_{\hat{i}}} \tilde{\varphi}^{j} \bigg|_{J=0} = G_c{}^{\hat{i}j}$$

which immediately follows from (4.3.4), one obtains

$$_{\hat{i},}\Gamma_{,j}[\langle\varphi\rangle] G_c{}^{j\hat{k}} = -\delta_{\hat{i}}{}^{\hat{k}}. \tag{4.3.9}$$

Comparing the relations (4.3.8) and (4.3.9) with (4.2.1) and (4.2.9), respectively, and also recalling that $\langle\varphi\rangle$ has been interpreted, due to relation (4.2.58), as the quantum analogue of φ_0, it seems reasonable to interpret $\Gamma[\tilde{\varphi}]$ as a quantum analogue of the classical action $S[\varphi]$.

Clearly, knowledge of $\Gamma[\tilde{\varphi}]$ makes it possible to restore $W[J]$, with the help of the inverse Legendre transform, and therefore to calculate arbitrary chronological averages, using the prescription (4.2.28). On the other hand, it is well known that the S-matrix is uniquely read off from $\Gamma[\tilde{\varphi}]$. As a result, all necessary information about quantum theory is encoded in the effective action. That is why the effective action is said to be the central object of quantum field theory.

$\Gamma[\tilde{\varphi}]$ is often called the generating functional for vertex functions, the n-point vertex function being defined as

$$\Gamma_{\hat{i}_1\hat{i}_2\ldots\hat{i}_n} = \Gamma_{,\hat{i}_1\hat{i}_2\ldots\hat{i}_n}[\tilde{\varphi}]|_{\tilde{\varphi}=\langle\varphi\rangle} \qquad n \geqslant 2. \tag{4.3.10}$$

These functions appear in a functional Taylor series for $\Gamma[\tilde{\varphi}]$:

$$\Gamma[\tilde{\varphi}] = \Gamma[\langle\varphi\rangle] + \sum_{n=2}^{\infty} \frac{1}{n!} \Gamma_{\hat{i}_1\hat{i}_2\ldots\hat{i}_n} \Delta\tilde{\varphi}^{\hat{i}_n} \ldots \Delta\tilde{\varphi}^{\hat{i}_2} \Delta\varphi^{\hat{i}_1}$$
$$\Delta\tilde{\varphi} = \tilde{\varphi} - \langle\varphi\rangle. \tag{4.3.11}$$

Every vertex function is known to be represented by one-particle irreducible Feynman diagrams without external lines (recall that a connected Feynman diagram is said to be one-particle irreducible if it cannot be transformed into two disconnected pieces by cutting one internal line).

Making use of the relations (4.2.26), (4.3.1) and (4.3.6, 7) leads to the functional integral representation for $\Gamma[\tilde{\varphi}]$:

$$e^{(i/\hbar)\Gamma[\tilde{\varphi}]} = \mathcal{N}_{\Phi} \int \mathcal{D}\varphi \, e^{(i/\hbar)(S[\varphi] - \Gamma_{,i}[\tilde{\varphi}](\varphi^i - \tilde{\varphi}^i))} \tag{4.3.12}$$

which proves to give a self-consistent way for perturbatively computing $\Gamma[\tilde{\varphi}]$. One starts by performing the change of integration variables $\varphi^i = \tilde{\varphi}^i + \hbar^{1/2}\chi^i$ (whose Jacobian is assumed to be unity, which is the case for supersymmetric theories) and expanding $S[\varphi] = S[\tilde{\varphi} + \hbar^{1/2}\chi]$ in a power series in $\hbar^{1/2}$. This gives

$$e^{(i/\hbar)\Delta\Gamma[\tilde{\varphi}]} = \mathcal{N}_{\Phi} \int \mathcal{D}\chi \exp\Bigg\{ i\bigg(\frac{1}{2}\chi^{\hat{i}}{}_{\hat{i},}S_{,\hat{j}}[\tilde{\varphi}]\chi^{\hat{j}}$$

$$+ \sum_{n=3}^{\infty} \frac{\hbar^{n/2-1}}{n!} S_{,\hat{i}_1 \ldots \hat{i}_n}[\tilde{\varphi}]\chi^{\hat{i}_n} \ldots \chi^{\hat{i}_1} - \hbar^{-1/2}\, \Delta\Gamma_{,\hat{i}}[\tilde{\varphi}]\chi^{\hat{i}} \Bigg\} \tag{4.3.13}$$

$$\Delta\Gamma[\tilde{\varphi}] = \Gamma[\tilde{\varphi}] - S[\tilde{\varphi}].$$

Next, adopting the ansatz

$$\Gamma[\tilde{\varphi}] = S[\tilde{\varphi}] + \sum_{n=1}^{\infty} \hbar^n \Gamma^{(n)}[\tilde{\varphi}] \tag{4.3.14}$$

the above integral relation makes it possible to determine $\Gamma^{(n+1)}$, provided all $\Gamma^{(i)}$, $i = 1, \ldots, n$, are known. $\Gamma^{(n)}$ represents the nth quantum correction to the effective action. It can also be shown that Feynman diagrams contributing to $\Gamma^{(n)}$ contain exactly n internal loops, so the expansion (4.3.14) is known as the 'loop expansion'.

As a simple example, let us calculate the one-loop contribution to $\Gamma[\tilde{\varphi}]$. For this purpose we keep only $\hbar$-independent terms in both sides of (4.3.13):

$$e^{i\Gamma^{(1)}[\tilde{\varphi}]} = \mathcal{N}_{\Phi} \int \mathcal{D}\chi \exp\left\{ i\frac{1}{2}\chi^{\hat{i}}{}_{\hat{i},}S_{,\hat{j}}[\tilde{\varphi}]\chi^{\hat{j}} \right\} = \mathrm{sDet}^{-1/2}(\eta^{\hat{i}\hat{k}}{}_{\hat{k},}S_{,\hat{j}}[\tilde{\varphi}]) \tag{4.3.15}$$

(compare this result with (4.2.31a)). Then we write

$$\Gamma^{(1)}[\tilde{\varphi}] = \frac{i}{2}\ln\Big(\mathrm{sDet}(\eta^{\hat{i}\hat{k}}{}_{\hat{k},}S_{,\hat{j}}[\tilde{\varphi}])\Big) = \frac{i}{2}\,\mathrm{sTr}\ln(\eta^{\hat{i}\hat{k}}{}_{\hat{k},}S_{,\hat{j}}[\tilde{\varphi}])$$

therefore the effective action up to second order in $\hbar$ is

$$\Gamma[\tilde{\varphi}] = S[\tilde{\varphi}] + \frac{i}{2}\hbar\,\mathrm{sTr}\ln(\eta^{\hat{i}\hat{k}}{}_{\hat{k},}S_{,\hat{j}}[\tilde{\varphi}]) + O(\hbar^2). \tag{4.3.16}$$

It should be remarked that relations (4.3.13) and (4.3.14) allow us to represent quantum corrections to $\Gamma[\tilde{\varphi}]$ as one-particle irreducible Feynman graphs without external lines arising in a perturbation theory in which the role of propagator is played by the Green's function of ${}_{\hat{i},}S_{,\hat{j}}[\tilde{\varphi}]$ under the Feynman boundary conditions, the functions $S_{,\hat{i}_1 \ldots \hat{i}_n}[\tilde{\varphi}]$ play the role of vertices, $\tilde{\varphi}$ being considered as an arbitrary background field.

In the case of a supersymmetric dynamical system, the effective action turns out to be the component form of a superfunctional defined in perfect analogy with $\Gamma[\tilde{\varphi}]$. Let us introduce the generating superfunctional for connected Green's functions

$$\mathbf{W}[\mathbf{J}] = \frac{\hbar}{i}\ln \mathbf{Z}[\mathbf{J}] \tag{4.3.17}$$

define the mean superfield in the presence of supersources

$$\tilde{v}^{\mathcal{I}} = \frac{\vec{\delta}}{\delta \mathbf{J}_{\mathcal{I}}} \mathbf{W}[\mathbf{J}] \tag{4.3.18}$$

and perform the Legendre transform of $\mathbf{W}[\mathbf{J}]$

$$\Gamma[\tilde{v}] = \mathbf{W}[\mathbf{J}[\tilde{v}]] - J_{\mathcal{I}}[\tilde{v}]\tilde{v}^{\mathcal{I}} \tag{4.3.19}$$

where **J**s should be expressed via $\tilde{v}$s with the help of equation (4.3.18). The superfunctional $\Gamma[\tilde{v}]$ will be called the 'effective action superfunctional'. Using the relations (4.2.50) and (4.2.53), one immediately has

$$\Gamma[\tilde{\varphi}] = \Gamma[\tilde{v}(\tilde{\varphi})] \qquad \tilde{v}^{\mathcal{I}}(\tilde{\varphi}) = Q^{\mathcal{I}}{}_{\hat{i}}\tilde{\varphi}^{\hat{i}}. \tag{4.3.20}$$

The trivial consequence of this relation is that the component loop expansion (4.3.14) implies the superfield loop expansion

$$\Gamma[\tilde{v}] = S[\tilde{v}] + \sum_{n=1}^{\infty} \hbar^n \Gamma^{(n)}[\tilde{v}] \tag{4.3.21}$$

and vice versa.

One can easily obtain the following superfunctional integral representation for $\Gamma[\tilde{v}]$:

$$\mathrm{e}^{(\mathrm{i}/\hbar)\Gamma[\tilde{v}]} = \mathcal{N}_{\mathrm{V}} \int \mathcal{D}v \, \mathrm{e}^{(\mathrm{i}/\hbar)(S[v] - \Gamma_{,\mathcal{I}}[\tilde{v}](v^{\mathcal{I}} - \tilde{v}^{\mathcal{I}}))}. \tag{4.3.22}$$

At the one-loop level, $\Gamma[\tilde{v}]$ is of the form

$$\Gamma[\tilde{v}] = S[\tilde{v}] + \frac{\mathrm{i}}{2}\hbar \, \mathbf{sTr} \ln(\eta^{\mathcal{I}\mathcal{K}}{}_{\mathcal{K},}S_{,\mathcal{J}}[\tilde{v}]) + O(\hbar^2). \tag{4.3.23}$$

Relation (4.3.20) tells us that all necessary information about the supersymmetric quantum theory is encoded in the effective action superfunctional.

4.3.2. Super Poincaré invariance of $\mathbf{W}[\mathbf{J}]$ *and* $\Gamma[\tilde{v}]$

Let us discuss the transformation properties of the superfunctionals $\mathbf{W}[\mathbf{J}]$ and $\Gamma[\tilde{v}]$ with respect to the super Poincaré group.

At a purely formal level, the question is resolved as follows. In a supersymmetric dynamical system, the classical action is invariant,

$$S[g \cdot v] = S[v] \tag{4.3.24}$$

under arbitrary super Poincaré transformations

$$v^{\mathcal{I}} \to g \cdot v^{\mathcal{I}} = \mathrm{e}^{\mathrm{i}(-b^a\mathbf{P}_a + [1/2]K^{ab}\mathbf{J}_{ab} + \epsilon\mathbf{Q} + \bar{\epsilon}\bar{\mathbf{Q}})} v^{\mathcal{I}} = (T(g))^{\mathcal{I}}{}_{\mathcal{J}} v^{\mathcal{J}} \tag{4.3.25}$$

the super Poincaré generators being given in (2.4.36). Any mapping (4.3.25), considered as a replacement of variables in a superfunctional integral, is characterized by the Jacobian

$$\mathbf{sDet}^{-1}\left((g\cdot v^{I})\frac{\overleftarrow{\delta}}{\delta v^{J}}\right)=\mathrm{sDet}^{-1}(T(g)).$$

Owing to the explicit structure of the super Poincaré generators, the operator in the exponential in expression (4.3.25) is of the type (4.1.28). Then, in accordance with Theorem 2 from subsection 4.1.2, we have

$$\mathbf{sDet}\left((g\cdot v^{I})\frac{\overleftarrow{\delta}}{\delta v^{J}}\right)=1 \qquad \forall\ g\in S\Pi. \tag{4.3.26}$$

As a result, all the super Poincaré transformations are volume-preserving:

$$\mathscr{D}(g\cdot v)=\mathscr{D}v \qquad \forall\ g\in S\Pi. \tag{4.3.27}$$

Now, setting the supersource $\mathbf{J}$ to transform contragradiently to v,

$$(g\cdot\mathbf{J})_{I}(g\cdot v)^{I}=\mathbf{J}_{I}v^{I} \qquad \forall\ g\in S\Pi \tag{4.3.28}$$

we obtain, using equations (4.3.24, 27, 28),

$$\begin{aligned}\mathrm{e}^{(\mathrm{i}/\hbar)\mathbf{W}[g\cdot\mathbf{J}]}&=\mathscr{N}_{\mathbb{V}}\int\mathscr{D}v\exp\left\{\frac{\mathrm{i}}{\hbar}(S[v]+(g\cdot\mathbf{J})_{I}v^{I})\right\}\\&=\mathscr{N}_{\mathbb{V}}\int\mathscr{D}(g^{-1}v)\exp\left\{\frac{\mathrm{i}}{\hbar}(S[g^{-1}\cdot v]+\mathbf{J}_{I}(g^{-1}\cdot v)^{I})\right\}\end{aligned}$$

therefore

$$\mathbf{W}[g\cdot\mathbf{J}]=\mathbf{W}[\mathbf{J}] \qquad \forall\ g\in S\Pi. \tag{4.3.39}$$

This implies, as a result of relations (4.3.18, 19), super Poincaré invariance of the effective action

$$\Gamma[g\cdot\tilde{v}]=\Gamma[\tilde{v}] \qquad \forall\ g\in S\Pi. \tag{4.3.40}$$

The final relation is the quantum analogue of equation (4.3.24) and expresses the fact that the dynamical system under consideration is supersymmetric at the quantum level.

The previous discussion is not completely strict and requires some comment. First, the superfunctional $\mathbf{W}[\mathbf{J}]$, as it has been constructed above, depends parametrically on the dynamical history $v_0\in\mathbb{V}$ chosen in the role of the classical ground state, that is, $\mathbf{W}[\mathbf{J}]=\mathbf{W}[\mathbf{J};v_0]$. Second, when obtaining $\mathbf{W}[\mathbf{J};v_0]$, we have integrated over some neighbourhood Σ_{v_0} of v_0, which is specified by the Feynman boundary conditions. So, the strict definition of

$\mathbf{W}[\mathbf{J}; v_0]$ is given by

$$\exp\left\{\frac{\mathrm{i}}{\hbar}(\mathbf{W}[\mathbf{J}; v_0] - S[v_0] - \mathbf{J}_I v_0^I)\right\}$$
$$= \mathcal{N}_{\mathbb{V}} \int_{\Sigma_{v_0}} \mathcal{D}(\Delta v) \exp\left\{\frac{\mathrm{i}}{\hbar}(S[v_0 + \Delta v] - S[v_0] - \mathbf{J}_I \Delta v^I)\right\}. \quad (4.3.41)$$

Let g be an element of the super Poincaré group. It moves v_0 into the dynamical history $g \cdot v_0 \in \mathbb{V}_0$ and the domain Σ_{v_0} onto $\Sigma_{g \cdot v_0}$. On the basis of arguments similar to those used to obtain relation (4.3.39), one now easily obtains

$$\mathbf{W}[g \cdot \mathbf{J}; g \cdot v_0] = \mathbf{W}[J; v_0] \qquad \forall\ g \in S\Pi \quad (4.3.42)$$

which represents the corrected version of (4.3.39).

Suppose the dynamical subspace $\mathbb{V}_0$ possesses a 'super Poincaré invariant dynamical history' v_0, as is the case in practice, so that

$$g \cdot v_0 = v_0 \qquad \forall\ g \in S\Pi. \quad (4.3.43)$$

Setting here $g = \exp\{\mathrm{i}(\epsilon^\alpha \mathbf{Q}_\alpha + \bar{\epsilon}_{\dot{\alpha}} \bar{\mathbf{Q}}^{\dot{\alpha}})\}$, $\mathbf{Q}_\alpha$ and $\bar{\mathbf{Q}}^{\dot{\alpha}}$ being the supersymmetry generators, we conclude that

$$\mathbf{Q}_\alpha v_0^I(z) = \bar{\mathbf{Q}}_{\dot{\alpha}} v_0^I(z) = 0 \Rightarrow \partial_a v_0^I(z) = 0 \Rightarrow v_0^I = \text{constant} \quad (4.3.44)$$

where the anticommutation relation

$$\{\mathbf{Q}_\alpha, \bar{\mathbf{Q}}_{\dot{\alpha}}\} = -2\mathrm{i}(\sigma^a)_{\alpha\dot{\alpha}} \partial_a$$

has been used. Next, setting $g = \exp([\mathrm{i}/2]K^{ab}\mathbf{J}_{ab}\}$ in equation (4.3.43), one finds that only scalar components of $\{v_0^I\}$ may have non-vanishing values. Under equation (4.3.32), the domain Σ_{v_0} turns out to be super Poincaré invariant, and the relation (4.3.42) takes the form

$$\mathbf{W}[g \cdot \mathbf{J}; v_0] = \mathbf{W}[\mathbf{J}; v_0] \qquad \forall\ g \in S\Pi. \quad (4.3.45)$$

Then, the mean superfield proves to be constant

$$\langle\langle v^I \rangle\rangle = \text{constant}. \quad (4.3.46)$$

In what follows, as a rule, perturbation theory will be based on the use of a super Poincaré invariant dynamical history in the role of v_0. By redefining the superfield variables $v^I(z) \rightarrow v^I(z) - v_0^I$ and the classical action $S[v] \rightarrow S[v] - S[v_0]$, we arrive at the typical situation

$$v_0 = 0 \qquad S[0] = 0 \qquad S_{,I}[0] = 0. \quad (4.3.47)$$

In addition, we shall adopt

$$\langle\langle v^I \rangle\rangle = 0 \quad (4.3.48)$$

which is really the case under reasonable physical assumptions.

4.3.3. Short excursion into renormalization theory

The effective action is usually calculated perturbatively within the framework of the loop expansion. It is quite typical that vertex functions are given by Feynman integrals divergent at large momenta. In this case the theory considered is said to contain ultraviolet divergences. Renormalization is a special procedure for reconstructing the theory, leading to finite vertex functions (that is, to a finite quantum theory).

Normally, the renormalization procedure consists of two parts. The first is regularization. Regularization is a prescription according to which all initially divergent Feynman integrals determining vertex functions, are replaced by other finite integrals depending on some (regularization) parameter Λ. These integrals are said to be regularized. It is required that the regularized integrals formally reduce to the initial ones when the regularization parameter goes to some definite value Λ'. We say the limit $\Lambda \to \Lambda'$ corresponds to the regularization being taken away.

There are a lot of popular regularization schemes in quantum field theory, Pauli–Villars regularization, dimensional regularization, higher derivatives regularization and so on. Each of them has advantages and disadvantages, the choice of scheme is a matter of convenience.

The second part of the renormalization procedure consists in constructing counterterms. In more detail, suppose we have a theory of fields $\check{\varphi}^i$ (in this subsection index i corresponds to $\hat{i}$ in our previous notation) described by an action $S[\check{\varphi}; \check{R}]$, where $\check{R}$ denotes all parameters of the theory (i.e. masses and couplings). The fields are assumed, for simplicity, to be bosonic. Introduce another theory of fields φ^i described by an action $S[\varphi; R, \Lambda]$, with R being the corresponding parameters, of the form

$$S[\varphi; R, \Lambda] = S[\varphi; R] + \Delta S[\varphi; R, \Lambda]. \tag{4.3.49}$$

Here $S[\varphi; R]$ is the original action expressed via new variables φ^i and R, and functional $\Delta S[\varphi; R, \Lambda]$ is of the form

$$\Delta S[\varphi; R, \Lambda] = \sum_{r=1}^{\infty} \hbar^r \, \Delta S_r[\varphi; R, \Lambda]. \tag{4.3.50}$$

The ΔS_rs are chosen in such a way that the effective action of the theory with the classical action (4.3.49, 50) is finite at each order of the loop expansion after taking away the regularization. The functional ΔS is called the action of counterterms (or simply counterterms). A fundamental result of renormalization theory is that in a local field theory counterterms ΔS_r exist and are local functionals of the fields φ^i.

It quite often turns out that the functional $\Delta S[\varphi; R, \Lambda]$ has a structure duplicating that of the action $S[\varphi; R]$ in the following sense. Let

$$S[\check{\varphi}; \check{R}] = \frac{1}{2} S^{(2)}_{i_1 i_2} \check{\varphi}^{i_1} \check{\varphi}^{i_2} + \sum_{n=3}^{\infty} \check{R}^{(n)}_{i_1 i_2 \ldots i_n} \check{\varphi}^{i_1} \check{\varphi}^{i_2} \ldots \check{\varphi}^{i_n} \tag{4.3.51}$$

then

$$S[\varphi; R] = \frac{1}{2} S^{(2)}_{i_1 i_2} \varphi^{i_1} \varphi^{i_2} + \sum_{n=3}^{\infty} R^{(n)}_{i_1 i_2 \ldots i_n} \varphi^{i_1} \varphi^{i_2} \ldots \varphi^{i_n} \tag{4.3.52}$$

with $R^{(n)}_{i_1 i_2 \ldots i_n}$ being parameters. $\Delta S[\varphi; R, \Lambda]$ is said to have a structure duplicating that of $S[\varphi; R]$ if it is of the form

$$\Delta S[\varphi; R, \Lambda] = \frac{1}{2} S^{(2)}_{j_1 j_2} (\Delta Z_1)^{j_1 j_2}_{i_1 i_2} \varphi^{i_1} \varphi^{i_2} + \sum_{n=3}^{\infty} R^{(n)}_{j_1 j_2 \ldots j_n} (\Delta Z_n)^{j_1 j_2 \ldots j_n}_{i_1 i_2 \ldots i_n} \varphi^{i_1} \varphi^{i_2} \ldots \varphi^{i_n}. \tag{4.3.53}$$

It is important to notice that the coefficients of $S^{(2)}$ and $R^{(n)}$ are the same in equation (4.3.52) and equation (4.3.53). In this case the functional (4.3.49) takes the form

$$S[\varphi; R, \Lambda] = \frac{1}{2} S^{(2)}_{j_1 j_2} (\delta^{j_1 j_2}_{i_1 i_2} + (\Delta Z_1)^{j_1 j_2}_{i_1 i_2}) \varphi^{i_1} \varphi^{i_2}$$
$$+ \sum_{n=3}^{\infty} R^{(n)}_{j_1 j_2 \ldots j_n} (\delta^{j_1 j_2 \ldots j_n}_{i_1 i_2 \ldots i_n} + (\Delta Z_n)^{j_1 j_2 \ldots j_n}_{i_1 i_2 \ldots i_n}) \varphi^{i_1} \varphi^{i_2} \ldots \varphi^{i_n}. \tag{4.3.54}$$

Here $\delta^{j_1 j_2 \ldots j_n}_{i_1 i_2 \ldots i_n} = \delta^{j_1}_{(i_1} \delta^{j_2}_{i_2} \ldots \delta^{j_n}_{i_n)}$ is a symmetrized product of delta symbols. Let us denote

$$\begin{aligned} (Z_1)^{j_1 j_2}_{i_1 i_2} &= \delta^{j_1 j_2}_{i_1 i_2} + (\Delta Z_1)^{j_1 j_2}_{i_1 i_2} \\ (\bar{Z}_n)^{j_1 j_2 \ldots j_n}_{i_1 i_2 \ldots i_n} &= \delta^{j_1 j_2 \ldots j_n}_{i_1 i_2 \ldots i_n} + (\Delta Z_n)^{j_1 j_2 \ldots j_n}_{i_1 i_2 \ldots i_n}. \end{aligned} \tag{4.3.55}$$

Then equation (4.3.54) can be rewritten in the manner

$$S[\varphi; R, \Lambda] = \tfrac{1}{2} S^{(2)}_{j_1 j_2} (Z_1)^{j_1 j_2}_{i_1 i_2} \varphi^{i_1} \varphi^{i_2} + \sum_{n=3}^{\infty} R^{(n)}_{j_1 j_2 \ldots j_n} (\bar{Z}_n)^{j_1 j_2 \ldots j_n}_{i_1 i_2 \ldots i_n} \varphi^{i_1} \varphi^{i_2} \ldots \varphi^{i_n}. \tag{4.3.56}$$

Suppose that $(Z_1)^{j_1 j_2}_{i_2 i_2}$ has the following structure

$$(Z_1)^{j_1 j_2}_{i_1 i_2} = (Z_1^{1/2})^{j_1}{}_{(i_1} (Z_1^{1/2})^{j_2}{}_{i_2)}. \tag{4.3.57}$$

This implies that the action $S[\varphi; R, \Lambda]$ can be obtained from the initial one $S[\check{\varphi}; \check{R}]$ with the help of the transformation

$$\begin{aligned} \check{\varphi}^i &= (Z_1^{1/2})^i_j \varphi^j \\ \check{R}^{(n)}_{i_1 i_2 \ldots i_n} &= R^{(n)}_{j_1 j_2 \ldots j_n} (Z_n)^{j_1 j_2 \ldots j_n}_{i_1 i_2 \ldots i_n} \end{aligned} \tag{4.3.58}$$

where

$$(Z_n)^{j_1 j_2 \ldots j_n}_{i_1 i_2 \ldots i_n} = (Z_1^{-1/2})^{j_1}_{k_1} (Z_1^{-1/2})^{j_2}_{k_2} \ldots (Z_1^{-1/2})^{j_n}_{k_n} (\bar{Z}_n)^{k_1 k_2 \ldots k_n}_{i_1 i_2 \ldots i_n}. \tag{4.3.59}$$

The relations (4.3.58) are known as the renormalization transformation, the quantities Z_1 (4.3.57) and Z_n (4.3.59) are said to be the renormalization constants. The following terminology is often used in practice: the initial objects $S[\breve{\varphi}; \breve{R}]$, $\breve{\varphi}$ and $\breve{R}$ are called the bare action, fields and parameters, respectively; the final ones $S[\varphi; R, \Lambda]$, φ and R are called the renormalized action, fields and parameters, respectively. As is seen from the above discussion, the renormalized action is obtained from the bare one with the help of the renormalization transformation. Every theory possessing this property is said to be multiplicatively renormalizable.

It has been pointed out that regularization is an important element of renormalization. A chief requirement of any regularization scheme, besides those stated above, is to preserve, as much as possible, symmetries of the theory at hand. Consider, for example, the most popular dimensional regularization. In this scheme a four-dimensional theory is continued to a d-dimensional space–time; the role of regularization parameter Λ is played by the space–time dimension d, Λ' coincides with $d=4$. Dimensionally regularized Feynman integrals are finite at $d\neq 4$ and may possess poles of the form $(d-4)^{-k}$, where $k=1, 2, \ldots$. Counterterms must cancel these poles resulting with finite Green's function in the limit $d\to 4$. Clearly, dimensional regularization breaks down those symmetries of the initial classical theory which can be formulated only in four dimensions. For example, $d=4$ massless theories are known to be scale invariant, but this invariance does not survive under continuation to $d\neq 4$. As a result, one can expect the appearance of scale anomalies after renormalization. Similarly, applying dimensional regularization to supersymmetric field theories creates an analogous problem. The point is that supersymmetry can be formulated only in some exceptional dimensions.

Regularization schemes suitable for supersymmetric field theories are those which preserve supersymmetry at every stage of the renormalization procedure. Such regularizations are said to be supersymmetric. Since only the superfield formulation explicitly displays supersymmetry, it seems evident that supersymmetric regularizations should be defined in terms of superfields.

4.3.4. Finite pathalogical supersymmetric theories

One of the most remarkable properties of supersymmetry is that it leads to a great suppression of ultraviolet divergencies. Moreover, some super Yang–Mills theories are known to be completely divergenceless. The basis of this phenomenon is the fact that some divergent contributions, coming from bosonic and fermionic fields in a supersymmetric theory, cancel each other. A simple example of such cancellations is given by equation (4.1.27), which is equivalent to

$$\mathrm{sTr}(\mathbb{1})=(-1)^{\varepsilon_I}\delta^I{}_I=0. \qquad (4.3.60a)$$

This can be rewritten, using equation (4.1.46), as

$$(-1)^{\varepsilon_i}\delta^{\hat{i}}_{\hat{i}}=(-1)^{\varepsilon_i}\delta^i{}_i\int d^4x\,\delta^4(0)=0. \qquad (4.3.60b)$$

In spite of the divergent factor $\int d^4x\,\delta^4(0)$, the full expression vanishes, since in every supersymmetric field theory the numbers of bosonic and fermionic fields coincide, hence

$$(-1)^{\varepsilon_i}\delta^i{}_i=0. \qquad (4.3.61)$$

It is worth noting that many divergent structures arise with the factor $(-1)^{\varepsilon_i}\delta^i{}_i$. As will be demonstrated here, there exists a wide class of completely finite supersymmetric models which are, however, physically pathological. Of course, to construct realistic supersymmetric theories free from ultraviolet divergencies is much harder, but possible.

Consider the model of a real scalar superfield $V(z)=\bar{V}(z)$ with the classical action

$$S[V]=\int d^8z\left\{-\frac{1}{2}V\Box V+\mathscr{P}(V)\right\} \qquad (4.3.62)$$

$\mathscr{P}$ being a real function of ordinary real variables. For simplicity, we assume

$$\mathscr{P}(0)=\mathscr{P}'(0)=0.$$

In our case, a super Poincaré invariant supermetric of the form (4.1.54) is uniquely fixed modulo a constant set to unity:

$$ds^2=\int d^8z\,\delta V(z)\,\delta V(z). \qquad (4.3.63)$$

By superfunctionally differentiating $S[V]$ twice, one obtains

$$\begin{aligned}\frac{\delta^2 S[V]}{\delta V(z)\,\delta V(z')}&=-(\Box-\mathscr{P}''(V))\delta^8(z-z')\\&=-(\theta-\theta')^2(\bar{\theta}-\bar{\theta}')^2(\Box-\mathscr{P}''(V))\delta^4(x-x'). \qquad (4.3.64)\end{aligned}$$

In accordance with equation (4.3.22), the effective action is

$$e^{(i/\hbar)\Gamma[U]}=\mathscr{N}\int\mathscr{D}V\,e^{(i/\hbar)(S[V]-\Gamma'[U](V-U))}$$

where

$$\Gamma'[U](V-U)\equiv\int d^8z\,\frac{\delta\Gamma[U]}{\delta U(z)}(V(z)-U(z))$$

U being a scalar superfield. In accordance with equations (4.3.23) and

(4.3.63, 64), the one-loop correction to $\Gamma[U]$ reads

$$\Gamma^{(1)}[U]=\frac{\mathrm{i}}{2}\,\mathbf{sTr}\ln\left(\frac{\delta^2 S[V]}{\delta V(z)\delta V(z')}\right).$$

Since the operator (4.3.64) is of the type (4.1.26), we conclude

$$\Gamma^{(1)}[U]=0$$

as a consequence of Theorem 1 from subsection 4.1.2. But this is not the end of the story. With the help of the supergraph technique described in the following sections, one can readily find that all higher-loop corrections to $\Gamma[U]$ also vanish,

$$\Gamma^{(n)}[U]=0 \qquad n=2,3,\ldots.$$

Therefore, the theory under consideration is finite and trivial as a quantum system: its effective action coincides with the classical one

$$\Gamma[U]=S[U] \tag{4.3.65}$$

quantum corrections are absent.

From the physical point of view, the theory with the action (4.3.62) is pathological, since the kinetic terms of all component fermionic fields include second-order differential operators. On the other hand, the corresponding dynamical equation

$$\Box V=\mathscr{P}'(V)$$

is non-trivial, due to the arbitrariness of $\mathscr{P}(V)$. In this respect it seems unexpected that the effective action is trivial (hence, so is the S-matrix).

The example described can be generalized, for instance, to the case of models of (anti)chiral superfields $\bar{\Phi}(z)$ and $\Phi(z)$, $\bar{D}_{\dot{\alpha}}\Phi=0$, of the form

$$S[\Phi,\bar{\Phi}]=\int \mathrm{d}^6 z\left\{-\frac{1}{2}\Phi\Box\Phi+\mathscr{L}_{\mathrm{c}}(\Phi)\right\}+\text{c.c.}$$

More generally, every supersymmetric theory with an action $S[V^{\underline{a}},\Phi^{\underline{m}},\bar{\Phi}^{\underline{\dot{m}}}]$, which does not possess V–Φ, V–$\bar{\Phi}$ or $\bar{\Phi}$–Φ mixings or does not contain terms with spinor covariant derivatives acting on superfields, proves to be finite and trivial, since in such cases the super Lagrangians will have a structure similar to that discussed above.

4.4. The Wess–Zumino model: perturbative analysis

4.4.1. Preliminary discussion

In this section the superfield perturbation theory described above is illustrated using the example of the standard Wess–Zumino model (3.2.11). As will be shown later, a general Wess–Zumino model (3.2.8) fails to be multiplicatively renormalizable unless the chiral superpotential takes the form

$$\mathscr{L}_c(\Phi)=\frac{1}{2}m\Phi^2+\frac{1}{3}\lambda\Phi^3 \tag{4.4.1}$$

where m and λ are constants. This is why our consideration is mainly concentrated on the standard case.

We begin by discussing some general aspects essential to every supersymmetric model (3.2.8). The space of histories $\mathbb{V}$ of such a system is parametrized by pairs of superfields: a chiral scalar $\Phi(z)$ and its conjugate antichiral scalar $\bar{\Phi}(z)$. So, the super Poincaré invariant supermetric of the form (4.1.54) is fixed, modulo a complex constant, as follows

$$\mathrm{d}s^2=\int \mathrm{d}^6z\, \delta\Phi(z)\,\delta\Phi(z)+\text{c.c.} \tag{4.4.2}$$

Its component form defind by (4.1.56) reads

$$\mathrm{d}s^2=\int \mathrm{d}^4x\left\{2\delta A(x)\delta F(x)-\frac{1}{2}\,\delta\psi^\alpha(x)\delta\psi_\alpha(x)\right\}+\text{c.c.}$$

where A, ψ_α and F are the component fields of Φ (see equations (2.8.4) and (4.1.41)).

By superfunctionally differentiating $S[\Phi,\bar{\Phi}]$ (equation (3.2.8)), with the help of the variational rules (3.1.46), one obtains

$$\frac{\delta S}{\delta\Phi(z)}=-\frac{1}{4}\bar{\mathrm{D}}^2\bar{\Phi}+\mathscr{L}'_c(\Phi)\qquad \frac{\delta S}{\delta\bar{\Phi}(z)}=-\frac{1}{4}\mathrm{D}^2\Phi+\bar{\mathscr{L}}'_c(\bar{\Phi}). \tag{4.4.3}$$

Taking the second superfunctional derivatives gives

$$\begin{pmatrix} \dfrac{\delta^2 S}{\delta\Phi(z)\delta\Phi(z')} & \dfrac{\delta^2 S}{\delta\Phi(z)\delta\bar{\Phi}(z')} \\[2ex] \dfrac{\delta^2 S}{\delta\bar{\Phi}(z)\delta\Phi(z')} & \dfrac{\delta^2 S}{\delta\bar{\Phi}(z)\delta\bar{\Phi}(z')} \end{pmatrix}=\begin{pmatrix} \mathscr{L}''_c(\Phi)\delta_+(z,z') & -\frac{1}{4}\bar{\mathrm{D}}^2\delta_-(z,z') \\[2ex] -\frac{1}{4}\mathrm{D}^2\delta_+(z,z') & \bar{\mathscr{L}}''_c(\bar{\Phi})\delta_-(z,z') \end{pmatrix}. \tag{4.4.4}$$

The final expression represents the superkernel of the linear operator

$$\mathbf{H}^{(\Psi)}=\begin{pmatrix} \Psi(z) & -\frac{1}{4}\bar{\mathrm{D}}^2 \\[2ex] -\frac{1}{4}\mathrm{D}^2 & \bar{\Psi}(z) \end{pmatrix} \tag{4.4.5}$$

$$\Psi\equiv\mathscr{L}''_c(\Phi)\qquad \bar{\mathrm{D}}_{\dot{\alpha}}\Psi=0$$

which maps the tangent space $\mathbb{T}_{(\Phi,\bar{\Phi})}$ to $\mathbb{V}$ at $(\Phi, \bar{\Phi})$ into its dual space. For a fixed Ψ, the $\mathbf{H}^{(\Psi)}$ can be considered as a linear operator acting on $\tilde{\mathbb{V}}$, due to the explicit form of the supermetric (4.4.2).

The operator $\mathbf{H}^{(\Psi)}$ will play an important role in subsequent considerations. Here it is worth remarking that $\mathbf{H}^{(\Psi)}$ arises in the quadratic model of (anti)chiral scalars $\bar{\chi}$ and χ coupled to external (anti)chiral scalars $\bar{\Psi}$ and Ψ with the action

$$S[\chi, \bar{\chi}; \Psi, \bar{\Psi}] = \int d^8z \bar{\chi}\chi + \left\{\frac{1}{2}\int d^6z\, \Psi\chi^2 + \text{c.c.}\right\} \tag{4.4.6}$$

$$\bar{D}_{\dot{\alpha}}\chi = \bar{D}_{\dot{\alpha}}\Psi = 0.$$

When $\Psi = \mathscr{L}''_c(\Phi_0)$, Φ_0 being a stationary point of expression (3.2.8), this superfunctional represents the linearized version (defined, in general, by equation (4.2.16)) of the Wess–Zumino action.

A Green's function

$$\mathbf{G}^{(\Psi)} = \begin{pmatrix} \mathbf{G}_{++}(z, z') & \mathbf{G}_{+-}(z, z') \\ \mathbf{G}_{-+}(z, z') & \mathbf{G}_{--}(z, z') \end{pmatrix} \tag{4.4.7}$$

of the operator $\mathbf{H}^{(\Psi)}$ is defined to satisfy the equation

$$\mathbf{H}^{(\Psi)}\mathbf{G}^{(\Psi)} = -\mathbb{I} \tag{4.4.8}$$

where

$$\mathbb{I}(z, z') = \begin{pmatrix} \delta_+(z, z') & 0 \\ 0 & \delta_-(z, z') \end{pmatrix}. \tag{4.4.9}$$

The signs $(\pm)$ indicate that $\mathbf{G}^{(\Psi)}$ is chiral $(+)$ or antichiral $(-)$ with respect to the corresponding argument; in particular,

$$\bar{D}_{\dot{\alpha}}\mathbf{G}_{++}(z, z') = \bar{D}'_{\dot{\alpha}}\mathbf{G}_{++}(z, z') = 0.$$

Below $\mathbf{G}^{(\Psi)}$ will always mean the unique Green's function of $\mathbf{H}^{(\Psi)}$ under the Feynman boundary conditions.

4.4.2. Feynman superpropagator

In the case of the standard Wess–Zumino model, there are two super Poincaré invariant dynamical histories determined, due to equations (3.2.9) and (4.3.44), by the equation

$$\mathscr{L}'_c(\Phi_0) = m\Phi_0 + \frac{1}{2}\lambda\Phi_0^2 = 0$$

hence

$$\Phi_0^{(A)} = 0 \qquad \text{or} \qquad \Phi_0^{(B)} = -2m/\lambda. \tag{4.4.10}$$

Both histories can equally well be taken in the role of classical ground state, since they are the only points minimizing the scalar potential (3.2.18). Because of the identity

$$\mathscr{L}_c(\Phi)=\frac{2}{3}\frac{m^3}{\lambda^2}-\frac{m}{2}(\Phi-\Phi_0^{(B)})^2+\frac{1}{3!}\lambda(\Phi-\Phi_0^{(B)})^3$$

the use of the latter solution in equation (4.4.10) is equivalent to the use of the former corresponding to the chiral superpotential (4.4.1) with $m\to -m$ (physically, only the combination $|m|^2$ proves relevant). Therefore, we choose $\Phi_0=0$, and the decomposition (4.2.15) for the Wess–Zumino model takes the form

$$S[\Phi,\bar{\Phi}]=S_0[\Phi,\bar{\Phi}]+S_{\mathrm{INT}}[\Phi,\bar{\Phi}]$$

$$S_0[\Phi,\bar{\Phi}]=\int \mathrm{d}^8z\bar{\Phi}\Phi+\left\{\frac{1}{2}m\int \mathrm{d}^6z\Phi^2+\mathrm{c.c.}\right\} \tag{4.4.11}$$

$$S_{\mathrm{INT}}[\Phi,\bar{\Phi}]=\frac{1}{3!}\lambda\int \mathrm{d}^6z\,\Phi^3+\mathrm{c.c.}$$

As is seen, the linearized action is determined by the operator $\mathbf{H}^{(m)}$.

To find Green's functions of $\mathbf{H}^{(m)}$, it is worth recalling the identity

$$\bar{\mathrm{D}}_{\dot{\alpha}}\Phi=0 \quad\Rightarrow\quad \bar{\mathrm{D}}^2\mathrm{D}^2\Phi=16\Box\Phi. \tag{4.4.12}$$

Now, one easily verifies

$$\mathbf{H}^{(m)}\mathbf{H}^{(-\bar{m})}=(\Box-|m|^2). \tag{4.4.13}$$

As a result, the formal solution of equation (4.4.8) with $\Psi=m$ reads

$$\mathbf{G}^{(m)}=-\mathbf{H}^{(-\bar{m})}\frac{1}{\Box-|m|^2}. \tag{4.4.14}$$

The Feynman superpropagator is specified by the well-known ε-prescription:

$$\mathbf{G}\equiv\mathbf{G}^{(m)}=-\mathbf{H}^{(-\bar{m})}\frac{1}{\Box-|m|^2+\mathrm{i}\varepsilon} \qquad \varepsilon\to+0. \tag{4.4.15}$$

From now on, we set m to be real. Then $\mathbf{G}$ can be explicitly written as

$$\mathbf{G}=\frac{1}{\Box-m^2+\mathrm{i}\varepsilon}\begin{pmatrix} m\delta_+(z,z') & \frac{1}{4}\bar{\mathrm{D}}^2\delta_-(z,z') \\ \frac{1}{4}\mathrm{D}^2\delta_+(z,z') & m\delta_-(z,z') \end{pmatrix}. \tag{4.4.16}$$

This relation shows that the elements of $\mathbf{G}$ are expressed via the 'chiral scalar superpropagator'

$$\mathbf{G}_c(z, z') = -\frac{1}{\Box - m^2 + i\varepsilon}\,\delta_+(z, z') \tag{4.4.17a}$$

and the 'antichiral scalar superpropagator'

$$\mathbf{G}_a(z, z') = -\frac{1}{\Box - m^2 + i\varepsilon}\,\delta_-(z, z'). \tag{4.4.17b}$$

Both $\mathbf{G}_c$ and $\mathbf{G}_a$ admit a representation similar to the well-known proper-time representation for the scalar propagator invented by J. Schwinger:

$$-\frac{1}{\Box - m^2 + i\varepsilon}\,\delta^4(x - x') = i\int_0^\infty ds\, U(x, x'; s) \tag{4.4.18}$$

where $U(x, x'; s)$ is the unique solution of the equation

$$\left(i\frac{\partial}{\partial s} + \Box - m^2\right)U(s) = 0 \tag{4.4.19a}$$

under the initial condition

$$U(x, x'; s \to +0) = \delta^4(x - x'). \tag{4.4.19b}$$

$U(x, x'; s)$ reads

$$U(x, x'; s) = -\frac{i}{(4\pi s)^2}\exp\left\{i\left[\frac{(x - x')^2}{4s} - m^2 s\right]\right\}. \tag{4.4.20}$$

Using the explicit form for $\delta_+(z, z')$ given by equation (4.1.17), one now has

$$\begin{gathered}\mathbf{G}_c(z, z') = i\int_0^\infty ds\, \mathbf{U}_c(z, z'; s) \\ \mathbf{U}_c(z, z'; s) = -\frac{i}{(4\pi s)^2}(\theta - \theta')^2 \exp\left\{i\left[\frac{(x_{(+)} - x'_{(+)})^2}{4s} - m^2 s\right]\right\}.\end{gathered} \tag{4.4.21}$$

Obviously, $\mathbf{U}_c(s)$ is the unique solution of equation (4.4.19a) under the initial condition

$$\mathbf{U}_c(z, z'; s \to +0) = \delta_+(z, z') \tag{4.4.22}$$

and is chiral in both superspace arguments.

4.4.3. Generating superfunctional

Introduce an external chiral–antichiral supersource

$$(\mathbf{J}(z), \bar{\mathbf{J}}(z)) \qquad \bar{\mathrm{D}}_{\dot\alpha}\mathbf{J} = \mathrm{D}_\alpha \bar{\mathbf{J}} = 0.$$

Using the notation

$$S_J = \int d^6z\, \mathbf{J}(z)\Phi(z) + \int d^6\bar{z}\, \bar{\mathbf{J}}(z)\bar{\Phi}(z) \equiv \mathbf{J}\Phi + \overline{\mathbf{J}\Phi} \qquad (4.4.23)$$

the generating superfunctional of the Wess–Zumino model is defined as

$$\mathbf{Z}[\mathbf{J}, \bar{\mathbf{J}}] = \mathscr{N} \int \mathscr{D}\bar{\Phi}\, \mathscr{D}\Phi\, e^{i(S_0 + S_{\mathrm{INT}} + S_J)}. \qquad (4.4.24)$$

After integration, this reduces to the general form (4.2.39, 42), the specialization to the case under consideration being as follows:

$$V_0 = (\Phi_0, \bar{\Phi}_0) = 0 \qquad H^I{}_{\bar{J}} = \mathbf{H}^{(m)} \qquad \mathbf{G}_{\mathrm{F}}{}^{I\bar{J}} = \mathbf{G}.$$

Note that $\mathbf{sDet}\,(\mathbf{H}^{(m)}) = \mathbf{sDet}^{-1}(\mathbf{G})$ is a constant independent of $\mathbf{J}$ and $\bar{\mathbf{J}}$. In fact, it can be shown that

$$\mathbf{sDet}(\mathbf{H}^{(m)}) = 1. \qquad (4.4.25)$$

This leads to

$$\begin{aligned} \mathbf{Z}[\mathbf{J}, \bar{\mathbf{J}}] &= \exp\left\{ iS_{\mathrm{INT}}\left[\frac{1}{i}\frac{\delta}{\delta \mathbf{J}}, \frac{1}{i}\frac{\delta}{\delta \bar{\mathbf{J}}}\right]\right\} \mathbf{Z}_0[\mathbf{J}, \bar{\mathbf{J}}] \\ \mathbf{Z}_0[\mathbf{J}, \bar{\mathbf{J}}] &= \exp\left\{\frac{i}{2}(\mathbf{J}, \bar{\mathbf{J}})\mathbf{G}\begin{pmatrix}\mathbf{J}\\ \bar{\mathbf{J}}\end{pmatrix}\right\} \end{aligned} \qquad (4.4.26)$$

where

$$\begin{aligned} (\mathbf{J}, \bar{\mathbf{J}})\mathbf{G}\begin{pmatrix}\mathbf{J}\\ \bar{\mathbf{J}}\end{pmatrix} &= \int d^6z\, d^6z'\, \mathbf{J}(z)\mathbf{G}_{++}(z, z')\mathbf{J}(z') \\ &\quad + 2\int d^6z\, d^6\bar{z}'\, \mathbf{J}(z)\mathbf{G}_{+-}(z, z')\bar{\mathbf{J}}(z') \\ &\quad + \int d^6\bar{z}\, d^6\bar{z}'\, \bar{\mathbf{J}}(z)\mathbf{G}_{--}(z, z')\bar{\mathbf{J}}(z'). \end{aligned} \qquad (4.4.27)$$

Using (4.4.16), the final expression can be rewritten in the manner

$$\int d^6z\, \mathbf{J}\frac{m}{\Box - m^2 + i\varepsilon}\mathbf{J} + 2\int d^6z\, \mathbf{J}\frac{1}{\Box - m^2 + i\varepsilon}\frac{1}{4}\bar{\mathrm{D}}^2\bar{\mathbf{J}} + \int d^6\bar{z}\, \bar{\mathbf{J}}\frac{m}{\Box - m^2 + i\varepsilon}\bar{\mathbf{J}}.$$

Since $\mathbf{J}$ is chiral, we have

$$\frac{1}{4}\int d^6z\, \mathbf{J}\frac{1}{\Box - m^2 + i\varepsilon}\bar{\mathrm{D}}^2\bar{\mathbf{J}} = -\int d^8z\, \mathbf{J}\frac{1}{\Box - m^2 + i\varepsilon}\bar{\mathbf{J}}.$$

As a result, we arrive at the expression

$$\mathbf{Z}_0[\mathbf{J}, \bar{\mathbf{J}}] = \exp\left\{i\left(-\int d^8z\, \mathbf{J}\frac{1}{\Box - m^2 + i\varepsilon}\bar{\mathbf{J}}\right.\right.$$

$$+\frac{m}{2}\int \mathrm{d}^6 z\, \mathbf{J}\frac{1}{\square - m^2 + \mathrm{i}\varepsilon}\mathbf{J}$$

$$+\frac{m}{2}\int \mathrm{d}^6 \bar{z}\, \bar{\mathbf{J}}\frac{1}{\square - m^2 + \mathrm{i}\varepsilon}\bar{\mathbf{J}}\bigg)\bigg\}. \qquad (4.4.28)$$

Let us introduce into consideration chronological averages for the linearized theory with the action S_0 (4.4.11) using the general rule:

$$\langle\langle A[\Phi, \bar{\Phi}]\rangle\rangle_0 \equiv \int \mathscr{D}\bar{\Phi}\mathscr{D}\Phi A[\Phi, \bar{\Phi}]\, \mathrm{e}^{\mathrm{i}S_0[\Phi,\bar{\Phi}]}$$

$$= A\left[\frac{1}{\mathrm{i}}\frac{\delta}{\delta \mathbf{J}}, \frac{1}{\mathrm{i}}\frac{\delta}{\delta \bar{\mathbf{J}}}\right]\mathbf{Z}_0[\mathbf{J}, \bar{\mathbf{J}}]\bigg|_{\mathbf{J}=\bar{\mathbf{J}}=0} \qquad (4.4.29)$$

A being a superfunctional. Then components of the Feynman superpropagator are given as the correlators:

$$\mathbf{G}_{+-}(z, z') = \mathbf{G}_{-+}(z', z) = \mathrm{i}\langle\langle \Phi(z)\bar{\Phi}(z')\rangle\rangle_0$$
$$\mathbf{G}_{++}(z, z') = \mathrm{i}\langle\langle \Phi(z)\Phi(z')\rangle\rangle_0 \qquad \mathbf{G}_{--}(z, z') = \mathrm{i}\langle\langle \bar{\Phi}(z)\bar{\Phi}(z')\rangle\rangle_0. \qquad (4.4.30)$$

It is clear that $\mathbf{Z}[\mathbf{J}, \bar{\mathbf{J}}]$ is expressible as a superfunctional Taylor series in $\mathbf{J}$ and $\bar{\mathbf{J}}$ with the coefficients being complicated linearized correlators of the form (4.4.29).

In practice, instead of using equations (4.4.27, 28) involving integrations over the (anti)chiral superspaces, it proves more convenient to express the components of $\mathbf{G}$ according to the law

$$\mathbf{G}_{++}(z, z') = \left(-\frac{1}{4}\right)^2 \bar{\mathrm{D}}^2\bar{\mathrm{D}}'^2\mathbf{K}_{++}(z, z')$$

$$\mathbf{G}_{+-}(z, z') = \left(-\frac{1}{4}\right)^2 \bar{\mathrm{D}}^2\mathrm{D}'^2\mathbf{K}_{+-}(z, z') \qquad (4.4.31)$$

$$\mathbf{G}_{--}(z, z') = \left(-\frac{1}{4}\right)^2 \mathrm{D}^2\mathrm{D}'^2\mathbf{K}_{--}(z, z')$$

where

$$\mathbf{K}_{+-}(z, z') = \mathbf{K}_{-+}(z, z') \equiv \mathbf{K}(z, z') = -\frac{1}{\square - m^2}\delta^8(z - z')$$

$$= -\delta^4(\theta - \theta')\frac{1}{\square - m^2}\delta^4(x - x') \qquad (4.4.32a)$$

$$\mathbf{K}_{++}(z, z') = m\frac{\mathrm{D}^2}{4\square}\mathbf{K}(z, z') \qquad \mathbf{K}_{--}(z, z') = m\frac{\bar{\mathrm{D}}^2}{4\square}\mathbf{K}(z, z'). \qquad (4.4.32b)$$

Here we have omitted the (iε)s in the denominators, which are assumed here and below. Note also that in deriving equations (4.4.32) we have used equations (2.5.30g) and the identities

$$\begin{aligned} \mathrm{D}_\alpha \delta^8(z-z') &= -\mathrm{D}'_\alpha \delta^8(z-z') \\ \bar{\mathrm{D}}_{\dot\alpha} \delta^8(z-z') &= -\bar{\mathrm{D}}'_{\dot\alpha} \delta^8(z-z'). \end{aligned} \tag{4.4.33}$$

Now, the relation (4.4.27) can be rewritten as

$$(\mathbf{J}, \bar{\mathbf{J}})\mathbf{G}\begin{pmatrix}\mathbf{J} \\ \bar{\mathbf{J}}\end{pmatrix} = \int \mathrm{d}^8 z \, \mathrm{d}^8 z' \Big\{ \mathbf{J}(z)\mathbf{K}_{++}(z, z')\mathbf{J}(z') + 2\mathbf{J}(z)\mathbf{K}_{+-}(z, z')\bar{\mathbf{J}}(z') + \bar{\mathbf{J}}(z)\mathbf{K}_{--}(z, z')\bar{\mathbf{J}}(z') \Big\} \tag{4.4.34}$$

and the free generating superfunctional takes the form

$$\mathbf{Z}_0[\mathbf{J}, \bar{\mathbf{J}}] = \exp\left\{ -\mathrm{i} \int \mathrm{d}^8 z \left(\bar{\mathbf{J}} \frac{1}{\Box - m^2} \mathbf{J} + \frac{m}{8} \mathbf{J} \frac{\mathrm{D}^2}{\Box(\Box - m^2)} \mathbf{J} + \frac{m}{8} \bar{\mathbf{J}} \frac{\bar{\mathrm{D}}^2}{\Box(\Box - m^2)} \bar{\mathbf{J}} \right) \right\}. \tag{4.4.35}$$

Both expressions involve only integrals over the full superspace.

Remark. The labels (±) carried by the **K**s do not indicate chirality or antichirality in the corresponding argument, as was the case of **G**s, but only the relationship between **K**s and **G**s (4.4.31).

The representations (4.4.27) and (4.4.34) lead to different supergraph techniques, which we now describe.

4.4.4. Standard Feynman rules

To obtain Feynman rules corresponding to the original representation for $\mathbf{Z}_0[\mathbf{J}, \bar{\mathbf{J}}]$ given by equations (4.4.26, 27), one must give the explicit structure of interaction (4.4.11), which implies

$$\mathrm{i}S_{\mathrm{INT}}\left[\frac{1}{\mathrm{i}}\frac{\delta}{\delta \mathbf{J}}, \frac{1}{\mathrm{i}}\frac{\delta}{\delta \bar{\mathbf{J}}}\right] = \mathrm{i}\frac{\lambda}{3!} \int \mathrm{d}^6 z \frac{\delta^3}{\delta(\mathrm{i}\mathbf{J}(z))^3} + \mathrm{i}\frac{\bar{\lambda}}{3!} \int \mathrm{d}^6 \bar{z} \frac{\delta^3}{\delta(\mathrm{i}\bar{\mathbf{J}}(z))^3}$$

and the variational identity

$$\frac{1}{3!}\frac{\delta^3}{\delta(\mathbf{J}(z))^3} \mathbf{J}(z_1)\mathbf{J}(z_2)\mathbf{J}(z_3) = \delta_+(z, z_1)\delta_+(z, z_2)\delta_+(z, z_3). \tag{4.4.36}$$

Then, the Feynman rules in coordinate space are:

1. Propagators:

$$\begin{aligned} &\Phi\Phi && -\mathrm{i}\mathbf{G}_{++}(z, z') \\ &\Phi\bar{\Phi} && -\mathrm{i}\mathbf{G}_{+-}(z, z') \\ &\bar{\Phi}\bar{\Phi} && -\mathrm{i}\mathbf{G}_{--}(z, z'). \end{aligned} \tag{4.4.37}$$

In the massless case, the $\Phi\Phi$- and $\bar{\Phi}\bar{\Phi}$-propagators vanish, as is seen from equation (4.4.16).

2. Vertices: There are chiral (Φ^3) and antichiral ($\bar{\Phi}^3$) vertices denoted by $\oplus$ and $\ominus$, respectively. Each chiral vertex is integrated over d^6z, each antichiral over $d^6\bar{z}$,

$$\oplus = i\lambda \int d^6z \qquad \ominus = i\bar{\lambda} \int d^6\bar{z}. \tag{4.4.38}$$

3. The usual symmetry factors for graphs should be taken into account.

Our chief goal consists in finding the effective action $\Gamma[\Phi, \bar{\Phi}]$ of the Wess–Zumino model defined by

$$\Gamma[\Phi, \bar{\Phi}] = -\frac{1}{i}\ln \mathbf{Z}[\mathbf{J}, \bar{\mathbf{J}}] - \mathbf{J}\Phi - \overline{\mathbf{J}\Phi} \tag{4.4.39}$$

where

$$\begin{pmatrix}\Phi \\ \bar{\Phi}\end{pmatrix} = \begin{pmatrix}\delta/\delta\mathbf{J} \\ \delta/\delta\bar{\mathbf{J}}\end{pmatrix}\frac{1}{i}\ln \mathbf{Z}[\mathbf{J}, \bar{\mathbf{J}}] = \begin{pmatrix}\mathbf{G}_{++} & \mathbf{G}_{+-} \\ \mathbf{G}_{-+} & \mathbf{G}_{--}\end{pmatrix}\begin{pmatrix}\mathbf{J} \\ \bar{\mathbf{J}}\end{pmatrix} + \text{loop corrections.}$$

To read off $\Gamma[\Phi, \bar{\Phi}]$ from $\mathbf{Z}[\mathbf{J}, \bar{\mathbf{J}}]$, one considers only the one-particle irreducible diagrams, amputate all the external lines and insert in their places Φ and $\bar{\Phi}$, according to the rule

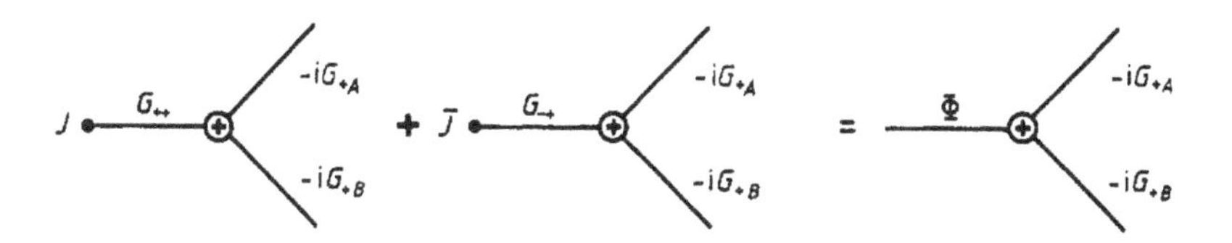

where $A, B = \pm$, and similarly for $\bar{\Phi}$-insertions. In addition, each diagram must be multiplied by $(-i)$ (see equation (4.4.39)).

The Feynman rules described here were originally developed by A. Salam and J. Strathdee. As a direct consequence of these rules, let us prove the following assertion.

Theorem 1. Every L-loop ($L \geqslant 1$) supergraph containing only chiral (antichiral) vertices vanishes.

Proof. Consider a connected L-loop supergraph of chiral type. It involves only $\mathbf{G}_{++}$-propagators, chiral in both arguments. On these grounds, it is convenient to use the chiral representation (see subsection 3.1.6), in which chiral superfields depend on x and θ only, to analyse the diagram. Using equation (3.1.66), in the chiral representation we have

$$\mathbf{G}_{++}(z,z') \equiv \frac{m}{\Box - m^2}\delta_+(z,z') = \delta^2(\theta-\theta')\frac{m}{\Box - m^2}\delta^4(x-x').$$

As is seen, the variables x and θ are separated. By assumption, the diagram contains a closed cycle formed by k points

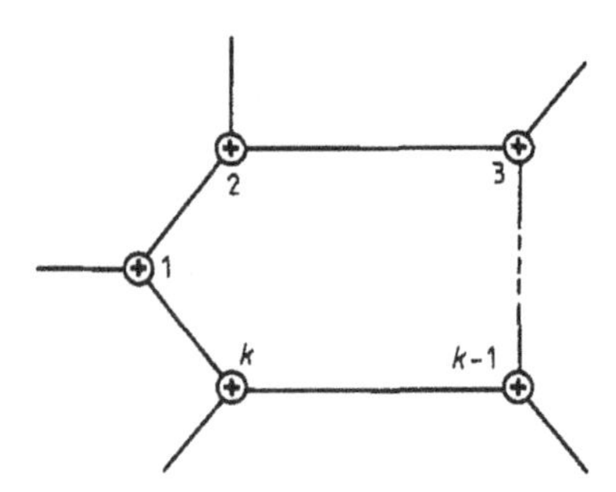

When $k>1$, its contribution to the diagram is proportional to

$$\begin{aligned}\mathbf{G}_{++}(z_1,z_2)\mathbf{G}_{++}(z_2,z_3)\ldots\mathbf{G}_{++}(z_{k-1},z_k)\mathbf{G}_{++}(z_k,z_1)\\ \sim\delta^2(\theta_1-\theta_2)\delta^2(\theta_2-\theta_3)\ldots\delta^2(\theta_{k-1}-\theta_k)\delta^2(\theta_k-\theta_1)\\ =\delta^2(\theta_1-\theta_2)\delta^2(\theta_1-\theta_3)\ldots\delta^2(\theta_1-\theta_k)\delta^2(\theta_k-\theta_1)=0\end{aligned}$$

where we have used the identities

$$\delta^2(\theta-\theta')f(\theta')=\delta^2(\theta-\theta')f(\theta) \qquad (\delta^2(\theta))^2=0. \tag{4.4.40}$$

The case $k=1$ corresponds to the tadpole

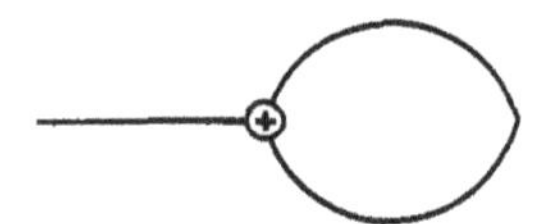

which is zero, owing to the identity

$$\delta^2(\theta-\theta)=0. \tag{4.4.41}$$

Remark. Just the same line of arguments can be used to prove equation (4.3.65).

Corollary. Every supergraph containing a closed purely chiral (antichiral) cycle vanishes.

For example, the following two-loop supergraph

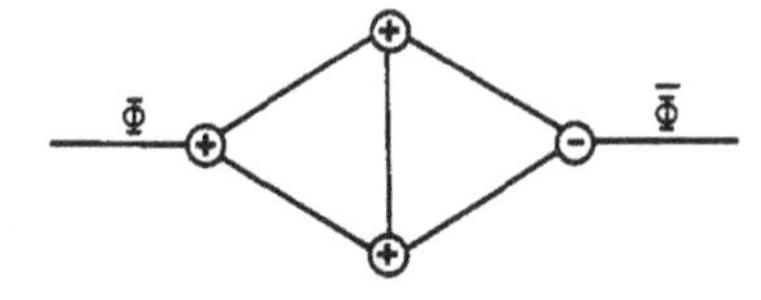

produces zero contribution to the effective action.

In accordance with Theorem 1, non-trivial supergraphs are those involving both chiral and antichiral vertices. From the viewpoint of practical calculations and general analysis, to handle such diagrams requires some improvement of the Feynman rules. The point is that neither chiral nor antichiral representations are helpful now, and one must treat $\delta_+(z, z')$ and $\delta_-(z, z')$ in the standard fashion

$$\delta_+(z, z') = -\frac{1}{4}\bar{\mathrm{D}}^2\delta^8(z-z') \qquad \delta_-(z, z') = -\frac{1}{4}\mathrm{D}^2\delta^8(z-z')$$

with all the superspace coordinates being on the same footing and the most reasonable separation being of the form $\delta^8(z) = \delta^4(x)\delta^4(\theta)$. Nevertheless, similarly to the proof of Theorem 1, it is reasonable to expect to get some simplifications caused by the properties of the Grassmann delta function $\delta^4(\theta)$ like those given in equations (4.4.40) and (4.4.41). To make use of such properties, one must convert the integration measures $\mathrm{d}^6z = \mathrm{d}^4x\,\mathrm{d}^2\theta$ and $\mathrm{d}^6\bar{z} = \mathrm{d}^4x\,\mathrm{d}^2\bar{\theta}$, associated with vertices, into $\mathrm{d}^8z = \mathrm{d}^4x\,\mathrm{d}^4\theta$ in accordance with the laws

$$-\frac{1}{4}\int \mathrm{d}^6z\,\bar{\mathrm{D}}^2U = -\frac{1}{4}\int \mathrm{d}^6\bar{z}\,\mathrm{D}^2U = \int \mathrm{d}^8z\,U \tag{4.4.42}$$

since the delta function $\delta^4(\theta)$ corresponds to the measure $\mathrm{d}^4\theta$.

4.4.5. Improved Feynman rules

Following M. Grisaru, M. Roček and W. Siegel, we now consider the reformulation of the above Feynman rules based on the use of the representation for superpropagators (4.4.42), the identity

$$(\bar{\mathrm{D}}^2U_1)(\bar{\mathrm{D}}^2U_2)\ldots(\bar{\mathrm{D}}^2U_n) = \bar{\mathrm{D}}^2\{U_1(\bar{\mathrm{D}}^2U_2)\ldots(\bar{\mathrm{D}}^2U_n)\}$$

and its conjugate, and the conversion rules (4.4.42). Together these imply that for each chiral vertex with three internal lines attached we have

$$\begin{aligned}\int \mathrm{d}^6z\, \mathbf{G}_{+A}(z, z_1)\mathbf{G}_{+B}(z, z_2)\mathbf{G}_{+C}(z, z_3) \\ = \int \mathrm{d}^8z\, \mathbf{K}_{+A}(z, z_1)\left(-\frac{1}{4}\bar{\mathscr{D}}^{(A)}\right)\left\{\left(-\frac{1}{4}\bar{\mathrm{D}}^2\right)\mathbf{K}_{+B}(z, z_2)\left(-\frac{1}{4}\bar{\mathscr{D}}^{(B)}\right)\right\} \\ \times\left\{\left(-\frac{1}{4}\bar{\mathrm{D}}^2\right)\mathbf{K}_{+C}(z, z_3)\left(-\frac{1}{4}\bar{\mathscr{D}}^{(C)}\right)\right\}\end{aligned}$$

where

$$A, B, C = \pm \qquad \mathscr{D}^{(A)} = \begin{cases}\bar{\mathrm{D}}^2 & A = + \\ \mathrm{D}^2 & A = -\end{cases}$$

and $\overleftarrow{\mathscr{D}}^{(A)}$ means that $\mathscr{D}^{(A)}$ acts on the second argument of $\mathbf{K}_{+A}$; for each chiral vertex with one external and two internal lines attached we have

$$\int d^6z\, \Phi(z)\mathbf{G}_{+A}(z, z_1)\mathbf{G}_{+B}(z, z_2)$$
$$= \int d^8z\, \Phi(z)\mathbf{K}_{+A}(z, z_1)\left(-\frac{1}{4}\overleftarrow{\mathscr{D}}^{(A)}\right)\left\{\left(-\frac{1}{4}\bar{D}^2\right)\mathbf{K}_{+B}(z, z_2)\left(-\frac{1}{4}\overleftarrow{\mathscr{D}}^{(B)}\right)\right\}$$

and so on. Equivalently, one could start with the representation for $\mathbf{Z}_0[\mathbf{J}, \bar{\mathbf{J}}]$ (4.4.35) and use the following integral version of the variational rule (4.4.36)

$$\frac{1}{3!}\int d^6z \frac{\delta^3}{\delta(\mathbf{J}(z))^3}\mathbf{J}(z_1)\mathbf{J}(z_2)\mathbf{J}(z_3)$$
$$= \int d^8z\, \delta^8(z-z_1)\left\{-\frac{1}{4}\bar{D}^2\delta^8(z-z_2)\right\}\left\{-\frac{1}{4}\bar{D}^2\delta^8(z-z_3)\right\}$$

for calculating $\mathbf{Z}[\mathbf{J}, \bar{\mathbf{J}}]$. As a result, we get the improved Feynman rules for the effective action in coordinate space.

1. Propagators:

$\Phi\bar{\Phi}$- line: z ——— z' $= \frac{i}{\Box - m^2}\delta^8(z-z')$ (4.4.43)

$\Phi\Phi$-line: z ▬▬▬ z' $=$ $(mD^2/4\Box)$ z —+— z' $= \frac{1}{4}D^2 \frac{im}{\Box(\Box - m^2)}\delta^8(z-z')$

$\bar{\Phi}\bar{\Phi}$- line: z ═══ z' $=$ $(m\bar{D}^2/4\Box)$ z —+— z' $= \frac{1}{4}\bar{D}^2 \frac{im}{\Box(\Box - m^2)}\delta^8(z-z')$.

2. Vertices: (4.4.44)

chiral (Φ^3) antichiral $(\bar{\Phi}^3)$

● $= i\lambda \int d^8z$ ○ $= i\bar{\lambda}\int d^8z$.

Associated with every chiral (antichiral) vertex with n internal lines attached are $n-1$ factors $-\frac{1}{4}\bar{D}^2$ $(-\frac{1}{4}D^2)$ acting on $n-1$ internal lines. All external lines (Φ- or $\bar{\Phi}$-ones) arise without any factors.

3. The standard combinatoric factors for graphs.

Remark. The following property of $\Phi\Phi$- and $\overline{\Phi\Phi}$-lines is often useful:

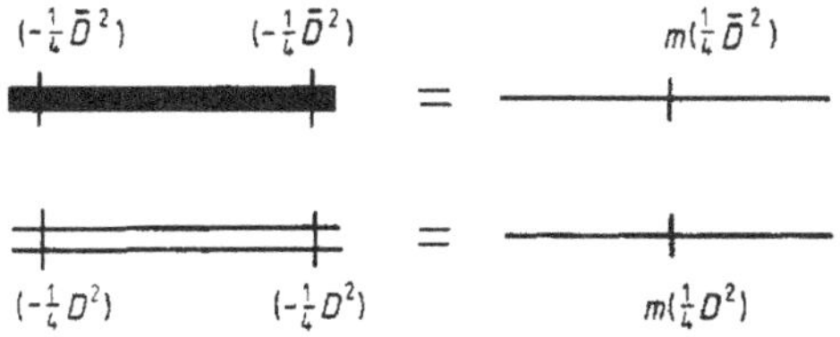

$(-\frac{1}{4}\bar{D}^2)$ $(-\frac{1}{4}\bar{D}^2)$ = $m(\frac{1}{4}\bar{D}^2)$

$(-\frac{1}{4}D^2)$ $(-\frac{1}{4}D^2)$ = $m(\frac{1}{4}D^2)$ (4.4.45)

One of the main advantages of the improved Feynman rules is the fact that for a graph with V vertices one can easily integrate $V-1$ vertices in $\theta, \bar{\theta}$. Each internal line represents itself as the superpropagator

$$\frac{\mathrm{i}}{\Box - m^2}\delta^8(z-z') = \delta^4(\theta-\theta')\frac{\mathrm{i}}{\Box - m^2}\delta^4(x-x')$$

or

$$\frac{im}{\Box(\Box - m^2)}\delta^8(z-z') = \delta^4(\theta-\theta')\frac{im}{\Box(\Box - m^2)}\delta^4(x-x')$$

with some number of $\mathbf{D}$ and $\bar{\mathbf{D}}$ factors acting on z or z'. All these factors can be transfered onto other lines (internal or external) with the help of integration by parts

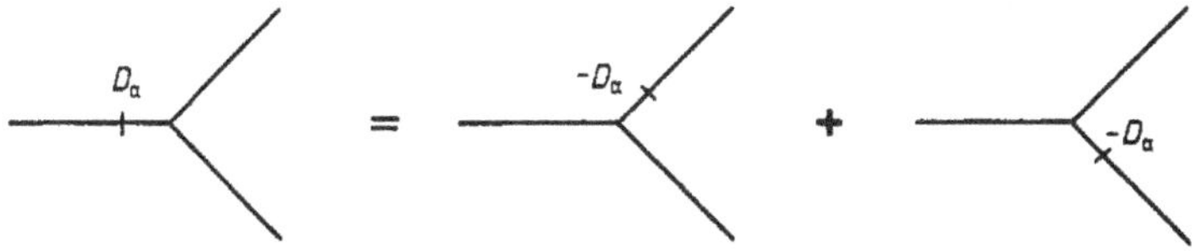

D_α = $-D_\alpha$ + $-D_\alpha$ (4.4.46)

After this, $\delta^4(\theta-\theta')$ factors out, and the integral over, for example, $\mathrm{d}^4\theta'$ is immediately done. The validity of integration by parts turns out to be obvious by rewriting the Feynman rules in momentum space.

Let us make the transformation to momentum space for the mean superfield $U=(\Phi, \bar{\Phi})$

$$U(z) = \int \frac{\mathrm{d}^4 p}{(2\pi)^4} \mathrm{e}^{-\mathrm{i}px} U(p, \theta)$$

the superpropagators $\mathscr{K} = (-\mathrm{i}\mathbf{K}_{+-}, -\mathrm{i}\mathbf{K}_{++}, -\mathrm{i}\mathbf{K}_{--})$

$$\mathscr{K}(z, z') = \int \frac{\mathrm{d}^4 p}{(2\pi)^4} \mathrm{e}^{-\mathrm{i}p(x-x')} \mathscr{K}(p; \theta, \theta')$$

and the delta function

$$\delta^8(z-z') = \int \frac{\mathrm{d}^4 p}{(2\pi)^4} \mathrm{e}^{-\mathrm{i}p(x-x')} \delta^4(\theta-\theta')$$

Here and below in this section we take the convention that θ means all the odd variables θ^α and $\bar{\theta}_{\dot{\alpha}}$. In momentum space the covariant derivatives $D_A = (\partial_a, \mathrm{D}_\alpha, \bar{\mathrm{D}}^{\dot{\alpha}})$ take the form

$$\begin{aligned} \partial_a &\to -\mathrm{i}p_a \\ \mathrm{D}_\alpha &\to \mathrm{D}_\alpha(p) = \partial_\alpha + (\sigma^a)_{\alpha\dot{\alpha}}\bar{\theta}^{\dot{\alpha}} p_a \\ \bar{\mathrm{D}}_{\dot{\alpha}} &\to \bar{\mathrm{D}}_{\dot{\alpha}}(p) = -\bar{\partial}_{\dot{\alpha}} - (\sigma^a)_{\alpha\dot{\alpha}}\theta^\alpha p_a \end{aligned} \tag{4.4.47}$$

and their algebra has the form

$$\begin{aligned} \{\mathrm{D}_\alpha(p), \mathrm{D}_\beta(p)\} &= \{\bar{\mathrm{D}}_{\dot{\alpha}}(p), \bar{\mathrm{D}}_{\dot{\beta}}(p)\} = 0 \\ \{\mathrm{D}_\alpha(p), \bar{\mathrm{D}}_{\dot{\alpha}}(p)\} &= -2(\sigma^a)_{\alpha\dot{\alpha}} p_a. \end{aligned} \tag{4.4.48}$$

The Feynman rules in momentum space read as follows:

1. Propagators:

$$\xrightarrow{p} \;=\; -\frac{\mathrm{i}}{p^2+m^2}\,\delta^4(\theta-\theta')$$

$$\xrightarrow{p}\ (\text{thick line}) \;=\; \frac{\mathrm{i}m}{p^2(p^2+m^2)}\left(\frac{1}{4}D^2(p)\right)\delta^4(\theta-\theta') \tag{4.4.49}$$

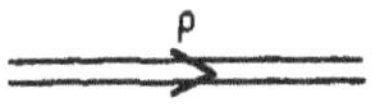

$$\xrightarrow{p}\ (\text{double line}) \;=\; \frac{\mathrm{i}m}{p^2(p^2+m^2)}\left(\frac{1}{4}\bar{D}^2(p)\right)\delta^4(\theta-\theta').$$

2. Each vertex is integrated over $\mathrm{d}^4\theta$.
3. Associated with every independent loop is the momentum integral $\int \mathrm{d}^4p/(2\pi)^4$.
4. Associated with the external momenta is the overall factor

$$\left[\prod_{p_{\text{ext}}} \int \frac{\mathrm{d}^4 p_{\text{ext}}}{(2\pi)^4}\right] (2\pi)^4 \delta^4\left(\sum_{\text{ext}} p_{\text{ext}}\right).$$

The other factors are the same as in coordinate space.

The operators $\mathrm{D}_\alpha(p)$ and $\bar{\mathrm{D}}_{\dot{\alpha}}(p)$ can be handled as ordinary covariant derivatives. It is worth remembering, however, that the coordinate space expressions

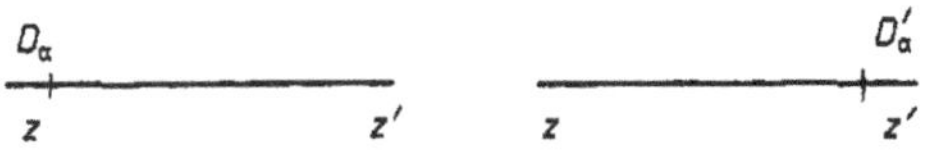

D_α (at z), line from z to z'; line from z to z', D'_α (at z')

correspond in momentum space to

$D_\alpha(p)$ θ θ′ θ θ′ $D'_\alpha(-p)$

respectively. The momentum space analogue of the relation

$$\mathrm{D}_\alpha \delta^8(z-z') = -\mathrm{D}'_\alpha \delta^8(z-z')$$

reads

$$\mathrm{D}_\alpha(p)\delta^4(\theta-\theta') = -\mathrm{D}'_\alpha(-p)\delta^4(\theta-\theta') \tag{4.4.50}$$

which implies

$D_\alpha(p)$ θ p θ′ = θ′ p θ′ $-D'_\alpha(-p)$ (4.4.51*a*)

hence

$D^2(p)$ θ p θ′ = θ p θ′ $D'^2(-p)$ (4.4.51*b*)

and so on. Using the basic property of the Berezin integral,

$$\int \mathrm{d}^4\theta\, \partial_\alpha f(\theta) = \int \mathrm{d}^4\theta\, \bar{\partial}_{\dot\alpha} f(\theta) = 0$$

f being arbitrary, we have the rules for integration by parts:

$D_\alpha(-p)$ p k p−k = $-D_\alpha(k)$ p k p−k + p k $-D_\alpha(p-k)$ p−k

(4.4.52)

$\bar{D}_{\dot\alpha}(-p)$ p k p−k = $-\bar{D}_{\dot\alpha}(k)$ p k p−k + p k $-\bar{D}_{\dot\alpha}(p-k)$ p−k.

The coordinate space version of the former coincides with equation (4.4.46). It is easy to see from the previous consideration that the momentum dependence of factors $\mathrm{D}(p)$ and $\bar{\mathrm{D}}(p)$ arising on internal or external lines can be omitted without any misunderstanding.

4.4.6. Example of supergraph calculations

Let us calculate the one-loop correction to the two-point vertex function. The relevant supergraphs are:

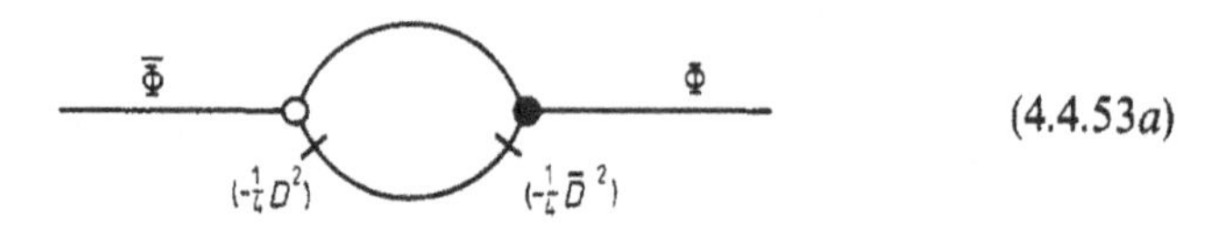

(4.4.53a)

(4.4.53b)

(4.4.53c)

In accordance with Theorem 1, the second and third graphs vanish. The contribution from the first graph is

$$\mathrm{i}\frac{1}{2}\bar{\lambda}\lambda\int \mathrm{d}^8z\,\mathrm{d}^8z'\,\bar{\Phi}(z)\left[\frac{1}{\Box-m^2}\delta^8(z-z')\right]\left[\frac{1}{16}\mathrm{D}^2\bar{\mathrm{D}}'^2\frac{1}{\Box-m^2}\delta^8(z-z')\right]\Phi(z').$$

Here $\frac{1}{2}$ is the combinatoric factor (it arises as $(3\times3\times2)/3!\,3!$). The propagator in the first set of brackets contains the free function $\delta^4(\theta-\theta')$, which implies that we must set $\theta=\theta'$ in the second brackets. The only non-zero contribution will arise when all the multipliers in $(\mathrm{D}^2\bar{\mathrm{D}}'^2)$ act on $\delta^4(\theta-\theta')$,

$$\frac{1}{16}(\mathrm{D}^2\bar{\mathrm{D}}'^2)\delta^4(\theta-\theta')\Big|_{\theta=\theta'}=1.$$

Finally, performing the integral in θ' gives

$$\mathrm{i}\frac{1}{2}\bar{\lambda}\lambda\int \mathrm{d}^4x\,\mathrm{d}^4x'\,\mathrm{d}^4\theta\,\bar{\Phi}(x,\theta)\Phi(x',\theta)[G(x,x')]^2$$
$$G(x,x')=-\frac{1}{\Box-m^2}\delta^4(x-x'). \qquad (4.4.54a)$$

Repeating the calculations in momentum space, where the graph (4.4.53a) reads

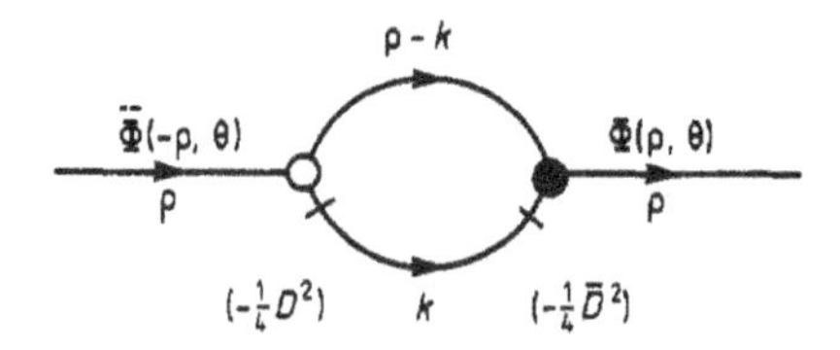

$$=\mathrm{i}\frac{1}{2}\bar{\lambda}\lambda\int\frac{\mathrm{d}^4p}{(2\pi)^4}\int\mathrm{d}^4\theta\,\mathrm{d}^4\theta'\,\bar{\Phi}(-p,\theta)\Phi(p,\theta')$$

$$\times\int\frac{\mathrm{d}^4p}{(2\pi)^4}\frac{1}{(p-k)^2+m^2}\delta^4(\theta-\theta')\frac{1}{k^2+m^2}\left[\frac{1}{16}\mathrm{D}^2(k)\bar{\mathrm{D}}'^2(-k)\delta^4(\theta-\theta')\right].$$

Applying the above arguments reduces this expression to

$$\frac{\bar{\lambda}\lambda}{2}\int\frac{\mathrm{d}^4p}{(2\pi)^4}A(p^2,m^2)\int\mathrm{d}^4\theta\,\bar{\Phi}(-p,\theta)\Phi(p,\theta) \tag{4.4.54b}$$

where

$$A(p^2,m^2)=\mathrm{i}\int\frac{\mathrm{d}^4k}{(2\pi)^4}\frac{1}{(p-k)^2+m^2}\frac{1}{k^2+m^2} \tag{4.4.55}$$

is an ordinary loop integral divergent at large momenta. Therefore, we must introduce a regularization scheme in order to make our diagram sensible.

4.4.7. Supersymmetric analytic regularization

We shall mainly use the so-called analytic regularization based on the replacement of each factor

$$\frac{1}{p^2+m^2-\mathrm{i}\varepsilon}=\mathrm{i}\int_0^\infty\mathrm{d}s\,\mathrm{e}^{-\mathrm{i}s(p^2+m^2)} \tag{4.4.56}$$

in equation (4.4.49) by its regularized version

$$\frac{\Gamma(1+\omega)\mu^{2\omega}}{(p^2+m^2-\mathrm{i}\varepsilon)^{1+\omega}}=\mathrm{i}\mu^{2\omega}\int_0^\infty\mathrm{d}s\,(\mathrm{i}s)^\omega\,\mathrm{e}^{-\mathrm{i}s(p^2+m^2)} \tag{4.4.57}$$

an analytic function of a complex variable ω (regularization parameter), $\omega\to 0$ at the end of calculations; μ is a real parameter with dimensions of mass (normalization point) introduced to fix the dimensions of propagators. There is no need to regularize $1/p^2$-factors in equation (4.4.49), since, in accordance with Theorem 1, neither $\Phi\Phi$- nor $\bar{\Phi}\bar{\Phi}$-lines form closed loops. It is obvious that every non-zero closed loop has to involve at least two $\Phi\bar{\Phi}$-lines. By simple power counting one can see that closed loops containing $\Phi\Phi$- or $\bar{\Phi}\bar{\Phi}$-lines produce finite contributions. Note also that the $1/p^2$-factors in superpropagators are often cancelled by rule (4.4.45), as is the case for the two-loop supergraph

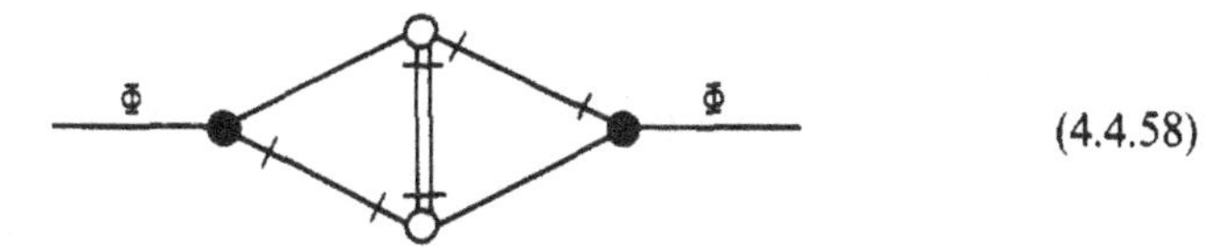

(4.4.58)

contributing to the two-point vertex function.

The Fourier transform of (4.4.57) reads

$$G(x, x'|\omega) = \mathrm{i}\mu^{2\omega}\int_0^\infty \mathrm{d}s\,(\mathrm{i}s)^\omega\,U(x, x'; s) \tag{4.4.59}$$

with $U(x, x'; s)$ being the kernel (4.4.20), and represents the regularized form of the propagator (4.4.18). Then, owing to the presence of $\delta^4(\theta-\theta')$ in expression (4.4.32*a*), the regularized form of $\mathbf{K}(z, z')$ can be written as

$$\mathbf{K}(z, z'|\omega) = \delta^4(\theta-\theta')\mu^{2\omega}\int_0^\infty \frac{\mathrm{d}s}{(4\pi s)^2}(\mathrm{i}s)^\omega\,\mathrm{e}^{\mathrm{i}[(w(z,z'))^2/4s - m^2 s]} \tag{4.4.60}$$

where

$$w^a(z, z') = (x-x')^a + \mathrm{i}\theta\sigma^a(\bar\theta-\bar\theta') - \mathrm{i}(\theta-\theta')\sigma^a\bar\theta.$$

It has been shown in Section 2.4 that the two-point function $w^a(z, z')$ is invariant under supersymmetry transformations. Therefore, the regularized superpropagator is explicitly super Poincaré invariant. As a result, the regularization scheme described preserves supersymmetry at every stage of perturbation theory.

Remark. Both expressions (4.4.59) and (4.4.60) are well behaved on the light cone when $\mathrm{Re}\,\omega > 2$.

In conclusion, let us point out that divergences arise within the framework of analytic regularization as poles in ω. For example, the regularized form of (4.4.55) reads

$$A(p^2, m^2|\omega) = -\mathrm{i}\mu^{4\omega}\int\frac{\mathrm{d}^4k}{(2\pi)^4}\int_0^\infty \mathrm{d}s_1\,\mathrm{d}s_2(\mathrm{i}s_1)^\omega(\mathrm{i}s_2)^\omega \times \mathrm{e}^{-\mathrm{i}s_1[(p-k)^2+m^2]}\,\mathrm{e}^{-\mathrm{i}s_2[k^2+m^2]}$$

and its direct calculation leads to

$$A(p^2, m^2|\omega) = \frac{1}{(4\pi)^2}\frac{1}{4\omega} + \text{finite terms.} \tag{4.4.61}$$

To cancel this divergence, we must introduce into the bare action the following divergent structure

$$-\frac{\bar\lambda\lambda}{(4\pi)^2}\frac{1}{2\omega}\int \mathrm{d}^2z\,\bar\Phi\Phi.$$

4.4.8. Non-renormalization theorem

Looking at the one-loop contribution to the two-point vertex function (4.4.54), one can see that the integrand is local in the odd variables θ (a function of only one θ^α and $\bar\theta_{\dot\alpha}$) and non-local in the odd variables x or p (a function of two xs or, equivalently, a non-polynomial function of p). The locality in θ turns out to be a general property of arbitrary supergraphs, in accordance with the theorem proved by M. Grisaru, M. Roček and W. Siegel.

Theorem 2 (Non-renormalization theorem). Each term in the effective action can be expressed as an integral over a single $d^4\theta$.

Proof. Consider a one-particle irreducible L-loop supergraph. Recall that each vertex is integrated over $d^4\theta$, each line represents $\delta^4(\theta-\theta')$ with some number of factors D and $\bar{D}$ acting on it (we suppress p-dependence). Choose in the graph a fixed loop involving, say, n vertices. Associated with this loop is a cycle $\delta^4(\theta_1-\theta_2)\delta^4(\theta_2-\theta_3)\ldots\delta^4(\theta_n-\theta_1)$ with derivatives Ds and $\bar{D}$s acting on delta functions. Integrating by parts, one can transfer all the covariant derivatives acting originally on $\delta^4(\theta_1-\theta_2)$ onto $\delta^4(\theta_2-\theta_3)$ or $\delta^4(\theta_n-\theta_1)$ or external lines to the loop. Now, performing the integral in θ_2 removes $\delta^4(\theta_1-\theta_2)$ and replaces θ_2 by θ_1 everywhere. Let us continue this procedure $n-1$ times. Then, the cycle is reduced to a single delta function, and one obtains an expression for the graph of the form

$$\prod_A \int d^4\theta_A \int d^4\theta_1 \, f(\theta_1, \theta_A)\{D\ldots D\delta^4(\theta_n-\theta_1)\}|_{\theta_n=\theta_1} \qquad (4.4.62)$$

where the label A enumerates the vertices external to the loop. The last term here is easily evaluated. As a result of the anticommutation relations for the covariant derivatives, any product of D and $\bar{D}$ factors can be reduced to an expression involving at the most four such factors. Since

$$\delta^4(\theta_n-\theta_1)=(\theta_n-\theta_1)^2(\bar{\theta}_n-\bar{\theta}_1)^2$$

one needs exactly two Ds and two $\bar{D}$s for expression (4.4.62) to be non-zero. In this case, we have

$$\frac{1}{16}D^2\bar{D}^2\,\delta^4(\theta_n-\theta_1)|_{\theta_n=\theta_1}=1.$$

As a result, the entire loop has been contracted to a point in θ-space. Continuing the above procedure loop-by-loop reduces the entire graph to a point in θ-space, and the total contribution takes the form

$$\prod_i \int d^4p_i \int d^4\theta \, F(p_i, \theta)$$

where p_i are independent external momenta.

Corollary 1. The general structure of the effective action of the Wess–Zumino model is as follows

$$\Gamma[\Phi, \bar{\Phi}]=\sum_n \int d^4x_1\ldots d^4x_n \quad \int d^4\theta \, \mathscr{T}_n(x_1, \ldots, x_n)F_1(x_1, \theta)\ldots F_n(x_n, \theta). \qquad (4.4.63)$$

Here $\mathcal{T}$s are translationally invariant functions on Minkowski space, Fs are local functions of $\Phi, \bar{\Phi}$ and their covariant derivatives,

$$F_i = F_i(\Phi, \bar{\Phi}, \mathrm{D}_A\Phi, \mathrm{D}_A\bar{\Phi}, \ldots). \tag{4.4.64}$$

Remark. Each term in structure (4.4.63) is represented by the integral over $\mathrm{d}^4\theta$, but not over $\mathrm{d}^2\theta$ or $\mathrm{d}^2\bar{\theta}$. Certainly, this does not mean that there are no purely chiral or antichiral contributions to the effective action. For example, the following term is possible:

$$m \int \frac{\mathrm{d}^4 p}{(2\pi)^4} B(p) \int \mathrm{d}^2\theta\, \mathrm{d}^2\bar{\theta}\, \Phi(-p, \theta, \bar{\theta}) \left[-\frac{1}{4} \mathrm{D}^2(p) \right] \Phi(p, \theta, \bar{\theta})$$

$$= m \int \frac{\mathrm{d}^4 p}{(2\pi)^4} p^2 B(p) \int \mathrm{d}^4\theta\, \Phi(-p, \theta)\Phi(p, \theta)$$

where we have indicated explicitly θ- and $\bar{\theta}$-dependencies and used the fact that in momentum space

$$\Phi(p, \theta, \bar{\theta}) = \mathrm{e}^{\theta\sigma^a\bar{\theta}p_a}\Phi(p, \theta).$$

The contribution from the graph (4.4.58) is exactly of the given type.

Corollary 2. All vacuum supergraphs vanish.

Proof. In accordance with Theorem 2, every vacuum supergraph can be represented as

$$\mathscr{A} \int \mathrm{d}^4\theta$$

where $\mathscr{A}$ is a loop momentum integral. The entire expression vanishes, since $\int \mathrm{d}^4\theta = 0$.

This last result implies that the generating superfunctional (4.4.21) is naturally normalized by

$$\mathbf{Z}[\mathbf{J} = 0, \bar{\mathbf{J}} = 0] = 1. \tag{4.4.65}$$

Other consequences of the non-renormalization theorem will be discussed below.

4.5. Note about gauge theories

The equivalence of Feynman rules in the component and superfield approaches has been established in the Section 4.2 for non-gauge supersymmetric theories. In principle, the proof can be extended to the case of gauge supersymmetric theories, but here it will depend on the class to which a given dynamical system belongs. As is known, there exist several different types of gauge theories (reducible or irreducible, possessing closed or open gauge algebra and so on), and for each of them there are specific features to be borne in mind when formulating Feynman rules. We find it reasonable to take the following line: to expose the main types of gauge theories, but to recall the Feynman rules only for irreducible theories with closed algebras. This is precisely the family to which the super Yang–Mills models considered in Section 3.5 belong. The Feynman rules for general gauge theories with linearly dependent generators and open algebras have been obtained by I. Batalin and G. Vilkovisky, and we refer the interested reader to the original publication[1].

4.5.1. *Gauge theories*

Let $\varphi^{\hat{i}} = \varphi^i(x)$ be the field variables, and $S[\varphi]$ be the classical action of a dynamical system. Suppose that on the space of histories $\mathbf{\Phi}$ there exists an infinite set of pure right supervector fields

$$R_{\hat{\alpha}} = \frac{\overleftarrow{\delta}}{\delta\varphi^{\hat{i}}} R^{\hat{i}}{}_{\hat{\alpha}}[\varphi] \qquad \varepsilon(R_{\hat{\alpha}}) = \varepsilon_\alpha \qquad \varepsilon(R^{\hat{i}}{}_{\hat{\alpha}}) = \varepsilon_i + \varepsilon_\alpha \tag{4.5.1}$$

non-vanishing along the dynamical subspace $\mathbf{\Phi}_0$,

$$R_{\hat{\alpha}}|_{\varphi=\varphi_0} \neq 0 \qquad \forall\ \varphi_0 \in \mathbf{\Phi}_0 \tag{4.5.2}$$

and satisfying the requirement

$$SR_{\hat{\alpha}} = S_{,\hat{i}}[\varphi]R^{\hat{i}}{}_{\hat{\alpha}}[\varphi] \equiv 0. \tag{4.5.3}$$

Here $\hat{\alpha} = (\alpha, x')$, the index α running over some finite number of values. Under the above conditions, the theory is said to be gauge. Associated with $R_{\hat{\alpha}}$ are the infinitesimal gauge transformations

$$\delta\varphi^{\hat{i}} = \varphi^{\hat{i}} R_{\hat{\alpha}}\xi^{\hat{\alpha}} = R^{\hat{i}}{}_{\hat{\alpha}}[\varphi]\xi^{\hat{\alpha}} \qquad \varepsilon(\xi^{\hat{\alpha}}) = \varepsilon_\alpha \tag{4.5.4}$$

leaving the action $S[\varphi]$ invariant, where the parameters $\xi^\alpha(x)$ are arbitrary φ-independent functions over space–time (but reasonably constrained at infinity). The $R_{\hat{\alpha}}$ or, equivalently, $R^{\hat{i}}{}_{\hat{\alpha}}[\varphi]$ are called generators of gauge transformations. Equation (4.5.2) implies that the gauge freedom does not reduce to a trivial invariance of the form

$$\delta\varphi^{\hat{i}} = S_{,\hat{j}}[\varphi]\Omega^{\hat{j}\hat{i}} \qquad \Omega^{\hat{j}\hat{i}} = -(-1)^{\varepsilon_i\varepsilon_j}\Omega^{\hat{i}\hat{j}}. \tag{4.5.5}$$

Obviously, every such transformation does not change $S[\varphi]$ and coincides with the identity map restricted to $\mathbf{\Phi}_0$.

[1]I.A. Batalin and G.A. Vilkovisky, *Phys. Rev.* **D 24** 2567, 1983

Owing to equations (4.5.2) and (4.5.3), the Hessian ${}_{i,}S_{,j}[\varphi_0]$, at each point φ_0 of $\mathbf{\Phi}_0$, is degenerate (that is, it possesses zero-eigenvalue eigenvectors of compact support in space–time) and has no Green functions, since

$${}_{i,}S_{,j}[\varphi_0]R^{j}{}_{\hat{\alpha}}[\varphi_0]=0. \tag{4.5.6}$$

We assume the set $\{R_{\hat{\alpha}}\}$ to be complete in the sense that each infinitesimal displacement $\delta\varphi^{\hat{i}}(x)$ leaving $S[\varphi]$ unchanged can be represented as a φ-dependent gauge transformation modulo the dynamical equations:

$$\begin{gathered} S_{,\hat{i}}[\varphi]\delta\varphi^{\hat{i}}=0 \quad \Rightarrow \\ \delta\varphi^{\hat{i}}=R^{\hat{i}}{}_{\hat{\alpha}}[\varphi]\xi^{\hat{\alpha}}[\varphi]+S_{,\hat{j}}[\varphi]\Omega^{\hat{j}\hat{i}}[\varphi] \end{gathered} \tag{4.5.7}$$

with the Ωs being constrained as in equation (4.5.5). This means that all the degeneracy of ${}_{i,}S_{,j}[\varphi_0]$ is due to the gauge invariance (4.5.4).

Introduce the super Lie bracket of generators

$$[R_{\hat{\alpha}}, R_{\hat{\beta}}\}=R_{\hat{\alpha}}R_{\hat{\beta}}-(-1)^{\varepsilon_\alpha\varepsilon_\beta}R_{\hat{\beta}}R_{\hat{\alpha}} \tag{4.5.8}$$

leading to a pure supervector field on $\mathbf{\Phi}_0$. For arbitrary infinitesimal parameters $\xi_1^{\hat{\alpha}}$ and $\xi_2^{\hat{\alpha}}$, $\varepsilon(\xi^{\hat{\alpha}}_{1,2})=\varepsilon_\alpha$, the transformation

$$\delta\varphi^{\hat{i}}=\varphi^{\hat{i}}[R_{\hat{\alpha}}, R_{\hat{\beta}}\}\xi_1^{\hat{\beta}}\xi_2^{\hat{\alpha}}$$

leaves $S[\varphi]$ invariant, as a consequence of (4.5.3). Therefore, equation (4.5.7) tells us that

$$[R_{\hat{\alpha}}, R_{\hat{\beta}}\}=R_{\hat{\gamma}}C^{\hat{\gamma}}{}_{\hat{\alpha}\hat{\beta}}+\frac{\overleftarrow{\delta}}{\delta\varphi^{\hat{i}}}S_{,\hat{j}}E^{\hat{j}\hat{i}}{}_{\hat{\alpha}\hat{\beta}} \tag{4.5.9}$$

where the Cs and Es satisfy the relations

$$\begin{gathered} C^{\hat{\gamma}}{}_{\hat{\alpha}\hat{\beta}}=-(-1)^{\varepsilon_\alpha\varepsilon_\beta}C^{\hat{\gamma}}{}_{\hat{\beta}\hat{\alpha}} \\ E^{\hat{j}\hat{i}}{}_{\hat{\alpha}\hat{\beta}}=-(-1)^{\varepsilon_\alpha\varepsilon_\beta}E^{\hat{i}\hat{j}}{}_{\hat{\alpha}\hat{\beta}}=-(-1)^{\varepsilon_i\varepsilon_j}E^{\hat{j}\hat{i}}{}_{\hat{\alpha}\hat{\beta}}. \end{gathered} \tag{4.5.10}$$

The coordinate form of equation (4.5.9) reads

$$R^{\hat{i}}{}_{\hat{\alpha},\hat{j}}[\varphi]R^{\hat{j}}{}_{\hat{\beta}}[\varphi]-(-1)^{\varepsilon_\alpha\varepsilon_\beta}R^{\hat{i}}{}_{\hat{\beta},\hat{j}}[\varphi]R^{\hat{j}}{}_{\hat{\alpha}}[\varphi]=R^{\hat{i}}{}_{\hat{\gamma}}[\varphi]C^{\hat{\gamma}}{}_{\hat{\alpha}\hat{\beta}}[\varphi]+S_{,\hat{j}}[\varphi]E^{\hat{j}\hat{i}}{}_{\hat{\alpha}\hat{\beta}}[\varphi]. \tag{4.5.11}$$

The equations (4.5.9) and (4.5.10) define the gauge algebra of the theory. The functionals $C^{\hat{\gamma}}{}_{\hat{\alpha}\hat{\beta}}[\varphi]$ are called the structure coefficients of the algebra. The gauge algebra is said to be closed when

$$E^{\hat{j}\hat{i}}{}_{\hat{\alpha}\hat{\beta}}[\varphi]=0 \tag{4.5.12}$$

otherwise, it is called open.

Remark. A famous example of a gauge theory with an open algebra is Einstein supergravity (see Chapters 5, 6). This theory is described by fields

$\varphi^i = \{e_a{}^m, \Psi^m{}_\alpha, \bar{\Psi}^{m\dot{\alpha}}, \mathbf{B}, \bar{\mathbf{B}}, \mathbf{A}^m\}$, where $e_a{}^m$ is a vierbein, $(\Psi^m{}_\alpha, \bar{\Psi}^{m\dot{\alpha}})$ is a spin-$\frac{3}{2}$ field called the gravitino, $(\mathbf{B}, \bar{\mathbf{B}}, \mathbf{A}^m)$ are supergravity auxiliary fields. The supergravity action functional (6.1.12) proves to be invariant under the general coordinate and local Lorentz transformations and the local supersymmetry transformations (5.1.45, 48). Before eliminating the auxiliary fields, with the help of their equations of motion (6.1.14), all the transformations form a closed algebra (with some field-dependent structure coefficients). However, after eliminating the auxiliary fields one discovers the presence of the terms $\delta S/\delta\Psi^m{}_\alpha$ and $\delta S/\delta\bar{\Psi}^{m\dot{\alpha}}$ in a commutator of two supersymmetry transformations.

The gauge theory is said to be irreducible if for every dynamical history $\varphi_0 \in \mathbf{\Phi}_0$ the generators $R^{\hat{j}}{}_{\hat{\alpha}}[\varphi_0]$ are linearly independent, that is, the equation

$$R^{\hat{j}}{}_{\hat{\alpha}}[\varphi_0]\xi^{\hat{\alpha}} = 0$$

has no non-trivial solution with compact support in space–time; otherwise, the theory is called reducible. In the case of a reducible gauge theory, there exists an infinite set of zero-eigenvalue eigenvectors $Z^{\hat{\alpha}}{}_{\hat{a}}[\varphi_0]$ for each functional supermatrix $R^{\hat{j}}{}_{\hat{\alpha}}[\varphi_0]$,

$$\begin{gathered} R^{\hat{j}}{}_{\hat{\alpha}}[\varphi_0]Z^{\hat{\alpha}}{}_{\hat{a}}[\varphi_0] = 0 \\ \varepsilon(Z^{\hat{\alpha}}{}_{\hat{a}}) = \varepsilon_\alpha + \varepsilon_a \end{gathered} \tag{4.5.13}$$

Where $\hat{a} = (a, x'')$, the index a running over a finite number of values. The off-mass-shell version of equation (4.5.13) is

$$\begin{gathered} R^{\hat{j}}{}_{\hat{\alpha}}[\varphi]Z^{\hat{\alpha}}{}_{\hat{a}}[\varphi] + S_{,\hat{i}}[\varphi]B^{\hat{j}\hat{i}}{}_{\hat{a}}[\varphi] = 0 \\ B^{\hat{j}\hat{i}}{}_{\hat{a}}[\varphi] = -(-1)^{\varepsilon_i\varepsilon_j}B^{\hat{i}\hat{j}}{}_{\hat{a}}[\varphi] \end{gathered} \tag{4.5.14}$$

and tells us that choosing in equation (4.5.4) the parameters

$$\xi^{\hat{\alpha}} = Z^{\hat{\alpha}}{}_{\hat{a}}[\varphi]\lambda^{\hat{a}} \qquad \varepsilon(\lambda^{\hat{a}}) = \varepsilon_a$$

$\lambda^{\hat{a}}$ being arbitrary, results in a trivial transformation of the form (4.5.5). When the equation

$$Z^{\hat{\alpha}}{}_{\hat{a}}[\varphi_0]\lambda^{\hat{a}} = 0$$

has no solution with compact support (other than $\lambda = 0$), the theory is said to have the first stage of reducibility. Otherwise, there are non-trivial zero-eigenvalue eigenvectors for $Z^{\hat{\alpha}}{}_{\hat{a}}[\varphi_0]$, which also may be linearly dependent, and so on. In other words, there can be gauge theories having any finite stage of reducibility and even infinitely reducible ones (the most famous example is the Green–Schwarz superstring[2]).

Of course, one can always choose in the set $\{R_{\hat{\alpha}}\}$ a linearly independent subset, hence making the theory irreducible. In actual theories, however,

[2]M.B. Green and J.H. Schwarz, *Phys. Lett.* **136B** 367, 1984.

the original generators $R^{\hat{j}}{}_{\hat{\alpha}}[\varphi]$ are local and covariant operators of the fields; after singling out a linearly independent subset, the locality or explicit covariance may be lost.

In our book we have met reducible gauge theories (both in components and superfields). Recall, for example, the antisymmetric tensor field model (1.8.34), which is invariant under the gauge transformations (1.8.35). Here we have

$$\varphi^{\hat{i}} = B_{ab}(x) \qquad R^{\hat{i}}{}_{\hat{\alpha}} = \frac{1}{2}(\partial_a \delta^c_b - \partial_b \delta^c_a)\delta^4(x - x')$$

where $\hat{i} = (ab, x)$ and $\hat{\alpha} = (c, x')$. The generators are linearly dependent, with zero-eigenvalue modes

$$Z^{\hat{\alpha}}{}_{\hat{a}} = \partial_c \delta^4(x' - x'')$$

where $\hat{a} = (x'')$. The reducibility is manifested in the fact that the gauge parameters in transformations (1.8.35) are defined modulo the transformations

$$\lambda_b \to \lambda_b + \partial_b \rho$$

ρ being arbitrary. Obviously, the theory has the first stage of reducibility. Furthermore, consider the chiral spinor superfield model (3.7.10) invariant under the gauge transformations (3.7.11). The theory has the first stage of reducibility, since the gauge parameter in transformations (3.7.11) is defined modulo the displacements

$$K \to K + \Lambda + \bar{\Lambda} \qquad \bar{D}_{\dot{\alpha}} \Lambda = 0$$

Λ being an arbitrary chiral scalar superfield. One more example is given by the superspin-0 model (3.7.24) invariant under the gauge transformations (3.7.22). The theory has the second stage of reducibility, since the gauge parameters in transformations (3.7.22) are defined modulo transformations of the form (3.7.11). Note also that the superfield supergravity (see Chapter 6) is a reducible gauge theory. The linearized supergravity action (6.2.5) is invariant under the gauge transformations (6.2.6), where L_α is defined modulo the redefinitions

$$L_\alpha \to L_\alpha + \eta_\alpha \qquad \bar{D}_{\dot{\alpha}} \eta_\alpha = 0$$

η_α being an arbitrary chiral spinor superfield.

4.5.2. Feynman rules for irreducible gauge theories with closed algebras

We now reproduce the Feynman rules for irreducible gauge theories with closed algebras. Together with the requirement of irreducibility and equation (4.5.12), we shall also assume the validity of the relation

$$(-1)^{\varepsilon_{\hat{i}}} {}_{\hat{i},} R^{\hat{i}}{}_{\hat{\alpha}}[\varphi] + (-1)^{\varepsilon_\beta} C^{\beta}{}_{\beta\hat{\alpha}}[\varphi] = 0 \tag{4.5.15}$$

whose role will soon become clear.

Under the above requirements, the in–out vacuum amplitude is known to be given as

$$\langle \text{out}|\text{in}\rangle = \mathcal{N}\int \mathcal{D}\varphi\, e^{iS[\varphi]}\, \delta[\chi^{\hat{\alpha}}[\varphi]]\, \text{sDet}(F^{\hat{\alpha}}{}_{\beta}[\varphi]) \tag{4.5.16}$$

where $\chi^{\hat{\alpha}}[\varphi]$, $\varepsilon(\chi^{\hat{\alpha}})=\varepsilon_{\alpha}$, are bosonic and fermionic functionals such that the functional supermatrix

$$F^{\hat{\alpha}}{}_{\beta}[\varphi] \equiv \chi^{\hat{\alpha}}{}_{,\hat{i}}[\varphi]R^{\hat{i}}{}_{\beta}[\varphi] \tag{4.5.17}$$

is non-singular at the point $\varphi_0 \in \mathbf{\Phi}_0$ chosen in the role of the classical ground state and hence in a neighbourhood of $\mathbf{\Phi}$. The $\chi^{\hat{\alpha}}[\varphi]$ are called gauge fixing functions. Note that their existence is due to irreducibility of the theory. The equation (4.5.16) represents the well-known Faddeev–Popov ansatz.

Remark. The delta function in equation (4.5.16) is to be understood as a functional analogue of the ordinary Fourier representation for $\delta^{p+2q}(z)$ (1.11.34)

The $\delta[\chi^{\hat{\alpha}}[\varphi]]$ is required in the right-hand side of (4.5.16) to break the degeneracy, owing to the gauge invariance, of the naive integral $\int \mathcal{D}\varphi \exp(iS[\varphi])$. On the other hand, its introduction results in arbitrariness having no relation to the dynamical content of the theory. Were the amplitude $\langle \text{out}|\text{in}\rangle$ well defined, it should not depend on $\chi[\varphi]$. Hence, the relation (4.5.16) may pretend to a physical correctness if its right-hand side, which will be denoted by $\langle \text{out}|\text{in}\rangle_{\chi}$, satisfies the identity

$$\langle \text{out}|\text{in}\rangle_{\chi} = \langle \text{out}|\text{in}\rangle_{\chi+\Delta\chi} \tag{4.5.18}$$

where $\Delta\chi^{\hat{\alpha}}[\varphi]$ is an arbitrary infinitesimal change of $\chi^{\hat{\alpha}}[\varphi]$. Let us recall the proof of identity (4.5.18) given by B. De Witt. In the functional integral

$$\langle \text{out}|\text{in}\rangle_{\chi+\Delta\chi} = \mathcal{N}\int \mathcal{D}\varphi\, e^{iS[\varphi]}\, \delta[\chi+\Delta\chi]\, \text{sDet}(F+\Delta F)$$

one makes the replacement of integration variables

$$\varphi^{i} = \tilde{\varphi}^{i} - R^{\hat{i}}{}_{\hat{\alpha}}[\tilde{\varphi}]\xi^{\hat{\alpha}}[\tilde{\varphi}]$$
$$\xi^{\hat{\alpha}}[\tilde{\varphi}] = (F^{-1}[\tilde{\varphi}])^{\hat{\alpha}}{}_{\beta}\, \Delta\chi^{\beta}[\tilde{\varphi}]$$

representing a field-dependent gauge transformation, hence

$$S[\varphi] = S[\tilde{\varphi}].$$

One also easily obtains

$$\delta[\chi[\varphi] + \Delta\chi[\varphi]] = \delta[\chi[\tilde{\varphi}]]$$

and

$$\mathrm{sDet}(F[\varphi]+\Delta F[\varphi])=\mathrm{sDet}(F[\tilde{\varphi}])\{1+\mathscr{A}[\tilde{\varphi}]\}$$

$$\mathscr{A}[\tilde{\varphi}]=(-1)^{\varepsilon_\alpha}(F^{-1}[\tilde{\varphi}])^{\hat{\alpha}}{}_{\beta}\{\chi^{\beta}{}_{,i}[\tilde{\varphi}]R^{i}{}_{\hat{\alpha}}[\tilde{\varphi}]-F^{\beta}{}_{\hat{\alpha},i}[\tilde{\varphi}]R^{i}{}_{\hat{\gamma}}[\tilde{\varphi}]\xi^{\hat{\gamma}}[\tilde{\varphi}]\}.$$

Next, the change in the integration measure must be calculated in accordance with rule (1.11.32) which gives

$$\mathscr{D}\varphi=\mathscr{D}\tilde{\varphi}\ \mathrm{sDet}\left(\varphi^{i}\frac{\overleftarrow{\delta}}{\delta\tilde{\varphi}^{j}}\right)=\mathscr{D}\tilde{\varphi}\{1-\mathscr{B}[\tilde{\varphi}]\}$$

$$\mathscr{B}[\tilde{\varphi}]=(-1)^{\varepsilon_i}{}_{\hat{i},}R^{\hat{i}}{}_{\hat{\alpha}}[\tilde{\varphi}]\xi^{\hat{\alpha}}[\tilde{\varphi}]$$

$$+(-1)^{\varepsilon_\alpha}(F^{-1}[\tilde{\varphi}])^{\hat{\alpha}}{}_{\beta}\left\{\Delta\chi^{\beta}{}_{,i}[\tilde{\varphi}]R^{\hat{i}}{}_{\hat{\alpha}}[\tilde{\varphi}]-(-1)^{\varepsilon_\alpha\varepsilon_\gamma}F^{\beta}{}_{\hat{\gamma},i}[\tilde{\varphi}]R^{\hat{i}}{}_{\hat{\alpha}}[\tilde{\varphi}]\xi^{\hat{\gamma}}[\tilde{\varphi}]\right\}.$$

Now, making use of equations (4.5.11, 12) and (4.5.17) leads to

$$\langle\mathrm{out}|\mathrm{in}\rangle_{\chi+\Delta\chi}=\mathscr{N}\int\mathscr{D}\tilde{\varphi}\ \mathrm{e}^{\mathrm{i}S[\tilde{\varphi}]}\ \delta[\chi^{\hat{\alpha}}[\tilde{\varphi}]]\ \mathrm{sDet}(F^{\hat{\alpha}}{}_{\beta}[\tilde{\varphi}])$$
$$\times\{1-((-1)^{\varepsilon_\alpha}{}_{\hat{i},}R^{\hat{i}}{}_{\hat{\alpha}}[\tilde{\varphi}]+(-1)^{\varepsilon_\beta}C^{\beta}{}_{\beta\hat{\alpha}}[\tilde{\varphi}])\xi^{\hat{\alpha}}[\tilde{\varphi}]\}\tag{4.5.19}$$

and this coincides with $\langle\mathrm{out}|\mathrm{in}\rangle_{\chi}$ under the fulfilment of equation (4.5.15).

Let $\Psi[\varphi]$ be a physical observable, that is, a gauge invariant functional of φ,

$$\Psi_{,i}[\varphi]R^{\hat{i}}{}_{\hat{\alpha}}[\varphi]=0.$$

On the basis of the arguments given above, its quantum average

$$\langle\mathrm{out}|T(\Psi[\varphi])|\mathrm{in}\rangle=\mathscr{N}\int\mathscr{D}\varphi\Psi[\varphi]\ \mathrm{e}^{\mathrm{i}S[\varphi]}\ \delta[\chi^{\hat{\alpha}}[\varphi]]\ \mathrm{sDet}(F^{\hat{\alpha}}{}_{\beta}[\varphi])\tag{4.5.20}$$

turns out to be independent of the gauge fixing,

$$\langle\mathrm{out}|T(\Psi[\varphi])|\mathrm{in}\rangle_{\chi+\Delta\chi}=\langle\mathrm{out}|T(\Psi[\varphi])|\mathrm{in}\rangle_{\chi}.\tag{4.5.21}$$

Obviously, this fact is necessary for physical correctness. At the same time, it serves as a starting point for representing each average (4.5.20) in a form completely similar to that which appeared in the non-gauge case (see equation (4.2.25)), namely

$$\langle\mathrm{out}|T(\Psi[\varphi])|\mathrm{in}\rangle=\mathrm{const}\int\mathscr{D}\varphi_{\mathrm{total}}\,\Psi[\varphi]\ \mathrm{e}^{\mathrm{i}S_{\mathrm{total}}[\varphi_{\mathrm{total}}]}$$

where the integration is performed over some extended space of histories parametrized by $\varphi_{\mathrm{total}}=(\varphi,\ldots)$, and S_{total} is a functional on this space. To obtain such a representation, the following two tricks may be applied. First, using equation (4.5.21), one can make in expression (4.5.20) the replacement

$$\begin{gathered}\chi^{\hat{\alpha}}[\varphi]\rightarrow\chi^{\hat{\alpha}}[\varphi]-f^{\hat{\alpha}}\\ \frac{\vec{\delta}}{\delta\varphi^{i}}f^{\hat{\alpha}}\equiv 0\qquad \varepsilon(f^{\hat{\alpha}})=\varepsilon_{\alpha}\end{gathered}\tag{4.5.22}$$

where the fs are arbitrary external bosonic and fermionic fields on the space–time, and then integrate the modified version of expression (4.5.20) over these fields with the weight

$$\exp\left(\frac{\mathrm{i}}{2}f^{\hat{\alpha}}B_{\hat{\alpha}\beta}[\varphi_0]f^{\beta}\right)\mathrm{sDet}^{1/2}(\eta^{\hat{\alpha}\beta}B_{\beta\hat{\gamma}}[\varphi_0])\tag{4.5.23a}$$

B being a supersymmetric functional supermatrix

$$\begin{gathered}\varepsilon(B_{\hat{\alpha}\beta})=\varepsilon_{\alpha}+\varepsilon_{\beta}\\ B_{\hat{\alpha}\beta}=(-1)^{\varepsilon_{\alpha}+\varepsilon_{\beta}+\varepsilon_{\alpha}\varepsilon_{\beta}}B_{\beta\hat{\alpha}}.\end{gathered}\tag{4.5.23b}$$

The integration over the fs is assumed to be normalized as

$$\begin{gathered}\mathcal{N}'\int\mathcal{D}f\exp\left(\frac{\mathrm{i}}{2}f^{\hat{\alpha}}\eta_{\hat{\alpha}\beta}f^{\beta}\right)=1\\ \varepsilon(\eta_{\hat{\alpha}\beta})=\varepsilon_{\alpha}+\varepsilon_{\beta}\qquad \eta_{\hat{\alpha}\beta}=(-1)^{\varepsilon_{\alpha}+\varepsilon_{\beta}+\varepsilon_{\alpha}\varepsilon_{\beta}}\eta_{\beta\hat{\alpha}}\end{gathered}\tag{4.5.23c}$$

for some fixed supermatrix η (compare with equation (4.2.32)). The second trick is due to the identities

$$\begin{gathered}\mathrm{sDet}^{-1}(F^{\hat{\alpha}}{}_{\beta}[\varphi])=\mathcal{N}''\int\mathcal{D}\xi'\,\mathcal{D}\xi\exp(\mathrm{i}\xi'_{\hat{\alpha}}F^{\hat{\alpha}}{}_{\beta}[\varphi]\xi^{\beta})\\ \varepsilon(\xi'_{\hat{\alpha}})=\varepsilon(\xi^{\hat{\alpha}})=\varepsilon_{\alpha}\end{gathered}\tag{4.5.24a}$$

$$\begin{gathered}\mathrm{sDet}(F^{\hat{\alpha}}{}_{\beta}[\varphi])=\mathcal{N}'''\int\mathcal{D}c'\,\mathcal{D}c\exp(\mathrm{i}c'_{\hat{\alpha}}F^{\hat{\alpha}}{}_{\beta}[\varphi]c^{\beta})\\ \varepsilon(c'_{\hat{\alpha}})=\varepsilon(c^{\hat{\alpha}})=\varepsilon_{\alpha}+1.\end{gathered}\tag{4.5.24b}$$

As may be seen, we can represent the superdeterminant in expression (4.5.20) by the Gaussian integral over fields $c'_{\hat{\alpha}}$ and $c^{\hat{\alpha}}$ of opposite statistics to the gauge parameters. As a result, the average (4.5.20) takes the form

$$\langle\mathrm{out}|T(\Psi[\varphi])|\mathrm{in}\rangle=\mathrm{sDet}^{1/2}(\eta^{\hat{\alpha}\beta}B_{\beta\hat{\gamma}}[\varphi_0])\tilde{\mathcal{N}}\int\mathcal{D}\varphi\mathcal{D}c'\mathcal{D}c\varphi\Psi[\varphi]\,\mathrm{e}^{\mathrm{i}S_{\mathrm{total}}[\varphi,c',c]}\tag{4.5.25}$$

where

$$S_{\mathrm{total}}[\varphi,c',c]=S[\varphi]+\frac{1}{2}\chi^{\hat{\alpha}}[\varphi]B_{\hat{\alpha}\beta}[\varphi_0]\chi^{\beta}[\varphi]+c'_{\alpha}\chi^{\hat{\alpha}}{}_{,\hat{i}}[\varphi]R^{\hat{i}}{}_{\beta}[\varphi]c^{\beta}.\tag{4.5.26}$$

The fields $c'_{\hat{\alpha}}$ and $c^{\hat{\alpha}}$ are called the Faddeev–Popov ghosts. It should be clear from the above consideration that the expression (4.5.25) does not depend on the explicit form of $\chi^{\hat{\alpha}}[\varphi]$ and $B_{\hat{\alpha}\hat{\beta}}[\varphi_0]$—external objects introduced for technical convenience.

Each quantum average (4.5.24) can be read off from the generating functional of Green's functions

$$Z_{(\chi,B)}[J] = \text{const} \int \mathscr{D}\varphi\, \mathscr{D}c'\, \mathscr{D}c\, \Psi[\varphi]\, e^{i(S_{\text{total}}[\varphi, c', c] + J_{\hat{i}}\varphi^{\hat{i}})} \tag{4.5.27}$$

$$\text{const} = \text{sDet}^{1/2}(\eta^{\hat{\alpha}\hat{\beta}} B_{\hat{\beta}\hat{\gamma}}[\varphi_0])\mathscr{N}$$

where $J_{\hat{i}}$ is an external source, $\varepsilon(J_{\hat{i}}) = \varepsilon_i$. As opposed to the physical correlators $\langle \text{out} | \text{in} \rangle$ and $\langle \text{out} | T(\Psi[\varphi]) | \text{in} \rangle$, $\Psi[\varphi]$ being a physical observable, the generating functional explicitly depends on the choice of $\chi^{\hat{\alpha}}[\varphi]$ and $B_{\hat{\alpha}\hat{\beta}}[\varphi_0]$. The same is true for the effective action $\Gamma_{(\chi,B)}[\tilde{\varphi}]$ constructed from $Z_{(\chi,B)}[J]$ according to the rule of Section 4.3. But the S-matrix derived from $Z_{(\chi,B)}[J]$ is known to be gauge independent.

From the viewpoint of practical perturbative calculations, it is useful to replace $Z_{(\chi,B)}[J]$ by its extension $Z_{(\chi,B)}[J, J_{c'}, J_c]$ corresponding to the case when sources for the ghosts are added in the exponential in expression (4.5.27). Next, one represents S_{total} as a sum of its free and coupled parts:

$$S_{\text{total}}[\varphi, c', c] = S[\varphi_0] + S_0[\Delta\varphi, c', c; \varphi_0] + S_{\text{INT}}[\Delta\varphi, c', c; \varphi_0]$$

$$S_0 = \frac{1}{2} \Delta\varphi^{\hat{i}} ({}_{\hat{i},}S_{,\hat{j}}[\varphi_0] + {}_{\hat{i},}\chi^{\hat{\alpha}}[\varphi_0] B_{\hat{\alpha}\hat{\beta}}[\varphi_0] \chi^{\hat{\beta}}{}_{,\hat{j}}[\varphi_0]) \Delta\varphi^{\hat{j}} + c'_{\hat{\alpha}} \chi^{\hat{\alpha}}{}_{,\hat{i}}[\varphi_0] R^{\hat{i}}{}_{\hat{\beta}}[\varphi_0] c^{\hat{\beta}} \tag{4.5.28}$$

where $\Delta\varphi = \varphi - \varphi_0$ and S_{INT} is at least cubic in $\Delta\varphi$, c' and c (assuming the normalization $\chi^{\hat{\alpha}}[\varphi_0] = 0$). Further manipulations are exactly the same as in the non-gauge case (see Section 4.2).

When establishing gauge independence, we have essentially used the validity of equation (4.5.15). Hence, the ansatz (4.5.16) proves not to be truly correct for gauge theories in which equation (4.5.15) does not take place. In this case, however, the modification needed turns out to consist merely in replacing the naive measure $\mathscr{D}\varphi$ by an invariant one, $\mathscr{D}\varphi\mu[\varphi]$, where $\mu[\varphi]$ is a functional transforming in such a way as to compensate the non-invariant terms in expression (4.5.19). Discussion of these theories is beyond the scope of our book. It should be remarked that both terms in expression (4.5.15), unless they vanish, are ill-defined in local field theories. The point here is that the generators $R^{\hat{i}}{}_{\hat{\alpha}}[\varphi]$, where $\hat{i} = (i, x)$ and $\hat{\alpha} = (\alpha, x')$, involve φ (and maybe its derivatives to a finite order) only at the point x, hence the quantity $(-1)^{\varepsilon_i}{}_{\hat{i},}R^{\hat{i}}{}_{\hat{\alpha}}[\varphi]$ contains delta functions with coincident arguments, and similarly for $(-1)^{\varepsilon_\beta} C^{\hat{\beta}}{}_{\hat{\beta}\hat{\alpha}}[\varphi]$. Fortunately, in many theories of interest both these objects vanish

$$(-1)^{\varepsilon_i} {}_{\hat{i},}R^{\hat{i}}{}_{\hat{\alpha}}[\varphi]=0 \tag{4.5.29a}$$

$$(-1)^{\varepsilon_\beta} C^{\beta}{}_{\beta\hat{\alpha}}[\varphi]=0. \tag{4.5.29b}$$

For example, in the case of a pure Yang–Mills theory, with $\varphi^{\hat{i}}=V_a{}^I(x)$, gauge transformations are given by the first expression in relation (3.5.33b), hence

$$R^{\hat{i}}{}_{\hat{\alpha}}[\varphi]=\delta^{IJ}\partial_a\delta^4(x-x')+f^{IKJ}V_a{}^K(x)\delta^4(x-x')$$

$${}_{\hat{k},}R^{\hat{i}}{}_{\hat{\alpha}}[\varphi]=\delta^b_a f^{IKJ}\delta^4(x-x'')\delta^4(x-x')$$

where $\hat{\alpha}=(J,x')$ and $\hat{k}=(Kb,x'')$. Here equations (4.5.29) are satisfied, since the structure constants of the gauge group are totally antisymmetric.

Remark. Equation (4.5.29a) implies the invariance of the functional measure $\mathscr{D}\varphi$ under the gauge transformations (4.5.4) with ξ being φ-independent.

4.5.3. Supersymmetric gauge theories

Consider a dynamical system, which is simultaneously supersymmetric and gauge. We shall assume that the full gauge freedom can be described by transformations of the corresponding superfield variables $v^{\hat{I}}=v^I(z)$ like

$$\begin{aligned} &\delta v^{\hat{I}}=\mathbf{R}^{\hat{I}}{}_{\hat{\mu}}[v]\zeta^{\hat{\mu}} \qquad S_{,\hat{I}}[v]\mathbf{R}^{\hat{I}}{}_{\hat{\mu}}[v]\equiv 0\\ &\varepsilon(\zeta^{\hat{\mu}})=\varepsilon_\mu \qquad \varepsilon(\mathbf{R}^{\hat{I}}{}_{\hat{\mu}})=\varepsilon_I+\varepsilon_\mu \end{aligned} \tag{4.5.30}$$

Here $\zeta^{\hat{\mu}}=\zeta^\mu(z')$ are tensor superfield parameters being, in general, unconstrained, chiral and antichiral; in the dependence upon superfield types of ζs; the summation over $\hat{\mu}$ in relation (4.5.30) includes the integration over d^8z', d^6z' or $d^6\bar{z}'$ (these conventions are exactly the same as for the superfield variables). The $\mathbf{R}^{\hat{I}}{}_{\hat{\mu}}[v]$ will be called supergenerators of gauge transformations. By supposition, the correspondence $\zeta^{\hat{\mu}}\to\delta v^{\hat{I}}$ maps tensor superfields into tensor fields, hence the operators $\mathbf{R}[v]$ are super Poincaré covariant; in particular, they (anti)commute with the supersymmetry generators. In general, the supergenerators constitute an algebra of the form

$$\mathbf{R}^{\hat{J}}{}_{\hat{\mu}}[v]\,\mathbf{R}^{\hat{J}}{}_{\hat{\nu}}[v]-(-1)^{\varepsilon_\mu\varepsilon_\nu}\mathbf{R}^{\hat{I}}{}_{\hat{\nu},\hat{J}}[v]\mathbf{R}^{\hat{J}}{}_{\hat{\mu}}\,[v]=\mathbf{R}^{\hat{I}}{}_{\hat{\lambda}}[v]\mathbf{C}^{\hat{\lambda}}{}_{\hat{\mu}\hat{\nu}}[v]+S_{,\hat{J}}[v]\mathbf{E}^{\hat{J}\hat{I}}{}_{\hat{\mu}\hat{\nu}}[v] \tag{4.5.31}$$

the **C**s and **E**s being superfunctionals under equations like (4.5.10).

Gauge transformations (4.5.30) can be rewritten in terms of components. For this purpose one must introduce not only the component fields $\varphi^i(x)$ of $v^I(z)$ in accordance with law (4.1.40), but also the component fields $\xi^\alpha(x)$ of $\zeta^\mu(z)$ defined with the help of proper picture-change operators $P^{\hat{\alpha}}{}_{\hat{\mu}}$ and $Q^{\hat{\mu}}{}_{\hat{\alpha}}$:

$$\begin{aligned} &\xi^{\hat{\alpha}}=P^{\hat{\alpha}}{}_{\hat{\mu}}\zeta^{\hat{\mu}} \qquad \zeta^{\hat{\mu}}=Q^{\hat{\mu}}{}_{\hat{\alpha}}\xi^{\hat{\alpha}}\\ &P^{\hat{\alpha}}{}_{\hat{\mu}}Q^{\hat{\mu}}{}_{\hat{\beta}}=\delta^{\hat{\alpha}}{}_{\hat{\beta}} \qquad Q^{\hat{\mu}}{}_{\hat{\alpha}}P^{\hat{\alpha}}{}_{\hat{\nu}}=\delta^{\hat{\mu}}{}_{\hat{\nu}}. \end{aligned} \tag{4.5.32}$$

Then, the component form of the gauge transformations (4.5.30) is given by equation (4.5.4), where

$$R^{\hat{i}}{}_{\hat{\alpha}}[\varphi]=P^{\hat{i}}{}_{\hat{I}}\mathbf{R}^{\hat{I}}{}_{\hat{\mu}}[v]Q^{\hat{\mu}}{}_{\hat{\alpha}}. \tag{4.5.33}$$

Note: we use letters from the beginning of Greek alphabet for component gauge parameters, and from the middle to denote superfield gauge parameters.

It follows from equation (4.5.33) that the superfield gauge algebra (4.5.31) induces the component version (4.5.11) with

$$\begin{aligned} C^{\hat{\gamma}}{}_{\hat{\alpha}\hat{\beta}}[\varphi]&=(-1)^{\varepsilon_\nu(\varepsilon_\mu+\varepsilon_\alpha)}P^{\hat{\gamma}}{}_{\hat{\lambda}}\mathbf{C}^{\hat{\lambda}}{}_{\hat{\mu}\hat{\nu}}[v]Q^{\hat{\mu}}{}_{\hat{\alpha}}Q^{\hat{\nu}}{}_{\hat{\beta}} \\ E^{\hat{i}\hat{j}}{}_{\hat{\alpha}\hat{\beta}}[\varphi]&=(-1)^{\varepsilon_I(\varepsilon_J+\varepsilon_i)+\varepsilon_\nu(\varepsilon_\mu+\varepsilon_\alpha)}P^{\hat{i}}{}_{\hat{I}}P^{\hat{j}}{}_{\hat{J}}\mathbf{E}^{\hat{I}\hat{J}}{}_{\hat{\mu}\hat{\nu}}[v]Q^{\hat{\mu}}{}_{\hat{\alpha}}Q^{\hat{\nu}}{}_{\hat{\beta}} \end{aligned} \tag{4.5.34}$$

As a result, the component gauge algebra is closed if and only if its superfield counterpart is closed. Furthermore, if the theory is irreducible in components, the same is true in terms of superfields. Finally, from the formal identities

$$(-1)^{\varepsilon_i}_{\hat{i},}R^{\hat{i}}{}_{\hat{\alpha}}[\varphi]=(-1)^{\varepsilon_I}{}_{\hat{I},}\mathbf{R}^{\hat{I}}{}_{\hat{\mu}}[v]Q^{\hat{\mu}}{}_{\hat{\alpha}}$$

$$(-1)^{\varepsilon_\beta}C^{\hat{\beta}}{}_{\hat{\beta}\hat{\alpha}}[\varphi]=(-1)^{\varepsilon_\nu}\mathbf{C}^{\hat{\nu}}{}_{\hat{\nu}\hat{\mu}}[v]Q^{\hat{\mu}}{}_{\hat{\alpha}}$$

we see that the relations (4.5.29) imply

$$(-1)^{\varepsilon_I}{}_{\hat{I},}\mathbf{R}^{\hat{I}}{}_{\hat{\mu}}[v]=0 \tag{4.5.35a}$$

$$(-1)^{\varepsilon_\nu}\mathbf{C}^{\hat{\nu}}{}_{\hat{\nu}\hat{\mu}}[v]=0. \tag{4.5.35b}$$

and vice versa. Therefore it proves unimportant which picture (superfield or component), is used for analysing gauge algebra.

Now, let us specialize our consideration to the case of irreducible gauge theories with closed algebras and under equations (4.5.35). Due to irreducibility, there exists a set of superfield gauge conditions $\kappa^{\hat{\mu}}[v]$ (having the same superfield types as the gauge parameters $\zeta^{\hat{\mu}}$, $\varepsilon(\zeta^{\hat{\mu}})=\varepsilon_\mu$, such that the supermatrix

$$\mathbf{F}^{\hat{\mu}}{}_{\hat{\nu}}[v]=\kappa^{\hat{\mu}}{}_{,\hat{I}}[v]\mathbf{R}^{\hat{I}}{}_{\hat{\nu}}[v] \tag{4.5.36}$$

is non-singular at some stationary point $v_0\in\mathbb{V}_0$ and hence, in its neighbourhood in $\mathbb{V}$. The $\kappa^{\hat{\mu}}[v]$ will be called gauge fixing superfunctions. The components of $\kappa^{\hat{\mu}}[v]$ defined by

$$\chi^{\hat{\alpha}}[\varphi]=P^{\hat{\alpha}}{}_{\hat{\mu}}\kappa^{\hat{\mu}}[v] \tag{4.5.37}$$

constitute an admissible set of gauge fixing functionals. Under the choice (4.5.37), the supermatrices (4.5.17) and (4.5.36) are connected as

$$F^{\hat{\alpha}}{}_{\hat{\beta}}[\varphi]=P^{\hat{\alpha}}{}_{\hat{\mu}}\mathbf{F}^{\hat{\mu}}{}_{\hat{\nu}}[v]Q^{\hat{\nu}}{}_{\hat{\beta}}. \tag{4.5.38}$$

Hence, non-singularity of $\mathbf{F}[v_0]$ implies non-singularity of $F[\varphi_0]$ at the point

$\varphi_0^{i} = P^{i}{}_{\hat{I}} v_0^{\hat{I}}$. As an immediate consequence of equation (4.5.38) we obtain

$$\mathrm{sDet}(F^{\hat{\alpha}}{}_{\beta}[\varphi]) = \mathbf{sDet}(\mathbf{F}^{\hat{\mu}}{}_{\hat{\nu}}[v]). \tag{4.5.39}$$

In the class of supersymmetric theories under consideration, the quantization scheme described in the previous subsection can be applied to superfields and components, and in both approaches it leads to equivalent physical results in the following sense. The generating functional (4.5.27) turns out to be the component form of its superfield counterpart

$$\begin{gathered}
\mathbf{Z}_{(\chi,\mathbf{B})}[\mathbf{J}] = \mathrm{const} \int \mathscr{D}v\, \mathscr{D}\mathbf{c}'\, \mathscr{D}\mathbf{c}\; \mathrm{e}^{\mathrm{i}(S_{\mathrm{total}}[v,\mathbf{c}',\mathbf{c}] + \mathbf{J}_{\hat{I}} v^{\hat{I}})} \\
\mathrm{const} = \mathbf{sDet}^{1/2}(\eta^{\hat{\mu}\hat{\nu}} \mathbf{B}_{\hat{\nu}\hat{\lambda}}[v_0]) \\
S_{\mathrm{total}}[v, \mathbf{c}', \mathbf{c}] = S[v] + \frac{1}{2} \kappa^{\hat{\mu}}[v] \mathbf{B}_{\hat{\mu}\hat{\nu}} \kappa^{\hat{\nu}}[v] + \mathbf{c}'_{\hat{\mu}} \kappa^{\hat{\mu}}{}_{,\hat{I}}[v] \mathbf{R}^{\hat{I}}{}_{\hat{\nu}}[v] \mathbf{c}^{\hat{\nu}}
\end{gathered} \tag{4.5.40}$$

provided that (1) the gauge fixing functions $\chi^{\hat{\alpha}}[\varphi]$ are chosen by the rule (4.5.37); (2) the functional supermatrices $B_{\hat{\alpha}\beta}[\varphi_0]$ and $\eta_{\hat{\alpha}\beta}$ represent the component versions of the superfunctional ones, that is

$$\mathbf{B}_{\hat{\mu}\hat{\nu}}[v_0] = P^{\mathrm{sT}}{}_{\hat{\mu}}{}^{\hat{\alpha}} B_{\hat{\alpha}\beta}[\varphi_0] P^{\beta}{}_{\hat{\nu}}.$$

$$\eta_{\hat{\mu}\hat{\nu}} = P^{\mathrm{sT}}{}_{\hat{\mu}}{}^{\hat{\alpha}} \eta_{\hat{\alpha}\beta} P^{\beta}{}_{\hat{\nu}}.$$

We suggest that the reader fills in the necessary details.

Before we conclude, let us make one important remark. The equations (4.5.35) are identically satisfied for a large family of local supersymmetric theories. In the local case, the supergenerators $\mathbf{R}^{\hat{I}}{}_{\hat{\mu}}[v]$, where $\hat{I} = (I, z)$ and $\hat{\mu} = (\mu, z')$, are functions of v and its covariant derivatives to a finite order at the point z. Suppose that the $\mathbf{R}^{\hat{I}}{}_{\hat{\mu}}[v]$ do not depend on covariant derivatives of v, then

$$_{\hat{J},}\mathbf{R}^{\hat{I}}{}_{\hat{\mu}}[v] \sim \delta^{(J)}(z'', z) \qquad \hat{J} = (J, z'')$$

the delta function being defined in equation (4.1.17). Since $\delta^{(J)}(z, z) = 0$, we obtain equation (4.5.35*a*). As an example, consider the general super Yang–Mills model of subsection 3.5.6. Here we have the superfield variables

$$v^{\hat{I}} = (V^I(z),\ \Phi^i(z),\ \bar{\Phi}_i(z))$$

and the gauge parameters

$$\zeta^{\hat{\mu}} = (\Lambda^J(z'),\ \bar{\Lambda}^J(z')) \equiv (\zeta^+, \zeta^-)$$

From equations (3.5.25) and (3.5.31) we find the supergenerators

$$\mathbf{R}^{\hat{I}}{}_{+}[v] = \begin{cases} -\dfrac{\mathrm{i}}{2} \delta^{IJ} \delta_+(z, z') + f^{IJ}(V(z)) \delta_+(z, z') \\ \mathrm{i}(T^J)^i{}_j \Phi^j(z) \delta_+(z, z') \\ 0 \end{cases}$$

$$\mathbf{R}^{\,\bar{I}}_{-}[v] = \begin{cases} \frac{i}{2}\delta^{IJ}\delta_{-}(z,z') + \bar{f}^{IJ}(V(z))\delta_{-}(z,z') \\ 0 \\ -i\Phi_j(z)(T^J)^j{}_i\delta_{-}(z,z'). \end{cases}$$

Both equations (4.5.35) are identically satisfied.

4.6. Feynman rules for super Yang–Mills theories

We proceed by obtaining superfield Feynman rules for supersymmetric Yang–Mills theories, based on the general considerations of Section 4.5.

4.6.1. Quantization of the pure super Yang–Mills model

We begin with superfield quantization of the pure super Yang–Mills theory (3.5.39). For later convenience, it is reasonable to rescale the coupling constant by $g \to \sqrt{2}g$ and the gauge superfield by $V \to gV$ thus setting the action

$$S_{\mathrm{SYM}} = \frac{1}{4g^2}\int d^6z\, \mathrm{tr}(W^\alpha W_\alpha) \tag{4.6.1}$$

where

$$\begin{aligned} W_\alpha &= -\frac{1}{8}\bar{D}^2(e^{-2gV} D_\alpha e^{2gV}) \\ V &= V^I(z)T^I \qquad \mathrm{tr}(T^I T^J) = \delta^{IJ}. \end{aligned} \tag{4.6.2}$$

The original normalization was useful when analysing the component content of the theory and the underlying geometry. The one above turns out to be convenient for perturbative superfield calculations. In particular, after the rescaling $V \to gV$, the action admits a well-defined limit $g \to 0$, since

$$S_{\mathrm{SYM}} = \frac{1}{16}\int d^8z\, \mathrm{tr}(VD^\alpha\bar{D}^2D_\alpha V) + O(g).$$

The action (4.6.1) is invariant with respect to the gauge transformations

$$\begin{aligned} \delta V &= L_{gV}[\Lambda - \bar{\Lambda} + \coth_{gV}(\Lambda + \bar{\Lambda})] \\ \bar{D}_{\dot{\alpha}}\Lambda &= 0 \qquad L_A B \equiv [A, B] \end{aligned} \tag{4.6.3}$$

obtained from the original transformations (3.5.31) by rescaling $V \to gV$ and $\Lambda \to 2ig\Lambda$. As we have seen in Section 3.5, the gauge transformations form a group, therefore the gauge algebra of the theory under consideration is closed. The corresponding supergenerators turn out to be linearly independent. To prove this statement, it is sufficient to consider the Abelian case, where equation (4.6.3) reads

$$\delta V = \Lambda + \bar{\Lambda}.$$

Setting here $\delta V = 0$ implies that

$$\mathrm{D}_\alpha \Lambda = 0 \quad \Rightarrow \quad \Lambda = -\bar{\Lambda} = \text{const}$$

hence the supergenerators have no zero-eigenvalue eigenfunction with compact support in space–time, other than $\Lambda = \bar{\Lambda} = 0$. Further, we have also seen that equations (4.5.35) are satisfied for the theory in question. So, it can be quantized by the scheme described in the previous section.

In the role of gauge fixing superfunctions we choose

$$\kappa[V] = -\frac{1}{4}\bar{\mathrm{D}}^2 V(z) + \mathbf{f}(z)$$

$$\bar{\kappa}[V] = -\frac{1}{4}\mathrm{D}^2 V(z) + \bar{\mathbf{f}}(z) \tag{4.6.4}$$

$$\mathbf{f} = \mathbf{f}^I T^I \qquad \bar{\mathrm{D}}_{\dot{\alpha}} \mathbf{f}^I = 0.$$

Here $\mathbf{f}^I$ are external chiral scalar superfields, inert under the gauge transformations (4.6.3). κ and $\bar{\kappa}$ constitute an admissible set of gauge fixing conditions, at least at the stationary point $V = 0$, where they transform under transformations (4.6.3) as

$$\delta\begin{pmatrix}\kappa[V]\\ \bar{\kappa}[V]\end{pmatrix}_{V=0} = \begin{pmatrix} 0 & -\frac{1}{4}\bar{\mathrm{D}}^2 \\ -\frac{1}{4}\mathrm{D}^2 & 0 \end{pmatrix}\begin{pmatrix}\Lambda\\ \bar{\Lambda}\end{pmatrix}.$$

The differential operator arising here is non-singular on the space of chiral–antichiral superfields, since its zero-eigenvalue eigenfunctions satisfy the Klein–Gordon equation and cannot be localized in Minkowski space.

The next step consists of deriving the ghost action that the gauge fixing (4.6.4) leads to. This is done most simply with the help of the standard observation that the ghost term in expression (4.5.40) can be represented in the manner

$$S_{\mathrm{GH}} = \mathbf{c}'_{\hat{\mu}} \delta\kappa^{\hat{\mu}}[v]$$

where $\delta\kappa$ is the variation of κ under the transformation (4.5.30) with $\zeta^{\hat{\mu}}$ being replaced by $\mathbf{c}^{\hat{\mu}}$. In our case, both gauge parameters and gauge conditions are (anti)chiral, and they give rise to (anti)chiral ghost superfields:

$$\zeta^{\hat{\mu}} = \begin{pmatrix}\Lambda\\ \bar{\Lambda}\end{pmatrix} \to \mathbf{c}^{\hat{\mu}} = \begin{pmatrix}\mathbf{c}\\ \bar{\mathbf{c}}\end{pmatrix} \qquad \mathbf{c} = \mathbf{c}^I(z) T^I \qquad \bar{\mathrm{D}}_{\dot{\alpha}} \mathbf{c}^I = 0$$

$$\kappa^{\hat{\mu}} = \begin{pmatrix} \kappa[V] \\ \bar{\kappa}[V] \end{pmatrix} \to \mathbf{c}'_{\hat{\mu}} = \begin{pmatrix} \mathbf{c}' \\ \bar{\mathbf{c}}' \end{pmatrix} \qquad \mathbf{c}' = \mathbf{c}'^{I}(z)T^{I} \qquad \bar{\mathrm{D}}_{\dot{\alpha}}\mathbf{c}'^{I} = 0.$$

The above observation leads to

$$\begin{aligned} S_{\mathrm{GH}} &= \int \mathrm{d}^6 z \,\mathrm{tr}\, \mathbf{c}' \frac{\delta\kappa}{\delta V} \delta V + \int \mathrm{d}^6 \bar{z} \,\mathrm{tr}\, \bar{\mathbf{c}}' \frac{\delta\bar{\kappa}}{\delta V} \delta V \\ &= \int \mathrm{d}^6 z \left(-\frac{1}{4} \bar{\mathrm{D}}^2 \right) \mathrm{tr}(\mathbf{c}' \,\delta V) + \int \mathrm{d}^6 \bar{z} \left(-\frac{1}{4} \mathrm{D}^2 \right) \mathrm{tr}(\bar{\mathbf{c}}' \,\delta V) \\ &= \int \mathrm{d}^8 z \,\mathrm{tr}((\mathbf{c}' + \bar{\mathbf{c}}')\delta V) \end{aligned}$$

and with the use of transformations (4.6.3) we finally obtain

$$S_{\mathrm{GH}} = \int \mathrm{d}^8 z \,\mathrm{tr}((\mathbf{c}' + \bar{\mathbf{c}}') L_{gV}[\mathbf{c} - \bar{\mathbf{c}} + \coth L_{gV}(\mathbf{c} + \bar{\mathbf{c}})]). \tag{4.6.5}$$

Now, in accordance with the results of subsection 4.5.3, the in–out vacuum amplitude reads

$$\begin{aligned} \langle \mathrm{out} | \mathrm{in} \rangle = \int & \mathcal{D}V \mathcal{D}\bar{\mathbf{c}}' \mathcal{D}\mathbf{c}' \mathcal{D}\bar{\mathbf{c}} \mathcal{D}\mathbf{c} \, \mathrm{e}^{\mathrm{i} S_{\mathrm{SYM}} + S_{\mathrm{GH}}} \\ & \times \delta_{+}\left[\mathbf{f} - \frac{1}{4} \bar{\mathrm{D}}^2 V \right] \delta_{-}\left[\bar{\mathbf{f}} - \frac{1}{4} \mathrm{D}^2 V \right] \end{aligned} \tag{4.6.6}$$

where

$$\delta_{+}\left[\mathbf{f} - \frac{1}{4} \bar{\mathrm{D}}^2 V \right] = \prod_{I=1}^{\dim G} \delta_{+}\left[\mathbf{f}^{I} - \frac{1}{4} \bar{\mathrm{D}}^2 V^{I} \right]$$

and does not depend on the external (anti)chiral superfields $\bar{\mathbf{f}}$ and $\mathbf{f}$. So, one can average over them with some appropriate weight. The standard choice is

$$\exp\left\{ \frac{\mathrm{i}}{\gamma} \int \mathrm{d}^8 z \,\mathrm{tr}(\bar{\mathbf{f}}\mathbf{f}) \right\} \tag{4.6.7}$$

γ being a real constant. As a result, one arrives at

$$\begin{aligned} \langle \mathrm{out} | \mathrm{in} \rangle &= \int \mathcal{D}V \mathcal{D}\bar{\mathbf{c}}' \mathcal{D}\mathbf{c}' \mathcal{D}\bar{\mathbf{c}} \mathcal{D}\mathbf{c} \, \mathrm{e}^{\mathrm{i} S_{\mathrm{total}}} \\ S_{\mathrm{total}} &= S_{\mathrm{SYM}} + S_{\mathrm{GB}} + S_{\mathrm{GH}} \end{aligned} \tag{4.6.8}$$

where

$$S_{\mathrm{GB}} = \frac{1}{16\gamma} \int \mathrm{d}^8 z \,\mathrm{tr}((\bar{\mathrm{D}}^2 V)(\mathrm{D}^2 V)). \tag{4.6.9}$$

Remark. Integrating the expression (4.6.6) over $\mathbf{f}$ and $\bar{\mathbf{f}}$ with the weight (4.6.7), we obtain exactly $\langle \text{out}|\text{in}\rangle$, due to the identity

$$\int \mathscr{D}\bar{\mathbf{f}}\mathscr{D}\mathbf{f} \exp\left\{\frac{\mathrm{i}}{\gamma}\int \mathrm{d}^8 z \, \mathrm{tr}(\bar{\mathbf{f}}\mathbf{f})\right\} = 1$$

which follows from equations (4.1.27), (4.4.25) and the relation

$$\int \mathscr{D}\bar{\chi}\mathscr{D}\chi \, \mathrm{e}^{\mathrm{i}S[\chi,\bar{\chi};\Psi,\bar{\Psi}]} = \mathbf{sDet}^{-1/2}(\mathbf{H}^{(\Psi)}) \tag{4.6.10}$$

with $S[\chi, \bar{\chi}; \Psi, \bar{\Psi}]$ and $\mathbf{H}^{(\Psi)}$ being defined by equations (4.4.6) and (4.4.5), respectively. Therefore, the integral (4.6.8) is independent of γ, and this constant can be fixed from considerations of convention.

The relation (4.6.8) motivates us to define the generating superfunctional for Green's functions in the manner

$$\mathbf{Z}[\mathscr{J}] = \int \mathscr{D}V\mathscr{D}\bar{\mathbf{c}}'\mathscr{D}\mathbf{c}'\mathscr{D}\bar{\mathbf{c}}\mathscr{D}\mathbf{c} \, \mathrm{e}^{\mathrm{i}(S_{\text{total}} + \int \mathrm{d}^8 z \, \mathrm{tr}(\mathscr{J}V))} \tag{4.6.11}$$

where $\mathscr{J} = \mathscr{J}^I(z)T^I$, $\mathscr{J}^I$ are real scalar superfields.

4.6.2. Propagators and vertices

We are now going to develop a perturbation theory to calculate $\mathbf{Z}[\mathscr{J}]$ (4.6.11). For this purpose, we choose the stationary point $V=0$ of S_{SYM} in the role of classical ground state and represent the action S_{total} as

$$S_{\text{total}} = S_0 + S_{\text{INT}}$$

where $S_0[V, \mathbf{c}', \mathbf{c}, \bar{\mathbf{c}}', \bar{\mathbf{c}}]$ is the quadratic part of S_{total} in the gauge and ghost superfields, while $S_{\text{INT}}[V, \mathbf{c}', \mathbf{c}, \bar{\mathbf{c}}', \bar{\mathbf{c}}]$ describes an interaction of the superfields. Then one can write

$$\mathbf{Z}[\mathscr{J}] = \exp\left(\mathrm{i}S_{\text{INT}}\left[\frac{\delta}{\delta \mathrm{i}\mathscr{J}}, \frac{\vec{\delta}}{\delta \mathrm{i}\eta'}, \frac{\vec{\delta}}{\delta \mathrm{i}\eta}, \frac{\vec{\delta}}{\delta \mathrm{i}\bar{\eta}'}, \frac{\vec{\delta}}{\delta \mathrm{i}\bar{\eta}}\right]\right)$$
$$\times \mathbf{Z}_0[\mathscr{J}, \eta', \eta, \bar{\eta}', \bar{\eta}]|_{\eta'=\eta=\bar{\eta}'=\bar{\eta}=0} \tag{4.6.12}$$

where $\mathbf{Z}_0$ is the free generating superfunctional defined by

$$\mathbf{Z}_0[\mathscr{J}, \eta', \eta, \bar{\eta}', \bar{\eta}] = \int \mathscr{D}V\mathscr{D}\bar{\mathbf{c}}'\mathscr{D}\mathbf{c}'\mathscr{D}\bar{\mathbf{c}}\mathscr{D}\mathbf{c} \exp \mathrm{i}\{S_0[V, \mathbf{c}', \mathbf{c}, \bar{\mathbf{c}}', \bar{\mathbf{c}}]$$
$$+ \mathrm{tr}\int \mathrm{d}^8 z \, \mathscr{J}V + \mathrm{tr}\int \mathrm{d}^6 z(\eta'\mathbf{c}' + \eta\mathbf{c}) + \mathrm{tr}\int \mathrm{d}^6 \bar{z}(\bar{\eta}'\bar{\mathbf{c}}' + \bar{\eta}\bar{\mathbf{c}})\}. \tag{4.6.13}$$

Here the supersources $\eta' = \eta'^I(z)T^I$ and $\eta = \eta^I(z)T^I$ are anticommuting chiral scalar superfields. The variational rules for supersources are:

$$\frac{\delta}{\delta \mathcal{J}^I(z)} \mathcal{J}^K(z') = \delta^{IK}\delta^8(z-z')$$

$$\frac{\vec{\delta}}{\delta \eta'^I(z)} \eta'^K(z') = \frac{\vec{\delta}}{\delta \eta^I(z)} \eta^K(z') = \delta^{IK}\left(-\frac{1}{4}\bar{\mathrm{D}}^2\right)\delta^8(z-z')$$

$$\frac{\vec{\delta}}{\delta \bar{\eta}'^I(z)} \bar{\eta}'^K(z') = \frac{\vec{\delta}}{\delta \bar{\eta}^I(z)} \bar{\eta}^K(z') = \delta^{IK}\left(-\frac{1}{4}\mathrm{D}^2\right)\delta^8(z-z').$$

Let us determine S_0 and S_{INT}. We start by discussing the gauge action (4.6.1), which can be rewritten in the form

$$S_{\mathrm{SYM}} = -\frac{1}{64g^2}\,\mathrm{tr}\int \mathrm{d}^8 z\, \mathrm{e}^{-2gV}\mathrm{D}^\alpha\, \mathrm{e}^{2gV}\bar{\mathrm{D}}^2(\mathrm{e}^{-2gV}\mathrm{D}_\alpha\, \mathrm{e}^{2gV}). \qquad (4.6.14)$$

Consider the identity

$$\begin{aligned}(\mathrm{e}^{-2gV}\mathrm{D}^\alpha\, \mathrm{e}^{2gV}) &= (\exp(-2gL_V)\mathrm{D}_\alpha)\cdot 1\\ &= -2g[V,\mathrm{D}_\alpha] + \frac{(2g)^2}{2!}[V,[V,\mathrm{D}_\alpha]] - \frac{(2g)^3}{3!}[V,[V,[V,\mathrm{D}_\alpha]]] + \ldots\\ &= 2g\mathrm{D}_\alpha V - 2g^2[V,(\mathrm{D}_\alpha V)] + \frac{4}{3}g^3[V,[V,(\mathrm{D}_\alpha V)]] + O(V^4)\end{aligned}$$

which we use to expand S_{SYM} in a power series in V. Up to fourth order, the action reads

$$\begin{aligned}S_{\mathrm{SYM}} = \int \mathrm{d}^8 z\, \mathrm{tr}\bigg\{&\frac{1}{16}V\mathrm{D}^\alpha\bar{\mathrm{D}}^2\mathrm{D}_\alpha V + \frac{1}{8}g(\bar{\mathrm{D}}^2\mathrm{D}^\alpha V)[V,(\mathrm{D}_\alpha V)]\\ &-\frac{1}{16}g^2[V,(\mathrm{D}^\alpha V)]\bar{\mathrm{D}}^2[V,(\mathrm{D}_\alpha V])-\frac{1}{12}g^2(\bar{\mathrm{D}}^2\mathrm{D}^\alpha V)[V,[V,(\mathrm{D}_\alpha V)]]\bigg\}. \qquad (4.6.15)\end{aligned}$$

Next, to the same order, the ghost action (4.6.5) reads

$$S_{\mathrm{GH}} = \int \mathrm{d}^8 z\, \mathrm{tr}\left\{\bar{\mathrm{c}}'\mathrm{c} - \bar{\mathrm{c}}\mathrm{c}' + g(\mathrm{c}'+\bar{\mathrm{c}}')[V,(\mathrm{c}-\bar{\mathrm{c}})] + \frac{1}{3}g^2(\mathrm{c}'+\bar{\mathrm{c}}')[V,[V,(\mathrm{c}+\bar{\mathrm{c}})]]\right\}. \qquad (4.6.16)$$

Notice that the terms $\mathrm{c}'\mathrm{c}$ and $\bar{\mathrm{c}}'\bar{\mathrm{c}}$ have been dropped due to their chirality and antichirality, respectively. The last step is to unify the quadratic part S_2 of action (4.6.15) with the gauge breaking equation (4.6.9):

$$S_2 + S_{\mathrm{GB}} = \frac{1}{2}\int \mathrm{d}^8 z\, \mathrm{tr}\left\{-V\Box V + \frac{1}{16}\left(1+\frac{1}{\gamma}\right)V\{\bar{\mathrm{D}}^2,\mathrm{D}^2\}V\right\}. \qquad (4.6.17)$$

Here we have used the identity (2.5.30*b*). The relations (4.6.15–17) lead to the following expression for S_{total}

$$S_{\text{total}} = S_0 + S_{\text{INT}} \tag{4.6.18a}$$

$$S_0 = \frac{1}{2}\int d^8z\, \text{tr}\left\{ -V\Box V + \frac{1}{16}\left(1+\frac{1}{\gamma}\right)V\{\bar{D}^2, D^2\}V\right\} + \int d^8z\, \text{tr}\{\bar{c}'c - \bar{c}c'\} \tag{4.6.18b}$$

$$\begin{aligned} S_{\text{INT}} = \int d^8z\, \text{tr}\bigg\{ & \frac{1}{8}g(\bar{D}^2D^\alpha V)[V, (D_\alpha V)] + g(c' + \bar{c}')[V, (c-\bar{c})] \\ & - \frac{1}{16}g^2[V, (D^\alpha V)]\bar{D}^2[V, (D_\alpha V)] - \frac{1}{12}g^2(\bar{D}^2D^\alpha V)[V, [V, (D_\alpha V)]] \\ & + \frac{1}{3}g^2(c' + \bar{c}')[V, [V, (c+\bar{c})]]\bigg\} + O(g^3). \end{aligned} \tag{4.6.18c}$$

As may be seen, the action S_0 involves fourth-order derivatives, for $\gamma \neq -1$, and takes the most simple form when $\gamma = -1$.

After establishing the explicit structure of S_0, the free generating superfunctional (4.6.13) is easily evaluated as:

$$\begin{aligned} \mathbf{Z}_0[\mathcal{J}, \eta', \eta, \bar{\eta}', \bar{\eta}] = \exp i\bigg\{ & -\frac{1}{2}\int d^8z\, d^8z'\, \mathcal{J}^I(z)\mathbf{G}_V^{IJ}(z, z')\mathcal{J}^J(z') \\ & - \int d^8z\, \text{tr}\left(\bar{\eta}'\frac{1}{\Box}\eta - \bar{\eta}\frac{1}{\Box}\eta'\right)\bigg\}. \end{aligned} \tag{4.6.19}$$

Here $\mathbf{G}_V^{IJ} = \delta^{IJ}\mathbf{G}_V$, and the Green's function $\mathbf{G}_V$ of a real scalar superfield satisfies the equation

$$\left[\Box - \frac{1}{16}\left(1+\frac{1}{\gamma}\right)\{\bar{D}^2, D^2\}\right]\mathbf{G}_V(z, z') = -\delta^8(z, z') \tag{4.6.20}$$

and the Feynmann boundary conditions. Note that the ghost sector of $\mathbf{Z}_0$ is calculated in complete analogy with the derivation of expression (4.4.28), but keeping in mind the anticommuting nature of ghosts. To find $\mathbf{G}_v$, let us introduce the orthogonal projectors

$$\mathscr{P}_{(0)} = -\frac{1}{8\Box}D^\alpha\bar{D}^2D_\alpha \qquad \mathscr{P}_{(+)} = \frac{1}{16\Box}\bar{D}^2D^2 \qquad \mathscr{P}_{(-)} = \frac{1}{16\Box}D^2\bar{D}^2$$

on spaces of linear, chiral and antichiral superfields, respectively. Their chief properties are given by the last line of expression (2.6.6) and equation (2.6.7). Now, equation (4.6.20) can be rewritten as

$$\left[\mathscr{P}_{(0)} - \frac{1}{\gamma}(\mathscr{P}_{(+)} + \mathscr{P}_{(-)})\right]\Box\mathbf{G}_V = -\mathbb{1}$$

and we will find its solution in the form

$$\mathbf{G}_V = \frac{1}{\Box}[\alpha\mathscr{P}_{(0)} + \beta(\mathscr{P}_{(+)} + \mathscr{P}_{(-)}].$$

Making use of the projectors' properties gives

$$\mathbf{G}_V = \frac{1}{\Box}[-\mathscr{P}_{(0)} + \gamma(\mathscr{P}_{(+)} + \mathscr{P}_{(-)})]. \tag{4.6.21}$$

In the case $\gamma = -1$, the Green's function takes the form

$$\mathbf{G}_V(z, z') = -\frac{1}{\Box + i\varepsilon}\delta^8(z, z') \qquad \varepsilon \to +0 \tag{4.6.22}$$

and the generating functional (4.6.19) reads

$$\mathbf{Z}_0[\mathscr{J}, \eta', \eta, \bar{\eta}', \bar{\eta}] = \exp\left[i\int d^8z\,\mathrm{tr}\left\{\frac{1}{2}\mathscr{J}\frac{1}{\Box}\mathscr{J} - \bar{\eta}'\frac{1}{\Box}\bar{\eta} - \eta'\frac{1}{\Box}\bar{\eta}\right\}\right]. \tag{4.6.23}$$

In what follows, we make the choice $\gamma = -1$.

Now we turn our attention to the vertices. Consider the coupling (4.6.18*c*). It describes the self-interaction of real scalar superfields V^I and their interaction with (anti)chiral ghost superfields $\bar{\mathbf{c}}'^I$, $\bar{\mathbf{c}}^I$, $\mathbf{c}'^I$ and $\mathbf{c}^I$. The explicit form of S_{INT} dictates, in standard fashion, the structure of the vertices. The only problem which may appear concerns vertices involving chiral and antichiral superfields. But that question has been discussed in detail in Section 4.4.

4.6.3. *Feynman rules for general super Yang–Mills models*

The previous consideration can be readily extended to the case of a general super Yang–Mills model describing interaction between gauge superfields V^I and matter chiral superfields Φ^k (and their conjugates $\bar{\Phi}_l = (\Phi^l)^*$) transforming in some representation R of the gauge group. We assume this representation to be irreducible and start with the classical gauge invariant action

$$S = S_{\mathrm{SYM}} + S_{\mathrm{MAT}} \tag{4.6.24}$$

where

$$S_{\mathrm{MAT}} = \int d^8z\,\bar{\Phi}_l(e^{2gV})^l{}_k\Phi^k + \int d^6z\left(\frac{1}{2}m\delta_{k_1k_2}\Phi^{k_1}\Phi^{k_2} + \frac{1}{3!}\lambda_{k_1k_2k_3}\Phi^{k_1}\Phi^{k_2}\Phi^{k_3}\right)$$
$$+ \int d^6\bar{z}\left(\frac{1}{2}m\delta^{l_1l_2}\bar{\Phi}_{l_1l_2} + \frac{1}{3}\bar{\lambda}^{l_1l_2l_3}\bar{\Phi}_{l_1}\bar{\Phi}_{l_2}\bar{\Phi}_{l_3}\right) \tag{4.6.25}$$

m and λ being mass and coupling constants, respectively ($\bar{\lambda}^{l_1l_2l_3} = (\lambda_{l_1l_2l_3})^*$).

The representation R should be real in the massive case, $m \neq 0$, and can be complex otherwise. Recall also that the coupling constants must form an invariant tensor of the gauge group in order that the action (4.6.25) be gauge invariant. Notice that S_{MAT} can be viewed as a multi-component Wess–Zumino model coupled to the gauge superfield.

The gauge theory under consideration is quantized in complete analogy with the pure super Yang–Mills one, and the corresponding generating superfunctional reads

$$\mathbf{Z}[\mathcal{J}, \mathbf{J}, \bar{\mathbf{J}}] = \int \mathscr{D}v \exp \mathrm{i}\left\{ S_{\text{total}} + \int \mathrm{d}^8 z \, \mathrm{tr}(\mathcal{J} V) + \int \mathrm{d}^6 z \, \mathbf{J}_k \Phi^k + \int \mathrm{d}^6 \bar{z} \, \bar{\mathbf{J}}^l \bar{\Phi}_l \right\} \tag{4.6.26}$$

$$v = (V, \Phi, \bar{\Phi}, \mathbf{c}', \mathbf{c}, \bar{\mathbf{c}}', \bar{\mathbf{c}}) \qquad \bar{\mathrm{D}}_{\dot{\alpha}} \mathbf{J}_k = \mathrm{D}_\alpha \bar{\mathbf{J}}^l = 0$$

where

$$S_{\text{total}} = S_{\text{SYM}} + S_{\text{GB}} + S_{\text{MAT}} + S_{\text{GH}} \tag{4.6.27}$$

with S_{GB} and S_{GH} being defined by equations (4.6.9) and (4.6.5), respectively. The next step is to represent S_{total} as a sum of its free and interacting parts. One thus finds

$$S_{\text{total}} = S_0 + S_{\text{INT}} \tag{4.6.28}$$

the free action being of the form

$$S_0 = S_0^{(V)} + S_0^{(\Phi)} + S_0^{(\text{gh})} \tag{4.6.29}$$

where

$$S_0^{(V)} = -\frac{1}{2} \int \mathrm{d}^8 z \, \mathrm{tr} \, V \Box V \qquad (\gamma = -1) \tag{4.6.30a}$$

$$S_0^{(\Phi)} = \int \mathrm{d}^8 z \, \bar{\Phi}_l \delta^l{}_k \Phi^k + \left\{ \frac{m}{2} \int \mathrm{d}^6 z \, \delta_{k_1 k_2} \Phi^{k_1} \Phi^{k_2} + \text{c.c.} \right\} \tag{4.6.30b}$$

$$S_0^{(\text{gh})} = \int \mathrm{d}^8 z \, \mathrm{tr}(\bar{\mathbf{c}}' \mathbf{c} - \bar{\mathbf{c}} \mathbf{c}'). \tag{4.6.30c}$$

The interaction reads

$$S_{\text{INT}} = S_{\text{INT}}^{(\Phi)} + S_{\text{INT}}^{(\Phi, V)} + S_{\text{INT}}^{(V)} + S_{\text{INT}}^{(V, \text{gh})} \tag{4.6.31}$$

where

$$S_{\text{INT}}^{(\Phi)} = \frac{1}{3!} \int \mathrm{d}^8 z \, \lambda_{k_1 k_2 k_3} \Phi^{k_1} \Phi^{k_2} \Phi^{k_3} + \frac{1}{3!} \int \mathrm{d}^8 \bar{z} \, \bar{\lambda}^{l_1 l_2 l_3} \bar{\Phi}_{l_1} \bar{\Phi}_{l_2} \bar{\Phi}_{l_3} \tag{4.6.32a}$$

$$S_{\text{INT}}^{(\Phi, V)} = \int \mathrm{d}^8 z \, \bar{\Phi}_l ((\mathrm{e}^{2gV})^l{}_k - \delta^l{}_k) \Phi^k \tag{4.6.32b}$$

$$S^{(V)}_{\mathrm{INT}} = \int d^8z \, \mathrm{tr}\left\{\frac{1}{8} g(\bar{D}^2 D^\alpha V)[V, (D_\alpha V)] - \frac{1}{16} g^2 [V, (D^\alpha V)]\bar{D}^2[V, (D_\alpha V)] - \frac{1}{12} g^2 (\bar{D}^2 D^\alpha V)[V, [V, (D_\alpha V)]]\right\} \quad (4.6.32c)$$

$$S^{(V,\mathrm{gh})}_{\mathrm{INT}} = \int d^8z \, \mathrm{tr}\left\{g(c' + \bar{c}')[V, (c - \bar{c})] + \frac{1}{3} g^2 (c' + \bar{c}')[V, [V, (c + \bar{c})]]\right\}. \quad (4.6.32d)$$

Here we have written out the expressions for $S^{(V)}_{\mathrm{INT}}$ and $S^{(V,\mathrm{gh})}_{\mathrm{INT}}$ up to fourth order in the superfields.

Based on the above relations, one now readily deduces superfield Feynman rules for calculating the generating superfunctional (4.6.26) and the effective action $\Gamma[V, \Phi, \bar{\Phi}]$ obtained as the Legendre transform of $(1/i)\ln \mathbf{Z}[\mathcal{J}, \mathbf{J}, \bar{\mathbf{J}}]$. The free generating superfunctional for the theory with action (4.6.27–32) is given as the product of the pure super Yang–Mills one (4.6.19) and that for matter

$$\mathbf{Z}_0[\mathbf{J}, \bar{\mathbf{J}}] = \exp\left\{-i \int d^8z \left(\bar{\mathbf{J}}^l \frac{\delta^k{}_l}{\Box - m^2} \mathbf{J}_k + \frac{m}{8} \mathbf{J}_{k_1} \frac{\delta^{k_1 k_2} D^2}{\Box(\Box - m^2)} \mathbf{J}_{k_2} + \frac{m}{8} \bar{\mathbf{J}}^{l_1} \frac{\delta_{l_1 l_2} \bar{D}^2}{\Box(\Box - m^2)} \bar{\mathbf{J}}^{l_2}\right)\right\} \quad (4.6.33)$$

(the latter has, in fact, been derived in Section 4.4). The structure of the vertices follows directly from the explicit form for interaction (4.6.21, 32). As in the case of the Wess–Zumino model, it is convenient to treat the ghost and matter (anti)chiral superfields within the framework of the improved supergraph technique described in subsection 4.4.5. Then all the vertices coming from expressions (4.6.32) will be integrated over d^8z in coordinate space or over $d^4\theta$ in momentum space. We leave the reader to deduce the rules for attaching the factors $(-\frac{1}{4}D^2)$ and $(-\frac{1}{4}\bar{D}^2)$ to the ends of the internal (anti)chiral ghost and matter lines. Notice that, up to sign, the ghost propagators coincide with the massless $\Phi\bar{\Phi}$-propagator (4.4.43). As for the VV-propagator, it reads in coordinate space

$$V^I V^J\text{-line: } \underset{z \qquad\qquad z'}{\sim\!\sim\!\sim\!\sim\!\sim} = -\delta^{IJ} \frac{1}{\Box} \delta^8(z - z') \quad (4.6.34a)$$

or, in momentum space,

$$\underset{p}{\sim\!\sim\!\blacktriangleright\!\sim\!\sim} = \delta^{IJ} \frac{1}{p^2} \delta^4(\theta - \theta') \quad (4.6.34b)$$

and has the opposite sign to the $\Phi\bar{\Phi}$-propagator.

As a simple example, let us calculate the one-loop gauge superfield contribution to the two-point $\Phi\bar{\Phi}$-vertex function. This contribution comes from the following part of $S_{\rm INT}$

$$S^{(\Phi,V)}_{\rm INT}=2g\int d^8z\,\bar{\Phi}_l(T^I)^l{}_k\Phi^k V^I+\ldots$$

where dots mean terms relevant for two-loop and higher corrections; $(T^I)^l{}_k$ are the generators of the representation R in which Φ^k transform. The corresponding supergraph is

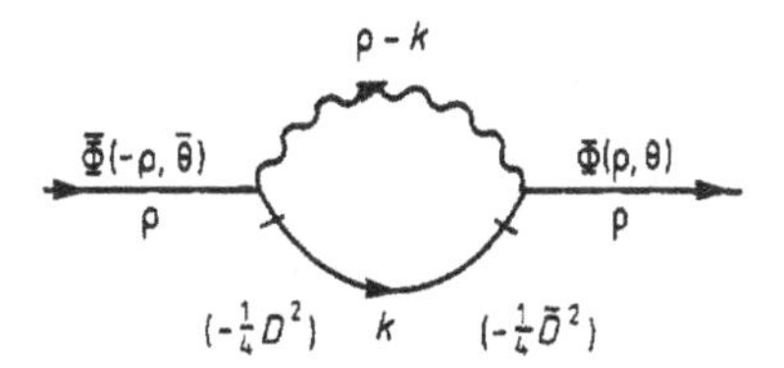

and leads to

$$-i4g^2\int\frac{d^4p}{(2\pi)^4}\int d^4\theta\, d^4\theta'\,\bar{\Phi}_{l_1}(-p,\theta)(T^I)^{l_1}{}_{k_1}$$
$$\times\int\frac{d^4k}{(2\pi)^4}\delta^{IJ}\frac{i}{(p-k)^2}\delta^4(\theta-\theta')\delta^{k_1}{}_{l_2}\frac{-i}{k^2+m^2}$$
$$\times\left[\frac{1}{16}D^2(k)\bar{D}'^2(-k)\delta^4(\theta-\theta')\right](T^J)^{l_2}{}_{k_2}\Phi^{k_2}(p,0). \qquad (4.6.35a)$$

Note that, in any representation, the operator

$$C_2=\sum_{I=1}^{\dim G}(T^I)^2$$

commutes with every generator T^J (since the structure constants in expression (3.5.24) are totally antisymmetric). Therefore, it is proportional to the unit operator, i.e. $C_2\propto\mathbb{1}$, in any irreducible representation. Since the matter representation R has been assumed to be irreducible, we have

$$(T^I)^l{}_{k_1}(T^I)^{k_1}{}_k=C_2(R)\delta^l{}_k$$

the constant $C_2(R)$ is known as the second Casimir of the representation. Using this fact and performing the integration over θ', the expression (4.6.35*a*) takes the form

$$-4g^2C_2(R)\int\frac{d^4p}{(2\pi)^4}A'(p^2,m^2)\int d^4\theta\,\bar{\Phi}_k(-p,\theta)\Phi^k(p,\theta) \qquad (4.6.35b)$$

where

$$A'(p^2,m^2)=i\int\frac{d^4p}{(2\pi)^4}\frac{1}{(p-k)^2}\frac{1}{k^2+m^2}. \qquad (4.6.36)$$

It is worth comparing the quantum corrections (4.4.54) and (4.6.35). In the massless case, the momentum functions (4.4.55) and (4.6.36) coincide, and the expressions (4.4.54*b*) and (4.6.35*b*) take similar structures but differ in sign. As a result, for some special values of the chiral coupling constants $\lambda_{k_1k_2k_3}$ and the gauge constant g, the one-loop corrections to the two-point $\Phi\bar{\Phi}$-vertex functions, which come from $S^{(\Phi)}_{\text{INT}}$ and $S^{(\Phi,V)}_{\text{INT}}$, cancel each other. Under such a choice, there are no one-loop $\Phi\bar{\Phi}$-divergences.

4.6.4. *Non-renormalization theorem*

The non-renormalization theorem described in subsection 4.4.8 turns out to preserve all its power in the case of the general super Yang–Mills theories just considered. Indeed, the only essential properties, which have been used in the proof of the theorem, were the facts that (1) each vertex is integrated over $d^4\theta$; (2) each propagator presents $\delta^4(\theta-\theta')$ multiplied by some θ-independent momentum function together with a number of D-factors acting on the delta-function. Both properties are implied by the Feynman rules given in the previous subsection. As a result, each supergraph forming the effective action of the super Yang–Mills theory can be represented by a single integral over $d^4\theta$. In other words, the effective action has the following structure

$$\Gamma[V,\Phi,\bar{\Phi}]=\sum_n\int d^4x_1\ldots d^4x_n\int d^4\theta\,\mathscr{T}_n(x_1,\ldots,x_n)F_1(x_1,\theta)\ldots F_n(x_n,\theta) \tag{4.6.37}$$

where the $\mathscr{T}$s are translationally invariant functions on Minkowski space, and the Fs are local functions of V, Φ, $\bar{\Phi}$ and their covariant derivatives

$$F_i=F_i(V,\Phi,\bar{\Phi},D_AV,D_A\Phi,D_A\bar{\Phi},\ldots). \tag{4.6.38}$$

4.7. Renormalization

As is well known, renormalization is one of the essential elements of quantum field theory. Formally speaking, renormalization consists in finding suitable counterterms and adding them to the initial action in order to substract the divergences loop-by-loop. A chief aim of this section is to investigate the general structure of the counterterms in super Yang–Mills theories.

4.7.1. *Superficial degree of divergence*

Let us consider some super Yang–Mills theory described by chiral Φ^k, antichiral $\bar{\Phi}_l$ and gauge $V=V^IT^I$ superfields interacting among themselves, the interaction determined from the classical action (4.6.24, 1, 25). As has been shown, its effective action is given entirely in terms of the supergraphs.

The corresponding Feynman rules were formulated in subsection 4.6.3 (see also subsection 4.4.5).

We pick out here some specific features of supergraphs. First, each vertex contains an integral over $d^4\theta$. Second, each vertex without external (anti)chiral lines attached includes four spinor covariant derivatives—D-factors; if an (anti)chiral external line is attached to the vertex, then the number of D-factors is reduced from four to two.

To secure finiteness of the supergraphs at all the intermediate stages of calculation, we have to introduce some regularization procedure. The problem of regularization in supersymmetric theories is a separate one. For the moment, we shall assume the existence of a regularization preserving supersymmetry. Then, at the end of this section we shall return to the discussion of supersymmetric regularizations.

The superficial degree of divergence of a Feynman diagram is defined to be the degree of homogeneity of the diagram in momenta. To calculate the superficial degree of divergence $w(G)$, G being an arbitrary supergraph, one has to take into account the momentum dimensions of the objects used to write the contribution of the given supergraph. These objects are propagators, vertices and momentum integrals.

Let us consider an arbitrary L-loop supergraph G contributing to the effective action, with V vertices, P propagators and E external lines. Let C of the P propagators be of $\Phi\Phi$- or $\overline{\Phi\Phi}$-type and E_c from these E external lines be (anti)chiral ones. According to the superfield Feynman rules (see subsections 4.4.5 and 4.6.3) the contribution of any such supergraph has the form

$$\int d^4p_1 \ldots d^4p_L \, d^4\theta_1 \ldots d^4\theta_V \, [\ldots] \tag{4.7.1}$$

where the square brackets include the above number of propagators and include some definite number of D-factors associated with vertices. It is obvious that only the integration over momenta may lead to divergences.

The momentum integrals (4.7.1) contribute the quantity $4L$ to $w(G)$. Taking into account the explicit dependence of the propagators on momenta, we have the contribution $-2(P+C)$. Recall that the $\Phi\Phi$- and $\overline{\Phi\Phi}$-propagators have the extra $1/p^2$ factor in comparison with the $\Phi\bar{\Phi}$- and VV-propagators (and the ghost ones). However, the quantity

$$4L - 2(P+C) \tag{4.7.2}$$

does not represent the final momentum dimension of the supergraph contribution (4.7.1), since the D-factors also depend on momentum.

Let us find the total number of D-factors depending on the internal momenta. The superfield Feynman rules tell us that each vertex without external (antichiral) lines includes four D-factors; only two D-factors are associated with a vertex possessing an external line of $\Phi\Phi$- or $\overline{\Phi\Phi}$-type. Therefore, the number of D-factors associated with the V vertices in expression (4.7.1) is equal to $4V-2E_c$. We also know that each propagator of $\Phi\Phi$- or $\overline{\Phi\Phi}$-type contains two D-factors. Hence, the total number of D-factors depending on the internal momenta in expression (4.7.1) is given by

$$4V-2E_c+2C \tag{4.7.3}$$

Now, the reader should recall the non-renormalization theorem (see subsections 4.4.8 and 4.6.4). According to this theorem, equation (4.7.1) can be transformed into an expression involving a single integral over $d^4\theta$. That is, $(V-1)$ θ-integrals can be taken explicitly, since each internal line contains a Grassmann delta-function $\delta^4(\theta_i-\theta_j)$. After taking $(V-1)$ θ-integrals, $(V-1)$ of the P delta-functions are cancelled, and we have $(P-V+1)$ remaining δ-functions. But owing to the well-known topological relation

$$V+L-P=1 \tag{4.7.4}$$

the number of such δ-functions is equal to L. As a result, after transforming equation (4.7.1) to the form with a single θ-integral, we have an expression including L Grassmann delta-functions. Since $\delta^4(\theta-\theta)=0$, this expression does not equal zero if all the δ-functions are completely reduced by means of relations like

$$\mathrm{D}^2\bar{\mathrm{D}}^2\delta^4(\theta-\theta')=16.$$

We see that $4L$ D-factors are required to cancel the remaining δ-functions.

In summary, after the transformation of equation (4.7.1) to the form dictated by the non-renormalization theorem we have the following number of D-factors depending on the internal momenta

$$4V-2E_c+2C-4L \tag{4.7.5}$$

What can we do with the remaining D-factors? Note that in the process of reducing equation (4.7.1) to the form with a single θ-integral one should perform integrations by parts, as in equation (4.4.52). Therefore, some D-factors can be transferred to the external lines. Those D-factors, which are not transferred to the external lines, must be converted into momenta via the relation $\{\mathrm{D},\bar{\mathrm{D}}\}\sim p$.

Suppose that all the remaining D-factors are converted into momenta by the law $\{\mathrm{D},\bar{\mathrm{D}}\}\sim p$ (no D-factors are transferred to the external lines). Then, they produce the following maximal number of internal momenta

$$\frac{1}{2}(4V-2E_c+2C-4L)=2(V-L)-E_c+C. \tag{4.7.6}$$

Then, the maximal superficial degree of divergence is given by the sum of equations (4.7.2) and (4.7.6). Therefore, we have

$$
\begin{aligned}
w(G) &= 4L - 2P - 2C + 2V - 2L - E_c + C \\
&= 2(L + V - P) - E_c - C = 2 - E_c - C
\end{aligned}
\tag{4.7.7}
$$

where equation (4.7.4) has been used.

Assuming that N_D D-factors are transferred to the external lines, the superficial degree of divergence is given as

$$
w(G) = 2 - \frac{1}{2} N_D - E_c - C. \tag{4.7.8}
$$

This is the final expression for the superficial degree of divergence.

4.7.2. Structure of counterterms

To ascertain the explicit structure of the counterterms, it is necessary to take into account the superficial degree of divergence, the non-renormalization theorem and the fact that the counterterms are local functionals in x-space. Also, we will assume that the regularization procedure chosen preserves supersymmetry. Therefore, the counterterm ΔS should be a local superfunctional of Φ, $\bar{\Phi}$ and V invariant under the supersymmetry transformations of these superfields. The locality of ΔS, supersymmetry and the non-renormalization theorem imply that the counterterm must have the form

$$
\Delta S = \int d^8z \, \Delta\mathscr{L}(\Phi, \bar{\Phi}, V, \ldots) \tag{4.7.9}
$$

where $\Delta\mathscr{L}(\Phi, \bar{\Phi}, V, \ldots)$ is a function of the basic superfields and their covariant derivatives to a finite order.

Let us consider some elementary consequences of equation (4.7.9). First, there are no counterterms to vacuum energies since

$$
\Delta S = \int d^8z \, \Delta\mathscr{L}|_{\Phi = \bar{\Phi} = V = 0} = 0.
$$

Therefore, there is no need to take into account the supergraphs for vertex functions without external superfield lines. In fact, all such supergraphs vanish, as a consequence of the non-renormalization theorem.

Second, the counterterms cannot contain terms having the structure of a chiral superpotential, such as

$$
\int d^6z \, f(\Phi)
$$

since their counterparts in the full superspace

$$
\int d^8z \left(-\frac{1}{4} \frac{D^2}{\Box} \right) f(\Phi)
$$

have nonlocal form.

Before we proceed to the study of the structure of counterterms, let us discuss the problem of gauge invariance of the counterterms. This problem is common to all gauge theories. The fact of the matter is that the quantization procedure demands that we introduce gauge fixing conditions which break the classical gauge invariance. On these grounds, the effective action turns out, in general, to be non-invariant under the initial gauge transformations. Certainly, the S-matrix is always gauge invariant. Thus it should not be a surprise that counterterms in the super Yang–Mills models may, in general, be non-invariant under supergauge transformations.

The standard way to classify counterterms in conventional gauge theories is based on the Slavnov–Taylor identities, which allow one to decrease the number of admissible counterterms. Analogous identities can also be formulated for superfield gauge theories. However, in many cases of interest it is possible to realize the quantization procedure in such a way that the corresponding Slavnov–Taylor identities imply invariance of the effective action under the classical gauge transformations. This formulation of quantum gauge theories is known as the background field method (it can be applied under the existence of gauge invariant regularization). Concrete realization of the above formulation is based on specific features of the given theory, and it is not clear from the outset that the super Yang–Mills theories are tractable within the framework of background field method. Fortunately, the background-field formulation of general super Yang–Mills models does exist, and it has been suggested by M. Grisaru, M. Roček and W. Siegel[1]. For our aims, it is essential to know only that there exists a formulation of the theory, in the framework of which the effective action is invariant under the initial supergauge transformations.

Now, let us find all the admissible counterterms. Divergent supergraphs are characterized by the condition $w(G)\geqslant 0$. In accordance with equation (4.7.8), there can be three possible basic types of divergent supergraphs:

$$\begin{aligned}
&1.\ E_c=2,\ C=0,\ N_D=0,\ w(G)=0\\
&2.\ E_c=0,\ C=2,\ N_D=0,\ w(G)=0\\
&3.\ E_c=0,\ C=0,\ N_D\leqslant 4,\ 0\leqslant w(G)\leqslant 2.
\end{aligned}\qquad(4.7.10)$$

Notice, the above conditions give no restrictions on the number of external V-lines.

Consider the first variant in expression (4.7.10). Divergent supergraphs of this type have exactly two (anti)chiral external lines and may carry an arbitrary number of external V-lines; their divergence is logarithmic. Hence, suitable counterterms should have the form

$$\Delta Z_1 \int d^8z\, \Phi F(V)\Phi \qquad (4.7.11)$$

[1]M. Grisaru, M. Roček and W. Siegel, *Nucl. Phys.* **B 159** 429, 1979.

all the indices being suppressed. Here $F(V)$ is some function of the gauge superfield V, ΔZ_1 is a series in couplings and the regularization parameter. We suppose that the calculation procedure is fulfilled within the framework of the background field method. Then, all the counterterms should be supergauge invariants. This implies that the only possible form for $F(V)$ is $\exp(2gV)$. Therefore, in the first variant of equations (4.7.10) the counterterm reads

$$\Delta Z_1 \int d^8z\, \bar{\Phi}\, e^{2gV} \Phi. \tag{4.7.12}$$

The supergraphs belonging to the second variant in equations (4.7.10) include only external V-lines. The corresponding counterterm should be proportional to

$$m^2 \int d^8z\, f(V) \tag{4.7.13}$$

$f(V)$ being some function of V. But if we carry out the calculations in the framework of the background field method, then such a counterterm is forbidden by supergauge invariance. More precisely, the only admissible choice is $f(V)=\text{const}$ and hence the expression (4.7.13) vanishes. As an example, the reader can explicitly check that the total divergent contribution produced by all possible one-loop diagrams of the form

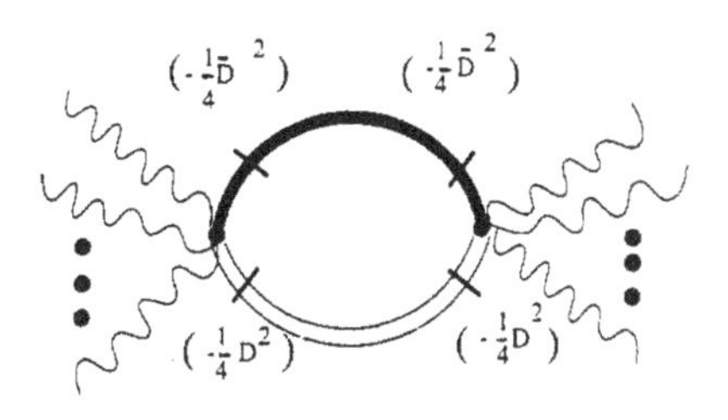

proves to be proportional to

$$m^2 \int d^8z\, \mathrm{tr}(e^{2gV}(e^{2gV})^{\mathrm{T}}) = m^2 \int d^8z\, \mathrm{tr}(e^{2gV}\, e^{-2gV}) = 0.$$

Finally, consider the third variant in equations (4.7.10). The most general form for the corresponding counterterm reads

$$\int d^8z\, G(V, D_A V, \ldots)$$

Here G is a function of the gauge superfield and its covariant derivatives to fourth order; in addition, G depends parametrically on the dimensionless coupling constants g and $\lambda_{k_1k_1k_3}$ and the regularization parameter. Note that the dimension of G is equal to two, while V is dimensionless. Obviously, one can write

$$\int \mathrm{d}^8z\, G = \int \mathrm{d}^8z\, V \mathscr{A} V + O(V^3) \tag{4.7.14}$$

where $\mathscr{A}$ is a V-independent differential operator of the fourth order in D_α and $\bar{\mathrm{D}}_{\dot{\alpha}}$. Assuming validity of the background-field method, the counterterm should be invariant under the gauge transformations (4.6.3). Then, its quadratic part in V must be invariant under the linearized transformations

$$\delta V = \Lambda + \bar{\Lambda} \qquad \bar{\mathrm{D}}_{\dot{\alpha}}\Lambda = 0.$$

Therefore, the most general form of the operator $\mathscr{A}$ is

$$\mathscr{A} = \frac{1}{16}\Delta Z_2 \mathrm{D}^\alpha \bar{\mathrm{D}}^2 \mathrm{D}_\alpha$$

ΔZ_2 being a series in the couplings and the regularization parameter. Now, the full counterterm can be recovered from the requirement of supergauge invariance to obtain

$$\Delta Z_2 \frac{1}{8g^2} \int \mathrm{d}^8z\, \mathrm{tr}(\mathrm{e}^{-2gV} \mathrm{D}^\alpha \mathrm{e}^{2gV} \mathrm{W}_\alpha) = \Delta Z_2 \frac{1}{4g^2} \int \mathrm{d}^6z\, \mathrm{tr}(W^\alpha W_\alpha). \tag{4.7.15}$$

In summary, we have shown that in the theory under consideration there are only two possible types of counterterm. They are given by the expressions (4.7.12) and (4.7.15).

Let the classical action be written as

$$S = \int \mathrm{d}^8z\, \bar{\Phi}\, \mathrm{e}^{2gV} \Phi + \frac{1}{4g^2} \int \mathrm{d}^6z\, \mathrm{tr}(W^\alpha W_\alpha)$$
$$+ \left\{ \int \mathrm{d}^6z \left(\frac{1}{2} m\Phi\Phi + \frac{1}{3!} \lambda\Phi\Phi\Phi \right) + \mathrm{c.c.} \right\} \tag{4.7.16}$$

with the group indices suppressed. The above analysis allows us to find the renormalized action S_R, Φ_R, V_R and parameters m_R, λ_R, g_R in the form

$$S_\mathrm{R} = Z_1 \int \mathrm{d}^8z\, \bar{\Phi}_\mathrm{R}\, \mathrm{e}^{2g_\mathrm{R} V_\mathrm{R}} \Phi_\mathrm{R} + Z_2 \frac{1}{4g_\mathrm{R}^2} \int \mathrm{d}^6z\, \mathrm{tr}(W_\mathrm{R}^\alpha W_{\mathrm{R}\alpha})$$
$$+ \left\{ \int \mathrm{d}^6z \left(\frac{1}{2} m_\mathrm{R} \Phi_\mathrm{R} \Phi_\mathrm{R} + \frac{1}{3!} \lambda_\mathrm{R} \Phi_\mathrm{R} \Phi_\mathrm{R} \Phi_\mathrm{R} \right) + \mathrm{c.c.} \right\} \tag{4.7.17}$$

where $Z_1 = 1 + \Delta Z_1$, $Z_2 = 1 + \Delta Z_2$ are the renormalization constants, W_R^α is the superfield strength expressed through the superfield V_R and the parameter g_R. Comparing the expressions (4.6.16) and (4.7.17), we see that

$$\Phi = (Z_1)^{1/2} \Phi_\mathrm{R} \qquad V = (Z_2)^{1/2} V_\mathrm{R}. \tag{4.7.18}$$

Hence, Z_1 is the wave function renormalization for the matter superfields and Z_2 that for the gauge superfield. One also finds

$$gV = g_R V_R \qquad m\Phi\Phi = m_R \Phi_R \Phi_R \qquad \lambda\Phi\Phi\Phi = \lambda_R \Phi_R \Phi_R \Phi_R. \qquad (4.7.19)$$

These relations are direct consequences of the non-renormalization theorem as well as of supergauge invariance.

Let us formally introduce renormalization constants for mass and couplings as follows

$$m = Z_m m_R \qquad \lambda = Z_\lambda \lambda_R \qquad g = Z_g g_R. \qquad (4.7.20)$$

Then, the relations (4.7.18) and (4.7.19) tell us that

$$Z_g (Z_2)^{1/2} = 1 \qquad Z_m Z_1 = 1 \qquad Z_\lambda (Z_1)^{3/2} = 1. \qquad (4.7.21)$$

We summarize the results. In super Yang–Mills theories, there are only two independent renormalization constants, Z_1 and Z_2. Renormalization is determined by the wave function divergences.

4.7.3. *Questions of regularization*

Previous considerations in this section were based on the assumption that there exists at least one regularization scheme preserving supersymmetry and supergauge invariance as well as being adapted for use in supergraph calculations. We review here several often-employed regularizations. In each case, fulfilment of the above requirements is not guaranteed by construction and has to be checked explicitly.

a. Supersymmetric Pauli-Villars regularization. Let us introduce a matrix which is formed from the propagators (4.4.49) of the Wess–Zumino model and defined as

$$\mathscr{K}(p) = -\mathrm{i}\begin{pmatrix} \mathbf{K}_{++}(p) & \mathbf{K}_{+-}(p) \\ \mathbf{K}_{-+}(p) & \mathbf{K}_{--}(p) \end{pmatrix}$$

$$= -\frac{\mathrm{i}}{p^2+m^2}\begin{pmatrix} \dfrac{m}{p^2}\left(-\dfrac{1}{4}\mathrm{D}^2(p)\right) & 1 \\ 1 & \dfrac{m}{p^2}\left(-\dfrac{1}{4}\bar{\mathrm{D}}^2(p)\right) \end{pmatrix}\delta^4(\theta-\theta'). \qquad (4.7.22)$$

As is known, the chief idea of the Pauli–Villars regularization consists in modifying, in each loop, the initial propagators in such a way as to obtain propagators possessing less singular behaviour on the light cone. This is achieved by adding to the initial propagators, in each loop, propagators of some auxiliary massive fields with suitable coefficients.

In our case, it seems reasonable to regularize the superpropagator (4.7.22) in the manner

$$\mathrm{i}\mathscr{K}^{(\mathrm{R})}(p)$$

$$=\begin{pmatrix} \dfrac{-\frac{1}{4}\mathrm{D}^2(p)}{p^2}\left(\dfrac{m}{p^2+m^2}+\sum\limits_i \dfrac{C_iM_i}{p^2+M_i^2}\right) & \dfrac{1}{p^2+m^2}+\sum\limits_i \dfrac{C_i}{p^2+M_i^2} \\ \dfrac{1}{p^2+m^2}+\sum\limits_i \dfrac{C_i}{p^2+M_i^2} & \dfrac{-\frac{1}{4}\bar{\mathrm{D}}^2(p)}{p^2}\left(\dfrac{m}{p^2+m^2}+\sum\limits_i \dfrac{C_iM_i}{p^2+M_i^2}\right) \end{pmatrix}$$

$$\times\,\delta^4(\theta-\theta') \tag{4.7.23}$$

where M_i are auxiliary masses, and C_i are dimensionless coefficients. To cancel the leading singularities of $\mathscr{K}$ on the light cone, it is sufficient to choose C_i under the conditions

$$\sum_i C_i=-1 \qquad \sum_i C_iM_i=-m. \tag{4.7.24}$$

All the auxiliary masses should be taken to infinity at the end of the supergraph calculations.

Remark. There is no need to regularize the factors $1/p^2$ in equation (4.7.22) since, as we have seen, their origin was purely kinematical: to convert the vertex integrals over $\mathrm{d}^2\theta$ or $\mathrm{d}^2\bar{\theta}$ into ones over $\mathrm{d}^4\theta$. In fact, one could even work with the unregularized propagators $\mathbf{K}_{++}$ and $\mathbf{K}_{--}$, because in the Wess–Zumino model the only divergent supergraphs with internal $\Phi\Phi$- or $\bar{\Phi}\bar{\Phi}$-lines are vacuum ones, due to equation (4.7.8), and they vanish. But this would lead to the breakdown of supersymmetry.

The explicit form of the regularized superpropagator (4.7.23) motivates us to associate with each mass M_i some auxiliary chiral superfield φ_i and its conjugate. This idea allows us to realize the Pauli–Villars regularization directly in the Lagrangian approach by adding to the classical action an action of auxiliary superfields as well as some coupling terms. A correct choice for the total action is of the form

$$S^{(\mathrm{R})}=\int \mathrm{d}^8z\left(\Phi\bar{\Phi}+\sum_i \frac{1}{C_i}\bar{\varphi}_i\varphi_i\right)$$
$$+\left\{\int \mathrm{d}^6z\left[\frac{m}{2}\Phi^2+\frac{1}{2}\sum_i \frac{1}{C_i}M_i\varphi_i^2+\frac{\lambda}{3!}\left(\Phi+\sum_i\varphi_i\right)^3\right]+\mathrm{c.c.}\right\} \tag{4.7.25}$$

Cs being solutions of equation (4.7.24). The action corresponds to a dynamical system of several (anti)chiral superfields. The corresponding perturbation consideration can be fulfilled in the standard way in terms of supergraphs.

Notice that supergraphs with external Φ- and $\bar{\Phi}$-lines only should be taken into account. The auxiliary superfields work in the loops. Obviously, in this regularization scheme the supersymmetry remains unbroken at all stages of the quantum calculations.

Unfortunately, Pauli–Villars regularization is not well adapted for gauge theories, in particular, for the super Yang–Mills theories. The problem is that the scheme demands that we introduce auxiliary massive (super)fields for all physical (super)fields, in particular, for the massless gauge ones. This can be accompanied by the breakdown of (super)gauge invariance.

b. Regularization by dimensional reduction. Any regularization procedure should answer the question: in what sense are the divergent Feynman integrals to be understood? The standard requirements of the regularization scheme consist in that it must lead to convergent Feynman integrals and preserve as many properties of the initial classical theory as possible. In supersymmetric theories it is natural to demand that the regularization scheme preserves supersymmetry. As we have noticed, the problem of existence of such a regularization should be discussed case by case. For instance, in the component approach one could use so-called dimensional regularization which is most popular in modern quantum field theory. Dimensional regularization demands the theory to be formulated in space–time of an arbitrary dimension from the very beginning. However, supersymmetry can exist only in space–times of special dimension. Therefore, it is conceivable that naive application of dimensional regularization will lead to a breakdown of supersymmetry.

Regularization by dimensional reduction invented by W. Siegel is a generalization of the standard dimensional regularization adapted to supergraph calculations.

Let us consider a divergent L-loop supergraph. The regularization consists in two steps. First, we fulfil all necessary manipulations with covariant derivatives (D-algebra) in four-dimensional space–time and integrate over θ, as in the proof of the non-renormalization theorem. As a result, we obtain a single integral over $d^4\theta$ and L momentum integrals. The integrand depends on external superfields and their covariant derivatives and therefore the full expression is manifestly supersymmetric. Second, to the remaining momentum integrals we apply standard dimensional regularization. All the momenta should be considered as $\mathscr{D}$-dimensional ones. The divergences appear only in the form of poles $1/(\mathscr{D}-4)^k$, $k=1, 2, \ldots$.

Regularization by dimensional reduction appears to preserve both supersymmetry and supergauge invariance. This scheme is well adapted to massless theories, since it does not demand the introduction of auxiliary masses.

To illustrate the use of regularization by dimensional reduction, let us consider the supergraph (4.4.53*a*). After fulfilment of the $\mathscr{D}$-algebra and

θ-integration, we arrive at equation (4.4.54*b*) in which the function $A(p^2, m^2)$ reads

$$A(p^2, m^2) = \mathrm{i} \int \frac{\mathrm{d}^4 k}{(2\pi)^2} \frac{1}{(k^2 + m^2)((p-k)^2 + m^2)}.$$

This integral is divergent. According to regularization by dimensional reduction, we have to replace the above expression by

$$A_{\mathscr{D}}(p^2, m^2) = \mathrm{i}\mu^{4-\mathscr{D}} \int \frac{\mathrm{d}^{\mathscr{D}} k}{(2\pi)^{\mathscr{D}}} \frac{1}{(k^2 + m^2)((p-k)^2 + m^2)} \tag{4.7.26}$$

μ being some parameters of mass dimension. It has been introduced in order to make $A_{\mathscr{D}}(p^2, m^2)$ dimensionless. Direct evaluation of the integral (4.7.26) leads to

$$A_{\mathscr{D}}(p^2, m^2) = -\frac{\mu^{4-\mathscr{D}}}{(4\pi)^{\mathscr{D}/2}} \Gamma\left(2 - \frac{\mathscr{D}}{2}\right) \int_0^1 \mathrm{d}\alpha \, [m^2 + p^2 \alpha(1-\alpha)]^{-2+\mathscr{D}/2}.$$

In the limit $\mathscr{D} \to 4$, one obtains

$$A_{\mathscr{D}}(p^2, m^2) = \frac{2}{(4\pi)^2(\mathscr{D}-4)} + \text{finite terms}.$$

To cancel this divergence, we have to introduce the one-loop counterterm

$$S_{\mathrm{COUNTR}} = -\frac{2\bar{\lambda}_{\mathrm{R}}\lambda_{\mathrm{R}}}{(4\pi)^2(\mathscr{D}-4)} \int \mathrm{d}^8 z \, \bar{\Phi}_{\mathrm{R}} \Phi_{\mathrm{R}}$$

expressed in terms of the regularized parameters λ_k, $\bar{\lambda}_k$ and superfields Φ_k, $\bar{\Phi}_k$.

Unfortunately, regularization by dimensional reduction leads to some ambiguities when applied to higher loop supergraph calculations. The problem is the following. To maintain supersymmetry, some objects which arise in the theory under consideration are to be defined directly in four dimensions. The others are to be continued, in the process of regularization, to $\mathscr{D}$ dimensions. However, there are many supergraphs which involve contractions of four-dimensional objects (spinors or tensors) with $\mathscr{D}$-dimensional ones. These contractions have an absolutely formal character and require special definitions. In principle, several reasonable definitions are available. But making the calculations in different ways may lead to different results. The differences among them disappear only at $\mathscr{D} = 4$, but at $\mathscr{D} = 4$ the supergraphs should be regularized. Nevertheless, in spite of possible ambiguities, regularization by dimensional reduction is the basic regularization scheme which is used in practice for supergraph calculations.

c. Other possibilities. As we have discussed, only the factor $1/(p^2 + m^2)$ in equation (4.7.22) is required to be regularized in order to make loop momentum integrals convergent. From this point of view, the situation is exactly the same as in conventional field theory. Hence we can use any of the standard regularization schemes applied to superfield theories.

For example, let us represent the scalar field propagator in the form

$$-\frac{\mathrm{i}}{p^2+m^2}=\int_0^\infty \mathrm{d}s\, \mathrm{e}^{-\mathrm{i}(p^2+m^2)s}.$$

To regularize the propagator, we have to modify the behaviour of the integrand at $s\to 0$. The regularized propagator can be defined as

$$\left(-\frac{\mathrm{i}}{p^2+m^2}\right)^{(\mathrm{R})}=\int_0^\infty \mathrm{d}s\, f_\omega(s)\, \mathrm{e}^{-\mathrm{i}(p^2+m^2)s} \tag{4.7.27}$$

f_ω being some smooth function chosen subject to the requirement that $f_\omega(s)$ and all its derivatives to a finite order vanish at $s=0$; also, $f_\omega(s)\to 1$ when the parameter ω tends to some fixed value. The choice $f_\omega(s)=(\mathrm{i}s\mu^2)^\omega$, where μ is a mass parameter, is often used in practice. The corresponding regularized propagator is given by equation (4.4.57); $\omega\to 0$ at the end of the calculations. The above choice defines so-called analytic regularization. It can be shown that the divergences of Feynman diagrams appear in the framework of analytic regularization only as poles of the type $1/\omega^k$, $k=1, 2, \ldots$.

When applied to supergraph calculations, analytic regularization has some features common with regularization by dimensional reduction. However, in analytic regularization all the calculations are performed in four-dimensional space. Unfortunately, analytic regularization can lead to the breakdown of Ward identities in gauge theories[2]. The example of supergraph calculations with use of analytic regularization has been presented in subsection 4.4.7.

4.8. Examples of counterterm calculations: an alternative technique

For given loop order, one must take into account infinitely many divergent supergraphs when trying to compute explicitly the relevant counterterms (4.7.12) and (4.7.15). There exists another approach to finding counterterms which makes it possible to account for all such supergraphs simultaneously. It is based on the Schwinger proper-time technique and its central object is a (super)propagator in an arbitrary external (super)field. This approach will be discussed in detail in Chapter 7. In the present section, the proper-time technique is used to calculate the one-loop counterterms for two supersymmetric models.

4.8.1. One-loop counterterms of matter in an external super Yang–Mills field

We consider a theory of matter chiral superfields $\Phi=\{\Phi^k(z)\}$ coupled to an external super Yang–Mills field described by $V=gV^I(z)T^I$, where g is the

[2]See, for example, R. Delbourgo, *Rep. Progr. Phys.* **39** 345, 1976.

coupling constant. The matter multiplet is assumed to transform in a real representation of the gauge group, hence the generators T^I are antisymmetric, $(T^I)^{\mathrm{T}} = -T^I$. Ignoring self-coupling of the matter superfields, the classical action reads

$$S[\Phi, \bar{\Phi}; V] = \int \mathrm{d}^8 z\, \bar{\Phi}\, \mathrm{e}^{2V} \Phi + \frac{m}{2} \left\{ \int \mathrm{d}^6 z\, \Phi^{\mathrm{T}} \Phi + \text{c.c.} \right\}. \tag{4.8.1}$$

Since $S[\Phi, \bar{\Phi}; V]$ involves no self-coupling of the dynamical superfields, the effective action $\Gamma[\Phi, \bar{\Phi}; V]$ of the theory is given only by its classical and one-loop parts:

$$\Gamma[\Phi, \bar{\Phi}; V] = S[\Phi, \bar{\Phi}; V] + \Gamma^{(1)}[V].$$

The one-loop correction $\Gamma^{(1)}[V]$, defined by the superfunctional integral

$$\mathrm{e}^{\Gamma^{(1)}}[V] = \int \mathscr{D}\bar{\Phi} \mathscr{D}\Phi\, \mathrm{e}^{\mathrm{i}S[\Phi, \bar{\Phi}; V]} \tag{4.8.2}$$

can be represented in the form

$$\Gamma^{(1)}[V] = \frac{\mathrm{i}}{2} \,\mathbf{sTr} \ln \mathbf{H}^{(V)} \tag{4.8.3}$$

where the operator

$$\mathbf{H}^{(V)} = \begin{pmatrix} m \mathbb{1}_{\Phi} & -\frac{1}{4} \bar{D}^2 \,\mathrm{e}^{-2V} \\ -\frac{1}{4} D^2 \,\mathrm{e}^{2V} & m \mathbb{1}_{\bar{\Phi}} \end{pmatrix} \equiv \begin{pmatrix} \mathbf{H}^{(V)}_{++} & \mathbf{H}^{(V)}_{+-} \\ \mathbf{H}^{(V)}_{-+} & \mathbf{H}^{(V)}_{--} \end{pmatrix} \tag{4.8.4}$$

determines the Hessian of $S[\Phi, \bar{\Phi}; V]$. It should be pointed out that $\Gamma^{(1)}[V]$ is a part of the one-loop effective action of the general super Yang–Mills theory considered earlier.

Remark. In equation (4.8.3) it is understood that $\mathbf{H}^{(V)}$ is a linear operator acting on the space of chiral-antichiral columns

$$\begin{pmatrix} \Phi \\ \bar{\Phi}^{\mathrm{T}} \end{pmatrix}.$$

More generally, given such an operator $\mathscr{A}$, its supertrace is defined with the help of the matrix superkernel

$$\mathscr{A}(z, z') = \begin{pmatrix} \mathscr{A}_{++} & \mathscr{A}_{+-} \\ \mathscr{A}_{-+} & \mathscr{A}_{--} \end{pmatrix} \begin{pmatrix} \mathbb{1}_{\Phi} \delta_+(z, z') & 0 \\ 0 & \mathbb{1}_{\bar{\Phi}} \delta_-(z, z') \end{pmatrix}$$

as follows:

$$\mathbf{sTr}\,\mathscr{A}=\int \mathrm{d}^6z\,\mathrm{tr}\,\mathscr{A}_{++}(z,z)+\int \mathrm{d}^6\bar{z}\,\mathrm{tr}\,\mathscr{A}_{--}(z,z)\equiv \mathbf{sTr}\,\mathscr{A}_{++}+\mathbf{sTr}\,\mathscr{A}_{--}$$

tr being the ordinary matrix trace over group indices.

In order to compute $\Gamma^{(1)}[V]$, we first apply the so-called doubling trick. Let us make in the integral (4.8.2) the change of variables

$$\Phi\to \mathrm{i}\Phi \qquad \bar{\Phi}\to -\mathrm{i}\bar{\Phi}$$

whose Jacobian is unity. This redefinition changes only the sign of the mass term (see equation (4.8.1)), in the exponential. Therefore, if we denote $\mathbf{H}(m)\equiv \mathbf{H}^{(V)}$, then

$$\mathbf{sDet}\,\mathbf{H}(m)=\mathbf{sDet}\,\mathbf{H}(-m).$$

One now obtains

$$\Gamma^{(1)}[V]=\frac{\mathrm{i}}{4}\mathbf{sTr}\ln\{\mathbf{H}(m)\mathbf{H}(-m)\}=\mathbf{sTr}\ln\begin{pmatrix}\mathscr{H}^{(V)}_{++}-m^2\mathbb{I}_{\Phi} & 0\\ 0 & \mathscr{H}^{(V)}_{--}-m^2\mathbb{I}_{\bar{\Phi}}\end{pmatrix} \tag{4.8.5}$$

where

$$\begin{aligned}\mathscr{H}^{(V)}_{++}&\equiv \mathbf{H}^{(V)}_{+-}\mathbf{H}^{(V)}_{-+}=\frac{1}{16}\bar{\mathrm{D}}^2\,\mathrm{e}^{-2V}\mathrm{D}^2\,\mathrm{e}^{2V}\\ \mathscr{H}^{(V)}_{--}&\equiv \mathbf{H}^{(V)}_{-+}\mathbf{H}^{(V)}_{+-}=\frac{1}{16}\mathrm{D}^2\,\mathrm{e}^{2V}\,\bar{\mathrm{D}}^2\,\mathrm{e}^{-2V}.\end{aligned} \tag{4.8.6}$$

Introduce chiral $\mathbf{G}^{(V)}_{++}$ and antichiral $\mathbf{G}^{(V)}_{--}$ superpropagators which are solutions under Feynman boundary conditions of the following equations

$$\begin{aligned}(\mathscr{H}^{(V)}_{++}-m^2)\mathbf{G}^{(V)}_{++}(z,z')&=-\mathbb{I}_{\Phi}\delta_+(z,z')\\ (\mathscr{H}^{(V)}_{--}-m^2)\mathbf{G}^{(V)}_{--}(z,z')&=-\mathbb{I}_{\bar{\Phi}}\delta_-(z,z').\end{aligned} \tag{4.8.7}$$

Then, the relation (4.8.5) is equivalent to

$$\Gamma^{(1)}[V]=-\frac{\mathrm{i}}{4}\mathbf{sTr}_+\ln\mathbf{G}^{(V)}_{++}-\frac{\mathrm{i}}{4}\mathbf{sTr}_-\ln\mathbf{G}^{(V)}_{--}. \tag{4.8.8}$$

As a result, the one-loop effective action is determined by the (anti)chiral superpropagators $\mathbf{G}^{(V)}_{--}$ and $\mathbf{G}^{(V)}_{++}$.

Remark. The trick applied above is the simplest way to prove equation (4.4.25).

Our next step in the treatment of $\Gamma^{(1)}[V]$ will be the introduction of the

Schwinger proper-time representation for the superpropagators

$$\mathbf{G}^{(V)}_{++}(z,z')=\mathrm{i}\int_0^\infty \mathrm{d}s\, U^{(V)}_{++}(z,z'|s)\,\mathrm{e}^{-\mathrm{i}m^2 s} \tag{4.8.9}$$

and similarly for $\mathbf{G}^{(V)}_{--}$. The superfield kernel $U^{(V)}_{++}(z,z'|s)$ is chiral in both arguments and satisfies the equation

$$\left(\mathrm{i}\frac{\partial}{\partial s}+\mathscr{H}^{(V)}_{++}\right)U^{(V)}_{++}(s)=0 \tag{4.8.10}$$

and the initial condition

$$U^{(V)}_{++}(z,z'|s)|_{s=0}=\mathbb{1}_\Phi\delta_+(z,z'). \tag{4.8.11}$$

With the help of purely formal manipulations described in detail in Chapter 7, one can show that the relation (4.8.8) is equivalent to

$$\Gamma^{(1)}[V]=-\frac{\mathrm{i}}{4}\int_0^\infty \frac{\mathrm{d}s}{s}\,\mathrm{e}^{-\mathrm{i}m^2 s}\{\mathbf{sTr}_+\,U^{(V)}_{++}(s)+\mathbf{sTr}_-\,U^{(V)}_{++}(s)\}. \tag{4.8.12}$$

Here the integral over s turns out to be divergent at $s\to 0$. We regularize $\Gamma^{(1)}[V]$ as follows:

$$\Gamma^{(1)}_\omega[V]=-\frac{\mathrm{i}}{4}\mu^{2\omega}\int_0^\infty \frac{\mathrm{d}(\mathrm{i}s)}{(\mathrm{i}s)^{1-\omega}}\,\mathrm{e}^{-\mathrm{i}m^2 s}\{\mathbf{sTr}_+\,U^{(V)}_{++}(s)+(+\to -)\} \tag{4.8.13}$$

μ, ω being the normalization point and regularization parameter, respectively; $\omega\to +0$ at the end of the calculations. Now, let us note the identity

$$\mathbf{sTr}_+\,U^{(V)}_{++}(s)=\mathbf{sTr}_-\,U^{(V)}_{--}(s) \tag{4.8.14}$$

which can be formally proved, using equations (4.8.6, 10, 11), as

$$\begin{aligned}\mathbf{sTr}_+\,U^{(V)}_{++}(s)&=\mathbf{sTr}_+\,\{\exp(\mathrm{i}s\mathbf{H}^{(V)}_{+-}\mathbf{H}^{(V)}_{-+})\}\\ &=\mathbf{sTr}_-\,\{\exp(\mathrm{i}s\mathbf{H}^{(V)}_{-+}\mathbf{H}^{(V)}_{+-})\}=\mathbf{sTr}_-\,U^{(V)}_{--}(s).\end{aligned}$$

Hence we can write

$$\Gamma^{(1)}_\omega[V]=-\frac{\mathrm{i}}{2}\mu^{2\omega}\int_0^\infty \frac{\mathrm{d}(\mathrm{i}s)}{(\mathrm{i}s)^{1-\omega}}\,\mathrm{e}^{-\mathrm{i}m^2 s}\,\mathbf{sTr}_+\,U^{(V)}_{++}(s). \tag{4.8.15}$$

It should be remarked that $\Gamma^{(1)}_\omega[V]$ is invariant under the super Yang–Mills gauge transformations (3.5.26); i.e. equations (4.8.6) and (3.5.26) show that

$$\mathscr{H}^{(V')}_{++}=\mathrm{e}^{\mathrm{i}\Lambda}\,\mathscr{H}^{(V)}_{++}\,\mathrm{e}^{-\mathrm{i}\Lambda}$$

Hence

$$\mathbf{sTr}_+\,U^{(V')}_{++}(s)=\mathbf{sTr}_+(\mathrm{e}^{\mathrm{i}\Lambda}\,U^{(V)}_{++}\,\mathrm{e}^{-\mathrm{i}\Lambda})=\mathbf{sTr}_+\,U^{(V)}_{++}(s).$$

The crucial observation is that the fourth-order differential operator $\mathscr{H}^{(V)}_{++}$

(4.8.6) is equivalent to a second-order one when acting on chiral superfields (by construction, $\mathscr{H}^{(V)}_{++}$ should act on chiral superfields only). In terms of the gauge covariant derivatives (3.6.40*a*) satisfying the algebra (3.6.45, 46*a*), we can write

$$\mathscr{H}^{(V)}_{++} = \mathscr{D}^{(+)a}\mathscr{D}^{(+)}_a - W^\alpha \mathscr{D}^{(+)}_\alpha - \frac{1}{2}(\mathscr{D}^{(+)\alpha}W_\alpha). \tag{4.8.16}$$

To simplify following expressions, we shall omit the label $(+)$ which the derivatives $\mathscr{D}^{(+)}_A$ carry. Now, the structure of $\mathscr{H}^{(V)}_{++}$ implies that the solution of equations (4.8.10, 11) can be looked for in the form (compare with equation (4.4.21))

$$U^{(V)}_{++}(z, z'|s) = -\frac{\mathrm{i}}{(4\pi s)^2}\exp\left[\frac{\mathrm{i}}{4s}(x_{(+)} - x'_{(+)})^2\right]\sum_{n=0}^{\infty} a^{(V)}_n(z, z')(\mathrm{i}s)^n \tag{4.8.17}$$

$$x^a_{(+)} = x^a + \mathrm{i}\theta\sigma^a\bar{\theta}.$$

Here the coefficients $a^{(v)}_n(z, z')$ are chiral in both arguments and satisfy the equations

$$(x_{(+)} - x'_{(+)})^a[\mathscr{D}_a - \mathrm{i}\bar{\theta}_{\dot{\alpha}}(\tilde{\sigma}_a)^{\dot{\alpha}\alpha}W_\alpha]a^{(V)}_0 = 0 \tag{4.8.18a}$$

$$(n+1)a^{(V)}_{n+1} + (x_{(+)} - x'_{(+)})^a[\mathscr{D}_a - \mathrm{i}\bar{\theta}\tilde{\sigma}_a W]a^{(V)}_{n+1} = \mathscr{H}^{(V)}_{++}a^{(V)}_n \qquad n = 0, 1, \ldots. \tag{4.8.18b}$$

These equations represent recurrence relations allowing us to determine the coefficients step by step. Note that the initial condition (4.8.11) and the equation (4.8.18*a*) lead to

$$a^{(V)}_0(z, z') = (\theta - \theta')^2\Omega(z, z') \tag{4.8.19}$$

Ω being the chiral two-point function subjected to equation (4.8.18*a*) and the boundary condition

$$\Omega(z, z) = \mathbb{1}_\Phi. \tag{4.8.20}$$

In accordance with the relations (4.8.15) and (4.8.17), we can now write

$$\Gamma^{(1)}_\omega[V] = \frac{\mu^{2\omega}}{2(4\pi)^2}\int_0^\infty \frac{\mathrm{d}(\mathrm{i}s)}{(\mathrm{i}s)^{3-\omega}}\,\mathrm{e}^{-\mathrm{i}m^2 s}\sum_{n=0}^{\infty}(\mathrm{i}s)^n\int \mathrm{d}^6x\,\mathrm{tr}\,a^{(V)}_n(z, z). \tag{4.8.21}$$

So, the effective action is determined by the coefficients $a^{(v)}_n$ at coincident points. As for the divergent part of $\Gamma^{(1)}_\omega[V]$, it is easy to show that

$$\lim_{\omega\to 0}\Gamma^{(1)}_\omega[V] = \frac{1}{\omega}\frac{1}{2(4\pi)^2}\int \mathrm{d}^6z\,\mathrm{tr}\left\{\frac{m^4}{2}a^{(V)}_0(z, z) - m^2 a^{(V)}_1(z, z) + a^{(V)}_2(z, z)\right\}$$
$$+\,\text{finite terms.}$$

On the other hand, it follows immediately from equations (4.8.18–20) that

$a_0^{(V)}(z,z)=a_1^{(V)}(z,z)=0$. Therefore, the divergent part of the effective action reads

$$\Gamma^{(1)}_{\mathrm{div}}[V]=\frac{1}{\omega}\frac{1}{2(4\pi)^2}\int d^6z\,\mathrm{tr}\,a_2^{(V)}(z,z). \tag{4.8.22}$$

As a consequence, all information about divergences of the theory is encoded in $a_2^{(V)}$.

Let us calculate $a_2^{(V)}$ at coincident points. Setting in equation (4.8.18*b*) $n=1$ and taking the limit $z\to z'$, one obtains

$$2a_2^{(V)}(z,z)=(\mathscr{D}^a\mathscr{D}_a-W^\alpha\bar{\mathrm{D}}_\alpha)a_1^{(V)}(z,z')|_{z=z} \tag{4.8.23}$$

where the identity $a_1^{(V)}(z,z)=0$ has been used. Other simple consequences of equations 4.8.18–20) are:

$$\begin{aligned}
\mathscr{D}_{a_1}\dots\mathscr{D}_{a_k}a_0^{(V)}(z,z')|_{z=z'}&=0 \qquad k=0,1,\dots\\
\mathscr{D}_\alpha\mathscr{D}_{a_1}\dots\mathscr{D}_{a_k}a_0^{(V)}(z,z')|_{z=z'}&=0\\
\mathscr{D}_{a_1}\dots\mathscr{D}_{a_k}a_1^{(V)}(z,z')|_{z=z'}&=0\\
\mathscr{D}^\alpha\mathscr{D}_\alpha a_0^{(V)}(z,z')|_{z=z'}&=-4\mathbb{1}_\Phi\\
\mathscr{D}_\alpha a_1^{(V)}(z,z')|_{z=z'}&=-2W_\alpha.
\end{aligned} \tag{4.8.24}$$

We arrive at

$$a_2^{(V)}(z,z)=W^\alpha W_\alpha. \tag{4.8.25}$$

Due to equation (4.8.22), the divergent part of the effective action is given by

$$\Gamma^{(1)}_{\mathrm{div}}[V]=\frac{1}{\omega}\frac{1}{2(4\pi)^2}\int d^6z\,\mathrm{tr}(W^\alpha W_\alpha) \tag{4.8.26}$$

and determines the relevant counterterm.

The results described were originally obtained by J. Honerkamp, M. Schlindwein, F. Krause and M. Scheunert in the case of supersymmetric electrodynamics and by I. Buchbinder in the case of general super Yang–Mills theories.

4.8.2. One-loop counterterms of the general Wess–Zumino model

Let us consider the general Wess–Zumino model (3.2.8) and determine the one-loop divergences of the corresponding effective action. In accordance with equation (4.3.15), the one-loop contribution to the effective action of the theory (3.2.8) is given as

$$\begin{gathered}
e^{i\Gamma^{(1)}}[\Phi,\bar{\Phi}]=\int\mathscr{D}\chi\,\mathscr{D}\bar{\chi}\,e^{iS[\chi,\bar{\chi};\Psi,\bar{\Psi}]}\\
\Psi\equiv\mathscr{L}''_c(\Phi).
\end{gathered} \tag{4.8.27}$$

Here the linearized action $S[\chi, \bar{\chi}; \Psi, \bar{\Psi}]$ is defined by equation (4.4.6). In terms of the operator (4.4.5), which determine the Hessian of $S[\chi, \bar{\chi}; \Psi, \bar{\Psi}]$, and its Green's function (4.4.7, 8) under the Feynman boundary conditions, $\Gamma^{(1)}[\Phi, \bar{\Phi}]$ is of the form

$$\Gamma^{(1)}[\Phi, \bar{\Phi}] = \frac{i}{2} \mathbf{sTr} \ln \mathbf{H}^{(\Psi)} = -\frac{i}{2} \mathbf{sTr} \ln \mathbf{G}^{(\Psi)} \tag{4.8.28}$$

where the operation of supertrace is defined in the manner:

$$\mathbf{sTr}\, \mathbf{G}^{(\Psi)} = \int d^6 z\, \mathbf{G}_{++}(z, z) + \int d^6 \bar{z}\, \mathbf{G}_{--}(z, z). \tag{4.8.29}$$

An important role in our subsequent consideration will be played by the superpropagator $\mathbf{G}_V^{(\Psi)}$ of real scalar superfield obeying the equation

$$\begin{gathered} \Delta \mathbf{G}_V^{(\Psi)}(z, z') = -\delta^8(z, z') \\ \Delta = \Box - \frac{1}{4} \Psi(z) \bar{D}^2 - \frac{1}{4} \bar{\Psi}(z) D^2. \end{gathered} \tag{4.8.30}$$

Its significance follows from the fact that $\mathbf{G}^{(\Psi)}$ can be expressed through $\mathbf{G}_V^{(\Psi)}$ by the rule:

$$\mathbf{G}^{(\Psi)}(z, z') = \frac{1}{16} \begin{pmatrix} \bar{D}_z^2 \bar{D}_{z'}^2 \mathbf{G}_V^{(\Psi)}(z, z') & \bar{D}_z^2 D_{z'}^2 \mathbf{G}_V^{(\Psi)}(z, z') \\ D_z^2 \bar{D}_{z'}^2 \mathbf{G}_V^{(\Psi)}(z, z') & D_z^2 D_{z'}^2 \mathbf{G}_V^{(\Psi)}(z, z') \end{pmatrix}. \tag{4.8.31}$$

In this connection, let us act with the operator

$$\mathbf{H}^{(0)} = \begin{pmatrix} 0 & -\frac{1}{4} \bar{D}^2 \\ -\frac{1}{4} D^2 & 0 \end{pmatrix}$$

on both sides of equation (4.4.8) from the left. Then one arrives at the equations

$$\begin{gathered} \Box_z \mathbf{G}_{++}(z, z') - \frac{1}{4} \bar{D}_z^2 (\Psi(z) \mathbf{G}_{-+}(z, z')) = 0 \\ \Box_z \mathbf{G}_{-+}(z, z') - \frac{1}{4} D_z^2 (\bar{\Psi}(z) \mathbf{G}_{++}(z, z')) = \frac{1}{16} D_z^2 \bar{D}_{z'}^2 \delta^8(z, z') \end{gathered} \tag{4.8.32a}$$

$$\begin{gathered} \Box_z \mathbf{G}_{--}(z, z') - \frac{1}{4} D_z^2 (\bar{\Psi}(z) \mathbf{G}_{+-}(z, z')) = 0 \\ \Box_z \mathbf{G}_{+-}(z, z') - \frac{1}{4} \bar{D}_z^2 (\Psi(z) \mathbf{G}_{--}(z, z')) = \frac{1}{16} \bar{D}_z^2 D_{z'}^2 \delta^8(z, z') \end{gathered} \tag{4.8.32b}$$

which the components of $\mathbf{G}^{(\Psi)}$, defined by equation (4.4.8), satisfy. On the other hand, let us introduce one more $\mathbf{G}^{(\Psi)}$ defined now by equation (4.8.31).

Its components turn out to satisfy the same equation (4.8.32). For example, acting with $\frac{1}{16}\bar{D}_{z'}^2 D_{z'}^2$ on both sides of equation (4.8.30) and using the chirality of Ψ, one obtains the first equation (4.8.32*a*), since

$$\bar{D}_z^2 D_{z'}^2 \delta^8(z, z') = \bar{D}_z^2 D_z^2 \delta^8(z, z') = 0.$$

The relation (4.8.31) leads to the following remarkable identity

$$\mathbf{sTr} \ln \mathbf{G}^{(\Psi)} = \mathbf{sTr} \ln \mathbf{G}_V^{(\Psi)} \tag{4.8.33}$$

where the operation of supertrace in the right-hand side is defined as

$$\mathbf{sTr}\, \mathbf{G}_V^{(\Psi)} = \int d^8z\, \mathbf{G}_V^{(\Psi)}(z, z) \tag{4.8.34}$$

(compare with equation (4.8.29)). To prove equation (4.8.33), we consider the variation of $\mathbf{sTr} \ln \mathbf{G}^{(\Psi)}$ with respect to arbitrary infinitesimal displacement $\Psi \to \Psi + \delta\Psi$ keeping, for simplicity, $\bar{\Psi}$ unchanged. We have

$$\begin{aligned}
\delta_\Psi\, \mathbf{sTr} \ln \mathbf{G}^{(\Psi)} &= \mathbf{sTr}\, (\delta_\Psi \mathbf{H}^{(\Psi)} \mathbf{G}^{(\Psi)}) = \mathbf{sTr}\left[\begin{pmatrix} \delta\Psi & 0 \\ 0 & 0 \end{pmatrix}\begin{pmatrix} \mathbf{G}_{++} & \mathbf{G}_{+-} \\ \mathbf{G}_{-+} & \mathbf{G}_{--} \end{pmatrix}\right] \\
&= \int d^6z\, d^6z'\, \delta\Psi(z) \mathbf{G}_{++}(z, z') \delta_+(z, z') \\
&= \int d^8z\, d^8z'\, \delta\Psi(z) \mathbf{G}_V^{(\Psi)}(z, z') \left(-\frac{1}{4}\bar{D}_z^2 \delta^8(z, z')\right) \\
&= \int d^8z\, \delta_\Psi \Delta_z\, \mathbf{G}_V^{(\Psi)}(z, z')|_{z=z'} = \delta_\Psi\, \mathbf{sTr} \ln \mathbf{G}_V^{(\Psi)}.
\end{aligned}$$

Since at $\Psi = 0$ the expressions in both sides of identity (4.8.33) vanish, the above analysis confirms equation (4.8.33).

Using equation (4.8.33), one can rewrite equation (4.8.28) in the form

$$\Gamma^{(1)}[\Phi, \bar{\Phi}] = -\frac{i}{2}\, \mathbf{sTr} \ln \mathbf{G}_V^{(\Psi)}. \tag{4.8.35}$$

This relation will be basic to our further investigations being, in fact, analogous to that described in the previous subsection. First, we introduce the proper-time representation for $\mathbf{G}_V^{(\Psi)}$:

$$\mathbf{G}_V^{(\Psi)}(z, z') = i\int_0^\infty ds\, U_V^{(\Psi)}(z, z'|s) \tag{4.8.36}$$

$U_V^{(\Psi)}$ being the unique solution of the equation

$$\left(i\frac{\partial}{\partial s} + \Delta\right) U_V^{(\Psi)}(s) = 0 \tag{4.8.37}$$

with the initial condition

$$U_V^{(\Psi)}(z, z'|s \to +0) = \delta^8(z, z'). \tag{4.8.38}$$

Second, we replace the ill-defined expression for $\Gamma^{(1)}$

$$\Gamma^{(1)}[\Phi, \bar{\Phi}] = -\frac{i}{2}\int_0^\infty \frac{ds}{s}\,\mathbf{sTr}\, U_V^{(\Psi)}(s)$$

which is equivalent to expression (4.8.35), by the regularized version

$$\Gamma_{(\omega)}^{(1)}[\Phi, \bar{\Phi}] = -\frac{i}{2}\mu^{2\omega}\int_0^\infty \frac{d(is)}{(is)^{1-\omega}}\,\mathbf{sTr}\, U_V^{(\Psi)}(s). \tag{4.8.39}$$

It now remains to analyse the kernel $U_V^{(\Psi)}$.

We look for the solution of equations (4.8.37, 38) using the ansatz

$$U_V^{(\Psi)}(z, z'|s) = -\frac{i}{(4\pi s)^2}\exp\left(i\frac{\sigma(z, z')}{2s}\right)\sum_{n=0}^{\infty} a_n^{(\Psi)}(z, z')(is)^n. \tag{4.8.40}$$

To fulfil equation (4.8.37), σ and $a^{(\Psi)}$ should satisfy the equations

$$2\sigma = \partial^a\sigma\,\partial_a\sigma - \frac{1}{4}\psi\bar{D}_{\dot{\alpha}}\sigma\bar{D}^{\dot{\alpha}}\sigma - \frac{1}{4}\bar{\psi}D^\alpha\sigma D_\alpha\sigma \tag{4.8.41a}$$

$$\left[(\partial^a\sigma)\partial_a - \frac{1}{4}\Psi(\bar{D}_{\dot{\alpha}}\sigma)\bar{D}^{\dot{\alpha}} - \frac{1}{4}\bar{\Psi}(D^\alpha\sigma)D_\alpha\right]a_0^{(\Psi)} = \frac{1}{2}(4 - (\Delta\sigma))a_0^{(\Psi)} \tag{4.8.41b}$$

$$(n+1)a_{n+1}^{(\Psi)} + \left[(\partial^a\sigma)\partial_a - \frac{1}{4}\Psi(\bar{D}_{\dot{\alpha}}\sigma)\bar{D}^{\dot{\alpha}} - \frac{1}{4}\bar{\Psi}(D^\alpha\sigma)D_\alpha\right]a_{n+1}^{(\Psi)}$$

$$= \frac{1}{2}(4 - (\Delta\sigma))a_{n+1}^{(\Psi)} + a_n^{(\Psi)} \qquad n = 0, 1, \ldots. \tag{4.8.41c}$$

To guarantee the initial condition (4.8.38), it proves sufficient to impose the following boundary conditions:

$$\sigma(z, z) = 0 \qquad D_A\sigma(z, z')|_{z=z'} = 0 \qquad D_\alpha D_A\sigma(z, z')|_{z=z'} = 0$$

$$D^2\partial_a\sigma(z, z')|_{z=z'} = 0 \qquad D_\alpha\partial_a\partial_b\sigma(z, z')|_{z=z'} = 0 \tag{4.8.42a}$$

$$D_{A_1}\ldots D_{A_k}a_0^{(\Psi)}(z, z')|_{z=z'} = 0 \qquad k = 0, 1, 2, 3 \tag{4.8.42b}$$

$$D^2\bar{D}^2 a_0^{(\Psi)}(z, z')|_{z=z'} = 16. \tag{4.8.42c}$$

Let us comment on these restrictions. At $\Psi = 0$ we have $\Delta = \Box$, hence

$$U_V(z, z|s) = U_V^{(\Psi=0)}(z, z|s) = -\frac{i}{(4\pi s)^2}(\theta - \theta')^2(\bar{\theta} - \bar{\theta}')^2\exp\left(i\frac{(x - x')^2}{4s}\right). \tag{4.8.43}$$

This implies that in the case $\Psi \neq 0$ the two-point functions from equation (4.8.40) read

$$\sigma(z, z') = \frac{1}{2}(x - x')^2 + O(\Psi)$$

$$a_0^{(\Psi)}(z, z') = \delta^4(\theta - \theta') + O(\Psi)$$

$$a_n^{(\Psi)}(z, z') = O(\Psi) \qquad n = 1, 2, 3, \ldots.$$

It can also be seen that any object such as

$$\mathrm{D}_{A_1} \ldots \mathrm{D}_{A_k} \sigma(z, z')|_{z=z'}$$

$$\mathrm{D}_{A_1} \ldots \mathrm{D}_{A_k} a_n^{(\Psi)}(z, z')|_{z=z'} \qquad k = 0, 1, 2, \ldots$$

is a tensor superfield possibly constructed from the Ψ-independent Lorentz-invariant tensors $\varepsilon_{\alpha\beta}$, $(\sigma_a)_{\alpha\dot{\alpha}}$, and so on, as well as depending analytically on Ψ, $\bar{\Psi}$ and their covariant derivatives to a finite order. The last observation, supplemented by considerations of dimension, tells us that all the expressions in the left-hand sides of equations (4.8.42*a*,*b*) should vanish. The boundary conditions (4.8.42) and the identity

$$\partial_a \partial_b \sigma(z, z')|_{z=z'} = \eta_{ab}$$

which follows from equations (4.8.41*a*) and (4.8.42*a*) show that as $s \to +0$ the kernel $U_V^{(\Psi)}$ behaves like the free one (4.8.43).

As can be easily checked, direct consequences of the equations (4.8.41) and of the boundary conditions (4.8.42) are the following relations:

$$\begin{aligned} a_1^{(\Psi)}(z, z) &= 0 \\ a_2^{(\Psi)}(z, z) &= \bar{\Psi}\Psi. \end{aligned} \tag{4.8.44}$$

Now, we are in a position to determine the divergent part of the one-loop effective action (4.8.39). By analogy with equation (4.8.22), in the present case we have

$$\Gamma_{\mathrm{div}}^{(1)}[\Phi, \bar{\Phi}] = \frac{1}{2\omega} \frac{1}{(4\pi)^2} \int \mathrm{d}^8 z \, a_2^{(\Psi)}(z, z) = \frac{1}{2\omega(4\pi)^2} \int \mathrm{d}^8 z \, \bar{\Psi}\Psi. \tag{4.8.45}$$

Therefore, the corresponding counterterm cancelling $\Gamma_{\mathrm{div}}^{(1)}[\Phi, \bar{\Phi}]$ can be written as

$$S_{\mathrm{COUNTR}} = \frac{1}{2(4\pi)^2 \omega} \int \mathrm{d}^8 z \, \bar{\mathscr{L}}_c''(\bar{\Phi}) \mathscr{L}_c''(\Phi) \tag{4.8.46}$$

where we have taken into account that $\Psi = \mathscr{L}_c''(\Phi)$.

As is seen from equation (4.8.46), the theory under consideration proves to be multiplicatively renormalizable only if $\mathscr{L}_c''(\Phi) \sim \Phi + \text{const}$. The final requirement means that $\mathscr{L}_c(\Phi)$ is a third-order polynomial in Φ.

4.9. Superfield effective potential

In conventional field theory the effective potential is defined to be the effective Lagrangian evaluated at constant values of the scalar fields, all other fields being taken to be zero. The effective potential $V_{\text{eff}}(\varphi)$ is an important tool for studying the questions of symmetry breaking and vacuum stability. In the present section we shall introduce a supersymmetric generalization of the effective potential leading to the so-called superfield effective potential.

4.9.1. Effective potential in quantum field theory (brief survey)

Let $\Gamma[\varphi]$ be the renormalized effective action of some scalar field theory. In general, $\Gamma[\varphi]$ is a nonlinear and non-local functional the calculation of which is usually performed in the framework of the loop expansion.

Suppose that φ is a slowly varying field. Then we can represent the effective action as a power series in the derivatives of field, that is

$$\Gamma[\varphi]=\int \mathrm{d}^4x\left(-V_{\text{eff}}(\varphi)-\frac{1}{2}Z(\varphi)\eta^{mn}\partial_m\varphi\partial_n\varphi+\ldots\right). \tag{4.9.1}$$

The function $V_{\text{eff}}(\varphi)$ is called the effective potential. To calculate the effective potential, it is sufficient to find the form of $\Gamma[\varphi]$ at $\varphi=\text{const}$.

Let $\varphi=\text{const}$ be a solution of the effective equation $\delta\Gamma[\varphi]/\delta\varphi=0$. Obviously, this is equivalent to

$$\frac{\partial V_{\text{eff}}(\varphi)}{\partial\varphi}=0. \tag{4.9.2}$$

The last equation is exact and allows one to determine those constant values of φ providing a minimum of the effective action. It is clear that the existence of such non-zero scalar fields can be associated with symmetry breaking.

According to the loop expansion, $V_{\text{eff}}(\varphi)$ is of the form

$$V_{\text{eff}}(\varphi)=V(\varphi)+\sum_{n=1}^{\infty}\hbar^n V^{(n)}(\varphi) \tag{4.9.3}$$

where $V(\varphi)$ is the classical potential of the theory under consideration, while $V^{(n)}(\varphi)$ is the n-loop quantum correction to the effective potential. The quantum corrections are usually evaluated using the standard diagram technique.

As an example, let us calculate the one-loop effective potential for a theory with the classical action

$$S[\varphi]=\int \mathrm{d}^4x\left(-\frac{1}{2}\eta^{mn}\partial_m\varphi\partial_n\varphi-V(\varphi)\right). \tag{4.9.4}$$

The one-loop contribution to the effective action is defined by the functional integral

$$e^{i\Gamma^{(1)}[\varphi]} = \int \mathscr{D}\chi \exp\left[\frac{i}{2}\int d^4x\, \chi(\Box - V''(\varphi))\chi\right]$$

and can be written in the form

$$\Gamma^{(1)}[\varphi] = -\frac{i}{2}\operatorname{Tr}\ln G^{(\varphi)} \tag{4.9.5}$$

where $G^{(\varphi)}(x, x')$ is a Green's function obeying the equation

$$(\Box_x - V''(\varphi))G^{(\varphi)}(x, x') = -\delta^4(x - x'). \tag{4.9.6}$$

Notice, we prefer here to use the notation Tr for the operation of functional supertrace (4.1.8), because the theory under consideration involves no fermionic fields. Introducing the proper-time representation

$$G^{(\varphi)}(x, x') = i\int_0^\infty ds\, U^{(\varphi)}(x, x'|s) \tag{4.9.7}$$

where kernel $U^{(\varphi)}$ satisfies the equation

$$\left[i\frac{\partial}{\partial s} + \Box_x - V''(\varphi(x))\right]U^{(\varphi)}(x, x'|s) = 0 \tag{4.9.8}$$

and the initial condition

$$U^{(\varphi)}(x, x'|s \to +0) = \delta^4(x - x') \tag{4.9.9}$$

we regularize $\Gamma^{(1)}[\varphi]$ as follows

$$\Gamma^{(1)}_\omega[\varphi] = -\frac{i}{2}\mu^{2\omega}\int_0^\infty \frac{d(is)}{(is)^{1-\omega}}\int d^4x\, U^{(\varphi)}(x, x|s) \tag{4.9.10}$$

$\omega \to 0$ at the end of the calculations (see the details in Chapter 7).

In order to find the effective potential, it is sufficient to solve the system (4.9.8, 9) only for $\varphi = \text{const}$. The relevant solution reads

$$U^{(\varphi=\text{const})}(x, x'|s) = \frac{i}{(4\pi i s)^2}\exp\left[\frac{i(x - x')^2}{4s} - isV''(\varphi)\right]. \tag{4.9.11}$$

Now, from equation (4.9.10) we deduce the regularized one-loop correction to $V_{\text{eff}}(\varphi)$:

$$\begin{aligned} V^{(1)}_\omega[\varphi] &= \frac{i}{2}\mu^{2\omega}\int_0^\infty \frac{d(is)}{(is)^{1-\omega}}\, U^{(\varphi=\text{const})}(x, x|s) \\ &= -\frac{1}{2(4\pi)^2(\omega - 1)(\omega - 2)}\Gamma(\omega)(V''(\varphi))^2\left(\frac{V''(\varphi)}{\mu^2}\right)^{-\omega}. \end{aligned} \tag{4.9.12}$$

In the limit $\omega \to 0$, we have $\Gamma(\omega) \approx 1/\omega + \gamma$, γ being the Euler constant,

$$\left(\frac{V''(\varphi)}{\mu^2}\right)^{-\omega} \approx 1 - \omega \ln \frac{V''(\varphi)}{\mu^2}.$$

Taking the same limit in equation (4.9.12) gives

$$V^{(1)}_{\omega} = V^{(1)}_{\text{div}} + V^{(1)}(\varphi)$$

where

$$V^{(1)}_{\text{div}} = -\frac{1}{64\pi^2\omega}(V''(\varphi))^2 \tag{4.9.13a}$$

$$V^{(1)}(\varphi) = \frac{1}{64\pi^2}(V''(\varphi))^2\left(\ln \frac{V''(\varphi)}{\mu^2} - \gamma - 3/2\right). \tag{4.9.13b}$$

$V^{(1)}_{\text{div}}$ is the divergent part of the effective potential. It should be cancelled by the counterterm

$$S_{\text{countr}} = \frac{1}{64\pi^2\omega}\int d^4x\,(V''(\varphi))^2.$$

Hence, the one-loop contribution to $V_{\text{eff}}(\varphi)$ is given by equation (4.9.13*b*). As a result, the one-loop effective potential reads

$$V_{\text{eff}}(\varphi) = V(\varphi) + V^{(1)}(\varphi) = V(\varphi) + \frac{\hbar}{64\pi^2}(V''(\varphi))^2\left(\ln \frac{V''(\varphi)}{\mu^2} - \gamma - 3/2\right). \tag{4.9.14}$$

The effective potential obtained depends on arbitrary mass parameter μ which reflects some inherent arbitrariness in the choice of renormalization scheme. This parameter should be fixed in practice by imposing so-called normalization conditions corresponding to a correct choice for observable parameters of the theory (masses, couplings). Let us consider, for instance, the theory with the following classical potential

$$V(\varphi) = \frac{\lambda_0}{4!}\varphi^4$$

where λ_0 is a bare coupling. One can choose suitable normalization conditions in the form

$$\left.\frac{d^2 V_{\text{eff}}}{d\varphi^2}\right|_{\varphi=0} = 0 \qquad \left.\frac{d^4 V_{\text{eff}}}{d\varphi^4}\right|_{\varphi=\varphi_0} = \lambda \tag{4.9.15}$$

where φ_0 is some non-zero constant. The quantity λ can be called the observable coupling. Now, from expressions (4.9.14) and (4.9.15) one finally obtains

$$V_{\text{eff}}(\varphi)=\frac{\lambda}{4!}\varphi^4+\frac{\hbar\lambda\varphi^4}{64\pi^2}\left(\ln\frac{\varphi^2}{\varphi_0^2}-\frac{25}{6}\right). \tag{4.9.16}$$

This result is known as the Coleman–Weinberg effective potential.

4.9.2. *Superfield effective potential*

Let $\Gamma[\Phi, \bar{\Phi}]$ be the renormalized effective action of some supersymmetric model described by chiral Φ and antichiral $\bar{\Phi}$ scalar superfields. Suppose that $\Phi(z)=\mathrm{e}^{\mathrm{i}\mathscr{H}}\Phi(x,\theta)$ and $\bar{\Phi}(z)=\mathrm{e}^{-\mathrm{i}\mathscr{H}}\bar{\Phi}(x,\bar{\theta})$, where $\mathscr{H}=\theta\sigma^a\bar{\theta}\partial_a$, vary slowly in space–time. Then we can represent the effective action as follows

$$\Gamma[\Phi,\bar{\Phi}]=\int \mathrm{d}^8z\, \mathscr{L}_{\text{eff}}(\Phi, \mathrm{D}_A\Phi, \mathrm{D}_A\mathrm{D}_B\Phi, \ldots, \bar{\Phi}, \mathrm{D}_A\bar{\Phi}, \mathrm{D}_A\mathrm{D}_B\bar{\Phi}, \ldots)$$
$$+\left\{\int \mathrm{d}^6z\, \mathscr{L}^{(\mathrm{c})}_{\text{eff}}(\Phi, \partial_a\Phi, \partial_a\partial_b\Phi, \ldots)+\text{c.c.}\right\}. \tag{4.9.17}$$

Here $\mathscr{L}_{\text{eff}}$ is a power series in the covariant derivatives $\mathrm{D}_A\Phi$, $\mathrm{D}_A\mathrm{D}_B\Phi, \ldots,$ $\bar{\Phi}$, $\mathrm{D}_A\bar{\Phi}$, $\mathrm{D}_A\mathrm{D}_B\bar{\Phi}, \ldots,$ while $\mathscr{L}^{(\mathrm{c})}_{\text{eff}}$ is a power series in the partial derivatives $\partial_a\Phi$, $\partial_a\partial_b\Phi, \ldots$. It seems natural to call $\mathscr{L}_{\text{eff}}$ an effective super Langrangian and $\mathscr{L}^{(\mathrm{c})}_{\text{eff}}$ an effective chiral super Lagrangian.

Before we proceed further, it is worth saying some words about the presented structure of the effective action (4.9.17). The effective action superfunctional is calculated in practice with the help of the supergraph technique described previously. In accordance with the non-renormalization theorem, the contribution to $\Gamma[\Phi,\bar{\Phi}]$ coming from an arbitrary supergraph can be represented as a single integral over $\mathrm{d}^4\theta$, but not over $\mathrm{d}^2\theta$ or $\mathrm{d}^2\bar{\theta}$. On these grounds, this is only the first term in equation (4.9.17) which is admissible on the basis of general principles. However, one can expect contributions to the effective action which really exist, of the type

$$\int \mathrm{d}^8z\left(-\frac{\mathrm{D}^2}{4\Box}\right)G=\int \mathrm{d}^6z\, G \tag{4.9.18}$$

where G is a chiral superfield. It is such corrections that form $\mathscr{L}^{(\mathrm{c})}_{\text{eff}}$.

The relation (4.9.17) defines the exact effective action as an expansion in superfield covariant derivatives. Now, let us look for a supersymmetric extension of the conventional effective potential. The theory under consideration is described by scalar and spinor fields, as is seen from the component expansion

$$\Phi(z)=\mathrm{e}^{\mathrm{i}\theta\sigma^a\bar{\theta}\partial_a}(A(x)+\theta^\alpha\psi_\alpha(x)+\theta^2F(x)).$$

To obtain the effective potential, we evaluate the effective Lagrangian at constant scalar fields, the remaining fields being switched off, that is, under the conditions

$$A = \text{const} \qquad F = \text{const} \qquad \psi_\alpha = 0. \tag{4.9.19}$$

Then, the effective potential is given as

$$-V_{\text{eff}} = \int d^4\theta \, \mathscr{L}_{\text{eff}} + \left\{ \int d^2\theta \, \mathscr{L}^{(c)}_{\text{eff}} + \text{c.c.} \right\} \Bigg|^{A,F=\text{const}}_{\psi_\alpha = 0} . \tag{4.9.20}$$

Note, however, that the conditions (4.9.19) are not supersymmetric, since applying a supersymmetry transformation makes ψ_α non-zero. Therefore, it is worth replacing equation (4.9.19) by the supersymmetric requirement

$$\partial_a \Phi = 0 \tag{4.9.21}$$

and equation (4.9.20) by

$$-\hat{V}_{\text{eff}} = \int d^4\theta \, \mathscr{L}_{\text{eff}} + \left\{ \int d^2\theta \, \mathscr{L}^{(c)}_{\text{eff}} + \text{c.c.} \right\} \Bigg|_{\partial_a \Phi = 0} . \tag{4.9.22}$$

The final object can be called a supersymmetric effective potential.

As is seen, V_{eff} and $\hat{V}_{\text{eff}}$ coincide when the component spinor field is zero. Certainly, such coincidence can always be achieved by applying a special supersymmetry transformation which switches off the spinor. However, the supersymmetric effective potential has one essential quality, namely, it can be calculated by purely superfield methods.

By virtue of equation (4.9.22), the supersymmetric effective potential is determined by the following superfield objects

$$\mathscr{V}_{\text{eff}} \equiv \mathscr{L}_{\text{eff}} |_{\partial_a \Phi = 0} \qquad \mathscr{V}^{(c)}_{\text{eff}}(\Phi) \equiv \mathscr{L}^{(c)}_{\text{eff}} |_{\partial_a \Phi = 0} \tag{4.9.23}$$

which will be called the real and chiral parts, respectively, of the superfield effective potential. We will also refer to $\mathscr{V}^{(c)}_{\text{eff}}$ as an effective chiral superpotential. It is easy to see that the most general form of $\mathscr{V}_{\text{eff}}$ reads

$$\mathscr{V}_{\text{eff}} = K(\Phi, \bar{\Phi}) + \mathscr{F}(D_\alpha \Phi, D^2 \Phi, \bar{D}_{\dot{\alpha}} \bar{\Phi}, \bar{D}^2 \bar{\Phi}; \Phi, \bar{\Phi}) \tag{4.9.24}$$

where

$$\mathscr{F} |_{D_\alpha \Phi = \bar{D}_{\dot{\alpha}} \bar{\Phi} = 0} = 0.$$

The function $K(\Phi, \bar{\Phi})$ is said to be the effective Kählerian potential, following the terminology of supersymmetric σ-models (see Section 3.3). As for the second term in equation (4.9.24), it seems reasonable to call $\mathscr{F}$ the auxiliary field effective potential. The point is that its contribution to the effective Lagrangian, $\int d^4\theta \, \mathscr{F}$, evaluated under the supersymmetric condition (4.9.21), turns out to be at least of third order in the auxiliary fields of Φ and $\bar{\Phi}$.

Within the framework of the loop expansion, K and $\mathscr{F}$ can be represented as

$$K(\Phi, \bar{\Phi}) = K_0(\Phi, \bar{\Phi}) + \sum_{n=1}^{\infty} \hbar^n K_n(\Phi, \bar{\Phi}) \tag{4.9.25a}$$

$$\mathscr{F} = \sum_{n=1}^{\infty} \hbar^n \mathscr{F}_n. \tag{4.9.25b}$$

Here $K_0(\Phi, \bar{\Phi})$ is the classical contribution. For example, for the Wess–Zumino model we have $K_0(\Phi, \bar{\Phi}) = \bar{\Phi}\Phi$. The $\hbar$-dependent terms in equation (4.9.25) mean quantum corrections to K and $\mathscr{F}$. Further, the effective chiral superpotential reads

$$\mathscr{V}^{(c)}_{\text{eff}}(\Phi) = \mathscr{L}_c(\Phi) + \sum_{n=1}^{\infty} \hbar^n \mathscr{V}^{(c)}_n(\Phi) \tag{4.9.26}$$

where $\mathscr{L}_c(\Phi)$ is the classical chiral superpotential, and $\mathscr{V}^{(c)}_n$ are quantum corrections. In the case of the Wess–Zumino model, the one-loop correction to $\mathscr{V}^{(c)}_{\text{eff}}$ proves to be zero.

4.9.3. Superfield effective potential in the Wess–Zumino model

Let us consider the effective action of the Wess–Zumino model (3.2.11). After performing standard manipulations, one obtains

$$\begin{aligned} e^{(i/\hbar)\tilde{\Gamma}[\Phi,\bar{\Phi}]} = \int \mathscr{D}\bar{\chi}\,\mathscr{D}\chi \exp i\bigg\{ S[\chi, \bar{\chi}; \Psi, \bar{\Psi}] \\ + \hbar^{1/2}\bigg[\int d^6z \left(\frac{\lambda}{3!}\chi^3 - \frac{1}{\hbar}\chi \frac{\delta\tilde{\Gamma}[\Phi, \bar{\Phi}]}{\delta\Phi}\right) + \text{c.c.}\bigg]\bigg\} \end{aligned} \tag{4.9.27}$$

where

$$\tilde{\Gamma}[\Phi, \bar{\Phi}] = \Gamma[\Phi, \bar{\Phi}] - S[\Phi, \bar{\Phi}]$$

$\Psi = \mathscr{L}''_c(\Phi)$, and the action $S[\chi, \bar{\chi}; \Psi, \bar{\Psi}]$ is given by equation (4.4.6). The representation (4.9.27) is the basis for the calculation of the effective action in the framework of the loop expansion. In order to find the superfield effective potential, we should determine $\tilde{\Gamma}[\Phi, \bar{\Phi}]$ under the condition (4.9.21).

Use of equation (4.9.27) allows us to search for the superfield effective potential on the basis of supergraph techniques with a superpropagator of the type $\mathbf{G}^{(\Psi)}$ (4.4.7), where $\Psi = \mathscr{L}''_c(\Phi)$, Φ being subject to condition (4.9.21). According to equation (4.8.31), $\mathbf{G}^{(\Psi)}$ can be expressed through the Green's function $\mathbf{G}^{(\Psi)}_V$ obeying equation (4.8.30). Hence to develop a perturbation theory for evaluating the superfield effective potential, we should find the Green's function $\mathbf{G}^{(\Psi)}_V$ in an external chiral superfield Ψ under the condition $\partial_a\Psi = 0$. In turn $\mathbf{G}^{(\Psi)}_V$ is determined, via the proper-time representation, by the corresponding kernel $U^{(\Psi)}_V$ (4.8.37, 38).

As a result, in order to find the superfield effective potential, we should solve the system (4.8.37, 38) for $U_V^{(\Psi)}$ in an external chiral superfield Ψ under the condition $\partial_a\Psi=0$ and then construct the propagator $\mathbf{G}_V^{(\Psi)}$. The final object is to be used in the loop calculations of K and $\mathscr{F}$. The solution of the system (4.9.37, 38) under the condition $\partial_a\Psi=0$ can be found explicitly[1].

Calculation of the contributions to $\mathscr{V}_{\text{eff}}^{(c)}$ demands additional comment. Because of equation (4.9.18), we can impose condition (4.9.21) only at the final stages of transformations. However, since $\mathscr{V}_{\text{eff}}^{(c)}$ depends only on Φ for its calculation in the massless case, $\Psi=\lambda\Phi$, it is necessary to keep in the kernel $U_V^{(\Psi)}$ and the propagator $\mathbf{G}_V^{(\Psi)}$ terms depending only on Ψ, but not on $\bar\Psi$. Taking into account the relations $\bar{\mathrm{D}}_{\dot\alpha}\Psi=0$ and $\bar{\mathrm{D}}^2\bar{\mathrm{D}}^2=0$, one notices from equations (4.8.30, 36, 37) that the $\bar\Psi$-independent parts of $U_V^{(\Psi)}$ and $\mathbf{G}_V^{(\Psi)}$ are at most linear in Ψ. Then one finds

$$\mathbf{G}_V^{(\Psi)}(z,z')=-\frac{1}{\Box}\,\delta^8(z,z')-\frac{1}{4\Box}\left[\Psi(z)\frac{\bar{\mathrm{D}}^2}{\Box}\,\delta^8(z,z')\right]+O(\bar\Psi). \qquad (4.9.28)$$

This ansatz can be used for loop calculations of $\mathscr{V}_{\text{eff}}^{(c)}$.

In the rest of the present section we shall discuss the superfield effective potential in the one-loop approximation. It is worth beginning with the consideration of some general properties of the one-loop effective action. Recall that the one-loop correction to $\Gamma[\Phi,\bar\Phi]$ is given by equations (4.8.27, 35).

The starting point of our analysis is the observation that the superfunctional integral (4.8.27) proves to be invariant under rigid transformations of the dynamical superfields

$$\chi\to \mathrm{e}^{\mathrm{i}\alpha}\chi \qquad \bar\chi\to\mathrm{e}^{-\mathrm{i}\alpha}\bar\chi \qquad (4.9.29a)$$

supplemented by the following displacements of the background superfields

$$\Psi\to\mathrm{e}^{-2\mathrm{i}\alpha}\Psi \qquad \bar\Psi\to\mathrm{e}^{2\mathrm{i}\alpha}\bar\Psi \qquad (4.9.29b)$$

α being a real constant. As a consequence, the unregularized one-loop correction $\Gamma^{(1)}$, defined by equation (4.8.35) and regarded as a functional of the superfields Ψ and $\bar\Psi$, is unchanged under the transformations (4.9.29*b*). Remarkably, this invariance turns out to be unbroken after introducing the regularization (4.8.39). Let us argue this assertion.

According to equations (4.8.37, 38), we can write $U_V^{(\Psi)}$ in the form

$$U_V^{(\Psi)}(s)=\mathrm{e}^{\mathrm{i}s\Delta} \qquad (4.9.30)$$

Δ being defined in expression (4.8.30). We represent the operator Δ in the manner

$$\Delta=\Box+\mathscr{P} \qquad \mathscr{P}=-\frac{1}{4}\Psi\bar{\mathrm{D}}^2-\frac{1}{4}\bar\Psi\mathrm{D}^2$$

[1]I.L. Buchbinder, S.M. Kuzenko and J.V. Yarevskaya, *Nucl. Phys.* **B411** 665, 1994.

and expand the exponential (4.9.30) in a power series in $\mathscr{P}$. One finds

$$U_V^{(\Psi)}(s) = \sum_{n=0}^{\infty} U_n(s)$$

$$U_0(s) = \mathrm{e}^{\mathrm{i}s\Box} \tag{4.9.31}$$

$$U_n(s) = \mathrm{i}\,\mathrm{e}^{\mathrm{i}s\Box} \int_0^s \mathrm{d}t\, \mathrm{e}^{-\mathrm{i}t\Box} \mathscr{P} U_{n-1}(t) \qquad n \geqslant 1.$$

As is seen, for $n \geqslant 1$ we have

$$U_n(s) = \mathrm{i}^n \int_0^s \mathrm{d}t_n \int_0^{t_n} \mathrm{d}t_{n-1} \ldots \int_0^{t_2} \mathrm{d}t_1\, A_n(s, t_1, \ldots, t_n) \tag{4.9.32}$$

$$A_n(s, t_1, \ldots, t_n) = U_0(s - t_n) \mathscr{P} U_0(t_n - t_{n-1}) \mathscr{P} \ldots U_0(t_2 - t_1) \mathscr{P} U_0(t_1).$$

Because of the identities

$$\bar{\mathrm{D}}_{\dot{\alpha}} \Psi = 0 \qquad \bar{\mathrm{D}}^2 \bar{\mathrm{D}}^2 = 0 \qquad [\bar{\mathrm{D}}_{\dot{\alpha}}, U_0(s)] = 0$$

and owing to the explicit form of $\mathscr{P}$, the As can be expressed as follows:

$$\begin{aligned} A_{2n}(s, t_1, \ldots, t_{2n}) = \left(\frac{1}{4}\right)^{2n} & U_0(s - t_{2n}) \Psi \bar{\mathrm{D}}^2 U_0(t_{2n} - t_{2n-1}) \\ & \times \Psi \mathrm{D}^2 U_0(t_{2n-1} - t_{2n-2}) \Psi \bar{\mathrm{D}}^2 \ldots U_0(t_2 - t_1) \Psi \mathrm{D}^2 U_0(t_1) \\ & + (\Psi \bar{\mathrm{D}}^2 \leftrightarrow \Psi \mathrm{D}^2) \end{aligned} \tag{4.9.33a}$$

$$\begin{aligned} A_{2n+1}(s, t_1, \ldots, t_{2n+1}) = -\left(\frac{1}{4}\right)^{2n+1} & U_0(s - t_{2n+1}) \Psi \bar{\mathrm{D}}^2 U_0(t_{2n+1} - t_{2n}) \\ & \times \Psi \mathrm{D}^2 U_0(t_{2n} - t_{2n-1}) \Psi \bar{\mathrm{D}}^2 \ldots U_0(t_2 - t_1) \Psi \bar{\mathrm{D}}^2 U_0(t_1) \\ & + (\Psi \bar{\mathrm{D}}^2 \leftrightarrow \Psi \mathrm{D}^2). \end{aligned} \tag{4.9.33b}$$

On the same grounds and due to the cyclic property of **sTr**, it follows from equation (4.9.33*b*) that all As carrying odd labels have zero supertrace,

$$\mathbf{sTr}\{A_{2n+1}(s, t_1, \ldots, t_{2n+1})\} = 0 \qquad n = 0, 1, \ldots. \tag{4.9.34}$$

In the even case, one has

$$\begin{aligned} \mathbf{sTr}\{A_{2n}(s, t_1, \ldots, t_{2n})\} = 2\left(\frac{1}{4}\right)^{2n} & \mathbf{sTr}\{U_0(s) \Psi \bar{\mathrm{D}}^2 U_0(t_{2n} - t_{2n-1}) \\ & \times \Psi \mathrm{D}^2 U_0(t_{2n-1} - t_{2n-2}) \Psi \bar{\mathrm{D}}^2 \ldots U_0(t_2 - t_1) \Psi \mathrm{D}^2 U_0(t_1 - t_{2n})\}. \end{aligned} \tag{4.9.35}$$

Obviously, this expression is invariant under the transformations (4.9.29*b*). It is worth also noting that

$$\mathbf{sTr}\, U_0(s) = 0$$

as a consequence of the explicit form of $U_0(s)=U_V(s)$ (4.8.43). Therefore, the integrand in equation (4.8.39) is of the form

$$\mathbf{sTr}\, U_V^{(\Psi)}(s)=\sum_{n=1}^{\infty} \mathbf{sTr}\, U_{2n}(s) \tag{4.9.36}$$

and each term on the right contains an equal number of Ψ- and $\bar{\Psi}$-factors.

We observe that $\Gamma_\omega^{(1)}$ possesses invariance with respect to the transformations (4.9.29*b*). The same is true for the one-loop corrections to $K(\Phi,\bar{\Phi})$ and $\mathscr{F}$. In particular, all these objects are even functionals of the variables Ψ and $\bar{\Psi}$. Another important consequence of the above analysis is that the one-loop correction to $\mathscr{V}_{\text{eff}}^{(c)}$ is zero.

4.9.4. *Calculation of the one-loop Kählerian effective potential*

In conclusion we would like to illustrate the general approach which has been described above, by evaluating one-loop Kählerian effective potential in the Wess–Zumino model.

As follows from the previous discussion, one must compute explicitly the kernel

$$U_V^{(\Psi)}(z,z'|s)=\mathrm{e}^{\mathrm{i}s\Delta}\,\delta^8(z,z') \tag{4.9.37}$$

the external chiral scalar Ψ being subject to the condition

$$\partial_a\Psi=0 \tag{4.9.38}$$

in order to find the one-loop correction to $\mathscr{V}_{\text{eff}}$. Then, the one-loop correction reads

$$\mathscr{V}_{\text{eff},\omega}^{\text{one-loop}}=-\frac{\mathrm{i}}{2}\mu^{2\omega}\int_0^\infty \frac{\mathrm{d}(\mathrm{i}s)}{(\mathrm{i}s)^{1-\omega}}\,U_V^{(\Psi)}(z,z|s) \tag{4.9.39}$$

where we have used the regularization prescription (4.8.39). On the other hand, if we are interested only in finding the one-loop correction to $K(\Phi,\bar{\Phi})$, we can replace the condition (4.9.38) by the stronger one

$$\Psi=\text{const} \tag{4.9.40}$$

and calculate the corresponding kernel (4.9.37). Then, the regularized correction $K_1(\Phi,\bar{\Phi})_\omega$ reads

$$K_1(\Phi,\bar{\Phi})_\omega=-\frac{\mathrm{i}}{2}\mu^{2\omega}\int_0^\infty \frac{\mathrm{d}(\mathrm{i}s)}{(\mathrm{i}s)^{1-\omega}}\,U_V^{(\Psi=\text{const})}(z,z|s). \tag{4.9.41}$$

Let us find the kernel $U_V^{(\Psi)}$ corresponding to the choice (4.9.40). We can now write

$$U_V^{(\Psi=\text{const})}(z,z'|s)=\mathrm{e}^{-(\mathrm{i}s/4)(\Psi\bar{D}^2+\bar{\Psi}D^2)}\,U_V(z,z'|s) \tag{4.9.42a}$$

where

$$U_V(z, z'|s) = e^{is\Box}\,\delta^8(z - z') = -\frac{i}{(4\pi s)^2}(\theta - \theta')^2(\bar\theta - \bar\theta')^2\, e^{(i/4s)(x - x')^2} \tag{4.9.42b}$$

and make use of the identities

$$\left(\frac{1}{16}\bar{D}^2 D^2\right)^n = \Box^{n-1}\left(\frac{1}{16}\bar{D}^2 D^2\right)$$

$$\left(\frac{1}{16}D^2\bar{D}^2\right)^n = \Box^{n-1}\left(\frac{1}{16}D^2\bar{D}^2\right) \qquad n = 1, 2, \ldots.$$

This leads to

$$U_V^{(\Psi=\mathrm{const})}(z, z'|s) = \left[1 + \frac{1}{16\Box}(\cosh(is\sqrt{\bar\Psi\Psi\Box}) - 1)\{D^2, \bar{D}^2\} + \frac{1}{4\sqrt{\bar\Psi\Psi\Box}}\sinh(is\sqrt{\bar\Psi\Psi\Box})(\Psi\bar{D}^2 + \bar\Psi D^2)\right]U_V(s). \tag{4.9.43}$$

The correction (4.9.41) is determined by the kernel at coincident points. From equations (4.9.42*b*) and (4.9.43) one immediately obtains

$$U_V^{(\Psi=\mathrm{const})}(z, z'|s) = \frac{2}{\Box}(\cosh(is\sqrt{\bar\Psi\Psi\Box}) - 1]U(x, x'|s)\Big|_{x=x'} \tag{4.9.44}$$

where

$$U(x, x'|s) = e^{is\Box}\,\delta^4(x - x') = -\frac{i}{(4\pi s)^2}e^{(i/4s)(x - x')^2}.$$

To compute the right-hand side of expression (4.9.44), we need to evaluate expressions of the form $\Box^n U(x, x|s)|_{x=x'}$. This can easily be done with the help of the equation

$$i\frac{\partial}{\partial s}U = -\Box U$$

which U satisfies, thus obtaining

$$\Box^n U(x, x'|s)\Big|_{x=x'} = \frac{\partial^n}{\partial(is)^n}U(x, x'|s)\Big| = \frac{i}{(4\pi)^2}(-1)^n\frac{(n+1)!}{(is)^{n+2}}.$$

As a result, the one-loop correction (4.9.41) takes the form

$$K_1(\Phi, \bar\Phi)_\omega = -\frac{\mu^{2\omega}}{(4\pi)^2}\int_0^\infty \frac{d(is)}{(is)^{2-\omega}}\sum_{n=1}^{\infty}\frac{(-1)^n n!(is\bar\Psi\Psi)^n}{(2n)!}$$

or, after introducing the new integration variable $\tau = \mathrm{i}s\bar{\Psi}\Psi$:

$$K_1(\Phi, \bar{\Phi})_\omega = -\frac{\mu^{2\omega}}{(4\pi)^2}(\bar{\Psi}\Psi)^{1-\omega}\int_0^\infty \frac{\mathrm{d}\tau}{\tau^{2-\omega}}\sum_{n=1}^{\infty}\frac{(-1)^n n!\tau^n}{(2n)!}. \qquad (4.9.45)$$

Let us use the well-known series representation for the so-called erfi-function

$$-\frac{\sqrt{\pi}}{2}x\mathrm{e}^{-x^2/4}\mathrm{erfi}(x/2) = \sum_{n=1}^{\infty}\frac{(-1)^n n! x^{2n}}{(2n)!}$$

where

$$\mathrm{erfi}(x) = \frac{2}{\sqrt{\pi}}\int_0^x \mathrm{e}^{t^2}\,\mathrm{d}t.$$

Then we obtain

$$K_1(\Phi, \bar{\Phi})_\omega = \frac{\mu^{2\omega}(\bar{\Psi}\Psi)^{1-\omega}}{2(4\pi)^2}\int_0^\infty \frac{\mathrm{d}\tau}{\tau^{1-\omega}}\int_0^1 \mathrm{d}t\,\mathrm{e}^{-\tau/4(1-t^2)}. \qquad (4.9.46)$$

Rewriting the integrals in the final expression as

$$\int_0^1 \frac{\mathrm{d}\tau}{\tau^{1-\omega}} + \int_0^1 \frac{\mathrm{d}\tau}{\tau^{1-\omega}}\left(\int_0^1 \mathrm{d}t\,\mathrm{e}^{-\tau/4(1-t^2)} - 1\right)$$
$$+ \int_1^\infty \frac{\mathrm{d}\tau}{\tau^{1-\omega}}\int_0^1 \mathrm{d}t\,\mathrm{e}^{-\tau/4(1-t^2)} \equiv \int_0^1 \frac{\mathrm{d}\tau}{\tau^{1-\omega}} + \zeta_\omega$$

we obtain

$$K_1(\Phi, \bar{\Phi})_\omega = \frac{\mu^{2\omega}(\bar{\Psi}\Psi)^{1-\omega}}{32\pi^2\omega} + \frac{\mu^{2\omega}(\bar{\Psi}\Psi)^{1-\omega}}{32\pi^2}\zeta_\omega$$

It is easy to see that $\zeta_\omega = \zeta + O(\omega)$, where

$$\zeta = \lim_{\omega\to 0}\zeta_\omega = \int_0^1 \frac{\mathrm{d}\tau}{\tau}\left(\int_0^1 \mathrm{d}t\,\mathrm{e}^{-\tau/4(1-t^2)} - 1\right) + \int_1^\infty \frac{\mathrm{d}\tau}{\tau}\int_0^1 \mathrm{d}t\,\mathrm{e}^{-\tau/4(1-t^2)}$$

is some finite constant, as follows from the asymptotics

$$\int_0^1 \mathrm{d}t\,\mathrm{e}^{-\tau/4(1-t^2)} \xrightarrow{\tau\to\infty} \frac{2}{\tau}.$$

Now, setting

$$\frac{\mu^{2\omega}(\bar{\Psi}\Psi)^{-\omega}}{\omega} = \frac{1}{\omega} - \ln\frac{\bar{\Psi}\Psi}{\mu^2} + O(\omega)$$

we obtain

$$K_1(\Phi, \bar{\Phi})_\omega = K_1(\Phi, \bar{\Phi})_{\mathrm{div}} + K_1(\Phi, \Phi)$$

where

$$K_1(\Phi, \bar{\Phi})_{\text{div}} = \frac{\Psi\Psi}{32\pi^2\omega} \tag{4.9.47}$$

$$K_1(\Phi, \bar{\Phi}) = -\frac{\Psi\Psi}{32\pi^2}\left(\ln\frac{\Psi\Psi}{\mu^2} - \zeta\right). \tag{4.9.48}$$

Here equation (4.9.47) defines the divergent part of the superfield effective action. This divergence has been calculated in subsection 4.8.2. The term $K_{1\text{div}}$ is cancelled by the counterterm (4.8.46). The relation (4.9.48) represents the one-loop correction to $K(\Phi, \bar{\Phi})$.

Thus the one-loop Kählerian potential has the form

$$K(\Phi, \bar{\Phi}) = \bar{\Phi}\Phi - \frac{\hbar}{32\pi^2}|\mathscr{L}''_c(\Phi)|^2\left(\ln\frac{|\mathscr{L}''_c(\Phi)|^2}{\mu^2} - \zeta\right). \tag{4.9.49}$$

To complete our work, we should fix μ by imposing some suitable renormalization condition. Consider, for simplicity, the massless Wess–Zumino model with $\mathscr{L}_c(\Phi) = (\lambda/3!)\Phi^3$ and choose the following renormaliation condition:

$$\frac{\partial^2 K(\Phi, \bar{\Phi})}{\partial\Phi\partial\bar{\Phi}}\Big|_{\Phi = \Phi_0} = 1 \tag{4.9.50}$$

where Φ_0 is a non-zero constant with dimensions of mass. Then one obtains

$$K(\Phi, \bar{\Phi}) = \bar{\Phi}\Phi - \frac{\hbar\lambda^2}{32\pi^2}\bar{\Phi}\Phi\left[\ln\frac{\bar{\Phi}\Phi}{\bar{\Phi}_0\Phi_0} - 2\right]. \tag{4.9.51}$$

The one-loop superfield effective potential $\mathscr{V}_{\text{eff}}$ includes not only the Kählerian part K, but also the auxiliary field effective potential $\mathscr{F}$. Calculation of the latter is a more tedious task, and we will not discuss it here, referring the reader to the paper by I. Buchbinder, S. Kuzenko and J. Yarevskaya.

4.9.5. Calculation of the two-loop effective chiral superpotential

As we know the one-loop contribution to the effective chiral superpotential is absent in the Wess–Zumino model (see for instance equation (4.9.35) at $\bar{\Phi} = 0$). Therefore one can expect that the quantum corrections to the chiral superpotential will arise starting at least with the two-loop approximation. A possibility of chiral quantum corrections to the effective potential was first noticed in the book by M.T. Grisaru, S.J. Gates, M. Roček and W. Siegel, and the relevant supergraphs producing such corrections were found by P. West[2].

[2]P. West, *Phys. Lett.* **258 B** 375, 1991.

For the first time the explicit two-loop correction to the chiral effective superpotential was found by I. Jack, D. R. T. Jones and P. West[3] in component form. We give here a completely superfield approach[4].

We start with the general equation (4.9.27) for the massless Wess–Zumino model. To find the chiral effective superpotential it is sufficient to set $\bar{\Psi} = 0$. In this case the two-loop contribution to effective action depending only on Ψ is of the form

$$\Gamma^{(2)} = \frac{\lambda^2}{12}\int d^6\bar{z}_1\, d^6\bar{z}_2 (\mathbf{G}_{--}(z_1, z_2))^3. \tag{4.9.52}$$

The Green function $\mathbf{G}_{--}(z_1, z_2)$ can be found from equation (4.8.31), where $G_V^{(\Psi)}$ is given by equation (4.9.28). Then

$$\mathbf{G}_{--}(z_1, z_2) = \frac{D_1^2 D_2^2}{16\Box_1}\left[\Psi(z_1)\frac{\bar{D}_1^2}{4\Box_1}\delta^8(z_1 - z_2)\right] \tag{4.9.53}$$

where $\Psi = -\lambda\Phi$. Converting here the integrals over $d^6\bar{z}$ into integrals over d^8z, in accordance with the rule $\int d^6\bar{z}(-\frac{1}{4}D^2) = \int d^8z$, and setting

$$\Phi(z_1)\frac{\bar{D}_1^2}{\Box_1}\delta^8(z_1 - z_2) = \int d^8z_3 \Phi(z_3)\delta^8(z_1 - z_2)\frac{\bar{D}_3^2}{\Box_2}\delta^8(z_1 - z_2)$$

we obtain

$$\begin{aligned}\Gamma^{(2)} = -\frac{\lambda^5}{12}\int &\left(\prod_{i=1}^{5} d^8z_i\right)\Phi(z_3)\Phi(z_4)\Phi(z_5)\left(\frac{1}{\Box_1}\delta^8(z_1 - z_2)\right)\\ &\times\left(\frac{D_2^2\bar{D}_3^2}{16\Box_2}\delta^8(z_1 - z_2)\right)\left(\frac{1}{\Box_2}\delta^8(z_1 - z_2)\right)\\ &\times\left(\frac{D_1^2\bar{D}_4^2}{16\Box_1}\delta^8(z_1 - z_2)\right)\left(\frac{D_1^2\bar{D}_5^2}{16\Box_1}\delta^8(z_1 - z_2)\right)\left(\frac{D_2^2}{4\Box_2}\delta^8(z_1 - z_2)\right).\end{aligned} \tag{4.9.54}$$

From the viewpoint of the supergraph technique, this expression corresponds to the supergraph given below, the external lines being chiral.

As the next step, we rewrite $\Gamma^{(2)}$ in a form that is suitable for extracting the correction to $\mathscr{V}^{(c)}_{\text{eff}}(\Phi)$ contained within it. Note first, that it is convenient to work in the chiral representation where

$$\Phi(z) = \Phi(x, \theta) \qquad \bar{D}_{\dot{\alpha}} = -\frac{\partial}{\partial\bar{\theta}^{\dot{\alpha}}}$$

since equation (4.9.54) involves only chiral superfields. Further, we apply standard D-algebra transformations (see subsection 4.7.1) to $\Gamma^{(2)}$ in order to reduce the number of integrals over $d^4\theta$ to a single one. Making use also of

[3]I. Jack, D.R.T. Jones and P. West, *Phys. Lett.* **258B** 382, 1991

[4]I.L. Buchbinder, S.M. Kuzenko and A.Yu. Petrov. *Phys. Lett.* **321B** 372, 1994.

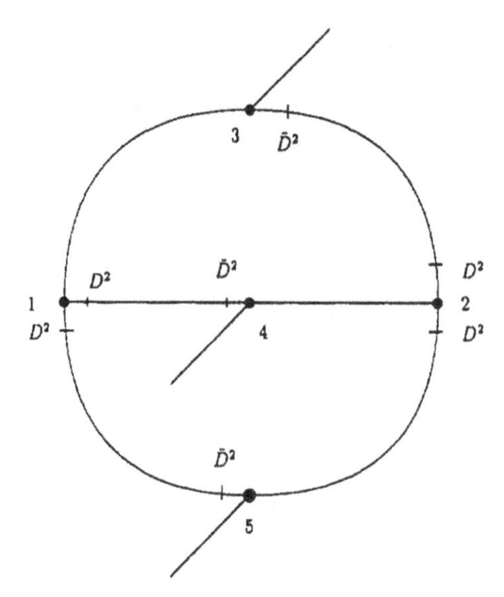

the obvious identities

$$\delta^4(x - x') = \int \frac{d^4k}{(2\pi)^4} e^{ik(x - x')} \qquad \Phi(x, \theta) = \int d^4y\, \delta^4(x - y)\Phi(y, \theta)$$

we arrive at

$$\begin{aligned}\Gamma^{(2)} = &-\frac{\lambda^5}{12}\int d^4x\, d^4\theta \int \frac{d^4k_1\, d^4k_2\, d^4p_1\, d^4p_2}{(2\pi)^{16}} \\ &\times \int d^4y_1\, d^4y_2 e^{ip_1(x - y_1) + ip_2(x - y_2)}\Phi(x, \theta) \\ &\times \left[k_1^2 p_1^2 \left(-\frac{D^2}{4\Box}\Phi(y_1, \theta)\right)\Phi(y_2, \theta) + (1 \leftrightarrow 2)\right. \\ &\left. + \frac{(k_1k_2)}{2}(D^\alpha\Phi(y_1, \theta))D_\alpha\Phi(y_2, \theta)\right]\Omega^{-1}(k, p)\end{aligned} \tag{4.9.55}$$

where

$$\Omega(k, p) = k_1^2k_2^2(k_1 + k_2)^2(k_1 + k_2 - p_1 - p_2)^2(k_1 - p_1)^2(k_2 - p_2)^2. \tag{4.9.56}$$

Finally, we integrate $\Gamma^{(2)}$ over $d^2\bar{\theta}$ in accordance with the rule $\int d^2\bar{\theta} = -\frac{1}{4}\bar{D}^2$. This leads to

$$\begin{aligned}\Gamma^{(2)} = &\frac{\lambda^5}{12}\int d^4x\, d^2\theta \int \frac{d^4p_1\, d^4p_2}{(2\pi)^8} \\ &\times \int d^4y_1\, d^4y_2 e^{ip_1(x - y_1) + ip_2(x - y_2)}\Phi(x, \theta)\Phi(y_1, \theta)\Phi(y_2, \theta)J(p_1, p_2)\end{aligned} \tag{4.9.57}$$

where

$$J(p_1, p_2) = \int \frac{d^4k_1\, d^4k_2}{(2\pi)^8} \frac{k_1^2p_1^2 + k_2^2p_2^2 - 2(k_1k_2)(p_1p_2)}{\Omega(k, p)}. \tag{4.9.58}$$

Up to now, the chiral suprfield $\Phi(x, \theta)$ was completely arbitrary. When computing the effective superpotential, we can assume Φ to be slowly varying in space–time. This assumption implies

$$\Phi(x, \theta)\Phi(y_1, \theta)\Phi(y_2, \theta) \approx \Phi^3(x, \theta). \tag{4.9.59}$$

Then equation (4.9.57) takes the form

$$\Gamma^{(2)} = \frac{\lambda^5}{12} J(p_1 \to 0, p_2 \to 0) \int d^6z\Phi^3. \tag{4.9.60}$$

It remains to use the known integral relation

$$J(p, 0) = \frac{6}{(4\pi)^4}\zeta(3) \tag{4.9.61}$$

$\zeta(\alpha)$ being the dzeta-function. As a result, the two-loop chiral effective superpotential read

$$\mathscr{V}^{(c)}_{\text{eff}}(\Phi) = \mathscr{L}_c + \hbar^2\mathscr{V}^{(c)}_{(2)} = \left(\frac{\lambda}{3!} + \frac{\lambda^5\zeta(3)\hbar^2}{2(4\pi)^4}\right)\Phi^3. \tag{4.9.62}$$

We note that the chiral effective superpotential is automatically finite. No renormalization is required for its calculation.

5 Superspace Geometry of Supergravity

To make a prairie it takes a clover and one bee,
One clover, and a bee,
And revery.
The revery alone will do,
If bees are few.

Emily Dickinson

5.1. Gauge group of supergravity and supergravity superfields

5.1.1. Curved superspace

In previous chapters, the supersymmetric generalization of Minkowski space as well as that of matter field theories in Minkowski space was given. Our next goal is to look for a supersymmetric generalization of Einstein gravity — for a supergravity. Ordinary gravity is treated as the theory of a curved space–time. Therefore it is tempting to think of supergravity as being theory of a curved superspace. But what is a curved superspace?

A space–time is said to be curved if the curvature tensor $\mathcal{R}^m{}_{npr}$, constructed from the metric $g_{mn}(x)$, is non-vanishing. It is a flat space–time when $\mathcal{R}^m{}_{npr} = 0$. This condition proves to be equivalent to the statement that there exist preferable coordinate systems in which the metric has the Minkowskian form,

$$g_{mn}(x) = \eta_{mn}.$$

Any two such coordinate systems are related by some Poincaré transformation. It is the Poincaré transformations which leave invariant the Minkowski metric.

DOI: 10.1201/9780367802530-5

In the case of a curved space–time, there is no natural way to choose preferable coordinate systems; all coordinate systems are on the same footing. Hence, a physical theory should be invariant under the group of general coordinate transformations

$$x^m \to x'^m = f^m(x)$$

or, in infinitesimal form,

$$x^m \to x'^m = x^m - K^m(x) \tag{5.1.1}$$

where $K^m(x)$ is an arbitrary infinitesimal vector field. The general coordinate transformation group is the gravity gauge group, and the metric, with the transformation law

$$\delta g_{mn}(x) = \nabla_m K_n + \nabla_n K_m \tag{5.1.2}$$

plays the role of the gravity gauge field.

Now, keeping in mind our space–time intuition, we return to superspace. Clearly, in order to define a notion of curved superspace, one should find a superfield 'metric' object. It was V. Ogievetsky and E. Sokatchev who found such an object.

In Chapter 2, subsection 2.5.3, we introduced the family of real surfaces $\mathbb{R}^{4|4}(\mathscr{H})$ in complex superspace $\mathbb{C}^{4|2}$ which were defined by equation (2.5.11). Every such surface can be understood as a real superspace $\mathbb{R}^{4|4}$ supplied by a real vector superfield $\mathscr{H}^m(x, \theta, \bar{\theta})$. Recall that the family $\{\mathbb{R}^{4|4}(\mathscr{H})\}$ contains a unique super-Poincaré invariant surface — $\mathbb{R}^{4|4}(\theta\sigma\bar{\theta})$, and that this surface was identified with flat global superspace $\mathbb{R}^{4|4}$. Obviously, we are faced with a situation analogous to the space–time one. We can look at space–time as a real space $\mathbb{R}^4$ provided with a second-rank tensor $g_{mn}(x)$. The family of all space–times $\{\mathbb{R}^4(g_{mn})\}$ contains a unique representative—flat global space–time $\mathbb{R}^4(\eta_{mn})$, which is characterized by the Poincaré invariant metric. Therefore, the family of superspaces $\{\mathbb{R}^{4|4}(\mathscr{H}^m)\}$ is a supersymmetric generalization of the family of space–times $\{\mathbb{R}^4(g_{mn})\}$.

Starting with a given complex coordinate system (y^m, θ^α) on $\mathbb{C}^{4|2}$, let us consider holomorphic coordinate transformations

$$y^m \to y'^m = f^m(y, \theta) \qquad \theta^\alpha \to \theta'^\alpha = f^\alpha(y, \theta)$$
$$\text{Ber}\left(\frac{\partial(y', \theta')}{\partial(y, \theta)}\right) \neq 0. \tag{5.1.3}$$

Infinitesimal holomorphic coordinate transformations are of the form

$$\begin{aligned} y^m &\to y'^m = y^m - \lambda^m(y, \theta) \\ \theta^\alpha &\to \theta'^\alpha = \theta^\alpha - \lambda^\alpha(y, \theta). \end{aligned} \tag{5.1.4}$$

Here λ^m and λ^α are infinitesimal vector and spinor holomorphic superfields.

The set of all holomorphic transformations (5.1.3) forms a supergroup (the 'holomorphic coordinate transformation supergroup'). Commuting two infinitesimal holomorphic transformations, generated by parameters $\lambda_1^m(y,\theta)$, $\lambda_1^\alpha(y,\theta)$ and $\lambda_2^m(y,\theta)$, $\lambda_2^\alpha(y,\theta)$, respectively, one obtains a transformation of the same type. The corresponding parameters are

$$\begin{aligned}\lambda_3^m(y,\theta) &= \lambda_2^n\partial\lambda_1^m/\partial y^n + \lambda_2^\beta\partial\lambda_1^m/\partial\theta^\beta - (1\leftrightarrow 2)\\ \lambda_3^\alpha(y,\theta) &= \lambda_2^n\partial\lambda_1^\alpha/\partial y^n + \lambda_2^\beta\partial\lambda_1^\alpha/\partial\theta^\beta - (1\leftrightarrow 2).\end{aligned} \tag{5.1.5}$$

Given a superspace $\mathbb{R}^{4|4}(\mathscr{H})$, every holomorphic transformation (5.1.4) induces a coordinate displacement on $\mathbb{R}^{4|4}(\mathscr{H})$ as well as changing the superfield $\mathscr{H}^m$ in a rather complicated nonlinear way. Explicitly, in accordance with equations (2.5.11) and (2.5.12), we have

$$\begin{aligned}x^m \to x'^m &= x^m - \frac{1}{2}\lambda^m(x^n + \mathrm{i}\mathscr{H}^n, \theta^\beta) - \frac{1}{2}\bar{\lambda}^m(x^n - \mathrm{i}\mathscr{H}^n, \bar{\theta}^{\dot\beta})\\ \theta^\alpha \to \theta'^\alpha &= \theta^\alpha - \lambda^\alpha(x^n + \mathrm{i}\mathscr{H}^n, \theta^\beta)\\ \bar{\theta}^{\dot\alpha} \to \bar{\theta}'^{\dot\alpha} &= \bar{\theta}^{\dot\alpha} - \bar{\lambda}^{\dot\alpha}(x^n - \mathrm{i}\mathscr{H}^n, \bar{\theta}^{\dot\beta})\end{aligned} \tag{5.1.6}$$

and

$$\mathscr{H}'^m(x',\theta',\bar{\theta}') = \mathscr{H}^m(x,\theta,\bar{\theta}) + \frac{\mathrm{i}}{2}\lambda^m(x^n + \mathrm{i}\mathscr{H}^n, \theta^\beta) - \frac{\mathrm{i}}{2}\bar{\lambda}^m(x^n - \mathrm{i}\mathscr{H}^n, \bar{\theta}^{\dot\beta}). \tag{5.1.7}$$

Introducing the superfield displacement

$$\delta\mathscr{H}^m(x,\theta,\bar{\theta}) = \mathscr{H}'^m(x,\theta,\bar{\theta}) - \mathscr{H}^m(x,\theta,\bar{\theta})$$

equation (5.1.7) leads to

$$\begin{aligned}\delta\mathscr{H}^m(x,\theta,\bar{\theta}) &= \frac{\mathrm{i}}{2}(\lambda^m - \bar{\lambda}^m) + \left(\frac{1}{2}(\lambda^n + \bar{\lambda}^n)\partial_n + \lambda^\alpha\partial_\alpha + \bar{\lambda}^{\dot\alpha}\bar{\partial}_{\dot\alpha}\right)\mathscr{H}^m(x,\theta,\bar{\theta})\\ \lambda^m &= \lambda^m(x^n + \mathrm{i}\mathscr{H}^n, \theta^\beta) \qquad \lambda^\alpha = \lambda^\alpha(x^n + \mathrm{i}\mathscr{H}^n, \theta^\beta).\end{aligned} \tag{5.1.8}$$

This transformation law defines a nonlinear superfield representation of the holomorphic coordinate transformation supergroup. It is natural to treat the holomorphic coordinate transformation supergroup as a gauge group and $\mathscr{H}^m(x,\theta,\bar{\theta})$ as the corresponding gauge superfield.

Geometrically, the superfields $\mathscr{H}^m$ and $\mathscr{H}^m + \delta\mathscr{H}^m$ are equivalent. They describe the same surface in $\mathbb{C}^{4|2}$ but written in two different complex coordinate systems on $\mathbb{C}^{4|2}$. Therefore, the space of all gauge superfields is decomposed into orbits of the holomorphic coordinate transformation supergroup. Gauge superfields from the same orbit determine equivalent geometries on $\mathbb{R}^{4|4}$. Gauge superfields from different orbits determine non-equivalent geometries on $\mathbb{R}^{4|4}$.

In accordance with the above consideration, we can make the following preliminary definitions. A superspace $\mathbb{R}^{4|4}(\mathscr{H})$ is said to be 'flat' if there exists a holomorphic coordinate transformation (5.1.3) which deforms $\mathscr{H}^m$ to the form

$$\mathscr{H}'^m(x', \theta', \bar{\theta}') = \delta_a{}^m \theta' \sigma^a \bar{\theta}'. \tag{5.1.9}$$

Otherwise, $\mathbb{R}^{4|4}(\mathscr{H})$ is said to be a 'curved superspace'. It is the superfield $\mathscr{H}^m(x, \theta, \bar{\theta})$ which plays the role of a superfield 'metric' object or 'gravitational superfield'. The transformation law (5.1.8) is the superspace analogue of equation (5.1.2).

At first sight, the holomorphic coordinate transformation supergroup of $\mathbb{C}^{4|2}$ is a candidate for the role of the superspace analogue of the general coordination transformation group of space–time. However, this is not quite correct. The point is that this supergroup includes only non-localizable transformations (a transformation is said to be localized near some point if it coincides with the identity mapping everywhere except for in a small region around this point). Indeed, whan taking the superfield parameters $\lambda^m(y, \theta)$ and $\lambda^\alpha(y, \theta)$ in equation (5.1.4) to be holomorphic on the whole superspace $\mathbb{C}^{4|2}$, one obtains non-localizable (in y^m) coordinate transformations only. Recall that one cannot make a holomorphic superfield on $\mathbb{C}^{4|2}$ non-vanishing in a small neighbourhood of some point y_0^m (see subsection 2.5.2). On the other hand, the general coordinate transformation group of space–time includes arbitrary, in particular localized, transformations. To resolve this difficulty, one must consider some 'extension' of the holomorphic coordinate transformation supergroup. Recall, flat global superspace $\mathbb{R}^{4|4}(\theta\sigma\bar{\theta})$ is embedded in the truncated superspace $\mathbb{C}_s^{4|2}$ defined by equation (2.5.9). The space $\mathbb{C}_s^{4|2}$ is mapped onto itself by every super-Poincaré transformation. Finally, the requirement of holomorphicity on $\mathbb{C}_s^{4|2}$ does not give any global restrictions on superfields. In particular, there are localized holomorphic superfields on $\mathbb{C}_s^{4|2}$.

Therefore, we can proceed as follows. We restrict our consideration to the case of superspaces $\mathbb{R}^{4|4}(\mathscr{H})$ embedded in $\mathbb{C}_s^{4|2}$ (by analogy with the flat global superspace). This means that $\mathscr{H}^m(x, \theta, \bar{\theta})$ has vanishing body,

$$(\mathscr{H}^m(x, \theta, \bar{\theta}))_B = 0. \tag{5.1.10}$$

Further we will consider only holomorphic coordinate transformations on $\mathbb{C}_s^{4|2}$. This restricts the superfield parameters in equation (5.1.4) in the manner

$$(\lambda^m(y, \theta) - \bar{\lambda}^m(\bar{y}, \bar{\theta}))_B = 0. \tag{5.1.11}$$

The requirements (5.1.10, 11) make subsequent results physically correct.

Let $\mathbb{R}^{4|4}(\mathscr{H})$ be a flat superspace. We choose a preferable coordinate system on $\mathbb{R}^{4|4}$ in which

$$\mathscr{H}^m = \delta_a{}^m \theta \sigma^a \bar{\theta}. \tag{5.1.12}$$

There exists a subset of transformations (5.1.4) preserving this superfield,

$$\delta\mathscr{H}^m = 0.$$

Many of them are given by the superfunctions

$$\begin{gathered}-\lambda^a(y,\theta) = b^a + 2i\theta\sigma^a\bar{\epsilon} + K^a{}_b y^b \\ -\lambda^\alpha(y,\theta) = \epsilon^\alpha - K^\alpha{}_\beta\theta^\beta \\ b^a = \bar{b}^a \qquad K^{ab} = (\sigma^{ab})_{\alpha\beta}K^{\alpha\beta} - (\tilde{\sigma}^{ab})_{\dot{\alpha}\dot{\beta}}\bar{K}^{\dot{\alpha}\dot{\beta}} \qquad K_{\alpha\beta} = K_{\beta\alpha}.\end{gathered} \tag{5.1.13}$$

Here b^a, $K^{\alpha\beta}$, ϵ^α are constant parameters. Obviously, transformations (5.1.4), induced by the given superfields, represent infinitesimal super-Poincaré transformations of $\mathbb{C}^{4|2}$ (see subsection 2.5.1). Therefore, every super-Poincaré transformation leaves invariant the flat gravitational superfield (5.1.12).

5.1.2. *Conformal supergravity*

Above we have interpreted $\mathscr{H}^m(x,\theta,\bar{\theta})$ as a gravitational superfield. In order for this interpretation to be non-contradictory, $\mathscr{H}^m$ should describe a multiplet of ordinary fields containing a gravitational field. Hence, it is necessary to investigate the component structure of $\mathscr{H}^m$ and to analyse the action of gauge transformations (5.1.8) on the component fields.

Decomposition of $\mathscr{H}^m$ in a power series in θ and $\bar{\theta}$ leads to

$$\begin{aligned}\mathscr{H}^m(x,\theta,\bar{\theta}) = {} & C^m(x) + i\theta^\alpha\chi^m{}_\alpha(x) - i\bar{\theta}_{\dot{\alpha}}\bar{\chi}^{m\dot{\alpha}}(x) \\ & + i\theta^2 S^m(x) - i\bar{\theta}^2\bar{S}^m(x) + \theta\sigma^a\bar{\theta}e_a{}^m(x) \\ & + i\bar{\theta}^2\theta^\alpha\Psi^m{}_\alpha(x) - i\theta^2\bar{\theta}_{\dot{\alpha}}\bar{\Psi}^{m\dot{\alpha}}(x) + \theta^2\bar{\theta}^2A^m(x).\end{aligned} \tag{5.1.14}$$

Constraint (5.1.10) is equivalent to

$$(C^m(x))_B = 0.$$

Among the component fields, there is a natural candidate for the role of the gravitational field. The $\theta\sigma^a\bar{\theta}$-coefficient $e_a{}^m(x)$ may be identified with the vierbein. Before clarifying this, we show that the component fields C^m, $\chi^m{}_\alpha$, $\bar{\chi}^{m\dot{\alpha}}$, S^m and $\bar{S}^m$ are purely gauge degrees of freedom.

The transformation law (5.1.8) can be rewritten in the form

$$\delta\mathscr{H}^m(x,\theta,\bar{\theta}) = \frac{i}{2}\lambda^m(x,\theta) - \frac{i}{2}\bar{\lambda}^m(x,\bar{\theta}) + O(\mathscr{H}) \tag{5.1.15}$$

where $O(\mathscr{H})$ denotes all terms depending on $\mathscr{H}$ and its derivatives. Further, decomposition of the superfield parameters in expression (5.1.4) in a power series in θ gives

$$\begin{aligned}\lambda^m(y,\theta) &= a^m(y) + \theta^\alpha\varphi^m{}_\alpha(y) + \theta^2 s^m(y) \\ \lambda^\alpha(y,\theta) &= \epsilon^\alpha(y) + \omega^\alpha{}_\beta(y)\theta^\beta + \theta^2\eta^\alpha(y).\end{aligned} \tag{5.1.16}$$

Constraint (5.1.11) is equivalent to

$$(a^m(y) - \bar{a}^m(y))_B = 0.$$

The remaining component fields in expression (5.1.16) are arbitrary. Then, equation (5.1.15) takes the form

$$\delta\mathscr{H}^m(x, \theta, \bar{\theta}) = \frac{i}{2}(a^m(x) - \bar{a}^m(x)) + \frac{i}{2}\theta^\alpha \varphi^m{}_\alpha(x) - \frac{i}{2}\bar{\theta}_{\dot{\alpha}}\bar{\varphi}^{m\dot{\alpha}}(x) + \frac{i}{2}\theta^2 s^m(x)$$

$$- \frac{i}{2}\bar{\theta}^2 \bar{s}^m(x) + O(\mathscr{H}).$$

We see that component fields C^m, $\chi^m{}_\alpha$ and S^m of $\mathscr{H}^m$ have arbitrary variations. Using gauge freedom (5.1.8), one can make these fields have any given values. In particular, there exists a gauge in which these fields are gauged out ('Wess–Zumino gauge'),

$$\mathscr{H}^m(x, \theta, \bar{\theta}) = \theta\sigma^a\bar{\theta}e_a{}^m(x) + i\bar{\theta}^2\theta^\alpha\Psi^m{}_\alpha(x) - i\theta^2\bar{\theta}_{\dot{\alpha}}\bar{\Psi}^{m\dot{\alpha}}(x) + \theta^2\bar{\theta}^2 A^m(x). \quad (5.1.17)$$

In the Wess–Zumino gauge, the transformation law (5.1.8) becomes polynomial.

Imposing the Wess–Zumino gauge does not fix the gauge freedom completely. There is a subset of transformations (5.1.4) preserving the Wess–Zumino gauge. Every such transformation is characterized by (field dependent) parameters:

$$\lambda^m(x, \theta) = b^m(x) + 2i\theta\sigma^a\bar{\epsilon}(x)e_a{}^m(x) - 2\theta^2\bar{\epsilon}(x)\bar{\Psi}^m(x)$$

$$\lambda^\alpha(x, \theta) = \epsilon^\alpha(x) + \frac{1}{2}(\sigma(x) + i\Omega(x))\theta^\alpha + K^\alpha{}_\beta(x)\theta^\beta + \theta^2\eta^\alpha(x) \quad (5.1.18)$$

$$K_{\alpha\beta} = K_{\beta\alpha}.$$

Here $b^m(x)$ and $\sigma(x)$, $\Omega(x)$ are arbitrary real vector and scalar functions, respectively, $\epsilon^\alpha(x)$ and $\eta^\alpha(x)$ are arbitrary spinor functions. Comparing expressions (5.1.13) and (5.1.18), we see that parameters $b^m(x)$ induce general coordinate transformations of a space–time, parameters $K^\alpha{}_\beta(x)$—local Lorentz transformations and parameters $\epsilon^\alpha(x)$—local supersymmetry transformations. Now we proceed to obtain the transformation laws of fields $e_a{}^m$, $\Psi^m{}_\alpha$, A^m.

We start with the general coordinate transformations which correspond to the choice

$$\lambda^m(x, \theta) = b^m(x) \qquad \lambda^\alpha(x, \theta) = 0.$$

In this case equation (5.1.8) takes the form

$$\delta_b\mathscr{H}^m = b^n\partial_n\mathscr{H}^m - \mathscr{H}^n\partial_n b^m$$

At the component level, one finds

$$\delta_b e_a{}^m = b^n \partial_n e_a{}^m - (\partial_n b^m) e_a{}^n$$
$$\delta_b \Psi^m{}_\alpha = b^n \partial_n \Psi^m{}_\alpha - (\partial_n b^m) \Psi^n{}_\alpha \tag{5.1.19}$$
$$\delta_b A^m = b^n \partial_n A^m - (\partial_n b^m) A^n.$$

Therefore, each of the component fields in expression (5.1.17) transforms as a world vector under the general coordinate transformations; hence the field index 'm' is to be understood as a curved-space one.

Let us consider the local Lorentz transformations. They are generated by the parameters

$$\lambda^m(x, \theta) = 0 \qquad \lambda^\alpha(x, \theta) = K^\alpha{}_\beta(x)\theta^\beta.$$

One can readily deduce from equations (5.1.8) that

$$\delta_K e_a{}^m = K_a{}^b e_b{}^m$$
$$\delta_K \Psi^m{}_\alpha = K_\alpha{}^\beta \Psi^m{}_\beta \tag{5.1.20}$$
$$\delta_K A^m = -\frac{1}{4} e_a{}^m e_b{}^n \varepsilon^{abcd} \partial_n K_{cd}$$

where

$$K^{ab} = (\sigma^{ab})_{\alpha\beta} K^{\alpha\beta} - (\tilde{\sigma}^{ab})_{\dot{\alpha}\dot{\beta}} \bar{K}^{\dot{\alpha}\dot{\beta}}.$$

It is seen that $e_a{}^m$ transforms as a covector and $\Psi^m{}_\alpha$ as an undotted spinor under the local Lorentz rotations; hence the field indices 'a' and 'α' are to be understood as tangent space (or flat) ones.

In accordance with equations (5.1.19, 20), the $\mathscr{H}^m$-component field $e_a{}^m(x)$ has the same behaviour under general coordinate and local Lorentz transformations as the vierbein. On these grounds, it is natural to identify $e_a{}^m(x)$ with the vierbein.

The last relation (5.1.20) shows that $A^m(x)$ changes non-trivially under local Lorentz transformations. It would be more desirable to work with a Lorentz scalar field. For this purpose, we redefine A^m as

$$A^m = \tilde{A}^m + \frac{1}{4} e_a{}^m \varepsilon^{abcd} \omega_{bcd} \tag{5.1.21}$$

where ω_{bcd} is the torsion-free spin connection (see equations (1.6.23)). $\tilde{A}^m$ proves to be a world vector and Lorentz scalar field,

$$\delta_b \tilde{A}^m = b^n \partial_n \tilde{A}^m - (\partial_n b^m) \tilde{A}^n \qquad \delta_K \tilde{A}^m = 0. \tag{5.1.22}$$

Before considering the local supersymmetry transformations, it is worth concentrating on one technical matter. Let us take two sets of the supergroup parameters:

$$\lambda_1^m = 2\mathrm{i}\theta\sigma^a\bar{\epsilon}e_a{}^m - 2\theta^2\bar{\epsilon}\bar{\Psi}^m$$
$$\lambda_1^\alpha = \epsilon^\alpha + \theta^2\eta^\alpha \qquad (5.1.23)$$

$$\lambda_2^m = 0 \qquad \lambda_2^\alpha = K^\alpha{}_\beta\theta^\beta$$

and calculate the corresponding bracket parameters (5.1.5). In the spinor case, one has

$$\lambda_3^\alpha = -K^\alpha{}_\beta\epsilon^\beta - \theta^2\left[K^\alpha{}_\beta\eta^\beta + \left(\frac{\mathrm{i}}{2}e_bK^{ba} + \frac{1}{4}\varepsilon^{abcd}e_bK_{cd}\right)(\bar{\epsilon}\tilde{\sigma}_a)^\alpha\right] \qquad (5.1.24)$$

where $e_b = e_b{}^m\partial_m$. It is seen that the local Lorentz transformations treat $\epsilon^\alpha(x)$ as an undotted spinor, but the transformation law of $\eta^\alpha(x)$ contains inhomogeneous $\bar{\epsilon}$-dependent terms. To avoid this complication, it is useful to make in expressions (5.1.23) the replacement

$$\eta^\alpha \to \eta^\alpha + (\bar{\epsilon}\tilde{\sigma}_a)^\alpha\left(\frac{\mathrm{i}}{2}\omega_b{}^{ba} + \frac{1}{4}\varepsilon^{abcd}\omega_{bcd}\right).$$

Owing to the local Lorentz transformation law of ω_{bcd}, one now has

$$\lambda_3^\alpha = -K^\alpha{}_\beta(\epsilon^\beta + \eta^\beta\theta^2)$$

instead of equation (5.1.24). On this basis, we change the parametrization (5.1.18) to

$$\lambda^m(x, \theta) = b^m + 2\mathrm{i}\theta\sigma^a\bar{\epsilon}e_a{}^m - 2\theta^2\bar{\epsilon}\bar{\Psi}^m$$
$$\lambda^\alpha(x, \theta) = \epsilon^\alpha + \frac{1}{2}(\sigma + \mathrm{i}\Omega)\theta^\alpha + K^\alpha{}_\beta\theta^\beta \qquad (5.1.25)$$
$$+ \theta^2\left[\eta^\alpha + (\bar{\epsilon}\tilde{\sigma}_a)^\alpha\left(\frac{\mathrm{i}}{2}\omega_b{}^{ba} + \frac{1}{4}\varepsilon^{abcd}\omega_{bcd}\right)\right].$$

Now, let us find the local supersymmetry transformations. Setting to zero all the parameters in expression (5.1.25), except ϵ^α and $\bar{\epsilon}_{\dot{\alpha}}$, and using equations (5.1.8) and (5.1.21), one obtains after long (requiring, maybe, some days) calculations

$$\delta_\epsilon e_a{}^m = \mathrm{i}\epsilon\sigma^a\bar{\Psi}^m - \mathrm{i}\Psi^m\sigma^a\bar{\epsilon}$$
$$\delta_\epsilon\Psi^m{}_\alpha = (\sigma^a\tilde{\sigma}^b\nabla_a\epsilon)_\alpha e_b{}^m - 2\mathrm{i}\tilde{A}^m\epsilon_\alpha$$
$$\delta_\epsilon\tilde{A}^m = \frac{\mathrm{i}}{4}e_a{}^m\varepsilon^{abcd}\nabla_b(\epsilon\sigma_c\bar{\Psi}^n - \Psi^n\sigma_c\bar{\epsilon})e_{nd} \qquad (5.1.26)$$
$$+ \frac{1}{2}\nabla_a(\epsilon\sigma^a\bar{\Psi}^m + \Psi^m\sigma^a\bar{\epsilon}) - e_a{}^m((\nabla_n\epsilon)\sigma^a\bar{\Psi}^n + \Psi^n\sigma^a\nabla_n\bar{\epsilon}).$$

Here ∇_n and $\nabla_a = e_a{}^n\nabla_n$ are the standard (torsion-free) covariant derivatives (see Section 1.6). As is seen, the Majorana spin-vector field

$$\boldsymbol{\Psi}^m = \begin{pmatrix} \Psi^m{}_\alpha \\ \bar{\Psi}^{m\dot{\alpha}} \end{pmatrix} \tag{5.1.27}$$

is a superpartner of the gravitational field. That is why $\boldsymbol{\Psi}^m$ is called the 'gravitino field'. It follows from the second relation (5.1.26) that the gravitino is a gauge field for the local supersymmetry transformations.

Our only remaining task is to consider transformations induced by the parameters σ, Ω and η^α in expression (5.1.25). Making the choice

$$\lambda^m(x, \theta) = 0 \qquad \lambda^\alpha(x, \theta) = \frac{1}{2}\sigma(x)\theta^\alpha$$

in equation (5.1.8), one finds

$$\begin{aligned} \delta_\sigma e_a{}^m &= \sigma e_a{}^m \\ \delta_\sigma \Psi^m{}_\alpha &= \frac{3}{2}\sigma\Psi^m{}_\alpha \\ \delta_\sigma \tilde{A}^m &= 2\sigma\tilde{A}^m. \end{aligned} \tag{5.1.28}$$

Comparing $\delta_\sigma e_a{}^m$ with equation (1.6.32) shows that the parameter $\sigma(x)$ corresponds to the Weyl transformations. Conformal weights of $\Psi^m{}_\alpha$ and $\tilde{A}^m$ are equal to $\frac{3}{2}$ and 2, respectively.

Further, making the choice

$$\lambda^m(x, \theta) = 0 \qquad \lambda^\alpha(x, \theta) = \frac{\mathrm{i}}{2}\Omega(x)\theta^\alpha$$

in equation (5.1.8), one arrives at the transformation

$$\begin{aligned} \delta_\Omega e_a{}^m &= 0 \\ \delta_\Omega \Psi^m{}_\alpha &= -\frac{\mathrm{i}}{2}\Omega\Psi^m{}_\alpha \\ \delta_\Omega \tilde{A}^m &= \frac{1}{2}g^{mn}\partial_n\Omega. \end{aligned} \tag{5.1.29}$$

This is a local chiral or γ_5-transformation. When rewritten in terms of the four-component spin-vector $\boldsymbol{\Psi}^m$, it has the form

$$\delta_\Omega \boldsymbol{\Psi}^m = -\frac{\mathrm{i}}{2}\Omega\gamma_5\boldsymbol{\Psi}^m.$$

In accordance with the last relation (5.1.29), $\tilde{A}^m$ is a gauge field for the local γ_5-transformations.

Finally, we set

$$\lambda^m(x, \theta) = 0 \qquad \lambda^\alpha(x, \theta) = \theta^2 \eta^\alpha(x)$$

in equation (5.1.8). A straightforward calculation gives

$$\begin{aligned} \delta_\eta e_a{}^m &= 0 \\ \delta_\eta \Psi^m{}_\alpha &= -\mathrm{i}(\sigma^a \bar{\eta})_\alpha e_a{}^m \\ \delta_\eta \tilde{A}^m &= \mathrm{i}\eta\Psi^m - \mathrm{i}\overline{\eta\Psi^m}. \end{aligned} \tag{5.1.30}$$

Evidently, this is a supersymmetry transformation mixing bosonic and fermionic fields. Therefore, we have two types of supersymmetry transformations generated by the parameters $(\epsilon^\alpha, \bar{\epsilon}_{\dot{\alpha}})$ and $(\eta^\alpha, \bar{\eta}_{\dot{\alpha}})$, respectively. To distinguish them, the former will be called Q-supersymmetry transformations and the latter — S-supersymmetry transformations.

The S-supersymmetry invariance can be used to make $\Psi^m{}_\alpha$ traceless. Introducing the gravitino field with a tangent-space index $\Psi^a{}_\alpha = e_m{}^a \Psi^m{}_\alpha$, and converting the vector index into a pair of spinor indices, $\Psi^{\beta\dot{\beta}}{}_\alpha = (\tilde{\sigma}_a)^{\dot{\beta}\beta} \Psi^a{}_\alpha$, the second relation (5.1.30) takes the form

$$\delta_\eta \Psi^{\beta\dot{\beta}}{}_\alpha = 2\mathrm{i}\delta^\beta{}_\alpha \bar{\eta}^{\dot{\beta}}.$$

So, one can choose the gauge $\Psi^{\alpha\dot{\alpha}}{}_\alpha = (\tilde{\sigma}_a \Psi^a)^{\dot{\alpha}} = 0$.

We have spent some time making exhaustive calculations, and now it is necessary to pause and think over the results. Our wish was to find a supersymmetric generalization of Einstein gravity. Such a generalization should contain three types of gauge transformations: general coordinate, local Lorentz and local Q-supersymmetry ones. However, the supergroup of holomorphic coordination transformations on $\mathbb{C}_s^{4|2}$, which were chosen in the role of gauge group, turned out to be too large. It includes, besides the mentioned transformations, local scale, local chiral and local S-supersymmetry transformations. Recall that the gauge group of Einstein gravity did not contain Weyl transformations and the vierbein transformation law was given by equation (1.6.48). Nevertheless, Weyl transformations were presented in the gauge group of conformal gravity, where the vierbein transformation law was given by equation (1.6.54). So, the formulation we have constructed is a 'conformal supergravity' — that is, a supersymmetric generalization of conformal gravity. The transformation law (5.1.8), with λ^m and λ^α arbitrary, represents a supersymmetric generalization of equation (1.6.54), but not of equation (1.6.48). In order to obtain a generalization of Einstein gravity, one must exclude from the superfield parameters λ^m and λ^α their independent components $\sigma(x)$, $\Omega(x)$ and $\eta^\alpha(x)$ (in a sense, $\Omega(x)$ and $\eta^\alpha(x)$ are supersymmetric partners of $\sigma(x)$). In other words, it is necessary to constrain the supergroup parameters or, equivalently, choose a subgroup in the supergroup of holomorphic coordinate transformations on $\mathbb{C}_s^{4|2}$.

5.1.3. Einstein supergravity

The supergroup of holomorphic coordinate transformations on $\mathbb{C}_s^{4|2}$ has a very natural subgroup — the supergroup of unimodular holomorphic transformations

$$y^m \to y'^m = f^m(y, \theta) \qquad \theta^\alpha \to \theta'^\alpha = f^\alpha(y, \theta)$$

$$\mathrm{Ber}\left(\frac{\partial(y', \theta')}{\partial(y, \theta)}\right) = 1. \tag{5.1.31}$$

Infinitesimal holomorphic transformations of the form (5.1.4) are unimodular, if the parameters are restricted as follows

$$\frac{\partial}{\partial y^m}\lambda^m - \frac{\partial}{\partial \theta^\alpha}\lambda^\alpha = 0 \tag{5.1.32}$$

It should be noted that any super Poincaré transformation (5.5.13) satisfies this equation. Rather beautifully, representing λ^m and λ^α as in equation (5.1.15), equation (5.1.32) can be uniquely resolved by expressing the parameters ω_α^α (note, $\omega_\alpha^\alpha = \sigma + \mathrm{i}\Omega$) and η^α in terms of other components. So, the supergroup of unimodular holomorphic transformations on $\mathbb{C}_s^{4|2}$ may be taken as a candidate for the role of gauge group in Einstein supergravity.

Now, following V. Ogievetsky and E. Sokatchev, *Einstein supergravity is postulated to be a gauge theory with the gravitational superfield $\mathscr{H}^m(x, \theta, \bar{\theta})$ being the dynamical object and the supergroup of unimodular holomorphic transformations beingthe gauge group.* The transformation law for $\mathscr{H}^m$ is given by equation (5.1.8), where the superfields λ^m and λ^α obey, in contrast to conformal supergravity, the constraint (5.1.32).

At this point, in principle, the descriptive part is over; one has all that is needed to proceed to extracting the consequences. However, it will be more convenient for us to employ a different, but equivalent, formulation of Einstein supergravity. The point is that usually working with fields under constraints is a more complicated technical problem then working with unconstrained fields. For example, it is a rather difficult task to define the path integral on the space of transverse vector fields $A_m^\perp(x)$, $\partial^m A_m^\perp = 0$. Fortunately, very often, it is possible to reformulate a theory of constrained fields as a theory of unconstrained fields with an additional gauge invariance. For example, electrodynamics may be treated as the theory $S[A^\perp] = \frac{1}{2}\int \mathrm{d}^4x A^{\perp m} \Box A_m^\perp$ of a transverse vector field $A_m^\perp(x)$ without gauge invariance or as the theory $S[A] = S[A^\perp]$ of an unconstrained vector field $A_m = A_m^\perp + A_m^\parallel$, where $A_m^\parallel$ is the longitudinal component of A_m, $\partial_{[m}A_{n]}^\parallel = 0$, but with the gauge invariance $A_m \to A_m + \partial_m \xi$. The auxiliary field $A_m^\parallel(x)$ is a compensating field for the gauge freedom.

5.1.4. Einstein supergravity (second formulation)

Our wish is to remove the constraint (5.1.32) and to handle arbitrary holomorphic transformations of $\mathbb{C}_s^{4|2}$ at the cost of the appearance of an auxiliary compensating superfield.

Following W. Siegel and S. J. Gates, let us consider, along with $\mathscr{H}^m$, one more superfield $\varphi(y, \theta)$ holomorphic on $\mathbb{C}_s^{4|2}$ and possessing the transformation law

$$\varphi'(y', \theta') = \left[\mathrm{Ber}\left(\frac{\partial(y, \theta)}{\partial(y', \theta')}\right)\right]^{1/3} \varphi(y, \theta) \tag{5.1.33}$$

under the supergroup of holomorphic coordinate transformations. Infinitesimally, this transformation law is written

$$\varphi'(y', \theta') = \varphi(y, \theta) + \frac{1}{3}\left(\frac{\partial \lambda^m}{\partial y^m} - \frac{\partial \lambda^\alpha}{\partial \theta^\alpha}\right)\varphi(y, \theta) \tag{5.1.34a}$$

or

$$\delta\varphi(y, \theta) = \varphi'(y, \theta) - \varphi(y, \theta) = (\lambda^m \partial_m + \lambda^\alpha \partial_\alpha)\varphi + \frac{1}{3}(\partial_m \lambda^m - \partial_\alpha \lambda^\alpha)\varphi. \tag{5.1.34b}$$

In accordance with equation (5.1.33), one can find a coordinate system on $\mathbb{C}_s^{4|2}$ in which

$$\varphi(y, \theta) = 1. \tag{5.1.35}$$

Evidently, all holomorphic coordinate transformations preserving this gauge choice are unimodular. As a result, we recover the gauge group of Einstein supergravity. It is the superfield $\varphi(y, \theta)$ which compensates for the additional gauge freedom. $\varphi(y, \theta)$ is called the 'chiral compensator'.

Now, one can take the following point of view. *Einstein supergravity is a gauge theory of two dynamical objects—the gravitational superfield $\mathscr{H}^m(x, \theta, \bar{\theta})$ and the chiral compensator $\varphi(y, \theta)$ transforming by the law (5.1.8) and (5.1.34b), respectively, under the supergroup of holomorphic coordinate transformations—the supergravity gauge group.* Of course, for physical applications it is necessary to restrict $\varphi(y, \theta)$ from $\mathbb{C}_s^{4|2}$ to the real superspace $\mathbb{R}^{4|4}(\mathscr{H})$ embedded in $\mathbb{C}_s^{4|2}$, resulting in the complex superfield $\varphi(x, \theta, \bar{\theta}) = \varphi(x^m + \mathrm{i}\mathscr{H}^m, \theta^\alpha)$.

The chiral compensator has a simple geometrical interpretation. One can develop integration theory over $\mathbb{C}_s^{4|2} = \mathbb{C}_s^{4|0} \times \mathbb{C}_a^2$ in complete analogy with the integration theory over $\mathbb{R}^{p|q}$ discussed in Section 1.10 (the only distinction between $\mathbb{C}_s^{4|0}$ and $\mathbb{R}^{4|0}$ is that variables used for parametrizing $\mathbb{C}_s^{4|0}$ and $\mathbb{R}^{4|0}$ have imaginary souls in the former case and have no souls in the latter case). Similarly to the rule for change of variables (1.11.22), one can prove that the measure $\mathrm{d}^4 y\, \mathrm{d}^2\theta$ on $\mathbb{C}_s^{4|2}$ transforms under the change of variables (5.1.3) as follows

$$d^4y'\,d^2\theta' = \mathrm{Ber}\left(\frac{\partial(y',\theta')}{\partial(y,\theta)}\right) d^4y\,d^2\theta.$$

Then, the transformation law (5.1.33) leads to

$$d^4y'\,d^2\theta'(\varphi'(y',\theta'))^3 = d^4y\,d^2\theta(\varphi(y,\theta))^3. \tag{5.1.36}$$

Therefore, the chiral compensator defines an invariant measure on $\mathbb{C}_s^{4|2}$. The field φ^3 is the superspace analogue of the density field e^{-1}, $e = \det(e_a{}^m)$, used in general relativity.

5.1.5. Einstein supergravity multiplet

In subsection 5.1.2, the conformal supergravity multiplet was obtained. It included the following fields: $e_a{}^m(x)$, $\Psi^m{}_\alpha(x)$, $\bar{\Psi}^{m\dot{\alpha}}(x)$ and $\tilde{A}^m(x)$. Now we proceed by finding a multiplet of fields corresponding to Einstein supergravity. Clearly, it should differ from the conformal supergravity multiplet, since Einstein supergravity is described by the gravitational superfield and the chiral compensator while all the dynamical content of conformal supergravity is encoded in $\mathscr{H}^m$ only.

A possible gauge choice in Einstein supergravity is $\varphi(y,\theta) = 1$. Instead of choosing this gauge, it is more convenient for us to impose the Wess–Zumino gauge (5.1.17) in which the gravitational superfield has the simplest form (in the gauge $\varphi = 1$, $\mathscr{H}^m$ will contain θ^2- and $\bar{\theta}^2$-terms as well). Every holomorphic transformation (5.1.4), preserving the Wess–Zumino gauge, is given by parameters (5.1.25).

Let us expand the chiral compensator in a power series in θ:

$$\varphi^3(x,\theta) = e^{-1}(x)\{F(x) + \theta^\alpha\chi_\alpha(x) + \theta^2 B(x)\}. \tag{5.1.37}$$

Here the factor $e^{-1}(x)$ is introduced to make $F(x)$ and $B(x)$ complex scalar fields and $\chi_\alpha(x)$ an undotted spinor field with respect to the general coordinate and local Lorentz transformations,

$$\begin{gathered}\delta_{b+K}F = b^m\partial_m F \qquad \delta_{b+K}B = b^m\partial_m B \\ \delta_{b+K}\chi_\alpha = b^m\partial_m\chi_\alpha + K_\alpha{}^\beta\chi_\beta\end{gathered} \tag{5.1.38}$$

in accordance with equation (5.1.34*b*). Making use once more of the transformation law (5.1.34*b*), one finds how the Weyl and local chiral transformations act on the fields:

$$\begin{aligned}\delta_{\sigma+\Omega}F &= (3\sigma - \mathrm{i}\Omega)F \\ \delta_{\sigma+\Omega}\chi_\alpha &= \left(\frac{7}{2}\sigma - \frac{\mathrm{i}}{2}\Omega\right)\chi_\alpha \\ \delta_{\sigma+\Omega}B &= 4\sigma B.\end{aligned} \tag{5.1.39}$$

As for the local S-supersymmetry transformations, they are of the form

$$\delta_\eta F = \delta_\eta B = 0$$
$$\delta_\eta \chi_\alpha = 2F\eta_\alpha \tag{5.1.40}$$

As may be seen, the Weyl and local chiral transformations can be used to gauge away F. The most useful gauge choice is $F = 1$. After this, the local S-supersymmetry transformations can be used to gauge away χ_α. A useful gauge fixing proves to be $\chi_\alpha = -2\mathrm{i}(\sigma_a)_{\alpha\dot\alpha}\bar\Psi^{a\dot\alpha}$. Then, one results with the gauge fixed chiral compensator

$$\varphi^3(x,\theta) = e^{-1}(x)\{1 - 2\mathrm{i}\theta\sigma_a\bar\Psi^a(x) + \theta^2 B(x)\}. \tag{5.1.41}$$

Finally, it remains to analyse the Q-supersymmetry transformations. Setting in expression (5.1.25) all the parameters, except ϵ_α and $\bar\epsilon^{\dot\alpha}$, to vanish, one obtains a transformation which breaks the gauge (5.1.41). To preserve our gauge choice, any Q-supersymmetry transformation must be accompanied by some ϵ-dependent local chiral $+$ S-supersymmetry transformation. Namely, instead of parameters (5.1.25), one has to consider the set of parameters

$$\lambda^m(x,\theta) = 2\mathrm{i}\theta\sigma^a\bar\epsilon e_a{}^m - 2\theta^2\epsilon\Psi^m$$

$$\lambda^\alpha(x,\theta) = \epsilon^\alpha + \frac{\mathrm{i}}{2}\Omega(\epsilon)\theta^\alpha + \theta^2\left[\eta^\alpha(\epsilon) + (\bar\epsilon\tilde\sigma_a)^\alpha\left(\frac{\mathrm{i}}{2}\omega_b{}^{ba} + \frac{1}{4}\varepsilon^{abcd}\omega_{bcd}\right)\right]$$

$$\Omega(\epsilon) = -\epsilon\sigma_a\bar\Psi^a - \Psi^a\sigma_a\bar\epsilon \tag{5.1.42}$$

$$\eta_\alpha(\epsilon) = \mathrm{i}(\sigma^a\nabla_a\bar\epsilon)_\alpha + \frac{2}{3}(\sigma_a\bar\Psi^a)_\alpha\epsilon\sigma_b\bar\Psi^b - \frac{2}{3}(\sigma_a\bar\epsilon)_\alpha\tilde A^a + \frac{1}{3}B\epsilon_\alpha$$
$$+ \frac{1}{3}(\sigma_a\bar\Psi_b)_\alpha(\epsilon\sigma^b\bar\Psi^a - \Psi^a\sigma^b\bar\epsilon).$$

It should be pointed that we have $\sigma(\epsilon) = 0$ due to the adopted gauge fixing for χ_α.

Let us denote the variation of any field under parameters (5.1.42) symbolically as $\hat\delta_\epsilon$. Clearly, this can be represented in the form

$$\hat\delta_\epsilon = \delta_\epsilon + \delta_{\Omega(\epsilon)} + \delta_{\eta(\epsilon)}.$$

The variations δ_ϵ, δ_Ω and δ_η of $\mathscr{H}^m$-component fields are given by equations (5.1.26, 29, 30). The variation $\delta_\epsilon B$ can easily be found using the transformation law (5.1.34b). Then, making use of equations (5.1.39, 40) gives

$$\hat\delta_\epsilon B = \mathrm{i}(\epsilon\sigma_a\bar\Psi^a - \Psi^a\sigma_a\bar\epsilon)B + 2\nabla_a(\overline{\epsilon\Psi^a}) + 2\nabla_a(\bar\Psi^b\tilde\sigma_b\sigma^a\bar\epsilon). \tag{5.1.43}$$

We do not present here expressions for $\hat\delta_\epsilon e_a{}^m$, $\hat\delta_\epsilon\Psi^m{}_\alpha$ and $\hat\delta_\epsilon\tilde A^m$. Instead, we consider the set of fields with lower curved-space indices

$$e_m{}^a \qquad \Psi_{m\alpha} = g_{mn}\Psi^n{}_\alpha \qquad \tilde{A}_m = g_{mn}\tilde{A}^n$$

and write down the transformation laws of these fields under the transformation (5.1.42) supplemented by the ϵ-dependent Lorentz rotation

$$K_{\alpha\beta}(\epsilon) = \frac{i}{2}[(\sigma_a\bar{\epsilon})_\alpha\Psi^a{}_\beta + (\sigma_a\bar{\epsilon})_\beta\Psi^a{}_\alpha + \epsilon_\alpha(\sigma_a\bar{\Psi}^a)_\beta + \epsilon_\beta(\sigma_a\bar{\Psi}^a)_\alpha]. \quad (5.1.44)$$

The corresponding field variations will be denoted by

$$\delta = \hat{\delta}_\epsilon + \delta_{K(\epsilon)}.$$

The gravitational field and the gravitino transform by the rule

$$\begin{aligned} \delta e_m{}^a &= i\Psi_m\sigma^a\bar{\epsilon} - i\epsilon\sigma^a\bar{\Psi}_m \\ \delta\Psi_{m\alpha} &= -2\hat{\nabla}_m\epsilon_\alpha - 2i\epsilon_\alpha\mathbf{A}_m - \frac{2i}{3}e_m{}^a(\sigma_a\tilde{\sigma}_b\epsilon)_\alpha\mathbf{A}^b - \frac{i}{3}e_m{}^a(\sigma_a\bar{\epsilon})_\alpha\bar{\mathbf{B}} \end{aligned} \quad (5.1.45)$$

where we have introduced the fields

$$\begin{aligned} \mathbf{B} &= B + \frac{1}{2}\bar{\Psi}^a\tilde{\sigma}_a\sigma_b\bar{\Psi}^b + \frac{1}{2}\Psi^a\Psi_a \\ \mathbf{A}_m &= \tilde{A}_m - \frac{1}{4}(\Psi^a\sigma_a\bar{\Psi}_m + \Psi_m\sigma_a\bar{\Psi}^a) - \frac{1}{2}e_m{}^a\left[\Psi^b\sigma_a\bar{\Psi}_b - \frac{i}{4}\varepsilon_{abcd}\Psi^b\sigma^c\bar{\Psi}^d\right]. \end{aligned} \quad (5.1.46)$$

$\hat{\nabla}_m$ denote covariant derivatives with torsion:

$$\begin{aligned} \hat{\nabla}_m\epsilon_\alpha &= \nabla_m\epsilon_\alpha + \frac{1}{2}\Phi_{mab}(\sigma^{ab}\epsilon)_\alpha \\ \Phi_{mab} &= e_m{}^c\Phi_{cab} = \frac{1}{2}e_m{}^c(\mathcal{T}_{abc} + \mathcal{T}_{acb} - \mathcal{T}_{bca}) \\ \mathcal{T}_{abc} &= \frac{i}{2}(\Psi_a\sigma_c\bar{\Psi}_b - \Psi_b\sigma_c\bar{\Psi}_a). \end{aligned} \quad (5.1.47)$$

The transformation laws of the fields just introduced are

$$\begin{aligned} \delta\mathbf{B} &= -2\epsilon\tilde{\sigma}^{ab}\Psi_{ab} - i\Psi^a\sigma_a\bar{\epsilon}\mathbf{B} + 2i\bar{\epsilon}\bar{\Psi}^m\mathbf{A}_m \\ \delta\mathbf{A}_m &= -\frac{1}{2}e_m^a(\epsilon\sigma^b\bar{\Psi}_{ab} + i\varepsilon_{abcd}\epsilon\sigma^b\bar{\Psi}^{cd}) - \frac{i}{4}\epsilon\Psi_m\mathbf{B} \\ &\quad - i\epsilon\sigma^a\bar{\Psi}_m\mathbf{A}_a + \frac{i}{2}\epsilon\sigma_a\bar{\Psi}^a\mathbf{A}_m + \frac{1}{4}e_m{}^a\varepsilon_{abcd}\epsilon\sigma^b\bar{\Psi}^c\mathbf{A}^d + \text{c.c.} \end{aligned} \quad (5.1.48)$$

where $\Psi_{ab}{}^{\dot{\alpha}}$ denotes the gravitino field strength

$$\Psi_{ab}{}^{\dot{\alpha}} = \hat{\nabla}_a \Psi_b{}^{\dot{\alpha}} - \hat{\nabla}_b \Psi_a{}^{\dot{\alpha}}$$
$$\hat{\nabla}_a \Psi_b{}^{\dot{\alpha}} \equiv \nabla_a \Psi_b{}^{\dot{\alpha}} + \frac{1}{2}\Phi_{acd}(\tilde{\sigma}^{cd}\Psi_b)^{\dot{\alpha}}. \qquad (5.1.49)$$

We summarize the results. The Einstein supergravity multiplet is given by the set of fields $\{e_m{}^a, \Psi_{m\alpha}, \Psi_m{}^{\dot{\alpha}}, \mathbf{A}_m, \mathbf{B}, \bar{\mathbf{B}}\}$. The component gauge (quasi) group includes the space–time general coordinate, local Lorentz and local supersymmetry transformation. The local supersymmetry transformation laws of the supergravity fields are given by equations (5.1.45, 48).

5.1.6. *Flat superspace (final definition) and conformally flat superspace*

In conclusion, we would like to correct some notation and definitions introduced in this section. The point is that conformal supergravity and Einstein supergravity, being gauge theories with the same gauge group, have different dynamical content. Conformal supergravity is described in terms of the gravitational superfield only, while Einstein supergravity needs one more dynamical superfield — the chiral compensator. The gravitational superfield defines the embedding of physical real superspace in $\mathbb{C}^{4|2}$, while the chiral compensator determines the invariant measure $\mathrm{d}\mu = \mathrm{d}^4y\,\mathrm{d}^2\theta\varphi^3$ on the physical surface. As a consequence, the more accurate symbol for denoting physical superspace is $\mathbb{R}^{4|4}(\mathscr{H}, \varphi)$, and not $\mathbb{R}^{4|4}(\mathscr{H})$. The symbol $\mathbb{R}^{4|4}(\mathscr{H})$ is reserved for denoting physical superspace in conformal supergravity.

In fact, conformal supergravity may also be treated as a theory of the gravitational superfield and the chiral compensator, but with the additional gauge invariance

$$\mathscr{H}^m \to \mathscr{H}'^m = \mathscr{H}^m$$
$$\varphi(y, \theta) \to \varphi'(y, \theta) = \mathrm{e}^{\sigma(y,\theta)}\varphi(y, \theta) \qquad (5.1.50)$$

where σ is an arbitrary superfield on $\mathbb{C}_s^{4|2}$. Equation (5.1.50) is a superspace analogue of space–time Weyl transformations (1.6.32). This is why the above transformations are called 'super Weyl transformations'.

Furthermore, the definition of flat superspace given in subsection 5.1.1 needs correction. In the literal sense of the word, we have defined a conformally flat superspace. To make this precise, let us remind ourselves of the usual meaning of conformally flat space. A space–time is said to be conformally flat if there exists a coordinate system in which the metric is, up to a factor, the Minkowski one,

$$g_{mn}(x) = \phi(x)\eta_{mn}. \qquad (5.1.51)$$

All transformations from the conformal group (see Section 1.7), and only these, preserve the conformally flat form of metric (5.1.51).

Now, we formulate the final definition of flat superspace. A superspace $\mathbb{R}^{4|4}(\mathscr{H}, \varphi)$ is said to be a 'flat superspace' if there exists a holomorphic coordinate transformation (5.1.3) such that

$$\mathscr{H}'^m(x', \theta', \bar{\theta}') = \delta_a{}^m \theta' \sigma^a \bar{\theta}' \qquad \varphi'(x', \theta') = 1. \tag{5.1.52}$$

The super Poincaré transformations (5.1.13), and only these, leave invariant flat superfields (5.1.52).

A superspace $\mathbb{R}^{4|4}(\mathscr{H}, \varphi)$ is said to be a 'conformally flat superspace' if there exists a coordinate system in which $\mathscr{H}^m$ has the flat form (5.1.12). It is not difficult to find the family of holomorphic coordination transformations of $\mathbb{C}_s^{4|2}$ preserving the flat gravitational superfield (5.1.12). Such transformations are generated by the superfield parameters

$$\begin{aligned} -\lambda^a(y, \theta) &= b^a + K^a{}_b y^b + \Delta y^a + f^a y^2 - 2y^a(f, y) \\ &\quad + 2i\theta\sigma^a\bar{\epsilon} - 2\theta\sigma^a\tilde{\sigma}_b\eta y^b \\ -\lambda^\alpha(y, \theta) &= \epsilon^\alpha - iy_a(\bar{\eta}\tilde{\sigma}^a)^\alpha + \left(\frac{1}{2}\Delta + \frac{i}{2}\Omega - (f, y)\right)\theta^\alpha \\ &\quad + (2y^a f^b(\sigma_{ab})^\alpha{}_\beta - K^\alpha{}_\beta)\theta^\beta + 2\eta^\alpha\theta^2 \\ K_{\alpha\beta} = K_{\beta\alpha} &\qquad K^{ab} = (\sigma^{ab})_{\alpha\beta}K^{\alpha\beta} - (\tilde{\sigma}^{ab})_{\dot{\alpha}\dot{\beta}}\bar{K}^{\dot{\alpha}\dot{\beta}} \end{aligned}$$

where Δ, Ω, b^a, f^a and K^{ab} are constant real c-number parameters and ϵ_α, η_α are constant complex a-number parameters. Transformations generated by the above parameters form a supergroup known as the 'superconformal group', which was studied in Chapter 2.

5.2. Superspace differential geometry

The supergravity gauge group has been realized in real superspace $\mathbb{R}^{4|4}$ in a rather unusual way, as the set of nonlinear transformations of special form (5.1.6). For mathematical beauty and to observe direct contact with standard general relativity, it would be desirable to have a supergravity formulation in which the gauge group is extended to include the full supergroup of general coordinate transformations on $\mathbb{R}^{4|4}$:

$$\begin{aligned} x^m &\to x'^m = f^m(x, \theta, \bar{\theta}) \\ \theta^\mu &\to \theta'^\mu = f^\mu(x, \theta, \bar{\theta}) \qquad \mathrm{Ber}\left(\frac{\partial(x', \theta', \bar{\theta}')}{\partial(x, \theta, \bar{\theta})}\right) \neq 0 \\ \bar{\theta}_{\dot{\mu}} &\to \bar{\theta}'_{\dot{\mu}} = \bar{f}_{\dot{\mu}}(x, \theta, \bar{\theta}) \end{aligned} \tag{5.2.1}$$

or, in infinitesimal form,

$$\begin{aligned} x^m \to x'^m &= x^m - K^m(x, \theta, \bar{\theta}) \\ \theta^\mu \to \theta'^\mu &= \theta^\mu - K^\mu(x, \theta, \bar{\theta}) \\ \bar{\theta}_{\dot{\mu}} \to \bar{\theta}'_{\dot{\mu}} &= \bar{\theta}_{\dot{\mu}} - \bar{K}_{\dot{\mu}}(x, \theta, \bar{\theta}) \end{aligned} \tag{5.2.2}$$

where K^m is an arbitrary real vector superfield and K^μ is an arbitrary undotted spinor superfield. In principle, such a formulation may be extracted from that constructed above by means of the introduction of superfluous compensating superfields. We take another course. For the time being we shall forget the results of the previous section and proceed to the development of a geometrical formalism in real superspace trying to keep, as far as possible, the analogy with ordinary differential geometry.

5.2.1. Superfield representations of the general coordinate transformation supergroup

We are going to describe some superfield representations of the supergroup of general coordinate transformations on $\mathbb{R}^{4|4}$. As in previous chapters, we adopt the compact notation $z^M = (x^m, \theta^\mu, \bar{\theta}_{\dot{\mu}})$ which makes it possible to rewrite (5.2.1) as

$$z^M \to z'^M = f^M(z) \qquad \mathrm{Ber}\left(\frac{\partial z'^N}{\partial z^M}\right) \neq 0. \tag{5.2.3}$$

Starting from this point, always when dealing with the general coordinate transformation supergroup, superspace indices will be denoted by letters from the middle of the Latin and Greek alphabets (letters from the beginning of each alphabet will be reserved for tangent-space indices).

The simplest representations of the general coordinate transformation supergroup are scalar and scalar densities. A 'scalar superfield' $\phi(z)$ is defined by the transformation law

$$\phi'(z') = \phi(z). \tag{5.2.4}$$

A 'scalar density superfield $\phi_{(w)}(z)$ of weight w' is characterized by the transformation law

$$\phi'_{(w)}(z') = \left(\mathrm{Ber}\left(\frac{\partial z'}{\partial z}\right)\right)^{-w} \phi_{(w)}(z). \tag{5.2.5}$$

Given a scalar density of weight one, $\phi_{(1)}(z)$, the integral over $\mathbb{R}^{4|4}$ of $\phi_{(1)}(z)$ does not depend on the choice of coordinate system on $\mathbb{R}^{4|4}$,

$$\int \mathrm{d}^8 z' \phi'_{(1)}(z') = \int \mathrm{d}^8 z \phi_{(1)}(z) \tag{5.2.6}$$

as a consequence of equation (1.11.32).

An invariant first-order differential operator on $\mathbb{R}^{4|4}$

$$V = V^M(z)\partial_M = V^m\partial_m + V^\mu\partial_\mu + V_{\dot\mu}\bar\partial^{\dot\mu} \tag{5.2.7}$$

where $V^M(z)$ are smooth superfunctions, is called a 'supervector field'. The supervector transformation law follows from the invariance requirement

$$V'^M(z')\partial'_M = V^M(z)\partial_M \tag{5.2.8a}$$

and it reads

$$V'^M(z') = V^N(z)\partial_N z'^M \tag{5.2.8b}$$

The value of the supervector field $V(p) = V^M\partial_M|_p$ at a given point p of superspace is said to be a 'supervector' at this point.

In accordance with the results of subsection 2.1.4, the set of all supervector fields on $\mathbb{R}^{4|4}$, denoted by $SVF(4, 4)$, forms an infinite-dimensional Berezin superalgebra with respect to the grading (2.1.46, 47) and the super Lie bracket (2.1.51). Based on equation (2.1.50), we introduce the operation of complex conjugation in $SVF(4, 4)$ as follows (see also equation (2.4.28)):

$$V \to \bar V = \bar V^M\partial_M \qquad (V\Psi)^* = (-1)^{\varepsilon(V)\varepsilon(\Psi)}\bar V\Psi^*$$

$$\Rightarrow \begin{cases} \bar V^m = (V^m)^* \\ \bar V^\mu = (-1)^{\varepsilon(V)}\varepsilon^{\mu\nu}(V_{\dot\nu})^* \\ \bar V_{\dot\mu} = (-1)^{\varepsilon(V)}\varepsilon_{\dot\mu\dot\nu}(V^\nu)^*. \end{cases} \tag{5.2.9}$$

where $\Psi(z)$ is a pure (bosonic or fermionic) superfield. One can explicitly check that $\bar V^M$ transforms according to the supervector law (5.2.8). A real c-type supervector field $K = \bar K$ is characterized by components of the structure

$$K^M = \bar K^M = (K^m, K^\mu, \bar K_{\dot\mu}) \qquad \varepsilon(K^M) = \varepsilon_M. \tag{5.2.10}$$

The set of complex (real) c-type supervector fields in $SVF(4, 4)$, denoted by ${}^0SVF(4, 4)$ (${}^0SVF_{\mathbb{R}}(4, 4)$), forms a complex (real) super Lie algebra with respect to the ordinary Lie bracket.

A 'supercovector field' (or super 1-form) $U_M(z)$ is defined to transform under the general coordinate transformation supergroup in the same way as a partial derivative of scalar a field, i.e.

$$U'_M(z') = (\partial'_M z^N)U_N(z). \tag{5.2.11}$$

Given a supervector field $V^M(z)$, the object

$$V^M U_M = V^m U_m + V^\mu U_\mu + V_{\dot\mu}U^{\dot\mu}$$

but not $U_M V^M$, is a scalar superfield. So, when working with supertensors, the relative order of factors turns out to be essential.

Higher-rank supertensor fields may be defined similarly. For example, a second-rank supertensor field $\Psi_M{}^N(z)$ is characterized by the transformation law

$$\Psi'_M{}^N(z') = (\partial'_M z^P)\Psi_P{}^R \partial_R z'^N. \tag{5.2.12}$$

Due to this transformation law, the superfield

$$(-1)^{\varepsilon_M}\Psi_M{}^M(z)$$

is a scalar if $(\Psi_M{}^N)$ is a c-type supermatrix.

5.2.2. The general coordinate transformation supergroup in exponential form

As in Section 1.11, here we restrict ourselves to the consideration of coordinate transformations on $\mathbb{R}^{4|4}$ expressible in the exponential form

$$z^M \to z'^M = \mathrm{e}^{-K} z^M \qquad K = K^M \partial_M = \bar{K} \tag{5.2.13}$$

with K being a real c-type supervector field. The set of all such transformations forms a connected subgroup of the general coordinate transformation supergroup. Recall that the Berezinian of transformation (5.2.13) is given by equation (1.11.17).

The exponential form for superspace reparametrizations is useful, since it makes it possible to write down superfield transformation laws in a similar manner to ordinary Yang–Mills transformations. Substituting equations (5.2.13) in the scalar transformation law (5.2.4) gives

$$\mathrm{e}^{-K}\phi'(z) = \phi(z)$$

therefore we have

$$\phi'(z) = \mathrm{e}^{K}\phi(z). \tag{5.2.14}$$

In the supervector case, making use of equation (1.11.19) leads to

$$V'^M(z')\partial'_M = (\mathrm{e}^{-K}V'^M(z))\mathrm{e}^{-K}\partial_M \mathrm{e}^{K} = \mathrm{e}^{-K}(V'^M(z)\partial_M)\mathrm{e}^{K}$$

therefore the transformation law (5.2.8) can be rewritten as

$$V' = V'^M(z)\partial_M = \mathrm{e}^{K} V \mathrm{e}^{-K}. \tag{5.2.15}$$

It is instructive also to consider the transformation law (5.2.5) in the special case $w = 1$. From equation (1.11.17), we have

$$\mathrm{e}^{-K}\phi'_{(1)}(z) = (1\cdot \mathrm{e}^{-\overleftarrow{K}})^{-1}\phi_{(1)}(z) = (\mathrm{e}^{-K}(1\cdot \mathrm{e}^{\overleftarrow{K}}))\phi_{(1)}(z)$$

where we have used equation (1.11.14). Recalling equation (1.11.13), we have

$$\phi'_{(1)}(z) = (1\cdot \mathrm{e}^{\overleftarrow{K}})\mathrm{e}^{K}\phi_{(1)}(z) = (\phi_{(1)}\mathrm{e}^{\overleftarrow{K}}). \tag{5.2.16}$$

It is obvious now that the integral of $\phi_{(1)}(z)$ over $\mathbb{R}^{4|4}$ is invariant with respect to general coordinate transformations.

5.2.3. *Tangent and cotangent supervector spaces*

The tangent space $T_p(\mathbb{R}^{4|4})$ at a point $p \in \mathbb{R}^{4|4}$ is defined as the supervector space of left first-order differential operators (supervectors) at this point,

$$T_p(\mathbb{R}^{4|4}) = \{V = V^M \partial_M|_p, \ V^M \in \Lambda_\infty\}. \tag{5.2.17}$$

The dimension of $T_p(\mathbb{R}^{4|4})$ is equal to (4, 4). Clearly, the set of supervectors $\{\partial_M|_p\}$ forms a pure basis of $T_p(\mathbb{R}^{4|4})$ which will be called the 'holonomic basis'. In accordance with equation (5.2.9), this basis is characterized by the following conjugation properties

$$\overline{\partial_m} = \partial_m \qquad \overline{\partial_\mu} = \varepsilon_{\dot{\mu}\dot{\nu}} \bar{\partial}^{\dot{\nu}} = -\bar{\partial}_{\dot{\mu}} \qquad \overline{\partial^{\mu}} = \varepsilon^{\dot{\mu}\dot{\nu}} \bar{\partial}_{\dot{\nu}} = -\bar{\partial}^{\dot{\mu}} \tag{5.2.18}$$

Every pure basis of $T_p(\mathbb{R}^{4|4})$ satisfying the same conjugation properties will be called a 'standard basis'. These are real c-type supervectors from $T_p(\mathbb{R}^{4|4})$ which can be defined as directional derivatives at point $p \in \mathbb{R}^{4|4}$.

The 'cotangent space' $T_p^*(\mathbb{R}^{4|4})$ at a point $p \in \mathbb{R}^{4|4}$ is defined to be the right dual of $T_p(\mathbb{R}^{4|4})$ (see subsection 1.9.5). Its element will be called super 1-forms at point p. Every super 1-form $\omega \in T_p^*(\mathbb{R}^{4|4})$ is a mapping

$$\omega: T_p(\mathbb{R}^{4|4}) \to \Lambda_\infty$$

possessing the property

$$(\alpha_2 V_1 + \alpha_2 V_2)\omega = \alpha_1(V_1)\omega + \alpha_2(V_2)\omega$$

for arbitrary $V_1, V_2 \in T_p(\mathbb{R}^{4|4})$ and $\alpha_1, \alpha_2 \in \Lambda_\infty$. The value of a super 1-form ω on a supervector V reads

$$(V)\omega = (V^M \partial_M)\omega = V^M(\partial_M)\omega. \tag{5.2.19}$$

The dimension of $T_p^*(\mathbb{R}^{4|4})$ is (4, 4).

Given a scalar superfield $\Psi(z)$ on $\mathbb{R}^{4|4}$, one can construct a super 1-form $\mathrm{d}\Psi|_p \in T_p^*(\mathbb{R}^{4|4})$ defined by

$$(V)\,\mathrm{d}\Psi|_p = (V\Psi)|_p \qquad \forall V \in T_p(\mathbb{R}^{4|4}) \tag{5.2.20}$$

and called the differential of Ψ at p. Choosing here $\Psi = z^N$ gives

$$(\partial_N)\,\mathrm{d}z^M = \delta_N{}^M.$$

Therefore, the set $\{\mathrm{d}z^M|_p\}$ forms a pure basis of $T_p^*(\mathbb{R}^{4|4})$, which is the right dual of the basis $\{\partial_M|_p\}$ on $T_p(\mathbb{R}^{4|4})$. For every superfield Ψ on $\mathbb{R}^{4|4}$, one has

$$\mathrm{d}\Psi|_p = \mathrm{d}z^M|_p \cdot \partial_M \Psi|_p. \tag{5.2.21}$$

Then, introducing the tangent bundle $T(\mathbb{R}^{4|4}) = \bigcup_{p \in \mathbb{R}^{4|4}} T_p(\mathbb{R}^{4|4})$ and the cotangent bundle $T^*(\mathbb{R}^{4|4}) = \bigcup_{p \in \mathbb{R}^{4|4}} T_p^*(\mathbb{R}^{4|4})$, every supervector field $V = V^M(z)\partial_M$ determines section V: $\mathbb{R}^{4|4} \to T(\mathbb{R}^{4|4})$ and every supercovector field $\omega = \mathrm{d}z^M \omega_M(z)$ determines section ω: $\mathbb{R}^{4|4} \to T^*(\mathbb{R}^{4|4})$.

5.2.4. Supervierbein

Supervector fields on $\mathbb{R}^{4|4}$

$$E_A = (E_a, E_\alpha, \bar{E}^{\dot\alpha}) = E_A{}^M(z)\partial_M$$

$$E_A{}^M = \begin{pmatrix} E_a{}^m & E_a{}^\mu & E_{a\dot\mu} \\ E_\alpha{}^m & E_\alpha{}^\mu & E_{\alpha\dot\mu} \\ \bar{E}^{\dot\alpha m} & \bar{E}^{\dot\alpha\mu} & \bar{E}^{\dot\alpha}{}_{\dot\mu} \end{pmatrix} \tag{5.2.22}$$

are said to form a 'supervierbein', if for any point $p \in \mathbb{R}^{4|4}$ the set $\{E_A|_p\}$ constitutes a standard basis in $T_p(\mathbb{R}^{4|4})$. Equivalently, the c-type supermatrix $E_A{}^M(z)$ must be non-singular at each point of the superspace, hence

$$E \equiv \mathrm{Ber}\,(E_A{}^M) \neq 0. \tag{5.2.23}$$

Note that the superfield $E(z)$ is real, $\bar{E} = E$. Since $E_A{}^M(z)$ is non-singular, it possesses the inverse supermatrix $E_M{}^A(z)$ ('inverse supervierbein') defined by

$$E_M{}^A E_A{}^N = \delta_M{}^N \qquad E_A{}^M E_M{}^B = \delta_A{}^B. \tag{5.2.24}$$

A general coordinate transformation on $\mathbb{R}^{4|4}$ acts on E_A^M and E_M^A according to the laws

$$E'_A{}^M(z') = E_A{}^N(z)\partial_N z'^M \tag{5.2.25a}$$

$$E'_M{}^A(z') = (\partial'_M z^N)E_N{}^A(z). \tag{5.2.25b}$$

It is clear that $E^A = \mathrm{d}z^M E_M{}^A(z)$ are supercovector fields and, due to the non-singularity of $E_M{}^A$, the set $\{E^A|_p\}$ forms a basis in $T_p^*(\mathbb{R}^{4|4})$. Equation (5.2.25*a*) leads to

$$E'(z') = \mathrm{Ber}\left(\frac{\partial z'}{\partial z}\right)E(z)$$

therefore $E^{-1}(z)$ is a scalar density superfield of weight one. In the case of the coordinate transformations (5.2.13), equation (5.2.25*a*) is equivalent to

$$E'_A = E'_A{}^M(z)\partial_M = \mathrm{e}^K E_A \mathrm{e}^{-K}. \tag{5.2.26}$$

Using equation (5.2.16), the density superfield $E^{-1}(z)$ transforms according to the law

$$(E^{-1})' = E^{-1}\mathrm{e}^{\overleftarrow{K}}. \tag{5.2.27}$$

Remark. We take the convention that letters from the beginning of the Latin or Greek alphabet denote tangent-space indices.

Any supervector field $V = V^M(z)\partial_M$ can be decomposed with respect to the supervierbein,

$$V = V^A(z)E_A \qquad V^A = V^M E_M{}^A.$$

Next, any supercovector field $\omega = dz^M \omega_M(z)$ can be decomposed with respect to the inverse supervierbein,

$$\omega = E^A \omega_A(z) \qquad \omega_A = E_A{}^M \omega_M.$$

It is clear that $V^A(z)$ and $\omega_A(z)$ are scalar superfields with respect to general coordinate transformations. More generally, using the supervierbein and its inverse, all superspace indices of a supertensor field can be converted into tangent-space indices. For example, in the case of the second-rank supertensor field $\Psi_M{}^N(z)$, the conversion reads as follows:

$$\Psi_M{}^N \rightarrow \Psi_A{}^B = E_A{}^M \Psi_M{}^N E_N{}^B.$$

Supertensor fields with tangent-space indices are very useful in practice, since they change as scalar superfields under general coordinate transformations.

In what follows, we will work only with supertensor fields carrying tangent-space indices.

5.2.5. *Superlocal Lorentz group*

To breathe life into our theory, it is necessary to choose a proper structure (super)group G of the superspace. By its very origion, the structure group does not act on points of the superspace. It merely determines, for each point $p \in \mathbb{R}^{4|4}$, physically equivalent bases (or frames) in $T_p(\mathbb{R}^{4|4})$. Two standard bases $\{\mathscr{E}_A\}$ and $\{\mathscr{E}'_A\}$ of $T_p(\mathbb{R}^{4|4})$ are said to be equivalent if and only if they are connected by a structure group transformation:

$$\mathscr{E}'_A = \Lambda_A{}^B(g)\mathscr{E}_B \qquad g \in G.$$

Here Λ is some linear representation of G. Next, two supervierbeins E_A and E'_A are equivalent if they define equivalent frames at each superspace point:

$$E'_A|_p = \Lambda_A{}^B(g(p))E_B|_p \qquad \forall p \in \mathbb{R}^{4|4}$$

or, expanding E_A and E'_A with respect to the holonomic basis,

$$E'_A{}^M(z) = \Lambda_A{}^B(g(z))E_B{}^M(z). \tag{5.2.28}$$

The corresponding inverse supervierbeins are connected by the rule

$$E'_M{}^A(z) = E_M{}^B(z)\Lambda_B{}^A((g(z))^{-1}). \tag{5.2.29}$$

The above expressions involve a supersmooth mapping f: $\mathbb{R}^{4|4} \rightarrow G$. The set of all such mappings forms an infinite-dimensional supergroup, with respect to the multiplication $(f_1 \cdot f_2)(z) = f_1(z)f_2(z)$, which will be called the 'superlocal structure group'. Any superlocal structure group transformation moves the supervierbein into a physically equivalent one. Therefore, the symmetry group of a physical dynamical system should include the product of the general coordinate transformation supergroup and the superlocal structure group.

Remark. Changing the supervierbein to an equivalent one induces superlocal structure group transformation of supertensor fields. In particular, the supervector transformation law is

$$V^A(z) \to V'^A(z) = V^B(z)\Lambda_B{}^A((g(z))^{-1}) \tag{5.2.30}$$

and the supercovector transformation law is

$$\omega_A(z) \to \omega'_A(z) = \Lambda_A{}^B(g(z))\omega_B(z). \tag{5.2.31}$$

Now, how do we choose the superspace structure group? In the space–time case, there was the physical principle requiring us to identify the structure group with the Lorentz group. Namely, two vierbeins are equivalent if they determine the same space–time metric. Unfortunately, there is no physically natural superspace generalization of the concept of metric. The most reasonable assumption one can make is to identify the zeroth-order terms in the $\theta, \bar{\theta}$-expansion of the supervierbein's component $E_a{}^m$ with the space–time vierbein,

$$E_a{}^m| = e_a{}^m(x).$$

This requires the Lorentz group to be contained in a possible superspace structure group. We simply postulate the Lorentz group to be the superspace structure group and choose the representation $(\frac{1}{2}, \frac{1}{2}) \oplus (\frac{1}{2}, 0) \oplus (0, \frac{1}{2})$ to constitute the equivalence relation between tangent space frames. Then, two physically equivalent supervierbeins are connected by a 'superlocal Lorentz transformation' of the form (5.2.29), where

$$\begin{gathered}\Lambda_A{}^B = (\mathrm{e}^K)_A{}^B \qquad K_A{}^B = \begin{pmatrix} K_a{}^b & & 0 \\ & K_\alpha{}^\beta & \\ 0 & & -\bar{K}^{\dot{\alpha}}{}_{\dot{\beta}} \end{pmatrix} \\ K^{ab}(z) = (\sigma^{ab})_{\alpha\beta}K^{\alpha\beta}(z) - (\tilde{\sigma}^{ab})_{\dot{\alpha}\dot{\beta}}\bar{K}^{\dot{\alpha}\dot{\beta}}(z)\end{gathered} \tag{5.2.32}$$

with $K_{\alpha\beta}(z) = K_{\beta\alpha}(z)$ being arbitrary.

Note that Ber $(\Lambda_A{}^B) = 1$. Therefore, the superlocal Lorentz transformations do not change the Berezinian of the supervierbein,

$$E' = E. \tag{5.2.33}$$

As is seen, the superlocal Lorentz group acts reducibly on the supervierbein. Each of the supervector fields E_a, E_α and $\bar{E}^{\dot{\alpha}}$ transforms independently. Then, any supervector field or any supertensor field can be invariantly decomposed into Lorentz irreducible components. On these grounds, it is possible to speak of Lorentz vector superfields, Lorentz spinor superfields (dotted and undotted) and, more generally, one can define arbitrary Lorentz tensor superfields. A tensor superfield of Lorentz type $(n/2, m/2)$ is defined to transform under the superlocal Lorentz group according to the law

$$V'_{\alpha_1\ldots\alpha_n\dot{\alpha}_1\ldots\dot{\alpha}_m}(z) = \mathrm{e}^{\frac{1}{2}K^{ab}(z)M_{ab}}V_{\alpha_1\ldots\alpha_n\dot{\alpha}_1\ldots\dot{\alpha}_m}(z). \tag{5.2.34}$$

Here the Lorentz generators M_{ab} act, as usual, on the external superfield indices. When the superspace general coordinate transformations are also taken into consideration, the above transformation law must be substituted by

$$V'_{\alpha_1\dots\alpha_n\dot\alpha_1\dots\dot\alpha_m}(z) = e^{\mathscr{K}} V_{\alpha_1\dots\alpha_n\dot\alpha_1\dots\dot\alpha_m}(z) \tag{5.2.35}$$

where

$$\mathscr{K} = K^M\partial_M + \frac{1}{2}K^{ab}M_{ab} = K^M\partial_M + K^{\alpha\beta}M_{\alpha\beta} + \bar{K}^{\dot\alpha\dot\beta}\bar{M}_{\dot\alpha\dot\beta}. \tag{5.2.36}$$

Remark. As in Chapter 2, we will assume every tensor superfield with an even (odd) number of spinor indices to be bosonic (fermionic).

5.2.6. Superconnection and covariant derivatives

Let us introduce a 'Lorentz superconnection' — that is, a real c-type super 1-form taking its values in the Lorentz algebra,

$$\Omega = dz^M\Omega_M = E^A\Omega_A$$
$$\Omega_A = \frac{1}{2}\Omega_A{}^{bc}(z)M_{bc} = \Omega_A{}^{bc}M_{\beta\gamma} + \Omega_A{}^{\dot\beta\dot\gamma}M_{\dot\beta\dot\gamma} \tag{5.2.37}$$
$$\Omega_A{}^{bc} = (\Omega_a{}^{bc}, \Omega_\alpha{}^{bc}, \bar\Omega^{\dot\alpha bc})$$

and transforming according to the law

$$\Omega' = dz^M\,\Omega'_M = -dg\cdot g^{-1} + g\Omega g^{-1}$$
$$g(z) = \exp\left(\frac{1}{2}K^{ab}(z)M_{ab}\right) \tag{5.2.38}$$

under the superlocal Lorentz group. Then, the operators

$$\mathscr{D}_A = (\mathscr{D}_a, \mathscr{D}_\alpha, \bar{\mathscr{D}}^{\dot\alpha}) \equiv E_A + \Omega_A \tag{5.2.39}$$

change covariantly with respect to the superspace general coordinate and superlocal Lorentz transformations,

$$\mathscr{D}'_A = e^{\mathscr{K}}\mathscr{D}_A e^{-\mathscr{K}}. \tag{5.2.40}$$

In the infinitesimal case we have

$$\mathscr{D}'_A = \mathscr{D}_A + \delta\mathscr{D}_A \qquad \delta\mathscr{D}_A = [\mathscr{K}, \mathscr{D}_A]$$

and, using equation (5.2.39), this transformation law is equivalent to the relations

$$\delta E_A{}^M = \mathscr{K} E_A{}^M - E_A K^M \tag{5.2.41a}$$
$$\delta\Omega_A{}^{bc} = \mathscr{K}\Omega_A{}^{bc} - E_A K^{bc} \tag{5.2.41b}$$

which represent the supervierbein and superconnection transformation laws, respectively.

Remark. Equation (5.2.41*a*) shows that $E_A{}^M$ is a gauge field for the general coordinate transformation supergroup. Similarly, equation (5.2.41*b*) means that $\Omega_A{}^{bc}$ is a gauge feld for the superlocal Lorentz group.

Let $V_{\alpha_1\ldots\alpha_n\dot\alpha_1\ldots\dot\alpha_m}(z)$ be a tensor superfield. In accordance with equations (5.2.35, 36, 40), the superfield $\mathscr{D}_B V_{\alpha_1\ldots\alpha_n\dot\alpha_1\ldots\dot\alpha_m}(z)$ transforms covariantly,

$$(\mathscr{D}_B V_{\alpha_1\ldots\alpha_n\dot\alpha_1\ldots\dot\alpha_m}(z))' = \mathrm{e}^{\mathscr{K}}\mathscr{D}_B V_{\alpha_1\ldots\alpha_n\dot\alpha_1\ldots\dot\alpha_m}(z).$$

Here the Lorentz generators involved in $\mathscr{K}$ act on all the indices of $\mathscr{D}_B V_{\alpha_1\ldots\alpha_n\dot\alpha_1\ldots\dot\alpha_m}$. Therefore, $\mathscr{D}_A$ moves each tensor superfield into a tensor one. On these grounds, the operators $\mathscr{D}_A$ are said to be 'covariant derivatives'.

Covariant derivatives act on super(co)vectors according to the rule

$$\begin{aligned}\mathscr{D}_B U_A &= E_B U_A + \Omega_{BA}{}^C U_C \\ \mathscr{D}_B V^A &= E_B V^A + \Omega_B{}^A{}_C V^C\end{aligned} \tag{5.2.42}$$

where

$$\Omega_{AB}{}^C = \begin{pmatrix} \Omega_{Ab}{}^c & & 0 \\ & \Omega_{A\beta}{}^\gamma & \\ 0 & & -\Omega_A{}^{\dot\beta}{}_{\dot\gamma}\end{pmatrix} = -\Omega_A{}^C{}_B. \tag{5.2.43}$$

(Anti)commuting the covariant derivatives gives

$$\begin{aligned}[\mathscr{D}_A, \mathscr{D}_B\} &= T_{AB}{}^C\mathscr{D}_C + R_{AB} \\ R_{AB} = \frac{1}{2}R_{AB}{}^{cd}M_{cd} &= R_{AB}{}^{\gamma\delta}M_{\gamma\delta} + R_{AB}{}^{\dot\gamma\dot\delta}\bar{M}_{\dot\gamma\dot\delta} \\ T_{AB}{}^C = -(-1)^{\varepsilon_A\varepsilon_B}T_{BA}{}^C &\qquad R_{AB} = -(-1)^{\varepsilon_A\varepsilon_B}R_{BA}.\end{aligned} \tag{5.2.44}$$

Here $T_{AB}{}^C$ is the 'supertorsion', and $R_{AB}{}^{cd}$ is the 'supercurvature'. As a result of equations (5.2.40), $T_{AB}{}^C$ and $R_{AB}{}^{cd}$ are supertensor fields. The supertorsion and supercurvatures are expressed in terms of the supervierbein and the superconnection as follows

$$T_{AB}{}^C = C_{AB}{}^C + \Omega_{AB}{}^C - (-1)^{\varepsilon_A\varepsilon_B}\Omega_{BA}{}^C \tag{5.2.45a}$$

$$R_{AB}{}^{cd} = E_A\Omega_B{}^{cd} + \Omega_A{}^{cl}\omega_{Bl}{}^d - (-1)^{\varepsilon_A\varepsilon_B}(A \leftrightarrow B) - C_{AB}{}^L\Omega_L{}^{cd}. \tag{5.2.45b}$$

Here $C_{AB}{}^C$ are the 'anholonomy supercoefficients':

$$\begin{aligned}[E_A, E_B\} &= C_{AB}{}^C E_C \\ C_{AB}{}^C &= (E_A E_B{}^M - (-1)^{\varepsilon_A\varepsilon_B}E_B E_A{}^M)E_M{}^C.\end{aligned} \tag{5.2.46}$$

Taking into account the block-diagonal structure of $\Omega_{AB}{}^C$, the expression

(5.2.45*a*) can be rewritten in the form

$$
\begin{aligned}
&T_{\alpha\dot\beta}{}^{\dot\gamma} = C_{\alpha\dot\beta}{}^{\dot\gamma} + \Omega_{\alpha\dot\beta}{}^{\dot\gamma} + \Omega_{\dot\beta\alpha}{}^{\dot\gamma} \\
&T_{\alpha\beta\dot\gamma} = C_{\alpha\beta\dot\gamma} \qquad T_{\alpha\dot\beta}{}^{c} = C_{\alpha\dot\beta}{}^{c} \\
&T_{\alpha}{}^{\dot\beta\gamma} = C_{\alpha}{}^{\dot\beta\gamma} + \Omega^{\dot\beta}{}_{\alpha}{}^{\gamma} \qquad T_{\alpha}{}^{\dot\beta c} = C_{\alpha}{}^{\dot\beta c} \\
&T_{\alpha b}{}^{\gamma} = C_{\alpha b}{}^{\gamma} - \Omega_{b\alpha}{}^{\gamma} \qquad T_{\alpha b\dot\gamma} = C_{\alpha b\dot\gamma} \\
&T_{\alpha b}{}^{c} = C_{\alpha b}{}^{c} + \Omega_{\alpha b}{}^{c} \\
&T_{ab}{}^{c} = C_{ab}{}^{c} + \Omega_{ab}{}^{c} - \Omega_{ba}{}^{c} \qquad T_{ab}{}^{\gamma} = C_{ab}{}^{\gamma}.
\end{aligned}
\tag{5.2.47}
$$

It is seen that the anholonomy coefficients $C_{\alpha\beta\dot\gamma}$, $C_{\alpha\dot\beta}{}^{c}$, $C_{\alpha b\dot\gamma}$ and $C_{ab}{}^{\gamma}$ are Lorentz tensor superfields.

5.2.7. Bianchi identities and the Dragon theorem

The super-Jacobi identies (2.1.8) are satisfied for any set of operators, in particular, for the covariant derivatives:

$$
\begin{aligned}
[\mathscr{D}_A, [\mathscr{D}_B, \mathscr{D}_C\}\} &+ (-1)^{\varepsilon_A(\varepsilon_B+\varepsilon_C)}[\mathscr{D}_B, [\mathscr{D}_C, \mathscr{D}_A\}\} \\
&+ (-1)^{\varepsilon_C(\varepsilon_A+\varepsilon_B)}[\mathscr{D}_C, [\mathscr{D}_A, \mathscr{D}_B\}\} = 0.
\end{aligned}
$$

Using equation (5.2.44), the left-hand side can be rewritten symbolically as

$$
\mathrm{S}_{ABC}{}^{E}\mathscr{D}_E + \frac{1}{2}\Psi_{ABC}{}^{dl}M_{dl}
$$

and each of $\mathrm{S}_{ABC}{}^{E}$ and $\Psi_{ABC}{}^{dl}$ must vanish. The condition $\mathrm{S}_{ABC}{}^{E} = 0$ reads explicitly as

$$
R_{ABC}{}^{D} + (-1)^{\varepsilon_A(\varepsilon_B+\varepsilon_C)}R_{BCA}{}^{D} + (-1)^{\varepsilon_C(\varepsilon_A+\varepsilon_B)}R_{CAB}{}^{D} = \Delta_{ABC}{}^{D} \tag{5.2.48}
$$

where we have introduced the notation

$$
\begin{aligned}
\Delta_{ABC}{}^{D} = \{\mathscr{D}_A T_{BC}{}^{D} - T_{AB}{}^{E}T_{EC}{}^{D}\} &+ (-1)^{\varepsilon_A(\varepsilon_B+\varepsilon_C)}\{\mathscr{D}_B T_{CA}{}^{D} - T_{BC}{}^{E}T_{EA}{}^{D}\} \\
&+ (-1)^{\varepsilon_C(\varepsilon_A+\varepsilon_B)}\{\mathscr{D}_C T_{AB}{}^{D} - T_{CA}{}^{E}T_{EB}{}^{D}\}.
\end{aligned}
\tag{5.2.49}
$$

Next, the condition $\Psi_{ABC} = \frac{1}{2}\Psi_{ABC}{}^{dl}M_{dl} = 0$ reads as

$$
\begin{aligned}
\{-\mathscr{D}_A R_{BC} + T_{AB}{}^{D}R_{DC}\} &+ (-1)^{\varepsilon_A(\varepsilon_B+\varepsilon_C)}\{-\mathscr{D}_B R_{CA} + T_{BC}{}^{D}R_{DA}\} \\
&+ (-1)^{\varepsilon_C(\varepsilon_A+\varepsilon_B)}\{-\mathscr{D}_C R_{AB} + T_{CA}{}^{D}R_{DB}\} = 0
\end{aligned}
\tag{5.2.50}
$$

The relations (5.2.48, 50) are said to be the 'Bianchi identities'. Given arbitrary supervierbein and superconnection, the corresponding supertorsion and supercurvature satisfy the Bianchi identities.

One of the most important consequences of the Bianchi identities is the fact that the supercurvature can be expressed completely in terms of the supertorsion (this statement is known as the Dragon theorem). The point is that the supercurvature takes its values in the Lorentz algebra,

$$R_{ABC}{}^{D} = \begin{pmatrix} R_{ABc}{}^{d} & & 0 \\ & R_{AB\gamma}{}^{\delta} & \\ 0 & & -R_{AB}{}^{\dot{\gamma}}{}_{\dot{\delta}} \end{pmatrix} \tag{5.2.51}$$

and the components $R_{ABc}{}^{d}$, $R_{AB\gamma}{}^{\delta}$ and $R_{AB}{}^{\dot{\gamma}}{}_{\dot{\delta}}$ are connected as follows

$$R_{AB}{}^{cd} = (\sigma^{cd})_{\gamma\delta} R_{AB}{}^{\gamma\delta} - (\tilde{\sigma}^{cd})_{\dot{\gamma}\dot{\delta}} R_{AB}{}^{\dot{\gamma}\dot{\delta}}. \tag{5.2.52}$$

Now, let us choose $A = \alpha$, $B = \beta$, $C = c$ and $D = d$ in equation (5.2.48). Then, by virtue of expressions (5.2.51), we have

$$R_{\alpha\beta c}{}^{d} = \Delta_{\alpha\beta c}{}^{d}. \tag{5.2.53}$$

Similarly, one obtains

$$R_{\dot{\alpha}\beta c}{}^{d} = \Delta_{\dot{\alpha}\beta c}{}^{d} \qquad R_{\alpha\beta c}{}^{d} = \Delta_{\alpha\beta c}{}^{d}. \tag{5.2.54}$$

Then, we choose $A = \alpha$, $B = b$, $C = c$ and $D = d$. The Bianchi identities (5.2.48) give

$$R_{\alpha bc}{}^{d} + R_{c\alpha b}{}^{d} = R_{\alpha bc}{}^{d} - R_{\alpha cb}{}^{d} = \Delta_{\alpha bc}{}^{d}.$$

Since $R_{\alpha bcd} = -R_{\alpha bdc}$, this equation can be resolved as follows

$$R_{\alpha bcd} = \frac{1}{2}(\Delta_{\alpha bcd} - \Delta_{\alpha cdb} + \Delta_{\alpha dbc}). \tag{5.2.55}$$

$R_{\dot{\alpha}bcd}$ can be found analogously. Finally, we choose $A = a$, $B = b$, $C = \gamma$ and $D = \delta$ (or $C = \dot{\gamma}$ and $D = \dot{\delta}$), obtaining

$$R_{ab\gamma\delta} = \Delta_{ab\gamma\delta} \qquad R_{ab\dot{\gamma}\dot{\delta}} = \Delta_{ab\dot{\gamma}\dot{\delta}}. \tag{5.2.56}$$

It only remains to apply equation (5.2.52).

Therefore, we have proved that the supercurvature is completely expressible in terms of the supertorsion, and hence, in a sense, the supercurvature is a redundant object. This fact had no place in general relativity, where the curvature and the torsion were independent, and one could switch off the torsion leaving a non-vanishing curvature. In superspace, the supertorsion is the main object determining superspace geometry.

5.2.8. Integration by parts

Theorem. Given a supervector field $V^{A}(z)$ under proper boundary conditions, the following identity

$$\int d^{8}z E^{-1}(-1)^{\varepsilon_A} \mathscr{D}_A V^{A} = \int d^{8}z E^{-1}(-1)^{\varepsilon_B} V^{A} T_{AB}{}^{B} \tag{5.2.57}$$

is fulfilled.

Proof. Making use of equations (5.2.42, 43) gives

$$(-1)^{\varepsilon_A}\mathscr{D}_A V^A = (-1)^{\varepsilon_A} E_A V^A - V^A \Omega_{BA}{}^B. \tag{5.2.58}$$

Further, equation (5.2.45*a*) leads to

$$(-1)^{\varepsilon_B} T_{AB}{}^B = (-1)^{\varepsilon_B} C_{AB}{}^B - \Omega_{BA}{}^B \tag{5.2.59}$$

since $(-1)^{\varepsilon_B}\Omega_{AB}{}^B = \Omega_{Ab}{}^b - \Omega_{A\beta}{}^\beta + \Omega_{A\dot\beta}{}^{\dot\beta} = 0$. Finally, recalling the definition of the anholonomy coefficients (5.2.46), one readily obtains

$$(-1)^{\varepsilon_B} C_{AB}{}^B = -E_A \ln E^{-1} - (1\cdot \overleftarrow{E}_A). \tag{5.2.60}$$

The relations (5.2.58–60) show that

$$(-1)^{\varepsilon_A} E^{-1}\mathscr{D}_A V^A = (E^{-1}V^A)\overleftarrow{E}_A + (-1)^{\varepsilon_B} E^{-1} V^A T_{AB}{}^B.$$

Here the first term in the r.h.s. is a total derivative. This completes the proof.

Remark. In the cases of vector or undotted spinor superfields equation (5.2.57) reads

$$\int \mathrm{d}^8 z E^{-1}\mathscr{D}_a V^a = \int \mathrm{d}^8 z E^{-1}(-1)^{\varepsilon_B} V^a T_{aB}{}^B$$
$$-\int \mathrm{d}^8 z E^{-1}\mathscr{D}_\alpha V^\alpha = \int \mathrm{d}^8 z E^{-1}(-1)^{\varepsilon_B} V^\alpha T_{\alpha B}{}^B. \tag{5.2.61}$$

Remark. The relation (5.2.57) takes the most simple form in the case when $(-1)^{\varepsilon_B} T_{AB}{}^B$ vanishes,

$$(-1)^{\varepsilon_B} T_{AB}{}^B = 0 \Rightarrow \int \mathrm{d}^8 z E^{-1}\mathscr{D}_A V^A = 0. \tag{5.2.62}$$

5.2.9. Flat superspace geometry

The approach developed above enables us to introduce the notion of 'superspace geometry' (or 'supergeometry') in literal analogy with the notion of space–time geometry. A supergeometry is determined by the set of superfields $\{E_A{}^M(z), \Omega_A{}^{bc}(z)\}$, where $E_A{}^M$ is a supervierbein and $\Omega_A{}^{bc}$ is a Lorentz superconnection. The $E_A{}^M$ and $\Omega_A{}^{bc}$ will be called the 'supergeometry potentials'. Two supergeometries are said to be equivalent if the corresponding potentials are connected by some superspace general coordinate and superlocal Lorentz transformation.

A 'flat superspace geometry' may be described as follows. A supergeometry is said to be flat if these exists a gauge choice with respect to the general coordinate transformation supergroup and the superlocal Lorentz group such that the covariant derivatives $\mathscr{D}_A$ coincide with the flat covariant derivatives D_A,

$$\mathscr{D}_A = D_A = \mathscr{E}_A{}^M \partial_M \qquad \Omega_A{}^{bc} = 0 \tag{5.2.63}$$

$$\mathscr{E}_A{}^M = \left(\begin{array}{c|c|c} \delta_a{}^m & 0 & 0 \\ \hline \mathrm{i}(\sigma^m)_{\alpha\dot{\nu}}\bar{\theta}^{\dot{\nu}} & \delta_\alpha{}^\mu & 0 \\ \hline \mathrm{i}(\tilde{\sigma}^m)^{\dot{\alpha}\nu}\theta_\nu & 0 & \delta^{\dot{\alpha}}{}_{\dot{\mu}} \end{array}\right).$$

These derivatives satisfy the algebra

$$[D_A, D_B\} = T_{AB}{}^C D_C \qquad R_{AB}{}^{cd} \equiv 0 \tag{5.2.64}$$

where the only non-vanishing supertorsion components are

$$T_{\alpha\dot{\beta}}{}^c = T_{\dot{\beta}\alpha}{}^c = -2\mathrm{i}(\sigma^c)_{\alpha\dot{\beta}}. \tag{5.2.65}$$

It is not difficult to find all gauge transformations (5.2.41) preserving the gauge (5.2.63). First, choosing $\mathscr{K} = \hat{K}^M \partial_M$ and demanding $\delta\mathscr{E}_A{}^M = 0$, one arrives at the equation

$$[D_A, \hat{K}^M \partial_M] = 0. \tag{5.2.66}$$

After the redefinition

$$\hat{K}^M \partial_M \equiv -b^m \partial_m + \mathrm{i}(\epsilon^\mu \mathbf{Q}_\mu + \bar{\epsilon}_{\dot{\mu}} \bar{\mathbf{Q}}^{\dot{\mu}}) \tag{5.2.67}$$

where $(\mathbf{Q}_\mu, \bar{\mathbf{Q}}^{\dot{\mu}})$ are the supersymmetry generators (2.4.37), the above equation leads to

$$D_A b^m = D_A \epsilon^\mu = 0 \Rightarrow b^m = \text{const} \qquad \epsilon^\mu = \text{const}. \tag{5.2.68}$$

Thus, we recover the space–time translations and the supersymmetry transformations. Further, we consider $\mathscr{K} = K^M \partial_M + \frac{1}{2} K^{ab} M_{ab}$ with $K_{ab} \neq 0$. In accordance with equation (5.2.41), the requirements $\delta\mathscr{E}_A{}^M = 0$, $\Omega'_A{}^{bc} = \Omega_A{}^{bc} = 0$ give two equations

$$\begin{aligned} D_A K^{bc} &= 0 \Rightarrow K^{bc} = \text{const} \\ K_A{}^B D_B &= [D_A, K^M \partial_M]. \end{aligned} \tag{5.2.69}$$

The general solution of the second equation is

$$K^M = z^N \delta_N{}^A K_A{}^B \delta_B{}^M + \hat{K}^M \tag{5.2.70}$$

and we come to the standard Lorentz transformations of flat global superspace. After this, the superspace and tanget-space incides can be identified.

5.3. Supergeometry with conformal supergravity constraints

The elegant supergeometrical formalism developed in previous section, turns out to be too general to correspond literally to superfield supergravity. To describe a superspace geometry, one must specify the superpotentials $E_A{}^M(z)$ and $\Omega_A{}^{bc}(z)$, which are arbitrary superfields in the general case. On the other hand, to describe a curved supergravity superspace, it is enough to introduce far lesser superfields: $\mathscr{H}^m(z)$, $\varphi(x + i\mathscr{H}, \theta)$ and $\bar{\varphi}(x - i\mathscr{H}, \bar{\theta})$. However it is not time to despair. We may hope that superfield supergravity corresponds to a subspace in the space of all supergeometries, where supervierbeins and superconnections satisfy some covariant constraints. Since the only covariant objects constructed from $E_A{}^M$ and $\Omega_A{}^{bc}$ are the supertorsion and supercurvature, any invariant subspace in the space of supergeometries (transforming into itself under the action of the general coordinate transformation supergroup and the superlocal Lorentz group) is selected out by imposing some constraints on $T_{AB}{}^C$ and $R_{AB}{}^{cd}$.

In principle, the situation is the same as in general relativity. Gravity is a gauge theory of vierbeins $e_a{}^m(x)$. From the other side, in order to specify a geometry it is necessary to accompany the vierbein by a Lorentz connection $\omega_a{}^{bc}(x)$. It is the torsion-free condition $\mathscr{T}_{ab}{}^c = 0$, which, expressing $\omega_a{}^{bc}$ uniquely through $e_a{}^m$, determines geometries corresponding to gravitation.

If we try to keep a direct analogy with general relativity, it is worth imposing the supertorsion-free condition

$$T_{AB}{}^C = 0. \tag{5.3.1}$$

This determines, due to equation (5.2.45*a*), the superconnection in terms of the supervierbein. However, this constraint is unsuitable for two reasons. First, flat global superspace is characterized by the non-vanishing supertorsion (5.2.65). Hence, the constraint (5.3.1) does not admit flat superspace as a particular solution. Secondly, it has been shown in subsection 5.2.7 that supercurvature is expressed via supertorsion. Then, the constraint (5.3.1) leads to the requirement

$$R_{AB} = 0$$

and as a consequence, to a trivial supergeometry.

We have reached the point where our space–time intuition is unable to help us. Now, in order to find reasonable restrictions on superspace geometry, some purely superspace arguments must be taken into account.

5.3.1. Conformal supergravity constraints

In global supersymmetry chiral scalar superfields, defined by the equation $\bar{D}_{\dot{\alpha}}\Phi = 0$, were of primary importance. Clearly, it would be very desirable to transfer this supersymmetry representation to the local case — that is, to introduce 'covariantly chiral scalar superfields' defined by

$$\bar{\mathscr{D}}_{\dot{\alpha}}\Phi = 0. \tag{5.3.2}$$

But this equation proves to be consistent only under non-trivial restrictions on the supertorsion. Indeed, demanding equation (5.3.2) and using the covariant derivative algebra (5.2.44), one obtains

$$0 = \{\bar{\mathscr{D}}_{\dot{\alpha}}, \bar{\mathscr{D}}_{\dot{\beta}}\}\Phi = T_{\dot{\alpha}\dot{\beta}}{}^{C}\mathscr{D}_{C}\Phi = T_{\dot{\alpha}\dot{\beta}}{}^{c}\mathscr{D}_{c}\Phi + T_{\dot{\alpha}\dot{\beta}}{}^{\gamma}\mathscr{D}_{\gamma}\Phi$$

and hence

$$T_{\dot{\alpha}\dot{\beta}}{}^{c} = T_{\dot{\alpha}\dot{\beta}}{}^{\gamma} = 0 \Rightarrow T_{\alpha\beta}{}^{c} = T_{\alpha\beta\dot{\gamma}} = 0. \tag{5.3.3}$$

Recalling that $T_{\alpha\beta}{}^{c} = C_{\alpha\beta}{}^{c}$ and $T_{\alpha\beta\dot{\gamma}} = C_{\alpha\beta\dot{\gamma}}$ (see equation (5.2.47)), we can rewrite equation (5.3.3) in the equivalent form

$$\{E_{\alpha}, E_{\beta}\} = C_{\alpha\beta}{}^{\gamma}E_{\gamma}. \tag{5.3.4}$$

We take the requirements (5.3.3) as constraints on superspace geometry.

In global supersymmetry, spinor covariant derivatives D_{α} and $\bar{\mathrm{D}}_{\dot{\alpha}}$ generated the full algebra of covariant derivatives $\mathrm{D}_{A} = \{\mathrm{D}_{a} = \partial_{a}, \mathrm{D}_{\alpha}, \bar{\mathrm{D}}^{\dot{\alpha}}\}$ since the vector covariant derivative was expressed as an anticommutator of spinor ones,

$$\{\mathrm{D}_{\alpha}, \bar{\mathrm{D}}_{\dot{\beta}}\} = -2\mathrm{i}(\sigma^{a})_{\alpha\dot{\alpha}}\mathrm{D}_{a} = -2\mathrm{i}\mathrm{D}_{\alpha\dot{\alpha}}.$$

It is very natural to conserve this basic property in the local case also, i.e. to consider only such supergeometries in which spinor derivatives $\mathscr{D}_{\alpha}$ and $\bar{\mathscr{D}}_{\dot{\alpha}}$ generate the algebra $\mathscr{D}_{A} = (\mathscr{D}_{a}, \mathscr{D}_{\alpha}, \bar{\mathscr{D}}^{\dot{\alpha}})$. This means that $\mathscr{D}_{a}$ can be extracted from $\{\mathscr{D}_{\alpha}, \bar{\mathscr{D}}_{\dot{\alpha}}\}$:

$$\{\mathscr{D}_{\alpha}, \bar{\mathscr{D}}_{\dot{\alpha}}\} = T_{\alpha\dot{\alpha}}{}^{b}\mathscr{D}_{b} + T_{\alpha\dot{\alpha}}{}^{\beta}\mathscr{D}_{\beta} + T_{\alpha\dot{\alpha}\dot{\beta}}\bar{\mathscr{D}}^{\dot{\beta}} + R_{\alpha\dot{\alpha}}$$

$$\det((\tilde{\sigma}_{a})^{\dot{\alpha}\alpha}T_{\alpha\dot{\alpha}}{}^{b}) \neq 0.$$

Now, one can make the redefinition

$$\mathscr{D}_{A} \to \tilde{\mathscr{D}}_{A} = (\tilde{\mathscr{D}}_{a}, \tilde{\mathscr{D}}_{\alpha}, \tilde{\bar{\mathscr{D}}}^{\dot{\alpha}}) \equiv \left(-\frac{\mathrm{i}}{4}(\tilde{\sigma}_{a})^{\dot{\beta}\beta}\{\mathscr{D}_{\beta}, \bar{\mathscr{D}}_{\dot{\beta}}\}, \mathscr{D}_{\alpha}, \bar{\mathscr{D}}^{\dot{\alpha}}\right)$$

changing only the vector supervierbein

$$E_{a}{}^{M} \to \tilde{E}_{a}{}^{M} = -\frac{\mathrm{i}}{4}(\tilde{\sigma}_{a})^{\dot{\alpha}\alpha}T_{\alpha\dot{\alpha}}{}^{B}E_{B}{}^{M}$$

and the vector superconnection

$$\Omega_{a}{}^{bc} \to \tilde{\Omega}_{a}{}^{bc} = -\frac{\mathrm{i}}{4}(\tilde{\sigma}_{a})^{\dot{\alpha}\alpha}\{T_{\alpha\dot{\alpha}}{}^{B}\Omega_{B}{}^{bc} + R_{\alpha\dot{\alpha}}{}^{cd}\}.$$

The operators $\tilde{\mathscr{D}}_A$ are characterized by the transformation laws:

$$\tilde{\mathscr{D}}'_A = e^{\mathscr{K}} \tilde{\mathscr{D}}_A e^{-\mathscr{K}} \qquad \mathscr{K} = K^M \partial_M + \frac{1}{2} K^{ab} M_{ab}$$

therefore they define a set of covariant derivatives. It is important that, after redefinition, the anticommutator of dotted and undotted spinor derivatives takes the same form as in global supersymmetry. As a result, the flat supergeometry arises as a particular solution.

Motivated by the above discussion, we postulate the constraints

$$\{\mathscr{D}_\alpha, \bar{\mathscr{D}}_{\dot\alpha}\} = -2i(\sigma^a)_{\alpha\dot\alpha}\mathscr{D}_a \equiv -2i\mathscr{D}_{\alpha\dot\alpha} \Leftrightarrow \begin{cases} T_{\alpha\dot\alpha}{}^a = -2i(\sigma^a)_{\alpha\dot\alpha} \\ T_{\alpha\dot\alpha}{}^\beta = T_{\alpha\dot\alpha\dot\beta} = 0 \\ R_{\alpha\dot\alpha}{}^{cd} = 0. \end{cases} \tag{5.3.5}$$

This means that the vector supervierbein E_a and superconnection Ω_a are expressed in terms of the spinor supervierbeins E_α and $\bar{E}_{\dot\alpha}$ and superconnections Ω_α and $\bar{\Omega}_{\dot\alpha}$. In particular, we have

$$E_{\alpha\dot\alpha} \equiv (\sigma^a)_{\alpha\dot\alpha} E_a = \frac{i}{2}\{E_\alpha, \bar{E}_{\dot\alpha}\} - \frac{i}{2}\Omega_{\alpha\dot\alpha\beta}\bar{E}^{\dot\beta} + \frac{i}{2}\bar{\Omega}_{\dot\alpha\alpha}{}^\beta E_\beta$$

$$\Rightarrow \{E_\alpha, \bar{E}_{\dot\alpha}\} = C_{\alpha\dot\alpha}{}^B E_B \qquad C_{\alpha\dot\alpha}{}^B = \begin{cases} -2i(\sigma^a)_{\alpha\dot\alpha} \\ -\bar{\Omega}_{\dot\alpha\alpha}{}^\beta \\ \Omega_{\alpha\dot\alpha\dot\beta}. \end{cases} \tag{5.3.6}$$

After imposing constraints (5.3.5), the only independent supergeometry variables are the spinor supervierbeins and superconnections. Similar to general relativity, it would be desirable to completely express the superconnection via the supervierbein. This requires additional constraints. Let us impose the following constraints

$$T_{\alpha\beta\gamma} = 0 \Rightarrow T_{\dot\alpha\dot\beta\dot\gamma} = 0 \tag{5.3.7}$$

and

$$T_{\alpha bc}(\tilde\sigma^{bc})_{\dot\beta\dot\gamma} = 0 \Rightarrow T_{\dot\alpha bc}(\sigma^{bc})_{\beta\gamma} = 0 \tag{5.3.8}$$

and show that they are sufficient to express Ω_α and $\bar\Omega_{\dot\alpha}$ via E_α and $\bar{E}_{\dot\alpha}$. First, by virtue of equation (5.2.47), the constraint $T_{\alpha\beta\gamma} = 0$ leads to

$$\Omega_{\alpha\beta\gamma} = -\frac{1}{2}(C_{\alpha\beta\gamma} + C_{\alpha\gamma\beta} - C_{\beta\gamma\alpha}). \tag{5.3.9}$$

Next, to resolve the constraint (5.3.8), we introduce the semi-covariant supervierbein (changing covariantly under the superspace general co-ordinate transformations and non-covariantly under the superlocal Lorentz transformations)

$$\check{E}_A = (\check{E}_a, \check{E}_\alpha, \check{\bar{E}}^{\dot\alpha}) \equiv \left(-\frac{i}{4}(\tilde\sigma_a)^{\dot\beta\beta}\{E_\beta, \bar{E}_{\dot\beta}\}, E_\alpha, \bar{E}^{\dot\alpha} \right)$$

$$[\check{E}_A, \check{E}_B\} = \check{C}_{AB}{}^C \check{E}_C \tag{5.3.10}$$

constructed in terms of E_α and $\bar{E}_{\dot\alpha}$ only. As opposed to $\check{E}_a$, the covariant vector supervierbein E_a depends, due to equation (5.3.6), on the spinor superconnections:

$$E_{\alpha\dot\alpha} = \check{E}_{\alpha\dot\alpha} + \frac{i}{2}\Omega_{\alpha\dot\alpha}{}^{\dot\beta}\bar{E}_{\dot\beta} + \frac{i}{2}\bar\Omega_{\dot\alpha\alpha}{}^{\beta}E_\beta$$

$$\Rightarrow C_{\alpha,\beta\dot\beta,}{}^{\gamma\dot\gamma} = \check{C}_{\alpha,\beta\dot\beta,}{}^{\gamma\dot\gamma} + 2\Omega_{\beta\dot\beta}{}^{\dot\gamma}\delta_\alpha{}^\gamma \tag{5.3.11}$$

where we have introduced the notation

$$C_{\alpha,\beta\dot\beta,\gamma\dot\gamma} = (\sigma^b)_{\beta\dot\beta}(\sigma^c)_{\gamma\dot\gamma}C_{\alpha bc}.$$

Now, demanding (5.3.8) and recalling (5.2.47), one obtains

$$\Omega_{\alpha\beta\dot\gamma} = \frac{1}{4}(\check{C}_{\alpha,\beta\beta,}{}^{\beta}{}_{\dot\gamma} + \check{C}_{\alpha,\beta\dot\gamma,}{}^{\beta}{}_{\beta}). \tag{5.3.12}$$

As a result, we have expressed Ω_α via E_α and $\bar{E}_{\dot\alpha}$.

Introducing the notation

$$T_{\alpha,\beta\dot\beta,\gamma\dot\gamma} = (\sigma^b)_{\beta\dot\beta}(\sigma^c)_{\gamma\dot\gamma}T_{\alpha bc} \tag{5.3.13}$$

the constraint (5.3.8) can be rewritten in the form

$$T_{\alpha,\beta\dot\beta,}{}^{\beta}{}_{\dot\gamma} = \varepsilon_{\dot\beta\dot\gamma}T_{\alpha b}{}^{b}. \tag{5.3.14}$$

We can summarize all the constraints on a superspace geometry:

representation-preserving constraints

$$T_{\dot\alpha\dot\beta}{}^{c} = T_{\dot\alpha\dot\beta}{}^{\gamma} = T_{\alpha\beta}{}^{c} = T_{\alpha\beta}{}^{\dot\gamma} = 0 \tag{5.3.15a}$$

conventional constraints (I)

$$T_{\alpha\dot\beta}{}^{\gamma} = T_{\alpha\dot\beta}{}^{\dot\gamma} = R_{\alpha\dot\beta}{}^{cd} = 0; \qquad T_{\alpha\dot\beta}{}^{c} = -2i(\sigma^c)_{\alpha\dot\beta}$$

$$\Leftrightarrow \mathfrak{D}_{\alpha\dot\alpha} = \frac{i}{2}\{\mathscr{D}_\alpha, \bar{\mathscr{D}}_{\dot\alpha}\} \tag{5.3.15b}$$

conventional constraints (II)

$$T_{\alpha\beta}{}^{\gamma} = T_{\dot\alpha\beta}{}^{\dot\gamma} = T_{\alpha,\beta(\dot\beta,}{}^{\beta}{}_{\dot\gamma)} = T_{\dot\alpha,(\beta\dot\beta,\gamma)}{}^{\dot\beta} = 0. \tag{5.3.15c}$$

Representation-preserving constraints make possible the existence of chiral scalar superfields; conventional constraints (I) determine the vector covariant derivative in terms of the spinor ones; conventional constraints (II) determine the spinor superconnections Ω_α and $\bar{\Omega}_{\dot\alpha}$ in terms of the spinor supervierbeins E_α and $\bar{E}_{\dot\alpha}$.

Constraints (5.3.15*a*) and (5.3.15*c*) mean that

$$T_{\alpha\beta}{}^C = T_{\dot\alpha\beta}{}^C = 0. \tag{5.3.16}$$

As will be shown later, constraints (5.3.15) correspond to conformal supergravity.

5.3.2. *The Bianchi identities*

For a given set of covariant derivatives $\mathscr{D}_A$, the Bianchi identities (5.2.48, 50) are satisfied identically. The Bianchi identities become non-trivial when the covariant derivatives are restricted by constraints. In this case the Bianchi identities play the role of consistency conditions and may be used to determine non-vanishing components of supertorsion and supercurvature. Now, our goal is to investigate the consequences to which the Bianchi identities lead when the constraints (5.3.15) are chosen.

We organize the analysis of the Bianchi identities according to their (mass) dimensions (recall that the spinor derivatives $\mathscr{D}_\alpha$ have dimension $\frac{1}{2}$, while the dimension of $\mathscr{D}_a$ is 1). Possible dimensions of superfield expressions in (5.2.48, 50) run from $\frac{1}{2}$ to 3. We reproduce below only the identities needed for our analysis, ordering them by dimension.

First of all, to simplify the analysis, it is useful to convert any vector index into a pair of spinor indices (dotted and undotted) in the standard fashion:

$$T_{\alpha,\beta\dot\beta,\gamma} = (\sigma^b)_{\beta\dot\beta} T_{\alpha b\gamma} \qquad T_{\alpha,\beta\dot\beta,\dot\gamma} = (\sigma^b)_{\beta\dot\beta} T_{\alpha b\dot\gamma}$$
$$T_{\alpha\dot\alpha,\beta\dot\beta,\gamma} = (\sigma^a)_{\alpha\dot\alpha}(\sigma^b)_{\beta\dot\beta} T_{ab\gamma} \tag{5.3.17}$$
$$T_{\alpha\dot\alpha,\beta\dot\beta,\gamma\dot\gamma} = (\sigma^a)_{\alpha\dot\alpha}(\sigma^b)_{\beta\dot\beta}(\sigma^c)_{\gamma\dot\gamma} T_{abc}$$

and analogously for complex conjugate quantities (see also equation (5.3.13)) and supercurvature. It is also necessary to keep in mind relation (5.2.51) along with the identity

$$R_{AB,\gamma\dot\gamma,\delta\dot\delta} \equiv (\sigma^c)_{\gamma\dot\gamma}(\sigma^d)_{\delta\dot\delta} R_{ABcd} = 2\varepsilon_{\dot\gamma\dot\delta} R_{AB\gamma\delta} + 2\varepsilon_{\gamma\delta} R_{AB\dot\gamma\dot\delta}. \tag{5.3.18}$$

All identities of dimension $\frac{1}{2}$ and 1 are contained in (5.2.48). Taking into account constraints (5.3.15), they can be written as:

Dim 1/2

Choose $A = \alpha$, $B = \beta$, $C = \dot\alpha$, $D = c$. Then one obtains

$$T_{\alpha,\beta\dot\alpha,\gamma\dot\gamma} + T_{\beta,\alpha\dot\alpha,\gamma\dot\gamma} = 0 \tag{5.3.19}$$

and its complex conjugate.

Dim 1

Choose $A = \alpha$, $B = \beta$, $C = \gamma$, $D = \delta$. Then one obtains

$$R_{\alpha\beta\gamma\delta} + R_{\beta\gamma\alpha\delta} + R_{\gamma\alpha\beta\delta} = 0 \tag{5.3.20}$$

and its complex conjugate.

Choose $A = \alpha$, $B = \beta$, $C = \dot{\gamma}$, $D = \dot{\delta}$. Then one obtains

$$R_{\alpha\beta\dot{\gamma}\dot{\delta}} = 2\mathrm{i}T_{\alpha,\beta\dot{\gamma},\dot{\delta}} + 2\mathrm{i}T_{\beta,\alpha\dot{\gamma},\dot{\delta}} \tag{5.3.21}$$

and its complex conjugate.

Choose $A = \alpha$, $B = \dot{\beta}$, $C = \beta$, $D = \gamma$. Then one obtains

$$T_{\alpha,\beta\dot{\beta},\gamma} + T_{\beta,\alpha\dot{\beta},\gamma} = 0 \tag{5.3.22}$$

as well as its complex conjugate.

Choose $A = \alpha$, $B = \beta$, $C = c$, $D = d$. This leads to

$$\begin{aligned} 2R_{\alpha\beta\gamma\delta}\varepsilon_{\dot{\gamma}\dot{\delta}} + 2R_{\alpha\beta\dot{\gamma}\dot{\delta}}\varepsilon_{\gamma\delta} = {} & \mathscr{D}_\alpha T_{\beta,\gamma\dot{\gamma},\delta\dot{\delta}} + \mathscr{D}_\beta T_{\alpha,\gamma\dot{\gamma},\delta\dot{\delta}} + 4\mathrm{i}\varepsilon_{\alpha\delta}T_{\beta,\gamma\dot{\gamma},\dot{\delta}} \\ & + 4\mathrm{i}\varepsilon_{\beta\delta}T_{\alpha,\gamma\dot{\gamma},\dot{\delta}} + \frac{1}{2}T_{\beta,\gamma\dot{\gamma},}{}^{\lambda\dot{\lambda}}T_{\alpha,\lambda\dot{\lambda},\delta\dot{\delta}} \\ & + \frac{1}{2}T_{\alpha,\gamma\dot{\gamma},}{}^{\lambda\dot{\lambda}}T_{\beta,\lambda\dot{\lambda},\delta\dot{\delta}} \end{aligned} \tag{5.3.23}$$

as well as its complex conjugate.

Choose $A = \alpha$, $B = \dot{\alpha}$, $C = b$, $D = d$. This gives

$$\begin{aligned} -2\mathrm{i}T_{\alpha\dot{\alpha},\beta\dot{\beta},\gamma\dot{\gamma}} = {} & \mathscr{D}_\alpha T_{\dot{\alpha},\beta\dot{\beta},\gamma\dot{\gamma}} + \bar{\mathscr{D}}_{\dot{\alpha}}T_{\alpha,\beta\dot{\beta},\gamma\dot{\gamma}} + 4\mathrm{i}\varepsilon_{\alpha\gamma}T_{\dot{\alpha},\beta\dot{\beta},\dot{\gamma}} - 4\mathrm{i}\varepsilon_{\dot{\alpha}\dot{\gamma}}T_{\alpha,\beta\dot{\beta},\gamma} \\ & + \frac{1}{2}T_{\dot{\alpha},\beta\dot{\beta},}{}^{\lambda\dot{\lambda}}T_{\alpha,\lambda\dot{\lambda},\gamma\dot{\gamma}} + \frac{1}{2}T_{\alpha,\beta\dot{\beta},}{}^{\lambda\dot{\lambda}}T_{\dot{\alpha},\lambda\dot{\lambda},\gamma\dot{\gamma}}. \end{aligned} \tag{5.3.24}$$

Dim 3/2

Choose $A = \alpha$, $B = \beta$, $C = \gamma$ in the second Bianchi identities (5.2.50). This gives

$$\mathscr{D}_\alpha R_{\beta\gamma} + \mathscr{D}_\beta R_{\gamma\alpha} + \mathscr{D}_\gamma R_{\alpha\beta} = 0. \tag{5.3.25}$$

Choose $A = \alpha$, $B = \dot{\beta}$, $C = \dot{\gamma}$ in (5.2.50). Then one obtains

$$\mathscr{D}_\alpha R_{\dot{\beta}\dot{\gamma}} = 2\mathrm{i}R_{\dot{\gamma},\alpha\dot{\beta}} + 2\mathrm{i}R_{\dot{\beta},\alpha\dot{\gamma}} \tag{5.3.26}$$

and its complex conjugate.

Choose $A = \dot{\alpha}, B = b, C = \gamma, D = \delta$ in expression (5.2.48). Then one obtains

$$R_{\dot{\alpha},\beta\dot{\beta},\gamma\delta} = -\bar{\mathscr{D}}_{\dot{\alpha}} T_{\gamma,\beta\dot{\beta},\delta} - \mathscr{D}_{\gamma} T_{\dot{\alpha},\beta\dot{\beta},\delta} - \frac{1}{2} T_{\dot{\alpha},\beta\dot{\beta},}{}^{\lambda\dot{\lambda}} T_{\gamma,\lambda\dot{\lambda},\delta} - \frac{1}{2} T_{\gamma,\beta\dot{\beta},}{}^{\lambda\dot{\lambda}} T_{\dot{\alpha},\lambda\dot{\lambda},\delta} - 2\mathrm{i} T_{\gamma\dot{\alpha},\beta\dot{\beta},\delta} \tag{5.3.27}$$

and its complex conjugate.

Finally, we choose $A = \dot{\alpha}$, $B = b$, $C = \dot{\gamma}$, $D = \delta$ in expression (5.2.48). Such choice leads to the identity

$$R_{\dot{\alpha},\beta\dot{\beta},\dot{\gamma}\delta} + R_{\dot{\gamma},\beta\dot{\beta},\dot{\alpha}\delta} = \bar{\mathscr{D}}_{\dot{\alpha}} T_{\dot{\gamma},\beta\dot{\beta},\delta} - \bar{\mathscr{D}}_{\dot{\gamma}} T_{\dot{\alpha},\beta\dot{\beta},\delta} + \frac{1}{2} T_{\dot{\alpha},\beta\dot{\beta},}{}^{\lambda\dot{\lambda}} T_{\dot{\gamma},\lambda\dot{\lambda},\delta} + \frac{1}{2} T_{\dot{\gamma},\beta\dot{\beta},}{}^{\lambda\dot{\lambda}} T_{\dot{\alpha},\lambda\dot{\lambda},\delta} \tag{5.3.28}$$

and its complex conjugate.

Dim 2

Choose $A = a, B = b, C = \gamma, D = \gamma$ in (5.2.48). Since $R_{ab\gamma}{}^{\gamma} = 0$, one obtains

$$\mathscr{D}_a T_{\gamma b}{}^{\gamma} - \mathscr{D}_b T_{\gamma a}{}^{\gamma} - T_{ab}{}^{c} T_{\gamma c}{}^{\gamma} + T_{\gamma a}{}^{E} T_{Eb}{}^{\gamma} - T_{\gamma b}{}^{E} T_{Ea}{}^{\gamma} + \mathscr{D}^{\gamma} T_{ab\gamma} = 0 \tag{5.3.29}$$

and its complex conjugate.

Choose $A = a$, $B = b$, $C = \dot{\gamma}$, $D = \delta$ in (5.2.48). One obtains

$$\bar{\mathscr{D}}_{\dot{\gamma}} T_{ab\delta} - \mathscr{D}_a T_{\dot{\gamma}b\delta} + \mathscr{D}_b T_{\dot{\gamma}a\delta} + T_{ab}{}^{E} T_{\dot{\gamma}E\delta} - T_{\dot{\gamma}a}{}^{E} T_{Eb\delta} + T_{\dot{\gamma}b}{}^{E} T_{Ea\delta} = 0 \tag{5.3.30}$$

and its complex conjugate.

The Bianchi identities of dimension $\frac{5}{2}$ and 3 will not be of help to us.

To solve the Bianchi identities, we shall decompose supertorsion and supercurvature into their Lorentz irreducible components and then apply the equations (5.3.19–5.3.30). The simplest situation occurs with the tensor $T_{\alpha,\beta\dot{\beta},\gamma\dot{\gamma}}$, of dimension $\frac{1}{2}$. In accordance with equation (5.3.19), it is antisymmetric in α and β. In accordance with constraint (5.3.14), it is antisymmetric in $\dot{\beta}$ and $\dot{\gamma}$. Hence, the result is

$$T_{\alpha,\beta\dot{\beta},\gamma\dot{\gamma}} = \varepsilon_{\alpha\beta}\varepsilon_{\dot{\beta}\dot{\gamma}} T_{\gamma} \Rightarrow T_{\dot{\alpha},\beta\dot{\beta},\gamma\dot{\gamma}} = \varepsilon_{\dot{\alpha}\dot{\beta}}\varepsilon_{\beta\gamma} \bar{T}_{\dot{\gamma}} \tag{5.3.31}$$

where T_{α} is a spinor superfield.

Recalling that $T_{\alpha\beta}{}^{\gamma} = T_{\alpha\dot{\beta}}{}^{\gamma} = 0$, equation (5.3.31) leads to

$$(-1)^{\varepsilon_B} T_{\alpha B}{}^{B} = T_{\alpha}. \tag{5.3.32}$$

5.3.3. Solution to the dim = 1 Bianchi identities

We begin with consideration of the curvature $R_{\alpha\beta\gamma\delta} = R_{(\alpha\beta)(\gamma\delta)}$ and decompose it into irreducible pieces:

$$R_{\alpha\beta\gamma\delta} = f^{1}(\varepsilon_{\alpha\gamma}\varepsilon_{\beta\delta} + \varepsilon_{\alpha\delta}\varepsilon_{\beta\gamma}) + f^{2}{}_{(\alpha\beta\gamma\delta)} + \varepsilon_{\alpha(\gamma}f^{3}{}_{\delta)\beta} + \varepsilon_{\beta(\gamma}f^{3}{}_{\delta)\alpha}$$

where $f^{2}{}_{\alpha\beta\gamma\delta}$ and $f^{3}{}_{\alpha\beta}$ are completely symmetric tensors. Substituting this decomposition into equation (5.3.20), one finds that $f^{2}{}_{\alpha\beta\gamma\delta} = 0$ and $f^{3}{}_{\alpha\beta} = 0$. If we denote $f^{1} \equiv -2\bar{R}$, then our result is

$$R_{\alpha\beta\gamma\delta} = -2\bar{R}(\varepsilon_{\alpha\gamma}\varepsilon_{\beta\delta} + \varepsilon_{\alpha\delta}\varepsilon_{\beta\gamma}) \Rightarrow$$
$$R_{\dot{\alpha}\dot{\beta}\dot{\gamma}\dot{\delta}} = 2R(\varepsilon_{\dot{\alpha}\dot{\gamma}}\varepsilon_{\dot{\beta}\dot{\delta}} + \varepsilon_{\dot{\alpha}\dot{\delta}}\varepsilon_{\dot{\beta}\dot{\gamma}}). \tag{5.3.33}$$

Next, let us analyse equations (5.3.21) and (5.3.23). By virtue of equation (5.3.31), we can rewrite equation (5.3.23) as

$$2\varepsilon_{\dot{\gamma}\dot{\delta}}R_{\alpha\beta\gamma\delta} + 2\varepsilon_{\gamma\delta}R_{\alpha\beta\dot{\gamma}\dot{\delta}} = -\frac{1}{4}\varepsilon_{\dot{\gamma}\dot{\delta}}(\varepsilon_{\alpha\gamma}\varepsilon_{\beta\delta} + \varepsilon_{\alpha\delta}\varepsilon_{\beta\gamma})T^{2}$$
$$+ \varepsilon_{\dot{\gamma}\dot{\delta}}(\varepsilon_{\beta\gamma}\mathscr{D}_{\alpha}T_{\delta} + \varepsilon_{\alpha\gamma}\mathscr{D}_{\beta}T_{\delta}) + 4i\varepsilon_{\alpha\delta}T_{\beta,\gamma\dot{\gamma},\dot{\delta}}$$
$$+ 4i\varepsilon_{\beta\delta}T_{\alpha,\gamma\dot{\gamma},\dot{\delta}} \tag{5.3.34}$$
$$T^{2} = T^{\alpha}T_{\alpha}.$$

$R_{\alpha\beta\dot{\gamma}\dot{\delta}}$ is symmetric in indices $\dot{\gamma}$ and $\dot{\delta}$, therefore equation (5.3.34) leads to

$$\varepsilon_{\gamma\delta}R_{\alpha\beta\dot{\gamma}\dot{\delta}} = i\varepsilon_{\alpha\delta}(T_{\beta,\gamma\dot{\gamma},\dot{\delta}} + T_{\beta,\gamma\dot{\delta},\dot{\gamma}}) + i\varepsilon_{\beta\delta}(T_{\alpha,\gamma\dot{\gamma},\dot{\delta}} + T_{\alpha,\gamma\dot{\delta},\dot{\gamma}}).$$

Comparing this result with equation (5.3.21), it is not difficult to deduce that

$$R_{\alpha\beta\dot{\gamma}\dot{\delta}} = 0 \Rightarrow R_{\dot{\alpha}\dot{\beta}\gamma\delta} = 0. \tag{5.3.35}$$

Then, equation (5.3.21) is re-solved by

$$T_{\alpha,\beta\dot{\beta},\dot{\gamma}} = \varepsilon_{\alpha\beta}f_{\dot{\beta}\dot{\gamma}}$$

for some tensor $f_{\dot{\beta}\dot{\gamma}}$. Now, the relations (5.3.33) and (5.3.34) are consistent only under the conditions:

$$\mathscr{D}_{\alpha}T_{\beta} + \mathscr{D}_{\beta}T_{\alpha} = 0 \tag{5.3.36}$$

as well as

$$T_{\alpha,\beta\dot{\beta},\dot{\gamma}} = i\varepsilon_{\alpha\beta}\varepsilon_{\dot{\beta}\dot{\gamma}}\left(\bar{R} - \frac{1}{16}T^{\gamma}T_{\gamma} + \frac{1}{8}\mathscr{D}^{\gamma}T_{\gamma}\right)$$
$$T_{\dot{\alpha},\beta\dot{\beta},\gamma} = i\varepsilon_{\dot{\alpha}\dot{\beta}}\varepsilon_{\beta\gamma}\left(R - \frac{1}{16}\bar{T}_{\dot{\gamma}}\bar{T}^{\dot{\gamma}} + \frac{1}{8}\bar{\mathscr{D}}_{\dot{\gamma}}\bar{T}^{\dot{\gamma}}\right). \tag{5.3.37}$$

It remains only to study equations (5.3.22) and (5.3.24). The first implies that

$$T_{\alpha,\beta\dot{\beta},\gamma} \equiv i\varepsilon_{\alpha\beta}\psi_{\gamma\dot{\beta}} \Rightarrow$$
$$T_{\dot{\alpha},\beta\dot{\beta},\dot{\gamma}} \equiv i\varepsilon_{\dot{\alpha}\dot{\beta}}\bar{\psi}_{\beta\dot{\gamma}} \tag{5.3.38}$$

where $\psi_{\gamma\dot\beta}$ is a complex vector superfield. Let us substitute equations (5.3.31) and (5.3.38) into equation (5.3.24), resulting in

$$\begin{aligned}T_{\alpha\dot\alpha,\beta\dot\beta,\gamma\dot\gamma} &= \mathrm{i}\varepsilon_{\alpha\beta}\left[\varepsilon_{\dot\alpha\dot\gamma}\left(\psi_{\gamma\dot\beta}+\frac{1}{4}\bar{\mathscr{D}}_{\dot\beta}T_{\gamma}\right)+\varepsilon_{\dot\beta\dot\gamma}\left(\psi_{\gamma\dot\alpha}+\frac{1}{4}\bar{\mathscr{D}}_{\dot\alpha}T_{\gamma}\right)\right]\\ &\quad -\mathrm{i}\varepsilon_{\dot\alpha\dot\beta}\left[\varepsilon_{\alpha\gamma}\left(\bar\psi_{\beta\dot\gamma}-\frac{1}{4}\mathscr{D}_{\beta}\bar T_{\dot\gamma}\right)+\varepsilon_{\beta\gamma}\left(\bar\psi_{\alpha\dot\gamma}-\frac{1}{4}\mathscr{D}_{\alpha}\bar T_{\dot\gamma}\right)\right]\\ &\quad +\mathrm{i}\varepsilon_{\alpha\beta}\varepsilon_{\dot\alpha\dot\beta}\left(\psi_{\gamma\dot\gamma}-\bar\psi_{\gamma\dot\gamma}-\frac{1}{4}\mathscr{D}_{\gamma}\bar T_{\dot\gamma}-\frac{1}{4}\bar{\mathscr{D}}_{\dot\gamma}T_{\gamma}\right).\end{aligned} \tag{5.3.39}$$

Supertorsion T_{abc} is antisymmetric in indices a and b, therefore

$$T_{\alpha\dot\alpha,\beta\dot\beta,\gamma\dot\gamma} = -T_{\beta\dot\beta,\alpha\dot\alpha,\gamma\dot\gamma}.$$

For consistency of this equation with equation (5.3.39), one demands that

$$\psi_{\gamma\dot\gamma}-\bar\psi_{\gamma\dot\gamma}=\frac{1}{4}\mathscr{D}_{\gamma}\bar T_{\dot\gamma}+\frac{1}{4}\bar{\mathscr{D}}_{\dot\gamma}T_{\gamma}. \tag{5.3.40}$$

Defining a real vector superfield G_a by the rule

$$\psi_a+\bar\psi_a\equiv 2G_a \tag{5.3.41}$$

we obtain

$$\begin{aligned}\psi_{\alpha\dot\alpha} &= G_{\alpha\dot\alpha}+\frac{1}{8}\mathscr{D}_{\alpha}\bar T_{\dot\alpha}+\frac{1}{8}\bar{\mathscr{D}}_{\dot\alpha}T_{\alpha}\\ \bar\psi_{\alpha\dot\alpha} &= G_{\alpha\dot\alpha}-\frac{1}{8}\mathscr{D}_{\alpha}\bar T_{\dot\alpha}-\frac{1}{8}\bar{\mathscr{D}}_{\dot\alpha}T_{\alpha}\end{aligned} \tag{5.3.42}$$

To summarize, we have resolved all the Bianchi identities of dimensions $\frac{1}{2}$ and 1 in terms of tensors T_α, $\bar T_{\dot\alpha}$, R, $\bar R$ and G_a. In fact, we have completely determined the anticommutators of spinor covariant derivatives, since the relations (5.3.16), (5.3.33) and (5.3.35) mean

$$\{\mathscr{D}_\alpha,\mathscr{D}_\beta\}=-4\bar R M_{\alpha\beta}\qquad \{\bar{\mathscr{D}}_{\dot\alpha},\bar{\mathscr{D}}_{\dot\beta}\}=4R\bar M_{\dot\alpha\dot\beta}. \tag{5.3.43}$$

5.3.4. Solution to the dim = $\frac{3}{2}$ Bianchi identities

We proceed by resolving the identities (5.3.25–28). Taking into account expressions (5.3.33), equation (5.3.25) means simply that

$$\mathscr{D}_\alpha\bar R=0\Rightarrow\bar{\mathscr{D}}_{\dot\alpha}R=0. \tag{5.3.44}$$

Therefore, R is a chiral scalar superfield. Further, since $R_{\beta\gamma\dot\beta\dot\gamma}=0$, equation (5.3.26) means that

$$\mathscr{D}_\alpha R_{\beta\dot\gamma\dot\alpha\dot\delta}=2\mathrm{i}R_{\dot\gamma,\alpha\beta,\dot\alpha\dot\delta}+2\mathrm{i}R_{\beta,\alpha\dot\gamma,\dot\alpha\dot\delta} \tag{5.3.45a}$$

and

$$R_{\dot{\gamma},\alpha\beta,\beta\gamma} + R_{\dot{\beta},\alpha\dot{\gamma},\beta\gamma} = 0. \tag{5.3.45b}$$

Recalling equation (5.3.33), we can rewrite the first relation (5.3.45), in the form

$$3\mathscr{D}_\alpha R = \mathrm{i}R_{\beta,\alpha\dot{\gamma},}{}^{\beta\dot{\gamma}}. \tag{5.3.46}$$

The identity (5.3.28) may be rewritten, by virtue of expression (5.3.31) and (5.3.38), as

$$R_{\dot{\alpha},\beta\dot{\beta},\dot{\gamma}\delta} + R_{\dot{\gamma},\beta\dot{\beta},\dot{\alpha}\delta} = \mathrm{i}\varepsilon_{\dot{\alpha}\dot{\beta}}\left(\bar{\mathscr{D}}_{\dot{\gamma}} - \frac{1}{2}\bar{T}_{\dot{\gamma}}\right)\bar{\psi}_{\beta\delta} + \mathrm{i}\varepsilon_{\dot{\gamma}\dot{\beta}}\left(\bar{\mathscr{D}}_{\dot{\alpha}} - \frac{1}{2}\bar{T}_{\dot{\alpha}}\right)\bar{\psi}_{\beta\delta}.$$

Since $R_{\dot{\alpha},\beta\dot{\beta},\dot{\gamma}\dot{\delta}}$ is symmetric in $\dot{\gamma}$ and $\dot{\delta}$, this relation is easily resolved:

$$R_{\dot{\alpha},\beta\dot{\beta},\dot{\gamma}\dot{\delta}} = \mathrm{i}\varepsilon_{\dot{\alpha}\dot{\beta}}\left(\bar{\mathscr{D}} - \frac{1}{2}\bar{T}\right)_{(\dot{\gamma}}\bar{\psi}_{\beta\dot{\delta})} - \frac{\mathrm{i}}{2}(\varepsilon_{\dot{\alpha}\dot{\gamma}}\varepsilon_{\dot{\beta}\dot{\delta}} + \varepsilon_{\dot{\alpha}\dot{\delta}}\varepsilon_{\dot{\beta}\dot{\gamma}})\left(\bar{\mathscr{D}}^{\dot{\lambda}} - \frac{1}{2}\bar{T}^{\dot{\lambda}}\right)\bar{\psi}_{\beta\dot{\lambda}}. \tag{5.3.47}$$

Now, equation (5.3.46) gives

$$\left(\bar{\mathscr{D}}^{\dot{\alpha}} - \frac{1}{2}\bar{T}^{\dot{\alpha}}\right)\bar{\psi}_{\alpha\dot{\alpha}} = \mathscr{D}_\alpha R. \tag{5.3.48}$$

This formula relates the tensors G_a and R.

Finally, we must consider the identity (5.3.27). First of all, we decompose the supertorsion $T_{\alpha\dot{\alpha},\beta\dot{\beta},\gamma}$, arising in the commutator of vector derivatives, into its irreducible components:

$$T_{\alpha\dot{\alpha},\beta\dot{\beta},\gamma} = \varepsilon_{\dot{\alpha}\dot{\beta}}W_{(\alpha\beta\gamma)} + \varepsilon_{\dot{\alpha}\dot{\beta}}(\varepsilon_{\alpha\gamma}f^1_\beta + \varepsilon_{\beta\gamma}f^1_\alpha) + \varepsilon_{\alpha\beta}f^2_{(\dot{\alpha}\dot{\beta})\gamma} \tag{5.3.49}$$

where $W_{\alpha\beta\gamma}$ is a totally symmetric tensor and $f^2_{\dot{\alpha}\dot{\beta}\gamma}$ is symmetric in its dotted indices. Inserting this decomposition into identity (5.3.27) and bearing in mind that $R_{\dot{\alpha},\beta\dot{\beta},\gamma\delta}$ is symmetric in γ and δ, one can express f^1_α and $f^2_{\dot{\alpha}\dot{\beta}\gamma}$ in terms of T_α, R and G_a. After doing this, one obtains

$$R_{\dot{\alpha},\beta\dot{\beta},\gamma\delta} = -\mathrm{i}\varepsilon_{\dot{\alpha}\dot{\beta}}(\varepsilon_{\beta\gamma}X_\delta + \varepsilon_{\beta\delta}X_\gamma) - 2\mathrm{i}\varepsilon_{\dot{\alpha}\dot{\beta}}W_{\beta\gamma\delta} \tag{5.3.50}$$

$$\begin{aligned} X_\alpha = &+\frac{1}{12}\left[\left(\bar{\mathscr{D}}_{\dot{\gamma}} - \frac{1}{2}\bar{T}_{\dot{\gamma}}\right)\left(\bar{\mathscr{D}}^{\dot{\gamma}} - \frac{1}{2}\bar{T}^{\dot{\gamma}}\right) - 4R\right]T_\alpha \\ &+\frac{1}{12}\left[2\psi_{\alpha\dot{\alpha}} + \left(\bar{\mathscr{D}}_{\dot{\alpha}} - \frac{1}{2}\bar{T}_{\dot{\alpha}}\right)\left(\mathscr{D}_\alpha - \frac{1}{2}T_\alpha\right)\right. \\ &\left.+\frac{1}{2}(\mathscr{D}_\alpha - T_\alpha)\left(\bar{\mathscr{D}}_{\dot{\alpha}} - \frac{1}{2}\bar{T}_{\dot{\alpha}}\right)\right]\bar{T}^{\dot{\alpha}}. \end{aligned}$$

Evidently, equation (5.3.45b) is satisfied.

5.3.5. Solution to the dim = 2 Bianchi identities

One can check that equation (5.3.29) is consistent under the requirement

$$(\mathscr{D}^{\gamma} - T^{\gamma})W_{\alpha\beta\gamma} = \frac{i}{2}\left(\mathscr{D}_{\alpha}{}^{\dot{\alpha}} - \frac{i}{2}(\mathscr{D}_{\alpha}\bar{T}^{\dot{\alpha}})\right)\psi_{\beta\dot{\alpha}} + \frac{i}{2}\left(\mathscr{D}_{\beta}{}^{\dot{\alpha}} - \frac{i}{2}(\mathscr{D}_{\beta}\bar{T}^{\dot{\alpha}})\right)\psi_{\alpha\dot{\alpha}}. \tag{5.3.51}$$

Next, equation (5.3.30) is satisfied if and only if the identity

$$\left(\bar{\mathscr{D}}_{\dot{\alpha}} - \frac{1}{2}\bar{T}_{\dot{\alpha}}\right)W_{\alpha\beta\gamma} = 0 \tag{5.3.52}$$

holds.

It may be shown that the other Bianchi identities, not considered above, are satisfied identically or play the role of relations determining the supercurvature R_{ab}.

5.3.6. Algebra of covariant derivatives

Now we are in a position to write down the algebra of covariant derivatives:

$$\{\mathscr{D}_{\alpha}, \bar{\mathscr{D}}_{\dot{\alpha}}\} = -2i\mathscr{D}_{\alpha\dot{\alpha}}$$

$$\{\mathscr{D}_{\alpha}, \mathscr{D}_{\beta}\} = -4\bar{R}M_{\alpha\beta}, \qquad \{\bar{\mathscr{D}}_{\dot{\alpha}}, \bar{\mathscr{D}}_{\dot{\beta}}\} = 4R\bar{M}_{\dot{\alpha}\dot{\beta}}$$

$$\begin{aligned}
[\bar{\mathscr{D}}_{\dot{\alpha}}, \mathscr{D}_{\beta\dot{\beta}}] &= \frac{1}{2}\varepsilon_{\dot{\alpha}\dot{\beta}}\bar{T}^{\dot{\gamma}}\mathscr{D}_{\beta\dot{\gamma}} - i\varepsilon_{\dot{\alpha}\dot{\beta}}\left(R + \frac{1}{8}\bar{\mathscr{D}}_{\dot{\gamma}}\bar{T}^{\dot{\gamma}} - \frac{1}{16}\bar{T}_{\dot{\gamma}}\bar{T}^{\dot{\gamma}}\right)\mathscr{D}_{\beta} \\
&\quad - i\varepsilon_{\dot{\alpha}\dot{\beta}}\left(G_{\beta}{}^{\dot{\gamma}} - \frac{1}{8}\mathscr{D}_{\beta}\bar{T}^{\dot{\gamma}} - \frac{1}{8}\bar{\mathscr{D}}^{\dot{\gamma}}T_{\beta}\right)\bar{\mathscr{D}}_{\dot{\gamma}} - i\mathscr{D}_{\beta}R\bar{M}_{\dot{\alpha}\dot{\beta}} \\
&\quad + i\varepsilon_{\dot{\alpha}\dot{\beta}}\left(\bar{\mathscr{D}}^{\dot{\delta}} - \frac{1}{2}\bar{T}^{\dot{\delta}}\right)\left(G_{\beta}{}^{\dot{\gamma}} - \frac{1}{8}\mathscr{D}_{\beta}\bar{T}^{\dot{\gamma}} - \frac{1}{8}\bar{\mathscr{D}}^{\dot{\gamma}}T_{\beta}\right)\bar{M}_{\dot{\delta}\dot{\gamma}} \\
&\quad + 2i\varepsilon_{\dot{\alpha}\dot{\beta}}X^{\gamma}M_{\beta\gamma} - 2i\varepsilon_{\dot{\alpha}\dot{\beta}}W_{\beta}{}^{\gamma\delta}M_{\gamma\delta} \\
[\mathscr{D}_{\alpha}, \mathscr{D}_{\beta\dot{\beta}}] &= \frac{1}{2}\varepsilon_{\alpha\beta}T^{\gamma}\mathscr{D}_{\gamma\dot{\beta}} - i\varepsilon_{\alpha\beta}\left(\bar{R} + \frac{1}{8}\mathscr{D}^{\gamma}T_{\gamma} - \frac{1}{16}T^{\gamma}T_{\gamma}\right)\bar{\mathscr{D}}_{\dot{\beta}} \\
&\quad + i\varepsilon_{\alpha\beta}\left(G^{\gamma}{}_{\dot{\beta}} + \frac{1}{8}\mathscr{D}^{\gamma}\bar{T}_{\dot{\beta}} + \frac{1}{8}\bar{\mathscr{D}}_{\dot{\beta}}T^{\gamma}\right)\mathscr{D}_{\gamma} + i\bar{\mathscr{D}}_{\dot{\beta}}\bar{R}M_{\alpha\beta} \\
&\quad - i\varepsilon_{\alpha\beta}\left(\mathscr{D}^{\delta} - \frac{1}{2}T^{\delta}\right)\left(G^{\gamma}{}_{\dot{\beta}} + \frac{1}{8}\mathscr{D}^{\gamma}\bar{T}_{\dot{\beta}} + \frac{1}{8}\bar{\mathscr{D}}_{\dot{\beta}}T^{\gamma}\right)M_{\delta\gamma} \\
&\quad - 2i\varepsilon_{\alpha\beta}\bar{X}^{\dot{\gamma}}\bar{M}_{\dot{\beta}\dot{\gamma}} + 2i\varepsilon_{\alpha\beta}\bar{W}_{\dot{\beta}}{}^{\dot{\gamma}\dot{\delta}}\bar{M}_{\dot{\gamma}\dot{\delta}}
\end{aligned} \tag{5.3.53}$$

where X_{α} is defined by equation (5.3.50). The supertorsions T_{α}, R, $G_{\alpha\dot{\alpha}}$ and

$W_{\alpha\beta\gamma}$ satisfy the following Bianchi identities:

$$\mathscr{D}_\alpha T_\beta + \mathscr{D}_\beta T_\alpha = 0 \qquad \bar{\mathscr{D}}_{\dot\alpha} R = 0$$
$$\bar{G}_a = G_a \qquad W_{\alpha\beta\gamma} = W_{(\alpha\beta\gamma)}$$
$$\mathscr{D}_\alpha R = \left(\bar{\mathscr{D}}^{\dot\alpha} - \frac{1}{2}\bar{T}^{\dot\alpha}\right)\left(G_{\alpha\dot\alpha} - \frac{1}{8}\mathscr{D}_\alpha \bar{T}_{\dot\alpha} - \frac{1}{8}\bar{\mathscr{D}}_{\dot\alpha} T_\alpha\right)$$
$$\left(\bar{\mathscr{D}}_{\dot\alpha} - \frac{1}{2}\bar{T}_{\dot\alpha}\right) W_{\alpha\beta\gamma} = 0 \tag{5.3.54}$$
$$(\mathscr{D}^\gamma - T^\gamma) W_{\alpha\beta\gamma} = \frac{i}{2}\left(\mathscr{D}_\alpha{}^{\dot\alpha} - \frac{i}{2}(\mathscr{D}_\alpha \bar{T}^{\dot\alpha})\right)\left(G_{\beta\dot\alpha} + \frac{1}{8}\mathscr{D}_\beta \bar{T}_{\dot\alpha} + \frac{1}{8}\bar{\mathscr{D}}_{\dot\alpha} T_\beta\right)$$
$$+ \frac{i}{2}\left(\mathscr{D}_\beta{}^{\dot\alpha} - \frac{i}{2}(\mathscr{D}_\beta \bar{T}^{\dot\alpha})\right)\left(G_{\alpha\dot\alpha} + \frac{1}{8}\mathscr{D}_\alpha \bar{T}_{\dot\alpha} + \frac{1}{8}\bar{\mathscr{D}}_{\dot\alpha} T_\alpha\right).$$

We did not give above the commutator of two-vector covariant derivatives. In accordance with equation (5.3.53), it can be calculated in accordance with the rule

$$[\mathscr{D}_{\alpha\dot\alpha}, \mathscr{D}_{\beta\dot\beta}] = \frac{i}{2}[\{\mathscr{D}_\alpha, \bar{\mathscr{D}}_{\dot\alpha}\}, \mathscr{D}_{\beta\dot\beta}] = \frac{i}{2}\{[\mathscr{D}_\alpha, \mathscr{D}_{\beta\dot\beta}], \bar{\mathscr{D}}_{\dot\alpha}\} + \frac{i}{2}\{\mathscr{D}_\alpha, [\bar{\mathscr{D}}_{\dot\alpha}, \mathscr{D}_{\beta\dot\beta}]\}.$$

5.3.7. Covariantly chiral tensor superfields

The set of representation-preserving constraints (5.3.15*a*) was sufficient to ensure the existence of covariantly chiral scalar superfields. The full set of constraints (5.3.15) makes possible the existence of *covariantly chiral tensor superfields with undotted indices only*. Indeed, the equation

$$\bar{\mathscr{D}}_{\dot\alpha} \Phi_{\alpha_1 \ldots \alpha_n} = 0 \tag{5.3.55}$$

is consistent with the anticommutator $\{\bar{\mathscr{D}}_{\dot\alpha}, \bar{\mathscr{D}}_{\dot\beta}\} = 4R\bar{M}_{\dot\alpha\dot\beta}$. As opposed to the flat-superspace case, *in a curved superspace there are no covariantly chiral superfields with dotted spinor indices unless $R \neq 0$.*

Note that, given a tensor superfield $U_{\alpha_1 \ldots \alpha_n}$, the tensor superfield $(\bar{\mathscr{D}}^2 - 4R)U_{\alpha_1 \ldots \alpha_n}$ is covariantly chiral

$$\bar{\mathscr{D}}_{\dot\alpha}(\bar{\mathscr{D}}^2 - 4R)U_{\alpha_1 \ldots \alpha_n} = 0. \tag{5.3.56}$$

5.3.8. Generalized super Weyl transformations

By construction, the algebra (5.3.53) is covariant under the general coordinate transformation supergroup and the superlocal Lorentz group acting by

$$\mathscr{D}_A \to \mathscr{D}'_A = e^{\mathscr{K}} \mathscr{D}_A e^{-\mathscr{K}}$$
$$\mathscr{K} = K^M \partial_M + \frac{1}{2} K^{bc} M_{bc}. \tag{5.3.57}$$

In fact, the algebra admits a richer symmetry group. It is not difficult to check that constraints (5.3.15) are invariant under the transformations

$$\mathscr{D}_A \to \mathscr{D}'_A \qquad \mathscr{D}'_{\alpha\dot\alpha} = \frac{\mathrm{i}}{2}\{\mathscr{D}'_\alpha, \bar{\mathscr{D}}'_{\dot\alpha}\}$$

$$\mathscr{D}'_\alpha = L\mathscr{D}_\alpha - 2(\mathscr{D}^\beta L)M_{\alpha\beta} \tag{5.3.58}$$
$$\bar{\mathscr{D}}'_{\dot\alpha} = \overline{L\mathscr{D}}_{\dot\alpha} - 2(\bar{\mathscr{D}}^{\dot\beta}\bar{L})\bar{M}_{\dot\alpha\dot\beta}$$

where $L(z)$ is an arbitrary complex scalar superfield (non-vanishing everywhere). In particular, one finds

$$T'_\alpha = L\bar{L}T_\alpha - \mathscr{D}'_\alpha \ln(L^4\bar{L}^2)$$
$$R' = -\frac{1}{4}(\bar{\mathscr{D}}^2 - 4R)\bar{L}^2. \tag{5.3.59}$$

Let us introduce real superfields $\Delta(z)$ and $\kappa(z)$ according to the rule

$$L\bar{L} = \mathrm{e}^\Delta \qquad L\bar{L}^{-1} = \mathrm{e}^{\mathrm{i}\kappa}. \tag{5.3.60}$$

Then, by virtue of equation (5.3.58), the supervierbein transforms according to the law

$$E_\alpha \to E'_\alpha = \mathrm{e}^{(\Delta + \mathrm{i}\kappa)/2}E_\alpha$$
$$\bar{E}_{\dot\alpha} \to \bar{E}'_{\dot\alpha} = \mathrm{e}^{(\Delta - \mathrm{i}\kappa)/2}\bar{E}_{\dot\alpha} \tag{5.3.61}$$
$$E_a \to E'_a = \mathrm{e}^\Delta E_a + \ldots.$$

Therefore, the parameter Δ induces superlocal scale transformations (or superspace Weyl transformations) and the parameter κ induces superlocal chiral or γ_5-transformations.

We will refer to the transformations (5.3.58) as the 'generalized super Weyl transformations'. The set of all such transformations forms a supergroup which represents a superspace extension of the Weyl group appearing in general relativity (see Section 1.6).

5.4. Prepotentials

After imposing constraints (5.3.15), the supergeometry potentials $E_A{}^M$ and $\Omega_A{}^{bc}$ are expressed in terms of the spinor vierbein superfields $E_\alpha{}^M$ and $\bar{E}_{\dot\alpha}{}^M$. The latter are constrained superfields, in accordance with equation (5.3.15*a*). In practice, operating with unconstrained objects is usually more convenient than with objects subject to constraints. On these grounds, it is worth resolving constraints (5.3.15*a*) in terms of unconstrained superfields ('supergeometry prepotentials'). The term 'prepotentials' is not incidental. Indeed, following the terminology of gauge theories, $T_{AB}{}^C$ and $R_{AB}{}^{cd}$ play the role of field strengths, being gauge covariant superfields, while the gauge superfields $E_A{}^M$ and $\Omega_A{}^{bc}$ play the role of potentials. So, it is perfectly reasonable to consider objects determining potentials as prepotentials.

The solution to the constraints described below was first given by W. Siegel.

5.4.1. *Solution to constraints (5.3.15a)*

Equation (5.3.4) says that the supervector fields E_α form a closed algebra. What usefulness does this fact contain for us? To answer this question, it is helpful to remember Frobenius' theorem from differential geometry (see, for example, Warner's book*). It may be formulated as follows.

Let $\mathcal{M}$ be a p-dimensional manifold with local coordinates x^m, and $\{V_i\}$, $i = 1,\ldots,q$, be a set of vector fields on $\mathcal{M}$, $V_i = V_i{}^m(x)(\partial/\partial x^m)$, linearly independent at each point of the manifold and forming a closed algebra:

$$[V_i, V_j] = C_{ij}{}^k(x)V_k.$$

Then, there exists a set of independent scalar functions $f^1(x),\ldots,f^{p-q}(x)$ such that the one-forms $\mathrm{d}f^\alpha$ vanish on vector fields V_i,

$$\mathrm{d}f^\alpha(V_i) = 0$$

Equivalently, if one considers submanifolds (surfaces) in $\mathcal{M}$, defined by the system of equations

$$f^\alpha(x) = \text{const} \qquad \alpha = 1,\ldots,p-q$$

then V_i are tangent to each surface:

$$V_i = A_i{}^j(y)\frac{\partial}{\partial y^j} \qquad \det(A_i{}^j) \neq 0$$

where y^i are local coordinates on the surface. The variables $\tilde{x}^m = (y^1,\ldots,y^q, f^1,\ldots,f^{p-q})$ define a local coordinate system on the manifold, and we have

$$\tilde{x}^m = \mathrm{e}^U x^m \qquad \frac{\partial}{\partial \tilde{x}^m} = \mathrm{e}^U \frac{\partial}{\partial x^m}\mathrm{e}^{-U} \qquad U = U^m(x)\frac{\partial}{\partial x^m}$$

for some vector field U. Finally, we obtain

$$V_i = A_i{}^j \mathrm{e}^U \frac{\partial}{\partial x^j}\mathrm{e}^{-U} \qquad i = 1,\ldots,q.$$

Returning to superspace, we deduce from equation (5.3.4) that

$$\begin{aligned} E_\alpha &= A_\alpha{}^\mu(z)\hat{E}_\mu \qquad \{\hat{E}_\mu, \hat{E}_\nu\} = 0 \\ \hat{E}_\mu &= \mathrm{e}^W \partial_\mu \mathrm{e}^{-W} \qquad W = W^M\partial_M \neq \bar{W} \end{aligned} \tag{5.4.1}$$

*Warner, F.W., *Foundations of Differentiable Manifolds and Lie groups.* — Graduate Texts in Mathematics, no. 94. New York, Berlin, Heidelberg, Tokyo: Springer-Verlag, 1983.

Here $W^M(z)$ are unconstrained complex superfields such that the operator W, being considered as a supervector field, is of c-type, $\varepsilon(W^M) = \varepsilon_M$. The bosonic superfields $A_\alpha{}^\mu(z)$ are subjected only to the requirement $\det(A_\alpha{}^\mu) \neq 0$. W is not restricted to be real because the submanifolds generated by E_α can be seen to be defined in a complex superspace, where dotted and undotted fermionic coordinates are not related by conjugation, in which our superspace is embedded as a real subspace.

Taking the complex conjugate of relations (5.4.1), one obtains

$$\bar{E}_{\dot\alpha} = \bar{A}_{\dot\alpha}{}^{\dot\mu}\hat{\bar{E}}_{\dot\mu} \qquad \hat{\bar{E}}_{\dot\mu} = -\mathrm{e}^{\bar{W}}\bar{\partial}_{\dot\mu}\mathrm{e}^{-\bar{W}}. \tag{5.4.2}$$

Recall that the superlocal Lorentz group acts on E_α as follows: $E'_\alpha = (\exp K)_\alpha{}^\beta E_\beta$, $\det(\exp K) = 1$. Therefore, it is useful to factor $A_\alpha{}^\mu$ into a Lorentz scalar and unimodular matrix,

$$\begin{gathered} E_\alpha = FN_\alpha{}^\mu\hat{E}_\mu \qquad \bar{E}_{\dot\alpha} = \overline{FN}_{\dot\alpha}{}^{\dot\mu}\hat{\bar{E}}_{\dot\mu} \\ \det(N_\alpha{}^\mu) = 1 \end{gathered} \tag{5.4.3}$$

Then, the superlocal Lorentz transformations act on F, $N_\alpha{}^\mu$ and W according to the rule:

$$F' = F \qquad W' = W \qquad N'_\alpha{}^\mu = (\mathrm{e}^K)_\alpha{}^\beta N_\beta{}^\mu. \tag{5.4.4}$$

With respect to the general coordinate transformation supergroup, the objects under consideration change in the manner:

$$\begin{gathered} E'_\alpha = \mathrm{e}^K E_\alpha \mathrm{e}^{-K} \qquad K = K^M\partial_M = \bar{K} \\ \Rightarrow F' = \mathrm{e}^K F \qquad N'_\alpha{}^\mu = \mathrm{e}^K N_\alpha{}^\mu \qquad \mathrm{e}^{W'} = \mathrm{e}^K\mathrm{e}^W. \end{gathered} \tag{5.4.5}$$

To summarize, after imposing constraints (5.3.15), all geometrical objects are expressed in terms of unconstrained complex superfields $N_\alpha{}^\mu$, F and W^M and their conjugates, which will be called the 'supergeometry prepotentials'.

5.4.2. *Useful gauges on the superlocal Lorentz group*

The prepotentials $N_\alpha{}^\mu$ prove to be compensating superfields for the superlocal Lorentz group. As is seen from rule (5.4.4), the superlocal Lorentz transformation $(\mathrm{e}^K)_\alpha{}^\beta = \delta_\alpha{}^\mu(N^{-1})_\mu{}^\beta$ moves $N_\alpha{}^\mu$ into the unit matrix. Therefore, the simplest gauge choice on the superlocal Lorentz group is

$$N_\alpha{}^\mu = \delta_\alpha{}^\mu \tag{5.4.6}$$

In this gauge we have

$$E_\alpha = F\hat{E}^\alpha \Rightarrow C_{\alpha\beta}{}^\gamma = E_\alpha \ln F\delta_\beta{}^\gamma + E_\beta \ln F\delta_\alpha{}^\gamma.$$

Then, equation (5.3.9) gives

$$\begin{aligned} \Omega_{\alpha\beta\gamma} &= \varepsilon_{\alpha\gamma} E_\beta \ln F + \varepsilon_{\alpha\beta} E_\gamma \ln F \Rightarrow \\ \bar{\Omega}_{\dot\alpha\dot\beta\dot\gamma} &= \varepsilon_{\dot\alpha\dot\gamma} \bar{E}_{\dot\beta} \ln \bar{F} + \varepsilon_{\dot\alpha\dot\beta} \bar{E}_{\dot\gamma} \ln \bar{F}. \end{aligned} \tag{5.4.7}$$

Another useful Lorentz gauge is of the form

$$\Omega_{\alpha\dot\beta\dot\gamma} = \bar{\Omega}_{\dot\alpha\beta\gamma} = 0. \tag{5.4.8}$$

Such a gauge choice is possible since the supercurvature $R_{\alpha\beta\dot\beta\dot\gamma}$ vanishes identically, in accordance with constraints (5.3.15). In fact, we have

$$R_{\alpha\beta\dot\beta\dot\gamma} = 0 \Leftrightarrow E_{(\alpha}\Omega_{\beta)\dot\beta\dot\gamma} = \frac{1}{2} C_{\alpha\beta}{}^{\gamma}\Omega_{\gamma\dot\beta\dot\gamma} - \Omega_{(\alpha\beta}{}^{\dot\delta}\Omega_{\beta)\dot\delta\dot\gamma}.$$

This equation means that the $SL(2, \mathbb{C})$ connection $\Omega_{\alpha\dot\beta\dot\gamma}$ is trivial:

$$\Omega_\alpha{}^{\dot\beta\dot\gamma}\bar{M}_{\dot\beta\dot\gamma} = \bar{g}^{-1}E_\alpha\bar{g} \qquad \bar{g} = \exp(\bar{L}^{\dot\beta\dot\gamma}\bar{M}_{\dot\beta\dot\gamma})$$

for some symmetric tensor superfield $\bar{L}_{\dot\beta\dot\gamma}$. Therefore, one can find a superlocal Lorentz transformation leading to the choice of gauge (5.4.8).

Gauge fixing (5.4.8) is helpful when working with covariantly chiral tensor superfields. To make our discussion clear, let us start with the scalar case. Consider a covariantly chiral scalar superfield $\Phi(z)$. By definition, it satisfies the equation

$$\bar{\mathscr{D}}_{\dot\alpha}\Phi = \bar{E}_{\dot\alpha}\Phi = 0.$$

By virtue of equations (5.4.2), we can represent Φ in the form

$$\Phi = e^{\bar{W}}\hat{\Phi} \qquad \bar{\partial}_{\dot\mu}\hat{\Phi} = 0. \tag{5.4.9}$$

Hence, every covariantly chiral scalar superfield is determined by a 'flat' chiral superfield $\hat{\Phi}(x, \theta)$, depending on x^m and θ^μ only. The exponential $e^{\bar{W}}$ plays the role of a picture-changing operator, since it transforms any 'flat' chiral superfield into a covariantly chiral scalar one. Note that $\hat{\Phi}$ is invariant under general coordinate and superlocal Lorentz transformations.

Next, consider a covariantly chiral tensor superfield $\chi_{\alpha_1\ldots\alpha_n}(z)$. We have

$$0 = \bar{\mathscr{D}}_{\dot\alpha}\chi_{\alpha_1\ldots\alpha_n} = \bar{E}_{\dot\alpha}\chi_{\alpha_1\ldots\alpha_n} + \bar{\Omega}_{\dot\alpha}{}^{\beta\gamma}M_{\beta\gamma}\chi_{\alpha_1\ldots\alpha_n} = \bar{E}_{\dot\alpha}\chi_{\alpha_1\ldots\alpha_n} + (g^{-1}\bar{E}_{\dot\alpha}g)\chi_{\alpha_1\ldots\alpha_n}$$

where $g = \exp(L^{\beta\gamma}M_{\beta\gamma})$. On these grounds, $\chi_{\alpha_1\ldots\alpha_n}$ can be represented in the form

$$\begin{gathered} \chi_{\alpha_1\ldots\alpha_n} = \exp(-L^{\beta\gamma}M_{\beta\gamma})e^{\bar{W}}\hat{\chi}_{\alpha_1\ldots\alpha_n} \\ \bar{\partial}_{\dot\mu}\hat{\chi}_{\alpha_1\ldots\alpha_n} = 0 \end{gathered} \tag{5.4.10}$$

which looks to be much more complicated than equation (5.4.9). However, in the gauge (5.4.8) both these representations take the same form.

5.4.3. The λ-supergroup

Often, when solving constraints in terms of unconstrained fields, we are faced with the appearance of a new gauge invariance. The most familiar example is electrodynamics. The first Maxwell equation reads $\varepsilon^{abcd}\partial_b F_{cd} = 0$, F_{cd} is an antisymmetric tensor unifying the electric and magnetic fields. Solving this equation using $F_{ab} = \partial_a A_b - \partial_b A_a$ introduces the gauge arbitrariness $\delta A_a = \partial_a \psi$.

So, we should scrutinize our theory for the appearance of an additional gauge invariance after solving constraints (5.3.15*a*) in the form (5.4.1). We shall look for transformations of the prepotentials $N_\alpha{}^\mu$, F and W^M leaving invariant the spinor supervierbein E_α and hence, the covariant derivatives.

The general coordinate transformation supergroup acts on e^W by the left shifts (5.4.5). Now, let us try to consider a right shift

$$\mathrm{e}^{W'} = \mathrm{e}^W \mathrm{e}^{-\bar{\lambda}} \qquad \mathrm{e}^{\bar{W}'} = \mathrm{e}^{\bar{W}} \mathrm{e}^{-\lambda} \qquad \lambda = \lambda^M \partial_M \tag{5.4.11}$$

with λ being a complex c-type supervector field. In the infinitesimal case, we have

$$\hat{E}'_\mu = \mathrm{e}^{W'} \partial_\mu \mathrm{e}^{-W'} = \hat{E}_\mu + \mathrm{e}^W (\partial_\mu \bar{\lambda}^N) \partial_N \mathrm{e}^{-W}.$$

To preserve the $\hat{E}_\mu$-form, one must demand

$$\begin{aligned} &\partial_\mu \bar{\lambda}^n = \partial_\mu \bar{\lambda}_{\dot{\nu}} = 0 \\ &\Rightarrow \bar{\partial}_{\dot{\mu}} \lambda^n = \bar{\partial}_{\dot{\mu}} \lambda^\nu = 0 \end{aligned} \tag{5.4.12}$$

resulting in

$$\delta \hat{E}_\mu = (\mathrm{e}^W \partial_\mu \bar{\lambda}^\nu) \hat{E}_\nu = (\mathrm{e}^W \partial_{(\mu} \bar{\lambda}^{\nu)}) \hat{E}_\nu + \frac{1}{2} (\mathrm{e}^W \partial_\nu \bar{\lambda}^\nu) \hat{E}_\mu.$$

If one supplements this transformation with

$$\begin{aligned} \delta F &= -\frac{1}{2} (\mathrm{e}^W \partial_\nu \bar{\lambda}^\nu) F \\ \delta N_\alpha{}^\mu &= -N_\alpha{}^\nu (\mathrm{e}^W \partial_{(\nu} \bar{\lambda}^{\mu)}) \end{aligned} \tag{5.4.13}$$

then E_α does not change. Rather beautifully, we do not obtain any restrictions on the parameter $\lambda_{\dot{\mu}}$, while the parameters λ^m and λ^μ satisfy chirality constraint (5.4.12).

Transformations (5.4.11) with the parameters restricted according to rule (5.4.12) form a supergroup (the 'λ-supergroup'). This follows from the fact that two supervector fields of the type (5.4.12) commute on a supervector field of the same type. The λ-supergroup can be realized as a group of triangular transformations

$$\begin{aligned} y'^m &= y'^m(y, \theta) \\ \theta'^\mu &= \theta'^\mu(y, \theta) \\ \bar{\rho}'_{\dot{\mu}} &= \bar{\rho}'_{\dot{\mu}}(y, \theta, \bar{\rho}) \end{aligned}$$

acting in complex superspace $\mathbb{C}^{4|4}$ with c-number coordinates y^m and a-number coordinates θ^μ and $\bar{\rho}_{\dot\mu}$. This is the geometrical interpretation of the λ-supergroup.

The covariant derivatives and tensor superfields are unaffected by the λ-supergroup. It acts on the prepotentials only. In addition, the prepotentials transform non-trivially with respect to the general coordinate transformation supergroup (the 'K-supergroup') and the superlocal Lorentz group. So, the full gauge group of the prepotentials is the product of three groups: the λ-supergroup, the K-supergroup and the superlocal Lorentz group. In accordance with equations (5.4.4, 5, 11, 13), the infinitesimal transformation laws of the prepotentials are

$$\begin{gathered}\delta \mathrm{e}^W = K\mathrm{e}^W - \mathrm{e}^W \bar{\lambda} \\ \delta F = KF - \frac{1}{2}(\mathrm{e}^W \partial_\nu \bar{\lambda}^\nu)F \\ \delta N_\alpha{}^\mu = KN_\alpha{}^\mu + K_\alpha{}^\beta N_\beta{}^\mu - N_\alpha{}^\nu(\mathrm{e}^W \partial_{(\nu}\bar{\lambda}^{\mu)}) \\ K = K^M\partial_M = \bar{K} \qquad \lambda = \lambda^M\partial_M \qquad \bar{\partial}_{\dot\mu}\lambda^n = \bar{\partial}_{\dot\mu}\lambda^\nu = 0.\end{gathered} \tag{5.4.14}$$

There is an interesting connection between the K- and λ-supergroups. Let Φ be a covariantly chiral scalar superfield. Then Φ is transformed by the K-supergroup and is inert with respect to the λ-supergroup. Next, we represent Φ in the form (5.4.9). Now, $\hat{\Phi}$ is inert with respect to the K-supergroup, but is transformed non-trivially by the λ-supergroup. In accordance with equation (5.4.11), we have

$$\Phi = \mathrm{e}^{\bar{W}}\hat{\Phi} = \mathrm{e}^{\bar{W}'}\hat{\Phi}'$$

therefore

$$\hat{\Phi}' = \mathrm{e}^{\lambda}\hat{\Phi} = \exp(\lambda^m\partial_m + \lambda^\mu\partial_\mu)\hat{\Phi} \qquad \bar{\partial}_{\dot\mu}\hat{\Phi}' = 0. \tag{5.4.15}$$

Note that constraints (5.4.12) could be obtained from the requirement $\bar{\partial}_{\dot\mu}\hat{\Phi}' = 0$.

Remark. In the Lorentz gauge (5.4.6) every λ-transformation should be supplemented by a λ-dependent Lorentz transformation with the parameters

$$K_{\alpha\beta} = \mathrm{e}^W \partial_{(\alpha}\bar{\lambda}_{\beta)} \qquad \bar{K}_{\dot\alpha\dot\beta} = \mathrm{e}^{\bar{W}}\bar{\partial}_{(\dot\alpha}\bar{\lambda}_{\dot\beta)} \tag{5.4.16}$$

(see equation (5.4.14)) where

$$\partial_\alpha = \delta_\alpha{}^\mu\partial_\mu \qquad \bar{\lambda}_\beta = \delta_\beta{}^\nu\bar{\lambda}_\nu.$$

5.4.4. *Expressions for E, T_α and R*

This subsection is devoted to obtaining expressions for the Berezinian $E = \mathrm{Ber}(E_A{}^M)$ and the supertorsion T_α and R in terms of the prepotentials. These results will be used later.

We begin by determining E. For this purpose we choose the Lorentz gauge (5.4.6) and introduce a semi-covariant supervierbein $\hat{E}_A$ (different from $\check{E}_A$ (5.3.10)) according to the rule:

$$\hat{E}_A = (\hat{E}_a, \hat{E}_\alpha, \hat{\bar{E}}^{\dot{\alpha}}) = \hat{E}_A{}^M \partial_M$$
$$\hat{E}_a = -\frac{i}{4}(\tilde{\sigma}_a)^{\dot{\alpha}\alpha}\{\hat{E}_\alpha, \hat{\bar{E}}_{\dot{\alpha}}\} \qquad \hat{E} = \mathrm{Ber}\,(\hat{E}_A{}^M). \tag{5.4.17}$$

Note that superfields $\hat{E}_A{}^M$, in particular Berezinian $\hat{E}$, are constructed in terms of the prepotentials W and $\bar{W}$ only. By virtue of equations (5.3.6) and (5.4.3), E_A and $\hat{E}_A$ are connected as follows:

$$E_\alpha = F\hat{E}_\alpha \qquad \bar{E}^{\dot{\alpha}} = \bar{F}\hat{\bar{E}}^{\dot{\alpha}}$$
$$E_a = \bar{F}F\hat{E}_a - \frac{i}{4}F(\tilde{\sigma}_a)^{\dot{\alpha}\alpha}(\Omega_{\dot{\alpha}\alpha}{}^\beta + \delta_\alpha{}^\beta \bar{E}_{\dot{\alpha}} \ln F)\hat{E}_\beta \tag{5.4.18}$$
$$- \frac{i}{4}\bar{F}(\tilde{\sigma}_a)^{\dot{\alpha}\alpha}(\bar{\Omega}_{\alpha\dot{\alpha}}{}^{\dot{\beta}} + \delta_{\dot{\alpha}}{}^{\dot{\beta}} E_\alpha \ln \bar{F})\hat{\bar{E}}_{\dot{\beta}}.$$

Then, one obtains

$$E = F^2\bar{F}^2\hat{E}. \tag{5.4.19}$$

Since E, $\hat{E}$ and F are Lorentz scalars, this result holds in the general case.

To find T_α, we again choose the Lorentz gauge $N_\alpha{}^\mu = \delta_\alpha{}^\mu$. Recalling equations (5.2.59, 60) and (5.3.32) leads to

$$T_\alpha = E_\alpha \ln E - (1\cdot\overleftarrow{E_\alpha}) - \Omega_{\beta\alpha}{}^\beta.$$

Then, making use of relation (5.4.7) reduces this expression to

$$T_\alpha = E_\alpha \ln E - (1\cdot\overleftarrow{E}_\alpha) + 3E_\alpha \ln F = E_\alpha \ln (EF^2) - (1\cdot\overleftarrow{\hat{E}}_\alpha)F.$$

Now, one can apply the technique of subsection 1.11.2:

$$(1\cdot\overleftarrow{\hat{E}}_\alpha) = (1\cdot\overleftarrow{\mathrm{e}^W \partial_\alpha \mathrm{e}^{-W}}) = (1\cdot\mathrm{e}^{-\overleftarrow{W}}\overleftarrow{\partial}_\alpha \mathrm{e}^{\overleftarrow{W}}) = \partial_\alpha(1\cdot\mathrm{e}^{-\overleftarrow{W}})\mathrm{e}^{\overleftarrow{W}}$$
$$= (1\cdot\mathrm{e}^{\overleftarrow{W}})(\mathrm{e}^W\partial_\alpha(1\cdot\mathrm{e}^{-\overleftarrow{W}})) = (1\cdot\mathrm{e}^{\overleftarrow{W}})\hat{E}_\alpha(\mathrm{e}^W(1\cdot\mathrm{e}^{-\overleftarrow{W}}))$$
$$= (1\cdot\mathrm{e}^{\overleftarrow{W}})\hat{E}_\alpha(1\cdot\mathrm{e}^{\overleftarrow{W}})^{-1}.$$

As a result, we have

$$T_\alpha = E_\alpha \ln (EF^2(1\cdot\mathrm{e}^{\overleftarrow{W}})). \tag{5.4.20}$$

Since T_α and E_α have the same Lorentz transformation laws, equation (5.4.20) holds in the general case. From equation (5.4.19), we read off the final expression

$$T_\alpha = E_\alpha T \qquad T = \ln [\bar{F}^2 F^4 \hat{E}(1 \cdot e^{\bar{W}})]$$
$$\bar{T}_{\dot{\alpha}} = \bar{E}_{\dot{\alpha}} \bar{T} \qquad T = \ln [F^2 \bar{F}^4 \hat{E}(1 \cdot e^{\bar{\bar{W}}})]. \tag{5.4.21}$$

To find R, we represent the anticommutator

$$\{\mathscr{D}_\alpha, \mathscr{D}_\beta\} = -4\bar{R} M_{\alpha\beta}$$

in terms of the spinor superconnection

$$\{\mathscr{D}_\alpha, \mathscr{D}_\beta\} = (E_\alpha \Omega_{\beta\gamma\delta} + E_\beta \Omega_{\alpha\gamma\delta} + 2\Omega_{\alpha\gamma}{}^{\lambda}\Omega_{\beta\lambda\delta}) M^{\gamma\delta}$$

and use equation (5.4.7). As a result, one obtains

$$\bar{R} = -\frac{1}{4} \hat{E}^\mu \hat{E}_\mu F^2 \Rightarrow R = -\frac{1}{4} \hat{\bar{E}}_{\dot{\mu}} \hat{\bar{E}}^{\dot{\mu}} \bar{F}^2. \tag{5.4.22}$$

Given a scalar superfield U, the superfield

$$(\bar{\mathscr{D}}^2 - 4R)U$$

is covariantly chiral and, by virtue of relation (5.4.22), we have

$$(\bar{\mathscr{D}}^2 - 4R)U = \hat{\bar{E}}_{\dot{\mu}} \hat{\bar{E}}^{\dot{\mu}} (\bar{F}^2 U). \tag{5.4.23}$$

The supertorsions G_a and $W_{\alpha\beta\gamma}$ can also be expressed in terms of the prepotentials. However, these expressions turn out to be highly complicated and we do not reproduce them.

5.4.5. *Gauge fixing for the K- and Λ-supergroups*

Let us analyse the W-transformation law (5.4.14). It can be rewritten in the form

$$\delta W = \delta W^M \partial_M = K - \bar{\lambda} + O(W)$$
$$= (K^m - \bar{\lambda}^m)\partial_m + (K^\mu - \bar{\lambda}^\mu)\partial_\mu + (\bar{K}_{\dot{\mu}} - \bar{\lambda}_{\dot{\mu}})\bar{\partial}^{\dot{\mu}} + O(W).$$

Recall that K^m is an arbitrary real vector superfield, while $\bar{\lambda}^\mu$ and $\bar{K}_{\dot{\mu}}$ are arbitrary spinor superfields. Using our freedom in the choice of K^m, we can kill the real part of W^m. On the other hand, the parameters $\bar{\lambda}^\mu$ and $\bar{K}_{\dot{\mu}}$ can be used to gauge away the spinor components W^μ and $W_{\dot{\mu}}$. After this, we obtain a purely imaginary vector superfield

$$W = -\mathrm{i}H \qquad H = H^m \partial_m = \bar{H}. \tag{5.4.24}$$

Therefore, the prepotentials $\mathrm{Re}\, W^m$, W^μ and $W_{\dot{\mu}}$ are compensating superfields for the K^M- and $\bar{\lambda}^\mu$-transformations.

Remark. Considering the K-transformations only, we can gauge away the real part of W resulting in a purely imaginary supervector field

$$W = -\mathrm{i}H \qquad H = H^M\partial_M = \bar{H} \tag{5.4.25}$$

Remark. After imposing the gauge (5.4.24), a covariantly chiral scalar superfield takes the form

$$\Phi(x, \theta, \bar{\theta}) = \mathrm{e}^{\mathrm{i}H}\hat{\Phi}(x, \theta) = \hat{\Phi}(\mathrm{e}^{\mathrm{i}H}x, \theta). \tag{5.4.26}$$

In the Lorentz gauge $N_\alpha{}^\mu = \delta_\alpha{}^\mu$, the spinor supervierbeins are of the form

$$E_\alpha = F\mathrm{e}^{-\mathrm{i}H}\partial_\alpha \mathrm{e}^{\mathrm{i}H} \qquad \bar{E}_{\dot{\alpha}} = \bar{F}\mathrm{e}^{\mathrm{i}H}(-\bar{\partial}_{\dot{\alpha}})\mathrm{e}^{-\mathrm{i}H}. \tag{5.4.27}$$

It is instructive to compare these expressions with the 'flat' ones (see (2.5.18–21)).

There are some residual K- and λ-transformations preserving the gauge choice (5.4.24). Making use of the W-transformation law

$$\begin{aligned}\delta \mathrm{e}^W &= K^M\partial_M \mathrm{e}^{-\mathrm{i}H} - \mathrm{e}^{-\mathrm{i}H}\bar{\lambda}^M\partial_M \\ &= \{K^M\partial_M - (\mathrm{e}^{-\mathrm{i}H}\bar{\lambda}^M)\mathrm{e}^{-\mathrm{i}H}\partial_M \mathrm{e}^{\mathrm{i}H}\}\mathrm{e}^{-\mathrm{i}H}\end{aligned}$$

and demanding δH to be real and to have no spinor pieces, $\delta H = \delta H^m\partial_m = \overline{\delta H}$, one arrives at the following restrictions:

$$\begin{gathered}K^m = K^m(\lambda, \bar{\lambda}) \qquad K^\mu = \mathrm{e}^{-\mathrm{i}H}\bar{\lambda}^\mu \qquad \bar{K}_{\dot{\mu}} = \mathrm{e}^{-\mathrm{i}H}\bar{\lambda}_{\dot{\mu}} \\ \Rightarrow \lambda^M = (\lambda^m(x, \theta), \lambda^\mu(x, \theta), \bar{\lambda}_{\dot{\mu}}(\mathrm{e}^{-2\mathrm{i}H}x, \theta)).\end{gathered} \tag{5.4.28}$$

Here $K^m(\lambda, \bar{\lambda})$ is a rather complicated function. It proves to be quite difficult to operate with this function in practical calculations. So, the gauge (5.4.24) is not a useful one. It is worth obtaining a gauge-fixed version in which the K-transformations are excluded by construction. As the key to the problem, let us consider the transformation law for $\exp(-2\mathrm{i}H)$:

$$\delta \mathrm{e}^{-2\mathrm{i}H} = \delta(\mathrm{e}^{-\bar{W}}\mathrm{e}^W) = \lambda \mathrm{e}^{-2\mathrm{i}H} - \mathrm{e}^{-\mathrm{i}2H}\bar{\lambda}. \tag{5.4.29}$$

We see that only λ-transformations act on $\exp(-2\mathrm{i}H)$. Hence, it is sufficient to find a reformulation in which H arises only in the combination $\exp(-2\mathrm{i}H)$. We shall turn to the chiral representation.

5.4.6. Chiral representation

For the time being, it is worth forgetting about any gauge fixing and starting from the beginning. Once again, we choose the Lorentz gauge $N_\alpha{}^\mu = \delta_\alpha{}^\mu$ and, then, perform the following picture-changing transformation: every superfield V is changed to

$$\tilde{V} \equiv \mathrm{e}^{-\bar{W}}V \tag{5.4.30a}$$

the covariant derivatives $\mathscr{D}_A$ are transformed into

$$\tilde{\mathscr{D}}_A \equiv \mathrm{e}^{-\bar{W}}\mathscr{D}_A\mathrm{e}^{\bar{W}} = \tilde{E}_A{}^M\partial_M + \frac{1}{2}\tilde{\Omega}_A{}^{bc}M_{bc}$$

$$\tilde{E}^{-1} \equiv (\mathrm{Ber}\,(\tilde{E}_A{}^M))^{-1} = (E^{-1}\mathrm{e}^{-\bar{W}}). \tag{5.4.30b}$$

The spinor supervierbeins take the form

$$\tilde{E}_\alpha = (\mathrm{e}^{-2\mathrm{i}H}F)\mathrm{e}^{-2\mathrm{i}H}\partial_\alpha\mathrm{e}^{2\mathrm{i}H} \qquad \tilde{\bar{E}}_{\dot{\alpha}} = -\bar{F}\bar{\partial}_{\dot{\alpha}}$$
$$\mathrm{e}^{-2\mathrm{i}H} \equiv \mathrm{e}^{-\bar{W}}\mathrm{e}^{W} \qquad H = H^M\partial_M = \bar{H} \tag{5.4.31}$$

where we have denoted $\tilde{\bar{F}}$ simply by $\bar{F}$. The representation obtained is said to be chiral, while the original representation is called real.

The chiral representation is well suited to the treatment of covariantly chiral scalar superfields:

$$\tilde{\Phi} = \hat{\Phi}(x, \theta) \equiv \Phi(x, \theta) \qquad \text{but } \tilde{\bar{\Phi}} = \mathrm{e}^{-2\mathrm{i}H}\bar{\Phi}(x, \theta). \tag{5.4.32}$$

The only complication this representation produces is that it modifies the complex conjugation rules: given a superfield V and its conjugate $\bar{V}$, for their chiral transforms we have

$$\bar{\tilde{V}} = (\mathrm{e}^{-\bar{W}}V)^* = \mathrm{e}^{-W}\,\bar{V} = \mathrm{e}^{2\mathrm{i}H}\tilde{\bar{V}}. \tag{5.4.33a}$$

In particular, if V is real, then

$$\bar{\tilde{V}} = \mathrm{e}^{2\mathrm{i}H}\tilde{V}. \tag{5.4.33b}$$

Now, let us analyse K- and λ-transformation laws. In the real representation, we have

$$\delta\mathscr{D}_A = [\mathscr{K}, \mathscr{D}_A] \qquad \delta V = \mathscr{K}V$$

$$\mathscr{K} = K^M\partial_M + (\mathrm{e}^{W}\,\partial_\beta\bar{\lambda}_\gamma)M^{\beta\gamma} + (\mathrm{e}^{\bar{W}}\bar{\partial}_{\dot{\beta}}\lambda_{\dot{\gamma}})\bar{M}^{\dot{\beta}\dot{\gamma}}$$

where V is a tensor superfield (indices are suppressed). Recall that in the Lorentz gauge $N_\alpha{}^\mu = \delta_\alpha{}^\mu$ every λ-transformation must be supplemented by the Lorentz transformation (5.4.16). Making the chiral transform, one obtains

$$\delta\tilde{\mathscr{D}}_A = [\lambda, \tilde{\mathscr{D}}_A] \qquad \delta\tilde{V} = \lambda\tilde{V}$$
$$\lambda = \lambda^M\partial_M + (\mathrm{e}^{-2\mathrm{i}H}\partial_\beta\bar{\lambda}_\gamma)M^{\beta\gamma} + (\bar{\partial}_{\dot{\beta}}\lambda_{\dot{\gamma}})\bar{M}^{\dot{\beta}\dot{\gamma}}. \tag{5.4.34}$$

In particular, the spinor supervierbeins change as follows:

$$\delta\tilde{E}_\alpha = (\mathrm{e}^{-2\mathrm{i}H}\partial_{(\alpha}\bar{\lambda}^{\beta)})\tilde{E}_\alpha + [\lambda^M\partial_M, \tilde{E}_\alpha]$$
$$\delta\tilde{\bar{E}}_{\dot{\alpha}} = (\partial_{(\dot{\alpha}}\lambda^{\dot{\beta})})\tilde{\bar{E}}_{\dot{\beta}} + [\lambda^M\partial_M, \tilde{\bar{E}}_{\dot{\alpha}}]. \tag{5.4.35}$$

As may be seen, the K-transformations never arise in the chiral representation. Of course, there is no mystery in this disappearance: the K-invariance can be used to gauge away the real part of W. There are no independent K-transformations in the gauge $\mathrm{Re}\, W = 0$. But in the chiral representation W enters into all quantities in the combination $\exp(-2iH) = \exp(-\bar{W}) \exp W$ only, and hence the real part of W is excluded by construction.

It should be pointed out that the spinor part of H can be gauged away by proper choice of gauge parameters $\lambda^{\dot{\mu}}$, resulting in the gauge (5.4.24). After this, we work with the restricted λ-transformations (5.4.28).

There is a simple recipe for transferring each expression in the real representation to the chiral representation. Namely, one is to make the following replacements: $N_{\alpha}{}^{\mu} \to \delta_{\alpha}{}^{\mu}$, $W \to -2iH$, $\bar{W} \to 0$, $\bar{F} \to \bar{F}$, $F \to e^{-2iH}F$ (and to place tildes on the remaining symbols). In particular, the chiral transformed supertorsions $\tilde{T}_{\dot{\alpha}}$ and $\tilde{R}$ read

$$\tilde{T}_{\dot{\alpha}} = \tilde{\bar{E}}_{\dot{\alpha}}\tilde{T} \qquad \tilde{T} = \ln(\bar{F}^2 \tilde{E})$$
$$\tilde{R} = \frac{1}{4}\bar{\partial}_{\dot{\mu}}\bar{\partial}^{\dot{\mu}}(\bar{F}^2) \tag{5.4.36}$$

where we have used equations (5.4.20, 22).

Let $\mathscr{L}$ be a scalar superfield. Equation (5.4.30b) then says that

$$\int d^8z E^{-1}\mathscr{L} = \int d^8z \tilde{E}^{-1}\tilde{\mathscr{L}} \tag{5.4.37}$$

anticipating suitable boundary conditions at infinity for $\mathscr{L}$. Therefore, there is no need to worry about representations (real or chiral) in which integrals over superspace are done.

5.4.7. *Gravitational superfield*

Instead of imposing the gauge (5.4.24) with purely imaginary vectorial W, we can retain some $\mathrm{Re}\, W^m$ if we wish. Then, the arbitrariness in choosing $\mathrm{Re}\, W^m$ can be used to make the gauge fixing

$$\exp(\bar{W}^n\partial_n)x^m = x^m + i\mathscr{H}^m(x, \theta, \bar{\theta}) \qquad \mathscr{H}^m = \bar{\mathscr{H}}^m. \tag{5.4.38a}$$

Understanding this relation as an equation in $\bar{W}^m$, its perturbative solution is

$$\bar{W}^m = i\mathscr{H}^m + \frac{1}{2}\mathscr{H}^n\partial_n\mathscr{H}^m + O(\mathscr{H}^3). \tag{5.4.38b}$$

Now, covariantly chiral scalar superfields have the form

$$\Phi(x, \theta, \bar{\theta}) = e^{\bar{W}}\hat{\Phi}(x, \theta) = \hat{\Phi}(x^m + i\mathscr{H}^m, \theta^{\mu}). \tag{5.4.39}$$

Let us find the residual K- and λ-transformations preserving the gauge (5.4.38). We have

$$\begin{aligned} i\delta\mathscr{H}^m &= \delta e^{\bar{W}}x^m = K^N\partial_N e^{\bar{W}}x^m - e^{\bar{W}}\lambda^N\partial_N x^m \\ &= K^N\partial_N(x^m + i\mathscr{H}^m) - e^{\bar{W}}\lambda^m = K^m - e^{\bar{W}}\lambda^m + iK^N\partial_N\mathscr{H}^m. \end{aligned}$$

This expression should be purely imaginary, hence

$$K^m = \frac{1}{2}e^{\bar{W}}\lambda^m + \frac{1}{2}e^{W}\bar{\lambda}^m = \frac{1}{2}\lambda^m(x^n + i\mathscr{H}^n, \theta^\nu) + \frac{1}{2}\bar{\lambda}^m(x^n - i\mathscr{H}^n, \bar{\theta}^{\dot\nu}) \quad (5.4.40)$$

and

$$\delta\mathscr{H}^m = K^N\partial_N\mathscr{H}^m + \frac{i}{2}(\lambda^m(x^n + i\mathscr{H}^n, \theta^\nu) - \bar{\lambda}^m(x^n - i\mathscr{H}^n, \bar{\theta}^{\dot\nu})). \quad (5.4.41)$$

It is also necessary to require two consistency conditions:

$$\delta e^{\bar{W}}\theta^\mu = 0 \qquad \delta e^{W}\bar{\theta}_{\dot\mu} = 0 \Rightarrow$$

$$K^\mu = \lambda^\mu(x^n + i\mathscr{H}^n, \theta^\nu), \qquad \bar{K}_{\dot\mu} = \bar{\lambda}_{\dot\mu}(x^n - i\mathscr{H}^n, \bar{\theta}^{\dot\nu}) \quad (5.4.42)$$

$$\lambda^M = (\lambda^m(x, \theta), \lambda^\mu(x, \theta), e^{-\bar{W}}e^{W}\bar{\lambda}_{\dot\mu}(x, \bar{\theta})).$$

Finally, let us substitute the expressions for K^m, K^μ and $\bar{K}_{\dot\mu}$ into expression (5.4.41) and compare the result with equation (5.1.8). Good Lord, we have recovered the transformation law of the gravitational superfield!

In the 'gravitational superfield gauge' (5.4.38), all the building blocks which have appeared in the expressions for the covariant derivatives can be readily expressed in terms of $\mathscr{H}^m$, F and $\bar{F}$. The semi-covariant supervierbein $\hat{E}_A$ is then given by

$$\hat{E}_\alpha = \partial_\alpha + i(\partial_\alpha\mathscr{H}^n)(1 - i\partial\mathscr{H})^{-1}{}_n{}^m\partial_m = \partial_\alpha + i(\hat{E}_\alpha\mathscr{H}^m)\partial_m$$

$$\hat{\bar{E}}_{\dot\alpha} = -\bar{\partial}_{\dot\alpha} + i(\bar{\partial}_{\dot\alpha}\mathscr{H}^n)(1 + i\partial\mathscr{H})^{-1}{}_n{}^m\partial_m = -\bar{\partial}_\alpha - i(\hat{\bar{E}}_\alpha\mathscr{H}^m)\partial_m \quad (5.4.43)$$

$$\hat{E}_a = -\frac{1}{4}(\tilde{\sigma}_a)^{\dot\alpha\alpha}([\hat{E}_\alpha, \hat{\bar{E}}_{\dot\alpha}]\mathscr{H}^m)\partial_m \equiv \hat{E}_a{}^m\partial_m$$

where we have introduced the notation

$$(1 + i\partial\mathscr{H})_n{}^m = \delta_n{}^m + i\frac{\partial\mathscr{H}^m}{\partial x^n}. \quad (5.4.44)$$

Introducing the anholonomy coefficients $\hat{C}_{AB}{}^C$ via

$$[\hat{E}_A, \hat{E}_B\} = \hat{C}_{AB}{}^C\hat{C}_C \qquad \hat{C}_{\alpha B}{}^\gamma = \hat{C}_{\alpha B\dot\gamma} = \hat{C}_{ab}{}^\gamma = 0$$

$$\hat{C}_{\alpha b}{}^c = \hat{E}_m{}^c\left(\frac{1}{4}\hat{E}^2(\sigma_b\hat{\bar{E}})_\alpha + i[\hat{E}_\alpha, \hat{E}_b]\right)\mathscr{H}^m$$

$$\hat{C}^{\dot\alpha}{}_b{}^c = -\hat{E}_m{}^c\left(\frac{1}{4}\hat{\bar{E}}^2(\tilde{\sigma}_b\hat{E})^{\dot\alpha} - i[\hat{\bar{E}}^{\dot\alpha}, \hat{E}_b]\right)\mathscr{H}^m \quad (5.4.45)$$

$$\hat{E}_a{}^m\hat{E}_m{}^b = \delta_a{}^b$$

the spinor superconnections prove to be

$$\Omega_{\alpha\beta\dot\gamma} = \varepsilon_{\alpha\beta}\hat{E}_{\dot\gamma}F + \varepsilon_{\alpha\dot\gamma}\hat{E}_{\beta}F$$
$$\Omega_{\alpha\dot\beta\dot\gamma} = \frac{1}{4}F(\hat{C}_{\alpha,\beta\dot\beta,}{}^{\beta}{}_{\dot\gamma} + \hat{C}_{\alpha,\beta\dot\gamma,}{}^{\beta}{}_{\dot\beta}) \tag{5.4.46a}$$

and

$$\bar{\Omega}_{\dot\alpha\dot\beta\gamma} = \varepsilon_{\dot\alpha\dot\beta}\hat{\bar{E}}_{\gamma}\bar{F} + \varepsilon_{\dot\alpha\gamma}\hat{\bar{E}}_{\dot\beta}\bar{F}$$
$$\bar{\Omega}_{\dot\alpha\beta\gamma} = \frac{1}{4}\bar{F}(\hat{C}_{\dot\alpha,\beta\dot\beta,\gamma}{}^{\dot\beta} + \hat{C}_{\dot\alpha,\beta\dot\gamma,\beta}{}^{\dot\beta}). \tag{(5.4.46b)}$$

Finally, the following identities

$$\hat{E} \equiv \operatorname{Ber}(\hat{E}_A{}^M) = \det(\hat{E}_a{}^m)$$
$$(1\cdot e^{\overleftarrow{W}}) = \det(1 + i\partial\mathscr{H}) \qquad (1\cdot e^{\overleftarrow{W}}) = \det(1 - i\partial\mathscr{H}) \tag{5.4.47}$$

hold.

Let us comment on the derivation of equations (5.4.43–47). Consider the change of variables

$$z^M_{(L)} = e^{\bar{W}}z^M = \begin{cases} x^m_{(L)} = x^m + i\mathscr{H}^m \\ \theta^\mu_{(L)} = \theta^\mu \\ \bar{\theta}_{\dot\mu(L)} = \bar{\theta}_{\dot\mu}. \end{cases} \tag{5.4.48}$$

The corresponding Berezinian is equal to

$$\operatorname{Ber}\left(\frac{\partial z_{(L)}}{\partial z}\right) = (1\cdot e^{\overleftarrow{W}}) = \det(1 + i\partial\mathscr{H})$$

which coincides with the second relation (5.4.47). Furthermore, we have

$$\partial^{(L)}_M \equiv \frac{\partial}{\partial z^M_{(L)}} = e^{\bar{W}}\partial_M e^{-\bar{W}} \Rightarrow \hat{\bar{E}}_{\dot\alpha} = -\bar{\partial}^{(L)}_{\dot\alpha}.$$

Now, making use of expressions (5.4.48) leads to the expression for $\hat{\bar{E}}_{\dot\alpha}$ (5.4.43). Finally, the expression for $\Omega_{\alpha\beta\dot\gamma}$ (5.4.46*a*) follows from equation (5.3.12).

In conclusion, it should be pointed out that the gauge transformations (5.4.40, 42) (preserving the gauge choice (5.4.38)) act on the covariant derivatives and tensor superfields in the manner:

$$\delta\mathscr{D}_A = [\mathscr{K}, \mathscr{D}_A] \qquad \delta V = \mathscr{K}V$$
$$\mathscr{K} = \mathscr{K}^M\partial_M + K^{\alpha\beta}M_{\alpha\beta} + \bar{K}^{\dot\alpha\dot\beta}\bar{M}_{\dot\alpha\dot\beta}$$
$$K^m = \frac{1}{2}\lambda^m(x + i\mathscr{H}, \theta) + \frac{1}{2}\bar{\lambda}^m(x - i\mathscr{H}, \bar{\theta}) \tag{5.4.49}$$

$$K^{\mu} = \lambda^{\mu}(x + \mathrm{i}\mathscr{H}, \theta)$$

$$K_{\alpha\beta} = \hat{E}_{(\alpha}\lambda_{\beta)}(x + \mathrm{i}\mathscr{H}, \theta)$$

since in the gauge $N_{\alpha}{}^{\mu} = \delta_{\alpha}{}^{\mu}$ every λ-transformation must be supplemented by the Lorentz transformation (5.4.16).

5.4.8. Gauge fixing on the generalized super Weyl group

We have nearly reached the aim formulated at the beginning of Section 5.2. Namely, we have found a supergeometrical description for the gravitational superfield. But what about the chiral compensator of Einstein supergravity? The point is that the set of prepotentials does not contain chiral superfields. Notice, however, that among the prepotentials $N_{\alpha}{}^{\mu}$, F and W^M the superfield F played a rather passive role in the above considerations. By construction, F is a complex superfield transforming as a scalar under the K-supergroup and as a density under the λ-supergroup (see equation (5.4.14)). In principle, there exists the possibility of obtaining a chiral superfield, if one manages to restrict F in a covariant way. Recall that in the supergeometrical approach every covariant constraint is a constraint on the supertorsion. Hence, in order to obtain a supergeometrical description of Einstein supergravity, one must supplement equations (5.3.15) by some additional constraints.

One can look on the prepotential F as a compensating superfield for the generalized super Weyl transformations (5.3.58). They act on the prepotentials as follows

$$F' = LF \qquad N'_{\alpha}{}^{\mu} = N_{\alpha}{}^{\mu} \qquad W'^{M} = W^{M}. \tag{5.4.50}$$

Now all the prepotentials acquire a clear interpretation. F, $N_{\alpha}{}^{\mu}$ and Re W are compensators for the generalized super Weyl group, the superlocal Lorentz group and the K-supergroup, respectively. W^{μ} and $\bar{W}_{\dot{\mu}}$ are compensators for the independent $\bar{\lambda}^{\mu}$- and $\lambda_{\dot{\mu}}$-transformations. Finally, Im W^m plays the role of gravitational superfield.

It has been shown in subsection 5.3.8 that the maximal symmetry group of constraints (5.3.15) is given by the product of three supergroups: the general coordinate transformation supergroup, the superlocal Lorentz group and the generalized super Weyl group. In a theory with such a symmetry group, the prepotential F can be gauged out by proper choice of super Weyl transformation. Then, in the gauge (5.4.38) we work with the gravitational superfield only, as in conformal supergravity. This is the reason why constraints (5.3.15) are said to be conformal supergravity constraints.

Since the prepotential F plays he role of super Weyl compensator, to impose a constraint on F is the same as fixing a gauge on the generalized super Weyl group. The simplest gauge choice is

$$F = 1.$$

However, this gauge fixing is not invariant under the Λ-supergroup (see equation (5.4.14)). To obtain a covariant gauge, let us recall the super Weyl transformation law for T_α (5.3.59). By virtue of equation (5.4.20), this can be rewritten in the form

$$T'_\alpha = \mathscr{D}'_\alpha T \qquad T' = T + \ln(L^4 \bar{L}^2). \tag{5.4.51}$$

Hence, one can take the super Weyl gauge

$$T_\alpha = 0.$$

Let us proceed to an analysis of this constraint.

5.5. Einstein supergravity

5.5.1. Einstein supergravity constraints

In the space of supergeometries (5.3.15) we consider the subspace characterized by the additional constraint

$$T_\alpha = \bar{T}_{\dot\alpha} = 0. \tag{5.5.1}$$

Now, the full set of constraints is

$$\begin{gathered} T_{\alpha\beta}{}^{C} = 0 \qquad T_{\dot\alpha\beta}{}^{C} = 0 \\ T_{\alpha\dot\alpha}{}^{B} + 2\mathrm{i}\delta_c^B(\sigma^c)_{\alpha\dot\alpha} = 0 \qquad R_{\alpha\dot\alpha}{}^{cd} = 0 \\ T_{\alpha b}{}^{c} = 0 \qquad T_{\dot\alpha b}{}^{c} = 0. \end{gathered} \tag{5.5.2}$$

These constraints (in a slightly different form) were suggested by J. Wess and B. Zumino.

We are going to show that the set of constraints (5.5.2) corresponds to Einstein supergravity.

5.5.2. Chiral compensator

The supertorsion $\bar{T}_{\dot\alpha}$ is expressed through the prepotentials in accordance with the rule (5.4.21). Setting $\bar{T}_{\dot\alpha} = 0$ gives

$$\begin{gathered} F^2 \bar{F}^4 \hat{E}(1 \cdot \mathrm{e}^{\overleftarrow{W}}) \equiv \varphi^{-3} \qquad \bar{E}_{\dot\alpha}\varphi = 0 \\ \Rightarrow \varphi = \mathrm{e}^{W}\phi \qquad \bar{\partial}_{\dot\mu}\phi = 0 \end{gathered} \tag{5.5.3}$$

where $\phi(x, \theta)$ is a flat chiral superfield. Adding the complex conjugate formula, one obtains

$$\begin{aligned} F &= \varphi^{1/2}\bar{\varphi}^{-1}(1 \cdot \mathrm{e}^{\overleftarrow{W}})^{-1/3}(1 \cdot \mathrm{e}^{\overleftarrow{\bar{W}}})^{1/6}\hat{E}^{-1/6} \\ \bar{F} &= \varphi^{-1}\bar{\varphi}^{1/2}(1 \cdot \mathrm{e}^{\overleftarrow{W}})^{1/6}(1 \cdot \mathrm{e}^{\overleftarrow{\bar{W}}})^{-1/3}\hat{E}^{-1/6}. \end{aligned} \tag{5.5.4}$$

As a result, all geometrical objects are expressed in terms of the old prepotentials $N_\alpha{}^\mu$ and W^M and the chiral superfield φ (and their conjugates).

To find the transformation law of φ, we represent it in the form

$$\varphi^3 = E^{-1}\bar{F}^{-2}(1\cdot \mathrm{e}^{\overleftarrow{\bar{W}}})^{-1}$$

using equation (5.4.19). E^{-1} changes only under the general coordinate transformations,

$$\delta E^{-1} = K^M\partial_M E^{-1} + (-1)^{\varepsilon_M}(\partial_M K^M)E^{-1}$$

(see equation (5.2.27)). The transformation law for $\bar{F}^{-2}$ follows from (5.4.14). Finally, using equations (1.11.21) and (5.4.14), one readily obtains

$$\begin{aligned}\delta(1\cdot \mathrm{e}^{\overleftarrow{\bar{W}}}) &= (1\cdot \mathrm{e}^{\overleftarrow{\bar{W}}}\overleftarrow{K}) - (1\cdot \overleftarrow{\lambda}\mathrm{e}^{\overleftarrow{\bar{W}}})\\ &= K^M\partial_M(1\cdot \mathrm{e}^{\overleftarrow{\bar{W}}}) + (-1)^{\varepsilon_M}(\partial_M K^M)(1\cdot \mathrm{e}^{\overleftarrow{\bar{W}}}) - (-1)^{\varepsilon_M}(\mathrm{e}^{\bar{W}}\,\partial_M\lambda^M)(1\cdot \mathrm{e}^{\overleftarrow{\bar{W}}})\end{aligned}$$

From these results, we deduce

$$\delta\varphi^3 = K^M\partial_M\varphi^3 + (\mathrm{e}^{\bar{W}}\,(\partial_m\lambda^m - \partial_\mu\lambda^\mu))\varphi^3 \qquad (5.5.5a)$$

or

$$\delta\phi^3 = \lambda^M\partial_M\phi^3 + (\partial_m\lambda^m - \partial_\mu\lambda^\mu)\varphi^3 = \partial_m(\lambda^m\phi^3) - \partial_\mu(\lambda^\mu\phi^3). \qquad (5.5.5b)$$

The second relation coincides with the transformation law of chiral compensator (5.1.34*b*) in Einstein supergravity! Hence, the prepotential φ can be identified with the chiral compensator. As a result, we have found the supergeometrical description for Einstein supergravity.

The supergravity formulation constructed in Section 5.1 and here, is usually called the minimal supergravity. Later we shall consider other versions of supergravity characterized by non-chiral compensating multiplets. The superfields $N_\alpha{}^\mu$, W^M and φ are said to be the 'minimal supergravity prepotentials'.

5.5.3. *Minimal algebra of covariant derivatives*

Under constraint (5.5.1), the covariant derivative algebra (5.3.53) is simplified drastically:

$$\begin{aligned}\{\mathcal{D}_\alpha, \bar{\mathcal{D}}_{\dot\alpha}\} &= -2\mathrm{i}\mathcal{D}_{\alpha\dot\alpha}\\ \{\mathcal{D}_\alpha, \mathcal{D}_\beta\} &= -4\bar{R}M_{\alpha\beta} \qquad \{\bar{\mathcal{D}}_{\dot\alpha}, \bar{\mathcal{D}}_{\dot\beta}\} = 4R\bar{M}_{\dot\alpha\dot\beta}\\ [\bar{\mathcal{D}}_{\dot\alpha}, \mathcal{D}_{\beta\dot\beta}] &= -\mathrm{i}\varepsilon_{\dot\alpha\dot\beta}(R\mathcal{D}_\beta + G_\beta{}^{\dot\gamma}\bar{\mathcal{D}}_{\dot\gamma}) - \mathrm{i}\mathcal{D}_\beta R\bar{M}_{\dot\alpha\dot\beta}\\ &\quad + \mathrm{i}\varepsilon_{\dot\alpha\dot\beta}\bar{\mathcal{D}}^{\dot\gamma}G_\beta{}^{\dot\delta}\bar{M}_{\dot\gamma\dot\delta} - 2\mathrm{i}\varepsilon_{\dot\alpha\dot\beta}W_\beta{}^{\gamma\delta}M_{\gamma\delta}\\ [\mathcal{D}_\alpha, \mathcal{D}_{\beta\dot\beta}] &= \mathrm{i}\varepsilon_{\alpha\beta}(\bar{R}\bar{\mathcal{D}}_{\dot\beta} + G^\gamma{}_{\dot\beta}\mathcal{D}_\gamma) + \mathrm{i}\bar{\mathcal{D}}_{\dot\beta}\bar{R}M_{\alpha\beta}\\ &\quad - \mathrm{i}\varepsilon_{\alpha\beta}\mathcal{D}^\gamma G^\delta{}_{\dot\beta}M_{\gamma\delta} + 2\mathrm{i}\varepsilon_{\alpha\beta}\bar{W}_{\dot\beta}{}^{\dot\gamma\dot\delta}\bar{M}_{\dot\gamma\dot\delta}.\end{aligned} \qquad (5.5.6)$$

The commutation relations for vector covariant derivatives are also quite simple:

$$[\mathscr{D}_{\alpha\dot{\alpha}}, \mathscr{D}_{\beta\dot{\beta}}] = \varepsilon_{\dot{\alpha}\dot{\beta}}\psi_{\alpha\beta} + \varepsilon_{\alpha\beta}\psi_{\dot{\alpha}\dot{\beta}} \tag{5.5.7}$$

where

$$\psi_{\alpha\beta} \equiv -\mathrm{i}G_{\beta}{}^{\dot{\gamma}}\mathscr{D}_{\alpha\dot{\gamma}} + \frac{1}{2}(\mathscr{D}_{\alpha}R)\mathscr{D}_{\beta} + \frac{1}{2}(\mathscr{D}_{\alpha}G_{\beta}{}^{\dot{\gamma}})\bar{\mathscr{D}}_{\dot{\gamma}} + W_{\alpha\beta}{}^{\gamma}\mathscr{D}_{\gamma}$$
$$+ \frac{1}{4}(\bar{\mathscr{D}}^2 - 8R)\bar{R}M_{\alpha\beta} + \mathscr{D}_{\alpha}W_{\beta}{}^{\gamma\delta}M_{\gamma\delta} - \frac{1}{2}\mathscr{D}_{\alpha}\bar{\mathscr{D}}^{\dot{\gamma}}G_{\beta}{}^{\dot{\delta}}\bar{M}_{\dot{\gamma}\dot{\delta}}$$

$$\psi_{\dot{\alpha}\dot{\beta}} \equiv \mathrm{i}G^{\gamma}{}_{\dot{\beta}}\mathscr{D}_{\gamma\dot{\alpha}} - \frac{1}{2}(\bar{\mathscr{D}}_{\dot{\alpha}}\bar{R})\bar{\mathscr{D}}_{\dot{\beta}} - \frac{1}{2}(\bar{\mathscr{D}}_{\dot{\alpha}}G^{\gamma}{}_{\dot{\beta}})\mathscr{D}_{\gamma} - \bar{W}_{\dot{\alpha}\dot{\beta}}{}^{\dot{\gamma}}\bar{\mathscr{D}}_{\dot{\gamma}}$$
$$+ \frac{1}{4}(\mathscr{D}^2 - 8\bar{R})R\bar{M}_{\dot{\alpha}\dot{\beta}} - \bar{\mathscr{D}}_{\dot{\alpha}}\bar{W}_{\dot{\beta}}{}^{\dot{\gamma}\dot{\delta}}\bar{M}_{\dot{\gamma}\dot{\delta}} + \frac{1}{2}\bar{\mathscr{D}}_{\dot{\alpha}}\mathscr{D}^{\gamma}G^{\delta}{}_{\dot{\beta}}M_{\gamma\delta}.$$

The supertorsions R, G_a and $W^{\alpha\beta\gamma}$ satisfy the Bianchi identities

$$\begin{gathered} G_a = \bar{G}_a \qquad W_{\alpha\beta\gamma} = W_{(\alpha\beta\gamma)} \\ \bar{\mathscr{D}}_{\dot{\alpha}}R = 0 \qquad \bar{\mathscr{D}}_{\dot{\alpha}}W_{\alpha\beta\gamma} = 0 \\ \bar{\mathscr{D}}^{\dot{\alpha}}G_{\alpha\dot{\alpha}} = \mathscr{D}_{\alpha}R \\ \mathscr{D}^{\gamma}W_{\alpha\beta\gamma} = \frac{\mathrm{i}}{2}\mathscr{D}_{\alpha}{}^{\dot{\alpha}}G_{\beta\dot{\alpha}} + \frac{\mathrm{i}}{2}\mathscr{D}_{\beta}{}^{\dot{\alpha}}G_{\alpha\dot{\alpha}}. \end{gathered} \tag{5.5.8}$$

We point out that in minimal supergravity $W_{\alpha\beta\gamma}$ is chiral.

From the covariant derivative algebra, one can obtain the following useful identities:

$$\begin{aligned} \mathscr{D}_{\alpha}\mathscr{D}_{\beta} &= \frac{1}{2}\varepsilon_{\alpha\beta}\mathscr{D}^2 - 2\bar{R}M_{\alpha\beta} \\ \bar{\mathscr{D}}_{\dot{\alpha}}\bar{\mathscr{D}}_{\dot{\beta}} &= -\frac{1}{2}\varepsilon_{\dot{\alpha}\dot{\beta}}\bar{\mathscr{D}}^2 + 2R\bar{M}_{\dot{\alpha}\dot{\beta}} \end{aligned} \tag{5.5.9a}$$

$$\begin{aligned} \mathscr{D}_{\alpha}\mathscr{D}^2 &= 4\bar{R}\mathscr{D}^{\beta}(\varepsilon_{\alpha\beta} + M_{\alpha\beta}) \\ \mathscr{D}^2\mathscr{D}_{\alpha} &= -2\bar{R}\mathscr{D}^{\beta}(\varepsilon_{\alpha\beta} + 2M_{\alpha\beta}) \end{aligned} \tag{5.5.9b}$$

$$\begin{aligned} \bar{\mathscr{D}}_{\dot{\alpha}}\bar{\mathscr{D}}^2 &= 4R\bar{\mathscr{D}}^{\dot{\beta}}(\varepsilon_{\dot{\alpha}\dot{\beta}} + \bar{M}_{\dot{\alpha}\dot{\beta}}) \\ \bar{\mathscr{D}}^2\bar{\mathscr{D}}_{\dot{\alpha}} &= -2R\bar{\mathscr{D}}^{\dot{\beta}}(\varepsilon_{\dot{\alpha}\dot{\beta}} + 2\bar{M}_{\dot{\alpha}\dot{\beta}}) \end{aligned} \tag{5.5.9c}$$

$$[\mathscr{D}^2, \bar{\mathscr{D}}_{\dot{\alpha}}] = -4(G_{\alpha\dot{\alpha}} + \mathrm{i}\mathscr{D}_{\alpha\dot{\alpha}})\mathscr{D}^{\alpha} + 4\overline{R\mathscr{D}}_{\dot{\alpha}} - 4(\mathscr{D}^{\gamma}G^{\delta}{}_{\dot{\alpha}})M_{\gamma\delta} + 8\bar{W}_{\dot{\alpha}}{}^{\dot{\gamma}\dot{\delta}}\bar{M}_{\dot{\gamma}\dot{\delta}} \tag{5.5.9d}$$

$$[\bar{\mathscr{D}}^2, \mathscr{D}_{\alpha}] = -4(G_{\alpha\dot{\alpha}} - \mathrm{i}\mathscr{D}_{\alpha\dot{\alpha}})\bar{\mathscr{D}}^{\dot{\alpha}} + 4R\mathscr{D}_{\alpha} - 4(\bar{\mathscr{D}}^{\dot{\gamma}}G_{\alpha}{}^{\dot{\delta}})\bar{M}_{\dot{\gamma}\dot{\delta}} + 8W_{\alpha}{}^{\gamma\delta}M_{\gamma\delta} \tag{5.5.9e}$$

$$\frac{1}{8}\mathscr{D}^{\alpha}(\bar{\mathscr{D}}^2 - 4R)\mathscr{D}_{\alpha} - \frac{1}{8}\bar{\mathscr{D}}_{\dot{\alpha}}(\mathscr{D}^2 - 4\bar{R})\bar{\mathscr{D}}^{\dot{\alpha}}$$

$$= \left\{\frac{1}{2}(\mathscr{D}_{\gamma}G_{\delta\dot{\alpha}})\bar{\mathscr{D}}^{\dot{\alpha}} - W_{\gamma\delta\alpha}\mathscr{D}^{\alpha} + \mathrm{i}(\mathscr{D}_{\gamma}{}^{\dot{\alpha}}G_{\delta\dot{\alpha}})\right\}M^{\gamma\delta} \qquad (5.5.9f)$$

$$+ \left\{\frac{1}{2}(\bar{\mathscr{D}}_{\dot{\gamma}}G_{\alpha\dot{\delta}})\mathscr{D}^{\alpha} - \bar{W}_{\dot{\gamma}\dot{\delta}\dot{\alpha}}\bar{\mathscr{D}}^{\dot{\alpha}} + \mathrm{i}(\mathscr{D}^{\alpha}{}_{\dot{\gamma}}G_{\alpha\dot{\delta}})\right\}\bar{M}^{\dot{\gamma}\dot{\delta}}.$$

It is worth noting that the identities (5.5.9*a–c*) hold for covariant derivatives under the conformal supergravity constraints, while the other identities are correct only in the Einstein supergravity case.

Let us give some simple applications of the above relations. If $\psi_{\alpha}(z)$ is an undotted spinor superfield, then equations (5.5.9*b*, *c*) lead to

$$(\mathscr{D}^2 - 4\bar{R})\mathscr{D}^{\alpha}\psi_{\alpha} = 0$$
$$(\bar{\mathscr{D}}^2 - 4R)\bar{\mathscr{D}}_{\dot{\alpha}}\bar{\psi}^{\dot{\alpha}} = 0. \qquad (5.5.10)$$

If $\eta_{\alpha}(z)$ is a covariantly chiral spinor superfield, then making use of equations (5.5.9*d*, *e*) gives

$$\bar{\mathscr{D}}_{\dot{\alpha}}\eta_{\alpha} = 0 \Rightarrow (\bar{\mathscr{D}}^2 - 4R)\mathscr{D}^{\alpha}\eta_{\alpha} = 0$$
$$(\mathscr{D}^2 - 4\bar{R})\bar{\mathscr{D}}_{\dot{\alpha}}\bar{\eta}^{\dot{\alpha}} = 0. \qquad (5.5.11)$$

For every scalar superfield $V(z)$, we have

$$\mathscr{D}^{\alpha}(\bar{\mathscr{D}}^2 - 4R)\mathscr{D}_{\alpha}V = \bar{\mathscr{D}}_{\dot{\alpha}}(\mathscr{D}^2 - 4\bar{R})\bar{\mathscr{D}}^{\dot{\alpha}}V \qquad (5.5.12)$$

in accordance with equation (5.5.9*f*).

5.5.4. *Super Weyl transformations*

It has been pointed out in subsection 5.4.7 that the Einstein supergravity constraint (5.5.1) can be treated as a gauge fixing condition for the generalized super Weyl group acting on the covariant derivatives according to the law (5.3.58). However, the gauge condition (5.5.1) does not completely fix the generalized super Weyl symmetry. There exist some residual transformations (5.3.58) preserving the gauge (5.5.1). In accordance with equation (5.4.51), the corresponding superfield parameters must satisfy the equation

$$\mathscr{D}_{\alpha}\ln(L^4\bar{L}^2) = 0 \Rightarrow$$
$$L = \exp\left(\frac{1}{2}\sigma - \bar{\sigma}\right) \qquad \bar{\mathscr{D}}_{\dot{\alpha}}\sigma = 0.$$

Here $\sigma(z)$ is an arbitrary covariantly chiral scalar superfield.

We conclude that the minimal supergravity algebra (5.5.6) is covariant under the transformations

$$\begin{aligned}
\mathscr{D}_\alpha \to \mathscr{D}'_\alpha &= L\mathscr{D}_\alpha - 2(\mathscr{D}^\beta L)M_{\beta\alpha} \\
\bar{\mathscr{D}}_{\dot\alpha} \to \bar{\mathscr{D}}'_{\dot\alpha} &= \overline{L\mathscr{D}}_{\dot\alpha} - 2(\bar{\mathscr{D}}^{\dot\beta}\bar{L})\bar{M}_{\dot\beta\dot\alpha} \\
\mathscr{D}_{\alpha\dot\alpha} \to \mathscr{D}'_{\alpha\dot\alpha} &= \frac{i}{2}\{\mathscr{D}'_\alpha, \bar{\mathscr{D}}'_{\dot\alpha}\} \\
L = \exp\left(\frac{1}{2}\sigma - \bar\sigma\right) &\qquad \bar{\mathscr{D}}_{\dot\alpha}\sigma = 0
\end{aligned} \tag{5.5.13}$$

which will be called the 'super Weyl transformations'.

It is not difficult to find how the supertorsions R, G_a and $W_{\alpha\beta\gamma}$ change under the transformations (5.5.13). The results are

$$\begin{aligned}
R' &= -\frac{1}{4}e^{-2\sigma}(\bar{\mathscr{D}}^2 - 4R)e^{\bar\sigma} \\
G'_{\alpha\dot\alpha} &= e^{-(\sigma-\bar\sigma)/2}\left\{G_{\alpha\dot\alpha} + \frac{1}{2}(\mathscr{D}_\alpha\sigma)(\bar{\mathscr{D}}_{\dot\alpha}\bar\sigma) + i\mathscr{D}_{\alpha\dot\alpha}(\bar\sigma - \sigma)\right\} \\
W'_{\alpha\beta\gamma} &= e^{-3\sigma/2}W_{\alpha\beta\gamma}.
\end{aligned} \tag{5.5.14}$$

As may be seen, $W_{\alpha\beta\gamma}$ changes homogeneously.

At the prepotential level, equation (5.5.13) means the following

$$N'_\alpha{}^\mu = N_\alpha{}^\mu \qquad W'^M = W^M \qquad \varphi' = e^\sigma\varphi. \tag{5.5.15}$$

Therefore, φ is a compensating superfield for the super Weyl transformations. The super Weyl invariance can be used to impose the gauge condition $\varphi = 1$.

5.5.5. *Integration by parts*

From the algebra (5.5.6, 7) one can obtain the relation

$$(-1)^{\varepsilon_B}T_{AB}{}^B = 0. \tag{5.5.16}$$

Therefore, in Einstein supergravity we have the rule for integration by parts (5.2.62). In particular, if $\psi_\alpha(z)$ and $V_a(z)$ are spinor and vector superfields under proper boundary conditions, then

$$\int d^8z\, E^{-1}\mathscr{D}^\alpha\psi_\alpha = 0 \qquad \int d^8z\, E^{-1}\mathscr{D}^a V_a = 0. \tag{5.5.17}$$

5.5.6. *Chiral integration rule*

Let $\mathscr{L}_c(z)$ be a covariantly chiral scalar superfield,

$$\mathscr{L}_c = e^{\bar W}\,\overset{\circ}{\mathscr{L}}_c \qquad \bar\partial_{\dot\mu}\overset{\circ}{\mathscr{L}}_c = 0$$

where $\mathscr{L}_c$ is characterized by the transformation law

$$\delta\mathscr{L}_c = (\lambda^m\partial_m + \lambda^\mu\partial_\mu)\mathscr{L}_c.$$

Equation (5.5.5*b*) tells us that the integral

$$S_c = \int d^4x\, d^2\theta \phi^3 \mathscr{L}_c \equiv \int d^6z\, \phi^3 \mathscr{L}_c. \tag{5.5.18}$$

is invariant with respect to all supergravity transformations (K, λ and superlocal Lorentz ones). Now, we are going to rewrite expression (5.5.18) as an integral over the superspace $\mathbb{R}^{4|4}$.

In the chiral representation (see subsection 5.4.6), equation (5.5.3) can be rewritten as

$$\phi^3 F^2 = \tilde{E}^{-1}. \tag{5.5.19}$$

Acting on both sides with the operator $\frac{1}{4}\bar{\partial}_{\dot\mu}\bar{\partial}^{\dot\mu}$ and recalling equation (5.4.36), one obtains

$$\phi^3 \tilde{R} = \frac{1}{4}\bar{\partial}_{\dot\mu}\bar{\partial}^{\dot\mu}\tilde{E}^{-1} \tag{5.5.20a}$$

or

$$\phi^3 = \frac{1}{4}\bar{\partial}_{\dot\mu}\bar{\partial}^{\dot\mu}(\tilde{E}^{-1}/\tilde{R}). \tag{5.5.20b}$$

Therefore, S_c can be represented in the form

$$S_c = \int d^4x\, d^2\theta\, d^2\bar{\theta}\,(\tilde{E}^{-1}/\tilde{R})\mathscr{L}_c.$$

Finally, due to the identity (5.4.37), we can write

$$S_c = \int d^6z\, \phi^3 \mathscr{L}_c = \int d^8z\,(E^{-1}/R)\mathscr{L}_c. \tag{5.5.21}$$

This relation, known as the 'chiral integration rule', was obtained by W. Siegel.

It should be stressed that the chiral integration rule becomes unacceptable in cases when the body of R vanishes at some superspace points. In particular, this prescription has no flat superspace limit implying $R \to 0$. Hence, the chiral integration rule illustrates itself the peculiarity of curved supergeometry.

Using the chiral integration rule, any integral over $\mathbb{R}^{4|4}$ can be reduced to an integral over $\mathbb{R}^4_c \times \mathbb{C}^2_a$. Given a scalar superfield $\mathscr{L}(z)$, two integrals

$$\int d^8z\, E^{-1}\mathscr{L}$$

and

$$\int d^8z\, E^{-1}\left(\mathscr{L} - \frac{1}{4}\bar{\mathscr{D}}^2\left(\frac{1}{R}\mathscr{L}\right)\right)$$

coincide, owing to property (5.5.17). Then, since R is chiral one can write

$$\begin{aligned}\int d^8z\, E^{-1}\mathscr{L} &= -\frac{1}{4}\int d^8z\,(E^{-1}/R)(\bar{\mathscr{D}}^2 - 4R)\mathscr{L} \\ &= -\frac{1}{4}\int d^6z\phi^3 e^{-\mathscr{W}}[(\bar{\mathscr{D}}^2 - 4R)\mathscr{L}].\end{aligned} \quad (5.5.22)$$

5.5.7. Matter dynamical systems in a supergravity background

We have completed the supergeometric description of curved superspace corresponding to Einstein supergravity. Now, one can easily develop matter superfield theory in a given (background) curved superspace.

By matter superfields we will understand unconstrained tensor superfields of arbitrary Lorentz types $(n/2, m/2)$, unconstrained chiral tensor superfields of Lorentz types $(n/2, 0)$ and unconstrained antichiral tensor superfields of Lorentz types $(0, m/2)$.

The main symmetry principle which underlies dynamical systems in a curved superspace is invariance with respect to the general coordinate superspace transformations and the superlocal Lorentz transformations (the general covariance principle):

$$\begin{aligned}&\mathscr{D}'_A = e^{\mathscr{K}}\mathscr{D}_A e^{-\mathscr{K}} \qquad \chi' = e^{\mathscr{K}}\chi \\ &\mathscr{K} = K^M\partial_M + \frac{1}{2}K^{ab}M_{ab}.\end{aligned} \quad (5.5.23)$$

Here χ is a set of matter superfields, in terms of which some dynamical system is described. By definition, a local dynamical system is characterized by an action superfunctional with the structure:

$$\begin{aligned}S[\chi; \mathscr{D}_A] &= \int d^8zE^{-1}\mathscr{L}(\chi; \mathscr{D}_A) \\ &\quad + \int d^6z\phi^3 e^{-\mathscr{W}}\mathscr{L}_c(\chi; \mathscr{D}_A) + \int d^6\bar{z}\bar{\phi}^3 e^{-\bar{\mathscr{W}}}\bar{\mathscr{L}}_c(\chi; \mathscr{D}_A) \quad (5.5.24) \\ &= \int d^8zE^{-1}\left\{\mathscr{L}(\chi; \mathscr{D}_A) + \left(\frac{1}{R}\mathscr{L}_c(\chi; \mathscr{D}_A) + \text{c.c.}\right)\right\}\end{aligned}$$

with $\mathscr{L}$ being a real scalar superfield and $\mathscr{L}_c$ a covariantly chiral scalar superfield; $\mathscr{L}$ and $\mathscr{L}_c$ are assumed to be functions of superfields χ and their covariant derivatives to a finite order. In addition, $\mathscr{L}$ and $\mathscr{L}_c$ may depend polynomially on the supertorsions R, G_a and $W_{\alpha\beta\gamma}$ in order to have a well-defined flat superspace limit.

Clearly, the dynamical superfield equations for the system (5.5.24) should be covariant under the transformations (5.5.23). In order to have an explicitly covariant form for these equations, it is worth obtaining covariant variational rules. We shall now find covariant variational rules in two particular cases: for dynamical systems described by a real scalar superfield $V(z)$ or by a covariantly chiral scalar superfield $\Phi(z)$ and its conjugate $\bar{\Phi}(z)$. Tensor cases may be treated similarly.

Having a dynamical system with an action superfunctional $S[V;\mathscr{D}]$, we represent an infinitesimal variation of $S[V;\mathscr{D}]$ in the form:

$$\delta S[V;\mathscr{D}] \equiv S[V+\delta V;\mathscr{D}] - S[V;\mathscr{D}]$$
$$= \int \mathrm{d}^8 z E^{-1} \delta V(z) \frac{\delta S[V;\mathscr{D}]}{\delta V(z)}. \tag{5.5.25}$$

The $\delta S[V;\mathscr{D}]/\delta V(z)$ will be called the 'left superfunctional (or variational) derivative' of $S[V;\mathscr{D}]$ with respect to $V(z)$. Obviously, $\delta S/\delta V(z)$ is a real scalar superfield. It follows from (5.5.25) that

$$\frac{\delta V(z')}{\delta V(z)} = E\delta^4(x-x')\delta^2(\theta-\theta')\delta^2(\bar{\theta}-\bar{\theta}') \equiv \delta^8(z,z'). \tag{5.5.26}$$

The bi-scalar $\delta^8(z,z')$ will be called the 'covariant delta-function'.

Having a dynamical system with an action superfunctional $S[\Phi,\bar{\Phi};\mathscr{D}]$, we represent an infinitesimal variation of $S[\Phi,\bar{\Phi};\mathscr{D}]$ as follows:

$$\delta S[\Phi,\bar{\Phi};\mathscr{D}] \equiv S[\Phi+\delta\Phi,\bar{\Phi}+\delta\bar{\Phi};\mathscr{D}] - S[\Phi,\bar{\Phi};\mathscr{D}]$$
$$= \int \mathrm{d}^6 z \phi^3 \mathrm{e}^{-\overleftarrow{W}} \left\{ \delta\Phi(z) \frac{\delta S[\Phi,\bar{\Phi};\mathscr{D}]}{\delta\Phi(z)} \right\} + \text{c.c.} \tag{5.5.27}$$
$$= \int \mathrm{d}^8 z (E^{-1}/R) \delta\Phi(z) \frac{\delta S[\Phi,\bar{\Phi};\mathscr{D}]}{\delta\Phi(z)} + \text{c.c.}$$

where $\delta S[\Phi,\bar{\Phi};\mathscr{D}]/\delta\Phi(z)$ is a covariantly chiral scalar superfield, which will be called the 'covariant superfunctional (or variational) derivative' of $S[\Phi,\bar{\Phi};\mathscr{D}]$ with respect to $\Phi(z)$. Owing to the identity

$$\delta\Phi(z') = \int \mathrm{d}^8 z E^{-1} \delta\Phi(z) \delta^8(z,z') = -\frac{1}{4} \int \mathrm{d}^8 z (E^{-1}/R) \delta\Phi(z) (\bar{\mathscr{D}}^2 - 4R) \delta^8(z,z')$$

we have

$$\frac{\delta\Phi(z')}{\delta\Phi(z)} = -\frac{1}{4}(\bar{\mathscr{D}}^2 - 4R)\delta^8(z,z') \equiv \delta_+(z,z'). \tag{5.5.28}$$

The bi-scalar $\delta_+(z, z')$ is covariantly chiral with respect to each argument,

$$\bar{\mathscr{D}}_{\dot{\alpha}}\delta_+(z, z') = \bar{\mathscr{D}}'_{\dot{\alpha}}\delta_+(z, z') = 0. \tag{5.5.29}$$

This can be readily seen in the chiral representation:

$$-(\tilde{\bar{\mathscr{D}}}^2 - 4\tilde{R})\psi = \bar{\partial}_{\dot{\mu}}\bar{\partial}^{\dot{\mu}}(F^2\psi) \qquad \forall\psi \Rightarrow$$

$$\begin{aligned}\tilde{\delta}_+(z, z') &= \frac{1}{4}\bar{\partial}_{\dot{\mu}}\bar{\partial}^{\dot{\mu}}\{F^2\tilde{E}\delta^4(x - x')\delta^2(\theta - \theta')\delta^2(\bar{\theta} - \bar{\theta}')\} \\ &= \frac{1}{4}\bar{\partial}_{\dot{\mu}}\bar{\partial}^{\dot{\mu}}\{\phi^{-3}\delta^4(x - x')\delta^2(\theta - \theta')\delta^2(\bar{\theta} - \bar{\theta}')\} \\ &= \phi^{-3}\delta^4(x - x')\delta^2(\theta - \theta')\end{aligned}$$

where we have used equation (5.5.19). $\delta_+(z, z')$ will be called the 'covariant chiral delta-function'.

5.6. Prepotential deformations

In this section we would like to discuss a covariant variational technique for the supergravity prepotentials. It will be necessary in order to obtain a supersymmetric generalization of the energy–momentum tensor and to find (in the following chapter) supergravity dynamical equations. To start with, we introduce a slightly modified parametrization of the supergravity prepotentials and the supergravity gauge group. The parametrization given below proves to be most convenient for perturbative calculations.

In the main body of subsections 5.7.1 and 5.7.2, the covariant derivatives under conformal supergravity constraints only will be considered. Each occasion that specification to the Einstein supergravity case is needed will be specifically mentioned.

5.6.1. Modified parametrization of prepotentials

In the prepotential parametrization of subsection 5.4.1, the flat superspace limit corresponds to the choice:

$$N_{\alpha}{}^{\mu} = \delta_{\alpha}{}^{\mu} \qquad F = 1 \qquad W = -\mathrm{i}\theta\sigma^a\bar{\theta}\partial_a \equiv -\mathrm{i}\mathscr{H}_0.$$

Hence, when making perturbative calculations with respect to a flat background, one must expand W around its flat non-vanishing value $(-\mathrm{i}\mathscr{H}_0)$. This is not so convenient. It would be preferable to have a parametrization in which the flat superspace limit corresponded to the choice $W = 0$. Such a parametrization is easily obtained from the old one by merely replacing

$$\mathrm{e}^W \rightarrow \mathrm{e}^W \mathrm{e}^{-\mathrm{i}\mathscr{H}_0}. \tag{5.6.1}$$

Then, the spinor covariant derivatives take the form

$$\mathscr{D}_\alpha = F N_\alpha{}^\mu e^W D_\mu e^{-W} + \frac{1}{2}\Omega_\alpha{}^{bc} M_{bc}$$
$$\bar{\mathscr{D}}_{\dot\alpha} = \overline{F N}_{\dot\alpha}{}^{\dot\mu} e^{\bar W} \bar{D}_{\dot\mu} e^{-\bar W} + \frac{1}{2}\bar\Omega_{\dot\alpha}{}^{bc} M_{bc} \tag{5.6.2}$$

where D_μ and $\bar{D}_{\dot\mu}$ are the flat spinor covariant derivatives (see equations (2.5.19, 21)). Then, it seems reasonable to write W as follows

$$W = W^M D_M \qquad D_M = (\partial_m, D_\mu, \bar{D}^{\dot\mu}). \tag{5.6.3}$$

Further, we exclude from our consideration the unimodular matrix $N_\alpha{}^\mu$ at the cost of introducing a complex superlocal Lorentz transformation. Namely, the covariant derivatives will be represented in the form:

$$\mathscr{D}_\alpha = e^{\mathscr{W}}\left(F D_\alpha + \frac{1}{2}\Sigma_\alpha{}^{bc} M_{bc}\right) e^{-\mathscr{W}}$$
$$\bar{\mathscr{D}}_{\dot\alpha} = e^{\bar{\mathscr{W}}}\left(\overline{F D}_{\dot\alpha} + \frac{1}{2}\bar\Sigma_{\dot\alpha}{}^{bc} M_{bc}\right) e^{-\bar{\mathscr{W}}} \tag{5.6.4}$$
$$\mathscr{W} = W^A D_A + W^{\alpha\beta} M_{\alpha\beta} + W^{\dot\alpha\dot\beta} \bar{M}_{\dot\alpha\dot\beta} \neq \bar{\mathscr{W}}$$

instead of the form (5.6.2). Here $W^{\alpha\beta}(z)$ and $W^{\dot\alpha\dot\beta}(z)$ are symmetric bi-spinor superfields. One can easily see that $W^{\alpha\beta}$ is in one-to-one correspondence with the old prepotential $N_\alpha{}^\mu$. A novel feature is the appearance of redundant superfields $W^{\dot\alpha\dot\beta}$ $(\neq (W^{\alpha\beta})^* = \bar{W}^{\dot\alpha\dot\beta})$ which we have introduced by hand. In principle, one can set $W^{\dot\alpha\dot\beta}$ to zero or uniquely fix their values by imposing the requirement

$$\Sigma_\alpha{}^{\dot\beta\dot\gamma} \bar{M}_{\dot\beta\dot\gamma} = 0 \tag{5.6.5}$$

which is possible owing to the results of subsection 5.4.2. However, we prefer to keep arbitrary $W^{\dot\alpha\dot\beta}$, similarly to the other superfields arising in $\mathscr{W}$. It will soon be shown that $W^{\dot\alpha\dot\beta}$ is a pure gauge degree of freedom associated with some auxiliary gauge invariance.

Considering the parametrization (5.6.4), it is convenient to represent $\mathscr{D}_A$ in the form $\mathscr{D}_A = E_A{}^M D_M + \frac{1}{2}\Omega_A{}^{bc} M_{bc}$ as well as to write the transformations (5.3.57) according to the rule

$$\mathscr{D}_A \to \mathscr{D}'_A = e^{\mathscr{K}} \mathscr{D}_A e^{-\mathscr{K}}$$
$$\mathscr{K} = K^A D_A + \frac{1}{2} K^{bc} M_{bc} = \bar{\mathscr{K}}. \tag{5.6.6}$$

These act on the prepotentials just introduced according to the law

$$e^{\mathscr{W}'} = e^{\mathscr{K}} e^{\mathscr{W}} \qquad F' = F \qquad \Sigma'_\alpha{}^{bc} = \Sigma_\alpha{}^{bc}. \tag{5.6.7}$$

To find an analogue of the λ-supergroup in the parametrization (5.6.4), we seek a transformation of the form

$$e^{\mathscr{W}''} = e^{\mathscr{W}'} e^{-\bar\Lambda} \qquad e^{\bar{\mathscr{W}}''} = e^{\bar{\mathscr{W}}'} e^{-\Lambda}$$
$$\Lambda = \Lambda^A D_A + \Lambda^{\alpha\beta} M_{\alpha\beta} + \Lambda^{\dot\alpha\dot\beta} \bar M_{\dot\alpha\dot\beta} \neq \bar\Lambda \tag{5.6.8}$$

supplemented by displacements of F and $\Sigma_{\dot\alpha}{}^{bc}$ such that the covariant derivatives remain unchanged. In the infinitesimal case, we have

$$0 = e^{-\bar{\mathscr{W}}'} \delta\bar{\mathscr{D}}_{\dot\alpha} e^{\bar{\mathscr{W}}'} = -\left[\Lambda, \overline{F D}_{\dot\alpha} + \frac{1}{2}\Sigma_{\dot\alpha}{}^{bc} M_{bc}\right] + \delta\overline{F D}_{\dot\alpha} + \frac{1}{2}\delta\Sigma_{\dot\alpha}{}^{bc} M_{bc}.$$

Due to the relation

$$[\bar D_{\dot\alpha}, \Lambda] = -\frac{1}{2}(\bar D_{\dot\alpha}\Lambda^{\beta\dot\beta})\partial_{\beta\dot\beta} + 2i\Lambda^{\beta}\partial_{\beta\dot\alpha} + (\bar D_{\dot\alpha}\Lambda^{\beta})D_{\beta}$$
$$- ((\bar D_{\dot\alpha}\Lambda^{\dot\beta}) + \Lambda_{\dot\alpha}{}^{\dot\beta})\bar D_{\dot\beta} + \frac{1}{2}(\bar D_{\dot\alpha}\Lambda^{bc})M_{bc}$$

the above requirement leads to restrictions on the parameters

$$\bar D_{\dot\alpha}\Lambda^{\beta\dot\beta} = 4i\Lambda^{\beta}\delta_{\dot\alpha}{}^{\dot\beta} \qquad \bar D_{\dot\alpha}\Lambda^{\beta} = 0$$
$$\Lambda_{\dot\alpha\dot\beta} = -\bar D_{(\dot\alpha}\Lambda_{\dot\beta)} \tag{5.6.9}$$
$$\Lambda_{\dot\alpha} \text{ and } \Lambda_{\alpha\beta} \text{ arbitrary}$$

and also the transformation laws

$$\delta\bar F = \Lambda\bar F + \frac{1}{2}(\bar D_{\dot\alpha}\Lambda^{\dot\alpha})\bar F$$
$$\delta\Sigma_{\dot\alpha}{}^{bc} = -\bar D_{\dot\alpha}\Lambda^{bc} + ([\Lambda, \Sigma_{\dot\alpha}])^{bc}. \tag{5.6.10}$$

Under the relations (5.6.9, 10), $\bar{\mathscr{D}}_{\dot\alpha}$ is invariant with respect to the transformation (5.6.8).

The first equation (5.6.9) is equivalent to the two identities:

$$\bar D_{\dot\alpha}\Lambda_{\beta\dot\beta} + \bar D_{\dot\beta}\Lambda_{\beta\dot\alpha} = 0 \qquad 8i\Lambda^{\alpha} = \bar D_{\dot\alpha}\Lambda^{\alpha\dot\alpha}.$$

They can be solved in terms of a spinor superfield L_α as follows:

$$\Lambda_{\alpha\dot\alpha} = -2i\bar D_{\dot\alpha}L_\alpha \qquad \Lambda_\alpha = -\frac{1}{4}\bar D^2 L_\alpha. \tag{5.6.11}$$

Then, the second equation (5.6.9) is satisfied automatically. We see that the Λ-transformations (5.6.8, 10) are generated by unconstrained parameters L_α, $\Lambda_{\dot\alpha}$ and $\Lambda_{\alpha\beta} = \Lambda_{\beta\alpha}$. This is one of the main advantages of the parametrization (5.6.4).

Let $\Phi(z)$ be a covariantly chiral scalar superfield, $\bar{\mathscr{D}}_{\dot\alpha}\Phi = 0$. It can be represented as

$$\Phi = e^{\mathscr{W}}\hat{\Phi} \qquad \bar{D}_{\dot\alpha}\hat{\Phi} = 0 \tag{5.6.12}$$

with $\hat{\Phi}$ being a flat chiral superfield. The Λ-transformations act on $\hat{\Phi}$ according to the law

$$\hat{\Phi}' = e^{\Lambda}\hat{\Phi} \tag{5.6.13}$$

or, in infinitesimal form,

$$\delta\hat{\Phi} = \Lambda\hat{\Phi} = -\frac{1}{4}\bar{D}^2(L^\alpha D_\alpha\hat{\Phi}) \tag{5.6.14}$$

where we have used equation (5.6.11). Clearly, $\delta\hat{\Phi}$ is chiral.

The $\mathscr{K}$-transformations (5.6.6) can be used to gauge away the real part of $\mathscr{W}$ resulting in

$$\mathscr{W} = -i\mathscr{H} \qquad \mathscr{H} = H^A D_A + H^{\alpha\beta}M_{\alpha\beta} + \bar{H}^{\dot\alpha\dot\beta}\bar{M}_{\dot\alpha\dot\beta} = \bar{\mathscr{H}}.$$

Such a gauge choice is equivalent to introducing a chiral representation defined by the rule: every tensor superfield V is to be changed by

$$\tilde{V} = e^{-\bar{\mathscr{W}}}V \tag{5.6.15a}$$

the covariant derivatives are to be transformed to

$$\tilde{\mathscr{D}}_A = e^{-\bar{\mathscr{W}}}\mathscr{D}_A e^{\bar{\mathscr{W}}}. \tag{5.6.15b}$$

There is one difference between the chiral representation just introduced and that considered in subsection 5.4.6. To obtain the representation of subsection 5.4.6, one has to make the complex coordinate transformation (5.4.30). In the present case, since

$$e^{-\bar{\mathscr{W}}} = \exp\left[-\bar{W}^A D_A - \frac{1}{2}\bar{W}^{ab}M_{ab}\right] = \exp\left[-\frac{1}{2}\bar{W}'^{ab}M_{ab}\right]\exp\left[-\bar{W}^A D_A\right]$$

one has to apply not only a complex coordinate transformation, but also to accompany it by some complex superlocal Lorentz transformation. Since Lorentz transformations (real or complex) do not change invariant functionals of the form $\int d^8z\, E^{-1}\mathscr{L}$, with $\mathscr{L}$ being a scalar superfield, the identity (5.4.37) holds in the present case also.

The covariant derivatives in the chiral representation are

$$\tilde{\mathscr{D}}_\alpha = e^{-2i\mathscr{H}}\left(F D_\alpha + \frac{1}{2}\Sigma^{bc}M_{bc}\right)e^{2i\mathscr{H}}$$

$$\tilde{\bar{\mathscr{D}}}_{\dot\alpha} = \overline{F D}_{\dot\alpha} + \frac{1}{2}\bar{\Sigma}^{bc}M_{bc} \qquad \mathscr{D}_{\alpha\dot\alpha} = \frac{i}{2}\{\mathscr{D}_\alpha, \bar{\mathscr{D}}_{\dot\alpha}\} \tag{5.6.16}$$

$$e^{-2i\mathscr{H}} \equiv e^{-\bar{\mathscr{W}}}e^{\mathscr{W}} \qquad \mathscr{H} = H^A D_A + \frac{1}{2}H^{ab}M_{ab} = \bar{\mathscr{H}}.$$

They transform according to the law:

$$\tilde{\mathscr{D}}'_A = e^{\Lambda}\mathscr{D}_A e^{-\Lambda} \qquad e^{-2i\mathscr{H}'} = e^{\Lambda}e^{-2i\mathscr{H}}e^{-\bar{\Lambda}} \tag{5.6.17}$$

where the parameters are defined as in equations (5.6.9) and (5.6.11). Since $(-2i\delta\mathscr{H}) = \Lambda_{\dot{\alpha}}\bar{D}^{\dot{\alpha}} - \bar{\Lambda}^{\alpha}D_{\alpha} + \Lambda^{\alpha\beta}M_{\alpha\beta} - \bar{\Lambda}^{\dot{\alpha}\dot{\beta}}\bar{M}_{\dot{\alpha}\dot{\beta}} + \ldots$, we can use $\Lambda_{\dot{\alpha}}$ and $\Lambda_{\alpha\beta}$ to gauge away H^{α}, $\bar{H}_{\dot{\alpha}}$, $H^{\alpha\beta}$ and $\bar{H}^{\dot{\alpha}\dot{\beta}}$ resulting in

$$\mathscr{H} = H^a\partial_a. \tag{5.6.18}$$

In this gauge, we have a restricted set of gauge transformations:

$$\delta e^{-2i\mathscr{H}} = \Lambda e^{-2i\mathscr{H}} - e^{-2i\mathscr{H}}\bar{\Lambda} \text{ has no spinor and Lorentz part} \Rightarrow$$

$$\Lambda_{\dot{\alpha}} = e^{-2i\mathscr{H}}\bar{\Lambda}_{\dot{\alpha}} \qquad \Lambda_{\alpha\beta} = e^{-2i\mathscr{H}}\bar{\Lambda}_{\alpha\beta}. \tag{5.6.19}$$

Remark. Instead of choosing the gauge (5.6.18), one can impose a weaker gauge condition

$$\mathscr{H} = H^A D_A. \tag{5.6.20}$$

In this case, only the second restriction (5.6.19) should be imposed.

In the chiral representation, one can readily repeat the analysis of subsection 5.4.4 obtaining explicit expressions for the geometrical objects:

$$\tilde{E} = \bar{F}^2(e^{-2i\mathscr{H}}F)^2\hat{E} \qquad \tilde{R} = -\frac{1}{4}\bar{D}^2(\bar{F}^2)$$

$$\tilde{\bar{T}}_{\dot{\alpha}} = \tilde{\bar{E}}_{\dot{\alpha}} \ln[\tilde{E}\bar{F}^2] \qquad \tilde{T}_{\alpha} = \tilde{E}_{\alpha} \ln[\tilde{E}(e^{-2i\mathscr{H}}F)^2(1\cdot e^{-2i\bar{\mathscr{H}}})] \tag{5.6.21}$$

where we have denoted:

$$\hat{E}_A = \hat{E}_A{}^M D_M = \left(-\frac{i}{4}(\tilde{\sigma}_a)^{\dot{\beta}\beta}\{\hat{E}_{\beta}, \hat{E}_{\dot{\beta}}\}, e^{-2i\mathscr{H}}D_{\alpha}e^{2i\mathscr{H}}, \bar{D}^{\dot{\alpha}}\right)$$

$$\hat{E} = \mathrm{Ber}(\hat{E}_A{}^M).$$

In the Einstein supergravity case, we have $T_{\alpha} = \bar{T}_{\dot{\alpha}} = 0$. Then it follows from the above relations that

$$\tilde{E}F^2 = \varphi^{-3} \qquad \bar{D}_{\dot{\alpha}}\varphi = 0. \tag{5.6.22}$$

The chiral compensator φ proves to have the transformation law

$$\delta\varphi^3 = \Lambda^A D_A\varphi^3 + (\partial_a\Lambda^a - D_{\alpha}\Lambda^{\alpha})\varphi^3 = \frac{1}{4}\bar{D}^2 D_{\alpha}(L^{\alpha}\varphi^3). \tag{5.6.23}$$

5.6.2. Background-quantum splitting

We go on by describing the covariant supergravity variational technique developed by M. Grisaru and W. Siegel.

Consider two sets of covariant derivatives under conformal supergravity constraints:

$$\mathscr{D}_A = E_A{}^M \mathrm{D}_M + \frac{1}{2}\Omega_A{}^{bc} M_{bc}$$

and

$$\nabla_A = (E + \delta E)_A{}^M \mathrm{D}_M + \frac{1}{2}(\Omega + \delta\Omega)_A{}^{bc} M_{bc} \equiv \mathscr{E}_A{}^M \mathrm{D}_M + \frac{1}{2}\Theta_A{}^{bc} M_{bc}. \tag{5.6.24}$$

Using the parametrization of previous subsections for $\mathscr{D}_A$, one can write

$$\nabla_\alpha = \mathrm{e}^{\mathscr{W} + \delta\mathscr{W}}\left[(F + \delta F)\mathrm{D}_\alpha + \frac{1}{2}(\Sigma + \delta\Sigma)_\alpha{}^{bc} M_{bc}\right]\mathrm{e}^{-\mathscr{W} - \delta\mathscr{W}} \tag{5.6.25}$$

where $\delta\mathscr{W} = \delta W^M \mathrm{D}_M + \frac{1}{2}\delta W^{bc} M_{bc}$ and δF are finite deformations ($\delta\Sigma_\alpha{}^{bc}$ is determined in terms of $\delta\mathscr{W}$ and δF because of the conformal supergravity constraints). The main problem one is faced with is that $\delta\mathscr{W}$ transforms highly nonlinearly: the $\mathscr{K}$- and Λ-transformations act on $\mathscr{W}$ and $(\mathscr{W} + \delta\mathscr{W})$ according to the laws

$$\mathrm{e}^{\mathscr{W}'} = \mathrm{e}^{\mathscr{K}}\mathrm{e}^{\mathscr{W}}\mathrm{e}^{-\bar{\Lambda}} \qquad \mathrm{e}^{\mathscr{W}' + \delta\mathscr{W}'} = \mathrm{e}^{\mathscr{K}}\mathrm{e}^{\mathscr{W} + \delta\mathscr{W}}\mathrm{e}^{-\bar{\Lambda}}. \tag{5.6.26}$$

To circumvent this problem, it is worth substituting the variation $\delta\mathscr{W}$ by the covariantized one, $\Delta\mathscr{W}$, defined by

$$\mathrm{e}^{\mathscr{W} + \delta\mathscr{W}} \equiv \mathrm{e}^{\Delta\mathscr{W}}\mathrm{e}^{\mathscr{W}}. \tag{5.6.27}$$

In accordance with equation (5.6.25), $\Delta\mathscr{W}$ transforms covariantly:

$$\mathrm{e}^{\Delta\mathscr{W}'} = \mathrm{e}^{\mathscr{K}}\mathrm{e}^{\Delta\mathscr{W}}\mathrm{e}^{-\mathscr{K}}. \tag{5.6.28}$$

Now, if one represents $\Delta\mathscr{W}$ in the form

$$\begin{aligned}\Delta\mathscr{W} &= \Delta W^M \mathrm{D}_M + \frac{1}{2}\Delta W^{bc} M_{bc} \\ &= \Delta W^M E_M{}^A \mathscr{D}_A + \frac{1}{2}(\Delta W^{bc} - \Delta W^M E_M{}^A \Omega_A{}^{bc}) M_{bc} \\ &\equiv \Delta\mathscr{W}^A \mathscr{D}_A + \frac{1}{2}\Delta\mathscr{W}^{bc} M_{bc}\end{aligned}$$

then $\Delta\mathscr{W}^A$ and $\Delta\mathscr{W}^{bc}$ will be Lorentz tensor superfields.

There is another argument in defence of variation (5.6.27). Namely, looking at equation (5.6.4), it is seen that $\mathscr{W}$ enters as a complex general coordinate and superlocal Lorentz transformation. Therefore, it seems reasonable to represent any change in $\mathrm{e}^{\mathscr{W}}$ as a (complex) supergroup shift. Similarly, the superfield $(\mathrm{e}^{\mathscr{W}} F)$ enters as a super Weyl transformation (5.3.58). Hence, it is tempting to represent its change as a supergroup shift, $F + \delta F = (e^{-\mathcal{W}'}\mathcal{F})F$. Finally, making a redefinition of $\delta\Sigma_\alpha{}^{bc}$ in equation (5.6.25), one arrives at the ∇ representation:

$$\nabla_\alpha = e^{\Delta\mathcal{W}'}\left[\mathcal{F}\mathcal{D}_\alpha + \frac{1}{2}\Delta\Omega_\alpha{}^{bc}M_{bc}\right]e^{-\Delta\mathcal{W}'}$$

$$\bar{\nabla}_{\dot\alpha} = e^{\Delta\bar{\mathcal{W}}'}\left[\overline{\mathcal{F}\mathcal{D}}_{\dot\alpha} + \frac{1}{2}\Delta\bar{\Omega}_{\dot\alpha}{}^{bc}M_{bc}\right]e^{-\Delta\bar{\mathcal{W}}'} \tag{5.6.29}$$

$$\nabla_{\alpha\dot\alpha} = \frac{i}{2}\{\nabla_\alpha, \bar{\nabla}_{\dot\alpha}\} \qquad \Delta\mathcal{W}' = \Delta\mathcal{W}'^A\mathcal{D}_A + \frac{1}{2}\Delta\mathcal{W}'^{bc}M_{bc} \neq \Delta\mathcal{W}.$$

This representation is known as 'background-quantum splitting' (originally, it was developed for covariant quantization of supergravity). Background-quantum splitting determines deformed covariant derivatives ∇_A in terms of the initial (background) derivatives $\mathcal{D}_A$. The derivatives ∇_A turn out to be coupled to the background derivatives $\mathcal{D}_A$. All objects $\Delta\mathcal{W}'^A$, $\Delta\mathcal{W}'^{bc}$, $\mathcal{F}$ and $\Delta\Omega_\alpha{}^{bc}$, which specify ∇_A, are tensor superfields.

There exists some inherent arbitrariness in the choice of $\Delta\mathcal{W}'$, $\mathcal{F}$ and $\Delta\Omega_\alpha^{bc}$. Let us seek a transformation of the form

$$\mathcal{D}'_A = \mathcal{D}_A \qquad e^{\Delta\bar{\mathcal{W}}''} = e^{\Delta\bar{\mathcal{W}}'}e^{-\Lambda}$$
$$\Lambda = \Lambda^A\mathcal{D}_A + \Lambda^{\alpha\beta}M_{\alpha\beta} + \Lambda^{\dot\alpha\dot\beta}\bar{M}_{\dot\alpha\dot\beta} \tag{5.6.30}$$

supplemented by displacements of $\mathcal{F}$ and $\Delta\Omega_\alpha{}^{bc}$ such that the operators ∇_A stay unchanged. Imposing the requirement

$$0 = e^{-\Delta\bar{\mathcal{W}}'}\delta\bar{\nabla}_{\dot\alpha}e^{\Delta\bar{\mathcal{W}}'} = -\left[\Lambda, \overline{\mathcal{F}\mathcal{D}}_{\dot\alpha} + \frac{1}{2}\Delta\bar{\Omega}_{\dot\alpha}{}^{bc}M_{bc}\right] + \delta\overline{\mathcal{F}\mathcal{D}}_{\dot\alpha} + \frac{1}{2}\delta(\Delta\bar{\Omega}_{\dot\alpha}{}^{bc}M_{bc})$$

and making use of the algebra (5.3.53), one finds the following restrictions on the parameters

$$\bar{\mathcal{D}}_{\dot\alpha}\Lambda_\alpha + \frac{i}{2}\Lambda_{\alpha\dot\alpha}\left(R + \frac{1}{8}\bar{\mathcal{D}}_{\dot\beta}\bar{T}^{\dot\beta} - \frac{1}{16}\bar{T}_{\dot\beta}\bar{T}^{\dot\beta}\right) = 0$$

$$2i\Lambda^\beta\delta^{\dot\beta}_{\dot\alpha} = \frac{1}{2}\bar{\mathcal{D}}_{\dot\alpha}\Lambda^{\dot\beta\dot\beta} + \frac{1}{4}\Lambda^\beta{}_{\dot\alpha}T^{\dot\beta}$$

$$\Lambda_{\dot\alpha\dot\beta} = -\bar{\mathcal{D}}_{(\dot\alpha}\Lambda_{\dot\beta)} + \frac{i}{2}\Lambda_{\beta(\dot\alpha}\left(G^\beta{}_{\dot\alpha)} - \frac{1}{8}\mathcal{D}^\beta\bar{T}_{\dot\beta)} - \frac{1}{8}\bar{\mathcal{D}}_{\dot\beta)}T^\beta\right) \tag{5.6.31}$$

$\Lambda_{\dot\alpha}$ and $\Lambda_{\alpha\beta}$ arbitrary.

The first two equations can be solved in terms of an unconstrained spinor superfield L_α:

$$\Lambda_{\alpha\dot\alpha} = -2\mathrm{i}\left(\bar{\mathscr{D}}_{\dot\alpha} + \frac{1}{2}\bar{T}_{\dot\alpha}\right)L_\alpha$$

$$\Lambda_\alpha = -\frac{1}{4}\left(\bar{\mathscr{D}}_{\dot\beta} - \frac{1}{2}\bar{T}_{\dot\beta}\right)\left(\bar{\mathscr{D}}^{\dot\beta} + \frac{1}{2}\bar{T}^{\dot\beta}\right)L_\alpha. \tag{5.6.32}$$

Here $\bar{T}_{\dot\alpha}$, R and $G_{\alpha\dot\alpha}$ are the supertorsions corresponding to $\mathscr{D}_A$. Further, the transformation law for $\mathscr{F}$ is

$$\delta\mathscr{F} = \Lambda\mathscr{F} + \frac{1}{2}(\bar{\mathscr{D}}_{\dot\alpha}\Lambda^{\dot\alpha})\mathscr{F} - \frac{\mathrm{i}}{4}\Lambda^{\alpha\dot\alpha}\left(G_{\alpha\dot\alpha} - \frac{1}{8}\mathscr{D}_\alpha\bar{T}_{\dot\alpha} - \frac{1}{8}\bar{\mathscr{D}}_{\dot\alpha}T_\alpha\right)\mathscr{F} \tag{5.6.33}$$

The explicit form of $\delta[\Delta\bar{\Omega}_{\dot\alpha}{}^{bc}]$ is not essential for later applications. In summary, the transformations (5.6.30–33) leave the derivatives ∇_A unchanged.

The background-quantum splitting proves to be very powerful when making covariant variations with respect to the supergravity prepotentials of invariant superfunctionals of the general structure

$$S_{(\mathscr{D})} = S[\chi_{(\mathscr{D})}; \mathscr{D}] = \int \mathrm{d}^8 z E^{-1}\mathscr{L}(\chi_{(\mathscr{D})}; \mathscr{D}] \tag{5.6.34}$$

with $\mathscr{L}$ being a scalar dependent on a set of matter superfields $\chi_{(\mathscr{D})}$ coupled to supergravity covariant derivatives $\mathscr{D}_A$. Here we mark the matter superfields by the label $(\mathscr{D})$, since covariantly defined superfields often depend on supergravity prepotentials. For example, if $\chi_{(\mathscr{D})}$ is a covariantly chiral scalar superfield, $\bar{\mathscr{D}}_{\dot\alpha}\chi_{(\mathscr{D})} = 0$, then one can write $\chi_{(\mathscr{D})} = \mathrm{e}^{\mathscr{W}}\hat{\chi}$, with $\hat{\chi}$ being an independent flat chiral superfield, $\bar{\mathrm{D}}_{\dot\alpha}\hat{\chi} = 0$. Hence, any supergeometry deformation $\mathscr{D}_A \to \nabla_A = \mathscr{D}_A + \delta\mathscr{D}_A$ will induce some change of the matter superfields, $\chi_{(\mathscr{D})} \to \chi_{(\nabla)} = \chi_{(\mathscr{D})} + \delta\chi$, as well as of the functional (5.6.34), $S_{(\mathscr{D})} \to S_{(\nabla)} = S_{(\mathscr{D})} + \delta S$, where

$$S_{(\nabla)} = S[\chi_{(\nabla)}; \nabla] = \int \mathrm{d}^8 z \mathscr{E}^{-1}\mathscr{L}(\chi_{(\nabla)}; \nabla). \tag{5.6.35}$$

Both functionals $S_{(\mathscr{D})}$ and $S_{(\nabla)}$ are invariant under the $\mathscr{K}$-transformations. In the former case, they take the form

$$\chi'_{(\mathscr{D})} = \mathrm{e}^{\mathscr{K}}\chi_{(\mathscr{D})} \qquad \mathscr{D}'_A = \mathrm{e}^{\mathscr{K}}\mathscr{D}_A\mathrm{e}^{-\mathscr{K}} \tag{5.6.36}$$

in the latter case, they read

$$\chi'_{(\nabla)} = \mathrm{e}^{\mathscr{K}}\chi_{(\nabla)} \qquad \nabla'_A = \mathrm{e}^{\mathscr{K}}\nabla_A\mathrm{e}^{-\mathscr{K}}. \tag{5.6.37}$$

However, after introducing background-quantum splitting (5.6.29), the transformations (5.6.37) can be realized in two different ways: (1) as 'background' transformations

$$\mathscr{D}'_A = e^{\mathscr{K}} \mathscr{D}_A e^{-\mathscr{K}} \qquad \nabla'_A = e^{\mathscr{K}} \nabla_A e^{-\mathscr{K}} \Rightarrow$$
$$\Delta\mathscr{W}' = e^{\mathscr{K}} \Delta\mathscr{W} e^{-\mathscr{K}} \qquad \mathscr{F}' = e^{\mathscr{K}} \mathscr{F} \qquad \Delta\Omega'_{\alpha bc} = e^{\mathscr{K}} \Delta\Omega_{\alpha bc} \tag{5.6.38a}$$

or (2) as 'quantum' transformations

$$\mathscr{D}'_A = \mathscr{D}_A \qquad \nabla'_A = e^{\mathscr{K}^{(Q)}} \nabla_A e^{-\mathscr{K}^{(Q)}} \Rightarrow$$
$$e^{\Delta\mathscr{W}'} = e^{\mathscr{K}^{(Q)}} e^{\Delta\mathscr{W}} \qquad \mathscr{F}' = \mathscr{F} \qquad \Delta\Omega'_{\alpha bc} = \Delta\Omega_{\alpha bc} \tag{5.6.38b}$$

where

$$\mathscr{K}^{(Q)} = K^A \mathscr{D}_A + \frac{1}{2} K^{bc} M_{bc} = \bar{\mathscr{K}}^{(Q)}.$$

It is worth noting that the existence of two types of $\mathscr{K}$-transformations is brought about by the splitting of a single object $(\mathscr{W} + \delta\mathscr{W})$ into two independent parts $\mathscr{W}$ and $\delta\mathscr{W}$ (a similar situation occurs, for example, in electrodynamics, where, after splitting $V_m = V_m^{(B)} + V_m^{(Q)}$, gauge transformations $\delta V_m = \partial_m \lambda$ can be realized as $\delta V_m^{(B)} = \partial_m \lambda$ and $\delta V_m^{(Q)} = 0$ or as $\delta V_m^{(B)} = 0$ and $\delta V_m^{(Q)} = \partial_m \lambda$). Note also that $S_{(\nabla)}$ is invariant under the Λ-transformations (5.6.30–33).

The real part of $\Delta\mathscr{W}$ is a purely gauge degree of freedom for the 'quantum' $\mathscr{K}$-invariance. This invariance can be used to gauge away $\mathrm{Re}\,\Delta\mathscr{W}$. Equivalently, $\mathrm{Re}\,\Delta\mathscr{W}$ is excluded automatically when working in the 'quantum' chiral representation introduced by

$$\chi_{(\nabla)} \to \tilde{\chi}_{(\nabla)} = e^{-\Delta\bar{\mathscr{W}}} \chi_{(\nabla)} \qquad \mathscr{D}_A \to \mathscr{D}_A$$
$$\nabla_A \to \tilde{\nabla}_A = e^{-\Delta\bar{\mathscr{W}}} \nabla_A e^{\Delta\bar{\mathscr{W}}} \equiv \tilde{\mathscr{E}}_A{}^M \mathrm{D}_M + \frac{1}{2} \tilde{\Theta}_A{}^{bc} M_{bc}. \tag{5.6.39}$$

Since $\tilde{\mathscr{E}}^{-1} = [\mathrm{Ber}(\tilde{\mathscr{E}}_A{}^M)]^{-1} = (\mathscr{E}^{-1} \cdot e^{-\Delta\overleftarrow{\bar{\mathscr{W}}}})$, the quantum chiral transform of $S_{(\nabla)}$ coincides with $S_{(\nabla)}$. In the quantum chiral representation, we have

$$\tilde{\nabla}_\alpha = e^{-2i\mathbf{H}} \left[\mathscr{F} \mathscr{D}_\alpha + \frac{1}{2} \Delta\Omega_\alpha{}^{bc} M_{bc} \right] e^{2i\mathbf{H}}$$

$$\tilde{\bar{\nabla}}_{\dot{\alpha}} = \overline{\mathscr{F} \mathscr{D}}_{\dot{\alpha}} + \frac{1}{2} \Delta\bar{\Omega}_{\dot{\alpha}}{}^{bc} M_{bc} \qquad \tilde{\nabla}_{\alpha\dot{\alpha}} = \frac{i}{2} \{\tilde{\nabla}_\alpha, \tilde{\bar{\nabla}}_{\dot{\alpha}}\}, \tag{5.6.40}$$

$$e^{-2i\mathbf{H}} \equiv e^{-\Delta\bar{\mathscr{W}}} e^{\Delta\mathscr{W}} \qquad \mathbf{H} = \mathbf{H}^A \mathscr{D}_A + \frac{1}{2} \mathbf{H}^{ab} M_{ab} = \bar{\mathbf{H}}$$

and the functional $S_{(\nabla)}$ is invariant under the transformations

$$\tilde{\chi}'_{(\nabla)} = e^{\Lambda} \tilde{\chi}_{(\nabla)} \qquad \tilde{\nabla}'_A = e^{\Lambda} \tilde{\nabla}_A e^{-\Lambda}$$
$$e^{-2i\mathbf{H}'} \equiv e^{\Lambda} e^{-2i\mathbf{H}} e^{-\bar{\Lambda}} \tag{5.6.41}$$

with Λ defined by equations (5.6.30–32). Using this invariance, one can impose the gauge

$$\mathbf{H} = \mathbf{H}^a \mathscr{D}_a. \tag{5.6.42}$$

After doing this, we work with a restricted set of gauge transformations (5.6.41) selected by the requirements

$$\Lambda_{\dot\alpha} = \bar\Lambda_{\dot\alpha} + O(\mathbf{H}) \qquad \Lambda_{\alpha\beta} = \bar\Lambda_{\alpha\beta} + O(\mathbf{H}). \tag{5.6.43}$$

Remark. If $\tilde\chi_{(\nabla)}$ is a covariantly chiral scalar superfield, $\tilde{\bar\nabla}_{\dot\alpha}\tilde\chi_{(\nabla)} = 0$, then its transformation law is

$$\delta\tilde\chi_{(\nabla)} = \Lambda\tilde\chi_{(\nabla)} = -\frac{1}{4}(\bar{\mathscr{D}}^2 - 4R)(L^\alpha \mathscr{D}_\alpha \tilde\chi_{(\nabla)}). \tag{5.6.44}$$

In conclusion, we give explicit expressions for some geometrical objects, constructed on the basis of $\tilde\nabla_A$, in terms of $\mathbf{H}$ and $\mathscr{F}$. We will denote the supercurvature and supertorsion tensors of $\tilde\nabla_A$ in bold type,

$$[\tilde\nabla_A, \tilde\nabla_B\} = \tilde{\mathbf{T}}_{AB}{}^C \tilde\nabla_C + \frac{1}{2}\tilde{\mathbf{R}}_{AB}{}^{cd} M_{cd}.$$

Demanding the anticommutator

$$\{\tilde{\bar\nabla}_{\dot\alpha}, \tilde{\bar\nabla}_{\dot\beta}\} = 4\tilde{\mathbf{R}}\bar M_{\dot\alpha\dot\beta}$$

one finds

$$\begin{aligned} \Delta\bar{\mathbf{\Omega}}_{\dot\alpha\dot\beta\dot\gamma} &= \varepsilon_{\dot\alpha\dot\beta}\bar{\mathscr{D}}_{\dot\gamma}\bar{\mathscr{F}} + \varepsilon_{\dot\alpha\dot\gamma}\bar{\mathscr{D}}_{\dot\beta}\bar{\mathscr{F}} \\ \tilde{\mathbf{R}} &= -\frac{1}{4}(\bar{\mathscr{D}}^2 - 4R)\bar{\mathscr{F}}^2. \end{aligned} \tag{5.6.45}$$

Next, similarly to the derivation of equation (5.4.20), one can obtain

$$\tilde{\bar{\mathbf{T}}}_{\dot\alpha} = \tilde{\bar\nabla}_{\dot\alpha} \ln [\bar{\mathscr{F}}^2 \tilde{\mathscr{E}} (\mathrm{e}^{\bar{\mathscr{W}}} \bar F)^2 (1 \cdot \mathrm{e}^{\overleftarrow{\bar{\mathscr{W}}}})].$$

On the other hand, the supertorsion $\bar T_{\dot\alpha}$, found from $\mathscr{D}_A$, takes the form

$$\bar T_{\dot\alpha} = \bar{\mathscr{D}}_{\dot\alpha} \ln [E(\mathrm{e}^{\bar{\mathscr{W}}} \bar F)^2 (1 \cdot \mathrm{e}^{\overleftarrow{\bar{\mathscr{W}}}})].$$

From these expressions we deduce that

$$\tilde{\bar{\mathbf{T}}}_{\dot\alpha} - \bar{\mathscr{F}} \bar T_{\dot\alpha} = \tilde{\bar\nabla}_{\dot\alpha} \ln [\bar{\mathscr{F}}^2 \tilde{\mathscr{E}} E^{-1}] \tag{5.6.46}$$

Finally, let us obtain a useful formula for $\tilde{\mathscr{E}}$. Introducing auxilliary derivatives

$$\hat\nabla_A \equiv \left(-\frac{\mathrm{i}}{4}(\tilde\sigma_a)^{\dot\beta\beta}\{\hat\nabla_\beta, \hat{\bar\nabla}_{\dot\beta}\}, \mathrm{e}^{-2\mathrm{i}\mathbf{H}} \mathscr{D}_\alpha \mathrm{e}^{2\mathrm{i}\mathbf{H}}, \bar{\mathscr{D}}^{\dot\alpha} \right) = U_A{}^B \mathscr{D}_B + \frac{1}{2}\hat\Omega_A{}^{bc} M_{bc} \tag{5.6.47}$$

and recalling equation (5.6.40), one readily obtains

$$\mathcal{E} = (\mathcal{F}e^{-2iH}\mathcal{F})^2 EU \qquad U = \mathrm{Ber}\,(U_A{}^B). \tag{5.6.48}$$

It is not difficult to check the following conjugation rules

$$(\mathcal{E}^{-1})^* = (\mathcal{E}^{-1}e^{2i\ddot{H}}) \qquad (U^{-1})^* = (U^{-1}\cdot e^{2i\overleftarrow{H}}). \tag{5.6.49}$$

5.6.3. Background-quantum splitting in Einstein supergravity

In the Einstein supergravity case, one must impose the constraints

$$T_{\dot\alpha} = \tilde{\mathbf{T}}_{\dot\alpha} = 0.$$

Then, the expressions (5.6.46) and (5.6.47) give

$$\mathcal{F}^4(e^{-2iH}\mathcal{F})^2 U = \varphi^{-3} \qquad \bar{\mathcal{D}}_{\dot\alpha}\varphi = 0 \tag{5.6.50}$$

where φ is a covariantly chiral scalar (with respect to $\mathcal{D}_A$) superfield. Using this relation together with equation (5.6.48), one can easily obtain

$$\begin{aligned} \mathcal{F} &= \varphi^{-1}(e^{-2iH}\bar\varphi)^{1/2}U^{-1/6}(1\cdot e^{-2i\overleftarrow{H}})^{1/6} \\ (e^{-2iH}\mathcal{F}) &= \varphi^{1/2}(e^{-2iH}\bar\varphi)^{-1}U^{-1/6}(1\cdot e^{-2i\overleftarrow{H}})^{-1/3}. \end{aligned} \tag{5.6.51}$$

Now the Berezinian $\mathcal{E}^{-1}$ takes the form

$$\mathcal{E}^{-1} = E^{-1}(\varphi e^{-2iH}\bar\varphi)U^{-1/3}(1\cdot e^{-2i\overleftarrow{H}})^{1/3}. \tag{5.6.52}$$

To complete out discussion, it is necessary to determine the transformation law of φ under the Λ-transformations (5.6.41). It is an easy task if one represents this superfield in the form: $\varphi^3 = \mathcal{F}^{-2}\mathcal{E}^{-1}E = \mathcal{F}^{-2}\check{\mathcal{E}}^{-1}$, where $\check{\mathcal{E}}$ is the Berezinian of the auxiliary supervierbein defined by

$$\check{\nabla}_A = \check{\mathcal{E}}_A{}^B\mathcal{D}_B + \ldots.$$

Here dots denote terms involving the Lorentz generators (superconnection terms). The transformations (5.6.41) change $\check{\mathcal{E}}_A^B$ as follows

$$\delta\check{\mathcal{E}}_A{}^B = \Lambda\check{\mathcal{E}}_A{}^B - \check{\nabla}_A\Lambda^B + \check{\mathcal{E}}_A{}^D\Lambda^C T_{CD}{}^B$$

and hence

$$\delta\check{\mathcal{E}}^{-1} = \Lambda\check{\mathcal{E}}^{-1} + (-1)^{\varepsilon_A}\check{\mathcal{E}}^{-1}\mathcal{D}_A\Lambda^A - (-1)^{\varepsilon_A}\check{\mathcal{E}}^{-1}\Lambda^B T_{BA}{}^A = \check{\mathcal{E}}^{-1}\overleftarrow{\Lambda}$$

where we have used the identity $(-1)^{\varepsilon_A}T_{BA}{}^A = 0$, which is fulfilled in Einstein supergravity. Then, the transformation law for $\mathcal{F}$ is given by equation (5.6.33) (in which one has to set $T_\alpha = \bar{T}_{\dot\alpha} = 0$). As a result, one obtains

$$\begin{aligned}\delta\varphi^3 &= \Lambda\varphi^3 + \varphi^3(\mathscr{D}_a\Lambda^a - \mathscr{D}_\alpha\Lambda^\alpha - \mathrm{i}G_a\Lambda^a) \\ &= \frac{1}{4}(\bar{\mathscr{D}}^2 - 4R)\mathscr{D}_\alpha(L^\alpha\varphi^3).\end{aligned} \tag{5.6.53}$$

The reader can compare this result with equation (5.6.44).

Note that, when $\mathbf{H} = 0$, the derivatives ∇_A coincide with those obtained from $\mathscr{D}_A$ after applying the super Weyl transformation (5.5.13) with $\sigma = \ln\varphi$.

5.6.4. First-order expressions

Now, we would like to calculate all geometrical objects constructed from $\tilde{\nabla}_A$ to first order in the quantum superfields $\mathbf{H}$ and σ ($\varphi = \mathrm{e}^\sigma$), i.e. considering the approximation

$$\mathrm{e}^{-2\mathrm{i}\mathbf{H}} \approx 1 - 2\mathrm{i}\mathbf{H}^a\mathscr{D}_a \qquad \varphi = \mathrm{e}^\sigma \approx 1 + \sigma. \tag{5.6.54}$$

An application will be given in the next subsection.

First of all, we calculate the Berezinian U of the supermatrix $U_A{}^B$ found from the derivatives $\hat{\nabla}_A$ (5.6.47). In the linear-in-$\mathbf{H}$ approximation we have

$$U_A{}^B = \delta_A{}^B + \Delta U_A{}^B \qquad U = (\mathrm{Ber}\, U_A{}^B) = 1 + (-1)^{\varepsilon_A}\Delta U_A{}^A.$$

From equation (5.6.47) we deduce that

$$\begin{aligned}\hat{\nabla}_\alpha &= \mathscr{D}_\alpha - 2\mathrm{i}[\mathbf{H}^b\mathscr{D}_b, \mathscr{D}_\alpha] \\ &= (\delta_\alpha{}^\beta - \mathbf{H}_{\alpha\dot\beta}G^{\beta\dot\beta})\mathscr{D}_\beta - \mathrm{i}(\mathscr{D}_\alpha\mathbf{H}^{\beta\dot\beta})\mathscr{D}_{\beta\dot\beta} + \mathbf{H}_\alpha{}^{\dot\beta}\bar{R}\bar{\mathscr{D}}_{\dot\beta} + \ldots \\ \hat{\nabla}_{\alpha\dot\alpha} &= \frac{\mathrm{i}}{2}\{\hat{\nabla}_\alpha, \hat{\bar{\nabla}}_{\dot\alpha}\} = \mathscr{D}_{\alpha\dot\alpha} - \mathbf{H}_{\alpha\dot\beta}G^{\beta\dot\beta}\mathscr{D}_{\beta\dot\alpha} + \frac{1}{2}(\bar{\mathscr{D}}_{\dot\alpha}\mathscr{D}_\alpha\mathbf{H}^{\beta\dot\beta})\mathscr{D}_{\beta\dot\beta} + \ldots\end{aligned}$$

where dots denote superconnection terms and off-diagonal terms. One readily obtains

$$U = 1 + 2G^a\mathbf{H}_a + \frac{1}{2}\bar{\mathscr{D}}^{\dot\alpha}\mathscr{D}^\alpha\mathbf{H}_{\alpha\dot\alpha}. \tag{5.6.55}$$

Analogously, we have

$$(1\cdot\mathrm{e}^{-2\mathrm{i}\overleftarrow{\mathbf{H}}}) = 1 - 2\mathrm{i}\mathscr{D}^a\mathbf{H}_a. \tag{5.6.56}$$

After plugging the upper two identities into expression (5.6.52), one obtains

$$\mathscr{E}^{-1} = E^{-1}\left(\sigma + \bar{\sigma} - \frac{2}{3}G^a\mathbf{H}_a + \frac{1}{12}[\mathscr{D}^\alpha, \bar{\mathscr{D}}^{\dot\alpha}]\mathbf{H}_{\alpha\dot\alpha} - \mathrm{i}\mathscr{D}^a\mathbf{H}_a\right). \tag{5.6.57}$$

Further, from equation (5.6.51) one has

$$\mathscr{F} \equiv 1 + \Delta\mathscr{F}$$
$$\Delta\mathscr{F} = \frac{1}{2}\bar{\sigma} - \sigma - \frac{1}{3}G^a\mathbf{H}_a - \frac{i}{3}\mathscr{D}^a\mathbf{H}_a - \frac{1}{12}\bar{\mathscr{D}}^{\dot{\alpha}}\mathscr{D}^{\alpha}\mathbf{H}_{\alpha\dot{\alpha}}. \tag{5.6.58}$$

Now, making use of equation (5.6.45) gives

$$\Delta\bar{\Omega}_{\dot{\alpha}\beta\dot{\gamma}} = \varepsilon_{\dot{\alpha}\beta}\bar{\mathscr{D}}_{\dot{\gamma}}\Delta\mathscr{F} + \varepsilon_{\dot{\alpha}\dot{\gamma}}\bar{\mathscr{D}}_{\beta}\Delta\mathscr{F}$$
$$\tilde{\mathbf{R}} = R - \frac{1}{2}(\bar{\mathscr{D}}^2 - 4R)\Delta\bar{\mathscr{F}}. \tag{5.6.59}$$

To finish the calculations, it is necessary to determine the superconnection $\Delta\bar{\Omega}_{\dot{\alpha}\beta\gamma}$ and the supertorsions $\tilde{\mathbf{G}}_a$ and $\tilde{\mathbf{W}}_{\alpha\beta\gamma}$. In order to find $\Delta\bar{\Omega}_{\dot{\alpha}\beta\gamma}$, one must calculate carefully the derivatives (5.6.47) and then impose the Einstein supergravity constraints on the covariant derivatives $\tilde{\nabla}_A$. After doing this, taking (anti)commutators $[\tilde{\nabla}_A, \tilde{\nabla}_B\}$ one may obtain explicit expressions for $\tilde{\mathbf{G}}_a$ and $\tilde{\mathbf{W}}_{\alpha\beta\gamma}$. It is a trivial but tedious task. Omitting the details, we reproduce the final results:

$$\Delta\bar{\Omega}_{\dot{\alpha}\alpha\beta} = \bar{\mathscr{D}}_{\dot{\alpha}}A_{\alpha\beta} \qquad A_{\alpha\beta} = \frac{1}{2}\bar{\mathscr{D}}^{\dot{\beta}}\mathscr{D}_{(\alpha}\mathbf{H}_{\beta)\dot{\beta}} + G_{(\alpha}{}^{\dot{\beta}}\mathbf{H}_{\beta)\dot{\beta}}$$

$$\tilde{\mathbf{W}}_{\alpha\beta\gamma} = W_{\alpha\beta\gamma} + \Delta W_{\alpha\beta\gamma}$$

$$\Delta W_{\alpha\beta\gamma} = -\frac{3}{2}\sigma W_{\alpha\beta\gamma} - A^{\rho\delta}M_{\rho\delta}W_{\alpha\beta\gamma}$$
$$-\frac{1}{8}(\bar{\mathscr{D}}^2 - 4R)[\mathscr{D}_{(\alpha}A_{\beta\gamma)} + \mathbf{H}_{(\alpha}{}^{\dot{\alpha}}\mathscr{D}_{\beta}G_{\gamma)\dot{\alpha}}]$$

$$\tilde{\mathbf{G}}_{\alpha\dot{\alpha}} = G_{\alpha\dot{\alpha}} + \Delta G_{\alpha\dot{\alpha}}$$

$$\Delta G_{\alpha\dot{\alpha}} = -i\mathbf{H}^b\mathscr{D}_bG_{\alpha\dot{\alpha}} - \frac{1}{2}(\sigma + \bar{\sigma})G_{\alpha\dot{\alpha}} + i\mathscr{D}_{\alpha\dot{\alpha}}(\bar{\sigma} - \sigma)$$
$$+\left\{\frac{1}{16}\bar{\mathscr{D}}_{\dot{\beta}}(\mathscr{D}^2 - 4\bar{R})\bar{\mathscr{D}}^{\dot{\beta}} + \frac{1}{16}\mathscr{D}^{\beta}(\bar{\mathscr{D}}^2 - 4R)\mathscr{D}_{\beta}\right. \tag{5.6.60}$$
$$\left. + \frac{1}{8}(\mathscr{D}^{\beta}R)\mathscr{D}_{\beta} + \frac{1}{8}(\bar{\mathscr{D}}_{\dot{\beta}}\bar{R})\bar{\mathscr{D}}^{\dot{\beta}} + \frac{2}{3}R\bar{R} + G^bG_b\right\}\mathbf{H}_{\alpha\dot{\alpha}}$$
$$+\left\{\frac{1}{3}\left(G_{\alpha\dot{\alpha}} + \frac{1}{2}[\bar{\mathscr{D}}_{\dot{\alpha}}, \mathscr{D}_{\alpha}]\right)\left(G^{\beta\dot{\beta}} - \frac{1}{2}[\bar{\mathscr{D}}^{\dot{\beta}}, \mathscr{D}^{\beta}]\right)\right.$$
$$\left. - \frac{1}{2}\mathscr{D}_{\alpha\dot{\alpha}}\mathscr{D}^{\beta\dot{\beta}} + \frac{1}{4}(\mathscr{D}_{\alpha}G^{\beta\dot{\beta}})\bar{\mathscr{D}}_{\dot{\alpha}} - \frac{1}{4}(\bar{\mathscr{D}}_{\dot{\alpha}}G^{\beta\dot{\beta}})\mathscr{D}_{\alpha}\right\}\mathbf{H}_{\beta\dot{\beta}}$$

$$+\frac{1}{2}\mathscr{D}^{\beta}(W_{\alpha\beta\gamma}\mathbf{H}_{\dot\alpha}{}^{\gamma}) - \frac{1}{2}\bar{\mathscr{D}}^{\dot\beta}(\bar{W}_{\dot\alpha\dot\beta\dot\gamma}\mathbf{H}_{\alpha}{}^{\dot\gamma})$$

$$+\frac{1}{4}\{-\mathbf{H}^{\beta\dot\beta}[\bar{\mathscr{D}}_{\dot\beta}, \mathscr{D}_{\beta}] + (\mathscr{D}_{\beta}\mathbf{H}^{\beta\dot\beta})\bar{\mathscr{D}}_{\dot\beta} - (\bar{\mathscr{D}}_{\dot\beta}\mathbf{H}^{\beta\dot\beta})\mathscr{D}_{\beta}\}G_{\alpha\dot\alpha}$$

$$+\frac{1}{4}\{(\mathscr{D}_{\alpha}R)\mathscr{D}^{\beta}\mathbf{H}_{\beta\dot\alpha} - (\bar{\mathscr{D}}_{\dot\alpha}\bar{R})\bar{\mathscr{D}}^{\dot\beta}\mathbf{H}_{\alpha\dot\beta}\}.$$

5.6.5. Topological invariants

As an application of the technique developed above, we show that the superfunctional

$$\mathscr{P} = \int \mathrm{d}^8 z\, E^{-1}\left\{\frac{1}{R}W^{\alpha\beta\gamma}W_{\alpha\beta\gamma} + G^a G_a + 2R\bar{R}\right\} \qquad (5.6.61)$$

is invariant with respect to arbitrary variations of the Einstein supergravity prepotentials. The superfunctional $\mathscr{P}$ turns out to be a supersymmetric generalization of the four-dimensional topological invariant $(P - \chi)$, where P is the Pontrjagin invariant and χ is the Euler invariant (see subsection 1.6.6).

First, we vary the superfunctional

$$I_1 = \int \mathrm{d}^8 z(E^{-1}/R)W^{\alpha\beta\gamma}W_{\alpha\beta\gamma}. \qquad (5.6.62)$$

For this purpose note the identity

$$\int \mathrm{d}^8 z(\mathbf{\mathscr{E}}^{-1}/\mathbf{R})\mathscr{L}_c = \int \mathrm{d}^8 z(E^{-1}/R)(1 + 3\sigma)\mathscr{L}_c \qquad \bar{\mathscr{D}}_{\dot\alpha}\mathscr{L}_c = 0 \quad (5.6.63)$$

which follows from the relations (5.6.57–59). Then, making use of equation (5.6.60) gives

$$\delta I_1 = \int \mathrm{d}^8 z(\mathbf{\mathscr{E}}^{-1}/\mathbf{R})\mathbf{W}^{\alpha\beta\gamma}\mathbf{W}_{\alpha\beta\gamma} - \int \mathrm{d}^8 z(E^{-1}/R)W^{\alpha\beta\gamma}W_{\alpha\beta\gamma}$$

$$= \int \mathrm{d}^8 z E^{-1}\mathbf{H}^{\alpha\dot\alpha}\Big\{-W_{\alpha}{}^{\beta\gamma}\mathscr{D}_{\beta}G_{\gamma\dot\alpha} + \bar{W}_{\dot\alpha}{}^{\dot\beta\dot\gamma}\bar{\mathscr{D}}_{\dot\beta}G_{\alpha\dot\gamma} - \mathscr{D}^b\mathscr{D}_b G_{\alpha\dot\alpha}$$

$$+\frac{1}{4}((\mathscr{D}^{\beta}R)\mathscr{D}_{\beta} + (\bar{\mathscr{D}}_{\dot\beta}\bar{R})\bar{\mathscr{D}}^{\dot\beta})G_{\alpha\dot\alpha} - (\bar{\mathscr{D}}_{\dot\alpha}G^b)\mathscr{D}_{\alpha}G_b - 3R\bar{R}G_{\alpha\dot\alpha} \qquad (5.6.64)$$

$$+\frac{1}{8}G_{\alpha\dot\alpha}(\bar{\mathscr{D}}^2\bar{R} + \mathscr{D}^2 R) + \frac{\mathrm{i}}{4}\mathscr{D}_{\alpha\dot\alpha}(\bar{\mathscr{D}}^2\bar{R} - \mathscr{D}^2 R)\Big\}$$

where we have used the relations (5.5.6, 8). It is seen that δI_1 does not involve variation of the chiral compensator, σ. Therefore, the superfunctional (5.6.62) does not depend on the chiral compensator φ. Equivalently, the above observation means that I_1 is invariant under the super Weyl transformations (5.5.13).

It is worth pointing out that the right-hand side of equation (5.6.64) is real. Hence, the superfunctional

$$\mathscr{P} - \bar{\mathscr{P}} = \int d^8z(E^{-1}/R)W^{\alpha\beta\gamma}W_{\alpha\beta\gamma} - \int d^8z(E^{-1}/\bar{R})\bar{W}_{\dot{\alpha}\dot{\beta}\dot{\gamma}}\bar{W}^{\dot{\alpha}\dot{\beta}\dot{\gamma}} \quad (5.6.65)$$

does not change under arbitrary variations of the Einstein supergravity prepotentials. This object represents the supersymmetric extension of the Pontrjagin invariant.

Secondly, let us vary the superfunctional

$$I_2 = \int d^8z\, E^{-1}G^aG_a. \quad (5.6.66)$$

Making use of the relations (5.6.57, 60) leads to

$$\begin{aligned}\delta I_2 &= \int d^8z\tilde{\mathscr{E}}^{-1}\tilde{\mathbf{G}}^a\tilde{\mathbf{G}}_a - \int d^8zE^{-1}G^aG_a \\ &= \frac{1}{2}\int d^8zE^{-1}\{\sigma\mathscr{D}^2R + \overline{\sigma\mathscr{D}^2R}\} \\ &+ \int d^8zE^{-1}\mathbf{H}^{\alpha\dot{\alpha}}\Big\{W_\alpha{}^{\beta\gamma}\mathscr{D}_\beta G_{\gamma\dot{\alpha}} + \bar{W}_{\dot{\alpha}}{}^{\dot{\beta}\dot{\gamma}}\bar{\mathscr{D}}_{\dot{\beta}}G_{\alpha\dot{\gamma}} + \mathscr{D}^b\mathscr{D}_bG_{\alpha\dot{\alpha}} \\ &- \frac{1}{4}((\mathscr{D}^\beta R)\mathscr{D}_\beta + (\bar{\mathscr{D}}_{\dot{\beta}}\bar{R})\bar{\mathscr{D}}^{\dot{\beta}})G_{\alpha\dot{\alpha}} + (\bar{\mathscr{D}}_{\dot{\alpha}}G^b)\mathscr{D}_\alpha G_b + R\bar{R}G_{\alpha\dot{\alpha}} \\ &- \frac{1}{2}G^b[\bar{\mathscr{D}}_{\dot{\alpha}}, \mathscr{D}_\alpha]G_b + \frac{1}{24}G_{\alpha\dot{\alpha}}(\bar{\mathscr{D}}^2\bar{R} + \mathscr{D}^2R) \\ &- \frac{5i}{12}\mathscr{D}_{\alpha\dot{\alpha}}(\bar{\mathscr{D}}^2\bar{R} - \mathscr{D}^2R) + \frac{2}{3}R\overleftrightarrow{\mathscr{D}}_{\alpha\dot{\alpha}}R + \frac{1}{3}(\bar{\mathscr{D}}_{\dot{\alpha}}\bar{R})\mathscr{D}_\alpha R\Big\}.\end{aligned} \quad (5.6.67)$$

Finally, we consider the superfunctional

$$I_3 = 2\int d^8z\, E^{-1}R\bar{R}. \quad (5.6.68)$$

Since in the quantum chiral representation

$$\tilde{\bar{\mathbf{R}}} = e^{-2i\mathbf{H}}(\mathbf{R})^*$$

making use of the relations (5.6.57, 59) leads to

$$\begin{aligned}\delta I_3 &= 2\int d^8z\tilde{\mathscr{E}}^{-1}\tilde{\mathbf{R}}\bar{\tilde{\mathbf{R}}} - 2\int d^8zE^{-1}R\bar{R} \\ &= -\frac{1}{2}\int d^8zE^{-1}\{\sigma\mathscr{D}^2R + \bar{\sigma}\bar{\mathscr{D}}^2\bar{R}\} \\ &+ \int d^8zE^{-1}\mathbf{H}^{\alpha\dot{\alpha}}\Big\{2R\bar{R}G_{\alpha\dot{\alpha}} - \frac{1}{6}G_{\alpha\dot{\alpha}}(\bar{\mathscr{D}}^2\bar{R} + \mathscr{D}^2R) \\ &+ \frac{i}{6}\mathscr{D}_{\alpha\dot{\alpha}}(\bar{\mathscr{D}}^2\bar{R} - \mathscr{D}^2R) - \frac{2}{3}\bar{R}\overleftrightarrow{\mathscr{D}}_{\alpha\dot{\alpha}}R - \frac{1}{3}(\bar{\mathscr{D}}_{\dot{\alpha}}\bar{R})\mathscr{D}_{\alpha}R\Big\}.\end{aligned} \tag{5.6.69}$$

From equations (5.6.64, 67, 69) we see that $\delta\mathscr{P} = 0$.

In conclusion, we point out that the superfunctional

$$\int d^8z\, E^{-1}(G^aG_a + 2R\bar{R}) \tag{5.6.70}$$

is invariant under the super Weyl transformations (5.5.13), similarly to the previously considered superfunctional (5.6.62).

5.7. Supercurrent and supertrace

In general relativity the energy–momentum tensor of a matter dynamical system coupled to a gravity background satisfies equation (1.6.60), owing to the general covariance. Our goal now is to find a superfield generalization of the energy-momentum tensor as well as to look for a supersymmetric version of equation (1.6.60).

5.7.1. Basic construction

Consider a theory of matter superfields χ living on a given (background) curved superspace. Its action

$$S = S[\chi; \mathscr{D}] = \int d^8z\, E^{-1}\mathscr{L}(\chi; \mathscr{D}) \tag{5.7.1}$$

$\mathscr{D}_A$ being the relevant covariant derivatives, is supposed to be invariant under the superspace general coodinate and superlocal Lorentz transformations. In the present section it will be useful for us to understand the action as a superfunctional of the matter superfields and of the background supergravity prepotentials $\mathscr{W} = W^M D_M + \frac{1}{2}W^{ab}M_{ab}$ and $\varphi(\bar{\mathscr{D}}_{\dot{\alpha}}\varphi = 0)$, $S[\chi; \mathscr{D}] = S[\chi; \mathscr{W}, \varphi]$.

Suppose the prepotentials suffer slight disturbances such that the covariant derivatives change in the manner

$$\mathscr{D}_A \to \nabla_A = \mathscr{E}_A{}^M \mathrm{D}_M + \frac{1}{2}\Theta_A{}^{bc}M_{bc}$$

$$\nabla_\alpha = \mathrm{e}^{-\mathrm{i}\mathbf{H}}\left(\mathscr{F}\mathscr{D}_\alpha + \frac{1}{2}\Delta\Omega_\alpha{}^{bc}M_{bc}\right)\mathrm{e}^{\mathrm{i}\mathbf{H}} \qquad \mathbf{H} = \mathbf{H}^a\mathscr{D}_a = \bar{\mathbf{H}} \tag{5.7.2}$$

$$\bar{\nabla}_{\dot\alpha} = \mathrm{e}^{\mathrm{i}\mathbf{H}}\left(\bar{\mathscr{F}}\bar{\mathscr{D}}_{\dot\alpha} + \frac{1}{2}\Delta\bar{\Omega}_{\dot\alpha}{}^{bc}M_{bc}\right)\mathrm{e}^{-\mathrm{i}\mathbf{H}} \qquad \nabla_{\alpha\dot\alpha} = \frac{\mathrm{i}}{2}\{\nabla_\alpha, \bar{\nabla}_{\dot\alpha}\}$$

with $\mathscr{F}$ being as in equation (5.6.51) and $\Delta\Omega_\alpha{}^{bc}$ being determined by supergravity constraints. This change in the supergeometry is accompanied by some disturbance of the matter superfields

$$\chi \to \chi_{(\nabla)} = f(\chi; \mathbf{H}, \varphi) \tag{5.7.3}$$

depending on their superfield types. The action changes as follows:

$$S[\chi; \mathscr{D}] \to S[\chi_{(\nabla)}; \nabla] \equiv S[\chi; \mathscr{D}|\mathbf{H}, \varphi] = \int \mathrm{d}^8 z\, E^{-1}\mathscr{L}(\chi_{(\nabla)}; \nabla). \tag{5.7.4}$$

In practice, it is useful to transform $S[\chi_{(\nabla)}; \nabla]$ into the 'quantum' chiral representation:

$$\tilde{\chi}_{(\nabla)} = \mathrm{e}^{-\mathrm{i}\mathbf{H}}\tilde{\chi}_{(\nabla)} \qquad \tilde{\nabla}_A = \mathrm{e}^{-\mathrm{i}\mathbf{H}}\nabla_A \mathrm{e}^{\mathrm{i}\mathbf{H}}$$
$$S[\chi; \mathscr{D}|\mathbf{H}, \varphi] = \int \mathrm{d}^8 z \tilde{\mathscr{E}}^{-1}\mathscr{L}(\tilde{\chi}_{(\nabla)}; \tilde{\nabla}). \tag{5.7.5}$$

Let us introduce two superfields

$$T_a \equiv \frac{\delta}{\delta \mathbf{H}^a} S[\chi; \mathscr{D}|\mathbf{H}, \varphi]|_{\mathbf{H}=0,\, \varphi=1}$$
$$T \equiv \frac{\delta}{\delta\varphi} S[\chi; \mathscr{D}|\mathbf{H}, \varphi]|_{\mathbf{H}=0,\, \varphi=1} \tag{5.7.6}$$

which will be called the 'supercurrent' and the 'supertrace' of the system, respectively. The supercurrent turns out to be a real vector superfield, since the same is true for $\mathbf{H}^a$. Similarly, since φ is covariantly chiral and scalar, the supertrace represents a covariantly chiral scalar superfield,

$$\bar{\mathscr{D}}_{\dot\alpha} T = 0. \tag{5.7.7}$$

Remark. For calculating the supercurrent and the supertrace, the following variational rules

$$\frac{\delta \mathbf{H}^b(z')}{\delta \mathbf{H}^a(z)} = \delta_a{}^b E \delta^8(z - z')$$
$$\frac{\delta \varphi(z')}{\delta \varphi(z)} = -\frac{1}{4}(\bar{\mathscr{D}}^2 - 4R) E \delta^8(z - z') \tag{5.7.8}$$

are helpful.

In accordance with the results of subsections 5.6.2 and 5.6.3, $S[\chi_{(\nabla)}; \nabla]$ is invariant under the transformations

$$\delta \mathrm{e}^{-2\mathrm{i}\mathbf{H}} = \Lambda \mathrm{e}^{-2\mathrm{i}\mathbf{H}} - \mathrm{e}^{-2\mathrm{i}\mathbf{H}} \bar{\Lambda}$$
$$\delta \varphi^3 = \frac{1}{4}(\bar{\mathscr{D}}^2 - 4R)\mathscr{D}_\alpha(L^\alpha \varphi^3) \qquad \delta \tilde{\chi}_{(\nabla)} = \Lambda \tilde{\chi}_{(\nabla)} \tag{5.7.9}$$

where

$$\Lambda = \Lambda^A \mathscr{D}_A + \Lambda^{\alpha\beta} M_{\alpha\beta} + \Lambda^{\dot{\alpha}\dot{\beta}} \bar{M}_{\dot{\alpha}\dot{\beta}}$$
$$\Lambda_{\alpha\dot{\alpha}} = (\sigma^a)_{\alpha\dot{\alpha}} \Lambda_a = -2\mathrm{i} \bar{\mathscr{D}}_{\dot{\alpha}} L_\alpha \qquad \Lambda_\alpha = -\frac{1}{4}\bar{\mathscr{D}}^2 L_\alpha, \qquad \Lambda_{\dot{\alpha}} = \bar{\Lambda}_{\dot{\alpha}} + O(\mathbf{H})$$
$$\Lambda_{\dot{\alpha}\dot{\beta}} = -\bar{\mathscr{D}}_{(\dot{\alpha}} \Lambda_{\dot{\beta})} + \frac{1}{2}\Lambda_{\beta(\dot{\alpha}} G^\beta_{\dot{\beta})} \qquad \Lambda_{\alpha\beta} = \bar{\Lambda}_{\alpha\beta} + O(\mathbf{H}). \tag{5.7.10}$$

Here L_α is an unconstrained spinor superfield. The transformation law for $\mathbf{H}_{\alpha\dot{\alpha}}$ can be rewritten in the form

$$\delta \mathbf{H}_{\alpha\dot{\alpha}} = \bar{\mathscr{D}}_{\dot{\alpha}} L_\alpha - \mathscr{D}_\alpha \bar{L}_{\dot{\alpha}} + O(\mathbf{H}) \tag{5.7.11}$$

where $O(\mathbf{H})$ denotes all $\mathbf{H}$-dependent terms.

Now, let us choose vanishing quantum superfields

$$\mathbf{H}^a = 0 \qquad \varphi = 1.$$

In this case, $\tilde{\chi}_{(\nabla)} = \chi$ and the transformation laws (5.7.9) reduce to

$$\delta \mathbf{H}_{\alpha\dot{\alpha}} = \bar{\mathscr{D}}_{\dot{\alpha}} L_\alpha - \mathscr{D}_\alpha \bar{L}_{\dot{\alpha}} \qquad \delta \tilde{\chi}_{(\nabla)} = \Lambda \chi$$
$$\delta \varphi^3 = \frac{1}{4}(\bar{\mathscr{D}}^2 - 4R)\mathscr{D}_\alpha L^\alpha \Rightarrow \delta \varphi = \frac{1}{12}(\bar{\mathscr{D}}^2 - 4R)\mathscr{D}_\alpha L^\alpha. \tag{5.7.12}$$

Further, the condition that $S[\chi_{(\nabla)}; \nabla]$ is invariant under the transformations (5.7.9) now reads

$$0 = \int \mathrm{d}^8 z E^{-1} \left\{ -\frac{1}{2}\delta \mathbf{H}^{\alpha\dot{\alpha}} T_{\alpha\dot{\alpha}} + \frac{1}{R}\delta\varphi T + \frac{1}{\bar{R}}\delta\bar{\varphi}\bar{T} \right\} + (\Lambda\chi)\frac{\delta S[\chi; \mathscr{D}]}{\delta \chi}. \tag{5.7.13}$$

Suppose the matter superfields satisfy their dynamical equations

$$\frac{\delta S[\chi;\mathcal{D}]}{\delta\chi}=0.$$

Then, equations (5.7.12) and (5.7.13) say

$$0=\int d^8z\,E^{-1}L^{\alpha}\left\{\frac{1}{2}\bar{\mathcal{D}}^{\dot\alpha}T_{\alpha\dot\alpha}+\frac{1}{3}\mathcal{D}_\alpha T\right\}+\text{c.c.}$$

Due to the arbitrariness of L^{α}, this relation is equivalent to the equation

$$\bar{\mathcal{D}}^{\dot\alpha}T_{\alpha\dot\alpha}=-\frac{2}{3}\mathcal{D}_\alpha T. \tag{5.7.14}$$

It is this equation which expresses the condition of invariance of the action superfunctional $S[\chi;\mathcal{D}]$ under superspace general coordinate and superlocal Lorentz transformations. One can look on equation (5.7.14) as a supersymmetric generalization of the conservation law (1.6.60).

Remark. There is a simple prescription for calculating the supertrace: considering the action $S[\chi;\mathcal{D}]$ as a superfunctional of the supergravity prepotentials, $S[\chi;\mathcal{D}]=S[\chi;\mathcal{W},\varphi]$, the supertrace proves to be given by

$$T=\frac{\delta}{\delta\sigma}S[\chi;\mathcal{W},e^{\sigma}\varphi]|_{\sigma=0}\qquad \bar{\mathcal{D}}_{\dot\alpha}\sigma=0 \tag{5.7.15}$$

with the variational derivative $\delta/\delta\sigma$ being defined as follows

$$\frac{\delta\sigma(z')}{\delta\sigma(z)}=-\frac{1}{4}(\bar{\mathcal{D}}^2-4R)E\delta^8(z-z'). \tag{5.7.16}$$

5.7.2. The relation with ordinary currents

Here we would like to discuss the component content of the supercurrent and the supertrace.

As we know, the invariance with respect to the supergravity gauge group can be fixed by imposing the gravitational superfield gauge (5.4.38). In this gauge, the matter action becomes a superfunctional of the gravitational superfield $\mathcal{H}^m(x,\theta,\bar\theta)$ and the chiral compensator $\phi(x,\theta)$, $S[\chi;\mathcal{D}]=S[\chi;\mathcal{H}^m,\phi]$. Clearly, making a slight disturbance of the gravitational superfield $\mathcal{H}^m\to\mathcal{H}^m+\delta\mathcal{H}^m$ will induce some change in the covariant derivatives. This change, supplemented by an auxiliary $\delta\mathcal{H}^m$-dependent general coordinate and superlocal Lorentz transformation, can be represented in the form (5.7.2). The two variations $\delta\mathcal{H}^m$ and $\mathbf{H}^a$ are in one-to-one correspondence. However, the former transforms in a nonlinear, complicated way, while the latter possesses the vector superfield transformation law.

Further, instead of considering $T_{\alpha\dot{\alpha}}$, one can introduce

$$\mathscr{T}_m \equiv \frac{\delta}{\delta \mathscr{H}^m} S[\chi; \quad \mathscr{H}^m, \hat{\varphi}] \tag{5.7.17}$$

which will be called the 'non-covariant supercurrent' of the system. The two objects T_a and $\mathscr{T}_m$ are uniquely connected to each other but, in contrast to T_a, $\mathscr{T}_m$ is characterized by a nonlinear complicated transformation law. We can look on the supercurrent T_a as a covariantized form of $\mathscr{T}_m$.

Expanding $\mathscr{T}_m$ in a power series in θ, $\bar{\theta}$, we obtain

$$\mathscr{T}_m = J^5_m - 2\mathrm{i}\theta^\alpha Q_{m\alpha} + 2\mathrm{i}\bar{\theta}_{\dot{\alpha}}\bar{Q}_m{}^{\dot{\alpha}} + 2\theta\sigma^a\bar{\theta}T_{ma} + \ldots \tag{5.7.18}$$

where

$$J^5_m = \frac{\delta S}{\delta A^m} \qquad Q_m{}^\alpha = \frac{\delta S}{\delta \Psi^m{}_\alpha} \qquad T_m{}^a = -\frac{\delta S}{\delta e_a{}^m}. \tag{5.7.19}$$

here $e_a{}^m$, $\Psi_\alpha{}^m$ and A^m are the component fields of $\mathscr{H}^m$ in the Wess–Zumino gauge (5.1.17). Obviously, $T_m{}^a$ is the energy–momentum tensor. Next, it was shown in section 5.1 that $\Psi^m{}_\alpha$ and A^m are gauge fields for local supersymmetry and local chiral transformations, respectively. Therefore, $Q_m{}^\alpha$ is the spinor supersymmetry current and J^5_m is the axial or γ_5-current. As a result, the supercurrent contains the multiplet of ordinary currents, including the energy–momentum tensor.

The leading term in the expansion of the supertrace in a power series in θ is

$$T(x, \theta) = 3G(x) + \ldots \qquad G = \delta S/\delta B \tag{5.7.20}$$

where B is the complex scalar field appearing in the power series expansion of φ^3 (see equations (5.1.41)). Higher-order component fields in the expansions (5.7.18) and (5.7.20) turn out to be expressed through the currents (5.7.19) and the field G. To argue this statement, let us analyse the conservation law (5.7.14) in a flat superspace.

5.7.3. *The supercurrent and the supertrace in flat superspace*

It is almost obvious that when background superspace is flat the supercurrents T_a and $\mathscr{T}_m$ coincide. Then, using equation (5.7.18), the leading component fields of $T_{\alpha\dot{\alpha}} = (\sigma^a)_{\alpha\dot{\alpha}}T_a$ are given by

$$\begin{gathered} T_{\alpha\dot{\alpha}}| = J^5_{\alpha\dot{\alpha}} \qquad \mathrm{D}_\alpha T_{\beta\dot{\beta}}| = -2\mathrm{i}Q_{\beta\dot{\beta},\alpha} = -2\mathrm{i}(\sigma^b)_{\beta\dot{\beta}}Q_{b\alpha} \\ \frac{1}{2}[\mathrm{D}_\alpha, \bar{\mathrm{D}}_{\dot{\alpha}}]T_{\beta\dot{\beta}}| = 2T_{\beta\dot{\beta},\alpha\dot{\alpha}} = 2(\sigma^b)_{\beta\dot{\beta}}(\sigma_a)_{\alpha\dot{\alpha}}T_b{}^a. \end{gathered} \tag{5.7.21}$$

Equation (5.7.14) takes the form

$$\bar{\mathrm{D}}^{\dot{\alpha}}T_{\alpha\dot{\alpha}} = -\frac{2}{3}\mathrm{D}_\alpha T \qquad \bar{\mathrm{D}}_{\dot{\alpha}}T = 0. \tag{5.7.22}$$

Recalling basic properties of the flat covariant derivatives, one can deduce from equation (5.7.22) some useful consequences:

$$\mathrm{D}^{\beta}\bar{\mathrm{D}}^{\dot{\alpha}}T_{\alpha\dot{\alpha}} = -\frac{1}{3}\delta^{\beta}{}_{\alpha}\mathrm{D}^2T \qquad \bar{\mathrm{D}}^{\dot{\beta}}\mathrm{D}^{\alpha}T_{\alpha\dot{\alpha}} = \frac{1}{3}\delta^{\dot{\beta}}{}_{\dot{\alpha}}\bar{\mathrm{D}}^2\bar{T} \tag{5.7.23a}$$

$$\bar{\mathrm{D}}^2T_{\alpha\dot{\alpha}} = \frac{8}{3}\mathrm{i}\partial_{\alpha\dot{\alpha}}T \qquad \mathrm{D}^2T_{\alpha\dot{\alpha}} = -\frac{8}{3}\mathrm{i}\partial_{\alpha\dot{\alpha}}\bar{T}. \tag{5.7.23b}$$

From relations (5.7.22) and (5.7.23*a*) we obtain

$$\begin{gathered}\mathrm{D}_{\alpha}T| = -3\mathrm{i}(\sigma^a\bar{Q}_a)_{\alpha} \\ -\frac{1}{4}\mathrm{D}^2T| = \frac{3}{8}\mathrm{D}^{\alpha}\bar{\mathrm{D}}^{\dot{\alpha}}T_{\alpha\dot{\alpha}}| = -\frac{3}{8}T_a{}^a + \frac{3}{4}\mathrm{i}\partial^aJ^5_a.\end{gathered} \tag{5.7.24}$$

Therefore, the supertrace is expanded in a power series in θ as follows:

$$\frac{1}{3}T = G - \mathrm{i}\theta^{\alpha}(\sigma^a\bar{Q}_a)_{\alpha} - \frac{1}{4}\theta^2(T_a{}^a - \mathrm{i}\partial^aJ^5_a). \tag{5.7.25}$$

We see that the supertrace contains the trace of the energy–momentum tensor, the trace of the supersymmetry current $(\sigma^a\bar{Q}_a)_{\alpha}$ and the axial current divergence $\partial^aJ^5_a$. On these grounds, the supertrace can be treated as the supersymmetric extension of the trace of the energy–momentum tensor.

The supercurrent component fields, omitted in the expansion (5.7.18), can be easily found with the help of equations (5.7.23, 24). The results are

$$\begin{gathered}-\frac{1}{4}\mathrm{D}^2T_{\alpha\dot{\alpha}}| = 2\mathrm{i}\partial_{\alpha\dot{\alpha}}G \\ -\frac{1}{4}\mathrm{D}_{\beta}\bar{\mathrm{D}}^2T_{\alpha\dot{\alpha}}| = 2\mathrm{i}\partial_{\alpha\dot{\alpha}}(\sigma^b\bar{Q}_b)_{\beta} \\ \frac{1}{32}\{D^2,\bar{D}^2\}T_{\alpha\dot{\alpha}}| = \frac{1}{2}\partial_{\alpha\dot{\alpha}}(\partial^bJ^5_b)_{\beta}.\end{gathered} \tag{5.7.26}$$

As we see, these components are expressed via those presented in equations (5.7.18, 19).

The conservation law (5.7.22) encodes information about the symmetry structure of the energy–momentum tensor as well as about divergences of the currents under consideration. It follows from equation (5.7.23*a*) that

$$\mathrm{D}_{\alpha}\bar{\mathrm{D}}^{\dot{\beta}}T_{\beta\dot{\beta}} + \mathrm{D}_{\beta}\bar{\mathrm{D}}^{\dot{\beta}}T_{\alpha\dot{\beta}} = 0 \tag{5.7.27}$$

$$\partial^aT_a = \frac{1}{6}\mathrm{i}(\mathrm{D}^2T - \bar{\mathrm{D}}^2\bar{T}). \tag{5.7.28}$$

Now, taking the space projection of equation (5.7.27), one readily obtains the relations

$$\partial^{\alpha\dot\beta} J^5_{\beta\dot\beta} = -\delta^\alpha{}_\beta \partial^c J^5_c$$

$$T_{\beta\dot\beta}{}^{\alpha\dot\beta} = -\delta_\beta{}^\alpha T^c_{c\cdot} \tag{5.7.29}$$

The second relation shows that the energy–momentum tensor is symmetric,

$$T_{ab} = T_{ba} \tag{5.7.30}$$

Further, equation (5.7.28) can be used to find

$$\begin{aligned} \partial^a Q_{a\beta} &= -\partial^a(\sigma_a \tilde\sigma^c Q_c)_\beta \\ \partial^a T_{ab} &= \frac{1}{2}\partial_b T^a_a \end{aligned} \tag{5.7.31}$$

Therefore, the currents $Q_{a\beta}$ and T_{ab}, obtained in accordance with rule (5.7.19), are not conserved. But the improved currents

$$\begin{aligned} Q'_{a\beta} &= Q_{a\beta} + (\sigma_a \tilde\sigma^c Q_c)_\beta \\ T'_{ab} &= T_{ab} - \frac{1}{2}\eta_{ab} T^c_c \end{aligned} \tag{5.7.32}$$

turn out to be conserved.

5.7.4. *Super Weyl invariant models*

Consider a dynamical system such that its action superfunctional $S[\chi; \mathscr{D}] = S[\chi; \mathscr{W}, \varphi]$ is invariant under super Weyl transformations of the general form

$$\chi' = \mathrm{e}^{-d_{(+)}\sigma - d_{(-)}\bar\sigma}\chi \qquad \mathscr{W}' = \mathscr{W} \qquad \varphi' = \mathrm{e}^\sigma \varphi \qquad \bar{\mathscr{D}}_{\dot\alpha}\sigma = 0 \tag{5.7.33}$$

with $d_{(+)}$ and $d_{(-)}$ being numbers (fixed for the system under consideration) and $\sigma(z)$ being an arbitrary covariantly chiral scalar parameter. In this case the dynamical system is called a 'super Weyl invariant model'. Recalling the definition of supertrace (5.7.15), we see that super Weyl invariance requires the supertrace to vanish when imposing the matter dynamical equations,

$$\mathcal{T}|_{\delta S/\delta\chi = 0} = 0. \tag{5.7.34}$$

Then, the conservation law (5.7.14) takes the form

$$\bar{\mathrm{D}}^{\dot\alpha} T_{\alpha\dot\alpha} = 0. \tag{5.7.35}$$

At the component level, equation (5.7.34) leads to the requirements

$$(\sigma^a \bar{Q}_a)_\alpha = 0 \qquad T^a_a = 0 \qquad \partial^a J^5_a = 0 \tag{5.7.36}$$

in accordance with equations (5.7.26).

5.7.5 Example

To obtain some experience in the calculation of the supercurrent and the supertrace, we are going to evaluate these objects in the case of a theory of a covariantly chiral scalar superfield $\chi, \bar{\mathscr{D}}_{\dot{\alpha}}\chi = 0$, and its conjugate $\bar{\chi}$. The action superfunctional reads

$$S[\chi, \bar{\chi}; \mathscr{D}] = \int \mathrm{d}^8 z\, E^{-1}\bar{\chi}\chi + \left\{ \int \mathrm{d}^8 z (E^{-1}/R)\mathscr{L}_c(\chi) + \text{c.c.} \right\}$$
$$\bar{\mathscr{D}}_{\dot{\alpha}}\mathscr{L}_c(\chi) = 0 \tag{5.7.37}$$

$\mathscr{L}_c$ being a chiral superpotential. This action generalizes the Wess–Zumino model action (3.2.8) to the case of a curved superspace. It is worth pointing out that if one represents the chiral compensator φ and the matter superfield χ in the form

$$\varphi = \mathrm{e}^{\bar{\mathscr{W}}}\phi \qquad \bar{\mathrm{D}}_{\dot{\alpha}}\phi = 0$$
$$\chi = \mathrm{e}^{\bar{\mathscr{W}}}\hat{\chi} \qquad \bar{\mathrm{D}}_{\dot{\alpha}}\hat{\chi} = 0 \tag{5.7.38}$$

then the chiral superpotential term can be rewritten as

$$\int \mathrm{d}^8 z\,(E^{-1}/R)\mathscr{L}_c(\chi) = \int \mathrm{d}^6 z\, \phi^3 \mathscr{L}_c(\hat{\chi}). \tag{5.7.39}$$

In accordance with the prescription of subsection 5.7.1, to find the supercurrent and the supertrace one must calculate the variation

$$\delta S = S[\tilde{\chi}_{(\mathbf{V})}, \tilde{\bar{\chi}}_{(\mathbf{V})}; \tilde{\nabla}] - S[\chi, \bar{\chi}; \mathscr{D}]$$

where

$$S[\tilde{\chi}_{(\mathbf{V})}, \tilde{\bar{\chi}}_{(\mathbf{V})}; \tilde{\nabla}] = \int \mathrm{d}^8 z\, \tilde{\mathscr{E}}^{-1}\tilde{\bar{\chi}}_{(\mathbf{V})}\tilde{\chi}_{(\mathbf{V})} + \left\{ \int \mathrm{d}^8 z (\tilde{\mathscr{E}}^{-1}/\tilde{\mathbf{R}})\mathscr{L}_c(\tilde{\chi}_{(\mathbf{V})}) + \text{c.c.} \right\}$$
$$\tilde{\chi}_{(\mathbf{V})} = \chi \qquad \tilde{\bar{\chi}}_{(\mathbf{V})} = \mathrm{e}^{-2\mathrm{i}\mathbf{H}}\bar{\chi}.$$

Using equations (5.6.57, 63), to first order in quantum superfields $\mathbf{H}^a$, σ and $\bar{\sigma}$ we have

$$\delta S = \int \mathrm{d}^8 z\, E^{-1} \left\{ -\frac{1}{2}\mathbf{H}^{\alpha\dot{\alpha}}T_{\alpha\dot{\alpha}} + \frac{1}{R}\sigma T + \frac{1}{R}\bar{\sigma}\bar{T} \right\}$$

where

$$T_{\alpha\dot{\alpha}} = \frac{1}{3}(\bar{\mathscr{D}}_{\dot{\alpha}}\bar{\chi})\mathscr{D}_\alpha \chi + \frac{2}{3}\mathrm{i}\bar{\chi}\overleftrightarrow{\mathscr{D}}_{\alpha\dot{\alpha}}\chi - \frac{2}{3}G_{\alpha\dot{\alpha}}\bar{\chi}\chi \tag{5.7.40}$$

and

$$T = -\frac{1}{4}\chi(\bar{\mathscr{D}}^2 - 4R)\bar{\chi} + 3\mathscr{L}_c(\chi). \tag{5.7.41}$$

Remarkably, the supercurrent does not involve $\mathscr{L}_c(\chi)$. This follows from the fact that the superpotential term (5.7.39) depends only on the chiral compensator.

Let us clarify whether or not the supertrace vanishes on-shell. Due to the variational rule

$$\frac{\delta\chi(z')}{\delta\chi(z)} = -\frac{1}{4}(\bar{\mathscr{D}}^2 - 4R)E\delta^8(z - z')$$

the dynamical equation for χ is given by

$$-\frac{1}{4}(\mathscr{D}^2 - 4R)\bar{\chi} + \mathscr{L}'_c(\chi) = 0. \tag{5.7.42}$$

The relations (5.7.41) and (5.7.42) show that the supertrace vanishes if and only if

$$\begin{aligned} 3\mathscr{L}_c(\chi) &= \chi\mathscr{L}'_c(\chi) \Rightarrow \\ \mathscr{L}_c(\chi) &= \frac{g}{3!}\chi^3 \end{aligned} \tag{5.7.43}$$

with g being a constant. In this case the action turns out to be invariant under the super Weyl transformations

$$\chi' = e^{-\sigma}\chi \qquad \mathscr{W}' = \mathscr{W} \qquad \varphi' = e^{\sigma}\varphi \qquad \bar{\mathscr{D}}_{\dot{\alpha}}\sigma = 0. \tag{5.7.44}$$

The reader can explicitly check that the supercurrent (5.7.40) and the supertrace (5.7.41) satisfy the conservation law (5.7.14) under imposition of the equation of motion (5.7.42).

5.8. Supergravity in components

In this section we intend to describe the technique of passing from superfields (supergravity or matter) to component fields. Explicit formulae expressing the component fields of the supertorsion tensor R, G_a and $W_{\alpha\beta\gamma}$ via the fields of the Einstein supergravity multiplet (see Section 5.1) will be given. It will be shown how to read off the action functional from the action superfunctional. Local supersymmetry transformation laws of matter component field will be discussed.

Throughout this section, we work in the gravitational superfield gauge (5.4.38), in which the covariant derivatives are built from the gravitational superfield $\mathscr{H}^m$ and the chiral compensator $\varphi(z) = e^{\bar{W}}\phi(x, \theta) = \phi(x + i\mathscr{H}, \theta)$ (and conjugate $\bar{\varphi}(z) = e^{W}\bar{\phi}(x, \bar{\theta}) = \bar{\phi}(x - i\mathscr{H}, \bar{\theta})$) by means of equations (5.4.43–47) and (5.5.4). As was shown in Section 5.1, the residual gauge invariance can be used to set the Wess–Zumino gauge

$$\begin{aligned} \mathscr{H}^m &= \theta\sigma^a\bar{\theta}e_a{}^m + i\bar{\theta}^2\theta^\alpha\Psi^m{}_\alpha - i\theta^2\bar{\theta}_{\dot{\alpha}}\bar{\Psi}^{m\dot{\alpha}} + \theta^2\bar{\theta}^2A^m \\ \phi^3 &= e^{-1}\{1 - 2i\theta\sigma_a\bar{\Psi}^a + \theta^2 B\} \\ \bar{\phi}^3 &= e^{-1}\{1 - 2i\bar{\theta}\tilde{\sigma}_a\Psi^a + \bar{\theta}^2\bar{B}\} \end{aligned} \tag{5.8.1}$$

where

$$\Psi^a{}_\alpha = e_m{}^a \Psi^m{}_\alpha \qquad \bar{\Psi}^{a\dot{\alpha}} = e_m{}^a \bar{\Psi}^{m\dot{\alpha}}. \tag{5.8.2}$$

We take the Wess–Zumino gauge as the starting point for investigating the supergravity component structure.

5.8.1. Space projections of covariant derivatives

In flat global superspace, superfield component fields were defined using the notion of space projection. It seems reasonable to follow the same line when working in a curved superspace. By definition, the space projection $V|$ of a superfield $V(x, \theta, \bar{\theta})$ coincides with the zero-order term in the power series expansion of V in $\theta, \bar{\theta}$:

$$V| = V(x, \theta = 0, \bar{\theta} = 0).$$

one can also define space projections of differential operators. Given a first-order differential operator of the general structure

$$X = X^M(z)\partial_M + X^{ab}(z)M_{ab}$$

its space projection is taken to be

$$X| \equiv X^M|\partial_M + X^{ab}|M_{ab}.$$

It is clear how to extend this definition to the case of operators of second and higher orders. We would like to point out, however, that when calculating space projections one should be careful and remember that the expression

$$(XV)| = X^M|(\partial_M V)| + X^{ab}|M_{ab}V|$$

does not coincide with

$$X|V| = X^m|\partial_m V| + X^{ab}|M_{ab}V|.$$

Now, we proceed to find the space projections of the covariant derivatives

$$\mathcal{D}_A| = E_A{}^M\Big|\partial_M + \frac{1}{2}\Omega_A{}^{bc}\Big|M_{bc}$$

in the Wess–Zumino gauges (5.8.1). For this purpose let us obtain some auxiliary relations.

Owing to the Wess–Zumino gauge, the semi-covariant supervierbein $\hat{E}_A$ (5.4.43) is characterized by the projections

$$\begin{aligned} \hat{E}_\alpha| &= \partial_\alpha \qquad \hat{\bar{E}}_{\dot{\alpha}}| = -\bar{\partial}_{\dot{\alpha}} \\ \hat{E}^2| &= -\partial^\alpha\partial_\alpha \qquad \hat{\bar{E}}^2| = -\bar{\partial}_{\dot{\alpha}}\bar{\partial}^{\dot{\alpha}} \\ \frac{1}{2}[\hat{E}_\alpha, \hat{\bar{E}}_{\dot{\alpha}}]| &= -\partial_\alpha\bar{\partial}_{\dot{\alpha}} \end{aligned} \tag{5.8.3}$$

and

$$\hat{E}_a| = e_a{}^m \partial_m \equiv e_a$$
$$\Rightarrow \hat{E}_a{}^m| = e_a{}^m \qquad \hat{E}_m{}^a| = e_m{}^a. \tag{5.8.4}$$

We see that $\hat{E}_a|$ coincides with the ordinary vierbein. Similarly, considering the superfield anholonomy coefficients $\hat{C}_{ab}{}^c$ appearing in the commutator

$$[\hat{E}_a, \hat{E}_b] = \hat{C}_{ab}{}^c \hat{E}_c$$

$\hat{C}_{ab}{}^c|$ proves to coincide with the ordinary anholonomy coefficients $\mathscr{C}_{ab}{}^c$,

$$[e_a, e_b] = \mathscr{C}_{ab}{}^c e_c$$

since $\hat{E}_a$ does not involve spinor partial derivatives.

The above identities make it possible to rewrite the expansions (5.8.1) in the manner

$$\mathscr{H}^m| = \hat{E}_\alpha \mathscr{H}^m| = \hat{E}^2 \mathscr{H}^m| = 0$$
$$-\frac{1}{2}[\hat{E}_\alpha, \hat{\bar{E}}_{\dot{\alpha}}]\mathscr{H}^m| = (\sigma^a)_{\alpha\dot{\alpha}} e_a{}^m$$
$$-\frac{1}{4}\hat{E}_\alpha \hat{\bar{E}}^2 \mathscr{H}^m| = \mathrm{i}\Psi^m{}_\alpha \tag{5.8.5a}$$
$$\frac{1}{32}\{\hat{E}^2, \hat{\bar{E}}^2\}\mathscr{H}^m| = A^m$$

and

$$\varphi| = e^{-1/3} \qquad \hat{E}_\alpha \varphi| = -\frac{2\mathrm{i}}{3} e^{-1/3}(\sigma^a \bar{\Psi}_a)_\alpha$$
$$-\frac{1}{4}\hat{E}^2 \varphi| = e^{-1/3}\left(B + \frac{2}{3}\Psi^a \tilde{\sigma}_a \sigma_b \bar{\Psi}^b\right) \tag{5.8.5b}$$

and similarly for $\bar{\varphi}$. Among the relations (5.8.5*a*), only the latter requires comment: due to equations (5.8.3) and (5.4.43), we have

$$\hat{\bar{E}}^2 \hat{E}^2 \mathscr{H}^m| = -\bar{\partial}_{\dot{\alpha}} \bar{\partial}^{\dot{\alpha}} \hat{E}^2 \mathscr{H}^m| = -\bar{\partial}_{\dot{\alpha}} \bar{\partial}^{\dot{\alpha}} (\partial_\alpha + 2\mathrm{i}(\sigma^a \bar{\theta})_\alpha e_a) \partial^\alpha \mathscr{H}^m|$$
$$= 16A^m + 4\mathrm{i}(\sigma^a)_{\alpha\dot{\alpha}} e_a \bar{\partial}^{\dot{\alpha}} \partial^\alpha \mathscr{H}^m|$$

which means

$$\frac{1}{16}\hat{\bar{E}}^2 \hat{E}^2 \mathscr{H}^m| = A^m - \frac{\mathrm{i}}{2} e_a e^{am}$$
$$\frac{1}{16}\hat{E}^2 \hat{\bar{E}}^2 \mathscr{H}^m| = A^m + \frac{\mathrm{i}}{2} e_a e^{am}. \tag{5.8.6}$$

To derive the relations (5.8.5*b*), it is sufficient to note that in the Wess–Zumino gauge φ is of the form

$$\varphi(z) = \left(1 + \mathrm{i}\mathscr{H}^m\partial_m - \frac{1}{2}\mathscr{H}^m\mathscr{H}^n\partial_m\partial_n\right)\phi(x, \theta)$$

and then to use the expansion of ϕ^3 (5.8.1).

Remark. Using equations (5.8.6), one can prove the relations

$$\begin{aligned}\frac{1}{8}\hat{E}_\alpha\hat{\bar{E}}^2\hat{E}_\beta\mathscr{H}^m| &= -\frac{\mathrm{i}}{2}(\sigma^{ab})_{\alpha\beta}\mathscr{C}_{ab}{}^c e_c{}^m + \varepsilon_{\alpha\beta}A^m \\ \frac{1}{8}\hat{\bar{E}}_{\dot\alpha}\hat{E}^2\hat{\bar{E}}_{\dot\beta}\mathscr{H}^m| &= \frac{\mathrm{i}}{2}(\sigma^{ab})_{\dot\alpha\dot\beta}\mathscr{C}_{ab}{}^c e_c{}^m - \varepsilon_{\dot\alpha\dot\beta}A^m.\end{aligned} \tag{5.8.7}$$

Now all the necesary ingredients are at our disposal in order to evaluate the projections $\mathscr{D}_\alpha|$ and $\bar{\mathscr{D}}_{\dot\alpha}|$. It is worth recalling that the spinor covariant derivatives read

$$\mathscr{D}_\alpha = F\hat{E}_\alpha + \Omega_{\alpha\beta\gamma}M^{\beta\gamma} + \Omega_{\alpha\dot\beta\dot\gamma}M^{\dot\beta\dot\gamma}$$

$$\bar{\mathscr{D}}_{\dot\alpha} = \overline{F\hat{E}_\alpha} + \bar{\Omega}_{\dot\alpha\beta\gamma}M^{\beta\gamma} + \bar{\Omega}_{\dot\alpha\dot\beta\dot\gamma}\bar{M}^{\dot\beta\dot\gamma}$$

Here the connection superfields are built from the F and $\bar{F}$ and anholonomy coefficients $\hat{C}_{\alpha b}{}^c$and $\hat{C}_{\dot\alpha b}{}^c$ (5.4.45) by the rule (5.4.46). In accordance with relations (5.4.47) and (5.5.4), F and $\bar{F}$ have the form

$$F = \varphi^{1/2}\bar{\varphi}^{-1}\det{}^{-1/3}(1 - \mathrm{i}\partial\mathscr{H})\det{}^{1/6}(1 + \mathrm{i}\partial\mathscr{H})\det{}^{-1/6}(\hat{E}_a{}^m)$$

$$\bar{F} = \varphi^{-1}\bar{\varphi}^{1/2}\det{}^{1/6}(1 - \mathrm{i}\partial\mathscr{H})\det{}^{-1/3}(1 + \mathrm{i}\partial\mathscr{H})\det{}^{-1/6}(\hat{E}_a{}^m)$$

where

$$(1 + \mathrm{i}\partial\mathscr{H})_n{}^m = \delta_n{}^m + \mathrm{i}\partial_n\mathscr{H}^m$$

$$\hat{E}_a{}^m = -\frac{1}{4}(\tilde{\sigma}_a)^{\dot\alpha\alpha}[\hat{E}_\alpha, \hat{\bar{E}}_{\dot\alpha}]\mathscr{H}^m.$$

Therefore, one must calculate $F|$, $\hat{E}_\alpha F|$, $\hat{\bar{E}}_{\dot\alpha}F|$ and $\hat{C}_{\alpha b}{}^c$ and their conjugate. Using identities (5.8.4, 5) and expressions (5.4.45) one obtains

$$\begin{aligned}\hat{C}_{\alpha,\beta\dot\beta,\gamma\dot\gamma}| &= 2\mathrm{i}\varepsilon_{\alpha\beta}\Psi_{\gamma\dot\gamma,\dot\beta} \\ \hat{C}_{\dot\alpha,\beta\dot\beta,\gamma\dot\gamma}| &= -2\mathrm{i}\varepsilon_{\dot\alpha\dot\beta}\Psi_{\gamma\dot\gamma,\beta}.\end{aligned} \tag{5.8.8}$$

Here we have converted the gravitino vector index into a pair of spinor indices,

$$\Psi_{\alpha\dot\alpha,\beta} = (\sigma_a)_{\alpha\dot\alpha}\Psi^a{}_\beta \qquad \Psi_{\alpha\dot\alpha,\dot\beta} = (\sigma_a)_{\alpha\dot\alpha}\Psi^a{}_{\dot\beta}. \tag{5.8.9}$$

Further, one readily obtains

$$\det(\hat{E}_a{}^m)| = e \qquad \hat{E}_\alpha\det(\hat{E}_a{}^m)| = \mathrm{i}e(\sigma_a\Psi^m)_\alpha. \tag{5.8.10}$$

Along with the relations (5.8.5*b*) these mean

$$F| = \hat{F}| = 1$$

$$\hat{E}_\alpha F| = -\hat{E}_\alpha \bar{F}| = -\frac{\mathrm{i}}{2}(\sigma_a \Psi^a)_\alpha \tag{5.8.11}$$

$$\hat{\bar{E}}_{\dot\alpha} F| = -\hat{\bar{E}}_{\dot\alpha} \bar{F}| = \frac{\mathrm{i}}{2}(\Psi^a \sigma_a)_{\dot\alpha}.$$

Now, in accordance with definitions (5.4.46), the space projections of the spinor superconnections are

$$\Omega_{\alpha\beta\gamma}| = -\frac{\mathrm{i}}{2}(\varepsilon_{\alpha\beta}\Psi_{\gamma\dot\gamma,}{}^{\dot\gamma} + \varepsilon_{\alpha\gamma}\Psi_{\beta\dot\gamma,}{}^{\dot\gamma})$$

$$\Omega_{\alpha\beta\dot\gamma}| = \frac{\mathrm{i}}{2}(\Psi_{\alpha\beta,\dot\gamma} + \Psi_{\alpha\dot\gamma,\beta}) \tag{5.8.12a}$$

and

$$\bar{\Omega}_{\dot\alpha\dot\beta\dot\gamma}| = \frac{\mathrm{i}}{2}(\varepsilon_{\dot\alpha\dot\beta}\Psi_{\gamma\dot\gamma,}{}^{\gamma} + \varepsilon_{\dot\alpha\dot\gamma}\Psi_{\gamma\dot\beta,}{}^{\gamma})$$

$$\bar{\Omega}_{\dot\alpha\dot\beta\gamma}| = -\frac{\mathrm{i}}{2}(\Psi_{\beta\dot\alpha,\gamma} + \Psi_{\gamma\dot\alpha,\beta}). \tag{5.8.12b}$$

Then, since $F| = 1$ and $\hat{E}_\alpha| = \partial_\alpha$, one can immediately arrive at the final results

$$\mathscr{D}_\alpha| = \partial_\alpha - \mathrm{i}(\sigma^a\Psi^b)_\alpha M_{ab}$$
$$\bar{\mathscr{D}}^{\dot\alpha}| = \bar\partial^{\dot\alpha} - \mathrm{i}(\tilde\sigma^a\Psi^b)^{\dot\alpha} M_{ab}. \tag{5.8.13}$$

It remains to determine $\mathscr{D}_a|$. Using the explicit form of the supervierbein vector components (5.4.18), it follows from the above identities that

$$E_a| = e_a + \frac{1}{2}\Psi_a{}^\beta\partial_\beta + \frac{1}{2}\Psi_{a\dot\beta}\bar\partial^{\dot\beta}. \tag{5.8.14}$$

Now, we find it convenient to represent $\mathscr{D}_a|$ as follows

$$\mathscr{D}_a| = \nabla_a + \frac{1}{2}\Psi_a{}^\beta\mathscr{D}_\beta| + \frac{1}{2}\Psi_{a\dot\beta}\bar{\mathscr{D}}^{\dot\beta}| \tag{5.8.15}$$

where ∇_a are ordinary space–time covariant derivatives,

$$\nabla_a = e_a + \frac{1}{2}\omega_a{}^{bc}M_{bc} \tag{5.8.16}$$

Here the Lorentz spin connection ω_{abc} is related to $\Omega_{abc}|$ by the rule

$$\omega_{abc} = \Omega_{abc}| + \frac{\mathrm{i}}{2}(\Psi_a\sigma_b\Psi_c - \Psi_a\sigma_c\Psi_b + \Psi_b\sigma_c\Psi_a - \Psi_c\sigma_b\Psi_a).$$

We shall determine the connection ω_{abc} (and hence $\Omega_{abc}|$) in subsection 5.8.3.

5.8.2. Space projections of R, $\bar{R}$ and G_a

As the next step, we evaluate the projections of the supertorsion tensors R, $\bar{R}$ and G_a.

Recall that $\bar{R}$ is given by

$$\bar{R} = -\frac{1}{4}\bar{\hat{E}}^2(F^2).$$

Then, making use of the explicit form of F gives

$$\bar{R}| = -\frac{1}{4}\bar{\varphi}^{-2}|\bar{\hat{E}}^2(\varphi \det{}^{-1/3}(\hat{E}_a{}^m))|.$$

After a short calculation, one obtains

$$\bar{R}| = \frac{1}{3}\mathbf{B}$$

$$\mathbf{B} = B + \frac{1}{2}\Psi^a\tilde{\sigma}_a\sigma_b\Psi^b + \frac{1}{2}\Psi^a\Psi_a \tag{5.8.17a}$$

hence

$$R| = \frac{1}{3}\bar{\mathbf{B}}$$

$$\bar{\mathbf{B}} = \bar{B} + \frac{1}{2}\Psi^a\sigma_a\tilde{\sigma}_b\Psi^b + \frac{1}{2}\Psi^a\Psi_a. \tag{5.8.17b}$$

We see that $3\bar{R}|$ and $3R|$ coincide with the scalar field $\mathbf{B}$ and $\bar{\mathbf{B}}$, respectively, from the Einstein supergravity multiplet.

To find $G_a|$, we note the relations

$$-\mathrm{i}G_a = T_{\beta a}{}^{\beta} = C_{\beta a}{}^{\beta} \tag{5.8.18}$$

$T_{\beta a}{}^{\gamma}$ being the components of the supertorsion,

$$[\mathscr{D}_\beta, \mathscr{D}_a] = T_{\beta a}{}^{\gamma}\mathscr{D}_\gamma + \ldots$$

and $C_{\beta a}{}^{\gamma}$ being the supervierbein's anholonomy coefficients,

$$[E_\beta, E_a] = C_{\beta a}{}^{\gamma}E_\gamma + \ldots.$$

In deriving relations (5.8.18), we have used the commutation relations (5.5.6) and the expressions for the supertorsion components (5.2.47).

Expanding the commutator

$$[E_\beta, E_{\alpha\dot{\alpha}}] = \Bigg[F\hat{E}_\beta, F\bar{F}\hat{E}_{\alpha\dot{\alpha}} + \frac{\mathrm{i}}{2}F(\hat{E}_\alpha\bar{F})\hat{\bar{E}}_{\dot{\alpha}} + \frac{\mathrm{i}}{2}\bar{F}(\hat{\bar{E}}_{\dot{\alpha}}F)\hat{E}_\beta$$
$$+ \frac{\mathrm{i}}{2}\Omega_{\alpha\dot{\alpha}}{}^{\dot{\gamma}}\overline{F\hat{E}}_{\dot{\gamma}} + \frac{\mathrm{i}}{2}\bar{\Omega}_{\dot{\alpha}\alpha}{}^{\gamma}F\hat{E}_\gamma\Bigg]$$

one arrives at

$$C_{\beta,\alpha\dot\alpha,}{}^{\beta}| = \frac{1}{4}\hat{C}_{\beta,\alpha\dot\alpha,}{}^{\gamma\dot\gamma}|\Psi_{\gamma\dot\gamma,}{}^{\beta} - \frac{\mathrm{i}}{2}(\hat{E}_\alpha\bar{F})|\hat{\bar{E}}_{\dot\alpha}F| + \frac{1}{2}(\hat{E}_\alpha\bar{F})|\Psi_{\beta\dot\alpha,}{}^{\beta} - (\hat{\bar{E}}_{\dot\alpha}F)|E_\alpha F|$$

$$-\mathrm{i}\bar{\Omega}_{\dot\alpha\alpha}{}^{\beta}|\hat{E}_\beta F| + \frac{\mathrm{i}}{2}\hat{E}_\alpha\hat{\bar{E}}_{\dot\alpha}F| + \frac{\mathrm{i}}{2}\hat{E}_\beta(F\bar{\Omega}_{\dot\alpha\alpha}{}^{\beta})|.$$

All the terms, except the last two, have been calculated above. When evaluating $\hat{E}_\alpha\hat{\bar{E}}_{\dot\alpha}F|$ and $\hat{E}_\beta(F\bar{\Omega}_{\dot\alpha\alpha}{}^{\beta})|$, the relations (5.8.6, 7) prove to be helpful. After some calculation, one arrives at

$$G_a| = \frac{4}{3}\mathbf{A}_a$$

$$\mathbf{A}^a = A^a + \frac{1}{8}\varepsilon^{abcd}\mathscr{C}_{bcd} - \frac{1}{4}(\Psi_a\sigma_b\bar{\Psi}^b + \Psi^b\sigma_b\bar{\Psi}_a) \tag{5.8.19}$$

$$-\frac{1}{2}\Psi^b\sigma_a\bar{\Psi}_b + \frac{\mathrm{i}}{8}\varepsilon^{abcd}\Psi_b\sigma_c\bar{\Psi}_d$$

We see that $\frac{3}{4}G_a$ coincides with the vector field from the Einstein supergravity multiplet.

5.8.3. *Basic construction*

The relations (5.8.13, 14, 17, 19) turn out to be sufficient to determine all the other quantities: the connection ω_{abc}, the projections $W_{\alpha\beta\gamma}|$, $\mathscr{D}_\alpha R|$, $\mathscr{D}_\alpha G_a|$, $\mathscr{D}^2 R|$ and so on. This is done as follows.

We are going to evaluate the space projection of commutator $[\mathscr{D}_{\alpha\dot\alpha}, \mathscr{D}_{\beta\dot\beta}]$ in two different ways: first, using the representation (5.8.15) and secondly, making use of the commutation relations (5.5.7). In the former case, we have

$$\mathscr{D}_{\alpha\dot\alpha}\mathscr{D}_{\beta\dot\beta}| = \nabla_{\alpha\dot\alpha}\mathscr{D}_{\beta\dot\beta}| + \frac{1}{2}\Psi_{\alpha\dot\alpha,}{}^{\gamma}\mathscr{D}_\gamma\mathscr{D}_{\beta\dot\beta}| + \frac{1}{2}\bar{\Psi}_{\alpha\dot\alpha,\dot\gamma}\bar{\mathscr{D}}^{\dot\gamma}\mathscr{D}_{\beta\dot\beta}|$$

$$= \nabla_{\alpha\dot\alpha}\left(\nabla_{\beta\dot\beta} + \frac{1}{2}\Psi_{\beta\dot\beta,}{}^{\gamma}\mathscr{D}_\gamma| + \frac{1}{2}\bar{\Psi}_{\beta\dot\beta,\dot\gamma}\bar{\mathscr{D}}^{\dot\gamma}|\right)$$

$$+\frac{1}{2}\Psi_{\alpha\dot\alpha,}{}^{\gamma}[\mathscr{D}_\gamma, \mathscr{D}_{\beta\dot\beta}]| + \frac{1}{2}\bar{\Psi}_{\alpha\dot\alpha,\dot\gamma}[\bar{\mathscr{D}}^{\dot\gamma}, \mathscr{D}_{\beta\dot\beta}]|$$

$$+\frac{1}{2}\Psi_{\alpha\dot\alpha,}{}^{\gamma}\left(\nabla_{\beta\dot\beta} + \frac{1}{2}\Psi_{\beta\dot\beta,}{}^{\delta}\mathscr{D}_\delta + \frac{1}{2}\bar{\Psi}_{\beta\dot\beta,\dot\delta}\bar{\mathscr{D}}^{\dot\delta}\right)\mathscr{D}_\gamma|$$

$$+\frac{1}{2}\bar{\Psi}_{\alpha\dot\alpha,\dot\gamma}\left(\nabla_{\beta\dot\beta} + \frac{1}{2}\Psi_{\beta\dot\beta,}{}^{\delta}\mathscr{D}_\delta + \frac{1}{2}\bar{\Psi}_{\beta\dot\beta,\dot\delta}\bar{\mathscr{D}}^{\dot\delta}\right)\bar{\mathscr{D}}^{\dot\gamma}|$$

which leads to

$$
\begin{aligned}
[\mathscr{D}_{\alpha\dot\alpha}, \mathscr{D}_{\beta\dot\beta}]| = {} & [\nabla_{\alpha\dot\alpha}, \nabla_{\beta\dot\beta}] \\
& + \frac{1}{2}(\nabla_{\alpha\dot\alpha}\Psi_{\beta\dot\beta,}{}^{\gamma} - \nabla_{\beta\dot\beta}\Psi_{\alpha\dot\alpha,}{}^{\gamma})\mathscr{D}_{\gamma}| + \frac{1}{2}(\nabla_{\alpha\dot\alpha}\bar\Psi_{\beta\dot\beta,\dot\gamma} - \nabla_{\beta\dot\beta}\bar\Psi_{\alpha\dot\alpha,\dot\gamma})\bar{\mathscr{D}}^{\dot\gamma}| \\
& + \frac{1}{2}\Psi_{\alpha\dot\alpha,}{}^{\gamma}[\mathscr{D}_{\gamma}, \mathscr{D}_{\beta\dot\beta}]| - \frac{1}{2}\Psi_{\beta\dot\beta,}{}^{\gamma}[\mathscr{D}_{\gamma}, \mathscr{D}_{\alpha\dot\alpha}]| \\
& + \frac{1}{2}\bar\Psi_{\alpha\dot\alpha,\dot\gamma}[\bar{\mathscr{D}}^{\dot\gamma}, \mathscr{D}_{\beta\dot\beta}]| - \frac{1}{2}\bar\Psi_{\beta\dot\beta,\dot\gamma}[\bar{\mathscr{D}}^{\dot\gamma}, \mathscr{D}_{\alpha\dot\alpha}]| \\
& + \frac{1}{4}\Psi_{\alpha\dot\alpha,}{}^{\gamma}\Psi_{\beta\dot\beta,}{}^{\delta}\{\mathscr{D}_{\gamma}, \mathscr{D}_{\delta}\}| + \frac{1}{4}\bar\Psi_{\alpha\dot\alpha,\dot\gamma}\bar\Psi_{\beta\dot\beta,\dot\delta}\{\bar{\mathscr{D}}^{\dot\gamma}, \bar{\mathscr{D}}^{\dot\delta}\}| \\
& - \frac{1}{4}(\Psi_{\alpha\dot\alpha,}{}^{\gamma}\bar\Psi_{\beta\dot\beta,}{}^{\dot\gamma} + \bar\Psi_{\alpha\dot\alpha,}{}^{\dot\gamma}\Psi_{\beta\dot\beta,}{}^{\gamma})\{\mathscr{D}_{\gamma}, \bar{\mathscr{D}}_{\dot\gamma}\}|.
\end{aligned}
$$

Now, introducing the torsion and curvature tensors associated with the derivatives ∇_a,

$$
\begin{aligned}
[\nabla_a, \nabla_b] &= \mathscr{T}_{ab}{}^{c}\nabla_c + \frac{1}{2}\mathscr{R}_{abcd}M^{cd} \\
&= \mathscr{T}_{ab}{}^{c}\nabla_c + \mathscr{R}_{ab\gamma\delta}M^{\gamma\delta} + \bar{\mathscr{R}}_{ab\dot\gamma\dot\delta}\bar M^{\dot\gamma\dot\delta}
\end{aligned}
\tag{5.8.20}
$$

and applying the (anti)commutation relations (5.5.6), the upper expression takes the form

$$
\begin{aligned}
[\mathscr{D}_{\alpha\dot\alpha}, \mathscr{D}_{\beta\dot\beta}]| = {} & \left\{\mathscr{T}_{\alpha\dot\alpha,\beta\dot\beta}{}^{c} + \frac{i}{2}(\Psi_{\alpha\dot\alpha}\sigma^{c}\bar\Psi_{\beta\dot\beta} - \Psi_{\beta\dot\beta}\sigma^{c}\bar\Psi_{\alpha\dot\alpha})\right\}\nabla_c \\
& + \left\{\frac{1}{2}(\nabla_{\alpha\dot\alpha}\Psi_{\beta\dot\beta,}{}^{\gamma} - \nabla_{\beta\dot\beta}\Psi_{\alpha\dot\alpha,}{}^{\gamma}) + \frac{i}{4}(\Psi_{\alpha\dot\alpha}\sigma^{c}\bar\Psi_{\beta\dot\beta} - \Psi_{\beta\dot\beta}\sigma^{c}\bar\Psi_{\alpha\dot\alpha})\Psi_{c}{}^{\gamma}\right. \\
& \left. - \frac{i}{2}(\Psi_{\alpha\dot\alpha,\beta}G^{\gamma}{}_{\dot\beta}| - \Psi_{\beta\dot\beta,\alpha}G^{\gamma}{}_{\dot\alpha}|) - \frac{i}{2}(\bar\Psi_{\alpha\dot\alpha,\beta}\delta^{\gamma}_{\beta} - \bar\Psi_{\beta\dot\beta,\dot\alpha}\delta^{\gamma}_{\alpha})R|\right\}\mathscr{D}_{\gamma}| \\
& + \left\{\mathscr{R}_{\alpha\dot\alpha,\beta\dot\beta,\gamma\delta} - \frac{i}{2}(\Psi_{\alpha\dot\alpha,\gamma}\varepsilon_{\beta\delta} + \Psi_{\alpha\dot\alpha,\delta}\varepsilon_{\beta\gamma})\bar{\mathscr{D}}_{\dot\beta}\bar R|\right. \\
& + \frac{i}{2}(\Psi_{\beta\dot\beta,\gamma}\varepsilon_{\alpha\delta} + \Psi_{\beta\dot\beta,\delta}\varepsilon_{\alpha\gamma})\bar{\mathscr{D}}_{\dot\alpha}\bar R| + \frac{i}{2}\bar\Psi_{\alpha\dot\alpha,\beta}\mathscr{D}_{(\gamma}G_{\delta)\dot\beta}| \\
& - \frac{i}{2}\bar\Psi_{\beta\dot\beta,\alpha}\mathscr{D}_{(\gamma}G_{\delta)\dot\alpha}| - i\bar\Psi_{\alpha\dot\alpha,\beta}W_{\beta\gamma\delta}| + i\bar\Psi_{\beta\dot\beta,\dot\alpha}W_{\alpha\gamma\delta}| \\
& \left. - \frac{1}{2}(\Psi_{\alpha\dot\alpha,\gamma}\Psi_{\beta\dot\beta,\delta} + \Psi_{\alpha\dot\alpha,\delta}\Psi_{\beta\dot\beta,\gamma})\bar R|\right\}M^{\gamma\delta} \\
& + \bar{\mathscr{D}}^{\dot\gamma}|\text{- and } \bar M^{\dot\gamma\dot\delta}\text{-terms.}
\end{aligned}
\tag{5.8.21}
$$

On the other hand, owing to the identity

$$(\sigma_a)_{\alpha\dot\alpha}(\sigma_b)_{\beta\dot\beta}\varepsilon^{abcd}G_c\mathscr{D}_d = -\mathrm{i}\varepsilon_{\dot\alpha\dot\beta}G_\beta{}^{\dot\gamma}\mathscr{D}_{\alpha\dot\gamma} + \mathrm{i}\varepsilon_{\alpha\beta}G^\gamma{}_{\dot\beta}\mathscr{D}_{\gamma\dot\alpha}$$

from relations (5.5.7) we deduce

$$\begin{aligned}
[\mathscr{D}_{\alpha\dot\alpha}, \mathscr{D}_{\beta\dot\beta}]| = & -(\sigma_a)_{\alpha\dot\alpha}(\sigma_b)_{\beta\dot\beta}\varepsilon^{abcd}G_d|\nabla_c \\
& + \left\{-\frac{1}{2}(\sigma_a)_{\alpha\dot\alpha}(\sigma_b)_{\beta\dot\beta}\varepsilon^{abcd}G_d|\Psi_c{}^\gamma + \frac{1}{2}\varepsilon_{\dot\alpha\dot\beta}\mathscr{D}_\alpha R|\delta^\gamma_\beta\right. \\
& \left. + \varepsilon_{\dot\alpha\dot\beta}W_{\alpha\beta}{}^\gamma| - \frac{1}{2}\varepsilon_{\alpha\beta}\bar{\mathscr{D}}_{\dot\alpha}G^\gamma{}_{\dot\beta}|\right\}\mathscr{D}_\gamma| \qquad (5.8.22) \\
& + \left\{\frac{1}{8}\varepsilon_{\dot\alpha\dot\beta}(\varepsilon_{\alpha\gamma}\varepsilon_{\beta\delta} + \varepsilon_{\alpha\delta}\varepsilon_{\beta\gamma})(\bar{\mathscr{D}}^2 - 8R)\bar{R}| + \varepsilon_{\dot\alpha\dot\beta}\mathscr{D}_\alpha W_{\beta\gamma\delta}|\right. \\
& \left. + \frac{1}{2}\varepsilon_{\alpha\beta}\bar{\mathscr{D}}_{\dot\alpha}\mathscr{D}_{(\gamma}G_{\delta)\dot\beta}|\right\}M^{\gamma\delta} + \bar{\mathscr{D}}^{\dot\gamma}| - \text{ and } \bar{M}^{\dot\gamma\dot\delta}\text{-terms.}
\end{aligned}$$

Now let us analyse the relations (5.8.21) and (5.8.22).

Since the ∇-terms in relations (5.8.21) snd (5.8.22) must coincide, the torsion $\mathscr{T}_{ab}{}^c$ is determined in terms of the gravitino and the vector field $\mathbf{A}_d$ as follows:

$$\mathscr{T}_{abc} = -\frac{\mathrm{i}}{2}(\Psi_a\sigma_c\bar\Psi_b - \Psi_b\sigma_c\bar\Psi_a) - \frac{4}{3}\varepsilon_{abcd}\mathbf{A}^d. \qquad (5.8.23)$$

Therefore, the ∇-connection reads

$$\omega_{abc} = \frac{1}{2}(\Gamma_{bca} + \Gamma_{acb} - \Gamma_{abc}) \equiv \omega_{abc}(e, \Psi) - \frac{2}{3}\varepsilon_{abcd}\mathbf{A}^d \qquad (5.8.24)$$

where

$$\Gamma_{abc} = \mathscr{C}_{abc} - \mathscr{T}_{abc} \qquad (5.8.25)$$

It is seen that the connection depends on the gravitino and the vector field $\mathbf{A}_d$. Extracting from ω_{abc} the term containing $\mathbf{A}^d$ gives the gravitino-dependent connection $\omega_{abc}(e, \Psi)$. The corresponding covariant derivatives

$$\begin{aligned}
\tilde\nabla_a &= e_a + \frac{1}{2}\omega_{abc}(e, \Psi)M^{bc} \\
[\tilde\nabla_a, \tilde\nabla_b] &= \tilde{\mathscr{T}}_{ab}{}^c\tilde\nabla_c + \frac{1}{2}\tilde{\mathscr{R}}_{abcd}M^{cd}
\end{aligned} \qquad (5.8.26)$$

will often be used later.

To simplify subsequent work, it is convenient to introduce the gravitino field strengths:

$$\Psi_{ab\gamma} \equiv \nabla_a \Psi_{b\gamma} - \nabla_b \Psi_{a\gamma} - \mathscr{T}_{ab}{}^c \Psi_{c\gamma} + \frac{4i}{3} \mathbf{A}^c \{(\sigma_{ca} \Psi_b)_\gamma - (\sigma_{cb} \Psi_a)_\gamma\}$$
$$+ \frac{2i}{3} \{\mathbf{A}_a \Psi_{b\gamma} - \mathbf{A}_b \Psi_{a\gamma}\}$$
$$\bar{\Psi}_{ab}{}^{\dot\gamma} \equiv \nabla_a \bar{\Psi}_b{}^{\dot\gamma} - \nabla_b \bar{\Psi}_a{}^{\dot\gamma} - \mathscr{T}_{ab}{}^c \bar{\Psi}_c{}^{\dot\gamma} - \frac{4i}{3} \mathbf{A}^c \{(\tilde\sigma_{ca} \bar{\Psi}_b)^{\dot\gamma} - (\tilde\sigma_{cb} \bar{\Psi}_a)^{\dot\gamma}\} \tag{5.8.27}$$
$$- \frac{2i}{3} \{\mathbf{A}_a \bar{\Psi}_b{}^{\dot\gamma} - \mathbf{A}_b \bar{\Psi}_a{}^{\dot\gamma}\}$$

and their spinor analogues

$$\Psi_{\alpha\beta,\gamma} = \frac{1}{2} (\sigma^{ab})_{\alpha\beta} \Psi_{ab\gamma} \qquad \Psi_{\dot\alpha\dot\beta,\gamma} = -\frac{1}{2} (\tilde\sigma^{ab})_{\dot\alpha\dot\beta} \Psi_{ab\gamma}$$
$$\bar\Psi_{\alpha\beta,\dot\gamma} = \frac{1}{2} (\sigma^{ab})_{\alpha\beta} \bar\Psi_{ab\dot\gamma} \qquad \bar\Psi_{\dot\alpha\dot\beta,\dot\gamma} = -\frac{1}{2} (\tilde\sigma^{ab})_{\dot\alpha\dot\beta} \bar\Psi_{ab\dot\gamma}. \tag{5.8.28}$$

In terms of the derivatives $\tilde\nabla_a$, then $\Psi_{ab\gamma}$ reads

$$\Psi_{ab\gamma} = \tilde\nabla_a \Psi_{b\gamma} - \tilde\nabla_b \Psi_{a\gamma} - \tilde{\mathscr{T}}_{ab}{}^c \Psi_{c\gamma} + \frac{2i}{3} \mathbf{A}^c \{(\sigma_{ca} \Psi_b)_\gamma - (\sigma_{cb} \Psi_a)_\gamma\}$$
$$+ \frac{2i}{3} \{\mathbf{A}_a \Psi_{b\gamma} - \mathbf{A}_b \Psi_{a\gamma}\} \tag{5.8.29a}$$

or

$$\Psi_{ab\gamma} = e_a{}^m e_b{}^n \left\{ \partial_m \Psi_{n\gamma} - \partial_n \Psi_{m\gamma} + \frac{1}{2} \tilde\omega_{mcd} (\sigma^{cd} \Psi_n)_\gamma - \frac{1}{2} \tilde\omega_{ncd} (\sigma^{cd} \Psi_m)_\gamma \right\}$$
$$+ \frac{2i}{3} \mathbf{A}^c \{(\sigma_{ca} \Psi_b)_\gamma - (\sigma_{cb} \Psi_a)_\gamma\} + \frac{2i}{3} \{\mathbf{A}_a \Psi_{b\gamma} - \mathbf{A}_b \Psi_{a\gamma}\} \tag{5.8.29b}$$

where $\tilde\omega_{mcd} = e_m{}^a \omega_{acd}(e, \Psi)$. It is worth pointing out that $\Psi_{ab\gamma}$ and $\bar\Psi_{ab}{}^{\dot\gamma}$ differ from similar quantities which arose in the transformation laws (5.1.47) only by the presence of **A**-dependent terms.

Let us return to the analysis of relations (5.8.21) and (5.8.22). Setting the $\mathscr{D}_\gamma|$-terms in both expressions to be equal, one obtains

$$\frac{1}{2} \varepsilon_{\dot\alpha\dot\beta} \mathscr{D}_\alpha R | \delta^\gamma_\beta + \varepsilon_{\dot\alpha\dot\beta} W_{\alpha\beta}{}^\gamma | - \frac{1}{2} \varepsilon_{\alpha\beta} \bar{\mathscr{D}}_{\dot\alpha} G^\gamma{}_{\dot\beta} |$$
$$= \varepsilon_{\dot\alpha\dot\beta} \Psi_{\alpha\beta,}{}^\gamma + \varepsilon_{\alpha\beta} \Psi_{\dot\alpha\dot\beta,}{}^\gamma - \frac{i}{6} \mathbf{B} (\Psi_{\alpha\dot\alpha,\beta\dot\beta} \delta^\gamma_\beta - \Psi_{\beta\dot\beta,\alpha\dot\alpha} \delta^\gamma_\alpha).$$

We decompose this identity into irreducible components resulting in

$$\mathscr{D}_\alpha R| = -\frac{2}{3}(\sigma^{ab}\Psi_{ab})_\alpha + \frac{i}{3}\bar{\mathbf{B}}(\sigma_a\bar{\Psi}^a)_\alpha = \frac{4}{3}\Psi_{\alpha\beta,}{}^{\beta} + \frac{i}{3}\bar{\mathbf{B}}\Psi_{\alpha\beta,}{}^{\beta} \quad (5.8.30a)$$

$$\bar{\mathscr{D}}_{(\dot\alpha}G^{\beta}{}_{\dot\beta)}| = -2\Psi_{\dot\alpha\dot\beta,}{}^{\beta} + \frac{i}{3}\mathbf{B}\bar{\Psi}^{\beta}{}_{(\dot\alpha,\dot\beta)} \quad (5.8.30b)$$

$$W_{\alpha\beta\gamma}| = \frac{1}{3}(\Psi_{\alpha\beta,\gamma} + \Psi_{\beta\gamma,\alpha} + \Psi_{\gamma\alpha,\beta}). \quad (5.8.30c)$$

Now, one can easily calculate the projection $\bar{\mathscr{D}}_{\dot\alpha}G_{\beta\dot\beta}|$, using the Bianchi identity $\bar{\mathscr{D}}^{\dot\alpha}G_{\alpha\dot\alpha} = \mathscr{D}_\alpha R$. Next, conjugating the relations (5.8.30), one obtains the projections $\bar{\mathscr{D}}_{\dot\alpha}\bar{R}|$, $\mathscr{D}_{(\alpha}G_{\beta)}{}^{\dot\beta}|$ and $\bar{W}_{\dot\alpha\dot\beta\dot\gamma}|$.

At present, it remains to investigate the $M^{\gamma\delta}$-terms in expressions (5.8.21) and (5.8.22). First, however, we would like to discuss the properties of the curvature tensor in the case of non-zero torsion.

5.8.4. *Algebraic structure of the curvature with torsion*

As is well known, in the torsion-free case the curvature tensor is characterized by the following algebraic properties:

$$\mathscr{R}_{abcd} = -\mathscr{R}_{bacd} = -\mathscr{R}_{abdc} \quad (5.8.31a)$$

$$\mathscr{R}_{abcd} + \mathscr{R}_{bcad} + \mathscr{R}_{cabd} = 0 \quad (5.8.31b)$$

$$\mathscr{R}_{abcd} = \mathscr{R}_{cdab}. \quad (5.8.31c)$$

The first property follows directly from the definition of the curvature. The second relation represents the torsion-free Bianchi identities. Finally, the third property is a direct consequence of the second one.

In the case of non-vanishing torsion, the Bianchi identities are modified in the manner

$$\begin{aligned}\mathscr{R}_{abc}{}^{d} + \mathscr{R}_{bca}{}^{d} + \mathscr{R}_{cab}{}^{d} &= \nabla_a\mathscr{T}_{bc}{}^{d} + \nabla_b\mathscr{T}_{ca}{}^{d} + \nabla_c\mathscr{T}_{ab}{}^{d} \\ &\quad -\mathscr{T}_{ab}{}^{e}\mathscr{T}_{ec}{}^{d} - \mathscr{T}_{bc}{}^{e}\mathscr{T}_{ea}{}^{d} - \mathscr{T}_{ca}{}^{e}\mathscr{T}_{eb}{}^{d}\end{aligned} \quad (5.8.32)$$

therefore neither equation (5.8.31*b*) nor equation (5.8.31*c*) hold. The curvature components induced by the torsion can easily be determined. Let us decompose the curvature into two parts:

$$\mathscr{R}_{abcd} = {}^{(S)}\mathscr{R}_{abcd} + {}^{(A)}\mathscr{R}_{abcd}$$

$${}^{(S)}\mathscr{R}_{abcd} = \frac{1}{2}\mathscr{R}_{abcd} + \frac{1}{2}\mathscr{R}_{cdab} = {}^{(S)}\mathscr{R}_{cdab}$$

$${}^{(A)}\mathscr{R}_{abcd} = \frac{1}{2}\mathscr{R}_{abcd} - \frac{1}{2}\mathscr{R}_{cdab} = -{}^{(A)}\mathscr{R}_{cdab}.$$

Since ${}^{(S)}\mathscr{R}_{abcd}$ is symmetric with respect to transposition of the first and second pairs of indices, one can readily see that

$$
{}^{(S)}\mathscr{R}_{abcd} + {}^{(S)}\mathscr{R}_{bcad} + {}^{(S)}\mathscr{R}_{cabd}
$$

is totally antisymmetric, therefore,

$$
{}^{(S)}\mathscr{R}_{abcd} = \check{\mathscr{R}}_{abcd} - \frac{1}{24}\varepsilon_{abcd}\varepsilon^{efhg}\mathscr{R}_{efhg} \tag{5.8.33}
$$

$\check{\mathscr{R}}_{abcd}$ being a curvature tensor under the torsion-free constraints (5.8.31*a–c*). As for ${}^{(A)}\mathscr{R}_{abcd}$, antisymmetric with respect to transposition of the first and second pairs of indices, it contains two irreducible components: antisymmetric

$$
{}^{(A)}\mathscr{R}^{c}{}_{acb} = \tfrac{1}{2}\mathscr{R}_{ab} - \tfrac{1}{2}\mathscr{R}_{ba} \tag{5.8.34}
$$

where

$$
\mathscr{R}_{ab} \equiv \mathscr{R}^{c}{}_{acb} \tag{5.8.35}
$$

and symmetric traceless

$$
\varepsilon^{acde(A)}\mathscr{R}_{cde}{}^{b} + \varepsilon^{bcde(A)}\mathscr{R}_{cde}{}^{a} = \varepsilon^{acde}\mathscr{R}_{cde}{}^{b} + \varepsilon^{bcde}\mathscr{R}_{cde}{}^{a} + \frac{1}{2}\eta^{ab}\varepsilon^{cdef}\mathscr{R}_{cdef}. \tag{5.8.36}
$$

It should be stressed that the tensors (5.8.34) and (5.8.36) and $\varepsilon^{abcd}\mathscr{R}_{abcd}$ are generated by the torsion, owing the equation (5.8.32).

For further analysis, it is useful to consider the curvature tensor with vector indices converted into spinor indices,

$$
\begin{aligned}
\mathscr{R}_{\alpha\dot\alpha,\beta\dot\beta,\gamma\dot\gamma,\delta\dot\delta} &= (\sigma^a)_{\alpha\dot\alpha}(\sigma^b)_{\beta\dot\beta}(\sigma^c)_{\gamma\dot\gamma}(\sigma^d)_{\delta\dot\delta}\mathscr{R}_{abcd} \\
&= 2\varepsilon_{\dot\gamma\dot\delta}\mathscr{R}_{\alpha\dot\alpha,\beta\dot\beta,\gamma\delta} + 2\varepsilon_{\gamma\delta}\bar{\mathscr{R}}_{\alpha\dot\alpha,\beta\dot\beta,\dot\gamma\dot\delta}
\end{aligned}
$$

where $\mathscr{R}_{\alpha\dot\alpha,\beta\dot\beta,\gamma\delta}$ proves to have the general form

$$
\begin{aligned}
\mathscr{R}_{\alpha\dot\alpha,\beta\dot\beta,\gamma\delta} &= \varepsilon_{\dot\alpha\dot\beta}C_{\alpha\beta\gamma\delta} + \varepsilon_{\alpha\beta}E_{\gamma\delta\dot\alpha\dot\beta} \\
&\quad + \varepsilon_{\dot\alpha\dot\beta}(\varepsilon_{\alpha\gamma}S_{\beta\delta} + \varepsilon_{\beta\gamma}S_{\alpha\delta} + \varepsilon_{\alpha\delta}S_{\beta\gamma} + \varepsilon_{\beta\delta}S_{\alpha\gamma}) \\
&\quad + \varepsilon_{\dot\alpha\dot\beta}(\varepsilon_{\alpha\gamma}\varepsilon_{\beta\delta} + \varepsilon_{\alpha\delta}\varepsilon_{\beta\gamma})F
\end{aligned} \tag{5.8.37}
$$

with $C_{\alpha\beta\gamma\delta}$ and $S_{\alpha\beta}$ being totally symmetric and $E_{\gamma\delta\dot\alpha\dot\beta}$ symmetric in its undotted and, independently, its dotted indices. We would like to recall that in the torsion-free case the scalar F and the symmetric traceless tensor

$$
E_{ab} = \frac{1}{4}(\tilde\sigma_a)^{\dot\alpha\alpha}(\tilde\sigma_b)^{\dot\beta\beta}E_{\alpha\beta\dot\alpha\dot\beta} \tag{5.8.38}
$$

were real and $S_{\alpha\beta}$ was absent (see equations 1.6.65, 66)). Another situation occurs in the general case. One can explicitly check the following relations:

$$F = \frac{1}{12}\left(\mathscr{R} + \frac{\mathrm{i}}{2}\varepsilon^{abcd}\mathscr{R}_{abcd}\right)$$

$$\begin{aligned} E_{ab} = \frac{1}{4}\Big\{ &\mathscr{R}_{ab} + \frac{\mathrm{i}}{2}\varepsilon_{acde}\mathscr{R}^{cde}{}_b + \mathscr{R}_{ba} + \frac{\mathrm{i}}{2}\varepsilon_{bcde}\mathscr{R}^{cde}{}_a \\ &- \frac{1}{2}\eta_{ab}\left(\mathscr{R} - \frac{\mathrm{i}}{2}\varepsilon^{cdef}\mathscr{R}_{cdef}\right)\Big\} \end{aligned} \tag{5.8.39}$$

$$S_{\alpha\beta} = -\frac{1}{4}(\sigma^{ab})_{\alpha\beta}\mathscr{R}_{ab}$$

where $\mathscr{R} = \eta^{ab}\mathscr{R}_{ab}$ is the scalar curvature.

Now, we are ready to complete the investigation of relations (5.8.21) and (5.8.22).

5.8.5. *Space projections of $\bar{\mathscr{D}}^2\bar{R}$, $\bar{\mathscr{D}}_{(\dot\alpha}\mathscr{D}_{(\gamma}G_{\delta)\beta)}$ and $\mathscr{D}_{(\alpha}W_{\beta\gamma\delta)}$*

Using the Bianchi identities (5.5.8), the $M^{\gamma\delta}$-terms in equation (5.8.22) can be rewritten as

$$\begin{aligned} &\frac{1}{8}\varepsilon_{\dot\alpha\dot\beta}(\varepsilon_{\alpha\gamma}\varepsilon_{\beta\delta} + \varepsilon_{\alpha\delta}\varepsilon_{\beta\gamma})\left(\left(\bar{\mathscr{D}}^2 - 8R\right)\bar{R}\,| + \varepsilon_{\dot\alpha\dot\beta}\mathscr{D}_{(\alpha}W_{\beta\gamma\rho)}|\right) \\ &+ \frac{1}{2}\varepsilon_{\alpha\beta}\bar{\mathscr{D}}_{(\dot\alpha}\mathscr{D}_{(\gamma}G_{\delta)\dot\beta)}| + \frac{1}{4}\varepsilon_{\dot\alpha\dot\beta}(\varepsilon_{\alpha\gamma}\mathscr{D}^\rho W_{\rho\beta\delta}| + \varepsilon_{\beta\gamma}\mathscr{D}^\rho W_{\rho\alpha\delta}| \\ &+ \varepsilon_{\alpha\delta}\mathscr{D}^\rho W_{\rho\beta\gamma}| + \varepsilon_{\beta\gamma}\mathscr{D}^\rho W_{\rho\alpha\gamma}|)) \end{aligned}$$

where dotted and undotted indices in $\bar{\mathscr{D}}_{(\dot\alpha}\mathscr{D}_{(\gamma}G_{\delta)\dot\beta)}$ (and everywhere below) are symmetrized independently. Further, we insert the expressions for $\mathscr{R}_{\alpha\dot\alpha,\beta\dot\beta,\gamma\delta}$ (5.8.37) in (5.8.21). Now, setting the $M^{\gamma\delta}$-terms in (5.8.21) and (5.8.22) to be equal leads to

$$\begin{aligned} \bar{\mathscr{D}}^2\bar{R}| = \frac{2}{3}&\left(\mathscr{R} + \frac{\mathrm{i}}{2}\varepsilon^{abcd}\mathscr{R}_{abcd}\right) + \frac{8}{9}\bar{\mathbf{B}}\mathbf{B} - \frac{2}{9}\mathbf{B}(\Psi^a\sigma_a\tilde\sigma_b\Psi^b + \Psi^a\Psi_a) \\ &+ i\bar{\mathscr{D}}_{\dot\alpha}\bar{R}|(\tilde\sigma_b\Psi^b)^{\dot\alpha} + \frac{2\mathrm{i}}{3}\Psi^{\alpha\dot\alpha,\beta}\mathscr{D}_{(\alpha}G_{\beta)\dot\alpha}| \end{aligned} \tag{5.8.40a}$$

$$\begin{aligned} \bar{\mathscr{D}}_{(\dot\alpha}\mathscr{D}^{(\gamma}G^{\delta)}{}_{\dot\beta)}| = 2E^{\gamma\delta}{}_{\dot\alpha\dot\beta} &+ 2\mathrm{i}\Psi^{(\gamma}{}_{(\dot\alpha,}{}^{\delta)}\bar{\mathscr{D}}_{\dot\beta)}\bar{R}| - \mathrm{i}\Psi_{\alpha(\dot\alpha,}{}^{\alpha}\mathscr{D}^{(\gamma}G^{\delta)}{}_{\dot\beta)}| \\ &+ 2\mathrm{i}\bar\Psi_{\alpha(\dot\alpha,\dot\beta)}W^{\alpha\gamma\delta}| + \frac{2}{3}\mathbf{B}(\tilde\sigma^{ab})_{\dot\alpha\dot\beta}\Psi_a{}^\gamma\Psi_b{}^\delta \end{aligned} \tag{5.8.40b}$$

$$\mathscr{D}_{(\alpha}W_{\beta\gamma\delta)}| = C_{\alpha\beta\gamma\delta} - \mathrm{i}\Psi_{(\alpha}{}^{\dot\alpha,}{}_{\beta}\Psi_{\gamma\delta),\dot\alpha} - \mathrm{i}\Psi_{(\alpha\beta,\gamma}\bar\Psi_{\delta)\dot\alpha,}{}^{\dot\alpha}. \tag{5.8.40c}$$

When deriving the last relation, we have used equations (5.8.30*b*, *c*).

Let us discuss the results obtained. From equation (5.8.40*a*) we see that the $\bar{\theta}^2$-component of the supertorsion $\bar{R}$ (and the θ^2-component of R) contains the scalar curvature. Therefore, one can look on R and $\bar{R}$ as supersymmetric extensions of the scalar curvature. Further, it follows from equation (4.8.40*b*) that the $\theta\sigma^b\bar{\theta}$-component of the supertorsion G_a contains the traceless Ricci tensor

$$\frac{1}{2}(\mathcal{R}_{ab} + \mathcal{R}_{ba}) - \frac{1}{4}\eta_{ab}\mathcal{R}$$

in accordance with equation (5.8.39). Therefore, G_a represents the supersymmetric extension of the Ricci tensor. Finally, equation (5.8.40*c*) shows that the leading term in the θ^α-component of the supertorsion $W_{\beta\gamma\delta}$ coincides with the anti-self-dual part of the Weyl tensor (see equation (1.6.66)). On these grounds, $W_{\beta\gamma\delta}$ can be called the 'super Weyl tensor'.

Above we have evaluated only a special set of supertorsion space projections. Beautifully, these results (combined with the commutation relations (5.5.6, 7) and the Bianchi identities (5.5.8)) turn out to be sufficient to determine $\bar{\mathcal{D}}_{\dot{\alpha}}\mathcal{D}_\gamma G_{\delta\beta}|$ and $\mathcal{D}_\alpha W_{\beta\gamma\delta}|$ and all the other space projections, such as $\mathcal{D}^2 G_{\alpha\dot{\alpha}}|$, $\mathcal{D}^2 W_{\alpha\beta\gamma}|$ and so on. For example, we have

$$\begin{aligned}
\mathcal{D}^2 G_{\alpha\dot{\alpha}}| &= \mathcal{D}^\beta\mathcal{D}_\alpha G_{\beta\dot{\alpha}}| - \mathcal{D}_\alpha\mathcal{D}^\beta G_{\beta\dot{\alpha}}| = \{\mathcal{D}^\beta, \mathcal{D}_\alpha\} G_{\beta\dot{\alpha}}| - 2\mathcal{D}_\alpha\mathcal{D}^\beta G_{\beta\dot{\alpha}}| \\
&= 4\bar{R}|M_{\alpha\beta}G^\beta{}_{\dot{\alpha}}| - 2\mathcal{D}_\alpha\bar{\mathcal{D}}_{\dot{\alpha}}\bar{\mathcal{R}}| = 6\bar{R}|G_{\alpha\dot{\alpha}}| + 4\mathrm{i}\mathcal{D}_{\alpha\dot{\alpha}}\bar{R}| \\
&= 8\mathbf{B}\mathbf{A}_{\alpha\dot{\alpha}}| + 4\mathrm{i}\nabla_{\alpha\dot{\alpha}}\bar{R}| + 2\mathrm{i}\Psi_{\alpha\dot{\alpha},\beta}\bar{\mathcal{D}}^\beta\bar{R}| \\
&= 8\mathbf{B}\left(\mathbf{A}_{\alpha\dot{\alpha}} - \frac{1}{12}\Psi_{\alpha\dot{\alpha}}\tilde{\sigma}_c\Psi^c\right) + \frac{4\mathrm{i}}{3}\nabla_{\alpha\dot{\alpha}}\mathbf{B} - \frac{4\mathrm{i}}{3}\Psi_{\alpha\dot{\alpha}}\tilde{\sigma}^{bc}\Psi_{bc}
\end{aligned}$$

where we have used equation (5.8.15). As a result, we have at our disposal the full component description of Einstein supergravity.

5.8.6. *Component fields and local supersymmetry transformation laws*

Our next goal is to give a proper definition of the component fields of tensor superfields and to discuss their transformation laws.

We begin by recalling that in the gravitational superfield gauge (5.4.38) we stay with the restricted set of gauge transformations of the general structure (5.4.49). The stronger gauge fixing (5.8.1) imposes severe constraints on the superfield parameters $\lambda^m(x,\theta)$ and $\lambda^\alpha(x,\theta)$. Namely, the only transformations preserving the Wess–Zumino gauge are: the general coordinate transformations described by

$$\lambda^m(x,\theta) = b^m(x) = \bar{b}^m(x) \qquad \lambda^\alpha = 0 \tag{5.8.41}$$

the local Lorentz transformations,

$$\lambda^m = 0 \qquad \lambda^\alpha(x,\theta) = K^\alpha{}_\beta(x)\theta^\beta \tag{5.8.42}$$

and the local supersymmetry transformations,

$$\begin{aligned} \lambda^m(x,\theta) &= 2\mathrm{i}\theta\sigma^a\bar{\epsilon}e_a{}^m - 2\theta^2\bar{\epsilon}\bar{\Psi}^m \\ \lambda^\alpha(x,\theta) &= \epsilon^\alpha(x) + \frac{\mathrm{i}}{2}\Omega(\epsilon)\theta^\alpha + K^\alpha{}_\beta(\epsilon)\theta^\alpha + \theta^2\text{-terms} \end{aligned} \tag{5.8.43}$$

with $\Omega(\epsilon)$ and $K^\alpha{}_\beta(\epsilon)$ being given in equations (5.1.42) and(5.1.44), respectively.

Consider a tensor superfield V (indices are suppressed). In accordance with relations (5.4.49), its space projection $V|$ transforms as follows

$$\begin{aligned} \delta V| = \frac{1}{2}(\lambda^m| + \bar{\lambda}^m|)\partial_m + \lambda^\alpha|\partial_\alpha V| + \bar{\lambda}_{\dot{\alpha}}|\bar{\partial}^{\dot{\alpha}}V| \\ + \partial_{(\alpha}\lambda_{\beta)}|M^{\alpha\beta}V| + \bar{\partial}_{(\dot{\alpha}}\bar{\lambda}_{\dot{\beta})}|\bar{M}^{\dot{\alpha}\dot{\beta}}V|. \end{aligned}$$

In the case of the general coordinate and local Lorentz transformations, this law is reduced to

$$\delta_{b+k}V| = b^m(x)\partial_m V| + \frac{1}{2}K^{ab}(x)M_{ab}V|. \tag{5.8.44}$$

Therefore, $V|$ is a tensor field. Similarly, since the covariant derivatives $\mathscr{D}_A$ move every tensor superfield to a tensor one, each of the fields $\mathscr{D}_A V|, \mathscr{D}_A\mathscr{D}_B V|, \ldots$, is a tensor.

Under the local supersymmetry transformations (5.8.43), $V|$ changes by

$$\delta V| = \epsilon^\alpha(x)\partial_\alpha V| + \bar{\epsilon}_{\dot{\alpha}}(x)\bar{\partial}^{\dot{\alpha}}V| + (K_{\alpha\beta}(\epsilon)M_{\alpha\beta} + \bar{K}_{\dot{\alpha}\dot{\beta}}(\epsilon)\bar{M}_{\dot{\alpha}\dot{\beta}})V|.$$

Now, comparing the explicit form of $K_{\alpha\beta}(\epsilon)$ and $\bar{K}_{\dot{\alpha}\dot{\beta}}(\epsilon)$ with the projections of the spinor superconnections (5.8.12) and using (5.8.13), one obtains

$$\delta_\epsilon V| = \epsilon^\alpha(x)\mathscr{D}_\alpha V| + \bar{\epsilon}_{\dot{\alpha}}(x)\bar{\mathscr{D}}^{\dot{\alpha}}V|. \tag{5.8.45}$$

This marvellous result means that the set of tensor fields $\{V|, \mathscr{D}_A V|, \mathscr{D}_A\mathscr{D}_B V|, \ldots\}$ is closed with respect to the supersymmetry transformations, and the relevant representation is linear, owing to the $\mathscr{D}$-transformation law (5.4.49). It can be readily seen that among the fields $\{V|, \mathscr{D}_A V|, \mathscr{D}_A\mathscr{D}_B V|, \ldots\}$ only a finite subset is functionally independent. In particular, if V is an unconstrained superfield, such a subset can be chosen as follows

$$\begin{aligned} &V| \quad \mathscr{D}_\alpha V| \quad \bar{\mathscr{D}}_{\dot{\alpha}}V| \quad -\frac{1}{4}\mathscr{D}^2 V| \quad -\frac{1}{4}\bar{\mathscr{D}}^2 V| \quad \frac{1}{2}[\mathscr{D}_\alpha, \bar{\mathscr{D}}_{\dot{\alpha}}]V| \\ &-\frac{1}{4}\mathscr{D}_\alpha\bar{\mathscr{D}}^2 V| \quad -\frac{1}{4}\bar{\mathscr{D}}_{\dot{\alpha}}\mathscr{D}^2 V| \quad \frac{1}{32}\{\mathscr{D}^2, \bar{\mathscr{D}}^2\}V|. \end{aligned} \tag{5.8.46}$$

Obviously, these fields are in one-to-one correspondence with those arising in the expansion of V in a power series in θ and $\bar{\theta}$. Therefore, we can identify tensor fields (5.8.46) with the component fields of V. In the case of a covariantly chiral tensor superfield χ, $\bar{\mathscr{D}}_{\dot{\alpha}}\chi = 0$, its component fields can be taken as

$$\chi| \qquad \mathscr{D}_\alpha\chi| \qquad -\frac{1}{4}\mathscr{D}^2\chi|.$$

As an example, let us find how the local supersymmetry transformations act on the component fields of a covariantly chiral scalar superfield χ. Defining the component fields by

$$C(x) = \chi| \qquad \lambda_\alpha(x) = \mathscr{D}_\alpha\chi| \qquad F(x) = -\frac{1}{4}\mathscr{D}^2\chi|$$

we obtain

$$\delta_\epsilon C = \epsilon^\alpha\mathscr{D}_\alpha\chi| = \epsilon\lambda$$

$$\delta_\epsilon\lambda_\alpha = (\epsilon^\beta\mathscr{D}_\beta + \bar{\epsilon}_{\dot{\beta}}\bar{\mathscr{D}}^{\dot{\beta}})\mathscr{D}_\alpha\chi| = -\frac{1}{2}\epsilon_\alpha\mathscr{D}^2\chi| + 2i\bar{\epsilon}^{\dot{\alpha}}\mathscr{D}_{\alpha\dot{\alpha}}\chi|$$

$$= 2\epsilon_\alpha F + 2i\bar{\epsilon}^{\dot{\alpha}}\nabla_{\alpha\dot{\alpha}}\chi| + i\bar{\epsilon}^{\dot{\alpha}}\Psi_{\alpha\dot{\alpha},}{}^\beta\mathscr{D}_\beta\chi|$$

$$= 2\epsilon_\alpha F + 2i\bar{\epsilon}^{\dot{\alpha}}\nabla_{\alpha\dot{\alpha}}C + i\bar{\epsilon}^{\dot{\alpha}}\Psi_{\alpha\dot{\alpha},}{}^\beta\lambda_\beta$$

$$\delta_\epsilon F = -\frac{1}{4}(\epsilon^\alpha\mathscr{D}_\alpha + \bar{\epsilon}_{\dot{\alpha}}\bar{\mathscr{D}}^{\dot{\alpha}})\mathscr{D}^2\chi| = -\epsilon^\alpha\bar{R}|\mathscr{D}_\alpha\chi| - \frac{1}{4}\bar{\epsilon}^{\dot{\alpha}}[\mathscr{D}^2, \bar{\mathscr{D}}_{\dot{\alpha}}]\chi|$$

$$= -\frac{1}{3}\mathbf{B}\epsilon\lambda + \bar{\epsilon}^{\dot{\alpha}}G_{\alpha\dot{\alpha}}|\mathscr{D}^\alpha\chi| + i\bar{\epsilon}^{\dot{\alpha}}\mathscr{D}_{\alpha\dot{\alpha}}\mathscr{D}^\alpha\chi|$$

$$= -\frac{1}{3}\mathbf{B}\epsilon\lambda - \frac{4}{3}\mathbf{A}^a\lambda\sigma_a\bar{\epsilon} - i(\nabla^a\lambda)\sigma_a\bar{\epsilon} - i\Psi^a\sigma_a\bar{\epsilon}F$$

where equations (5.5.9*b*, *d*) have been used.

We hope the reader has already been convinced that finding local supersymmetry transformation laws is a simple procedure.

Remark. The transformation laws (5.1.48)can easily be reproduced utilizing the fact that the fields $\mathbf{B}$ and $\mathbf{A}_a$ represent the space projections of tensor superfields $\bar{R}$ and G_a, respectively (and keeping in mind the difference between the definitions of gravitino field strengths (5.1.49) and (5.8.27)).

5.8.7. *From superfield action to component action*

Before we conclude, one more important question must be discussed — how to represent an action superfunctional of the general structure

$$S = \int d^8z \, E^{-1} \mathscr{L}$$

in terms of the relevant component fields.

Owing to the well-known identity

$$\int d^8z E^{-1} \mathscr{L} = -\frac{1}{4} \int d^8z \frac{E^{-1}}{R} (\bar{\mathscr{D}} - 4R) \mathscr{L}$$

it is sufficient to solve the problem for chiral action superfunctionals of the type

$$S_c = \int d^8z \frac{E^{-1}}{R} \mathscr{L}_c \qquad \bar{\mathscr{D}}_{\dot{\alpha}} \mathscr{L}_c = 0. \tag{5.8.47}$$

Recall that one can equivalently write S_c as

$$S_c = \int d^6z \, \phi^3 \hat{\mathscr{L}}_c \tag{5.8.48}$$

where $\mathscr{L}_c$ and its chiral transform $\hat{\mathscr{L}}_c$ are connected in the Wess–Zumino gauge (5.8.1) in the manner

$$\mathscr{L}_c(z) = \left(1 + i\mathscr{H}^m \partial_m - \frac{1}{2} \mathscr{H}^m \mathscr{H}^n \partial_m \partial_n\right) \hat{\mathscr{L}}_c(x, \theta).$$

Next, performing the integral over θ reduces S_c to

$$S_c = \frac{1}{4} \int d^4x \partial^\alpha \partial_\alpha (\phi^3 \hat{\mathscr{L}}_c).$$

Now, since $\hat{E}^2| = -\partial^\alpha \partial_\alpha$ and $\mathscr{H}^m \sim \bar{\theta}$, we can rewrite the last expression in the form

$$\begin{aligned} S_c &= \int d^4x \left\{ -\frac{1}{4} \hat{E}^\alpha \hat{E}_\alpha (\varphi^3 \mathscr{L}_c) \right\} | \\ &= \int d^4x \left\{ -\frac{1}{4} \varphi^3 | \hat{E}^2 \mathscr{L}_c | - \frac{1}{4} \mathscr{L}_c | \hat{E}^2 \varphi^3 | - \frac{1}{2} \hat{E}^\alpha \varphi^3 | \hat{E}_\alpha \mathscr{L}_c | \right\}. \end{aligned} \tag{5.8.49}$$

On the other hand, we have the simple relations

$$\mathscr{D}_\alpha \mathscr{L}_c | = F | \hat{E}_\alpha \mathscr{L}_c | = \hat{E}_\alpha \mathscr{L}_c |$$

$$\begin{aligned} -\frac{1}{4} \mathscr{D}^2 \mathscr{L}_c | &= -\frac{1}{4} \mathscr{D}^\alpha (F \hat{E}_\alpha \mathscr{L}_c) | = -\frac{1}{4} F | \hat{E}^\alpha (F \hat{E}_\alpha \mathscr{L}_c) | - \frac{1}{4} \Omega^\beta{}_\beta{}^\alpha | F | \hat{E}_\alpha \mathscr{L}_c | \\ &= -\frac{1}{4} (\hat{E}^\alpha F | + \Omega^\beta{}_\beta{}^\alpha |) \hat{E}_\alpha \mathscr{L}_c) | - \frac{1}{4} \hat{E}^2 \mathscr{L}_c | \end{aligned}$$

since $F| = 1$. Using equations (5.8.11, 12), one obtains the relation

$$-\frac{1}{4}\hat{\mathscr{E}}^2\mathscr{L}_c| = -\frac{1}{4}\mathscr{D}^2\mathscr{L}_c| + \frac{i}{2}(\Psi^a\tilde{\sigma}_a)^\alpha\mathscr{D}_\alpha\mathscr{L}_c|$$

which must be inserted in expression (5.8.49). Finally, after using equations (5.8.5b), S_c takes the form

$$S_c \equiv \int d^8z\frac{E^{-1}}{R}\mathscr{L}_c = \int d^4x\, e^{-1}\left\{-\frac{1}{4}\bar{\mathscr{D}}^2\mathscr{L}_c| - \frac{i}{2}(\Psi^a\tilde{\sigma}_a)^\alpha\mathscr{D}_\alpha\mathscr{L}_c| + \left(\mathbf{B} - \frac{1}{2}\Psi^a\tilde{\sigma}_a\sigma_b\Psi^b - \frac{1}{2}\Psi^a\Psi_a\right)\mathscr{L}_c|\right\}. \tag{5.8.50}$$

This is the working formula for calculating the action functional.

6 Dynamics in Supergravity

To invest existence with a stately air
Needs but to remember
That the acorn there
Is the egg of forests
For the upper air!

Emily Dickinson:
To Venerate the Simple Days

6.1. Pure supergravity dynamics

We have already succeeded in developing a deep understanding of curved superspace geometry. Naturally, the formalism developed inevitably leads those familiar with Einstein general relativity to new hypothetical worlds—supersymmetric universes. By a supersymmetric universe we shall understand a curved superspace, empty or inhabited by some matter tensor superfields denoted χ, with a joint evolution law (action principle). It is reasonable to take an action superfunctional determining physically admissible supergeometries and dynamical matter superfield histories in the form

$$S = S_{SG}[\mathscr{D}_A] + S_M[\chi; \mathscr{D}_A]$$

where S_{SG} is a pure supergravity action, depending on the Einstein supergravity prepotentials only, and S_M is a matter action. Both terms in S are required to be invariant under the general coordinate transformation supergroup and the superlocal Lorentz group. In this section we consider the simplest candidates for the role of pure supergravity action.

DOI: 10.1201/9780367802530-6

6.1.1. Einstein supergravity action superfunctional

We begin by finding a supergravity action such that the corresponding dynamical equations reduce to ordinary Einstein equations after switching off all the fields, except the vierbein, of the Einstein supergravity multiplet. In other words, the dynamical equation should admit the following particular solution

$$\mathscr{R}_{ab}(e)=0 \qquad \Psi^{m\alpha}=\mathbf{A}^m=\mathbf{B}=0$$

$\mathscr{R}_{ab}(e)$ being the Ricci tensor constructed from the torsion-free covariant derivatives. This implies that the purely vierbein-dependent part of supergravity action must coincide with the ordinary gravity action

$$S_G=\frac{1}{2\kappa^2}\int \mathrm{d}^4x\, e^{-1}\mathscr{R}(e) \tag{6.1.1}$$

where κ is the graviational coupling constant.

It seems natural to look for a supergravity action in the class of superfunctionals of the general structure

$$\frac{1}{\kappa^2}\int \mathrm{d}^8x\,\frac{E^{-1}}{R}\,\mathscr{L}_c(T_{BC}{}^D,\mathscr{D}_AT_{BC}{}^D,\ldots)+\text{c.c.} \tag{6.1.2}$$

$$\bar{\mathscr{D}}_{\dot\alpha}\mathscr{L}_c=0$$

where $\mathscr{L}_c$ is a covariantly chiral scalar depending polynomially on the supertorsion and its covariant derivatives up to a finite order.

Remark. Recall that owing to the identity

$$\int \mathrm{d}^8z\,E^{-1}\mathscr{L}=-\frac{1}{4}\int \mathrm{d}^8z\,\frac{E^{-1}}{R}(\bar{\mathscr{D}}^2-4R)\mathscr{L}$$

each integral over $\mathbb{R}^{4|4}$ can be reduced to a chiral-like integral.

Let us show, guided by considerations of dimension, that the supergravity action can be restored uniquely modulo a constant. The dimensions of quantities appearing in (6.1.2) are

$$[1/\kappa^2]=2 \qquad [\mathrm{d}^8z]=-2 \qquad [E]=0 \qquad [R]=1.$$

Since the action is dimensionless, we must have

$$[\mathscr{L}_c]=1.$$

Next, we suppose that $\mathscr{L}_c$ has no explicit dependence on the gravitational coupling constant. Then, by virtue of the relations

$$[G_a]=1 \qquad [W_{\alpha\beta\gamma}]=3/2 \qquad [\mathscr{D}_AT_{BC}{}^D]\geqslant 3/2$$

which follow from the commutation relations (5.5.6), $\mathscr{L}_c$ may be at most a linear combination of R and $\bar{R}$. Finally, the requirement of chirality means

$\mathscr{L}_c \sim R$. We therefore obtain

$$S_{SG} = -\frac{3}{\kappa^2}\int d^8z\, E^{-1}. \tag{6.1.3}$$

The normalization constant (-3) is needed to produce the correct vierbein contribution (6.1.1) at the component level (see below).

As may be seen, the supergravity action has a simple geometrical interpretation. Namely, S_{SG} is proportional to the supervolume of the curved superspace. Using relations (5.4.19) and (5.5.4), the action can be rewritten in terms of the prepotentials as follows:

$$S_{SG} = -\frac{3}{\kappa^2}\int d^8z\, \bar{\varphi}\varphi\{\hat{E}^{-1}(1\cdot e^{\overleftarrow{W}})(1\cdot e^{\overleftarrow{\bar{W}}})\}^{1/3}. \tag{6.1.4}$$

In the chiral representation, the action reads

$$S_{SG} = -\frac{3}{\kappa^3}\int d^6z\, \hat{\varphi}^3\hat{R} \qquad R = e^{\bar{W}}\hat{R} \tag{6.1.5}$$

and this form of the supergravity action is just like the gravity action (6.1.1).

6.1.2. Supergravity dynamical equations

To vary the supergravity action, we follow the prescription of background-quantum splitting described in Section 5.6 (see also subsection 5.7.1). Making use of equation (5.6.57) gives

$$\begin{aligned}\delta S_{SG} &= S_{SG}[\tilde{\nabla}_A] - S_{SG}[\mathscr{D}_A]\\ &= -\frac{3}{\kappa^2}\int d^8z\, E^{-1}\left\{\sigma + \bar{\sigma} - \frac{2}{3}G^a\mathbf{H}_a + \frac{1}{2}[\mathscr{D}^\alpha, \bar{\mathscr{D}}^{\dot\alpha}]\mathbf{H}_{\alpha\dot\alpha} - i\mathscr{D}^a\mathbf{H}_a\right\}\\ &= -\frac{3}{\kappa^2}\int d^8z\, E^{-1}\left\{\sigma + \bar{\sigma} - \frac{2}{3}G^a\mathbf{H}_a\right\}\end{aligned}$$

where we have omitted total derivative terms. Recalling the variational rules (5.7.8), with $\varphi = 1 + \sigma$ being the quantum chiral compensator, one obtains

$$\frac{\delta S_{SG}}{\delta \mathbf{H}^a} = \frac{2}{\kappa^2}G_a \qquad \frac{\delta S_{SG}}{\delta\varphi} = -\frac{3}{\kappa^2}R. \tag{6.1.7}$$

Therefore, the supergravity dynamical equations look like

$$G_a = 0 \qquad R = 0 \tag{6.1.8}$$

We see that on-shell, only the super Weyl tensor $W_{\alpha\beta\gamma}$ remains non-vanishing, and it satisfies the equation

$$\mathscr{D}^\alpha W_{\alpha\beta\gamma} = 0 \tag{6.1.9}$$

in accordance with the Bianchi indentities (5.5.8).

6.1.3. Einstein supergravity action functional

We would like to describe how the action (6.1.3) looks in terms of components. For this purpose we apply the component technique developed in Section 5.8.

To convert S_{SG} into components, we make use of the reduction formula (5.8.50):

$$S_{SG} = -\frac{3}{\kappa^2}\int d^8z \frac{E^{-1}}{R} R = -\frac{3}{\kappa^2}\int d^4x\, e^{-1}\left\{-\frac{1}{4}\mathcal{D}^2 R| - \frac{i}{2}(\Psi^a\tilde{\sigma}_a)^\alpha \mathcal{D}_\alpha R| + \left(\mathbf{B} - \frac{1}{2}\Psi^a\tilde{\sigma}_a\sigma_b\Psi^b - \frac{1}{2}\Psi^a\Psi_a\right)R|\right\}.$$

Next, recalling that $R| = \frac{1}{3}\bar{\mathbf{B}}$ and making use of equation (5.8.40a) one obtains

$$S_{SG} = -\frac{3}{\kappa^2}\int d^4x\, e^{-1}\left\{-\frac{1}{6}\left(\mathcal{R} - \frac{i}{2}\varepsilon^{abcd}\mathcal{R}_{abcd}\right) + \frac{1}{9}\bar{\mathbf{B}}\mathbf{B} - \frac{1}{9}\bar{\mathbf{B}}(\Psi^a\tilde{\sigma}_a\sigma_b\Psi^b + \frac{1}{2}\Psi^a\Psi_a) - \frac{i}{4}(\Psi^a\tilde{\sigma}_a)^\alpha\mathcal{D}_\alpha R| - \frac{i}{6}\Psi^{\alpha\dot{\alpha},\beta}\bar{\mathcal{D}}_{(\dot{\alpha}}G_{\alpha\beta)}|\right\}.$$

The last two terms can be written, with the help of equations (5.8.30a,b), in the form

$$-\frac{1}{6}\varepsilon^{abcd}\Psi_a\tilde{\sigma}_b\Psi_{cd} + \frac{1}{9}\bar{\mathbf{B}}\left(\Psi^a\tilde{\sigma}_a\sigma_b\Psi^b + \frac{1}{2}\Psi^a\Psi_a\right)$$

with $\Psi_{cd\alpha}$ being the gravitino field strength (5.8.27). Therefore, the action functional reads

$$S_{SG} = \frac{1}{\kappa^2}\int d^4x\, e^{-1}\left\{\frac{1}{2}\mathcal{R} - \frac{i}{4}\varepsilon^{abcd}\mathcal{R}_{abcd} + \frac{1}{4}\varepsilon^{abcd}\Psi_a\tilde{\sigma}_b\Psi_{cd} - \frac{1}{3}\bar{\mathbf{B}}\mathbf{B}\right\}. \quad (6.1.10)$$

Note that the second term in S_{SG} is purely imaginary. In addition, the third term also has an imaginary part. These constributions must form a total derivative, since the original action (6.1.3) is real by construction. Hence, we can rewrite S_{SG} in the explicitly real form:

$$S_{SG} = \frac{1}{\kappa^2}\int d^4x\, e^{-1}\left\{\frac{1}{2}\mathcal{R} - \frac{1}{3}\bar{\mathbf{B}}\mathbf{B} + \frac{1}{4}\varepsilon^{abcd}(\Psi_a\tilde{\sigma}_b\Psi_{cd} - \Psi_a\sigma_b\bar{\Psi}_{cd})\right\}. \quad (6.1.11)$$

The action has hidden dependence on the vector field $\mathbf{A}^d$ via the connection ω_{abc} (5.8.24) and the gravitino field strengths (see expressions (5.8.29)). Extracting all $\mathbf{A}$-dependent terms, one obtains

$$S_{SG} = \frac{1}{\kappa^2}\int d^4x\, e^{-1}\left\{\frac{1}{2}\mathcal{R}(e, \Psi) - \frac{1}{3}\bar{\mathbf{B}}\mathbf{B} + \frac{4}{3}\mathbf{A}^d\mathbf{A}_d + \frac{1}{4}\varepsilon^{abcd}(\Psi_a\tilde{\sigma}_b\tilde{\Psi}_{cd} - \Psi_a\sigma_b\bar{\tilde{\Psi}}_{cd})\right\}. \quad (6.1.12)$$

Here $\mathscr{R}(e, \Psi) = \tilde{\mathscr{R}}^{ab}{}_{ab}$ is the scalar curvature generated by the covariant derivatives $\tilde{\nabla}_a$ (5.8.26), and

$$\tilde{\Psi}_{ab\gamma} = \tilde{\nabla}_a \Psi_{b\gamma} - \tilde{\nabla}_b \Psi_{a\gamma} - \tilde{\mathscr{T}}_{ab}{}^c \Psi_{c\gamma}. \tag{6.1.13}$$

From equation (6.1.12) we see the part of S_{SG} determined by the vierbein simply coincides with the gravitational action (6.1.1). It is seen also that S_{SG} leads to trivial dynamics for the fields $\mathbf{B}$ and $\mathbf{A}_d$, which vanish on-shell,

$$\mathbf{B} = \mathbf{A}_d = 0. \tag{6.1.14}$$

On these grounds, $\mathbf{B}$ and $\mathbf{A}_d$ are said to be the 'supergravity auxiliary fields'. As for the last term in (6.1.12), it turns out to be a curved space nonlinear generalization of the Rarita–Schwinger action of a massless spin-$\frac{3}{2}$ particle (1.8.42). Therefore, we can consider the supergravity theory as describing interacting massless spin-2 and spin-$\frac{3}{2}$ particles. It should be noted that there exists a wide class of models of interacting spin-2 and spin-$\frac{3}{2}$ fields, with Lagrangians of the form

$$\frac{1}{2}\mathscr{R}(e) + \varepsilon^{abcd}\bar{\Psi}_a \tilde{\sigma}_b \nabla_c(e) \Psi_d + O(\Psi^4)$$

where $\nabla_a(e)$ are the torsion-free covariant derivatives. But it is the supergravity action which describes consistent interaction in the sense that S_{SG} possesses invariance under local supersymmetry transformations $\delta\Psi_{m\alpha} = \partial_m \varepsilon_\alpha + \ldots$, which generalize the linearized gauge invariance. Let us stress also that a naive curved-space generalization of the Rarita–Schwinger action

$$S = \int d^4x\, e^{-1} \varepsilon^{abcd} \bar{\Psi}_a \tilde{\sigma}_b \nabla_c(e) \Psi_d$$

proves to be contradictory, since the dynamical equations

$$\varepsilon^{abcd} \tilde{\sigma}_b \nabla_c(e) \Psi_d = 0$$

are consistent only if

$$0 = \varepsilon^{abcd} \tilde{\sigma}_a \nabla_b(e) \nabla_c(e) \Psi_d = \frac{1}{2} \varepsilon^{abcd} \tilde{\sigma}_a [\nabla_b(e), \nabla_c(e)] \Psi_d = \frac{1}{4} \varepsilon^{abcd} \mathscr{R}_{ab}{}^{kl} \tilde{\sigma}_c \sigma_{kl} \Psi_d$$

which implies a strong constraint on the curvature of curved space. It is supergravity which gives us the only satisfactory way to describe massless spin-$\frac{3}{2}$ particles in curved space.

In conclusion, we would like to comment about the component form of the supergravity equations of motion (6.1.8). First, setting $\bar{R}| = G_a| = 0$ gives the auxiliary field equations (6.1.14). These mean, in particular, that the connections ω_{abc} and $\omega_{abc}(e, \Psi)$, defined by equation (5.8.24), coincide on-shell. Then the requirements $\mathscr{D}_\alpha R| = \bar{\mathscr{D}}_{(\dot{\alpha}} G^\beta{}_{\beta)}| = 0$ are equivalent, due to equations (5.8.30a,b), and to the gravitino field equations which can be written in several equivalent forms:

$$\Psi_{\alpha\beta,}{}^{\beta}=0 \qquad \Psi_{\dot{\alpha}\beta,\gamma}=0 \tag{6.1.15a}$$

$$\varepsilon^{abcd}\tilde{\sigma}_b\Psi_{cd}=0 \tag{6.2.15b}$$

$$\tilde{\sigma}_a\Psi_{bc}+\frac{1}{2}\tilde{\sigma}_c\Psi_{ab}=\tilde{\sigma}_b\Psi_{ac}+\frac{1}{2}\tilde{\sigma}_c\Psi_{ba} \tag{6.1.15c}$$

$$\frac{1}{2}\varepsilon^{abcd}\Psi_{cd}=-\mathrm{i}\Psi^{ab} \qquad \tilde{\sigma}^b\Psi_{ab}=0. \tag{6.1.15d}$$

Finally, setting the left-hand sides of the relations (5.8.40*a*) and (5.8.40*b*) and their conjugate to vanish, one arrives at

$$\mathscr{R}_{ab}+\mathscr{R}_{ba}=\mathrm{i}\Psi^c\left(\tilde{\sigma}_a\Psi_{bc}+\frac{1}{2}\tilde{\sigma}_c\Psi_{ab}\right)+\mathrm{i}\Psi^c\left(\sigma_a\bar{\Psi}_{bc}+\frac{1}{2}\sigma_c\bar{\Psi}_{ab}\right) \tag{6.1.16}$$

where we have used the second relation (5.8.39). These equations constitute the supersymmetric extension of ordinary Einstein equations.

6.1.4. Supergravity with a cosmological term

It has been pointed out that S_{SG} is actually given by the invariant volume of curved superspace $\mathbb{R}^{4|4}$. Besides the supervolume $\int \mathrm{d}^8z\, E^{-1}$, we have at our disposal two other invariant 'volumes'

$$\int \mathrm{d}^6z\, \varphi^3=\int \mathrm{d}^8z\, E^{-1}R^{-1}$$

and

$$\int \mathrm{d}^6\bar{z}\, \bar{\varphi}^3=\int \mathrm{d}^8z\, E^{-1}\bar{R}^{-1}$$

corresponding to the superspaces on which chiral and antichiral superfields respectively, reside. Hence, one can consider an action superfunctional which combines all the supervolumes:

$$\begin{aligned} S^{(\mu)}_{SG}&=-\frac{3}{\kappa^2}\int \mathrm{d}^8z\, E^{-1}+\frac{\mu}{\kappa^2}\int \mathrm{d}^6z\, \varphi^3+\frac{\bar{\mu}}{\kappa^2}\int \mathrm{d}^6\bar{z}\, \bar{\varphi}^3\\ &=\frac{1}{\kappa^2}\int \mathrm{d}^8z\, E^{-1}\left\{-3+\frac{\mu}{R}+\frac{\bar{\mu}}{\bar{R}}\right\} \end{aligned} \tag{6.1.17}$$

with μ being a complex constant. What physics does this action lead to?

In accordance with the reduction formula (5.8.50) at the component level we have

$$\int \mathrm{d}^8z\,\frac{E^{-1}}{R}=\int \mathrm{d}^4x\, e^{-1}\left(\mathbf{B}-\frac{1}{2}\Psi^a\tilde{\sigma}_a\sigma_b\Psi^b-\frac{1}{2}\Psi^a\Psi_a\right)$$

therefore the total action functional reads

$$S_{SG}^{(\mu)} = \frac{1}{\kappa^2} \int d^4x \left\{ \frac{1}{2} \mathscr{R}(e, \Psi) - \frac{1}{3} \bar{\mathbf{B}}\mathbf{B} + \mu \mathbf{B} + \overline{\mu \mathbf{B}} + \frac{4}{3} \mathbf{A}^d \mathbf{A}_d \right.$$

$$\left. + \frac{1}{2} \left[\Psi_a \left(\frac{1}{2} \varepsilon^{abcd} \tilde{\sigma}_b \Psi_{cd} - \mu \tilde{\sigma}^a \sigma^b \bar{\Psi}_b - \mu \bar{\Psi}^a \right) + \text{c.c.} \right] \right\}. \tag{6.1.18}$$

As is seen, the fields $\mathbf{B}$, $\bar{\mathbf{B}}$ and $\mathbf{A}_d$ do not propagate, since their dynamical equations are

$$\bar{\mathbf{B}} = 3\mu \qquad \mathbf{A}_d = 0. \tag{6.1.19}$$

However, in contrast to S_{SG}, $S_{SG}^{(\mu)}$ leads to non-zero constant values for the scalar fields. Eliminating $\mathbf{B}$ and $\bar{\mathbf{B}}$ with the help of their equations of motion produces the following contribution to the action

$$3 \frac{\mu \bar{\mu}}{\kappa^2} \int d^4x \, e^{-1} \tag{6.1.20}$$

which can easily be recognized as an ordinary cosmological term. As a result, we can interpret the superfunctional

$$S_{S.cosm} = \frac{\mu}{\kappa^2} \int d^6z \, \hat{\varphi}^3 + \text{c.c.} \tag{6.1.21}$$

as the 'supersymmetric cosmological term'. The action (6.1.17) describes supergravity with a cosmological term.

The dynamical equations of supergravity with a cosmological term can be readily found with the help of equations (6.1.7) and (5.6.63):

$$G_a = 0 \qquad R = \mu. \tag{6.1.22}$$

Clearly, their space projections coincide with the auxiliary field equations (6.1.19). Further, setting $\mathscr{D}_\alpha R|$, $\bar{\mathscr{D}}_{(\dot{\alpha}} G^\beta{}_{\dot{\beta})}|$ and $\bar{\mathscr{D}}^2 \bar{R}|$ to vanish, one obtains the gravitino field equations

$$\frac{1}{2} \varepsilon^{abcd} \tilde{\sigma}_b \Psi_{cd} - \mu \tilde{\sigma}^a \sigma_b \bar{\Psi}^b - \mu \bar{\Psi}^a = 0 \tag{6.1.23}$$

and the expression for the scalar curvature

$$\mathscr{R} = -12|\mu|^2 + \frac{1}{2} \{ \bar{\mu}(\Psi^a \sigma_a \tilde{\sigma}_b \Psi^b + \Psi^a \Psi_a) + \text{c.c.} \}. \tag{6.1.24}$$

Equations (6.1.23) show that the gravitino becomes effectively massive in the presence of the cosmological term. As for the scalar curvature, it acquires a constant negative value modulo gravitino excitations.

6.1.5. Conformal supergravity

We have seen that the full covariance group of the commutator algebra (5.5.6, 7) also includes, on a level with the superspace general coordinate and superlocal Lorentz transformations, the super Weyl transformations (5.5.13). Clearly, the Einstein supergravity action is not super Weyl invariant. However, it is not difficult to find a supergravity action possessing such an invariance. In accordance with the results of subsection 5.6.5, the superfunctional

$$S_{SC}=\frac{1}{\gamma}\int d^8z\,\frac{E^{-1}}{R}\,W^{\alpha\beta\gamma}W_{\alpha\beta\gamma} \qquad (6.1.25)$$

does not depend on the chiral compensator φ, and, hence it turns out to be super Weyl invariant. Note, S_{SC} is real modulo total derivative terms (see subsection 5.6.5). In components, the purely vierbein-dependent part of S_{SC} coincides, due to equation (5.8.40*c*), with the integral from the squared Weyl tensor. Hence, the superfunctional (6.1.25) represents the supersymmetric extension of the conformal gravity action (1.6.53). On these grounds and since the super Weyl invariance is inherent in conformal supergravity (see subsection 5.1.6), the S_{SC} can be identified with the action superfunctional of conformal supergravity.

Arbitrary variation of S_{SC} can be represented in the form (5.6.64). Therefore, conformal supergravity is characterized by the dynamical equations

$$\begin{aligned} &W_\alpha{}^{\beta\gamma}\mathscr{D}_\beta G_{\gamma\dot\alpha}-\bar W_{\dot\alpha}{}^{\dot\beta\dot\gamma}\bar{\mathscr{D}}_{\dot\beta}G_{\alpha\dot\gamma}+(\bar{\mathscr{D}}_{\dot\alpha}G^b)\mathscr{D}_\alpha G_b\\ &+\left\{\mathscr{D}^b\mathscr{D}_b-\frac{1}{4}(\mathscr{D}^\beta R)\mathscr{D}_\beta-\frac{1}{4}(\bar{\mathscr{D}}_{\dot\beta}\bar R)\bar{\mathscr{D}}^{\dot\beta}+3R\bar R-\frac{1}{8}(\bar{\mathscr{D}}^2\bar R)-\frac{1}{8}(\mathscr{D}^2R)\right\}G_{\alpha\dot\alpha}\\ &+\frac{i}{4}\mathscr{D}_{\alpha\dot\alpha}(\mathscr{D}^2R-\bar{\mathscr{D}}^2\bar R)=0. \end{aligned} \qquad (6.1.26)$$

It follows from these equations that the component vector field $\mathbf{A}_a$, being auxiliary in Einstein supergravity, becomes dynamical in conformal supergravity.

Remark. Owing to the fact that the integrand in (5.6.61) forms a total derivative, the conformal supergravity action can be rewritten in the manner

$$S_{SC}=-\frac{1}{\gamma}\int d^8z\,E^{-1}(G^aG_a+2R\bar R). \qquad (6.1.27)$$

6.1.6. Renormalizable supergravity models

As is well known, the sufficient condition for a quantum field theory to be renormalizable is the appearance of dimensionless coupling constants in the classical action. Since the gravitational coupling constant κ has non-zero dimension, Einstein gravity turns out to be non-renormalizable, and this fact has proved to be one of the most serious obstacles on the way to constructing quantum gravity. Renormalizable gravity models exist, but they are described by higher-derivative actions of the general structure

$$S=\int d^4x\, e^{-1}\left\{\frac{1}{2\kappa^2}\mathscr{R}(e)+\frac{\lambda}{\kappa^2}+\xi\mathscr{R}^{ab}(e)\mathscr{R}_{ab}(e)+\zeta\mathscr{R}^2(e)\right\} \tag{6.1.28}$$

with ξ and ζ being dimensionless constants. These models are known as R^2-gravity theories. It is worth pointing out that we could add to the expression in braces in (6.1.28) one more term—the squared Weyl tensor. However, there is no necessity to do so because of the topological nature of the functional (1.6.44).

In superspace, the most general supergravity action with dimensionless coupling constants reads

$$S_{SR^2}=\int d^8z\, E^{-1}\{\xi G^aG_a+\zeta R\bar{R}+qR^2+\overline{qR^2}\} \tag{6.1.29}$$

and renormalizable supergravity models are described by classical actions of the form

$$S=S_{SG}+S_{S.cosm}+S_{SR^2}. \tag{6.1.30}$$

Using the relations (5.8.40*a*,*b*), one can readily see that at the component level the first and second terms in (6.1.29) contain the squared Ricci tensor and the squared scalar curvature, respectively. So, they are supersymmetric analogues of the structures presented in the R^2-gravity action (6.1.28). The last two terms in (6.1.29), however, prove to have no direct space-time analogues in the sense that they do not produce purely vierbein-dependent contributions in components.

In conclusion, we note the action (6.1.30) leads to non-trivial dynamics for the component fields $\mathbf{B}$ and $\mathbf{A}_a$.

6.1.7. *Pathological supergravity model*

The naive superspace generalization of the gravity action (6.1.1) would be of the form

$$S=\int d^8z\, E^{-1}(\eta R+\overline{\eta R}) \tag{6.1.31}$$

where η is a complex constant. But it turns out to be pathological from a physical point of view. To argue, this point we take for simplicity η to be real, $\eta=\bar{\eta}\equiv-9\kappa^{-1}$. In components, one obtains

$$S=\frac{1}{\kappa}\int d^4x\, e^{-1}(\mathbf{B}+\bar{\mathbf{B}})\left(\mathscr{R}+\frac{1}{3}\bar{\mathbf{B}}\mathbf{B}\right)+\dots$$

where dots mean gravitino-dependent terms. Setting the gravitino to vanish, the **B**-equation of motion reads

$$\mathscr{R}+\frac{2}{3}\bar{\mathbf{B}}\mathbf{B}+\frac{1}{3}\bar{\mathbf{B}}^2=0$$

which has the solution

$$\mathbf{B}=\bar{\mathbf{B}}=\pm\sqrt{-\mathscr{R}}.$$

For consistency, the scalar curvature should be non-positive. Now, eliminating **B** from the action gives

$$S=\pm\frac{4}{3}\frac{1}{\kappa}\int \mathrm{d}^4x\, e^{-1}(-\mathscr{R})^{3/2}.$$

We see that the action becomes non-polynomial in the curvature after elimination of the scalar auxiliary fields!

6.2. Linearized supergravity

We are going to show that at the linearized level the Einstein supergravity theory, constructed in the previous section, is reduced to describing two massless super Poincaré states of superhelicities (-2) and $\frac{3}{2}$ (or, equivalently, four massless Poincaré states of helicities $\pm\frac{3}{2}$ and ± 2). We also consider linearized conformal supergravity.

6.2.1. Linearized Einstein supergravity action

To obtain a linearized form of the supergravity action (6.1.3), we shall use the prepotential parametrization of subsection 5.6.1. It will also be convenient for us to work in the chiral representation defined by equations (5.6.15, 16). After imposing the gauge (5.6.18), the action S_{SG} reads in terms of the prepotentials $H^a=\bar{H}^a$, φ and $\bar{\varphi}$, with φ being a flat chiral superfield, $\bar{\mathrm{D}}_{\dot{\alpha}}\varphi=0$, as follows

$$S_{\mathrm{SG}}=-\frac{3}{\kappa^2}\int \mathrm{d}^8z\,(\varphi\,\mathrm{e}^{-2\mathrm{i}H}\,\bar{\varphi})(1\cdot\mathrm{e}^{-2\mathrm{i}\overleftarrow{H}})\hat{E}^{-1/3} \tag{6.2.1}$$

where

$$\begin{gathered} H=H^a\partial_a \qquad \hat{E}=\mathrm{Ber}(\hat{E}_A{}^B) \\ \hat{E}_A=\hat{E}_A{}^B\mathrm{D}_B=\left(-\frac{\mathrm{i}}{4}(\tilde{\sigma}_a)^{\dot{\gamma}\gamma}\{\hat{E}_\gamma,\hat{E}_{\dot{\gamma}}\},\ \mathrm{e}^{-2\mathrm{i}H}\mathrm{D}_\alpha\mathrm{e}^{2\mathrm{i}H},\bar{\mathrm{D}}^{\dot{\alpha}}\right) \end{gathered} \tag{6.2.2}$$

with D_A being the flat covariant derivatives. Here we have used the relations

(5.6.21, 22). The action is invariant under the gauge transformations

$$\delta e^{-2iH} = \Lambda e^{-2iH} - e^{-2iH}\bar{\Lambda}$$
$$\delta\varphi^3 = -\frac{1}{4}\bar{D}^2 D^\alpha(L_\alpha\varphi^3) \tag{6.2.3}$$

where

$$\Lambda = \Lambda^A D_A \qquad \Lambda_\alpha = -\frac{1}{4}\bar{D}^2 L_\alpha$$
$$\Lambda_{\alpha\dot{\alpha}} = -2i\bar{D}_{\dot{\alpha}} L_\alpha \qquad \Lambda_{\dot{\alpha}} = -\frac{1}{4} e^{-2iH} D^2 \bar{L}_{\dot{\alpha}}$$

with L_α being an arbitrary spinor superfield.

Now, let us set

$$\varphi = e^\sigma \qquad \bar{\varphi} = e^{\bar{\sigma}} \qquad \bar{D}_{\dot{\alpha}}\sigma = 0 \tag{6.2.4}$$

and decompose the supergravity action to second order in the superfields H^a, σ and $\bar{\sigma}$. First, we note that $\hat{E} = \det(\hat{E}_a{}^b)$, because H does not involve spinor pieces. From (6.2.2) one readily obtains

$$\hat{E}_a = \partial_a + U_a{}^b\partial_b$$

$$U_a{}^b = \frac{1}{2}(\tilde{\sigma}_a)^{\dot{\alpha}\alpha}\bar{D}_{\dot{\alpha}}D_\alpha H^b - \frac{i}{2}(\tilde{\sigma}_a)^{\dot{\alpha}\alpha}\bar{D}_{\dot{\alpha}}\{(D_\alpha H^c)\partial_c H^b - H^c D_\alpha\partial_c H^b\} + O(H^3).$$

Due to the formula $\det e^A = e^{\operatorname{tr} A}$, we have

$$\hat{E}^{-1/3} = \det{}^{-1/3}(\mathbb{1} + U) = 1 - \frac{1}{3}\operatorname{tr} U + \frac{1}{6}\operatorname{tr} U^2 + \frac{1}{18}(\operatorname{tr} U)^2 + O(U^3)$$

which leads to

$$\hat{E}^{-1/3} = 1 - \frac{1}{6}\bar{D}_{\dot{\alpha}}D_\alpha H^{\alpha\dot{\alpha}} + \frac{1}{72}(\bar{D}_{\dot{\alpha}}D_\alpha H^{\alpha\dot{\alpha}})^2 + \frac{1}{24}(\bar{D}_{\dot{\alpha}}D_\alpha H^{\beta\dot{\beta}})(\bar{D}_{\dot{\beta}}D_\beta H^{\alpha\dot{\alpha}})$$
$$+ H^2\text{-total derivative terms} + O(H^3).$$

Up to the same accuracy, we have

$$(1\cdot e^{-2i\overleftarrow{H}})^{1/3} \approx 1 - \frac{2i}{3}\partial_a H^a + \frac{4}{9}(\partial_a H^a)^2$$

and also

$$\varphi\, e^{-2iH}\,\bar{\varphi} \approx 1 + \sigma + \bar{\sigma} + \frac{1}{2}(\sigma + \bar{\sigma})^2 - 2iH^a\partial_a\bar{\sigma}.$$

Finally, making use of the identities

$$\bar{D}_{\dot{\alpha}}D_{\alpha}H^{\alpha\dot{\alpha}} = -\frac{1}{2}[D_{\alpha}, \bar{D}_{\dot{\alpha}}]H^{\alpha\dot{\alpha}} + 2i\partial_a H^a$$

$$\int d^8z\,(\bar{D}_{\dot{\alpha}}D_{\alpha}H^{\beta\dot{\beta}})(\bar{D}_{\dot{\beta}}D_{\beta}H^{\alpha\dot{\alpha}}) = -\int d^8z\{(\bar{D}_{\dot{\alpha}}D_{\alpha}H^{\alpha\dot{\alpha}})^2 + H^a D^{\alpha}\bar{D}^2 D_{\alpha}H_a\}$$

one arrives, after the rescaling $H^a \to \kappa H^a$ and $\sigma \to \kappa\sigma$, at the superfunctional

$$S^{(2)}_{SG} = \int d^8z\left\{\frac{1}{8}H^a D^{\alpha}\bar{D}^2 D_{\alpha}H_a - 3\bar{\sigma}\sigma + \frac{1}{48}([D_{\alpha}, \bar{D}_{\dot{\alpha}}]H^{\alpha\dot{\alpha}})^2 - (\partial_a H^a)^2 + 2i(\sigma - \bar{\sigma})\partial_a H^a\right\} \tag{6.2.5}$$

where we have dropped numerous total derivative terms. It is instructive to compare the first term in equation (6.2.5) with the massless vector multiplet action (3.4.9).

Looking at $S^{(2)}_{SG}$, we see that the (anti)chiral superfields are characterized by a kinetic term with the wrong sign (compare with equation (3.2.5)). Nevertheless, this is not a defect of the theory, since σ and $\bar{\sigma}$ turn out to be purely gauge degrees of freedom with respect to the transformations

$$\delta H_{\alpha\dot{\alpha}} = \bar{D}_{\dot{\alpha}}L_{\alpha} - D_{\alpha}\bar{L}_{\dot{\alpha}} \qquad \delta\sigma = -\frac{1}{12}\bar{D}^2 D^{\alpha}L_{\alpha} \tag{6.2.6}$$

which leave invariant $S^{(2)}_{SG}$ and represent the linearized version of the original nonlinear transformations (6.2.3). The above gauge invariance can be used to impose the guage condition

$$\sigma = \bar{\sigma} = 0. \tag{6.2.7}$$

Another useful gauge is

$$D^{\alpha}H_{\alpha\dot{\alpha}} = \bar{D}^{\dot{\alpha}}H_{\alpha\dot{\alpha}} = 0. \tag{6.2.8}$$

In this gauge $S^{(2)}_{SG}$ proves to take the form

$$S^{(2)}_{SG} = -\int d^8z\,\{H^a\Box H_a + 3\bar{\sigma}\sigma\} \qquad \Box = \partial^a\partial_a. \tag{6.2.9}$$

Remark. All we have done above is, in fact, the expansion of S_{SG} around a flat superspace background. To obtain an expansion of S_{SG} around a curved superspace background described by covariant derivatives $\mathscr{D}_A$, one starts with the following ansatz for S_{SG}:

$$S_{SG} = -\frac{3}{\kappa^2}\int d^8z\,E^{-1}(\varphi\,e^{-2i\mathbf{H}}\,\varphi)(1\cdot e^{-2i\overleftarrow{\mathbf{H}}})U^{-1/3} \tag{6.2.10}$$

where

$$\mathbf{H}=\mathbf{H}^a\mathscr{D}_a=\bar{\mathbf{H}} \qquad \varphi=\mathrm{e}^\sigma \qquad \bar{\mathscr{D}}_{\dot{\alpha}}\sigma=0 \qquad U=\mathrm{Ber}(U_A{}^B)$$

with the supermatrix $U_A{}^B$ being defined by equation (5.6.47). In accordance with the results of Section 5.6, the action is invariant under the background transformations

$$\delta\mathscr{D}_A=[\mathscr{K},\mathscr{D}_A] \qquad \delta\mathbf{H}^a=\mathscr{K}\mathbf{H}^a \qquad \delta\varphi=\mathscr{K}\varphi$$

$$\mathscr{K}=K^A\mathrm{D}_A+\frac{1}{2}K^{ab}M_{ab}=\bar{\mathscr{K}}$$

and the quantum gauge transformations

$$\delta\mathbf{H}_{\alpha\dot{\alpha}}=\bar{\mathscr{D}}_{\dot{\alpha}}L_\alpha-\mathscr{D}_\alpha\bar{L}_{\dot{\alpha}}+O(\mathbf{H}) \qquad \delta\varphi^3=-\frac{1}{4}(\bar{\mathscr{D}}^2-4R)\mathscr{D}^\alpha(L_\alpha\varphi^3)$$

with L_α being arbitrary.

6.2.2. *Linearized superfield strengths and dynamical equations*

Our next goal is to analyse which supersymmetry representation the theory (6.2.5) describes on-shell. For this purpose, we must investigate relevant superfield strengths—that is, descendants of H^a, σ and $\bar{\sigma}$ which are invariants of the gauge transformations (6.2.6). We can take in the role of such objects linearlzed versions of the supertorsion tensors R, G_a and $W_{\alpha\beta\gamma}$. Indeed, since R, G_a and $W_{\alpha\beta\gamma}$ transform homogeneously with respect to the supergravity gauge group, their linearized counterparts, which will be denoted by $\mathbf{R}$, $\mathbf{G}_a$ and $\mathbf{W}_{\alpha\beta\gamma}$, respectively, should be invariants of the linearized gauge group. Explicit expressions for linearized supertorsions follow from the relations (5.6.59, 60) by setting the covariant derivatives $\mathscr{D}_A$ to be flat. This leads to the superfields

$$\mathbf{R}=-\frac{1}{4}\bar{\mathrm{D}}^2\bar{\sigma}+\frac{\mathrm{i}}{6}\bar{\mathrm{D}}^2\partial^a H_a$$

$$\mathbf{G}_{\alpha\dot{\alpha}}=\mathrm{i}\partial_{\alpha\dot{\alpha}}(\bar{\sigma}-\sigma)+\frac{1}{8}\mathrm{D}^\beta\bar{\mathrm{D}}^2\mathrm{D}_\beta H_{\alpha\dot{\alpha}}-\frac{1}{24}[\mathrm{D}_\alpha,\bar{\mathrm{D}}_{\dot{\alpha}}][\mathrm{D}^\beta,\bar{\mathrm{D}}^{\dot{\beta}}]H_{\beta\dot{\beta}}+\partial_{\alpha\dot{\alpha}}\partial^b H_b \tag{6.2.11}$$

$$\mathbf{W}_{\alpha\beta\gamma}=\frac{\mathrm{i}}{8}\bar{\mathrm{D}}^2\partial_{(\alpha}{}^{\dot{\gamma}}\mathrm{D}_\beta H_{\gamma)\dot{\gamma}}.$$

It is instructive to explicitly check their invariance under the transformations (6.2.6). The Bianchi identities (5.5.8) are reduced at the linearized level to

$$\begin{gathered}\bar{\mathrm{D}}_{\dot{\alpha}}\mathbf{R}=\bar{\mathrm{D}}_{\dot{\alpha}}\mathbf{W}_{\alpha\beta\gamma}=0 \qquad \bar{\mathrm{D}}^{\dot{\alpha}}\mathbf{G}_{\alpha\dot{\alpha}}=\mathrm{D}_\alpha\mathbf{R}\\ \mathrm{D}^\alpha\mathbf{W}_{\alpha\beta\gamma}=\mathrm{i}\partial_{(\beta}{}^{\dot{\gamma}}\mathbf{G}_{\gamma)\dot{\gamma}}.\end{gathered} \tag{6.2.12}$$

Remark. In the gauge (6.2.8), the expressions (6.2.11) are simplified drastically

$$\mathbf{R} = -\frac{1}{4}\bar{\mathrm{D}}^2\bar{\sigma} \qquad \mathbf{G}_{\alpha\dot{\alpha}} = \mathrm{i}\partial_{\alpha\dot{\alpha}}(\bar{\sigma}-\sigma) - \Box H_{\alpha\dot{\alpha}}$$
$$\mathbf{W}_{\alpha\beta\gamma} = \frac{\mathrm{i}}{8}\bar{\mathrm{D}}^2\partial_\alpha{}^{\dot{\gamma}}\mathrm{D}_\beta H_{\gamma\dot{\gamma}}. \tag{6.2.13}$$

The dynamical system (6.2.5) is characterized by the equations of motion

$$\mathbf{G}_a = 0 \qquad \mathbf{R} = 0 \tag{6.2.14}$$

which represent the linearized form of equations (6.1.8). We see that only the (anti)chiral superfield strengths $\bar{\mathbf{W}}_{\dot{\alpha}\dot{\beta}\dot{\gamma}}$ and $\mathbf{W}_{\alpha\beta\gamma}$ do not vanish on-shell, and they satisfy, due to identities (6.2.12), the equations

$$\mathrm{D}^\alpha\mathbf{W}_{\alpha\beta\gamma} = \bar{\mathrm{D}}^{\dot{\alpha}}\bar{\mathbf{W}}_{\dot{\alpha}\dot{\beta}\dot{\gamma}} = 0. \tag{6.2.15}$$

In accordance with the results of Section 2.7, these equations define massless on-shell superfields, and the corresponding superhelicities are

$$\kappa(\mathbf{W}_{\alpha\beta\gamma}) = \frac{3}{2} \qquad \kappa(\bar{\mathbf{W}}_{\dot{\alpha}\dot{\beta}\dot{\gamma}}) = -2. \tag{6.2.16}$$

Therefore, the Einstein supergravity theory at the linearized level describes two massless super Poincaré states of superhelicities (-2) and $\frac{3}{2}$.

Similar on-shell analysis can be done in components. The component supergravity action (6.1.12) reduces at the linearized level to the sum of the linearized gravity action (1.8.47) and the Rarita–Schwinger action (1.8.41), modulo auxiliary fields. As we have seen, the former theory describes two massless states of helicities (± 2) and the latter, two massless states of helicities $(\pm\frac{3}{2})$.

6.2.3. Linearized conformal supergravity

The action describing linearized conformal supergravity is given in terms of the linearized super Weyl tensor $\mathbf{W}_{\alpha\beta\gamma}$ (6.2.11) as follows

$$S^{(2)}_{\mathrm{SC}} = \int \mathrm{d}^6 z\, \mathbf{W}^{\alpha\beta\gamma}\mathbf{W}_{\alpha\beta\gamma}. \tag{6.2.16}$$

It is evidently invariant under the gauge transformations

$$\delta H_{\alpha\dot{\alpha}} = \bar{\mathrm{D}}_{\dot{\alpha}} L_\alpha - \mathrm{D}_\alpha \bar{L}_{\dot{\alpha}} \tag{6.2.17}$$

with L_α being arbitrary. The dynamical equations are of the form

$$\Box\mathbf{G}_a + \frac{\mathrm{i}}{4}\partial_a(\mathrm{D}^2\mathbf{R} - \bar{\mathrm{D}}^2\bar{\mathbf{R}}) = 0 \tag{6.2.18}$$

and represent the linearized form of the conformal supergravity equations (6.1.26).

Let us rewrite $S^{(2)}_{\mathrm{SC}}$ in terms of H^a. Using equation (6.2.11), we obtain

$$S^{(2)}_{\mathrm{SC}} = \frac{1}{64}\int \mathrm{d}^6 z\, (\bar{\mathrm{D}}^2 \partial^{(\alpha}{}_{\dot\alpha} \mathrm{D}^\beta H^{\gamma)\dot\alpha})\bar{\mathrm{D}}^2 \partial_{(\alpha}{}^{\dot\beta} \mathrm{D}_\beta H_{\gamma)\dot\beta}$$

$$-\frac{1}{16}\int \mathrm{d}^8 z (\partial^{(\alpha}{}_{\dot\alpha} \mathrm{D}^\beta H^{\gamma)\dot\alpha})\bar{\mathrm{D}}^2 \partial_{(\alpha}{}^{\dot\beta} \mathrm{D}_\beta H_{\gamma)\dot\beta}$$

$$= -\frac{1}{16}\int \mathrm{d}^8 z (\partial^{\alpha}{}_{\dot\alpha} \mathrm{D}^\beta H^{\gamma\dot\alpha})\bar{\mathrm{D}}^2 \partial_{(\alpha}{}^{\dot\beta} \mathrm{D}_\beta H_{\gamma)\dot\beta}.$$

Then, after some calculation, one arrives at

$$S^{(2)2}_{\mathrm{SC}} = \int \mathrm{d}^8 z \left\{ -H^a \Box\Box H_a - \frac{4}{3}(\partial_a H^a)\Box \partial_b H^b + \frac{1}{12} H^a \bar{\mathrm{D}}^2 \mathrm{D}^2 \Box H_a \right.$$

$$\left. + \frac{1}{12}(\partial_a H^a)\bar{\mathrm{D}}^2 \mathrm{D}^2 \partial_b H^b + \frac{\mathrm{i}}{6}(\mathrm{D}^\alpha H_{\alpha\dot\alpha})\Box \partial^{\beta\dot\alpha} \bar{\mathrm{D}}^{\dot\beta} H_{\beta\dot\beta} \right\}. \qquad (6.2.19)$$

In the gauge (6.2.8), $S^{(2)}_{\mathrm{SC}}$ reduces to

$$S^{(2)}_{\mathrm{SC}} = -\int \mathrm{d}^8 z\, H^a \Box\Box \mathrm{H}_a. \qquad (6.2.20)$$

We see that conformal supergravity leads to higher derivative dynamical equations.

6.3. Supergravity–matter dynamical systems

Einstein supergravity action S_{SG} is suitable only for describing empty supersymmetric universes. In the presence of matter, it must be substituted by an action superfunctional of the general structure

$$S = S_{\mathrm{SG}} + S_{\mathrm{M}}[\chi; \mathscr{D}] \qquad (6.3.1)$$

where S_{M} is a matter action and χ denote matter superfields under consideration. Such a dynamical system is characterized by equations of motion which naturally fall into two groups: supergravity equations and matter ones. The matter dynamical equations read simply as

$$\delta S_{\mathrm{M}}/\delta\chi = 0. \qquad (6.3.2)$$

Introducing the supercurrent

$$T_a = \delta S_{\mathrm{M}}/\delta \mathbf{H}^a \qquad T_a = \bar{T}_a \qquad (6.3.3)$$

and the supertrace

$$T = \delta S_{\mathrm{M}}/\delta\varphi \qquad \bar{\mathscr{D}}_{\dot\alpha} T = 0 \qquad (6.3.4)$$

of the matter superfields (see Section 5.7), the supergravity equations of motion are given by

$$\frac{2}{\kappa^2} G_a + T_a = 0 \qquad -\frac{3}{\kappa^2} R + T = 0 \tag{6.3.5}$$

where we have used the relations (6.1.7). These equations generalize the ordinary Einstein equations with matter. They show that the supercurrent and supertrace characterize the coupling of matter to supergravity.

Since the supertorsions G_a and R satisfy the Bianchi identity $\bar{\mathscr{D}}^{\dot\alpha} G_{\alpha\dot\alpha} = \mathscr{D}_\alpha R$, the system (6.3.5) is consistent under the supercurrent conservation law

$$\bar{\mathscr{D}}^{\dot\alpha} T_{\alpha\dot\alpha} = -\frac{2}{3} \mathscr{D}_\alpha T \tag{6.3.6}$$

which, at the same time, expresses the fact that S_M is gauge invariant, in accordance with the results of Section 5.7.

Now, we consider some specific matter models in a curved superspace.

6.3.1. *Chiral scalar models*

We commence by discussing simple models of a covariantly chiral scalar superfield Φ, $\bar{\mathscr{D}}_{\dot\alpha}\Phi = 0$, and its conjugate $\bar\Phi$, embraced by the common action

$$S = \int d^8z\, E^{-1} \bar\Phi \Phi + \left\{ \int d^8z \frac{E^{-1}}{R} \mathscr{L}_c(\Phi, R) + \text{c.c.} \right\} \tag{6.3.7}$$

where

$$2\mathscr{L}_c(\Phi, R) = m\Phi^2 + \xi R \Phi^2 + \frac{1}{3}\lambda \Phi^3 \equiv 2\mathscr{L}_c(\Phi) + \xi R \Phi^2$$

with m, ξ and λ being coupling constants. The system with this action can be considered as a curved-superspace extension of the standard Wess–Zumino model (3.2.11).

Using the technique of Section 5.8, the above action can be easily reduced to components. We leave this as an exercise for the reader. Let us comment, however, about the coupling of the scalar component fields

$$C(x) = \Phi| \qquad \bar C(x) = \bar\Phi|$$

to the vierbein. Setting the gravitino and supergravity auxiliary fields $\mathbf{B}$ and $\mathbf{A}_a$ to zero and keeping only C and $\bar C$ among the component fields of Φ and $\bar\Phi$, from equations (5.8.40, 50) we deduce

$$\int d^8z\, E^{-1} \bar\Phi \Phi \sim \int d^4x\, e^{-1} \bar C \left(\nabla^a \nabla_a - \frac{1}{6} \mathscr{R} \right) C \tag{6.3.8a}$$

$$\int d^8z\, E^{-1} \Phi^2 \sim -\frac{1}{6} \int d^4x\, e^{-1} C^2 \mathscr{R}. \tag{6.3.8b}$$

We see that the kinetic term in action (6.3.7) leads to Weyl invariant coupling of the scalar fields to gravity (see also subsection 1.7.7). This is not incidental, since the kinetic term is super Weyl invariant (see subsection 5.7.5). The mass term and the ξ-term in action (6.3.7) break this invariance. Decomposing C into real and imaginary parts,

$$C=\frac{1}{\sqrt{2}}(\mathscr{F}+\mathrm{i}\mathscr{G})$$

we observe from expressions (6.3.8) that

$$\int \mathrm{d}^8 z\, E^{-1}\left\{\bar{\Phi}\Phi+\frac{1}{2}\xi\Phi^2+\frac{1}{2}\bar{\xi}\bar{\Phi}^2\right\}$$

$$\sim\frac{1}{2}\int \mathrm{d}^4 x\, e^{-1}\left\{\mathscr{F}\left(\nabla^a\nabla_a-\frac{1}{6}(\xi+1)\mathscr{R}\right)\mathscr{F}+\mathscr{G}\left(\nabla^a\nabla_a+\frac{1}{6}(\xi-1)\mathscr{R}\right)\mathscr{G}\right\} \qquad (6.3.9)$$

where we have taken ξ to be real. It is seen that in the particular cases $\xi=\pm 1$ one of the scalar fields does not couple to the scalar curvature (minimal coupling with gravity).

Now, let us determine the supercurrent and the supertrace of the system with action (6.3.7). When $\xi=0$, their expressions are given by equations (5.7.40) and (5.7.41), respectively. Hence, it remains to vary the superfunctional

$$I=\frac{1}{2}\int \mathrm{d}^8 z\, E^{-1}\{\xi\Phi^2+\overline{\xi\Phi^2}\}$$

with respect to the supergravity prepotentials by the rule given in subsection 5.7.1. Namely, we are to calculate the variation

$$\delta I=\frac{1}{2}\int \mathrm{d}^8 z\, \tilde{\mathscr{E}}^{-1}\{\xi\tilde{\Phi}^2_{(\mathrm{V})}+\bar{\xi}\tilde{\bar{\Phi}}^2_{(\mathrm{V})}\}-\frac{1}{2}\int \mathrm{d}^8 z\, E^{-1}\{\xi\Phi^2+\overline{\xi\Phi^2}\}$$

$$\tilde{\Phi}_{(\mathrm{V})}=\Phi \qquad \tilde{\bar{\Phi}}_{(\mathrm{V})}=\mathrm{e}^{-2\mathrm{i}\mathbf{H}^a\mathscr{D}_a}\bar{\Phi}$$

to first order in the superfields $\mathbf{H}^a$ and σ, where $\tilde{\mathscr{E}}^{-1}$ reads as in equation (5.6.57). After simple calculation, one obtains the supercurrent

$$T_{\alpha\dot{\alpha}}=\frac{1}{3}(\bar{\mathscr{D}}_{\dot{\alpha}}\bar{\Phi})\mathscr{D}_\alpha\Phi+\frac{2}{3}\mathrm{i}\bar{\Phi}\overleftrightarrow{\mathscr{D}}_{\alpha\dot{\alpha}}\Phi+\frac{1}{3}\mathrm{i}\mathscr{D}_{\alpha\dot{\alpha}}(\xi\Phi^2-\overline{\xi\Phi^2})$$

$$-\frac{2}{3}G_{\alpha\dot{\alpha}}\left(\bar{\Phi}\Phi+\frac{1}{2}\xi\Phi^2+\frac{1}{2}\overline{\xi\Phi^2}\right) \qquad (6.3.10)$$

and the supertrace

$$T=-\frac{1}{4}(\bar{\mathscr{D}}^2-4R)\left(\bar{\Phi}\Phi+\frac{1}{2}\xi\Phi^2+\frac{1}{2}\overline{\xi\Phi^2}\right)+3\mathscr{L}_{\mathrm{c}}(\Phi). \qquad (6.3.11)$$

Remark. Analysing the above expressions, we note an important property of the supercurrent and supertrace in flat global superspace. In the flat-superspace limit, the action (6.3.7) reduces to the standard Wess–Zumino model action (3.2.11), which is ξ-independent, but the corresponding supercurrent and supertrace reduce to

$$T^{(\xi)}_{\alpha\dot{\alpha}} = \frac{1}{3}(\bar{D}_{\dot{\alpha}}\bar{\Phi})D_{\alpha}\Phi + \frac{2}{3}i\bar{\Phi}\overleftrightarrow{\partial}_{\alpha\dot{\alpha}}\Phi + \frac{1}{3}i\partial_{\alpha\dot{\alpha}}(\xi\Phi^2 - \overline{\xi\Phi^2}) \tag{6.3.12}$$

and

$$T^{(\xi)} = -\frac{1}{4}\bar{D}^2\left(\bar{\Phi}\Phi + \frac{1}{2}\overline{\xi\Phi^2}\right) + 3\mathscr{L}_c(\Phi) \tag{6.3.13}$$

respectively, where some ξ-dependence is present. For every ξ, we can take $T^{(\xi)}_{\alpha\dot{\alpha}}$ and $T^{(\xi)}$ in the roles of supercurrent and supertrace of the Wess–Zumino model. Different choices correspond to different possible ways of extending them to curved superspace. On the other hand, all choices are physically equivalent due to the fact that if $T^{(\xi)}_{\alpha\dot{\alpha}}$ and $T^{(\xi)}$ satisfy the conservation law

$$\bar{D}^{\dot{\alpha}}T^{(\xi)}_{\alpha\dot{\alpha}} = -\frac{2}{3}D_{\alpha}T^{(\xi)} \tag{6.3.14}$$

then so do the improved supercurrent and supertrace

$$\begin{aligned} \hat{T}^{(\xi)}_{\alpha\dot{\alpha}} &= T^{(\xi)}_{\alpha\dot{\alpha}} + \frac{1}{3}i\partial_{\alpha\dot{\alpha}}(\Omega - \bar{\Omega}) \qquad \bar{D}_{\dot{\alpha}}\Omega = 0 \\ \hat{T}^{(\xi)} &= T^{(\xi)} - \frac{1}{8}\bar{D}^2\bar{\Omega} \end{aligned} \tag{6.3.15}$$

for an arbitrary chiral scalar Ω. This is a fundamental arbitrariness in the choice of supercurrent and supertrace in flat global superspace.

The action (6.3.7) leads to the following dynamical equations

$$\begin{aligned} -\frac{1}{4}(\bar{\mathscr{D}}^2 - 4R)\bar{\Phi} + m\Phi + \xi R\Phi + \frac{1}{2}\lambda\Phi^2 &= 0 \\ -\frac{1}{4}(\mathscr{D}^2 - 4\bar{R})\Phi + m\bar{\Phi} + \bar{\xi}\bar{R}\bar{\Phi} + \frac{1}{2}\bar{\lambda}\bar{\Phi}^2 &= 0 \end{aligned} \tag{6.3.16}$$

where we have taken m real. If $\xi = \lambda = 0$, this system is easily separated:

$$\frac{1}{16}(\bar{\mathscr{D}}^2 - 4R)(\mathscr{D}^2 - 4\bar{R})\Phi = m^2\Phi \tag{6.3.17a}$$

$$\frac{1}{16}(\mathscr{D}^2 - 4\bar{R})(\bar{\mathscr{D}}^2 - 4R)\bar{\Phi} = m^2\bar{\Phi}. \tag{6.3.17b}$$

Owing to the chirality of Φ, the fourth-order operation in the l.h.s. of equation (6.3.17*a*) can be reduced, with the help of the commutator algebra (5.5.6), to second-order. Then, the equation (6.3.17*b*) takes the form

$$\left(\Box_+ - \frac{1}{4}(\bar{\mathscr{D}}^2\bar{R}) + R\bar{R} - m^2\right)\Phi = 0$$
$$\Box_+ = \mathscr{D}^a\mathscr{D}_a + \frac{1}{4}R\bar{\mathscr{D}}^2 + \mathrm{i}G^a\mathscr{D}_a + \frac{1}{4}(\mathscr{D}^\alpha R)\mathscr{D}_\alpha . \tag{6.3.18}$$

Similarly, the equation in $\bar{\Phi}$ takes the form

$$\left(\Box_- - \frac{1}{4}(\mathscr{D}^2 R) + R\bar{R} - m^2\right)\bar{\Phi} = 0$$
$$\Box_- = \mathscr{D}^a\mathscr{D}_a + \frac{1}{4}\overline{R\mathscr{D}^2} - \mathrm{i}G^a\mathscr{D}_a + \frac{1}{4}(\bar{\mathscr{D}}_{\dot\alpha}\bar{R})\bar{\mathscr{D}}^{\dot\alpha}. \tag{6.3.19}$$

The operator $\Box_+$ ($\Box_-$) transforms every covariantly chiral (antichiral) scalar superfield into a covariantly chiral (antichiral) one. For this reason $\Box_+$ ($\Box_-$) may be called the 'chiral (antichiral) d'Alembertian'. The operator $\Box_+$ is unique in the sense that any second-order differential operator constructed from the covariant derivatives, with the leading term $\mathscr{D}^a\mathscr{D}_a$ acting on the space of covariantly chiral scalar superfields turns out to have the form

$$\Box_+ + \Psi \qquad \bar{\mathscr{D}}_{\dot\alpha}\Psi = 0$$

where Ψ is a covariantly chiral scalar superfield.

6.3.2. *Vector multiplet models*

The massless vector multiplet model (3.4.9) is uniquely continued to a curved superspace when taking the superfield variable $V(z)$ to be a real scalar superfield (with respect to the general coordinate transformation supergroup and the superlocal Lorentz group) and requiring the invariance of the curved-superspace action under the gauge transformations

$$V \to V' = V + \frac{\mathrm{i}}{2}(\bar{\Lambda} - \Lambda) \qquad \bar{\mathscr{D}}_{\dot\alpha}\Lambda = 0 \tag{6.3.20}$$

with $\Lambda(z)$ being an arbitrary covariantly chiral scalar. Namely, the gauge invariant action superfunctional is of the form

$$S = \frac{1}{8}\int \mathrm{d}^8z\, E^{-1} V\mathscr{D}^\alpha(\bar{\mathscr{D}}^2 - 4R)\mathscr{D}_\alpha V = \frac{1}{2}\int \mathrm{d}^8z \frac{E^{-1}}{R} W^\alpha W_\alpha \tag{6.3.21}$$

where

$$W_\alpha = -\frac{1}{4}(\bar{\mathscr{D}}^2 - 4R)\mathscr{D}_\alpha V \qquad \bar{\mathscr{D}}_{\dot\alpha} W_\alpha = 0. \tag{6.3.22}$$

As a result of the identity (5.5.12), the action is real. Its invariance under transformation (6.3.20) follows from the observation

$$\bar{\mathscr{D}}_{\dot{\alpha}}\Lambda=0 \;\Rightarrow\; (\bar{\mathscr{D}}^2-4R)\mathscr{D}_\alpha\Lambda=0 \tag{6.3.23}$$

which is a direct consequence of the identity (5.5.9*e*). Therefore, W_α and its conjugate $\bar{W}_{\dot{\alpha}}$ are invariant superfield strengths.

The action superfunctional proves to be invariant under the super Weyl transformations (5.5.13) if we take V to be super Weyl inert. This can be easily checked using the transformation laws for R (5.5.14) and for E^{-1}, $(E^{-1})'=e^{\sigma+\bar{\sigma}}E^{-1}$. Therefore, S does not depend on the chiral compensator, which means that the supertrace vanishes identically.

To find the supercurrent, we must, following the prescription of Section 5.7, calculate the variation

$$\delta S=\frac{1}{2}\int d^8z\,\frac{\tilde{\mathscr{E}}^{-1}}{\tilde{\mathbf{R}}}\,\tilde{W}^{\alpha}_{(\nabla)}\tilde{W}_{(\nabla)\alpha}-\frac{1}{2}\int d^8z\,\frac{E^{-1}}{R}\,W^{\alpha}W_{\alpha}$$

where

$$\tilde{W}_{(\nabla)\alpha}=-\frac{1}{4}(\tilde{\bar{\nabla}}^2-4\tilde{\bar{\mathbf{R}}})\tilde{\nabla}_\alpha\tilde{V}_{(\nabla)}\qquad \tilde{V}_{(\nabla)}=e^{-i\mathbf{H}^a\mathscr{D}_a}V$$

to first order in $\mathbf{H}^a$. With the help of equations (5.6.40, 58–60), one can derive the identity

$$(\tilde{\bar{\nabla}}-4\tilde{\bar{\mathbf{R}}})\lambda_\alpha=(\bar{\mathscr{D}}^2-4R)\lambda_\alpha-A_\alpha{}^\beta(\bar{\mathscr{D}}^2-4R)\lambda_\beta+(\bar{\mathscr{D}}^2-4R)[A_\alpha{}^\beta\lambda_\beta+2\Delta\bar{\mathscr{F}}\lambda_\alpha] \tag{6.3.24}$$

which is convenient when performing variations. It is also necessary to recall relation (5.6.63). After some calculation, one arrives at the supercurrent

$$T_{\alpha\dot{\alpha}}=2W_\alpha\bar{W}_{\dot{\alpha}}+\frac{1}{2}(\mathscr{D}^\beta W_\beta)[\mathscr{D}_\alpha,\bar{\mathscr{D}}_{\dot{\alpha}}]V. \tag{6.3.25}$$

Making use of the dynamical equation $\mathscr{D}^\alpha W_\alpha=0$, we can change the above expression to

$$T_{\alpha\dot{\alpha}}=2W_\alpha\bar{W}_{\dot{\alpha}}. \tag{6.3.26}$$

As can be seen, the supercurrent is not gauge invariant off-shell. The point is that the gauge parameters in expression (6.3.20) are covariantly chiral, hence they depend on supergravity prepotentials.

For completeness, let us also consider the massive vector multiplet model in a curved superspace with the action

$$S=\frac{1}{2}\int d^8z\,\frac{E^{-1}}{R}\,W^\alpha W_\alpha+m^2\int d^8z\,E^{-1}V^2 \tag{6.3.27}$$

One can readily verify that this system is characterized by the supercurrent

$$T_{\alpha\dot\alpha}=2W_\alpha\bar W_{\dot\alpha}+\frac{1}{2}(\mathscr{D}^\beta W_\beta)[\mathscr{D}_\alpha,\bar{\mathscr{D}}_{\dot\alpha}]V$$
$$+\frac{2}{3}m^2\left\{\bar{\mathscr{D}}_{\dot\alpha}V\mathscr{D}_\alpha V-G_{\alpha\dot\alpha}V^2-\frac{1}{2}V[\mathscr{D}_\alpha,\bar{\mathscr{D}}_{\dot\alpha}]V\right\} \tag{6.3.28}$$

and the supertrace

$$T=-\frac{1}{4}m^2(\bar{\mathscr{D}}^2-4R)V^2. \tag{6.3.29}$$

On-shell, the supercurrent takes the form

$$T_{\alpha\dot\alpha}=2W_\alpha\bar W_{\dot\alpha}+\frac{2}{3}m^2\{\bar{\mathscr{D}}_{\dot\alpha}V\mathscr{D}_\alpha V-G_{\alpha\dot\alpha}V^2+V[\mathscr{D}_\alpha,\bar{\mathscr{D}}_{\dot\alpha}]V\}\,. \tag{6.3.30}$$

6.3.3. Super Yang–Mills models

Now we are going to obtain a curved-superspace extension of the super Yang–Mills dynamical systems (3.5.46). In a curved superspace, superfields Φ^i and $\bar\Phi_i$ forming the matter multiplet should be treated as covariantly chiral and antichiral scalars, respectively. Superfields V^I forming Yang–Mills superfield $V=V^IT^I=V^+$, where T^I are Hermitian generators of the gauge group, should be treated as real scalars. The most natural generalization of the super Yang–Mills gauge transformations (3.5.25, 26) reads

$$\Phi'=\mathrm{e}^{\mathrm{i}\Lambda}\Phi \qquad \Lambda=\Lambda^IT^I \qquad \bar{\mathscr{D}}_{\dot\alpha}\Lambda^I=0 \tag{6.3.31a}$$

$$\bar\Phi'=\bar\Phi\mathrm{e}^{-\mathrm{i}\bar\Lambda} \qquad \bar\Lambda=\bar\Lambda^IT^I \qquad \mathscr{D}_\alpha\bar\Lambda^I=0 \tag{6.3.31b}$$

$$\mathrm{e}^{2V'}=\mathrm{e}^{\mathrm{i}\bar\Lambda}\,\mathrm{e}^{2V}\,\mathrm{e}^{-\mathrm{i}\Lambda} \tag{6.3.31c}$$

with the parameters Λ^I being covariantly chiral scalar superfields. Invariant under these transformations the action superfunctional, with the flat-superspace limit (3.5.46), is of the form

$$S=\int \mathrm{d}^8z\,E^{-1}\bar\Phi_i(\mathrm{e}^{2V})^i{}_j\Phi^j+\left\{\int \mathrm{d}^8z\,\frac{E^{-1}}{R}\,\mathscr{L}_\mathrm{c}(\Phi^i)+\mathrm{c.c.}\right\}$$
$$+\frac{1}{2g^2}\int \mathrm{d}^8z\,\frac{E^{-1}}{R}\,\mathrm{tr}\,W^\alpha W_\alpha \tag{6.3.32}$$

where

$$W_\alpha=-\frac{1}{8}(\bar{\mathscr{D}}^2-4R)(\mathrm{e}^{-2V}\mathscr{D}_\alpha\mathrm{e}^{2V}) \qquad \bar{\mathscr{D}}_{\dot\alpha}W_\alpha=0$$
$$\bar W_{\dot\alpha}=\frac{1}{8}(\mathscr{D}^2-4\bar R)(e^{2V}\bar{\mathscr{D}}_{\dot\alpha}\mathrm{e}^{-2V}) \qquad \mathscr{D}_\alpha\bar W_{\dot\alpha}=0 \tag{6.3.33}$$

are gauge covariant superfield strengths possessing the transformation laws

$$W'_\alpha = e^{i\Lambda} W_\alpha e^{-i\Lambda} \qquad \bar{W}'_{\dot\alpha} = e^{i\bar\Lambda} \bar{W}_{\dot\alpha} e^{-i\bar\Lambda}. \tag{6.3.34}$$

Note that the last term in equation (6.3.32) turns out to be real modulo total derivatives, see below.

In order to describe the dynamical properties of the system with action (6.3.32), it is convenient to introduce two sets of gauge covariant derivatives:

$$\begin{aligned} &\mathbb{D}_A^{(+)} = \mathscr{D}_A + i\Gamma_A^{(+)} \qquad \Gamma_A^{(+)} = \Gamma_A^{I(+)}(z) T^I \\ &\mathbb{D}_\alpha^{(+)} = e^{-2V} \mathscr{D}_\alpha e^{2V} \qquad \bar{\mathbb{D}}_{\dot\alpha}^{(+)} = \bar{\mathscr{D}}_{\dot\alpha} \qquad \mathbb{D}_{\alpha\dot\alpha}^{(+)} = \frac{i}{2}\{\mathbb{D}_\alpha^{(+)}, \bar{\mathbb{D}}_{\dot\alpha}^{(+)}\} \end{aligned} \tag{6.3.35a)}$$

and

$$\begin{aligned} &\mathbb{D}_A^{(-)} = \mathscr{D}_A + i\Gamma_A^{(-)} \qquad \Gamma_A^{(-)} = \Gamma_A^{I(-)}(z) T^I \\ &\mathbb{D}_\alpha^{(-)} = \bar{\mathscr{D}}_\alpha \qquad \bar{\mathbb{D}}_{\dot\alpha}^{(-)} = e^{2V} \bar{\mathscr{D}}_{\dot\alpha} e^{-2V} \qquad \mathbb{D}_{\alpha\dot\alpha}^{(-)} = \frac{i}{2}\{\mathbb{D}_\alpha^{(-)}, \bar{\mathbb{D}}_{\dot\alpha}^{(-)}\} \end{aligned} \tag{6.3.35b}$$

with the transformation laws

$$\mathbb{D}'^{(+)}_A = e^{i\Lambda} \mathbb{D}_A^{(+)} e^{-i\Lambda} \qquad \mathbb{D}'^{(-)}_A = e^{i\bar\Lambda} \mathbb{D}_A^{(-)} e^{-i\bar\Lambda} \tag{6.3.36}$$

connected to each other in the manner

$$\mathbb{D}_A^{(+)} = e^{-2V} \mathbb{D}_A^{(-)} e^{2V}. \tag{6.3.37}$$

The operators $\mathbb{D}_A^{(+)}$ and $\mathbb{D}_A^{(-)}$ are curved-superspace versions of the flat ones (3.6.40*a*) and (3.6.40*b*), respectively. They satisfy commutator algebras of the form

$$[\mathbb{D}_A^{(\pm)}, \mathbb{D}_B^{(\pm)}\} = T_{AB}{}^C \mathbb{D}_C^{(\pm)} + R_{AB} + iF_{AB}^{(\pm)}$$

where $T_{AB}{}^C$ and $R_{AB} = \frac{1}{2} R_{AB}{}^{cd} M_{cd}$ are the Einstein supergravity supertorsion and supercurvature, $F_{AB}^{(\pm)} = F_{AB}^{I(\pm)} T^I$ is the Yang–Mills supercurvature. Explicitly, we have

$$\begin{aligned} &\{\mathbb{D}_\alpha^{(\pm)}, \mathbb{D}_\beta^{(\pm)}\} = -4\bar{R} M_{\alpha\beta} \qquad \{\bar{\mathbb{D}}_{\dot\alpha}^{(\pm)}, \bar{\mathbb{D}}_{\dot\beta}^{(\pm)}\} = 4R\bar{M}_{\dot\alpha\dot\beta} \\ &[\bar{\mathbb{D}}_{\dot\alpha}^{(\pm)}, \mathbb{D}_{\beta\dot\beta}^{(\pm)}] = -i\varepsilon_{\dot\alpha\dot\beta}(R\mathbb{D}_\beta^{(\pm)} + G_\beta{}^{\dot\gamma} \bar{\mathbb{D}}_{\dot\gamma}^{(\pm)}) - i\mathscr{D}_\beta R \bar{M}_{\dot\alpha\dot\beta} \\ &\qquad + i\varepsilon_{\dot\alpha\dot\beta}(\mathscr{D}^\gamma G_\beta{}^{\dot\delta} \bar{M}_{\dot\gamma\dot\delta} - 2W_\beta{}^{\gamma\delta} M_{\gamma\delta}) + 2i\varepsilon_{\dot\alpha\dot\beta} W_\beta^{(\pm)} \\ &[\mathbb{D}_\alpha^{(\pm)}, \mathbb{D}_{\beta\dot\beta}^{(\pm)}] = i\varepsilon_{\alpha\beta}(\overline{R\mathbb{D}_{\dot\beta}^{(\pm)}} + G^\gamma{}_{\dot\beta} \mathbb{D}_\gamma^{(\pm)}) + i\bar{\mathscr{D}}_{\dot\beta} \bar{R} M_{\alpha\beta} \\ &\qquad - i\varepsilon_{\alpha\beta}(\bar{\mathscr{D}}^{\dot\gamma} G^\delta{}_{\dot\beta} M_{\gamma\delta} - 2\bar{W}_{\dot\beta}{}^{\dot\gamma\dot\delta} \bar{M}_{\dot\gamma\dot\delta}) + 2i\varepsilon_{\alpha\beta} \bar{W}_{\dot\beta}^{(\pm)} \end{aligned} \tag{6.3.38}$$

where $W_\alpha^{(\pm)}$ and $\bar{W}_{\dot\alpha}^{(\pm)}$ are expressed in terms of the superfield strengths (6.3.33) as follows

$$\begin{aligned} &W_\alpha^{(+)} = W_\alpha \qquad \bar{W}_{\dot\alpha}^{(+)} = e^{-2V} \bar{W}_{\dot\alpha} e^{2V} \\ &W_\alpha^{(-)} = e^{2V} W_\alpha e^{-2V} \qquad \bar{W}_{\dot\alpha}^{(-)} = \bar{W}_{\dot\alpha}. \end{aligned} \tag{6.3.39}$$

Together with the Bianchi identities (5.5.8), we also have

$$\bar{\mathbb{D}}^{(\pm)}_{\dot\alpha} W^{(\pm)}_\alpha = 0 \qquad \mathbb{D}^{(\pm)}_\alpha \bar{W}^{(\pm)}_{\dot\alpha} = 0$$
$$\mathbb{D}^{\alpha(\pm)} W^{(\pm)}_\alpha = \bar{\mathbb{D}}^{(\pm)}_{\dot\alpha} \bar{W}^{\dot\alpha(\pm)}. \tag{6.3.40}$$

To derive equations (6.3.38–40), one simply repeats the steps performed in subsection 3.6.4 when obtaining (3.6.45–48).

Now, let us consider the dynamical equations of the system. Varying the action (6.3.32) with respect to Φ_i gives the equations

$$-\frac{1}{4}(\bar{\mathscr{D}}^2 - 4\bar{R})(\mathrm{e}^{2V}\Phi)^i + \frac{\partial \mathscr{L}_c(\Phi)}{\partial \Phi_i} = 0. \tag{6.3.41}$$

Next, varying the action with respect to V leads to

$$2(\bar{\Phi}\mathrm{e}^{2V} T^I \Phi) T^I = \frac{1}{g^2} \mathbb{D}^{\alpha(+)} W^{(+)}_\alpha \tag{6.3.42a}$$

or

$$2(\bar{\Phi} T^I \mathrm{e}^{2V} \Phi) T^I = \frac{1}{g^2} \mathbb{D}^{\alpha(-)} W^{(-)}_\alpha \tag{6.3.42b}$$

depending on which variation, $\Delta^{(+)}V$ or $\Delta^{(-)}V$ (3.6.57), was used. Clearly, the equations (6.3.42*a*) and (6.3.42*b*) are equivalent.

Remark. Reality of the last term in equation (6.3.32) follows from the fact that its arbitrary variation over V can be represented as

$$-\frac{1}{g^2}\mathrm{tr}\int \mathrm{d}^8 z\, E^{-1}\Delta^{(+)}V(\mathbb{D}^{\alpha(+)}W_\alpha) = -\frac{1}{g^2}\mathrm{tr}\int \mathrm{d}^8 z\, E^{-1}\Delta^{(-)}V(\mathbb{D}^{\alpha(-)}W^{(-)}_\alpha)$$
$$= -\frac{1}{g^2}\mathrm{tr}\int \mathrm{d}^8 z\, E^{-1}\Delta^{(-)}V(\bar{\mathbb{D}}^{(-)}_{\dot\alpha}\bar{W}^{\dot\alpha})$$

and this expression is real, owing to the Hermitian conjugation rules

$$(\Delta^{(+)}V)^+ = \Delta^{(-)}V \qquad (\mathbb{D}^{\alpha(+)}W_\alpha)^+ = \bar{\mathbb{D}}^{(-)}_{\dot\alpha}\bar{W}^{\dot\alpha}.$$

Let us set the chiral superpotential $\mathscr{L}_c(\Phi)$ to zero. Then, equation (6.3.41) has the consequence

$$\frac{1}{16}(\mathscr{D}^2 - 4R)\mathrm{e}^{-2V}(\bar{\mathscr{D}}^2 - 4\bar{R})\mathrm{e}^{2V}\Phi = \frac{1}{16}\bar{\mathbb{D}}^{2(+)}\mathbb{D}^{2(+)}\Phi = 0.$$

Since Φ is gauge covariantly chiral, $\bar{\mathbb{D}}^{(+)}_{\dot\alpha}\Phi = 0$, making use of the algebra

(6.3.38) reduces the above system to

$$\{\mathbb{D}^{a(+)}\mathbb{D}_a^{(+)}+\frac{1}{4}R\mathbb{D}^{2(+)}+\mathrm{i}G^a\mathbb{D}_a^{(+)}+\frac{1}{4}(\mathscr{D}^\alpha R)\mathbb{D}_\alpha^{(+)}-W^\alpha\mathbb{D}_\alpha^{(+)}$$

$$-\frac{1}{2}(\mathbb{D}^{\alpha(+)}W_\alpha)-\frac{1}{4}(\bar{\mathscr{D}}^2\bar{R})+R\bar{R}\}\Phi=0. \quad (6.3.43)$$

These equations represent the Yang–Mills covariantized form of equation (6.3.18).

To calculate the supercurrent and supertrace of the system with action (6.3.32), it is helpful to operate with the identify (6.3.24). Tedious calculations lead to the supercurrent

$$T_{\alpha\dot{\alpha}}=-\frac{1}{6}[\mathscr{D}_\alpha,\bar{\mathscr{D}}_{\dot{\alpha}}](\bar{\Phi}\mathrm{e}^{2V}\Phi)+\mathrm{i}\bar{\Phi}\mathrm{e}^{2V}\mathscr{D}_{\alpha\dot{\alpha}}\Phi-\mathrm{i}(\mathscr{D}_{\alpha\dot{\alpha}}\bar{\Phi})\mathrm{e}^{2V}\Phi-\frac{2}{3}G_{\alpha\dot{\alpha}}\bar{\Phi}\mathrm{e}^{2V}\Phi$$

$$-\frac{1}{g^2}\mathrm{tr}\left\{2W_\alpha^{(+)}\bar{W}_{\dot{\alpha}}^{(+)}+\frac{1}{4}(\mathbb{D}^{\beta(+)}W_\beta^{(+)})(\mathrm{e}^{-2V}[\mathscr{D}_\alpha,\bar{\mathscr{D}}_{\dot{\alpha}}]\mathrm{e}^{2V}\right.$$

$$\left.+2(\mathrm{e}^{-2V}\bar{\mathscr{D}}_{\dot{\alpha}}\mathrm{e}^{2V})(\mathrm{e}^{-2V}\mathscr{D}_\alpha\mathrm{e}^{2V}))\right\} \quad (6.3.44)$$

and the supertrace

$$T=-\frac{1}{4}(\bar{\mathscr{D}}^2-4R)(\bar{\Phi}\mathrm{e}^{2V}\Phi)+3\mathscr{L}_c(\Phi). \quad (6.3.45)$$

As may be seen, the supercurrent is not gauge invariant off-shell. However, using the dynamical equations (6.3.42), one can easily rewrite $T_{\alpha\dot{\alpha}}$ in the gauge invariant form:

$$T_{\alpha\dot{\alpha}}=-\frac{1}{6}[\mathscr{D}_\alpha,\bar{\mathscr{D}}_{\dot{\alpha}}](\bar{\Phi}\mathrm{e}^{2V}\Phi)-\frac{2}{3}G_{\alpha\dot{\alpha}}\bar{\Phi}\mathrm{e}^{2V}\Phi+\mathrm{i}(\bar{\Phi}\mathrm{e}^{2V})\overleftrightarrow{\mathbb{D}}_{\alpha\dot{\alpha}}{}^{(+)}\Phi$$

$$+\frac{2}{g^2}\mathrm{tr}(W_\alpha^{(+)}\bar{W}_{\dot{\alpha}}^{(+)}). \quad (6.3.46)$$

Further, due to the dynamical equations (6.3.41), the on-shell supertrace reads

$$T=-\frac{\partial\mathscr{L}_c(\Phi)}{\partial\Phi^i}\Phi^i+3\mathscr{L}_c(\Phi). \quad (6.3.47)$$

Therefore, the supertrace vanishes if and only if

$$\mathscr{L}_c(\Phi)=\frac{1}{3!}\lambda_{ijk}\Phi^i\Phi^j\Phi^k. \quad (6.3.48)$$

In this case the action (6.3.32) turns out to be super Weyl invariant.

6.3.4. Chiral spinor model

In conclusion, we would like to consider the model of a covariantly chiral spinor superfield χ_α, $\bar{\mathscr{D}}_{\dot\alpha}\chi_\alpha=0$, and its conjugate $\bar\chi_{\dot\alpha}$ with the action superfunctional

$$S=-\int \mathrm{d}^8z\, E^{-1}\mathbb{L}^2 \tag{6.3.49}$$

where

$$\mathbb{L}=\frac{1}{2}(\mathscr{D}^\alpha\chi_\alpha+\bar{\mathscr{D}}_{\dot\alpha}\bar\chi^{\dot\alpha})=\bar{\mathbb{L}}.$$

In the flat-superspace limit, the theory reduces to the massless chiral spinor model (3.7.10).

Owing to the identity (5.5.12), the above action is invariant under the gauge transformations

$$\delta\chi_\alpha=\mathrm{i}(\bar{\mathscr{D}}^2-4R)\mathscr{D}_\alpha K \qquad K=\bar K \tag{6.3.50}$$

with K being an arbitrary real scalar superfield, hence $\mathbb{L}$ is a gauge invariant superfield strength. It is worth also pointing out that, by virtue of relation (5.5.11), $\mathbb{L}$ is characterized by

$$(\bar{\mathscr{D}}^2-4R)\mathbb{L}=(\mathscr{D}^2-4\bar R)\mathbb{L}=0. \tag{6.3.51}$$

Each real scalar superfield under this constraint will be said to be covariantly linear.

To obtain dynamical equations, one uses the variational rules

$$\begin{aligned}\frac{\delta\chi_\alpha(z')}{\delta\chi_\beta(z)}&=-\frac{1}{4}\delta_\alpha^\beta(\bar{\mathscr{D}}^2-4R)E\delta^8(z-z')\\ \frac{\delta\bar\chi_{\dot\alpha}(z')}{\delta\bar\chi_{\dot\beta}(z)}&=-\frac{1}{4}\delta_{\dot\alpha}^{\dot\beta}(\mathscr{D}^2-4\bar R)E\delta^8(z-z').\end{aligned} \tag{6.3.52}$$

Then one obtains

$$(\bar{\mathscr{D}}^2-4R)\mathscr{D}_\alpha\mathbb{L}=(\mathscr{D}^2-4\bar R)\bar{\mathscr{D}}_{\dot\alpha}\mathbb{L}=0. \tag{6.3.53}$$

To calculate the supercurrent and supertrace of the theory, we must evaluate the variation

$$\delta S=-\frac{1}{4}\int \mathrm{d}^8z\,\tilde{\mathscr{E}}^{-1}(\tilde\nabla^\alpha\tilde\chi_{(\nabla)\alpha}+\tilde{\bar\nabla}_{\dot\alpha}\tilde{\bar\chi}^{\dot\alpha}_{(\nabla)})^2-S$$

where the derivatives $\tilde\nabla_A$ are given by equation (5.6.40), to first order in the superfields $\mathbf{H}^a$ and σ. Note, $\tilde\chi_{(\nabla)\alpha}$ should be a covariantly chiral spinor superfield with respect to $\tilde\nabla_A$, $\tilde{\bar\nabla}_{\dot\alpha}\tilde\chi_{(\nabla)\alpha}=0$. In the linear approximation in $\mathbf{H}^a$,

we have

$$\tilde{\chi}_{(\nabla)\alpha} = \chi_\alpha - A_\alpha{}^\beta \chi_\beta$$
$$\tilde{\bar{\chi}}_{(\nabla)\dot{\alpha}} = \bar{\chi}_{\dot{\alpha}} - \bar{A}_{\dot{\alpha}}{}^{\dot{\beta}} \bar{\chi}_{\dot{\beta}} - 2\mathrm{i}\mathbf{H}^b \mathscr{D}_b \chi_{\dot{\alpha}}$$

with $A_{\alpha\beta}$ being defined by equation (5.6.60). After some effort, one arrives at the supercurrent

$$T_{\alpha\dot{\alpha}} = \frac{2}{3}(\bar{\mathscr{D}}_{\dot{\alpha}}\mathbb{L})\mathscr{D}_\alpha\mathbb{L} + \frac{2}{3}\mathbb{L}[\mathscr{D}_\alpha, \bar{\mathscr{D}}_{\dot{\alpha}}]\mathbb{L} - \frac{2}{3}G_{\alpha\dot{\alpha}}\mathbb{L}^2 + \left\{\frac{1}{2}\bar{\chi}_{\dot{\alpha}}(\bar{\mathscr{D}}^2 - 4R)\mathscr{D}_\alpha\mathbb{L} + \text{c.c.}\right\} \tag{6.3.54}$$

and the supertrace

$$T = -\frac{1}{4}(\bar{\mathscr{D}}^2 - 4R)\mathbb{L}^2 - \frac{3}{8}\chi^\alpha(\bar{\mathscr{D}}^2 - 4R)\mathscr{D}_\alpha\mathbb{L}. \tag{6.3.55}$$

On-shell, the supercurrent and the supertrace become gauge invariant:

$$T_{\alpha\dot{\alpha}} = \frac{2}{3}(\bar{\mathscr{D}}_{\dot{\alpha}}\mathbb{L})\mathscr{D}_\alpha\mathbb{L} + \frac{2}{3}\mathbb{L}[\mathscr{D}_\alpha, \bar{\mathscr{D}}_{\dot{\alpha}}]\mathbb{L} - \frac{2}{3}G_{\alpha\dot{\alpha}}\mathbb{L}^2$$
$$T = -\frac{1}{4}(\bar{\mathscr{D}}^2 - 4R)\mathbb{L}^2. \tag{6.3.56}$$

It is instructive to compare the expressions obtained here with the supercurrent and supertrace of the chiral scalar model with action

$$S = \frac{1}{2}\int \mathrm{d}^8 z\, E^{-1}(\Phi + \bar{\Phi})^2 \qquad \bar{\mathscr{D}}_{\dot{\alpha}}\Phi = 0 \tag{6.3.57}$$

which is obtained from action (6.3.7) by setting $\xi = 1$, $\mathscr{L}_c(\Phi) = 0$. Making this choice in equations (6.3.10) and (6.3.11) gives

$$T_{\alpha\dot{\alpha}} = \frac{2}{3}(\bar{\mathscr{D}}_{\dot{\alpha}}\mathbb{M})\mathscr{D}_\alpha\mathbb{M} + \frac{2}{3}\mathbb{M}[\mathscr{D}_\alpha, \bar{\mathscr{D}}_{\dot{\alpha}}]\mathbb{M} - \frac{2}{3}G_{\alpha\dot{\alpha}}\mathbb{M}^2$$
$$T = -\frac{1}{4}(\bar{\mathscr{D}}^2 - 4R)\mathbb{M}^2 \tag{6.3.58}$$

where we have introduced the notation

$$\mathbb{M} = \frac{1}{\sqrt{2}}(\Phi + \bar{\Phi}). \tag{6.3.59}$$

We observe that the expressions (6.3.56) and (6.3.58) have the same form. Remarkably, this is not a formal coincidence. We have seen that $\mathbb{L}$ was submitted to the kinematical constraints (6.3.51) and the dynamical equations

(6.3.53). As for $\mathbb{M}$, from its definition we directly deduce

$$(\bar{\mathscr{D}}^2-4R)\mathscr{D}_\alpha\mathbb{M}=(\mathscr{D}^2-4\bar{R})\bar{\mathscr{D}}_{\dot{\alpha}}\mathbb{M} \tag{6.3.60}$$

with the help of identities (5.5.9*d,e*). On the other hand, the dynamical equations of the theory (6.3.57) are

$$(\bar{\mathscr{D}}^2-4R)\mathbb{M}=(\mathscr{D}^2-4\bar{R})\mathbb{M}=0 \tag{6.3.61}$$

hence $\mathbb{M}$ is covariantly linear on-shell. As a result, $\mathbb{L}$ and $\mathbb{M}$ satisfy the same differential equations on-shell. Therefore, under a proper choice of initial conditions for the theories (6.3.49) and (6.3.57) (and gauge conditions for the former theory) their supercurrents (supertraces) coincide on-shell. This implies classical equivalence of the theories under consideration.

The equivalence can be also seen with the help of the first-order model with action

$$S[V,\Phi,\bar{\Phi}]=-\int d^8z\,E^{-1}\left\{\frac{1}{2}V^2+V(\Phi+\bar{\Phi})\right\} \qquad V=\bar{V} \qquad \bar{\mathscr{D}}_{\dot{\alpha}}\Phi=0 \tag{6.3.62}$$

which constitutes duality of the above theories. Indeed, varying $S[V,\Phi,\bar{\Phi}]$ over V gives the equation

$$-V=\Phi+\bar{\Phi}$$

which can be used to eliminate V resulting in the action (6.3.57). On the other hand, varying $S[V,\Phi,\bar{\Phi}]$ with respect to Φ and $\bar{\Phi}$ gives

$$(\bar{\mathscr{D}}^2-4R)V=(\mathscr{D}^2-4\bar{R})V=0$$

which has the following solution:

$$V=\frac{1}{\sqrt{2}}(\mathscr{D}^\alpha\chi_\alpha+\bar{\mathscr{D}}_{\dot{\alpha}}\bar{\chi}^{\dot{\alpha}}) \qquad \bar{\mathscr{D}}_{\dot{\alpha}}\chi_\alpha=0.$$

Substituting this expression into $S[V,\Phi,\bar{\Phi}]$ and using the identity

$$(\bar{\mathscr{D}}^2-4R)V=0 \qquad \bar{\mathscr{D}}_{\dot{\alpha}}\Phi=0 \quad\Rightarrow\quad \int d^8z\,E^{-1}V\Phi=0 \tag{6.3.63}$$

one obtains the action (6.3.49).

6.4. (Conformal) Killing supervectors. Superconformal models

We now turn our attention to looking for superspace analogues of conformal Killing vector fields—the objects discussed in Section 1.7. We also consider superconformal dynamical systems and superconformal massless superfields.

6.4.1. (Conformal) Killing supervector fields

To begin with, let us introduce the notion of conformal Killing supervector fields.

Consider a curved superspace described by covariant derivatives

$$\mathcal{D}_A = E_A + \Omega_A = E_A{}^M(z)\partial_M + \frac{1}{2}\Omega_A{}^{bc}(z)M_{bc}.$$

Let $\xi = \xi^A(z)E_A$ be a real c-type supervector field and $K^{ab}(z)$ be a real antisymmetric tensor superfield. ξ^A and K^{ab} generate an infinitesimal general coordinate and superlocal Lorentz transformation acting on the covariant derivatives and tensor superfields as follows:

$$\delta_{\mathscr{K}}\mathcal{D}_A = [\mathscr{K}, \mathcal{D}_A] \qquad \delta_{\mathscr{K}}\chi = \mathscr{K}\chi \tag{6.4.1}$$

where

$$\mathscr{K} = \xi^A\mathcal{D}_A + K^{\alpha\beta}M_{\alpha\beta} + \bar{K}^{\dot\alpha\dot\beta}\bar{M}_{\dot\alpha\dot\beta}$$

with χ being a tensor superfield (with suppressed indices). Using the (anti)commutation relations (5.5.6, 7), it is easy to calculate explicitly the variations $\delta_{\mathscr{K}}\mathcal{D}_A$. In particular, in the undotted spinor case one obtains

$$\begin{aligned}\delta\mathcal{D}_\alpha = &\left(K_\alpha{}^\beta - \mathcal{D}_\alpha\xi^\beta - \frac{i}{2}\xi_{\alpha\dot\beta}G^{\beta\dot\beta}\right)\mathcal{D}_\beta + \left(\mathcal{D}_\alpha\bar{\xi}^{\dot\beta} + \frac{i}{2}\xi_\alpha{}^{\dot\beta}\bar{R}\right)\bar{\mathcal{D}}_{\dot\beta}\\ &+2i\left(\delta_\alpha{}^\beta\bar{\xi}^{\dot\beta} - \frac{i}{4}\mathcal{D}_\alpha\xi^{\beta\dot\beta}\right)\mathcal{D}_{\beta\dot\beta}\\ &-\left(\mathcal{D}_\alpha K^{\beta\gamma} + 4\delta_\alpha{}^{(\beta}\xi^{\gamma)}\bar{R} - \frac{i}{2}\delta_\alpha{}^{(\beta}\xi^{\gamma)\dot\gamma}\bar{\mathcal{D}}_{\dot\gamma}\bar{R} - \frac{i}{2}\xi_{\alpha\dot\alpha}\mathcal{D}^{(\beta}G^{\gamma)\dot\alpha}\right)M_{\beta\gamma}\\ &-\left(\mathcal{D}_\alpha\bar{K}^{\dot\beta\dot\gamma} + i\xi_{\alpha\dot\alpha}\bar{W}^{\dot\alpha\dot\beta\dot\gamma}\right)\bar{M}_{\dot\beta\dot\gamma}.\end{aligned} \tag{6.4.2}$$

Let us also recall how infinitesimal super Weyl transformations act on the covariant derivatives:

$$\begin{aligned}&\delta_\sigma\mathcal{D}_\alpha = \left(\frac{1}{2}\sigma - \bar\sigma\right)\mathcal{D}_\alpha - \mathcal{D}^\beta\sigma M_{\beta\alpha} \qquad \delta_\sigma\bar{\mathcal{D}}_{\dot\alpha} = \left(\frac{1}{2}\bar\sigma - \sigma\right)\bar{\mathcal{D}}_{\dot\alpha} - \bar{\mathcal{D}}^{\dot\beta}\bar\sigma\bar{M}_{\dot\beta\dot\alpha}\\ &\delta_\sigma\mathcal{D}_{\alpha\dot\alpha} = \frac{i}{2}\{\delta_\sigma\mathcal{D}_\alpha, \bar{\mathcal{D}}_{\dot\alpha}\} + \frac{i}{2}\{\mathcal{D}_\alpha, \delta_\sigma\bar{\mathcal{D}}_{\dot\alpha}\} \qquad \bar{\mathcal{D}}_{\dot\alpha}\sigma = 0.\end{aligned} \tag{6.4.3}$$

Here σ is a covariantly chiral scalar superfield.

Now we proceed by finding a set of transformation parameters $\{\xi^A, K^{ab}, \sigma\}$ such that their combined action on the covariant derivatives is trivial,

$$(\delta_{\mathscr{K}} + \delta_\sigma)\mathcal{D}_A = 0.$$

In accordance with the relations (6.4.2) and (6.4.3), this requirement means the parameters ζ_α, $\bar{\xi}_{\dot\alpha}$, $K_{\alpha\beta}$, $\bar{K}_{\dot\alpha\dot\beta}$, σ and $\bar\sigma$ should be expressed via ξ^a in the manner:

$$\delta_\alpha{}^\beta \bar{\xi}^{\dot\beta} = \frac{i}{4}\mathscr{D}_\alpha \xi^{\beta\dot\beta} \quad \Rightarrow \quad \bar{\xi}^{\dot\alpha} = \frac{i}{8}\mathscr{D}_\alpha \xi^{\alpha\dot\alpha} \tag{6.4.4a}$$

$$K[\xi]_{\alpha\beta} = \mathscr{D}_{(\alpha}\xi_{\beta)} - \frac{i}{2}\xi_{(\alpha}{}^{\dot\beta} G_{\beta)\dot\beta} \tag{6.4.4b}$$

$$\sigma[\xi] = \frac{1}{3}(\mathscr{D}^\alpha \xi_\alpha + 2\bar{\mathscr{D}}_{\dot\alpha}\bar{\xi}^{\dot\alpha} - i\xi^a G_a). \tag{6.4.4c}$$

In addition, the following consistency conditions

$$\mathscr{D}_\alpha \bar{\xi}_{\dot\alpha} = -\frac{i}{2}\xi_{\alpha\dot\alpha}\bar{R} \tag{6.4.5a}$$

$$\bar{\mathscr{D}}_{\dot\alpha} K[\xi]^{\beta\gamma} = i\xi_{\alpha\dot\alpha} W^{\alpha\beta\gamma} \tag{6.4.5b}$$

$$\mathscr{D}_\alpha K[\xi]^{\beta\gamma} = -\delta_\alpha{}^{(\beta}\mathscr{D}^{\gamma)}\sigma[\xi] - 4\delta_\alpha{}^{(\beta}\xi^{\gamma)}\bar{R} + \frac{i}{2}\delta_\alpha{}^{(\beta}\xi^{\gamma)\dot\gamma}\bar{\mathscr{D}}_{\dot\gamma}\bar{R} + \frac{i}{2}\xi_{\alpha\dot\alpha}\mathscr{D}^{(\beta}G^{\gamma)\dot\alpha} \tag{6.4.5c}$$

$$\bar{\mathscr{D}}_{\dot\alpha}\sigma[\xi] = 0 \tag{6.4.5d}$$

should hold.

From relation (6.4.4*a*) it is easy to read off the closed equation on the vector superfield ξ^a:

$$\mathscr{D}_\alpha \xi_{\beta\dot\beta} + \mathscr{D}_\beta \xi_{\alpha\dot\beta} = 0 \quad \Rightarrow \quad \bar{\mathscr{D}}_{\dot\alpha}\xi_{\beta\dot\beta} + \bar{\mathscr{D}}_{\dot\beta}\xi_{\beta\dot\alpha} = 0. \tag{6.4.6}$$

Then, making use of the covariant derivatives algebra (5.5.6), one immediately obtains the important consequences

$$(\mathscr{D}^2 + 2\bar{R})\xi_a = 0 \qquad (\bar{\mathscr{D}}^2 + 2R)\xi_a = 0 \tag{6.4.7}$$

and

$$\mathscr{D}_a \xi_b + \mathscr{D}_b \xi_a = \frac{1}{2}\eta_{ab}\mathscr{D}^c \xi_c. \tag{6.4.8}$$

Remarkably, the last relation coincides in form with equation (1.7.5) defining conformal Killing vector fields of a space–time. But the highly important fact for us is that *the consistency conditions (6.4.5a–d) are satisfied identically under equation (6.4.6)*. For example, let us check the requirement (6.4.5*d*). We have

$$3\bar{\mathscr{D}}_{\dot\alpha}\sigma[\xi] = \bar{\mathscr{D}}_{\dot\alpha}\left(\frac{1}{2}\mathscr{D}_{\beta\dot\beta}\xi^{\beta\dot\beta} - \frac{i}{8}\mathscr{D}_\beta\bar{\mathscr{D}}_{\dot\beta}\xi^{\beta\dot\beta} + \frac{i}{2}\xi^{\beta\dot\beta}G_{\beta\dot\beta}\right)$$

$$=\frac{1}{2}[\bar{\mathscr{D}}_{\dot{\alpha}}, \mathscr{D}_{\beta\dot{\beta}}]\xi^{\beta\dot{\beta}}+\frac{i}{8}\mathscr{D}_{\beta}\bar{\mathscr{D}}_{\dot{\alpha}}\bar{\mathscr{D}}_{\dot{\beta}}\xi^{\beta\dot{\beta}}+\frac{i}{2}\bar{\mathscr{D}}_{\dot{\alpha}}(\xi^{\beta\dot{\beta}}G_{\beta\dot{\beta}})$$

$$=\frac{1}{2}[\bar{\mathscr{D}}_{\dot{\alpha}}, \mathscr{D}_{\beta\dot{\beta}}]\xi^{\beta\dot{\beta}}+\frac{i}{2}\mathscr{D}_{\beta}(R\xi^{\beta}{}_{\dot{\alpha}})+\frac{i}{2}\bar{\mathscr{D}}_{\dot{\alpha}}(\xi^{\beta\dot{\beta}}G_{\beta\dot{\beta}})$$

where we have used equations (6.4.6, 7). Expanding the commutator in this expression by the rule (5.5.6), one obtains $\bar{\mathscr{D}}_{\dot{\alpha}}\sigma[\xi]=0$. Slightly more work is necessary to check the conditions (6.4.5*b*,*c*). As a result, we come to the conclusion that the only way to achieve the goal formulated above is to choose a solution of the equations (6.4.6) in the role of ξ^a and to take the other parameters as in (6.4.4).

A real *c*-type supervector field

$$\xi=\xi^A E_A \qquad \xi^A=\left(\xi^a, -\frac{i}{8}\bar{\mathscr{D}}_{\dot{\beta}}\xi^{\alpha\dot{\beta}}, -\frac{i}{8}\mathscr{D}^{\beta}\xi_{\beta\dot{\alpha}}\right) \tag{6.4.9}$$

is said to be a 'conformal Killing supervector' (for the supergeometry determined by $\mathscr{D}_A$) of ξ^a satisfies the master equation (6.4.6). The basic property of conformal Killing supervectors is that they act on the covariant derivatives like superlocal Lorentz and super Weyl transformations:

$$\delta_{\mathscr{K}[\xi]}\mathscr{D}_A=[\mathscr{K}[\xi], \mathscr{D}_A]=-\delta_{\sigma[\xi]}\mathscr{D}_A \tag{6.4.10}$$

where

$$\mathscr{K}[\xi]=\xi^A\mathscr{D}_A+K[\xi]^{\alpha\beta}M_{\alpha\beta}+\bar{K}[\xi]^{\dot{\alpha}\dot{\beta}}\bar{M}_{\dot{\alpha}\dot{\beta}} \tag{6.4.11}$$

with $K[\xi]_{\alpha\beta}$ and $\sigma[\xi]$ being defined as in expressions (6.4.4).

A conformal Killing supervector ξ is said to be a 'Killing supervector' if it satisfies the requirement:

$$\sigma[\xi]=\bar{\sigma}[\xi]=0 \quad\Leftrightarrow\quad \mathscr{D}^{\alpha}\xi_{\alpha}+i\xi^a G_a=0. \tag{6.4.12}$$

The basic property of Killing supervectors is that they define symmetry transformations of the curved superspace: transformations generated by $\mathscr{K}[\xi]$ leave the covariant derivatives unchanged,

$$\delta_{\mathscr{K}[\xi]}\mathscr{D}_A=[\mathscr{K}[\xi], \mathscr{D}_A]=0. \tag{6.4.13}$$

From the point of view of supersymmetric field theory in curved superspace, Killing supervectors are of great importance. Indeed, consider a theory of matter superfields χ with an action superfunctional $S[\chi;\mathscr{D}]$ invariant, as usual, under general coordinate and superlocal Lorentz transformations 6.4.1). Then equation (6.4.13) shows that $S[\chi;\mathscr{D}]$ is unchanged under the transformations

$$\delta\chi=\mathscr{K}[\xi]\chi \tag{6.4.14}$$

for every Killing supervector ξ. So, Killing supervectors constitute symmetries of any dynamical system. As for arbitrary conformal Killing supervectors, they generate symmetry transformations of super Weyl invariant models only. If $S[\chi; \mathscr{D}]$ is a scalar with respect to super Weyl transformations of the form (5.7.33), then it does not change under the transformations

$$\delta\chi = \mathscr{K}[\xi]\chi - \mathrm{d}_{(+)}\sigma[\xi]\chi - \mathrm{d}_{(-)}\bar{\sigma}[\xi]\chi \qquad (6.4.15)$$

for every conformal Killing supervector ξ. This statement is a trivial consequence of transformation (6.4.10).

For a given supergeometry, the set of all conformal Killing supervectors and the set of all Killing supervectors form real super Lie algebras. The latter assertion is almost obvious. If ξ_1 and ξ_2 are two Killing supervectors,

$$[\mathscr{K}[\xi_{1,2}], \mathscr{D}_A] = 0$$

then

$$[[\mathscr{K}[\xi_1], \mathscr{K}[\xi_2]], \mathscr{D}_A] = [[\mathscr{K}[\xi_1], \mathscr{D}_A], \mathscr{K}[\xi_2]] + [\mathscr{K}[\xi_1], [\mathscr{K}[\xi_2], \mathscr{D}_A]] = 0$$

therefore

$$[\mathscr{K}[\xi_1], \mathscr{K}[\xi_2]] = \mathscr{K}[\xi_3]$$

for some Killing supervector ξ_3. One can readily see that

$$\xi_3 = [\xi_1, \xi_2]$$

which completes the proof. The fact that the set of conformal Killing supervectors forms a super Lie algebra follows from the theorem below. Before formulating the theorem, let us give one more definition.

Two supergeometries determined by covariant derivatives $\mathscr{D}_A$ and $\tilde{\mathscr{D}}_A$ are said to be Weyl equivalent if $\mathscr{D}_A$ and $\tilde{\mathscr{D}}_A$ are connected by some super Weyl transformation

$$\begin{aligned} &\tilde{\mathscr{D}}_\alpha = L\mathscr{D}_\alpha - 2\mathscr{D}^\beta L M_{\alpha\beta} \qquad \tilde{\bar{\mathscr{D}}}_{\dot\alpha} = \overline{L\mathscr{D}}_{\dot\alpha} - 2\bar{\mathscr{D}}^{\dot\beta}\overline{LM}_{\dot\alpha\dot\beta} \\ &\tilde{\mathscr{D}}_{\alpha\dot\alpha} = \frac{\mathrm{i}}{2}\{\tilde{\mathscr{D}}_\alpha, \tilde{\bar{\mathscr{D}}}_{\dot\alpha}\} \qquad L = \exp\left(\frac{1}{2}\omega - \bar{\omega}\right) \qquad \bar{\mathscr{D}}_{\dot\alpha}\omega = 0. \end{aligned} \qquad (6.4.16)$$

Theorem. Weyl equivalent supergeometries have the same conformal Killing supervectors.

Proof. Let ξ be a conformal Killing supervector with respect to $\mathscr{D}_A$. ξ is given in the form (6.4.9), where ξ^a obeys the equation (6.4.6). Decompose ξ with respect to the supervierbein $\tilde{E}_A$ ($\tilde{\mathscr{D}}_A = \tilde{E}_A + \tilde{\Omega}_A$):

$$\begin{aligned} &\xi = \xi^A E_A = \tilde{\xi}^A \tilde{E}_A \qquad \tilde{\xi}^A = (\tilde{\xi}^a, \tilde{\xi}^\alpha, \tilde{\bar{\xi}}_{\dot\alpha}) \\ &\tilde{\xi}^a = (L\bar{L})^{-1}\xi^a \qquad \tilde{\xi}^\alpha = L^{-1}\left(\xi^\alpha + \frac{\mathrm{i}}{4}\xi^{\alpha\dot\alpha}\bar{\mathscr{D}}_{\dot\alpha}\ln L\right) \end{aligned} \qquad (6.4.17)$$

where we have used relations (6.4.16). It is not difficult to check that $\tilde{\xi}^A$ can be represented as

$$\tilde{\xi}^A = \left(\tilde{\xi}^a, -\frac{i}{8} \bar{\tilde{\mathcal{D}}}_{\dot\beta} \tilde{\xi}^{\alpha\dot\beta}, -\frac{i}{8} \tilde{\mathcal{D}}^{\beta} \tilde{\xi}_{\beta\dot\alpha} \right) \tag{6.4.18}$$

and $\tilde{\xi}^a$ turns out to satisfy the equation

$$\tilde{\mathcal{D}}_{\alpha} \tilde{\xi}_{b\dot\beta} + \tilde{\mathcal{D}}_{\beta} \tilde{\xi}_{\alpha\dot\beta} = 0.$$

Therefore, ξ is a conformal Killing supervector with respect to $\tilde{\mathcal{D}}_A$. One can also prove the important relations:

$$\mathcal{K}[\tilde{\xi}] \equiv \tilde{\xi}^A \tilde{\mathcal{D}}_A + K[\tilde{\xi}]^{\alpha\beta} M_{\alpha\beta} + \bar{K}[\tilde{\xi}]^{\dot\alpha\dot\beta} \bar{M}_{\dot\alpha\dot\beta} = \mathcal{K}[\xi] \tag{6.4.19a}$$

$$\sigma[\tilde{\xi}] = \sigma[\xi] - \xi^A \mathcal{D}_A \omega \tag{6.4.19b}$$

where

$$K[\tilde{\xi}]_{\alpha\beta} = \tilde{\mathcal{D}}_{(\alpha} \tilde{\xi}_{\beta)} - \frac{i}{2} \tilde{\xi}_{(\alpha}{}^{\dot\beta} \tilde{G}_{\beta)\dot\beta}$$

$$\sigma[\tilde{\xi}] = \frac{1}{3} (\tilde{\mathcal{D}}^{\alpha} \tilde{\xi}_{\alpha} + 2 \bar{\tilde{\mathcal{D}}}_{\dot\alpha} \bar{\tilde{\xi}}^{\dot\alpha} - i \tilde{\xi}^a \tilde{G}_a)$$

and $\tilde{G}_a$ is the supertorsion component corresponding to $\tilde{\mathcal{D}}_A$. The relation (6.4.19*a*) shows that the operator $\mathcal{K}[\xi]$ is super Weyl invariant. The relation (6.4.19*b*) tells us that the property of a conformal Killing supervector ξ to be a Killing supervector is not super Weyl invariant.

6.4.2. *The gravitational superfield and conformal Killing supervectors*

Here we would like to look at conformal Killing supervectors from the point of view of their action on the supergravity prepotentials.

In the gravitational superfield gauge (5.4.38), Einstein supergravity covariant derivatives $\mathcal{D}_A$ are experessed in terms of the superfields $\mathcal{H}^m$, φ and $\bar\varphi$, with φ being the chiral compensator, $\bar{\mathcal{D}}_{\dot\alpha}\varphi = 0$. Residual gauge transformations, generated by parameters $\lambda^m(x, \theta)$ and $\lambda^\mu(x, \theta)$, act on $\mathcal{H}^m$, $\mathcal{D}_A$ and φ in accordance with the laws (5.4.41), (5.4.49) and (5.5.5), respectively. Let us choose (if possible) a set of superfields $\{\lambda^m(x, \theta), \lambda^\mu(x, \theta)\}$ such that

$$\delta \mathcal{H}^m = 0 \tag{6.4.20}$$

and only φ may change. Owing to (5.4.41), λ^m and λ^μ should satisfy the equations

$$\lambda^m(x + i\mathcal{H}, \theta) - \bar{\lambda}^m(x - i\mathcal{H}, \bar\theta) = 2i\xi \mathcal{H}^m \tag{6.4.21}$$

where

$$\xi = K^M \partial_M \qquad K^m = \frac{1}{2}\lambda^m(x+\mathrm{i}\mathscr{H},\theta) + \frac{1}{2}\bar{\lambda}^m(x-\mathrm{i}\mathscr{H},\bar{\theta})$$
$$K^\mu = \lambda^\mu(x+\mathrm{i}\mathscr{H},\theta) \qquad \bar{K}_{\dot{\mu}} = \bar{\lambda}_{\dot{\mu}}(x-\mathrm{i}\mathscr{H},\bar{\theta}). \tag{6.4.22}$$

Under these conditions, the covariant derivatives vary by the rule (6.4.3) with $\sigma = \delta \ln \varphi$. On the other hand, their change is given by equation (5.4.49). Therefore the supervector field (6.4.22) is a conformal Killing supervector! Remarkably, the reverse assertion is also true. Namely, it can be shown, using the formulae (5.4.43–47) and (5.5.4) expressing $\mathscr{D}_A$ via $\mathscr{H}^m$, φ and $\bar{\varphi}$, that every conformal Killing supervector has the form (6.4.22). Therefore, the basic property of conformal Killing supervectors is that they leave $\mathscr{H}^m$ unchanged.

Given a Killing supervector ξ, it leaves $\mathscr{D}_A$ unchanged. In accordance with (5.5.5), this means that the parameters $\lambda^m(x,\theta)$ and $\lambda^\mu(x,\theta)$ should satisfy the equation

$$\xi \ln \varphi = -\frac{1}{3} \mathrm{e}^{\overleftarrow{W}}(\partial_m \lambda^m - \partial_\mu \lambda^\mu) \tag{6.4.23a}$$

in addition to equations (6.4.21). Since $\varphi(z) = \varphi((x+\mathrm{i}\mathscr{H},\theta) = \mathrm{e}^{\overleftarrow{W}}\varphi(x,\theta)$, this equation can be rewritten, with the help of conditions (6.4.22), in the following useful form:

$$(\lambda^m \partial_m + \lambda^\mu \partial_\mu) \ln \varphi(x,\theta) = -\frac{1}{3}(\partial_m \lambda^m - \partial_\mu \lambda^\mu). \tag{6.4.23b}$$

6.4.3. *(Conformal) Killing supervectors in flat global superspace*

We are going to find (conformal) Killing supervectors of flat global superspace.

In flat global superspace, a conformal Killing supervector ξ has the form

$$\xi = \xi^A(z) \mathrm{D}_A \qquad \xi^A = \bar{\xi}^A = \left(\xi^a, -\frac{\mathrm{i}}{8}\bar{\mathrm{D}}_{\dot{\beta}}\xi^{\alpha\dot{\beta}}, -\frac{\mathrm{i}}{8}\mathrm{D}^\beta \xi_{\beta\dot{\alpha}} \right) \tag{6.4.24}$$

where ξ^a satisfy the master equation

$$\mathrm{D}_\alpha \xi_{\beta\dot{\beta}} + \mathrm{D}_\beta \xi_{\alpha\dot{\beta}} = 0 \quad \Leftrightarrow \quad \bar{\mathrm{D}}_{\dot{\alpha}} \xi_{\beta\dot{\beta}} + \bar{\mathrm{D}}_{\dot{\beta}} \xi_{\beta\dot{\alpha}} = 0. \tag{6.4.25}$$

The relations (6.4.7) and (6.4.8) take the form

$$\mathrm{D}^2 \xi_a = \bar{\mathrm{D}}^2 \xi_a = 0 \tag{6.4.26}$$

and

$$\partial_a \xi_b + \partial_b \xi_a = \frac{1}{2}\eta_{ab}(\partial^c \xi_c). \tag{6.4.27}$$

From relations (6.4.26) we see that ξ^a is a linear vector superfield, and $\xi^\alpha = -(\mathrm{i}/8)\bar{D}_{\dot\beta}\beta^{\alpha\dot\beta}$ is a chiral spinor superfield,

$$\bar{D}_{\dot\alpha}\xi^\alpha = 0 \tag{6.4.28}$$

Equation (6.4.27) has been investigated in Section 1.7. In particular, it was shown that every solution ξ^a of this equation must be at most quadratic in space–time coordinates,

$$\partial_a\partial_b\partial_c\xi^d(x,\theta,\bar\theta) = 0. \tag{6.4.29}$$

Owing to this observation, the general solution of the master equation (6.4.25) can be found quite easily. The result reads

$$\begin{aligned}\xi^a(x,\theta,\bar\theta) = {} & b^a + \Delta x^a + K^a{}_b x^b + f^a x^2 - 2x^a(x,f) + 2\mathrm{i}(\theta\sigma^a\bar\epsilon - \epsilon\sigma^a\bar\theta) \\ & - 2(\theta\sigma^a\tilde\sigma^b\eta + \bar\eta\tilde\sigma^b\sigma^a\bar\theta)x_b - \theta\sigma^a\bar\theta\Omega + \varepsilon^{abcd}\left(\frac{1}{2}K_{bc} + 2f_b x_c\right)\theta\sigma_d\bar\theta \\ & + 2\mathrm{i}(\bar\theta^2\theta\sigma^a\bar\eta - \theta^2\eta\sigma^a\bar\theta) - \theta^2\bar\theta^2 f^a\end{aligned} \tag{6.4.30a}$$

or, in spinor notation,

$$\begin{aligned}\xi_{\alpha\dot\alpha}(x,\theta,\bar\theta) = {} & b_{\alpha\dot\alpha} + \Delta x_{\alpha\dot\alpha} + K_\alpha{}^\beta x_{\beta\dot\alpha} + \bar K_{\dot\alpha}{}^{\dot\beta} x_{\alpha\dot\beta} + x_{\alpha\dot\beta}x_{\beta\dot\alpha}f^{\beta\dot\beta} \\ & - 4\theta_\alpha(\mathrm{i}\bar\epsilon_{\dot\alpha} + \eta^\beta x_{\beta\dot\alpha}) - 4\bar\theta_{\dot\alpha}(\mathrm{i}\epsilon_\alpha - x_{\alpha\dot\beta}\bar\eta^{\dot\beta}) + 2\theta_\alpha\bar\theta_{\dot\alpha}\Omega \\ & + 2\mathrm{i}\theta_\alpha\bar\theta^{\dot\beta}(\bar K_{\dot\alpha\dot\beta} - f_{\beta(\dot\alpha}x^\beta{}_{\dot\beta)}) + 2\mathrm{i}\bar\theta_{\dot\alpha}\theta^\beta(K_{\alpha\beta} + f_{(\alpha}{}^{\dot\beta}x_{\beta)\dot\beta}) \\ & - 4\mathrm{i}\theta_\alpha\bar\theta^2\bar\eta_{\dot\alpha} - 4\mathrm{i}\bar\theta_{\dot\alpha}\theta^2\eta_\alpha - \theta^2\bar\theta^2 f_{\alpha\dot\alpha}.\end{aligned} \tag{6.4.30b}$$

Here Δ, Ω, b^a, f^a and $K^{ab} = -K^{ba}$ are constant real *c*-number parameters, ε_α and η_α are constant *a*-number parameters. It is instructive to compare expression (6.4.30*a*) with the expression (1.7.12) determining conformal Killing vectors of Minkowski space.

Fom (6.4.30*b*) one immediately obtains

$$\xi_\alpha = \frac{\mathrm{i}}{8}\bar D^{\dot\alpha}\xi_{\alpha\dot\alpha} = \mathrm{e}^{\mathrm{i}\mathscr{H}}\lambda_\alpha(x,\theta) \qquad \mathscr{H} = \theta\sigma^a\bar\theta\partial_a \tag{6.4.31}$$

with $\lambda_\alpha(x,\theta)$ being defined by equation (2.9.16) (recall that the parameters (2.9.16) generate infinitesimal superconformal transformations (2.9.17)). Moreover, it is a simple exercise to check that

$$\begin{aligned}\xi = \xi^A D_A & = (\xi^a + \mathrm{i}\xi\sigma^a\bar\theta - \mathrm{i}\theta\sigma^a\bar\xi)\partial_a + \xi^\alpha\partial_\alpha + \bar\xi_{\dot\alpha}\bar\partial^{\dot\alpha} \\ & = \frac{1}{2}(\mathrm{e}^{\mathrm{i}\mathscr{H}}\lambda^a(x,\theta) + \mathrm{e}^{-\mathrm{i}\mathscr{H}}\bar\lambda^a(x,\bar\theta))\partial_a + (\mathrm{e}^{\mathrm{i}\mathscr{H}}\lambda^\alpha(x,\theta))\partial_\alpha + (\mathrm{e}^{-\mathrm{i}\mathscr{H}}\bar\lambda_{\dot\alpha}(x,\bar\theta))\bar\partial^{\dot\alpha}\end{aligned} \tag{6.4.32}$$

with $\lambda^a(x,\theta)$ defined as in equation (2.9.16). Therefore, infinitesimal coordinate transformations

$$z^A \to z'^A = z^A + \xi^B D_B z^A \tag{6.4.33}$$

generated by conformal Killing supervectors are exactly the superconformal transformations (2.9.17).

Now, let us find Killing supervectors of flat global superspace. In accordance with equation (6.4.12), Killing supervectors are conformal Killing supervectors that obey the constraint

$$D^\alpha \bar{D}^{\dot\alpha} \xi_{\alpha\dot\alpha} = 0 \quad \Leftrightarrow \quad \bar{D}^{\dot\alpha} D^\alpha \xi_{\alpha\dot\alpha} = 0. \tag{6.4.34}$$

Those superfields (6.4.30*b*) which respect this constraint are given by

$$\xi_{\alpha\dot\alpha}(x, \theta, \bar\theta) = b_{\alpha\dot\alpha} + K_\alpha{}^\beta x_{\beta\dot\alpha} + \bar{K}_{\dot\alpha}{}^{\dot\beta} x_{\alpha\dot\beta} - 4\mathrm{i}(\theta_\alpha \bar\epsilon_{\dot\alpha} + \bar\theta_{\dot\alpha} \epsilon_\alpha) + 2\mathrm{i}(\theta_\alpha \bar\theta^{\dot\beta} \bar{K}_{\dot\alpha\dot\beta} + \bar\theta_{\dot\alpha} \theta^\beta K_{\alpha\beta}). \tag{6.4.35}$$

Clearly, the transformations induced by Killing supervectors are exactly the super Poincaré transformations.

6.4.4. *Superconformal models*

As is well known, the conformal group is the space–time symmetry group of massless relativistic field theories (see Section 1.7). Similarly, there exist a variety of massless supersymmetric field theories possessing an invariance under the superconformal group. Such theories are called 'superconformal'. More precisely, a supersymmetric dynamical system with an action superfunctional $S[\chi]$, where χ are tensor superfield dynamical variables, is said to be superconformal if (1) the space of superfield histories $\mathbb{V}$ is endowed with an action of the superformal group

$$T(g)\colon \chi(z) \to (T(g)\chi)(z) \equiv \chi_g(z) \qquad \forall\, \chi \in \mathbb{V}$$

$$T(g_1 g_2) = T(g_1) T(g_2)$$

for every elements g_1 and g_2 from the superconformal group; and (2) the action superfunctional is invariant with respect to arbitrary superconformal transformations,

$$S[\chi_g] = S[\chi] \qquad \forall\, x \in \mathbb{V}.$$

In all known superconformal models, we have linear superconformal representations, and they can be described as follows. Let g be a group element of the form

$$g = \exp\left[\mathrm{i}\left(-b^a \mathbf{p}_a - f^a \mathbf{v}_a + \frac{1}{2} K^{ab} \mathbf{j}_{ab} + \Delta \mathbf{d} + \Omega \mathbf{a} + \varepsilon \mathbf{q} + \bar\varepsilon \bar{\mathbf{q}} + \eta \mathbf{s} + \bar\eta \bar{\mathbf{s}}\right)\right] \tag{6.4.36}$$

and let $\xi = \xi^A D_A$ be the conformal Killing supervectors having in the expansion (6.4.30*a*) the same coefficient as in expression (6.4.36). Denote by

$$\mathscr{J}_g(z) = \mathrm{Ber}\left(\frac{\partial(g\cdot y^b, g\cdot\theta^\beta)}{\partial(y^a, \theta^\alpha)}\right)\Bigg|_{y=x+i\theta\sigma\bar\theta} \tag{6.4.37}$$

the restriction on $\mathbb{R}^{4|4}(\theta\sigma\bar\theta)$ of the Berezinian of the superconformal transformation $y^a \to g\cdot y^a$, $\theta^\alpha \to g\cdot\theta^\alpha$ on $\mathbb{C}^{4|2}$ (constructed in Section 2.9), and introduce the operator

$$\begin{aligned}\mathscr{K}[\xi] = \xi^A \mathrm{D}_A + K[\xi]^{\alpha\beta}M_{\alpha\beta} + \bar K[\xi]^{\dot\alpha\dot\beta}\bar M_{\dot\alpha\dot\beta}\\ K[\xi]^{\alpha\beta} = \mathrm{D}^{(\alpha}\xi^{\beta)}\end{aligned} \tag{6.4.38}$$

associated with ξ (in accordance with the general prescription (6.4.4, 11)). Now, for arbitrary numbers $d_{(+)}$ and $d_{(-)}$, on the linear space of tensor superfields we define the following superconformal representation:

$$T(g)\colon \chi(z) \to \chi_g(z) = (\mathscr{J}_g)^{-(1/3)d_{(+)}}(\bar{\mathscr{J}}_g)^{-(1/3)d_{(-)}}\, \mathrm{e}^{-\mathscr{K}[\xi]}\xi(z). \tag{6.4.39}$$

In the infinitesimal case, the above transformation reads (compare with equation (6.4.15))

$$\delta\chi = -\mathscr{K}[\xi]\chi + \{d_{(+)}\sigma[\xi]\chi + d_{(-)}\bar\sigma[\xi]\}\chi \tag{6.4.40}$$

where

$$\begin{aligned}\sigma[\xi] &= \frac{1}{3}(\mathrm{D}^\alpha\xi_\alpha + 2\bar{\mathrm{D}}_{\dot\alpha}\bar\xi^{\dot\alpha}) = -\frac{1}{3}(\partial_a\xi^a - \mathrm{D}_\alpha\xi^\alpha)\\ &= -\frac{1}{3}(\partial_a\lambda^a(y,\theta) - \partial_\alpha\lambda^\alpha(y,\theta))\Big|_{y=x+i\theta\sigma\bar\theta}\end{aligned} \tag{6.4.41}$$

with $\lambda^a(y,\theta)$ and $\lambda^\alpha(y,\theta)$ defined as in relation (2.9.16). To derive equation (6.4.41), one can use the relation (6.4.32).

When choosing in representation (6.4.39) the dilatations $g = \exp(\mathrm{i}\Delta\mathbf{d})$ and the axial rotations $g = \exp(\mathrm{i}\Delta\mathbf{a})$, one obtains

$$\chi'(x,\theta,\bar\theta) = \mathrm{e}^{-(d_{(+)}+d_{(-)})\Delta}\,\chi(\mathrm{e}^{-\Delta}x, \mathrm{e}^{-(1/2)\Delta}\theta, \mathrm{e}^{-(1/2)\Delta}\bar\theta) \tag{6.4.42}$$

and

$$\chi'(x,\theta,\bar\theta) = \mathrm{e}^{-(\mathrm{i}/3)(d_{(+)}-d_{(-)})\Omega}\chi(x, \mathrm{e}^{(\mathrm{i}/2)\Omega}\theta, \mathrm{e}^{-(\mathrm{i}/2)\Omega}\bar\theta) \tag{6.4.43}$$

respectively. Hence $(d_{(+)}+d_{(-)})$ is the dimension of χ, and $(d_{(+)}-d_{(-)})$ is proportional to the axial charge of χ. $d_{(+)}$ and $d_{(-)}$ will be called the 'superconformal weights' of χ.

To find generators of the representation under consideration, we rewrite, with the help of relations (6.4.32, 41), the expression (6.4.40) in the manner

$$-\delta\chi=\frac{1}{2}\{\lambda^a(x_{(+)},\theta)+\bar{\lambda}^a(x_{(-)},\bar{\theta})\}\partial_a\chi+\{\lambda^\alpha(x_{(+)},\theta)\partial_\alpha+\bar{\lambda}^{\dot{\alpha}}(x_{(-)},\bar{\theta})\bar{\partial}_{\dot{\alpha}}\}\chi$$
$$+\{\mathrm{D}^\alpha\lambda^\beta(x_{(+)},\theta)M_{\alpha\beta}-\bar{\mathrm{D}}^{\dot{\alpha}}\bar{\lambda}^{\dot{\beta}}(x_{(-)},\bar{\theta})\}\bar{M}_{\dot{\alpha}\dot{\beta}}\}\chi$$
$$+\frac{1}{3}\left\{d_{(+)}(\partial_a\lambda^a(y,\theta)-\partial_\alpha\lambda^\alpha(y,\theta))\Big|_{y=x_{(+)}}\right.$$
$$\left.+d_{(-)}(\partial_a\bar{\lambda}^a(\bar{y},\bar{\theta})-\bar{\partial}_{\dot{\alpha}}\bar{\lambda}^{\dot{\alpha}}(\bar{y},\bar{\theta}))\Big|_{\bar{y}=x_{(-)}}\right\}\chi$$

where $x^a_{(\pm)}=x^a\pm\mathrm{i}\theta\sigma^a\bar{\theta}$, λ^a and λ^α are given in expression (2.9.16). Then, one obtains the super Poincaré generators (2.4.37), the dilatation generator

$$\mathbf{D}=\frac{\mathrm{i}}{2}\{(x^a_{(+)}+x^a_{(-)})\partial_a+\theta^\alpha\partial_\alpha+\bar{\theta}^{\dot{\alpha}}\bar{\partial}_{\dot{\alpha}}\}+\mathrm{i}(d_{(+)}+d_{(-)}) \qquad (6.4.44a)$$

the axial generator

$$\mathbf{A}=\frac{1}{2}(\theta^\alpha\partial_\alpha-\bar{\theta}^{\dot{\alpha}}\bar{\partial}_{\dot{\alpha}})-\frac{1}{3}(d_{(+)}-d_{(-)}) \qquad (6.4.44b)$$

the special conformal generators

$$\mathbf{V}_{\alpha\dot{\alpha}}=(\sigma^a)_{\alpha\dot{\alpha}}\mathbf{V}_a=-\frac{\mathrm{i}}{2}(x_{(+)\alpha}{}^{\dot{\beta}}x_{(+)}{}^{\beta}{}_{\dot{\alpha}}+x_{(-)\alpha}{}^{\dot{\beta}}x_{(-)}{}^{\beta}{}_{\dot{\alpha}})\partial_{\beta\dot{\beta}}$$
$$+2\mathrm{i}(\theta_\alpha x_{(+)}{}^{\beta}{}_{\dot{\alpha}}\partial_\beta+\bar{\theta}_{\dot{\alpha}}x_{(-)\alpha}{}^{\dot{\beta}}\bar{\partial}_{\dot{\beta}})-2\mathrm{i}(x_{(+)}{}^{\beta}{}_{\dot{\alpha}}M_{\alpha\beta}+x_{(-)\alpha}{}^{\dot{\beta}}\bar{M}_{\dot{\alpha}\dot{\beta}})$$
$$+2\mathrm{i}(d_{(+)}x_{(+)\alpha\dot{\alpha}}+d_{(-)}x_{(-)\alpha\dot{\alpha}}) \qquad (6.4.44c)$$

and the S-supersymmetry generators

$$\mathbf{S}_\alpha=-\mathrm{i}x_{(+)\alpha}{}^{\dot{\beta}}h^b\partial_{\beta\dot{\beta}}+2\mathrm{i}\theta^2\partial_\alpha+x_{(-)\alpha}{}^{\dot{\beta}}\bar{\partial}_{\dot{\beta}}-4\mathrm{i}\theta^\beta M_{\alpha\beta}+4\mathrm{i}d_{(+)}\theta_\alpha$$
$$\bar{\mathbf{S}}_{\dot{\alpha}}=-\mathrm{i}x_{(-)}{}^{\beta}{}_{\dot{\alpha}}\bar{\theta}^{\dot{\beta}}\partial_{\beta\dot{\beta}}+x_{(+)}{}^{\beta}{}_{\dot{\alpha}}\partial_\beta-2\mathrm{i}\bar{\theta}^2\bar{\partial}_{\dot{\alpha}}-4\mathrm{i}\bar{\theta}^{\dot{\beta}}\bar{M}_{\dot{\alpha}\dot{\beta}}+4\mathrm{i}d_{(-)}\bar{\theta}_{\dot{\alpha}}. \qquad (6.4.44d))$$

It is not difficult to check that the generators satisfy the (amti)commutation relations (2.9.21).

Now, consider two examples of superconformal models. Our first example is the super Yang–Mills theory described by the action superfunctional

$$S[\Phi,\bar{\Phi},V]=\int \mathrm{d}^8z\,\bar{\Phi}_i(\mathrm{e}^{2V})^i{}_j\Phi^j+\left\{\frac{1}{3}\lambda_{ijk}\int \mathrm{d}^6z\,\Phi^i\Phi^j\Phi^k+\mathrm{c.c.}\right\}$$
$$+\frac{1}{2g^2}\int \mathrm{d}^6z\,\mathrm{tr}\,W^\alpha W_\alpha \qquad (6.4.45)$$

$$\bar{\mathrm{D}}_{\dot{\alpha}}\Phi^i=0 \qquad W_\alpha=-\frac{1}{8}\bar{\mathrm{D}}^2(\mathrm{e}^{-2V}\mathrm{D}_\alpha\mathrm{e}^{2V}).$$

The action is invariant under the superconformal transformations

$$\delta\Phi^i = -\xi^A \mathrm{D}_A \Phi^i + \sigma[\xi]\Phi^i \qquad \delta V = -\xi^A \mathrm{D}_A V \tag{6.4.46}$$

with $\xi = \xi^A \mathrm{D}_A$ being an arbitrary conformal Killing supervector. This follows from the fact that the curved-superspace analogue (6.3.32) (with the same chiral superpotential) of the above action turns out to be invariant under the super Weyl transformations (6.4.3) supplemented by

$$\delta\Phi^i = -\sigma\Phi^i \qquad \delta V = 0.$$

The second example is the chiral spinor model with the action superfunctional

$$S[\chi, \bar{\chi}] = -\frac{1}{k^2}\int \mathrm{d}^8 z\, G \ln G$$
$$G = \frac{1}{2}(\mathrm{D}^\alpha \chi_\alpha + \bar{\mathrm{D}}_{\dot{\alpha}} \bar{\chi}^{\dot{\alpha}}) \qquad \bar{\mathrm{D}}_{\dot{\alpha}} \chi_\alpha = 0 \tag{6.4.47}$$

k being a constant. Here χ_α is a chiral spinor superfield. The action is invariant under the superconformal transformations

$$\delta\chi_\alpha = -\xi^A \mathrm{D}_A \chi_\alpha - K[\xi]_\alpha{}^\beta \chi_\beta + \frac{3}{2}\sigma[\xi]\chi_\alpha \tag{6.4.48}$$

for an arbitrary conformal Killing supervector ξ. The statement follows from the fact that $S[\chi, \bar{\chi}]$ can be continued to a curved superspace by the rule

$$S[\chi, \bar{\chi}; \mathcal{D}] = -\frac{1}{k^2}\int \mathrm{d}^8 z\, E^{-1} G \ln G$$
$$G = \frac{1}{2}(\mathcal{D}^\alpha \chi_\alpha + \bar{\mathcal{D}}_{\dot{\alpha}} \bar{\chi}^{\dot{\alpha}}) \qquad \bar{\mathcal{D}}_{\dot{\alpha}} \chi_\alpha = 0 \tag{6.4.49}$$

with χ_α being a covariantly chiral spinor superfield. The superfunctional obtained proves to be invariant under the super Weyl transformations (6.4.3) supplemented by

$$\delta\chi_\alpha = -\frac{3}{2}\sigma\chi_\alpha. \tag{6.4.50}$$

6.4.5. *On-shell massless conformal superfields*

In Section 2.7, we studied on-shell massless superfields. Recall that a tensor superfield of Lorentz type $(A/2, B/2)$ (totally symmetric in its undotted indices and, independently, in its dotted indices) $G_{\alpha_1 \ldots \alpha_A \dot{\alpha}_1 \ldots \dot{\alpha}_B}(z)$ is said to be on-shell massless if it satisfies the system of supersymmetric massless equations

$$\mathrm{D}^2 G = \bar{\mathrm{D}}^2 G = 0 \tag{6.4.51a}$$

$$\partial_{\gamma\dot{\gamma}} \mathrm{D}^\gamma G = \partial_{\gamma\dot{\gamma}} \bar{\mathrm{D}}^{\dot{\gamma}} G = 0 \tag{6.4.51b}$$

where superfield indices are suppressed, and the supplementary conditions

$$\partial^{\alpha_1\dot{\gamma}} G_{\alpha_1\ldots\alpha_A\dot{\alpha}_1\ldots\dot{\alpha}_B} = 0 \qquad (6.4.52a)$$

$$\partial^{\gamma\dot{\alpha}_1} G_{\alpha_1\ldots\alpha_A\dot{\alpha}_1\ldots\dot{\alpha}_B} = 0 \qquad (6.4.52b)$$

of which the former (latter) must be imposed when $A \neq 0$ $(B \neq 0)$. Now, we would like to obtain an answer to the following problem. Suppose, G transforms according to some representation of the superconformal group,

$$\delta G = -\mathscr{K}[\xi]G + \{d_{(+)}\sigma[\xi] + d_{(-)}\bar{\sigma}[\xi]\}G \qquad (6.4.53)$$

for any conformal Killing supervector ξ. The question is: in what cases is this transformation law consistent with the equations (6.4.51, 52)? In other words, under which conditions does δG remain on-shell massless? Let us remind ourselves that the similar problem for on-shell massless fields in Minkowski space was solved in Section 1.8.

To begin with, consider the case $A \neq 0$. Under the special conformal transformations, G changes according to the rule

$$\delta_f G_{\alpha_1\ldots\alpha_A\dot{\alpha}_1\ldots\dot{\alpha}_B} = \frac{\mathrm{i}}{2} f^{\beta\dot{\beta}} \mathbf{V}_{\beta\dot{\beta}} G_{\alpha_1\ldots\alpha_A\dot{\alpha}_1\ldots\dot{\alpha}_B}$$

with $\mathbf{V}_{\alpha\dot{\alpha}}$ written as in expression (6.4.44c). We demand the requirement

$$\partial^{\alpha_1\dot{\gamma}} \delta_f G_{\alpha_1\ldots\alpha_A\dot{\alpha}_1\ldots\dot{\alpha}_B} = 0. \qquad (6.4.54)$$

With the help of condition (6.4.52a), it is not difficult to verify that

$$\begin{aligned}
\partial^{\alpha_1\dot{\gamma}} \delta_f G_{\alpha_1\ldots\alpha_A\dot{\alpha}_1\ldots\dot{\alpha}_B} = 2f^{\beta\dot{\beta}} \Big\{ & \delta_{\dot{\beta}}{}^{\dot{\gamma}} \Big(d_{(+)} + d_{(-)} - \frac{1}{2}A - 1 \Big) G_{\beta\alpha_2\ldots\alpha_A\dot{\alpha}_1\ldots\dot{\alpha}_B} \\
& + \bar{\theta}_{\dot{\beta}} \bar{\mathrm{D}}^{\dot{\gamma}} G_{\beta\alpha_2\ldots\alpha_A\dot{\alpha}_1\ldots\dot{\alpha}_B} - \sum_{k=1}^{B} G_{\beta\alpha_2\ldots\alpha_A\dot{\alpha}_1\ldots\hat{\dot{\alpha}}_k\ldots\dot{\alpha}_B(\dot{\beta}}\delta^{\dot{\gamma})}{}_{\dot{\alpha}_k} \\
& - \delta_{\dot{\beta}}{}^{\dot{\gamma}} \theta_\beta \mathrm{D}^{\alpha_1} G_{\alpha_1\alpha_2\ldots\alpha_A\dot{\alpha}_1\ldots\dot{\alpha}_B} \Big\}. \qquad (6.4.55)
\end{aligned}$$

Let us point out that the second equation (6.4.51) and the supplementary condition (6.4.52a) lead to the consequence

$$\mathrm{D}^{\alpha_1} G_{\alpha_1\alpha_2\ldots\alpha_A\dot{\alpha}_1\ldots\dot{\alpha}_B} = 0 \qquad (6.4.56)$$

(for more details see Section 2.7). Hence the last term in relation (6.4.55) vanishes. To cancel the other terms in relation (6.4.55), one must demand the system of conditions:

$$\begin{aligned}
&(1)\ d_{(+)} + d_{(-)} = \frac{1}{2}A + 1 \\
&(2)\ B = 0 \qquad (6.4.57) \\
&(3)\ \bar{\mathrm{D}}_{\dot{\gamma}} G_{\alpha_1\ldots\alpha_A} = 0
\end{aligned}$$

So, our superfield must be chiral with undotted indices only. Finally, let us recall that $\sigma[\xi]$ is chiral. Then the variation (6.4.53) proves to be chiral, $\bar{D}_{\dot{\alpha}}\delta G=0$, when

$$d_{(-)}=0. \tag{6.4.58}$$

As a result, the superconformal weights $d_{(+)}$ and $d_{(-)}$ of G are fixed completely. Further examination of the compatibility of the superconformal law (6.4.53) with the massless equations (6.4.51, 52) can be shown to give no additional restrictions on G.

The case $B\neq 0$ is treated similarly.

Finally, let us examine the case $A=B=0$. We begin by observing that equations (6.4.51) lead to the on-shell equation $\Box G=0$. Then, imposing the requirement

$$\Box\delta_f G=0 \tag{6.4.59}$$

proves to be equivalent to

$$d_{(+)}+d_{(-)}=1. \tag{6.4.60}$$

Next, imposing the conditions

$$D^2\delta_f G=\bar{D}^2\delta_f G=0 \tag{6.4.61}$$

leads to the equations

$$\begin{aligned}(1-d_{(+)})D_\alpha G&=0\\(1-d_{(-)})\bar{D}_{\dot{\alpha}}G&=0.\end{aligned} \tag{6.4.62}$$

The system of equations (6.4.60, 62) has the solutions:

$$d_{(+)}=1 \qquad d_{(-)}=0 \qquad \bar{D}_{\dot{\alpha}}G=0 \tag{6.4.64a}$$

or

$$d_{(+)}=0 \qquad d_{(-)}=1 \qquad D_\alpha G=0. \tag{6.4.63b}$$

From the above considerations we see that only undotted (or scalar) chiral and dotted (or scalar) antichiral on-shell massless superfields constitute representations of the superconformal group. Such superfields are said to be 'on-shell massless conformal superfields'.

In conclusion, let us summarize all the conditions determining on-shell massless conformal superfields:

The scalar case ($A=B=0$)

$$\begin{aligned}\bar{D}_{\dot{\alpha}}G&=0 \qquad D^2G=0\\ d_{(+)}&=1 \qquad d_{(-)}=0\end{aligned} \tag{6.4.64a}$$

or

$$\mathrm{D}_\alpha \bar{G} = 0 \qquad \bar{\mathrm{D}}^2 \bar{G} = 0$$
$$d_{(+)} = 0 \qquad d_{(-)} = 1. \tag{6.4.64b}$$

The undotted case $(A \neq 0,\ B = 0)$

$$\bar{\mathrm{D}}_{\dot\gamma} G_{\alpha_1 \ldots \alpha_A} = 0 \qquad \mathrm{D}^{\alpha_1} G_{\alpha_1 \alpha_2 \ldots \alpha_A} = 0$$
$$d_{(+)} = A/2 + 1 = \kappa(G) + 1 \qquad d_{(-)} = 0. \tag{6.4.64c}$$

The dotted case $(A = 0,\ B \neq 0)$

$$\mathrm{D}_\alpha \bar{G}_{\dot\alpha \ldots \dot\alpha_B} = 0 \qquad \bar{\mathrm{D}}^{\dot\alpha_1} \bar{G}_{\dot\alpha_1 \dot\alpha_2 \ldots \dot\alpha_B} = 0$$
$$d_{(+)} = 0 \qquad d_{(-)} = B/2 + 1 = -\kappa(\bar{G}) + 1/2. \tag{6.4.64d)}$$

Here $\kappa(G)$ or $\kappa(\bar{G})$ denotes the superhelicity of the superfield under consideration.

6.5. Conformally flat superspaces, anti-de Sitter superspace

In Section 5.1, we introduced the notions of flat superspace and conformally flat superspace. Recall that a superspace is said to be flat if there exists a coordinate system in which the gravitational superfield and the chiral compensator have the flat form

$$\mathscr{H}^m = \delta_a{}^m \theta \sigma^a \bar\theta \qquad \varphi = 1 \tag{6.5.1}$$

a superspace is said to be conformally flat if in some coordinate system the gravitational superfield has the flat form

$$\mathscr{H}^m = \delta_a{}^m \theta \sigma^a \bar\theta \tag{6.5.2}$$

with φ being unrestricted. Now, our goal is to reformulate these definitions in terms of the supertorsion tensors R, G_a and $W_{\alpha\beta\gamma}$, that is, in a covariant fashion. We also consider andi-de Sitter superspace—the simplest solution of the Einstein supergravity equations with cosmological term (6.1.22).

6.5.1. Flat superspace

To begin with, let us obtain necessary and sufficient conditions for a superspace to be flat. In spite of the simplicity of the problem, it deserves some discussion.

In the case of a supergeometry determined (in the gravitational superfield gauge) by the prepotentials (6.5.1), the corresponding supertorsion tensors R, G_a and $W_{\alpha\beta\gamma}$ are zero,

$$R = G_a = W_{\alpha\beta\gamma} = 0. \tag{6.5.3}$$

So, every flat superspace should be characterized by the conditions (6.5.3), they are necessary conditions. Remarkably, they also turn out to be sufficient. Let us argue this statement.

We choose the gravitational superfield gauge (5.4.38) supplemented by the Wess–Zumino gauge (5.8.1) and analyse the consequences of equations (6.5.3). First, from relations (5.8.17, 19) we see that the supergravity auxiliary fields vanish,

$$\mathbf{B}=\mathbf{A}^a=0. \tag{6.5.4}$$

Secondly, from relations (5.8.21, 22) we deduce

$$\mathscr{R}_{abcd}=0 \tag{6.5.5}$$

where $\mathscr{R}_{abcd}$ is the curvature tensor associated with the space–time covariant derivatives ∇_a defined by equations (5.8.16, 23–25). Finally, owing to relation (5.8.30), the gravitino field strengths (5.8.27) vanish,

$$\Psi_{ab\gamma}=\bar{\Psi}_{ab}{}^{\dot{\gamma}}=0. \tag{6.5.6}$$

The system (6.5.4)–(6.5.6) represents the total set of component requirements which equations (6.5.3) lead to. Recalling relations (5.8.20) and (5.8.27), we can rewrite expressions (6.5.5) and (6.5.6) as follows:

$$[\nabla_a, \nabla_b]=\mathscr{T}_{ab}{}^c\nabla_c \qquad \mathscr{T}_{ab}{}^c=-\mathrm{i}\Psi_{[a}\sigma^c\bar{\Psi}_{b]} \tag{6.5.7}$$

and

$$2\nabla_{[a}\Psi_{b]\gamma}=\mathscr{T}_{ab}{}^c\Psi_{c\gamma} \qquad 2\nabla_{[a}\bar{\Psi}_{b]}{}^{\dot{\gamma}}=\mathscr{T}_{ab}{}^c\bar{\Psi}_c{}^{\dot{\gamma}}. \tag{6.5.8}$$

As may be seen, space–time may possess non-trivial torsion even if all the superfield strengths characterizing the supergeometry are zero. However, the torsion turns out to be gauge dependent and, under the conditions (6.5.4–6), it can be switched off using the local supersymmetry invariance (recall that in the Wess–Zumino gauge, the transformations (5.8.43) are also at our disposal). Indeed, the relations (6.5.7) and (6.5.8) show that

$$\Psi_{a\gamma}=\nabla_a\xi_\gamma \tag{6.5.9}$$

for some undotted spinor field $\xi_\gamma(x)$. On the other hand, under equation (6.5.4) the gravitino transformation law (5.1.45) takes the form

$$\delta\Psi_{m\alpha}=-2\nabla_m\varepsilon_\alpha.$$

Therefore, the local supersymmetry invariance can be used to impose the gauge

$$\Psi_{a\gamma}=\bar{\Psi}_a{}^{\dot{\gamma}}=0 \quad\Rightarrow\quad \mathscr{T}_{ab}{}^c=0. \tag{6.5.10}$$

Note that the system of relations (6.5.4)–(6.5.6) is locally supersymmetric invariant.

In the gauge (6.5.10), the only non-zero field in the expressions (5.8.1) is the vierbein, and the covariant derivatives ∇_a satisfy the algebra

$$[\nabla_a, \nabla_b] = 0.$$

Hence, the space-time general coordinate and local Lorentz invariances can be used to take the gauge

$$e_a{}^m = \delta_a{}^m. \tag{6.5.11}$$

This completes the proof.

In summary, a supergeometry (under the Einstein supergravity constraints) is flat if and only if the corresponding supertorsion tensors R, G_a and $W_{\alpha\beta\gamma}$ vanish.

6.5.2. Conformally flat superspace

Consider a supergeometry admitting the gauge fixing (6.5.2). Evidently, the corresponding covariant derivatives $\mathcal{D}_a$ are connected to flat global derivatives D_A by means of some super Weyl transformation:

$$\begin{gathered}\mathcal{D}_\alpha = F D_\alpha - 2(D^\beta F) M_{\alpha\beta} \qquad \bar{\mathcal{D}}_{\dot\alpha} = \bar{F} \bar{D}_{\dot\alpha} - 2(\bar{D}^{\dot\beta} \bar{F}) \bar{M}_{\dot\alpha\dot\beta} \\ \mathcal{D}_{\alpha\dot\alpha} = \bar{F} F \partial_{\alpha\dot\alpha} + \frac{i}{2} F(D_\alpha \bar{F}) \bar{D}_{\dot\alpha} + \frac{i}{2} \bar{F}(\bar{D}_{\dot\alpha} F) D_\alpha - iF(D_\alpha \bar{D}^{\dot\beta} \bar{F}) \bar{M}_{\dot\alpha\dot\beta} - i\bar{F}(\bar{D}_{\dot\alpha} D^\beta F) M_{\alpha\beta}\end{gathered} \tag{6.5.12}$$

where

$$F = \varphi^{1/2} \bar{\varphi}^{-1} \qquad \bar{D}_{\dot\alpha} \varphi = 0.$$

Expressions for the supertorsion tensors can be read off from relations (5.5.14):

$$\begin{gathered}R = -\frac{1}{4} \varphi^{-2} \bar{D}^2 \bar{\varphi} \\ G_{\alpha\dot\alpha} = (\bar{\varphi}\varphi)^{-1/2} \left\{ \frac{1}{2} (D_\alpha \ln \varphi) \bar{D}_{\dot\alpha} \ln \bar{\varphi} + i \partial_{\alpha\dot\alpha} \ln(\bar{\varphi}/\varphi) \right\}\end{gathered} \tag{6.5.13}$$

and

$$W_{\alpha\beta\gamma} = 0. \tag{6.5.14}$$

We see that the super Weyl tensor of any conformally flat superspace is zero, Now, we are going to argue that the requirement (6.5.14) turns out to be the sufficient condition for conformal flatness, for a wide class of supergeometries.

Let $\mathcal{D}_A$ be a set of covariant derivatives which satisfy equation (6.5.14). If the corresponding chiral scalar $R \neq 0$, we change $\mathcal{D}_A$ by some super Weyl transformed derivatives $\mathcal{D}'_A$, according to rule (5.5.13), such that $R' = 0$. To achieve this, we must choose, due to relation (5.5.14), a covariantly chiral scalar superfield σ satisfying the equation

$$-\frac{1}{4} (\mathcal{D}^2 - 4\bar{R}) e^\sigma = 0 \qquad \bar{\mathcal{D}}_{\dot\alpha} \sigma = 0. \tag{6.5.15}$$

As a particular solution, one can take

$$e^{\sigma}=1+\frac{1}{4}\left(\Box_{+}-\frac{1}{4}(\bar{\mathscr{D}}^2\bar{R})+R\bar{R}\right)^{-1}(\bar{\mathscr{D}}^2-4R)\bar{R} \tag{6.5.16}$$

where $\Box_{+}$ is the chiral d'Alembertian (6.3.18). In order that the expression (6.5.16) is well defined, all supergeometry superfields should obey appropriate boundary conditions (for instance, be rapidly vanishing at infinity). Note that $W'_{\alpha\beta\gamma}=0$, due to relation (5.5.14). Suppose that $G'_a=0$ also. Then, the supergeometry determined by $\mathscr{D}'_A$ is flat, and, as a result, the supergeometry determined by $\mathscr{D}_A$ is conformally flat. Thus, it remains to show that the relations

$$R=W_{\alpha\beta\gamma}=0 \tag{6.5.17}$$

imply

$$G_a=0. \tag{6.5.18}$$

Our proof will not pretend to strictness, we simply sketch the basic steps. Let $\mathscr{D}_A$ be a set of covariant derivatives characterized by conditions (6.5.17). We choose the Lorentz gauge (5.4.8) resulting in

$$\mathscr{D}_\alpha=E_\alpha+\Omega_{\alpha\beta\gamma}M^{\beta\gamma} \qquad \bar{\mathscr{D}}_{\dot\alpha}=\bar{E}_{\dot\alpha}+\bar{\Omega}_{\dot\alpha\dot\beta\dot\gamma}\bar{M}^{\dot\beta\dot\gamma}. \tag{6.5.19}$$

Such a gauge fixing leaves the possibility of making some residual Lorentz transformations of the form

$$\begin{gathered}\mathscr{D}_A\to\mathscr{D}'_A=e^{l}\mathscr{D}_A e^{-l}\\ l=l_{\beta\gamma}M^{\beta\gamma}+\bar{l}_{\dot\beta\dot\gamma}\bar{M}^{\dot\beta\dot\gamma} \qquad \bar{\mathscr{D}}_{\dot\alpha}l_{\beta\gamma}=0.\end{gathered} \tag{6.5.20}$$

Next, since $\{\mathscr{D}_\alpha,\mathscr{D}_\beta\}=0$, the superconnection $\Omega_{\alpha\beta\gamma}$ is trivial and we can write

$$\begin{gathered}\mathscr{D}_\alpha=e^{U}\hat{E}_\alpha e^{-U} \qquad \bar{\mathscr{D}}_{\dot\alpha}=e^{\bar{U}}\hat{\bar{E}}_{\dot\alpha}e^{-\bar{U}}\\ U=U_{\beta\gamma}M^{\beta\gamma} \qquad \hat{E}_\alpha=\hat{E}_\alpha{}^M\partial_M \qquad \{\hat{E}_\alpha,\hat{E}_\beta\}=0.\end{gathered} \tag{6.5.21}$$

There is some inherent arbitrariness in choosing $U_{\alpha\beta}$. Indeed, changing e^U to

$$e^{U'}=e^{U}e^{h} \qquad h=h_{\beta\gamma}M^{\beta\gamma} \qquad \hat{E}_\alpha h_{\beta\gamma}=0 \tag{6.5.22}$$

leaves $\mathscr{D}_A$ unchanged. The gauge freedom (6.5.20, 22) can be used to impose the gauge

$$\hat{E}^2U_{\beta\gamma}=\hat{\bar{E}}^2U_{\beta\gamma}=0. \tag{6.5.23}$$

Further, due to relation (6.5.17), the commutator $[\bar{\mathscr{D}}_{\dot\alpha},\mathscr{D}_{\beta\dot\beta}]$ does not contain $M^{\beta\gamma}$-terms (see expressions (5.5.6)). In the gauge (6.5.19), this requirement is equivalent to

$$\bar{\mathscr{D}}^2\Omega_{\alpha\beta\gamma}=0. \tag{6.5.24}$$

Under reasonable boundary conditions, the system of equations (6.5.23) and

(6.5.24) has only trivial solutions, so

$$\Omega_{\alpha\beta\gamma}=0.$$

As a result, the commutator $[\bar{\mathscr{D}}_{\dot{\alpha}}, \mathscr{D}_{\beta\dot{\beta}}]$ does not involve $\bar{M}^{\dot{\beta}\dot{\gamma}}$-terms, which leads to

$$\bar{\mathscr{D}}_{\dot{\beta}}G_a=0 \quad \Rightarrow \quad G_a=0.$$

In summary, it is the super Weyl tensor that measures whether or not a superspace is conformally flat. Superspace is conformally flat if and only if the corresponding super Weyl tensor is identically zero.

6.5.3. Physical sense of conformal flatness

We have seen that every curved superspace admits the choice of the Wess–Zumino gauge (5.8.1). What are the peculiar features of conformally flat superspaces?

In the Wess–Zumino gauge, the residual gauge freedom consists of the space–time general coordinate, local Lorentz and local supersymmetry transformations. When our superspace is conformally flat, the above gauge freedom can be used to impose the gauge

$$\begin{aligned} e_a{}^m &= e^{1/4}\delta_a{}^m \\ \Psi^m{}_\alpha &= e_a{}^m(\sigma^a\bar{\eta})_\alpha \\ A^m &= g^{mn}\partial_n\Omega+O(\Psi,\bar{\Psi}) \end{aligned} \tag{6.5.25}$$

for fields $\bar{\eta}^{\dot{\alpha}}(x)$ and $\Omega(x)$. Equivalently, the fields $e_a{}^m$, $\Psi^m{}_\alpha$ and A^m propagate only those modes that are generated by the Weyl transformations (5.1.28), the local S-supersymmetry transformations (5.1.30) and the local chiral transformations (5.1.29), respectively. In this and only this case, we are in a position to move from the Wess–Zumino gauge (5.8.1) to the gauge (6.5.2), applying a proper Weyl+local S-supersymmetry+local chiral transformation. As a result of equations (5.1.39, 40), the chiral compensator, transformed from the gauge (5.8.1) to the gauge (6.5.2), is of the form

$$\varphi^3(x,\theta)=\mathrm{e}^{-3/4}\{\mathrm{e}^{\mathrm{i}\hat{\Omega}(x)}+\theta^\alpha\hat{\eta}_\alpha(x)+\theta^2e^{-1/4}(x)B(x)\}. \tag{6.5.26}$$

Here $\hat{\Omega}=\bar{\hat{\Omega}}$ and $\hat{\eta}_\alpha$ are fields uniquely connected to the similar ones in gauge (6.5.25).

Given a conformally flat superspace admitting the gauge choice

$$e_a{}^m=e^{1/4}\delta_a{}^m \qquad \Psi^m{}_\alpha=A^m=0 \tag{6.5.27}$$

the corresponding supergeometry can be described by the superfields

$$\begin{aligned} &\mathscr{H}^m=\delta_a{}^m\theta\sigma^a\bar{\theta} \qquad \varphi(z)=\varphi(x+\mathrm{i}\theta\sigma\bar{\theta},\theta) \\ &\varphi(x,\theta)=e^{-1/4}\left\{1+\frac{1}{3}\theta^2e^{-1/4}(x)B(x)\right\}=\left[e^{1/4}(x)-\frac{1}{3}\theta^2B(x)\right]^{-1}. \end{aligned} \tag{6.5.28}$$

6.5.4. Anti-de Sitter superspace

We are going to construct a supergeometry determined by

$$W_{\alpha\beta\gamma} = G_a = 0$$
$$R = \mu = \text{const} \neq 0 \tag{6.5.29}$$

or, equivalently, possessing the covariant derivatives algebra

$$\{\mathscr{D}_\alpha, \bar{\mathscr{D}}_{\dot\alpha}\} = -2\mathrm{i}\mathscr{D}_{\alpha\dot\alpha}$$
$$\{\mathscr{D}_\alpha, \mathscr{D}_\beta\} = -4\bar\mu M_{\alpha\beta} \qquad \{\bar{\mathscr{D}}_{\dot\alpha}, \bar{\mathscr{D}}_{\dot\beta}\} = 4\mu \bar M_{\dot\alpha\dot\beta}$$
$$[\bar{\mathscr{D}}_{\dot\alpha}, \mathscr{D}_{\beta\dot\beta}] = -\mathrm{i}\mu\varepsilon_{\dot\alpha\dot\beta}\mathscr{D}_\beta \qquad [\mathscr{D}_\alpha, \mathscr{D}_{\beta\dot\beta}] = \mathrm{i}\bar\mu\varepsilon_{\alpha\beta}\bar{\mathscr{D}}_{\dot\beta} \tag{6.5.30}$$
$$[\mathscr{D}_{\alpha\dot\alpha}, \mathscr{D}_{\beta\dot\beta}] = -2\bar\mu\mu(\varepsilon_{\dot\alpha\dot\beta}M_{\alpha\beta} + \varepsilon_{\alpha\beta}\bar M_{\dot\alpha\dot\beta}).$$

It will be called the 'anti-de Sitter supergeometry'. Clearly, anti-de Sitter supergeometry is conformally flat.

One can verify that the supergeometry under consideration admits the gauge fixing (6.5.27). Therefore, it is within our power to take the prepotentials in the form (6.5.28). In accordance with the relations (6.5.13) and (6.5.29), flat (anti)chiral superfields $\bar\varphi(z)$ and $\varphi(z)$, $\bar{\mathrm{D}}_{\dot\alpha}\varphi = 0$, should satisfy the equations

$$-\frac{1}{4}\bar{\mathrm{D}}^2\bar\varphi = \mu\varphi^2 \qquad -\frac{1}{4}\mathrm{D}^2\varphi = \bar\mu\bar\varphi^2 \tag{6.5.31a}$$

$$\frac{1}{2}(\mathrm{D}_\alpha \ln \varphi)\bar{\mathrm{D}}_{\dot\alpha} \ln \bar\varphi + \mathrm{i}\partial_{\alpha\dot\alpha} \ln(\bar\varphi/\varphi) = 0. \tag{6.5.31b}$$

Note that the equations (6.5.31a) coincide in form with the dynamical equations (3.2.9) of the Wess–Zumino model with chiral superpotentials $\mathscr{L}_c(\Phi) = -(\mu/3)\Phi^3$. A particular solution of the above problem is

$$\varphi(x, \theta) = \left[1 - \frac{1}{4}\bar\mu\mu x^2 - \bar\mu\theta^2\right]^{-1} \tag{6.5.32}$$

where $x^2 = \eta_{mn}x^m x^n$.

From solution (6.5.32) we may read off the space–time vierbein

$$e_a{}^m = \left(1 - \frac{1}{4}\bar\mu\mu x^2\right)\delta_a{}^m \tag{6.5.33}$$

and the space–time metric

$$\mathrm{d}s^2 = \frac{1}{\left(1 - \frac{1}{4}\bar\mu\mu x^2\right)^2}\eta_{mn}\,\mathrm{d}x^m\,\mathrm{d}x^n. \tag{6.5.34}$$

This is the metric of the well-known space from general relativity, the anti-de Sitter space. As a manifold, anti-de Sitter space represents locally the domain of $\mathbb{R}^4$ contained inside the hyperboloid

$$x^2 = 4/\bar{\mu}\mu. \tag{6.5.35}$$

The Lorentz covariant derivatives associated with the vierbein (6.5.33) read

$$\begin{aligned} \nabla_a &= e_a + \frac{1}{2}\omega_a{}^{bc}M_{bc} \\ \omega_a{}^{bc} &= \frac{1}{2}\bar{\mu}\mu(\delta_a{}^b x^c - \delta_a{}^c x^b) \end{aligned} \tag{6.5.36}$$

where $x^a \equiv \delta^a_m x^m$, and satisfy the algebra

$$\begin{aligned} [\nabla_a, \nabla_b] &= -\bar{\mu}\mu M_{ab} \quad \Rightarrow \\ \mathcal{R}_{abcd} &= -\bar{\mu}\mu(\eta_{ab}\eta_{cd} - \eta_{ad}\eta_{bc}) \quad \Rightarrow \quad \mathcal{R} = -12\bar{\mu}\mu. \end{aligned} \tag{6.5.37}$$

Before going further, we would like to make one comment. In general relativity, there are known to be two (maximally symmetric) space–times of constant non-zero curvature, that is, of the form

$$C_{abcd} = 0 \qquad \mathcal{R}_{ab} = \frac{1}{4}g_{ab}\mathcal{R} \qquad \mathcal{R} = \text{const} \neq 0. \tag{6.5.38}$$

The choice $\mathcal{R} > 0$ corresponds to de Sitter space, and the choice $\mathcal{R} < 0$ corresponds to anti-de Sitter space. Clearly, the set of requirements (6.5.29) represents the supersymmetric analogue of that in (6.5.38). So, every supergeometry of the type (6.5.30) can be called a 'supergeometry of constant supertorsion' (recall that supertorsion determines supercurvature). Our consideration shows that, for any complex μ, the supergeometry described leads to a space–time geometry of constant negative curvature. Therefore, it is anti-de Sitter geometry which admits supersymmetrization. The de Sitter geometry cannot be extended to a supergeometry of constant supertorsion.

Now, we formulate what will be understood by 'anti-de Sitter superspace'. As a supermanifold, it is locally identified with the domain of $\mathbb{R}^{4|4}$ such that the bodies of c-number variables x^m lie inside the hyperboloid (6.5.35). The relevant supergeometry is determined by the prepotentials (6.5.28), with $\varphi(x, \theta)$ defined as in expression (6.5.32).

6.5.5. *Killing supervectors of anti-de Sitter superspace*

To develop supersymmetric field theory in anti-de Sitter superspace, it is necessary to determine all possible Killing supervectors. Recall that for every curved superspace, the corresponding Killing supervectors (if they exist) generate symmetry transformations, that is, ones leaving the covariant derivative unchanged.

The simplest way to find the Killing supervectors of anti-de Sitter superspace is to use the observation that Weyl equivalent supergeometries possess the same set of conformal Killing supervectors (see subsection 6.4.1).

Evidently, anti-de Sitter and flat supergeometries are Weyl equivalent. Above we have found all the conformal Killing supervectors of flat global superspace, hence, of anti–de Sitter superspace too. They are given by the expression (6.4.32), with $\lambda^a(x, \theta)$ and $\lambda^\alpha(x, \theta)$ written as in expression (2.9.16). Killing supervectors of anti–de Sitter superspace are those conformal Killing supervectors which satisfy the equation (6.4.23), with $\varphi(x, \theta)$ being the relevant chiral compensator (6.5.32). Solving equation (6.4.23) restricts the parameters in expression (2.9.16) as follows:

$$\Delta=\Omega=0 \qquad f^a=\frac{1}{4}\bar{\mu}\mu b^a \qquad \eta^\alpha=-\frac{1}{2}\bar{\mu}\epsilon^\alpha \qquad \bar{\eta}_{\dot{\alpha}}=-\frac{1}{2}\mu\bar{\epsilon}_{\dot{\alpha}}. \tag{6.5.39}$$

Notice that no restrictions on the Lorentz parameters $K_{\alpha\beta}$ appear, since the chiral compensator (6.5.32) is Lorentz invariant. Inserting the above expressions into relations (2.9.16) gives

$$\lambda^a(x, \theta)=K^a{}_b x^b+b^a\left(1+\frac{1}{4}\bar{\mu}\mu x^a\right)-\frac{1}{2}\bar{\mu}\mu x^a(b, x)+2\mathrm{i}\theta\sigma^a\bar{\epsilon}+\bar{\mu}\theta\sigma^a\tilde{\sigma}_b\epsilon x^b$$

$$\lambda^\alpha(x, \theta)=-K^\alpha{}_\beta\theta^\beta+\frac{1}{4}\bar{\mu}\mu b^a x^b(\theta\sigma^a\tilde{\sigma}_b)^\alpha+\epsilon^\alpha(1-\bar{\mu}\theta^2)+\frac{\mathrm{i}}{2}\mu(\bar{\epsilon}\tilde{\sigma}_b)^\alpha x^b. \tag{6.5.40}$$

The Killing supervector associated with these superparameters can be read off from expression (6.4.27).

Given a Killing supervector $\xi=\xi^A\mathrm{D}_A$ of anti-de Sitter superspace, it generates the operator $\mathscr{K}[\xi]$, defined by expression(6.4.38), which commutes with the anti-de Sitter covariant derivatives (6.5.12, 32). $\mathscr{K}[\xi]$ can be represented in the form

$$\mathscr{K}[\xi]=-\mathrm{i}\left[\frac{1}{2}b^{\alpha\dot{\alpha}}\mathbb{P}_{\alpha\dot{\alpha}}+K^{\alpha\beta}\mathbb{J}_{\alpha\beta}+\bar{K}^{\dot{\alpha}\dot{\beta}}\bar{\mathbb{J}}_{\dot{\alpha}\dot{\beta}}+\epsilon^\alpha\mathbb{Q}_\alpha+\bar{\epsilon}_{\dot{\alpha}}\bar{\mathbb{Q}}^{\dot{\alpha}}\right\}$$

$$\mathbb{P}_{\alpha\dot{\alpha}}=\mathbf{P}_{\alpha\dot{\alpha}}+\frac{1}{4}\bar{\mu}\mu\mathbf{V}_{\alpha\dot{\alpha}} \qquad \mathbb{J}_{\alpha\beta}=\mathbf{J}_{\alpha\beta} \tag{6.5.41}$$

$$\mathbb{Q}_\alpha=\mathbf{Q}_\alpha-\frac{1}{2}\bar{\mu}\mathbf{S}_\alpha \qquad \bar{\mathbb{Q}}_{\dot{\alpha}}=\bar{\mathbf{Q}}_{\dot{\alpha}}-\frac{1}{2}\mu\bar{\mathbf{S}}_{\dot{\alpha}}$$

where $\mathbf{P}_{\alpha\dot{\alpha}}$, $\mathbf{J}_{\alpha\beta}$, $\bar{\mathbf{J}}_{\dot{\alpha}\dot{\beta}}$, $\mathbf{Q}_\alpha$ and $\bar{\mathbf{Q}}_{\dot{\alpha}}$ are the super Poincaré generators (2.4.37), and $\mathbf{V}_{\alpha\dot{\alpha}}$, $\mathbf{S}_\alpha$ and $\bar{\mathbf{S}}_{\dot{\alpha}}$ are the same as in expressions (6.4.37) but with $d_{(+)}=d_{(-)}=0$. The operators just introduced satisfy, due to relation (2.9.21), the following (anti)commutation relations

$$
\begin{gathered}
\{\mathbb{Q}_\alpha, \bar{\mathbb{Q}}_{\dot{\alpha}}\} = 2\mathbb{P}_{\alpha\dot{\alpha}} \\
\{\mathbb{Q}_\alpha, \mathbb{Q}_\beta\} = -4\mathrm{i}\bar{\mu}\mathbb{J}_{\alpha\beta} \qquad [\mathbb{Q}_\alpha, \mathbb{P}_{\beta\dot{\beta}}] = -\varepsilon_{\alpha\beta}\mu\overline{\mathbb{Q}}_{\dot{\beta}} \\
[\mathbb{P}_{\alpha\dot{\alpha}}, \mathbb{P}_{\beta\dot{\beta}}] = -2\mathrm{i}\bar{\mu}\mu(\varepsilon_{\dot{\alpha}\dot{\alpha}}\mathbb{J}_{\alpha\beta} + \varepsilon_{\alpha\beta}\bar{\mathbb{J}}_{\dot{\alpha}\dot{\beta}}) \qquad (6.5.42) \\
[\mathbb{J}_{\alpha\beta}, \mathbb{Q}_\gamma] = \mathrm{i}\varepsilon_{\gamma(\alpha}\mathbb{Q}_{\beta)} \qquad [\mathbb{J}_{\alpha\beta}, \mathbb{P}_{\gamma\dot{\gamma}}] = \mathrm{i}\varepsilon_{\gamma(\alpha}\mathbb{P}_{\beta)\dot{\gamma}} \\
[\mathbb{J}_{\alpha\beta}, \mathbb{J}_{\gamma\delta}] = \mathrm{i}\varepsilon_{\gamma(\alpha}\mathbb{J}_{\beta)\delta} + \mathrm{i}\varepsilon_{\delta(\alpha}\mathbb{J}_{\beta)\gamma}.
\end{gathered}
$$

The remaining (anti)commutators vanish or can be found by Hermitian conjugation. The relations (6.5.42) define the superalgebra known as 'super de Sitter algebra'.

6.6. Non-minimal supergravity

6.6.1. Preliminary discussion

This section is devoted to consideration of a family of supergravity–matter dynamical systems classically equivalent to the pure supergravity theory with the action (6.1.3). The existence of such systems implies that there are several off-shell supergravity formulations realized in terms of different sets of superfields, but possessing the same dynamics on-shell.

One naturally comes to the results given below, when trying to solve the following dynamical problem. Suppose, we have a super Weyl invariant theory of supergravity coupled to a (constrained) scalar superfield Ψ. As usual, by super Weyl invariance we understand that the total supergravity–matter action superfunctional $S[\Psi, \mathscr{D}_A]$ is unchanged under transformations of the form

$$\varphi \to \varphi' = \mathrm{e}^\sigma \varphi \qquad \Psi \to \Psi' = \mathrm{e}^{-d_{(+)}\sigma - d_{(-)}\bar{\sigma}}\Psi \qquad \bar{\mathscr{D}}_{\dot{\alpha}}\sigma = 0 \qquad (6.6.1)$$

where φ is the supergravity chiral compensator, $d_{(+)}$ and $d_{(-)}$ are fixed numbers and σ is an arbitrary covariantly chiral scalar superfield. Suppose also that our system is dynamically equivalent to the pure supergravity theory. By equivalence we understand that one can choose a super Weyl gauge in which the dynamical equations of the system under consideration

$$\frac{\delta}{\delta\Psi} S[\Psi, \mathscr{D}_A] = 0 \Rightarrow T = 0 \qquad T_a = 0 \qquad (6.6.2)$$

imply the Einstein supergravity equations

$$R = G_a = 0 \qquad (6.6.3)$$

and vice versa (in equation (6.6.2) T and T_a are the supertrace and supercurrent, respectively). As a result, $S[\Psi, \mathscr{D}_A]$ can be taken to describe the supergravity dynamics, instead of the original supergravity action.

Off-mass-shell, the super Weyl invariance (6.6.1) can be used to choose the gauge

$$\varphi = 1. \tag{6.6.4}$$

Then, due to relation (5.5.5), every K- and λ-transformation should be supplemented by a special super Weyl one to preserve the gauge, and the transformation law of Ψ takes the form

$$\delta\Psi = K^M \partial_M \Psi + \frac{1}{3} d_{(+)}(\mathrm{e}^{\overline{W}}(\partial_m \lambda^m - \partial_\mu \lambda^\mu))\Psi + \frac{1}{3} d_{(-)}(\mathrm{e}^W(\partial_m \bar{\lambda}^m - \bar{\partial}_{\dot\mu} \bar{\lambda}^{\dot\mu}))\Psi \tag{6.6.5}$$

instead of the original law

$$\delta\Psi = K^M \partial_M \Psi. \tag{6.6.6}$$

Therefore, the superfield Ψ loses its scalar nature and turns into a density. In the case of non-chiral Ψ, we then obtain a supergravity formulation with non-chiral compensator.

The above dynamical problem is trivially solved when setting Ψ to be covariantly chiral. One introduces the action

$$S[\Phi, \mathcal{D}_A] = -\frac{3}{\kappa^2} \int \mathrm{d}^8 z\, E^{-1} \bar{\Phi}\Phi \qquad \bar{\mathcal{D}}_{\dot\alpha}\Phi = 0 \tag{6.6.7}$$

obtained from the supergravity action (6.1.3) by the replacement

$$\varphi \to \varphi\Phi. \tag{6.6.8}$$

Evidently, $S[\Phi, \mathcal{D}_A]$ possesses the super Weyl invariance

$$\varphi \to \mathrm{e}^{\sigma}\varphi \qquad \Phi \to \mathrm{e}^{-\sigma}\Phi \qquad \bar{\mathcal{D}}_{\dot\alpha}\sigma = 0. \tag{6.6.9}$$

The corresponding matter and supergravity dynamical equations are

$$\begin{gathered} (\bar{\mathcal{D}}^2 - 4R)\bar{\Phi} = 0 \\ (\bar{\mathcal{D}}_{\dot\alpha}\bar{\Phi})\mathcal{D}_\alpha\Phi + 2\mathrm{i}\bar{\Phi}\overleftrightarrow{\mathcal{D}}_{\alpha\dot\alpha}\Phi - 2G_{\alpha\dot\alpha}\bar{\Phi}\Phi = 0 \end{gathered} \tag{6.6.10}$$

where we have used equation (6.3.10). Now, imposing the super Weyl gauge

$$\Phi = 1 \tag{6.6.11}$$

reduces the action (6.6.7) to the supergravity one (6.1.3), and the dynamical equations (6.6.10) to the supergravity ones (6.6.3).

Remark. The theory considered represents the supersymmetric analogue of the Weyl invariant version of the Einstein gravity described by the action

$$S[e_a{}^m, C] = \frac{3}{\kappa^2} \int \mathrm{d}^4 x\, \mathrm{e}^{-1} \left\{ \nabla^a C \nabla_a C + \frac{1}{6} \mathcal{R} C^2 \right\} \tag{6.6.12}$$

where C is a scalar field. This action is obtained from the gravity action (6.6.1) by the replacement

$$e_a{}^m \to e_a{}^m C^{-1} \tag{6.6.13}$$

so it is explicitly Weyl invariant.

In the remainder of the section we shall analyse another, non-chiral, choice with Ψ being a 'complex covariantly linear scalar superfield' defined by

$$(\bar{\mathscr{D}}^2 - 4R)\Psi = 0 \qquad \bar{\Psi} \neq \Psi \qquad ((\mathscr{D}^2 - 4\bar{R})\Psi \neq 0). \tag{6.6.14}$$

Wishing to work with unconstrained superfields only, one must represent, due to relations (5.5.10), the superfields Ψ and $\bar{\Psi}$ in the manner

$$\Psi = \bar{\mathscr{D}}_{\dot{\alpha}}\bar{\chi}^{\dot{\alpha}} \qquad \bar{\Psi} = \mathscr{D}^{\alpha}\chi_{\alpha} \tag{6.6.15}$$

with χ_α being a spinor superfield. Then, any theory of the superfields Ψ and $\bar{\Psi}$ can be equivalently treated as a gauge theory of the superfields χ_α and $\bar{\chi}_{\dot{\alpha}}$ possessing the gauge invariance

$$\delta\chi_\alpha = \mathscr{D}^{\beta}\Lambda_{\alpha\beta} \qquad \Lambda_{\alpha\beta} = \Lambda_{\beta\alpha}. \tag{6.6.16}$$

In such an approach, Ψ and $\bar{\Psi}$ appear as gauge invariant superfield strengths. One more useful representation for Ψ follows from the relations (5.4.2) and (5.4.23). The requirement (6.6.14) means that

$$\begin{gathered} \Psi = \bar{F}^{-2}\gamma \qquad \gamma = e^{\bar{W}}\hat{\gamma} \\ \hat{\bar{E}}^2\gamma = \bar{\partial}^2\hat{\gamma} = 0. \end{gathered} \tag{6.6.17}$$

It is seen that every complex covariantly linear superfield Ψ is determined by a flat complex linear superfield $\hat{\gamma}$.

6.6.2. *Complex linear compensator*

We begin by examining the question: what are admissible superconformal weights $d_{(+)}$ and $d_{(-)}$ for which the transformation law (6.6.1) is consistent with the constraint $(\bar{\mathscr{D}}^2 - 4R)\Psi = 0$? Under the super Weyl transformations (5.5.13), the operator $(\bar{\mathscr{D}}^2 - 4R)$ changes as

$$(\bar{\mathscr{D}}^2 - 4R) \to e^{-2\sigma}(\bar{\mathscr{D}}^2 - 4R)e^{\bar{\sigma}} + \bar{M}_{\dot{\alpha}\dot{\beta}}\text{-dependent terms.} \tag{6.6.18}$$

Therefore, the most general transformation law for Ψ is

$$\Psi \to \Psi' = \exp[-d_{(+)}\sigma - \bar{\sigma}]\Psi \tag{6.6.19}$$

with $d_{(+)}$ being arbitrary. The same result also follows from relation (6.6.17). Indeed, the relation (6.6.17) shows that Ψ depends explicitly on $\bar{\varphi}$ as $\Psi \propto \bar{\varphi}^{-1}$ (recall, $\bar{F} \propto \bar{\varphi}^{1/2}\varphi^{-1}$) and can be taken to depend on φ as $\Psi \propto \varphi^{-d_{(+)}}$ by means of the redefinition $\gamma \to \varphi^{-d_{(+)}}\gamma$.

Remark. It is not difficult to check that the spinor superfield $\bar{\chi}_{\dot{\alpha}}$, which arose in expression (6.6.15), should transform according to the rule

$$\bar{\chi}_{\dot{\alpha}} \to \bar{\chi}'_{\dot{\alpha}} = \exp\left[(1 - d_{(+)})\sigma - \frac{3}{2}\bar{\sigma}\right]\bar{\chi}_{\dot{\alpha}} \qquad (6.6.20)$$

to produce the law (6.6.19).

Further, we restrict ourselves to the consideration of cases with real $d_{(+)} \neq 1$. The choice $d_{(+)} = 1$ seems to be very special, since here the representation defined by the law (6.6.19) becomes reducible, and the constraint $\bar{\Psi} = \Psi$ can be imposed. The case $d_{(+)} = 1$ will be studied in the next section. By tradition, we will write $d_{(+)}$ in the form

$$d_{(+)} = \frac{1 - 3n}{3n + 1} \qquad n \neq -1/3,\ 0. \qquad (6.6.21)$$

The action superfunctional invariant under the super Weyl transformations

$$\varphi \to \varphi' = e^{\sigma}\varphi \qquad \Psi \to \Psi' = \exp\left[\frac{3n-1}{3n+1}\sigma - \bar{\sigma}\right]\Psi \qquad \bar{\mathscr{D}}_{\dot{\alpha}}\sigma = 0 \quad (6.6.22)$$

reads

$$S_{(n)} = \frac{1}{n\kappa^2}\int d^8z\, E^{-1}(\bar{\Psi}\Psi)^{(3n+1)/2}. \qquad (6.6.23a)$$

Note that the superfield Ψ^{ξ}, $\xi \neq 1$, is not linear and so the choice of the super Lagrangian in the form $(\bar{\Psi}\Psi)^{\xi}$ seems to be natural. With the help of the relations (5.4.19) and (5.5.4), the action can be rewritten in terms of the supergravity prepotentials and the superfield γ (6.6.17):

$$S_{(n)} = \frac{1}{n\kappa^2}\int d^8z\, \hat{E}^n[(\bar{\Phi}\varphi)^3(1\cdot e^{\overleftarrow{W}})(1\cdot e^{\overleftarrow{\bar{W}}})]^{(n+1)/2}(\bar{\gamma}\gamma)^{(3n+1)/2}. \qquad (6.6.23b)$$

The constant $(1/n)$ has been introduced into the action to normalize $\hat{E}^n$ ($\hat{E}^n \approx (1 + \theta^2\bar{\theta}^2\mathscr{L})^n = 1 + n\theta^2\bar{\theta}^2\mathscr{L}$). The action (6.6.23) was studied by W. Siegel and S. J. Gates. Below we shall show that it describes supergravity. This supergravity formulation is known as 'non-minimal supergravity'.

Remark. Let us set $n = -1/3$ in equation (6.6.23*b*). Then, the superfields γ and $\bar{\gamma}$ drop out of the action and we recover the supergravity action (6.1.4). So, the choice $n = -1/3$ is said to correspond to the minimal supergravity described in detail above.

Owing to the super Weyl invariance of $S_{(n)}$, we are in a position to impose the super Weyl gauge (6.6.4). After doing this, the supergravity gauge group acts on Ψ according to the law (6.6.5), where $d_{(-)} = 1$ and $d_{(+)}$ is given in equation (6.6.21). Now, let us determine the transformation law of the superfield $\hat{\gamma}$ (6.6.17). $\bar{F}$ changes according to the rule

$$\delta\bar{F}^2 = K^M \partial_M \bar{F}^2 - (e^{\bar{W}}\bar{\partial}_{\dot{\mu}}\lambda^{\dot{\mu}})\bar{F}^2 + \frac{2}{3}(e^{\bar{W}}(\partial_m\lambda^m - \partial_\mu\lambda^\mu))\bar{F}^2 - \frac{1}{3}(e^{W}(\partial_m\bar{\lambda}^m - \partial_\mu\bar{\lambda}^\mu))\bar{F}^2$$

where we have used equation (5.4.14) and the fact that in the gauge (6.6.4) every λ-transformation should be supplemented by the super Weyl transformation with $\sigma = -\frac{1}{3}e^{\bar{W}}(\partial_m\lambda^m - \partial_\mu\lambda^\mu)$. Next, the operator $e^{\bar{W}}$ changes as in relation (5.4.14). As a result, one arrives at

$$\delta\hat{\gamma} = (\lambda^m\partial_m + \lambda^\mu\partial_\mu)\hat{\gamma} - \bar{\partial}^{\dot{\mu}}(\lambda_{\dot{\mu}}\hat{\gamma}) + \frac{n+1}{3n+1}(\partial_m\lambda^m - \partial_\mu\lambda^\mu)\hat{\gamma}. \tag{6.6.24}$$

It is worth recalling that here the parameters λ^m and λ^μ are $\bar{\theta}$-independent while $\lambda_{\dot{\mu}}$ are arbitrary. On these grounds, the variation $\delta\hat{\gamma}$ turns out to be linear,

$$\bar{\partial}^2\hat{\gamma} = 0 \quad \Rightarrow \quad \bar{\partial}^2\delta\hat{\gamma} = 0.$$

A complex linear superfield $\hat{\gamma}$, $\bar{\partial}^2\hat{\gamma} = 0$, transforming by the law (6.6.24) under the supergravity gauge group, is said to be a 'non-minimal supergravity compensator' or, simply a 'complex linear compensator'.

The arbitrariness in the choice of $\lambda_{\dot{\mu}}$ can be used to eliminate the compensator, that is to impose the gauge

$$\hat{\gamma} = 1. \tag{6.6.25}$$

Then, the parameters of residual gauge transformations should obey the constraint

$$(3n+1)\bar{\partial}^{\dot{\mu}}\lambda_{\dot{\mu}} = (n+1)(\partial_m\lambda^m - \partial_\mu\lambda^\mu). \tag{6.6.26}$$

6.6.3. Non-minimal supergeometry

Up to a factor, the action (6.6.23) is obtained from the minimal supergravity action (6.1.3) by changing the chiral compensator φ to the super Weyl invariant combination

$$\varphi[\Psi^{3n-1}\bar{\Psi}^{3n+1}]^{(3n-1)/12n}. \tag{6.6.27}$$

This observation can be used for constructing super Weyl invariant and generally covariant derivatives. Namely, let us introduce the set of operators $\mathbb{D}_A = (\mathbb{D}_a, \mathbb{D}_\alpha, \bar{\mathbb{D}}^{\dot{\alpha}})$ as

$$\begin{aligned}
&\mathbb{D}_\alpha = \mathbb{U}\mathscr{D}_\alpha - 2(\mathscr{D}^\beta\mathbb{U})M_{\alpha\beta} \qquad && \mathbb{U} = [\Psi^{n+1}\bar{\Psi}^{n-1}]^{-(3n+1)/8n} \\
&\bar{\mathbb{D}}_{\dot{\alpha}} = \overline{\mathbb{U}\mathscr{D}}_{\dot{\alpha}} - 2(\bar{\mathscr{D}}^{\dot{\beta}}\bar{\mathbb{U}})\bar{M}_{\dot{\alpha}\dot{\beta}} \qquad && \mathbb{D}_a = -\frac{i}{4}(\tilde{\sigma}_a)^{\dot{\alpha}\alpha}\{\mathbb{D}_\alpha, \bar{\mathbb{D}}_{\dot{\alpha}}\}.
\end{aligned} \tag{6.6.28}$$

Owing to the scalar nature of Ψ, the operators $\mathbb{D}_A$ change covariantly under the supergravity gauge group:

$$\mathbb{D}_A \to \mathbb{D}'_A = e^{\mathscr{K}} \mathbb{D}_A e^{-\mathscr{K}} \qquad \mathscr{K} = K^M \partial_M + \frac{1}{2} K^{ab} M_{ab}. \tag{6.6.29}$$

Since $\mathbb{D}_\alpha \ln(\mathbb{U}^4 \bar{\mathbb{U}}^2) \neq 0$, $\mathbb{D}_A$ are obtained from the minimal covariant derivatives $\mathscr{D}_A$ (possessing the algebra (5.5.6)) by applying a special generalized super Weyl transformation (compare with equation (5.5.13)). It is a simple exercise to verify that the derivative $\mathbb{D}_A$ are invariant under the minimal super Weyl transformations (6.6.22). That is why one can impose the gauge $\varphi = 1$ and this does not spoil the transformation properties (6.6.29).

Covariant derivatives $\mathbb{D}_A$ obey (anti)commutation relations of the form (5.3.53), and the corresponding supertorsion tensors will be denoted as $\mathbb{T}_\alpha$, $\mathbb{R}$, $\mathbb{G}_a$ and $\mathbb{W}_{\alpha\beta\gamma}$. With the help of straightforward but tedious calculations, one can check that they are expressed in terms of the supertorsions R, G_a and $W_{\alpha\beta\gamma}$, associated with the minimal covariant derivatives $\mathscr{D}_A$, as follows:

$$\mathbb{T}_\alpha = \mathbb{D}_\alpha \mathbb{T} \qquad \mathbb{T} = \ln(\mathbb{U}^4 \bar{\mathbb{U}}^2)$$

$$\mathbb{R} = -\frac{1}{4}(\bar{\mathscr{D}}^2 - 4R)\bar{\mathbb{U}}^2 \qquad \mathbb{W}_{\alpha\beta\gamma} = \bar{\mathbb{U}}^2 \mathbb{U} W_{\alpha\beta\gamma} \tag{6.6.30}$$

$$\mathbb{G}_{\alpha\dot\alpha} = \bar{\mathbb{U}}\mathbb{U} G_{\alpha\dot\alpha} + \frac{1}{2}(\bar{\mathbb{D}}_{\dot\alpha} \ln \mathbb{U})\mathbb{D}_\alpha \ln \mathbb{U} + \frac{1}{4}\bar{\mathbb{D}}_{\dot\alpha}\mathbb{D}_\alpha \ln(\mathbb{U}^2 \bar{\mathbb{U}}^{-1}) - \frac{1}{4}\mathbb{D}_\alpha \bar{\mathbb{D}}_{\dot\alpha} \ln(\bar{\mathbb{U}}^2 \mathbb{U}^{-1}).$$

Finally, the linearity condition (6.6.14) proves to be equivalent to the following relation

$$4\mathbb{R} = -\frac{n+1}{3n+1}\bar{\mathbb{D}}_{\dot\alpha}\bar{\mathbb{T}}^{\dot\alpha} + \left(\frac{n+1}{3n+1}\right)^2 \bar{\mathbb{T}}_{\dot\alpha}\bar{\mathbb{T}}^{\dot\alpha} \tag{6.6.31}$$

which is known as the 'non-minimal supergravity constraint'. The covariant derivatives algebra (5.3.53) under the constraint (6.6.31) defines the 'non-minimal supergeometry'.

Remark. Representing $\mathbb{D}_A$ in the form

$$\mathbb{D}_A = \mathbb{E}_A{}^M \partial_M + \frac{1}{2} \mathbb{Y}_A{}^{bc} M_{bc} \tag{6.6.32}$$

and defining the Berezinian of the non-minimal supervierbein

$$\mathbb{E} = \mathrm{Ber}(\mathbb{E}_A{}^M)$$

one readily obtains

$$\mathbb{E} = E(\bar{\mathbb{U}}\mathbb{U})^2 = E(\bar{\Psi}\Psi)^{-(3n+1)/2}. \tag{6.6.33}$$

Therefore, the action (6.6.23) can be rewritten as

$$S_{(n)}=\frac{1}{n\kappa^2}\int \mathrm{d}^8z\, \mathbb{E}^{-1}. \qquad (6.6.34)$$

In such a form, $S_{(n)}$ looks completely similar to the minimal supergravity action (6.1.3).

6.6.4. Dynamics in non-minimal supergravity

We are going to show that the supergravity–matter theory with the action (6.6.23) is dynamically equivalent to the Einstein supergravity.

Dynamical equations corresponding to the action $S_{(n)}$ are

$$\mathbb{T}_\alpha=\bar{\mathbb{T}}_{\dot\alpha}=0 \qquad (6.6.35)$$

and

$$\mathbb{G}_a=0. \qquad (6.6.36)$$

Note that the equations (6.6.35) lead to

$$\mathbb{R}=\bar{\mathbb{R}}=0 \qquad (6.6.37)$$

owing to the non-minimal constraint (6.6.31). Let us comment on the derivation of equations (6.6.35, 36). First, it is necessary to vary $S_{(n)}$ with respect to Ψ and $\bar\Psi$. Any variation $\delta\Psi$ satisfies the constraint $(\bar{\mathscr{D}}^2-4R)\delta\Psi=0$, but it can be represented, due to (6.6.15), in the form $\delta\Psi=\bar{\mathscr{D}}_{\dot\alpha}\delta\bar\chi^{\dot\alpha}$, with $\delta\bar\chi^{\dot\alpha}$ being arbitrary. Then, simple calculations give equation (6.6.35). Secondly, it is necessary to find the supercurrent and to set it to zero. To determine the supercurrent, one must calculate the variation

$$\delta_{\mathbf{H}}S_{(n)}=\frac{1}{n\kappa^2}\int \mathrm{d}^8z\, \tilde{\mathscr{E}}^{-1}[(\tilde\nabla^\alpha\tilde\chi_\alpha)\tilde{\bar\nabla}_{\dot\alpha}\tilde{\bar\chi}^{\dot\alpha}]^{(3n+1)/2}-S_{(n)}$$

$$\tilde\chi_\alpha=\mathrm{e}^{-\mathrm{i}\mathbf{H}^a\mathscr{D}_a}\chi_\alpha \qquad \tilde{\bar\chi}_{\dot\alpha}=\mathrm{e}^{-\mathrm{i}\mathbf{H}^a\mathscr{D}_a}\bar\chi_{\dot\alpha}$$

where $\tilde{\mathscr{E}}^{-1}$, $\tilde\nabla_\alpha$ and $\tilde{\bar\nabla}_{\dot\alpha}$ are defined by equations (5.6.40, 57–59) with $\sigma=\bar\sigma=0$. Then, one obtains (6.6.36).

Consider the consequences of the dynamical equations. From relations (6.6.30) and (6.6.35) we deduce

$$\mathbb{U}=\exp\left[\bar\rho-\frac{1}{2}\rho\right] \qquad \bar{\mathscr{D}}_{\dot\alpha}\rho=0 \quad\Rightarrow\quad \Psi=\exp\left[\frac{3n-1}{3n+1}\rho-\bar\rho\right]$$

for some covariantly chiral scalar superfield ρ. Hence, due to the super Weyl invariance (6.6.22), the super Weyl gauge

$$\Psi=1 \qquad (6.6.38)$$

is admissible on-shell. Under this gauge fixing, the equations (6.6.36) and (6.6.37) reduce to the Einstein supergravity equations (6.6.3), as is seen from relations (6.6.30). Thus, the action $S_{(n)}$ describes the Einstein supergravity, for any $n \neq -\frac{1}{3}, 0$. The $S_{(n)}$ is known as the 'non-minimal supergravity action'.

The equivalence of the two theories with classical actions S_{SG} (6.1.3) and $S_{(n)}$, respectively, can also be seen as follows. Consider the auxiliary supergravity–matter model with the action

$$S[V, \Phi, \mathscr{D}_A] = \frac{1}{n\kappa^2} \int d^8z\, E^{-1} \left\{ (\bar{V}V)^{(3n+1)/2} - \frac{3n+1}{2} V\Phi^{6n/(3n+1)} - \frac{3n+1}{2} \overline{V\Phi^{6n/(3n+1)}} \right\} \tag{6.6.39}$$

where V is an unconstrained complex scalar superfield and Φ is a covariantly chiral scalar superfield, $\bar{\mathscr{D}}_{\dot{\alpha}}\Phi = 0$. The action is invariant under the super Weyl transformations

$$\begin{gathered} \varphi \to e^{\sigma}\varphi \qquad \Phi \to e^{-\sigma}\Phi \\ V \to \exp\left[\frac{3n-1}{3n+1}\sigma - \bar{\sigma}\right] V \qquad \bar{\mathscr{D}}_{\dot{\alpha}}\sigma = 0. \end{gathered} \tag{6.6.40}$$

Varying $S[V, \Phi, \mathscr{D}_A]$ with respect to Φ and $\bar{\Phi}$ gives

$$(\bar{\mathscr{D}}^2 - 4R)V = (\mathscr{D}^2 - 4\bar{R})\bar{V} = 0 \quad \Rightarrow \quad V = \Psi$$

with Ψ a complex covariantly linear scalar superfield. Then we have

$$\int d^8z\, E^{-1} V\Phi^{6n/(3n+1)} = -\frac{1}{4} \int d^8z \frac{E^{-1}}{R} \Phi^{6n/(3n+1)} (\bar{\mathscr{D}}^2 - 4R)V = 0$$

and the above superfunctional reduces to the non-minimal supergravity action. On the other hand, imposing the dynamical equations for V and $\bar{V}$ makes it possible to eliminate these superfields. As a result $S[V, \Phi, \mathscr{D}_A]$ reduces to the action (6.6.7), which represents the super Weyl invariant version of the minimal supergravity action. Therefore, the minimal and non-minimal supergravities are dually equivalent.

6.6.5. *Prepotentials and field content in non-minimal supergravity*

Now it is time to discuss the results obtained. We commence by recalling that imposing the conformal supergravity constraints (5.3.15) leads to the covariant derivatives algebra (5.3.53) and expresses all the supergeometry objects in terms of the unconstrained superfields $N_\alpha{}^\mu$, W^M and F (with $N_\alpha{}^\mu$, W^μ and $W_{\dot{\mu}}$ purely gauge degrees of freedom). The minimal supergeometry formulation is characterized by one more constraint (5.5.1) manifesting the fact that F becomes a function of W^M, $\bar{W}^M$, $\hat{\varphi}$ and $\bar{\hat{\varphi}}$, where $\hat{\varphi}$ is a $\bar{\theta}$-independent superfield, $\bar{\partial}_{\dot{\mu}}\hat{\varphi} = 0$. The superfields $N_\alpha{}^\mu$, W^M and $\hat{\varphi}$ are the prepotentials of minimal supergravity. The non-minimal supergeometry formulation is characterized by the set of constraints (5.3.15) and (6.6.31), where $n \neq -\frac{1}{3}$, 0. The last constraints leads to the fact that F becomes a function of W^M, $\bar{W}^M$, $\hat{\gamma}$ and $\hat{\bar{\gamma}}$, where $\hat{\gamma}$ is a superfield under the restriction $\bar{\partial}^2\hat{\gamma}=0$. The superfields $N_\alpha{}^\mu$, W^M and $\hat{\gamma}$ are the prepotentials of non-minimal supergravity. Clearly, the requirement $\bar{\partial}_{\dot{\mu}}\hat{\varphi}=0$ is stronger than $\bar{\partial}_{\dot{\mu}}\bar{\partial}^{\dot{\mu}}\hat{\gamma}=0$. As a consequence, $\hat{\gamma}$ describes more field degrees of freedom than $\hat{\varphi}$, and the non-minimal algebra (5.3.53) looks much more complicated than the minimal algebra (5.5.6). That is why we use the term 'non-minimal'.

With respect to the supergravity gauge group, the prepotentials W^M, $\hat{\varphi}$ and $\hat{\gamma}$ are characterized by the transformation laws (5.4.14), (5.5.5) and (6.6.24), respectively. In both minimal and non-minimal supergravities, one can gauge away the compensators resulting in $\hat{\varphi}=1$ and $\hat{\gamma}=1$. Then we end up with residual gauge transformations constrained by equation (5.1.32) in the minimal case and by equation (6.6.26) in the non-minimal case. Such a gauge fixing does not seem to be the most useful. The point is that equation (6.6.26) restricts the parameters $\lambda_{\dot{\mu}}$, and this makes it impossible to gauge away all W^μ and $W_{\dot{\mu}}$. The most useful gauge choice is the following. In both minimal and non-minimal supergravities, it is in our power to choose the gravitational superfield gauge (5.4.38) supplemented by the Wess–Zumino gauge (5.1.17). Residual gauge invariance is described by the superparameters (5.4.42), (5.1.18). Then, the local scale$+\gamma_5$-transformations and the local S-supersymmetry transformations can be used to bring the minimal and non-minimal compensators into the form:

$$\hat{\varphi}(x,\theta)=1+\theta^2 B(x) \tag{6.6.41a}$$

and

$$\hat{\gamma}(x,\theta,\bar{\theta})=1+\theta^\alpha\rho_\alpha(x)+\theta^2 B(x)+\theta\sigma^a\bar{\theta}v_a(x)+\theta^2\bar{\theta}_{\dot{\alpha}}\bar{v}^{\dot{\alpha}}(x). \tag{6.6.41b}$$

Here B and v^a are complex scalar and vector fields, respectively, ρ_α and $\bar{v}^{\dot{\alpha}}$ are spinor fields. The real vector field A^m from (5.1.17) and the fields presented in expressions (6.6.41a) and (6.6.41b) do not propagate on-shell. So, the minimal and non-minimal supergravity formulations have the same dynamical fields, the graviton and the gravitino, but different sets of auxiliary fields.

6.6.6. Geometrical approach to non-minimal supergravity

Non-minimal supergravity has been introduced above in the dynamical approach as the supergravity–matter system classically equivalent to Einstein supergravity. Remarkably, it possesses a purely geometric realization analogous to that developed for minimal supergravity in Section 5.1. Here we describe the approach suggested by E. Sokatchev.

Consider complex superspace $\mathbb{C}^{4|4}$ parametrized by four complex c-number coordinates y^m and four complex a-number coordinates: undotted θ^μ and dotted $\bar{\rho}_{\dot{\mu}}$, $\bar{\rho}_{\dot{\mu}} \neq \varepsilon_{\dot{\mu}\dot{\nu}}(\theta^\nu)^* = \varepsilon_{\dot{\mu}\dot{\nu}}\bar{\theta}^{\dot{\nu}}$. Coordinates will be unified in

$$\zeta^M=(y^m,\theta^\mu,\bar{\rho}_{\dot{\mu}}) \quad \text{and} \quad \xi^{\underline{M}}=(y^m,\theta^\mu) \tag{6.6.42}$$

looking on $\xi^{\underline{M}}$ as the coordinates of some complex superspace $\mathbb{C}^{4|2}$. We take the point of view that $\mathbb{C}^{4|4}$ is the trivial fibre bundle over base $\mathbb{C}^{4|2}$ with fibre $\mathbb{C}^{0|2}$. Then, natural coordinate transformations on $\mathbb{C}^{4|4}$ are 'triangular' transformations of the general structure:

$$y'^m = f^m(y, \theta) \qquad \theta'^\mu = f^\mu(y, \theta) \qquad \bar{\rho}'_{\dot{\mu}} = f_{\dot{\mu}}(y, \theta, \bar{\rho}) \tag{6.6.43}$$

with

$$\mathrm{Ber}\left(\frac{\partial \zeta'^N}{\partial \zeta^M}\right) \neq 0 \quad \Rightarrow \quad \mathrm{Ber}\left(\frac{\partial \xi'^{\underline{N}}}{\partial \xi^{\underline{M}}}\right) \neq 0.$$

The infinitesimal form for such transformations is

$$\begin{aligned} y'^m &= y^m - \lambda^m(y, \theta) \\ \theta'^\mu &= \theta^\mu - \lambda^\mu(y, \theta) \\ \bar{\rho}'_{\dot{\mu}} &= \bar{\rho}_{\dot{\mu}} - \lambda_{\dot{\mu}}(y, \theta, \bar{\rho}) \end{aligned} \tag{6.6.44}$$

with λ^m, λ^μ and $\lambda_{\dot{\mu}}$ being arbitrary superfunctions of the proper arguments.

The supergroup of coordinate transformations (6.6.43) has a natural family of subgroups, each of which is singled out by a restriction of the form

$$\left[\mathrm{Ber}\left(\frac{\partial \zeta'^N}{\partial \zeta^M}\right)\right]^{3n+1} = \left[\mathrm{Ber}\left(\frac{\partial \xi'^{\underline{N}}}{\partial \xi^{\underline{M}}}\right)\right]^{2n} \tag{6.6.45}$$

for some real number n. In the infinitesimal case, this restriction coincides with equation (6.6.26).

Further, let us define an embedding of real superspace $\mathbb{R}^{4|4}$, parametrized by coordinates x^m, θ^μ and $\bar{\theta}_{\dot{\mu}}$, into our superspace $\mathbb{C}^{4|4}$ in the manner:

$$\begin{aligned} & y^m + \bar{y}^m = 2x^m \qquad y^m - \bar{y}^m = 2\mathrm{i}\mathscr{H}^m(x, \theta, \bar{\theta}) \\ & \rho^\mu - \theta^\mu = \mathscr{H}^\mu(x, \theta, \bar{\theta}) \quad \Rightarrow \quad \bar{\rho}_{\dot{\mu}} - \bar{\theta}_{\dot{\mu}} = \bar{\mathscr{H}}_{\dot{\mu}}(x, \theta, \bar{\theta}). \end{aligned} \tag{6.6.47}$$

Here $\mathscr{H}^m$ and $\mathscr{H}^\mu$ are superfields on $\mathbb{R}^{4|4}$ which determine the embedding. Obviously, every coordinate transformation (6.6.44) induces, first, the coordinate transformation on $\mathbb{R}^{4|4}$:

$$\begin{aligned} x'^m &= x^m - \frac{1}{2}\lambda^m(x + \mathrm{i}\mathscr{H}, \theta) - \frac{1}{2}\bar{\lambda}^m(x - \mathrm{i}\mathscr{H}, \bar{\theta}) \\ \theta'^\mu &= \theta^\mu - \lambda^\mu(x + \mathrm{i}\mathscr{H}, \theta) \\ \bar{\theta}'_{\dot{\mu}} &= \bar{\theta}_{\dot{\mu}} - \bar{\lambda}_{\dot{\mu}}(x - \mathrm{i}\mathscr{H}, \bar{\theta}) \end{aligned} \tag{6.6.47}$$

and secondly, the change of superfields $\mathscr{H}^m$ and $\bar{\mathscr{H}}_{\dot{\mu}}$:

$$\mathscr{H}'^m(x', \theta', \bar{\theta}') = \mathscr{H}^m(x, \theta, \bar{\theta}) + \frac{\mathrm{i}}{2}\lambda^m(x + \mathrm{i}\mathscr{H}, \theta) - \frac{\mathrm{i}}{2}\bar{\lambda}^m(x - \mathrm{i}\mathscr{H}, \bar{\theta}) \tag{6.6.48a}$$

$$\bar{\mathscr{H}}'_{\dot{\mu}}(x', \theta', \bar{\theta}') = \bar{\mathscr{H}}_{\dot{\mu}}(x, \theta, \bar{\theta}) + \bar{\lambda}_{\dot{\mu}}(x - \mathrm{i}\mathscr{H}, \bar{\theta}) - \lambda_{\dot{\mu}}(x + \mathrm{i}\mathscr{H}, \theta, \bar{\theta} + \bar{\mathscr{H}}) \tag{6.6.48b}$$

Here $(x \pm \mathrm{i}\mathscr{H})$ and $(\bar{\theta} + \bar{\mathscr{H}})$ mean $(x^n \pm \mathrm{i}\mathscr{H}^n)$ and $(\bar{\theta}_{\dot{\nu}} + \bar{\mathscr{H}}_{\dot{\nu}})$, respectively.

Based on the above consideration, one can make the following proposal. For a given real $n \neq 0$, the supergravity gauge group is the supergroup of coordinate transformations (6.6.43) constrained by (6.6.45). A curved real superspace is described by a 'generalized gravitational superfield'

$$\mathscr{H}^M(z) = \bar{\mathscr{H}}^M(z) = (\mathscr{H}^m(z), \mathscr{H}^\mu(z), \bar{\mathscr{H}}_{\dot{\mu}}(z)). \tag{6.6.49}$$

The supergravity gauge group acts on $\mathscr{H}^M(z)$ according to the law (6.6.48).

Let us argue the given proposal. To start with, consider the case $n = -\frac{1}{3}$. Then, the left-hand side in restriction (6.6.45) is identically unity, and the superparameters $\lambda_{\dot{\mu}}$ in expressions (6.6.44) and (6.6.48*b*) are completely arbitrary. On these grounds, we are able to impose the gauge

$$\mathscr{H}^M = (\mathscr{H}^m, 0, 0)$$

After this, we arrive at the Einstein supergravity formulation described in subsection 5.1.3.

Next, consider the case $n \neq -\frac{1}{3}, 0$. We intend to show that there is a special gauge in non-minimal supergravity under which the situation described above is realized explicitly. In non-minimal supergravity, the prepotentials are N^μ_α, W^M and $\hat{\gamma}$. The invariance under the superlocal Lorentz group can be used to set $N_\alpha{}^\mu = \delta_\alpha{}^\mu$. The invariance under the general coordinate transformation supergroup can be used to gauge away Re W^m (leaving some part of Re W^m, if we wish) and W^μ. As a result, the following gauge

$$\begin{aligned} &\mathrm{e}^{W} x^m = x^m + \mathrm{i}\mathscr{H}^m(x, \theta, \bar{\theta}) \qquad \bar{\mathscr{H}}^m = \mathscr{H}^m \\ &\mathrm{e}^{W}\theta^\mu = \theta^\mu \qquad \mathrm{e}^{W}\bar{\theta}_{\dot{\mu}} = \bar{\theta}_{\dot{\mu}} + \bar{\mathscr{H}}_{\dot{\mu}}(x, \theta, \bar{\theta}) \end{aligned} \tag{6.6.50}$$

is admissible. Evidently, this gauge fixing is weaker than the gravitational superfield gauge (5.4.38). So, the choice (6.6.50) will be called the 'generalized gravitational superfield gauge'. Under equation (6.6.50), every λ-transformation should be supplemented by some K-transformation to preserve the gauge. Similarly to the derivation of equations (5.4.40–42), one now finds

$$\delta \mathrm{e}^{W} = K^M \partial_M \mathrm{e}^{W} - \mathrm{e}^{W} \lambda^M \partial_M$$

where

$$\begin{aligned} &\lambda^M = (\lambda^m(x, \theta), \lambda^\mu(x, \theta), \lambda_{\dot{\mu}}(x, \theta, \bar{\theta})) \\ &K^m = \frac{1}{2}\lambda^m(x + \mathrm{i}\mathscr{H}, \theta) + \frac{1}{2}\bar{\lambda}^m(x - \mathrm{i}\mathscr{H}, \bar{\theta}) \\ &K^\mu = \lambda^\mu(x + \mathrm{i}\mathscr{H}, \theta) \qquad \bar{K}_{\dot{\mu}} = \bar{\lambda}_{\dot{\mu}}(x - \mathrm{i}\mathscr{H}, \bar{\theta}) \end{aligned} \tag{6.6.51}$$

and the transformation laws

$$\begin{aligned} &\delta\mathscr{H}^m = K^N \partial_N \mathscr{H}^m + \frac{\mathrm{i}}{2}\lambda^m(x + \mathrm{i}\mathscr{H}, \theta) - \frac{\mathrm{i}}{2}\bar{\lambda}^m(x - \mathrm{i}\mathscr{H}, \bar{\theta}) \\ &\delta\bar{\mathscr{H}}_{\dot{\mu}} = K^N \partial_N \bar{\mathscr{H}}_{\dot{\mu}} + \bar{\lambda}_{\dot{\mu}}(x - \mathrm{i}\mathscr{H}, \bar{\theta}) - \lambda_{\dot{\mu}}(x + \mathrm{i}\mathscr{H}, \theta, \bar{\theta} + \bar{\mathscr{H}}). \end{aligned} \tag{6.6.52}$$

It is seen that the relations (6.6.52) coincide with those in (6.6.48). Finally, imposing the gauge $\hat{\gamma}=1$ restricts superparameters λ^M in accordance with equation (6.6.26).

6.6.7. *Linearized non-minimal supergravity*

In conclusion, let us describe a linearized version of non-minimal supergravity. Recall that the same construction for minimal supergravity was described in Section 6.2. Similarly to the minimal case, it is now convenient to use the prepotential parametrization described in subsection 5.6.1 and the chiral representation defined by equations (5.6.15, 16). Upon imposing the Λ-supergroup gauge (5.6.18) and the super Weyl gauge $\varphi=1$, the non-minimal supergravity action (6.6.23*b*) turns out to be a superfunctional of the prepotentials $H^a=\bar{H}^a$, γ and $\bar{\gamma}$, with γ being a flat linear complex superfield, $\bar{\mathrm{D}}^2\gamma=0$ $(\mathrm{D}^2\gamma\neq 0)$. The action reads

$$S_{(n)}=\frac{1}{n\kappa^2}\int \mathrm{d}^8z\,(\gamma \mathrm{e}^{-2\mathrm{i}H}\bar{\gamma})^{(3n+1)/2}(1\cdot \mathrm{e}^{-2\mathrm{i}\overleftarrow{H}})^{(n+1)/2}\hat{E}^n. \tag{6.6.53}$$

Here the objects H and $\hat{E}$ are exactly the same as in equation (6.2.2). The action is invariant under the gauge transformations

$$\begin{gathered}\delta \mathrm{e}^{-2\mathrm{i}H}=\Lambda \mathrm{e}^{-2\mathrm{i}H}-\mathrm{e}^{-2\mathrm{i}H}\bar{\Lambda}\\ \delta\gamma=(\Lambda^a\partial_a+\Lambda^\alpha \mathrm{D}_\alpha)\gamma+\bar{\mathrm{D}}_{\dot{\alpha}}(\Lambda^{\dot{\alpha}}\gamma)+\frac{n+1}{3n+1}(\partial_a\Lambda^a-\mathrm{D}_\alpha\Lambda^\alpha)\gamma.\end{gathered} \tag{6.6.54}$$

The explicit form of the gauge parameters $\Lambda=\Lambda^A\mathrm{D}_A$ was given in equation (6.2.3).

The set of superfields $\{H^a=0, \gamma=1)$ constitutes a stationary point of (6.6.53) corresponding to flat global superspace. To linearize at this stationary point, we must expand the action up to third order in the superfields H^a, Γ and $\bar{\Gamma}$, Γ being defined by

$$\gamma=1+\Gamma \qquad \bar{\mathrm{D}}^2\Gamma=0 \qquad (\mathrm{D}^2\Gamma\neq 0). \tag{6.6.55}$$

Since the relevant computational work has been done in subsection 6.2.1, here we reproduce only the final result:

$$\begin{aligned}S^{(2)}_{\text{NON-MIN.SG}}=\int \mathrm{d}^8z\bigg\{&-\frac{1}{16}H^{\alpha\dot{\alpha}}\mathrm{D}^\beta\bar{\mathrm{D}}^2\mathrm{D}_\beta H_{\alpha\dot{\alpha}}+\frac{n+1}{8n}(\partial_{\alpha\dot{\alpha}}H^{\alpha\dot{\alpha}})^2\\ &+\frac{n+1}{32}([\mathrm{D}_\alpha,\bar{\mathrm{D}}_{\dot{\alpha}}]H^{\alpha\dot{\alpha}})^2-\frac{(n+1)(3n+1)}{4n}\,\mathrm{i}\,H^{\alpha\dot{\alpha}}\partial_{\alpha\dot{\alpha}}(\Gamma-\bar{\Gamma})\\ &-\frac{3n+1}{4}H^{\alpha\dot{\alpha}}(\mathrm{D}_\alpha\bar{\mathrm{D}}_{\dot{\alpha}}\Gamma-\bar{\mathrm{D}}_{\dot{\alpha}}\mathrm{D}_\alpha\bar{\Gamma})\\ &+\frac{(3n+1)^2}{4n}\bar{\Gamma}\Gamma+\frac{9n^2-1}{8n}(\Gamma^2+\bar{\Gamma}^2)\bigg\}.\end{aligned} \tag{6.6.56}$$

Note that we have made the rescaling $H^a \to \kappa H^a$, $\Gamma \to \kappa\Gamma$. The action obtained is invariant under the gauge transformations

$$\begin{aligned}\delta H_{\alpha\dot\alpha} &= \bar{D}_{\dot\alpha} L_\alpha - D_\alpha \bar{L}_{\dot\alpha} \\ \delta\Gamma &= -\frac{1}{4}\frac{n+1}{3n+1}\bar{D}^2 D^\alpha L_\alpha + \frac{1}{4}\bar{D}^{\dot\alpha} D^2 \bar{L}_{\dot\alpha}\end{aligned} \qquad (6.6.57)$$

L_α being arbitrary. The equation (6.6.57) presents the linearized version of equation (6.6.54).

One can readily see that the theories (6.2.5) and (6.6.56) are equivalent. It is sufficient to construct an auxiliary model realizing the dual transform for these theories. In the role of such a model, one can choose

$$\begin{aligned}S[H,\sigma,V] = \int d^8z \bigg\{ &-\frac{1}{16} H^{\alpha\dot\alpha} D^\beta \bar{D}^2 D_\beta H_{\alpha\dot\alpha} + \frac{n+1}{8n}(\partial_{\alpha\dot\alpha} H^{\alpha\dot\alpha})^2 \\ &+ \frac{n+1}{32}([D_\alpha, \bar{D}_{\dot\alpha}] H^{\alpha\dot\alpha})^2 - \frac{(n+1)(3n+1)}{4n}\, i\, H^{\alpha\dot\alpha}\partial_{\alpha\dot\alpha}(V - \bar{V}) \\ &- \frac{3n+1}{4} H^{\alpha\dot\alpha}(D_\alpha \bar{D}_{\dot\alpha} V - \bar{D}_{\dot\alpha} D_\alpha \bar{V}) + \frac{(3n+1)^2}{4n}\bar{V}V \\ &+ \frac{9n^2-1}{8n}(V^2 + \bar{V}^2) - 3(V\sigma + \bar{V}\bar{\sigma}) \bigg\}\end{aligned} \qquad (6.6.58)$$

V being an arbitrary complex scalar superfield. As a result, the theories (6.2.5) and (6.6.56) describe two on-shell massless super Poincaré states of superhelicities (-2) and $\frac{3}{2}$.

6.7. New minimal supergravity

In the previous section, we have carefully avoided dealing with the value $n=0$. This case, similar to $n=-\frac{1}{3}$, turns out to be very special, for at least three reasons. First, the Ψ-transformation law (6.6.22) is consistent with the constraint $\Psi = \bar{\Psi}$ if and only if $n=0$. Secondly, the action superfunctional (6.6.23) becomes ill-defined for $n=0$. Thirdly, the prescription for the construction of super Weyl invariant and generally covariant derivatives, given in subsection 6.6.3, does not work here. Now, we get down to $n=0$.

6.7.1. Real linear compensator

The chief hero of our subsequent considerations will be a real covariantly linear scalar superfield $\mathbb{L}$:

$$(\bar{\mathcal{D}}^2 - 4R)\mathbb{L} = (\mathcal{D}^2 - 4\bar{R})\mathbb{L} = 0 \qquad \mathbb{L} = \bar{\mathbb{L}} \qquad (6.7.1)$$

There exists a covariantly chiral spinor superfield η_α such that

$$\mathbb{L}=\frac{1}{2}(\mathscr{D}^\alpha\eta_\alpha+\bar{\mathscr{D}}_{\dot\alpha}\bar\eta^{\dot\alpha}) \qquad \bar{\mathscr{D}}_{\dot\alpha}\eta_\alpha=0. \tag{6.7.2}$$

Clearly, η_α is defined modulo redefinitions of the form:

$$\eta_\alpha\rightarrow\eta'_\alpha=\eta_\alpha+\mathrm{i}(\bar{\mathscr{D}}^2-4R)\mathscr{D}_\alpha K \qquad K=\bar K \tag{6.7.3}$$

with K being an arbitrary real scalar superfield. So, every dynamical theory of $\mathbb{L}$ can be equivalently treated as a gauge theory of the superfields η_α and $\bar\eta_{\dot\alpha}$ with the gauge invariance (6.7.3).

Owing to the identities (5.4.23) and (5.4.2), the conditions (6.7.1) lead to

$$\mathbb{L}=\bar F^{-2}\gamma=F^{-2}\hat\gamma \qquad \begin{cases}\gamma=\mathrm{e}^{\bar W}\hat\gamma & \bar\partial_\mu\bar\partial^\mu\hat\gamma=0\\ \bar\gamma=\mathrm{e}^{W}\hat{\hat\gamma} & \partial^\mu\partial_\mu\hat{\hat\gamma}=0.\end{cases} \tag{6.7.4}$$

Therefore, $\mathbb{L}$ is determined by some superfield $\hat\gamma$ satisfying two equations:

$$\bar\partial^2\hat\gamma=0 \qquad \hat{\hat\gamma}=(\hat\gamma)^*=\mathrm{e}^{-W}[F^2\bar F^2\mathrm{e}^{\bar W}\hat\gamma]. \tag{6.7.5}$$

Note that in the case of a complex covariantly linear scalar superfield Ψ, the superfield $\hat\gamma$ defined by (6.6.17) and its conjugate $\hat{\hat\gamma}$ are independent. In accordance with equation (5.5.4), the reality condition (6.7.5) can be rewritten as

$$\hat{\hat\varphi}^3\hat{\hat\gamma}=\mathrm{e}^{-W}\left[\frac{(1\cdot\mathrm{e}^{\overleftarrow{\bar W}})}{(1\cdot\mathrm{e}^{\overleftarrow{W}})}\mathrm{e}^{\bar W}(\hat\varphi^3\hat\gamma)\right] \tag{6.7.6}$$

where $\varphi=\mathrm{e}^{\bar W}\hat\varphi$ and $\bar\varphi=\mathrm{e}^{W}\hat{\hat\varphi}$. Then, making use of the identities (1.1.13, 14) leads to the very elegant formula:

$$\hat{\hat\varphi}^3\hat{\hat\gamma}=((\hat\varphi^3\hat\gamma)\mathrm{e}^{\overleftarrow{\bar W}}\mathrm{e}^{-\overleftarrow{W}}). \tag{6.7.7}$$

Suppose, the supergravity chiral compensator φ acquires some displacement. To preserve the linearity condition (6.7.1), $\mathbb{L}$ should also acquire some change. Remarkably, the transformation law of $\mathbb{L}$ is uniquely fixed. Due to relation (6.6.18), one finds

$$\varphi\rightarrow\varphi'=\mathrm{e}^\sigma\varphi \qquad \mathbb{L}\rightarrow\mathbb{L}'=\mathrm{e}^{-\sigma-\bar\sigma}\mathbb{L} \qquad \bar{\mathscr{D}}_{\dot\alpha}\sigma=0. \tag{6.7.8a}$$

Remark. To produce the relations (6.7.8*a*), the chiral superfield η_α (6.7.2) should be transformed according to the law

$$\eta_\alpha\rightarrow\eta'_\alpha=\mathrm{e}^{-3\sigma/2}\eta_\alpha. \tag{6.7.8b}$$

An action superfunctional possessing invariance under the super Weyl transformations (6.7.8) can be obtained from the non-minimal supergravity action (6.6.23), if one suitably defines the limit $n\rightarrow 0$. Namely, let us set

$\Psi = \bar{\Psi} = \mathbb{L}$ in $S_{(n)}$ and take the limit $n \to 0$:

$$S_{(n)} = \frac{1}{n\kappa^2} \int d^8z\, E^{-1}\mathbb{L} + \frac{3}{\kappa^2} \int d^8z\, E^{-1}\mathbb{L} \ln \mathbb{L}.$$

The first term here vanishes identically, since

$$\int d^8z\, E^{-1}\mathbb{L} = -\frac{1}{4} \int d^8z \frac{E^{-1}}{R} (\bar{\mathscr{D}}^2 - 4R)\mathbb{L} = 0 \tag{6.7.9}$$

Thus we obtain the action

$$S = \frac{3}{\kappa^2} \int d^8z\, E^{-1}\mathbb{L} \ln \mathbb{L}. \tag{6.7.10}$$

It is invariant under the super Weyl transformations (6.7.8), since

$$(E^{-1}\mathbb{L})' = E^{-1}\mathbb{L}$$

and, therefore,

$$\begin{aligned} \delta S &= -\frac{3}{\kappa^2} \int d^8z\, E^{-1}\mathbb{L}\{\sigma + \bar{\sigma}\} \\ &= \frac{3}{4\kappa^2} \int d^8z\, E^{-1} \left\{ \frac{\sigma}{R} (\bar{\mathscr{D}}^2 - 4R)\mathbb{L} + \frac{\bar{\sigma}}{\bar{R}} (\mathscr{D}^2 - 4\bar{R})\mathbb{L} \right\} = 0. \end{aligned}$$

The superfunctional (6.7.10), known as 'the action of new minimal supergravity', was invented by P.S. Howe, K.S. Stelle and P. Townsend and independently by S. J. Gates, M. Roček and W. Siegel.

Owing to the super Weyl invariance (6.7.8), one can impose the gauge $\varphi = 1$. Then, the transformation laws of $\mathbb{L}$ and $\hat{\gamma}$ under the supergravity gauge group take the form

$$\delta\mathbb{L} = K^M \partial_M \mathbb{L} + \frac{1}{3}(e^{\overleftarrow{W}}(\partial_m \lambda^m - \partial_\mu \lambda^\mu))\mathbb{L} + \frac{1}{3}(e^{W}(\partial_m \bar{\lambda}^m - \bar{\partial}_{\dot{\mu}} \bar{\lambda}^{\dot{\mu}}))\mathbb{L} \tag{6.7.11}$$

and

$$\delta\hat{\gamma} = \hat{\gamma}\overleftarrow{\lambda} = \partial_m(\lambda^m \hat{\gamma}) - \partial_\mu(\lambda^\mu \hat{\gamma}) - \bar{\partial}^{\dot{\mu}}(\lambda_{\dot{\mu}} \hat{\gamma}) \tag{6.7.12}$$

respectively. The derivation of equation (6.7.12) is the same as of equation (6.6.24) (formally, equation (6.7.12) is obtained from equation (6.6.24) by setting $n = 0$). The superfield $\hat{\gamma}$ constrained by equations (6.7.5) and transforming by the law (6.7.12) is said to be 'real linear compensator'.

Due to arbitrariness of the superparameters $\lambda_{\dot{\mu}}$ in equation (6.7.12), it is possible to choose the gauge

$$\hat{\gamma} = 1. \tag{6.7.13}$$

In this gauge, the residual gauge transformations are generated by superparameters λ^M constrained as

$$(1\cdot\overleftarrow{\lambda})=0. \tag{6.7.14}$$

Next, from equation (6.7.6) we deduce

$$(1\cdot e^{\overleftarrow{W}})=(1\cdot e^{\overleftarrow{\bar{W}}}). \tag{6.7.15}$$

Therefore, imposing the gauge (6.7.13) restricts the prepotentials W^M according to rule (6.7.15).

6.7.2. Dynamics in new minimal supergravity

Our next goal is to show that the supergravity–matter system with action (6.7.10) is dynamically equivalent to Einstein supergravity.

Using relations (6.7.2), we represent S in terms of the (anti)chiral superfields $\bar{\eta}_{\dot{\alpha}}$ and η_α,

$$S=\frac{3}{2\kappa^2}\int d^8z\, E^{-1}(\mathscr{D}^\alpha\eta_\alpha+\bar{\mathscr{D}}_{\dot{\alpha}}\bar{\eta}^{\dot{\alpha}})\ln(\mathscr{D}^\beta\eta_\beta+\bar{\mathscr{D}}_{\dot{\beta}}\bar{\eta}^{\dot{\beta}}).$$

This model, considered as the dynamical system of the supergravity prepotentials and the superfields η_α and $\bar{\eta}_{\dot{\alpha}}$, is characterized by the dynamical equations

$$\frac{\delta S}{\delta\eta_\alpha}=\frac{\delta S}{\delta\bar{\eta}_{\dot{\alpha}}}=0\Rightarrow T=0 \qquad T_{\alpha\dot{\alpha}}=0 \tag{6.7.16}$$

with T and $T_{\alpha\dot{\alpha}}$ being the supertrace and supercurrent, respectively. In accordance with the variational rules for (anti)chiral spinor superfields (6.3.52), the equations of motion for η_α and $\bar{\eta}_{\dot{\alpha}}$ are

$$(\bar{\mathscr{D}}^2-4R)\mathscr{D}_\alpha\ln\mathbb{L}=(\mathscr{D}^2-4\bar{R})\bar{\mathscr{D}}_{\dot{\alpha}}\ln\mathbb{L}=0. \tag{6.7.17}$$

Here we have used equation (6.7.9) and the fact that variations $\delta\eta_\alpha$ and $\delta\bar{\eta}_{\dot{\alpha}}$ must be covariantly chiral and antichiral, respectively, hence $\delta\mathbb{L}=\frac{1}{2}(\mathscr{D}^\alpha\delta\eta_\alpha+\bar{\mathscr{D}}_{\dot{\alpha}}\delta\bar{\eta}^{\dot{\alpha}})$ is covariantly linear. The supercurrent of the theory under consideration can be found in the same way as was done in subsection 6.3.3. for the theory (6.3.49). The result reads

$$\begin{aligned}\kappa^2T_{\alpha\dot{\alpha}}&=\mathbb{L}\{2G_{\alpha\dot{\alpha}}+(\mathscr{D}_\alpha\ln\mathbb{L})\bar{\mathscr{D}}_{\dot{\alpha}}\ln\mathbb{L}-[\mathscr{D}_\alpha,\bar{\mathscr{D}}_{\dot{\alpha}}]\ln\mathbb{L}\}\\&\quad+\frac{3}{2}\eta_\alpha(\mathscr{D}^2-4\bar{R})\bar{\mathscr{D}}_{\dot{\alpha}}\ln\mathbb{L}-\frac{3}{2}\bar{\eta}_{\dot{\alpha}}(\bar{\mathscr{D}}^2-4R)\mathscr{D}_\alpha\ln\mathbb{L}\end{aligned} \tag{6.7.18}$$

or, after taking into account equations (6.7.17), as

$$\kappa^2T_{\alpha\dot{\alpha}}=\mathbb{L}\{2G_{\alpha\dot{\alpha}}+(\mathscr{D}_\alpha\ln\mathbb{L})\bar{\mathscr{D}}_{\dot{\alpha}}\ln\mathbb{L}-[\mathscr{D}_\alpha,\bar{\mathscr{D}}_{\dot{\alpha}}]\ln\mathbb{L}\}. \tag{6.7.19}$$

The general solution of equations (6.7.17) is

$$\mathbb{L}=e^{-\rho-\bar{\rho}} \qquad \bar{\mathscr{D}}_{\dot{\alpha}}\rho=0 \tag{6.7.20}$$

for some covariantly chiral scalar superfield ρ. Then, the super Weyl invariance (6.7.8) can be used to gauge away ρ. In other words, the gauge choice

$$\mathbb{L}=1 \tag{6.7.21}$$

is admissible on-shell. After this, the linearity condition (6.7.1) means $R=0$, and the dynamical equation $T_{\alpha\dot{\alpha}}=0$ implies $G_{\alpha\dot{\alpha}}=0$. As a result, we obtain the Einstein supergravity equations.

In summary, we have shown that the action (6.7.10) leads to the same dynamics as the supergravity action (6.1.3). So, the action (6.7.10) can be taken in the role of the Einstein supergravity action. This supergravity formulation is known as new minimal supergravity. The term 'new minimal' is used due to the fact that the chiral superfield (the compensator of the minimal formulation is chiral) and the real linear superfield (the compensator of new minimal formulation is real linear) carry equal numbers of bosonic and fermionic field degrees of freedom, 4 real bosonic ones + 4 real fermionic ones.

There is one more way to establish the equivalence of the supergravity theories (6.1.3) and (6.7.10). Consider the auxiliary model with the action superfunctional

$$S[V, \mathbb{L}, \mathcal{D}_A]=\frac{3}{\kappa^2}\int \mathrm{d}^8z\, E^{-1}(V\mathbb{L}-\mathrm{e}^V) \tag{6.7.22}$$

where V is an arbitrary real scalar superfield and $\mathbb{L}$ is a real covariantly linear scalar superfield. The action turns out to be invariant under the super Weyl transformations

$$\varphi'=\mathrm{e}^{\sigma}\varphi \qquad V'=V-\sigma-\bar{\sigma} \qquad \mathbb{L}'=\mathrm{e}^{-\sigma-\bar{\sigma}}\mathbb{L} \qquad \bar{\mathcal{D}}_{\dot{\alpha}}\sigma=0. \tag{6.7.23}$$

It may be seen that V can be interpreted as a gauge superfield or the super Weyl group. Varying the action with respect to V gives the equation $V=\ln \mathbb{L}$, which can be used to eliminate V. Then one obtains the new minimal supergravity action. On the other hand, by varying $S[V, \mathbb{L}; \mathcal{D}_A]$ with respect to the (anti)chiral superfields $\bar{\eta}_{\dot{\alpha}}$ and η_{α}, from which $\mathbb{L}$ is composed, we obtain the equations

$$(\bar{\mathcal{D}}^2-4R)\mathcal{D}_{\alpha}V=(\mathcal{D}^2-4\bar{R})\bar{\mathcal{D}}_{\dot{\alpha}}V=0.$$

which have solutions

$$V=\ln \Phi+\ln \bar{\Phi} \qquad \bar{\mathcal{D}}_{\dot{\alpha}}\Phi=0$$

for some covariantly chiral scalar superfield. After doing this, our action reduces to the action (6.6.7) representing the super Weyl invariant form of the Einstein supergravity action (6.1.3). As a result, the minimal and new minimal supergravity formulations turn out dually equivalent.

6.7.3. Gauge fixing and field content in new minimal supergravity

Upon fixing the superlocal Lorentz invariance (by setting $N_\alpha{}^\mu = \delta_\alpha{}^\mu$) and the super Weyl invariance (by setting $\varphi = 1$), new minimal supergravity is described by the prepotentials W^M and $\hat{\gamma}$, with W^M being arbitrary and $\hat{\gamma}$ subject to constraint (6.7.5). These superfields possess the transformation laws (5.4.14) and (6.7.12), respectively, under the K- and λ-supergroups. Above we have pointed out that the λ-invariance can be partly fixed by imposing the condition (6.7.13), which requires the prepotentials W^M to be constrained as in (6.7.15). Then, one ends up with residual λ-transformations under equation (6.7.14) and with arbitrary K-transformations. The latter can be used to take the generalized gravitational superfield gauge (6.6.50). In this gauge, we have

$$(1\cdot \mathrm{e}^{\overleftarrow{W}}) = \frac{\det(\delta_m{}^n + \mathrm{i}\partial_m \mathscr{H}^n)}{\det(\delta^{\dot\mu}{}_{\dot\nu} + \hat{\bar{E}}^{\dot\mu} \bar{\mathscr{H}}_{\dot\nu})} \tag{6.7.24}$$

where $\hat{E}_\mu$ and $\hat{\bar{E}}^{\dot\mu}$ are defined by equation (5.4.43). So, the requirement (6.7.15) takes the form

$$\frac{\det(\delta_m{}^n + \mathrm{i}\partial_m \mathscr{H}^n)\det(\delta_\mu{}^\nu + \hat{E}_\mu \mathscr{H}^\nu)}{\det(\delta_m{}^n - \mathrm{i}\partial_m \mathscr{H}^n)\det(\delta^{\dot\mu}{}_{\dot\nu} + \hat{\bar{E}}^{\dot\mu} \bar{\mathscr{H}}_{\dot\nu})} = 1. \tag{6.7.25}$$

Remark. To derive equation (6.7.24), one can consider the change of variables

$$z^M_{(L)} = \mathrm{e}^W z^M = \begin{cases} x^m_{(L)} = x^m + \mathrm{i}\mathscr{H}^m \\ \theta^\mu_{(L)} = \theta^\mu \\ \bar\theta_{\dot\mu(L)} = \bar\theta_{\dot\mu} + \bar{\mathscr{H}}_{\dot\mu} \end{cases} \Rightarrow \frac{\partial z^N_{(L)}}{\partial z^M} = \begin{pmatrix} \delta_m{}^n + \mathrm{i}\partial_m \mathscr{H}^n & 0 & \partial_m \bar{\mathscr{H}}_{\dot\nu} \\ \mathrm{i}\partial_\mu \mathscr{H}^n & \delta_\mu{}^\nu & 0 \\ \mathrm{i}\bar\partial^{\dot\mu} \mathscr{H}^n & 0 & \delta^{\dot\mu}{}_{\dot\nu} + \bar\partial^{\dot\mu} \bar{\mathscr{H}}_{\dot\nu} \end{pmatrix} \tag{6.7.26}$$

and use the relations (1.11.17) and (1.10.64).

In subsection 6.6.6, we saw that minimal ($n = -\frac{1}{3}$) and non-minimal ($n \neq -\frac{1}{3}, 0$) supergravities can be treated as gauge theories of an (unconstrained) generalized gravitational superfield $\mathscr{H}^M = \bar{\mathscr{H}}^M$ with the gauge group being the supergroup of coordinate transformations (6.6.44, 45). It follows from the above consideration that in the case $n = 0$ such a treatment leads to a theory differing from new minimal supergravity. New minimal supergravity is to be understood as follows. It is the gauge theory of a generalized gravitational superfield $\mathscr{H}^M = \bar{\mathscr{H}}^M$ satisfying the constraint (6.7.25). The gauge group is the supergroup of coordinate transformations (6.6.44) under the unimodularity condition

$$\mathrm{Ber}\left(\frac{\partial \zeta'^N}{\partial \zeta^M}\right) = 1 \tag{6.7.27}$$

which coincides with constraint (6.6.45) for $n = 0$. The gauge group acts on

$\mathcal{H}^M$ according to law (6.6.48). Note that requirement (6.7.14) gives the infinitesimal form of requirement (6.7.27).

Instead of imposing the gauge $\hat{\gamma}=1$, let us take another, more convenient, course. The K- and λ-transformations can be used to choose the gravitational superfield gauge (5.4.38). In this approach, the gravitational superfield $\mathcal{H}^M=\bar{\mathcal{H}}^M$ is unconstrained, but the compensator $\hat{\gamma}$ is governed by the equations

$$\bar{\partial}^2\hat{\gamma}=0 \qquad \bar{\gamma}=(\hat{\gamma}\mathrm{e}^{\overleftarrow{\bar{W}}}\mathrm{e}^{-\overleftarrow{W}}) \tag{6.7.28}$$

where $W=W^M\partial_m$ is some functional of $\mathcal{H}^M$ (see equations (5.4.38)).

In the gravitational superfield gauge, the residual gauge invariance is described by the superparameters (5.4.42). It is in our power to choose the Wess–Zumino (5.1.17) for our exploration. Since now $\mathcal{H}^3=0$, we have

$$\bar{W}^M=\mathrm{i}\mathcal{H}^M+\frac{1}{2}\mathcal{H}^n\partial_n\mathcal{H}^M$$

and the requirements for $\hat{\gamma}$ take the form

$$\bar{\partial}^2\hat{\gamma}=0 \qquad \hat{\bar{\gamma}}=\hat{\gamma}+2\mathrm{i}(\hat{\gamma}\overleftarrow{\mathcal{H}})-2(\hat{\gamma}\overleftarrow{\mathcal{H}}\overleftarrow{\mathcal{H}}) \tag{6.7.29}$$

where $\mathcal{H}=\mathcal{H}^m\partial_m$. It follows from here that $\hat{\gamma}$ has the structure:

$$\hat{\gamma}(x,\theta,\bar{\theta})=C(x)+\theta^\alpha\varphi_\alpha(x)+\bar{\theta}_{\dot{\alpha}}\bar{\varphi}^{\dot{\alpha}}(x)+\theta\sigma^a\bar{\theta}v_a(x)+\theta^2\bar{\theta}_{\dot{\alpha}}\bar{\zeta}^{\dot{\alpha}}(x) \tag{6.7.30}$$

with $C(x)$ being real. Further, the transformations surviving in the Wess–Zumino gauge are described by the superparameters (5.1.18). In accordance with the law (6.7.12), the Weyl and local S-supersymmetry transformations (corresponding to the parameters σ and η^α in (5.1.18)) can be used to gauge away the components C and φ_α to give

$$\hat{\gamma}(x,\theta,\bar{\theta})=e^{-1}(x)(1+\theta\sigma^a\bar{\theta}v_a(x)+\theta^2\bar{\theta}_{\dot{\alpha}}\bar{\zeta}^{\dot{\alpha}}(x)). \tag{6.7.31}$$

Here the factor e^{-1} is introduced to make the fields v_a and $\bar{\zeta}^{\dot{\alpha}}$ scalars with respect to the space–time general coordinate transformations (setting $\lambda^m=b^m(x)$ and $\lambda^\mu=\lambda_{\dot{\mu}}=0$ in equation (6.7.12) gives $\delta\hat{\gamma}=\partial_m(b^m\hat{\gamma})$). Now, the reality condition (6.7.29) means

$$\begin{gathered}\bar{v}_a=v_a+2\mathrm{i}e\partial_m(e^{-1}e_a{}^m)\\ \partial_m\left[e^{-1}\left(A^m+\frac{1}{4}(v^a+\bar{v}^a)e_a{}^m\right)\right]=0\end{gathered} \tag{6.7.32a}$$

and

$$\zeta^\alpha=-2e\partial_m(e^{-1}\Psi^{m\alpha}). \tag{6.7.32b}$$

The system (6.7.32a) has the solution

$$v_a=w_a+2A_a-\mathrm{i}e\partial_m(e^{-1}e_a{}^m) \tag{6.7.33}$$

where

$$w^m = e_a{}^m w^a = \frac{1}{2}\varepsilon^{mnkl}\partial_n B_{kl} \qquad (\varepsilon^{0123} = e)$$
$$B_{kl} = -B_{lk} = \bar{B}_{kl}. \tag{6.7.34}$$

We see that the only independent component of $\hat{\gamma}$ is an antisymmetric tensor field B_{kl} possessing the gauge invariance

$$\delta B_{kl} = \partial_m f_l - \partial_l f_k. \tag{6.7.35}$$

There is no mystery in the appearance of B_{kl} and corresponding gauge invariance. One can easily recognize B_{kl} as a component field of the (anti)chiral superfields $\bar{\eta}_{\dot{\alpha}}$ and η_α from which $\mathbb{L}$ (and $\hat{\gamma}$) is composed. Next, the invariance (6.7.35) represents a fragment of invariance (6.7.3).

The most important feature of new minimal supergravity is that its component gauge group includes not only the space–time general coordinate, local Lorentz and local supersymmetry transformations (as in the minimal and non-minimal versions), but the local chiral ones also. It is worth reminding ourselves that the Weyl and local S-supersymmetry transformations have been used to achieve the gauge (6.7.31). If one chooses in the law (6.7.12)

$$\lambda^m = 0 \qquad \lambda^\alpha = \frac{i}{2}\Omega\theta^\alpha$$

$$\lambda_{\dot{\alpha}} = e^{-\bar{W}} e^{W} \bar{\lambda}_{\dot{\alpha}} = -\frac{i}{2}\Omega\bar{\theta}_{\dot{\alpha}} - \bar{\theta}_{\dot{\alpha}}\mathscr{H}^m\partial_m\Omega$$

which corresponds to some γ_5-transformation, then the result will be

$$\delta v_a = e_a\Omega \qquad \delta\bar{\zeta}^{\dot{\alpha}} = -i\partial_m(\Omega\Psi^{m\dot{\alpha}})$$

hence

$$\delta A_a = \frac{1}{2}\partial_a\Omega \qquad \delta\Psi^{m\dot{\alpha}} = \frac{i}{2}\Omega\Psi^{m\dot{\alpha}}.$$

These are exactly the γ_5-transformation laws (5.1.29).

Owing to the equivalence of minimal and new minimal supergravities, the fields A_m and B_{mn} do not propagate on-shell. They are the auxiliary fields of new minimal supergravity. At the same time, A_m and B_{mn} are gauge fields for the local chiral invariance and the invariance (6.7.35), respectively. One can check that A_m and B_{mn} enters the action (6.7.10) as

$$\int d^4x\, e^{-1}\{C_1 w^m w_m + C_2 w^m A_m\} \tag{6.7.36}$$

C_1 and C_2 being non-zero constants. Here w^m is defined in equation (6.7.34). The non-dynamical nature of A_m and B_{mn} is now explicit.

6.7.4. Linearized new minimal supergravity

A linearized version of the new minimal supergravity can be obtained by means of direct examination of its action superfunctional (6.7.10) or by finding a dual transform for the linearized minimal supergravity action (6.2.5). We choose the second course.

Let us consider a globally supersymmetric theory which is described by unconstrained real vector and scalar superfields H^a and U as well as by a chiral scalar superfield σ, $\bar{D}_{\dot{\alpha}}\sigma=0$, and its conjugate $\bar{\sigma}$. The action superfunctional is given by

$$S[H,U,\sigma]=\int d^8z\left\{-\frac{1}{16}H^{\alpha\dot{\alpha}}D^\beta\bar{D}^2D_\beta H_{\alpha\dot{\alpha}}-\frac{1}{4}(\partial_{\alpha\dot{\alpha}}H^{\alpha\dot{\alpha}})^2+\frac{1}{16}([D_\alpha,\bar{D}_{\dot{\alpha}}]H^{\alpha\dot{\alpha}})^2\right.$$
$$\left.+U\left(\frac{1}{2}[D_\alpha,\bar{D}_{\dot{\alpha}}]H^{\alpha\dot{\alpha}}-3\sigma-3\bar{\sigma}\right)+\frac{3}{2}U^2\right\} \qquad (6.7.37)$$

and turns out to be invariant under the gauge transformations

$$\delta H_{\alpha\dot{\alpha}}=\bar{D}_{\dot{\alpha}}L_\alpha-D_\alpha\bar{L}_{\dot{\alpha}}$$
$$\delta U=\frac{1}{4}(D^\alpha\bar{D}^2L_\alpha+\bar{D}_{\dot{\alpha}}D^2\bar{L}^{\dot{\alpha}}) \qquad (6.7.38)$$
$$\delta\sigma=-\frac{1}{12}\bar{D}^2D^\alpha L_\alpha$$

L_α being arbitrary. Note that the variation δU is linear, $\bar{D}^2\delta U=0$.

The equation of motion for U makes it possible to express U via the other superfields. Then, the action (6.7.37) is reduced to the linearized supergravity action (6.2.5). On the other hand, the equation of motion for σ means nothing more than the linearity of U,

$$U=\mathbb{U}=\bar{\mathbb{U}} \qquad \bar{D}^2\mathbb{U}=0. \qquad (6.7.39)$$

Then, σ and $\bar{\sigma}$ drop out of the action and it takes the form:

$$S^{(2)}_{\text{NEW MIN.SG}}=\int d^8z\left\{-\frac{1}{16}H^{\alpha\dot{\alpha}}D^\beta\bar{D}^2D_\beta H_{\alpha\dot{\alpha}}-\frac{1}{4}(\partial_{\alpha\dot{\alpha}}H^{\alpha\dot{\alpha}})^2\right.$$
$$\left.+\frac{1}{16}([D_\alpha,\bar{D}_{\dot{\alpha}}]H^{\alpha\dot{\alpha}})^2+\frac{1}{2}\mathbb{U}[D_\alpha,\bar{D}_{\dot{\alpha}}]H^{\alpha\dot{\alpha}}+\frac{3}{2}\mathbb{U}^2\right\}. \qquad (6.7.40)$$

This superfunctional represents the linearized version of the new minimal supergravity action. It is obviously invariant under the transformations

$$\delta H_{\alpha\dot{\alpha}}=\bar{D}_{\dot{\alpha}}L_\alpha-D_\alpha\bar{L}_{\dot{\alpha}}$$
$$\delta\mathbb{U}=\frac{1}{4}(D^\alpha\bar{D}^2L_\alpha+\bar{D}_{\dot{\alpha}}D^2\bar{L}^{\dot{\alpha}}) \qquad (6.7.41)$$

where L_α is an arbitrary spinor superfield.

Using the invariance (6.7.41), one can set the gauge

$$\mathbb{U}=0. \tag{6.7.42}$$

Parameters of the residual gauge transformations are constrained by

$$\mathrm{D}^{\alpha}\bar{\mathrm{D}}^{2}L_{\alpha}+\bar{\mathrm{D}}_{\dot{\alpha}}\mathrm{D}^{2}\bar{L}^{\dot{\alpha}}=0$$

which has the solution

$$L_{\alpha}=\mathrm{i}\mathrm{D}_{\alpha}K+\mathrm{i}\bar{\mathrm{D}}^{\dot{\alpha}}\zeta_{\alpha\dot{\alpha}} \qquad K=\bar{K}.$$

As a result, in the gauge (6.7.42) we end up with an invariance of the form

$$\begin{gathered}\delta H_{a}=\partial_{a}K+\mathrm{i}(\Lambda_{a}-\bar{\Lambda}_{a})\\ K=\bar{K} \qquad \bar{\mathrm{D}}_{\dot{\beta}}\Lambda_{a}=0.\end{gathered} \tag{6.7.43}$$

Remarkably, one can look upon this law as a simultaneous generalization of ordinary Yang–Mills and super Yang–Mills linearized transformations.

6.8. Matter coupling in non-minimal and new minimal supergravities

Owing to the existence of several supergravity formulations discussed in detail in Sections 6.6 and 6.7, it seems reasonable to investigate peculiarities of matter coupling in the non-minimal and new minimal supergravities. Strictly speaking, this problem is not a fundamental one. By their very origin, each of the $n\neq -\frac{1}{3}$, 0 and $n=0$ versions represents a theory of (minimal) supergravity coupled to some special matter. On these grounds, every dynamical system describing an interaction, for example, of non-minimal supergravity with some matter superfields $\{V^I\}$ can be equivalently reformulated as a theory of minimal supergravity coupled to the extended matter $\{V^I, \Psi, \bar{\Psi}\}$, where Ψ is a complex covariantly linear scalar superfield. In this sense, minimal supergravity is unique. Nevertheless, sometimes use of the new minimal formulation (or, less often, the non-minimal one) may turn out to be preferable.

6.8.1. Non-minimal chiral compensator

We are going to introduce a chiral superfield, constructed from the non-minimal prepotentials W^M and γ, playing the same role in non-minimal supergravity as the chiral compensator φ in minimal supergravity. Our consideration will be based on the non-minimal covariant derivatives (6.6.28) and the corresponding supertorsion tensors (6.6.30, 31).

Let us introduce the following complex scalar superfield:

$$\begin{gathered}\Sigma=\frac{n+1}{3n+1}(4\mathbb{R})^{-1}\bar{\mathbb{T}}_{\dot{\alpha}}\bar{\mathbb{T}}^{\dot{\alpha}}=\frac{3n+1}{n+1}+(4\mathbb{R})^{-1}\bar{\mathbb{D}}_{\dot{\alpha}}\bar{\mathbb{T}}^{\dot{\alpha}} \qquad n\neq -1\\ \Sigma=-(\bar{\mathbb{D}}_{\dot{\beta}}\bar{\mathbb{T}}^{\dot{\beta}})^{-1}\bar{\mathbb{T}}_{\dot{\alpha}}\bar{\mathbb{T}}^{\dot{\alpha}} \qquad n=-1.\end{gathered} \tag{6.8.1}$$

Its basic properties are:

$$\bar{\Sigma}^2=0 \tag{6.8.2a}$$

$$\bar{\mathbb{D}}_{\dot{\alpha}}\bar{\Sigma}=\bar{\mathbb{T}}_{\dot{\alpha}} \tag{6.8.2b}$$

$$\frac{1}{4}(\bar{\mathbb{D}}^2-4\mathbb{R})\,\mathrm{e}^{-\bar{\Sigma}}=\frac{2n}{n+1}\mathbb{R} \qquad n\neq -1. \tag{6.8.2c}$$

In deriving equations (6.8.2), we have used the identities

$$\{\bar{\mathbb{D}}_{\dot{\alpha}},\bar{\mathbb{D}}_{\dot{\beta}}\}=4\mathbb{R}\bar{M}_{\dot{\alpha}\dot{\beta}} \qquad \bar{\mathbb{D}}_{\dot{\alpha}}\mathbb{R}=0 \qquad \bar{\mathbb{T}}_{\dot{\alpha}}=\bar{\mathbb{D}}_{\dot{\alpha}}\bar{\mathbb{T}}$$

and the fact that for $n=-1$ we have $\mathbb{R}=0$ and $\bar{\mathbb{D}}_{\dot{\alpha}}\bar{\mathbb{T}}^{\dot{\alpha}}$ is chiral. In what follows, we choose $n\neq -1$. The case $n=-1$ can be treated similarly, with evident modifications.

Remark. $\bar{\Sigma}$ contains the chiral superfield $\mathbb{R}$ in the denominator. It is not difficult to see that the leading term in the power series expansion in θ for $\mathbb{R}$ is proportional to the auxiliary field $\bar{B}$ (see equation (6.6.41*b*) modulo bi-linear fermionic combinations. So, $\bar{\Sigma}$ turns out to be well-defined only for some special non-minimal supergeometries. In particular, $\bar{\Sigma}$ becomes singular in the flat superspace limit.

Since $\bar{\mathbb{T}}_{\dot{\alpha}}=\bar{\mathbb{D}}_{\dot{\alpha}}\bar{\mathbb{T}}$, from equation (6.8.2*b*) we deduce that

$$\bar{\Sigma}-\bar{\mathbb{T}}=\ln\bar{\Delta}^3 \qquad \bar{\mathbb{D}}_{\dot{\alpha}}\Delta=\bar{\mathscr{D}}_{\dot{\alpha}}\Delta=0 \tag{6.8.3}$$

for some covariantly chiral scalar superfield Δ. One can check that the transformation (6.6.22) changes Δ according to the rule

$$\Delta\to\Delta'=\mathrm{e}^{-\sigma}\Delta \tag{6.8.4}$$

coinciding with the transformation of Φ in equation (6.6.9). Further, the non-minimal derivatives (6.6.28) admit the form

$$\begin{gathered}\mathbb{D}_\alpha=\mathrm{e}^{\rho}[\tilde{\mathscr{D}}_\alpha-2(\tilde{\mathscr{D}}^\beta\rho)M_{\alpha\beta}] \qquad \rho=-\frac{1}{6}\bar{\Sigma}+\frac{1}{3}\Sigma \\ \bar{\mathbb{D}}_{\dot{\alpha}}=\mathrm{e}^{\bar{\rho}}[\bar{\tilde{\mathscr{D}}}_{\dot{\alpha}}-2(\bar{\tilde{\mathscr{D}}}^{\dot{\beta}}\bar{\rho})\bar{M}_{\dot{\alpha}\dot{\beta}}].\end{gathered} \tag{6.8.5}$$

Here $\tilde{\mathscr{D}}_A=(\tilde{\mathscr{D}}_a,\tilde{\mathscr{D}}_\alpha,\bar{\tilde{\mathscr{D}}}^{\dot{\alpha}})$ are the minimal covariant derivatives obtained from $\mathscr{D}_A$ by replacing the compensator φ by the super Weyl invariant combination $\varphi\Delta$. (In other words, if one indicates the explicit dependence of $\mathscr{D}_A$ on the prepotentials as $\mathscr{D}_A=\mathscr{D}_A[W^M,\varphi]$, then $\tilde{\mathscr{D}}_A=\mathscr{D}_A[W^M,\varphi\Delta]$. Similarly, the non-minimal supergravity action (6.6.34) can be rewritten as

$$S_{(n)}=\frac{1}{n\kappa^2}\int\mathrm{d}^8z\,\tilde{E}^{-1}\exp\left[-\frac{1}{3}(\Sigma+\bar{\Sigma})\right] \qquad \tilde{E}^{-1}=E^{-1}\bar{\Delta}\Delta. \tag{6.8.6}$$

It follows from equations (6.8.5) and (6.8.6) that the minimal compensator φ

enters into all non-minimal quantities only in the super Weyl invariant combination $\varphi\Delta$.

The chiral superfield Δ transforms as a scalar under the supergravity gauge group as long as φ is not eliminated. However, in the super Weyl gauge $\varphi=1$ every λ-transformation must be supplemented by the super Weyl one with $\sigma=-\frac{1}{3}e^{\bar{W}}(\partial_m\lambda^m-\partial_\mu\lambda^\mu)$, and the transformation law for Δ takes the form

$$\delta\hat{\Delta}=K^M\partial_M\hat{\Delta}+\frac{1}{3}(\partial_m\lambda^m-\partial_\mu\lambda^\mu)\hat{\Delta} \qquad \Delta=e^{\bar{W}}\hat{\Delta}. \tag{6.8.7}$$

So, non-minimal supergravity possesses its own chiral compensator. But it is necessary to remember that Δ, similarly to $\bar{\Sigma}$ from which Δ is constructed, turns out to be well defined only for some special non-minimal supergeometries. Δ is known as the 'non-minimal chiral compensator'.

With the help of Δ, we are in a position to introduce chiral superpotentials into non-minimal supergravity. Given a covariantly chiral scalar superfield $\mathscr{L}_c=e^{\bar{W}}\hat{\mathscr{L}}_c$, $\bar{\mathscr{D}}_{\dot{\mu}}\mathscr{L}_c=0$, we define the integral of $\mathscr{L}_c$ over superspace in the three equivalent forms:

$$I=\int d^6z\,(\hat{\varphi}\hat{\Delta})^3\hat{\mathscr{L}}_c \tag{6.6.8a}$$

$$=\int d^8z\,\frac{\mathbb{E}^{-1}}{\mathbb{R}}\,e^{\bar{\Sigma}}\mathscr{L}_c \tag{6.8.8b}$$

$$=-\frac{n+1}{2n}\int d^8z\,\frac{\mathbb{E}^{-1}}{\mathbb{R}}\,\mathscr{L}_c. \tag{6.8.8c}$$

Let us comment on the derivation of expressions (6.8.8*b*) and (6.8.8*c*). Using the chiral integration rule (5.5.21), we can write

$$I=\int d^8z\,\frac{\tilde{E}^{-1}}{\tilde{R}}\,\mathscr{L}_c=-\frac{1}{4}\int d^8z\,\frac{E^{-1}U}{[(\tilde{\bar{\mathscr{D}}}^2-4\tilde{R})U]}\,\mathscr{L}_c.$$

Choosing here $U=\exp(\frac{1}{6}\bar{\Sigma}-\frac{1}{3}\Sigma)$ and noting that

$$\mathbb{R}=-\frac{1}{4}(\tilde{\bar{\mathscr{D}}}^2-4\tilde{R})\exp\left(\frac{1}{6}\bar{\Sigma}-\frac{1}{3}\Sigma\right) \tag{6.8.9}$$

one obtains equation (6.8.8*b*). Next, the relation (6.8.2*c*) can be used to write

$$I=\frac{n+1}{8n}\int d^8z\,\frac{\mathbb{E}^{-1}}{\mathbb{R}^2}\,e^{\bar{\Sigma}}(\bar{\mathbb{D}}-4\mathbb{R})(e^{-\bar{\Sigma}}\mathscr{L}_c).$$

Then making use of the rule for integration by parts

$$\int d^8z\,\mathbb{E}^{-1}V(\bar{\mathbb{D}}-4\mathbb{R})U=\int d^8z\,\mathbb{E}^{-1}U\,e^{\bar{\Sigma}}(\bar{\mathbb{D}}-4\mathbb{R})(e^{-\bar{\Sigma}}V) \tag{6.8.10}$$

leads to equation (6.8.8*c*).

6.8.2. Matter dynamical systems in a non-minimal supergravity background
Now it is easy to to generalize the results of subsection 5.5.7, where we have discussed general properties of matter systems coupled to a minimal supergravity background, to the non-minimal case. Generalization consists merely in choosing the structure

$$S[\chi; \mathbb{D}_A] = \int d^8z\, \mathbb{E}^{-1}\left\{\mathscr{L}(\chi; \mathbb{D}_A) + \frac{e^{\bar{\Sigma}}}{\mathbb{R}}\mathscr{L}_c(\chi; \mathbb{D}_A) + \frac{e^{\Sigma}}{\bar{\mathbb{R}}}\bar{\mathscr{L}}_c(\chi; \mathbb{D}_A)\right\} \qquad \bar{\mathbb{D}}_{\dot\alpha}\mathscr{L}_c = 0 \tag{6.8.11}$$

in the role of action superfunctional describing the dynamics of matter superfields χ in some non-minimal supergravity background, instead of (5.5.24). In addition, it seems reasonable to admit a non-polynomial dependence on the supertorsions $\mathbb{T}_\alpha$ and $\mathbb{T}_{\dot\alpha}$ for $\mathscr{L}$ and $\mathscr{L}_c$.

All models of matter superfields coupled to non-minimal supergravity can be naturally divided onto two classes: 'effectively minimal' and 'essentially non-minimal'. A matter model is said to be effectively minimal (essentially non-minimal) if after expressing the non-minimal covariant derivatives $\mathbb{D}_A$ via the induced minimal ones $\tilde{\mathscr{D}}_A$ (6.8.5) the action $S[\chi; \mathbb{D}_A]$ turns out to be independent (explicitly dependent) on the superfields Σ and $\bar{\Sigma}$. Examples of effectively minimal models are the chiral scalar superfield model with the action

$$S[\chi, \bar\chi; \mathbb{D}_A] = \int d^8z\, \mathbb{E}^{-1}\bar\chi\, e^{(1/3)(\Sigma+\bar\Sigma)}\chi + \left\{\int d^8z \frac{\mathbb{E}^{-1}}{\mathbb{R}} e^{\bar\Sigma}\mathscr{L}_c(\chi) + \text{c.c.}\right\} \qquad \bar{\mathbb{D}}_{\dot\alpha}\chi = 0 \tag{6.8.12}$$

and the vector multiplet model with the action

$$\begin{aligned} S[V; \mathbb{D}_A] &= \frac{1}{2}\int d^8z \frac{\mathbb{E}^{-1}}{\mathbb{R}} e^{\bar\Sigma}\, W^\alpha W_\alpha \\ W_\alpha &= -\frac{1}{4}(\bar{\mathbb{D}} - 4\mathbb{R})(e^{-\bar\Sigma/2}\mathbb{D}_\alpha V) \qquad V = \bar{V}. \end{aligned} \tag{6.8.13}$$

Making use of the representation (6.8.5) reduces the former to the model (5.7.37), and the latter to the model (6.3.21). As an example of essentially non-minimal systems, one can consider the chiral scalar superfield model with the action

$$S[\chi, \bar\chi; \mathbb{D}_A] = \int d^8z\, \mathbb{E}^{-1}\bar\chi\, e^{\xi(\Sigma+\bar\Sigma)}\chi + \left\{\int d^8z \frac{\mathbb{E}^{-1}}{\mathbb{R}} e^{\bar\Sigma}\mathscr{L}_c(\chi) + \text{c.c.}\right\} \tag{6.8.14}$$

where $\xi \neq \frac{1}{3}$.

Many essentially non-minimal theories are potentially pathalogical due to non-analyticity in the supergravity auxiliary fields.

6.8.3. New minimal supergravity and supersymmetric σ-models

In conclusion, we would like to elucidate one of the most important applications of the new minimal formulation—describing locally supersymmetric nonlinear σ-models.

Supersymmetric σ-models have been studied in Section 3.3. The underlying σ-model symmetry principle is invariance under the Kähler transformations (3.3.11). The σ-model action superfunctional (3.3.10) possesses this invariance, since the operation of integration over flat global superspace is characterized by the property

$$\bar{\mathrm{D}}_{\dot{\alpha}}\Lambda=0 \quad \Rightarrow \quad \int \mathrm{d}^8 z\, \Lambda=0.$$

When trying to extend the σ-model to a curved superspace, we must treat the σ-model variables Φ^i as covariantly chiral scalar superfields, $\bar{\mathscr{D}}_{\dot{\alpha}}\Phi^i=0$, as well as covariantize the integration measure. However, the naive covariantization

$$\int \mathrm{d}^8 z\, E^{-1}\, K(\Phi^i, \bar{\Phi}^{\underline{j}}) \tag{6.8.15}$$

proves to be inconsistent with the Kähler invariance (3.3.11), since

$$\bar{\mathrm{D}}_{\dot{\alpha}}\Lambda=0 \quad \Rightarrow \quad \int \mathrm{d}^8 z\, E^{-1}\Lambda=\int \mathrm{d}^6 z\, \hat{\varphi}^3\, \mathrm{e}^{-\bar{W}}(R\Lambda)\neq 0$$

as long as $R\neq 0$. On the other hand, the relations (6.7.1) imply

$$\bar{\mathrm{D}}_{\dot{\alpha}}\Lambda=0 \quad \Rightarrow \quad \int \mathrm{d}^8 z\, E^{-1}\mathbb{L}\Lambda=\int \mathrm{d}^8 z\, E^{-1}\mathbb{L}\bar{\Lambda}=0.$$

As a result, the action

$$\int \mathrm{d}^8 z\, E^{-1}\mathbb{L}\, K(\Phi^i, \bar{\Phi}^{\underline{j}}) \tag{6.8.16}$$

is explicitly invariant under the Kähler transformations (3.3.11). Remarkably, it is also invariant under the super Weyl transformations (6.7.8). We see that the Kähler invariance appears to be consistent with the new minimal supergravity formulation.

By adding the pure supergravity term (6.7.10) to expression (6.8.16), one obtains the action

$$S[\Phi^i, \mathbb{L}, \mathscr{D}_A]=\int \mathrm{d}^8 z\, E^{-1}\mathbb{L}\left\{\frac{3}{\kappa^2}\ln \mathbb{L}+K(\Phi^i, \bar{\Phi}^{\underline{j}})\right\} \tag{6.8.17}$$

defining the new minimal supergravity–σ-model dynamical system.

It is interesting to notice that the action (6.8.17) can be obtained from the auxiliary action

$$S[\Phi^i, \mathbb{L}, V, \mathscr{D}_A] = \frac{3}{\kappa^2} \int d^8z\, E^{-1} \left\{ V\mathbb{L} - \exp\left(V - \frac{\kappa^2}{3} K(\Phi^i, \bar{\Phi}^{\underline{i}}) \right) \right\} \qquad (6.8.18)$$

which generalizes action (6.7.22), by eliminating the Lagrange multiplet V. The action introduced proves to be invariant under the super Weyl transformations (6.7.23) and the Kähler transformations

$$\begin{aligned} K(\Phi, \bar{\Phi}) &\rightarrow K(\Phi, \bar{\Phi}) + \Lambda(\Phi) + \bar{\Lambda}(\bar{\Phi}) \\ V &\rightarrow V + \frac{\kappa^2}{3} (\Lambda(\Phi) + \bar{\Lambda}(\bar{\Phi})). \end{aligned} \qquad (6.8.19)$$

If one now varies action (6.8.18) with respect to $\mathbb{L}$ (obtaining $(\bar{\mathscr{D}} - 4R)\mathscr{D}_\alpha V = 0$) and then imposes the super Weyl gauge $V = 0$, the action reduces to

$$\begin{aligned} S[\Phi^i, \mathscr{D}_A] &= -\frac{3}{\kappa^2} \int d^8z\, E^{-1} \exp\left(-\frac{\kappa^2}{3} K(\Phi^i, \bar{\Phi}^{\underline{i}}) \right) \\ &= -\frac{3}{\kappa^2} \int d^8z\, E^{-1} + \int d^8z\, E^{-1} K(\Phi^i, \bar{\Phi}^{\underline{i}}) + O(\kappa^2). \end{aligned} \qquad (6.8.20)$$

Here the first term represents the Einstein supergravity action and the second term describes the coupling of the σ-model superfields to minimal supergravity. In the flat superspace limit and $\kappa \rightarrow 0$, $S[\Phi^i, \mathscr{D}_A]$ takes the flat form (3.3.10).

In the super Weyl gauge $V = 0$ we have adopted, every Kähler transformation (6.8.19) must be accompanied by a special super Weyl transformation with $\sigma = (\kappa^2/3)(\Lambda(\Phi) + \bar{\Lambda}(\bar{\Phi}))$. On these grounds, the σ-model action possesses the following super Weyl–Kähler invariance:

$$\begin{aligned} \varphi &\rightarrow \exp\left[\frac{\kappa^2}{3} \Lambda(\Phi) \right] \varphi \\ K(\Phi, \bar{\Phi}) &\rightarrow K(\Phi, \bar{\Phi}) + \Lambda(\Phi) + \bar{\Lambda}(\bar{\Phi}). \end{aligned} \qquad (6.8.21)$$

Clearly, supergravity–matter models (6.8.17) and (6.8.20) are equivalent.

6.9. Free massless higher superspin theories

As has been shown, linearized Einstein supergravity describes free massless particles of spin $\frac{3}{2}$ and 2. So, Einstein supergravity can be treated as a supersymmetric field theory of interacting massless spin-$\frac{3}{2}$ and spin-2 particles. It is quite natural to ask: Do supersymmetric field theories exist which consistently describe an interaction of massless particles having higher spins $s > 2$?. A consistent system of equations for interacting massless fields of all spins, including the gravitational field, was constructed by M. Vasiliev[1]. It is not clear, however, whether there exists an action functional leading to such equations of motion. Description of Vasiliev's construction is out of the scope of this book. The chief goal of the present section is to describe free supersymmetric field theories of massless higher spin particles. These models may happen to be linearized versions of some fundamental higher spin theories extending, in a nontrivial way, Einstein supergravity.

We begin with two definitions. A free relativistic field theory is said to describe a massless particle of spin s if the Poincaré representation which acts in the physical dynamical subspace $\mathbf{\Phi}_0^{(\text{phys})}$, is given as a direct sum of two irreducible massless representations carrying the helicies $(-s)$ and s. By $\mathbf{\Phi}_0^{(\text{phys})}$ we mean the dynamical subspace $\mathbf{\Phi}_0$ factorized over all the purely gauge histories (massless theories turn out to be gauge when $s \geqslant 1$). It is clear that $\mathbf{\Phi}_0^{(\text{phys})}$ may be parametrized by (functionally independent) gauge invariant field strengths which remain non-vanishing on-shell.

A free supersymmetric (super)field theory is said to describe a massless multiplet (superparticle) of superspin Y if the super Poincaré representation acting on the physical dynamical subspace is equivalent to a direct sum of two irreducible massless representations carrying the superhelicities $(-Y-\frac{1}{2})$ and Y. A massless multiplet of superspin Y describes four massless particle states of the helicities $\pm Y$ and $\pm(Y+\frac{1}{2})$ or, equivalently, two massless particles having spins Y and $(Y+\frac{1}{2})$. This is why the massless superspin-Y multiplet is often denoted as $(Y, Y+\frac{1}{2})$.

Previously, we have discussed a lot of supersymmetric models realizing multiplets with lower superspins $Y=0, \frac{1}{2}$, and $\frac{3}{2}$. In particular, the cases $Y=0$ and $Y=\frac{1}{2}$ were realized by the models (3.2.5) and (3.4.9), respectively. Three equivalent formulations were found, given by equations (6.2.5), (6.6.56) and (6.7.40), for a superspin-$\frac{3}{2}$ multiplet. Below we shall present the superfield realizations for massless multiplets of higher superspins $Y=2, \frac{5}{2}, 3, \ldots$, as well as for the so-called gravitino multiplet $(Y=1)$.

6.9.1. Free massless theories of higher integer spins

In this subsection we discuss free field theories describing massless particles of higher integer spins $s=3, 4, 5, \ldots$. These theories were constructed by C. Fronsdal[2].

Let us fix some integer $s>2$ and introduce into consideration a set of two real bosonic tensor fields over Minkowski space of the form

$$\varphi^{\hat{i}}=\{h_{\alpha_1\ldots\alpha_s\dot{\alpha}_1\ldots\dot{\alpha}_s}(x),\ h_{\alpha_1\ldots\alpha_{s-2}\dot{\alpha}_1\ldots\dot{\alpha}_{s-2}}(x)\}. \tag{6.9.1}$$

These fields are assumed to be totally symmetric in their undotted indices and, independently, in their dotted indices.

Remark. Throughout the section, excepting one specially mentioned case,

[1] M. A. Vasiliev *Phys. Lett.* **243B** 378, 1990; *Class. Quantum. Grav.* **8** 1387, 1991.
[2] C. Fronsdal *Phys. Rev.* **18D** 3624, 1978.

all (super)fields which appear will be totally symmetric in their undotted indices and, independently, in the dotted ones. To simplify expressions involving numerous indices, we will often use compact notation of the type

$$F_{\alpha(A)\dot\alpha(B)} \equiv F_{\alpha_1\ldots\alpha_A\dot\alpha_1\ldots\dot\alpha_B} = F_{(\alpha_1\ldots\alpha_A)(\dot\alpha_1\ldots\dot\alpha_B)}$$

indicating in brackets the number of indices. We will also use the following summation convention

$$F^{\alpha(A)\dot\alpha(B)}G_{\alpha(A)\dot\alpha(B)} \equiv F^{\alpha_1\ldots\alpha_A\dot\alpha_1\ldots\dot\alpha_B}G_{\alpha_1\ldots\alpha_A\dot\alpha_1\ldots\dot\alpha_B}.$$

Remark. Spinor indices of the fields (6.9.1) can be converted into vector ones according to the rule

$$h_{a_1\ldots a_s} = (-1/2)^s(\tilde\sigma_{a_1})^{\dot\alpha_1\alpha_1}\ldots(\tilde\sigma_{a_s})^{\dot\alpha_s\alpha_s}h_{\alpha_1\ldots\alpha_s\dot\alpha_1\ldots\dot\alpha_s}$$

and similarly for $h_{\alpha(s-2)\dot\alpha(s-2)}$. The resultant real tensor fields $h_{a_1\ldots a_s}$ and $h_{a_1\ldots a_{s-2}}$ are obviously symmetric and traceless. They can be unified in the symmetric tensor field

$$\Phi_{a_1\ldots a_s} = h_{a_1\ldots a_s} + \eta_{(a_1a_2}h_{a_3\ldots a_s)}$$

which has vanishing second trace,

$$\Phi^{bc}{}_{bca_1\ldots a_{s-4}} = 0.$$

The fields (6.9.1) play the role of dynamical variables in the theory with action

$$\begin{aligned}S_{(s)} = \frac{1}{2}\left(-\frac{1}{2}\right)^s \int \mathrm{d}^4x \Big\{ & h^{\alpha(s)\dot\alpha(s)}\Box h_{\alpha(s)\dot\alpha(s)} - \frac{s}{2}\partial_{\beta\dot\beta}h^{\beta\alpha(s-1)\dot\beta\dot\alpha(s-1)}\partial^{\gamma\dot\gamma}h_{\gamma\alpha(s-1)\dot\gamma\dot\alpha(s-1)} \\ & - s(s-1)h^{\alpha(s-2)\dot\alpha(s-2)}\partial^{\beta\dot\beta}\partial^{\gamma\dot\gamma}h_{\beta\gamma\alpha(s-2)\dot\beta\dot\gamma\dot\alpha(s-2)} \\ & - s(2s-1)h^{\alpha(s-2)\dot\alpha(s-2)}\Box h_{\alpha(s-2)\dot\alpha(s-2)} \\ & - \frac{1}{2}s(s-2)^2\,\partial_{\beta\dot\beta}h^{\beta\alpha(s-3)\dot\beta\dot\alpha(s-3)}\partial^{\gamma\dot\gamma}h_{\gamma\alpha(s-3)\dot\gamma\dot\alpha(s-3)}\Big\}.\end{aligned} \tag{6.9.2}$$

The action remains invariant under the following gauge transformations

$$\begin{aligned}\delta h_{\alpha_1\ldots\alpha_s\dot\beta_1\ldots\dot\beta_s} &= \partial_{(\alpha_1(\dot\beta_1}\zeta_{\alpha_2\ldots\alpha_s)\dot\beta_2\ldots\dot\beta_s)} \\ \delta h_{\alpha_1\ldots\alpha_{s-2}\dot\alpha_1\ldots\dot\alpha_{s-2}} &= \frac{s-1}{s^2}\partial^{\beta\dot\beta}\zeta_{\beta\alpha_1\ldots\alpha_{s-2}\dot\beta\dot\alpha_1\ldots\dot\alpha_{s-2}}\end{aligned} \tag{6.9.3}$$

where the real tensor parameters $\zeta_{\alpha(s-1)\dot\alpha(s-1)}$ are completely arbitrary functions over space–time. Note that the operations of symmetrization over undotted and dotted indices are always done independently.

We are going to show that the theory (6.9.2) describes a massless particle of spin s. It is worth beginning with the following simple table.

Table 6.9.1. Number of fields and gauge parameters.

Fields		Gauge parameters
$h_{\alpha(s)\dot{\alpha}(s)}$: $(s+1)^2$	$\Big\}\, 2s^2+2$	$\zeta_{\alpha(s-1)\dot{\alpha}(s-1)}$: s^2
$h_{\alpha(s-2)\dot{\alpha}(s-2)}$: $(s-1)^2$		

Next, let us prove an important theorem.

Theorem 1. On-shell, among the components of $h_{\alpha(s)\dot{\alpha}(s)}$ and $h_{\alpha(s-2)\dot{\alpha}(s-2)}$ there are only two independent components modulo gauge transformations.

Proof. Suppose that $h_{\alpha(s)\dot{\alpha}(s)}$ and $h_{\alpha(s-2)\dot{\alpha}(s-2)}$ satisfy the dynamical equations determined by the action (6.9.2). Using the gauge freedom, one is in a position to impose the gauge condition

$$h_{\alpha(s-2)\dot{\alpha}(s-2)}=0 \tag{6.9.4}$$

which constrains, due to transformations (6.9.3), the residual gauge invariance via

$$\partial^{\beta\dot{\beta}}\zeta_{\beta\alpha(s-2)\dot{\beta}\dot{\alpha}(s-2)}=0. \tag{6.9.5}$$

Note also that the dynamical equation for $h_{\alpha(s-2)\dot{\alpha}(s-2)}$ implies

$$\partial^{\beta\dot{\beta}}\partial^{\gamma\dot{\gamma}}h_{\beta\gamma\alpha(s-2)\dot{\beta}\dot{\gamma}\dot{\alpha}(s-2)}=0. \tag{6.9.6}$$

If equation (6.9.5) is satisfied, the divergence $\partial^{\beta\dot{\beta}}h_{\beta\alpha(s-1)\dot{\beta}\dot{\alpha}(s-1)}$ transforms in the manner

$$\delta(\partial^{\beta\dot{\beta}}h_{\beta\alpha(s-1)\dot{\beta}\dot{\alpha}(s-1)})\approx\Box\zeta_{\alpha(s-1)\dot{\alpha}(s-1)}$$

and, owing to relation (6.9.6), can be gauged away

$$\partial^{\beta\dot{\beta}}h_{\beta\alpha(s-1)\dot{\beta}\dot{\alpha}(s-1)}=0. \tag{6.9.7}$$

Under the gauge conditions (6.9.4) and (6.9.7), $h_{\alpha(s)\dot{\alpha}(s)}$ satisfies the equation

$$\Box h_{\alpha(s)\dot{\alpha}(s)}=0 \tag{6.9.8}$$

and the residual gauge invariance is constrained by

$$\Box\zeta_{\alpha(s-1)\dot{\alpha}(s-1)}=0 \tag{6.9.9}$$

in addition to equation (6.9.5).

For further analysis, it is convenient to transform to momentum space both the fields and residual gauge parameters. In accordance with equation

(6.9.8), we have

$$h_{\alpha(s)\dot{\alpha}(s)}(x)=\int\frac{\mathrm{d}^3\vec{p}}{p^0}\,\mathrm{e}^{\mathrm{i}px}h_{\alpha(s)\dot{\alpha}(s)}(p)+\text{c.c.}$$

$$p^a=(p^0,\vec{p})\qquad p^0=|\vec{p}|$$

and similarly for the residual gauge parameters. It is also useful to choose a standard reference system in which

$$p^a=(E,0,0,E)\quad\Rightarrow$$
$$p_{\alpha\dot{\alpha}}=\begin{pmatrix}p_{1\dot{1}} & 0\\ 0 & 0\end{pmatrix}\qquad p^{\alpha\dot{\alpha}}=\begin{pmatrix}0 & 0\\ 0 & p^{2\dot{2}}\end{pmatrix}\tag{6.9.10}$$

where $p_{1\dot{1}}=p^{2\dot{2}}=2E$. Now, the relation (6.9.7) means

$$h_{2\alpha(s-1)\dot{2}\dot{\alpha}(s-1)}(p)=0.\tag{6.9.11}$$

Analogously, equation (6.9.5) leads to

$$\zeta_{2\alpha(s-2)\dot{2}\dot{\alpha}(s-2)}(p)=0.$$

Making the gauge transformation generated by such a parameter, one has

$$\delta h_{1\alpha(s-1)\dot{1}\dot{\alpha}(s-1)}(p)\approx p_{1\dot{1}}\zeta_{\alpha(s-1)\dot{\alpha}(s-1)}(p).$$

Therefore, we can impose the gauge condition

$$h_{1\alpha(s-1)\dot{1}\dot{\alpha}(s-1)}(p)=0\tag{6.9.12}$$

which turns out to completely fix the gauge freedom. The relations (6.9.11) and (6.9.12) tell us that the only non-zero components of $h_{\alpha(s)\dot{\alpha}(s)}(p)$ are the following:

$$h_{\underbrace{1\ldots1}_{s}\underbrace{\dot{2}\ldots\dot{2}}_{s}}(p),\qquad h_{\underbrace{2\ldots2}_{s}\underbrace{\dot{1}\ldots\dot{1}}_{s}}(p).\tag{6.9.13}$$

This completes the proof.

The next important observation is the fact that the theory under consideration possesses the following gauge invariant field strengths:

$$C_{\alpha_1\ldots\alpha_{2s}}=\partial_{(\alpha_1}{}^{\dot{\beta}_1}\ldots\partial_{\alpha_s}{}^{\dot{\beta}_s}h_{\alpha_{s+1}\ldots\alpha_{2s})\dot{\beta}_1\ldots\dot{\beta}_s}$$
$$\bar{C}_{\dot{\alpha}_1\ldots\dot{\alpha}_{2s}}=\partial^{\beta_1}{}_{(\dot{\alpha}_1}\ldots\partial^{\beta_s}{}_{\dot{\alpha}_s}h_{\beta_1\ldots\beta_s\dot{\alpha}_{s+1}\ldots\dot{\alpha}_{2s})}.\tag{6.9.14}$$

On-shell, these field strengths do not vanish and satisfy the equations

$$\partial^{\beta\dot{\gamma}}C_{\beta\alpha(2s-1)}=0\qquad\partial^{\gamma\dot{\beta}}\bar{C}_{\dot{\beta}\dot{\alpha}(2s-1)}=0.\tag{6.9.15}$$

The last statement can be immediately proved by choosing the gauge (6.9.4, 7). It follows from equation (6.9.15) that $C_{\alpha(2s)}$ and $\bar{C}_{\dot{\alpha}(2s)}$ are on-shell massless fields (see Section 1.8) and their helicities are s and $(-s)$, respectively.

Now, let us recall that each on-shell massless field has only one independent component. If one fixes the reference system in the manner (6.9.10), then the relevant components $C_{\alpha(2s)}$ and $\bar{C}_{\dot{\alpha}(2s)}$ are the following:

$$C_{1\ldots 1}(p) \qquad \bar{C}_{\dot{1}\ldots\dot{1}}(p).$$

In the guage (6.9.4, 7, 12), these components are uniquely connected with the fields (6.9.13) according to the rule:

$$C_{1\ldots 1}(p) \sim (E)^s h_{1\ldots 1\dot{2}\ldots\dot{2}}(p)$$

$$\bar{C}_{\dot{1}\ldots\dot{1}}(p) \sim (E)^s h_{2\ldots 2\dot{1}\ldots\dot{1}}(p).$$

Therefore, $C_{\alpha(2s)}$ and $\bar{C}_{\dot{\alpha}(2s)}$ are the only (functionally independent) field strengths surviving on-shell. As a result, the theory (6.9.2) describes a massless particle of spin s.

6.9.2. Free massless theories of higher half-integer spins

In this subsection we discuss free field theories of massless particles having half-integer spins $(s+\frac{1}{2})$, where $s=2, 3, 4, \ldots$. These theories were constructed by J. Fang and C. Fronsdal[2].

Let us fix some integer $s \geqslant 2$ and introduce into consideration a set of fermionic tensor fields over Minkowski space with the structure

$$\varphi^{i} = \{\Psi_{\alpha(s+1)\dot{\alpha}(s)}(x),\ \Psi_{\alpha(s-1)\dot{\alpha}(s)}(x),\ \Psi_{\alpha(s-1)\dot{\alpha}(s-2)}(x)$$
$$\Psi_{\alpha(s)\dot{\alpha}(s+1)}(x),\ \Psi_{\alpha(s)\dot{\alpha}(s-1)}(x),\ \Psi_{\alpha(s-2)\dot{\alpha}(s-1)}(x)\}. \quad (6.9.16)$$

Remark. The fields can be unified in a four-component Majorana spin-tensor field as follows. First, let us combine $\Psi_{\alpha(s+1)\dot{\alpha}(s)}$ and $\Psi_{\alpha(s-1)\dot{\alpha}(s)}$ in the field

$$\lambda_{\beta\alpha_1\ldots\alpha_s\dot{\alpha}_1\ldots\dot{\alpha}_s} = \lambda_{\beta(\alpha_1\ldots\alpha_s)\dot{\alpha}_1\ldots\dot{\alpha}_s} = \Psi_{\beta\alpha_1\ldots\alpha_s\dot{\alpha}_1\ldots\dot{\alpha}_s} + \varepsilon_{\beta(\alpha_1}\Psi_{\alpha_2\ldots\alpha_s)\dot{\alpha}_1\ldots\dot{\alpha}_s}$$

which is not completely symmetric in its undotted indices. Secondly, introduce the two-component spin-tensors

$$\lambda_{a_1\ldots a_s\beta} = (-1/2)^s(\tilde{\sigma}_{a_1})^{\dot{\alpha}_1\alpha_1}\ldots(\tilde{\sigma}_{a_s})^{\dot{\alpha}_s\alpha_s}\lambda_{\beta\alpha_1\ldots\alpha_s\dot{\alpha}_1\ldots\dot{\alpha}_s}$$

$$\Psi_{a_1\ldots a_{s-2}\beta} = (-1/2)^{s-2}(\tilde{\sigma}_{a_1})^{\dot{\alpha}_1\alpha_1}\ldots(\tilde{\sigma}_{a_{s-2}})^{\dot{\alpha}_{s-2}\alpha_{s-2}}\Psi_{\beta\alpha_1\ldots\alpha_{s-2}\dot{\alpha}_1\ldots\dot{\alpha}_{s-2}}$$

and then construct the four-component Majorana spin-tensors

$$\Lambda_{a_1\ldots a_s} = \begin{pmatrix}\lambda_{a_1\ldots a_s\beta}\\ \bar{\lambda}_{a_1\ldots a_s}{}^{\dot{\beta}}\end{pmatrix} \qquad \Psi_{a_1\ldots a_{s-2}} = \begin{pmatrix}\Psi_{a_1\ldots a_{s-2}\beta}\\ \bar{\Psi}_{a_1\ldots a_{s-2}}{}^{\dot{\beta}}\end{pmatrix}.$$

Both spin-tensors are symmetric and traceless, with respect to the vector indices, and $\Psi_{a_1\ldots a_{s-2}}$ satisfies the relation

$$\gamma^b \Psi_{ba_1\ldots a_{s-3}} = 0.$$

[2]J. Fang and C. Fronsdal, *Phys. Rev.* **D 18** 3630, 1978.

Finally, we construct the spin-tensor

$$\Theta_{a_1 \ldots a_s} = \Lambda_{a_1 \ldots a_s} + \eta_{(a_1 a_2} \Psi_{a_3 \ldots a_s)}$$

which is symmetric with respect to the vector indices and satisfies the equation

$$\gamma^b \gamma^c \gamma^d \Theta_{bcda_1 \ldots a_{s-3}} = 0.$$

The fields (6.9.16) play the role of dynamical variables in the theory with action

$$\begin{aligned} S_{(s+1/2)} = \mathrm{i}\left(-\frac{1}{2}\right)^s \int \mathrm{d}^4x \{ & \bar{\Psi}^{\alpha(s)\dot{\beta}\dot{\alpha}(s)} \partial^\beta{}_{\dot{\beta}} \Psi_{\beta\alpha(s)\dot{\alpha}(s)} \\ & + s(\bar{\Psi}^{\alpha(s)\dot{\alpha}(s-1)} \partial^{\beta\dot{\beta}} \Psi_{\beta\alpha(s)\dot{\beta}\dot{\alpha}(s-1)} + \partial_{\beta\dot{\beta}} \bar{\Psi}^{\beta\alpha(s-1)\dot{\beta}\dot{\alpha}(s)} \Psi_{\alpha(s-1)\dot{\alpha}(s)}) \\ & + (2s+1)\Psi^{\alpha(s-1)\dot{\beta}\dot{\alpha}(s-1)} \partial^\beta{}_{\dot{\beta}} \bar{\Psi}_{\beta\alpha(s-1)\dot{\alpha}(s-1)} \\ & + (s+1)(\bar{\Psi}^{\alpha(s-2)\dot{\alpha}(s-1)} \partial^{\beta\dot{\beta}} \Psi_{\beta\alpha(s-2)\dot{\beta}\dot{\alpha}(s-1)} \\ & + \partial_{\beta\dot{\beta}} \bar{\Psi}^{\beta\alpha(s-1)\dot{\beta}\dot{\alpha}(s-2)} \Psi_{\alpha(s-1)\dot{\alpha}(s-2)}) \\ & - \bar{\Psi}^{\alpha(s-2)\dot{\beta}\dot{\alpha}(s-2)} \partial^\beta{}_{\dot{\beta}} \Psi_{\beta\alpha(s-2)\dot{\alpha}(s-2)} \}. \end{aligned} \tag{6.9.17}$$

The action turns out to be invariant under the gauge transformations

$$\begin{aligned} \delta\Psi_{\alpha_1 \ldots \alpha_{s+1} \dot{\alpha}_1 \ldots \dot{\alpha}_s} &= \partial_{(\alpha_1 (\dot{\alpha}_1} \xi_{\alpha_2 \ldots \alpha_{s+1}) \dot{\alpha}_2 \ldots \dot{\alpha}_s)} \\ \delta\Psi_{\alpha_1 \ldots \alpha_{s-1} \dot{\alpha}_1 \ldots \dot{\alpha}_s} &= \frac{1}{s+1} \partial^\beta{}_{(\dot{\alpha}_1} \xi_{\beta\alpha_1 \ldots \alpha_{s-1} \dot{\alpha}_2 \ldots \dot{\alpha}_s)} \\ \delta\Psi_{\alpha_1 \ldots \alpha_{s-1} \dot{\alpha}_1 \ldots \dot{\alpha}_{s-2}} &= \frac{s-1}{s} \partial^{\beta\dot{\beta}} \xi_{\beta\alpha_1 \ldots \alpha_{s-1} \dot{\beta}\dot{\alpha}_1 \ldots \dot{\alpha}_{s-1}} \end{aligned} \tag{6.9.18}$$

with arbitrary tensor parameters $\xi_{\alpha(s)\dot{\alpha}(s-1)}$.

The theory (6.9.17) describes a massless particle of spin $(s+1/2)$. To verify this statement, we take the same course as in the previous subsection. It is worth beginning with the following table.

Table 6.9.2. Number of fields and gauge parameters.

Fields			Gauge parameters
$\Psi_{\alpha(s+1)\dot{\alpha}(s)}$:	$(s+2)(s+1)$	$3s(s+1)+2$	$\xi_{\alpha(s)\dot{\alpha}(s-1)}$: $s(s+1)$
$\Psi_{\alpha(s-1)\dot{\alpha}(s)}$:	$s(s+1)$		
$\Psi_{\alpha(s-1)\dot{\alpha}(s-2)}$:	$s(s-1)$		

Now we prove an important assertion useful for our analysis.

Theorem 2. On-shell, among the components of $\Psi_{\alpha(s+1)\dot{\alpha}(s)}$, $\Psi_{\alpha(s-1)\dot{\alpha}(s)}$ and $\Psi_{\alpha(s-1)\dot{\alpha}(s-2)}$ there is only one independent component modulo gauge transformations.

Proof. Let $\Psi_{\alpha(s+1)\dot\alpha(s)}$, $\Psi_{\alpha(s-1)\dot\alpha(s)}$ and $\Psi_{\alpha(s-1)\dot\alpha(s-2)}$ be solutions of the dynamical equations determined by the action (6.9.17). Using the gauge freedom, we can annihilate $\Psi_{\alpha(s-1)\dot\alpha(s-2)}$, that is, impose the gauge condition

$$\Psi_{\alpha(s-1)\dot\alpha(s-2)}=0. \tag{6.9.19}$$

Then, the residual gauge freedom is constrained, due to transformations (6.9.18), by

$$\partial^{\beta\dot\beta}\xi_{\beta\alpha(s-1)\dot\beta\dot\alpha(s-2)}=0. \tag{6.9.20}$$

Next, owing to the $\Psi_{\alpha(s-2)\dot\alpha(s-1)}$ dynamical equation, the field $\Psi_{\alpha(s-1)\dot\alpha(s)}$ satisfies the equation

$$\partial^{\beta\dot\beta}\Psi_{\beta\alpha(s-2)\dot\beta\dot\alpha(s-1)}=0. \tag{6.9.21}$$

The relations (6.9.20, 21) and the $\Psi_{\alpha(s-1)\dot\alpha(s)}$ transformation law (6.9.18) show that $\Psi_{\alpha(s-1)\dot\alpha(s)}$ can be completely gauged away

$$\Psi_{\alpha(s-1)\dot\alpha(s)}=0. \tag{6.9.22}$$

This leads to additional restrictions on the gauge parameters of the form

$$\partial^{\beta}{}_{\dot\gamma}\xi_{\beta\alpha(s-1)\dot\alpha(s-1)}=0 \tag{6.9.23}$$

hence $\Box\xi_{\beta\alpha(s)\dot\alpha(s-1)}=0$. In the gauge (6.9.19, 22), $\Psi_{\alpha(s+1)\dot\alpha(s)}$ satisfies the equation

$$\partial^{\beta}{}_{\dot\gamma}\Psi_{\beta\alpha(s)\dot\alpha(s)}=0. \tag{6.9.24}$$

Similarly to the bosonic case, we can now make the transform to momentum space. Choosing our reference system in the manner (6.9.10), one immediately finds that equation (6.9.24) means

$$\Psi_{2\alpha(s)\dot\alpha(s)}(p)=0. \tag{6.9.25}$$

Further, the residual gauge invariance described by equations (6.9.20, 23) can be used to impose the condition

$$\Psi_{1\alpha(s)\dot 1\dot\alpha(s-1)}(p)=0 \tag{6.9.26}$$

which completely fixes the gauge freedom. Now, the expressions (6.9.25) and (6.9.26) imply that the only non-vanishing component of $\Psi_{\alpha(s+1)\dot\alpha(s)}(p)$ is

$$\Psi_{\underbrace{1\ldots1}_{s+1}\underbrace{\dot2\ldots\dot2}_{s}}(p). \tag{6.9.27}$$

Thus, the theorem has been proved.

Another crucial observation is the fact that the theory under consideration has the following gauge invariant field strengths:

$$\begin{aligned}C_{\alpha_1\ldots\alpha_{2s+1}}&=\partial_{(\alpha_1}{}^{\dot\beta_1}\ldots\partial_{\alpha_s}{}^{\dot\beta_s}\Psi_{\alpha_{s+1}\ldots\alpha_{2s+1})\dot\beta_1\ldots\dot\beta_s}\\ \bar C_{\dot\alpha_1\ldots\dot\alpha_{2s+1}}&=\partial^{\beta_1}{}_{(\dot\alpha_1}\ldots\partial^{\beta_s}{}_{\dot\alpha_s}\Psi_{\beta_1\ldots\beta_s\dot\alpha_{s+1}\ldots\dot\alpha_{2s+1})}.\end{aligned} \tag{6.9.28}$$

On-shell, they turn out to satisfy the equations

$$\partial^{\beta\dot{\gamma}}C_{\beta\alpha(2s)}=0 \qquad \partial^{\gamma\dot{\beta}}\bar{C}_{\dot{\beta}\dot{\alpha}(2s)}=0. \tag{6.9.29}$$

Therefore, $C_{\alpha(2s+1)}$ and $\bar{C}_{\dot{\alpha}(2s+1)}$ carry the helicities $(s+\frac{1}{2})$ and $-(s+\frac{1}{2})$, respectively. It is easy to see that, in the gauge (6.9.19, 22, 26), the only non-zero component of $C_{\alpha(2s+1)}$ is expressed via the field component (6.9.27) and vice versa. As a result, $C_{\alpha(2s+1)}$ and $\bar{C}_{\dot{\alpha}(2s+1)}$ prove to be the only independent field strengths surviving on-shell. We see that the theory (6.9.17) describes a massless particle of spin $(s+\frac{1}{2})$.

6.9.3. *Free massless theories of higher half-integer superspins*

In this subsection we are going to discuss free theories describing massless multiplets of half-integer superspins $Y=(s+\frac{1}{2})$, where $s=2, 3, 4, \ldots$. These theories were constructed by S. Kuzenko, V. Postnikov and A. Sibiryakov.

Let us fix some integer $s\geqslant 2$ and consider two sets of bosonic superfields over flat global superspace:

$${}_{\perp}v^{I}=\{H_{\alpha_1\ldots\alpha_s\dot{\alpha}_1\ldots\dot{\alpha}_s}(z),\ \Gamma_{\alpha_1\ldots\alpha_{s-1}\dot{\alpha}_1\ldots\dot{\alpha}_{s-1}}(z),\ \bar{\Gamma}_{\alpha_1\ldots\alpha_{s-1}\dot{\alpha}_1\ldots\dot{\alpha}_{s-1}}(z)\} \tag{6.9.30}$$

and

$${}_{\parallel}v^{I}=\{H_{\alpha_1\ldots\alpha_s\dot{\alpha}_1\ldots\dot{\alpha}_s}(z),\ G_{\alpha_1\ldots\alpha_{s-1}\dot{\alpha}_1\ldots\dot{\alpha}_{s-1}}(z),\ \bar{G}_{\alpha_1\ldots\alpha_{s-1}\dot{\alpha}_1\ldots\dot{\alpha}_{s-1}}(z)\} \tag{6.9.31}$$

which will play the role of dynamical variables in two different, but equivalent, supersymmetric theories. In both cases, $H_{\alpha(s)\dot{\alpha}(s)}$ is an unconstrained real scalar superfield which transforms in the representation $(s/2, s/2)$ of the Lorentz group. The complex superfields $\Gamma_{\alpha(s-1)\dot{\alpha}(s-1)}$ and $G_{\alpha(s-1)\dot{\alpha}(s-1)}$ transform in the same representation $((s-1)/2, (s-1)/2))$ of the Lorentz group. However, they are subject to the following, essentially different, constraints:

$$\bar{\mathrm{D}}^{\dot{\beta}}\Gamma_{\alpha_1\ldots\alpha_{s-1}\dot{\beta}\dot{\alpha}_1\ldots\dot{\alpha}_{s-2}}=0 \tag{6.9.32}$$

$$\bar{\mathrm{D}}_{(\dot{\alpha}_1}G_{\beta_1\ldots\beta_{s-1}\dot{\alpha}_2\ldots\dot{\alpha}_{s-1})}=0. \tag{6.9.33}$$

Obviously, the equations (6.9.32) and (6.9.33) imply that Γ and G are linear superfields,

$$\bar{\mathrm{D}}^2\Gamma_{\alpha(s-1)\dot{\alpha}(s-1)}=0 \tag{6.9.34}$$

$$\bar{\mathrm{D}}^2G_{\alpha(s-1)\dot{\alpha}(s-1)}=0. \tag{6.9.35}$$

It seems reasonable to call $\Gamma_{\alpha(s-1)\dot{\alpha}(s-1)}$ a 'transversally linear superfield'. Analogously, $G_{\alpha(s-1)\dot{\alpha}(s-1)}$ is said to be a 'longitudinally linear superfield'.

It is worth pointing out that the general solution of equation (6.9.32) reads

$$\Gamma_{\alpha_1\ldots\alpha_{s-1}\dot{\alpha}_1\ldots\dot{\alpha}_{s-1}}=\bar{\mathrm{D}}^{\dot{\alpha}_s}\xi_{\alpha_1\ldots\alpha_{s-1}(\dot{\alpha}_1\ldots\dot{\alpha}_{s-1}\dot{\alpha}_s)} \tag{6.9.36}$$

where $\xi_{\alpha(s-1)\dot{\alpha}(s)}$ is an unconstrained fermionic superfield of the Lorentz type $((s-1)/2, s/2)$. Such a superfield is defined modulo transformations of the form

$$\delta\xi_{\alpha_1\ldots\alpha_{s-1}\dot{\alpha}_1\ldots\dot{\alpha}_s}=\bar{\mathrm{D}}^{\dot{\alpha}_{s+1}}\Lambda_{\alpha_1\ldots\alpha_{s-1}(\dot{\alpha}_1\ldots\dot{\alpha}_s\dot{\alpha}_{s+1})} \tag{6.9.37}$$

with arbitrary bosonic tensor parameters $\Lambda_{\alpha(s-1)\dot{\alpha}(s+1)}$. Further, the general solution of equation (6.9.33) is

$$G_{\alpha_1\ldots\alpha_{s-1}\dot{\alpha}_1\ldots\dot{\alpha}_{s-1}} = \bar{\mathrm{D}}_{(\dot{\alpha}_1}\zeta_{\alpha_1\ldots\alpha_{s-1}\dot{\alpha}_2\ldots\dot{\alpha}_{s-1})}. \tag{6.9.38}$$

Here $\zeta_{\alpha(s-1)\dot{\alpha}(s-2)}$ is unconstrained fermionic tensor superfield defined modulo shifts of the form

$$\delta\zeta_{\alpha_1\ldots\alpha_{s-1}\dot{\alpha}_1\ldots\dot{\alpha}_{s-2}} = \bar{\mathrm{D}}_{(\dot{\alpha}_1}\kappa_{\alpha_1\ldots\alpha_{s-1}\dot{\alpha}_2\ldots\dot{\alpha}_{s-2})} \tag{6.9.39}$$

with arbitrary bosonic tensor parameters $\kappa_{\alpha(s-1)\dot{\alpha}(s-3)}$. We see that the superfield $\Gamma_{\alpha(s-1)\dot{\alpha}(s-1)}$ constrained by equation (6.9.32) can be treated as a gauge invariant superfield strength of the superpotentials $\xi_{\alpha(s-1)\dot{\alpha}(s)}$, with respect to the gauge invariance (6.9.37) (of infinite range of reducibility). Similarly, the superfield $G_{\alpha(s-1)\dot{\alpha}(s-1)}$ subject to equation (6.9.33) can be treated as a gauge invariant superfield strength of the superpotentials $\zeta_{\alpha(s-1)\dot{\alpha}(s-2)}$, with respect to the gauge invariance (6.9.39) (of finite range of reducibility).

Let us define gauge transformations of the superfields $H_{\alpha(s)\dot{\alpha}(s)}$, $\Gamma_{\alpha(s-1)\dot{\alpha}(s-1)}$ and $G_{\alpha(s-1)\dot{\alpha}(s-1)}$ according to the law:

$$\delta H_{\alpha_1\ldots\alpha_s\dot{\alpha}_1\ldots\dot{\alpha}_s} = \bar{\mathrm{D}}_{(\dot{\alpha}_1}L_{\alpha_1\ldots\alpha_s\dot{\alpha}_2\ldots\dot{\alpha}_s)} - \mathrm{D}_{(\alpha_1}\bar{L}_{\alpha_2\ldots\alpha_s)\dot{\alpha}_1\ldots\dot{\alpha}_s} \tag{6.9.40}$$

$$\delta\Gamma_{\alpha_1\ldots\alpha_{s-1}\dot{\alpha}_1\ldots\dot{\alpha}_{s-1}} = -\frac{1}{4}\bar{\mathrm{D}}^{\beta}\mathrm{D}^2\bar{L}_{\alpha_1\ldots\alpha_{s-1}\dot{\beta}\dot{\alpha}_1\ldots\dot{\alpha}_{s-1}} \tag{6.9.41}$$

$$\delta G_{\alpha_1\ldots\alpha_{s-1}\dot{\alpha}_1\ldots\dot{\alpha}_{s-1}} = -\frac{1}{4}\bar{\mathrm{D}}^2\mathrm{D}^{\beta}L_{\beta\alpha_1\ldots\alpha_{s-1}\dot{\alpha}_1\ldots\dot{\alpha}_{s-1}}$$
$$+\mathrm{i}(s-1)\partial^{\beta\dot{\beta}}\bar{\mathrm{D}}_{(\dot{\alpha}_1}L_{\beta\alpha_1\ldots\alpha_{s-1}\dot{\alpha}_2\ldots\dot{\alpha}_{s-1})\dot{\beta}}. \tag{6.9.42}$$

Here $L_{\alpha(s)\dot{\alpha}(s-1)}$ is an arbitrary fermionic tensor superfield of the Lorentz type $(s/2, (s-1)/2)$.

There exists a quadratic superfunctional of the superfields $H_{\alpha(s)\dot{\alpha}(s)}$, $\Gamma_{\alpha(s-1)\dot{\alpha}(s-1)}$ and $\bar{\Gamma}_{\alpha(s-1)\dot{\alpha}(s-1)}$ which is invariant under the transformations (6.9.40) and (6.9.41):

$$S^{\perp}_{(s+1/2,s+1)} = \left(-\frac{1}{2}\right)^s\int \mathrm{d}^8z\,\Bigg\{\frac{1}{8}H^{\alpha(s)\dot{\alpha}(s)}\mathrm{D}^{\beta}\bar{\mathrm{D}}^2\mathrm{D}_{\beta}H_{\alpha(s)\dot{\alpha}(s)}$$
$$+H^{\beta\alpha(s-1)\dot{\beta}\dot{\alpha}(s-1)}(\mathrm{D}_{\beta}\bar{\mathrm{D}}_{\dot{\beta}}\Gamma_{\alpha(s-1)\dot{\alpha}(s-1)} - \bar{\mathrm{D}}_{\dot{\beta}}\mathrm{D}_{\beta}\bar{\Gamma}_{\alpha(s-1)\dot{\alpha}(s-1)})$$
$$+(\bar{\Gamma}^{\alpha(s-1)\dot{\alpha}(s-1)}\Gamma_{\alpha(s-1)\dot{\alpha}(s-1)}$$
$$+\frac{s+1}{s}\Gamma^{\alpha(s-1)\dot{\alpha}(s-1)}\Gamma_{\alpha(s-1)\dot{\alpha}(s-1)} + \mathrm{c.c.})\Bigg\}. \tag{6.9.43}$$

It turns out that this action superfunctional describes a massless multiplet of superspin $(s+\frac{1}{2})$. Explicitly, the gauge freedom (6.9.40, 41) makes it possible to choose the following Wess–Zumino gauge:

$$H_{\alpha_1\ldots\alpha_s\dot\alpha_1\ldots\dot\alpha_s}=\theta^\beta\bar\theta^{\dot\beta}h_{(\beta\alpha_1\ldots\alpha_s)(\dot\beta\dot\alpha_1\ldots\dot\alpha_s)}+\bar\theta^2\theta^\beta\Psi_{(\beta\alpha_1\ldots\alpha_s)\dot\alpha_1\ldots\dot\alpha_s}-\theta^2\bar\theta^{\dot\beta}\bar\Psi_{\alpha_1\ldots\alpha_s(\dot\beta\dot\alpha_1\ldots\dot\alpha_s)}$$
$$+\theta^2\bar\theta^2A_{\alpha_1\ldots\alpha_s\dot\alpha_1\ldots\dot\alpha_s} \qquad (6.9.44)$$
$$\Gamma_{\alpha_1\ldots\alpha_{s-1}\dot\alpha_1\ldots\dot\alpha_{s-1}}=\mathrm{e}^{\mathrm{i}\theta\sigma^a\bar\theta\partial_a}[h_{\alpha_1\ldots\alpha_{s-1}\dot\alpha_1\ldots\dot\alpha_{s-1}}+\theta^\beta\Psi_{(\beta\alpha_1\ldots\alpha_{s-1})\dot\alpha_1\ldots\dot\alpha_{s-1}}$$
$$+\theta_{(\alpha_1}\Psi_{\alpha_2\ldots\alpha_{s-1})\dot\alpha_1\ldots\dot\alpha_{s-1}}+\bar\theta^{\dot\beta}\lambda_{\alpha_1\ldots\alpha_{s-1}(\dot\beta\dot\alpha_1\ldots\dot\alpha_{s-1})}$$
$$+\theta^2B_{\alpha_1\ldots\alpha_{s-1}\dot\alpha_1\ldots\dot\alpha_{s-1}}+\theta^\beta\bar\theta^{\dot\beta}U_{(\beta\alpha_1\ldots\alpha_{s-1})(\dot\beta\dot\alpha_1\ldots\dot\alpha_{s-1})}$$
$$+\bar\theta^{\dot\beta}\theta_{(\alpha_1}F_{\alpha_2\ldots\alpha_{s-1})(\dot\beta\dot\alpha_1\ldots\dot\alpha_{s-1})}+\theta^2\bar\theta^{\dot\beta}\rho_{\alpha_1\ldots\alpha_{s-1}(\dot\beta\dot\alpha_1\ldots\dot\alpha_{s-1})}]$$

all the component fields being totally symmetric in their undotted indices and independently in the dotted indices. Here the bosonic fields $h_{\alpha(s+1)\dot\alpha(s+1)}$, $h_{\alpha(s-1)\dot\alpha(s-1)}$ and $A_{\alpha(s)\dot\alpha(s)}$ are real, the fields $B_{\alpha(s-1)\dot\alpha(s-1)}$ and $U_{\alpha(s)\dot\alpha(s)}$ are complex. One can readily see that the bosonic fields $A_{\alpha(s)\dot\alpha(s)}$, $B_{\alpha(s-1)\dot\alpha(s-1)}$, $U_{\alpha(s)\dot\alpha(s)}$ and $F_{\alpha(s-2)\dot\alpha(s)}$ as well as the fermionic fields $\lambda_{\alpha(s-1)\dot\alpha(s)}$, $\rho_{\alpha(s-1)\dot\alpha(s)}$ are auxiliary. Performing the integration over θ, $\bar\theta$ in equation (6.9.43) and eliminating all the auxiliary fields, we obtain a special theory of bosonic fields

$$h_{\alpha(s+1)\dot\alpha(s+1)} \qquad h_{\alpha(s-1)\alpha(s-1)}$$

and fermionic fields

$$\Psi_{\alpha(s+1)\dot\alpha(s)} \qquad \Psi_{\alpha(s-1)\dot\alpha(s)} \qquad \Psi_{\alpha(s-1)\dot\alpha(s-2)}+\text{c.c.}$$

Then upon trivial rescalings, the bosonic part of the action proves to coincide with the massless action $S_{(s+1)}$ (see equation (6.9.2)), while the fermionic part coincides with the action $S_{(s+1/2)}$ (6.9.17). Therefore, the dynamical system describes a massless multiplet of superspin $(s+\frac{1}{2})$.

Now, we define an action superfunctional of the superfields (6.9.31) which is invariant under the gauge transformations (6.9.40) and (6.9.42):

$$S^{\parallel}_{(s+1/2,s+1)}=\left(-\frac{1}{2}\right)^s\int\mathrm{d}z\left\{\frac{1}{8}H^{\alpha(s)\dot\alpha(s)}\mathrm{D}^\beta\bar{\mathrm{D}}^2\mathrm{D}_\beta H_{\alpha(s)\dot\alpha(s)}\right.$$
$$-\frac{1}{8}\frac{s}{2s+1}[\mathrm{D}_\beta,\bar{\mathrm{D}}_{\dot\beta}]H^{\beta\alpha(s-1)\dot\beta\dot\alpha(s-1)}[\mathrm{D}^\gamma,\bar{\mathrm{D}}^{\dot\gamma}]H_{\gamma\alpha(s-1)\dot\gamma\dot\alpha(s-1)}$$
$$+\frac{s}{2}\partial_{\beta\dot\beta}H^{\beta\alpha(s-1)\dot\beta\dot\alpha(s-1)}\partial^{\gamma\dot\gamma}H_{\gamma\alpha(s-1)\dot\gamma\dot\alpha(s-1)}$$
$$+\frac{2\mathrm{i}s}{2s+1}\partial_{\beta\dot\beta}H^{\beta\alpha(s-1)\dot\beta\dot\alpha(s-1)}(G_{\alpha(s-1)\dot\alpha(s-1)}-\bar G_{\alpha(s-1)\dot\alpha(s-1)})$$
$$+\frac{1}{2s+1}(\bar G^{\alpha(s-1)\dot\alpha(s-1)}G_{\alpha(s-1)\dot\alpha(s-1)}$$
$$\left.-\frac{s+1}{s}G^{\alpha(s-1)\dot\alpha(s-1)}G_{\alpha(s-1)\dot\alpha(s-1)}+\text{c.c.}\ )\right\}. \qquad (6.9.45)$$

This model also describes a massless multiplet of superspin $(s+\frac{1}{2})$. Indeed, the theories (6.9.43) and (6.9.45) are connected by duality transformation via the auxiliary model with the action

$$S[H, G, V] = \left(-\frac{1}{2}\right)^s \int d^8z \left\{ \frac{1}{8} H^{\alpha(s)\dot\alpha(s)} D^\beta \bar{D}^2 D_\beta H_{\alpha(s)\dot\alpha(s)} \right.$$
$$+ \left(H^{\beta\alpha(s-1)\dot\alpha(s-1)} D_\beta \bar{D}_{\dot\beta} V_{\alpha(s-1)\dot\alpha(s-1)} \right.$$
$$- \frac{2}{s} G^{\alpha(s-1)\dot\alpha(s-1)} V_{\alpha(s-1)\dot\alpha(s-1)} + \bar{V}^{\alpha(s-1)\dot\alpha(s-1)} V_{\alpha(s-1)\dot\alpha(s-1)}$$
$$\left.\left. + \frac{s+1}{s} V^{\alpha(s-1)\dot\alpha(s-1)} V_{\alpha(s-1)\dot\alpha(s-1)} + \text{c.c.} \right) \right\} \tag{6.9.46}$$

$V_{\alpha(s-1)\dot\alpha(s-1)}$ being an unconstrained complex tensor superfield of the Lorentz type $((s-1)/2, (s-1)/2)$. Thus, the theories (6.9.43) and (6.9.45) are equivalent.

The two superfield formulations for the massless superspin-$(s+\frac{1}{2})$ multiplet, which are determined by the actions (6.9.43) and (6.9.45), can be called transversal and longitudinal, respectively.

There is a purely superfield way to prove that both the theories (6.9.43) and (6.9.45) describe a massless multiplet of superspin $(s+\frac{1}{2})$. Namely, they possess the following gauge invariant superfield strengths:

$$\begin{aligned} W_{\alpha_1 \ldots \alpha_{2s+1}} &= \bar{D}^2 \partial_{(\alpha_1}{}^{\dot\beta_1} \ldots \partial_{\alpha_s}{}^{\dot\beta_s} D_{\alpha_{s+1}} H_{\alpha_{s+2} \ldots \alpha_{2s+1})\dot\beta_1 \ldots \dot\beta_s} \\ \bar{W}_{\dot\alpha_1 \ldots \dot\alpha_{2s+1}} &= D^2 \partial^{\beta_1}{}_{(\dot\alpha_1} \ldots \partial^{\beta_s}{}_{\dot\alpha_s} \bar{D}_{\dot\alpha_{s+1}} H_{\beta_1 \ldots \beta_s \dot\alpha_{s+2} \ldots \dot\alpha_{2s+1})} \end{aligned} \tag{6.9.47}$$

$W_{\alpha(2s+1)}$ and $\bar{W}_{\dot\alpha(2s+1)}$ being chiral and antichiral, respectively. On-shell, they satisfy the equations

$$D^\beta W_{\beta\alpha(2s)} = 0 \qquad \bar{D}^{\dot\beta} \bar{W}_{\dot\beta\dot\alpha(2s)} = 0 \tag{6.9.48}$$

which can be readily checked, for example, by applying the gauge conditions

$$D^\beta H_{\beta\alpha(s-1)\dot\alpha(s)} = \Gamma_{\alpha(s-1)\dot\alpha(s)} = 0$$

which the theory (6.9.43) admits on-shell. Therefore, $W_{\alpha(2s+1)}$ and $\bar{W}_{\dot\alpha(2s+1)}$ carry the superhelicities $(s+\frac{1}{2})$ and $-(s+1)$, respectively. It is easy to see that $W_{\alpha(2s+1)}$ and $\bar{W}_{\dot\alpha(2s+1)}$ are the only independent superfield strengths surviving on-shell.

In conclusion, let us turn our attention to the case $s=1$, which was excluded above and, in fact, should correspond to linearized supergravity. This choice is seen to be exceptional for both formulations developed. Concerning the transversal formulation, the point is that the constraint (6.9.32) cannot be applied when $s=1$. However, one can still impose constraint (6.9.34), a consequence of equation (6.9.32). Clearly, this is the most natural reduction of the transversal formulation to the case $s=1$. If one now sets $s=1$ and takes Γ to be a complex linear superfield, $\bar{D}^2\Gamma=0$ and $\Gamma \neq \bar{\Gamma}$, then the action (6.9.43) proves to coincide (after the redefinition $\Gamma \to -\Gamma$) with the linearized action of non-minimal $n=-1$ supergravity (6.6.56). As for the longitudinal formulation, upon the choice $s=1$, and only in this case, the equation (6.9.33) implies chirality, $\bar{D}_{\dot{\alpha}}G=0$; the right-hand side in the transformation law (6.9.42) also becomes chiral. The corresponding action (6.9.45) coincides (up to a trivial rescaling) with the linearized action of minimal supergravity (6.2.5). It is quite possible that the new minimal formulation is a specific feature of supergravity (superspin $\frac{3}{2}$) and cannot surve as a basis for extensions to higher superspins.

6.9.4. *Free massless theories of higher integer superspins*

We now present free superfield theories describing massless multiplets of higher integer superspins $Y=s$, where $s=2, 3, 4, \ldots$. These theories were constructed by S. Kuzenko and A. Sibiryakov.

Let us fix some integer $s \geqslant 2$ and consider two sets of bosonic superfields over flat global superspace:

$$_{\perp}v^I=\{H_{\alpha_1\ldots\alpha_{s-1}\dot{\alpha}_1\ldots\dot{\alpha}_{s-1}}(z),\ \Gamma_{\alpha_1\ldots\alpha_s\dot{\alpha}_1\ldots\dot{\alpha}_s}(z),\ \bar{\Gamma}_{\alpha_1\ldots\alpha_s\dot{\alpha}_1\ldots\dot{\alpha}_s}(z)\} \tag{6.9.49}$$

and

$$_{\parallel}v^I=\{H_{\alpha_1\ldots\alpha_{s-1}\dot{\alpha}_1\ldots\dot{\alpha}_{s-1}}(z),\ G_{\alpha_1\ldots\alpha_s\dot{\alpha}_1\ldots\dot{\alpha}_s}(z),\ \bar{G}_{\alpha_1\ldots\alpha_s\dot{\alpha}_1\ldots\dot{\alpha}_s}(z)\}. \tag{6.9.50}$$

In both cases, the tensor superfield $H_{\alpha(s-1)\dot{\alpha}(s-1)}$ is real and unconstrained. The complex tensor superfields $\Gamma_{\alpha(s)\dot{\alpha}(s)}$ and $G_{\alpha(s)\dot{\alpha}(s)}$ are transversally linear and longitudinally linear, respectively,

$$\bar{D}^{\dot{\beta}}\Gamma_{\alpha_1\ldots\alpha_s\dot{\beta}\dot{\alpha}_1\ldots\dot{\alpha}_{s-1}}=0 \tag{6.9.51}$$

$$\bar{D}_{(\dot{\alpha}_1}G_{\alpha_1\ldots\alpha_s\dot{\alpha}_2\ldots\dot{\alpha}_{s+1})}=0. \tag{6.9.52}$$

We define gauge transformations of the superfields $H_{\alpha(s-1)\dot{\alpha}(s-1)}$, $\Gamma_{\alpha(s)\dot{\alpha}(s)}$ and $G_{\alpha(s)\dot{\alpha}(s)}$ according to the law:

$$\delta H_{\alpha_1\ldots\alpha_{s-1}\dot{\alpha}_1\ldots\dot{\alpha}_{s-1}}=D^{\beta}L_{\beta\alpha_1\ldots\alpha_{s-1}\dot{\alpha}_1\ldots\dot{\alpha}_{s-1}}-\bar{D}^{\dot{\beta}}\bar{L}_{\alpha_1\ldots\alpha_{s-1}\dot{\beta}\dot{\alpha}_1\ldots\dot{\alpha}_{s-1}} \tag{6.9.53}$$

$$\begin{aligned}\delta\Gamma_{\alpha_1\ldots\alpha_s\dot{\alpha}_1\ldots\dot{\alpha}_s}=&-\frac{1}{4}\bar{D}^2D_{(\alpha_1}\bar{L}_{\alpha_2\ldots\alpha_s)\dot{\alpha}_1\ldots\dot{\alpha}_s}\\&+\mathrm{i}(s+1)\bar{D}^{\dot{\beta}}\partial_{(\alpha_1(\dot{\beta}}\bar{L}_{\alpha_2\ldots\alpha_s)\dot{\alpha}_1\ldots\dot{\alpha}_s)}\end{aligned} \tag{6.9.54}$$

$$\delta G_{\alpha_1\ldots\alpha_s\dot{\alpha}_1\ldots\dot{\alpha}_s}=-\frac{1}{4}\bar{D}_{(\dot{\alpha}_1}D^2L_{\alpha_1\ldots\alpha_s\dot{\alpha}_2\ldots\dot{\alpha}_s)}. \tag{6.9.55}$$

Here $L_{\alpha(s)\dot{\alpha}(s-1)}$ is an arbitrary bosonic tensor superfield of the Lorentz type $(s/2, (s-1)/2)$.

An action superfunctional for the superfields (6.9.49), which is invariant with respect to the transformations (6.9.53) and (6.9.54), reads

$$S^{\perp}_{(s,s+1/2)} = -\left(-\frac{1}{2}\right)^s \int d^8z \left\{ -\frac{1}{8} H^{\alpha(s-1)\dot{\alpha}(s-1)} D^\beta \bar{D}^2 D_\beta H_{\alpha(s-1)\dot{\alpha}(s-1)} \right.$$

$$+ \frac{1}{8} \frac{s^2}{(s+1)(2s+1)} [D^\beta, \bar{D}^{\dot{\beta}}] H^{\alpha_1 \ldots \alpha_{s-1} \dot{\alpha}_1 \ldots \dot{\alpha}_{s-1}} [D_{(\beta}, \bar{D}_{(\dot{\beta}}] H_{\alpha_1 \ldots \alpha_{s-1})\dot{\alpha}_1 \ldots \dot{\alpha}_{s-1})}$$

$$+ \frac{1}{2} \frac{s^2}{s+1} \partial^{\beta\dot{\beta}} H^{\alpha_1 \ldots \alpha_{s-1} \dot{\alpha}_1 \ldots \dot{\alpha}_{s-1}} \partial_{(\beta(\dot{\beta}} H_{\alpha_1 \ldots \alpha_{s-1})\dot{\alpha}_1 \ldots \dot{\alpha}_{s-1})}$$

$$+ \frac{2is}{2s+1} H^{\alpha(s-1)\dot{\alpha}(s-1)} \partial^{\beta\dot{\beta}} (\Gamma_{\beta\alpha(s-1)\dot{\beta}\dot{\alpha}(s-1)} - \bar{\Gamma}_{\beta\alpha(s-1)\dot{\beta}\dot{\alpha}(s-1)})$$

$$\left. + \frac{1}{2s+1} \left(\bar{\Gamma}^{\alpha(s)\dot{\alpha}(s)} \Gamma_{\alpha(s)\dot{\alpha}(s)} - \frac{s}{s+1} \Gamma^{\alpha(s)\dot{\alpha}(s)} \Gamma_{\alpha(s)\dot{\alpha}(s)} + \text{c.c.} \right) \right\}. \tag{6.9.56}$$

We are going to show that the dynamical system determined by $S^{\perp}_{(s,s+1/2)}$ describes a massless multiplet of superspin s. It turns out that the gauge freedom (6.9.53, 54) makes it possible to choose the following Wess–Zumino gauge:

$$\begin{aligned} H_{\alpha_1 \ldots \alpha_{s-1} \dot{\alpha}_1 \ldots \dot{\alpha}_{s-1}} = {} & \theta_{(\alpha_1} \bar{\theta}_{(\dot{\alpha}_1} h_{\alpha_2 \ldots \alpha_{s-1})\dot{\alpha}_2 \ldots \dot{\alpha}_{s-1})} \\ & + \theta^2 \bar{\theta}_{(\dot{\alpha}_1} \Psi_{\alpha_1 \ldots \alpha_{s-1} \dot{\alpha}_2 \ldots \dot{\alpha}_{s-1})} - \bar{\theta}^2 \theta_{(\alpha_1} \bar{\Psi}_{\alpha_2 \ldots \alpha_{s-1})\dot{\alpha}_1 \ldots \dot{\alpha}_{s-1}} \\ & + \theta^2 \bar{\theta}^2 A_{\alpha_1 \ldots \alpha_{s-1} \dot{\alpha}_1 \ldots \dot{\alpha}_{s-1}} \end{aligned} \tag{6.9.57}$$

$$\begin{aligned} \Gamma_{\alpha_1 \ldots \alpha_s \dot{\alpha}_1 \ldots \dot{\alpha}_s} = e^{i\theta\sigma^a\bar{\theta}\partial_a} \Big[& h_{\alpha_1 \ldots \alpha_s \dot{\alpha}_1 \ldots \dot{\alpha}_s} + \theta^\beta \Psi_{(\beta\alpha_1 \ldots \alpha_s)\dot{\alpha}_1 \ldots \dot{\alpha}_s} + \theta_{(\alpha_1} \Psi_{\alpha_2 \ldots \alpha_s)\dot{\alpha}_1 \ldots \dot{\alpha}_s} \\ & + \bar{\theta}^{\dot{\beta}} \lambda_{\alpha_1 \ldots \alpha_s (\dot{\beta}\dot{\alpha}_1 \ldots \dot{\alpha}_s)} + \theta^2 B_{\alpha_1 \ldots \alpha_s \dot{\alpha}_1 \ldots \dot{\alpha}_s} + \theta^\beta \bar{\theta}^{\dot{\beta}} U_{(\beta\alpha_1 \ldots \alpha_s)(\dot{\beta}\dot{\alpha}_1 \ldots \dot{\alpha}_s)} \\ & + \bar{\theta}^{\dot{\beta}} \theta_{(\alpha_1} F_{\alpha_2 \ldots \alpha_s)(\dot{\beta}\dot{\alpha}_1 \ldots \dot{\alpha}_s)} + \theta^2 \bar{\theta}^{\dot{\beta}} \rho_{\alpha_1 \ldots \alpha_s (\dot{\beta}\dot{\alpha}_1 \ldots \dot{\alpha}_s)} \Big]. \end{aligned}$$

Here the bosonic fields $h_{\alpha(s)\dot{\alpha}(s)}$, and $A_{\alpha(s-1)\dot{\alpha}(s-1)}$ are real and the fields $B_{\alpha(s)\dot{\alpha}(s)}$ and $U_{\alpha(s+1)\dot{\alpha}(s+1)}$ are complex. It follows from considerations of dimension that the bosonic fields $A_{\alpha(s-1)\dot{\alpha}(s-1)}$, $B_{\alpha(s)\dot{\alpha}(s)}$, $U_{\alpha(s+1)\dot{\alpha}(s+1)}$, $F_{\alpha(s-1)\dot{\alpha}(s+1)}$ and the fermionic fields $\lambda_{\alpha(s)\dot{\alpha}(s+1)}$," $\rho_{\alpha(s)\dot{\alpha}(s+1)}$ are auxiliary. Integrating over θ, $\bar{\theta}$ in equation (9.6.56) and eliminating all the auxiliary fields, one arrives at a special theory of bosonic fields

$$h_{\alpha(s)\dot{\alpha}(s)} \qquad h_{\alpha(s-2)\dot{\alpha}(s-2)}$$

and fermionic fields

$$\Psi_{\alpha(s+1)\dot{\alpha}(s)} \qquad \Psi_{\alpha(s-1)\dot{\alpha}(s)} \qquad \Psi_{\alpha(s-1)\dot{\alpha}(s-2)} + \text{c.c.}$$

It proves that, up to a simple rescaling of fields, the bosonic part of the action coincides with the action (6.9.2), while the fermionic part coincides with (6.9.17). As a result, our dynamical system describes a massless multiplet of superspin s.

An action superfunctional for the superfields (6.9.50), which is invariant under arbitrary transformations (6.9.53) and (6.9.55), reads

$$S^{\|}_{(s,s+1/2)}=\left(-\frac{1}{2}\right)^s\int d^8z\left\{\frac{1}{8}H^{\alpha(s-1)\dot\alpha(s-1)}D^\beta\bar D^2D_\beta H_{\alpha(s-1)\dot\alpha(s-1)}\right.$$
$$+\frac{s}{s+1}H^{\alpha(s-1)\dot\alpha(s-1)}(D^\beta\bar D^{\dot\beta}G_{\beta\alpha(s-1)\dot\beta\dot\alpha(s-1)}-\bar D^{\dot\beta}D^\beta\bar G_{\beta\alpha(s-1)\dot\beta\dot\alpha(s-1)})$$
$$\left.+\left(\bar G^{\alpha(s)\dot\alpha(s)}G_{\alpha(s)\dot\alpha(s)}+\frac{s}{s+1}G^{\alpha(s)\dot\alpha(s)}G_{\alpha(s)\dot\alpha(s)}+\text{c.c.}\right)\right\}. \tag{6.9.58}$$

This model also describes a massless multiplet of superspin s. The point is that the theories (6.9.56) and (6.9.58) are equivalent, since they are connected by duality transformation via the auxiliary action

$$S[H,\Gamma,V]=\left(-\frac{1}{2}\right)^s\int d^8z\left\{\frac{1}{8}H^{\alpha(s-1)\dot\alpha(s-1)}D^\beta\bar D^2D_\beta H_{\alpha(s-1)\dot\alpha(s-1)}\right.$$
$$+\frac{1}{s+1}(sH^{\alpha(s-1)\dot\alpha(s-1)}D^\beta\bar D^{\dot\beta}V_{\beta\alpha(s-1)\dot\beta\dot\alpha(s-1)}+2\Gamma^{\alpha(s)\dot\alpha(s)}V_{\alpha(s)\dot\alpha(s)}$$
$$\left.+(s+1)\bar V^{\alpha(s)\dot\alpha(s)}V_{\alpha(s)\dot\alpha(s)}+sV^{\alpha(s)\dot\alpha(s)}V_{\alpha(s)\dot\alpha(s)}+\text{c.c.})\right\} \tag{6.9.59}$$

$V_{\alpha(s)\dot\alpha(s)}$ being an unconstrained complex tensor superfield of the Lorentz type $(s/2, s/2)$.

Let us present one more proof of the fact that the theory (6.9.58) describes a massless multiplet of superspin s. The chiral tensor superfield

$$W_{\alpha_1\ldots\alpha_{2s}}=\frac{\mathrm{i}s}{2}\,\partial_{(\alpha_1}{}^{\dot\beta_1}\ldots\partial_{\alpha_{s-1}}{}^{\dot\beta_{s-1}}\bar D^{\dot\beta_s}D_{\alpha_s}G_{\alpha_{s-1}\ldots\alpha_{2s})\dot\beta_1\ldots\dot\beta_s}+\partial_{(\alpha_1}{}^{\dot\beta_1}\ldots\partial_{\alpha_s}{}^{\dot\beta_s}G_{\alpha_{s-1}\ldots\alpha_{2s})\dot\beta_1\ldots\dot\beta_s} \tag{6.9.60}$$

and its conjugate are gauge invariant superfield strengths (in fact, the only independent ones on-shell). When restricted to the dynamical subspace, they satisfy the equations

$$D^\beta W_{\beta\alpha(2s-1)}=0\qquad \bar D^{\dot\beta}\bar W_{\dot\beta\dot\alpha(2s-1)}=0.$$

Therefore, $W_{\alpha(2s)}$ and $\bar W_{\dot\alpha(2s)}$ carry the superhelicities s and $-(s+\frac{1}{2})$, respectively.

Remark. If one expresses $G_{\alpha(s)\dot\alpha(s)}$ via an unconstrained fermionic tensor superfield according to the rule

$$G_{\alpha(s)\dot\alpha_1\ldots\dot\alpha_s}=\bar D_{(\dot\alpha_1}\zeta_{\alpha(s)\dot\alpha_2\ldots\dot\alpha_s)}$$

then the expression (6.9.60) can be rewritten in the manner

$$W_{\alpha_1\ldots\alpha_{2s}}=\frac{\mathrm{i}}{4}(s+1)\bar D^2\partial_{(\alpha_1}{}^{\dot\beta_1}\ldots\partial_{\alpha_{s-1}}{}^{\dot\beta_{s-1}}D_{\alpha_s}\zeta_{\alpha_{s+1}\ldots\alpha_{2s})\dot\beta_1\ldots\dot\beta_{s-1}}.$$

6.9.5. Massless gravitino multiplet

In the case $s=1$, it turns out that both the formulations (6.9.56) and (6.9.58) describe a massless multiplet of superspin $Y=1$ (gravitino multiplet). To argue this assertion, it is sufficient to consider the transversal model in which the dynamical variables are

$$\{H, \Gamma_{\alpha\dot{\alpha}} \bar{\Gamma}_{\alpha\dot{\alpha}}\}$$

where H is a real scalar superfield, while $\Gamma_{\alpha\dot{\alpha}}$ is subject to the constraint

$$\bar{\mathrm{D}}^{\dot{\alpha}}\Gamma_{\alpha\dot{\alpha}}=0. \tag{6.9.62}$$

The corresponding action (6.9.56) is invariant under the gauge transformations

$$\begin{aligned}\delta H &= \mathrm{D}^{\alpha}L_{\alpha}+\bar{\mathrm{D}}_{\dot{\alpha}}\bar{L}^{\dot{\alpha}} \\ \delta\Gamma_{\alpha\dot{\alpha}} &= -\frac{1}{4}\bar{\mathrm{D}}^2\mathrm{D}_{\alpha}\bar{L}_{\dot{\alpha}}+2\mathrm{i}\bar{\mathrm{D}}^{\dot{\beta}}\partial_{\alpha(\dot{\beta}}\bar{L}_{\dot{\alpha})}\end{aligned} \tag{6.9.63}$$

L_{α} being arbitrary. This gauge freedom makes it possible to choose the following Wess–Zumino gauge (compare with (6.9.57))

$$H=\theta^2\bar{\theta}^2 A \tag{6.9.64}$$

$$\begin{aligned}\Gamma_{\alpha\dot{\alpha}}=\mathrm{e}^{\mathrm{i}\theta\sigma^a\bar{\theta}\partial_a}\Big[& h_{\alpha\dot{\alpha}}+\theta^{\beta}\Psi_{(\beta\alpha)\dot{\alpha}}+\theta_{\alpha}\Psi_{\dot{\alpha}}+\bar{\theta}^{\dot{\beta}}\lambda_{\alpha(\dot{\beta}\dot{\alpha})}+\theta^2 B_{\alpha\dot{\alpha}} \\ &+\theta^{\beta}\bar{\theta}^{\dot{\beta}}U_{(\beta\alpha)(\dot{\beta}\dot{\alpha})}+\bar{\theta}^{\dot{\beta}}\theta_{\alpha}F_{(\dot{\beta}\dot{\alpha})}+\theta^2\bar{\theta}^{\dot{\beta}}\rho_{\alpha(\dot{\beta}\dot{\alpha})}\Big]\end{aligned}$$

$h_{\alpha\dot{\alpha}}$ being real. All component fields, except $h_{\alpha\dot{\alpha}}$, $\Psi_{\alpha\beta\dot{\alpha}}$, $\bar{\Psi}_{\alpha\dot{\alpha}\dot{\beta}}$, $\Psi_{\dot{\alpha}}$ and $\bar{\Psi}_{\alpha}$, are auxiliary. If one inserts the above expansions into equation (6.9.56), integrates $S^{\perp}_{(1,3/2)}$ over θ, $\bar{\theta}$ and eliminates the auxiliary fields, then the resultant action will coincide with a sum of two action functionals describing massless particles of spin 1 and $\frac{3}{2}$.

Let us consider the $s=1$ longitudinal formulation (6.9.58) developed originally by a number of authors[3]. It can be described by a set of unconstrained superfields of the form

$$\{H, \Psi_{\alpha}, \bar{\Psi}_{\dot{\alpha}}\} \qquad H=\bar{H}. \tag{6.9.65}$$

Here Ψ_{α} is connected to the longitudinally linear superfield $G_{\alpha\dot{\alpha}}$ which arose in equation (6.9.58) in the manner

$$G_{\alpha\dot{\alpha}}=\bar{\mathrm{D}}_{\dot{\alpha}}\Psi_{\alpha}. \tag{6.9.66}$$

[3]E.S. Fradkin and M.A. Vasiliev, *Nuovo Cimento Lett.* **25** 79, 1979; B. de Wit and J.W. van Holten, *Nucl. Phys.* **B 155** 530, 1979; S.J. Gates and W. Siegel, *Nucl. Phys.* **B 164** 484, 1980.

Using the vector multiplet strength

$$W_\alpha = -\frac{1}{4}\bar{D}^2 D_\alpha H \tag{6.9.67}$$

one can rewrite $S^{\parallel}_{(1,3/2)}$ as follows:

$$S^{\parallel}_{(1,3/2)} = \hat{S}_{(1,3/2)} - \frac{1}{4}\int d^6z\, W^\alpha W_\alpha - \int d^8z\,(W^\alpha \Psi_\alpha + \bar{W}_{\dot\alpha}\bar{\Psi}^{\dot\alpha}) \tag{6.9.68}$$

where

$$\hat{S}_{(1,3/2)} = \int d^8z \left(D^\alpha \bar{\Psi}^{\dot\alpha} \bar{D}_{\dot\alpha} \Psi_\alpha - \frac{1}{4} D^\alpha \bar{\Psi}^{\dot\alpha} D_\alpha \bar{\Psi}_{\dot\alpha} - \frac{1}{4} \bar{D}^{\dot\alpha} \Psi^\alpha \bar{D}_{\dot\alpha} \Psi_\alpha \right). \tag{6.9.69}$$

As a result of conditions (6.9.53, 55) and (6.9.66), the action (6.9.68) is invariant under the transformations

$$\begin{aligned} \delta H &= D^\alpha L_\alpha + \bar{D}_{\dot\alpha} \bar{L}^{\dot\alpha} \\ \delta \Psi_\alpha &= \Lambda_\alpha + \frac{1}{2} D_\alpha D^\beta L_\beta \qquad \bar{D}_{\dot\alpha} \Lambda_\alpha = 0 \end{aligned} \tag{6.9.70}$$

L_α being arbitrary. As may be seen, it is in our power to impose the gauge fixing

$$H = 0. \tag{6.9.71}$$

In this gauge, the action (6.9.68) coincides with $\hat{S}_{(1,3/2)}$, while the invariance (6.9.70) is reduced to

$$\begin{aligned} \delta \Psi_\alpha &= \Lambda_\alpha + \partial_{\alpha\dot\alpha} D^2 \bar{D}^{\dot\alpha} K \\ \bar{D}_{\dot\alpha} \Lambda_\alpha &= 0 \qquad K = \bar{K}. \end{aligned} \tag{6.9.72}$$

We conclude that the massless gravitino multiplet admits the formulation in terms of unconstrained spinor superfields Ψ_α and $\bar{\Psi}_{\dot\alpha}$, whose dynamics is determined by the action $\hat{S}_{(1,3/2)}$, invariant with respect to the transformations (6.9.72).

There exists another formulation for the gravitino multiplet, in terms of unconstrained spinor superfields Ψ_α and $\bar{\Psi}_{\dot\alpha}$, suggested by V. Ogievetsky and E. Sokatchev[4]. It can be obtained using the following scheme. The longitudinal formulation for the gravitino multiplet proves to be equivalent to that described by the action

$$S[\Psi, H, \Phi] = S^{\parallel}_{(1,3/2)} - \frac{1}{2}\int d^8z\,(\bar{\Phi}\Phi + \Phi D^\alpha \Psi_\alpha + \overline{\Phi \bar{D}_{\dot\alpha}} \bar{\Psi}^{\dot\alpha}) \tag{6.9.73}$$

[4]V.I. Ogievetsky and E. Sokatchev, *J. Phys. A: Math. Gen.* **10** 2021, 1977.

where Φ is a chiral scalar superfield. This action is derived from $S^{\|}_{(1,3/2)}$ by the replacement

$$\Psi_\alpha \to \Psi_\alpha + \frac{1}{16} D_\alpha \frac{\bar{D}^2}{\square} \bar{\Phi}.$$

The action $S[\Psi, H, \Phi]$ is invariant under the gauge transformations

$$\begin{aligned} &\delta H = U + \bar{U} \qquad \delta\Phi = -\frac{1}{2}\bar{D}^2\bar{U} \\ &\delta\Psi_\alpha = \Lambda_\alpha + \frac{1}{2} D_\alpha U \qquad \bar{D}_{\dot{\alpha}}\Lambda_\alpha = 0. \end{aligned} \tag{6.9.74}$$

Here U is a complex scalar superfield. The above invariance can easily be deduced from expressions (6.9.70) by making the replacement $L_\alpha \to L_\alpha + D_\alpha T$ and identifying $U = D^\alpha L_\alpha + \bar{D}^2 \bar{T}$. Next, setting the gauge (6.9.71) reduces the action (6.9.73) to

$$S[\Psi, \Phi] = \hat{S}_{(1,3/2)} - \frac{1}{2}\int d^8z\, (\bar{\Phi}\Phi + \Phi D^\alpha \Psi_\alpha + \overline{\Phi D_{\dot{\alpha}}} \bar{\Psi}^{\dot{\alpha}}) \tag{6.9.75}$$

and the invariance (6.9.74) to the following:

$$\begin{aligned} &\delta\Psi_\alpha = \Lambda_\alpha + \mathrm{i} D_\alpha K \qquad \delta\Phi = \mathrm{i}\bar{D}^2 K \\ &K = \bar{K} \qquad \bar{D}_{\dot{\alpha}}\Lambda_\alpha = 0. \end{aligned} \tag{6.9.76}$$

The model (6.9.75) turns out to be equivalent to the dynamical system with action superfunctional

$$\begin{aligned} S[\Psi, \Phi, V] = \hat{S}_{(1,3/2)} &+ \frac{1}{4}\int d^8z\, (D^\alpha\Psi_\alpha + \bar{D}_{\dot{\alpha}}\bar{\Psi}^{\dot{\alpha}})^2 \\ &+ \int d^8z\, \{V^2 + V(\bar{\Phi} + \Phi + D^\alpha\Psi_\alpha + \bar{D}_{\dot{\alpha}}\bar{\Psi}^{\dot{\alpha}})\} \end{aligned} \tag{6.9.77}$$

V being a real scalar superfield. The final action remains invariant under the transformations (6.9.76) supplemented by

$$\delta V = -\frac{1}{2}(D^\alpha\Lambda_\alpha + \bar{D}_{\dot{\alpha}}\bar{\Lambda}^{\dot{\alpha}}). \tag{6.9.78}$$

On using the equations of motion for Φ and $\bar{\Phi}$, the action (6.9.77) is reduced to

$$S[\Psi, \mathscr{F}] = \hat{S}_{(1,3/2)} + \frac{1}{4}\int d^8z\, (D^\alpha\Psi_\alpha + \bar{D}_{\dot{\alpha}}\bar{\Psi}^{\dot{\alpha}})^2 + \int d^8z\, \{\mathscr{F}^2 + \mathscr{F}(D^\alpha\Psi_\alpha + \bar{D}_{\dot{\alpha}}\bar{\Psi}^{\dot{\alpha}})\} \tag{6.9.79}$$

where $\mathscr{F}$ is a real scalar superfield subject to the linear constraint, that is,

$$\mathscr{F}=\bar{\mathscr{F}} \qquad \bar{\mathrm{D}}^2\mathscr{F}=0. \tag{6.9.80}$$

Obviously, $S[\Psi,\mathscr{F}]$ possesses an invariance of the form

$$\delta\Psi_\alpha \rightarrow \Lambda_\alpha + \mathrm{i}\mathrm{D}_\alpha K \qquad \delta\mathscr{F} = -\frac{1}{2}(\mathrm{D}^\alpha\Lambda_\alpha + \bar{\mathrm{D}}_{\dot{\alpha}}\bar{\Lambda}^{\dot{\alpha}}) \tag{6.9.81}$$

with Λ_α and K defined as in relations (6.9.76). This gauge freedom admits the gauge choice $\mathscr{F}=0$, which reduces $S[\Psi,\mathscr{F}]$ to the superfunctional

$$\hat{\hat{S}}_{(1,3/2)}=\hat{S}_{(1,3/2)}+\frac{1}{4}\int \mathrm{d}^8z\,(\mathrm{D}^\alpha\Psi_\alpha+\bar{\mathrm{D}}_{\dot{\alpha}}\bar{\Psi}^{\dot{\alpha}})^2 \tag{6.9.82}$$

which is invariant under the transformations

$$\delta\Psi_\alpha=\mathrm{i}\bar{\mathrm{D}}^2\mathrm{D}_\alpha K_1+\mathrm{i}\mathrm{D}_\alpha K_2 \qquad K_i=\bar{K}_i. \tag{6.9.83}$$

Our considerations show that the dynamical system with action (6.9.82) describes the massless gravitino multiplet. This is the model proposed by V. Ogievetsky and E. Sokatchev.

7 Effective Action in Curved Superspace

Time present and time past
Are both perhaps present in time future
And time future contained in time past
If all time is eternally present
All time is unredeemable

T.S. Eliot:
'*Four Quartets*' *Burnt Norton*

7.1. The Schwinger–De Witt technique

The Schwinger–De Witt technique (or the proper-time technique) is a powerful method for studying the effective action in theories with external fields. This technique was first introduced by J. Schwinger and later formulated in a general-covariant manner by B.S. De Witt. The Schwinger–De Witt technique is known to be well adapted to calculations of divergences and anomalies in the framework of the background field method.

In the present chapter we describe a superspace extension of the Schwinger–De Witt technique and explore it for investigating the structure of effective action superfunctional, including the analysis of divergences, anomalies and so on, in simple locally supersymmetric theories in a curved superspace. To make the account complete and independent, we find it pertinent to give here a brief review of the standard Schwinger–De Witt technique.

7.1.1. When the proper-time technique can be applied

Let us consider a non-gauge theory of bosonic and fermionic fields $\varphi^{\hat{i}} \equiv \varphi^{i}(x)$ which live on a curved background space–time to be described by a vierbein

DOI: 10.1201/9780367802530-7

$e_a{}^m(x)$. Let $S[\varphi]$ be the classical action of the theory. The one-loop correction to the effective action was shown in Chapter 4 to be formally given by

$$\Gamma^{(1)}[\tilde{\varphi}] \approx \frac{\mathrm{i}}{2}\,\mathrm{sTr}\,\ln({}_{i,}S_{,j}[\tilde{\varphi}]) \tag{7.1.1}$$

where the operator $S'' \equiv {}_{i,}S_{,j}[\tilde{\varphi}]$ determines the quadratic part of $S[\varphi]$ with respect to the mean field $\tilde{\varphi}$.

The standard Schwinger–De Witt technique allows one to obtain information about the explicit structure of $\Gamma^{(1)}[\tilde{\varphi}]$ only when $\mathrm{sDet}(S'')$ can be represented as a product of functional determinants from second-order differential operators of the general form

$$H = \nabla^a\nabla_a - m^2\mathbb{1} + 2\mathscr{A}^a(x)\nabla_a + \mathscr{Q}(x) \tag{7.1.2}$$

where ∇ denote torsion-free covariant derivatives. Each operator H is assumed to act on a space of purely bosonic fields or purely fermionic fields, $\mathscr{A}^a$ and $\mathscr{Q}$ being matrix-valued functions over the space–time. The chief advantage of the Schwinger–De Witt technique is that it supplies us with a universal prescription to determine the short-distance behaviour of the Feynman propagator associated with any operator (7.1.2).

It is worth recalling two simple theories meeting the above requirement. We first consider the model of a scalar field φ with the action

$$S[\varphi] = -\frac{1}{2}\int \mathrm{d}^4x\, e^{-1}\{\nabla^a\varphi\nabla_a\varphi + m^2\varphi^2 + \xi\mathscr{R}\varphi^2 + \frac{\lambda}{6}\varphi^4\} \tag{7.1.3}$$

where m, ξ and λ are coupling constants and $\mathscr{R}$ is the scalar curvature of the background space–time. Here the one-loop quantum correction is

$$\Gamma^{(1)}[\tilde{\varphi}] = \frac{\mathrm{i}}{2}\,\mathrm{Tr}\,\ln(\nabla^a\nabla_a - m^2 - \xi\mathscr{R} - \lambda\tilde{\varphi}^2). \tag{7.1.4}$$

Another model describes a charged massive field Ψ_{D}, a four-component Dirac spinor, and its conjugate $\bar{\Psi}_{\mathrm{D}}$ coupled to external gravitational and electromagnetic fields:

$$S[\Psi_{\mathrm{D}}] = -\int \mathrm{d}^4x\, e^{-1}\bar{\Psi}_{\mathrm{D}}\{\mathrm{i}\gamma^a(\nabla_a - \mathrm{i}qA_a) + m\}\Psi_{\mathrm{D}}. \tag{7.1.5}$$

The quantum correction to the classical action is now given as

$$\Gamma = -\mathrm{i}\,\mathrm{Tr}\ln h^{(m)} = -\frac{\mathrm{i}}{2}\,\mathrm{Tr}\ln H^{(m)} \tag{7.1.6}$$

where

$$\begin{aligned} h^{(m)} &= \mathrm{i}\gamma^a(\nabla_a - \mathrm{i}qA_a) + m\mathbb{1} \equiv \mathrm{i}\gamma^a\tilde{\nabla}_a + m\mathbb{1} \\ H^{(m)} &= h^{(m)}h^{(-m)} = \tilde{\nabla}^a\tilde{\nabla}_a - \left(m^2 + \frac{1}{4}\mathscr{R}\right)\mathbb{1} + \frac{\mathrm{i}}{4}qF_{ab}[\gamma^a, \gamma^b] \end{aligned} \tag{7.1.7}$$

$\tilde{\nabla}$ being the gauge-covariant derivatives. In equations (7.1.4) and (7.1.6) we have taken into account that sTr coincides with (differs in sign from) ordinary functional trace denoted by Tr in the purely bosonic (purely fermionic) case.

The second example considered illustrates the observation that any operator (7.1.2) can be brought to the form

$$H=\tilde{\nabla}^a\tilde{\nabla}_a-m^2\mathbb{1}+\tilde{\mathscr{Q}}(x) \tag{7.1.8}$$

by redefining ∇_a in the manner $\nabla_a\to\tilde{\nabla}_a=\nabla_a+\mathscr{A}_a$. In what follows, the generalized covariant derivatives will be assumed, for the sake of simplicity, to have no torsion, that is, to satisfy the commutation relations

$$[\tilde{\nabla}_a, \tilde{\nabla}_b]=\tilde{\mathscr{R}}_{ab} \tag{7.1.9a}$$

with some generalized curvature $\tilde{\mathscr{R}}_{ab}$, given by

$$\tilde{\mathscr{R}}_{ab}\varphi^i=\tilde{\mathscr{R}}_{ab}{}^i{}_j\varphi^j. \tag{7.1.9b}$$

In contrast to the definitions of Section 4.1, in curved space–time it is useful to define formal operations, like variational derivatives and sTr, with the help of the covariant delta-function

$$\delta^4(x, y)\equiv e(x)\delta^4(x-y)=\delta^4(y, x)$$

which is bi-scalar under general coordinate transformations. Thus we have

$$\frac{\vec{\delta}}{\delta\varphi^i(x)}\varphi^j(y)=\delta_i{}^j\delta^4(x, y)$$

$$\text{sTr}\, F=\int d^4x\, e^{-1}(-1)^{\varepsilon_i}F^i{}_i(x, x)\equiv\int d^4x\, e^{-1}\,\text{str}\, F(x, x).$$

Here F denotes some operator, and its kernel $F^i_j(x, y)$ is determined by

$$F\varphi^i(x)=\int d^4y\, e^{-1}(y)F^i{}_j(x, y)\,\varphi^j(y)$$

for any φ^i with compact support.

7.1.2. Schwinger's kernel

We introduce a Green's function $G^i{}_j(x, y)$ of the operator (7.1.2) as a solution to the equation

$$H^i{}_j(x)G^j{}_k(x, y)=-\delta^i{}_k\delta^4(x, y) \tag{7.1.10}$$

To pick out a single solution of this equation, one must take into account suitable boundary conditions. We choose those boundary conditions which specify the Feynman propagator by giving to the operator H an infinitesimal imaginary part $(H+i\varepsilon)$, $\varepsilon\to+0$. Then in the limit of turning off all the external fields, in particular when the background metric $g_{mn}(x)$ takes its flat value

η_{mn}, the Green's function G reduces to the Feynman propagator in flat space–time.

To solve equations (7.1.10) under the boundary conditions chosen, we use the integral representation

$$G(x, y) = \mathrm{i} \int_0^\infty \mathrm{d}s \, U(x, y|s) \tag{7.1.11}$$

where $U(x, y|s)$ satisfies the Schrödinger-type equation

$$\mathrm{i} \frac{\partial U(x, y|s)}{\partial s} = -H(x) U(x, y|s). \tag{7.1.12}$$

This representation is known as the Fock–Schwinger representation, with parameter s the proper time and operator $U(x, y|s)$ the Schwinger's kernel. It is worth noting that the ε-prescription guarantees convergence of the integral in (7.1.11) at the upper limit.

Substituting (7.1.11) into equation (7.1.10) and making use of equation (7.1.12), we obtain

$$H(x) G(x, y) = -U(x, y|0) = -\mathbb{1} \delta(x, y).$$

Hence the Schwinger's kernel must be subjected to the initial condition

$$\lim_{s \to +0} U(x, y|s) = \mathbb{1} \delta(x, y). \tag{7.1.13}$$

Therefore, finding G is equivalent to solving equation (7.1.12) under the initial condition (7.1.13), but the latter problem turns out to be simpler in practice than a direct examination of the Green's function.

Before discussing the general structure of the Schwinger's kernel, we first compute this object in the case of Minkowski space and for vanishing background fields, i.e. when the operator (7.1.8) takes the simplest form

$$H_0 = (\eta^{mn} \partial_m \partial_n - m^2) \mathbb{1}. \tag{7.1.14}$$

Here, use of the Fourier transform

$$U_0(x, y|s) = \int \frac{\mathrm{d}^4 p}{(2\pi)^4} U_0(p|s) \mathrm{e}^{\mathrm{i}p(x-y)}$$

reduces the evolution problem

$$\left(\mathrm{i} \frac{\partial}{\partial s} + H_0 \right) U_0(x, y|s) = 0 \qquad U_0(x, y|0) = \mathbb{1} \delta^4(x - y)$$

to the following one

$$\mathrm{i} \frac{\partial U_0(p|s)}{\partial s} = (p^2 + m^2) U_0(p|s) \qquad U_0(p|0) = \mathbb{1}.$$

Then we obtain

$$U_0(x, y|s) = \int \frac{d^4p}{(2\pi)^4} e^{ip(x-y) - ip^2 s - im^2 s} \mathbb{1}. \qquad (7.1.15)$$

Using the integral identity

$$\int_{-\infty}^{\infty} d\tau\, e^{ia\tau^2 + ib\tau} = \begin{cases} \dfrac{1+i}{\sqrt{2}} \sqrt{\dfrac{\pi}{a}} \exp\left(-\dfrac{ib^2}{4a}\right) & a > 0 \\[2ex] \dfrac{1-i}{\sqrt{2}} \sqrt{\dfrac{\pi}{|a|}} \exp\left(\dfrac{ib^2}{4|a|}\right) & a < 0 \end{cases}$$

where a and b are real numbers, $a \neq 0$, it is easy to calculate the integral (7.1.15) resulting in

$$U_0(x, y|s) = \frac{i}{(4\pi i s)^2} e^{i(x-y)^2/4s - im^2 s} \mathbb{1}. \qquad (7.1.16)$$

We note that U_0 involves the Poincaré-invariant two-point function

$$\sigma_0(x, y) = \frac{1}{2} \eta_{mn} (x-y)^m (x-y)^n \qquad (7.1.17)$$

which is the half squared interval between two points x and y in Minkowski space and satisfies the identity

$$\partial^m \sigma_0 \partial_m \sigma_0 = 2\sigma_0. \qquad (7.1.18)$$

Returning to a curved space–time, we will search, following De Witt[1], $U(x, y|s)$ in the form

$$U^i{}_j(x, y|s) = \frac{i}{(4\pi i s)^2} \Omega^{1/2}(x, y) e^{i\sigma(x,y)/2s - im^2 s} F^i{}_j(x, y|s). \qquad (7.1.19)$$

Here $\sigma(x, y)$ and $\Omega(x, y)$ are bi-scalar functions to be specified below, $F^i{}_j(x, y|s)$ is supposed to be analytic in s in a neighbourhood of $s = 0$,

$$F^i{}_j(x, y|s) = \sum_{n=0}^{\infty} a_n{}^i{}_j(x, y)(is)^n. \qquad (7.1.20)$$

Comparison with equation (7.1.16) shows that each two-point function $a_n(x, y)$, for $n = 1, 2, \ldots$, is created by the background fields and disappears in the limit of turning off all these fields. By virtue of equation (7.1.20), one can see the ansatz (7.1.19) is consistent with equation (7.1.12) only if $\sigma(x, y)$ is subjected to the requirement

$$\nabla^a \sigma \nabla_a \sigma = 2\sigma. \qquad (7.1.21)$$

[1]B.S. De Witt, *Dynamical Theory of Groups and Fields* (Gordon and Breach, New York, 1965).

This is the very equation which identically satisfies an important geometric object known as the 'geodesic distance', which is half of the squared distance along the geodesic line between two space–time points x and y. Next, the function σ should possess the boundary behaviour

$$\sigma(x, x) = \nabla_a \sigma(x, x')|_{x=x'} = 0 \tag{7.1.22a}$$

and

$$\det(\partial_m \partial'_n \sigma(x, x'))|_{x=x'} = -e^{-2}(x) \tag{7.1.22b}$$

in order to satisfy the initial condition (7.1.13) and to have the correct limit in flat space. Such boundary conditions prove to pick out a unique solution of equation (7.1.21). We note also that for near points the geodesic interval is

$$\frac{1}{2} g_{mn}(x)(x-y)^m(x-y)^n + O((x-y)^3) \tag{7.1.23}$$

and, hence is consistent with equation (7.1.22). Therefore, we identify σ with the geodesic interval.

The introduction of $\Omega(x, y)$ in expression (7.1.19) is a matter of convenience; the role of this object is to make maximally simple the equation for $F(x, y|s)$. It is useful to choose Ω in the form of the Van Vleck–Morette determinant

$$\Omega(x, x') = -\det(-\nabla_a \nabla'_b \sigma(x, x')) \tag{7.1.24}$$

obeying the equation

$$\nabla^a(\Omega \nabla_a \sigma) = 4\Omega \tag{7.1.25}$$

and the boundary relation

$$\Omega(x, x) = 1. \tag{7.1.26}$$

Then equation (7.1.12) tells us that

$$\frac{\partial F}{\partial s} + \frac{\mathrm{i}}{s} \nabla^a \sigma \tilde{\nabla}_a F = \mathrm{i}\Omega^{-1/2} \tilde{\nabla}^a \tilde{\nabla}_a (\Omega^{1/2} F) + \mathrm{i}\mathscr{Q} F. \tag{7.1.27}$$

Using the decomposition (7.1.20), we obtain the system of equations

$$\nabla^a \sigma \tilde{\nabla}_a a_0 = 0 \tag{7.1.28a}$$

$$(\nabla^a \sigma \tilde{\nabla}_a + n + 1) a_{n+1} = \Omega^{-1/2} \tilde{\nabla}^a \tilde{\nabla}_a (\Omega^{1/2} a_n) + \mathscr{Q} a_n. \tag{7.1.28b}$$

The zeroth coefficient must be subjected to the boundary condition

$$a_0{}^i{}_j(x, x) = \delta^i{}_j \tag{7.1.29}$$

which guarantees, together with equations (7.1.22, 23) and (7.1.26), fulfilment of the initial condition (7.1.13). The two-point functions a_n, $n = 0, 1, 2, \ldots$, under equations (7.1.28) and (7.1.29) are known in the literature as the

De Witt coefficients. Equations (7.1.28) are recurrence relations allowing one to calculate the De Witt coefficients step by step.

In accordance with equations (7.1.21) and (7.1.25), the bi-scalars σ and Ω entering the Schwinger's kernel are completely determined by the pseudo-Riemannian manifold under consideration and do not depend on the background fields except the gravitational field. In other words, these objects are universal ones for field theories in a fixed curved space–time. A different situation will be shown to take place in curved superspace.

Remark. In equations (7.1.21) and (7.1.25) we have taken into account the fact that the derivatives ∇_a and $\tilde{\nabla}_a$ differ only in an internal connection which does not act on Lorentzian indices.

7.1.3. One-loop divergences of effective action

We turn to the analysis of the first quantum correction to $S[\varphi]$ in a bosonic field theory

$$\Gamma^{(1)}[\tilde{\varphi}]=\frac{\mathrm{i}}{2}\operatorname{Tr}\ln H=-\frac{\mathrm{i}}{2}\operatorname{Tr}\ln G. \tag{7.1.30}$$

This definition of $\Gamma^{(1)}[\tilde{\varphi}]$ is rather formal. Rewritten with the aid of the proper-time representations for G, $\Gamma^{(1)}[\tilde{\varphi}]$ turns out to acquire a more transparent form.

Let us consider an arbitrary variation δH of the operator H. Then $\Gamma^{(1)}[\tilde{\varphi}]$ has the displacement

$$\delta\Gamma^{(1)}[\tilde{\varphi}]=\frac{\mathrm{i}}{2}\operatorname{Tr}(H^{-1}\delta H)$$

in which we replace $H^{-1}=-G$ by its explicit expression (7.1.11),

$$\delta\Gamma^{(1)}[\tilde{\varphi}]=\frac{1}{2}\int_0^\infty \mathrm{d}s\,\operatorname{Tr}(U(s)\delta H).$$

Equation (7.1.12) tells us that $\operatorname{Tr}(U(s)\delta H))=(\mathrm{i}s)^{-1}\operatorname{Tr}(\delta U(s))$, hence

$$\delta\Gamma^{(1)}[\tilde{\varphi}]=\delta\left(-\frac{\mathrm{i}}{2}\int_0^\infty \frac{\mathrm{d}s}{s}\operatorname{Tr}U(s)\right).$$

Therefore, up to some additive constant, $\Gamma^{(1)}[\tilde{\varphi}]$ has the form

$$\Gamma^{(1)}[\tilde{\varphi}]=-\frac{\mathrm{i}}{2}\int_0^\infty \frac{\mathrm{d}s}{s}\int \mathrm{d}^4x\, e^{-1}(x)\operatorname{tr}U(x,x|s) \tag{7.1.31}$$

where tr denotes the trace over matrix indices. As a result, the one-loop effective action is determined by the Schwinger's kernel at coincident points. We come to the conclusion that the Schwinger's kernel should be a central object in one-loop quantum field theory.

Use of the explicit form (7.1.19) for $U(x, y|s)$ in equation (7.1.31) leads to

$$\Gamma^{(1)}[\tilde{\varphi}] = \frac{1}{2}\int \mathrm{d}^4x\, e^{-1} \int_0^\infty \frac{\mathrm{d(is)}}{\mathrm{is}(4\pi \mathrm{is})^2}\, \mathrm{e}^{-\mathrm{i}(m^2 - \mathrm{i}\varepsilon)s}\, \mathrm{tr}\, F(x, x|s). \qquad (7.1.32)$$

Now, equation (7.1.20) shows that the first three terms of the series (7.1.20) cause the above integral to be divergent at the lower limit. Hence, one must regularize the integral over proper time in practical calculations. Below we list several popular regularizations.

In the framework of the dimensional regularization scheme, the Schwinger's kernel, corresponding to the four-dimensional theory, should be replaced by its analogue for a d-dimensional theory

$$U_\omega(x, y|s) = \frac{\mathrm{i}\mu^{2\omega}}{(4\pi \mathrm{is})^{2-\omega}}\, \Omega_\omega^{1/2}(x, y)\mathrm{e}^{\mathrm{i}\sigma_\omega(x,y) - \mathrm{i}m^2 s} F_\omega(x, y|s). \qquad (7.1.33)$$

Here $\omega = (4-d)/2$, and the objects σ_ω, Ω_ω and F_ω correspond to d-dimensional space–time. The mass parameter μ has been introduced into (7.1.33) to keep fixed dimension of U_ω. At $\omega = 0$ the kernel U_ω turns into the original one (7.1.19). The dimensionally continued effective action is then given by

$$\Gamma_\omega^{(1)}[\tilde{\varphi}] = \frac{\mu^{2\omega}}{2(4\pi)^{2-\omega}} \int \mathrm{d}^d x\, e_\omega^{-1}(x) \int_0^\infty \frac{\mathrm{d(is)}}{(\mathrm{is})^{3-\omega}}\, \mathrm{e}^{-\mathrm{i}m^2 s} F_\omega(x, x|s). \quad (7.1.34)$$

Treating ω as a complex variable, the integral will be convergent for $\mathrm{Re}\,\omega > 2$ and can be readily computed in this region. Making use of the decomposition

$$F_\omega(x, y|s) = \sum_{n=0}^{\infty} a_n^{(\omega)}(x, y)(\mathrm{i}s)^n \qquad (7.1.35)$$

$a^{(\omega)}$ being the De Witt coefficients in d dimensions, and introducing the notation

$$A_n^{(\omega)} = \int \mathrm{d}^d x\, e_\omega^{-1}\, \mathrm{tr}\, a_n^{(\omega)}(x, x) \qquad A_n = A_n^{(0)} \qquad (7.1.36)$$

one finds in the massive case

$$\Gamma_\omega^{(1)}[\varphi] = \frac{\mu^{2\omega}}{2(4\pi)^{2-\omega}} \sum_{n=0}^{\infty} A_n^{(\omega)} (m^2)^{2-\omega-n}\Gamma(n+\omega-2) \qquad (7.1.37)$$

where $\Gamma(z)$ is the gamma-function. We separate the part of $\Gamma_\omega^{(1)}[\tilde{\varphi}]$ leading to divergences when $\omega \to 0$:

$$\Gamma_\omega^{(1)}[\tilde{\varphi}]_{\mathrm{div}} = \frac{\Gamma(\omega)}{2(4\pi)^{2-\omega}} \left(\frac{m}{\mu}\right)^{-2\omega} \left\{ \frac{m^4}{(2-\omega)(1-\omega)} A_0^{(\omega)} - \frac{m^2}{(1-\omega)} A_1^{(\omega)} + A_2^{(\omega)} \right\}.$$

In the limit $\omega\to 0$ this expression reduces to

$$\Gamma^{(1)}_{\omega}[\tilde{\varphi}]_{\text{div}}=\frac{1}{32\pi^2\omega}\left\{A_2-m^2A_1+\frac{m^4}{2}A_0\right\}+\text{finite terms.} \qquad (7.1.38)$$

Therefore, we must calculate the coefficients $a_1(z,z)$ and $a_2(z,z)$ to determine the one-loop divergences of the theory.

The relation (7.1.37) constitutes an asymptotic expansion of the effective action in inverse powers of m^2. It is obvious that such an expansion is inadmissible in the massless case, where one must apply some other technique to sum an infinite number of terms coming from the Schwinger's kernel. In practice, one usually introduces an auxiliary mass M at intermediate stages of the calculation to improve the behaviour of the integral over proper time at the upper limit. This mass parameter should be eliminated only after completing all manipulations with ultravioletly regularlized expressions. Any polynomial-in-M term in the effective action can be freely removed. In particular, the one-loop divergence of $\Gamma^{(1)}[\tilde{\varphi}]$ in the massless theory is determined by the first term in the figure brackets of equation (7.1.38). The corresponding counterterm in the minimal substractions scheme reads

$$S^{(1)}_{\text{count}}=-\frac{1}{32\pi^2\omega}\int d^4x\, e^{-1}\operatorname{tr} a_2(x,x). \qquad (7.1.39)$$

To find the counterterms, one can use a scheme which is technically analogous to dimensional regularization but less sophisticated. Such a regularization consists in simple insertion of an additional factor $(\mathrm{i}s)^{\omega}$ into the integral over proper time in equation (7.1.32), without any continuation of the Schwinger's kernel in d-dimensional space. Then we obtain

$$\Gamma^{(1)}_{\omega}[\tilde{\varphi}]=\frac{1}{2}\mu^{2\omega}\int_0^{\infty}\frac{ds}{(\mathrm{i}s)^{1-\omega}}\int d^4x\, e^{-1}(x)\operatorname{tr} U(x,x|s). \qquad (7.1.40)$$

This prescription will be called the ω-regularization.

One more way to regularize the integral over proper time is as follows

$$\begin{aligned}\Gamma^{(1)}_{L}[\tilde{\varphi}]&=-\frac{\mathrm{i}}{2}\int_{-\mathrm{i}L^2}^{\infty}\frac{ds}{s}\operatorname{Tr} U(s)\\ &=\frac{1}{32\pi^2}\int_{-\mathrm{i}L^2}^{\infty}\frac{d(\mathrm{i}s)}{(\mathrm{i}s)^3}\mathrm{e}^{-\mathrm{i}m^2s}\int d^4x\, e^{-1}(x)\operatorname{tr} F(x,x|s)\end{aligned} \qquad (7.1.41)$$

L being a real regularization parameter of inverse mass dimension, $L\to +0$ at the end of calculations. Here the integration over s is performed along the positive part of the real axis, except a small region near $s=0$ where the integration contour goes to the domain $\operatorname{Im} s<0$ to end at $s=-\mathrm{i}L^2$. The point is that everyone computes, starting from Schwinger, proper-time integrals by rotating the integration contour into the negative imaginary axis or, which is the same, by making the replacement $s=-i\tau$ with real non-negative τ. Hence, the above prescription is equivalent to cutting off time-rotated integrals at the lower limit.

For $s\approx 0$ the decomposition (7.1.20) is applicable, and one can readily calculate the divergent part of (7.1.41). One finds

$$\Gamma_L^{(1)}[\tilde{\varphi}]_{\text{div}}=\frac{1}{32\pi^2}\left\{\frac{1}{2}L^{-4}A_0+L^{-2}(A_1-m^2A_0)\right.$$
$$\left.-\ln(L^2\mu^2)\left(A_2-m^2A_1+\frac{m^4}{2}A_0\right)\right\} \tag{7.1.42}$$

where μ is a parameter of mass dimension. Dropping here all m-dependent terms, we obtain the one-loop divergences for the massless theory. Then counterterms can be chosen in the manner

$$S_{\text{count}}^{(1)}=-\frac{1}{32\pi^2}\left\{\frac{1}{2}L^{-4}A_0+L^{-2}A_1-\ln(L^2\mu^2)A_2\right\}. \tag{7.1.43}$$

7.1.4. Conformal anomaly

The Schwinger–De Witt technique is an ideal tool for the study of breaking local symmetries by quantum corrections. In this subsection we demonstrate its power by giving explicit calculations of the so-called conformal (or Weyl) anomaly.

As has been discussed in Section 1.7, certain field theories in curved space-time possess a local scale (or Weyl) invariance of the form

$$e_a'^{\ m}(x)=e^{\sigma(x)}e_a^{\ m}(x) \qquad \varphi'^i(x)=e^{p_{(i)}\sigma(x)}\varphi^i(x) \tag{7.1.44}$$

$p_{(i)}$ being fixed constants determined by the field types under consideration. Such a symmetry is inconsistent with the presence of massive parameters in the classical action, hence it can originate only in massless theories. The Weyl invariance appears to be potentially anomalous at the quantum level. The problem is that one inevitably introduces some massive parameter in the process of regularization. As a result, there is room for anomalies in any classically Weyl invariant theory.

We will consider two models describing bosonic fields $\varphi^i(x)$ and fermionic fields $\Psi^I(x)$, respectively, on a curved gravitational background. Their action functionals read as follows:

$$S_{\text{B}}=\frac{1}{2}\int d^4x\, e^{-1}\eta_{ij}\varphi^i(H\varphi)^j$$
$$H=\nabla^a\nabla_a-\xi\mathscr{R}\mathbb{1} \tag{7.1.45}$$

ξ being a constant, and

$$S_F = -\int d^4x\, e^{-1}\eta_{IJ}\Psi^I(h\Psi)^J$$
$$h = i\gamma^a\nabla_a. \tag{7.1.46}$$

Here η_{ij} and η_{IJ} are constant Lorentz-invariant matrices. It is easy to see that S_B may be invariant under transformations of the form (7.1.44) only if $p_{(i)}=1$, for each index i, and the operator H changes according to the rule

$$H' = e^{q_B\sigma}He^{-p_B\sigma}$$
$$q_B = 3 \qquad p_B = 1. \tag{7.1.47}$$

Owing to the first and fourth relations (1.6.33), such a transformation law proves to be admissible only under the conditions that $\xi = \frac{1}{6}$ and the field family $\{\varphi^i\}$ is a set of scalars. Analogously, S_F turns out to be Weyl invariant only under the conditions: (1) $p_I = \frac{3}{2}$ for each I; (2) the field family $\{\Psi^I\}$ is a set of Majorana spinors. Then, the operator h is characterized by the transformation law

$$h' = e^{q_F\sigma}he^{-p_F\sigma}$$
$$q_F = \tfrac{5}{2} \qquad p_F = \tfrac{3}{2}. \tag{7.1.48}$$

The squared operator $H_F \equiv hh$, which is used to define the fermionic effective action, transforms in more complicate way than the bosonic operator H

$$H'_F = e^{q_F\sigma}he^{(q_F - p_F)\sigma}he^{-p_F\sigma}. \tag{7.1.49}$$

Now, we are going to investigate the behaviour of renormalized effective action for a massless conformal scalar field under the Weyl transformations (the case of a massless spinor can be handled similarly). Naive effective action

$$\Gamma = \frac{i}{2}\operatorname{Tr}\ln H \qquad H = \nabla^a\nabla_a - \frac{1}{6}\mathcal{R} \tag{7.1.50}$$

is formally invariant with respect to the redefinitions (7.1.47) provided Det is defined such that $\operatorname{Det}(\sigma(x)\delta(x,y)) = 1$ for any σ. However, here we operate with infinite expressions. To make our manipulations sensible, we introduce some regularization. Let us first consider the effective action in the ω-regularization scheme:

$$\Gamma_\omega = \frac{1}{2}\mu^{2\omega}\int_0^\infty \frac{ds}{(is)^{1-\omega}}\operatorname{Tr} U(s) \tag{7.1.51}$$

where the kernel $U(s)$ corresponds to the conformally covariant operator H given in equation (7.1.50). In accordance with equation (7.1.47), an infinitesimal Weyl transformation changes H according to the rule

$$\delta H = q_B \sigma H - p_B H \sigma$$

hence

$$\delta \operatorname{Tr} U(s) = is(q_B - p_B) \operatorname{Tr}(\sigma H U(s)) = (q_B - p_B) s \frac{\partial}{\partial s} \operatorname{Tr}(\sigma U(s)). \quad (7.1.52)$$

This leads to

$$\delta \Gamma_\omega = -\omega \mu^{2\omega} \int_0^\infty \frac{is}{(is)^{1-\omega}} \operatorname{Tr}(\sigma U(s)). \quad (7.1.53)$$

Since $\delta\Gamma_\omega$ is proportional to ω, only the divergent part of the proper-time integral, appearing in equation (7.1.53), contributes to $\delta\Gamma_\omega$ in the limit $\omega \to 0$. This part can be found in complete analogy with the derivation of equation (7.1.38), that is, by introducing the factor $\exp(-iM^2 s)$, M being an infrared regulator, into the proper-time integral. Then, after separating the divergent part, M should be removed. As a result, one obtains

$$\delta \Gamma_\omega = -\frac{1}{(4\pi)^2} \int d^4x \, e^{-1} \sigma a_2(x, x) + O(\omega). \quad (7.1.54)$$

Let us define the renormalized effective action

$$\Gamma_{ren} = \lim_{\omega \to 0} (\Gamma_\omega + S_{count})$$

with S_{count} given as in equation (7.1.39). Equation (7.1.54) shows that Γ_{ren} is not invariant under the Weyl transformations:

$$\left. \frac{\delta \Gamma_{ren}[e^\sigma e_a{}^m]}{\delta \sigma(x)} \right|_{\sigma=0} = -\frac{1}{(4\pi)^2} a_2(x, x). \quad (7.1.55)$$

This phenomenon is known as the Weyl anomaly.

Equation (7.1.54) allows one to observe an interesting property of counterterms in a classically Weyl invariant theory. The point is that $\delta\Gamma_\omega$ has no divergent part. Since Γ_ω is of the form

$$\Gamma_\omega = -S_{count} + \Gamma_{ren} + O(\omega)$$

we conclude that the functional S_{count} is Weyl invariant. Therefore, the combination $e^{-1}(x) a_2(x, x)$ varies in total derivative under the Weyl transformations.

It is instructive also to re-obtain the Weyl anomaly in the framework of the L-regularization scheme. Starting with the regularized effective action

$$\Gamma_L = -\frac{i}{2} \int_{-iL^2}^\infty \frac{ds}{s} \operatorname{Tr} U(s) \quad (7.1.56)$$

we consider its variation with respect to an infinitesimal Weyl transformation. Making use of equation (7.1.52) gives

$$\delta\Gamma_L = -\frac{\mathrm{i}}{2}(q_B - p_B)\int_{-\mathrm{i}L^2}^{\infty} \mathrm{d}s \frac{\partial}{\partial s}\mathrm{Tr}(\sigma U(s)) = \frac{\mathrm{i}}{2}(q_B - p_B)\,\mathrm{Tr}(\sigma U(s))|_{s=-\mathrm{i}L^2}.$$

Here and below we do not at this stage set explicit values for q_B and p_B. For $s \approx 0$ the decomposition (7.1.20) is powerful, and the above expression can be rewritten as follows:

$$\delta\Gamma_L = -\frac{1}{32\pi^2}(q_B - p_B)\int \mathrm{d}^4x\, e^{-1}\sigma\{L^{-4} + L^{-2}a_1(x,x) + a_2(x,x)\} + O(L^2) \tag{7.1.57}$$

where we have used condition (7.1.29). Then, the finite functional $(\Gamma_L + S_{\text{count}})$, for S_{count} given as in equation (7.1.43), is

$$\begin{aligned}\delta(\Gamma_L + S_{\text{count}}) = -\frac{1}{32\pi^2}\int \mathrm{d}^4x \Big\{ & L^{-4}\Big[(q_B - p_B)\sigma\, e^{-1} + \frac{1}{2}\delta e^{-1}\Big] \\ & + L^{-2}[(q_B - p_B)\sigma e^{-1}a_1(x,x) + \delta(e^{-1}a_1(x,x))] \\ & - \ln(L^2\mu^2)\delta(e^{-1}a_2(x,x)) + (q_B - p_B)\sigma e^{-1}a_2(x,x)\Big\} + O(L^2).\end{aligned}$$

Since the variation must be finite as $L \to 0$, there appear the consistency conditions:

$$q_B - p_B = 2 \tag{7.1.58a}$$

$$\delta(e^{-1}a_1(x,x)) = -2\sigma e^{-1}a_1(x,x) + \text{total deriv.} \tag{7.1.58b}$$

$$\delta(e^{-1}a_2(x,x)) = \text{total deriv.} \tag{7.1.58c}$$

In addition, the renormalized effective action $\Gamma_{\text{ren}} = (\Gamma_L + S_{\text{count}})|_{L\to 0}$ satisfies the same relation (7.1.55) which arose in the ω-regularization scheme. Therefore, the Weyl anomaly is a physical effect independent of the regularization chosen (in contrast with counterterms).

Let us discuss the relations (7.1.58). Equation (7.1.58*a*) arose as a requirement necessary for Weyl invariance of the classical action. We note further that equation (7.1.58*b*) is satisfied identically if $a_1(x,x) \sim \mathscr{R}$ (see equation (1.6.33)). As will be shown below, $a_1(x,x) = 0$ for a massless conformal scalar field, but this is not the case for a massless spinor field where $a_1(x,x)$ coincides with the scalar curvature. Finally, equation (7.1.58*c*) we met when describing the Weyl anomaly in the framework of ω-regularization. The correctness of equation (7.1.58*c*) can be easily checked on the basis of explicit results concerning the general structure of $a_2(x,x)$ and given in the next subsection.

In conclusion, let us return to the relation (7.1.55). We recall that any Weyl invariant field theory is characterized by the condition that its energy–momentum tensor is traceless (see subsection 1.6.8). The breakdown of Weyl invariance by quantum corrections appears to be equivalent to the fact that the averaged energy–momentum tensor $\langle T^{mn}\rangle \sim \delta\Gamma_{\text{ren}}/\delta g_{mn}$ acquires a non-zero trace $\langle T^m_m\rangle$, the latter coinciding with the expression in the left-hand side of equation (7.1.55). Therefore equation (7.1.55) shows that the trace of the averaged energy–momentum tensor is given by the a_2-coefficient.

7.1.5. The coefficients $a_1(x,x)$ and $a_2(x,x)$

We have already seen that one-loop counterterms and anomalies are determined by De Witt coefficients a_1 and a_2 at coincident points, thus it is necessary to be able to calculate these objects. In principle, the calculation of De Witt coefficients at coincident points is a simple, but tedious task. It consists of numerous differentiations of the master equations (7.1.21), (7.1.25) and (7.1.28) together with the use of the covariant derivatives algebra (7.1.9) and the boundary conditions (7.1.22), (7.1.26) and (7.1.29).

To find $a_1(x,x)$ and $a_2(x,x)$, it is necessary first of all to calculate purely geometric quantities of the form $\nabla_{a_1}\ldots\nabla_{a_k}\sigma(x,x')|_{x=x'}$ and $\nabla_{a_1}\ldots\nabla_{a_l}\Omega^{1/2}(x,x')|_{x=x'}$ to finite orders k and l. Direct calculations lead to the relations:

$$\nabla_a\nabla_b\sigma(x,x')|_{x=x'}=\eta_{ab} \tag{7.1.59a}$$

$$\nabla_a\nabla_b\nabla_c\sigma(x,x')|_{x=x'}=0 \tag{7.1.59b}$$

$$\nabla_a\nabla_b\nabla_c\nabla_d\sigma(x,x')|_{x=x'}=\frac{2}{3}\mathscr{R}_{a(cd)b} \tag{7.1.59c}$$

$$\nabla_a\Box\Box\sigma(x,x')|_{x=x'}=-\nabla_a\mathscr{R} \tag{7.1.59d}$$

$$\Box\Box\Box\sigma(x,x')|_{x=x'}=-\frac{4}{15}\mathscr{R}^{abcd}\mathscr{R}_{abcd}+\frac{4}{15}\mathscr{R}^{ab}\mathscr{R}_{ab}-\frac{8}{5}\Box\mathscr{R} \tag{7.1.59e}$$

and

$$\nabla_a\Omega^{1/2}(x,x')|_{x=x'}=0 \tag{7.1.60a}$$

$$\nabla_a\nabla_b\Omega^{1/2}(x,x')|_{x=x'}=\frac{1}{6}\mathscr{R}_{ab} \tag{7.1.60b}$$

$$\nabla_a\Box\Omega^{1/2}(x,x')|_{x=x'}=\frac{1}{6}\nabla_a\mathscr{R} \tag{7.1.60c}$$

$$\Box\Box\Omega^{1/2}(x,x')|_{x=x'}=\frac{1}{30}\mathscr{R}^{abcd}\mathscr{R}_{abcd}-\frac{1}{30}\mathscr{R}^{ab}\mathscr{R}_{ab}+\frac{1}{36}\mathscr{R}^2+\frac{1}{5}\Box\mathscr{R} \tag{7.1.60d}$$

where $\Box = \nabla^a \nabla_a$. For example, let us derive the important relation (7.1.59*a*). Acting with $\nabla_b \nabla_c$ on both sides of equation (7.1.21), we then put $x = x'$ in the resulting identity:

$$\nabla_b \nabla^a \sigma \nabla_c \nabla_a \sigma + \nabla^a \sigma \nabla_b \nabla_c \nabla_a \sigma = \nabla_b \nabla_c \sigma$$

and use equation (7.1.22*a*). This leads to

$$F_{ba} \eta^{ad} F_{dc} = F_{bc}$$

where $F_{bc} = \nabla_b \nabla_c \sigma(x, x')|_{x=x'}$. Since F_{bc} is a non-singular matrix as a result of equation (7.1.22*b*), the final relation is equivalent to relation (7.1.59*a*).

Other information which is necessary for computing $a_1(x, x)$ and $a_2(x, x)$, involves multiple derivatives of a_0 at coincident points:

$$\tilde{\nabla}_a a_0(x, x')|_{x=x'} = 0 \tag{7.1.61a}$$

$$\tilde{\nabla}_a \tilde{\nabla}_b a_0(x, x')|_{x=x'} = \frac{1}{2} \tilde{\mathscr{R}}_{ab} \tag{7.1.61b}$$

$$\tilde{\Box} \tilde{\Box} a_0(x, x')|_{x=x'} = \frac{1}{2} \tilde{\mathscr{R}}^{ab} \tilde{\mathscr{R}}_{ab} \tag{7.1.61c}$$

where $\tilde{\Box} = \tilde{\nabla}^a \tilde{\nabla}_a$. Let us describe the derivation of equation (7.1.61*b*). Acting with $\tilde{\nabla}_b \tilde{\nabla}_c$ on the left-hand side of equation (7.1.28*a*) gives

$$\nabla_b \nabla_c \nabla^a \sigma \tilde{\nabla}_a a_0 + \{\nabla_b \nabla^a \sigma \tilde{\nabla}_c \tilde{\nabla}_a a_0 + (b \leftrightarrow c)\} + \nabla^a \sigma \tilde{\nabla}_b \tilde{\nabla}_c \tilde{\nabla}_a a_0 = 0.$$

We put $x = x'$ and make use of equations (7.1.22*a*), (7.1.59*a*) and (7.1.59*b*) resulting in

$$(\tilde{\nabla}_b \tilde{\nabla}_c a_0 + \tilde{\nabla}_c \tilde{\nabla}_b a_0)|_{x=x'} = 0$$

which can be rewritten as follows:

$$2\tilde{\nabla}_b \tilde{\nabla}_c a_0|_{x=x'} = [\tilde{\nabla}_b, \tilde{\nabla}_c] a_0|_{x=x'} = \tilde{\mathscr{R}}_{bc} a_0(x, x).$$

Then, the boundary condition (7.1.29) leads to equation (7.1.61*b*).

Remark. In the above expressions, $\mathscr{R}_{ab} = \mathscr{R}_{ba}$ denotes the Riccci tensor, while $\tilde{\mathscr{R}}_{ab} = -\tilde{\mathscr{R}}_{ba}$ denotes the generalized curvature tensor taking its values in some matrix algebra (see equation (7.1.9*b*)).

Now, we are ready to calculate $a_1(x, x)$ and $a_2(x, x)$. In accordance with equation (7.1.28*b*), the equation determining $a_1(x, x')$ reads

$$\nabla^a \sigma \tilde{\nabla}_a a_1 + a_1 = \Omega^{-1/2} \tilde{\nabla}^a \tilde{\nabla}_a (\Omega^{1/2} a_0) + \tilde{\mathscr{Q}} a_0. \tag{7.1.62}$$

We set here $x = x'$ and make use of equations (7.1.22*a*), (7.1.26), (7.1.29), (7.1.60*a*,*b*) and (7.1.61*b*). This immediately gives

$$a_1(x, x) = \tilde{\mathscr{Q}} + \frac{1}{6} \mathscr{R} \mathbb{1}. \tag{7.1.63}$$

Next, the equation determining a_2 reads

$$\nabla^a \sigma \tilde{\nabla}_a a_2 + 2a_2 = \Omega^{-1/2} \tilde{\nabla}^a \tilde{\nabla}_a (\Omega^{1/2} a_1) + \tilde{\mathscr{Q}} a_1$$

and reduces, for $x = x'$, to

$$a_2(x, x) = \frac{1}{2}\left(\tilde{\mathscr{Q}} + \frac{1}{6}\mathscr{R}\mathbb{1}\right)\left(\tilde{\mathscr{Q}} + \frac{1}{6}\mathscr{R}\mathbb{1}\right) + \frac{1}{2}\tilde{\Box}(x, x')a_1(x, x')|_{x=x'}.$$

The quantity $\tilde{\Box} a_1(x, x')|_{x=x'}$ can be easily computed with the help of equation (7.1.62) and the relations (7.1.59–61). One obtains as a result:

$$a_2(x, x) = \frac{1}{180}(\mathscr{R}^{abcd}\mathscr{R}_{abcd} - \mathscr{R}^{ab}\mathscr{R}_{ab} + \Box\mathscr{R})\mathbb{1} + \frac{1}{12}\tilde{\mathscr{R}}^{ab}\tilde{\mathscr{R}}_{ab}$$
$$+ \frac{1}{6}\tilde{\Box}\left(\tilde{\mathscr{Q}} + \frac{1}{6}\mathscr{R}\mathbb{1}\right) + \frac{1}{2}\left(\tilde{\mathscr{Q}} + \frac{1}{6}\mathscr{R}\mathbb{1}\right)\left(\tilde{\mathscr{Q}} + \frac{1}{6}\mathscr{R}\mathbb{1}\right). \qquad (7.1.64)$$

The relations (7.1.63, 64) solve the problem of finding the De Witt coefficients a_1 and a_2 at coincident points.

For applications it is sometimes useful to re-express $(\mathscr{R}_{abcd})^2$ in equation (7.1.64) via the squared Weyl tensor. Making use of the identity

$$C^{abcd}C_{abcd} = \mathscr{R}^{abcd}\mathscr{R}_{abcd} - 2\mathscr{R}^{ab}\mathscr{R}_{ab} + \frac{1}{3}\mathscr{R}^2 \qquad (7.1.65)$$

which follows from equation (1.6.17), we obtain

$$a_2(x, x) = \frac{1}{180}\left(C^{abcd}C_{abcd} + \mathscr{R}^{ab}\mathscr{R}_{ab} - \frac{1}{3}\mathscr{R}^2 + \Box\mathscr{R}\right)\mathbb{1}$$
$$+ \frac{1}{12}\tilde{\mathscr{R}}^{ab}\tilde{\mathscr{R}}_{ab} + \frac{1}{6}\tilde{\Box}\left(\tilde{\mathscr{Q}} + \frac{1}{6}\mathscr{R}\mathbb{1}\right) + \frac{1}{2}\left(\tilde{\mathscr{Q}} + \frac{1}{6}\mathscr{R}\mathbb{1}\right)\left(\tilde{\mathscr{Q}} + \frac{1}{6}\mathscr{R}\mathbb{1}\right).$$
$$(7.1.66)$$

As examples, let us determine the coefficients a_1 and a_2 associated with the conformally covariant operators

$$H_{sc} = \Box - \frac{1}{6}\mathscr{R} \qquad H_{sp} = hh = \Box - \frac{1}{4}\mathscr{R}\mathbb{1}$$

for h given as in expression (7.1.46). The H_{sc} and H_{sp} are assumed to act on spaces of scalar and four-component spinor fields, respectively. In the former case we have $\tilde{\mathscr{Q}} = -\frac{1}{6}\mathscr{R}$ and $\tilde{\mathscr{R}}_{ab} = 0$, hence

$$a_1^{(sc)}(x, x) = 0 \qquad (7.1.67a)$$

$$a_2^{(sc)}(x, x) = \frac{1}{180}\left(C^{abcd}C_{abcd} + \mathscr{R}^{ab}\mathscr{R}_{ab} - \frac{1}{3}\mathscr{R}^2 + \Box\mathscr{R}\right). \qquad (7.1.67b)$$

In the latter case we have $\tilde{\mathscr{Q}} = -\frac{1}{4}\mathscr{R}\mathbb{1}$ and $\tilde{\mathscr{R}}_{ab} = \frac{1}{2}\mathscr{R}_{abcd}M^{cd}$, where $M^{cd} = -\frac{1}{4}[\gamma^c, \gamma^d]$, hence

$$\mathrm{tr}\, a_1^{(\mathrm{sp})}(x,x) = -\frac{1}{3}\mathscr{R} \tag{7.1.68a}$$

$$\mathrm{tr}\, a_2^{(\mathrm{sp})}(x,x) = 4a_2^{(\mathrm{sc})}(x,x) - \frac{1}{24}\left(C^{abcd}C_{abcd} + 2\left(\mathscr{R}^{ab}\mathscr{R}_{ab} - \frac{1}{3}\mathscr{R}^2\right)\right) - \frac{1}{18}\,\Box\mathscr{R}. \tag{7.1.68b}$$

The results of subsection 1.6.6 imply that the functionals

$$A_2^{(\mathrm{sc})} = \int \mathrm{d}^4x\, e^{-1} a_2^{(\mathrm{sc})}(x,x)$$

and

$$A_2^{(\mathrm{sp})} = \int \mathrm{d}^4x\, e^{-1}\, \mathrm{tr}\, a_2^{(\mathrm{sp})}(x,x)$$

are invariant under arbitrary Weyl transformations.

In summary, we have described the standard Schwinger–De Witt technique established in its final form in the 1960s, and still remaining the most efficient method for studying functional determinants of operators of the form (7.1.2). For more complicated differential operators there exists a generalized Schwinger–De Witt technique developed later by A. Barvinsky and G. Vilkovisky[2]. The interested reader can find the necessary details in the original paper.

7.2. Proper-time representation for covariantly chiral scalar superpropagator

We proceed to investigating the one-loop effective action in theories of matter superfields on a curved superspace. Among the specific features of quantum theory in curved superspace (or space–time) is the fact that the matter effective action begins to depend here not only on the matter mean (super)fields, but also on the supergravity prepotentials (or gravitational field). Analysis of the latter dependence will be our primary goal. Throughout this chapter, only the minimal formulation for Einstein supergravity will be used to describe the background supergeometry. In the present section, we often make use of the prepotentials parametrization adopted in Sections 5.4 and 5.5.

As mentioned above, the standard Schwinger–De Witt technique is most suitable for studying functional determinants of second-order differential

[2]A.O. Barvinsky and G.A. Vilkovisky, *Phys. Rep.* **119** 1, 1985.

operators with the universal second-order term $\nabla^a\nabla_a$. Most differential operators which arise in superfield theories involve spinor partial derivatives along with ordinary ones. It turns out that the Schwinger–De Witt technique preserves its power for superspace second-order operators of the general form

$$\mathscr{D}^a\mathscr{D}_a + f^{\alpha\dot{\alpha}}(z)[\mathscr{D}_\alpha, \bar{\mathscr{D}}_{\dot{\alpha}}] + f_1(z)\mathscr{D}^2 + f_2(z)\bar{\mathscr{D}}^2 + \ldots \qquad (7.2.1)$$

where the dots mean all terms are at most linear in covariant derivatives. A mixed term with spinor and vector derivatives can, in principle, be included into expression (7.2.1), but such structures do not appear in practice. When considering the one-loop quantum correction in a superfield theory, we will try to re-express it via determinants of operators having the form (7.2.1).

7.2.1. *Basic chiral model*

Let us begin with the curved-superspace action

$$S_{(\Psi)}[\Phi] = \int \mathrm{d}^8z\, E^{-1}\bar{\Phi}\Phi + \left\{\frac{1}{2}\int \mathrm{d}^8z \frac{E^{-1}}{R}\Psi\Phi^2 + \text{c.c.}\right\} \qquad (7.2.2)$$

describing the interaction between dynamical (Φ) and external (Ψ) covariantly chiral scalar superfields, $\bar{\mathscr{D}}_{\dot{\alpha}}\Phi = \bar{\mathscr{D}}_{\dot{\alpha}}\Psi = 0$. A set of chiral models, which will be discussed below, can be read from $S_{(\Psi)}$. For $\Psi = m$ we obtain the massive chiral action

$$S_{(m)}[\Phi] = \int \mathrm{d}^8z\, E^{-1}\bar{\Phi}\Phi + \left\{\frac{m}{2}\int \mathrm{d}^8 \frac{E^{-1}}{R}\Phi^2 + \text{c.c.}\right\}. \qquad (7.2.3)$$

Setting $\Psi = 0$ gives the massless super Weyl invariant model with the action

$$S[\Phi] = \int \mathrm{d}^8z\, E^{-1}\bar{\Phi}\Phi. \qquad (7.2.4)$$

For $\Psi = R$ we obtain another interesting massless action

$$S_{(R)}[\Phi] = \frac{1}{2}\int \mathrm{d}^8z\, E^{-1}(\Phi + \bar{\Phi})^2. \qquad (7.2.5)$$

In curved superspace, the variational derivative for a covariantly chiral scalar superfield is given by equation (5.5.28). Hence the Hessian of $S_{(\Psi)}[\Phi]$ reads

$$\begin{pmatrix} \dfrac{\delta^2 S_{(\Psi)}}{\delta\Phi(z)\delta\Phi(z')} & \dfrac{\delta^2 S_{(\Psi)}}{\delta\Phi(z)\delta\bar{\Phi}(z')} \\ \dfrac{\delta^2 S_{(\Psi)}}{\delta\bar{\Phi}(z)\delta\Phi(z')} & \dfrac{\delta^2 S_{(\Psi)}}{\delta\bar{\Phi}(z)\delta\bar{\Phi}(z')} \end{pmatrix} = \mathbf{H}^{(\Psi)}\begin{pmatrix} \delta_+(z,z') & 0 \\ 0 & \delta_-(z,z') \end{pmatrix} \qquad (7.2.6)$$

where

$$\mathbf{H}^{(\Psi)}=\begin{pmatrix} \Psi(z) & -\frac{1}{4}(\bar{\mathscr{D}}^2-4R) \\ -\frac{1}{4}(\mathscr{D}^2-4\bar{R}) & \bar{\Psi}(z) \end{pmatrix}. \tag{7.2.7}$$

For $\Psi=0$ this operator reduces to

$$\mathbf{H}=\begin{pmatrix} 0 & -\frac{1}{4}(\bar{\mathscr{D}}^2-4R) \\ -\frac{1}{4}(\mathscr{D}^2-4\bar{R}) & 0 \end{pmatrix}\equiv\begin{pmatrix} 0 & \mathbf{H}_{+-} \\ \mathbf{H}_{-+} & 0 \end{pmatrix}. \tag{7.2.8}$$

Let us note that $\mathbf{H}^{(\Psi)}$ turns into the operator (4.4.5) in the flat superspace limit. We denote both flat and curved operators by the same symbol, but only the latter will appear in the present chapter and hence no confusion should arise.

The quantum correction $\Gamma_{(\Psi)}$ to $S_{(\Psi)}$ is defined by the superfunctional integral

$$e^{i\Gamma_{(\Psi)}}=\int \mathscr{D}\Phi\,\mathscr{D}\bar{\Phi}\,e^{iS_{(\Psi)}[\Phi]} \tag{7.2.9}$$

and can be represented in the form

$$\Gamma_{(\Psi)}=\frac{i}{2}\,\mathbf{Tr}\ln\mathbf{H}^{(\Psi)}. \tag{7.2.10}$$

Remark. Throughout this chapter, the operations **sTr** and **sDet** are denoted by **Tr**(−**Tr**) and **Det**($\mathbf{Det}^{-1}$), respectively, in the purely bosonic (fermionic) case.

Remark. We define the superkernel of an operator

$$\mathbf{A}=\begin{pmatrix} \mathbf{A}_{++} & \mathbf{A}_{+-} \\ \mathbf{A}_{-+} & \mathbf{A}_{--} \end{pmatrix}$$

which acts on a space of covariantly chiral-antichiral scalar columns

$$\begin{pmatrix} \Phi(z) \\ \bar{\Phi}(z) \end{pmatrix}=\begin{pmatrix} e^{\bar{W}}\hat{\Phi}(x,\theta) \\ e^{W}\hat{\bar{\Phi}}(x,\bar{\theta}) \end{pmatrix}$$

with the help of the covariant chiral delta-function

$$\delta_+(z,z')=-\frac{1}{4}(\bar{\mathscr{D}}^2-4R)E\delta^8(z-z')=\varphi^{-3}\,e^{\bar{W}}\,e^{\bar{W}'}\,\delta^4(x-x')\delta^2(\theta-\theta')$$

and its conjugate $\delta_-(z,z')$ in accordance with the rule

$$\begin{pmatrix} \mathbf{A}_{++}(z,z') & \mathbf{A}_{+-}(z,z') \\ \mathbf{A}_{-+}(z,z') & \mathbf{A}_{--}(z,z') \end{pmatrix}=\mathbf{A}\begin{pmatrix} \delta_+(z,z') & 0 \\ 0 & \delta_-(z,z') \end{pmatrix}. \tag{7.2.11}$$

This definition is convenient in practice since the components of the superkernel are bi-scalar and covariantly chiral or antichiral according to which label, + or −, is attached to the corresponding argument (for example, $\bar{\mathscr{D}}_{\dot\alpha}\mathbf{A}_{++}(z,z')=\bar{\mathscr{D}}'_{\dot\alpha}\mathbf{A}_{++}(z,z')=0$). Then $\mathbf{Tr\,A}$ reads as follows:

$$\begin{aligned}\mathbf{Tr\,A}&=\int \mathrm{d}^6z\,\mathrm{e}^{-\bar{W}}(\varphi^3\mathbf{A}_{++}(z,z))+\int \mathrm{d}^6\bar{z}\,\mathrm{e}^{-W}(\bar\varphi^3\mathbf{A}_{--}(z,z))\\ &=\int \mathrm{d}^8z\,\frac{E^{-1}}{R}\mathbf{A}_{++}(z,z)+\int \mathrm{d}^8z\,\frac{E^{-1}}{R}\mathbf{A}_{--}(z,z)\\ &\equiv \mathbf{Tr}_+\mathbf{A}_{++}+\mathbf{Tr}_-\mathbf{A}_{--}.\end{aligned}\tag{7.2.12}$$

The action (7.2.2) is the minimal extension of the flat action (4.4.6) in a curved superspace. As we have seen, the flat action was used to define the one-loop quantum correction in the Wess–Zumino model (see equation (4.8.27)). Analogously, the one-loop correction $\Gamma^{(1)}[\Phi]$ in the generalized Wess–Zumino model (6.3.7) is connected with the superfunctional (7.2.10) by the rule

$$\Gamma^{(1)}[\Phi]=\Gamma_{(\Psi)}|_{\Psi=m+\xi R+\lambda\Phi}.\tag{7.2.13}$$

This is why $\Gamma_{(\Psi)}$ deserves consideration for arbitrary Ψ.

When $\Psi\neq$const the operator $\mathbf{H}^{(\Psi)}$ has such a structure that the Schwinger–De Witt technique proves to be inappropriate to handle directly the Feynman superpropagator of $\mathbf{H}^{(\Psi)}$. There are two approaches, however, to reduce the problem of computing $\mathbf{Det\,H}^{(\Psi)}$ to similar problems, but for operators which meet the chief requirement of the proper-time technique. One of these approaches can be applied in the case when the body of $\Psi(z)$ is non-zero at each point z of superspace. Under this assumption, we have the relation

$$\begin{pmatrix}\mathrm{e}^{-2\sigma} & 0\\ 0 & \mathrm{e}^{-2\bar\sigma}\end{pmatrix}\mathbf{H}^{(\Psi)}\begin{pmatrix}\mathrm{e}^{\sigma} & 0\\ 0 & \mathrm{e}^{\bar\sigma}\end{pmatrix}=\mathbf{H}'^{(m)}\tag{7.2.14}$$

where

$$\mathrm{e}^{\sigma}=\Psi/m \qquad m=\text{const}\neq 0\tag{7.2.15}$$

and m is a real parameter and $\mathbf{H}'^{(m)}$ is just the operator $\mathbf{H}^{(m)}$ constructed on the base of the super Weyl transformed covariant derivatives (5.5.13) with σ given by conditions (7.2.15). Equation (7.2.14) leads to the formal relation

$$\mathbf{Det\,H}^{(\Psi)}=\mathbf{Det\,H}'^{(m)}$$

which shows that $\Gamma_{(\Psi)}$ can be read from $\Gamma_{(m)}$. If we explicitly indicate the dependence of $\Gamma_{(m)}$ on the supergravity prepotentials W^M and φ, $\Gamma_{(m)}=\Gamma_{(m)}[W,\varphi]$, then $\Gamma_{(\Psi)}$ is given by

$$\Gamma_{(\Psi)}=\Gamma_{(m)}\left[W,\frac{\Psi}{m}\varphi\right].\tag{7.2.16}$$

Another approach to computing $\Gamma_{(\Psi)}$, which needs no assumption about Ψ, will be described in the following section.

7.2.2. Covariantly chiral Feynman superpropagator

Let us consider the Feynman superpropagator

$$\mathbf{G}^{(m)}=\begin{pmatrix} \mathbf{G}^{(m)}_{++} & \mathbf{G}^{(m)}_{+-} \\ \mathbf{G}^{(m)}_{-+} & \mathbf{G}^{(m)}_{--} \end{pmatrix}$$

associated with the operator $\mathbf{H}^{(m)}$. By definition, $\mathbf{G}^{(m)}$ satisfies the equation

$$\begin{pmatrix} m & -\frac{1}{4}(\bar{\mathscr{D}}^2-4R) \\ -\frac{1}{4}(\mathscr{D}^2-4\bar{R}) & m \end{pmatrix}\begin{pmatrix} \mathbf{G}^{(m)}_{++} & \mathbf{G}^{(m)}_{+-} \\ \mathbf{G}^{(m)}_{-+} & \mathbf{G}^{(m)}_{--} \end{pmatrix}=-\begin{pmatrix} \delta_+ & 0 \\ 0 & \delta_- \end{pmatrix}. \tag{7.2.17}$$

From this equation we obtain

$$\mathbf{G}^{(m)}=\mathbf{H}^{(-m)}\begin{pmatrix} \mathbf{G}^{(m)}_{\mathrm{c}} & 0 \\ 0 & \mathbf{G}^{(m)}_{\mathrm{a}} \end{pmatrix} \tag{7.2.18}$$

where

$$\mathbf{G}^{(m)}_{\mathrm{c}}\equiv-\frac{1}{m}\mathbf{G}^{(m)}_{++} \qquad \mathbf{G}^{(m)}_{\mathrm{a}}\equiv-\frac{1}{m}\mathbf{G}^{(m)}_{--}.$$

The bi-scalar Green's function $\mathbf{G}^{(m)}_{\mathrm{c}}(z,z')$ is covariantly chiral in both arguments and because of equation (7.2.17) satisfies the equation

$$(\mathscr{H}_{\mathrm{c}}-m^2)\mathbf{G}^{(m)}_{\mathrm{c}}(z,z')=-\delta_+(z,z') \tag{7.2.19}$$

where

$$\mathscr{H}_{\mathrm{c}}=\frac{1}{16}(\bar{\mathscr{D}}^2-4R)(\mathscr{D}^2-4\bar{R}). \tag{7.2.20}$$

Similarly, the antichiral superpropagator $\mathbf{G}^{(m)}_{\mathrm{a}}(z,z')$ obeys the equation

$$(\mathscr{H}_{\mathrm{a}}-m^2)\mathbf{G}^{(m)}_{\mathrm{a}}(z,z')=-\delta_-(z,z') \tag{7.2.21}$$

where

$$\mathscr{H}_{\mathrm{a}}=\frac{1}{16}(\mathscr{D}^2-4\bar{R})(\bar{\mathscr{D}}^2-4R). \tag{7.2.22}$$

As pointed out in subsection 6.3.2, the fourth-order differential operator $\mathscr{H}_{\mathrm{c}}$ ($\mathscr{H}_{\mathrm{a}}$) turns into a second-order operator when acting on the full space of covariantly chiral (antichiral) scalar superfields; that is

$$\mathscr{H}_{\mathrm{c}}\Big|_{\substack{\text{chiral}\\ \text{scalar space}}}=\square_+-\frac{1}{4}(\bar{\mathscr{D}}^2\bar{R})+R\bar{R}$$

$$\mathscr{H}_{\mathrm{a}}\Big|_{\substack{\text{antichiral}\\ \text{scalar space}}}=\square_--\frac{1}{4}(\mathscr{D}^2R)+R\bar{R}.$$

Here $\square_+$ and $\square_-$ are the chiral and anticharal d'Alembertians, respectively, defined in equations (6.3.18) and (6.3.19). Both operators $\square_+$ and $\square_-$ have the form (7.2.1).

The effective action $\Gamma_{(m)}$ can be readily expressed in terms of the superpropagators $\mathbf{G}_c^{(m)}$ and $\mathbf{G}_a^{(m)}$. With the aid of the formal relation

$$\mathbf{Det}\,\mathbf{H}^{(m)} = \mathbf{Det}^{1/2}(\mathbf{H}^{(m)}\mathbf{H}^{(-m)})$$

one obtains

$$\Gamma_{(m)} = -\frac{i}{4}\mathbf{Tr}_+ \ln \mathbf{G}_c^{(m)} - \frac{i}{4}\mathbf{Tr}_- \ln \mathbf{G}_a^{(m)} \qquad (7.2.23)$$

for $\mathbf{Tr}_\pm$ defined as in equation (7.2.12). In accordance with equations (7.2.18) and (7.2.23), the Green's functions $\mathbf{G}_c^{(m)}$ and $\mathbf{G}_a^{(m)}$ are central objects in the quantum theory.

Both terms in the right-hand side of equation (7.2.23) produce in fact the same contribution modulo boundary terms at infinity, because of the formal equality

$$\delta\mathbf{Tr}_- \ln \mathbf{G}_c^{(m)} = \delta\mathbf{Tr}_- \ln \mathbf{G}_a^{(m)} \qquad (7.2.24)$$

which holds for an arbitrary variation of the supergravity prepotentials. Equation (7.2.24) is a simple consequence of the operator relations

$$\mathbf{H}_{-+}\mathbf{G}_c^{(m)} = \mathbf{G}_a^{(m)}\mathbf{H}_{-+} \qquad (7.2.25)$$
$$\mathbf{G}_c^{(m)}\mathbf{H}_{+-} = \mathbf{H}_{+-}\mathbf{G}_a^{(m)}$$

for $\mathbf{H}_{+-}$ and $\mathbf{H}_{-+}$ given in identity (7.2.8), together with the identity

$$\mathbf{Tr}_+(\mathbf{A}_{+-}\mathbf{B}_{-+}) = \mathbf{Tr}_-(\mathbf{B}_{-+}\mathbf{A}_{+-}). \qquad (7.2.26)$$

To prove relation (7.2.25), it is sufficient to take account of the composite nature of $\mathscr{H}_c$ and $\mathscr{H}_a$,

$$\mathscr{H}_c = \mathbf{H}_{+-}\mathbf{H}_{-+} \qquad \mathscr{H}_a = \mathbf{H}_{-+}\mathbf{H}_{+-} \qquad (7.2.27)$$

which implies

$$\mathbf{H}_{-+}\mathscr{H}_c = \mathscr{H}_a\mathbf{H}_{-+} \qquad \mathscr{H}_c\mathbf{H}_{+-} = \mathbf{H}_{+-}\mathscr{H}_a. \qquad (7.2.28)$$

As a result, we can rewrite $\Gamma_{(m)}$ in the form

$$\Gamma_{(m)} = -\frac{i}{2}\mathbf{Tr}_+ \ln \mathbf{G}_c^{(m)}. \qquad (7.2.29)$$

In perfect analogy to the massive case, the quantum correction Γ in the massless theory (7.2.4) can be represented as

$$\Gamma = -\frac{i}{4}\mathbf{Tr}_+ \ln \mathbf{G}_c - \frac{i}{4}\mathbf{Tr}_- \ln \mathbf{G}_a = -\frac{i}{2}\mathbf{Tr}_+ \ln \mathbf{G}_c \qquad (7.2.30)$$

where the chiral ($\mathbf{G}_c$) and antichiral ($\mathbf{G}_a$) Green's functions constitute the Feynman superpropagator $\mathbf{G}$ associated with the operator (7.2.8)

$$\mathbf{G}=\mathbf{H}\begin{pmatrix} \mathbf{G}_c & 0 \\ 0 & \mathbf{G}_a \end{pmatrix} \tag{7.2.31}$$

and satisfy the equations

$$\mathscr{H}_c\mathbf{G}_c(z,z')=-\delta_+(z,z') \tag{7.2.32a}$$

$$\mathscr{H}_a\mathbf{G}_a(z,z')=-\delta_-(z,z'). \tag{7.2.32b}$$

7.2.3. The chiral d'Alembertian

Before proceeding further, we would like to discuss the chiral d'Alembertian

$$\Box_+=\mathscr{D}^a\mathscr{D}_a+\frac{1}{4}\mathscr{R}\mathscr{D}^2+\mathrm{i}G^a\mathscr{D}_a+\frac{1}{4}(\mathscr{D}^\alpha R)\mathscr{D}_\alpha \tag{7.2.33}$$

entering the equation for $\mathbf{G}_c^{(m)}$, and related geometric objects. As has been already mentioned in subsection 6.3.1, the chiral d'Alembertian is unique in the sense that any operator of the form

$$\mathscr{D}^a\mathscr{D}_a+f^{\alpha a}(z)\mathscr{D}_\alpha\mathscr{D}_a+f^a(z)\mathscr{D}_a+f^\alpha(z)\mathscr{D}_\alpha$$

which maps the full space of covariantly chiral scalar superfields into itself, coincide with $\Box_+$. Closely related to $\Box_+$ is an operation of multiplication on the chiral space. This operation is defined by associating with arbitrary chiral scalars Φ_1 and Φ_2 another scalar superfield

$$\Phi_1*\Phi_2=\Phi_2*\Phi_1\equiv\frac{1}{16}(\bar{\mathscr{D}}^2-4R)(\mathscr{D}^\alpha\Phi_1\mathscr{D}_\alpha\Phi_2)=\mathscr{D}^a\Phi_1\mathscr{D}_a\Phi_2+\frac{1}{4}R\mathscr{D}^\alpha\Phi_1\mathscr{D}_\alpha\Phi_2 \tag{7.2.34}$$

which is covariantly chiral by construction. The connection between $\Box_+$ and $*$ reads

$$\int \mathrm{d}^8z\,\frac{E^{-1}}{R}\Phi_1\Box_+\Phi_2=-\int \mathrm{d}^8z\,\frac{E^{-1}}{R}\Phi_1*\Phi_2 \tag{7.2.35}$$

for arbitrary covariantly chiral scalars Φ_1 and Φ_2.

Similarly to $\Box_+$, the $*$-product is uniquely defined in the sense that any binary operation

$$(\Phi_1,\Phi_2)=\mathscr{D}^a\Phi_1\mathscr{D}_a\Phi_2+f(z)\mathscr{D}^\alpha\Phi_1\mathscr{D}_\alpha\Phi_2+f^{\alpha a}(z)(\mathscr{D}_\alpha\Phi_1\mathscr{D}_a\Phi_2+\mathscr{D}_a\Phi_1\mathscr{D}_\alpha\Phi_2)$$

which maps arbitrary chiral scalars Φ_1 and Φ_2 into a chiral scalar, must coincide with $\Phi_1*\Phi_2$. This observation has important consequences. First, chiral generalizations of the equations (7.1.21) and (7.1.25) for the geodesic interval and the Van Vleck–Morette determinant, respectively, are of the form

$$\sigma * \sigma \equiv \mathscr{D}^a\sigma\mathscr{D}_a\sigma + \frac{1}{4}R\mathscr{D}^\alpha\sigma\mathscr{D}_\alpha\sigma = 2\sigma \tag{7.2.36}$$

and

$$\mathscr{D}^a\sigma\mathscr{D}_a\Omega + \frac{1}{4}R\mathscr{D}^\alpha\sigma\mathscr{D}_\alpha\Omega = (4 - \Box_+\sigma)\Omega. \tag{7.2.37}$$

Analogously, the equation for an antichiral extension of the geodesic interval should have the form

$$\mathscr{D}^a\bar{\sigma}\mathscr{D}_a\bar{\sigma} + \frac{1}{4}\bar{R}\bar{\mathscr{D}}_{\dot\alpha}\bar{\sigma}\bar{\mathscr{D}}^{\dot\alpha}\bar{\sigma} = 2\bar{\sigma}. \tag{7.2.38}$$

From equations (7.2.36) and (7.2.38) we see that there is no universal extension of the geodesic interval in curved superspace. Different Green's functions will involve different super-intervals.

Chirality of the final expression (7.2.34) implies that there is a natural chiral supermetric. Let us introduce the complex coordinates $z^M_{(L)} = e^{\bar{W}} z^M$ most suitable for operating with covariantly chiral scalars $\Phi(z) = e^{\bar{W}}\Phi(x, \theta) = \Phi(y^{\underline{M}})$, where $y^{\underline{M}} = (x^m_{(L)}, \theta^\mu_{(L)})$. Next we re-express the supervierbein in the coordinate system introduced, $E_A{}^M\partial_M = E^{(L)M}_A\,\partial^{(L)}_M$. Because of the relations (5.4.2, 3), in the gauge (5.4.6) we have

$$E^{(L)M}_A = \begin{pmatrix} E^{(L)\underline{M}}_{\underline{A}} & | & E^{(L)}_{\underline{A}}{}_{\dot\mu} \\ --- & - & -- \\ 0 & | & \delta^{\dot\alpha}{}_{\dot\mu}\bar{F} \end{pmatrix} \equiv \begin{pmatrix} l^{\underline{M}}_{\underline{A}} & | & E^{(L)}_{\underline{A}}{}_{\dot\mu} \\ -- & - & -- \\ 0 & | & \delta^{\dot\alpha}{}_{\dot\mu}\bar{F} \end{pmatrix}. \tag{7.2.39}$$

$$\underline{A} = (a, \alpha) \qquad \underline{M} = (m, \mu)$$

Now equation (7.2.34) can be rewritten as follows

$$\Phi_1 * \Phi_2 = g^{MN}(y)\frac{\partial\hat{\Phi}_1}{\partial y^{\underline{N}}}\frac{\partial\hat{\Phi}_2}{\partial y^{\underline{M}}} \tag{7.2.40}$$

where the supermetric is given by

$$g^{\underline{MN}} = (l^{\mathrm{sT}})^{\underline{M}}{}_{\underline{A}}\,\eta^{\underline{AB}}\,l_{\underline{B}}{}^{\underline{N}} \tag{7.2.41}$$

$$\eta^{\underline{AB}} = \begin{pmatrix} \eta^{ab} & 0 \\ 0 & \frac{1}{4}R\varepsilon^{\alpha\beta} \end{pmatrix}$$

l^{sT} being the supertranspose of the supermatrix l (see equation (1.9.59). The sypermetric is seen to be a function of the chiral variables $y^{\underline{M}}$ only; though $g^{\underline{MN}}$ is built in terms of the reduced supervierbein $l_{\underline{A}}{}^{\underline{M}}$ with non-chiral components. We note, however, that the Berezinian of $l_{\underline{A}}{}^{\underline{M}}$ is chiral. With the aid of equations (1.11.17), (5.4.19), (5.5.3) and (7.2.39), one can readily prove

$$l = \mathrm{Ber}(l_{\underline{A}}{}^{\underline{M}}) = \varphi^{-3}. \tag{7.2.42}$$

Because of the relations (7.2.35) and (7.2.40), the chiral d'Alembertian can be written in the form

$$\Box_+ = (-1)^{\varepsilon_{\underline{M}}} \varphi^{-3} \frac{\partial}{\partial y^{\underline{M}}} \left(\varphi^3 g^{\underline{MN}} \frac{\partial}{\partial y^{\underline{N}}} \right) \tag{7.2.43}$$

completely analogous to the standard representation

$$\nabla^a \nabla_a = e \partial_m (e^{-1} g^{mn} \partial_n)$$

for the scalar d'Alembertian.

7.2.4. Covariantly chiral Schwinger's superkernel

The structure of the chiral superpropagator $\mathbf{G}_c^{(m)}$ can be described in the framework of the proper-time technique. As in the case of propagators in curved space-time, we introduce the integral representation

$$\mathbf{G}_c^{(m)}(z, z') = \mathrm{i} \int_0^\infty \mathrm{d}s\, \mathrm{e}^{-\mathrm{i}(m^2 - \mathrm{i}\varepsilon)s}\, \mathbf{U}_c(z, z'|s) \tag{7.2.44}$$

where $\mathbf{U}_c(z, z'|s)$ is now required to satisfy the equation

$$\mathrm{i}\frac{\partial \mathbf{U}_c}{\partial s} = -\mathscr{H}_c \mathbf{U}_c = -\left(\Box_+ - \frac{1}{4}(\bar{\mathscr{D}}^2 \bar{R}) + R\bar{R} \right) \mathbf{U}_c \tag{7.2.45}$$

and the initial condition

$$\mathbf{U}_c(z, z'|s \to +0) = \delta_+(z, z'). \tag{7.2.46}$$

Similarly to $\mathbf{G}_c^{(m)}$, the superkernel $\mathbf{U}_c(z, z'|s)$ must be covariantly chiral and scalar in both the superspace arguments.

Due to the explicit form of $\Box_+$, it is possible to search for a solution of equation (7.2.45) by applying the ansatz

$$\mathbf{U}_c(z, z'|s) = \frac{\mathrm{i}}{(4\pi \mathrm{i}s)^2} \exp\left(\frac{\mathrm{i}\sigma(z, z')}{2s} \right) \sum_{n=0}^{\infty} \mathbf{a}_n^c(z, z')(\mathrm{i}s)^n \tag{7.2.47}$$

where σ and $\mathbf{a}^c$'s are chiral bi-scalars. To ensure the fufilment of equation (7.2.45), σ must be a solution of equation (7.2.36), whereas the remaining bi-scalars must obey the system of equations

$$\sigma * \mathbf{a}_0^c + \frac{1}{2}(\Box_+ \sigma - 4)\mathbf{a}_0^c = 0 \tag{7.2.48a}$$

$$(n+1)\mathbf{a}_{n+1}^c + \sigma * \mathbf{a}_{n+1}^c + \frac{1}{2}(\Box_+ \sigma - 4)\mathbf{a}_{n+1}^c = \mathscr{H}_c \mathbf{a}_n^c \tag{7.2.48b}$$

the $*$-product being defined in identity (7.2.34). The initial condition (7.2.46) proves to be fulfilled under the following boundary conditions:

$$\sigma(z,z)=\mathscr{D}_A\sigma(z,z')|_{z=z'}=\mathscr{D}_\alpha\mathscr{D}_A\sigma(z,z')|_{z=z'}=0 \qquad (7.2.49a)$$

$$\mathscr{D}_\alpha\mathscr{D}_a\mathscr{D}_b\sigma(z,z')|_{z=z'}=\mathscr{D}^2\mathscr{D}_a\sigma(z,z')|_{z=z'}=0$$

$$\mathbf{a}_0^c(z,z)=\mathscr{D}_A\mathbf{a}_0^c(z,z')|_{z=z'}=\mathscr{D}_\alpha\mathscr{D}_a\mathbf{a}_0^c(z,z')|_{z=z'}=0 \qquad (7.2.49b)$$

and

$$\mathscr{D}_a\mathscr{D}_b\sigma(z,z')|_{z=z'}=\eta_{ab} \qquad (7.2.50a)$$

$$\mathscr{D}_\alpha\mathscr{D}_\beta\mathbf{a}_0^c(z,z')|_{z=z'}=-2\varepsilon_{\alpha\beta}. \qquad (7.2.50b)$$

Equations (7.2.49–50) completely specify all the objects entering the right-hand side in equation (7.2.47). Rather than demanding condition (7.2.50a), it is sufficient in fact to require $\det(-\mathscr{D}_a\mathscr{D}_b\sigma(z))|_{z=z'}\neq 0$; then equation (7.2.50$a$) is a simple consequence of the master equation (7.2.36).

There are simple observations which lead directly to the above boundary conditions even without careful analysis concerning their compatibility with condition (7.2.46). We first note that in the flat superspace limit $\mathbf{U}_c(z,z'|s)$ should take the form (4.4.21), hence the flat values of σ and a_0^c read

$$\sigma(z,z')_{\text{flat}}=\frac{1}{2}(x_{(+)}-x'_{(+)})^2 \qquad \mathbf{a}_0^c(z,z')_{\text{flat}}=(\theta-\theta')^2$$

and are seen to be consistent with the flat version of equations (7.2.49, 50). Another remark is that all tensor superfields $\mathscr{D}_{A_1}\ldots\mathscr{D}_{A_k}\sigma(z,z')|_{z=z'}$ and $\mathscr{D}_{A_1}\ldots\mathscr{D}_{A_k}\mathbf{a}_n^c(z,z')|_{z=z'}$, where $k, n=0, 1, 2, \ldots$, should have purely geometric origin by their very construction. What is more, they should be polynomials of the supertorsion tensors R, $\bar{R}$, G_a, $W_{\alpha\beta\gamma}$ and $\bar{W}_{\dot\alpha\dot\beta\dot\gamma}$ and their covariant derivatives in order to make possible the existence of a flat limit. Then, the equalities (7.2.49) follow from considerations of dimension. Dimensions of the basic geometric structures are

$$[R]=[G_a]=1 \qquad [W_{\alpha\beta\gamma}]=3/2 \qquad [\mathscr{D}_A]\geqslant 1/2$$

but all objects entering equalities (7.2.49) have lesser dimensions ($[\sigma]=-2$, $[\mathbf{a}_0^c]=-1$) or their tensor form differs from that of the invariant tensors $\varepsilon_{\alpha\beta}$, $\varepsilon_{\dot\alpha\dot\beta}$ and $(\sigma_a)_{\alpha\dot\alpha}$. Finally, the relations (7.2.50) are dictated by the flat limit.

Equations (7.2.48) can be rewritten in a simpler form if one redefines $\mathbf{a}_n^c$ in the manner

$$\mathbf{a}_n^c(z,z')=\mathbf{\Omega}^{1/2}(z,z')\tilde{\mathbf{a}}_n^c(z,z') \qquad n=0, 1, 2, \ldots \qquad (7.2.51)$$

where the chiral bi-scalar $\mathbf{\Omega}$ is subject to equation (7.2.37) and the boundary conditions

$$\mathbf{\Omega}(z,z)=1 \qquad \mathscr{D}_\alpha\mathbf{\Omega}(z,z')|_{z=z'}=0 \qquad (7.2.52)$$

Then $\tilde{\mathbf{a}}^{\mathrm{c}}$s satisfy the equations

$$\sigma * \tilde{\mathbf{a}}_0^{\mathrm{c}} = 0 \tag{7.2.53a}$$

$$(n+1)\tilde{\mathbf{a}}_{n+1}^{\mathrm{c}} + \sigma * \tilde{\mathbf{a}}_{n+1}^{\mathrm{c}} = \Omega^{-1/2}\mathscr{H}_{\mathrm{c}}(\Omega^{1/2}\tilde{\mathbf{a}}_n^{\mathrm{c}}). \tag{7.2.53b}$$

By analogy with the terminology of Section 7.1, it is natural to call $\mathbf{a}_n^{\mathrm{c}}$ or $\tilde{\mathbf{a}}_n^{\mathrm{c}}$ the De Witt supercoefficients.

7.2.5. $\mathbf{a}_1^{\mathrm{c}}(z, z)$ *and* $\mathbf{a}_2^{\mathrm{c}}(z, z)$

From the viewpoint of renormalization, of exceptional interest are the coefficients $\mathbf{a}_1^{\mathrm{c}}$ and $\mathbf{a}_2^{\mathrm{c}}$ at coincident points. To calculate these coefficients, one follows the same scheme described in the previous section. Namely, it is necessary to make numerous covariant differentiations of the master equations (7.2.36) and (7.2.48), with the use of the algebra of covariant derivatives (5.5.6), and take account of the boundary conditions (7.2.49–50). The immediate consequences of equations (7.2.48–50) are:

$$\mathbf{a}_1^{\mathrm{c}}(z, z) = -R \tag{7.2.54}$$

and

$$2\mathbf{a}_2^{\mathrm{c}}(z, z) = \Box_+ \mathbf{a}_1^{\mathrm{c}}(z, z')|_{z=z'} + \frac{1}{4}(\bar{\mathscr{D}}^2 - 4R)R\bar{R}. \tag{7.2.55}$$

To find $\Box_+ \mathbf{a}_1^{\mathrm{c}}(z, z')|_{z=z'}$ it is necessary to obtain a number of auxiliary relations. Straightforward but tedious calculation leads to

$$\begin{gathered}
\mathscr{D}_a\mathscr{D}_b\mathscr{D}_c\sigma(z, z')|_{z=z'} = \mathscr{D}^2\mathscr{D}^a\mathscr{D}_a\sigma(z, z')|_{z=z'} = 0 \\
\mathscr{D}^a\mathscr{D}_a\mathscr{D}^2\sigma(z, z')|_{z=z'} = -16\bar{R} \\
\mathscr{D}^a\mathscr{D}_a\mathscr{D}^b\mathscr{D}_b\sigma(z, z')|_{z=z'} = 8R\bar{R} + G^aG_a - \frac{1}{2}\bar{\mathscr{D}}^2\bar{R}^2 - \frac{1}{2}\mathscr{D}^2 R \\
\mathscr{D}^a\mathscr{D}_a\mathscr{D}_\alpha\mathbf{a}_0^{\mathrm{c}}(z, z')|_{z=z'} = \mathscr{D}_a\mathscr{D}_b\mathscr{D}_c\mathbf{a}_0^{\mathrm{c}}(z, z')|_{z=z'} = 0 \\
\mathscr{D}^2\mathscr{D}_a\mathbf{a}_0^{\mathrm{c}}(z, z')|_{z=z'} = -\mathscr{D}_a\mathscr{D}^2\mathbf{a}_0^{\mathrm{c}}(z, z')|_{z=z'} = 2\mathrm{i}G_a \\
\mathscr{D}^a\mathscr{D}_a\mathscr{D}^2\mathbf{a}_0^{\mathrm{c}}(z, z')|_{z=z'} = 12R\bar{R} + 2G^aG_a - \mathscr{D}^2 R \\
\mathscr{D}^2\mathscr{D}^a\mathscr{D}_a\mathbf{a}_0^{\mathrm{c}}(z, z')|_{z=z'} = 4R\bar{R} + 2G^aG_a - \bar{\mathscr{D}}^2\bar{R} \\
\mathscr{D}^a\mathscr{D}_a\mathscr{D}^b\mathscr{D}_b\mathbf{a}_0^{\mathrm{c}}(z, z')|_{z=z'} = \frac{1}{2}W^{\alpha\beta\gamma}W_{\alpha\beta\gamma} + \frac{1}{8}\mathscr{D}^\alpha R\mathscr{D}_\alpha R + \frac{1}{4}\bar{\mathscr{D}}_{\dot\alpha}G^a\bar{\mathscr{D}}^{\dot\alpha}G_a.
\end{gathered} \tag{7.2.56}$$

With the help of these identities one readily obtains the final relation

$$\mathbf{a}_2^{\mathrm{c}}(z, z) = \frac{1}{12}W^{\alpha\beta\gamma}W_{\alpha\beta\gamma} + \frac{1}{48}(\bar{\mathscr{D}}^2 - 4R)G^aG_a - \frac{1}{96}(\bar{\mathscr{D}}^2 - 4R)(\mathscr{D}^2 - 4\bar{R})R. \tag{7.2.57}$$

At intermediate stages of computing $\mathbf{a}_2^c(z,z)$ there appear non-chiral expressions, but the final result is explicitly chiral.

It is worth noting that the superfunctional

$$\int d^8z \frac{E^{-1}}{R} \mathbf{a}_2^c(z,z) = \frac{1}{12} \int d^8z\, E^{-1} \left\{ \frac{1}{R} W^{\alpha\beta\gamma} W_{\alpha\beta\gamma} - (G^a G_a + 2R\bar{R}) \right\} \tag{7.2.58}$$

is invariant under arbitrary super Weyl transformations (5.5.13), in accordance with the results of subsection 5.6.5.

7.2.6. One-loop divergences

Once the proper-time representation for $\mathbf{G}_c^{(m)}$ has been established, the effective action (7.2.29) can be handled by the rules given in the previous section. We rewrite $\Gamma_{(m)}$ in the form

$$\Gamma_{(m)} = -\frac{i}{2} \int_0^\infty \frac{ds}{s} e^{-im^2 s} \mathbf{Tr}_+ \mathbf{U}_c(s) \tag{7.2.59}$$

and then improve the behaviour of the proper-time integral at the lower limit by applying the prescription of ω-regularization. Since $\mathbf{a}_0^c(z,z)=0$, the divergent part of the effective action reads

$$\begin{aligned} \Gamma_{(m),\mathrm{div}} &= \frac{1}{32\pi^2\omega} \int d^8z \frac{E^{-1}}{R} \{\mathbf{a}_2^c(z,z) - m^2 \mathbf{a}_1^c(z,z)\} \\ &= \frac{1}{32\pi^2\omega} \int d^8z\, E^{-1} \left\{ \frac{1}{R} \mathbf{a}_2^c(z,z) + m^2 \right\}. \end{aligned} \tag{7.2.60}$$

Because of the explicit form of $\mathbf{a}_2^c(z,z)$, the integral (7.2.60) is real modulo total derivatives.

With the aid of equation (7.2.60) it is easy to obtain divergences of the effective action (7.2.10). Since $\Gamma_{(\Psi)}$ and $\Gamma_{(m)}$ are connected as in equation (7.2.16), we have

$$\Gamma_{(\Psi),\mathrm{div}} = \frac{1}{32\pi^2\omega} \int d^8z\, E^{-1} \left\{ \frac{1}{R} \mathbf{a}_2^c(z,z) + \bar{\Psi}\Psi \right\} \tag{7.2.61}$$

where we have accounted for the fact that the functional (7.2.58) is super Weyl invariant. Now, equation (7.2.13) tells us that the one-loop divergences in the generalized Wess–Zumino model (6.3.7) are

$$\begin{aligned} \Gamma^{(1)}[\Phi]_{,\mathrm{div}} &= \frac{\lambda^2}{32\pi^2\omega} \int d^8z\, E^{-1}\bar{\Phi}\Phi + \frac{\lambda}{32\pi^2\omega} \int d^8z\, E^{-1}\{\Phi(m+\xi\bar{R}) + \bar{\Phi}(m+\xi R)\} \\ &+ \frac{m\xi}{32\pi^2\omega} \int d^8z\, E^{-1}\{R+\bar{R}\} \\ &+ \frac{1}{32\pi^2\omega} \int d^8z\, E^{-1} \left\{ m^2 + \xi^2 R\bar{R} + \frac{1}{R} \mathbf{a}_2^c(z,z) \right\} \end{aligned} \tag{7.2.62}$$

with λ and ξ being chosen real for simplicity. To eliminate these divergences one must introduce counterterm $S_{\text{count}} = -\Gamma^{(1)}[\Phi]_{,\text{div}}$ into the classical action (6.3.7).

Let us discuss the final result. As is clear, in the flat limit only the first term in equation (7.2.62) survives. In a curved superspace with $R \neq 0$ integrals like $\int \mathrm{d}^8 z\, E^{-1}\Phi$ do not vanish, and there arise linear-in-Φ divergences. As a result, the Wess–Zumino model turns out to be unstable in curved superspace. Both the third and fourth terms in equation (7.2.62) show vacuum divergences which appear due to non-trivial background supergeometry. Functional

$$S_{\text{vac.count}} = -\frac{m\xi}{32\pi^2\omega}\int \mathrm{d}^8 z\, E^{-1}\{R+\bar{R}\}$$
$$-\frac{m}{32\pi^2\omega}\int \mathrm{d}^8 z\, E^{-1}\left\{m^2+\xi^2 R\bar{R}+\frac{1}{R}\mathbf{a}_2^{\mathrm{c}}(z,z)\right\}. \qquad (7.2.63)$$

is to be understood as a one-loop counterterm to the classical supergravity action in a full supergravity–matter quantum theory. The explicit form of the second term in equation (7.2.63) implies that the supergravity action should include the standard Einstein term (6.1.3) along with higher-derivative R^2-terms (6.1.19). Because of the first term in (7.2.63), the supergravity action should also involve pathalogical structures (6.1.31) as long as $\xi \neq 0$.

7.2.7. *Switching on an external Yang–Mills superfield*

The above results admit a natural extension to the case of a system of covariantly chiral scalars $\Phi = \{\Phi^i(z)\}$ coupled to an external Yang–Mills gauge superfields $V = V^I(z)T^I$, with T^i the generators of a real representation of the gauge group. The quantum correction in the chiral model

$$S_{(m)}[\Phi; V] = \int \mathrm{d}^8 z\, E^{-1}\bar{\Phi}\,\mathrm{e}^{2V}\Phi + \frac{m}{2}\left\{\int \mathrm{d}^8 z \frac{E^{-1}}{R}\Phi^{\mathrm{T}}\Phi + \mathrm{c.c.}\right\} \qquad (7.2.64)$$

can be represented in the form

$$\Gamma_{(m)}[V] = -\frac{\mathrm{i}}{2}\mathbf{Tr}_+ \ln \hat{\mathbf{G}}_{\mathrm{c}}^{(m)} \qquad (7.2.65)$$

where the matrix superpropagator $\hat{\mathbf{G}}_{\mathrm{c}}^{(m)}(z,z')$ is covariantly chiral in both arguments and subjected to the equation

$$\left(\hat{\Box}_+ - \frac{1}{4}(\bar{\mathscr{D}}^2\bar{R}) + R\bar{R} - m^2\right)\hat{\mathbf{G}}_{\mathrm{c}}^{(m)}(z,z') = -\mathbb{1}_\Phi \delta_+(z,z'). \qquad (7.2.66)$$

Here $\hat{\Box}_+$ is the Yang–Mills covariantized chiral d'Alembertian,

$$\hat{\Box}_+ = \mathbb{D}^a\mathbb{D}_a + \frac{1}{4}R\bar{\mathbb{D}}^2 + \mathrm{i}G^a\mathbb{D}_a + \frac{1}{4}(\mathscr{D}^\alpha R)\mathbb{D}_\alpha - W^\alpha\mathbb{D}_\alpha - \frac{1}{2}(\mathbb{D}^\alpha W_\alpha) \qquad (7.2.67)$$

where $\mathbb{D}$ denote the gauge-covariant derivatives (6.3.35*a*), and W^α is the corresponding superfield strength defined in equation (6.3.33). In comparison with equation (7.2.47), the Schwinger's superkernel for $\hat{\mathbf{G}}^{(m)}_{\rm c}$ involves the same chiral super-interval $\boldsymbol{\sigma}$, but all coefficients $\hat{\mathbf{a}}^{\rm c}_n$ are now matrix-valued. The coefficients $\hat{\mathbf{a}}^{\rm c}_1(z, z)$ and $\hat{\mathbf{a}}^{\rm c}_2(z, z)$ prove to be given simply by combining equations (4.8.25) and (7.2.54, 57):

$$\hat{\mathbf{a}}^{\rm c}_2(z, z) = \mathbf{a}^{\rm c}_1(z, z)\mathbb{1} \tag{7.2.68a}$$

$$\hat{\mathbf{a}}^{\rm c}_2(z, z) = \mathbf{a}^{\rm c}_2(z, z)\mathbb{1} + W^\alpha W_\alpha. \tag{7.2.68b}$$

In conclusion, it is necessary to note that the coefficients $\mathbf{a}^{\rm c}_1(z, z)$ and $\mathbf{a}^{\rm c}_2(z, z)$ were originally computed by I.N. McArthur with the aid of a normal coordinate frame in curved superspace[1].

7.3. Proper-time representation for scalar superpropagators

In the present section we give an approach to the computation of the quantum correction $\Gamma_{(\Psi)}$ defined by equation (7.2.10). This approach does not require us to make any assumption concerning Ψ and is based on a remarkable link between superpropagators arising in special chiral models and in the massless vector multiplet model. For the scalar superpropagators of interest we obtain the relevant proper-time decompositions.

7.3.1. Quantization of the massless vector multiplet model

The massless vector multiplet in a curved superspace is described by the action

$$S[V] = \frac{1}{16}\int \mathrm{d}^8 z\, E^{-1} V \mathscr{D}^\alpha(\bar{\mathscr{D}}^2 - 4R)\mathscr{D}_\alpha V \tag{7.3.1}$$

where $V(z)$ is a real scalar superfield. The classical properties of this model have been discussed in subsection 6.3.2 (in comparison with (6.3.21), $S[V]$ contains an additional factor $\frac{1}{2}$ for later convenience). In particular, it was shown that $S[V]$ remains invariant under the following gauge transformations

$$V \to V^\Lambda = V + \Lambda + \bar{\Lambda} \qquad \bar{\mathscr{D}}_{\dot\alpha}\Lambda = 0 \tag{7.3.2}$$

Λ being an arbitrary covariantly chiral scalar superfield. As is obvious, the gauge generators are Abelian and linearly independent, so the theory can be quantized with the help of the Faddeev–Popov prescription (see Section 4.5).

[1]I.N. McArthur, *Class. Quantum Grav.* **1** 233; 245, 1984.

We fix the invariance (7.3.2) by imposing the gauge conditions

$$\kappa[V] \equiv -\frac{1}{4}(\bar{\mathscr{D}}^2 - 4R)V + a = 0$$
$$\bar{\kappa}[V] \equiv -\frac{1}{4}(\mathscr{D}^2 - 4\bar{R})V + \bar{a} = 0 \tag{7.3.3}$$

where $a(z)$ is an external covariantly chiral scalar superfield. The functions $\kappa[V]$ and $\bar{\kappa}[V]$ have been chosen to be covariantly chiral and antichiral, respectively, since they should belong to the same functional space where the gauge parameters Λ and $\bar{\Lambda}$ live. Because of the chiral variational rule (5.5.28), the Faddeev–Popov matrix reads

$$\begin{pmatrix} \dfrac{\delta\kappa[V^\Lambda]}{\delta\Lambda} & \dfrac{\delta\kappa[V^\Lambda]}{\delta\bar{\Lambda}} \\ \dfrac{\delta\bar{\kappa}[V^\Lambda]}{\delta\Lambda} & \dfrac{\delta\bar{\kappa}[V^\Lambda]}{\delta\bar{\Lambda}} \end{pmatrix} = \mathbf{H}^{(R)}. \tag{7.3.4}$$

Here $\mathbf{H}^{(R)}$ is defined by equation (7.2.7) for $\Psi = R$ and is the Hessian of the chiral model (7.2.5). As a result, the in–out vacuum amplitude is given by

$$e^{i\Gamma_v} = \int \mathscr{D}V\, \delta_+\left[-\frac{1}{4}(\bar{\mathscr{D}}^2 - 4R)V + a\right]\delta_-\left[-\frac{1}{4}(\mathscr{D}^2 - 4\bar{R})V + \bar{a}\right]\mathbf{Det}(\mathbf{H}^{(R)})\, e^{iS[V]}. \tag{7.3.5}$$

In accordance with the results of subsection 4.5.2, Γ_v does not depend on the external superfields a and $\bar{a}$. Therefore, one can freely integrate the right-hand side of equation (7.3.5) over a and $\bar{a}$ with some weight $\rho(a, \bar{a})$ such that

$$\int \mathscr{D}a\mathscr{D}\bar{a}\, \rho(a, \bar{a}) = 1.$$

It is useful to choose $\rho(a, \bar{a})$ in the form

$$\mathbf{Det}^{1/2}(\mathbf{H}^{(\Psi)})\, e^{-iS_{(\Psi)}[a]}$$

the action $S_{(\Psi)}$ being defined by equation (7.2.2). This leads to the following representation

$$e^{i\Gamma_v} = \mathbf{Det}(\mathbf{H}^{(R)})\, \mathbf{Det}^{1/2}((\mathbf{H}^{(\Psi)})\, e^{-(i/2)\int d^8z\, E^{-1} V \Box_v^{(\Psi)} V} \tag{7.3.6}$$

where

$$\Box_v^{(\Psi)} = -\frac{1}{8}\mathscr{D}^\alpha(\bar{\mathscr{D}}^2 - 4R)\mathscr{D}_\alpha + \frac{1}{16}\{(\bar{\mathscr{D}}^2 - 4R), (\mathscr{D}^2 - 4\bar{R})\} - \frac{1}{4}\Psi(\bar{\mathscr{D}}^2 - 4R)$$
$$-\frac{1}{4}\bar{\Psi}(\mathscr{D}^2 - 4\bar{R}). \tag{7.3.7}$$

When acting on scalar superfields, $\Box_{\rm v}^{(\Psi)}$ reads

$$\Box_{\rm v}^{(\Psi)}=\mathscr{D}^a\mathscr{D}_a-\frac{1}{4}G^{\alpha\dot\alpha}[\mathscr{D}_\alpha,\bar{\mathscr{D}}_{\dot\alpha}]-\frac{1}{4}\Psi(\bar{\mathscr{D}}^2-4R)-\frac{1}{4}\bar\Psi(\mathscr{D}^2-4\bar R)$$

$$-\frac{1}{4}(\bar{\mathscr{D}}_{\dot\alpha}\bar R)\bar{\mathscr{D}}^{\dot\alpha}-\frac{1}{4}(\mathscr{D}^\alpha R)\mathscr{D}_\alpha-\frac{1}{4}(\bar{\mathscr{D}}^2\bar R)-\frac{1}{4}(\mathscr{D}^2R)+2R\bar R \qquad (7.3.8)$$

as a consequence of the covariant derivatives algebra (5.5.6), and is a curved-superspace extension of the operator Δ (4.8.30). Let $\mathbf{G}_{\rm v}^{(\Psi)}$ be the Feynman superpropagator associated with $\Box_{\rm v}^{(\Psi)}$ and defined by

$$\Box_{\rm v}^{(\Psi)}\mathbf{G}_{\rm v}^{(\Psi)}(z,z')=-\delta^8(z,z'). \qquad (7.3.9)$$

Then, equation (7.3.6) can be rewritten in the form

$$\Gamma_{\rm v}=-\frac{\mathrm{i}}{2}\mathbf{Tr}\ln\mathbf{G}_{\rm v}^{(\Psi)}-\frac{\mathrm{i}}{2}\mathbf{Tr}\ln\mathbf{H}^{(\Psi)}-\mathrm{i}\mathbf{Tr}\ln\mathbf{H}^{(R)}. \qquad (7.3.10)$$

Remark. Given an operator $\mathbf{A}$ acting on some space of unconstrained scalar superfields, it is useful to define its superkernel to be bi-scalar:

$$\mathbf{A}(z,z')=\mathbf{A}\delta^8(z,z')=\mathbf{A}E\delta^8(z-z'). \qquad (7.3.11)$$

Then $\mathbf{Tr\,A}$ is given as

$$\mathbf{Tr\,A}=\int\mathrm{d}^8z\,E^{-1}\mathbf{A}(z,z). \qquad (7.3.12)$$

Equation (7.3.10) contains more interesting information than may appear at first sight. The point is that the right-hand side in equation (7.3.10) does not depend on Ψ, which leads to important consequences. Let us first choose $\Psi=0$. Then equation (7.3.10) takes the form

$$\Gamma_{\rm v}=-\frac{\mathrm{i}}{2}\mathbf{Tr}\ln\mathbf{G}_{\rm v}-\frac{\mathrm{i}}{2}\mathbf{Tr}\ln\mathbf{H}-\mathrm{i}\mathbf{Tr}\ln\mathbf{H}^{(R)}. \qquad (7.3.13)$$

where $\mathbf{G}_{\rm v}\equiv\mathbf{G}_{\rm v}^{(0)}$ and $\mathbf{H}\equiv\mathbf{H}^{(0)}$. Choosing $\Psi=R$, we obtain

$$\Gamma_{\rm v}=-\frac{\mathrm{i}}{2}\mathbf{Tr}\ln\mathbf{G}_{\rm v}^{(R)}-\frac{3\mathrm{i}}{2}\mathbf{Tr}\ln\mathbf{H}^{(R)}. \qquad (7.3.14)$$

Equations (7.3.10, 13, 14) allow us to express $\Gamma_{\rm v}$ and $\mathbf{Tr}\ln\mathbf{H}^{(\Psi)}$ as follows:

$$\Gamma_{\rm v}=-\frac{3\mathrm{i}}{2}\mathbf{Tr}\ln\mathbf{G}_{\rm v}+\mathrm{i}\mathbf{Tr}\ln\mathbf{G}_{\rm v}^{(R)}-\frac{3\mathrm{i}}{2}\mathbf{Tr}\ln\mathbf{H} \qquad (7.3.15)$$

$$\mathbf{Tr}\ln\mathbf{H}^{(\Psi)}=\mathbf{Tr}\ln\mathbf{G}_{\rm v}-\mathbf{Tr}\ln\mathbf{G}_{\rm v}^{(\Psi)}+\mathbf{Tr}\ln\mathbf{H}. \qquad (7.3.16)$$

In the previous section we saw that $\mathbf{Tr}\ln\mathbf{H}$ can be represented according to the rule

$$\mathbf{Tr}\ln\mathbf{H}=-\mathbf{Tr}_+\ln\mathbf{G}_c$$

$\mathbf{G}_c$ being the massless chiral superpropagator defined by equation (7.2.32*a*). As a result, we arrive at the final relations:

$$\Gamma_v=-\frac{3i}{2}\mathbf{Tr}\ln\mathbf{G}_v+i\mathbf{Tr}\ln\mathbf{G}_v^{(R)}+\frac{3i}{2}\mathbf{Tr}_+\ln\mathbf{G}_c \tag{7.3.17}$$

$$\mathbf{Tr}\ln\mathbf{H}^{(\Psi)}=\mathbf{Tr}\ln\mathbf{G}_v-\mathbf{Tr}\ln\mathbf{G}_v^{(\Psi)}-\mathbf{Tr}_+\ln\mathbf{G}_c. \tag{7.3.18}$$

The latter is a by-product of our consideration, and it supplies us with a simple expression for the quantum correction (7.2.10):

$$\Gamma_{(\Psi)}=-\frac{i}{2}(\mathbf{Tr}\ln\mathbf{G}_v^{(\Psi)}-\mathbf{Tr}\ln\mathbf{G}_v)+\Gamma \tag{7.3.19}$$

where $\Gamma\equiv\Gamma_{(0)}$ is the quantum correction in the massless chiral model (7.2.4).

The above relations make it possible to compute Γ_v and $\Gamma_{(\Psi)}$ in the framework of the proper-time technique. For any Ψ the operator $\Box_v^{(\Psi)}$ has exactly the form (7.2.1), hence the corresponding Green's functions $\mathbf{G}_v^{(\Psi)}$ possess a well-defined proper time decomposition. The proper-time representation for $\mathbf{G}_c$ follows from equations (7.2.44, 47) by setting $m=0$.

7.3.2. Connection between $\mathbf{G}_v^{(\Psi)}$ and $\mathbf{G}^{(\Psi)}$

Let us consider the Feynman superpropagator

$$\mathbf{G}^{(\Psi)}=\begin{pmatrix}\mathbf{G}_{++}^{(\Psi)} & \mathbf{G}_{+-}^{(\Psi)}\\ \mathbf{G}_{-+}^{(\Psi)} & \mathbf{G}_{--}^{(\Psi)}\end{pmatrix} \tag{7.3.20}$$

associated with $\mathbf{H}^{(\Psi)}$,

$$\mathbf{H}^{(\Psi)}\mathbf{G}^{(\Psi)}(z,z')=-\begin{pmatrix}\delta_+(z,z') & 0\\ 0 & \delta_-(z,z')\end{pmatrix}. \tag{7.3.21}$$

It turns out that $\mathbf{G}^{(\Psi)}$ is expressed via the scalar superpropagator $\mathbf{G}_v^{(\Psi)}$, defined by equation (7.3.9), in the manner

$$\mathbf{G}^{(\Psi)}(z,z')=\begin{pmatrix}\mathbf{h}\mathbf{h}'\mathbf{G}_v^{(\Psi)}(z,z') & \mathbf{h}\bar{\mathbf{h}}'\mathbf{G}_v^{(\Psi)}(z,z')\\ \bar{\mathbf{h}}\mathbf{h}'\mathbf{G}_v^{(\Psi)}(z,z') & \bar{\mathbf{h}}\bar{\mathbf{h}}'\mathbf{G}_v^{(\Psi)}(z,z')\end{pmatrix} \tag{7.3.22}$$

where

$$\mathbf{h}=-\frac{1}{4}(\bar{\mathscr{D}}^2-4R)\qquad \bar{\mathbf{h}}=-\frac{1}{4}(\mathscr{D}^2-4\bar{R}).$$

This relation generalizes equation (4.8.31) to the case of curved superspace. It can be proved as follows. First, we act with $\mathbf{H}^{(R)}$ from the left on both sides in equation (7.3.21) resulting in

$$\mathbf{H}^{(R)}\mathbf{H}^{(\Psi)}\mathbf{G}^{(\Psi)}(z,z') = -\mathbf{H}^{(R)}\begin{pmatrix} \mathbf{h} & 0 \\ 0 & \bar{\mathbf{h}} \end{pmatrix}\delta^8(z,z').$$

Further, we apply $\mathbf{H}=\mathbf{H}^{(0)}$ from the left to both sides of the relation

$$\Box_{\mathrm{v}}^{(\Psi)}\begin{pmatrix} \mathbf{h}' & \bar{\mathbf{h}}' \\ \mathbf{h}' & \bar{\mathbf{h}}' \end{pmatrix}\mathbf{G}_{\mathrm{v}}^{(\Psi)}(z,z') = -\begin{pmatrix} \mathbf{h} & \bar{\mathbf{h}} \\ \mathbf{h} & \bar{\mathbf{h}} \end{pmatrix}\delta^8(z,z')$$

which is a consequence of equation (7.3.9). The result can be presented in the form

$$\mathbf{H}^{(R)}\mathbf{H}^{(\Psi)}\hat{\mathbf{G}}^{(\Psi)}(z,z') = -\mathbf{H}\begin{pmatrix} \mathbf{h} & \bar{\mathbf{h}} \\ \mathbf{h} & \bar{\mathbf{h}} \end{pmatrix}\delta^8(z,z')$$

where $\hat{\mathbf{G}}^{(\Psi)}$ is the operator defined by equation (7.3.22). Here we have used the identities

$$\mathbf{h}\mathscr{D}^\alpha(\bar{\mathscr{D}}^2-4R)\mathscr{D}_\alpha = \bar{\mathbf{h}}\mathscr{D}^\alpha(\bar{\mathscr{D}}^2-4R)\mathscr{D}_\alpha = 0$$

$$\mathbf{h}^2 = =R\mathbf{h} \qquad \bar{\mathbf{h}}^2 = \overline{R\mathbf{h}}$$

which are correct when acting on scalars. Since

$$\mathbf{H}^{(R)}\begin{pmatrix} \mathbf{h} & 0 \\ 0 & \bar{\mathbf{h}} \end{pmatrix} = \mathbf{H}\begin{pmatrix} \mathbf{h} & \bar{\mathbf{h}} \\ \mathbf{h} & \bar{\mathbf{h}} \end{pmatrix}$$

both $\mathbf{G}^{(\Psi)}$ and $\hat{\mathbf{G}}^{(\Psi)}$ satisfy the same equation.

Using equation (7.3.22), it is not difficult to obtain an independent proof of the relation (7.3.19).

7.3.3. Scalar Schwinger's superkernel

The proper-time representation for $\mathbf{G}_{\mathrm{v}}^{(\Psi)}(z,z')$ reads

$$\mathbf{G}_{\mathrm{v}}^{(\Psi)}(z,z') = \mathrm{i}\int_0^\infty \mathrm{d}s\, \mathbf{U}_{\mathrm{v}}^{(\Psi)}(z,z'|s) \tag{7.3.23}$$

where the bi-scalar superkernel $\mathbf{U}_{\mathrm{v}}^{(\Psi)}(z,z')$ satisfies the evolution equation

$$\left(\mathrm{i}\frac{\partial}{\partial s} + \Box_{\mathrm{v}}^{(\Psi)}(z,z')\right)\mathbf{U}_{\mathrm{v}}^{(\Psi)}(z,z') = 0 \tag{7.3.24}$$

and the initial condition

$$\mathbf{U}_{\mathrm{v}}^{(\Psi)}(z,z'|s\to+0) = \delta^8(z,z'). \tag{7.3.25}$$

We search for a solution of equation (7.3.24) in the form

$$\mathbf{U}_{\mathrm{v}}^{(\Psi)}(z,z'|s) = \frac{\mathrm{i}}{(4\pi\mathrm{i}s)}\exp\left(\frac{\mathrm{i}\sigma_{\mathrm{v}}^{(\Psi)}(z,z')}{2s}\right)\sum_{n=0}^{\infty}\mathbf{a}_n^{(\Psi)}(z,z')(\mathrm{i}s)^n \tag{7.3.26}$$

$\boldsymbol{\sigma}^{(\Psi)}_{\mathrm{v}}$ and $\mathbf{a}^{(\Psi)}$ being real bi-scalar superfields. It is easy to see that $\boldsymbol{\sigma}^{(\Psi)}_{\mathrm{v}}$ should obey the equation

$$\mathscr{D}^a\sigma^{(\Psi)}_{\mathrm{v}}\mathscr{D}_a\sigma^{(\Psi)}_{\mathrm{v}}-\frac{1}{2}G^{\alpha\dot\alpha}\mathscr{D}_\alpha\boldsymbol{\sigma}^{(\Psi)}_{\mathrm{v}}\bar{\mathscr{D}}_{\dot\alpha}\boldsymbol{\sigma}^{(\Psi)}_{\mathrm{v}}-\frac{1}{4}\Psi\bar{\mathscr{D}}_{\dot\alpha}\boldsymbol{\sigma}^{(\Psi)}_{\mathrm{v}}\bar{\mathscr{D}}^{\dot\alpha}\boldsymbol{\sigma}^{(\Psi)}_{\mathrm{v}}$$

$$-\frac{1}{4}\Psi\mathscr{D}^\alpha\boldsymbol{\sigma}^{(\Psi)}_{\mathrm{v}}\mathscr{D}_\alpha\boldsymbol{\sigma}^{(\Psi)}_{\mathrm{v}}=2\boldsymbol{\sigma}^{(\Psi)}_{\mathrm{v}} \qquad (7.3.27)$$

whereas the recurrence relations for $\mathbf{a}^{(\Psi)}$ are

$$\langle\sigma^{(\Psi)}_{\mathrm{v}},\mathbf{a}^{(\Psi)}_0\rangle+\frac{1}{2}(\Delta^{(\Psi)}\sigma^{(\Psi)}_{\mathrm{v}}-4)\mathbf{a}^{(\Psi)}_0=0 \qquad (7.3.28)$$

$$(n+1)\mathbf{a}^{(\Psi)}_{n+1}+\langle\boldsymbol{\sigma}^{(\Psi)}_{\mathrm{v}},\mathbf{a}^{(\Psi)}_{n+1}\rangle+\frac{1}{2}(\Delta^{(\Psi)}\boldsymbol{\sigma}^{(\Psi)}_{\mathrm{v}}-4)\mathbf{a}^{(\Psi)}_{n+1}=\Box^{(\Psi)}_{\mathrm{v}}\mathbf{a}^{(\Psi)}_n$$

where we have introduced a bracket $\langle\ ,\ \rangle$ for real scalar superfields, defined by

$$\langle V_1,V_2\rangle=\mathscr{D}^aV_1\mathscr{D}_aV_2-\frac{1}{4}G^{\alpha\dot\alpha}(\mathscr{D}_\alpha V_1\bar{\mathscr{D}}_{\dot\alpha}V_2-\bar{\mathscr{D}}_{\dot\alpha}V_1\mathscr{D}_\alpha V_2)$$

$$-\frac{1}{4}\Psi\bar{\mathscr{D}}_{\dot\alpha}V_1\bar{\mathscr{D}}^{\dot\alpha}V_2-\frac{1}{4}\Psi\mathscr{D}^\alpha V_1\mathscr{D}_\alpha V_2. \qquad (7.3.29)$$

The operator $\Delta^{(\Psi)}$ reads

$$\Delta^{(\Psi)}=\mathscr{D}^a\mathscr{D}_a-\frac{1}{4}G^{\alpha\dot\alpha}[\mathscr{D}_\alpha,\bar{\mathscr{D}}_{\dot\alpha}]-\frac{1}{4}\Psi\bar{\mathscr{D}}^2-\frac{1}{4}\Psi\mathscr{D}^2-\frac{1}{4}(\bar{\mathscr{D}}_{\dot\alpha}\bar{R})\bar{\mathscr{D}}^{\dot\alpha}-\frac{1}{4}(\mathscr{D}^\alpha R)\mathscr{D}_\alpha.$$

$$(7.3.30)$$

Equation (7.3.27) shows once again that in superspace there is no universal generalization of the geodesic interval.

The initial condition for $\mathbf{U}^{(\Psi)}_{\mathrm{v}}$ implies that the system of equations (7.3.27, 28) must be solved under the following boundary conditions:

$$\boldsymbol{\sigma}^{(\Psi)}_{\mathrm{v}}(z,z')=\mathscr{D}_A\boldsymbol{\sigma}^{(\Psi)}_{\mathrm{v}}(z,z')|_{z=z'}=\mathscr{D}_\alpha\mathscr{D}_A\boldsymbol{\sigma}^{(\Psi)}_{\mathrm{v}}(z,z')|_{z=z'}=0$$

$$\mathscr{D}_a\mathscr{D}_b\boldsymbol{\sigma}^{(\Psi)}_{\mathrm{v}}(z,z')|_{z=z'}=\eta_{ab} \qquad (7.3.31)$$

$$\bar{\mathscr{D}}_{\dot\alpha}\mathscr{D}^2\boldsymbol{\sigma}^{(\Psi)}_{\mathrm{v}}(z,z')|_{z=z'}=\mathscr{D}_a\mathscr{D}^2\boldsymbol{\sigma}^{(\Psi)}_{\mathrm{v}}(z,z')|_{z=z'}=\mathscr{D}_a[\mathscr{D}_\alpha,\bar{\mathscr{D}}_{\dot\alpha}]\boldsymbol{\sigma}^{(\Psi)}_{\mathrm{v}}(z,z')|_{z=z'}=0$$

$$\mathscr{D}^2\bar{\mathscr{D}}^2\boldsymbol{\sigma}^{(\Psi)}_{\mathrm{v}}(z,z')|_{z=z'}=16$$

and

$$\mathbf{a}^{(\Psi)}_0(z,z)=\mathscr{D}_A\mathbf{a}^{(\Psi)}_0(z,z')|_{z=z'}=\mathscr{D}_A\mathscr{D}_B\mathbf{a}^{(\Psi)}_0(z,z')|_{z=z'}=0$$

$$\mathscr{D}_\alpha\mathscr{D}_A\mathscr{D}_B\mathbf{a}^{(\Psi)}_0(z,z')|_{z=z'}=0 \qquad (7.3.32)$$

$$\mathscr{D}_\alpha\mathscr{D}_\beta\bar{\mathscr{D}}_{\dot\alpha}\bar{\mathscr{D}}_{\dot\beta}\mathbf{a}^{(\Psi)}_0(z,z')|_{z=z'}=-4\varepsilon_{\alpha\beta}\varepsilon_{\dot\alpha\dot\beta}.$$

These requirements are grounded in the same considerations as were made in subsection 7.2.4. In the flat-superspace limit $\mathbf{U}_v \equiv \mathbf{U}_v^{(0)}$ takes the form (4.8.43).

Using equations (7.3.27, 28, 31, 32), one readily obtains

$$\begin{aligned} \mathbf{a}_1^{(\Psi)}(z, z) &= 0 \\ \mathbf{a}_2^{(\Psi)}(z, z) &= -G^a G_a + \bar{\Psi}\Psi. \end{aligned} \tag{7.3.33}$$

7.3.4. Divergences of effective action

Now, it is a trivial task to obtain divergent parts of the superfunctionals (7.3.17) and (7.3.19) with the help of the general algorithm given in subsection 7.1.3. We rewrite Γ_v in the form

$$\Gamma_v = -\frac{i}{2}\int_0^\infty \frac{ds}{s}\{\mathbf{Tr}(3\mathbf{U}_v(s) - 2\mathbf{U}_v^{(R)}(s)) - 3\mathbf{Tr}_+\mathbf{U}_c(s)\} \tag{7.3.34}$$

and modify the integral over proper time by applying, for instance, the prescription for ω-regularization. This leads to the following divergent part of Γ_v:

$$\Gamma_{v,\mathrm{div}} = \frac{1}{32\pi^2\omega}\int d^8z\, E^{-1}\left\{3\mathbf{a}_2(z,z) - 2\mathbf{a}_2^{(R)}(z,z) - 3\frac{\mathbf{a}_2^c(z,z)}{R}\right\}. \tag{7.3.35}$$

Similarly, the divergent part of $\Gamma^{(\Psi)}$ reads

$$\Gamma_{(\Psi),\mathrm{div}} = \frac{1}{32\pi^2\omega}\int d^8z\, E^{-1}\left\{\mathbf{a}_2^{(\Psi)}(z,z) - \mathbf{a}_2(z,z) + \frac{\mathbf{a}_2^c(z,z)}{R}\right\} \tag{7.3.36}$$

which coincides, as a result of equation (7.3.33), with our previous result (7.2.61).

In conclusion let us point out that the superfunctional (7.3.35) is invariant under arbitrary super Weyl transformations (5.5.13). This is a consequence of super Weyl invariance of the corresponding classical theory (7.3.1).

7.4. Super Weyl anomaly

In Chapter 6 we met a number of super Weyl invariant models in curved superspace. The classical action $S[\chi; \mathscr{W}, \varphi]$ in such a theory remains unchanged under transformations of the form (5.7.33), the parameter σ being an arbitrary covariantly chiral scalar. This implies vanishing of the classical supertrace T, defined by equation (5.7.15), on the mass shell $\delta S/\delta\chi = 0$.

Since equation (5.7.33) presents a supersymmetric extension of ordinary Weyl transformations, the super Weyl invariance is potentially anomalous at the quantum level (see also subsection 7.1.4). Renormalized effective action $\Gamma[\tilde{\chi}; \mathscr{W}, \varphi]$ can be chosen to be invariant only under the supergravity gauge transformations, but not with respect to the super Weyl ones. Both symmetries cannot be preserved in general, which means that the effective supertrace

$$\langle T\rangle_{\tilde{\chi}} = \frac{\delta\Gamma[\tilde{\chi}; \mathscr{W}, e^{\sigma}\varphi]}{\delta\sigma}\bigg|_{\sigma=0}$$

remains non-zero when imposing the effective equation of motion $\delta\Gamma/\delta\tilde{\chi}=0$. As a result, the condition of general covariance in quantum theory is given by the following equation for the effective supercurrent $\langle T_{\alpha\dot{\alpha}}\rangle_{\tilde{\chi}}$

$$\bar{\mathscr{D}}^{\dot{\alpha}}\langle T_{\alpha\dot{\alpha}}\rangle_{\tilde{\chi}} = -\frac{2}{3}\mathscr{D}_{\alpha}\langle T\rangle_{\tilde{\chi}}$$

which replaces the classical equation (5.7.35).

In the case when $\tilde{\chi}=0$ is a solution to the effective equations of motion, the occurrence of the super Weyl anomaly can be studied by analysing the vacuum part Γ of the effective action, which is related to the in–out amplitude $\langle \text{out}|\text{in}\rangle = \exp(\mathrm{i}\Gamma)$. The breakdown of super Weyl invariance is expressed by the fact that Γ possesses some dependence on the chiral compensator,

$$\langle T\rangle = \frac{\delta\Gamma[\mathscr{W}, e^{\sigma}\varphi]}{\delta\sigma}\bigg|_{\sigma=0}$$

7.4.1. Super Weyl anomaly in a massless chiral scalar model

We are going to calculate anomalous supertrace $\langle T\rangle$ in the simplest super Weyl invariant model with classical action

$$S[\Phi] = \int d^8z\, E^{-1}\bar{\Phi}\Phi = \int d^8 z\,(E^{-1}/R)\Phi\mathbf{H}_{+-}\bar{\Phi} = \int d^8 z\,(E^{-1}/\bar{R})\bar{\Phi}\mathbf{H}_{-+}\Phi \tag{7.4.1}$$

Φ being a covariantly chiral scalar superfield. The operators $\mathbf{H}_{+-}$ and $\mathbf{H}_{-+}$ are given by equation (7.2.8). Since Φ changes as $\Phi' = e^{-\sigma}\Phi$ with respect to local rescaling of the chiral compensator, the super Weyl transformation laws for $\mathbf{H}_{+-}$ and $\mathbf{H}_{-+}$ read

$$\mathbf{H}'_{+-} = e^{-2\sigma}\mathbf{H}_{+-}e^{\bar{\sigma}} \qquad \mathbf{H}'_{-+} = e^{-2\bar{\sigma}}\mathbf{H}_{-+}e^{\sigma}. \tag{7.4.2}$$

Therefore, the transformation law for $\mathscr{H}_c = \mathbf{H}_{+-}\mathbf{H}_{-+}$ is given by

$$\mathscr{H}'_c = e^{-2\sigma}\mathscr{H}_c e^{\sigma-\bar{\sigma}}. \tag{7.4.3}$$

In the framework of the L-regularization scheme, the regularized effective action is given by

$$\Gamma_L = -\frac{\mathrm{i}}{2}\int_{-\mathrm{i}L^2}^{\infty}\frac{ds}{s}\mathbf{Tr}_+\mathbf{U}_c(s) \tag{7.4.4}$$

where

$$\mathbf{U}_c(z, z'|s) = \exp(\mathrm{i}s\mathscr{H}_c)\delta_+(z, z')$$

is the chiral Schwinger's superkernel described in Section 7.2. Then, the renormalized effective action reads

$$\Gamma_{\text{ren}} = \lim_{L\to 0} (\Gamma_L + S_{\text{count}}) \tag{7.4.5}$$

the counterterm being chosen as

$$S_{\text{count}} = -\frac{1}{32\pi^2}\left\{L^{-2}\int \mathrm{d}^8z\,\frac{E^{-1}}{R}\,a_1^c(z, z) - \ln(L^2\mu^2)\int \mathrm{d}^8z\,\frac{E^{-1}}{R}\,a_2^c(z, z)\right\}. \tag{7.4.6}$$

Our goal is to find the variation of Γ_{ren} under an infinitesimal super Weyl transformation.

Let us vary Γ_L. In accordance with equation (7.4.3), we have

$$\delta\mathbf{Tr}_+\mathbf{U}_c(s) = -\mathrm{i}s\,\mathbf{Tr}_+(\sigma\mathscr{H}_c\mathbf{U}_c(s)) - \mathrm{i}s\,\mathbf{Tr}_+(\mathscr{H}_c\bar{\sigma}\mathbf{U}_c(s)). \tag{7.4.7}$$

The first term in this variation can be rewritten with the aid of equation (7.2.45), in the form

$$-\mathrm{i}s\,\mathbf{Tr}_+(\sigma\mathscr{H}_c\mathbf{U}_c(s)) = -s\frac{\partial}{\partial s}\mathbf{Tr}_+(\sigma\mathbf{U}_c(s)). \tag{7.4.8}$$

To transform the second term in equation (7.4.7), we make use of the operator identities

$$\mathscr{H}_c\bar{\sigma} \equiv \mathbf{H}_{+-}\mathbf{H}_{-+}\bar{\sigma} = \mathbf{H}_{+-}\bar{\sigma}\mathbf{H}_{-+}$$

and

$$\mathbf{H}_{-+}\mathbf{U}_c(s) \equiv \mathbf{H}_{-+}\exp(\mathrm{i}s\mathbf{H}_{+-}\mathbf{H}_{-+}) = \exp(\mathrm{i}s\mathbf{H}_{-+}\mathbf{H}_{+-})\mathbf{H}_{-+} \equiv \mathbf{U}_a(s)\mathbf{H}_{-+}.$$

Taking into account equation (7.2.26) also, we can write

$$\mathbf{Tr}_+(\mathscr{H}_c\bar{\sigma}\mathbf{U}_c(s)) = \mathbf{Tr}_+(\mathbf{H}_{+-}\bar{\sigma}\mathbf{U}_a(s)\mathbf{H}_{-+}) = \mathbf{Tr}_-(\bar{\sigma}\mathbf{U}_a(s)\mathscr{H}_a)$$

which leads to the relation

$$\mathrm{i}s\mathbf{Tr}_+(\mathscr{H}_c\bar{\sigma}\mathbf{U}_c(s)) = -s\frac{\partial}{\partial s}\mathbf{Tr}_-(\bar{\sigma}\mathbf{U}_a(s)). \tag{7.4.9}$$

Equations (7.4.47–9) show that Γ_L varies as follows:

$$\delta\Gamma_L = -\frac{\mathrm{i}}{2}\mathbf{Tr}_+(\sigma\mathbf{U}_c(-iL^2)) - \frac{\mathrm{i}}{2}\mathbf{Tr}_+(\bar{\sigma}\mathbf{U}_a(-iL^2)). \tag{7.4.10}$$

The proper-time decomposition for $\mathbf{U}_c$ was described in Section 7.2. As is easily seen, an analogous representation for the antichiral Schwinger's superkernel

$$\mathbf{U}_{\mathrm{a}}(z, z'|s) = \exp(\mathrm{i}s\mathcal{H}_{\mathrm{a}})\delta_{-}(z, z') \tag{7.4.11}$$

is of the form

$$\mathbf{U}_{\mathrm{a}}(z, z'|s) = \frac{\mathrm{i}}{(4\pi \mathrm{i}s)^2} \exp\left(\frac{\mathrm{i}\bar{\sigma}(z, z')}{2s}\right) \sum_{n=0}^{\infty} \bar{\mathbf{a}}_n^{\mathrm{c}}(z, z')(\mathrm{i}s)^n \tag{7.4.12}$$

with the antichiral bi-scalars $\bar{\sigma}$ and $\bar{\mathbf{a}}^{\mathrm{c}}$ being conjugates of those presented in equation (7.2.47). As a result, equation (7.4.10) can be rewritten in the form

$$\delta\Gamma_L = \frac{1}{32\pi^2}\left\{-L^2 \int \mathrm{d}^8 z\, E^{-1}\sigma + \int \mathrm{d}^8 z \frac{E^{-1}}{R} \sigma \mathbf{a}_2^{\mathrm{c}}(z, z)\right\} + \mathrm{c.c.} + O(L^2). \tag{7.4.13}$$

where we have used the fact that $\mathbf{a}_1^{\mathrm{c}}(z, z) = -R$.

Now, let us turn to the counterterm S_{count}. As was noted above, the second term in the right-hand side of equation (7.4.6) is invariant with respect to arbitrary super Weyl transformations. Hence only the first term contributes to the super Weyl variation of S_{count}:

$$\delta S_{\mathrm{count}} = \frac{L^{-2}}{32\pi^2} \int \mathrm{d}^8 z\, E^{-1}\sigma + \mathrm{c.c.} \tag{7.4.14}$$

From equations (7.4.13) and (7.4.14) we obtain

$$\delta\Gamma_{\mathrm{ren}} = \frac{1}{32\pi^2} \int \mathrm{d}^8 z \frac{E^{-1}}{R} \sigma \mathbf{a}_2^{\mathrm{c}}(z, z) + \mathrm{c.c.} \tag{7.4.15}$$

Therefore, the anomalous supertrace in the theory under consideration reads

$$\langle T \rangle = \frac{1}{32\pi^2} \mathbf{a}_2^{\mathrm{c}}(z, z). \tag{7.4.16}$$

Remark. Another example of classically super Weyl invariant theories gives the vector multiplet model (7.3.1). Calculation of the anomalous supertrace in this theory requires more subtle consideration[1]. We give only the final result:

$$\langle T \rangle = -\frac{1}{32\pi^2}\left\{3\mathbf{a}_2^{\mathrm{c}}(z, z) + \frac{1}{4}(\bar{\mathcal{D}}^2 - 4R)(3\mathbf{a}_2(z, z) - 2\mathbf{a}_2^{(R)}(z, z))\right\}. \tag{7.4.17}$$

Here the coefficients $\mathbf{a}_2(z, z)$ and $\mathbf{a}_2^{(R)}(z, z)$ are obtained from $\mathbf{a}_2^{(\Psi)}(z, z)$ (7.3.33) by setting $\Psi = 0$ and $\Psi = R$, respectively.

[1] I.L. Buchbinder and S.M. Kuzenko, *Nucl. Phys.* **B 274** 653, 1986.

7.4.2. Anomalous effective action

The relation defining anomalous supertrace

$$\frac{\Delta\Gamma}{\Delta\varphi} \equiv \left.\frac{\delta\Gamma[\mathscr{W}, e^{\sigma}\varphi]}{\delta\sigma}\right|_{\sigma=0} = \langle T \rangle \tag{7.4.18}$$

can be understood as an equation for the renormalized effective action provided we have at our disposal an exact expression for $\langle T \rangle$ obtained by some indirect method. This is just the situation under consideration.

Substituting the explicit values of $\mathbf{a}_2^c(z, z)$, $\mathbf{a}_2(z, z)$ and $\mathbf{a}_2^{(R)}(z, z)$ into equations (7.4.16) and (7.4.17), the anomalous supertrace can be written in the following general form:

$$\langle T \rangle = aW^2 + bP + \frac{c}{16}(\bar{\mathscr{D}}^2 - 4R)\mathscr{D}^2 R \tag{7.4.19}$$

where

$$W^2 = W^{\alpha\beta\gamma} W_{\alpha\beta\gamma}$$

$$P = W^2 - \frac{1}{4}(\bar{\mathscr{D}}^2 - 4R)(G^a G_a + 2R\bar{R}).$$

These are only the numerical coefficients a, b, c which reflect the specific features of the initial theory. The chiral superfield P is originated as a density for the topological invariant (5.6.61),

$$\mathscr{P} \equiv \int d^8z\, E^{-1}\left\{\frac{1}{R}W^2 + G^a G_a + 2R\bar{R}\right\} = \int d^8z \frac{E^{-1}}{R} P. \tag{7.4.20}$$

Now, our goal is to obtain a solution of equation (7.4.18). As is clear, the renormalized effective action consists, in general, of two parts:

$$\Gamma = \Gamma_{\mathrm{A}} + \Gamma_{\mathrm{I}}$$

where Γ_{A} generates the anomaly (7.4.19):

$$\frac{\Delta\Gamma_{\mathrm{A}}}{\Delta\varphi} = \langle T \rangle$$

whereas Γ_{I} is a super Weyl invariant functional. Only the anomalous part of Γ can be restored by solving the above equation.

Let us proceed by solving the equation

$$\frac{\Delta\Gamma_{\mathrm{A}}}{\Delta\varphi} = aW^2 + bP + \frac{c}{16}(\bar{\mathscr{D}}^2 - 4R)\mathscr{D}^2 R. \tag{7.4.21}$$

An important role in our considerations will be played by the covariantly chiral scalar superfield Ω and its conguate $\bar{\Omega}$ defined by

$$\begin{pmatrix}\Omega(z)\\ \bar{\Omega}(z)\end{pmatrix}=\begin{pmatrix}1\\ 1\end{pmatrix}+\mathbf{G}\begin{pmatrix}R\\ \bar{R}\end{pmatrix}=\begin{pmatrix}1+\int d^8z'\,E^{-1}(z')\mathbf{G}_{+-}(z,z')\\ 1+\int d^8z'\,E^{-1}(z')\mathbf{G}_{-+}(z,z')\end{pmatrix}. \tag{7.4.22}$$

Here $\mathbf{G}$ is the Feynman superpropagator for the massless theory (7.4.1). The superfield Ω satisfies the massless equation of motion

$$(\bar{\mathscr{D}}^2-4R)\Omega=0 \tag{7.4.23}$$

and proves to obey the transformation law

$$\Omega'=e^{-\sigma}\Omega \tag{7.4.24}$$

with respect to the super Weyl transformations (5.5.13). As an instructive exercise, the reader can explicity verify equation (7.4.24). It is worth pointing out that the flat-superspace limit for Ω is unity, and its dependence on the supergravity prepotentials takes the form

$$\Omega=\varphi^{-1}+O(\mathscr{W}). \tag{7.4.25}$$

In accordance with definition (7.4.22), Ω is a non-local functional of $\mathscr{W}$ and φ. However, in the case of a conformally flat superspace, Ω turns into the inverse of φ in the gauge $\mathscr{W}=0$. It is necessary to note that the chiral scalar Ω was introduced by Grisaru, Nielsen, Siegel and Zanon[2] and is a supersymmetric generalization of the conformal scalar field $1-\frac{1}{6}(-\Box+\mathscr{R}/6)^{-1}\mathscr{R}$ of Fradkin and Vilkovisky[3].

Using the definition of the variational derivative $\Delta/\Delta\varphi$ (7.4.18) and the super Weyl transformation laws (5.5.14) and (7.4.24), it is extremely simple to obtain the following identities:

$$\frac{\Delta}{\Delta\varphi}\int d^8z\,\frac{E^{-1}}{R}\,W^2\ln\Omega=-W^2 \qquad \frac{\Delta}{\Delta\varphi}\int d^8z\,\frac{E^{-1}}{R}\,\bar{W}^2\ln\bar{\Omega}=0 \tag{7.4.26a}$$

$$\begin{aligned}\frac{\Delta}{\Delta\varphi}\int d^8z\,E^{-1}(G^aG_a+2R\bar{R})\ln(\bar{\Omega}\Omega)&=\frac{1}{4}(\bar{\mathscr{D}}^2-4R)(G^aG_a+2R\bar{R})\\ &\quad-\frac{1}{8}(\bar{\mathscr{D}}^2-4R)\mathscr{D}^\alpha(G_{\alpha\dot{\alpha}}\bar{\mathscr{D}}^{\dot{\alpha}}\ln\bar{\Omega}-R\mathscr{D}_\alpha\ln\Omega)\end{aligned} \tag{7.4.26b}$$

[2] M. T. Grisaru, N. K. Nielsen, W. Siegel and D. Zanon *Nucl. Phys.* **B 247** 157, 1984.
[3] E. S. Fradkin and G. A. Vilkovisky *Phys. Lett.* **73B** 209, 1978.

$$\frac{\Delta}{\Delta\varphi}\int d^8z\, E^{-1}G^{\alpha\dot{\alpha}}\bar{\mathscr{D}}_{\dot{\alpha}}(\ln\bar{\Omega})\mathscr{D}_\alpha(\ln\Omega)$$

$$= -\frac{1}{4}(\bar{\mathscr{D}}^2-4R)\{-\mathscr{D}^\alpha(G_{\alpha\dot{\alpha}}\bar{\mathscr{D}}^{\dot{\alpha}}\ln\bar{\Omega})+\mathrm{i}\mathscr{D}_{\alpha\dot{\alpha}}(\bar{\mathscr{D}}^{\dot{\alpha}}(\ln\bar{\Omega})\mathscr{D}^\alpha\ln\Omega)\} \qquad (7.4.26c)$$

$$\frac{\Delta}{\Delta\varphi}\int d^8z\, E^{-1}\mathscr{D}^\alpha(\ln\Omega)\mathscr{D}_\alpha(\ln\Omega)\bar{\mathscr{D}}_{\dot{\alpha}}(\ln\bar{\Omega})\bar{\mathscr{D}}^{\dot{\alpha}}(\ln\bar{\Omega})$$

$$=2(\bar{\mathscr{D}}^2-4R)\{\mathrm{i}\bar{\mathscr{D}}_{\alpha\dot{\alpha}}(\bar{\mathscr{D}}^{\dot{\alpha}}(\ln\bar{\Omega})\mathscr{D}^\alpha(\ln\Omega))-\mathscr{D}^\alpha(R\mathscr{D}_\alpha\ln\Omega)\} \qquad (7.4.26d)$$

$$\frac{\Delta}{\Delta\varphi}\int d^8z\, E^{-1}R\bar{R}=\frac{1}{16}(\bar{\mathscr{D}}^2-4R)\mathscr{D}^2R. \qquad (7.4.26e)$$

In deriving these relations we have also made use of the equation of motion (7.4.23).

Let us consider the functional

$$\Gamma_A^{(1)}=-a\int d^8z\, E^{-1}\left\{\frac{\ln\Omega}{R}W^2+\frac{\ln\bar{\Omega}}{\bar{R}}\bar{W}^2\right\}. \qquad (7.4.27)$$

From equation (7.4.26*a*) we obtain

$$\frac{\Delta\Gamma_A^{(1)}}{\Delta\varphi}=aW^2. \qquad (7.4.28)$$

Introduce one more functional

$$\Gamma_A^{(2)}=c\int d^8z\, E^{-1}R\bar{R}. \qquad (7.4.29)$$

Equation (7.4.26*e*) leads to

$$\frac{\Delta\Gamma_A^{(2)}}{\Delta\varphi}=\frac{c}{16}(\bar{\mathscr{D}}^2-4R)\mathscr{D}^2R. \qquad (7.4.30)$$

It remains to look for a superfunctional generating the term bP in equation (7.4.19). With the aid of equations (7.4.26*b–d*) one finds that the functional

$$\Gamma_A^{(3)}=-b\int d^8z\, E^{-1}\left\{\frac{\ln\Omega}{R}P+\frac{\ln\bar{\Omega}}{\bar{R}}\bar{P}+\frac{1}{2}\left(G^{\alpha\dot{\alpha}}+\frac{1}{8}\bar{\mathscr{D}}^{\dot{\alpha}}(\ln\bar{\Omega})\mathscr{D}^\alpha(\ln\Omega)\right)\right.$$

$$\left.\times\bar{\mathscr{D}}_{\dot{\alpha}}(\ln\bar{\Omega})\mathscr{D}_\alpha(\ln\Omega)\right\} \qquad (7.4.31)$$

solves the problem,

$$\frac{\Delta\Gamma_A^{(3)}}{\Delta\varphi}=bP. \qquad (7.4.32)$$

Equations (7.4.27–32) lead to the anomalous effective action

$$\Gamma_A=\Gamma_A^{(1)}+\Gamma_A^{(2)}+\Gamma_A^{(3)}=\int d^8z\,E^{-1}\left\{-\frac{\ln\Omega}{R}(aW^2+bP)-\frac{\ln\bar\Omega}{\bar R}(a\bar W^2+b\bar P)\right.$$
$$\left.-\frac{b}{2}\left(G^{\alpha\dot\alpha}+\frac{1}{8}\bar{\mathscr{D}}^{\dot\alpha}(\ln\bar\Omega)\mathscr{D}^{\alpha}(\ln\Omega)\right)\bar{\mathscr{D}}_{\dot\alpha}(\ln\bar\Omega)\mathscr{D}_{\alpha}(\ln\Omega)+cR\bar R\right\}. \quad (7.4.33)$$

One more interesting representation for the anomalous effective action, which is based on the use of a natural generalization for Ω, can be found in the original publication[4].

The non-local functional Γ_A turns into a local one in the case of a conformally flat superspace, i.e. when $W_{\alpha\beta\gamma}=0$. Then, in the gauge $\mathscr{W}=0$ the covariant derivatives and the supertorsions R, G_a are given by equations (6.5.12, 13), whereas Ω coincides with φ^{-1}. As a result, Γ_A takes the form

$$\Gamma_A=-\int d^8z\left\{c\partial^a(\ln\bar\varphi)\partial_a(\ln\varphi)+\frac{1}{4}(c-b)\bar{\mathrm{D}}^{\dot\alpha}(\ln\bar\varphi)\mathrm{D}^{\alpha}(\ln\varphi)\right.$$
$$\left.\times\left([\bar{\mathrm{D}}_{\dot\alpha},\mathrm{D}_\alpha](\ln\bar\varphi\varphi)-\frac{1}{2}\bar{\mathrm{D}}_{\dot\alpha}(\ln\bar\varphi)\mathrm{D}_\alpha(\ln\varphi)\right)\right\} \quad (7.4.34)$$

where $\mathrm{D}_A=(\partial_a,\mathrm{D}_\alpha,\bar{\mathrm{D}}^{\dot\alpha})$ are the flat global covariant derivatives. This functional can be considered as the classical action of some supersymmetric theory in flat superspace.

7.4.3. Solution of effective equations of motion in conformally flat superspace

It is natural to understand the renormalized effective action $\Gamma[\mathscr{W},\varphi]$ as a quantum correction, due to vacuum effects, to the classical supergravity action S_{SG} which can be taken in one of two possible forms (6.1.3) or (6.1.17). In general, the effective gravity equation of motion

$$\frac{\delta(S_{SG}+\Gamma)}{\delta\mathbf{H}^a}=0 \qquad \frac{\Delta(S_{SG}+\Gamma)}{\Delta\varphi}=0 \quad (7.4.35)$$

cannot be written explicitly since the calculation of Γ is an extraordinary technical problem. However, the above results supply us with a remarkable possibility to obtain exact effective equations in the case of classically super Weyl invariant theories. In such a theory the effective action is $\Gamma=\Gamma_A+\Gamma_I$ where Γ_A is given by equation (7.4.33) and Γ_I is some unknown superfunctional. Because of the super Weyl invariance of Γ_I, the effective equations read

$$\frac{\delta}{\delta\mathbf{H}}(S_{SG}+\Gamma_A+\Gamma_I)=0 \quad (7.4.36a)$$

[4] I.L. Buchbinder and S.M. Kuzenko, *Phys. Lett.* **202B** 233, 1988.

$$\frac{\Delta}{\Delta\varphi}(S_{SG}+\Gamma_A)=\frac{\Delta S_{SG}}{\Delta\varphi}+\langle T\rangle=0. \tag{7.4.36b}$$

Here the first equation is still unknown. We may suppose, nevertheless, that there are conformally flat solutions ($W_{\alpha\beta\gamma}=0$) for the full system of equations. Then, only the second equation remains essential.

So, let us consider the equation for the chiral compensator choosing the supergravity action with cosmological term (6.1.17) in the role of S_{SG}. With $W_{\alpha\beta\gamma}=0$, equation (7.4.36b) reads

$$\frac{3}{\kappa^2}(R-\mu)+\frac{1}{4}(\bar{\mathscr{D}}^2-4R)\left\{b(G^aG_a+2R\bar{R})-\frac{c}{4}\mathscr{D}^2R\right\}=0. \tag{7.4.37}$$

A particular solution is obtained by setting $G_a=0$ and $R=\text{const}$, R being subject to the constraint

$$R-\mu-\frac{\kappa^2}{12}R^2\bar{R}=0 \tag{7.4.38}$$

and corresponds to the anti-de Sitter supergeometry.

7.5. Quantum equivalence in superspace

7.5.1. Problem of quantum equivalence

One of the specific features of superspace, as compared to ordinary space–time, is that any on-shell supersymmetry representation admits, in general, several off-shell superfield realizations. Of course, there exist examples of different field theories possessing equivalent dynamics, but superspace provides far more possibilities. The most impressive example is the existence of infinitely many superfield formulations for Einstein supergravity (minimal ($n=-\frac{1}{3}$), new minimal ($n=0$), non-mininal ($n\neq-\frac{1}{3},0$)) instead of the unique generally accepted formulation for Einstein gravity. All these formulations were shown in Chapter 6 to be equivalent at the classical level. So, in supersymmetric field theory we inevitably run into a problem of quantum equivalence for classically equivalent realizations. By quantum equivalence is meant equivalence of S-matrices. For theories in external (super)fields, quantum equivalence can be understood as the coincidence of in–out vacuum amplitudes modulo an irrelevant factor expressed via boundary values of background (super)fields at infinity. It is clear that the problem of quantum equivalence is non-trivial in general and deserves special consideration in any concrete case.

Of principal importance is the study of quantum equivalence between different supergravity formulations. Unfortunately this is still an open problem because of the non-renormalizability of (super)gravity. In the simpler case of renormalizable classically equivalent theories the study of their quantum equivalence can, in principle, be carried out in the framework of perturbation theory.

In the present section we investigate three superfield realizations for the so-called non-conformal scalar multiplet which is normally described by a covariantly chiral scalar superfield Φ on the basis of the action superfunctional

$$S_{(R)}[\Phi]=\frac{1}{2}\int \mathrm{d}^8z\, E^{-1}(\Phi+\bar{\Phi})^2. \tag{7.5.1}$$

This model is said to be non-conformal because the action involves the term

$$\frac{1}{2}\int \mathrm{d}^8z\, E^{-1}(\Phi^2+\bar{\Phi}^2)$$

which breaks the super Weyl invariance present in the action (7.4.1). Among other chiral models based on classical actions of the form

$$\int \mathrm{d}^8z\, E^{-1}\left\{\bar{\Phi}\Phi+\frac{\xi}{2}(\Phi^2+\bar{\Phi}^2)\right\}$$

with $\xi\neq\pm 1$, the above model is selected because of its equivalence to the gauge theory of a covariantly chiral spinor superfield χ_α, $\bar{\mathscr{D}}_{\dot\alpha}\chi_\alpha=0$, which is described by the action

$$S[\chi]=-\frac{1}{4}\int \mathrm{d}^8z\, E^{-1}(\mathscr{D}^\alpha\chi_\alpha+\bar{\mathscr{D}}_{\dot\alpha}\bar{\chi}^{\dot\alpha})^2. \tag{7.5.2}$$

The classical equivalence of these theories has been established in subsection 6.3.4. One more realization for the non-conformal scalar multiplet follows from equation (7.5.1) by expressing Φ via a real scalar superfield U in accordance with the rule

$$\Phi=-\frac{1}{4}(\bar{\mathscr{D}}^2-4R)U \qquad U=\bar{U} \tag{7.5.3}$$

Then we obtain the action

$$S[U]=\frac{1}{32}\int \mathrm{d}^8z\, E^{-1}U((\bar{\mathscr{D}}^2-4R)+(\mathscr{D}^2-4\bar{R}))^2U. \tag{7.5.4}$$

Thus we have three classically equivalent models realized in terms of (i) chiral scalar (CSC), (ii) chiral spinor (CSP), and (iii) real scalar (RSC) superfields.

Let us introduce into considerations the in–out vacuum amplitudes and effective actions for the above classically equivalent theories:

$$\langle\text{out}|\text{in}\rangle_{\text{CSC}}=\exp(\mathrm{i}\Gamma_{\text{CSC}}[\mathscr{W},\varphi])$$

$$\langle\text{out}|\text{in}\rangle_{\text{CSP}}=\exp(\mathrm{i}\Gamma_{\text{CSP}}[\mathscr{W},\varphi])$$

$$\langle\text{out}|\text{in}\rangle_{\text{RSC}}=\exp(\mathrm{i}\Gamma_{\text{RSC}}[\mathscr{W},\varphi]).$$

To have equivalence at the quantum level, the renormalized effective actions must possess the same local dependence on the supergravity prepotentials. In other words, the average supercurrents as well as the average supertraces arising in the theories under examination must coincide.

The CSC model is non-gauge, and the corresponding effective action $\Gamma_{\text{CSC}} \equiv \Gamma_{(R)}$ is formally given by

$$\Gamma_{\text{CSC}} = \frac{\mathrm{i}}{2} \mathbf{Tr} \ln \mathbf{H}^{(R)} \tag{7.5.5}$$

(see also Sections 7.2 and 7.3). As for the models (7.5.2) and (7.5.4), they are gauge theories with linearly dependent generators (see subsection 4.5.1). The action (7.5.2) is invariant under the gauge transformations

$$\chi_\alpha \to \chi_\alpha^V = \chi_\alpha + \frac{\mathrm{i}}{8}(\bar{\mathscr{D}}^2 - 4R)\mathscr{D}_\alpha V \qquad V = \bar{V} \tag{7.5.6}$$

the gauge parameter being an unconstrained real scalar superfield defined modulo arbitrary shifts of the form

$$V \to V^\Lambda \to V + \Lambda + \bar{\Lambda} \qquad \bar{\mathscr{D}}_{\dot{\alpha}} \Lambda = 0. \tag{7.5.7}$$

Next, the action (7.5.4) remains unchanged under gauge transformations

$$U \to U^\eta = U + \frac{1}{2}(\mathscr{D}^\alpha \eta_\alpha + \bar{\mathscr{D}}_{\dot{\alpha}} \bar{\eta}^{\dot{\alpha}}) \qquad \bar{\mathscr{D}}_{\dot{\alpha}} \eta_\alpha = 0 \tag{7.5.8}$$

as a consequence of equation (5.5.11). Here the gauge parameters η_α is a covariantly chiral spinor superfield defined modulo arbitrary shifts of the form (7.5.6), and in its turn V is defined modulo arbitrary shifts (7.4.7). So, the CSP and RSC models are reducible gauge theories with the first and second, respectively, stages of reducibility. Hence, it is required to give proper definitions for Γ_{CSP} and Γ_{RSC} before proceeding further.

As is known, the Faddeev–Popov approach is inapplicable to quantize reducible gauge theories. It is the Batalin–Vilkovisky quantization method[1] which leads to correct Feynman rules in the Lagrangian approach for arbitrary finitely reducible gauge theories. In the case of Abelian finitely reducible gauge theories, the Batalin–Vilkovisky method turns out to be equivalent to the quantization procedure of A. Schwarz[2], the latter being the first known technique developed to quantize reducible theories. Since the CSP and RSP models are Abelian gauge theories, their quantization can be fulfilled in the framework of the Schwarz procedure. In order to illustrate this technique, we first discuss the quantization of the more familiar gauge theory of a second-rank antisymmetric tensor field in a curved space–time.

[1]I.A. Batalin and G.A. Vilkovisky, *Phys. Lett.* **120B** 166, 1983; *Phys. Rev.* D **28** 2567, 1983.
[2]A.S. Schwarz, *Lett. Math. Phys.* **2** 247, 1976; *Commun. Math. Phys.* **67** 1, 1979.

7.5.2. Gauge antisymmetric tensor field

The gauge theory of antisymmetric tensor field $A_{ab}(x)$ coupled to an external gravitational field is determined by the action

$$A[S]=\frac{1}{2}\int \mathrm{d}^4x\, e^{-1}L^a(A)L_a(A) \tag{7.5.9}$$

$$L^a(A)=\frac{1}{2}\varepsilon^{abcd}\nabla_b A_{cd}.$$

This theory is known to be classically equivalent to the theory of a scalar field $\varphi(x)$ minimally coupled to gravity

$$S[\varphi]=-\frac{1}{2}\int d^4x\, e^{-1}\nabla^a\varphi\nabla_a\varphi. \tag{7.5.10}$$

Both models are of interest to us since the former arises at the component level in the model (7.5.2) and the latter in the model (7.5.1).

The action $S[A]$ is invariant under the following gauge transformations

$$A_{ab}\to A^{\lambda}_{ab}=A_{ab}+\nabla_{[a}\lambda_{b]} \tag{7.5.11}$$

the gauge parameter being defined modulo arbitrary shifts of the form $\delta\lambda_a=\nabla_a\Psi$. So, the model under consideration is a reducible gauge theory with the first stage of reducibility. In the role of gauge-fixing functions we choose

$$K_a(A)=\nabla^b A_{ab} \tag{7.5.12}$$

which is a transversal vector field,

$$\nabla^a K_a(A)=0 \tag{7.5.13}$$

and transforms under (7.5.11) according to the law

$$\delta K_a(A)=\nabla^b\nabla_{[b}\lambda_{a]}\equiv \mathscr{F}_a{}^b\lambda_b. \tag{7.5.14}$$

Adding to $S[A]$ the gauge-breaking term

$$S_{\mathrm{GB}}[A]=-\frac{1}{2}\int \mathrm{d}^4x\, e^{-1}K^a(A)K_a(A) \tag{7.5.15}$$

we arrive at the non-gauge action

$$S[A]+S_{\mathrm{GB}}[A]=\frac{1}{4}\int \mathrm{d}^4x\, e^{-1}A^{ab}\Box_2 A_{ab} \tag{7.5.16}$$

where

$$\Box_2 A_{ab}=\nabla^c\nabla_c A_{ab}+2\mathscr{R}^c{}_{[a}A_{b]c}-2\mathscr{R}^c{}_{[ab]}{}^d A_{cd} \tag{7.5.17}$$

$\Box_2$ being an antisymmetric tensor field d'Alembertian.

Now, it is worth making some general remarks concerning the Schwarz quantization prescription. Let us recall that the central idea in the Faddeev–Popov construction was separation of the gauge group volume from the naive path integral. From the technical point view, this was done by attracting the delta-function of a gauge-fixing function. The chief observation made by Schwarz was that basic elements of the Faddeev–Popov construction can be preserved, but must be properly modified, in the case of Abelian reducible gauge theories. First, it is necessary to modify the notion of gauge group volume, since some of the gauge parameters do not appear in the transformation law of the gauge fields. Second, one must extend the notion of functional delta-function to the case when its argument is a constrained field. The point is that admissible gauge-fixing functions in reducible gauge theories should obey some constraints.

We proceed to the quantization of the theory (7.5.9). Because of the constraint (7.5.13), the delta-function $\delta^4[K_a(A)]$ is ill-defined ($\delta^4[K_a(A)] \sim \delta[0]$, where $\delta^4[\ldots]$ and $\delta[\ldots]$ denote the vector and scalar functional delta-functions). As a result, the standard Faddeev–Popov construction does not work.

Let us try to define a generalized delta-function $\tilde{\delta}^4[B_a]$, where $B_a(x)$ is an arbitrary transverse vector field, $\nabla^a B_a = 0$. We start with the formal Fourier-representation

$$\delta^4[B_a] = \int \mathscr{D}V \exp\left(\mathrm{i} \int \mathrm{d}^4x \, e^{-1} V^a B_a \right). \tag{7.5.18}$$

Owing to the transversality of B_a, the exponential in equation (7.5.18) is invariant under the transformations

$$V_a \to V_a^\Psi = V_a + \nabla_a \Psi \tag{7.5.19}$$

with arbitrary Ψ, hence the path integral is ill-defined. However, making use of the Faddeev–Popov prescription allows us to extract from (7.5.18) the integral over the gauge group (7.5.19) in the same fashion as in electrodynamics. Then one arrives at the following well-defined object

$$\tilde{\delta}^4[B_a] \equiv \int \mathscr{D}V \exp\left(\mathrm{i} \int \mathrm{d}^4x \, e^{-1} V^a B_a \right) \delta[\nabla^a V_a] \mathrm{Det}(\Box_0) \tag{7.5.20}$$

where $\Box_0 = \nabla^a \nabla_a$ is a scalar field d'Alembertian. With the help of the identity

$$\delta[\nabla^a V_a] = \int \mathscr{D}\varphi \exp\left(\mathrm{i} \int \mathrm{d}^4x \, e^{-1} V^a \nabla_a \varphi \right)$$

$\tilde{\delta}^4[B_a]$ can be rewritten in the form

$$\tilde{\delta}^4[B_a] = \int \mathscr{D}\varphi \, \delta^4[B_a + \nabla_a \varphi] \mathrm{Det}(\Box_0). \tag{7.5.21}$$

Next, we turn to finding a proper integration measure $\mathcal{D}\mu_\lambda$ for the group of gauge transformations (7.5.11). Let $\Gamma[A]$ be a functional of the gauge field. Naive path integral $I[A]=\int \mathcal{D}\lambda\, F[A^\lambda]$ over the gauge group is seen to be degenerate, since $A^\lambda = A^{\lambda'}$ for $\lambda'_a = \lambda_a + \nabla_a \Psi$, Ψ being arbitrary. In order to extract the Ψ-group volume from $I[A]$, we can again make use of the Faddeev–Popov prescription. This leads to the following definition of integrals over the λ-group:

$$I[A]=\int \mathcal{D}\mu_\lambda F[A^\lambda] \qquad \mathcal{D}\mu_\lambda \equiv \delta[\nabla^a \lambda_a] \mathrm{Det}(\Box_0) \mathcal{D}\lambda\,. \tag{7.5.22}$$

Once $\delta^4[K_a(A)]$ and $\mathcal{D}\mu_\lambda$ have been defined, we can follow the standard Faddeev–Popov procedure to quantize the theory (7.5.9). First, we introduce the Faddeev–Popov determinant Δ:

$$\Delta^{-1}=\int \mathcal{D}\mu_\lambda\, \delta^4[K_a(A^\lambda)]. \tag{7.5.23}$$

This can be expressed in the form

$$\Delta = \mathrm{Det}(\Box_1)\mathrm{Det}^{-2}(\Box_0) \tag{7.5.24}$$

where $\Box_1$ is the vector field d'Alembertian defined by

$$\Box_1 V_a = \nabla^b \nabla_b V_a - \mathcal{R}_a{}^b V_b. \tag{7.5.25}$$

To prove equation (7.5.24), we rewrite the right-hand side of equation (7.5.23) as follows:

$$\int \mathcal{D}\mu_\lambda \delta^4[K_a(A^\lambda)] = \int \mathcal{D}V \mathcal{D}\lambda \delta[\nabla^a V_a - \rho]\delta[\nabla^a \lambda_a - \gamma]$$
$$\times \mathrm{Det}^2(\Box_0) \exp\left(\mathrm{i} \int \mathrm{d}^4x\, e^{-1} V^a (K_a(a) + \mathcal{F}_a{}^b \lambda_b)\right) \tag{7.5.26}$$

for $\mathcal{F}_a{}^b$ the operator given in (7.5.14). By construction, this expression does not depend on external scalar fields ρ and γ, so we can integrate the right-hand side of equation (7.5.26) over ρ and γ with the weight $\exp(-\mathrm{i}\int \mathrm{d}^4x\, e^{-1}\rho\gamma)$. Making also the shift $\lambda_a \to \lambda_a - (1/\Box_1)K_a(A)$ in the path integral, one obtains (7.5.24).

With the aid of the representation (7.5.21), it is possible to rewrite equation (7.5.23) in one more useful form:

$$\Delta^{-1} = \int \mathcal{D}\mu_\lambda \mathcal{D}\varphi\, \delta^4[K_a(A^\lambda) + \nabla_a \varphi - \zeta_a] \mathrm{Det}(\Box_0) \tag{7.5.27}$$

where ζ_a is an arbitrary external vector field. This expression does not depend on the transversal part of ζ_a by its very construction. The longitudinal part of ζ_a arises in expression (7.5.27) as a result of the shift $\varphi \to \varphi - \nabla^a(1/\Box_1)\zeta_a$.

The next step in the Faddeev–Popov construction is to extract the gauge group volume. In our case this consists of inserting into the naive path integral

$$\int \mathscr{D}A\, e^{iS[A]}$$

the following unit

$$1=\Delta\int \mathscr{D}\mu_\lambda \mathscr{D}\varphi\; \delta^4[K_a(A^\lambda)+\nabla_a\varphi-\zeta_a]\det(\Box_0)$$

along with the replacement of the integration variable $A\to A^{-\lambda}$. This leads to the effective action

$$e^{i\Gamma_A}=\int \mathscr{D}A\mathscr{D}\varphi\; \delta^4[K_a(A)+\nabla_a\varphi-\zeta_a]\Delta \mathrm{Det}(\Box_0)\exp(iS[A]). \quad (7.5.28)$$

Since Γ_A does not depend on ζ_a, it is in our power to integrate over ζ_a with the weight $\exp(-\frac{1}{2}\int d^4x\, e^{-1}\zeta^2)$. Then we obtain

$$e^{i\Gamma_A}=\int \mathscr{D}A\mathscr{D}\varphi\; \Delta\, \mathrm{Det}(\Box_0)\exp\left\{i(S[A]+S_{GB}[A])-\frac{i}{2}\int d^4x\, e^{-1}\nabla^a\varphi\nabla_a\varphi\right\}. \quad (7.5.29)$$

Now, equations (7.5.15) and (7.5.23) allow us to obtain the final expression for the effective action

$$\Gamma_A=\Gamma_\varphi+\frac{i}{2}\{\mathrm{Tr}\ln\Box_2-2\,\mathrm{Tr}\ln\Box_1+2\,\mathrm{Tr}\ln\Box_0\} \quad (7.5.30)$$

where

$$\Gamma_\varphi=\frac{i}{2}\mathrm{Tr}\ln\Box_0 \quad (7.5.31)$$

is the effective action of the scalar field model (7.5.10).

7.5.3. Quantization of the chiral spinor model

We proceed by constructing the effective action for the gauge superfield theory with classical action (7.5.2) on the basis of the quantization technique described above.

An admissible gauge-fixing function in the theory under consideration can be chosen in the form

$$\mathbb{K}(\chi)=\frac{i}{2}(\mathscr{D}^\alpha\chi_\alpha-\bar{\mathscr{D}}_{\dot\alpha}\bar\chi^{\dot\alpha}) \quad (7.5.32)$$

$\mathbb{K}(\chi)$ being a covariantly linear real scalar superfield,

$$(\bar{\mathscr{D}}^2-4R)\mathbb{K}(\chi)=0 \qquad \overline{\mathbb{K}(\chi)}=\mathbb{K}(\chi). \tag{7.5.33}$$

The transformation law

$$\mathbb{K}(\chi^V)=\mathbb{K}(\chi)-\frac{1}{8}\mathscr{D}^\alpha(\bar{\mathscr{D}}^2-4R)\mathscr{D}_\alpha V \tag{7.5.34}$$

tells us that the requirement $\mathbb{K}(\chi^V)=0$ fixes V modulo aribitrary shifts (7.5.7). Adding the gauge-breaking term

$$S_{\mathrm{SB}}[\chi]=\int \mathrm{d}^8x\, E^{-1}(\mathbb{K}(\chi))^2 \tag{7.5.35}$$

to the classical action $S[\chi]$, we obtain the following non-gauge superfunctional

$$S[\chi]+S_{\mathrm{GB}}[\chi]=-\frac{1}{4}\int \mathrm{d}^8\, E^{-1}\chi^\alpha(\mathscr{D}^2-6\bar{R})\chi_\alpha+\text{c.c.}$$

$$=\int \mathrm{d}^8\,\frac{E^{-1}}{R}\,\chi^\alpha(\mathscr{H}_{\mathrm{c}}\chi)_\alpha+\int \mathrm{d}^8\,\frac{E^{-1}}{\bar{R}}(\bar{\mathscr{H}}_{\mathrm{a}}\bar{\chi})_{\dot\alpha}\bar{\chi}^{\dot\alpha} \tag{7.5.36}$$

where

$$(\mathscr{H}_{\mathrm{c}}\chi)_\alpha\equiv\frac{1}{16}(\bar{\mathscr{D}}^2-4R)(\mathscr{D}^2-6\bar{R})\chi_\alpha$$

$$=\left\{\mathscr{D}^b\mathscr{D}_b+\frac{1}{4}R\mathscr{D}^2+\mathrm{i}G^b\mathscr{D}_b+\frac{1}{4}(\mathscr{D}^\beta R)\mathscr{D}_\beta-\frac{3}{8}(\bar{\mathscr{D}}^2-4R)\bar{R}\right\}\chi_\alpha$$

$$-\left\{W^\beta{}_{\alpha\gamma}\mathscr{D}_\beta+\frac{1}{2}(\mathscr{D}^\beta W_{\alpha\beta\gamma})\right\}\chi^\gamma. \tag{7.5.37}$$

The operator $\mathscr{H}_{\mathrm{c}}$ is a d'Alembertian on the space of covariantly chiral spinor superfields.

Let $\delta[\ldots]$, $\delta_+[\ldots]$ and $\delta_-[\ldots]$ denote the functional delta-functions on spaces of unconstrained real scalar, covariantly chiral and antichiral scalar superfields, respectively. Because of constraint (7.5.33), the expression $\delta[\mathbb{K}(\chi)]$ turns out to be ill-defined, $\delta[\mathbb{K}[\chi]]\sim\delta_+[0]\delta_-[0]$. Following the Schwarz approach, we must look for generalized delta-functions $\tilde{\delta}[\ldots]$ for covariantly linear real scalars. Let $\mathbb{L}$ be such a superfield, $(\bar{\mathscr{D}}^2-4R)\mathbb{L}=0$, $\mathbb{L}=\bar{\mathbb{L}}$. Then the path integral

$$\delta[\mathbb{L}]=\int \mathscr{D}\Pi\exp\left(\mathrm{i}\int \mathrm{d}^8z\, E^{-1}\Pi\,\mathbb{L}\right) \tag{7.5.38}$$

is seen to be degenerate, since the exponential remains unchanged under the gauge transformations

$$\Pi \to \Pi^{\Lambda} = \Pi + \Lambda + \bar{\Lambda} \qquad \bar{\mathscr{D}}_{\dot{\alpha}}\Lambda = 0. \tag{7.5.39}$$

This is exactly the gauge invariance of the massless vector multiplet model. Therefore, the gauge group volume can be extracted from equation (7.5.38) in the same fashion as was done in subsection 7.3.1. As a result, we obtain the following well-defined superfunctional

$$\begin{aligned}\delta[\mathbb{L}] = \int \mathscr{D}\Pi\, \delta_+\left[\frac{1}{4}(\bar{\mathscr{D}}^2 - 4R)\Pi\right]\delta_-\left[\frac{1}{4}(\mathscr{D}^2 - 4\bar{R})\Pi\right] \mathbf{Det}(\mathbf{H}^{(R)})\\ \times \exp\left(\mathrm{i}\int \mathrm{d}^8 z\, E^{-1}\Pi\mathbb{L}\right)\end{aligned} \tag{7.5.40}$$

which can be called a functional delta-function of a covariantly linear scalar superfield. Using the identity

$$\delta_+\left[\frac{1}{4}(\bar{\mathscr{D}}^2 - 4R)\Pi\right]\delta_-\left[\frac{1}{4}(\mathscr{D}^2 - 4\bar{R})\Pi\right] = \int \mathscr{D}\bar{\Phi}\mathscr{D}\Phi\, \mathrm{e}^{\mathrm{i}\int \mathrm{d}^8 z\, E^{-1}\Pi(\Phi + \bar{\Phi})}$$

where the integration is performed over chiral (Φ) and antichiral ($\bar{\Phi}$) variables, equation (7.5.40) can be rewritten in the form

$$\delta[\mathbb{L}] = \int \mathscr{D}\bar{\Phi}\mathscr{D}\Phi\, \delta[\mathbb{L} + \Phi + \bar{\Phi}]\, \mathbf{Det}(\mathbf{H}^{(R)}). \tag{7.5.41}$$

The next major step is to define an integration measure $\mathscr{D}\mu_V$ over a group of gauge transformations (7.5.6). Let $\mathscr{F}[\chi]$ be a functional of the gauge superfields χ_α, $\bar{\chi}_{\dot{\alpha}}$. The naive path integral $\int \mathscr{D}V \mathscr{F}[\chi^V]$ is degenerate, since $\mathscr{F}[\chi^V]$ does not change under arbitrary Λ-transformations (7.5.7). Extracting the Λ-group volume from the above path integral, with the aid of the results of subsection 7.3.1, we arrive at

$$\int \mathscr{D}\mu_V\, \mathscr{F}[\chi^V] \tag{7.5.42}$$

$$\mathscr{D}\mu_V = \delta_+\left[\frac{1}{4}(\bar{\mathscr{D}}^2 - 4R)V\right]\delta_-\left[\frac{1}{4}(\mathscr{D}^2 - 4\bar{R})V\right]\mathbf{Det}(\mathbf{H}^{(R)})\mathscr{D}V.$$

We set up equation (7.5.42) as the definition of integration over the gauge group (7.5.6).

Further consideration is a simple repetition of the Faddeev–Popov prescription. One introduces the Faddeev–Popov determinant Δ defined by

$$\Delta^{-1} \equiv \int \mathscr{D}\mu_V\, \delta[\mathbb{K}(\chi^V)] = \int \mathscr{D}\mu_V \mathscr{D}\bar{\Phi}\mathscr{D}\Phi\, \delta[\mathbb{K}(\chi^V) + \Phi + \bar{\Phi} - U]\, \mathbf{Det}(\mathbf{H}^{(R)}) \tag{7.5.43}$$

U being an external real scalar superfield (as is readily seen, Δ does not depend on U). Then one inserts the unit

$$1=\Delta^{-1}\int \mathscr{D}\mu_V\,\delta[\mathbb{K}(\chi^V)]$$

into the naive path integral

$$\int \mathscr{D}\bar{\chi}\mathscr{D}\chi\, e^{iS[\chi]}$$

and separates the gauge group volume $\int \mathscr{D}\mu_V$. This leads to the effective action

$$e^{i\Gamma_{CSP}}=\int \mathscr{D}\bar{\chi}\mathscr{D}\chi\mathscr{D}\bar{\Phi}\mathscr{D}\Phi\,\Delta\,\mathbf{Det}(\mathbf{H}^{(R)})\delta[\mathbb{K}(\chi)+\Phi+\bar{\Phi}-U]e^{iS[\chi]}. \tag{7.5.44}$$

It remains to integrate the right-hand side over U with the weight $\exp(i\int d^8z\,E^{-1}U^2)$. As a result, one obtains

$$e^{i\Gamma_{CSP}}=\Delta\,\mathbf{Det}(\mathbf{H}^{(R)})\int \mathscr{D}\bar{\chi}\mathscr{D}\chi\bar{\Phi}\mathscr{D}\Phi\, e^{i(S[\chi]+S_{GB}[\chi]+2S_{(R)}[\Phi])}. \tag{7.5.45}$$

Direct calculation of the Faddeev–Popov determinant gives

$$\Delta=\mathbf{Det}(\Box_v^{(R)})\,\mathbf{Det}^{-3}(\mathbf{H}^{(R)}) \tag{7.5.46}$$

$\Box_v^{(R)}$ being the scalar superfield d'Alembertian (7.3.8) with $\Psi=R$. Now, using equations (7.5.1) and (7.5.36), the effective action can be expressed in the form

$$\Gamma_{CSP}=\Gamma_{CSC}+X \tag{7.5.47}$$

where

$$X=-\frac{i}{2}(\mathbf{Tr}_+\ln\mathscr{H}_c+\mathbf{Tr}_-\ln\mathscr{H}_a+2\,\mathbf{Tr}\ln\Box_v^{(R)}-4\,\mathbf{Tr}\ln\mathbf{H}^{(R)}) \tag{7.5.48}$$

Γ_{CSC} being the effective action (7.5.5) of the chiral scalar model (7.5.1).

As for the real scalar model (7.5.4), its quantization requires more tedious consideration. Omitting details, which can be found in our paper[3], we present the final expression for the effective action

$$\Gamma_{RSP}=\Gamma_{CSC}-2X \tag{7.5.49}$$

where X is exactly the superfunctional given by equation (7.5.48).

[3] I.L. Buchbinder and S.M. Kuzenko, *Nucl. Phys.* B **308** 162, 1988.

7.5.4. Analysis of quantum equivalence

Comparing equations (7.5.47) and (7.5.49) we see that the effective actions Γ_{CSC}, Γ_{CSP} abd Γ_{RSP} do not coincide. However, this fact does not yet mean absence of equivalence at the quantum level. It may appear that X is a constant locally independent of the supergravity prepotentials. If this is the case,

$$\frac{\delta X}{\delta \mathscr{W}} = \frac{\delta X}{\delta \varphi} = 0 \tag{7.5.50}$$

then the models under consideration will remain equivalent in the quantum theory. In this subsection, we are going to give a proof of equation (7.5.50). We show first of all that X has no divergent part.

We begin by rewriting equation (7.5.48) in terms of Green's functions. Let us introduce the Feynman superpropagator $\tilde{\mathbf{G}}_{c\alpha}{}^{\alpha'}$,

$$\mathscr{D}_\beta \tilde{\mathbf{G}}_{c\alpha}{}^{\alpha'}(z, z') = \mathscr{D}'_\beta \tilde{\mathbf{G}}_{c\alpha}{}^{\alpha'}(z, z') = 0$$

associated with the chiral spinor d'Alembertian $\tilde{\mathscr{H}}_c$ (7.5.37),

$$\tilde{\mathscr{H}}_c \tilde{\mathbf{G}}_c(z, z') = -\mathbb{I}\delta_+(z, z'). \tag{7.5.51}$$

Its antichiral analogue will be denoted by $\tilde{\mathbf{G}}_a$. Taking into account equations (7.2.30) and (7.3.18), the superfunctional X can be rewritten as follows:

$$X = \frac{i}{2}(\mathbf{Tr}_+ \ln \tilde{\mathbf{G}}_c - 2\mathbf{Tr}_+ \ln \mathbf{G}_c + \mathbf{Tr}_- \ln \tilde{\mathbf{G}}_a - 2\mathbf{Tr}_- \ln \mathbf{G}_a$$
$$+ 4\mathbf{Tr} \ln \mathbf{G}_v - 2\mathbf{Tr} \ln \mathbf{G}_v^{(R)}). \tag{7.5.52}$$

The structure of the Green's functions $\mathbf{G}_c$ and $\mathbf{G}_v^{(\Psi)}$ were discussed in Sections 7.2 and 7.3. In complete analogy with our previous consideration of $\mathbf{G}_c$, one can develop a proper-time representation for the chiral spinor superprogator $\tilde{\mathbf{G}}_c$ and calculate the corresponding coefficients $\tilde{\mathbf{a}}_1^c$ and $\tilde{\mathbf{a}}_1^c$ at coincident points.

Then one obtains

$$\operatorname{tr} \tilde{\mathbf{a}}_1^c(z, z) = 2\mathbf{a}_1^c(z, z) \tag{7.5.53}$$

$$\operatorname{tr} \tilde{\mathbf{a}}_2^c(z, z) = 2\mathbf{a}_2^c(z, z) - W^{\alpha\beta\gamma} W_{\alpha\beta\gamma} + \frac{1}{4}(\bar{\mathscr{D}}^2 - 4R)R\bar{R}.$$

In the framework of the L-regularization scheme, the divergent part of X

reads

$$X_{\text{div}}=\frac{L^{-2}}{32\pi^2}\left\{\int \mathrm{d}^8 \frac{E^{-1}}{R}[2\mathbf{a}_1^{\mathrm{c}}(z,z)-\mathrm{tr}\,\tilde{\mathbf{a}}_1^{\mathrm{c}}(z,z)]+\mathrm{c.c.}\right\}$$

$$+\frac{\ln(L^2\mu^2)}{32\pi^2}\int \mathrm{d}^8z\, E^{-1}\left\{\frac{1}{R}[\mathrm{tr}\,\tilde{\mathbf{a}}_2^{\mathrm{c}}(z,z)-2\mathbf{a}_2^{\mathrm{c}}(z,z)]+\mathrm{c.c.}\right.$$

$$\left.+4\mathbf{a}_2(z,z)-2\mathbf{a}_2^{(R)}(z,z)\right\}. \tag{7.5.54}$$

Making use of equations (7.3.33), (7.5.53) and omitting obvious total derivatives, one arrives at

$$X_{\text{div}}=-\frac{\ln(L^2\mu^2)}{32\pi^2}(\mathscr{P}+\bar{\mathscr{P}}) \tag{7.5.55}$$

$\mathscr{P}$ being the topological invariant (5.6.61). As a result, X_{div} is a constant with respect to arbitrary infinitesimal displacements of the supergravity prepotentials.

Now, we give a formal proof of equation (7.5.50). Let us consider the local non-degenerate change of variables

$$U=\Phi+\bar{\Phi}+\frac{1}{2}(\mathscr{D}^\alpha\chi_\alpha+\bar{\mathscr{D}}_{\dot{\alpha}}\bar{\chi}^{\dot{\alpha}}) \tag{7.5.56}$$

$$V=\Psi+\bar{\Psi}+\frac{\mathrm{i}}{2}(\mathscr{D}^\alpha\chi_\alpha-\bar{\mathscr{D}}_{\dot{\alpha}}\bar{\chi}^{\dot{\alpha}})$$

where U and V are real scalar superfields, Φ and Ψ covariantly chiral scalar superfields and χ_α a covariantly chiral spinor superfield. It turns out that the corresponding Jacobian is equal to

$$J(U,V|\Phi,\Psi,\chi)=\mathbf{Det}(\mathbf{H}^{(R)})\mathbf{Det}_+^{-1/2}(\mathscr{H}_{\mathrm{c}})\mathbf{Det}_-^{-1/2}(\mathscr{H}_{\mathrm{a}}). \tag{7.5.57}$$

This relation can be proved by making the above change of variables in the path integral

$$\int \mathscr{D}U\,\mathscr{D}V\exp\left(\frac{\mathrm{i}}{2}\int \mathrm{d}^8z\,E^{-1}(U^2-V^2)\right).$$

Then we consider the inverse change of variables

$$\Phi=\frac{1}{16}(\bar{\mathscr{D}}^2-4R)[(\bar{\mathscr{D}}^2-4R)+(\mathscr{D}^2-4\bar{R})]\frac{1}{\Box_{\mathrm{v}}^{(R)}}U$$

$$\Psi=\frac{1}{16}(\bar{\mathscr{D}}^2-4R)[(\bar{\mathscr{D}}^2-4R)+(\mathscr{D}^2-4\bar{R})]\frac{1}{\Box_{\mathrm{v}}^{(R)}}V \tag{7.5.58}$$

$$\chi_\alpha=-\frac{1}{8}(\bar{\mathscr{D}}^2-4R)\mathscr{D}_\alpha\frac{1}{\Box_{\mathrm{v}}^{(R)}}(U-\mathrm{i}V).$$

The corresponding Jacobian reads

$$J(\Phi,\ \Psi,\ \chi|U,\ V)=\mathbf{Det}(\mathbf{H}^{(R)})\mathbf{Det}^{-1}(\Box_{\mathrm{v}}^{(R)}) \tag{7.5.59}$$

which can be proved by carrying out the replacement of variables (7.5.58) in the path integral

$$\mathbf{Det}(\mathbf{H}^{(R)})=\int \mathscr{D}\bar{\Phi}\mathscr{D}\Phi\mathscr{D}\bar{\Psi}\mathscr{D}\Psi\mathscr{D}\bar{\chi}\mathscr{D}\chi \exp\left\{\frac{\mathrm{i}}{2}\int \mathrm{d}^8 z\, E^{-1}\left[-(\Phi+\bar{\Phi})\frac{1}{\Box_{\mathrm{v}}^{(R)}}(\Phi+\bar{\Phi})\right.\right.$$
$$\left.\left.+(\Psi+\bar{\Psi})\frac{1}{\Box_{\mathrm{v}}^{(R)}}(\Psi+\bar{\Psi})+\frac{1}{R}\chi^{\alpha}\chi_{\alpha}+\frac{1}{\bar{R}}\bar{\chi}_{\dot{\alpha}}\bar{\chi}^{\dot{\alpha}}\right]\right\}.$$

It follows from equations (7.5.57) and (7.5.59) that the superfunctional (7.5.48) must vanish modulo boundary terms at infinity, the latter being always ignored in (super)field redefinitions.

The above analysis gives strong grounds in defence of equation (7.5.50) and, hence, quantum equivalence of the three superfield models under consideration.

As an instructive exercise, we suggest that the reader find a field analogue of the superfield definition (7.5.56), which makes it possible to establish formal coincidence of the effective actions (7.5.30) and (7.5.31).

Bibliography

Pioneering papers on supersymmetry

Yu. A. Gol'fand and E. P. Likhtman. Extension of the algebra of Poincaré group generators and violation of P-invariance. *JETP Lett.* **13** 323, 1971.

D. V. Volkov and V. P. Akulov. Possible universal neutrino interaction. *JETP Lett.* **16** 438, 1972.

J. Wess and B. Zumino. Supergauge transformations in four dimensions. *Nucl. Phys.* **B 70** 39, 1974.

J. Wess and B. Zumino. A Lagrangian model invariant under supergauge transformations. *Phys. Lett.* **49B** 52, 1974.

Pioneering papers on supergravity

D. Z. Freedman, P. van Nieuwenhuizen and S. Ferrara. Progress towards a theory of supergravity. *Phys. Rev.* **D 13** 3214, 1976.

S. Deser and B. Zumino. Consistent supergravity. *Phys. Lett.* **62B** 335, 1976.

Basic reviews and books on supersymmetry and supergravity

V. I. Ogievetsky and L. Mezincescu. Boson-fermion symmetries and superfields. *Sov. Phys.-Usp.* **18** 960, 1976.

P. Fayet and S. Ferrara. Supersymmetry. *Phys. Rep.* **32** 249, 1977.

A. Salam and J. Strathdee. Supersymmetry and Superfields. *Fortschr. Phys.* **26** 57, 1978.

P. van Nieuwenhuizen. Supergravity. *Phys. Rep.* **68** 189, 1981.

H. P. Nilles. Supersymmetry, supergravity and particle physics. *Phys. Rep.* **110** 1, 1984.

H. E. Haber and G. L. Kane. The search for supersymmetry: probing physics beyond the standard model. *Phys. Rep.* **117** 75, 1985.

E. S. Fradkin and A. A. Tseytlin. Conformal supergravity. *Phys. Rep.* **119** 233, 1985.

M. F. Sohnius. Introducing supersymmetry. *Phys. Rep.* **128** 39, 1985.

M. J. Duff, B. E. W. Nilsson and C. N. Pope. Kaluza-Klein supergravity. *Phys.*

Rep. **130** 1, 1986.
N. Dragon, U. Ellwanger and M. Schmidt. Supersymmetry and Supergravity. *Prog. Particle Nucl. Phys.* **18** 1, 1987.
S. J. Gates, M. T. Grisaru, M. Roček and W. Siegel. *Superspace or One Thousand and One Lessons in Supersymmetry.* Benjamin Cummings, Reading, MA, 1983.
J. Wess and J. Bagger. *Supersymmetry and Supergravity.* Princeton University Press, Princeton, NJ, 1991 (Second Edition, Revised and Corrected).
P. West. *Introduction to Supersymmetry and Supergravity.* World Scientific, Singapore, 1992 (Extended Second Edition).
P. Freund. *Introduction to Supersymmetry.* Cambridge University Press, Cambridge, 1986.
R. N. Mohapatra, *Unification and Supersymmetry.* Springer-Verlag, New York, 1986.
O. Piquet and K. Sibold. *Renormalized Supersymmetry: The Perturbation Theory of $N = 1$ supersymmetric Theories in Flat Space-Time.* Birkhäuser, Boston, 1986.
H. J. W. Müller-Kirsten and A. Wiedemann. *Supersymmetry: An Introduction with Conceptual and Calculational Details.* World Scientific, Singapore, 1987.
M. Müller. *Consistent Classical Supergravity Theories.* Springer-Verlag, New York, 1989.
L. Castellani, R. D' Auria and P. Fre. *Supergravity and Superstrings: A Geometric Perspective.* World Scientific, Singapore, 1991.
P. Fre and P. Soriani. *The $N = 2$ Wonderland: From Calabi-Yau Manifolds to Topological Field Theories.* World Scientific, Singapore, 1995.

Field theory in superspace

A. Salam and J. Strathdee. Super-gauge transformations. *Nucl. Phys.* **B 76** 477, 1974.
S. Ferrara, J. Wess and B. Zumino. Supergauge multiplets and superfields. *Phys. Lett.* **51B** 239, 1974.
J. Wess and B. Zumino. Supergauge invariant extension of quantum electrodynamics. *Nucl. Phys.* **B 78** 1, 1974.
A. Salam and J. Strathdee. Super-symmetry and non-Abelian gauges. *Phys. Lett.* **51B** 353, 1974.
S. Ferrara and B. Zumino. Supergauge invariant Yang-Mills theories. *Nucl. Phys.* **B 79** 413, 1974.
A. Salam and J. Strathdee. Superfields and Fermi-Bose symmetry. *Phys. Rev.* **D 11** 1521, 1975.
S. Ferrara and B. Zumino. Transformation properties of the supercurrent. *Nucl. Phys.* **B 87** 207, 1975.
E. Sokatchev. Projection operators and supplementary conditions for superfields

with arbitrary spin. *Nucl. Phys.* **B 99** 96, 1975.
V. I. Ogievetsky and E. S. Sokatchev. The supercurrent. *Sov. J. Nucl. Phys.* **28** 423, 1978.
W. Siegel. Gauge spinor superfield as scalar multiplet. *Phys. Lett.* **85B** 333, 1979.
M. T. Grisaru, M. Roček and W. Siegel. Improved methods for supergraphs. *Nucl. Phys.* **B 159** 429, 1979.
M. T. Grisaru and W. Siegel. Supergraphity (I). Background field formalism. *Nucl. Phys.* **B 187** 149, 1981.
S. J. Gates and W. Siegel. Variant superfield representations. *Nucl. Phys.* **B 187** 389, 1981.
M. T. Grisaru and W. Siegel. Supergraphity (II). Manifestly covariant rules and higher-loop finiteness. *Nucl. Phys.* **B 201** 292, 1982.
B. de Wit and M. Roček. Improved tensor multiplets. *Phys. Lett.* **109B** 439, 1982.
E. A. Ivanov. Intristic geometry of the $N = 1$ supersymmetric Yang-Mills theory. *J. Phys. A: Math. Gen.* **16** 2571, 1983.
B. B. Deo and S. J. Gates. Comments on nonminimal $N = 1$ scalar multiplets. *Nucl. Phys.* **B 254** 187, 1985.

Superfield supergravity

J. Wess and B. Zumino. Superspace formulation of supergarvity. *Phys. Lett.* **66B** 361, 1977.
J. Wess and B. Zumino. Superfield Lagrangian for supergarvity. *Phys. Lett.* **74B** 51, 1978.
V. Ogievetsky and E. Sokatchev. Structure of supergravity group. *Phys. Lett.* **79B** 222, 1978.
P. S. Howe and R. W. Tucker. Scale invariance in superspace. *Phys. Lett.* **80B** 138, 1978.
R. Grimm, J. Wess and B. Zumino. A complete solution of the Bianchi identities in superspace with supergravity constraints. *Nucl. Phys.* **B 152** 255, 1979.
W. Siegel and S. J. Gates. Superfield supergravity. *Nucl. Phys.* **B 147** 77, 1979.
V. I. Ogievetsky and E. S. Sokatchev. The simplest Einstein supergarvity group. *Sov. J. Nucl. Phys.* **31** 140, 1980.
V. I. Ogievetsky and E. S. Sokatchev. The gravitational axial-vector superfield and the formalism of differential geometry. *Sov. J. Nucl. Phys.* **31** 424, 1980.
V. I. Ogievetsky and E. S. Sokatchev. The normal gauge in supergarvity. *Sov. J. Nucl. Phys.* **32** 443, 1980.
V. I. Ogievetsky and E. S. Sokatchev. Torsion and curvature in terms of the axial superfield. *Sov. J. Nucl. Phys.* **32** 447, 1980.
V. I. Ogievetsky and E. S. Sokatchev. Equations of motion for the axial gravitational superfield. *Sov. J. Nucl. Phys.* **32** 589, 1980.

E. A. Ivanov and A. S. Sorin. Superfield formulation of OSp(1,4) supersymmetry. *J. Phys. A: Math. Gen.* **13** 1159, 1980.
S. J. Gates and W. Siegel. Understanding constraints in superspace formulations of supergravity. *Nucl. Phys.* **B 163** 519, 1980.
S. J. Gates, K. S. Stelle and P. C. West. Algebraic origins of superspace constraints in supergravity. *Nucl. Phys.* **B 169** 347, 1980.
P. S. Howe, K. S. Stelle and P. K. Townsend. The vanishing volume of $N = 1$ superspace. *Phys. Lett.* **107B** 420, 1981.
S. J. Gates, M. Roček and W. Siegel. Solution to constraints for $n = 0$ supergravity. *Nucl. Phys.* **B 198** 113, 1982.

Supermathematics

F. A. Berezin. *The Method of Second Quantization.* Academic Press, New York, 1966.
F. A. Berezin. *Introduction to Superanalysis.* Reidel, Dordrecht, 1987.
B. S. De Witt. *Supermanifolds.* Cambridge University Press, Cambridge, 1986.

Index

For Product Safety Concerns and Information please contact our EU representative GPSR@taylorandfrancis.com Taylor & Francis Verlag GmbH, Kaufingerstraße 24, 80331 München, Germany

Printed and bound by CPI Group (UK) Ltd, Croydon, CR0 4YY
01/05/2025
01858515-0001